Commercial Ornamental Crops

CUT FLOWERS

The Editors

Dr. R. L. Misra served the cause of floricultural science for more than 16 years in temperate regions of Himachal Pradesh and about 24 years in sub-tropical conditions of New Delhi. After his joining at IARI, New Delhi in 1984, he was associated with all the 7-8 floricultural courses being offered to the M.Sc. and Ph.D. students, apart from looking his mandated research programmes. Out of his whole official career, as a chairman, he guided 16 Ph.D. and 6 M.Sc. students apart from being members of the advisory committees of many more students, and developed 30 gladiolus varieties which have high demand countrywide. Now all these students are serving various institutions as Professors and Associate Professors and in other departments holding a very high position. As a teacher as well as a researcher, his contribution is well known as he has in all some 500 publications including 6 books. He is the founder member of Indian Society of Ornamental Horticulture. From 1993 to 2004 he edited its regular publication *Journal of Ornamental Horticulture*. He is Fellow of Indian Society of Genetics and Plant Breeding since 80's, and Indian Society of Ornamental Horticulture since 2010. For his most significant contribution in the field of floriculture, he was honoured with 'Gold Medal' of Horticultural Society of India in 2013. He is life member in 11 professional societies. Joined as a regular Project Coordinator of All India Coordinated Research Project on Floriculture of ICAR at Division of Floriculture and Landscaping, IARI, New Delhi where he coordinated some 21 centres across 17 states. He served IARI for 40 years and retired as Project Coordinator cum Principal Scientist in May 2008. Only because of his highly significant contributions in the field of teaching and research, after his retirement he was inducted as adjunct Faculty of IARI in Floriculture.

Dr. Sanyat Misra is a graduate from UAS, Bangalore; postgraduate from GBP University of Agriculture and Technology, Pantnagar; and Ph.D. from Dr. YSP University of Horticulture and Forestry, Solan (H.P.), and then after completing his Ph.D. degree served for 3 years as Research Associate at NBPGR, New Delhi and Shimla, and in 2004 joined as regular Scientist cum Assistant Professor (Floriculture) in the Department of Horticulture, Birsa Agricultural University, Kanke, Ranchi (Jharkhand). So far he has guided Ten M.Sc. students for their thesis work. Presently he is Head, Department of Horticulture, BAU, Ranchi. He is a very active member of the horticulture faculty looking after research, teaching and extension. He has to his credit about 45 publications, two bulletins and Five books of international repute.

Commercial Ornamental Crops

CUT FLOWERS

– Editors –

R.L. Misra

Sanyat Misra

2023

Daya Publishing House®

A Division of

Astral International Pvt. Ltd.

New Delhi – 110002

ISBN: 9789359191034 (HB)

Published by : **Daya Publishing House®**
A Division of
Astral International Pvt. Ltd.
– ISO 9001:2015 Certified Company –
4736/23, Ansari Road, Darya Ganj
New Delhi-110 002
Ph. 011-43549197, 23278134
E-mail: info@astralint.com
Website: www.astralint.com

Dr. K.L CHADHA
Ph.D (Hort.), D.Sc. (Honoris causa), FNAAS, FHSI, FISHS
(Padma Shri Awardee)
President
The Horticultural Society of India
F-1, Societies Block, NASC Complex,
New Delhi-110 012
Tel.: (011-25842127
E-mail.:-hsi42@rediffmail.com

Formerly
National Professor (Hort.) ICAR, New Delhi
Deputy Director General (Hort.) ICAR, New Delhi
Horticultural Commissioner, Min. of Agri., GoI, New Delhi
Executive Director, NHB, Min. of Agri., GoI, New Delhi
Director, IIHR (ICAR), Bangalore
Project Coordinator (Fruits) ICAR, Lucknow

Foreword

Traditional flower culture forms the backbone of Indian Floriculture and is mostly in the hands of small and marginal farmers. In recent years emphasis has also been laid on growing flowers under protected cultivation both for domestic market and exports. Lot of efforts has also been made to promote R&D in floriculture. As a result floriculture is now recognized as a lucrative profession with much higher potential for returns per unit area. India along with China are leading countries in the world in terms of area and production. While the traditional flower sector in India has registered an impressive growth during the past few plans, India's contribution to global floriculture is still insignificant.

The interest in education in floriculture has been increasing of late. However, authentic scientific literature in floriculture is not adequate. Most publication in India are not very recent or are from foreign countries. I am happy that Drs. R.L. Misra and Sayant Misra a father-son team, both experts in the field of floriculture and landscaping have made an effort to put their life long experience and expertise in bringing out this book titled, Commercial Ornamentals dealing worth 36 Cut Flower Crops. Each Crop gives details about origin, nomenclature, production, protection and post-harvest technology. It is hoped that the publication will be a valuable reference for all those engaged in research, education, development and management for floriculture industry. The publication will also act as a critical guide book for both undergraduate and post graduate students in floriculture and landscape gardening. I compliment the authors for their effort on bringing out this publication and wish them success in this and future endeavours.

K.L. Chadha

Dr. K.L. Chadha

Preface

Since the time immemorial, the flowers in the worldwide system, especially in Asian countries are being used traditionally irrespective of the caste, creed and religions. There are many plants and trees which are therefore attached with our day-to-day life and many a plants have been associated with gods and goddesses. All over India, people from different walks of life, irrespective of their sex, faith, creed and caste grow some ornamentals near their homes to be used in needs. Also in the villages, mostly people have a patch where local flowers are allowed to grow on their own from where villagers used to pluck them for ceremonial occasions and for offering to gods and goddesses. The trees or shrubs emitting pleasant fragrances especially by the sunset used to be planted in villages. Then there was no flower market, no flower business or any other thing alike and the people used to collect them unhesitantly from anywhere, irrespective of the fact that who is growing these, so had been the system in India. For making the look of the house and its surrounding beautiful, people were and are using ornamentals by planting around. Even in the mango orchard one can see such plantings and there had been no restrictions that who will use them but most of such plants were *Amaranthus, Amaryllis, Barleria, Bougainvillea, Calotropis, Canna, Catharanthus, Celosia, Clerodendron, Codiaeum, Crossandra, Dahlia, Dendranthema, Eranthemum, Ervatamia, Gaillardia, Helianthus, Heliconia, Hibiscus, Impatiens, Ixora, Jasminum, Narcissus, Nerium, Pandanus, Polianthes, Rosa, Tagetes, Thevetia* and so on, which people used to grow and enjoy, and out of which a few were being used even for extraction of essential oils. The cut flower concept is of the western countries which Asian people also found quite attractive, therefore, now realizing its importance they have started growing these in greenhouses to earn their livelihood and to generate more money to raise their status even by export of such flowers to international markets. Apart from the internationally recognized cut flower markets in Holland, Germany and elsewhere, every country has its own national and local centres for such markets where all types of cut flowers reach every day from many countries in this country. Here in this book, we have put our all efforts to include almost all the important **Cut Flowers** which not only in India are being grown but world over. We have tried to furnish here their 'A to Z' proper cultivation practices in a scientific manner.

Indian Council of Agricultural Research, New Delhi has formulated an uniform course programme in Floriculture for Bachelor's, Master's and Doctoral programmes, therefore, this book **Commercial Ornamental Crops: Cut Flowers** has been written to avoid any confusion that what are important cut flower crops. This is the first publication of the kind, dealing exclusively with cut flowers. This is being followed by two other publications, *viz.* first **Commercial Ornamental Crops: Traditional and Loose Flowers**, and second **Ornamental Plants (Commonly Used)**. However, one more book on **Soilless Culture** is also ready. We assure that not only to the students and their professors this book will be useful, but will also serve an useful guide to the scientific community of floriculture, the landcapers and horticulturists, the growers and other garden enthusiasts.

R.L. Misra

Sanyat Misra

Contents

List of Contributors

(Chapters written in **bold** show that author is senior contributor)

Ajit Kumar

(*Antirrhinum, Callistephus chinensis,* ***Centaurea, Helianthus, Iberis,*** *Lilium, Scabiosa, Strelitzia,* ***Tulipa***)

Department of Horticulture, G.B. Pant University of Agriculture & Technology, Pantnagar-263145, Distt. Udham Singh Nagar (Uttarakhand).

Anuradha Sane

(*Heliconia*)

Division of Plant Genetic Resources,IIHR, Hessaraghatta Lake P.O., Bangalore-560089.

Atul Batra

(***Gaillardia***)

Floriculture section, National Botanical Research Institute, Rana Pratap Marg, Lucknow-226001.

Bag, M.K.

(*Zinnia*)

Germplasm Evaluation Division, NBPGR, Pusa Campus, New Delhi 110012.

Banerji, B.K.

(*Gaillardia*)

Floriculture section, National Botanical Research Institute, Rana Pratap Marg, Lucknow-226001.

E-mail: banerjibk@yahoo.co.in

Bhavya Bhargava

(*Gladiolus*)

Institute of Himalayan Bioresource Technology, CSIR, Palampur-176061 (H.P.).

Bini Sundar, S.T.

(*Eustoma, Solidago*)

Department of Floriculture & Landscaping, H.C.R.I., TNAU, Coimbatore-641 003.

Dhiman, S.R.

(Proteaceous Ornamentals)

Department of Floriculture and Landscaping, Dr Y S Parmar University of Horticulture and Forestry, Nauni, Distt.- Solan (H.P.).

Dubey, R.K.

(*Strelitzia*)

Department of Floriculture and Landscaping, P.A.U., Ludhiana-141004 (Punjab).

Gupta, Y.C.

(*Callistephus chinensis,* ***Dianthus caryophyllus, Gladiolus,*** Proteaceous Ornamentals)

Department of Floriculture and Landscaping, Dr Y S Parmar University of Horticulture and Forestry, Nauni, Distt.- Solan (H.P.).

Himanshu Trivedi

(*Helianthus*)

Department of Horticulture, G.B. Pant University of Agriculture & Technology, Pantnagar-263145, Distt. Udham Singh Nagar (Uttarakhand).

Janakiram, T.

(***Heliconia***)

Division of Floriculture and Landscaping, Indian Agricultural Research Institute, New Delhi-110012.

Jyoti Bajeli

(*Centaurea, Tulipa*)

Department of Horticulture, G.B. Pant University of Agriculture & Technology, Pantnagar-263145, Distt. Udham Singh Nagar (Uttarakhand).

Kalkame Ch. Momin

(*Dianthus caryophyllus,* **Proteaceous Ornamentals**)

Department of Floriculture and Landscaping, Dr Y S Parmar Univ. of Horticulture and Forestry, Nauni, Distt.- Solan (H.P.).

Kannan, M.

(*Eustoma,* ***Solidago***)

Department of Floriculture & Landscaping, H.C.R.I., TNAU, Coimbatore-641 003.

Khushboo Kathayat

(*Centaurea*)

Department of Horticulture, G.B. Pant University of Agriculture & Technology, Pantnagar-263145, Distt. Udham Singh Nagar (Uttarakhand).

Kiran Kumari

(*Iberis*)

Department of Horticulture, G.B. Pant University of Agriculture & Technology, Pantnagar-263145, Distt. Udham Singh Nagar (Uttarakhand).

Lakshmi Durga M.

(*Achillea*)

Division of Floriculture and Landscaping, I.A.R.I., New Delhi-110 012.

Mamta Bohra

(*Helianthus*)

Department of Horticulture, College of Agriculture, G.B. Pant University of Agriculture & Technology, Pantnagar-263145, Distt. Udham Singh Nagar (Uttarakhand).

Manish Kapoor

(*Helianthus, Iberis, Tulipa*)

Department of Horticulture, G.B. Pant University of Agriculture & Technology, Pantnagar-263145, Distt. Udham Singh Nagar (Uttarakhand).

Meenu Kumari

(*Digitalis* & *Helianthus*)

Division of Floriculture and Landscaping, I.A.R.I., New Delhi-110 012.

Misra, R.L.

(***Alstroemeria,*** *Amaryllis* & *Hippeastrum, Anemone,* ***Anthurium, Antirrhinum,*** *Callistephus chinensis, Centaurea,* Cut Foliage/Cut Greens and Other Fillers, ***Dahlia,*** *Delphinium,* ***Dendranthema,*** *Eustoma, Gerbera,Gladiolus, Gypsophila, Helianthus, Heliconia, Iberis, Lilium, Limonium, Matthiola, Narcissus,* Orchids, *Polianthes tuberosa,* Proteaceous Ornamentals, ***Rosa,*** *Salpiglossis, Scabiosa, Solidago, Strelitzia, Tulipa, Zantedeschia*)

(Ex-Project Coordinator, AICRP on Floriculture, ICAR; & Former Officer-in-Charge, Indo-Israel Project on R&D, IARI, New Delhi-110 012)

C-4, Brahma Apartments, Plot 7, Sector 7, Dwarka, New Delhi-110 075 (Mob.: 9968287841).

Namita

(***Digitalis***)

Division of Floriculture and Landscaping, I.A.R.I., New Delhi-110 012.

Narendra Bhandari

(*Lilium*)

Department of Horticulture, G.B. Pant University of Agriculture & Technology, Pantnagar-263145, Distt. Udham Singh Nagar (Uttarakhand).

Naveen Kumar

(***Strelitzia***)

(Division of Floriculture and Landscaping, I.A.R.I., New Delhi-110 012)

Present address: 280, Chhoti Baradari, Path 2, Jalandhar-144001 (Pun jab).

Nomita Laishram

(*Gladiolus*)

Department of Floriculture and Landscaping, Dr Y S Parmar University of Horticulture and Forestry, Nauni, Distt.- Solan (H.P.).

Poonam Kumari

(*Salpiglossis*)

Division of Floriculture and Landscaping, I.A.R.I., New Delhi-110 012.

Pragya Ranjan

(*Callistephus chinensis,* ***Gypsophila,*** *Lilium, Strelitzia,* ***Zinnia***)

Indian Institute of Vegetable Research, P.O. Jakhani, Varanasi-221 305.

Priyanka Sharma

(*Gladiolus*)

Department of Floriculture and Landscaping, Dr Y S Parmar University of Horticulture and Forestry, Nauni, Distt.- Solan (H.P.).

Priyanka Thakur

(***Narcissus***)

Department of Floriculture and Landscaping, Dr. Y. S. Parmar University of Horticulture and Forestry, Nauni, Solan-173230 (H.P.). E.mail address: priyanka.thakur@gmail.com

Raju, D.V.S.

(**Cut Foliage/Cut Greens and Other Fillers,** *Rosa*)

Division of Floriculture and Landscaping, I.A.R.I., New Delhi-110012.

Ranchana, P.

(*Eustoma, Solidago*)

Department of Floriculture & Landscaping, H.C.R.I., TNAU, Coimbatore-641 003.

Ranjan, J.K.

(*Antirrhinum, Callistephus chinensis, Gypsophila, Strelitzia, Zinnia*)

Indian Institute of Vegetable Research, P.O. Jakhani, Varanasi-221 305.

Ranjan Srivastava

(*Iberis, Lilium*)

Department of Horticulture, G.B. Pant University of Agriculture & Technology, Pantnagar-263145, Distt. Udham Singh Nagar (Uttarakhand).

Ritu Jain

(***Achillea, Campanula***)

Division of Floriculture and Landscaping, I.A.R.I., New Delhi-110 012.

Sanjay Kumar

(*Limonium, Matthiola, Scabiosa, Zantedeschia*)

CSIR-Institute of Himalayan Bioresource Technology (IHBT), Palampur-176061 (H.P.).

Sanyat Misra

(*Achillea, Alstroemeria,* ***Amaryllis & Hippeastrum, Anemone,*** *Anthurium, Antirrhinum,* ***Callistephus chinensis,*** *Campanula, Centaurea,* Cut Foliage/Cut Greens and Other Fillers, *Dahlia,* **Delphinium,** *Dendranthema, Digitalis,* ***Eustoma, Gerbera,*** *Gladiolus, Gypsophila, Helianthus, Iberis,* ***Lilium,*** *Limonium, Matthiola, Narcissus,* **Orchids,** ***Polianthes tuberosa,*** Proteaceous Ornamentals, *Rosa, Scabiosa, Solidago, Strelitzia, Tulipa, Zantedeschia, Zinnia*)

Department of Horticulture, Birsa Agricultural University, Kanke, Ranchi-834 006 (Jharkhand).

Sapna Panwar

(*Digitalis,* ***Salpiglossis***)

Division of Floriculture and Landscaping, I.A.R.I., New Delhi-110 012.

Shisa Ullas P.

(*Digitalis*)

Division of Agricultural Engineering, I.A.R.I., New Delhi-110 012.

Singh, M.K.

(***Limonium, Matthiola, Scabiosa, Zantedeschia***)

Division of Floriculture and Landscaping, I.A.R.I., New Delhi-110 012.

Sunder Pal

(*Helianthus, Tulipa*)

Department of Horticulture, College of Agriculture, G.B. Pant University of Agriculture & Technology, Pantnagar-263145, Distt. Udham Singh Nagar (Uttarakhand).

Thaneswari

(*Salpiglossis*)

Division of Ornamental Crops, I.I.H.R., Hessaraghatta Lake P.O., Bengaluru-560 089.

Tripti

(*Tulipa*)

Department of Horticulture, College of Agriculture, G.B. Pant University of Agriculture & Technology, Pantnagar-263145, Distt. Udham Singh Nagar (Uttarakhand).

Varun M. Hiremath

(*Campanula*)

Division of Floriculture and Landscaping, I.A.R.I., New Delhi-110 012.

Yalek Messar

(*Lilium*)

Department of Horticulture, G.B. Pant University of Agriculture & Technology, Pantnagar-263145, Distt. Udham Singh Nagar (Uttarakhand).

1

Achillea (Family: Asteraceae)

Ritu Jain, Lakshmi Durga M. and Sanyat Misra

[**Common names**: Black yarrow (*Achillea atrata*), Fernleaf yarrow (*Achillea filipendulina*), Bood wort/Capenter's grass/Milfoil/Nose-bleed/Soldier's woundwort/Stanch weed/Yarrow (*A. millefolium*), Sneezewort (*A. ptarmica*)]

Introduction

Achillea acquired its name after the mythical Greek character, Homer's hero 'Achilles', who discovered the medicinal wonders of this plant (Thieret, 2001) and distributed yarrow among his soldiers to stanch the blood from their wounds. Its common name yarrow originated from Scots language meaning a 'rough stream', owing to the Scotland's Yarrow River. *Achillea millefolium* roughly translates to 'thousand-leaved'. It is native to Eurasia and widely naturalized in temperate and boreal zones of the Northern Hemisphere, and to a lesser extent in southern regions. These plants are widespread, and have a tendency to become weedy and invasive (Armitage, 1987). However, many ornamental cultivars have been introduced and selections of the species are now considered as an important garden plants for cut flowers, container-growing, and xeriscapic landscapes. Yarrow is also marketed as fresh fillers and dry flowers in flower arrangements. The foliage of *Achillea filipendulina* is unique in its airy, ferny appearance and is useful as an alternative to the standard foliage. In some areas, it is recommended as a groundcover to control soil erosion on slopes and hillsides as it spreads by rhizomes. It is also recommended as a low-maintenance, infrequently mowed lawn, where low-growing eco-types are capable of blooming under the lawnmower blades (Small and Catling, 1999). *Achillea millefolium* plants are characterized as heavy metal accumulators which aid in phytoremediation (Radanovic *et al.*, 2000).

Yarrow has a wide array of medicinal uses like anti-inflammatory, anti-rheumatic, antiseptic, antispasmodic, analgesic, carminative, diaphoretic, digestive, expectorant, hypotensive, hemostatic and vulnerary properties (Balbir *et al.*, 2012). In addition, it is used to treat coughs, ease the women's menstrual cramps and regenerates the cardio-vascular system by lowering blood pressure and slowing one's heartbeat. *A. millefolium* is used in beer, wine, and soft drink flavouring. It is an insect repellent that dissuades visits from some ants, flies, and beetles (Schwartz, 2006). The Cherokee tribes drank Yarrow Tea steep to reduce fever and aid in sleep. Yarrow contains achilleine, apigenin and azulene, inulin, menthol, and a neurotoxin, thujone (Plants Database, 2011). The dried root of *Achillea ptarmica* (sneezewort) was once ground and used for snuff, which resulted in a runny nose. Yarrow is considered to be a minor essential oil crop.

Botany

Achillea commonly represents *Achillea millefolium*, a long-lived and long-blooming perennial, with pleasant aromatic scent and distinct morphology. It grows 30 cm to 1 m tall. Stem is simple or somewhat forking above, and very sparsely to rather densely villous. Leaves appear in basal rosettes and alternate along stems, and are lanceolate, 2-4 times pinnately-dissected with lanceolate-subulate segments, the blade 3-15 cm long to 2.5 cm wide, basal leaves long-petiolate, all but the lowermost are cauline and sessile. Heads numerous in a flat or round-topped, short and broad, paniculate corymbiform inflorescence; the disc is 2-4 mm wide. Involucral bracts about 20, imbricate in several series, short oblong, 1.5-2.5 mm long, blunt to pointed, glabrous to villous, with broad hyaline to brown or blackish scarious margins. Receptacle is convex and chaffy. Ray florets are mostly five, three toothed, pistillate and fertile, white or occasionally pink or deep rose-magenta, 2-4 mm long. Disc florets are 10-30, bisexual and fertile, corolla 2.5-3.0 mm long, lobes five, white or cream-coloured; style branches flattened, truncate, penicillate; and anthers with ovate-obtuse tips. Achenes oblong, 1.5-2.0 mm long, compressed parallel to the involucral bracts, finely lined lengthwise with thick wing-margins, glabrous and pappus is absent unlike Asteraceaous members. The basic chromosome number often reported in different species of *Achillea* is x= 9; however, *Achillea millefolium* is a very complex polyploid group with its diversity in chromosome numbers and ploidy levels frequently occurring in the genus (Ebrahimi *et al.*, 2012). *A. filipendulina* and *A. tenuifolia*, are diploid (2n = 2x = 18) and three species, *A. bieberstinii*, *A. pachycephala* and *A. aucheri* are tetraploid (2n = 4x = 36). Two ploidy levels of 2n = 6x = 54 and 2n = 8x = 72 in *A. millefolium* and two ploidy levels of 2n = 2x = 18 and 2n = 4x = 36 occur in *A. santolina* species. Several *Achillea* species show high morphological variability and worthy diversity in different populations due to its polyploidy (Sheidai *et al.*, 2009).

Achillea yields sufficient nectar to entice a variety of insects for pollination to its obligate outcrossed (self incompatible) flowers. Plenty of honey bees also visit these flowers which help in honey production. The light-weight achenes do not have specialized hairs or plumes for wind dispersal but are easily captured by wind as they detach from the inflorescence that rise above the leaves. Differentiated populations can be crossed, and the hybrids of crosses among cytotypes can be fertile, but crosses among tetraploid and hexaploid forms are less fertile. Guo *et al.* (2005) in Europe studied 27 taxa from 66 populations where they recorded molecular DNA evidence for reticulate evolution in the species *Achillea millefolium* which is very complex. Evidences suggest that many new taxa have formed after hybridization between species followed by polyploidization.

Propagation

Achillea species can be propagated from seed, divisions, or tip cuttings whereas hybrids mostly by divisions or tip cuttings to preserve the intensity of flower colours and maintain plant uniformity (Still and Steven, 1994).

The storage life of **seed** is up to three years but afterwards the viability drops gradually (Stevens *et al.*, 1996). Seeds can be sown throughout the year where germination occurs easily in light, warm and humid conditions. However, to have flowers within the first year, seeds should be sown in plug trays or flats containing a moist mixture of equal parts of peat and vermiculite, barely covering with vermiculite, in a heated greenhouse in late February to early March. After germination, the trays are covered with plastic or placed under a mist system. Germination of *A. filipendulina* takes 7 to 14 days at 18.3 to 21 °C and 90 per cent humidity. Approximately 0.4g of seed yields 1,000 seedlings (Nau, 1999). Direct sowing is not recommended. Transplants are grown further for another 2-3 weeks at 13 to 15 °C in full sunlight. Seedlings are fed fortnightly with 100 ppm nitrogen until plants are ready to be placed in the field but afterwards feeding of nitrogen is stopped as this produces weak stems and poor quality blooms. Transplants are normally placed in the production beds approximately 8-10 weeks after sowing.

Rhizomes can be **divided** any time, but preferably in early spring or immediately after flowering (Armitage, 1993). *A. millefolium* and its hybrids can be divided after every 2-3 years and *A. filipendulina* selections every 3-5 years to maintain plant vigour. As plants become crowded and/or pot-bound, disease problems increase and flower size and quality decrease. After grading of the divisions, the large crowns are directly planted in the production beds but smaller ones in pots until they attain a desirable size.

The fastest method of propagating *A. ptarmica* and *A. filipendulina* cultivars is by **soft tip cuttings** having 3-4 nodes, from late spring to early summer in the greenhouse having 10-15 °C temperature. Rooting hormones are generally not needed if adequate bottom heating (21-23 °C) is used during propagation. However, 0.1 per cent (1,000 ppm) IBA concentration is best for faster and efficient rooting.

Plants acclimatize well in outdoor conditions if terminal cuttings of *A. millefolium* are used as explants for **micropropagation** on half-strength MS medium and *in vitro* rooting without hormones (Marija *et al.*, 2015). As an alternative to seed propagation, an efficient *in vitro* propagation and conservation method of *A. millefolium* developed by Shatnawi (2012) indicated that maximum root number is obtained from media containing 1.2 mg l^{-1} of IBA and maximum shoots per microshoot on MS medium supplemented with 0.9 mg l^{-1} of BAP. *In vitro* shoots can be stored up to 32 weeks on MS medium supplemented with different sugar solutions, out of which maximum number of shoots survive on the medium supplemented with 3 per cent sucrose. Cryopreservation through vitrification is successfully achieved when shoot tips are precultured on a medium supplemented with 0.4M sorbitol and 0.1M sucrose for 1 day, followed by loading shoot tips with concentrated plant vitrification solution 2 (PVS2) for 20 minutes, then being dehydrated with PVS2 for 60 minutes at 0 °C prior to storage in liquid nitrogen. Plant regeneration system *via* shoot multiplication from shoot-tips and adventitious shoot regeneration from root segments of *A. millefolium* was developed by Turker *et al.* (2009). Higher numbers of shoots were obtained when shoot-tips were cultured on MSMO (Murashige and Skoog medium with minimal organics) supplemented with 3.0 mg l^{-1} BA and 0.5 mg l^{-1} IAA, or 5.0 mg l^{-1} kinetin and 1.0 mg l^{-1} IBA, producing 17.3 and 17.0 shoots per explant at 100 per cent frequency, respectively. For adventitious shoot regeneration, only root segments developed shoots when cultured on medium containing a combination of 1 mg l^{-1} TDZ, 0.5 mg l^{-1} IAA and 0.5 mg l^{-1} GA_3 (18.9 shoots per explant at 100 per cent frequency). Regenerated shoots rooted readily on MSMO medium containing different concentrations of IAA, IBA, NAA or 2,4-D. Ninety eight per cent of regenerated plants survived through the hardening process when the rooted plantlets were kept at 55-65 per cent relative humidity for 2 weeks, and then planted in pots containing potting soil and kept at 25-35 per cent humidity. To establish a clean stock devoid of fungal and bacterial infection, Evanor and Reuveni (2004) gave a method for large scale propagation of *A. filipendulina* cv. 'Parker' through clean meristem culture. The best conditions for propagating *A. filipendulina* was found to be MS medium supplemented with 3 per cent sucrose and 1mg l^{-1} IAA plus 2mg l^{-1} BA under 16 h of cool fluorescent light.

Species and Cultivars

The genus *Achillea* consists of 200 species of perennial herbs distributed throughout Europe, North and West Asia, and North Africa (Gabbrielli *et al.*, 1988). Three species such as *Achillea* 'Coronation Gold', *A. filipendulinam, A. millefolium* and *A. ptarmica* are used as cut flowers in world flower markets (Armitage, 1992). Following are the important ornamental species as defined by Hay and Beckett (1971), Pizzetti and Cocker (1975), Beckett (1985) and Brickell (1994).

Achillea atrata. A native to the moist calcareous soils mostly of western Alps at and above 1,800 m altitude, growing up to 20 cm high with erect, unbranched and leafy stems, and the leaves being smooth to slightly hairy, pinnate to bipinnate. Flowerhead comprises of 5-15 short terminal trusses of white-petalled short-lived flowers with off-white disc which appear late. The scales of the involucre on the unopened buds are initially green but afterwards turn black hence its specific name *atrata* and common name 'black yarrow'. It is not easy to cultivate as its requirement is always of moist location and away from hottest summer. It is propagated through seeds or by division of clumps.

Achillea chrysocoma. A native of Greece, this species grows to a height of only 10 cm but with 25 cm spread. The leaves are grey-green, deeply lobed and form a woolly carpet. Flowers of mustard-yellow appear during summer in 5.0-7.5 cm heads.

Achillea clavennae (*A. clavnae, A. argentea*)**.** A native to rocky limestone meadows of W & SW alpine Alps, it is a short bushy plant growing only 10-25 cm tall with all the parts of the plant being silver-grey and hairy. Its stems are little branched, leaves are alternate, irregular and pinnatifid where upper ones are stalkless. Flowerhead umbrella-shaped with several spreading clusters of terminal flowers, where petals are white and the disc off-white and flowering is for about 3-4 months. It has several varieties with improved characters. It is propagated by division and seeds.

Achillea clypeolata. It is a semi-evergreen upright perennial native to the Balkans which grows up to the height of 45 cm. Its plants are hairy in all its parts, foliage is silvery, feathery, hairy, linear-lanceolate, upper ones stalkless but lower ones short-stalked, and the inflorescence is umbrella-shaped with densely packed masses of tiny flowers. Petals are small and lemon-yellow but disc is yellow. It flowers profusely almost for three months and its cultivation is quite easy. It is propagated through division and seeds. It requires fertile acidic or alkaline soil which is moderately heavy. Its noteworthy variety is 'Moonshine'. Its hybrids with *A. taygetea* have produced feathery silver-grey foliage and flat bright yellow flowerheads 5.0-7.5 cm wide but the hybrids do not tolerate wet soil and become susceptible to foliar diseases and root rot during hot weathers.

Achillea filipendulina (*A.eupatorium*)**.** This is the common lemon-yellow yarrow of the florists for outdoor markets as cut flowers, native to Caucasus and Central and SW Asia though naturalised in parts of Europe and North America, and is quite easy in cultivation. It grows up to 1.5 m high bearing erect, stout and leafy stems with alternate, pinnatifid with 18-25 lanceolate segments, and feathery fern-like foliage which is linear, irregularly serrated, hairy and rough. The bright lemon-yellow flowers are arranged in large and dense umbrella-shaped corymbs some 15 cm in diameter on strong tall stems. Lemon-yellow petals are small and rounded. Flowering duration is for about three months. Its requirement is for heavy fertile soil at a sunny situation. It is propagated by division or by seeds. 'Altgold' is 60 cm tall and bears deep yellow flowers. Plants often bloom twice a year. Its 'Cloth of Gold' is a popular gold-flowered cultivar most suitable for cutting. Its stems are 60–90 cm long and flowerheads are 8–10 cm wide. 'Coronation Gold' (*A. filipendulina* × *A. clypeolata*) is compact, clump-forming, upright growing up to 90 cm high with about 60 cm spread and carries flat heads of bright mustard-yellow flowers in dense, plate-like, compound corymbs on stiff and erect stems. It is noted for its silvery to grey-green, deeply-dissected and fern-like foliage similar to *A. filipendulina*, which are strongly spicy-aromatic whose aroma persists even in dry arrangements. 'Gold Plate' produces long stems (1.2-1.5 m) with about 60 cm spread and large deep golden-yellow flowers. 'Moonwalker' produces 10-13 cm wide heads with yellow blooms on up to 90 cm tall stems. The flowers of 'Schwellenberg' are bright yellow. 'Parker's Variety' has 60–105 cm long stems and deep yellow flowers some 8–10 cm wide. *A. filipendulina* × *A. ptarmica* bears compact, well-branched, erect and strong stems bearing feathery and silvery foliage and deep gold flowerheads, and this was introduced to the United States in 1987 by Longwood Gardens and the National Arboretum.

A.× lewisii 'King Edward'**.** This perennial hybrid is semi-evergreen, compact, rounded and with woody base which grows up to 10 cm high with spreads of up to 23 cm. Its grey-green leaves are soft and feathery. Buff-yellow compact flowerheads appear during summer with minute flowers. It can suitably be planted along side the banks, on walls and in the rock gardens.

Achillea millefolium. A native to Great Britain, entire Europe eastwards to Siberia, western Himalayan Asia, and naturalized in Australia and New Zealand, it is a mat-forming species growing up to 75 cm high with spread of about 40 cm which flowers for about four months and its flowers are used for cutting though plants of var. 'Kelwayi' also for making unorthodox lawn which appears better than *Dichondra repens*. It is in cultivation for more than six centuries. Its alternate dark green leaves are deeply cut in various lobes (18-25 segments). The inflorescence is compact and regular. Tiny white to cerise flowers appear on some 10 cm wide flattened heads. Its varieties have large colour selections (white, cream, pink, rose, mauve and bicolours) which make this most popular yarrow for cut flowers. It can be grown along the sea shores, in arid and dry alkaline or acidic soils, in sand dunes, *vis-à-vis* in semi-wild garden. It is also resistant to cold and drought. It is highly productive and is readily propagated by seeds as well as through division. 'Galaxy hybrids' (*A. millefolium* × *A. taygetea*) has resulted in numerous selections with leaves quite similar to its first parent in shape and texture though flowerheads much larger and colourful. Its most popular varieties are 'Cerise Queen' (cherry red), 'Fire King' (deep red), 'Heidi' (dark violet), 'Kelwayi' (clear dark red), 'Lavender Beauty' (light lavender), 'Lilac Beauty'(lilac), 'Paprika' (red with yellow centre), 'Red and Gold' (bright red with gold centre), 'Rose Beauty' (pink), 'Summer Pastels' (pastel colour mix), 'Weserandstein' (light rose), 'White Beauty' (creamy-white), *etc.* Stem length and propensity to fade earlier are the limitations of the varieties for use as cut flowers.

Achillea moschata **(*A. genipi*).** It is found wild in the poor rocky soils having only traces of lime, of the Alps screes at altitudes above 1,350 m, growing up to 20 cm in height with erect, spreading, non-branching and downy stems. Green and pointed leaves are stalkless, pinnatifid, little downy and with a moss-like aroma. Inflorescence appear in open rounded umbrella- or semi-circular form with flowers having white petals. Flowering duration is for about two months. Though a alpine plant, it is not a long-lived one and is very difficult to cultivate. It is propagated through division and by seeds.

Achillea nobilis **(*A. camphorata*).** A plant difficult to grow, is native to Southern Europe, the Balkans, southern Russia to southern Siberia, Persia and Asia Minor which grows bushy up to 90 cm in height when in bloom. It prefers dry and sunny positions, and calcareous alkaline soil though it maintains vigour even in non-alkaline soils. Whole plant is pubescent, almost white and emits the camphor aroma. Stems are hairy, upper part branched, and leafy. Upper leaves are stalkless, lower stalked, feathery and lanceolate-acuminate. Inflorescences are regular, thickly branched and umbrella-shaped, bearing flower trusses where petals are white or yellowish and recurved. It is propagated through division or from seeds.

Achillea ptarmica. It is widespread across Europe, the Caucasus, Siberia, Asia Minor and naturalized in places of North America, and is less commercially exploited but its flowers are worth cutting. It is in cultivation since 1542. It survives in moist and damp soils. Its dried powdered leaves and roots are used as an inexpensive substitute for snuff. It is a creeping, spreading and glabrous shrubby plant attaining the height of up to 75 cm with the spread of about 40 cm. Stems grow erect, are branched and leafy. Its stalkless leaves are mid-green, alternate, narrowly lanceolate, serrated and little hairy. Daisy-like creamy-white flowers with ovate petals are held in terminal umbrella-shaped corymbose clusters up to 10 cm wide. It is high-yielding with good vase life hence most suitable as cut flowers. Its most popular varieties are 'Angels' Breath' (flowers white and suitable as filler), 'Ballerina' (suitable for drying), 'Globe' (small button-like flowers), 'La Schneeball' (suitable for growing in acidic or alkaline soils), 'Perry's White' (white button-shaped flowers appear in loose heads about a week earlier than the type species), 'The Pearl' (a most popular variety producing a profusion of double creamy-white flowers), *etc.*

Achillea taygetea. A native of Middle East, this species is suitable for cutting and border planting, growing to a height of 45 cm and spread of 12.5 cm. It bears silver-grey pinnate leaves. It produces pale-yellow flowers in flat heads some 5.0-10.0 cm wide. Its var. 'Moonshine' is little taller than the type species and bears pale-sulphur-yellow flowers.

Achillea tomentosa **(*A. aurea*).** A native to the southern Alps, the Balkans, central France, Spain and central Russia to western Siberia on dry and sunny slopes in sandy and stony soils, this attains the height of up to 15 cm when in bloom and flowers from June to September. Plants on bruising emit an aroma. Its golden-yellow flowers can be dried for arrangement. This is an unbranched and spreading silver-grey downy plant with leafy stems. Leaves are short-stalked, bipinnate and linear-lanceolate.

Cultural Practices

With its capability for regional adaptation and ability to grow in various moisture regimes, yarrow is a relatively undemanding crop that can grow in an array of environments (Small and Catling, 1999). Because of its hardy nature (thick stem cuticle, and a vast root system) and drought tolerance, it has the ability to be successful in nutrient-poor and dried-out soil. *Achillea millefolium* is involved in mutualistic relationship with members of the fungi phylum Glomeromycota. These fungi form endomycorrhizae with yarrow, allows the plant to receive more water, gases, and other nutrients from the fungus. In return, the fungus receives sugars that are stored within the plant cells. Yield and stem length are better with a soil pH of 6.4. *Achillea millefolium* possesses extreme tolerance and drought resistance which allows it to grow in areas that other ornamentals cannot hope to inhabit.

Achillea depends on plant maturity rather than **photoperiod** for flowering. Seedlings grown at 12 to 15 °C perform better. While cold temperatures are not necessary for flowering in plants with *A. filipendulina* in their parentage, yarrows seem to have more uniform growth and better vigour when grown in winter **temperatures** below 4 °C for about four weeks. During summer when the flower stalks are elongating, *A. millefolium* cultivars will produce strong stalks and intense flower colouration if night temperatures fall below 18 °C. Optimum duration of cold is normally less than four weeks. In general, plant height increases rapidly one week before the flower buds are visible, and the rapid growth phase ends two weeks after flower buds are visible. Photoperiod is a key factor in flower formation. *Achillea millefolium* 'Summer Pastels' is a qualitative long-day plant with a critical photoperiod between 12-16 hours at 18 °C, and under 16 h photoperiod in growth chambers, these flower after 57, 45 and 37 days when grown under 100, 200, or 300 μmol m^{-2} s^{-1}, respectively. A per day increase in mean dry weight gain from 0.32 to 1.02g d^{-1} occurred as irradiance increased from 100 to 300 μmol m^{-2} s^{-1} in *A. millefolium* (Zhang *et al.*, 1996). Plants are not tolerant of dense shade according to Aleksoff (1999). Engle *et al.* (1993) found that *A. filipendulina* 'Cloth of Gold' produces terminal and lateral buds under short days but only terminal buds under long days. Temperatures >10 °C and moderate to higher light levels (200-300 μmol m^{-2} s^{-1}) in combination with long days may permit producing this cultivar as a greenhouse crop.

Rows in production beds can range from 30-90 cm apart. Close **spacing** produces a high yield per square metre but per plant the yield is low. In one experiment using *A.* × 'Coronation Gold', plants spaced at 30 cm yielded 46 stems per plant and 46.5 stems per square foot, while those spaced 60 cm apart yielded 74 stems per plant and 18.6 stems per 30 cm^2. In another experiment with the same variety, Armitage (1987) reported stem length increasing over a 4-year period. In a comparison, *A. filipendulina* produces fewer but longer stems than *A.* × 'Coronation Gold'. The stems are staked as they tend to fall over in hot and humid conditions.

A soft **pinch** to 2-3 nodes shortly after transplanting will encourage more uniform branching. Plants can also be pinched during the later stages of propagation and this is useful for quick cropping in smaller pot sizes. A second pinch can be given on plants grown in very large containers with longer cropping times. Achillea is a vigorous grower and will require **growth regulator** treatments under most growing conditions. One to two sprays of B-9 at 2,500 ppm will help keep the plants more compact. Under warm growing conditions, a Bonzi (paclobutrozol) drench at 0.5-1.0 ppm during mid-late season will keep the plant growth under control.

Overhead **watering** is not recommended as it damages the flowers, causes petal spotting, splashing of the soil onto the foliage and promotes the spread of disease. Yarrows do better if kept moderately moist. They are not much demanding with regard to **nutrients** and these do well in moderately fertile soil.

Postharvest

Flowers are harvested with stems from the base of the plant after visibility of the pollen in the flower. After harvest, plants are cut back to remove necrotic foliage and to promote new growth. 'Coronation Gold' yarrow can remain productive for at least four years without any decline in its flower yield. *A.filipendulina* and its hybrids last 7-12 days in plain water. Flowers tolerate up to one week when stored at 1.67 °C. *Achillea ptarmica* cultivars remain fresh for 5-8 days if harvested at full flower open stage. *Achillea millefolium* and its cultivars have a vase life of 3-4 days.

Insect-Pests and Diseases

Yarrows are relatively pest-free. In wet springs and summers, **aphids** (*Aphis gossypii, Myzus persicae*), **leaf hoppers**, **thrips** and **spider mites** (*Tetranychus urticae*) are the most common insects found on new growth. Several species of the **root-knot nematode** (*Meloidogyne* spp.) can attack yarrows and cause stunting, decline, and reduction of flower yield, and their heavy infestation can kill the plants. Root-knot infection is recognized by the lumpy appearance of the roots. Always check roots carefully when dividing plants. Chemical nematicides can be applied only to fallow ground as they are toxic to plants. Control of plant-parasitic nematodes must be done prior to planting.

Foliar fungal diseases, including **powdery mildew**, **downy mildew** and **rust** may infect yarrows. Powdery mildew is distinguished by white spots on both sides of the leaves. Downy mildew is distinguished by small yellow spots on the top of the leaves and white mould on the bottom. If mildew is a recurring problem, it requires slight increase of the spacing between plants so that air circulation around the foliage is increased. Also, irrigating early in the day will lower the relative humidity in the microclimate around the crop before nightfall, which will help reduce the incidence of foliar fungal diseases. Rust is characterized by raised spots (pustules) on the undersides of leaves and stems. The pustules usually appear reddish in colour, but may range from bright yellow to black. If the infection is severe, the pustules will become enlarged and grow together, destroying leaves and occasionally the entire plant. Removal and destroying rust-infected leaves is the best measure. At the end of the season, infected plants are cut at the ground and the whole plant debris is removed. **Stem rot** (*Rhizoctinia solani*) infection results into decay of the stem base. Fungicides and crop rotation help to alleviate the problem. Allowing the soil to dry between watering will help to retard the development of soil pathogens. **Root rots** are also caused due to infection of *Botrytis* and *Pythium*, by which apart from rotting of the roots, seedling blight in the greenhouse also occurs. These diseases are promoted by wet conditions and can be prevented through good sanitation and ventilation. *Botrytis* can damage flowers and foliage outdoors during extended periods of wet weather. Removal of blighted plant parts and old flowers from production beds will reduce further spread of such pathogens.

References

Aleksoff, K. C. 1999. *Achillea millefolium*. In: *Fire Effects Information System* (online). U.S. Department of Agriculture, Forest Service, Rocky Mountain Research Station, Fire Sciences Laboratory (http://www.fs.fed.us/database/feis).

Armitage, A.M. 1987. The influence of spacing on field-grown perennial crops. *HortSci.*, **22**: 904-907.

Armitage, A.M. 1992. Field studies of *Achillea* as a cut flower: longevity, spacing and cultivar response. *J. Amer. Soc. hort. Sci.*, **117**(1): 65-67.

Armitage, A.M. 1993. *Specialty Cut Flowers*. Timber Press, Portland, Oregon, USA.

Balbir, S., P. Kaur, N. Kumari and R. Kumar, 2012. Review on *Achillea millefolium*, important medicinal plant. *Intern. J. appl. Res. Nat. Prod.* (Spl. Issue), **1**: 176.

Beckett, K.A. 1985. The Concise Encyclopedia of Garden Plants, pp. 10-11. Orbis Publishing Ltd, London.

Brickell, C. 1994. *Th Royal Horticultural Society Gardeners' Encyclopedia of Plants and Flowers*, p. 428. Dorling Kindersley, London.

Ebrahimi, M., M. Farajpour and Beigmohammadi, 2012. Genetic relationships among yarrow based on random amplified polymorphic DNA markers. *Indian J. biotech. pharm. Res.*, **3**(4): 69-73(10.1016/j.bse.2012.02.017).)

Engle, B.E., A.C. Cameron and R.D. Heins, 1993. The effects of long days and short days on growth of several species of herbaceous perennials. *HortSci.*, **28**:573.

Evenor, D and M. Reuveni, 2004. Micropropagation of *Achillea filipendulina* cv. Parker. *Pl. Cell, Tissue & Org. Cult.*, **79**: 91-93.

Gabbrielli, G., G. Loggini, P. Cioni, B. Giannaccini and E. Mancuso, 1988. Activity of lavandino essential oil against non-tubercular opportunistic rapid grown mycobacteria. *Pharmacol. Res. Commun.*, **20** (5): 434-439.

Guo, Y.P., J. Saukel, R. Mittermayr and F. Ehrendorfer, 2005. AFLP analyses demonstrate genetic divergence, hybridization and multiple polyploidization in the evolution of *Achillea* (Asteraceae-Anthemideae). *New Phytologist*, **166**: 273-290.

Hay, R. and K.A. Beckett, 1971. *Reader's Digest Encyclopedia of Garden Plants and Flowers*, pp. 16-17. The Reader's Digest Association Ltd., London.

Marija, M., S. Dragana, G. Mihailo, D. Matilda, O.P. Dragica, D.B. Danijela and B. Milica, 2015. Micropropagation of *Achillea millefolium* L. on half-strength MS medium and direct rooting and acclimatization of microshoots in hydroponic culture. *Glasnik Sumarskog Faculteta*, **111**: 99-1112 (doi:10.2298/GSF1511099M).)

Nau, J. 1999. Cultivar testing-America's Trial Gardens. In: *Issues, Challenges and Opportunities* (ed. Neil O. Anderson), pp. 6-14. NRC Research Press, Springer Publications, Ottawa, Ontario, Canada.

Pizzetti, I. and H. Cocker, 1975. *Flowers ~ A Guide for your Garden*, pp. 12-18. Harry N. Abrams Inc., Pulishers, New York.

Radanovic, D., M. Jakovljevic and S.A. Mladenovic, 2000. Content of potentially toxic trace elements in some medicinal plants. In: *Proc. of the First Conf. on Medic. & Aromatic Plants of SE European Countries* (ed. Josif Pancic). VI Meeting: "Day of Medicinal Plants - 2000", IMPR, Belgrade, pp. 481–488.

Shatnawi, M. 2012. Multiplication and cryopreservation of yarrow (*Achillea millefolium* L., Asteraceae). *J. agric. Sci. & Techn.*, **15**(1): 163-173.

Sheidai, M., N. Azanei and F. Attar, 2009. New chromosome number and unreduced pollen formation in *Achillea* species (Asteraceae). *Acta Biol. Szegediensis*, **53**: 39-43.

Small, E. and P.M. Catling, 1999. *Achillea millefoilum*. In. *Canadian Medicinal Crops* (pp. 6–14).NRC Research Press, Ottawa, Ontario, Canada.

Stevens, R, K.R. Jorgensen, S.A. Young and S. B. Monsen, 1996. *Forb and Shrub Seed Production Guide for Utah*. Utah State University Extension Service, Logan, Utah, USA.

Still, L. and M. Steven, 1994. *Manual of Herbaceous Ornamental Plants* (4th ed.). Stipes Publishing Company, Champaign, Illinois, USA.

Thieret, J.W. 2001. *Yarrow Encyclopedia Americana*. Grolier International Inc., Danbury, Connecticut, USA.

Turker, A.U., B. Yucesan and E. Gurel, 2009. *In vitro* regeneration of *Achillea millefolium*. L from shoot-tips and root segments of seedlings. *J. Pl. Biochem. & Biotechn.*, **18**(1): 65-69.

Zhang, D., A.M. Armitage, J.M. Affolter and M.A. Dirr, 1996. Environmental control of flowering and growth of *Achillea millefolium* L. 'Summer Pastels'. *HortSci.*, **31**(3): 364-365.

2

Alstroemeria (Family: Alstroemeriaceae)

R.L. Misra and Sanyat Misra

[**Common names**: Brazilian parrot lily, Chilian lily (*A. chiliensis*), Golden Peruvian lily/Peruvian lily (*A. aurea* or *A. aurantiaca*), Herb lily/Inca lily (*A. pelegrina*), Lily of Peru, Lily of the Incas, Parrot lily (*A. pulchella*), St. Martin's flower (*A. ligtu*)]

Introduction and Origin

Alstroemeria was named so by Linnaeus in honour of his friend cum pupil Baron Claus von Alstroemer (1736-94), a Swedish naturalist. About 50 species of *Alstroemeria* are known, most being native to Chile, Brazil, Bolivia, Ecuador, Peru, Paraguay, Venezuela and Argentina, Chile being the centre of distribution. Species are found from dry and warm desert-like regions to the moist and cool elevations and from 26° to 40° south latitude and this type of diversity will certainly contribute new traits in the future hybridization programme.

Alstroemeria is of recent introduction to the floriculture world scenario and has most recently entered and has started capturing Indian markets. It produces beautifully marked flowers, the hardier species being suited to the herbaceous border and the more tender ones for pot cultivation in the cool greenhouse or conservatory. This can ideally be planted on slopes where moisture during summer is not a limitation provided the area is not waterlogged and there if once established will continue to multiply. The colour of the flower is white to dark yellow, pink, violet, purple and red. Yellow throat and black dots at the base of the throat and petals make them very attractive. Under protected condition they are grown for cut flowers. It is a crop of the temperate and sub-temperate regions, particularly the higher reaches of the Western Ghats, the Nilgiri hills (Ootacamund and Kodaikanal), higher altitudes of North-East Hill regions, Darjeeling hills of West Bengal, the higher reaches in Orissa, *i.e.* Koraput regions, Mt. Abu in Rajasthan, sub-temperate regions of Uttarakhand, and from 1,000-2,000 metres height of Himachal Pradesh and Jammu & Kashmir.

Botany

It is an herbaceous perennial with clustered, creeping and sympodial rhizomes having white fleshy, very brittle and fibrous roots, dried roots and rhizomes of *A. haemantha* yielding a farinaceous flour (farina), the shoots either vegetative or reproductive based on the environmental conditions, the leaves are entire, thin, linear with parallel veins, lanceolate, glaucous, twisted close to stem and alternate or scattered along the stem, shoots a whorled cymose inflorescence, each cyme sympodially branched with up to 4 florets per cyme, opening one after the other, flowers zygomorphic, non-tubular, brightly coloured and in a simple or compound terminal umbel (corymbose) though a few bear solitary flowers, tepals 6 and free, 3 inner (lower) ones narrower and often blotched, streaked or suffused contrasting with the base colour and longer than the outer three, stamens 6 and attached to the base of tepals, anthers basifixed, pollen dehiscing before stigma receptivity (a dichogamous stage), style slender with downward curving, stigma trifid, ovary inferior and 3-chambered, and fruit a capsule with many seeds. It possesses very large chromosomes and the basic chromosome number is n = 8. Tsuchiya and Hangi (1987) made cytological investigations of nine species and 25 alstroemeria cultivars, and reported that all the species had 2n = 2x = 16 chromosomes with different karyotypes, four cultivars including 'Walter Flaming' showed 2n = 2x = 16 chromsomes, 12 with 2n = 3x = 24, cv. 'Orange Beauty' with 2n = 3x+1 = 25, and eight were tetraploid with 4x-1 = 31, 4x = 32, 4x+1 = 33, & the six plants showed 2n = 32. They recorded abnormal meiotic behaviour of chromosomes in all the cultivars including the diploids. Tsuchiya *et al.* (1987) made cytological studies in nine new alstroemeria cultivars and two plants designated as *A. ligtu* hybrids and reported 'Orchid' and 'Canaria' as diploid with 2n = 2x = 16; 'Campfire', 'Carmen', 'Marina', 'Pink Perfection', 'Red surprise' and 'Regina' as triploid with 2n = 3x = 24; 'Rosario' as hypertetraploid with 2n = 4x+1 = 33; and *A. ligtu* hybrids as diploids with 2n = 2x = 16.

Mikkelsen and Mikkelsen (1989) reported that in Denmark, a 35-year old lady working as a florist developed dermatitis after becoming sensitized to *Alstroemeria*. Slob *et al.* (1975) reported that *Alstroemeria* contains tuliposide A (a known allergin). Saito *et al.* (1985) reported anthocyanidin glycosides 6-hydroxycyanidin 3-rutinoside as the major pigment and 6-hydroxycyanidin 3-glucoside as the minor pigment occurring in the red flowers of 'King Cardinal', 'Red Sunrise' and 'Red Sunset', and the minor component in the flowers of 'Red Star'. In a further experiment, Saito *et al.* (1988) reported that 6-hydroxycyanidin 3-rutinoside was abundant in orange, pink and red flowers of *A. aurantiaca, A. pulchella* (*A. versicolor*), *A. ligtu* var. *angustifolia* x *A. haemantha* and 11 cultivars; a novel anthocyanidin glycoside 6-hydroxydelphinidin 3-rutinoside was found only in five cultivars bearing pink-purple flowers. They stated that cyanidine 3-rutinoside was observed in all coloured flowers while delphinidin 3-rutinoside was found in 10 cultivars and one species (*A. violacea*) bearing pink and pink-purple flowers.

Classification, Species and Varieties

There are about 50 known species of *Alstroemeria*. Most of the *Alstroemeria* species are native to Chile, Brazil, Bolivia, Peru, Paraguay, Venezuela and Argentina. Assis (2003) reported two new species of *Alstroemeria* from Brazil and these are ***A. amabilis*** from Paranta and Santa Catarina and ***A. capixaba*** from Expírito Santo. *Alstroemeria* can be classified into two groups: i) '**Orchid types**', which are developed by using two Chilean species, one of them being *Alstroemneria aurea*. These are characterized by tall growing habits (2.5-3.0 metres), producing a large number of flowers in a short duration (3-5 months in spring) after passing through a 10-15°C winter growing temperature but no flowering in summer (when these remain vegetative and are divided for next spring flowering) and other months and have little or no photoperiodic responsiveness; and ii) '**Butterfly types**', which are hybrids of Chilean and Brazilian species, and are characterized by shorter growing habits (good for pot plant production and also for cut flowers) with larger and more open flowers and a different foliage arrangement, where flowering is induced and maintained at 13-16° C temperature and long photoperiods for 9-12 months each year depending upon the cultivar and environmental conditions. These can be planted in any month, but summer planting is best, and when the plants are given long photoperiods, high light intensities and cool temperatures, these flower within 3-4 months. Garton and Peiser (1989) advocated breeding objectives of alstroemeria for (a) dwarf habit with short flower stems, (b) many flowers per stem, (c) many flowering stems per plant, (d) high floral longevity, (e) bright attractive colours, (f) year-round flowering, (g) predictable flowering and (h) sterility or reduced seed set, and Baeure and Bakken (1997) stressed the development of cultivars which flower during high temperatures and maintain flower production during winter when there is low light stress. Moreover, there should also be genetically dwarf cultivars most suitable for pot cultivation (Dole and Wilkins, 1999).

The most common **species** are: ***Alstroemeria aurea*** (syn. *A. aurantiaca*), native to Chile, with yellow, orange or deep orange-red flowers, growing to a height of 90 cm and has cultivars like 'Angustifolia', 'Dover Orange', 'Flava', 'Lutea', 'Major; 'Moerheim's Orange', 'Rubra' and 'Splendors'; ***A. brasiliensis***, native to Brazil, with reddish-yellow flowers having brown spots, and height 90 cm; ***A. caryophyllaea***, native to Brazil, with sepals and lower petals rose while upper petal white in the centre, fragrant and grows to a height of 45 cm; ***A. chilensis*** with pink to blood-red flowers, and height 60-90 cm; ***A. haemantha*** (syn. *A. simsii*), native to Chile, having red flowers with yellow streaks, and 75-90 cm in height, and the 'Roseae' is a listed cultivar; ***A. hookeri***, native to Andes, with pink flowers, 15 cm in height, and summer and early autumn flowering; ***A. ligtu***, native to Chile, itself a good cut flower species, flower colour pastel in shades of light yellow, pink, red, ivory, and peach and grows to a 40 cm height; ***A. pelegrina***, native to Chile, bears largest flowers with rose to lilac to yellow and outer petals have large green point while upper inner petals have yellow centre and maroon flecks throughout, and height 30-35 cm; ***A. psittacina*** (syn. *A. pulchella*), native to Brazil, has dark red flowers, tipped green and spotted brown, and height 90 cm; ***A. pulchra*** (syn. *A. tricolor*), native to Chile, has white to light greyish pink or soft lilac flowers with yellow, red or purple blotches, and height 45 cm; ***A. versicolor***, native to Chile, which resembles ***A. psittacina*** but smaller, flowers yellow or orange, flecked purple and height 20-60 cm; and ***A. violacea***, native to Chile, with large lavender flowers, two upper petals with yellow centre, flecked dark purple throughout and the height is up to 70 cm. The other **varieties** are:

- ☆ **White**-Alba, Alaska, Amanda, Apollo, Bianca, Friendship, Ice Cream, Labelle, Casablanca, Maria, Mona Lisa, Nevada, Orchid Flower, Paloma (*A. aurantiaca* cv. Lutea x cv. Pulchra), Snow Flake, Snow Queen, White Dream, White Swan, etc.
- ☆ **Yellow**-Barcelona, Bolivia, Butterscotch, Canaria, Eleanor, La Paz (an induced mutant of Rio), Minas, Moonshine, Rio, Walter Fleming (Orchid), Yellow Crown, Yellow King, Yellow Libelle, etc.
- ☆ **Orange/salmon**-Afterglow, Bogota, Comet, Delight, Diana, Dover Orange, Flaming Star, Helios, Moerheim Orange, Monarch, Orange, Orange Beauty, Orange King, Valiant, Victoria, etc.
- ☆ **Pink**-Ballerina, Cinderella, Carola, Elisabeth, Fanfare, Fiona, Leticia, Lorena, Marie, Parigo Charm, Pink Jewel, Olivia, Pink Triumph, Rose Marie, Rosita, Samora, Serena, Sonata, Valparaisa, Veronica, etc.
- ☆ **Red**-Carmen, King Cardinal, Liliana, Marina, Margaret, Parade, Rosea, Sangria, Tiara, Tucumano, Vanitas, Wilhelmina, etc.
- ☆ **Lavender**-Alicia, etc.

☆ **Purple/lilac**-Azula, Barbara, Bingo, Bonito, Brocado, Chimbotina, Ice Cream, Jubilee, Jupiter, La Paza, Quitona, etc.

Lin and Molnar (1983) in Canada developed 'Alsaan' (orange) by crossing *A. aurantiaca* with cv. 'Orange Beauty'. Konst Alstroemeria (1990) developed the vars 'Serena' (orange-white) and 'Wilhemina' (grey-purple) by crossing *A. aurantiaca* with a butterfly-type alstroemeria in Holland. Broertjes and Verboom (1974) subjected the actively growing young rhizomes having multicellular apices to X-irradiation and reported optimum dose for diploid cultivars as 400 rads while for triploids as 500-600 rads, without any sign of chimaerism. They introduced five mutants into commerce.

Milde (1981a) in FRG when grew six *A. ligtu* cultivars in greenhouse from October 1977 to July 1979, observed each cultivar flowering during this period thrice for roughly the same length of time, second flowering occurring about $3^1/_2$ months after the first flowering was over though 'Orchid' and 'Yellow Tiger' flowered 4-5 weeks before others at first and third blooming. Third blooming produced highest numbers of stems while second the lowest. 'Orange Beauty' and 'Orchid' produced 477 and 628 number of stems/m^2, respectively, and all the cultivars except 'Orange Beauty' produced over 40 per cent top quality stems. In a further glasshouse trial, Milde (1981b) tried 10 *A. ligtu* hybrids from September 1979 to December 1980 which flowered from the February end or mid-May, and again four months later. The varieties flowering earliest were 'Orchid', 'Canaria', 'Zebra' and 'Red Sunset', producing highest number of blooms were 'Orchid' and 'King Cardinal' (431.5 and 420.2 stems/m^2, respectively), and producing highest percentage of best quality blooms were 'Zebra' and 'Canaria'. Milde (1979) tried six cultivars in greenhouse during three flowering periods, *i.e.* April-June, 1976, September-December 1976 and March-July 1977, and reported that 'Orchid', followed by 'Stavita Canaria' gave earliest and highest yields though 'Starosa' and 'Beauty' did not produce an autumn crop. Kwiatkowska and Brzozowska (1980) evaluated four culivars, *viz.* 'Canaria', 'Orchid', 'Regina' and 'Harmony' in the glasshouse of Poland and reported 'Harmony' followed by 'Regina' producing highest flower yield in the first year though quality-wise the varieties were 'Canaria' and 'Orchid', production of inflorescence in second year was alright but in the third year it declined drastically. In Holland, Leliveld (1973) reported that early planting in the glasshouse in October increases the inflorescence yield with about three weeks advance flowering than December and January plantings but such crops require supplementary lighting during January-February. Baersch (1983) in Germany tried five cultivars (Canaria, Pink Triumph, Orange Beauty, Orchid Flower and Red Sunset) by planting in September at 23,725 plants/ha and started harvesting the flowers from February to December. Jesand and Waithaka (1988) in Kenya reported that propagated rhizomes of the cvs 'Carmen', 'Marina' and 'Pink Perfection' sprouted 9-10 days after planting and started flowering after 132, 245 and 215 days after planting, respectively, highest shoots appearing in 'Carmen' but with smallest floral stems, *i.e.* 47.02 cm, followed by 'Pink Perfection' with tallest, *i.e.* 94.32 cm, and 'Marina' with 82.93 cm stems. In Holland, Nijssen and Hoogeveen (1990) while working with six new alstroemeria varieties (Apollo, Bianca, Bingo, Cavalier, Tiara and Westland) in the glasshouse reported that in a crop duration from September 8, 1987 to April 1989, heaviest total crops in 'Bingo' (322 stems/m^2) and 'Cavalier' (313 stems/m^2) were recorded though winter yield (November 7, 1988 to February 1989) was very poor. Healy and Lang (1984) in USA tried single rhizomes of nine cultivars into 10-cm pots in October at 16°C continuous temperature, and then the pots were transferred to raised gravel or soil benches at 6 plants/7^{ft-1} at 16°C day and 13°C night temperatures where soil benches were found better.

Propagation

Alstroemeria plants consist of a sympodial, fleshy, multistemmed rhizome growing subterranean from which shoots and roots (fibrous) arise, and with the development of the plant this becomes thickened storage roots storing a tremendous amount of starch (Cox and MacMasters, 1947) and this is called as 'radices medullosae'. These are white, fleshy, very brittle and densely haired below. As per the environmental conditions, the shoot can be reproductive or vegetative. Aerial shoots develop from a sympodially branching rhizome, and the new aerial shoots can also arise from axillary buds at the base of these shoots and these buds can develop into rhizomes (Vonk Noordegraaf, 1975; Healy and Wilkins, 1985). The terminal growing point of the rhizome has natural tendency growing downward by a few degrees which with a pace of time results into soil exhaustion, hence, these rhizomes every 2-4 years are replanted in fortified soil (Wilkins and Heins, 1976). The **rhizomes** are divided continuously at every 10-12 weeks which cause them to remain vegetative to make the division efficient or these should be divided during dormancy, *i.e.* when they have stopped flowering. The dormant rootstocks should be removed gently and with care as this is a complex mass of storage roots, fibrous roots and rhizomes. The division is made with as much storage and supporting material as possible, kept moist and replanted immediately. For the first nine months the growth is slow but afterwards these may be left in place for several years. While making the division care should be taken that each division contains at least three vegetative shoots. Until the emergence of the new shoot, the growth is subterranean and aerial shoots emerge from a sympodially branching rhizome as well as also from axillary buds located at the base of these shoots, and these buds can develop into rhizomes. Through division, the rate of rhizome multiplication is not rapid. Lisiecka and Szczepaniak (1989) in Poland selected 2-year old plants of nine alstroemeria cultivars so that each plant part possesses a 3-5 cm rhizome section having a bud and at least three storage roots and grew them in pots from August 5 to October 24. They reported coefficient of propagation depending on the variety and positively correlating with the weight of the underground part. Powell and Bunt (1986) in England tried for propagation two cultivars every two months throughout the year at a minimum temperature of 9°C and reported that apices of aerial parts of the plants propagated in January and March showed signs of flower initiation within

3-4 months. The main rhizome in 'Red Sunset' continued floral-stem production throughout the year but in 'Campfire' the flower initiation was found delayed during autumn and about two shoots on the main rhizome of each plant remained vegetative until their senescence. In both the cultivars, those propagated in July, September and November, the shoots remained vegetative until spring when flower initiation occurred in both new and previously vegetative shoots.

Micropropagation is adopted for commercial cultivation through which 4-7 fold increase in plants can occur every four weeks which will also have an additional advantage of getting material free from viruses and other diseases. For this modified MS medium, 30 g sucrose, 1-5 mg benzylaminopurine, 0.01 mg NAA and 1.2 g Gel Gro per litre of medium are added. High levels of cytokinin which hastens vegetative production can also cause *in vitro* mutation. Ziv *et al.* (1973) cultured inflorescence explants taken from 1-2 mm below the apex on White's medium as well as on MS medium which produced roots and buds directly from the tissue and not from the callus. They found a higher ratio of auxin to cytokinin resulting in root regeneration while the reverse ratio promoted bud differentiation. Through bud subculturing, plantlets were obtained on a low sucrose medium supplemented with IAA acid but without kinetin. Gabryszewska and Hempel (1985) obtained stimulated axillary rhizome initial growth and formation of erect shoots in the cv. 'Zebra' with cytokinin such as BA at 0.5-4.0 mg/l which inhibited rooting while auxins like NAA at 1.0-16.0 mg/l stimulated rooting. Lin and Monette (1987) got regeneration through intact (not longitudinally sliced) rhizome tips cultured on solid and liquid MS salt formulations initially at 8°C and further establishment of the plantlets to a tune of 90 per cent. Pierik *et al.* (1988) cultured terminal and lateral tips from fleshy rhizomes of cv. 'Toledo' on MS macro- and micro-salts at full strength (except Fe), NaFeEDTA (25 mg/l), sucrose (3 per cent), BA (2-4 mg/l), vitamin B_1 (0.4 mg/l) and Difco Bacto-agar (0.7 per cent) at 21°C and 8 h fluorescent light (16 h darkness) for forming new rhizomes but for rooting the medium used had increased level of sucrose (5 per cent), no BA and 0.5 per cent mg/l NAA. They stated that optimum rhizome multiplication required 3-week cycles of subculturing, better being in liquid medium.

Propagation through **seeds** causes genetic variability hence growers should not opt for this system. However, one year old seeds will germinate within 8-10 weeks if are subjected to four week's moist and warm (18-25° C) environment, followed by four week's of moist and cool (7° C) conditions. Fresh seeds will germinate *in vitro* at 18°C within 10-14 days. Thompson and Newman (1979) reported that seeed germination in alstroemeria is very erratic if sown without any treatment. However, a constant and fluctuating temperature between 6 and 31°C improves germination. They stated that a conditioning treatment by sowing the seeds initially for 4-5 weeks at 25°C then subjecting to 10°C gives 40 per cent germination, chipping the testa with a sharp knife to expose the embryo gives 20 per cent germination, while combination of these both gave 98 per cent germination when first sown at 25°C for one month then chipped and re-sown at 6-11°C where these germinated within 2-3 weeks. Thompson *et al.* (1979) reported that seeds of *A. aurantiaca, A. gayana, A. haemantha, A. pulchra, A. revoluta, A. spathulata, A. versicolor* and *A. violacea* germinated well when a patch of seed coat and the underlying endosperm was removed and sown at 26-31°C where these germinated after several weeks though without chipping only a few seeds germinated at 11°C but after a considerable time. However, the conditioning treatment at 26°C for one month, followed by seed chipping and then exposing at 6°C caused seed germination within 10-20 days. King and Bridgen (1990) gave pre-sowing treatments with various temperature regimes to the seeds of certain *Alstroemeria* hybrids 3-12 months after harvest and five selections from 20 interspecific F_1 hybrids 3-7 or 8-12 weeks after harvest. Seeds were chipped of the seed coat above the embryo, were given abraision treatments for whole of the seed coat, soaked in distilled water, GA_3 (0.029, 0.29 or 2.9 mM) or KNO_3 (0.5 or 1.0 M) and then were evaluated under three environmental regimes (8 weeks at a constant warm temperature of 18-25°C, 4 weeks at 18-25°C followed by 4 weeks at 7°C warm-cold temperature or 4 weeks at 7°C followed by 4 weeks at 18-25°C cold-warm temperature). They recorded water soaking or soaking in GA_3 in 0.29 or 2.9 mM giving high percentage of seed germination, *vis-à-vis* warm-cold environment, however, cold-warm environment took longest time for germination though there had been genotypic variation and age of the seed.

Soils, Preparation of Land, Planting and Weed Control

Alstroemeria can be grown in a variety of **soils** which is loose, well-drained, pH 5.5 to 6.5, and rich in organic matter, *i.e.* to a tune of 3 per cent. A sandy loam soil with adequate humus is ideal for its growing. Bridgen and Soriano (1986) tried 1 part soil : 1 part peatmoss : 1 part perlite or 80 per cent topsoil and 20 per cent peatmoss both at pH 5.8 ± 0.2 using dolomitic limestone, and recorded the former medium giving best results. It likes cool sub-tropical climatic conditions (10-12 °C in winter and 14-16 °C later in summer, *i.e.* sometimes in May-June; or a range of 18/13 °C day/night temperatures) but where summers are not to the extremes, as a very hot persisting condition coupled with dry spell kills it. For their growing, alstroemerias require full light (15 W/m^2 in winter to 60 W/m^2 during vigorous growth), while plants are shaded only to maintain a temperature range of 13-17 °C, which is optimum for their growing, and an RH of 80-85 per cent. Bridgen (1993) has given a full account of its cultivation. CO_2 concentration around 0.12 per cent in the glasshouse has been found increasing inflorescence yield and its effect was quite pronounced at higher temperature (Verboom and Staaveren, 1978). Blomme and Dambre (1984) stated that planting alstroemerias during autumn in Belgium in containers under glasshouse and raising the temperature to 13-14 °C for one month from November 15 versus a constant temperature of 5 °C increased the number of young shoots and weight of rhizomes, and advanced flowering by 4-6 weeks. Since it perennializes up to three years after planting once, it would be better to have a deep (30-40cm) and rich soil for its growing so that it may sustain producing quality blooms season after season. For **preparation of land**, the soil should be worked into 40 cm depth 2-3 times with cultivators followed by levelling each time, the rootstocks of perennial weeds including the rhizomes of the nut-grass should thoroughly be taken out with fork or through hands, and then finally levelled and beds are prepared. Milde (1985) tried for 23 months cropping through planting of seedlings of cvs 'Canaria', 'King Cardinal' and 'Orchid' on November 13 in FRG into a substrate mixture of 15 litre peat + 150 g Oscornahum + 80 g Oscorna Animalin at minimum winter temperature of 6 °C and summer temperature of 15 °C and obtained flowering of cultivars twice each year between April and mid-June and again between August and November, highest stem yield being in the first flowering of the second year, and the yields over four harvests were 286 stems in 'Canaria', 369 stems in 'King Cardinal' and 272 stems/m^2 in 'Orchid'. It would be better if the bed size is sufficiently lengthy but the width may be restricted to two metres to facilitate proper field operations through the clearance left in between the beds. The soil should have sufficient moisture at the time of planting, and the **planting** is done 10-15 cm deep in sandy-loam soil but shallower in heavy soils with a precaution that growing points should not be, in any case, more than 10 cm deep or appropriately to 5-7 cm only, and 45-60 cm apart (2-4 plants/m^2), depending on the cultivars and the cropping years. The rhizomes are divided and planted at the time when these have finished the flowering, and this may either be after spring or in November looking into the type of cultivars being grown, or at any time as per convenience of the grower but avoiding the peak flowering period. At this time support lines of galvanized or plastic wire are also set up with 20 × 15 or 20 × 20 openings, first two layers at every 20-30 cm above the soil line and subsequent at 50 cm in height. For small plantings, bamboo or *Salix* sticks or string support may be opted.

Pinching and **disbudding** are normally not required but since a majority of stems that emerge are vegetative so these are removed regularly to maintain production and to stimulate axillary shoot elongation. Older shoots no longer unfolding leaves as well as weak and short-stemmed shoots not dark green are also thinned out (Wilkins and Heins, 1976). Depending on the cultivar, the cycles of tall shoots (up to 1.5 m) and thin and short shoots (too short for commercial sale) develop. The tall ones should be cut to the marketing length and remaining stems are left intact for carbohydrate synthesis through photosynthesis to help prevent next cycle of short stems. During the next routine thinning (pulling of the stems gently from the rhizomes which have a natural abscission layer at the point of shoot attachment and not by cutting) or when leaf senescence has commenced the flower heads of the short stems are sacrificed, again leaving the shoot for carbohydrate synthesis (Wilkins and Heins, 1976; Healy and Wilkins, 1985). Thinning stimulates rhizome branching. In 'orchid-types', flowering is over by the end of summer, so the plants are headed back in July-September, and in winter the thinning may continue by removing 15-25 per cent of the vegetative shoots until commencement of flowering. 'Butterfly types' are thinned out each month to a tune of 15-25 per cent during autumn and winter months, till commencement of flowering. In spring or summer plantings, some 10 stems per plant are pinched to build up the crop, and later in the winter these pinched stems are pulled out. Planting in spring and summer produces flowers within 10-12 weeks.

For **potted plants,** rhizomes are planted shallower with growing points 2.5-3.0 cm from the surface which will permit more branching as compared to deep planting. For 6-8 weeks of planting, the newly planted rhizomes should not receive long days of 13-16 hours so that sufficient root system is developed before flower induction, and when planting liners directly in pots, the vegetative shoots arising from the liners should not be cut back as this will delay flowering by 2-3 weeks. To get quick flower production, liners of rhizomes are planted directly in their final pots. When *Alstroemeria* is forced as potted plant, the height control becomes a major concern. Though many of the retardants fail to express any good effect but daminozide at 200-2,500 ppm exerts favourable effects. For pots, in fact, only genetically dwarf cultivars should be planted where there may not be any need of applying retardants. The standard pot size is 15-18 cm so that developing roots may not face any problem. These plants flower within 90-120 days. In potted plants, some 15-20 per cent of vegetative shoots should be removed during winter till commencement of flowering (Bridgen, 1997) and afterwards only damaged and dead ones are removed (pulled and not cut) every 2-3 weeks, however, the operation should be gentle not to pull the plant out of the pot, especially in case of young plants.

In small holdings, **weeding** is carried out manually but in large holdings, it is done through an effective weedicide. Decamba as a pre-emergence is quite effective weed-killer with no any adverse effect.

Irrigation, Manures and Fertilizers

Alstroemeria does not like wet feet. Too much water encourages root rot, especially when the plants are initially planted. However, when plants are initially being established, overwatering should be avoided until emergence of the new shoots (Wilkins and Heins, 1976; Verboom, 1980). In poorly drained soils this crop does not come up well though it requires moist and cool situation. It would be better if this crop is being provided with peripheral system of irrigation with spray nozzles pointed into the centre of the beds or with trickle tubes down the centre of the bed. Frequent but semi-heavy irrigation is required for *Alstroemeria* crop. It is sensitive to salts. Water EC should not be more than 1.3 dS/m (Sonneveld, 1988) and soluble salt (especially Na and Cl) contents not more than 1.2-1.5 mmho/cm.

Farmyard manure or compost at the rate of 50-60 tonnes/ha should be incorporated in the soil thoroughly at the time of land preparation. Since it is grown necessarily as perennial cropping, hence fertigation will be more effective system of fertilization, but before applying any fertilizer to the crop, the optimum quantity of each and every nutrient should be calculated on the basis of soil and leaf analysis. In usual cases, 200-400 ppm N through calcium nitrate and potassium nitrate should be applied. Nitrogen in nitrate form (200-280 ppm per week) has been found to increase flower production and hastening flowering. Bridgen and Soriano (1986) recorded soil, peatmoss and perlite medium at 1:1:1 quite good over 80 per cent topsoil + 20 per cent peatmoss for alstroemeria cultivation and advocated weekly application of 450 ppm N. Milde (1983) reported that alstroemerias are salt-sensitive and suggested application of K at flowering for improving the vase life. Meij (1976) recommended top dressing of K only in the second year for better flower yield though N decreased the yield. Blomme and Dambre (1979) recommended 25 g K_2O/m^2 as 50 per cent potassium phosphate as basal as well as top-dressing. He also stated that Plantosan 4D (20:10:16) @ 30 g/m^2 and four top-dressings per year, each at 60 g/m^2 provided highest yield of flowers. Bik and van der Berg (1982) during winter at media pH of 5.5 recommended N and K_2O in the ratio of 1:1 or 1:1.4 weekly at 300 mg/12.5 litre plastic pot filled with sphagnum peatmoss, and recorded good flower production. Lisiecka and Szczepaniak (1983) in a 3-year trials where plants of 'Regina' were grown singly in 10-litre containers on a peat substrate, reported highest flower yields at NPK levels of 300:150-300:600 mg/l in the first year and at 300-450:300:900 mg/l in the second year. Hermann (1986) in plastic house trials used 15N:5P:20K:2Ca (Nitrophoska perfect) or 15N:9P:15K:2Ca (Nitrophoska permanent, slow release fertilizer) at 30 or 40 g N/m^2, and reported that though there was no variation in flower yield at any levels of fertilizers in the cv. 'Vasa Rosita' which produced 510-531 stems/m^2 in 2-year cropping while cv. 'Vasa Canaria' gave highest cut flower yield (731 stems/m^2 over 2 years) with the higher rate of Nitrophoska perfect.

Growth, Development, Flowering and Flower Forcing

Powell and Bunt (1984) stated the phenomenon of growth, floral initiation and further development a complex one, due to varietal differencess. The flowering in *Alstroemeria* is controlled by the **temperature** of the rhizome getting from the medium or soil (Vonk Noordengraaf, 1975; Heins and Wilkins, 1979; Wilkins *et al.*, 1980; Healy and Wilkins, 1981) and if the temperature of the medium is maintained at 16° C the flowering duration is extended, irrespective of the temperature in the atmosphere. For forcing, they are kept for six weeks at 5° C and then continuously at 13° C (Healy and Wilkins, 1981, 1982b) or throughout at 13° C over a longer period (16 weeks) though increase in the duration of 5°C temperature than usual decreases number of flowering shoots. Also forcing from 9-13° C and then continuously up to 17° C produces flowering shoots while at 21° C the rhizomes are devernalized and only vegetative shoots are produced. Greenhouse temperature is maintained at 16-18° C during the day time and 10-13° C during nights for the growth of the rhizomes and for production of the flowers (Wilkins and Heins, 1976; Healy and Wilkins, 1981; Bridgen and Bartok, 1990) though response to light and temperature, as per study conducted in Belgium, *i.e.* NW Europe, varies greatly due to cultivars (Labeke and Dambre, 1993) as has been observed in cvs 'King Kardinal' and 'Amanda' where flowering was higher at 14 °C media temperature, in 'Helios' it was more at 18 °C whereas in 'Cinderella' the flower production was higher at 22 °C (Baeure and Bakken, 1997). In Columbia the alstroemerias are produced essentially on the equator with 12 h ± 30 minutes natural photoperiods (Vonk Noordegraaf, 1975; Wilkins *et al.*, 1980; Healy and Wilkins, 1985). During summer, mulches in the greenhouse are used with intermittent misting so that through evaporation the medium may remain cool. Uematsu (1990) stated that up to December in Japan in Ligtu hybrids, the elongation of blind shoots and formation of flowering shoots and rhizomes depended on nutrient reserves in the tuberous roots used as mother plants, in January the elongation of blind shoots and growth of feeder roots depended on the photosynthesis of rosette shoots, and from February onwards the development of flowering shoots and new tuberous roots depended on the photosynthesis of blind shoots. He described that a low temperature treatment was essential for elongation of flowering shoots, and the low temperatures occurring naturally from late November to mid-December were sufficient.

Spieser (1987) in the wooden greenhouse with double polyethylene air-inflated cladding circulated the cold water at a rate of approximately 3 litres/second during spring and summer through the polyethylene pipes buried 15 cm below the soil in Canada to maintain the soil temperature from 12-15 °C and reported successful cropping. Hurka (1989) reported that in cv. 'Regina', a soil temperature of 11-15 °C produced more flowering stems and fewer non-flowering stems than when grown at a soil temperature of more than 18 °C though air temperature did not show any direct effect on flowering. There were varietial differences as flowering stem production ranged from 80 to 400/m^2, 'Jubilee' producing 80-100 per cent of the flowering stems than 50-70 per cent of 'Appelbloesem'. Keil-Gunderson *et al.* (1989) tried cvs 'Atlas' and 'Monika' in cooled (12 to 16 °C) and non-cooled (15 to 18 °C) gravel substrate at four temperature-controlled greenhouse compartments having mean day/night air temperatures of 20/14, 20/13, 22/13 and 24/14 °C and reported 1.6 times greater stem yield in 'Atlas' than in 'Monika'. They summed out that it was 20 °C average daytime air temperature and 12 to 14 °C the root substrate temperatures which favoured year-round production. Bridgen and Bartok (1990) recorded two months longer flowering with higher yield in alstroemeria by placing of cooling polythene tubes 10 cm below the surface and circulation of well water at 10-15 ° C. Healy and Wilkins (1982a) stated that number of days to flowering is inversely related to the number of weeks (up to 8, usually 6 weeks) the plants are vernalized at 5°C. These are vernalized further at 13 °C for forcing. In a further experiment, when Healy and Wilkins (1982b) subjected the plants at 5 °C for 6-8 weeks, and then at 18 °C night temperatures which though hastened the flowering but with reduced production and quality of flowers as compared to usual 13 °C. They found decreased stem production by increased duration of 5 °C temperature than usual six weeks. Schubert (1989) in Germany tried 'Appelbloesem', 'King Kardinal', 'Manon', 'Mona Lisa', 'Ramona', 'Red Bird' 'Samora' and 'Yellow King' by planting in late September at a spacing of 0.60 × 0.55 m in 2 parts Floraton 1 + 1 part bark humus (low-nutrient substrate) or 2 parts steam-sterilized compost + 1 part limed peat containing 10 per cent Styromul (expanded polystyrene) in tiered benches under plastic at 16 °C for three years, and reported better results under low-nutrient substrate, maximum production being under 'Appelbloesem', *i.e.* 400

stems/m²/year with 170 stems of class I size. Wiel (1989) mulched and/or cold-treated from June to October in Holland through subsoil circulation of cold water in polyethylene tubes in the plantings of alstroemeria cvs 'Flamengo' and 'Jubilee', and recorded 16-30 per cent more stems due to extension of flowering time in certain varieties under combined treatment of mulching and cold water circulation.

Incandescent **lighting** for natural day + day continuation (or night interruption, *i.e.* 10 minutes on and 20 minutes off being equally effective) for 14-16 hours induces earlier flowering but this treatment from April to September results in reduced flower production (Verboom, 1980) in Holland and during this period the required light intensity is at least 3 fc through incandescent lights (Vonk Noordegraaf, 1975; Heins and Wilkins, 1979; Healy and Wilkins, 1981, 1982b, 1985). Short days (8 h) delay the flowering with reduced flower production (Healy *et al.*, 1982). However, photosynthetic lighting, in Minnesota which is at 45° North latitude, has been recorded to be ineffective as here during winters, it is sufficiently cold but the days are bright and sunny though short, in the Pacific NW (USA) at 50° North latitude only high intensity lighting is effective as this region has poor winter light, while during winters in Norway, *i.e.* Northern Europe, supplemental lighting of 900 fc from high-pressure sodium lamps is used and this lighting is necessary only for winter production of *Alstroemeria*. Hurka (1989) stated that a photoperiod of at least 13 h was required for flowering, and supplementary lighting (100-200 lx) in spring in Germany hastened flowering by two months, *i.e.* from May-June to March-April though stems falling under various quality classes were affected differently in the three cultivars used (Applebloesem, Eleanor and Jubilee) due to supplementary lighting. Molnar (1975) in Canada planted in November the young plants of cvs 'Regina' and 'Orchid' under glasshouse with 10- or 16-h days for two successive years and recorded heavier yield with 20 per cent taller plants under 16-h day, 'Regina' outyielding 40 per cent more than 'Orchid'. In Canada, he recorded only one flowering period in a year, *i.e.* from February to August end in 'Regina' while from February to December in 'Orchid'. Waithaka and Chepkairor (1988) while studying with three *Alstroemeria* cvs 'Carmen', 'Marina' and 'Pink Perfection' in Kenya, stated that alstroemerias start flowering at 130 to 245 days of planting. In England, Powell and Bunt (1984) stated that with a September planting, the first flowers were initiated in February or March but the extension of shoots was inhibited for about two weeks in 'Campfire', four weeks in 'Red Sunset' and over two months in 'Zebra', a case of dormancy but after this a second flush of flowering shoots was produced. They reported that flower initiation was progressively delayed in successive shoots until November but in 'Red Sunset' the vegetative phase was confined to shoots growing from branch rhizomes, and when flower initiation was resumed, many of the shoots produced during the non-inductive phase initiated weak inflorescences most of which aborted. Normally, alstroemerias produce 1.00 to 2.25 metres of plant height with about 45.0 to 100.0 cm of inflorescence length. A 12-h day (long photoperiods) length with a light intensity of 40 lx has been found advancing flowering in cv. 'Walter Fleming' (syn. Orchid) but beyond this it caused fewer shoots, flower initiation occurring earliest at the lowest of temperatures tested (ranging from 9 to 25 °C) but further flower development occurred only at higher temperatures though practically no flowering at 25 °C (Noordegraaf, 1972). In low light areas the supplementary lighting (a minimum of 600 fc for 16 h) of potted plants will induce earlier flowering with increase in flower numbers while shading of the plants is carried out to maintain 13-17 °C temperature (Wilkins and Heins, 1976; Bridgen, 1997). Verboom (1980) further reported that use of CO_2 in the altroemeria greenhouses increases the yield of flower stems slightly.

Postharvest

Leaf chlorosis and yellowing, poor flower opening, *vis-a-vis* poor pigment development (Woltering and Harkema, 1981; Dai and Paull, 1991; Hicklenton, 1991; Sacalis, 1993), loss of turgidity, flower desiccation, and petal shedding are the problems associated with *Alstroemeria* cut flowers. *Alstroemeria* cut flowers are harvested when the first florets on the cyme are opening or are fully openend which will alleviate the flower opening and pigment development to some extent as this is almost associated with premature harvesting, but while transporting to long distances open flowers are not preferred because of the obvious reasons. Cytokinins and GA_3 reduce chlorosis to some extent (Dole and Wilkins, 1999). The flowers are harvested by gentle pulling of the stems from the rhizome which has natural abscission layer at the point of shoot attachment, and not by cutting but while putting the cut ends in water or vase solution the white portion of the cut ends are removed otherwise the stems will not absorb the solution and collapse. Water absorption is optimum when stems are cut at tepal reflexing stage and are placed in high quality water after removing the white portion at the cut end. Leaves are also stripped from the lower portion of the stem. Thick and taller stems are preferred over weak and smaller ones. Class A stems are 80 cm long, turgid and thick while class B stems have 60-70 cm length, each having three or more trusses of flowers. If the stems are removed from the plant prematurely, there is a problem in opening of the flowers in the cyme as well as pigment development on the petals is also hampered, even after putting such flowers in the vase solution.

Sucrose, 8-HQC and GA_3 in vase solution at pH 3.5 increase the cut flower life (Dai and Paull, 1991; Hicklenton, 1991; Michalczuk *et al.*, 1992; Sacalis, 1993). Stems pretreated (pulsed) in silver thiosulphate (STS) 4 mM and sucrose 200 g/l solution from 12- 24 hours prolong the vase life for about four days and decrease flower abscission, and this can also be stored up to two weeks in water at 2-4 °C as this reduces ethylene production. Cut flowers held at 4.5 °C for 6 days last for 7-8 days in water whereas for 9-10 days in a vase-life preservative solution at room temperature (Molnar, 1975). STS treated stems should be kept in high quality water before being transported. Leaf chlorosis (yellowing) can also be reduced by using cytokinin (BA) and GA_3 (1×10^{-4} M) immediately after harvest. *Alstroemeria* is an ethylene sensitive cut flower so pretreatment with low doses of 1-MCP gives results similar to STS in negating ethylene effects and prolongs the vase life in an ethylene-free environment and in presence of ethylene at 1 μl/l, the vase life of 1-MCP-treated flowers increases up to four times than control (Serek *et al.*, 1995). The effects of STS and 1-MCP were studied by Dole *et al.* (2005) on *Alstroemeria* cv. 'Rebecca' by treating the unpacked cut stems and placing them either in deionized water (DI) and subjected to 1-MCP (740 nl l^{-1}) or ambient air for 4 h or DI + STS at 0.1 mM for 4 h, and then the stems were removed and stored in polythene sleeves either wet in DI water or dry in plastic-lined floral boxes at 5 °C in the dark for 4 days, then afterwards these bunches were placed in DI water under 12 h light (76-100 μmol $m^{-2}s^{-1}$) per day where it was found that STS increased the vase life though 1-MCP did not. Hibma (1988) through ^{14}C-labelled solution of GA_3 at 20 mg/l for 16 h (as found optimum) to the freshly cut stems of cvs 'Annabel', 'Ontario' and 'Veronica' and then storing these in dark or in red light and GA_3's distribution in the stems, leaves and flowers as assessed through ELISA test (maximum distribution being observed in leaves) revealed that leaf yellowing in the dark following GA treatment was delayed by 13 days in 'Veronica' and 7 days in 'Annabel' though it was delayed by 9 and 5 days, respectively, without such treatment. Under red light, GA_3 treatment postponed leaf yellowing for 22 and 15 days, respectively, compared with 17 and 9 days, respectively, without GA treatment. Milde (1989) suggested picking of stems in the morning with one flower open and removal of the lower leaves coming in contact of vase water whose initial temperature should be 20-25 °C so that its O_2 content is minimized, immediately after harvest treating with Chrysal SVB and then at 10 °C temperature in the storage room or during transport causes lasting of flowers and buds a week longer than the foliage, and the leaf chlorosis in Chrysal SVB has been noted at 13th day than 6-7 days in untreated ones. Hickelenton (1991) when used BA or GA_3 at 50 mg/l to cut stems, foliage colour did not change even after storing the stems for 14 days in tap water. Jacquier (1983) recommended placing of cut stems in water for 12 h and then in opening solution (containing 5.0-7.5 per cent sugar and silver nitrate at 20 mg/l) before transport. Chepkairor and Waithaka (1988) recommended storing the stems harvested at bud stage in deionized water at 20-30°C and then placing of 15 cm of cut ends in deionized water, 4 per cent sucrose, STS complex, 2.5 per cent Chrysal, STS + 4 per cent sucrose or 2.5 per cent Chrysal + 4 per cent sucrose as 1-h pulsing and then into fresh deionized water containing a biocide which caused lasting of flowers in control for 21 days in 'Carmen', 19 days in 'Pink Perfection' and 16 days in 'Marina', however, the vase life of 'Carmen' by 8 days, of 'Pink Perfection' by 7 days and of 'Marina' by 6 days was found increased when flower stems were pulsed in STS solution. Since exudes from the injured portion of *Alstroemeria* cause dermatitis hence while harvesting the flowers, gloves should be used. Macnish *et al.* (2008) when added aqueous chlorine dioxide (ClO_2) at 2 or 10 μl l^{-1} in clean deionized water, the build up of bacteria in the vase solution was prevented and vase life was found extended in *Altroemeria peruviana* cv. 'Senna'. While transporting, 10 stems are bunched together with bottom leaves up to 10-15 cm removed, then packed in plastic sleeves. Before packing, these are stored at 5-7°C. Potted plants are ready for shipping when two flowers open and these can be shipped at 10 °C to sustain up to four days while at 4 °C up to six days. Under ordinary room conditions the flowers in potted plants last up to 19 days.

Lifting of Rhizomes and Storage

Since growing of the *Alstroemeria* is a continuous process and for propagation or for further cropping these are lifted from the soil, divided and planted when flowering in the crop is over. Even these can be lifted and planted throughout the year. However, if by any means field is to be vacated or for marketing or for self growing the bare rooted rhizomes may also be stored though bare rooted storage is not as ideal as in peat moss. The rhizomes can be stored for several months at moderately moist condition at 1-3° C. Thick grade vermiculite and peat moss are ideal to maintain optimum moisture level. Any other light and well-drained compound is also suitable for its storage. The rhizomes can also be wrapped in plastic but this also requires regular checking as rhizomes will spoil under too wet condition for long periods. The moisture level of the stored rhizomes is regularly monitored to ensure that the rhizomes do not dry out or become too wet as both these conditions will spoil them. Under wet condition prevailing for long, root rot may occur. Root development continues during storage.

Insect-Pests, Diseases and Physiological Disorders

Leaf aphid deforms the leaves and shoot, and with increase of temperature this becomes very active and builds up the population rapidly. First this attacks the young leaves, and then the flowers when emerge. **Thrips** also feed on the tender parts of the plant, especially the young leaves and flowers and the shoot tops are crumpled, and flowers malformed and streaked. **Leaf miners** also damage the crop, adults leave a trail of white dots on the young foliage and the larvae tunnel the foliage irregularly. **Caterpillars** are normally active during summers and autumn and feed on whole plant, first eating the leaves and flowers and then after stems also. Aphids can be controlled through spraying with nicotine sulphate, thrips with 0.1 per cent spraying of methyl parathion and caterpillars through 0.2 per cent spraying of methyl parathion or Rogor. Leaf miners are controlled with 0.1 per cent spraying of Vertimec at 5-days intervals. **Red spider mites** attack the plants that are under shaded location and are left uncared. They feed on the underside of the leaves by positioning themselves in the plant cells and suck them until the cells are empty. Their presence may be noticed through the small webs these have made. In their serious attack, foliage turns yellow and ultimately plant dies. Vertimec (abamectine) at 0.1 per cent sprays at 10-days interval will control the mites. When the crop is being grown close to some marshy places or when humid weather is persisting or if field is not clean, **slugs** and **snails** attack the crop from the underside of leaves during the dark or in nights and batter whole of the plant. Adult females lay its eggs in the soil which hatch and start feeding on the crops. These can be controlled through metaldehyde pellets. **Nematodes** (*Pratylenchus penetrans* and *P. bolivianus*) attack roots and cause root red spots or lesions on roots, stunting and premature yellowing of foliage. Use of soil fumigation, HWT with formaldehyde or soil application of Furadan will control the nematodes.

Alstroemerias have relatively only a few disease problems. However, these do have *Pythium* and *Rhizoctonia* root rot problems but if newly planted rhizome divisions are drenched with Captan or Thiride along with good sanitation of the field, no such problems shall be encountered. These fungi develop under warm humid conditions and stems show rotting just at the soil level. *Botrytis* infects during cool weathers under high humidity conditions and due to overcrowding of the crop which hinders aeration in between the plants. The stubble of the crop also encourages its infection and the disease can attack even the flowers. A great many viruses attack *Alstroemeria,* out of which 'tomato spotted wilt virus' (TSWV) is most devastating and kills the plants. Virus infected plants should be rogued out and burnt.

Abortion or blasting of flowers occurs during periods of low irradiance or when roots are damaged due to excessive watering or water-logging or due to excessive salts in the soil. In this case fully developed buds senesce before opening.

References

Assis, M.C. De, 2003. Two new species of *Alstroemeria* L. (Alstroemeriaceae) from Brazil. *Acta Botanica Brasilica,* **17**(2): 179-182.

Baersch, V. 1983. Alstroemeria : an enrichment of the cut flower market supply (German). *Deutscher Gartenbau,* **37**(29): 1330-1331.

Baeure, O.A. and A.K. Bakken, 1997. Light intensive production of *Alstroemeria* under different combinations of air and soil temperature. *Sci. Hort.,* **68**:137-143.

Bik, R.A. and T.J.M. van der Berg, 1982. Nitrogen and potassium fertilization of the *Alstrioemeria* cvs Orchid and Carmen grown on peat. *Acta Hort.,* No. 126, pp. 287-292.

Blomme, R. and P. Dambre, 1979. The culture of alstroemerias and their response to fertilizer (Dutch). *Verbondsnieuws voor de Belgische Sierteelt,* **23**(14): 561-565.

Blomme, R. and P. Dambre, 1984. Advancing the spring flowering of alstroemeria (Dutch). *Verbondsnieuws Belgische Sierteelt,* **28**(5): 221, 223, 225, 227.

Bridgen, M.P. 1993. *Alstroemeria.* In: The Physiology of Flower Bulbs (eds De Hertogh, A.A. and Le Nard, M.), pp. 201-209. Elsevier Science Publishers A.V., Amsterdam, The Netherlands.

Bridgen, M.P. 1997. *Alstroemeria.* In: *Tips on Growing Specialty Potted Plants*(eds Gasten, M.L., S.A. Carver, C.A. Irwin and R.A. Larson), pp. 13-16. Ohio Florists' Association, Columbus, Ohio.

Bridgen, M.P. and J. Bartok, 1990. Evaluation of a growing medium cooling system and its effects on the flowering of *Alstroemeria. HortSci.,* **25**(12): 1592-1594.

Bridgen, M.P. and C. Soriano, 1986. A comparison of two growing media for *Alstroemeria. Connecticut Greenhouse Newsletter,* No. 132, pp. 5-6.

Broertjes, C. and H. Verboom, 1974. Mutation breeding of *Alstroemeria. Euphytica,* **23**(1): 39-44.

Chepkairor, M.J. and K. Waithaka, 1988. The effect of floral preservatives on the vase life of *Alstroemeria* cut flowers. *E. Afr. Agric. and For. J.,* **53**(4): 229-231.

Cox, M.J. and M.M. MacMasters, 1947. Starch from *Alstroemeria. Plant Life,* **14**: 71-80.

Dai, J.W. and R.E. Paull, 1991. Postharvest handling of *Alstroemeria. HortSci.,* **26**:314.

Dole, J.M. and H.F. Wilkins, 1999. Alstroemeria. In: *Floriculture ~ Principles and Species,* pp. 179-184. Prentice Hall, Upper Saddle River, New Jersey.

Dole, J.M., W.C. Fonteno and S.L. Blankenship, 2005. Comparison of silver thiosulfate with 1-methylcyclopropene on 19 cut flower taxa. *Acta Hort.* (*Proceedings of the Fifth International Postharvest Symposium,* eds Mencarelli, F. and P. Tonutti), held at Verona, Italy, on 6-11 June, 2004; No. 682 (vol. 2), pp. 949-956.

Gabryszewska, E. and M. Hempel, 1985. The influence of cytokinins and auxins on *Alstroemeria* in tissue culture. *Acta Hort.,* No. 167, pp. 295-300.

Garton, S. and G. Peiser, 1989. Breeding and propagation of *Alstroemeria* for potted flowering plant production. *Combined Proc. Intern. Pl. Propagator's Soc.,* **39**: 166-169.

Healy, W.E. and D. Lang, 1984. Alstroemeria cultivar trials – 1984. *Res. Bull., Colorado Greenhouse Gr's Assocn,* No. 414, pp. 1-2.

Healy, W.E. and H.F. Wilkins, 1981. Interaction of soil temperature, air temperature and photoperiod on growth and flowering of *Alstroemeria* 'Regina'. *HortSci.,* **16**: 459.

Healy, W.E. and H.F. Wilkins, 1982a. Responses of *Alstroemeria* 'Regina' to temperature treatments prior to flower-inducing temperatures. *Scientia Hort.,* **17**(4): 383-390.

Healy, W.E. and H.F. Wilkins, 1982b. The interaction of temperature on flowering of *Alstroemeria* "Regina". *J. Amer. Soc. hort. Sci.,* **107**(2): 248-251.

Healy, W.E. and H.F. Wilkins, 1985. *Alstroemeria.* In: *Handbook of Flowering,* vol. I (ed. Halevy, A.H.), pp. 415-424. CRC Press, Boca Raton, Fla.

Healy, W.E., H.F. Wilkins and M. Celusta, 1982. Role of light quality, photoperiod and high intensity supplemental lighting on flowering of *Alstroemeria* 'Regina'. *J. Amer. Soc. hort. Sci.,* **107**: 1046-1049.

Heins, R.D. and H.F. Wilkins, 1979. Effect of soil temperature and photoperiod on vegetative and reproductive growth of *Alstroemeria* 'Regina'. *J. Amer. Soc. hort. Sci.,* **104**: 359-365.

Hermann, P. 1986. Fertilizer experiments with *Alstroemeria* (German). *Zierpflanzenbau,* **26**(6): 228-229.

Hibma, J.T. 1988. Ontwikkeling van een toets ter controle op het gebruik van voorbehandelings middelen tegen bladvergeling bij *Alstroemeria*. *Verslag - Centrum voor Agrobiologisch Onderzoek*, No. 91, 26 pp.

Hicklenton, P.R. 1991. GA_3 and benzylaminopurine delay leaf yellowing in cut *Alstroemeria* stems. *HortSci.*, **26**((): 1198-1199.

Hurka, W. 1989. Blütenbildung bei *Alstroemeria* - Hybriden. *DeutscherGartenbau*, **43**(9): 570-572.

Jacquier, C. 1983. A study of the possibilities of early cutting of *Alstroemeria* flowers (French). *Revue Horticole Suiss*, **56**(10) 306-310.

Jesang, C.M. and K. Waithaka, 1988. Growth and flowering of *Alstroemeria*. *Acta Hort.*, No. 218, pp. 115-120.

Keil-Gunderson, L.S., K.L. Goldsberry and P.L. Chapman, 1989. Air and substrate temperatures for 'Atlas' and 'Monika' alstroemeria. *HortSci.*, **24**(4): 613-616.

King, J. J. and M.P. Bridgen, 1990. Environmental and genotypic regulation of *Alstroemeria* seed germination. *HortSci.*, **25**(12): 1607-1609.

Konst Alstroemeria, B.V. 1990. *Alstroemeria* (*Alstroemeria* hybrid). *Plant Varieties J.*, **3**(3): 6-8.

Kwiatkowska, B. and J. Brzozowska, 1980. Studies on the evaluation of 4 *Alstroermeria hybrida* cultivars (Polish). *Prace Instytutu Sadownictwa I Kwiaciarstwa w Skierniewicach, B*, **5**: 31-40.

Labeke, van M.C. and P. Dambre, 19933. Response of five *Alstroemeria* cultivars to soil cooling and supplemental lighting. *Sci. Hort.*, **56**: 135-145.

Leliveld, H.P.J. 1973. Early planting of alstroemerias is necessary (Dutch). *Vakblad Bloemisterij*, **28**(23): 11.

Lin, W.C. and J.M. Molnar, 1983. *Alstroemeria* Alsaan. *Canadian J. Pl. Sci.*, **63**(2): 565-566.

Lin, W.C. and P.L. Monette, 1987. *In vitro* propagation of *Alstroemeria* 'Alsaan'. *Plant Cell, Tissue Organ Cult.*, **9**(1): 29-35.

Lisiecka, A. and S. Szczepaniak, 1983. Effect of NPK levels on flower yield in *Alstroemeria* (*Alstroemeria* x *hybrida* Hort.) (Polish). *Prace Instytutu Sadownictwa I Kwiaciarstwa w Skierniewicach, B* (Roœliny Ozdobne), **8**: 83-99.

Lisiecka, A. and S. Szczepaniak, 1989. Porównanie wspólczynnika rozmna žania kilku od mian alstremerii. *Prace Komisji Nauk Rolniczych I Komisji Nauk Leœnych*, No. 67, pp. 89-93.

Macnish, A.J., R.T. Leonard and T.A. Nell, 2008. Treatment with chlorine dioxide extends the vase life of selected cut flowers. *Postharv. Biol. & Tech.*, **50**(2/3): 197-207.

Meij, J. 1976. Alstroemeria growing can be heading for a good future. *Vakblad Bloemisterij*, **31**(14): 14-15.

Michalczuk, B., A. Przybyla, D.M. Goszczyñska aand R.M. Rudnicki, 1992. Effect of postharvest chemical treatment on longevity of different cultivars of cut *Alstroemeria* flowers. *Acta Hort.*, No. 325: 199-205.

Mikkelsen, S. and F. Mikkelsen, 1989. Erhvervbetinget eksem efter *Alstroemeria* (Inkalilje). *Ugeskrift for Laeger*, **151**(14): 898.

Milde, H. 1979. Alstroemerias (German). *Gb+Gw*, **79**(18): 434-437.

Milde, H. 1981a. *Alstroemeria ligtu* hybrids (German). *Gb+Gw*, **81**(7): 156-158.

Milde, H. 1981b. *Alstroemeria ligtu* hybrids (German). *Gb+Gw*, **81**(34): 778-780.

Milde, H. 1983. *Alstroemeria*. *Gb+Gw*, **83**(4): 81-89.

Milde, H. 1985. 'King Cardinal' gave the highest *Alstroemeria* yield (German). *Gb+Gw*, **85**(35): 1330-1331.

Milde, H. 1989. Haltbarkeit von *Alstroemeria* - Blätter und Blüten. *Deutscher Gartenbau*, **43**(9): 576-579.

Molnar, J. 1975. *Alstroemeria* - a promising new cut flower. *Ohio Florists' Assocn Bull.*, No. 553, pp. 3-5.

Nijssen, H.M.C. and M.G. Hoogeveen, 1990. Gebruikswaar-deonderzoek zes cultivars *Alstroemeria*. Cultivars leveren geen bijdrage aan produktiespreiding. *Vakblad Bloemist.*, **45**(15) 26-28.

Noordegraaf, C.V. 1972. Advancing flowering of alstroemerias (Dutch). *Vakblad voor de Bloemisterij*, **27**(40): 13.

Pierik, R.L.M., A. Van Voorst, G. Booy, C.A.M. van Acker, C.L.C. Lelivelt and J.C. De Wit, 1988. Vegetative propagation of *Alstroemeria* hybrids *in vitro*. *Acta Hort.*, No. 226, Vol. I, pp. 81-89.

Powell, M.C. and and A.C. Bunt, 1984. Periodicity of growth in the alstroemeria cultivars 'Campfire', 'Red Sunset' and 'Zebra'. *Scient. Hort.*, **24**(3/4): 359-367.

Powell, M.C. and A.C. Bunt, 1986. The effects of propagation date on flower production in alstroemeria 'Campfire' and 'Red Sunset'. *Sci. Hort.*, **28**(1): 147-157.

Sacalis, J.N. 1993. *Alstroemeria* hybrids. In: *Cut Flowers ~ Prolonging Freshness*, 2nd ed. (ed. Seals, J.L.), pp. 23-24. Ball Publishing, Batavia, Illinois.

Saito, N., M. Yokoi, M. Ogawa, M. Kamijo and T. Honda, 1988. 6-Hydroxyanthocyanidin glycosides in the flowers of *Alstroemeria*. *Phytochem.*, **27**(5): 1399-1401.

Saito, N., M. Yokoi, M. Yamaka and T. Honda, 1985. Anthocyanidin glycosides from the flowers of *Alstroemeria*. *Phytochem.*, **24**(9): 2125-2126.

Schubert, W. 1989. Alstroemerien in Bankbeeten, im Folienhaus und im Freiland. *Deutscher Gartenbau*, **43**(9): 579-581.

Serek, M., E.C. Sisler and M.S. Reid, 1995. Effects of 1-MCP on the vase life and ethylene response of cut flowers. *Pl. Gr. Reg.*, **16**(1): 93-97.

Slob, A., B. Jekel, B. De Jong and E. Schlatmann, 1975. On the occurrence of tuliposides in the Liliflorae. *Phytochem.*, **14**(9): 1997-2005.

Sonneveld, C. 1988. The salt tolerance of Alstroemeria (*Alstroemeria* ×). *Acta Hort.*, No. 228, pp. 317-326.

Spieser, H. 1987. Water source heat pumps for greenhouse soil cooling (Final Report). *Ontario Min. of Agric. and Food, Agricultural Energy Centre,*Canada, 17 pp.

Thompson, P.A. and P. Newman, 1979. Germination of alstroemerias. *Garden* (UK), **104**(2): 75-76.

Thompson, P.A., P. Newman and P.D. Keefe, 1979. Germination of species of *Alstroemeria* L. *Gartenbauwissenschaft*, **44**(3): 97-102.

Tsuchiya, T. and A. Hang, 1987. Chromosome studies in genus *Alstroemeria. Acta Hort.*, No. 205, pp. 281-287.

Tsuchiya, T., A. Hang, W.E. Jr. Healy and H. Hughes, 1987. Chromosome studies in the genus *Alstroemeria*. I. Chromosome numbers in 10 cultivars. *Botanical Gaz.*, **148**(4): 519-524.

Uematsu, J. 1990. Regulation of flowering in *Alstroemeria* Ligtu hybrid. I. Growth and differentiation (Japanese). *Bull. Saitama hort. Exp. Stn*, No. 17, pp. 11-18.

Verboom, H. 1980. *Alstroemeria* and some other flower crops for the future. *Scientific Hort.*, **30**: 33-42.

Verboom, H. and M.C. van Staaveren, 1978. The application of CO_2 to alstroemerias (Dutch). *Vakblad Bloemisterij*, **33**(23): 61.

Vonk Noordegraaf, C. 1975. Temperature and daylength requirements of *Alstroemeria. Acta Hort.*, No. 51, pp. 267-274.

Waithaka, K. and M.J. Chepkairor, 1988. Outdoor growth and flowering patterns of *Alstroemeria* in Kenya. *East African Agrricultural and Forestry J.*, **53**(4): 213-220.

Wiel, A. van De, 1989. Tussentijds verslag *Alstroemeria* – proef in Venlo. Grondkoeling en Styromul maken oogstspreiding mogelijk. *Vakblad Bloemisterij*, **44**(6): 32-33.

Wilkins, H.F. and R.D. Heins, 1976. *Alstroemeria* general culture. *Flor. Rev.*, **159**(4121): 30-31,78-80.

Wilkins, H.F., W.E. Healy and T.L. Gilbertson-Ferriss, 1980. Comparing and contrasting the control of flowering in *Alstroemeria* 'Regina', *Freesia × hybrida* and *Lilium longiflorum*. In: *Petaloid Monocotyledons, hort.& bot. Res.* (eds Brickell, C., D.F. Cutler and M. Gregory), pp. 51-63. Academic Press, London.

Wotering, E.J. and H. Harkema, 1981. Ethylene damage to cut flowers. *Bedriifsontwikkeling*, **12**: 193-196. Sprenger Institute, Wageningen, Nethertlands.

Ziv, M., R. Kanterovitz and A.H. Halevy, 1973. Vegetative propagation of *Alstroemeria in vitro. Scientia Hort.*, **1**(3): 271-277.

3

Amaryllis & *Hippeastrum* (Family: Amaryllidaceae)

Sanyat Misra and R.L. Misra

[**Common names**: Amaryllis/Belladonna lily/South African amaryllis (*Amaryllis belladonna*); Barbados lily/Equestrian star-flower (*Hippeastrum equestre, H. puniceum*), Butterfly amaryllis (*H. papilio*), Dutch amaryllis, Green amaryllis (*H. calyptratum*), Knight's star lily, Lily of the palace (*H. aulicum*), Mexican lily (*H. reginae*)]

Introduction and Origin

Amaryllis is a classical name, or probably named so after the beautiful Greek shepherdess of the same name, and has only one species, *A. belladonna,* being native of SW Cape of South Africa (coastal hills and stream banks on the southwestern seaboard of the Cape Province). *Hippeastrum* takes its name from the *hippeus* meaning a 'knight', 'rider' or 'horse' and *astron* meaning a 'star' due to some fancied resemblance in *H. equestre*, perhaps of the equitant leaves and the star-shaped corolla opening. *Hippeastrum* is native to subtropical Central and South America, centred from eastern Brazil to the south-central Andes of Peru via Paraguay, and distributed from Mexico to Chile via Surinam, Bolivia and Argentina.

Amaryllis is most suitable for planting in a sunny border, for massing in patches of the gardens, as a flowering potted plant, and also for cutting. Amaryllis and hippeastrums are excellent plants for growing indoors as pot plants and also make excellent cut flowers, colour ranging from white to purple via green, yellow, orange, pink, red (scarlet, crimson, vermillion, and maroon), mauve and purple. Certain varieties are fragrant, some less though others with more intense fragrance. Flowers are trumpet shaped and quite large. These flower in March and April in the sub-tropical climates of India though during late-summer in the temperate areas.

Botany

Amaryllis differs from the *Hippeastrum* in having solid scape and the absence of scales between the filaments. It has brown tunicated, perrennial, large and rounded bulbs with contractile roots. The leaves are hysteranthous, deciduous, strap-shaped, and in two ranks; the scape is un-branched, solid, stout, flushed red-purple, leafless, 60-90 cm long and inflorescence is an umbel of 6 or more, white to deep pink but mostly white towards the base, trumpet-shaped, short stalked and highly fragrant large flowers with 6-10 cm perianth, flowers subtended by 2 large and equal spathes which enclose the whole umbel in the bud, tube funnel-form and somewhat bilaterally symmetric, lobes 6 and spreading. Style and stamens deflexed and then curving towards apex, stamens 6 and ovary 3-chambered, and the fruit a few seeded capsule with large, fleshy and round seeds. On the hills it is late-summer flowering as in Holland also though under subtropical conditions it is spring flowering. In Holland, its foliage appears in spring and dies off at the end of July, flowers appear in September-October, the bulbs consist of dried scales, sound scales, a dried flower bud, several series of phyllomes, the remains of floral stalks and a flower bud (Hartsema and Leupen, 1942), and when last foliage has been split off the growing point proceeds to form the flowers. The inflorescence consists of two botryces, each botryx of 8-9 flowers though all may not develop. Out of both the botryces, only 9-12 flowers develop. Flowers start initiating once in a year and that too before the inflorescence of the previous season has appeared and the formation of the flower is quickest and best when the bulbs are kept at 23°C. The basic chromosome numbers are n = 11, 12. It has been crossed with *Brunsvigia, Clivia, Crinum* and *Nerine* (Coertze and Louw, 1990).

Balina (1976) found the temperature range of 23-30°C as optimum for *Amaryllis* pollen tube growth while at 35°C pollen tube deformation was noted. Stanley (1961) tried phenyle borate, tributyle borate, sodium tetraborate, boric acid, potassium metaborate and potassium tetraborate at 5 μg/ml in the pollen of *Amaryllis hybrida* where though phenyl borate inhibited pollen growth but others stimulated the germination, largest growth being under tributyl borate. In a further experiment, Stanley and Lichtenberg (1963) at 10 μg/ml of butyl borate and sodium tetraborate (borax) found good pollen germination though in latter case the per cent germination was half of the former. At and above 100 μg/ml of butyl borate and boric acid, at and above 50 μg/ml of potassium tetraborate and at and above 4 μg/ml of phenyl borate, there was inhibition of growth and germination. Growth stimulation was also observed under 10 μg/ml of boric acid, potassium tetraborate and potassium metaborate. In fact, amaryllis requires both B and exogenous sugar for pollen germination.

Sealy (1968) described about the intergeneric hybrid *Amarine tubergenii* 'Zwanenburg Beauty' which was developed through *Nerine bowdenii* and *Amaryllis belladonna* with initial raiser's name of *Nerinodonna tubergenii* 'Zwanenburg'. This produces up to five stems of large purplish pink flowers that last well when cut (Dix, 1977). Bell (1977) elaborated the inheritance of double flowered trait as a method of introducing new genes into tetraploid populations though most of the available germplasm is diploid, and partially fertile triploids allowed new traits to be added but at a very slow rate. They further elaborated that selected diploids can be treated for polyploidy induction. Hannibal (1955) stated that when *Amaryllis belladonna* was crossed with *Crinum moorei*, three types of seeds, *viz.* large pulpy green seeds, medium-sized coral red seeds, and small coral red seeds were obtained, the latter two being parthenocarpic arising from spontaneous growth of the female germ cells.

Hippeastrum has tunicated bulb with hairy and branched roots, *vis-à-vis* contractile roots, bulbs bearing linear or strap-shaped basal leaves with enlarged leaf bases. Bulbs exhibit sympodial growth where the vegetative growth occurs from a lateral bud at the base of the inflorescence (Dole and Wilkins, 1999), the floral stalk always being formed at the side of the four leaves that emerge, thus in the ready to be planted bulb two sets of four leaves and two floral scapes are formed at different stages of development (Rees, 1972, 1985; Okubo, 1993). Its scapes are robust but hollow and leafless, flowers white, red, orange-pink to deep crimson or their striped combinations, shape tubular to funnelform or salverform, horizontal or drooping, perianth lobes 6 erect and spreading, filaments 6 often with small scales in between, fruit a 3-chambered capsule with about 80 black, flattened or compressed seeds. In ***Hippeastrum hybridum***, the growing point ordinarily changes into a flower bud after four leaves have been formed, and then a new growing point from the base of the lowest leaf starts (Okada and Yamado, 1953). Jana (1983) reported that in the cv. 'Fire Dance', the onset of flower initiation occurs at 120 days after planting. Otto (1967) stated that a type of hippeastrum has been recorded producing several flowering stems (2.3 to 2.5) during the season (January to August) in Germany without showing any dormant period when in the second season bulbs had been left as such in the greenhouse beds over winter. Wahi and Bhattacharjee (1986) studied 45 genotypes of hybrid hippeastrums for 11 quantitative characters and recorded genotypic correlations greater in magnitude than phenotypic ones. They found high heritabilities for length, diameter and longevity of flowers and number of bulbs per plant, the latter being significantly associated with number of shoots, *vis-à-vis* flowers per stalk at genotypic level.

Pettit *et al.* (1984) recorded acetylcaranine, ambelline and anhydrolycorinium chloride from bulbs which were active against P-388 lymphocytic leukemia *in vitro.* Rao ***et al.*** (1971) found lycorine (0.002 per cent), tazettine (0.003 per cent) and pseudolycorine (0.04 per cent) in the bulbs of *Hippeastrum johnsonii*. Amico et al. (1977) recorded lycorine in the bulbs with innermost fleshy scales which extend into foliage leaves containing 0.09 per cent, middle scales 0.16 per cent and the outer one with 0.028 per cent after flowering in *Amaryllis belladonna* grown in Apulia (Italy). The lycorine in traces was also recorded in the foliage leaves and roots. Tamahashi *et al.* (1990) found galanthamine alkaloid in the bulbs of *Amarcrinum* (*Amaryllis belladonna* x *Crinum moorei*). Mohgazi *et al.* (1975) isolated lycorine, hippeastrine, tazettine, vittatine, hippacine, hippadine, hippagine, hippafine and five others from Egyptian grown *H. vittatum*. Ali *et al.* (1984) extracted pancracine, the first 5, 1-methanomorphanthridine alkaloid (structure of hippagine) from hippeastrum. Tomoda *et al.* (1985) isolated a mucous polysaccharide, hippeastrum-H-glucomannan from the bulbs of *H. hybridum*.

Hippeastrums have basic chromosome count of n = 9 in *H. advenum*, *H. chiliense* and *H. pretense*; n = 11 in triploid *H. argentinum* (syn. *H. candidum*), *H. aulicum* (syn. *H. robustum*), *H. elegans* (syn. *H. ambiguum, H. solandriflorum*), *H. puniceum* (syn. *H. equestre*), triploid *H. reginae, H. stylosum*, and tetraploid or 4x-1 in *H. vittatum*. Arroyo (1982) recorded 2n = 22 in most of the nine *Hippeastrum* species tested though *H. blumenavia* was found with 2n = 20. Karihaloo (1984) recorded 2n = 4x+1 = 45 chromosomes in the cv. 'Dawn'. Nwankiti (1983) from root tip cells of *H. equestre* recorded 2n = 33 and suggested that this species is a natural aneuploid with complete infertility, and the odd chromosome was not to be a B-chromosome as its length differed from any of the other 16 homologues. The basic chromosome number is said to be n (x) = 11, out of which two consist of median, five submedian and four subterminal as has been reported by Khaleel and Siemsen (1989) while studying the four *Hippeastrum* hybrids, *viz.* 'Apple Blossom', 'Basuto', 'Lucky Strike' and 'Red Stripe'. Singh *et al.* (1980) found greater pollen tube growth of *Amaryllis vittata* (*Hippeastrum vittatum*) in suspension culture medium containing a sugar (lactose, sucrose, raffinose, glucose, maltose, trehalose or fructose), maximum being in lactose and sucrose while minimum in fructose, than in a medium containing the non-metabolizable compound pentaerythritol. Sharma *et al.* (1981a) did not observe any significant change in the enzyme activities (α-glucosidase, α-galactosidase, β-glucosidase and β-galactosidase) at different stages of tube growth during *in vitro* germination, and no increase in β-glucosidase activity in a suspension of intact germinating pollen was observed. Sharma *et al.* (1981b) reported that sodium cyanide at 0.01M or sodium fluoride at 0.002 M inhibited respiration in pollen of *A. vittata* whereas 2, 4-dinitrophenol at 10 μM stimulated it. Complete inhibition of pollen tube growth occurred in 2, 4-dinitrophenol and partial inhibition in sodium cyanide or in sodium fluoride. Sidhu *et al.* (1986) stated that ABA at 5 and 10 μg/ml promoted tube elongation, $^{14}CO_2$ fixation and phophoenolpyruvate carboxylase activity during *in vitro* germination of *A. vittata* pollen though IAA, BA, GA and ethephon did not influence pollen tube elongation at 1 μg/ml and instead suppressed it at 10 μg/ml. Bhandal and Bala (1989) cultured the pollen of *A. vittata* in a medium containing 3 per cent sucrose and 10 μg boric acid/ml along with $Cd(NO_3)_2$ at 50 μM, $NiCl_2$ at 100 μM, $CuSO_4$ at 150 μM, $Pb(NO_3)_2$ at 200 μM, $Co(NO_3)_2$ at 200 μM and $Zn(NO_3)_2$ at 250 μM at 28°C. Co_2+ did not cause complete inhibition of pollen germination. Pollen germination was recorded to a tune of 51 per cent at 200 μM and only 25 per cent at 500 μM of $Co(NO_3)_2$. Suzuki and Tamura (1979) tried for breeding 13 hippeastrum cultivars through cross combinations and reported highest fertility in cross pollinations between the Japanese cultivars though seting of seeds per fruit was recorded highest when 'Ludwig' cultivars of USA were intercrossed. In Tadjikistan, Davydova and Kozlova (1983) tried some most attractive cultivars such asa 'Beautiful Lady', 'Dutch Bell', 'Lucky Strike', 'Pink Perfection', 'Royal Dutch', 'United Nations' and 'Zanzibar' and obtained certain interesting hybrids. Merrow (1988) used *Hippeastrum cardenasianum, H. fragrantissimum, H. lapacense, H. papilio, H. reticulatum* var. *striatifolium* and an unidentified Bolivian species, out of which he got success through reciprocal crosses of F_3 progeny with *H. lapacense, H. cardenasianum* and an unidentified Bolivian species. Through intervarietal crosses involving many cultivars, Jana (1989) recorded the hybrids of 'Bouquet' with 'White Favorite' having a typical plant form and flower shape with a different pigment distribution pattern on the perianth, stamens and carpel which could be used as a new commercial cultivar.

Species and Varieties

Amaryllis and *Hippeastrum*, both are separate genera, but through intergeneric hybridization both are so merged with each other that both the genera are taken as synonymous, and the modern cultivars are referred to as *Hippeastrum* hybrids. *Hippeastrum* contains some 75 species; most prominent one *H.* × *ackermannii* is of garden origin, probably between *H. aulicum* and *H. psittacinum*, scape some 60 cm long, flowers about 15 cm long, green in throat, white at centre and rest scarlet with white margins; *H. aulicum* (*H. robustum*) is native to Brazil, Paraguay, bulb ovoid, 10 cm diameter, blunt-tipped leaves are 6-9 and some 60 cm long, scape up to 60 cm, stout and terete, flowers are crimson with green throat, usually 2, 15 cm long, and its variety *platypetalum* more vigorous than the parent; *H. candidum* is indigenous to Tucuman Province in Argentina, bulb black-purple, globose and 7.5 cm diameter, leaves ligulate, hysteranthous, shallowly channelled and more than 30 cm in length, scape up to 70 cm, dark purple below and glaucous above, flowers 6, fragrant white with greenish base, pendant, funnel-form and 20 cm long; *H. equestre* (*H. puniceum*) is native to Mexico to Chile and Brazil, bulb globular with brown scales, some 5 cm in diameter, producing bulblets freely, leaves 6-8, 30-50 cm long, scape stout and 30-60 cm long and terete, flowers 2-4, 10-13 cm across, bright red with green at the base and tube green, and its varieties have orange with white, orange with green, etc. colours and also the double forms; *H. reticulatum* is native to S. Brazil, bulbs subglobose, leaves and flowers appear together, leaves and scapes 30 cm long, flowers 3-6, mauve or red-purple with crimson stripes and cross lines and drooping; *H. rutilum* is native to Brazil and Venezuela, bulbs almost globular, 5-7.5 cm in diameter and stoloniferous, leaves 6-8, 30 cm long and elongating after flowering, scapes also 30 cm long with 2-3 red flowers, perianth-tube green, and its varieties are *fulgidum* (deep crimson), *crocatum* (saffron), *citrinum* (bright yellow), *acuminatum* (pink), etc., *H. vittatum* is native to Peru, bulbs globose and 7.5 cm in diameter, leaves 6-8 and 60 cm long, scape 90 cm long, with 3-6 flowers which are white striped red and the flowers are 12.5 cm across, etc. Europe, North America and Japan have a long history of breeding and producing numerous cultivars (Meerow and Kane, 1991), the first being European hybrids produced in 1799 by using *H. reginae* and *H. vittatum*.

Though there are more than 300 cultivars but the most important ones are 'American Express' (crimson-scarlet), 'Apple Blossom' (tetraploid, white flushed pink), 'Athos' (dark wine-red), 'Bambara' (deep red with light green or white midrib), 'Basato' or 'Basuto' (tetraploid, darkest red), 'Beautiful Lady' (pale mandarin-red), 'Belinda' (dark red), 'Best Seller' (cerise), 'Blushing Bride' (rose, large blooms), 'Bouquet' (salmon), 'Byjou' (soft burnt apricot), 'Candy Cane' (red striped white), 'Cantate' (deep red), 'Christmas Gift' (white), 'Cicero' (orange red), 'Cinderella' (orange striped white), 'Dazzler' (white), 'Desert Dawn' (light salmon-orange, leaves and flowers appear together), 'Double Record' (white-red), 'Dutch Belle' (opal rose), 'Fair Lady' (salmon), 'Fire Dance' (deep red), 'Helsinki' (white), 'Jewel' (white, double), 'Kalahari' (rose), 'Lady Jane' (salmon, double), 'Lucky Strike' (tetraploid), 'Ludwig's Goliath' (large, bright scarlet), 'Lydia' (pale-salmon), 'Masai' (white flushed with pink stripes), 'Minerva' (red, flamed white), 'Nivalis' (white, throat flushed yellow), 'Orange Sovereign' (orange), 'Picottee' (white rimmed red), 'Orange Sun' (orange), 'Oskar' (deep red), 'Picotee' (white with a narrow red margin), 'Pink Favourite' (pink), 'Prima Donna' (crimson), 'Purple Sensation' (lipstick pink), 'Queen of Sheba' (pink), 'Red Lion' (dark red), 'Red Master' (red), 'Red Peacock' (red, double), 'Red Strike' (tetraploid), 'Rose Marie' (pink) 'Royal Velvet' (deep velvety red), 'Safari' (red, petals bearded), 'Salmon Tower' (salmon), 'Samrat' (salmon), 'Springtime' (light rose, white centre and white tips to petals), 'Star of Holland' (scarlet with white star at throat), 'Susan' (large, pink), 'Sweet Heart' (red), 'United Nations' (white striped vermillion), 'Valentine' (white, veins pink), 'Vera' (pale-pink), 'White Christmas' (white), 'White Dazzler' (white), 'Zanzibar' (dark red, midrib green), 'Zenith' (pink with a white stripe), etc. Loeser (1979) stated 'Athos' as the best wine-red cultivar, 'Rose Marie' the best pink and 'Belinda' the best dark red.

Other noteworthy varieties are 'Berry Beauty', 'Bhagiraj', 'Blazing Star', 'Bleeding Heart', 'Brilliance', 'Brilliant Red', 'Canary', 'Candidum', 'Cardinal', 'Carmine King', 'Carmine Red', 'Carrara', 'Cocktail', 'Courtesy', 'Daintiness', 'Dawn', 'Day Light', 'Enchantress', 'Fairy Queen', 'Faust', 'Fire Fly', 'Fire Dance', 'Five Star General', 'Forget-Me-Not', 'Gay Lady', 'Geest Glory', 'Honeymoon', 'Jessica', 'Love Desire', 'Love Song', 'Lyric', 'Maiden's Blush', 'Melody', 'Mevrouw Warmenhoven', 'Moon Mirage', 'Morning Star', 'My Love', 'Orange Gleam', 'Orange King', 'Orange Majesty', 'Peppermint', 'Pink Pearl', 'Rainbow', 'Red Matador', 'Red Riding Hood', 'Salmon King', 'Scarleto', 'Royal Dutch', 'Scarlet Beauty', 'Scarlet Pimpernel', 'Sensational Giant', 'Silver Jubilee', 'Snow Princess', 'Star Dust', 'Striking Strikes', 'Vermillion Record', 'Wedding Wells', 'White Queen', 'Zebrina', etc.

Bhattacharjee (1981) when studied the performance of 45 *Hippeastrum* cultivars, reported 'Bhagiraj', 'Canary Bird', 'Candidum', 'Carmine King', 'Christmasx Gift', 'Dutch Belle', 'Honey Moon', 'Lyric', 'Pink Pearl', 'Peppermint' and 'Sophia' as outstanding with regard to attractiveness of flowers and flower colours, lasting quality and bulb production. Through the glasshouse performance trial of the 11 new hippeastrum cultivars along with five standard cultivars, Gelder and Ettema (1990) reported that 'Bestseller', 'Piquant' and 'No. 263' as best pot plants, followed by 'Cinderella', whereas 'London' proved best for cut flowers, followed by 'Liberty' and 'No. 263'.

Propagation

Amaryllis and hippeastrums are multiplied through vegetative means (offsets, fractional scale stems, twin-scaling and tissue culture) and sexual means (seeds). When grown vegetatively, these maintain their characters in the off-springs. Generally, the vegetative propagation is by natural increase of offsets from the existing mother bulbs which is very slow, normally 0-4, depending on the cultivar. Production of commercial bulbs from offsets takes about 1-2 years to attain blooming stage. Fully grown bulbs (minimum commercial bulb size is normally 20 cm in circumference) can produce up to three floral scapes per year (Boyle and Stimart, 1987). Commercially the bulbs are being produced by Israel, South Africa and the Netherlands (Okubo, 1993) and the size of the bulbs for forcing is usually 24-26 cm.

Miss Ida Luyten (1935, 1936) suggested removal of any leaves present on the bulbs having active growth and then **scooping** of basal plate is done. Now the bulbs are halved lengthwise and the scales from the centre are removed one by one and are inserted in slanting position with curved portion down in an earthenware seed pan filled with moist fine sand and then again the same type of sand is filled in the pan in the fashion that only the ends of the exceptionally long scales protrude. These pans covered with glass panes are now placed at a constant temperature of 30°C and these pans are very lighly sprayed with water at alternate days with fine nozzle water can. There should not be direct sunlight if these are kept outdoors. Inspection of the pans may be done every 3-4 weeks and around two months tiny bulblet formation is found which may be removed after three months with a sharp knife and planted in fertile soil, better in glasshouse at a temperature range of 25-28°C. By '**fractional scale stems**', the bulbs are vertically fractionated into 2-16 pieces by a sharp sterilized knife and these pieces are planted in sprouting medium *i.e.* equal parts of sphagnum peat and coarse sand in the month of February. In this case, only 1.25 cm upper part may be left uncovered. These may be watered daily but lightly through a fine mist. By this method sprouts may appear within 6 weeks and the leaves within three months whereas flowers within two years. Everett (1936) described sectioning of the bulbs vertically into 8 or 16 parts without scooping, each part further divided vertically (depending on the variety) and then each fraction is planted in greenhouse in flats containing peat moss where one or more bulbs develop between the scales. Fujioka (1964) stated that to prevent decay and to promote bulblet formation, a 4-day period at 25°C and R.H. 90-92 per cent for curing bulb scales before planting, proved most effective. Rooting was increased by increased length of treatment at 30°C and when treatments at 25° and 30°C were alternated every 12 hours, bulblet formation took place 10 days earlier. Gyenes (1975) apart from using fractional scale stems (vertically 8 or 16 parts) in laboratory or in glasshouse in Hungary or those cut into 8 parts were further split into single or twin-scale cuttings and set in 1:1 perlite and peat mixture, found bulblet formation more readily in the glasshouse than the laboratory and in June-July than in December. Using $^{1}/_{16}$ of a bulb was most suitable for commercial propagation. Bose *et al.* (1980a) propagated *Hippeastrum hybridum* cv. 'Fire Dance' bulbs by cutting horizontally retaining only the top scale-stem portions (1.5 cm in diameter) which were again cut longitudinally into 8-40 sections and planted in vermiculite, sand, leaf mould or sand/leaf mould (1:1) mixture. They obtained 100 per cent plantlets in vermiculite when scale-stem portions were cut into 28 sections and 37 per cent plantlets through 32 sections. In sand, leaf mould and in sand-leaf mould, the highest number of plantlets was obtained from scale-stem portions cut into 24, 20 and 20 sections, respectively.

Cultivars that produce few or no offsets are multiplied through **twin-scaling** and now this is considered a standard method of propagation. In twin-scales, the thickness and the length of the outer scale affect the rate of bulblet formation *vis-à-vis* subsequent bulb development but inner scale does not. Two scales together are required for propagation system because vascular bundles of the swellings initiated at abaxial surface of the inner scale connect only with the vascular system of the outer scale and not of the inner scale. Twin-scales (50-80 bulblets in number from one large mother bulb) are prepared in November and incubated in sand beds with moist vermiculite at 25-28°C. After three months of incubation, these bulblets are planted @ 260/m^2 and harvested after eight months and then again planted within two months and harvested after eight months to get flowering size bulbs. The sizes of bulbs for commercial use are 20-22, 22-24, 26-28, 30-32 and >32 cm in circumference. Alkema *et al.* (1977) suggested storage of twin-scales of *Amaryllis belladonna* at 25°C rather than at lower temperatures. Sakinishi and Yanagawa (1979) also recommended storage of the twin-scale pieces of *Hippeastrum hybridum* at 20 or 25°C than at 15 or 30°C for a period of four months though effectiveness of the storage was dependent on the species. They also found that for evergreen species the optimal temperature was higher than those which normally remain dormant in summer. Hanks (1983, 1986) elaborated the making of twin-scales (pieces of two adjacent scales joined by a more-or-less triangular piece of basal plate) and their incubation in polythene bag at 15-22°C for up to 8 weeks, and then after formation of the bulbils on the twin-scales, their planting in sterilized compost. The bulbs produced by this method require 3-4 years to reach flowering size. Huang *et al.* (1990a) and Okubo *et al.* (1990) stated that in twin-scaling of *H. hybridum* cv. 'Akamaruben', the thickness and length of the outer scale affected the rate of bulblet formation and leaf development, while those of inner scale did not. They observed that protuberance initiation occurred on the abaxial surface of the inner scale but the vascular bundles of the protuberances became united with the vascular system of the outer scale and not with those of the inner scale. They further added that instead of protuberance initiation, protocorm-like bodies were regenerated from single-scales, and the bulblets were formed by the protocorm-like bodies similar to orchids morphologically as well as physiologically.

Bases of the scales are normally used for quick multiplication through ***in vitro* culture** technique though young elongating scape tissues and shoot apices (for getting virus-free plants) are also capable for such culture. Bapat and Narayanaswamy (1976) established callus cultures successfully from the explants of bulbs, floral peduncles and anther tissues on the medium in the presence of a synergistic combination of coconut milk and 2, 4-D.They observed shoot and root redifferentiation leading to the development of complete plantlets occurring when the tissue was transferred from an auxin-based medium to a medium lacking 2, 4-D. Mii *et al.* (1974) stated that excised bulb scales of *Hippeastrum hybridum* cultured on MS medium supplemented with combinations of NAA and kinetin developed buds or roots or both together, the pattern of organogenesis depending upon the NAA concentration in the medium as kinetin had little effect and was toxic at high concentrations. Huang *et al.* (1990b) compared the bulb formation through tissue culturing of single scales and twin scales *in vitro*. They obtained protocorm-like bodies similar to orchid protocorm from single scales though twin scales produced the bulblets directly. They suggested multiplication of protocorms in liquid MS medium with 1.0 mg/l zeatin, followed by bulblet and shoot formation in liquid medium with 1.0 mg/l IAA and 1.0 mg/l zeatin. Pierik *et al.* (1990) used floral explants in a high sucrose medium along with an auxin and cytokinin in the medium, in continuous darkness and at a relatively high temperature and then the regenerated bulblets are subcultured in light on basal media without growth regulators for root and leaf formation. After reaching the bulb size of at least 0.8-1.2 cm diameter, the bulbs are quartered and subcultured again like to those of bulblets and then planted directly in the prepared field. Wang *et al.* (1989) cultured the explants from floral organs or bulb segments on MS medium supplemented with 0.1-1.0 mg IAA or 0.01-0.5 mg NAA and 0.5-2.0 mg BAP per litre and obtained 5-7 times the multiplication rate of a 6.19 cm diameter bulb compared to usual type of vegetative propagation. However, a greater size of bulb poroduces a greater number of bulblets with larger size. Yanagawa and Sakanishi (1977) used segments containing i) both tissues of the scale base and the basal plate which produced bulblets from the proximal end of the scale and roots from the basal plate tissue at a high rate, ii) segments of the extreme base of a scale which also produced bulblets but rooting later, and iii) segments of a single scale base with or without basal plate tissue where no difference was observed whether segments were excised from the thicker or thinner parts of a scale. Some explants with basal plate tissue planted with the adaxial surface upwards regenerated bulblets from the adaxial side of the scale base though most explants regenerated bulblets from basal abaxial areas alone irrespective of the orientation of segments. Bose and Jana (1977) took scale-stem pieces of the cv. 'Red' and grew them on modified Knudson C medium with added IAA, NAA or GA_3, each at 2 or 5 mg/l, or 2, 4-D at 0.2 or 0.5 mg/l and recorded best shoot and root growth in medium containing NAA at 5 mg/l. 2, 4-D proved phytotoxic, GA_3 delayed regeneration and inhibited plantlet growth while IAA gave intermediate results.

Sexual propagation is by means of **seeds** but one is not sure to get the same colour of the flowers and plant type as to that of its mother or parents hence this means is applied only by the breeders to generate new varieties. Seeds have short viability so should be sown immediately after harvesting and grown at 20-25°C temperature. However, seeds stored at 11 per cent or 52 per cent relative humidity at 5°C or 15°C, respectively, remain viable up to 12 months. Seeds germinate in 1-2 weeks but for the emergence of first leaf it still takes many more days. After emergence, the seedlings are fed regularly with a complete fertilizer. If the seedlings attain good size, *i.e.* 10-12 cm, these are transplanted by gentle lifting. Within two years from germination, the full flowering size bulbs are formed.

Soils, Preparation of Land, Planting and Weed Control

Light sandy **loam** soil, with medium humus is suitable for its growing but in no case the medium should contain the bark. Though the optimum soil reaction for its growing is at 6.0-6.5 pH but can easily be grown up to 8.0 pH range without any much deterioration. Clay soil may be improved by addition of coarse sand. The soil should be well drained. Fresh manure should be avoided. At early stage of growth, it does not require high nutrition. It can be grown in partial shade but not too much, for spring and summer crop. Too much shade encourages appearance of leaves but abortion of growing inflorescence. Since these are tender perennials, therefore, these can be grown best at temperature range of 16-27 °C. In case of hot summers, these may be sheltered.

The soil to a depth of 30 cm is dug up and properly pulverized, and fortified with 200 quintals/ha well rotten farmyard manure, followed by flooding to check the nitrogen loss. At proper stage, again the soil is dug up and the beds of convenient sizes are made. Planting is followed in November when there is sufficient moisture in the soil, at a distance of

15 cm in the rows spaced at 20-25 cm, by burying two-third bottom of the bulb in the soil keeping its old viable roots intact (De Hertogh and Tilley, 1991; Okubo, 1993; De Hertogh, 1996), and one-third peeping out of the medium, pressing firmly from all the sides. Haegeman and Onsem (1970) planted 1-year pips (bulbs) in 4 size-grades (1-2, 2-3, 3-4 and 4-5 cm in circumference) and 2-year buttons (bulbs) in 3 grades (2-3, 3-4 and 4-5 cm) in mid-March in Holland and flowering appeared between June 2 and August 4. They obtained earliest high yields with 3-4 and 4-5 size buttons but in the end of the cropping season the pips cropped more heavily than buttons. The pips and buttons <3 cm grade-size though provided fewer flowers up to July 14 but thereafter there was little difference with larger ones. Largest bulbs (26-28 cm circumference) flower earlier (about 17 days) with about 5 cm longer stems and more uniform plants whereas smaller ones (24-26 cm) flower about 9-10 days later than largest ones (Loeser, 1979). They are nice pot plants and look very attractive when in flowering. For pot planting, the pots should be 20-30 cm size having three parts rich fibrous garden loam, one part leaf mould, half part peat, little ground charcoal, 15 g bone meal and enough coarse sand to ensure proper drainage.

In India, this is spring to summer crop so summer weeds are aplenty which are weeded out manually. Certain herbicides such as chloroxuron @ 50 g in 10 litres of water/100 m^2 or chloropropham are quite effective for **controlling weeds** in hippeastrum plantings. Simazine or diuron is also recommended. Jana and Bose (1981) tried five herbicides at different rates as pre- or post-emergence and obtained good results with Weedone (2, 4, 5-T) at 5 litres/ha as pre-emergence.

Irrigation and Nutrition

Potted hippeastrums require light showering daily from sprouting to flowering. They are left in the pots for 3-4 years as these do not require frequent disturbances. They are not heavy feeder but to sustain proper flowering every year, these should be provided with liquid manure fortnightly during growing period. Field grown crops are watered at weekly intervals just to keep the soil moist throughout the growing season. However, watering is withheld when the leaves start turning yellow. Boyle and Stimart (1987) through imposing irrigation interruptions of 0, 2, 4 and 8 weeks on actively growing 'Red Lion' plants in 12.5-cm plastic pots, recorded first scape flowering on all plants by 160 days in 2-week interruptions, 140 days in 4 week interruptions and 60 days under 8 week's interruptions, after resumption of irrigation.

Half quintal of calcium nitrate, 50 kg of potassium sulphate (Indian soils are normally not deficient in potassium so this should be applied only when the soil is really deficient) and two quintals of single superphosphate may be applied per hectare area of planting at the time when leaf growth begins. Apart from this dose, half of this may be applied once in March and again in May to encourage quicker bulb formation. After the flowering, the scapes are cut off from the ground. Jana and Bose (1980) reported that *Hippeastrum hybridum* cv. 'Fire Dance' plants gave maximum yields of bulbs, bulblets and flowers when were supplied with 20 g N, 40 g P_2O_5 and 20 g K_2O/m^2. For bulb production under glass in three hippeastrum cultivars, Roorda van Esynga (1984) compared the effects of different soil levels of N (0-8 mmol/l soil extract) and K (0-4 mmol/l soil extract). He reported that though rising N levels had little effect on bulb weight at lifting but K at 1 mmol/l extract (higher level being no more effective) increased it, and when forcing, control bulbs gave poorer number of flowers per stem and the number and length of flower stems were found reduced as compared to those grown in fertilized plots. Tsukamoto and Fujioka (1956) obtained best results in case of bulblets for production of bulbs when treated either with 300 ppm each of N, P and K; or 225 ppm N, 150 ppm P and 300 ppm K. They further reported that plants given high rates of N and K absorbed more N and K, respectively, than plants given low rates. Sakanishi and Fukuzumi (1964) provided with either 1, 3 or 9 g of N, P_2O_5 or K_2O to the bulbils of an amaryllis clone grown in 5-gallon glazed crocks for 4 years and obtained largest bulbs under 3:3:3 g and 3:3:1 g. They found high levels of N retarding the bulb growth, and N and K contents of bulbs were increased with increase in N and K levels of nutrition but P application did not affect the P content. They recorded lowest rate of N resulting into slightly more bulbils, and the largest number of bulbils was produced by plants receiving 1:1:1 g but their subsequent growth and leaf development were greatest when the parent bulb had received the highest rate of N. Shoushan *et al.* (1978ab) applied N, NP, NK, NPK or Nutrin (commercial foliage fertilizer) to amaryllis grown in peat moss in the greenhouse and recorded that NPK, NP and Nutrin increased stem length and diameter, flower diameter, fresh weight of cut spikes, leaf and bulb N contents and the sugar content of the leaves but not of the bulbs though flowering was delayed in N but accelerated by NPK and Nutrin. The fertilizers containing K were most effective. Nautiyal and Bajpai (1979) applied N at 40-80 g/m^2, P_2O_5 at 30 or 60 g/m^2 and K_2O at 30 or 60 g/m^2 and reported that 60:30:30 g/m^2 N:P:K provided best plant growth and maximum flower production, however, highest N rates (60 and 80 g/m^2) slightly delayed the flowering. Bhattacharjee *et al.* (1982) recorded 20:40:40 g/m^2 N:P_2O_5:K_2O, the highest doses provided, giving the best results with respect to bulb and bulblet weight and diameter, the number of flowers & flower stalk, length of floral stalk & flower diameter and flowering duration.

Growth, Development and Flowering

Because of their tropical origin, there is no real dormant period in hippeastrums, but in subtropical and temperate regions environmental factors impose a rest period, hence, there is only one flowering period. Throughout the entire growth period, flower initiation alternates with leaf formation. Leaf emergence and flowering require 3-5 weeks. The time span between flower initiation and emergence is 11-14 months and for leaves it is 7-10 months (Okubo, 1993) that is why flowering precedes the leaf emergence. However, to obtain synchronous emergence of foliage and flowers, only mature bulbs having greater than 2 cm long floral buds inside should be harvested (De Hertogh, 1996) and on flowering the leaves should be to the half of the length of the spike when forcing. Cultivars may have two to eight flowers, normally four. The two main flowers develop with the two spathe-leaves, and each with one bract. Normally, the inflorescence branches into a double helicoids cyme and in axils third and fourth flower may originate, likewise fifth and sixth or seventh and eighth flowers may also appear, depending on the varieties. A flowering scape emerges with a set of four leaves. The time span between the initiation of each set of vegetative and reproductive structures is four months. Daylength has no qualitative effect on growth though slight growth reduction in short days appeared to result directly from poor photosynthesis (Hayashi, 1974). Hong (1970) reported that warm-water treatment to the bulbs at 30°C for 10 days accelerated flowering and flowering occurred in 50.5 ± 7.8 days and the treatment at 40°C caused root damage. Hayashi and Suzuki (1970) recorded fastest bulb growth at 18/13°C and 13/8°C day/night temperatures though growth inhibition was found hampered at 23/8°C. Hayashi (1972) recorded best bulb thickening at 20°C, and after 28 weeks of planting the bulbs formed 1-2 flower buds at 20 or 15°C, and when the treatment of 10°C was given the bulbs having 18 or less bulb scales also developed flower buds. Basu and Bose (1980) found increased number of flowers per plant from 2.6 in the control to 3.4 and advanced flowering by 18 days when *Hippeastrum hybridum* cv. 'Red' was treated with 200 ppm ancymidol applied at 2-3 leaf stage. Bose *et al.* (1980b) obtained increased bulb weight and flower diameter when the bulbs of *H. hybridum* cv. 'Fire Dance' were soaked for 24 h in GA_3, however, two foliar sprays 30 days apart of GA_3 at 10, 100 or 1000 mg/l raised bulblet production and increased flower size and number. Bulb soaking in IAA at 1-100 ppm increased number and weight of bulblet, two foliar sprayings 30 days apart at 1-100 ppm increased number and size of flowers and 100 mg/l increased bulblet production. Bulb soaking for 24 h in chlormequat (Cycocel) at 1,000 mg a.i./l increased flower number, chlormequat sprayings 30 days apart at 1,000 mg/l increased weight of bulblets and number and size of flowers or ethephon (Ethrel) at 1,000 mg/l sprayings twice at 30 days apart improved bulblet weight. Bhattacharjee (1983) recorded induced bulb sprouting, delayed flowering, increased flower size and longevity, accelerated bulb production and improved bulb size and weight by soaking the hippeastrum bulbs in CCC at 1000 or 2500 ppm and B-nine (daminozide) at 1000-5000 ppm. B-nine even increased the floral stalk. Ethrel and MH both at 500-2000 ppm delayed sprouting but induced early appearance of floral buds and increased flower longevity. Ethrel also increased bulb formation though MH, TIBA and NAA proved ineffective. IAA and GA_3, each at 10-1000 ppm promoted vegetative growth, induced early flowering, increased number of flowers per stalk, flower size, stalk length, flower longevity and bulb production. Maiko and Aks'onova (1983) found suppressed bulb development and flowering when applied GA at early stages of morphogenesis, but advanced flower stem growth and induced early and uniform flowering when applied at the last stages of morphogenesis after formation of the reproductive initials. Though height control in this crop is not required but Adriansen (1985) reported Cycocel and Bonzi marginally effective.

Flower Forcing

For programming and forcing techniques, selection of bulbs produced in specific countries is done. After lifting, bulbs should be dried in shade under proper ventilation for two weeks at 23-25 °C, followed by subsequent storage at 13 °C and 80 per cent R.H. for 8-10 weeks, and for longer storage at 5-9 °C, by keeping the old root system viable. During storage, the bulbs are held at 5 °C if the buds or leaves are visible but held at 9 °C when there is no sprouting. If the bulbs are still dormant they may be stored at 13 °C if need be but if they have begun sprouting, they may be stored at 5 °C. For forcing, normally one bulb is planted per pot, pot size depending upon the size of the bulb to be planted, *e.g.* a 12 cm pot for 24-26 cm circumference, a 15 cm pot for 28-30 and a 20 cm pot for >32 cm bulb circumference. For cut flower forcing, the planting density of the bulb per square metre is 40. Light intensity of 1,000-5,000 foot candles *i.e.* low to medium, is favourable in greenhouse but potted plants require higher light intensities. The optimal temperature for forcing is 24-27 °C, though varying from 21-27 °C (Okubo, 1993; De Hertogh, 1996). For bulb production, CO_2 concentration in the greenhouses is kept at 500-600 ppm. Normally, no feeding is required for forcing, and watering is done just to keep the soil moist. Boyle and Stimart (1987) when subjected the actively growing potted 'Red Lion' bulbs to 0, 2, 4 or 8 weeks drying (no watering), a water stress of 4 to 8 weeks resulted in cent per cent flowering in 140 or 60 days, respectively, when compared to 160 days for a 2-week stress, and only 83 per cent bulbs were observed flowering in control (0 week) in 160 days. Luyten (1946) advocated lifting of hippeastrum bulbs in September in Holland, then storing at 15° or 17°C for $4^1/_2$ weeks, followed by 23°C for 4 weeks and then placing in the greenhouse at 17°-24°C for getting the blooms around Christmas. Sheehan and Howe (1957) stated that amaryllis bulbs having a circumference over 23.75 cm flowered when forced under glass. Kurki (1958) stated that flower buds initiate from February to October and those stored dry and warm conditions from September to mid-January will flower in February and March, however, for flowering during Christmas, the bulbs should be kept dry during August, should be then kept at 15-17°C for $4^1/_2$ weeks and at 23°C for 4 weeks. Dijkhuizen and Leeuwen (1975) recommended lifting the bulbs of cv. 'Scarlet Glow' in Holland on July 10 provided flower initials have formed, and then immediate storage (or after a short period of drying) at 13° or 17°C for 10-12 weeks provided emergence of most flower stems under shortest forcing period.

Postharvest

For cut flower production, the spikes are harvested when flowers have started showing colour but not yet open. Out-rolling and splitting of the base of the cut stem after harvesting can be prevented by 24 h soaking in 0.125 M sucrose solution at 22 °C (Halevy and Kofranek, 1984) which also enhances vase life and promotes opening of all the flower buds. It is advisable not to cold store the cut flowers but in case of any urgent need, these may be dry stored at 5-10 °C, and when taken out the cut ends should be recut and placed in a preservative solution (Secalis, 1993). Vase life of cut flower is normally up to 9 days. Halevy and Kofranek (1984) stated that splitting and out-rolling of the cut stem bases of *H.* x *hybridum* can be prevented by 1-day pulsing with 0.125 M sucrose which together also promotes opening of all the flower buds and increases the vase life slightly. Dijkhuizen (1980a) observed a very little difference in the duration of flowering, *vis-à-vis* on the vase life in water among the 75 hippeastrum cultivars tested though the differences among many cultivars between longest and the shortest flowering period were greater. Dijkhuizen (1980b) recorded that an individual flower on the same stem lasted for almost the same duration though there were varietal differences. Mehlhorn (1984) stated that harvesting in hippeastrum flowers is effected when buds have started showing colour, these are packed dry into plastic containers and stored at 4-6°C for up to 10 days. Carow and Rober (1979) reported 'cut floral stalk topple' due to more K being present in the sap, and the tissues in the damaged region were found containing less bound Ca than at the base of the affected stalk. The inhibition of water uptake due to CCC aggravated the situation than use of water alone to the cut stems.

The potted hippeastrum is marketed when the flower stalk length is 30 cm tall and the leaves are 15-30 cm long (Nell, 1993). Potted flowering plants can be stored at various stages of development for 7-10 days at 7-10 °C temperature. The postharvest life of the potted amaryllis is 14-21 days. With proper care and treatments the amaryllis plants at home can repeatedly be flowered (Dole and Wilkins, 1999).

Insect-Pests, Diseases and Physiological Disorders

In India, no any major **insect-pest** problem has been noticed in hippeastrum plantings but spider mite, though very rare, may sometimes pose a problem while forcing the bulbs which can be controlled by use of sulphur vaporizer. Aphids, mealy bugs and thrips may also attack which may be controlled by spraying with 0.2 per cent Rogor. Bulb scale mite (*Steneotarsonemus laticeps*) has been found attacking bulbs which can be controlled by exposing the infested bulbs to frost, drenching with endrin or endosulfan, dipping bulbs in a thionazin solution and through HWT (Anon., 1971). De Figueiredo and Pereira (1944) recorded the caterpillars of *Xanthopastis timais* moth infesting amaryllis whose control they suggested through spraying with a mixture of lead arsenate and Paris green. Snails also attack the crop seldom which can be controlled through formaldehyde poison baits. Goodey (1951) observed that though roots of hippeastrums were badly affected with the nematode *Rotylenchus* but undamaged parts of the roots were covered with root hairs of secondary origin arising from cortical cells right up to the base of the bulb. Johansson and Niedieck (1964) recorded a nematode (*Scutellonema brachyurum*) infesting amaryllis in Sweden which they controlled through soil treatment with Terracur but five months later again there was population build up, however, soil steaming appeared to keep populations at a low level at least for four months under observation. Nong and Weber (1965) when inoculated the amaryllis seedlings with *Scutellonema brachyurum,* no root symptoms were apparently noticed after two months but after four months only slight infestation was noticed but with no discernible foliage symptoms and no nematode was observed to penetrate below the endodermis of live roots. Also inoculation of root lesion nematode (*Pratylenchus scribneri*) showed no apparent foliage symptoms but after four months most of roots had been destroyed and the bulbs were also found infested. Andersson (1971) recorded *Pratylenchus scribneri* infesting rotted hippeastrum roots and for its control he recommended HWT most effective.

Rhizopus, *Sclerotium*, *Botrytis* and *Pythium* species cause **soft rot** of the bulbs and roots, and *Aecidium amaryllidis* and *Cercospora* species **leaf spots** which may be controlled by the use of 0.2 per cent Dithane M-45 for leaf spot diseases and Captaf alternate with Bavistin for rots, spraying fortnightly. Abo-El-Dahab and El-Goorani (1970) observed in Egypt a dry spathe rot and in severe cases covering the entire flowers in *Amaryllis vittata*. Though they isolated a grey sterile fungus but this could not be identified. Pag (1964) reported leaf spot of hippeastrum due to attack of *Colletotrichum crassipes* which he found associated with wounding, different *Fusarium* species, and *Stagonospora curtisii*; Tapio (1966) isolated 11 species of fungi (*Stagonospora cutisii, Colletotrichum crassipes, Fusarium bulbigenum, F. moniliforme, Botrytis cinerea, Chaetonium* sp., *Popularia sphaerosperma, Trichothecium roseum, Rhizoctonium tuliparum, Penicillium nigricans* and *Melanospora fallax*) and found good control of various pathogens individually or in combination at various stages with Captan, thiram, maneb, zineb, ferbam, PCNB, and mercuric compounds which sometimes impaired seed germination; van Leeuwen (1974) reported *Stagonospora curtisii* attacking flower stems for which he suggested benomyl 0.2 per cent + trichlorophenol (as Aaglitan 1 per cent) cold bulb dip for 0.5 h on dormant bulbs or HWT and against *Fusarium* infection he suggested steaming of soil, cutting off the roots immediately and to provide HWT using 0.5 per cent benomyl followed by a cold dip in 1 per cent Aaglitan for 20-30 minutes; and Conijn (1983, 1984) found good control with 0.4 per cent Sportek (prochloraz) by dipping for 15 minutes the dormant bulbs of cv. 'Red Lion' showing symptoms of infection with *Stagonospora* sp. at lifting, and through Sportak 0.4 per cent + Daconil 2787 (chlorothalonil) 0.5 per cent dip for 15 minutes of infected bulbs of the cvs 'Diana', 'Apple Blossom' and 'Masai' severely infected with root rot (*Fusarium* sp.). Schenk (1954), Bonifacio (1958), Gill (1959), Assawah (1968) and Tarabeih *et al.* (1980) also reported prevalence of **'hippeastrum fire'**, **'leaf scorch'** or **'red blotch'** disease (*Stagonospora curtisii*) appearing on the flower stem as rows of spreading red to brown spots or cracks. Its control measures are spraying the stams with 0.75 per cent copper oxychloride or HWT of the bulbs at 43.5°C for two hours (Schenk, 1954), as pathogen remains latent in the bulb so such bulbs should be treated with 1 per cent formalin at 40-50°C, however, spraying the shoots with Bordeaux mixture along with some sticker is also recommended (Bonifacio, 1958), spraying six times from small leaf stage to flowering with zineb or ferbam (900 g/100 gallons of water) (Gill,

1959), and benomyl and thiocur treatment is recommended as protective fungicides (Tarabeih *et al.*, 1980). Horst (1990) advised an appropriate prevention spray on the cut surface (the point of scape cutting) of the bulbs or through microtube or sub-irrigation.

Tomato spotted wilt virus causes the leaves with numerous white or yellow spots. Certain blood red spots may also appear on leaf edges. Amaryllis mosaic exhibits irregular light and darker green areas on the leaves so such plants should be burnt or buried, and the insect vectors feeding on the leaves should be controlled by spraying with 0.2 per cent Metacid-50 (methyl parathion). Brants and van Den Heuvel (1965) reported that sap of infected hippeastrums showing hippeastrum mosaic symptoms (HMV) which may also be transmitted through seeds, infected other hippeastrum plants through inoculation though it was not transmitted through aphids. This virus is found in all parts of the plants except spathe. Iwaki (1967) in Japan found this virus widespread where one-third of the plants were also noted with cucumber mosaic virus (CMV). CMV causes mosaic-like patterns on the foliage which may vary from a symptomless mottling to a very conspicuous mosaic pattern, and is transmitted mechanically and through *Myzus persicae* but not through seeds (Stouffer, 1963). He suggested isolation of propagating stock especially during vegetative propagation and control of insects. Brölman-Hupkes (1975) stated about two non-seed-borne viruses, one causing leaf mosaic and the other latent which were suggested to be controlled by controlling the thrips, *Frankliniella fusca*. De Leeuw (1972b) stated *Hyoscymus niger* as a host plant of mosaic virus of hippeastrum. Cevat (1967) described about two viruses infecting hippeastrums, one mosaic virus and the other *Lycopersicon* virus 3, and he suggested their control measures as controlling of aphid vectors by soaking of infected bulbs after lifting in a cold solution pf Parathion @ 200 g of 25 per cent w.p. per 100 litres of water for 2-3 hours. Holmes (1965) suggested that hippeastrum bulbs infected with mosaic during vegetative propagation can be controlled first by massive trimming of the roots and most of the bulb scales leaving a little more than the stem of each bulb. This stem is now cut into sectors and planted in boxes of sand putting upper surfaces up and these sectors regenerate small bulblets on the surface which are removed when still small and planted in 10-cm pots filled with soil. Out of whole survived ones, some $^{1}/_{7}^{th}$ show free of virus. De Leeuw (1972a) reported that certain hippeastrum plants appearing to have been infected with mosaic are, in fact, infected with a virus which closely resembles to that of TRV though this could not be re-inoculated possibly due to presence of virus inhibitors.

Flower abortion can occur which may be due to high levels of ethephon used during forcing, and high temperature may cause competition for water, nutrition, hormones, etc. during growth or storage which may be one of the factors causing abortion. Improper stage of bulb harvesting, *i.e.* when floral bud inside the bulb is less than 2 cm long also causes abortion. Carow and Rober (1979) reported 'flower stalk topple' due to more K in the sap of harvested flower stems. They stated that tissues in the damaged region had less bound Ca than at the base of the affected stems though there also it was low than healthy stems. When water uptake of cut stems was inhibited by CCC, topple was more prevalent than when water alone was used.

Lifting, Dormancy and Storage of Bulbs

Since bulbs form the flower initials in the previous season, hence bulb lifting should be delayed so that in this case at least >2 cm length of the floral bud has been formed, therefore the bulbs should be harvested not before August, and preferably at the time of planting itself, *i.e.* in the month of October-November. They are lifted gently without damaging the root zone. At least for two weeks they are dried at 23-25 °C under proper ventilation, and then stored for 8-10 weeks with moist sphagnum moss at 13-15 °C and 80 per cent R.H., followed by two weeks at 21-23 °C for leaf and floral stalk growth and development. There should not be any break during 8-10 weeks storage at 13-15 °C temperature otherwise there is maximum possibility for development of floral stalk prematurely but no leaves (Okubo, 1993; De Hertogh, 1996). However, for late forcing, bulbs are subjected for two weeks to 23-25 °C for flower development, followed by eight weeks at 13 °C and then these bulbs become ready for immediate planting but for further storage these can be stored for up to nine months at 5 °C (Okubo, 1993; De Hertogh, 1996). There is no absolute dormancy in amaryllis bulbs. Sheehan and Howe (1957) in USA started lifting the bulbs from October 19 to January 14 and stored them at room temperature until January 14 when all were potted for forcing. There had been little effect of lifting time, and earlier the lifting delayed was flowering though differences were not so marked, however, delayed lifting gave taller spikes. Bose *et al.* (1979) when stored the bulbs of cv. 'Fire Dance' at 5°, 10°, 17° or 30°C for 30, 60 or 90 days, observed that those stored for 30°C for shorter duration developed maximum number of daughter bulbs and leaves, *vis-à-vis* increased number of flower stalks and flowers, stalk length and flower size, however, the duration of storage had very little effect.

References

Abo-El-Dahab, M.K. and M.A. El-Goorani, 1970. A dry rot of amaryllis floral parts in Egypt (U.A.R.). *Phytopathologia Mediterranea*, **9**(2/3): 184-186.

Adriansen, E. 1985. Kemisk vaekstregulering (Danish). In: *Potteplanter 1-Produktion, Metoder, Midler* (eds Christensen, O.V., A. Klougart, I.S. Pedersen and K. Wikesjö), pp. 142-162. GartnerINFO, København, Denmark.

Ali, A.A., M.K. Mesbah and A.W. Frahm, 1984. Phytochemical investigation of *Hippeastrum vittatum*. Part IV. Stereochemistry of pancracine, the first 5, 11-methanomorphanthridine alkaloid from *Hippeastrum* – structure of "hippagine". *Planta Medica*, **50**(2): 188-189.

Alkema, H.Y. and C.J.M. van Leenwen, 1977. Propagation of some minor crops by double scales (Dutch). *Vakblad Bloemisterij*, **32**(36): 24-25.

Andersson, S. 1971. *Pratylenchus scribneri*, a root-damaging nematode in *Hippeastrum* (Swedish). *Växtskyddsnotiser*, **35**(4): 43-47.

Amico, A., L. Stefanizzi, S. Bruno and V. Bonvino, 1977. Localizaztion of lycorine in *Amaryllis belladonna* L. bulbs (Italian). *Fitoterapia*, **48**(2): 60-67.

Anon. 1971. Bulb scale mite. Advisory Leaflet from Min. Agric. Fish. Food, No. 456, 5 pp.

Arroyo, S. 1982. The chromosomes of *Hippeastrum, Amaryllis* and *Phycella* (Amaryllidaceae). *Kew Bull.*, **37**(2): 211-216.

Assawah, M.W. 1968. *Stagonospora curtisii* (Besk.) Sacc. on amaryllis and trumpet daffodil in Egypt (U.A.R.). *Phytopath. Medit.*, **7**: 21-27.

Balina, N.V. 1976. The effect of high temperatures on the growth of pollen tubes (Russian). *Fiziologiya Rastenii*, **23**(4): 805-811.

Bapat, V.A. and S. Narayanaswamy, 1976. Growth and organogenesis in explanted tissues of *Amaryllis* in culture. *Bull. Torrey bot. Club*, **103**(2): 53-56.

Basu, T.K. and T.K. Bose, 1980. Uses of ancymidol on flowering bulbous plants. *Sci. Cult.*, **46**(9): 323-325.

Bell, W.D. 1977. Double flowered *Amaryllis. Proc. Fla St. hort. Soc.*, **90**: 121-122.

Bhandal, I.S. and R. Bala, 1989. Heavy metal inhibition of *in vitro* pollen germination and pollen tube growth in *Amaryllis vittata* (Ait). *Curr. Sci.*, **58**(7): 379-380.

Bhattacharjee, S.K. 1981. Horticultural description and performance studies of some cultivars of *Hippeastrum hybridum* Hort. *Singapore J. Prim. Industries*, **9**(2): 127-135.

Bhattacharjee, S.K. 1982. Effect of nitrogen, phosphorus and potash fertilization on *Hippeastrum. Indian Agriculturist*, **26**(3): 193-197.

Bhattacharjee, S.K. 1983. Influence of growth-regulating chemicals on *Hippeastrum hybridum* hort. *Gardens' Bull.*, **36**(2): 2347-242.

Bonifacio, A. 1958. A new disease of hippeastrum (Italian). *Riv. Ortoflorofruttic. Ital.*, **42**: 362-367.

Bose, T.K. and B.K. Jana, 1977. Reageneration of plantlets in *Hippeastrum in vitro. Indian J. Hort.*, **34**(4): 446-447.

Bose, T.K., B.K. Jana and T.P. Chattopadhyay, 1979. Effect of temperature and duration of storage of bulbs on growth and flowering in *Hippeastrum. Pb. Hort. J.*, **19**(3/4): 205-207.

Bose, T.K., B.K. Jana and T.P. Mukhopadhyay, 1980a. A note on the propagation of *Hippeastrum* from scale-stem sections. *Pb. Hort. J.*, **20**(3/4): 222-226.

Bose, T.K., B.K. Jana and T.P. Chattopadhyay, 1980b. Effects of growth regulators on growth and flowering in *Hippeastrum hybridum* Hort. *Scientia Hort.*, **12**(2): 195-200.

Boyle, T.H. and D.P. Stimart, 1987. Influence of irrigation interruptions on flowering of *Hippeastrum* x *hybridum* 'Red Lion'. *HortSci.*, **22**(6): 1290-1292.

Brölman-Hupkes, J.E. 1975. A virus-complex in *Hippeastrum hybridum* (abstr.). *Acta Botanica Neerlandica*, **24**(2): 253.

Brants, D.H. and J. van den Heuvel, 1965. Investigation of *Hippeastrum* mosaic virus in *Hippeastrum hybridum. Neth. J. Plant Path.*, **71**: 145-151.

Carow, B. and R. Rober, 1979. Flower stalk topple in *Hippeastrum* (German). *Gartenbauwissenschaft*, **44**(2): 67-70.

Cevat, H.N. 1967. Hippeastrum growers: clean up now. *Vakblad Bloemisterij*, **22**: 547.

Coertze, A.F. and E. Louw, 1990. The breeding of interspecies and intergenera hybrids in the Amaryllidaceae. *Acta Hort. (Fifth intern. Symp. Flowerbulbs*, held on July 10-14, 1989 in Seatle, Washington), No. 266, pp. 349-352.

Conijn, C.G.M. 1983. New material for controlling *Stagonospora* in *Hippeastrum* ("amaryllis") (Dutch). *Bloembollencultuur*, **93**(40): 1039.

Conijn, C.G.M. 1984. Considered advice on hippeastrum bulb disinfection (Dutch). *Vakblad voor de Bloemisterij*, **39**(6): 40-41.

Davydova, A.A. and A.G. Kozlova, 1983. Some data on varietal studies and breeding of *Hippeastrum* (Russian). *Introduktsiya I Ekologiya Rastenii*, No. 8, pp. 55-78.

De Figueiredo, E.R., Jr. and H.F. Pereira, 1944. Notas sôbre *Xanthopastis timais* Cram. (Lep. Noct.), praga das Amarylidáceas (*Xanthopastis timais*, a pest of amaryllids). *Arq. Inst. Boil.*, **15**: 289-298.

De Hertogh, A.A. 1996. *Amaryllis*. In: *Holland Bulb Forcer's Guide* (5th ed), pp. C7-16. International Flower Bulb Centre, Hillegom, The Netherlands.

De Hertogh, A.A. and M. Tilley, 1991. Planting medium effects on forced Swaziland and Dutch grown *Hippeastrum* hybrids. *HortSci.*, **29**: 1168-1170.

De Leeuw, G.T.N. 1972a. Tobacco mosaic virus in *Hippeastrum hybridum. Netherlands J. Pl. Path.*, **78**(2): 69-71.

De Leeuw, G.T.N. 1972b. *Hyocymus niger*, a useful local lesion host for a mosaic virus in *Hippeastrum. Netherlands J. Pl. Path.*, **78**(3): 107-109.

Dijkhuizen, T. 1980a. Keeping quality of hippeastrum flowers (Dutch). *Bloembollencultuur*, **90**(33): 890-891.

Dijkhuizen, T. 1980b. The lasting quality of *Hippeastrum* ("amaryllis") is important (Dutch). *Vakblad voor de Bloemisterij*, **35**(18): 34-35.

Dijkhuizen, I. and A.J.M. van Leeuwen, 1975. Amaryllis (Dutch). *Vakblad Bloemisj.*, **30**(4):12-13.

Dix, J.F.C. 1977. *Amarine tubergenii* Zwanenburg Beauty : a valuable bulbous plant. *Bloembollenbcultuur*, **88**(18): 391.

Dole, J.M. and H.F. Wilkins, 1999. *Hippeastrum*. In: *Floriculture Principles and Species*. Prentice Hall, Upper Saddle River, New Jersey, USA.

Everett, T.H. 1936. The vegetative propagation of amaryllis. *Gdnrs' Chron.*, **99**: 235.

Fujioka, S. 1964. Studies on bulblet formation on bulb scales used for propagation of amaryllis (*Hippeastrum hybridum* Hort.) (Japanese). *J. Jap. Soc. hort. Sci.*, **33**: 159-170.

Gelder, A. De and J. Ettema, 1990. Gebruikswaarde – onderzoek met nieuwe amaryllis – cultivars. *Vakblad Bloemist.*, **45**(27): 40-45.

Gill, D.L. 1959. Reducing amaryllis leaf spot by spraying. *Plant Dis. Reptr*, **43**: 1272-1273.

Goodey, J.B. 1951. A secondary piliferous layer on the roots of *Hippeastrum*. *Nature*, **167**: 822-823.

Gyenes, L. 1975. Vegetative propagation of *Amaryllis* (Hungarian). *Kertészet és Szölészet*, **24**(24): 7.

Haegeman, J. and J.G. Onsem, 1970. The relationship between corm size, corm age and cut flower yield in anemones. *Meded. Rijksstat. SierpfTeelt Melle*, No. 20, pp. 1-21.

Halevy, A.H. and A.M. Kofranek, 1984. Prevention of stem-base splitting in cut *Hippeastrum* flowers. *HortSci.*, **19**(1): 113-114.

Hanks, G.R. 1983. Propagation of bulbs by twin-scaling. *Garden* (U.K.), **108**(10): 402-405.

Hanks, G.R. 1986. The effect of temperature and duration of incubation on twin-scale propagation of *Narcissus* and other bulbs. *Crop Res.*, **25**(2): 143-152.

Hannibal, L.S. 1955. Unusual seeding habits of *Amaryllis belladonna*. *J. roy. Hort. Soc.*, **80**: 130-131.

Hartsema, A.M. and F.F. Leupen, 1942. Orgaanvorming en periodiciteit van *Amaryllis belladonna* L. *Meded. LandbHoogesch. Wageningen*, dl. **46,** verh. 4, 30 pp.

Hayashi, I. 1972. Studies on the growth and flowering of *Hippeastrum hybridum*. II. The effect of temperature and planting density on growth and flower bud differentiation (Japanese). *Bull. Kanagawa hort. Expt Stn*, No. 20, pp. 95-102.

Hayashi, I. 1974. Studies on the growth and flowering of *Hippeastrum hybridum*. III. The effect of daylength on growth and flower bud development (Japanese). *Bull. Kanagawa hort. Exp. Stn*, No. 22, pp. 116-119.

Hayashi, I. and M. Suzuki, 1970. Studies on the growth and flowering of *Hippeastrum hybridum*. 1(I). Effect of temperature on the growth of young seedlings and bulbs. (II). Growth and flowering out-of-doors and the effects of autumn temperature on growth (Japanese). *Bull. Kanagawa hort. Exp. Stn*, No. 18, pp. 171-188.

Hofmann, W. 1986. Anemonen 'Mona Lisa' – ein Hit? *Deutscher Gartenbau*, **40**(26): 1212-1213.

Holmes, F.O. 1965. Elimination of mosaic disease from a clone of *Hippeastrum* (abstr.). *Phytopath.*, **55**: 504.

Hong, Y.P. 1970. The effect of temperature treatments on flowering and growth in amaryllis, *Hippeastrum hybridum* (Korean). *Res. Repts of the Office of Rural Development, Horticulture*, **13**: 57- 63.

Horst, R.K. 1990. *Amaryllis*. In: *Westcott's Plant Disease Handbook* (5th ed.), p. 526. Van Nostrand Reinhold, New York, USA.

Huang, C.W., H. Okubo and S. Uemoto, 1990a. Importance of two scales in propagating *Hippeastrum hybridum* by twin scaling. *Scientia Hort.*, **42**(1-2): 141-149.

Huang, C.W., H. Okubo and S. Uemoto, 1990b. Comparison of the bulblet formation from twin scales and single scales in *Hippeastrum hybridum* cultured *in vitro*. *Scientia Hort.*, **42**(1-2): 151-160.

Iwaki, M. 1967. Viruses causing mosaic diseases of amaryllis in Japan (Japanese). *Ann. Phytopath. Soc., Japan*, **33**: 237-243.

Jana, B.K. 1983. Studies on the flower bud differentiation and its development in *Hippeastrum hybridum* Hort. *South Indian Hort.*, **31**(2/3): 127-128.

Jana, B.K. 1989. A new flower-colour pattern for Royal Dutch amaryllis (*Hippeastrum hybridum*). *Indian Agriculturist*, **33**(2): 111-114.

Jana, B.K. and T.K. Bose, 1980. Effects of fertilizers on growth and flowering of *Hippeastrum*. *Indian Agriculturist*, **24**(1): 23-30.

Jana, B.K. and T.K. Bose, 1981. Effect of weedicides on growth and flowering of hippeastrum. *Indian Agriculturist*, **25**(1) 39-48.

Johansson, E. and H. Niedieck, 1964. *Scutellonema brachyurum*, a nematode pest new to Sweden (Swedish). *Växtskyddsnotiser*, **28**: 94-98.

Karihaloo, J.L. 1984. A 45 chromosome variety of *Hippeastrum* hybrid. *Curr. Sci.*, **53**(16): 865-866.

Khaleel, T.F. and D. Siemsen, 1989. Cytoembryology of *Amaryllis* hybrids. *Canadian J. Bot.*, **67**(3): 839-847.

Kurki, L. 1958. Retarinkukan kukkimisen taustaa (Conditions for flowering in hippeastrum). *Puutarha*, **61**: 424.

Loeser, H. 1979. Comparison of amaryllis cultivars in 1979. *Zierpflanzenbau*, **19**(25): 1102-1103.

Luyten, Ida, 1935. Vegetative propagation of hippeastrums. I. *Yearbk Amer. Amaryllis Soc.*, **2**: 115-122.

Luyten, Ida, 1936. Vegetative propagation of hippeastrums. II. *Meded. Lab. Pl. Physiol. Wageningen*, **46**: 252-260.

Luyten, Ida, 1946. Over goeden en vervroegden bloei van *Hippeastrum*. *Meded. Lab. Plphysiol. Onderz., Wageningen*, No. 70, 31 pp.

Maiko, T.K. and L.M. Aks'onova, 1983. Effect of gibberellin on *Hippeastrum hybridum* development in relation to application date (Russian). *Introdukts)ya ta Aklimatizats)ya Roslin na Ukraînî*, No. 22, pp. 76-78.

Merrow, A.W. 1988. New trends in amaryllis (*Hippeastrum*) breeding. *Proc. Annual Meeting Fla St. hort. Soc.*, **101**: 285-288.

Merrow, A.W. and K.E. Kane, 1991. *Hippeastrum* breeding at the University of Florida. *Herbertia*, **47**: 4-10.

Mehlhorn, M. 1984. Production of *Hippeastrum* cut flowers at the VEG Unit of Ornamental Plant Seed Production, Barth (German). *Gartenbau,* **31**(10): 311-312.

Mii, M., T. Mori and N. Iwase, 1974. Organ formation from the excised bulb scales of *Hippeastrum hybridum in vitro. J. hort. Sci.*, **49**(3): 241-244.

Mohgazi, A.M.El, A.A. Ali and M.K. Hesbah, 1975. Phytochemical investigation of *Hippeastrum vittatum* growing in Egypt. Part II. Isolation and identification of new alkaloids. *Planta Medica*, **28**(4): 336-342.

Nautiyal, M.C. and P.N. Bajpai, 1979. Studies on nutritional requirement of amaryllis (*Amaryllis belladonna* Linn.) variety Royal Dutch. *Plant Sci.*, **11**: 75-83.

Nell, T.A. 1993. *Hippeastrum* × *cu*. In: *Flowering Potted Plants, Prolonging Shelf Performance*, pp. 51-52. Ball Publishing, Batavia, Illinois, USA.

Nong, L. and G.F. Weber, 1965. Pathological effects of *Pratylenchus scribneri* and *Scutellonema brachyurum* on amaryllis. *Phytopath.*, **55**: 228-230.

Nwankiti, O.C. 1983. Cytotaxonomy of the Harmattan lily. *Indian J. Genet.*, **43**(2): 272-275.

Okada, M. and T. Yamada, 1953. On flower bud differentiation and its development in amaryllis (*Hippeastrum hybridum*) (Japanese). *J. hort. Ass. Japan,* **22**: 119-122.

Okubo, H. 1993. *Hippeastrum* (*Amaryllis*). In: *The Physiology of Flowering Bulbs* (eds De Hertogh, A.A. and M.Le Nard), pp. 321-334. Elsevier Science Publishers, Amsterdam, Holland.

Okubo, H., C.W. Huang and S. Uemoto, 1990. Role of outer scale in twin-scale propagation of *Hippeastrum hybridum* and comparison of bulblet formation from single- and twin-scales. *Acta Hort.*, No. 266, pp. 59-66.

Otto, A. 1967. Hippeastrum flowers several times in the year (German). *Gartenwelt*, **67**: 479-480.

Pag, H. 1964. *Colletotrichum crassipes* (Speg.) v. Arx as a pathogen on *Hippeastrum vittatum. Phytopath. Z.*, **51**: 317-323.

Pettit, G.R., V. Gaddamidi, A. Goswami and G.M. Gragg, 1984. Antineoplastic agents, 99. *Amaryllis belladonna. J. Natural Products*, **47**(5): 796-801.

Pierik, R.L.M., J.S. Blokker, M.W.C. Dekker, H. De Does, A.C. Kuip, T.A. van der Made, Y.M.J. Menten and N.C. M.H. De Vetten, 1990. Micropropagation of *Hippeastrum* hybrids. In: *Integration of in vitro Techniques in Ornamental Plant Breeding. EUCARPIA Proceedings* (Wageningen). Symp. held on November 10-14, 1990, pp. 21-27.

Rao, R.V.K., N. Ali and Vimaladevi, 1971. Phytochemical studies on *Hippeastrum johnsonii* bulbs. *Indian J. Pharmacy*, **33**(3): 56-58.

Rees, A.M. 1972. *The Growth of Bulbs*. Academic Press, London, United Kingdom.

Rees, A.M. 1985. *Hippeastrum*. In: *Handbook of Flowering*, vol. I (ed. Halevy, A.H.), pp. 294-296. CRC Press, Boca Raton, Florida, USA.

Reimherr, P. 1987. Was tut sich bei 'Mona Lisa'? *Deutscher Gartenbau*, **41**(25): 1480-1481.

Roorda van Eysinga, J.P.N.L. 1984. Research on optimal fertilization. High artificial manure application is essential before planting hippeastrum (Dutch). *Vakblad Bloemisterij*, **39**(4): 36-37.

Sakanishi, Y. and H. Fukuzumi, 1964. Investigation into the ratio and amount of nitrogen, phosphorus and potassium fertilizers required for bulbil formation on bulb scales of amaryllis (Japanese). *J. Jap. Soc. hort. Sci.*, **33**: 259-264.

Sakanishi, Y. and T. Yanagawa, 1979. Bulblet formation on scale pieces of various bulbous ornamentals (Japanese). *Studies from the Instt. of Hort.*, Kyoto Univ., **9**: 100-107.

Schenk, P.J. 1954. Het vuur van *Hippeastrum. Cult. Hand.*, **20**: 279-280.

Sealy, R. 1968. X*Amarine tubergenii. J. roy. Hort. Soc.*, **93**: 430-433.

Secalis. J.N. 1993. *Hippeastrum* hybrids. In: *Handbook of Flowering*, vol. I (ed. Halevy, A.H.), pp. 294-296. CRC Press, Boca Raton, Florida, USA.

Sheehan, T.J. and K.J. Howe, 1957. A study of some factors affecting amaryllis flowering. *Proc. Fla St. hort. Soc.*, **70**: 387-389.

Sharma, S., M.B. Singh and C.P. Malik, 1981a. Relation of glycosidases to *Amaryllis vittata* pollen tube growth. *Plant Cell Physiol.*, **22**(5): 927-931.

Sharma, S., C.P. Malik and M.B. Singh, 1981b. Effect of some metabolic inhibition on germination, tube growth and O_2 uptake by *in vitro* cultures of *Amaryllis vittata* Ait. pollen grains. *Acta Bot. Indica*, **9**(1): 154-157.

Shoushan, A.M., A.M. AbouDahab, R. Eldabh and M. Auda, 1978a. Mineral nutrition of amaryllis (*Hippeastrum vittatum*). I. Effect of fertilizers on growth and flowering. *Res. Bull., Fac. Agric., Ain Shams Univ.*, No. 965, 11 p.

Shoushan, A.M., A.M. AbouDahab, R. S. Eldabh and M. Auda, 1978b. Mineral nutrition of amaryllis (*Hippeastrum vittatum* Herb.). II. Effect of fertilizers on bulb growth, bulblet production and nitrogen and sugar contents of leaf and bulb. *Res. Bull., Fac. Agric., Ain Shams Univ.*, No. 966, 11 p.

Sidhu, R.K., A.S. Basra and C.P. Malik, 1986. Hormonal effects on tube elongation, $^{14}CO_2$ fixation and phosphoenolpyruvate carboxylase activity in *Amaryllis* pollen: Promotion by abscisic acid. *Plant Growth Regulation*, **4**(3): 293-298.

Singh, M.B., S. Sharma and C.P. Malik, 1980. Uptake and utilization of exogenous sugars by *Amaryllis vittata* pollen suspension cultures. *Indian J. exp. Biol.*, **18**(8): 828-831.

Stanley, R.G. 1961. Effects of various forms of boron on germinating pollen (abstr.). *Plant Physiol.* (Suppl.), **36**: xvi.

Stanley, R.G. and E.A. Lichtenberg, 1963. The effect of various boron compounds on *in vitro* germination of pollen. *Physiol. Plant.*, **16**: 337-346.

Stouffer, R.F. 1963. A mosaic disease of hybrid amaryllis caused by cucumber mosaic virus. *Proc. Fla St. hort. Soc.*, **76**: 462-466.

Suzuki, M. and T. Tamura, 1979. Studies on the breeding of *Hippeastrum hybridum*. Cross-compatibility of cultivars and characteristics of F_1 seedlings (Japanese). *Studies from the Instt. Hort.*, Kyoto Univ., **9**: 108-118.

Tanahashi, T., A. Poulev and M.H. Zenk, 1990. Radio-immunoassay for the quantitative determination of galanthamine. *Planta Medica*, **56**(1): 77-81.

Tapio, E. 1966. Red spot of amaryllis caused by fungi. *Ann. Agric. Fenn.*, **5**: 26-47.

Tarabeih, A.M., A.J. Al-Zarari and S.H. Michail, 1980. Leaf scorch of *Amaryllis*, new to Iraq. *Phytopathologia Mediterranea*, **19**(2/3): 150-152.

Tomoda, M., N. Shimizu, K. Shimada and R. Gonda, 1985. Plant mucilages. XXXV. Isolation and characterization of a mucous polysaccharide, hippeastrum-H-glucomannan, from the bulbs of *Hippeastrum hybridum. Chemical pharmaceut. Bull.*, **33**(1): 48-51.

Tsukamoto, Y. and S. Fujioka, 1956. studies on fertilizers for florist crops. I. Fertilizers for amaryllis. *J. hort. Ass. Japan*, **25**: 208-212.

Van Leeuwen, A.J. 1974. Bulb disinfection for hippeastrums (Dutch). *Bloembollencultuur*, **84**(30): 708, 711.

Wahi, S.D. and S.K. Bhattacharjee, 1986. Correlation and path-coefficient analysis in *Hippeastrum hybridum. South Indian Hort.*, **34**(4): 244-251.

Wang, J.F., C.L. Jia and B. Jin, 1989. An investigation on the growth habit of tissue cultured plantlets of Barbados lily (*Amaryllis vittata* Ait.) (Chinese). *Scientia Agric. Sinica.*, **22**(1): 53-56.

Yanagawa, T. and Y. Sakanishi, 1977. Regeneration of bulblets on *Hippeastrum* bulb segments excised from various parts of a parent bulb. *J. Jap. Soc. hort. Sci.*, **46**(2): 250-260.

4

Anemone (Family: Ranunculaceae)

Sanyat Misra and R.L. Misra

[**Common names**: Anemone/Blue wood anemone/Poppy anemone/Poppy flowered anemone (*A. coronaria*), Broad-leaved garden anemone (*A. hortensis*), Buttercup anemone (*A. ranunculoides*), Double snowdrop anemone (*A. sylvestris* var. *flore-pleno*), Grecian windflower, Greek anemone, Greek windflower, Honorine Jobert/The Bride/Whirlwind (*A. japonica* var. *alba*), Japanese anemone (*A. hupehensis, A. japonica*), Lady Ardilaun (*A. japonica* var. *rubra*), Lily-of-the-field (a Biblical name for *A. coronaria*), Poppy scarlet anemone (*A. fulgens*), Pasque flower (*A. pulsatilla*), Snowdrop anemone/Snowdrop windflower (*A. sylvestris*), Windflower/Spring anemone (*A. quinquefolia*), Wood anemone (*A. nemorosa*), Yellow wood anemone (*A. ranunculoides*)]

Intrtoduction and Origin

The genus was first named by Theophrastus and is thought to have been derived from the Greek *anemos* means 'wind'. However, some authorities now consider it deriving from the Syrian *nama'an,* the cry for the dead Adonis whose blood returned to life as scarlet anemone, as per the legend, or is a Greek version of the Semetic Naamen, or Adonis, whose blood as per the legend, resurrected into *Anemone coronaria,* or *Adonis* when Adonis died in battle. Anemone is a subject only of the temperate regions and not of the tropical ones. Its cultivation dates back to 1936 when Mr. R. Pitcher at Rhinebeck (USA) grew it through seeds, and soon after its introduction into Europe, in 1948 breeding of giant F_1 hybrids began and in 1982 cv. 'Mona Lisa' was introduced for cut flower production (Hegele, 1986a).

It makes an excellent cut flower as well as excellent pot plant. The summer and autumn flowering species are fibrous-rooted, suitable for growing in a border, and have branching stems. Spring-flowering species are generally tuberous-rooted (rhizomes), suitable for rock gardens, for naturalizing in open woodlands and for providing cut flowers, especially the 'de Caen' and 'St. Brigid' strains. Many species and varieties are also used as garden, container or landscape plants and for mass effect. These have violet, dark blue, scarlet, pink and white colour forms. These can effectively be planted preferably on shady banks beneath high-crowned trees. Plants form tuberous rhizome which can last up to 10 years. The majority of the species come from the temperate and mountainous regions of the Northern Hemisphere.

Botany

Anemone is a dicotyledonous geophyte which can be grown only in the temperate regions and not in the tropical regions. Its rootstock is rhizomatous, roots fleshy or fibrous and sometimes woody, basal leaves 3- or 5-lobed, deeply dissected or compound and sometimes entire, stem leaves often in whorls below the inflorescence in cluster of three, flowers petalless, composed of 5-20 coloured petaloid sepals (perianth segments), actinomorphic, hermaphrodite, solitary or in cymes, shallow dish-shaped, white, yellow, purple, blue, red or pink, stamens free, numerous and shorter than sepals, carpels numerous, 1-ovuled, ovule pendulous, and fruit of many single-seeded achenes. Its flowers are cup- or bowl-shaped. As per the values obtained for photosynthesis, respiration and dry matter content in relation to leaf area of *Anemone nemorosa*, Löhr (1952) found that neither it is a decided shade nor a decided light loving plant. Petiet (1958) reported that self-pollination in *Anemone coronaria* cvs 'Hollandia' and 'Sylphide' was successful in open. Gill (1977) reported that anemone strain 'St. Piran' has been bred for high flower quality and winter hardiness in England. Wide colour range of the *Anemone coronaria* flowers is a result of the anthocyanin pigments. Pollen has pelargonidin, cyanidin and malvidin-3-

glucoside and scarlet petals contain pelargonidin-3-lathyroside pigments. Cyanidin from purple and delphinidin from blue and violet petals have been isolated. Harvey (1971) isolated three anthocyanidin (aglycone) moieties of the anthocyanins present in petal tissue of different *A. coronaria* phenotypes as pelargonidin, delphinidin and cyanidin. Figurkin and Ogurtsova (1975) isolated triterpenic glycosides from the rhizomes of *A. ranunculoides*. The haploid chromosome number in *Anemone* is 7 or 8, depending on the species. Madahar (1967) experimented with many taxa and observed that most were diploid with 2n = 16 though *A. palmata* is tetraploid. Sushila Bhattarai (1989) studied the mitosis and meiosis of the Himalayan species *A. obtusiloba* and *A. elongata* (both 2n = 14), and *A. vitifolia* and *A. rivularis* (both 2n = 16) and concluded that since all these species had symmetrical karyotypes, these all derived from a common ancestral stock.

Classification, Species and Varieties

There are some 150 species under this genus and all are herbaceous perennials. Darnell (1935, 1936ab) described some 106 species. The most important ones are *Anemone altaica*, a native to Arctic Russia and Japan, growing to 20 cm, rhizomes cylindrical, thick and creeping, basal leaf solitary, petiolate and often cut, stem leaves tripartite in whorls of 3 and mostly divided, flower solitary, white and veined violet inside and sometimes flushed violet outside, perianth segments 8-9 or even up to 12; *A. apennina*, native to S. Europe, grows to 15 cm height, rhizomes almost tuberous and creeping, leaves basal and also frequently in whorls on the flower stalk just below the inflorescence, normally 3-lobed, each further deeply lobed, flowers solitary and blue, rarely flushed pink or white, segments 8-23, and its varieties are *A. a. alba* or *albiflora* (white), 'Petrovac' (rich blue, multi-petalled), 'Purpurea' (purple-rose), etc.; *A. baldensis*, native to Alps mountains, rhizomes tuberous, grows to 15 cm, leaves biternate with lobed leaflets, flowers white and segments 6-10; *A. biflora*, native to Afghanistan, Iran, India (Kashmir) and Pakistan, grows up to 20 cm, rhizomes tuberous and congested, leaves 3-lobed each with further deep lobing, flowers 2-3, nodding and crimson, rarely orange to yellow, and perianth seagments 5; *A. blanda*, native to SE Europe, Turkey, Cyprus and Caucasus (USSR), very similar to *A. apennina*, grows to 18 cm, rhizomes tuberous, thick and congested, basal leaf 1 or absent, 3-parted and each part further divided, stem leaves 2-3 and deeply parted into 3, flowers solitary, white, pink, mauve or blue with 9-15 segments, and its varieties are 'Atrocaerulea' ('Ingrami', blue), 'Blue Shades' (pale to deep blue), 'Blue Star' (pale-blue), 'Bridesmaid' (large white), 'Bright Star' (pink), 'Charmer' (pink), 'Fairy' (pure white), 'Pink Star' (large, pale-pink), 'Radar' (magenta with white centre), 'Rosea' (pale-pink), 'Violet Star' (large, violet with white exterior), 'White Splendour' (large, white with pink exterior), var. *alba* (white), var. *scythinica* (recorded from N. Turkestan, white with blue exterior).

A.coronaria, a native of Southern Europe, Mediterranean region and Turkey which grows up to 60 cm in height. Its solitary flowers appear during spring in white or in various shades of blue, pink, scarlet, red and bicolours. Its perianth segments are 5-8. Rootstock is a brown tuber where one year old is rounded on the top and the older tubers often present stoloniferous extensions with misshapen and knobby roots, and the tubers have contractile roots. It is the primary species used in commerce for cut flower production and as potted plant, and its varieties are single to semi-double such as 'de Caen' group [collective name of single-flowered cultivar, *viz.* 'His Excellency' (flowers large and scarlet with white eye, and its strain as a separate variety 'Hollandia'), 'Mr. Fokker' (blue), 'The Bride' (white), etc. and the varieties most popular commercially are 'Jerusalem' (Israel) from wild *A. coronaria* × 'Wicabri' hybrids, 'San Piran' (England), 'Tetranemone' (France), 'Wicabri' (the Netherlands), etc.], 'Mona Lisa' group [F_1 hybrids propagated through seeds (full colour range), the var. 'Cleopatra' is also grown through seeds, 'Sylphide' (mauve)], 'St. Brigid' group [collective name for semi-double cultivars, having pastel and bicolours, *viz.* 'Lord Lieutenant' (blue), 'Mt. Everest' (semi-double form of 'The Bride', flower white), 'The Admiral' (deep pink), 'The Governor' (vermillion-scarlet)], and the forms, *viz.*, var. *coronaria* (scarlet), var. *alba* cv. 'Burnat' (white), var. *cyanea* (blue), var. *rosea* (pink), etc. The species has produced robust hybrids with *A. hortensis*, *A. pavonina* and *A.* × *fulgens* which can flower throughout the year through judicious planting or by growing in winter under cloches.

Other cut flower species are *A.*× *fulgens*, {syn. *A. stellata*), recorded from S. France as a natural hybrid, flowers solitary of scarlet colour with narrow sepals, spring-flowering, rhizomatous, and developed through *A. pavonina* (rhizomatous, flowers solitary of violet, scarlet, purple or pink, rarely white or yellow, and the basal portion white, spring-flowering) × *A. hortensis* (flowers solitary of pink-mauve colour, spring-flowering, rhizomatous)} which includes many varieties and forms such as 'Antulata Grandiflora' (large flowers with yellow centre), 'Multipetala' (same as the species but with semi-double flowers), 'St. Babo Hybrids' (developed through hybridization with *A. coronaria*, the hybrids are also known as 'Peacock Eye' hybrids, and colours ranging from pink to salmon, brick red and dark red to violet-blue); *A. hortensis* and *A. hupehensis* (autumn flowering with mauve-pink flowers, and one of the parents of *A. hybrida*, rhizomatous); *A. hortensis*, native to C. Mediterranean, grows to 30 cm, rhizomes congested and resembling a tuber, basal leaves lobed and stem leaves unlobed, flowers solitary and pink-mauve, and segments 12-20; *A. nemorosa*, native to whole Europe, grows to 30 cm, rhizomes slender and creeping, leaves trilobed and sub-lobes further divided, flowers solitary and white which may turn flushed purple or pink with age, segments 6-8, rarely 5-12, and its forms and varieties are 'Alba Plena' (double white), 'Allenii' (lilac or pale-blue exterior and lavender interior), 'Blue Beauty' (pale-blue), 'Blue Bonnet' (large blue), 'Bracteata' (large double white, outer segments wholly or in part green), 'Grandiflora' (large white), 'Green Fingers' (flowers red), 'Leeds Variety' (white but with age flushed pink), 'Lychette' (large white), 'Monstrosa' (greenish), 'Plena' (double), 'Robinsoniana' (grey exterior, lavender interior), 'Rosea' (opening white but changing to deep pink with age), 'Royal Blue' (blue), 'Vestal' (a petaloid button in the centre of the flowers), 'Vindobonensis' (cream), 'Virescens' (a pyramid of leaf-like bracts on the place of flowers), etc.; *A. quinquefolia*, similar to *A. nemorosa*, native to USA (Washington to N. Carolina), grows to 30 cm, rhizomes tough and horizontal, leaves solitary, and 3-parted but stem leaves smaller, flowers solitary, flowers white often flushed pink, and segments 4-9, mostly 5; *A. ranunculoides*, native to entire Europe except Mediterranean, grows to 15 cm, rhizomes slender and creeping, basal leaves may be absent or solitary and both the types 3-parted, each part deeply lobed, flowers solitary and yellow, segments 5-6 but sometimes more, and its varieties are 'Flore Pleno' (yellow, semi-double, perianth segments >12), 'Grandiflora' (large), 'Superba' (yellow), etc.; *A. rivularis*, native to N. India and SW China, grows to 90 cm, rhizome tuberous, leaves trifoliate and leaflets trifid, flowers several, segments 5-8, colour white tinged metallic-blue below; *A. trifolia*, native to Balkans, Italy, Hungary, Portugal and Spain, grows to 15-20 cm, rhizome thin and creeping, basal leaves usually absent, stem leaves in whorls of 3, each deeply cut to form 3 leaflets, flowers solitary and white, rarely flushed pink, segments 5-12, and its ssp. *albida* (white) was recorded from Portugal and NW Spain; *A. sylvestris*, native to Europe and W. Asia, grows to 45 cm, spreading by suckers from the roots, clump-forming, leaves 3- to 5-lobed, each deeply toothed, flowers fragrant white, and segments 5; *A. vitifolia*, native to N. India, Myanmar and China, grows to 60 cm, clump-forming, leaves 5-lobed, and flowers white with 5 segments; etc.

Other species of economic importance are: (i) **summer and autumn-flowering**: - *A.* × *hybrida* (sometimes listed as *A. japonica* as well as *A.* × *fulgens*), is a group of free-flowering hybrids once established (normally after one year) and includes the varieties 'Honorine Jobert' (white) popularly known as *A. japonica, 'Lorelei'* (rose-pink), 'Louise Uhink' (white), 'Max Vogel' (pink), 'Queen Charlotte' (semi-double, pink), 'September Charm' (pink), etc.; *A.* × *lesseri* (summer flowering *i.e.* in May and June, rose-red, purple, yellow and white forms); *A. glauciifolia* (syn. *A. glaucophylla*), rootstock neither rhizomatous nor fibrous but thick, flowers solitary and mauve in colour; *A. lancifolia*, rootstocks neither fibrous nor rhizomatous but fleshy, flowers solitary and white; *A. narcissiflora*, flowers white in umbels and clump-forming; etc.; and (ii) **spring and winter-flowering** (tuberous rooted, mostly rhizomatous):- *A. apenina* and *A. nemorosa* (both are described above), and *A. ranunculoides* (also described above) has produced hybrids with *A. nemorosa* which are called as *A. lipsiensis* (sulphur-yellow flowers), etc. Other spring-flowering species are *A. altaica* (already described above), *A. biflora* (also described above), *A. bucharica* (rhizomatous, flowers usually in pairs but sometimes solitary or in threes and the colour red or violet-red), *A. caroliniana* (tuberous, flowers purple or white-purple), *A. caucasica* (syn. *A. blanda* var. *parvula*, rootstock small tuber, flowers blue or white), *A. eranthoides* (rhizomatous, flowers golden-yellow inside, yellow-green outside), *A. flaccida* (rhizomatous, flowers 1-3, cream-white sometimes flushed pink), *A. gortschakowii* (rhizomatous, flowers in pairs but sometimes solitary, pale-yellow but seldom flushed red), *A. palmata* (rhizomatous, flowers solitary but sometimes in pairs, yellow but often flushed outside), *A. petiolulosa* (rhizomatous, in clusters of 2-4 but sometimes solitary, yellow with flushed red exterior), *A. quinquefolia* (described above), *A. trifolia* (described above), *A. tschernjacwii* (rhizomatous, flowers 2-3 together, white basal portion flushed purple-pink), *A. tuberosa* (tuberous rhizome, flowers solitary white or pink), etc.

Hansen and Sieber (1965) evaluated autumn anemone varieties in Germany and recommended 'Honorine Jobert', 'Königin Charlotte', 'Praecox', 'September Charm', 'Robustissima', 'Prinze', 'Heinrich', 'Wirbelwind (Whirlwind)', 'Splendens', 'Albadura', 'Géante des Blanches' and 'Kriemhilde' for commercial cultivation. 'Bressingham Glow' and 20 other varieties were found inferior. Clausen (1972) tried 14 Japanese *Anemone hupehensis* and others in Denmark and recommended the cultivation of 'Honorine Jobert', 'Konigin Charlotte', 'September Charm' and 'Rubra'. In Holland, Hensen (1979) recommended *A. tomentosa* cv. 'Superba', *A.* x *hybrida* cvs 'Honorine Jobert', 'Frau Marie Maushard', 'Aureole' and 'Whirlwind', and *A. hupehensis* cv. 'Prinz Heinrich'.

Propagation

Summer and autumn flowering species are planted from October to March by dividing the clumps or through root cuttings. The cuttings are planted in pots or boxes containing peat and sand (both in equal quantity by volume and placed in a cold frame. When three leaves have developed, these are planted out in the nursery and in the ensuing October these are planted out for flowering. **Spring or winter flowering species** are planted through separation of the rhizomes or tubers, as the case may be, in the end of the summer when the growth has completely ceased.

Since many rhizome transmitted diseases are transmitted by vegetative means and also as multiplication rate being utterly poor, anemones are best produced from seeds which come up, at the most, in two seasons to produce flowers but in one year, seedlings form rhizomes of commercial size which are known as 'pips'. De Winter (1974) stated one of the causes for non-viability of anemone 'pips' (rhizomes or tubers obtained by ripening-off seedling plants a few months after flowering) due to harvesting at improper time. Through experiment with 'De Caen' and 'St. Brigid' types, he found the best time for lifting in Holland around start of September though it depends on the extent to which the crop foliage has died down. Growing in mild climates or under protection, seeds go on flowering continuously. However, *A. blanda* is propagated through division of the rhizomes. Controlled parental lines are used for producing hybrid seeds and rhizomes or tubers, especially in case of *A. coronaria*. Seeds germinate at 10-14 °C or at 15-21 °C. With French cultivars, the seeds harvested in the autumn are immediately germinated at 23 °C whereas winter and spring harvested seeds are first dried for 10 days at 12 °C and then sown for quick germination. Most of the anemones produce abundance of seeds which are very light and weigh some 70-1200 seeds per gramme. One flower may produce up to 200-300 seeds. Seeds germinate from 7-14 days at 15 °C. Certain parents are identified for F_1 seed production. One such F_1 hybrid is available for commercial production with the name of 'Mona Lisa'. Matured seeds are sown in pots or pans of seed compost filled deeply, *i.e.* not less than 15 cm deep as anemone develops the taproot of about 10 cm length. The medium where seeds are to be sown should be well drained. The seed-raised rhizomes are discarded after taking flower for one season. Piskornik (1984) found best seed germination at 15 or 20 °C when sown below or at 0.5 cm deep though 25 °C as well as 1.0 or 2.0 cm deep sowing inhibited it markedly, irrespective of darkness or light. Hassan *et al.* (1985a) studied the effects of temperature (24 or 8 °C) and supplementary lighting (2000 lux for 0, 2, 4 or 6 h/day) on *Anemone coronaria* seed germination and found best results with 2h light at 8°C. Hegele (1986) stated that at a low temperature of <15°C the anemone 'Mona Lisa' seeds can be grown easily with good percentage of seedlings. Jones (1986) studied the seed germination and corm sprouting in anemone 'St. Piran' and recorded best and rapid germination of seeds at 15°C and the corm sprouting at 18°C. Jones *et al.* (1986) stated that defluffing of anemone seeds (single seeded achenes are covered with unicellular hairs which hinder handling) is possible through sulphuric acid treatment for 3 minutes where seeds of 'St. Piran' germinated up to 39 per cent and for 10 minutes where seeds germinated only 1 per cent while in the control it was 69-74 per cent though defluffing through high air pressure for 0.5 to 4 minutes by spinning the seeds in a cylinder where the hairs were getting separated due to abrasive action, the percentage germination of seeds was from 36 to 58 per cent while 'Mona Lisa' performed slightly better than 'St. Piran'. Piskornik (1986/1987) severed the fruiting parts + stems at three stages (green, white and yellow) and placed it in water to ripen, and after shedding, the achenes were collected and sown to see their viability in comparison to those collected after full ripening on the mother plant. Also the usually ripened achenes on the mother plants were collected and stored in plastic bags at 20°C for up to 36 months or at 4°C for up to 72 months. Achenes harvested prematurely and then ripening them in water had no any adverse effect on germination potential but those harvested after usual ripening and stored at 20°C gave <45 per cent germination after 24 months while those stored at 4°C gave <45 per cent germination after 48 months. Piskornik (1989) in Poland had sown graded (0.83 g/1000 seeds) and ungraded (0.64 g/1000 seeds) seeds of *A. coronaria* 'de Caen' race of varied colours on March 1 at 20°C in a greenhouse and transplanted in field after 82 days, *i.e.* May 22, the flowers were harvested every 2-3 days, and tubers were lifted at 265 days of sowing, *i.e.* November 22. He obtained almost twice as many flowers, *vis-à-vis* 25 per cent greater tuber weight under graded seeds than from ungraded seeds. Horovitz *et al.* (1975) in Israel obtained germination of achenes in dark in *A. coronaria* 'de Caen' type at 15-20°C and in wild forms growing there at 10-15°C.

Micropropagation is the quickest method for multiplying its elite germplasm in shortest duration of time. Maia and Pellergrin (1974) cultured mature *Anemone coronaria* achenes collected from mother plants from November to May in France. They obtained easy germination of those achenes at 23°C which matured in November and December though only a low percentage of achenes germinated those maturing from January to May. They further observed that a 10-day pre-treeatment at 12°C enhanced germination at 23°C as did dry storage at room temperature. Anther culture is well adopted method of micropropagation for quick multiplication in anemone. Johansson and Eriksson (1977) induced embryo formation through anther cultures of *Anemone virginiana* and three other species. Through addition of activated charcoal to the Nitsch and Nitsch medium, a higher percentage of embryos was recorded than cytokinin 2ip at 10^{-6} M. Johansson *et al.* (1982) selected anthers of *anemone* and certain other ornamentals with pollen at the uninucleate stage and cultured by floating on a liquid medium which overlay a solid agar medium with charcoal. They found stimulated embryogenesis in *A. canadensis, A. radicatum, A. setigerum* and *A. vitifolia.* Cold treatment of *A. canadensis* anthers in the culture medium for 20 days at 7°C stimulated embryogenesis if the anthers were cultured in air but inhibited embryogenesis in anther cultures in 2 per cent CO_2. Cold treatment reduced ABA concentration in *A. canadensis* anthers from 2.2 x 10^{-6} g/g in untreated anthers to 0.6 x 10^{-6} g/g in treated anthers. Johansson (1983) reported that inhibition of embryogenesis in *A. canadensis* due to added ABA to the medium was reduced to a tune of 80 per cent through addition of activated charcoal (AC). They stated that embryos from anther culture produce phenolic substances in higher quantity in absence of AC so addition of AC reduces the concentration of such substances. Johansson and Eriksson (1984) reported that in *A. canadensis* cold treatment increased the number of proembryos but incubation in 2 per cent CO_2 had no effect though in case of *A. dichotoma* it increased embryo production and incubation in 5 per cent CO_2 increased embryo production in *A. hupehensis.* In a further trial with *A. canadensis,* Johansson *et al.* (1990) found that embryogenesis was independent of the presence in the culture medium of glycine, nicotinic acid, pyridoxine, thiamine, folic acid, D-biotin or myoinositol, absence of Fe-EDTA totally inhibited embryogenesis, and AC absorbed Fe-EDTA, pyridoxine, folic acid and nicotinic acid in a double-layer medium almost completely within 24 h. AC has also been found releasing some substance(s) that stimulates embryogenesis. Addition of polyvinylpolypyrrolidone (PVPP) in the medium stimulated embryogenesis but was not as effective as AC. There was complete inhibition of embryogenesis when both PVPP and AC were added together but in presence of AC addition of ethanol to the culture medium also stimulated embryogenesis.

Soils, Preparation of Land, Planting and Weed Control

Summer and autumn flowering types may be grown from October to March in partial shade in a well-drained but moisture retentive soil rich in humus. Jeff (1958) stated that acid conditions affect anemones severely. At a pH of 3.5-5.0, no emergence of anemones took place, and 5.0-6.5 pH resulted in a very slow emergence and growth, growth was found improved at 6.5-7.0, and pH above 7.0 the growth was normal, whereas above pH 7.5 there was no any further improvement, so it is concluded that anemones are comfortable at pH range of 7.0-7.5. These plants, especially *A. x hybrida* do not like disturbance for several years after once planted. After flowering is over, the plants are headed back to ground level. Winter and spring flowering types may be grown under sun or partial shade in water retentive but well-drained soil.

A. coronaria and *A. x fulgens* varieties prefer being grown under full sun. Large rhizomes may take about three months and smaller ones up to six months to come into blooms. Both the types of varieties deteriorate quickly so should be replaced at every 2-3 years. They are planted 4.0-7.5 cm deep in September or October. Jeff (1956a, 1957a, 1958) in England planted anemones on May 29 or May 20, June 19, and July 10 and 31 and recorded highest total yields through earliest planting with peak cropping occurring in October but in January and February when flowers are in great demand the production particularly from largest corms ($^3/_4$ cm) fell down considerably. Jeff (1959) recorded highest yields and the earliest peak crops by planting largest ($^3/_4$ cm) corms on May 29, June 19 and July 10 & 31 and for cropping between November and January, he suggested planting on June 19 for continuous cropping. Meredith and Sheehan (1965) tried 30, 20 and 15 cm spacing intervals in 'Wicabri' anemones in open in Florida and recorded increase in flower production and decrease in flower diameter with every reduction in spacing interval, and they obtained excellent crop of flowers from January to late March. Gradner and Reimherr (1986) planted the seedlings of 'Mona Lisa' anemones at 13-14 day's intervals starting from July 17 to September 9 and recorded highest yields with better quality blooms with early plantings, even better being in the mixed colour forms, followed by blue and then red ones. Reimherr (1987) in a greenhouse trial in Germany with 'Mona Lisa' anemones planted on August 11 found that planting on ridges increased the flower yield by 8 per cent. Trickle irrigation increased the stem length while heating during winter (8°/5°C day/night with a ventilation temperature of 13°C) decreased it though yield was not affected by either of the treatments. He also observed decreased yield with poor stem length by delayed planting, usual planting being mid-August. However, he recorded more yield in scarlet, orchid and blue types than the white types. Piskornik (1989) obtained better flowering when seeds were sown in Poland on March 7 than on March 21and the tuber weight was found 30 per cent less when the plants were allowed to form seeds than those used for cut flower production and 60 per cent less than those from de-flowered plants. Petiet (1959) advocated cultivation of anemones in Holland from seed to flower in one year by sowing the seeds of *Anemone coronaria* 'Sylphide' in mid-January, transplanting of seedlings on May 1 at 15 x 15 cm apart, harvesting of flowers in mid-June and continuing till Octonber end with yield of flowers averaging 6 per plant but when the seeds were sown from late February to mid-March, seedlings were planted out on May 6 at 25 x 25 cm apart, the flowering started from early July until mid-October and the yield of flowers averaged 8 per plant. Yasuda and Yokoyama (1957) by planting one lot of anemone tubers at normal time of planting, *i.e.* October, while other lots were stored either at room temperature or at 0°C and planted every month from November until March and found 6 week's difference in the time of flowering between tubers planted in October and those planted in March. In Morocco, Moret (1952) stated that anemones are planted there in September-October at 0.4 x 0.4 m spacing and 5 cm depth after steeping in warm water for 24 hours where first shoots appear after 15-20 days and flowering occurs from January to April with about 50-100 marketable flowers per plant. He reported that 4-500 pea-size corms/m^2 are obtained. Kamel and Nabih (1978) at monthly intervals from July 15 to October 15 planted out in pots the room-stored tubers of *A. coronaria* and reported that earlier planting took more time (July planting taking 227 days) for flowering though later planting required less (October planting 146 days), and they suggested to plant the anemones in half shade rather than full sun. Scheu and Reimherr (1986) with anemone 'Mona Lisa', planted at 14, 16, 18, 20 or 25 plants/m^2 on raised beds in greenhouse with a 5-cm mesh netting for support and recorded best yields having longest stems and no *Botrytis* with high density planting though red-flowering forms produced the lowest yields. Hofmann (1986) tried greenhouse cultivation of 'Mona Lisa' anemones and reported that 7°C temperature is optimum for its growing in the greenhouse. Fischer and Kalthoff (1987) with cv. 'Mona Lisa' planted in large troughs on September 25 in a clay-peat medium provided with a basal dressing of 1 g of Flory 3 Grün fertilizer per litre of substrate (representing 24.0 g N, 17.6 g P_2O_5 and 24.0 g K_2O/m^2) and 3 rates of supplementary fertilization and obtained best vegetative growth with highest quality cut flowers and the longest vase life through weekly fertilization totalling over 24 week's period as 32.0 g N, 17.6 g P_2O_5 and 24.0 g K_2O/m^2. In a further studies with seedlings of white, scarlet and dark blue F_1 hybrid variants of anemone cv. 'Mona Lisa' planted on August 11 and fertilized with 1 g Flory 3 Grün and 50 mg Flory 10 per litre of substrate and with supplemenatary fertilization at various concentrations from September 1987 to April 1988, Fischer and Kalthoff (1989) harvested the flowers from December 16 to April 14 in Germany, maximum yield being in basal fertilizer with 1 g Flory 3 Grün complete fertilizer per litre of substrate (180 g/m^2) followed by weekly application of 0.2 per cent fertilizer solution (total being 187 g/m^2 over 28 weeks including 28 g N, 21 g P_2O_5 and 28 g K_2O/m^2). Van Leeuwen and Dop (1990) lifted the tubers of *A. blanda* cvs 'White Splendour' and 'Blue Shades' in the end of June or July, stored dry at 9, 13 or 17°C and planted in 11-cm pots in August-September, then stored at 2, 5, 9 or 13°C for 11, 13, 15 or 17 weeks and then pots were moved into a greenhouse at 12 or 18°C at the end of December. They observed increase in flowering percentage and number of flowers per plant under 5°C pot-storage and 12°C greenhouse temperature as growing at 18°C greenhouse temperature there was high percentage of flower abortion. Those stored for 7 weeks produced more flowers with longer stems than those stored for 3 or 12 weeks, and the storage at room temperature was found better than at 7.2°C (Radspinner and Sheehan, 1963). It is essential to protect the crop from both the sides and overhead during severe winters (Jeff, 1956a, 1957a), however, the best yields were obtained under flat-topped cloches with the top glass in position only during bad weather.

For anemone growing, the soil should be porous, rich in organic matter (3 per cent or more), preferably sandy loam and should have good water holding capacity but well-drained. Poor drainage is the major cause of rhizome or tuber rotting. The soil structure and texture should be to the extent that a constant moisture level in the soil is maintained during the germination of seeds and during warmer months. At the time of field preparation, 20-30 tonnes of compost or farmyard manure should be incorporated in the soil, and no any fertilizer should be used at the time of planting. Anemones, especially those producing rhizomes or tubers require deep soil, at least 30 cm so while preparing the land three deep ploughings are necessary followed by planking, and then finally the beds are prepared of convenient sizes. The seed sowing density that provides the optimum yield of flowering size rhizomes or tubers is 5 g/m^2 or 3000 seeds/m^2. Before sowing of seeds, these should be held for 10 days at 12° C. While sowing the seeds, these can be mixed with moist sand for equal distribution on the soil surface and the crop from the seed is harvested when leaves have senesced. For flower production, the rhizomes are graded into 6-7 cm, 5-6 cm, 4-5 cm and 3-4 cm circumference when grown from seeds as these will certainly flower in the ensuing season, *i.e.* the second season of seed sowing. The size influences the flower quality and quantity. The 20 x 20 cm spacing for planting of the rhizomes or tubers, as well as the fibrous-rooted plants is optimum. Jeff and Fuller (1961) advocated a wider spacing which favours better performance and growth though they said that increases in number of blooms cut are proportional to increases in the number of corms planted per hectare, and in general, they advocated between 2,00,000 and 2,12,500 corms per hectare. Petiet (1962) while studying 25 x5 cm, 35 x 5 cm and 35 x 10 cm spacings recommended wider spacing for increasing flowers though with delayed flowering, increasing seed heads, and their weight per plant though the yield of flowers and seeds was found poor per unit area. Garibaldi (1986) while working with 1-cm circumference imported anemone corms from Holland by planting in northern Italy in July, August and September at 250 corms/m^2 planting density, obtained more corms of 3 cm and above in circumference after 9 months of planting, September planting giving 148 corms/m^2 though July or August plantings 68 and 65 corms/m^2, respectively. However, he recorded slightly more flower production from corms multiplied in Italy than from Dutch corms grown in Italy.

Throughout the growing period the field should be kept weed-free to ensure proper growth of the plants. It would be better to hand-pick the weeds whenever these are found growing. Regular weeding of the field will ensure that whole of the nutrients available in the soil will be consumed only by the main crop. Jeff (1958) recommended PCP at 13.5 kg/ha four days before emergence as effective and safe. He suggested that weed control is better under dry than under wet conditions. Jeff (1959) stated that simazine and monuron both control weeds as highest flower yields were obtained with simazine at 560 g/ha or monuron at 1.125 kg/ha. Jeff (1958, 1959) recommended black polythene films for mulching to eliminate weeds. Mulching the beds with straw may be done to keep the soil cool during summer on the hills. Jeff (1960) applied five residual herbicides six days after planting anemone corms and recorded highest yield of marketable flowers with simazine at 0.56 kg/ha and monuron 1.12 kg/ha while by applying chlorpropham at 0.56 kg/ha + fenuron at 1.12 kg/ha and monuron 1.12 kg/ha 10 days after planting in the very dry season he obtained best residual weed control. Fouchard (1961) when applied triazines as pre-panting found good control and when applied ametryne and prometryne as post-planting to rooted corms these were found less injurious. Jeff (1964) found 1.4 litre (measure) of 40 per cent CIPC + 4.5 kg of 25 per cent fenuron or 1.125 to 2.25 kg of 50 per cent simazine per hectare as quite effective residual herbicides in anemone fields. De Rooy (1966) recommended linuron as most promising herbicide to be applied a few days before emergence and if the corms of anemones are planted below the normal 1-2 cm then the injury caused by this herbicide is drastically reduced, however, it is never recommended for use when the crop is raised from seeds. Gill (1971) stated that though all the weedicides in one way or the other reduce the production but labour cost saved while manual weeding compensates it. He recommended simazine at 840 g/ha as the best all round herbicide, and chlorpropham + fenuron at 11.25 kg formulated product per hectare or terbacil at 560 g/ha quite effective. Brosh *et al.* (1972) found pre-emergence application of chlorthal – dimethyl at 7.5 kg/ha quite safe for anemone 'de Caen'.

Irrigation, Manures and Fertilizers

The plants should be kept uniformly moist through peripheral irrigation as water at the crowns of the plants results in crown rot. Over watering should strictly be avoided as this will cause rot of the rootstock and ultimate collapse of the plant. Compost or farmyard manure at the rate of 20-30 tonnes per hectare should be incorporated in the soil at the time of bed preparation. A 200 ppm N regime at every 10 days to the crop at 'run off' stage will be quite effective. The crop is not heavy feeder; however, use of fertilizer should be based on soil and leaf analysis. Anemones never tolerate soluble salt levels more than 2g/l. $Ca(NO_3)_2$ encourages good growth. Since this crop is grown under low temperatures, the uses of urea and ammonium fertilizers are avoided. Jeff (1956a, 1957a, 1958) stated that plots in which a good balance of P and K had been maintained showed highest yields together with plots supplied with dung manure. He noted that N had a depressing effect on total yields and the two plots which received only K also provided poor yield. Jeff (1959) suggested balanced doses of phosphate and potash and of farmyard manure giving good results, and the nitrogenous top-dressings in mid-August gave better leaf colour, longer stems and sometimes higher flower yields. Puccini (1961) stated that $N:P_2O_5:K_2O$ requirement of anemone was in the ratio of 1:0.35:2 with minor element absorption of Cu, Mn, Zn, B, Va and Ti being considerable, Mo and Co being moderate and Li, Ba, Be and Zr requirement being in very minute quantity. Radspinner and Sheehan (1963) with *Anemone coronaria* 'Wacabri' when tried for winter and spring cut flower in North Central Florida, recorded highest yield with greatest stem length and flower diameter at the highest fertilizer level of 2.4 kg of coated slow-release 9-9-9 fertilizer per 100 square feet, and the largest and greatest number of flowers were produced with non-stored corms.

McDowell (1985) transplanted 'Mona Lisa' F_1 hybrid anemones of 7 colours on August 21 in USA in 15-cm pot containing peat-soil-vermiculite (2:1:2) mixture. The substrate was supplied with a 12-36-14 ($N-P_2O_5-K_2O$) fertilizer having 550 ppm N one week after planting and after the establishment of the plants a regular application of 25-5-20 at 550 ppm N used to be given. Flowering appeared after 51 days of transplanting in white, blue and orchid types, a few days later in bicolours, wine and pink types while in red-flowered types 62 days after transplanting. 'Purple Splash' was transplanted six weeks later than others and flowered at 91 days. The mean flower yield per plant for the 7 colours was 6-7, flower diameter >8.0 and with the mean stem height of 40 cm. Meredith and Sheehan (1965) used 8-8-8 fertilizer at 4160-6185 kg/ha with 'Wicabri' anemones and obtained enhanced flower production, however, increase in the levels had no effect.

Growth, Development, Flowering, Flower Forcing and Height Control

In nature, rhizomes or tubers and seeds go into rest during the dry and hot summers, a condition akin to Mediterranean climate where most of the *Anemone* species originated, adapting summer drought conditions. This condition is necessary for their survival. Sprouting of the clonal part or germination of the seeds occurs only when cool conditions with adequate moisture reappear in autumn and winter. Climatic factors have strong influence on anemone flowering. The cooler and moistened the growing area, the later and longer is the flowering time. The areas which experience cool and wet summers, the seedlings of certain varieties, *e.g.* 'Mona Lisa' can be used for production of winter and spring cut flowers. *A. coronaria*, a species native to Mediterranean regions (high summer temperatures and cool winters) requires two seasons to flower through seeds and flower initiation occurs during the winter. The minimum tuber size for flowering of *A. coronaria* is 3-4 cm in circumference for diploid forms flowering earlier and 4-5 cm for tetraploid forms flowering later than the diploids and this size is attained in one year through seeds, and on replanting these small tubers normally grow up >7 cm in circumference. The increase in size of the tubers or rhizomes above the minimum required for flowering has nothing to do with the time taken for flowering but for producing quantity and quality blooms. The plants can remain alive up to 10 years. For early autumns and winter flowerings in Mediterranean regions, it is only tubers which are required. In case of *A. blanda*, it requires usually two years to grow commercial size tubers *i.e.* >5 cm.

Domestication of *A. coronaria* and continuous selections for late spring and summer flowering types at a place quite away from its native haunts have led to shortening of the juvenile period of the seedlings, and therefore, the selections are more tolerant to heat on the cost of their winter hardiness. *A. coronaria* flowers naturally in the cool winters and springs. After flowering, the foliage matures and senesces and the tuber goes into high temperature induced (summer) dormancy and with the onset of cooler autumn temperatures the root and shoot growth re-initiate. *A. coronaria* tubers can be dry stored for 2-3 years at 15-25°C in an ethylene- or naphthalene-free room if properly dried, and on planting these still produce flowers satisfactorily. However, storage periods and temperatures of tubers or rhizomes have greater influence on time taken for flowering and flower productivity. In France, these are stored at 2°C for 4-6 weeks, in Japan these are stored at 5° or 10°C for four weeks for getting 100 per cent flowering after planting but the quality being better under 10°C treatment, while in U.S.A. (Florida) they have found 6°C for seven weeks as optimum for getting maximum number of flowers. The hydrated tubers of *A. coronaria* var. 'Wicabri' show inhibited root growth and leaf elongation but accelerated differentiation of the reproductive meristems and hastened flowering when stored at 1°C for 0-7 weeks, however, 1°C treatment for 4-6 weeks is most efficient with tetraploid varieties. Long days and high temperatures with or without moisture stress induce tuber dormancy. Long day conditions at low temperature below 20°C do not induce dormancy. A 5-10°C night and 14-18°C day temperature after planting results into highest number of flowers with quality blooms but a higher temperature regime suppresses growth and delays flowering or speed up flower production but flower size and stem lengths are found reduced. A healthy plant of *A. coronaria* produces 10-15 marketable blooms under favourable weather conditions. Too high temperature may induce summer dormancy. Ohkawa (1987) found earliest emergence in tubers held at 20°C, followed by 15, 10, 5 and 0°C, and floral primordia development was recorded 100 per cent four weeks after planting when tubers were held at 5°C for four weeks though in non-cooled tubers this process was slower which consequently delayed the anthesis by about one month. He recorded highest flower weight and diameter at growing temperature of 5°C though number of flowers per plant and stem lengths were found highest at a growing temperature of 10°C. In the present trial, photoperiod treatments had no any significant effect though long days treatment hastened flowering. Zimmer and Girmen (1987) in Germany recorded high root growth in *A. blanda* at 17°C (low temperatures reduced it) but the number of buds was found highest at 8-14°C while elongation was promoted at 5-11°C temperature. Ben-Hod *et al.* (1989) reported that in a hybrid cv. 'Hollandia x Israeli wild type' anemone possessing summer dormancy inductible by long days and/or high temperatures, corm differentiation did not require photothermal induction and occurred very early in seedling development but the corm size at the end of the growth cycle was the net result of the differential effects of daylength and temperature on its growth rate and growth period. They further added that LD of 16 h promoted corm growth rate as compared to 8 h SD but high temperature and LD reduced its final size owing to the earlier induction of dormancy. Larger corms were found producing more shoots and flowers per corm but number of flowers produced per shoot was unaffected by initial corm weight.

Gundersen (1959) reported that GA at 10 or 100 ppm applied to dormant rhizomes of *Anemone nemorosa* was effective in breaking dormancy. Hassan *et al.* (1985c) tried three applications of GA_3 from 30 days after planting and recorded improved flowering and earliest flowering at 100 ppm in *A. coronaria*, and 150 ppm gave greatest number of flowers per plant, flower diameter and dry weight. The flower production was best under three applications of GA_3 at 150 ppm. For advancing the flowering in 'De Caen' anemones, Talia and Ferrari (1989) sprayed the: a) dry rhizomes at emergence with 100 ppm GA_3, b) rhizomes pre-germinated in water moistened peat and sand for 15 days and then vernalized at 5°C for 30 days, c) rhizomes pre-germinated in 100 ppm-moistened peat and sand for 15 days and then vernalized at 5°C for 30 days, d) pre-germinated rhizomes planted in water-moistened peat and sand for 15 days and then vernalized at 5°C for 30 days and treated with 100 ppm GA_3 20 days after planting, and e) dry rhizomes as control, and obtained advanced flowering by 40-50 days in a), by 30 days in c) & d), and by 15 days in b) without any quality variation though number of flowers was found reduced in the first year.

In *A. coronaria,* seed germination is also responsive to temperature. For germination, optimum temperature regime for Israeli wild types is 10-15°C and for 'de Caen' varieties 15-20°C. Winter and spring produced seeds require dry storage for 10 days at 12°C while autumn produced seeds germinate at 23°C if sown immediately after collecting. Seedling development in *A. coronaria*, tuberization occurs very early. Juvenile stage through seeds is much more than the tubers. Before planting the tubers are soaked in water for 36 hours and then for 12 hours in mancozeb + carbendazim at 20°C, and then placed for 5-6 weeks in an aerated, cool (2°C) and moist conditions (preferably in moist perlite) in plastic bags, and then after sprouting the tubers are planted and the field watered. To provide required coolest temperature at this time, moisture, shading and ventilation should be controlled. When sprouted tubers dehydrate in storage before planting, suddenly roots stop growing and the plants wilt or enter into dormancy for several months. Optimum daily irradiance and at 8-10°C night temperature after planting causes early flowering with more in number and better in quality. Hassan *et al.* (1985b) exposed the plants of anemone to full sunlight (12000 lux) for 5, 6, 8 or 11 h per day or in the lath houses at low light intensity (2800-3000 lux) for 11 h per day, from 23 days after planting, and obtained greatest vegetative growth, *i.e.* number of leaves per plant and fresh & dry weights, early flowering, high flower fresh and dry weights, high flower diameter, and greatest number of tubers per plant, *vis-à-vis* highest leaf N, P and K concentrations while the plants in lath house exhibited greatest stem lengths and diameters, fresh weights, and the highest leaf nucleic acid concentrations.

Being a cold temperature crop, anemones do not require glasshouse cultivation but the provision of the moveable cover may protect the plants from excess winter moisture and may improve crop health. In glasshouse growing, fan and pad cooling during summer is essential. *A. coronaria* plants are moderately cold hardy and can tolerate -3°C temperature but -5°C is lethal. In *A. blanda*, the potted plants require 5°C temperature for 15-17 weeks for rhizome storage in a rooting room but the storage for shorter times will abort the flower buds, however, these may be forced at 12°C temperature in a glasshouse, and after the roots have grown out the bottom of the pot the temperature is lowered to 5°C until the shoots become 2.5 cm long and then the pots are placed at 0-2°C for a total cold treatment of 16 weeks. *A. fulgens* and *A. blanda* are stored at 9-17°C with half hourly air changes; however, excessive atmosphere drying can reduce sprouting. Pre-germinated tubers should be avoided to drying and warm temperatures, and can be transported in moist perlite at 15°C. *A. blanda* can tolerate -9°C soil temperature. The rhizomes of *A. hupehensis* can be forced at 10°C after being cold stored at 4°C for six weeks where flower production may last for more than 13 weeks producing around 45 cm long stems. Twisk and De Rooy (1962) described that if corms of *Anemone coronaria* are kept in moist peat at 4-9°C, it causes production of shoots and roots which may be damaged during transit, so preparation in polythene bags at the same temperature produces a small amount of shoot growth but no root growth which facilitates its safe transportation. Flowering normally starts after 5-6 weeks after planting out. Maia and Venard (1974) advocated soaking of corms in running water for 36 h, followed by dipping for 12 h in captan at 12 g/l and storage for 5-6 weeks at 1°C in perforated polyethylene bags for getting early flowering in anemones. They further stated that this treatment worked satisfactorily with Dutch corms harvested the previous year and with corms from the south of France harvested in June and planted in the following August in France.

When the flowering is over and the leaves have senesced in *A. coronaria*, the tubers are lifted and dried at warm temperature (25-30°C) for one week so that tubers shrink to $^1/_3$ of their original size, *i.e.* more than 15 per cent reduction in their moisture level. Delayed lifting of the tubers will make them unfit for flower production because root emergence takes place which are damaged when lifting. In Israel, in the *A. coronaria* growing in the wild, the summer dormancy can be induced through long photoperiods even during mid-winter when it is normal flowering time, and the dormancy induced by long days causes early cessation of flowering though with longer scapes (Kadman-Zahavi *et al.*, 1984).

Height control is not required in the field grown plants grown for cut flower production but in potted plants. Ancymidol at 0.25 to 0.5 mg/plant application produces compact plant in the pots. A 21 days second tuber dip in 100 mg a.i./l of ancymidol (A-rest) reduces peduncle length drastically in the pots. Van Leeuwen and van Der Lans (1989) with cvs 'Splendour' and 'Blue Shades' obtained compact plants when 13°C stored tubers were treated with drenching of 11-cm diameter pot with 75 ml of ancymidol (Reducynol) per pot after pot-planted tubers were further stored at 5°C for 15 weeks and then forced in a glasshouse temperature of about 12°C before dwarfing treatment. Van Leeuwen and Dop (1990) controlled excessive stem elongation under warm greenhouse conditions by using ancymidol or paclobutrazol as a soil drench at the time when plants were moved in the greenhouse. Other retardants are not effective in this crop.

Postharvest

Hofmann (1986) reported that 'Mona Lisa' anemones produce 200-300 cut stems/m^2 per year. Harvesting only of the mature flowers is done. Only those flowers are matured where one or two petals have gone through the opening and closing cycle once. This provides an optimum vase life. However, Moret (1952) from Morocco reported that flowers should be harvested before opening by pulling and not by cutting. While harvesting, the stems should be detached by hand at a low temperature so that flowers may remain closed. While putting in water, the stems are recut to ensure good water absorption. Hanke and Reimherr (1986) harvested the flowers of anemone at 13 stages of floral development, *i.e.* 10 days after to 2 days before the right bud stage and found that stems cut at the bud stage required 3-4 days for opening at 20°C while in winter the stems cut at quite large bud stage with 2-cm necks (stem from upper leaf to base of flower) and in spring stems with smaller buds had the greatest vase life. They further stated that the older the flowers were at cutting the longer the petals and necks became at flower opening and addition of Chrysal (15 g/l water) further increased the petal length though blue and red flowers showed some marginal discoloration and distortion and the blue flowers turned to poor violet. Piskornik (1983) recorded longest vase life of 8.2 days of the flowers cut at bud stage when kept in vase containing 40 g/l sucrose + 40 mg/l silver thiosulphate + 100 mg/l 8-HQS at 22°C and 65-75 per cent RH than 5.6 days in control (plain water). The stems wrapped tightly in paper are kept straight for transportation to prevent crooking of some varieties *e.g.* 'Mona Lisa'. The cut flowers can be stored at 1-2° C in an ethylene-free room for a maximum of 7 days before marketing. The flowers can be put in vase solution containing 2-4 per cent sucrose, silver thiosulphate and 8-hydroxyquinoline. The flowers have 6-10 days vase life.

Lifting and Storage

The tubers or rhizomes are lifted when after flowering the leaves have senesced. Significant delay in lifting may cause root emergence which may be injured when lifting. The root injury may force the rhizomes or tubers to die down or may go in dormancy for several months. After lifting the tubers should be dried to about 15 per cent moisture content at 25-30° C. The dried *A. coronaria* tubers can be stored for 2-3 years at 15-25° C in an ethylene or naphthalene free room. *A. fulgens* and *A. blanda* tubers can be stored at 9-17° C. While transporting, the tubers should be packed in moist perlite in alkathene bags and transported at 15° C temperature.

Insect-Pests, Diseases and Physiological Disorders

In United Kingdom, Woodville (1953) observed a wood mouse (*Apodemus sylvaticus sylvaticus*) damaging and feeding on anemone plantings for which he recommended 1.25-cm mesh wire-netting dug in at the bottom and curved over and down at the top. Aphids infest the stems and leaves, making the plants sticky and sooty. Thrips also suck the sap from flowers and stems. Whitefly (*Trialeurodes vaporariorum*) is a 1.5 mm long and white insect resembling a tiny moth. Their tiny yellow nymphs feed on the under side of the leaves and make the leaves chlototic. Caterpillars and various cutworms also feed on anemone plants. Garden swift moth (*Hepialus lupulinus*) is effectively controlled by soil drenching of 20 per cent DDT emulsion. Perret (1949) found 10 per cent benzene hexachloride for complete control of anemone flower eating beetle (*Tropinota squalida*) and lead arsenate, calcium arsenate, 1 per cent dintrophenol, 10 per cent DDT, 10 per cent benzene hexachloride for complete control of another flower eating beetle, *Hoplia aulica*. All other insects can be controlled by spraying with 0.2 per cent Rogor or 0.1 per cent Nuvacron. Springtails also infest anemone which can be controlled by lindane and aldrin (Jeff, 1956a).

Vovlas *et al.* (1973) reported root-knot nematode, *Meloidogyne incognita* infesting *Anemone coronaria* in Italy for which they suggested soil sterilization with methyl bromide and fumigation with organic halogenates before planting for its control.

The CMV, TNV and TSWV viruses infect the crop showing various symptoms. CMV is transmitted through aphids, TNV through a soil fungus, *Olpidium brassicae* and TSWV through thrips. Such plants should be rogued out. Hollings (1957) in U.K. recorded an already existing viral disease of *Anemone coronaria* which he named as 'anemone mosiac' (AMV). In this disease the leaves are mottled, the flowers are broken and distorted, and causes winter browning and crinkling. It is transmitted mechanically and through aphids. This disease was found along with CMV and TNV in imported corms. Hollings (1965) recorded a strain of tobacco ringspot virus (TRSV) from severely necrotic *A. coronaria* plants and named it anemone necrosis virus (ANV) which is transmitted through sap. Kochman *et al.* (1973) stated that CMV causes general chlorosis and reduces the size of leaf blades and of whole plant. Gera *et al.* (2006) recorded phytoplasma and/or spiroplasma disease on *Anemone* spp. in Israel. Field sanitation, roguing and control of vectors are some of the measures adopted in keeping the viral diseases under check.

A Peronospora anemones causes leaf curling, discoloration and rapid destruction of the foliage under high humidity and cool temperatures, *P. ficariae* and *Plasmopara pygmeaea* cause downy mildew which is a major problem in England during summer (Gregory, 1949; Jeff, 1956b 1958). Against *Peronospora* species, Boerema and Silver (1960) recommended steam sterilization as an effective measure. Jeff (1956b, 1968) found metham-sodium at rates as low as 150 gal per hectare quite effective against soil-borne infection by *Perosnospora* sp. Dithane Z-78 application maintained good health of the crop and treatment with 2 per cent formalin solution at 5 gallons per square yard reduced the primary infection considerably. Campbell (1957) recommended treatment of the infected soil with sodium-m-dithiocarbamate at 56 g per square yard (applied in water at 2.25 kg/30 gal/20 square yard) which recorded almost complete control of *P. ficariae* in the field. In case of *Peronospora anemones*, Tramier (1963, 1965) in France stated that optimum temperature for germination of its conidia is 12°C and for its development at least for 8 hours water droplets on the leaves should remain, and its spread is chiefly found during flower harvesting coinciding with extra humid conditions prevailing during August and September. He suggested planting of anemone in the same field only after a gap of 4 years, and sterilization of contaminated soil with steam or N-methyldithiocarbamate to prevent the disease. Garibaldi and Gulline (1972) recommended 3-4 year crop rotation, soil fumigation with methyl bromide, Vapam or methyl isothiocyanate and treatment of plants during conditions favouring infection (rainy periods from September to December) with dithiocarbamate fungicides or phthalimide derivatives. *Phytophthora cactorum* causes tuber rot. Garibaldi and Gullino (1974) stated that severe symptoms of *P. cactorum* infection develop only at soil temperatures above 18°C. *Rhizoctonia solani* and *Sclerotinia sclerotiorum* cause plant rot. *Erysiphe polygoni* (powdery mildew) also attacks this crop. *Tranzschelia pruni-spinosae* rust comes from the tubers to the leaves and causes leaf distortions, and its alternate hosts are *Prunus* species so anemone should be grown quite away from *Prunus* species. *Botrytis cinerea* causes grey mould in anemones and its control measures are pre-plant soil dusting with 2, 6-dichloro-4-nitroaniline and its occasional applications afterwards to prevent the disease though this fungicide is not able to eradicate the disease once this enters the plant. Skilled picking, balancing of N aplications through sufficient K_2O supplies and wide spacing (Brenchley and Johnstone, 1955); and zineb (65 per cent), captan (50 per cent) and thiram, all at 2.25 kg/250 gallons/ha, as well as maneb and dicloran are also effective in controlling the disease (Jeff, 1961). *Botrytis* causes leaf, stem and flower blight. Against *Botrytis*, water is kept off the foliage and crown of the plant, old leaves are removed to enhance ventilation and flowers are harvested without leaving the stump of the stem. During crop season, regular fortnightly spraying with benomyl 0.1 per cent alternate with Captan 0.2 per cent, and rhizome dipping in 0.2 per cent carbendazim before storing and before planting will keep all these diseases under check. Jeff (1957b) suggested soil fumigation with either formalin or chloropicrin in May in U.K. for reducing the infection of *Peronospora ficariae* until the autumn but as a result of secondary infection no difference in disease incidence by mid-winter was noticed. Timing of spray against *Botrytis* is most important and whenever there occurs any damage to the crop due to frost or wind, immediately Dithane Z-78 should be applied.Against powdery mildew, sulphur dusting may prove quite beneficial. Gregory (1949) reported occurrence of leaf diseases of anemone in England, *viz.* winter browning cluster cut rust (*Puccinia pruni-spinosae*), powdery mildew (*Oidium* sp.), downy mildew (*Plasmopara pygmaea* and *Perosnospora ficariae*), and black leaf spot (*Septoria anemones* var. *coronariae*).

Colletotrichum acutatum and *C. gloeosporioides* (*Glomerella cingulata*) have been reported causing shrivelling and curling of the leaves and leaf necrosis, twisting of the flower stems and often plant death. *Colletotrichum* can infect seeds, corms and plants of *Anemone coronaria*. Mevel (1980) stated that development of *C. acutatum* is encouraged by 25-30°C temperatures and high humidity, more prominently being in early plantings in France. The control measures suggested by different workers are dipping of pips in either 2 per cent captan for 1.5 h or 2 per cent captafol for 15 minutes immediately before planting in late November-early December, or after pre-soaking the pips may be dipped for 30 minutes in a solution of benomyl, carbendazim + captafol or captan, and for checking the spread of the infection in the field the plants may be sprayed either with the solution of 150 g benomyl or carbendazim + either 1 kg captan or 1-2 kg captafol per hectare whereas for crops grown from seeds the amount of captan or captafol must be reduced to 0.5 kg to avoid root damage, and seed treatment with a systemic fungicide such as benomyl, carbendazim or thiophanate-methyl + captafol or captan in Holland (Weststeijn, 1979, 1980); corm soaking in captan at 800 g/hl, autumn planting in frost-free regions, foliar spraying with mancozeb at 200 g/hl and corm harvesting in May-June before appearance of hot weather in France (Mevel, 1980); corm soaking for 15 h in maneb at 800 g/hl, zineb at 800 g/hl, captan at 839 g/hl or dichlofluanid at 500 g/hl in France (Tramier and Bettachini, 1980); seed harvesting from the stocks that have been disinfected before planting and had been disease-free, pips disinfection in captan or captafol in combination with a systemic fungicide and immediately after harvesting the washing of the pips (small corms) in running water (De Winter, 1980); dipping of corms in captan or dichlofluanid at 1 g/l for 60 minutes and transplanting them 24 or 48 h after treatment, or HWT at 40°C for 30 minutes for forced corms or at 50° for 30 minutes for dormant corms combined with low rates, *i.e.* 0.25 to 1.0 g/l of fungicides, and spraying in the field with captafol (80 g/100 litres) + benomyl (30 g/100 litres) or captafol (57 g/100 litres) + folpet (135 g/100 litres) (Gullino and Garibaldi, 1981); disinfection of seeds and pips in solutions of captan or captafol, each with benomyl, carbendazim or thiphanate-methyl before sowing or planting and spraying of the same chemicals at weekly intervals to the crop from emergence, or zineb/maneb + a systemic fungicide and watering the soil at 3-leaf stage with PCNB (quintozene), followed by above treatments (De Winter, 1981); post-planting drench of Sanspor (48 per cent captafol) at 4 ml/l and 100 ml solution/plant, followed by fortnightly spraying with Sanspor 93 litres in 1000 litres water/ha (O'Neill, 1983); treating corms in HWT at 47.5°C for 1.5 h or 50°C for 1 hour, followed by storage of treated corms in moist vermiculite at 20°C for 4 days (Doornik, 1990); and storage of corms with increased duration of storage at 17 or 20°C, both at 45 per cent RH also decreased necrosis in plants after planting (Doornik and Booden, 1990).

References

Ben-Hod, G., J. Kigel and B. Steinitz, 1989. Photothermal effects on corm and flower development in *Anemone coronaria* L. *Scientia Hort.*, **40**(3): 247-258.

Boerema, G.H. and C.N. Silver, 1960. Downy mildew of anemones (Dutch). *Meded. PlZiekt. Dienst*, No. 134, pp. 155-157.

Brenchley, G.H. and K.H. Johnstone, 1955. Grey mould of anemones in Cornwall. *Plant Path.*, **4**: 54-57.

Brosh, S., E. Dubitski and Z. Safran, 1972. Chemical weed control in anemones. In : *Proc. 4th Israeli Weed Control Conf., Rehovot, 1970,* under the auspices of Weed Science Society, pp. 49.

Campbell, W. 1957. Anemone downy mildew (*Peronospora ficariae*). *A.R. Rosewarne exp. Hort. Stat.*, pp. 82-83.

Clausen, G. 1972. Variety trials with Japanese anemones 1965-70 (Danish). *Tidsskrift for Planteavl*, **76**(1): 13-21.

Darnell, A.W. 1935. The genus *Anemone*. *Gdnrs' Chron.*, **98**: 386-388, 408-410 and 444.

Darnell, A.W. 1936a. The genus *Anemone*. *Gdnrs' Chron.*, **99**: 10-11, 28, 89-90, 251, 283-284, 314, 353, 374 and 405-406.

Darnell, A.W. 1936b. The genus *Anemone*. *Gdnrs' Chron.*, **100**: 12-13, 28-29 and 47-48.

De Rooy, M. 1966. Anemones and weed control (Dutch). *Weekbl. BloembollCult.*, **76**: 718.

De Winter, J.A.T. 1974. Oogt en verwerking van anemonepitten. *Bloembollencultuur*, **85**(8): 178.

De Winter, J.A.T. 1980. Naar een export van anemonen vrij van krulbladziekte. *Bloembollencultuur*, **91**(3): 72.

De Winter, J.A.T. 1981. Bestrijding van krulbladziekte in anaemonen. *Bloembollencultuur*, **91**(33): 912.

Doornik, A.W. 1990. Hot-water treatment to control *Colletotrichum aculatum* on corms of *Anemone coronaria*. *Acta Hort.*, No. 266, pp. 491-494.

Doornik, A.W. and E.M.C. Booden, 1990. Decrease in viability of *Colletotrichum aculatum* in corms of *Anemone coronaria* during storage. *Acta Hort.*, No. 266, pp. 505-507.

Edwards, C.A. and E.B. Dennis, 1960. Observations on the biology and control of the garden swift moth. *Plant Path.*, **9**: 95-99.

Figurkin, B.A. and L.N. Ogyrtsova, 1975. Triterpenic glycosides of *Anemone ranunculoides* (Russian). *Khimiya Prirodnykh Soedinenii*, **11**(1): 101.

Fischer, P. and F. Kalthoff, 1987. Die Düngung von Anemone 'Mona Lisa'. *Deutscher Gartenbau*, **41**(25): 1482-1484.

Fischer, P. and F. Kalthoff, 1989. Mittlerer Nährstoffbedarf für 'Mona Lisa'. *Deutscher Gartenbau*, **43**(5): 263-266.

Fouchard, D. 1961. Possible new uses of triazines in flower and market-garden crops (French). *1er Conf. franc. Mauv. Herbes (COLUMA)*, pp. 4.

Garibaldi, E.A. 1986. Prove di ingrossamento dei rizomi di anemone nel nostro paese. *Coluture Protette*, **15**(11): 31-33.

Garibaldi, A. and G. Gullino, 1972. A serious outbreak of *Peronospora* on anemones (Italian). *Informatore Fitopatologica*, **22**(9): 9-11.

Garibaldi, A. and G. Gullino, 1974. Preliminary observations on a new disease of *Anemone coronaria* L. caused by *Phytophthora cactorum* (Leb. Et Cohn) Schroet. *Mededelingen van de Facultiet Landbouwwetenschappen, Rijksuniversiteit Gent*, **39**(2): 951-956.

Gera, A., L. Maslenin, A. Rosner, M. Zeidan and P.G. Weintraub, 2006. Phytoplasma diseases in ornamental crops in Israel. *Acta Hort.* (*Proceedings of the XIth Int. Symp. on Virus Dis. of Ornam. Plants*, held at Taichung, Taiwan, on March 9-14, 2004; ed. Chang, C.A.), No. 722, pp. 155-161.

Gill, L.M. 1971. Anemones : weed control on corms. *Exptl Hort.*, No. 22, pp. 312-37.

Gill. L.M. 1977. Rosewarne hope new anemone strain will halt decline. *Grower*, **88**(22): 1138.

Gradner, U. and P. Reimherr, 1986. Pflanzzeiten bei Anemone 'Mona Lisa'. *Deutscher Gartenbau*, **40**(26): 1202-1203.

Gregory, P.H. 1949. Leaf diseases of *Anemone coronaria* in Cornwall. *Trans. Brit. Mycol. Soc.*, **32**: 241-244.

Gullino, M.L. and A. Garibaldi, 1981. Results of experimental trials for controlling leaf curling of anemone caused by *Colletotrichum gloeosporioides*. *Mededelingen van de Faculteit Landbouwwetenschappen, Rijksuniversiteit Gent*, **46**(3): 873-879.

Gundersen, K. 1959. Some experiments with gibberellic acid. *Acta Hort. Gotoburg.*, **22**: 87-110.

Hanke, H. and P. Reimherr, 1986. Zur Haltbarkeit von Anemone 'Mona Lisa' (German). *Deutscher Gartenbau*, **40**(26): 1208-1210.

Hansen, R. and J. Sieber, 1965. The evaluation of autumn anemone varieties. *Dtsche Gärtnerbörse*, **65**: 367-368.

Hassan, H.A., E.A. Agina, I.O. Baz and S.M. Mohamed, 1985a. Some physiological studies on seed germination of some ornamental plants. *Ann. Agric. Sci., Moshtohor*, **22**(2): 617-630.

Hassan, H.A., E.A. Agina, E.M. Koriesh and S.M. Mohamed, 1985b. Physiological studies in *Anemone coronaria* L. and *Ranunculus asiaticus* L. I. Effect of light (intensity and duration). *Ann. Agric. Sci., Moshtohor*, **22**(2): 571-582.

Hassan, H.A., E.A. Agina, E.M. Koriesh and S.M. Mohamed, 1985c. Physiological studies in *Anemone coronaria* L. and *Ranunculus asiaticus* L. II. Effect of gibberellic acid. *Ann. Agric. Sci., Moshtohor*, **22**(2): 583-595.

Hensen, K.J.W. 1979. Voortgezet onderzoek van het sortiment Japanese anemonen. *Groen*, No. 9, pp. 363-370.

Harvey, D.M. 1971. Phenotypic variation in flower colour within the *Anemone coronaria* cultivars. *Ann. Bot.*, **35**: 1-8.

Hassan, H.A., E.A. Agina, I.O. Baz and S.M. Mohamed, 1985. Some physiological studies on seed germination of some ornamental plants. *Ann. Agric. Sci., Moshtohor*, **22**(2): 617-630.

Hegele, A. 1986a. Die Geschichte der Anemone 'Mona Lisa'. Eine aussergewöhnliche Schnittblume. *Deutscher Gartenbau*, **40**(26): 1200.

Hegele, A. 1986b. Anemonen 'Mona Lisa' immer beliebter. Jungpflanzen – Produktion nicht ganz einfach. *Deutscher Gartenbau*, **40**(26): 1201-1202.

Hollings, M. 1957. Anemone mosaic – a virus disease. *Ann. Appl. Biol.*, **45**: 44-61.

Hollings, M. 1965. Anemone necrosis, a disease caused by a strain of tobacco ringspot virus. *Ann. Appl. Biol.*, **55**: 447-457.

Horovitz, A., S. Bullowa and M. Negbi, 1975. Germination characters in wild and cultivated *Anemone coronaria* L. *Euphytica*, **24**(1): 213-220.

Jeff, A.E. 1956a. Anemones. *Rep. Rosewarne Exp. Hort. Stat.1952-55*, pp. 11-18.

Jeff, A.E. 1956b. Anemones. *A.R. Rosewarne exp. Hort. Stat.*, pp. 6-13.

Jeff, A.E. 1957a. Report from Rosewarne-1. Observations on anemones. *Comm. Gr*, No. 3186, pp. 179.

Jeff, A.E. 1957b. Anemones. *A.R. Rosewarne exp. Hort. Stat.*, pp. 4-11.

Jeff, A.E. 1958. Anemones. *A.R. Rosewarne exp. Hort. Stat.*, pp. 6-16.

Jeff. A.E. 1959. Anemones. *A.R. Rosewarne exp. Hort. Stat.*, pp. 7-17.

Jeff, A.E. 1960. Observation studies on the use of residual herbicides on anemones during 1958-59 and 1959-60. *Proc. 5th Brit. Weed Control conf.*, pp. 6.

Jeff, A.E. 1961. Observations on the control of *Botrytis* in anemones. *Proc. Brit. Insectic. Fungic. Conf., Brighton*, pp. 321-326.

Jeff, A.E. 1968. The effects of soil treatments on anemone downy mildew infection. *Exp. Hort.*, No. 18, pp. 101-109.

Jeff, A.E. and D.J. Fuller, 1961. Anemone spacing experiments in the South-West. *Exp. Hort.*, No. 5, pp. 31-36.

Jeff, E. 1964. Weed control in anemones. *British Fmr*, No. 331, p. 68.

Johansson, L. 1983. Effects of activated charcoal in anther cultures. *Physiologia Plantarum*, **59**(3): 397-403.

Johansson, L., B. Andersson and T. Eriksson, 1982. Improvement of anther culture technique : Activated charcoal bound in agar medium in combination with liquid medium and elevated CO_2 concentration. *Physiologia Plantarum*, **54**(1): 24-30.

Johansson, L., E. Calleberg and A. Gedin, 1990. Correlation between activated charcoal, Fe-EDTA and other organic media ingredients in cultured anthers of *Anemone canadensis*. *Physiologia Plantarum*, **80**(2): 243-249.

Johansson, L. and T. Eriksson, 1977. Induced embryo formation in anther cultures of several *Anemone* species. *Physiologia Plantarum*, **40**(3): 172-174.

Johansson, L. and T. Eriksson, 1984. Effects of carbon dioxide in anther cultures. *Physiologia Plantarum*, **60**(1): 26-30.

Jones, S., P. Atkkey and J. Pegler, 1986. De-fluffed-courtesy of GCRI craftsmen. *Grower*, **105**(25): 24-26.

Jones, S.K. 1986. The germination of *Anemone* 'St. Piran' seed and corms. *Acta Hort.*, No. 177, Vol. II, pp. 675-679.

Kadman-Zahavi, A., A. Horovitz and Y. Ozeri, 1984. Long-day induced dormancy in *Anemone coronaria* L. *Ann. Bot.*, **53**(2): 213-217.

Kamel, H.A. and A. Nabih, 1978. Effect of storage and planting dates on *Anemone coronaria* flowering. *Agric. Res. Rev.*, **56**(3): 135-137.

Kochman, J., A. Kowalska and R. Krasuska, 1973. Cucumber mosaic virus on *Anemone coronaria* L. (Polish). *Acta Agrobotanica*, **26**(2): 237-249.

Löhr, E. 1952. Photosynthese von *Anemone nemorosa*. *Physiol. Plant.*, **5**:221-227.

Madahar, C. 1967. Mediterranean and Asian taxa of anemone (section *Eriocephalus*) with tuberous rootstocks. *Canad. J. Bot.*, **45**: 725-755.

Maia, N. and M.C. Pellergrin, 1974. Caractéristiques de la germination des akènes d'*Anemone coronaria* L. *Bull. Soc. bot. France*, **121**(3/4): 79-88.

Maia, N. and P. Venard, 1974. Vernalization of anemone corms (French). *Revue Horticole*, **146**(2322): 17-21.

McDowell, J. 1985. Greenhouse culture of 'Mona Lisa' anemones. *Bull., Pennsylvania Flower Grs*, No. 362, pp. 5-6.

Meredith, W.C. and T.J. Sheehan, 1965. Anemone fertilization and spacing. *Proc. Fla St. hort. Soc.*, **78**: 404-405.

Mevel, A. 1980. Le point sur la frisure de l'anémone. *Horticulture Française*, No. 114, pp. 3-5.

Moret, 1952. La culture des anemones. *Terre Maroc.*, **26**: 134-135.

Ohkawa, K. 1987. Growth and flowering of *Anemone coronaria* L. 'de Caen'. *Acta Hort.*, No. 205, pp. 159-167.

O'Neill, T. 1983. Curl up and die. *GC & HTS*, **194**(10): 24-25.

Perret, J.E. 1949. Essais sur deux insectes floricoles. *Rev. Hort. Paris*, **121**: 138-139.

Petiet, J. 1958. Self-pollination of anemones. *Meded. Dir. Tuinb.*, **21**: 31-34.

Petiet, J. 1959. Anemones in one year from seed to flower. *Weekbl. BloembollCult.*, **69**: 520.

Petiet, J. 1962. The effect of plant spacing on anemones (Dutch). *Meded. Dir. Tuinb.*, **25**: 151-153.

Piskornik, M. 1983. The longevity and water relations of cut poppy anemone flowers (*Anemone coronaria* L.). *Prace Instytutu Sadownictwa I Kwiaciarstwa w Skierniewicach, B, (Roœliny Ozdobne)*, **8**: 191-198.

Piskornik, M. 1984. Kielkowanie nasion zawilca wiencowego (*Anemone coronaria* L.). *Zeszyty Naukowe Akademii Rolniczej w Krakowie, Ogrodnictwo*, No. 12/179, pp. 51-59.

Piskornik, M. 1986/1987. Wpływ stadium dojrza łości owocostanu oraz czasu przechowywania niełupek na żywotność zawilca wieńcowatego (*Anemone coronaria* L.). *Prace Instytutu Sadownictwa I Kwiaciarstwa w Skierniewicach. Seria B, Rośliny Ozdobne*, **11**: 33-39.

Piskornik, M. 1989. Generative reproduction of poppy anemone (*Anemone coronaria* L.). *Folia Horticulturae*, **1**(2): 65-74.

Puccini, G. 1961. Studies on the nutrition of anemones grown for cutting. The commercial growing of anemones in Italy (Italian). *Ann. Sper. Agrar.*, **15**: 513-530.

Radspinner, A.L. and T.J. Sheehan, 1963. Effects of fertilization and storage treatments on growth and flowering of Wacabri anemones. *Proc. Fla St. hort. Soc.*, **76**: 428-431.

Scheu, M. and P. Reimherr, 1986. Hügelbeet lässt *Botrutus* auch bei enger Pflanzung wenig Chancen. *Deutscher Gartenbau*, **40**(26): 1205-1207.

Sushila Bhattarai, 1989. Karyomorphological studies in four species of *Anemone*. *Cytologia*, **54**(4): 709-713.

Talia, M.C. and I. Ferrari, 1989. Effetti della gibberelina sulla fioritura dell'anemone. *Colture Protette*, **18**(8-9): 97-100.

Tramier, R. 1963. A study on mildew of *Anemone coronaria* in southern France (French). *Ann.Éuphyt.*, **14**: 311-323.

Tramier, R. 1965. Epiphytology and treatment of *Anemone coronaria* L. mildew (*Peronospora anemones* Tram.) in the south of France (French). *Phytiat.- Phytopharm.*, **14**: 49-56.

Tramier, R. and A. Bettachini, 1980. *Colletotrichum aculatum* : Maladie nouvelle de l'anémone en France. Biologie et méthodes de lutte. *Phytiatrie -Phytopharmacie*, **29**(2): 121-124.

Twisk, D. and M. de Rooy, 1962. Is flower forcing possible with anemones and ranunculus (Dutch). *Weékbl. BloembollCult.*, **73**: 484.

Van Leeuwen, P.J. and A.J. Dop, 1990. Effects of storage, cooling and greenhouse conditions on *Anemone blanda, Fritillaria meleagris* and *Oxalis adenophylla* for use as pot plant. *Acta Hort.*, No. 266, pp. 101-107.

Van Leeuwen, P.J. and A.M. van Der Lans, 1989. Mogdijkheden voor teelt nadir onderzocht. *Anemone blanda* geeft op pot rijkbloeiende gewas. *Bloembollencultuur*, **100**(5): 26-27.

Vovlas, N., A. Avgelis and M. D'Urso, 1973. Nematodes and viruses, causes of serious damage in anemone cultivation in southern Italy (Italian). *Informatore Fitopatologica*, **23**(8): 19-22.

Weststeijn, G. 1979. Behandeling van anemonenpitten ter bestrijding van Krulbladziekte. *Bloembollencultuur*, **90**(21): 499.

Weststeijn, G. 1980. Krulbladziekte bij anemonen. *Bloembollencultuur*, **90**(35): 960-961.

Woodville, H.C. 1953. Damage to anemones by field mice. *Plant Path.*, **2**: 21-24.

Yasuda, I. and N. Yokoyama, 1957. Research on the planting in spring of bulbs normally planted in the autumn (Japanese). *Sci. Rep. Fac. Agric. Okayama Univ.*, No. 10, pp. 51-66.

Zimmer, K. and M. Girmen, 1987. Temperature dependency of the development of *Anemone blanda and Eranthis hiemalis* (Dutch). *Gartenbauwissenschaft*, **52**(6): 263-265.

5

Anthurium (Family: Araceae, and sub-family: Pothoideae)

R.L. Misra and Sanyat Misra

[**Common names**: Flame plant/Flamingo flower (*A. scherzerianum*) and Oil cloth flower/Painter's palette/Tail flower (*A. andraeanum*)]

Introduction and Botany

Anthurium takes its name from the Greek word *anthos* meaning flower, and *aura* meaning tail, referring to the spadix. Anthuriums are evergreen tropical ornamentals, valued for their most beautiful flowers having peculiar but delicate look and for their unusually attractive foliage. Its cut flowers are long lasting and give bold effect in the floral arrangement. The major countries importing anthurium are USA, Germany and Japan. South India (Kerala, Tamil Nadu and Karnataka) is the hub of anthurium growing in India, Kerala being its intensive growing zone, followed by Coorg in Karnataka. Quality anthuriums are also being grown in Sikkim, West Bengal and Maharashtra and here many export-oriented anthurium based units have also started functioning. From Kerala, Rajeevan *et al.* (2002) have brought a publication named 'Anthurium' which has been published through AICRPF, New Delhi. Bhatt and Desai (1989) and Devinder Prakash *et al.* (2006) also contributed chapters on this crop in two different books. Though ideal growing areas in the country are NEH region (tropical and sub-tropical regions, especially Sikkim, Meghalaya and Arunanchal Pradesh), Assam, West Bengal (Kalimpong and Darjeeling areas), Orissa (Koraput region and certain other adjoining areas), Tripura, Nagaland, Mizoram, Manipur, Andaman & Nicobar Islands, lower part (foot hills) of Jammu & Kashmir as well Himachal Pradesh and Uttarakhand, Maharashtra and Gujarat, but presently it is being cultivated only in entire Kerala, Yercaud and Ootacamund, *i.e.* Nilgiri hills in Tamil Nadu, Coorg in Karnataka and in a limited part of Maharashtra for commercial use. In fact, there are three major consumer markets in the world, *viz.* Europe with 47.8 per cent per capita consumption, USA with 21.8 per cent and Japan with 16.8 per cent.

Anthuriums are tropical, evergreen, perennial but herbaceous aroids, having creeping, climbing or arborescent stems. They can be found growing as an epiphyte on trees not as a parasite but use them as an anchoring foundation, as lithophyte growing on rocks or as terrestrial on the ground. They are native to Central America, Colombia, Brazil, Guatemala, Peru and Venezuela and were introduced into Great Britain during early nineteenth century. Leaves have a thick midrib with some well defined lateral veins radiating from the place where midrib originates and all such veins are also connected with veinlets. Flower is composed of spadix enclosed in a spathe. The true flower is protogynous, and the fruit is a berry.

Leena Ravidas (2003) selected six commercially important *A. andraeanum* varieties, *viz.* 'Agnihotri', 'Candy Queen', 'Eureka Red', 'Lima', 'Nitta' and 'Red Dragon' and three species, *viz. A. amnicola, A. crystallinum* and *A. ornatum* for their morphological characterization and genetic variability, crossability and ^{60}Co gamma rays mutagenesis. All the varieties were found perpetual bloomer with straight peduncle except in 'Candy Queen'. 'Candy Queen' produced largest spathe, 'Red Dragon' shortest spadix, and the vars 'Eureka Red' & 'Red Dragon' more number of spikes per year. Spike emergence to unfurling ranged from 19.00 days in 'Lima' and 'Candy Queen' to 25.33 days in 'Agnihotri', female receptivity from 12.00 days in 'Red Dragon' to 20.33 days in 'Eureka Red', an interface of 3 to 20 days was found between female phase and male phase, male phase ranging from 9.33 days in 'Lima' to 20.67 days in

'Red Dragon' and early ripening of seeds in 'Red Dragon' (145 days) to late in 'Agnihotri' (208 days). For postharvest qualities, 'Eureka Red', 'Red Dragon' and 'Lima' were superior to others. Heritability was of moderate to high magnitude for most of the characters. PCV values were slightly higher than those of GCV, and the plant height recorded highest PCV (70.57) and GCV (69.96). Renu (1999) studied 10 varieties and through analysis of variance, she found significant variation among the 12 quantitative characters. She found 42 per cent pollen fertility in var. 'Liver Red' and 13.7 per cent in 'Mauritius Orange', however, the number of fruits per candle ranged from 5 to 183, and the seed germination varied from 3 to 12 days. The percentage of fruit set was below 50 per cent in all the crosses except 'Pompon Red' × 'Liver Red', and the best female parents were identified as 'Nitta Orange', 'Liver Red', 'Pompon Red and 'Ceylon Red' whereas best male parents as 'Ceylon Red', 'Merengue White' and 'Liver Red'. Maya Devi (2001) studied 100 anthurium genotypes for 10 characters where all the characters were highly influenced by genotypic variation, and high heritability with a good genetic advance was found for all the characters except the number of spadices per plant per year which exhibited medium heritability and low genetic advance. Leaf area and suckering abilities were found to be the two potential contributing characters for divergence. She studied gene action for 24 characters in a diallel fashion using five parents and 10 hybrids and reported that for almost all the characters (except for plant height) the selected parents possessed most of the dominant genes with positive effects. Based on gca and sca effects, vars 'Honeymoon Red', 'Kalympong Red', 'Chilli Red' and 'Liver Red' were selected as general combiners. Most of the hybrids were observed with negative heterosis for plant height, leaf blade length & width, days for emergence to maturity of leaves, leaf area, days for flowering and inclination of candle. Through the variability studies in *A. andraeanum*, Binodh (2002) reported high phenotypic and genotypic coefficients of variation for total anthocyanin content, pollen fertility, inclination of candle to spathe and duration of interphase. Phenotypic variation highly influenced all the characters except leaf area and number of flowers per candle. High heritability with a good genetic advance was found for all the 13 characters except suckering ability and the number of spadices per plant per year as these exhibited medium heritability and high genetic advance. Plant height was found phenotypically positively correlated with leaf area, internodal length and days from emergence to maturity of inflorescence. Candle length showed significant positive correlation with leaf area, flower number per candle, spadix life and duration of female phase.For most of the characters it showed highly positive genotypic correlations. Out of five varieties ('Album' 'White', 'Chilli Red', 'Honeymoon Red', 'Lady Jane' and 'Pink') of *Anthurium andraeanum* studied for their pollen characters, Bindu (1992) obtained highest amount of pollen in vars 'Honeymoon Red' and 'Pink', largest pollen in var. 'Lady Jane' though the size difference among the varieties was not significant, and the pollen fertility very low in all the varieties studied. Sindhu (1995) while studying six anthurium varieties ('Album', 'Chilli Red', 'Honeymoon Red', 'Kalympong Orange', 'Kalympong Red' and 'Pink') for their morphological characters and crossability, recorded significant correlation between leaf size and spathe size, number of leaves per plant found were from 3.5 to 6.5, largest plants in 'Pink' and shortest in 'Kalympong Red', lowest leaf plastochron duration in 'Album' while longest in 'Chilli Red', maximum number of flowers were in 'Pink and 'Honeymoon Red', annual number of spadices per plant ranged from 4 to 8, 'Pink' and 'Kalympong Red' produced super large flowers while 'Album' the smallest, duration between the emergence of two successive spadices ranged from 43-51 days, maximum angle between the spadix and plane of the spathe was in 'Honeymoon Red' while smallest in 'Chilli Red', percentage of candles bearing fruits being maximum in 'Album' and lowest in 'Kalympong Red', fruit maturity period ranged from 4.8 to 8.0 months, maximum percentage of fruits (52.3 per cent) harvested were in the cross 'Pink' × 'Honeymoon Red', larger size seeds in 'Pink' and 'Honeymoon Red', and seed germination was observed in all the 23 combinations with maximum (63.4 per cent) being in a cross where 'Album' was used as female parent and minimum where 'Kalympong Orange' was used. Mini Balachandran (1998) observed 150 Gy radiation dose lethal to callus as well as to shoot tips, however, lower dose of 50 Gy gave maximum positive response. Leena Ravidas (2003) reported reduced germination and those germinated with reduced growth in seeds irradiated with gamma rays. Leena Ravidas (2003) tried 42 combinations but succeeded in 17 as all the selfs and interspecific crosses failed, and out of the 17 crosses, it was 'Lima' which produced largest number of compatible crosses, high seed set and germination percentage, followed by 'Candy Queen', 'Red Dragon' and 'Eureka Red' whereas 'Lima', followed by 'Red Dragon' and 'Eureka Red' were best performer as male parents. She found flowering in three hybrids, *viz.* 'Lima' × 'Eureka Red' 11 months after planting, and 'Lima' × 'Red dragon' & 'Candy Queen' × 'Lima' 12 months after planting out.

Basic chromosome numbers in anthuriums are n = 15 (diploid being *Anthurium andraeanum and A. hookeri*, triploid being *A. scandens* and tetraploid being *A. digitatum* and *A. wallisii*), n = 16 (*A. magnificum*) and n = 22. Somatic chromosome number in *A. andraeanum* has been found 2n = 30+2B and the species having a high percentage of meiotic abnormalities and the karyotype difference suggests it of hybrid origin. Bindu and Mercy (1997) stated that the five of its varieties studied have been grouped into karyotype asymmetry category 3 which indicated that the species is highly evolved. This species is a secondary polyploid with a basic chromosome number of probably x = 6. Cotias-de-Oliveira *et al.* (1999) from Brazil again found out the same already recorded chromosome count of 2n = 30 and 2n = 60 for *A. pentaphyllum* var. *pentaphyllum* and for the first time reported 2n = 30 in *A. longipes* and *A. affine*, latter species possessing 1-4 B chromosomes. Chi-GekLan *et al.* (1998) stated that highest transient expression of GUS gene in *A. andraeanum* callus tissues were obtained using pUGCI, indicating that the promoter from maize, a monocot, is more efficient in expressing the GUS gene in anthurium than dicot promoters. Kuchnle *et al.* (1995) got genetically transformed against resistance to bacterial pathogens, two cvs 'Rudolph' and

'UH 1060' of anthurium with vectors containing antibacterial genes and synthetic derivatives from *Hyalophora cecropia* and bacteriophages and the regenerated plants showed a delay in disease symptom development than controls.

The plant is native to Central America, Colombia, Peru, Brazil, Guatemala and Venezuela. Hawaii's main cut flower for export is anthurium. USA (Hawaii, Florida and California), Mauritius and the Netherlands are the major countries for commercial cultivation of anthuriums. Apart from these, Denmark, Italy, Germany, Jamaica, Philippines, Sri Lanka, Trinidad and Tobago also deal with anthuriums as there are many commercial nurseries dealing with anthuriums for export. The Netherlands occupies top position in anthurium production under glass. The most important market for Dutch anthuriums is Western Europe, Japan, Hong Kong and Singapore, Germany being the largest importer. Trinidad is the leading supplier of anthurium cut flowers to United States, Dominican Republic and Jamaica. Among Asian countries, Singaporean growers produce anthuriums in Malaysia for export to other Asian and Middle East countries. Mauritius anthuriums are being supplied mainly to French market.

Classification, Species and Varieties

There are more than 700 species but only some 50 species are in cultivation out of which only 10-15 species are in trade. They are grown either for their very showy flowers, *i.e.* 'spathes and spadices' or for their showy velvety leaves. Hence, they fall under two natural groups, *viz.* i) flowering such as *Anthurium andraeanum, A. bakeri, A. brownii, A.×ferrierense, A. ornatum, A. regale, A. regnellianum, A. robustum* and *A. scherzerianum*, and ii) foliage such as *A. clarinervium, A. corrugatum, A. crystallinum, A. holtonianum, A. leuconerum, A. magnificum, A. panduratum, A. papilionensis, A. splendidum, A. veitchii* and *A. warocqueanum*. Though all the anthuriums flower but those having showy and large flowers (spathes and spadices) are the flowering anthuriums such as ***A. andraeanum*** (an epiphyte grown as terrestrial) with heart-shaped, waxy red, reddish orange, scarlet or white spathes and yellow and white straight spadix [vars *album* (white), *amoenum* (rose-red), *gameri* (spathe shining red), *Lawrenciae* (pure white, spathe large), *rhodochlorum* (rose and green), *roseum* (spathes shining rose-pink), *salmoneum* (spathes salmon), *sanguineum* (spathes dark crimson), etc.], ***A brownie*** (an epiphyte) with spathe greenish, rose tinted, 15-20 cm long and spadix 25-40 cm long, ***A. regnellianum*** (an epiphyte or terrestrial) with spathe narrow green, some 4 cm long and 0.6 to 1.0 cm wide, spadix dark green and 4-5 cm long, ***A. spathiphyllum*** (an epiphyte) having spathe up to 5 cm long and 2.5-4.0 cm wide, narrow, pale green or whitish and spadix 2.5 cm long and pale yellow, and ***A. scherzerianum*** (flamingo flower, a popular house plant of the genus is either epiphytic or terrestrial) with long, waxy and palette-shaped spathes of brilliant scarlet colour and spiral (coiled) spadix of orange-red to golden-yellow colour [vars *album* (white), *album magnificum* (white), *andegavense* (scarlet on the back and white & scarlet spotted above), *bennettii* (sharp pointed leaves and spathes), *giganteum* (very large spathe), *lacteum* (white), *maximum* (very large spathe), *maximum album* (white), *mutabile* (white bordered), *nebulosum* (double white spotted rose), *parisiense* (rose salmon spathe and orange spadix), *pygmaeum* (quite dwarf), *roseum* (rose coloured spathes), *rothschildianum* (creamy white), *sanguineum* (deep blood-red spathes), *wardianum* (scarlet with extra large bracts), *wardii* (very large spathe), *warocqueanum* not the species but a variety of *A. scherzerianum* (white spotted red), *williamsii* (white), *woodbridgei* (very large spathe), *verbaeneum* (white), etc.].

Anthuriums where flowers are insignificant and the foliage is large, handsome and velvety are foliage anthuriums such as ***Anthurium crystallinum*** (an epiphyte) with large heart-shaped deep green leaves (violet when young) having midribs and the veins ivory above and pale pink beneath, ***A. magnificum*** (terrestrial) with some 60 cm long, deep cordate and oval leaves where upper surface is olive-green with white nerves, ***A. regale*** (epiphyte) with cordate-oblong, long-cuspidate, up to 90 cm long leaf blades which are dull green (young ones tinged rose) with white veins, ***A. splendidum*** (terrestrial) with coriaceous, sea-green, glaucous leaves having depressions and nerves brownish, ***A. veitchii*** (terrestrial) having pendent leaf blades, 90-120 cm long and metallic green but marked by deep-sunk nerves, ***A. warocqueanum,*** an epiphyte having a climbing stem is very vigorous with 60-120 cm long, oblong-lanceolate, hanging, tapering and deep velvety green leaves with midrib and other veins of contrast lighter colour, etc.

These all the 11 anthuriums are the most important ones but under flowering anthuriums *A. andreanum* and *A. scherzerianum*, and under foliage anthuriums *A. crystallinum* and *A. warocqueanum* are more admired. Apart from these, *A. aemulum, A. digitatum, A. x ferrierense* (*A. andraeanum* x *A. nymphaeiplium*), *A. pentaphyllum, A. radicans* (a terrestrial creeper), *A. scandens* (an epiphytic climber), *A. subsignatum, A. undatum*, etc. are creeping or climbing anthuriums. *A. acutangulum* is an epiphyte (rarely terrestrial), *A. acutifolium* is terrestrial (often epilithic), *A. amnicola* is an epiphyte, *A. andicola* is either epiphytic or epilithic, *A. bakeri* is an epiphyte, *A. bogotense* is epiphytic and terrestrial, *A. clarinervium* and *A. corrugatum* are terrestrial, *A. hacumense* is an epiphyte, *A. hoffmannii* is an epiphyte or terrestrial, *A. hookeri* is an epiphyte or lithophyte, *A. kyburzii* and *A. pedatoradiatum* are terrestrial, *A. pendulifolium* is an epiphyte (rarely terrestrial), *A. pentaphyllum, A. pottery* and *A. spectabile* are epiphytes, *A. watermaliense* is a terrestrial species, etc. Some of the new species recorded recently are *A. faustomirandae* sp. nov. (Perez-Farrera and Croat, 2001), *A. minarum* (Sakuragui and Mayo, 1999) and *A. thompsonii* (Arias-Granda, 1996). The species cross easily among themselves.

The present day anthurium cultivars are mostly hybrids of various species involving mainly *A. andraeanum* and *A. scherzerianum*. The cultivars are dwarf types suitable as pot plant, large types with large flowers suitable for arrangement and foliage types as pot plant for indoor decoration.

The pot plant var. 'Red Hot' developed through cv. 'Southern Blush' (an F_1 hybrid of *A. andraeanum* × *A. amnicola*) x 'Lady Jane' (a pink miniature) has medium red (at anthesis) to lighter red (prior to senescence) spathes which are

6-7 cm long and 4-5 cm wide, and spadix red to orange red from base to apex. It attains full marketable quality growth in a 1.6 litre pot in 11 months. 'Showbiz' is another interspecific pot plant hybrid released from Florida that has compact branched growth and produces numerous showy light red spathes. Other pot plant varieties popular in USA are 'Allura' (white), 'Champion' (white obake), 'Dutch Treat' (*A. scherzerianum*, red), 'Kohara Double' (red), 'Leilani' (light lavender), 'Lollipop' (reddish pink and cream), 'Maize' (red and white), 'Mary Jean' (white and pink), 'Pink Aristocrat' (pink), 'Pink Frost' (pink and white), 'Pixie Pink' (pink), 'Red Hot' (red and pink), 'Shazzam' (*A. scherzerianum*, red and pink). University of Hawaii has released some outstanding pot cultivars, *viz.*, *A.* x *atroquiense* (*A. anthrophyoides* x *A. antioquiense*, mini, spathe white, minty scent), *A.* x *amnioquiense* (*A. antioquiense* x *A. amnicola*, mini, spathe light lavender, minty scent), 'Kalapana' (red obake, blight tolerant, cut flower use), 'Tropic Fire' (red, blight tolerant), 'Tropic Ice' (white obake, blight tolerant), etc. *A. andraeanum* x *A. amnicola* hybrids, *A. antioquiens* hybrids and *A. amnicola* hybrids are compact, highly floriferous and make very good pot blooming anthuriums with attractive foliage.

Other cultivars for cut flower use are i) **reds** such as 'Arizona', 'Asahi', 'Aumana', 'Avo-Claudia', 'Avo-Ingrid', 'Avo-Netta', 'Avo-Red', 'Avo-Rosette', 'Avo-Serge', 'Ayashii', 'Calypso', 'Cancan', 'Carre', 'Cherry Red', 'Chilli Red', 'Claudia', 'Dragon's Tongue', 'Eureka Red', 'Fla King', 'Fla Red', 'Fla Success', 'Hawaiian Red', 'Hayashi', 'Honduras', 'Honeymoon Red', 'Ingrid', 'Inka', 'Jacqueline', 'Jertrood', 'Kalimpong Red', 'Kaumana', 'Kozohara', 'Kansako No.1', 'Liver Red', 'Madame Butterfly', 'Magic Red', 'Mauritius Red', 'Mickey Mouce', 'Mirjam', 'Nova-Aurora', 'Ozaki', 'Pha', 'Pohoa', 'Pompon Red', 'Pronto', 'Red Dragon', 'Red Elf', 'Rio', 'Scarlette', 'Scarlet Red', 'Sikkim Red', 'Splish Splash', 'Sweet Heart', 'Tanake', 'Temptation', 'Tina Red', 'Tropical Red', 'Tyam', 'Violetta', etc., ii) **roses** such as 'Marian Seefurth' (rose pink), 'Rico', 'Sarina' (white and rose), 'Tropical', etc., iii) **pinks** such as 'Abe', 'Abe Pink', 'Agnihotri', 'Aneunue' (green and coral pink), 'Avo-Anneke', 'Bettine', 'Bl'ush', 'Marian Seefurth', 'Calypso' (dark pink inside and lighter outside), 'Candy Queen', 'Candy Stripe', 'Cheers', 'Hoenette', 'Lady Jane' (miniature), 'Launette', 'Magic Pink', 'Paradise Pink', 'Passion', 'Sarina', 'Sonata', 'Spirit', 'Surprise', etc., iv) **oranges** such as 'Avo-Gino', 'Casino', 'Diamond Jubilee', 'Favoriet', 'Fla Orange', 'Horning Orange', 'Horning Rubin', 'Kalimpong Orange', 'Mauritius Orange', 'Nitta', 'Nitta Orange', 'Orangeeth' (spathe orange red and spadix yellowish orange), 'Orange Glory', 'Ordinary Orange', 'Sun Burst', 'Sunset Orange', etc., v) **whites** such as 'Acropolis', 'Avo-Jose', 'Avo-Margarette', 'Bianco', 'Chameleon', 'Cuba', 'De Weese', 'Fla Exotic', 'Gaisha', 'Haga White', 'Hidden Treasure', 'Jamaica', 'Lima White', 'Manoa Mist', 'Mauna Kea' (margin green), 'Mauritius White', 'Meringue White', 'Morocco', 'Myron Moori', 'Pierrot', 'Suchiro', 'Trinidad' (off white), 'Uniwai', 'Uranus', etc., and vi) **miscellaneous** such as 'Amigo' (red obake), 'Aneunue', 'Blush' (spathe red-veined), 'Candy Queen' (peach spathe and yellow spadix), 'Carnival' (pink margined white spathe), 'Caroline Simmon' (purple), 'Choco' (chocolate brown), 'Double' (various colours), 'Fantasia' (spathe cream veined pink), 'Farao' (orange-green obake), 'Fla Rose' (obake-peach), 'Lambada' (white-green-obake), 'Madonna' (cream obake), 'Midori' (obake-green), 'Midori-Green' (green), 'Pistache' (green), 'President' (pink obake), 'Red Dragon' (red obake), 'Rico' (obake-rose), 'Sultan' (pink obake), etc.

Propagation

Vegetative propagation of anthurium is through rootstock division, stem cuttings, suckers, and through leaf axillary bud. After top cutting, the **rootstock** is removed from the containers in February-March and cleaned off the compost. The fibrous roots are divided each piece containing a growing point (node). These pieces are dipped in 0.2 per cent Captan or Thiride, planted in fresh compost or 1:3 organic matter and sand medium, and are kept in humid conditions until fully established. The most common method of propagation is through **stem cuttings**. Terminal cuttings preferably from older plants with 1, 2 or 3 nodes or leaves and treated with Seradix 1 or IBA 500 ppm, provides good rooting if kept under intermittent mist which accelerates rooting and increases survival. Performance of the cuttings taken with 2 or 3 nodes is better than 1-node cutting. However, if such cuttings have intact aerial roots, the success is quite assured under intermittent mist. Hata *et al.* (1994) suggested hot water treatment of cuttings at 49°C for 10 minutes + basal application of 0.8 per cent IBA which apart from enhancing rooting also increased number of shoots per cutting in *Anthurium andraeanum* cv. 'Marian Seefurth'. After taking terminal cuttings, remaining part of the stem develops **side shoots** (**suckers** or **offshoots**) coming up from the base which with aerial roots should also be removed gently and planted in the medium and such plants flower earlier than any other method adopted for its multiplication. However, such offshoots (suckers) when attain 4-5 leaf stage (not the larger size) and 2-3 roots, only then are separated gently and planted in the medium (1 part sand: 3 parts compost, by volume). One normal plant may produce 10-20 offshoots in one year, more being in *A. scherzerianum* and less in case of *A. andraeanum*. The dormant **axillary buds** arising from the leaves are gently removed from the plant along with the leaf and root and planted as for suckers. These buds develop into a new plant.

Tissue culture, the non-conventional technique is prevalent in anthurium multiplication through which higher production than those from seedlings is obtained. Leaf, spadix and root segments, stem sections, vegetative buds, etc. are used as explants for callus formation on MS or Nitsch medium. Except apical or the axillary bud meristem explants, others will develop into plants through callus where genetic variability may be observed, though apical or axillary bud meristem explants produce true to type plants. The explant from unfolding leaves takes 11 months from leaf explants to complete plantlet formation, *i.e.* sprout induction and leaf development and further two months for root initiation. Leena Ravidas (2003) reported that when immature seeds of *A. andreanum* varieties 40-45 days before maturity were cultured *in vitro* in ½ MS medium supplemented with 1 mg l^{-1} BA, there were good germination and further development, for callus initiation ½ MS medium with 6 mg l^{-1} BA and 3 mg l^{-1} NAA and for rooting and growth promotion ½ MS + 0.5 mg l^{-1} BA + 1 mg l^{-1} IAA were quite effective. This way received plantlets could be transferred to the field (1:1:1 sand, *Trichoderma*-treated cow dung and husk pieces) by 6 to 7 months, which start flowering by 12 months and thus it will require only 22 to 24 months from crossing to flowering instead of usual 30-35 months. Rooted tissue-cultured plants may be planted in egg-trays filled with washed fine and coarse sand 1 part and leaf mould 1 part or in soilrite. Weekly spraying with 0.2 per cent NPK (3:1:1) mixed with 0.1 per cent mancozeb fungicide keeps the plants healthy. In a 10-13 cm pot also some 10-20 plants can be planted in the soil mixture incorporated with VAM and *Glomus* sp. for better growth of the plantlets *ex vitro.* Synseed (synthetic seed) micropropagation can help multiplying the desired genotype in large scale in short duration at a reduced cost of production.

Sreelatha (1992) reported segments of leaf, petiole, spathe, spike and inflorescence stalk as explants for callus initiation in continuous darkness, and in *A. andreanum* 0.08 mg l^{-1} 2, 4-D and 1.0 mg l^{-1} BA, in *A. veitchii* 0.2 mg l^{-1} 2, 4-D and 1.0 mg l^{-1} BA and in *A. grande* 0.5 mg l^{-1} 2, 4-D and 1.0 mg l^{-1} BA were found optimum for callusing, though MS medium with ¼th strength of major nutrients was found ideal for callus multiplication. Shoot tips from *in vitro* grown seedlings were used as explants for enhanced release of axillary buds. Maximum number of shoots (4.5) was observed with 2.0 mg l^{-1} kinetin and 1.0 mg l^{-1} BA and there was no callusing when kinetin was used. BA and 2ip treatment produced callus growth at the base of the explant. MS major nutrients at ¼th strength and full-strength micro-nutrients were observed ideal for multiple shoot induction in light. The shoots rooted naturally without any treatment and for potting, sand was found as best medium. For survival and further growth of the plantlets, VAM (*Glomus constrictum* and *G. etunicatum*) were found quite useful. Krishnan (1997) tried GA_3 and BAP both at 250, 500, 750 and 1,000 mg per litre on usual and topped plants of *A. andreanum* and obtained callusing in explants taken from leaf, petiole and spadix of such plants, with good callusing being from spadix explants in ½ MS medium supplemented with 2 mg 2, 4-D and 0.3 mg kinetin, each per litre of culture, though addition of casein hydrolysate in the medium improved callusing in leaf explants, however, the calli did not respond to somatic embryogenesis induction treatments. Thomas (1996) obtained regeneration in anthurium on MS medium supplemented with 0.5 mg BA, 2.0 mg IAA, 30.0 g sucrose and 6.0 g agar, each per litre but in presence of light; maximum shoot proliferation (13.49) and rooting on MS medium supplemented with 1.5 mg kinetin, 3.0 mg IAA, 150.0 mg casein hydrolysate, 30.0 g sucrose and 6.0 g agar, each per litre of medium; and growth improvement of shoots in MS medium supplemented with activated charcoal 1.0 g l^{-1} and further subculturing on MS medium supplemented with 0.5 mg kinetin and 16.0 mg IAA, each per litre of culture and this way rooted plantlets were when planted out, she obtained 60 per cent survival rate. Hamidah *et al.* (1997) studied somatic embryogenesis in *A. scherzerianum* and reported that leaf pieces from micropropagated plants were the best explants on a medium containing 18 μM 2, 4-D and 6 per cent sucrose concentration, and 0.46 μM kinetin, solidified with Gelrite where somatic embryos regenerated entire plants. Cen *et al.* (1993) recommended full strength MS medium + 0.1 ppm BA which have the best inductive effect on bud formation while half-strength MS medium + 0.1 ppm was found best for inducting root formation. Teng WheiLan and Teng (1997) used adventitious shoots of *A. andreanum* on MS medium supplemented with 0.5 mg thiamin-HCl, 0.4 mg pyridoxine-HCl, 0.5 mg nicotinic acid, 20 mg Na-Fe EDTA/litre, 15 per cent (v/v) coconut water, 2 per cent sucrose, 2.2-4.4 μM BA and 0.9 μM 2, 4-D where regeneration rates and frequency were found affected due to inoculum size and all the regenerated shoots were also observed normal. Sreelatha *et al.* (1998) used MS medium supplemented with various combinations of 2,4-D, BA, NAA, kinetin, IAA and 2,4,5-T and found best callus initiation in leaf explants from basal portions of leaves of four anthurium species in combinations of 2,4-D and BA under continuous darkness. Modified MS medium containing 2.2 μM BA produced multiple shoots under weak light in root explants of inter-specific anthurium hybrid 'UH1060' and such regenerated plants grew normally and flowered within 16 months (Chen-Fure Chyi, 1997). Yu-KwangJin *et al.* (1995a,b) cultured shoot tips of *A. andreanum* cvs 'Hazrija' and 'Ingrid' (for cut flowers) and *A. scherzerianum* cv. 'Belinda' (for pot plants) on MS medium supplemented with macronutrients, and recorded enhanced shoot growth when medium was added with 0.3-0.8x strength of macronutrients though under prolonged culture duration symptoms of nutrient deficiencies were observed. They found IBA more effective for rooting than IAA or NAA, and survival rate was observed above 90 per cent when plantlets were grown in the medium of fine bark than on peat moss + perlite. Ajithkumar and Nair (1998) recorded good success with *A. andreanum* when cultured leaf explants. Malhotra *et al.* (1998) cultured leaf explants of *A. andreanum* cvs 'Anouchka', 'Nitta' and 'Osaki' on modified MS or Nitsch medium to determine the optimum concentrations of 2, 4-D, BA and ammonium nitrate for callus induction and plantlet regeneration and recorded Nitsch medium better for callus induction when supplemented with BA at 1 mg/l and 2, 4-D at 0.1 mg/l with a reduced concentration of ammonium nitrate, *i.e.* 200 mg/l, the occurrence of shoot formation was found under reduced BA concentration (0.5 mg/l) and increased ammonium nitrate concentration (720 mg/l), and the regenerated shoots were found rooting easily on Nitsch medium supplied with 1 mg/l of IBA and 0.04 per cent AC. Mini Balachandran (1998) for surface disinfection of leaf explants recorded 70 per cent ethyl alcohol wipe followed by 0.1 per cent $HgCl_2$ for 8 minutes as the best treatment up to 99.74 per cent survival of cultures, and of spadix explants 70 per cent ethyl alcohol wipe + 0.1 per cent Emisan dip for 3 minutes followed by 0.1 per cent $HgCl_2$ for 10 minutes as the best treatment up to 87.56 per cent culture survival. Among these 4 explants tried, callus induction was maximum in case of *in vivo* leaf explants where callusing occurred within 51 days when cultured in darkness on to Nitsch-White medium with kinetin 0.5 mg l^{-1} +2, 4-D 0.3 mg l^{-1} + sucrose 20 g l^{-1} + glucose 10 g l^{-1} + agar 6 g l^{-1}. Spadix explants and *in vitro* derived leaf, nodes, petiole and root explants exhibited good response to callus induction, while among the *in vitro* derived explants, roots explants showed maximum callus multiplication. She found spadix explants showing better callus induction in half strength MS medium supplemented with 0.3 mg l^{-1} 2, 4-D + 0.5 mg l^{-1} kinetin + 30 g l^{-1} sucrose + 6 g l^{-1} agar. Nitsch medium supplemented with BAP 0.5 mg l^{-1} gave 92 per cent shoot regeneration with leaf callus. For somatic embryogenesis, *in vitro* derived leaves, petioles and immature seeds showed positive response of 53.0 per cent, 18.9 per cent and 8.33 per cent, respectively. Devinder Prakash *et al.* (2001) cultured 5 mm long petiole explants on MS medium + 2 per cent sucrose + 0.8 per cent (w/v) agar supplemented alone or in different combinations of 2, 4-D, kinetin and BA in dark at 25±2°C for 2-3 months, explants excised from 8-week old seedlings grown on agarified Nitsch medium (2,000 lux and 25°C) through selfed seeds of *A. andreanum* cv. 'Maurititius Orange', and obtained mass callusing in 6-8 weeks in MS+1.0 or 0.5 mg 2, 4-D/l medium, and these calluses were subcultured every six weeks, after 8-10 months of callus initiation the callus pieces were subcultured on MS basal medium, and after four weeks of subculture green spots appeared which subsequently developed into shoot primordia and then to shoots.

Atta Alla *et al.* (1998) cultured the excised seedlings of *A. parvispathum* (a foliage species) on a multiplication medium containing 2 mg BA and 0.2 mg NAA/l where they got 4-fold increase in the shoot number, these shoots were then placed in elongation medium containing 20 mg kinetin/l and when these attained 2.3 cm length apart from further multiplication, these were placed in various rooting media and then these produced 3.6 roots per plantlet with 94 per cent success rate in the medium which contained 0.25 mg IBA/l. This whole process from seed to root took 24 weeks. Now these plantlets after being treated with 0.2 per cent benomyl were placed in acclimatization trays containing peat, bark and soil (!:1:1) for four weeks in a mist chamber with 30°C bottom heat and then these were shifted to the greenhouse where for next three months these were watered twice a week along with every second week spraying with 1:500 dilution of a seaweed concentrate (Kelpak) and obtained 98 per cent success. Somaya *et al.* (1998) cultured seedling tips and nodes (seeds sterilized in 3 per cent NaOCl for 30 minutes gave highest percentage of germination) on MS medium supplemented with 1.0 mg BA and the medium containing 0.25 mg NAA/l accelerated root formation. Montes *et al.* (1999) after disinfecting the ripe seeds of *A. cubense* with 70 per cent alcohol for 1 second, then in sodium hypochlorite for 5 minutes followed by thrice rinsing with distilled water and then again with sodium hypochlorite for 8 minutes followed by rinsing with distilled water thrice and recorded 15 adventitious sprouts per explant within 60 days when grown on MS medium supplemented with BA at 0.22-0.88 µM at a temperature of 27±1°C and a photoperiod of 16 hours. These plantlets after attaining 3- to 4-leaf stage when planted in pots containing 1:1 soil and organic matter the rate of survival was 90 per cent.

For developing new varieties and even otherwise, the anthuriums are raised through **seeds**. The seeds raised through crossing or otherwise (open pollinated ones) are sown in peaty compost preferably in a propagating chamber at 21-24 °C (Hay, 1971) where these germinate within 10 days. These are transplanted within 4-6 months in the proper medium where these will flower within $1^1/_2$-2 years. In *A. scherzerianum*, if berries are picked at orange-red stage (not the overripe, *i.e.* reddish brown) and fermented only for 4-5 days in water at 22 °C to separate out the seeds from the pulp results in good germination within 4-5 days if sown at 20-25 °C temperature range though *A. andraeanum* seeds require 28 °C temperature, high peat substrate or FYM + peat + sphagnum moss having 4-5 pH range and continuous lighting for seed germination, and other species require 21.1-23.9 °C temperature. In case of *A. andraeanum* varieties, Leena Ravidas (2003) advocated 1:1:1 sand, *Trichoderma*-treated cow dung, and husk pieces for plantlets grown through seeds. Berries in the same inflorescence ripe at different times so as soon as one ripes, may be collected and stored in water but storage beyond 5 days results in poor germination. Bose and Yadav (1989) stated that storage of fermented seeds in water for beyond five days reduced germination from 99 to 80 per cent though storage for 10 days reduced it to 53 per cent. Seeds extracted through crushed mature fruits by soaking in 13 per cent crystalline sodium carbonate for 2.5 h at 20 °C or in 6 per cent pectinase solution for 5h at 26-30 °C results in perfect cleaning of the seeds for sowing singly (Bose and Yadav, 1989). Thiram dusting of the shade-dried seeds before storage at 10 °C and stored for not more than 12 weeks results in 95-100 per cent germination.

Climate, Media and Cultural Practices

Anthuriums are tropical aroids which come up well under high humid conditions, and at constant temperature these continue flowering throughout the year if planted in proper rooting medium. Anthuriums produce aerial roots around the moss-covered tree trunks and branches and draw nutrients and absorb moisture from the air and dead barks. In fact, it produces a flower from the axil of each leaf and together below a thick and succulent root which toughens with age and if this root is not buried at this stage in the compost it will become hard and redundant. Too high a plant produces a few flowers of poor quality until the medium is built up around the stem to facilitate the roots, or the plant is cut over, rooted afresh in sand and planted anew in a pot filled with coarsely chopped sphagnum moss 1 part, coarsely chopped fern roots 1 part, and sand + well decomposed manure 1 part, all by volume, and these plants may require repotting every two or three years but should be top-dressed every year before start of the new growth. The life of one anthurium plant is considered only 6-7 years. Their proper growth depends upon optimum conditions, *viz.* soil and media, nutrients, temperature (day temperature of 25-28 °C), relative humidity (80 per cent), light and shade regulation (75 per cent shade though varying with species and cultivars, age and climate), rainfall and water management, wind management, and CO_2 concentration (900 ml/m^3) if planted in greenhouse environment. Salvi (1997) with anthurium recorded 80 per cent shade levels the best treatment for plant height, though superior combination with respect to plant height was found as 70 per cent shade + Knop's solution + 750 ppm BA. Sucker production (1.35) was better thsn control under 80 per cent shade and 750 ppm BA though it was even better (2.5) when 80 per cent shade was combined with Ohio solution and 750 ppm BA. Decrease in shade intensity decreased aerial root production, and at 13 months of plant-age, inflorescence with normal size appeared first under 80 per cent shade treatment and subsequently in other shade levels in descending order, *i.e.* 70, 60 and 50 per cent.

Anthurium cultivation requires sufficiently porous **media** for adequate aeration as it is a necessity because of the roots which are in natural contact with the outside air (the aerial roots). The media for its growing may be any thing from coco-husk to oasis (soil polyphenol foam as is there in the Netherlands) via coco-peat, rice husk, tree bark, fern root, sphagnum moss, bagasse (in Mauritius), wood shaving, rock wool, lava, charcoal, sand and gravels, brick pieces, etc. with certain amendments. Medium should conform to certain qualities, *viz.* should offer sufficient support to the plants (anchoring), should be highly porous for adequate root aeration, should be able to hold required water, fertilizers and other necessary nutrients and drain out easily, should have a pH range of 5.7 to 6.2 with Ec 0.7-1.3 dS/m, and should be capable to degenerate quite slowly. Rapid degenerating medium will become messy soon which will cause water-logging by which its porosity is marred and the scope of aeration is gone which will not permit the roots to function properly for its growth and for drawing nutrients and instead will invite various harmful fungi and algae to develop. The soil will also become sour due to non-leaching of salts. Consequently, the plants will also turn yellow and die. *A. andraeanum* does quite well in gravel than in peat or in a mixture of sphagnum moss and coniferous forest soil. However, 1:1 mixture of soil and wood shavings or 5:1 mixture of wood shavings and cow manure + tree fern fibre, sugarcane bagasse, coffee leaf mould, spent ground coffee, cured coffee pulp/coffee parchment, tree bark, powdered bark, black cinder (less nematode damage), chicken manure, top soil + filter press cake, pumice, peat moss or bark-based compost, are equally good for production of anthurium cut flowers. Any substrate described here when is mixed with high peat, the flower yield is found increased with better quality. Tai-Chun and Ricketts (1992) recommended gravel medium as the best one as this is non-biodegradable, easily manageable, has minimal disease problems and possesses required physical and chemical properties for anthurium growing but Simon *et al.* (1999) reported gravel giving poorest yield in *A. andreanum*, and together recommended coconut husks and coconut fibroblast for higher yield, followed by gravel and fibroblast, brick medium and the brick + fibroblast medium.

Peat + pine-bark + perlite (2:2:1), peat + superphosphate + perlite (1:1:1), high peat + perlite + sphagnum moss (2:1:1), powdered bark + perlite + high peat (1:1:2), coir + compost (1:1) or a well-drained spongy compost made of peat and sphagnum moss (1:2) is ideal for most of anthuriums. Coarse leaf mould and sphagnum moss 2-3 parts, turfy loam 1 part and coarse sand 1 part, along with little amount of charcoal also makes an ideal medium for anthurium. Matysiak *et al.* (1996) in a growth chamber found greatest growth in organic medium of peat 3 parts + perlite 1 part or in Jiffy-7 peat pots or in mineral medium of rockwool + perlite with CO_2 enrichment (350 or 1,200 μmol/mol) under normal atmospheric conditions in rooted microcuttings of *A.* x *cultorum* (*A.* x *ferrierense*). These rooted microcuttings were grown for two weeks at 50 μmol m^{-2} s^{-1} PPFD (photosynthetic photon flux density), then afterwards for six weeks at 150 μmol m^{-2} s^{-1} PPFD. Jawaharlal *et al.* (2001a) with *A. andreanum* cv. 'Temptation' in shadenet-house recorded highest number of suckers, branches, *vis-à-vis* flowers with increased inflorescence longevity and reduced juvenile period. Holcroft *et al.* (1995) reported greatest flower production in medium containing coarse potting mix of pine bark fortified with water absorbent gel or wetting agents. Welsch and Rober (1995) recommended substrates with large proportion of broken aggregates for growing *Anthurium scherzerianum*. Pawar *et al.* (2002) tried coconut coir pieces, brick pieces, wooden charcoal, and garden soil, all alone and coconut coir pieces+brick pieces, coconut coir pieces+wooden charcoal and coconut coir pieces+brick pieces+wooden charcoal all in equal volume and recorded largest flower size in coconut coir pieces+wooden charcoal, longest floral stalk in wooden charcoal though longest vase life was obtained in coconut coir pieces. For anthurium cultivation, the presence of salts in the medium hampers the plant growth, soil chloride being more detrimental. A mixture of coarse sand and dried cow-dung (1:1) or coarse sand, coconut husk, cow dung and charcoal/brick pieces (2:2:1:$^1/_2$) is being used in Kerala Agricultural University while the combination of leaf mould and coco-peat in Tamil Nadu Agricultural University as the cheapest and best media for better sucker production and cut flower yield. A 25-30 cm pot (preferably earthen which permits adequate breathing by the roots) with a minimum of two holes is first filled with a 2-3 cm layer of coarse sand over the crock pieces, followed by some 3-5 cm thick layer of charcoal or brick pieces upon which root ball is placed by pressing with pieces of coconut husk in the sides of the root ball in the pots. Dried cow dung + coarse sand mixture is evenly spread over the root ball, in a way that sand, brick/charcoal pieces and mixture cover only some $^1/_3{}^{rd}$ - $^1/_2$ depth of the pot, watered and kept under high humid conditions, preferably under intermittent mist. *Trichoderma viride* for better health may be mixed in dried powder of cow dung in 1:100, immediately moistened and is kept under shade by moistening and stirring it daily for 10 days and then it is filled in the pot while planting. Once in every three months, the potting mix is built up around the stem to cover the aerial roots before being redundant, which apart from encouraging the growth will also provide good anchorage to the growing plant. Basheer (2005) stated sand + coir pith compost an ideal medium for juvenile *A. andreanum* plants for obtaining greater plant height, leaf area, leaf duration, petiole length at 3rd and 4th week after emergence, shortest phyllachron, the high fresh and dry weights of leaves, their N and K contents, earlier flowering, greater number of flowers per plant, increased spathe size, greater length and thickness of floral stalk and enhanced vase life.

In the large scale, the anthuriums are grown directly on the raised or even on flat **beds** in the open or in the **polyhouse** equipped with drip system of irrigation and foggers/humidifiers to manage water and humidity requirements. Polyhouse protects plants against adverse weather conditions, especially temperature and to some extent rainfall. The sophisticated polyhouse is equipped with a climate control computer to regulate temperature, humidity, ventilation and CO_2. It would be even better if the polyhouse is also equipped with fertigation system. To protect the plants from a very high temperature and low humidity, the closed polyhouses are constructed even with 'pad and fan' cooling system, pad (cladding material or mattress) with water flowing over it at one end and an exhaust fan (ventilator) to another end, *i.e.* opposite direction so that hot air can be blown outside by sucking it through the pads, which will ultimately keep the greenhouse cool. If there be further need, shade installation may also be added but in extreme conditions. Making such greenhouses in sub-temperate areas or in the forest will save the energy to a greater extent. Femina (2006) recommended an integrated air temperature – relative humidity – light intensity regime to maximize growth, production and quality of anthurium, and the reduction of temperature could be accomplished through the use of water spray/mist system to increase the humidity in the shade-houses or polyhouses if that is not hi-tech. The medium used in this case is direct soil or gravel mixed with FYM or leaf mould some 15 cm in depth as is being practiced by commercial growers in the forest near Yercaud (Tamil Nadu). People suggest more depth, even 25-30 cm but that is of no use as root spread of anthurium is not so deep in the medium. The size of the beds is ordinarily 1.5 m in width while it may be of any length from 1.5-5.0 m, as per the convenience of the grower and the size of the polyhouse. However, the sides of the beds are slightly raised, *i.e.* from 20 to 30 cm. The bed-width is more important as while standing on the paths on both sides of the bed one can attend to all sorts of field operations including flower cutting. The width of the path is normally 80 cm. The **planting distance** in the beds is kept 30-60 cm apart, generally 30 × 45 cm for most of the varieties, depending upon the growth and the height of the individual varieties. Thick planting hinders the air circulation among the plants and even spraying of insecticide/fungicide in the beds; though to some extent this problem may be overcome through timely thinning of old or even otherwise normal leaves, as more than five leaves are not good for health of a plant. Thinning is also done to maintain plant equilibrium. Healthy and good looking thinned leaves may also be marketed as cut foliage. Suckering in flowering plants will cost flower production, both in terms of quantity and quality, therefore, when a sucker in a plant appears it should immediately be removed gently by hand pulling. In temperature controlled polyhouses there may not be any problem but in Kerala where winter is very mild, October **planting** af *A. andreanum* is recommended (Femina, 2006).

For proper growth and flowering, anthuriums also need adequate amount of **nutrients**, especially nitrogen, potassium and calcium. Inadequate levels of N and K reduce flower yield and quality. It is suggested that 126 mg N and 225 mg K application in a 12.5 litre container per week improves flower yield and quality in *A. andraeanum*. Annual dressing with 29 g N and 30 g K_2O/m^2 is said to be optimum in *A. andraeanum*, where mature leaves show 2 per cent N and 3 per cent K though *A. scherzerianum* provides best results with 21.6 mg N/pot/week. Foliar application of even 0.05 per cent urea is detrimental to anthuriums, however, earthworm (*Eudrilus eugeniae*) wash 50 per cent application has been found quite encouraging with cv. 'Crinkle Red' where plants remained healthy with better quality flowers and more sucker production (Karuna *et al.*, 1999). Cv. 'Temptation' responded well to 0.2 per cent 30-10-10 NPK +100 ppm GA_3 which reduced pre-blooming period and provided highest number of suckers (Jawaharlal *et al.*, 2001b). Under 80 per cent shade, 20-20-40 NPK at 0.25 per cent at weekly intervals when applied to *A. andraeanum* resulted in longest vase life (Valsalakumari *et al.*, 2001). Basheer (2005) reported that 4 g l^{-1} cowdung extract applied to the plants at 165 and 225 days after transplanting maximized fresh as well as dry weight of leaves, increased spathe size and induced earlier flowering with maximum length of floral stalk & vase life along with greater N and K content while P content was greater during the period in plants receiving 6 g l^{-1} fresh cowdung extract. Through application of 25 g organic manure mix, she recorded greater plant height, leaf area and duration, and petiole length in first and second week after leaf emergence, higher spathe size, length of the floral stalk and vase life along with greater N and K content in the leaves, greater P content in plants receiving 50 g mix though fresh and dry weights of the leaves were greater in plants receiving 75 g mix. Ca deficiency results in colour breakdown, collapse of the proximal part of spathe and instability of the middle lamella and the situation aggravates when pH of substrate is in between 3-4. The optimum Ca concentration in the leaf tissues and lobe should be 0.54-0.16 per cent, respectively. In *A. scherzerianum*, 4 g $CaCO_3$ and 2 mg boron produce largest number of flowers with best quality. Dufour *et al.* (2001) studied optimization of mineral nutrition in soilless culture under tropical conditions with *A. andreanum* and reported best growth and yields of plants in the solution having lowest calcium (2.25 meq Ca litre^{-1}) and the highest ammonium (2.43 meq NH_4 litre^{-1}) concentrations. Sonneveld *et al.* (1993) used nutrient solution with the irrigation water by adjusting the EC values of the solution in a way that EC values ranged from 0.6 to 3.0 dS/m at 25°C while studying the two *A. andreanum* cvs 'Tropical' and 'Cuba' and recorded poor vegetative growth, *vis-à-vis* flower production, decreasing with every increase in the nutrient levels. They recorded Ec of the drainage water ranging from 0.8 to 4.7 dS/m and suggested that the solution of nutrients should have Ec of only 0.7 dS/m for optimum production.

Anthurium being a tropical crop does well at constant **temperature** in between 18-28°C. The minimum winter temperature for *A. andraeanum* and *A. scherzerianum* is 13°C, and for *A. crystallinum* 16°C though only for a short duration these may bear 3°C lower temperatures. The optimum temperature for vegetative growth in both the species is 18.3°C and the constant lower temperatures cause spotting on leaves and marked reduction in the growth of plants. A range of temperatures from 21.1 to 23.9°C is favourable for initiation of flower buds in *A. scherzerianum* but in case of hybrid anthurium cultivars a temperature of 30°C is ideal for getting higher flower count of good quality (Wang Yin Tung and Wang, 1999). It can tolerate as high as 35°C temperature provided atmospheric humidity is very high, *i.e.* 80-85 per cent.

Anthurium loves filtered **sunlight** as it is a plant of semi-shady conditions. Too much light will burn and scorch the flowers as well as leaves. It thrives well at 1500 to 2500 fc light but *A. scherzerianum* requires 1000 to 1500 fc light for good growth and flowering. Direct sunlight can cause severe burn of the leaves as it increases the temperature in anthurium leaves up to 10°C above the air temperature. During summer, it should be given 80 per cent shade, or even more in case of *A. crystallinum.* Under lower percentage of shade stem lengths are found reduced and leaves may also become chlorotic as under 27 per cent shade such abnormalities have been observed though quality flowers have been reported under 75 per cent shade, largest number of flowers have been observed under 25 per cent shade, however, under 55 per cent shading the number of flower was found increased from 5 to 12 per plant per year (Bose and Yadav, 1989). Photoperiod and light intensity requirement of the plant though is cultivar dependent but by and large all can do well between 1500-2500 fc light.

Whenever temperature rises above 16°C, the plants are freely **watered.** However, during the growing season these are moderately watered. Watering is dependent on season, the greenhouse climatic conditions, size of the plant and pot and the medium used, plant growth stage, and type of container. During spring and summer the water demand by the plants is much more when plants should be watered twice daily but in winters it reduces and should be watered only when surface of the medium seems to be dry. Optimum value of water EC is 0.6 m mhos/cm and in no case it should rise above 1.5 m mhos/cm. Sodium chloride present in water is highly detrimental to the crop. Likewise, Na_2SO_4 or $MgCl_2$ in the irrigation water also lowers flower production. In commercial cultivation, the irrigation water is economical through sprinkler system but in the polyhouse it should be through drip system. It would be more economical to give them fertigation rather than only irrigation. In fertigation, only a very light concentration of the nutrients should be given. Rains damage the flowers and increase the chance of fungal infection. Therefore, anthurium should be grown under protection. The **relative humidity** requirement by the anthurium is also very high, *i.e.* above 70 per cent but during bright sunlight it may be as high as 85 per cent. High humidity may be maintained by damping down the greenhouse twice a day and by syringing the leaves daily during hot weathers. Paths and benches should also be kept moist. For maintaining high humidity levels inside the greenhouse, foggers or humidifiers may be used.

Wind and storms also damage this crop, hence right type of greenhouse should be erected to limit the risk of damage. CO_2 is already present in the air, normally at 300 ppm concentration but the requirement is up to 800 ppm which can be maintained easily in the greenhouse. The concentration above 800 ppm is detrimental.

Growth and Development

Plant growth regulators have been used in anthurium for increasing growth, sucker production, quality and yield of flowers and to reduce juvenile phase (when only vegetative buds are formed in the axils and not the generative ones). Anthuriums undergo two distinct developmental stages, a juvenile phase and a generative phase. The flower buds forming in the axils of leaves during generative phase remain dormant and only after breaking of dormancy. Environmental conditions not favourable at the time of flower development may either prolong the dormant phase or may damage the flower buds depending on the stage at which stress condition has been encountered.

Monthly application of BA at 750 mg/l to tissue cultured plants, starting from second month of planting, results into precocious and healthy suckering, each sucker having independent shoot and root system. BA at 100 mg/l induces 3.6 buds (shoots) per plant followed by PBA 1500 mg/l with 2.2 buds and ethephon at 1000 mg/l producing 1.8 buds as compared to nil in untreated ones (Bose and Yadav, 1989). GA_3 at 1000 mg/l application during vegetative stage induces branching while at flowering when applied every month increases yield of the flowers. Krishnan (1997) tried topping, *vis-à-vis* GA_3 and BAP at 250, 500, 750 and 1,000 mg per litre on usual and topped plants of *A. andreanum* and reported that topping alone could induce lateral shoot production with good size. Further studies revealed that GA_3 750 mg per litre on topped plants generated highest number of lateral shoots after 8 months of treatment and its 500 mg l^{-1} produced flowers with maximum angle between apathe and spadix because of increase in the content of Ca and Mg in the spadix, whereas BAP 250 mg l^{-1} was found to be more effective only on intact plants and that too after 5 months of treatment. Boschi *et al.* (1998) applied 500 ppm GA_3 at potting and then again to the plants after one month of potting in *A. Scherzerianum* cv. 'Amazonas' which resulted into largest floral buds without reducing the leaf number. GA_3 at 375 or 500 mg/l and light intensity of 280 µmol m^{-2} s^{-1} resulted into more than three blooms per plant in cv. 'Renate' (Henny and Hamilton, 1992). Abdussamed (1999) obtained more than three times suckering in anthurium 'Hawaiian Red' with 1,000 ppm GA_3 when applied at monthly intervals, highest number of flowers per plant (2.34) with BA and GA_3 treatment at 250 each under ground planting compared to only one plant in control, and the combined effect of BA and GA_3 each at 500 ppm produced better anthocyanin content of flowers. Beena (2000) with three cultivars of anthurium, *viz.* 'Ceylon Red', 'Kalympong Orange' and 'Liver Red' used GA_3, TIBA and kinetin each at 100, 300 and 500 ppm concentrations and recorded maximum vegetative growth, number of suckers, length of spadix and size of spathe with 500 ppm GA_3, minimum plant spread with 100 ppm TIBA, and significantly superior longevity of spadix under 500 ppm kinetin.

Flower Harvesting and Postharvest Management

Anthuriums are **harvested** with long stalks leaving only 3 cm of stem on the plant to prevent rotting of the stem. Under proper management conditions, an anthurium plant produces some 8-12 flowers per year. When unfolding of the spathe is complete and the spadix is just fully developed with about $^{1}/_{3}{}^{rd}$ true flowers on the spadix fully open, the flowers are harvested with a sharp knife. Flowers wilt quickly if harvested too early. Another criterion for flower harvesting is that when one-third of the true flowers on spadix are fully developed in acropetal succession, as is the practice in Hawaii. Salvi (1997) also recommended right stage of flower harvesting when $^1/_3$ of flowers open on the spadix. After harvesting, the cut ends of the flowers are placed in lukewarm water (38°C) overnight and then **graded** as per the colour, the length of stems, and the size of spathes and spadices. The flowers are graded as per Mauritius system or as per Dutch system. Mauritius system is based on average length and width of the spathe while Dutch system is based as per diameter of the bract and stalk length, as are furnished in the Table 5.1.

The flowers of anthurium can be **stored** successfully at 13-17 °C for up to two weeks which may last for 2-4 more weeks in an arrangement. The storage temperature below 13 °C may cause blueing of flowers, especially the red ones though some varieties tolerate even 5 °C storage temperature. These can also be stored at 2-10 per cent O_2 and at ambient temperature of 24-25 °C where cool chamber is not available. The most commonly used corrugated box sizes for **packing** anthurium flowers are 21.6 x 50.8 x 91.4 cm or 27.9 x 43.2 x 101.6 cm which can accommodate some 120 flowers (Salunkhe *et al.*, 1990). A box with 60 × 30 × 22.5 cm dimension will accommodate some 70-100 flowers. Cartons are lined with polythene sheets and insulated with moist paper to maintain proper humidity during transit. The flowers having blemishes, black spots, short stalks, deformity and discolouration are discarded while grading. Anthurium cut flowers after grading are packed with cut ends inserted either in water-filled rubber balloon or in water-filled plastic tubes, mouths of the balloons or tubes being securely tied to the stem above the cut ends so that there may not be water leakage. Cotton swabs or pads saturated with water and enclosed in wax paper are placed above the cut ends and securely tied around the cut portion to prevent desiccation and then the flowers are placed into a polythene cover of proper size after putting some soft protective material in between the spathe and spadix to prevent bruising. The open end of the polythene is stapled so that the flower inside the polythene remains in a position. The modern practice is different as individual flowers are packed in individual polythene bags or the spadices are dipped in melted paraffin to check the moisture loss as is the practice in whole of North America including Hawaii. In the Netherlands, the flower stems are placed in flasks containing water and pack them in moist boxes after putting a soft protective material between the spathe and spadix to save the flowers from bruising during transit. The flowers are placed in cartons by providing foam plastic lining (to help maintaining high humidity) and supports in the box and the flowers are securely taped there.

Wang and Wang (1999) when stored the anthurium cut flowers at 4, 14 and 20 °C for 10-15 days, observed lower sugar content under 14 °C than at 4 °C as under higher temperature the respiration rate is also higher but at lower temperature (4 °C) there had been flower discoloration (even under those conditions where flowers had received pre-storage treatments with 2.5 per cent DMSO) which could not be improved when subsequently the storage temperaturtes were increased. Normally, the flowers of *Anthurium andreanum* are insensitive to the problems of water uptake possibly due to low transpiration rate. Sankat *et al.* (1994) studied the water balance (water uptake, loss and accumulation) in the cut flowers of *Anthurium* cv. 'Trinidad Pink' placed singly in bottles filled with distilled water and kept for 30 days at ambient temperatures of 8, 13, 18 and 28 °C, and recorded a good balance between transpiration and water uptake, higher

Table 5.1

1. Mauritius System of Grading			*2. Dutch System of Grading*			
Grade	*Average Length and Width (cm)*	*Length and Width (cm)*	*Grade*	*Flower Width (cm)*	*Minimum Stalk Length (cm)*	*Number per Box*
Extra super	>	35.6	Peewee	<6.25	25	21
Super	+15	30.4-35.6	Mini	6.25-7.5	25	18
Large	12.5-15	25.4-30.4	Small	7.5-10.0	30	15
Medium	10-12.5	20.4-25.4	Medium	10.0-12.5	35	11
Small	7.5-10	15.2-20.4	Large	12.5-13.75	40	16
Mini	7.5	14.6-15.2	Extra Large	13.75-15.0	45	12
Peewee	7.5	14.6	Premium	>15	50	10

The VBN (Federation of Dutch Flower Auctions) grading specifications for Class I flowers are:

~Absolute freshness

~Pest and disease free

~Free of residue of protective substances or other matter which may affect the product's appearance

~Fault free

~Green petal margins are permitted provided the flowers are graded separately as indicated at the auction

~The stalks must be straight and sufficiently firm to carry the inflorescence

Cut anthuriums with slight deviation from the above requirements are auctioned as Class II though a 5 per cent deviation is permitted within each lot (Anonymous, 1997).

fresh weight, and higher moisture content in the flowers stored at 13 °C though those flowers stored at 8 °C showed spathe bluing and chilling injury.

The growing-condition's light-intensity does not affect cut flower quality in anthuriums, however, under the climatic conditions of the Netherlands the flowers cut at the stage when spadix is almost completely white last for 20 days (in winter) to 25 days (in summer), though there is great varietal variation. Vascular blockage in the cut flower stem causes flower senescence. While senescing, there is weight loss of the flower, glowlessness and blueing of spathe, spadix necrosis, stem collapse and abscission of spathe and spadix from the stem. **Pulsing** in 2.25 per cent 7-Up (a carbonated beverage), 500 ppm benzoic acid, 7.3 ppm of sodium hypochlorite, BA at 50 ppm for 12 h, triadimefon at 25 ppm for 8 h or silver nitrate at 4 mM for 10-60 minutes dip (within 12 h of harvesting) before storage or shipping improves the life of cut flowers significantly but to these treatments varieties differ in their response. Salvi (1997) obtained BA 50 ppm pulsing treatment for 12 h with maximum vase life (20 days). Abdussamed (1999) recorded highest vase life of 19.5 days compared to 9 days in control in case of anthurium var. 'Hawaiian Red' when pulsed the flowers with BA 150 ppm for 8 h. He recorded extended vase life of 13.5 days compared to 10.5 days in control when for packing polythene covers with waxing and $KMnO_4$ in packing cases were used. The response of anthurium var. 'Hawaiian Red' to BA at 25 ppm and 8-HQ at 30 ppm **holding solution** results in delayed spadix necrosis, spathe blueing and increases vase life up to 27 days as per the findings of Kerala Agricultural University. Salvi *et al.* (1997) while working with cv. 'Agnihotri' by subjecting the cut flowers in tap water and the holding solutions of sucrose, $AgNO_3$, CCC and BA, recorded no effect on vase life of sucrose or $AgNO_3$, CCC extending the life only by 1-2 days though in water these lasted for 15 days while BA at 25 ppm prolonged the life by 22 days, and observed that water uptake and electrolyte leakage were directly correlated with vase life. Abdussamed (1999) obtained earlier and highest number of flowers with maximum vase life of 19 days as compared to 14-15 days in control through weekly fertigation with 0.25 per cent NPK (20:20:40) in anthurium var. 'Hawaiian Red'. Yapa *et al.* (2000) used different concentrations of anti-ethylene compounds such as $AgNO_3$, STS, BAP, $KMnO_4$, glycerine, hot water and control (distilled water) against the cut flowers of *Anthurium andreanum* and obtained longest vase life of 14 days with 2 mg/l $KMnO_4$, followed by 12.87 days in STS, 12.43 days in 10 mg/l BAP, 12.37 days in 5 mg/l BAP and 12.25 days in 1,000 mg/l $AgNO_3$ though hot water and 15 mg/l BAP did not have any effect, however, glycerine proved even detrimental. BA 100 mg l^{-1} dip or spray to different *A. andreanum* cultivars showed a great variation from a 20 per cent reduction to 2.5-fold increase in the vase life and those varieties responding positively showed 20 days more vase life than control.

Insect-Pests and Diseases

Anthuriums being highly sensitive to certain pesticides, it is not an easy task to control all these pests effectively.

Green peach **aphid** (*Myzus persicae* and *M. circumflexus* on *A. scherzerianum*), *Aphis gossypii, Aphis nicotiana, Aulacorthum solani* and *Macrosiphum euphorbiae* infest on anthurium by sucking plant saps from various parts of the plant, inject substances that are toxic to plants, produce honey dew which attracts ants and certain fungal infection and transmit viruses. Being viviparous, these build up their population quickly. Aphids may be light green, yellow, pink or red in colour. Their control on anthurium will not serve the purpose so long other vegetation is nearby where these find an alternate host. Therefore there should be proper sanitation and then these should be controlled through 0.2 per cent spraying of some systemic insecticides such as dimethoate or malathion. Pyrethrum extract gives complete control in greenhouses. **Mealy bugs** suck sap of the leaves from the lower surface producing honey dew which attracts sooty mould. Drenching the media with systemic insecticide like dimethoate will control this pest also. Youngs and adult **thrips** (*Chaetanaphorothrips orchidii*, Ranjith, 2013) are light yellow-brown in colour and for sucking the sap bore the cells by feeding on leaves and flowers (spathe and spadix) causing brown stripes on the affected part and mottle them. Control of aphids will control these insects also. The adults and nymphs of **anthurium whiteflies** (*Crenidorsum aroidephagus* and *Aleurotulus anthuricola*; Ranjith, 2013), smaller than common glasshouse whiteflies (*Trialeurodes anthuricola* and *A. vaporariorum*) where adults have white waxy coating, colonise on the under surface of foliage and petiole sheath making the plants weak with localized necrosis and chlorotic spotting, honeydew deposition and the growth of sooty mould. **Mites** [cyclamen mite, *Phytonemus pallidus*; and red spider mites, *Tetranychus urticae, T. telarius* & the privet or false spider mite, *Brevipalpus obovatus*; Ranjith, 2013) are tiny, spider-like, 8-legged, white green (sometimes brown-red) and transparent insects which suck up the contents by boring through the plant cells especially on the underside of the leaves or flowers which make the infested part (flowers, *i.e.* spathe and spadix, and leaves) discoloured to silver-white (red spider mites cause yellow discolouration by feeding on older leaves), being more serious during summer months and later on make the webs. Spraying of the underside of the leaves and other affected parts with vertimec 0.4 per cent or Kelthane 0.08 per cent will control this pest. **Scale insects** suck the sap from stems and leaves which can be controlled by wiping the affected parts with methylated spirit soaked cotton-wool swab or spraying with 0.2 per cent Malathion. **Caterpillars** of the tomato looper, the beet army worms and certain other lepidopterous insects chew various parts of flowers and leaves which may be controlled through catching and killing or through lannate 0.1 per cent or decis at 0.5 per cent sprayings. **Slugs and snails** chew over the root tips, leaves, buds as well as flowers, especially during night and batter the leaves which should effectively be controlled through metaldehyde baits. **Nematodes** [*Meloidogyne incognita* (Cadet *et al.*, 1993), *Aphelenchoides fragariae, Pratylenchus infestans, Radopholus citrophilus* (Goo and Sipes, 1997) *and R. similes* (Bala and Hosein, 1996), etc.) infest on anthurium. *Meloidogyne* galls on the roots show brown lesions. *Aphelenchoides* feeds in leaves causing necrotic areas between leaf veins. *Pratylenchus* attacks only the roots by which leaves yellow and plant may die. *Radopholus* also infests on roots and show dark necrotic lesions which causes stunting and poor growth of the plant with fewer and smaller flowers and premature leaf yellowing. These pests can be put under check by planting comparatively more tolerant species or varieties such as *A. pittieri* and *A. ravenii* instead of *A. andraeanum* cv. 'Midori', by planting healthy stock in disinfected media through steam or methyl bromide, by treating the cuttings for 10 minutes at 50°C temperature, by separating planting media from the natural soil of the bed, through proper sanitation, and through precautionary measures. The nematodes can also be controlled through recirculation of water by ozone and activated hydrogen peroxide @ 1 h to 20 g ozone. With its new formulation, only 200 ppm for 24 h is required for their complete elimination (Runia and Amsing, 1996).

Viral (mosaic and malformation of leaves and spathes, mostly in white cultivars) diseases are transmitted through *Bemisia tabaci* and through mechanical tools so controlling the vector will control these problems, and such plants should be uprooted and destroyed. Dasheen mosaic potyvirus has also been observed infecting this plant in Taiwan (Liang *et al.*, 1994). Control of insect-pests controls even the viruses as insect-vectors spreading the viruses are eliminated.

Bacterial blight is caused by *Xanthomonas axonopodis* (*campestris*) pv. *diffenbachiae*, initially causing small and angular water-soaked spots on the leaf margins where afterwards after killing the tissues bright yellow halo around the spot will be formed and spadix also starts rotting from tip downwards. The disease is favoured by warm and wet weather. This disease has been found recording heavy loss to anthurium plantings to the growers in Kerala (Dhanya, 2000). *Aglaonema robelini, Colocasia esculenta, Dieffenbachia* sp., *Philodendron oxycardium* and *Syngonium podophyllum* are the carriers of the bacterium which can survive up to 45 days on plant debris and up to 60 days in the soil (Dhanya, 2000). Under *in vitro* evaluation, Dhanya (2000) found 100 ppm streptocycline and 0.3 per cent Captan more effective in inhibiting its growth, however, under *in vivo* condition 0.15 per cent turmeric powder + sodium bicarbonate (10:1) and its weekly 5 sprayings controlled the pathogen similar to streptocycline 100 ppm. Sabitha (2002) reported that in Thiruvananthapuram, through surveying some 30 anthurium plantings, the pink varieties were found tolerant though reds susceptible. After evaluating 15 botanicals, she advocated crude extract and 2:1 dilution of *Allium sativum* and *Tagetes erecta*, crude and 2 per cent *neem* cake extract and 1-2 per cent *neem* oil and coconut oil quite effective under *in vitro* control of the pathogen. Under field conditions in seven month old tissue-cultured seedlings, she found pre- and post-inoculation sprayings with crude extract of *neem* cake and *T. erecta* reducing the disease to a tune of 85 per cent, and when the same spraying schedule with crude extract of *neem* cake and *T. erecta* as well as 100 ppm streptocycline were used on flowering plants, *neem* cake was found most effective. Nair (2005) with 10-month old greenhouse plants of anthurium var. 'Cancan' observed equally effective the prophylactic root dip and foliar spray of *Bacillus* sp. or turmeric powder + sodium bicarbonate (10:1) at 1.5 per cent for controlling the bacterial blight. However, as a precautionary measure the contamination should be checked through tools, water and flower cutting, and by restricting nitrogenous fertilizers, especially the ammonical forms completely. Apart from Hawaii, this disease has also been reported from France by Soustrade *et al.* (2000) and from Italy by Zoina *et al.* (2000). **Bacterial wilt** reported in Taiwan by Su and Leu (1995) is caused by *Pseudomonas solanacearum* which initially yellows the leaf margins but afterwards covers entire leaf which ultimately results into wilting of the plant. Its occurrence is normally noticed where unfermented bark has been used as mulch. Strict sanitation of the growing area, removal of the affected part or the plants and spraying with streptomycin sulphate or oxytetracycline will keep the bacterial diseases under check.

Fungal diseases that attack anthuriums are anthracnose (black nose or spadix rot) caused by *Colletotrichum gleosporioides* (*Glomerella cingulata*) which is highly prevalent in high rainfall areas initially with tiny dark spots on the individual flowers of the spadix which afterwords spread in angular or triangular shapes even on the leaves and flower stalks where its infection causes damage of spadices, and typical anthracnose symptoms on leaves, flower stalks and spikes, and it can effectively be controlled through 0.2 per cent maneb sprayings (Deshmukh *et al.*, 1999); root rot caused by *Pythium splendens, P. spinosum, P. vexans, Rhizoctonia* sp. [Pasini *et al.* (1996) reported *R. solani* from Italy causing collar or foot rot of anthurium; and Guo and Ko (1994, 1996) reported that incidence of *Rhizoctonia* sp. increases with use of Diuron herbicide], *Calonectria crotolariae, Phytophthora* sp. and *Fusarium* sp. and in such cases of root rot the symptoms appear as plant stunting, size reduction of leaves and flowers, yellowing and hanging of leaves at the edges and the browning of roots and stems, and their control measures are drenching the plants with 0.1 per cent furalaxyl, mancozeb 0.25 per cent or thiram 0.25 per cent or spraying twice at 10 days intervals with aliette 0.15 per cent ; and leaf spots caused by *Septoria anthurii* which can be controlled as to that of anthracnose and powdery mildew caused by *Erysiphe communis* (Orlikowski, 1999) can be controlled through 0.1 per cent benomyl spraying. Benomyl or bavistin at 0.15 per cent alternate with captan 0.2 per cent spraying will also control almost all the diseases.

Physiological disorders, *viz.*, 'flower abortion', 'flower deformation' and 'rosette formation' due to excessive root activity or pressure not in line with the activity of the aboveground plant part which is quite slow and the problem is mainly genetic but to some extent can be overcome by use of restricted amount of media, balance of climatic factors and use of gibberellins; 'folder ears' where basal lobes of flowers do not open properly which may also be genetic; 'sticking' also probably a genetic cause is that where while opening the spathe is stuck which can either be loosened by hand or the lower RH may overcome this; 'jamming' also a genetic problem being encountered under arid conditions in long sheathed cultivars where flowers are jammed too tightly inside the unfolding sheaths and get damaged; and 'cracks' on either or on both sides of the spathe may occur due to active growth of flower when RH is higher for a short duration which may be overcome to some extent by lowering the RH during the night.

References

Abdussamed, K.P. 1999. Regulation of Flowering and Postharvest Behaviour of *Anthurium andraeanum* cv. 'Hawaiian Red' (M.Sc. Thesis). Kerala Agril. University, Vellanikkara, Thrissur.

Ajithkumar, P.V. and S.R. Nair, 1998. Establishment and hardening of *in vitro* derived plantlets of *Anthurium andreanum*. *J. Ornam. Hort.*, **4**(1-2): 9-12.

Anonymous, 1997. *Production Manual on Anthurium andreanum*. A publication by APEDA, Min. of Commerce, Govt. of India.

Arias-Granda, I. 1996. A new species of *Anthurium* Schott (Araceae) from Cuba. *Revista Jard. Bot. Nac.*, No. 17-18, pp. 5-7.

Atta Alla, H., B.G. McAlister, J. Van Staden and JH. Van staden, 1998. *In vitro* culture and establishment of *Anthurium parvispathum*. *S. Afr. J. Bot.*, **64**(5): 296-298.

Bala, G. and F. Hosein, 1996. Plant parasitic nematodes associated with anthuriums and other tropical ornamentals. *Nematropica*, **26**(1): 9-14.

Basheer, S.N. 2005. Standardization of Growing Media and Organic Nutrition for Juvenile Anthurium Plants (*Anthurium andreanum* Lind.). M.Sc. Thesis. College of Agriculture, Kerala Agricultural University, Vellayani, Kerala.

Beena, R. 2000. Effect of Growth Regulators on the Growth and Flowering of Anthurium (*Anthurium andreanum* Linden). M.Sc. Thesis. College of Agriculture (K.A.U.), Vellayani, Kerala.

Bhatt, N.R. and B.B. Desai, 1989. Anthurium. In: Commercial Flowers (Eds Bose, T.K. Yadav, L.P.), pp.623-641. Naya Prokash, Calcutta.

Bindu, M.R. 1992. Chromosome Behavior and Pollen Analysis in *Anthurium* sp. (M.Sc. Thesis). College of Agriculture, Kerala Agricultural University, Vellayani, Kerala.

Bindu, M.R. and S.T. Mercy, 1997. Chromosome behaviour and karyotype analysis in *Anthurium andreanum* Lind. *South Indian Hort.*, **45**(3-4): 134-138.

Binodh, A.K. 2002. Genetic Variability and Character Associations in *Anthurium andreanum* Linden (M.Sc. Thesis). College of Agriculture, Kerala Agril. University, Vellayani, Kerala.

Boschi, C., D. Benedicto, J. Molinari, and A. di Benedetto, 1998. Gibberellic acid application to *Anthurium scherzerianum*. Foliage and inflorescence responses. *Rev. Fac. Agron.*, **18**(1-2): 89-92.

Bose, T.K. and L.P. Yadav, 1989. Anthurium. In: Commercial Flowers, pp. 623-641. Naya Prokash, Kolkata.

Cadet, P., E. van den Berg, L. Nema and E. van den Berg, 1993. Parasitic nematodes of flower crops in Martinique. *PHM Rev. Hortic.*, No. 341, pp. 53-58.

Cen, Y.Q., R.M. Jiang, Z.L. Deng and D.X. Ni, 1993. *In vitro* propagation of *Anthurium* andraeanum. Morphogenesis and effects of physical and chemical factors. Acta *Horticulturae Sinica*, **20**(2): 187-192.

Chen-Fure Chyi, A.R. Kuehnle and N. Sugii, 1997. Anthurium roots for micropropagation and *Agrobacterium tumefaciens*-mediated gene transfer. *Plant Cell, Tissue and Organ Culture*, **49**(1): 71-74.

Chi-GekLan, H.K.L. Goh, Yee Hoo-Kuwan, T. Legavre, G.L. Chi, K.Y. Hoo and R.A. Drew, 1998. GUS gene expression in *Anthurium andreanum, Oncidium* Grower Ramsey and *Brassilaeliocattleya* Orange Glory Empress after particle bombardment. *Acta Hort.*, No. 461, pp. 379-383.

Cotias-de-Oliveira, A.L.P., M.L.S. Guedes and E.C. Barreto, 1999. Chromosome numbers for *Anthurium* and *Philodendron* sp. (Araceae) occurring in Bahia, Brazil. *Genet. Mol. Biol.*, **22** (2): 237-242.

Deshmukh, H. V., D.R. Pawar and M.S. Joshi, 1999. Symptomatology and host range studies of anthracnose disease of anthurium incited by *Colletotrichum gloeosporiodes*. *Orissa J. Hort.*, **27**(1): 26-28.

Devinder Prakash, K. Sujatha, and Sangama, 2006. Anthurium. In: Advances in Ornamental Horticulture, Vol. 2 (ed. Bhattacharjee, S.K.), pp.109-129. Pointer Publishers, Jaipur.

Devinder Prakash, M.L. Choudhary, K.V. Prasad, N. Nagesh and D. Prakash, 2001. Regeneration of plantlets from petiole explants of *Anthurium andraeanum* Lind. Cv. 'Mauritius Orange'. *Phytomorphology*, **51**(1): 83-85.

Dhanya, M.K. 2000. Etiology and Management of Bacterial Blight of Anthurium (*Anthurium andreanum* Linden). M.Sc. Thesis. College of Agriculture (K.A.U.), Vellayani, Kerala.

Dufour, L., W.J. Horst (Ed.), M.K. Schenk (Ed.), A. Burkert (Ed.), N. Claassen (Ed.), H. Flessa (Ed.), W.B. Frommer (Ed.), H. Goldbach (Ed.), H.W. Olfs (Ed.) and V. Romheld, 2001. Optimisation of *Anthurium andreanum* mineral nutrition in soil-less culture under tropical conditions. *Fourteenth intern. Plant Nutrition Colloquium*, Hanover (Germany), pp. 784-785.

Femina, 2006. Microclimatic Relations on the Growth, Yield and Quality of Anthurium (Linden) under Different Growing Systems (M.Sc. Thesis). K.A.U., Vellanikkara, Thrissur, Kerala.

Goo, M.Y.C. and B.S. Sipes, 1997. Host preference of *Radopholus citrophilus* from Hawaiian anthurium among selected tropical ornamentals. *HortSci.*, **32**(7): 1237-1238.

Guo, L.Y. and W.H. Ko, 1994. Survey of root rot of anthurium. *Plant Path. Bull.*, **3**(1): 18-23.

Guo, L.Y. and W.H. Ko, 1996. Nature of enhanced severity of anthurium root rot by diuron treatment for weed control. *J. Phytopathol.*, **144**(1): 7-11.

Hamidah, M., A.G.A. Karim and P. Debergh, 1997. Somatic embryogenesis and plant regeneration in *Anthurium scherzerianum*. *Plant Cell, Tissue and Organ Culture*, **48**(3): 189-193.

Hata, T.Y., A.H. Hara, M.A. Nagao, and B.K.S. Hu, 1994. Hot water treatment and indole-3- butyric acid stimulates rooting and shoot development of tropical ornamental cuttings. *Hort. Tech.*, **4**(2): 159-162.

Hay, Roy 1971. Anthurium. Reader's Digest Encyclopaedia of Garden Plants and Flowers, pp. 41-42. Reader's Digest Asscn, London.

Henny, R.J. and R.L. Hamilton, 1992. Flowering of anthurium following treatment with gibberellic acid. *HortScience*, **27**(12): 1328.

Holcroft, D.M., M.D. Laing, P. adams (Ed.), A.P. Hidding (Ed.), J.A. Kipp (Ed.), C. Sonneveld (Ed.) and C. de Kreij, 1995. Evaluation of pine bark as substrate for anthurium production in South Africa. *Acta Hort.*, No. 401, pp. 177-184.

Jawaharlal, M., J.P. Joshua, T. Arumugam, S. Subramanian and M. Vijayakumar, 2001a. Standardisation of growing media for anthurium (*Anthurium andreanum*) cv. Temptation under shade net house. *South Indian Hort.* (Special Issue), **49**: 323-325.

Jawaharlal, M., J.P. Joshua, T. Arumugam, S. Subramanian and M. Vijayakumar, 2001b. Standardisation of nutrients and growth regulators to reduce pre-blooming period and to promote growth and flowering in anthurium (*Anthurium andreanum*) under protected shade net house. *South Indian Hort.* (Special Issue), **49**: 342-344.

Karuna, K., C.R. Patil, P. Narayanaswamy and R.D. Kale, 1999. Stimulatory effect of earthworm body fluid (vermiwash) on Crinkle Red variety of *Anthurium andreanum* Lind. *Crop Res. Hisar*, **17**(2): 253-257.

Krishnan, A.G. 1997. Improvement of Propagation Efficiency of *Anthurium andreanum* Andre. (M.Sc. Thesis). College of Horticulture, Kerala Agril. University, Vellanikkara, Thrissur.

Kuchnle, A.R., F.C. Chen and N. Sugii, 1995. Novel approaches for genetic resistance to bacterial pathogens in flower crops. *HortSci.*, **30**(3): 456-461.

Leena Ravidas, 2003. Improvement of *Anthurium andreanum* Lind. by *in vivo* and *in vitro* Methods (Ph.D. Thesis). College of Horticulture, Kerala Agril. University, Vellanikkara, Thrissur.

Liang, Y.G., T.C. Yang and C.T. Chen, 1994. Identification of dasheen mosaic virus in Taiwan by immunolabelling. *Report of the Taiwan Sugar Res. Institute*, No. 143, pp. 41-50.

Malhotra, S., D. Puchooa and K. Goofoolye, 1998. Callus induction and plantlet regeneration in three varieties of *Anthurium andreanum*. *Rev. Agric. Sucr. Ile Maurice*, **77**(1): 25-32.

Matysiak, B., J. Nowak, K. Tozai (Ed.), C. Kubota (Ed.), K. Fujiwara (Ed.), Y. Ibaraki (Ed.) and S. Sase, 1996. The effects of root media and CO_2 on acclimatization and growth of microcuttings of *Anthurium, Dieffenbachia, Homalomena* and *Spathiphyllum*. *Acta Hort.*, No. 440: 628-632.

Mayadevi, P. 2001. Genetic Divergence in *Anthurium andraeanum* Linden (Ph.D. Thesis). College of Agriculture, Kerala agricultural University, Vellayani, Kerala.

Mini Balachandran, 1998. Improvement of *Anthurium andraeanum* Lind. *in vitro* (Ph.D. Thesis). College of Agriculture, Kerala agricultural University, Vellayani, Kerala.

Montes, S., M.M. Hernandez and M. Varela, 1999. Organogenesis in *Anthurium cubense*. *Cultivos Tropicales*, **20**(1): 51-54.

Nair, A.R. 2005. Microbial Antagonists and Resistance Inducers for the Management of Bacterial Blight of Anthurium (*Anthurium andreanum* Linden). M.Sc. Thesis. College of Agriculture, Kerala Agricultural University, Vellayani, Kerala.

Orlikowski, L. 1999. Fungi from genus *Phytophthora* - an increasing risk to ornamental plants (Polish). *Ochr. Rostl.*, **43**(3): 18.

Pasini, C., T. Berio, P. Curir and F. D'Aquila, 1996. Further characterization of *Rhizoctonia solani* isolated from carnation and other ornamental plants. *Informatore Fitopatologico*, **46**(6): 33-36.

Pawar, G.M., M.T. Patil and A.M. Gaikwad, 2002. Effect of substrates on anthurium culture. In: *Floriculture Research Trend in India* (eds Misra, R.L. and Sanyat Misra), pp. 326-327. Indian Society of Ornamental Horticulture. Division of Floriculture and landscaping, I.A.R.I., New Delhi.

Paull, R.E., T. Chantrachit and Theeranuch Chantrachit, 2001. Benzyladenine and the vase life of tropical ornamentals. *Postharvest Biol. and Techn.*, **21**(3): 303-310.

Perez-Farrera, M.A. and T.B. Croat, 2001. A new species of *Anthurium* (Araceae) from Chiapas, Mexico. *Novon.*, **11**(1): 88-91.

Rajeevan, P.K., P.K. Valsalakumari, C.K. Geetha, Leena Ravidas, Vinod Kumar and S.K. Bhattacharjee, 2002. Anthurium, 42 p. AICRPF, IARI, New Delhi.

Ranjith, A.M. 2013. Anthurium. In: *Identification & Management of Horticultural Pests*, pp. 125-127. New India Publishing Agency, Pitam Pura, Neew Delhi-110 088.

Renu, R.S. 1999. Intervarietal Hybridization in *Anthurium andreanum* Linden (M.Sc. Thesis). College of Agriculture, Kerala Agricultural University, Vellayani, Kerala.

Runia, W. Th. and J.J. Amsing, 1996. Disinfestation of nematode-infested recirculation water by ozone and activated hydrogen peroxide. *Intern. Soc. Soilless Cult (Proc. 9th intern. Congr. Soilless Culture)*. St. Helier, Jersey, Channel Islands, April 12-19, pp. 381-393.

Sabitha, S.R. 2002. Management of Bacterial Blight of Anthurium (*Anthurium andreanum* Linden) using Botanicals (M.Sc. Thesis). College of Agriculture (K.A.U.), Vellayani, Kerala.

Sakuragui, C.M. and S.J. Mayo, 1999. A new species of *Anthurium* (Araceae) from south-eastern Brazil. *Feddes Report*, **110**(7-8): 535-539.

Salunkhe,D.K., N.R. Bhatt and B.B. Desai, 1990. Minor cut flower crops - Anthurium. In: *Postharvest Biotechnology of Flowers and Ornamental Plants*, pp. 242-252. Naya Prokash, Kolkata.

Salvi, B.R. 1997. Optimization of Shade, Nutrients and Growth Regulators for Cut Flower Production in Anthurium (Ph.D. Thesis). Kerala Agril. University, Vellanikkara, Thrissur.

Salvi, B.R., P.K. Valsalakumari, P.K. Rajeevan and C.K. Geetha, 1997. Effect of holding solutions on cut anthurium flowers. *Hort. J.*, **10**(2): 65-72.

Sankat, C.K., S. Mujaffar and P. Sass, 1994. Water balance in cut anthurium flowers in storage and its effect on quality. *Acta Hort.*, No. 368, pp. 723-732.

Simon, G., G. Bala (Ed.) and M. Fortune, 1999. The effect of different growing media on *Anthurium andraeanum* production and most effective rate of fertilizer application. *Proc. Res. Div., Min. Agric, Land and Marine Resources. Res. Seminar series* **3 & 4**: 25-29.

Sindhu, K. 1995. Cross Compatibility in *Anthurium andraeanum* Lind. (M.Sc. Thesis). College of Agriculture, Kerala Agricultural University, Vellayani, Kerala.

Somaya, K.U., P. Narayanaswamy and K.V. Jayaprasad, 1998. Micropropagation studies in *Anthurium andreanum* Lind. *Karnataka J. agric. Sci.*, **11**(2): 466-470.

Sonneveld, C., W. Voogt and M. Tattini, 1993. The concentration of nutrients for growing *Anthurium andreanum* in substrate. *Acta Hort.*, No. 342, pp. 61-67.

Soustrade, I., L. Gagnevin, P. Roumagnac, O. Gambin, D. Guillaumin and E. Jeoffrault, 2000. First report of anthurium blight caused by *Xanthomonas axonopodis* pv. *dieffenbachiae* in Reunion Ialand. *Plant Dis.*, **84**(12): 1343-1344.

Sreelatha, U. 1992. Improvement of Propagation Efficiency of *Anthurium* Species *in vitro* (Ph.D. Thesis). College of Agriculture (K.A.U.), Vellayani, Kerala.

Sreelatha, U., S.R. Nair and K. Rajmohan, 1998. Factors affecting somatic organogenesis from leaf explants of *Anthurium* species. *J. Ornam. Hort.* (New Series), **1**(2): 48-54,

Su, C.C. and L.S. Leu, 1995. Bacterial wilt of anthurium caused by *Pseudomonas solanacearum*. *Plant Path. Bull.*, **4**(1): 34-38.

Tai-Chun, P.A. and M.R. Ricketts, 1992. Developing techniques for the production of anthuriums in a gravel type medium in Jamaica. *Proc. Interamerican Soc. Trop. Hort.*, **35:** 141-144.

Teng WheiLan and W.L. Teng, 1997. Regeneration of *Anthurium* adventitious shoots using liquid or raft culture. *Plant Cell, Tissue and Organ Culture*, **49**(2): 153-156.

Thomas, A.S. 1996. Micropropagation in Selected Varieties of *Anthurium andreanum* Lind. (M.Sc. Thesis). College of Agriculture, Kerala Agricultural University, Vellayani, Kerala.

Valsalakumari, P.K., K.P. Abdussamed, P.K. Rajeevan and C.K. Geetha, 2001. Shade and nutrient management in *Anthurium andreanum*. *South Indian Hort.* (Speccial Issue), **49**: 326-331.

Wang, HeLi and H.L. Wang, 1999. Effect of low temperature on sugar content, respiration rate and quality of anthurium flowers. *Plant Physiol. Communications*, **35**(6): 458-460.

Wang, Yin Tung and Y.T. Wang, 1999. Greenhouse performance of six potted anthurium cultivars in a subtropical area. *Hort. Technology*, **9**(3): 409-412.

Welsch, J. and R. Rober, 1995. Raising plants in soil, further culture in expanded clay. *Gartenbaumagazin*, **4**(8): 19-20, 22-23.

Yapa, S.S., B.C.N. Peiris and S.E. Peiris, 2000. Potential low cost treatments for extending the vase-life of anthurium (*Anthurium andreanum* Lind.) flowers. *Trop. agric. Res.*, **12**: 334-343.

Yu-KwangJin, Yoeup Paek-Kee, K.J. Yu and K.Y. Paek, 1995a. Micropropagation of *Anthurium* sp. through shoot tip and callus culture. *J. Korean Soc. hort. Sci.*, **36**(5): 684-694.

Yu-KwangJin, Yoeup Paek-Kee, K.J. Yu and K.Y. Paek, 1995b. Effect of macroelement levels in the media on shoot tip culture of *Anthurium* sp. and re-establishment of plantlets in soil. *J. Korean Soc. hort. Sci.*, **36**(6): 893-899.

Zoina, A., A. Raio and A. Spasiano, 2000. First report of anthurium bacterial blight in Italy. *J. Plant Path.*, **82**(1): 65.

6

Antirrhinum (Family: Scrophulariaceae)

R.L. Misra, Sanyat Misra, Ajit Kumar and J.K. Ranjan

[**Common name**: Dog flower, Common or Large snapdragon (*A. majus*), Small snapdragon (*A. orontium*), Snout flower]

Introduction and Origin

Antirrhinum takes its name from the Greek *anti* = like and *rhin* (or *rhinos*) = nose or snout, alluding to the curious appearance (shaped as snout of the rhinoceros) of its flowers that resemble the snout of a dragon. The genus *Antirrhinum* comprises 42 species of annuals, perennials and sub-shrubs native to W. Mediterranean and WN America. It is believed that it reached Britain through the Romans at an early date where it got naturalized in the mountainous regions and from here it reached to other parts of the world. *Antirrhinum majus*, a short-lived woody-based hardy perennial (though usually treated as an annual) which belongs to Mediterranean region, is a progenitor of all the modern varieties and these varieties have almost completely replaced the type. The varieties are now available in a wide range of colours (white, yellow, orange, pink, rose, scarlet, red, crimson, lilac, purple, bicolours & multicolours, and shapes - single, double and open-faced = butterfly) and stature, and are divided into four groups according to height. Taller forms are suitable for cut flowers while dwarf types for rock gardens and for edging borders. The flowers are fragrant.

It is a versatile flower which can be grown in any part of the world from tropical to temperate areas, albeit during winter season in the tropical areas and during summer in the temperate, as a principal cut flower crop in many of the countries, for bedding to landscape the gardens, in rockeries, in herbaceous borders, for edging, in the greenhouses or in pots. In the language of flowers, it stands for presumption. It was also considered, in the long time past, that plants of antirrhinum save from being bewitched. Russians used to grow it in a large scale also for its seeds to extract oil which is similar to olive oil. To restore beauty and youthfulness, the women in the past used to wash the face with antirrhinum seeds-soaked water. During renaissance period, people used to wear its flowers on sleeves to welcome the king as then wearing its flowers was considered a matter of honour. The plants contain bitter and stimulant properties. They are diuretic and are used against scurvy, in tumour, as detergent and as an astringent. The first cultivars of antirrhinum were first grown in United States during long day periods but around 1926 Frank Volz of Chevoit (Ohio) introduced the cv. 'Chevoit Maid' which produced flowers in winter, and in Columbus (Ohio) in 1938 Fred and Helen Windmiller introduced the first F_1 hybrid 'Christmas Cheer', followed by others like George J. Ball, Inc. and Yoder Brothers, Inc. who introduced F_1 hybrids such as 'Maryland Pink', 'Mary Ellen' and 'Dorcas Jane' around 1950 (Arora, 1989). Now F_1 hybrids suitable for growing in early and late winters, and rust resistant cultivars and F_2 tall and semi-tall hybrids with attractive colours are also available (Arora, 1989).

Antirrhinums are prostrate to erect or climbing herbaceous plants or even half-shrubs, having usually opposite paired to alternate, lanceolate to ovate, entire, and dark green to purplish leaves (Bailey and Bailey, 1976), and inflorescence a terminal raceme of numerous individual tubular flowers, individual perfect flowers being borne on a short pedicel along the stem, each flower pinched together to form a closed mouth, and then flaring into 5 waved lobes. Corolla is saccate or gibbous at base but not spurred, and is fused into a tube. Three lower lips spread out but the two upper ones remain erect and form a mouth, hence it is named commonly as snapdragon. Stamens are four and non-winged. Though there are many important genera in Scrophulariaceae such as *Calceolaria*,

Castilleja, Digitalis, Hebe, Mimulus, Nemesia, Pedicularis, Penstemon, Verbascum, Veronica, etc. but *Antirrhinum* is closely allied to *Linaria*, differing only in spurless flowers. The diploid number of chromosomes in antirrhinum is 2n = 16 and tetraploid forms are also available and some 200 genes have been found in its various species and interspecific hybrids [Baur (1910, 1919, 1924, 1930), cited from Arora (1989)]. On crossing red with ivory-flowered antirrhinums the F_1 population gives pink flowers but in F_2 these segregate into 1 red, 2 pink and 1 ivory and further in F_3 the pink F_2s segregate similarly. It was further found that red flowers differed from ivory flowers by single gene (*F*) and from yellow flowers by two independent genes (*F* and *C*). Baur identified the genes for colours as *C* for ivory or yellow, *F* for red, *G* for modifying red and *D* for controlling tube-colour, and these genes were redesignated by Baur [(1930, cited from Arora (1989)] as *N* for albino, *P* for pink, *M* for modifier and *Y* for yellow though later Stubbe (1966, cited from Arora, 1989) identified some other genes in the list and changed the gene nomenclature as nivea (*Niv*), sulfurea (*Sulf* = *C*), incolorata (*Inc* = *F*), eosinea (*Eos*), pallida (*Pal*), diluta (*Dil*) and eluta (*El*) which determine different colours due to synthesis of different pigments, the main seven colours being white, yellow, ivory, bronze, pink, crimson and magenta. He further stated that homozygous recessive *niv*/*niv* block pigment formation and therefore results in white flowers while dominant *niv*+ allele is necessary for anthocyanin formation.

Mahal (1972) intercrossed *Antirrhinum siculum, A. linkianum, A. tortuosum, A. cirrhigerum, A. molle, A. glutinosum, A. majus* and *A. orontium* in nearly all the directions. *A. majus* × *A. linkianum, A. majus* × *A. siculum, A. molle* × *A. tortuosum* and *A. tortuosum* × *A. linkianum* were successful only in one direction; excepting *A. siculum* with *A. majus*, other crosses involving *A. orontium* and *A. siculum* failed; and the crosses *A. majus* × *A. cirrhigerum, A. majus* × *A. tortuosum, A. majus* × *A. glutinosum* and *A. cirrhigerum* × *A. linkianum* were successful in both the directions. The hybrids of all the successful crosses were fertile and intermediate with only a few exceptions where one parent showed dominance for some of the characters, and at meiosis, 16 chromosomes were found to form bivalents showing genetic homology of the participating species. In F_2, some of the disharmonious gene combinations were recorded manifesting into abnormal growth and flower forms quite different from either of the parents which suggests evolutionary potentialities through transgressive gene action. Goldsmith [(1968), cited from Arora (1989)] reported an apetalous pollen sterile form with a thicker fasciated pistil appearing quite commonly which though sets seeds but with poor yield. *A. majus* inbred lines were when crossed with *A. glutinosum* (a dwarf species) resulted into uniform and semi-dwarf F_1 hybrid having more tillers with closely arranged large flowers in inflorescences on them (Mahal, 1972). Malik (1979) recorded appreciable heterosis in F_1 hybrids when crossed the diploid and tetraploid progenies. Arora (1989) stated that interspecific hybrids between *A. majus* and *A. molle* gave variation in plant height, branching habit and corolla form, and a new class of antirrhinum with creeping habit suitable as carpet for ground covering or for rock gardens has been developed and this group is named as miniature or magic carpet.

With the treatment of 0.4 per cent colchicine on the seeds of antirrhinum cv. 'Katrain Local' (diploid), Jain *et al.* (1962) obtained a tetraploid form with increased size of individual flowers on compact and tall inflorescences with more ruffled corolla marked by presence of small papillae-like structure not found in the diploids. Mahal *et al.* (1968) and Mahal (1972) also raised colchiploids of *A. linkianum, A. majus, A. siculum* and two varieties of *A. orontium* which were stout, hardier, shorter and slower in growth but with larger flowers, pollen grains, fruits and seeds and with quite high pollen and seed fertility, and a plant was also recorded bearing varying number of chromosomes in the pollen cells from the enucleate condition to a very high number (2n = 72) where pollen and seeds were both fertile and when this was crossed with diploid and tetraplopid forms in both the directions, succeeded slighty only with tetraploid form when tetraploid form was used as a seed parent and the resultant hybrid was a vigorous and better triploid with faster growth and longer duration of flowering. Out of this work, NBRI, Lucknow released a tetraploid variety 'Tetra Giant'.

Classification, Species and Cultivars

In antirrhinum, the varieties are in plenty in a wide range of colours and stature, which on the basis of their height are basically divided into three groups, the taller ones (90-180 cm tall) suitable as cut flowers, semi-tall (50-70 cm tall) for bedding and also sometimes as cut flowers while dwarf types (15-30 cm tall) for edging, bedding, rock gardens and for planting in pots or bowls.

A. Under taller group (90-180 cm tall) the types included are:

   (i) *Antirrhinum majus* 'Maximum': This group is used as cut flowers and for bedding and has one or different shades of colour, *viz.* white, yellow, orange-scarlet, scarlet, rose, lilac, crimson-purple, coppery, etc.

   (ii) Tip Top: Good for cut flower where branches appear at base and flower colour range from pure white to dark red via canary-yellow, orange-rose, rose-pink and scarlet.

   (iii) Rocket hybrids (F_1): Good for cut flowers and bedding, spikes are long and are produced on stout stems. The colour ranges pure white to bronze via golden yellow, pink, rose-red, cherry and red.

   (iv) Butterfly type F_1 hybrid (Penstemon flowered): This type has open-faced flowers some 70-75 cm tall plants with two distinct strains, single flowered 'Bright Butterflies' and double azalea flowered 'Madame Butterflies'.

The cut flower cultivars are further classified into four flowering response groups, each suiting to different production conditions (Cockshull, 1985), *viz.*:

- ✰ Group I (winter series): Requiring short days, low light and 7-10°C night temperatures.
- ✰ Group II (late winter to early spring series): Requiring short days but longer than Group I, with a moderate light and 10-13°C night temperatures.
- ✰ Group III (spring series): Requiring medium to long days, moderate to high light and 13-16°C night temperatures.
- ✰ Group IV (summer series): Requiring high light, long days and night temperatures higher than 16°C.

All the cultivars are individually placed into these four groups on the basis of their vegetative and reproductive responses to seasonal changes of temperature and photoperiod in North America and Europe (Cockshull, 1985).

B. Under semi-tall group (50-70 cm tall) the types included are:

(i) *A. majus* 'Nanum': A most popular type suitable for both, cut flower and bedding and the colours of the varieties being pure white, dark yellow, orange-yellow, orange-scarlet, bright scarlet, rose, fiery red and dark red.

(ii) Tetra Ruffled: Tetraploid snapdragons having stout stems with long and compact spikes and the flowers are large and ruffled in many attractive colours.

(iii) F_1 Carioca: Branching at base, vigorous, uniform, ideal for bedding with various colours.

(iv) F_1 Coronette: Growth sturdy, ideal for bedding.

C. Under dwarf group (15-30 cm tall). They are bushy plants with large number of spikes and are most suitable for edging, for bedding, for rockeries and for boxes and pots. This includes:

(i) F_1 Floral Carpet: Dwarf, ball-shaped plants bearing masses of flowers.

(ii) Little Darling: Bushy in nature and flowers facing upward.

(iii) F_1 Kolibre: Early flowering and in many colours.

(iv) Tom Thumb: Dwarf and slight spreading, and suitable for borders.

(v) Magic Carpet: Very dwarf, suitable for edging and for use in rockeries.

The florist's classification of the cut flowers as given by the 'Society of American Specifications for Snapdragon Standard Grades' (Dole and Wilkins, 1999) is as shown in Table 6.1

The genus *Antirrhinum* comprises of 42 species, the major cultivated ones being *A. glandulosum* which is an annual growing to about 60 cm height, branches reddish or glandular, flowers are rose with whitish-yellow at lower portion and are borne on leafy racemes; *A.glutinosum* (*A. hispanicum*) native to SE Spain is a mat-forming sub-shrub, flowers white to pale-yellow with pale-red or purplish striping, at time suffused all over, flowers 2 cm long, and flowering in summer. *A. molle* (*A. hispanicum mollissimum*) native to Pyrenees possesses large hairs on stems and leaves, dwarf and shrubby, branches decumbent and the racemes bear a few flowers at the end of the branches and petals pale-pink or white. *A. majus* native to Mediterranean region (SW Europe), though treated as an annual, in fact, it is a herbaceous perennial species, plants erect and branching from the base, plant height 20-120 cm and the flowers 2.5-5.0 cm long in various colours from pink to red-purple, arranged on racemes of variable length. This species has many cultivars in commerce which have a wide range of genetically controlled heights such as *A. majus* 'Maximum', *A. majus* 'Nanum', and *A. majus* 'Pumilum' (syn. *A. majus* 'Nanum Compactum'). *A. majus* 'Maximum' a **tall cut flower** group which grows 90-120 cm tall and has many forms: 'Tip Top' a base-branching larger strain, the pink 'Apple Blossom', and the rust-resistant varieties including 'Harrison's Rust Resistant' of mixed colour with compact spikes; the 'Rocket Hybrids' group produces quite vigorous plants while 'Supreme F_1 Doubles' including the mixture 'Glamour Shades' are good cut flowers from among the F_1 hybrids, and the F_1 varieties bred for cut flowers include 'Lavender Lady' and 'Pink Ice'; the 'Sentinel' group produces 90 cm tall, well-shaped and stout inflorescence and this includes the varieties 'Cavalier' (orange flushed rose), 'Majorette' (pink), 'Sunlight' (bright pale-yellow) and 'White Spire' (pure white); the mutant 'Tetraploids' or 'Tetra Snaps' most suitable for cutting, being the next addition to the 'Penstemon-flowered group', are outstanding for their colour combinations (white, yellow, orange, pink and red shades), the size of the flowers and the spike, and the most suitable cut flower F_1's are 'Bright Butterflies' mixture and 'Innovation Mixed'. *A. majus* 'Nanum' grows 45 cm tall, most suitable for summer **bedding** and falling under intermediate or **semi-dwarf** varieties which includes 'Black Prince' (leaves dark bronze and flowers deep crimson), 'Nelrose' (pink), 'Welcome' (brilliant crimson), rust-resistant varieties – 'Orange Glow', 'Scarlet Monarch' and 'Yellow Monarch', the F_1 hybrids available in mixtures, *vis-à-vis* in separate colours such as

Table 6.1

Grade (No. of Stems 12/bunch)	*Label (Colour)*	*Individual Flower Stem Weight (g)*		*Minimum Flowers Open per Stem*	*Minimum Stem Length (cm)*
		Minimum	*Maximum*		
Special	Blue	71	113	15	91
Fancy	Red	43	70	12	76
Extra	Green	29	42	9	61
First	Yellow	14	28	9	46

'Carioca Hybrids' (40-45 cm tall), 'Regal Hybrids' (45 cm tall) and 'Sprite Hybrids' (35-40 cm tall), and the other group the 'Hyacinth-flowered' (45 cm tall) ones branching freely from the base and carrying large flowers on dense spikes are most suitable for bedding. *A. majus* 'Pumilum' (syn. *A. majus* 'Nanum Compactum') is a **dwarf** group growing 15 cm tall which includes 'Tom Thumb Mixed' (20 cm tall), 'Magic Carpet' (10 cm tall) and 'F_1 Floral Carpet Mixed' (20 cm tall). Other important cultivated species are *A. asarina,* a native of SW Europe which is greyish-clammy and procumbent, leaves 5-lobed, flowers axillary & solitary and white sometimes tinged red, the palate yellow; *A. blanquetti,* a glabrous perennial though inflorescence glandular and pubescent, 120 cm high, inflorescence up to 20 cm bearing 5-30 flowers, and corolla 3-4 cm long and yellow; *A. calycium,* a glabrous annual growing 45 cm high; *A. coulterianum,* an erect or scandent annual growing some 100 cm high where lateral branches are tendril-like, raceme pubescent and corolla is white with pale-yellow palate and some 9.5-13.0 mm long; *A. filipes,* a glabrous and scandent annual attaining 100 cm height, pedicels 10 cm and tendril-like and flowers yellow with golden at lower lips and palate speckled maroon; *A latifolium,* an erect perennial species growing up to 100 cm high, and flowers pale-yellow; *A. maurandioides* (syn. *Maurandia antirrhiniflora*), a climbing species native of Texas to California growing up to 180 cm through its coiling petioles and peduncles, most suitable for basket growing, on the windows or in cool greenhouses, leaves 3-lobed sweet-potato type and flowers axillary and more than 2.5 cm with attractive violet to purple colour; *A. multiflorum,* an erect and pubescent perennial growing up to 150 cm high, corolla 13-18 mm long with pale-pink to carmine flowers with white palate but with opening of the flowers the palate turns brown to tan; *A. muttallinum,* an erect and pubescent anuual some 200 cm tall bearing tendril-like branchlets, flowers solitary, axillary or on branchlets, oblique-ovoid, glandular to pubescent and lavender to pale-purple while tube and palate are white; *A. orcuttianum,* a native of lower part of South California, grows to a height of 60-120 cm, flower clusters bear tendril-like branchlets, and flowers are white or violet and some 8 mm long; *A. orontium* (small snapdragon), normally not in cultivation, grows to 30 cm, an annual with linear leaves and small purple or white 8 mm long flowers in the axils; *A. ovatum,* an annual with numerous branches, height 60 cm, corolla 17-20 mm long, flowers oblique to ovoid and cream flushed pink on upper lip and tube; *A. pulverulentum,* though quite similar to *A. sempervirens* but differs only in hairs which are larger, upper leaves alternate and flowers pale-yellow; *A. sempervirens* native to Pyrenees and EC Spain is a prostrate and woolly sub-shrub, 1-3 cm long, flowers white or cream veined or lined purple and splashed violet on upper lip and is suitable for summer flowering in temperate regions; *A. siculum,* a glabrous and erect perennial herb freely branching, growing up to 50 cm, and corolla 1.7-2.5 cm long with yellow flower colour; and *A. speciosum,* a native to islands of South California growing 90-120 cm tall, and the plant is somewhat pubescent with flowers of scarlet or pink-red.

Apart from those cultivars described above, other important additional ones are 'Bells Mix' (butterfly type, dwarf 20-25 cm), 'Chimes' series (usually available as mix, dwarf 15-20 cm), 'Dwarf Trumpet Serenade' (flowers trumpet shaped open, dwarf 30 cm), 'Floral Showers' series (short day, dwarf 15-20 cm, 11 colours), 'Giant Forerunner Mix' (basal branching, tall 105 cm), 'Jack Pot' (pink), 'Lavender Bicolor' (lavender and white bicolour, dwarf 30 cm), 'Liberty' series (9 colours, medium 45-55 cm), 'Lipstick Gold' (gold bicolour, throat red, medium 45 cm), 'Madame Butterfly' (azalea-type, double florets, mixed colours, tall 60-90 cm), 'Monarch Mix' (mixed, medium 40-45), 'Orchid Rocket' (rose), 'Peaches & Cream' (peach and cream bicolour, dwarf 20-25 cm), 'Pink Ice' (pink), 'Pixie Mix' (florets butterfly shaped, mixed colour, dwarf 15-20 cm), 'Popette' (white and purple-rose, throat bicolour, medium 45 cm), 'Princess White' (white, eye purple, medium 40-45 cm), 'Purple King' (lilac-purple, medium 45 cm), 'Royal Carpet Mix' (rust resistant, 10 colours, dwarf 30 cm), 'Sonnet' series (spike sturdy and branched, 9 colours, medium 50-60 cm), 'Sweetheart Mix' (double azalea type, dwarf 30 cm), 'Tahiti' series (13 colours, early type, dwarf 18-32 cm), 'White Wonder' (white, medium 45 cm), etc.

Propagation

For antirrhnums, the seed propagation is the best commercial proposition. Seed germinate within 10 days under light and at 18-20°C temperature. In India, seeds are sown thinly in September on the raised nursery beds or in the pots or trays in the soil medium containing coarse sand and sufficient organic matter. The seed beds are watered lightly daily in the evening so that seedlings do not encounter water-stress. For plug production (Ball, 1991; Laughner and Corr, 1996), seeds are left uncovered in the plugs for 5-7 days at 21-24°C, medium (water:medium 2:1) of plugs having 5.5-5.8 pH and less than 0.75 dS/m^2 EC and the light levels 450-1500 fc; then these are kept at 18-21°C for 14 days with the application of 50-75 ppm calcium or potassium nitrate; then for 14 days these are again kept at 16-18°C and at light levels 1000-2500 fc with application of 100-150 ppm N once a week along with the supplement of magnesium sulphate or magnesium nitrate at 1.2 g/l in way to maintain 3:2:1 ratio of K:Ca:Mg; and then finally these are given 16-17°C for 7-10 days with adequate fertilizers (no ammonical) if need be and then these are promptly transplanted. Damping-off of seedlings is often caused by ammonia which becomes a problem in seedling raising. Plants kept in the plugs for too long will cause stunting. For marketing or for delayed transplanting, these plugs can be stored for 6-7 weeks at 2-4°C and at 250 fc light from fluorescent bulbs. Ordinarily, on raised beds the seeds are sown after dressing them with thiram (0.3 per cent) or captan (0.5 per cent) as seedlings are highly susceptible to *Pythium* and *Rhizoctonia.* To avoid the ammonical damping-off, seeds are sown thinly in a compost of equal parts sand and peat moss with no nutrients, and only after germination watering with dilute liquid feed is done. When the seedlings develop two leaves or have attained about 3 cm height, are transplanted.

Though cuttings root easily but easy and quick method is only through seeds. Snapdragons are also multiplied from cultures initiated from root and shoot tissues. Full plant development is achieved by placing the roots in the soil or medium. It is also propagated asexually by tissue culture (Pfister and Widholm, 1984; Sangwan and Harada, 1975).

Cultural Practices

Sandy-loam soil which has sufficient aeration (Larsen *et al.*, 1985), with a pH of 6.0 to 7.5 and rich in organic matter is quite suitable for growing of snapdragons provided it is at the sunny location. Well aerated medium does not permit water-logging and instead there is free movement of air, *vis-a-vis* root spreading is facilitated. However, apart from soil there are various other media (with or without soil) which can effectively be used for its cultivation. Peat:perlite (1:1), peat:soil:perlite or sand (1:1:1) or soil:sand (1:1) has been used for taking bumper crop of antirrhinum (Hanan and Langhans, 1963) whereas Lee *et al.* (1986) used only rockwool for its successful growing. In an ordinary polyhouse when Loeser (1991) grew China aster in two batches, it gave most uniform results and the flowers produced were of good quality in a cellular potting system with 60, 90 or 96 plants/m^2.

While preparing the land, there should be one deep ploughing some 25 cm deep, followed by planking and then second ploughing when all the rootstocks of perennial weeds, all the slow degrading plant remnants, polythene pieces, pebbles and brick pieces, etc. should be removed and then planked. This is the time when well rotten farmyard manure, compost or dung manure @ 30,000 kg per hectare should be broadcast in the soil and then final ploughing is carried out, followed by planking. At this time the soil should be properly levelled. The soil should not be too dry and it would be better if there is sufficient moisture so that there may be efficient working and which also facilitates the bed preparation. As per the lay out, now the main bunds and channels are made and then side bunding of the beds as per the convenience is done so that these may facilitate cultural operations. Normally the width of the beds is 160 cm and the length to any level as per the convenience looking into the topographical level of the land. Now the beds are properly levelled so that while watering no water may stand anywhere.

Seedlings are transplanted after these have developed two leaves or have attained about 3 cm height. This stage is attained within 35 to 45 days of sowing. The distances apart depends on the purpose for which these are grown, on the varieties looking into their response group and the prevailing climatic conditions. Broadly, these are planted at 12.5 x 7.5 cm to 12.5 x 10 cm apart for winter cut flower production, and 10 x 7.5 cm to 12.5 x 7.5 cm apart for summer cut flower production (Stefanis and Langhans, 1982). However, the closer spacing increases the yield of flowers or cut flowers but quality is hampered. In a 15-cm pot, some 4-5 seedlings are planted. Pot to pot distance is kept as close as possible but when plants are so grown that shade of one plant is falling on the other then these are spaced properly. For cut flower production, these require a two-tier support. The nets having 12.5 x 10 cm, 15 x 15 cm or 20 x 15 cm openings are quite suitable for supporting the plants. For cut flowers these are grown single stemmed. For enhancing the number of branches per plant or per pot, the plants are pinched. This encourages greater number of flowers though flowering is delayed and plant height is also reduced. Early stage pinching when plants have attained 10 cm height or when they are established in their flowering position is more preferable as branches become quite strong through early stage pinching. However, pinching is the regular practice to make the potted plants floriferous, in shape and attractive. Under glass, the plants may be flowered any time of the year by varying the sowing dates though to some extent this can be attained even to open-grown crops by manipulating the sowing dates. In India, its seeds are normally sown in the nursery beds from September to November and planting is done from mid-October to December end. On the hills (temperate areas) also though sowing time is the same but seeds sown in January-February in pots, pans, flats, bowls or boxes inside the greenhouse, cloches or tunnels may be transplanrted in March in the open for harvesting a bumper crop.

If the area is small, manually the weeds should be taken out as soon as these are seen in the field. Weed control is absolutely necessary to take a vigorous crop. Controlling the weeds at an early stage of the crop season will help the crop branching more and after some time the beds will be full of the main crop, and then afterwards these will not allow growing of the weeds. Until the spread of the crop is full, the weeding is must so it requires atleast four weedings per crop. Alachlor, chloramben, or oxidiazon at 4.5 kg/ha after transplanting have been found quite effective in controlling the weeds. Oryzalin also controls the weeds completely in the antirrhinum fields. Crag 1 at 4.5 kg/ha or simazine 1.1 kg/ha controls the weeds in snapdragon field when the plant is in full growth and no injury to the plants is observed.

Nutrition and Water Management

Right from young seedlings and plugs the care should be taken that these get nutrition in proper quantity so that quality plants are produced with repeated flowering (Larson *et al.*, 1985; Rogers, 1951). Nitrogen levels should be sufficient at the initial stage of growth for developing a strong stem (Boodley, 1962) as in its deficiency plants are stunted and leaves yellow. Arora (1989) described the deficiency symptoms of various elements in snapdragon plants. Deficiency of phosphorus suppresses growth, young leaves become dark green and the leaf-tips may curve downward and inward towards the main stem. Potassium deficiency causes interveinal chlorosis in youg leaves similar to iron deficiency and afterwards necrosis of tips and margins of older leaves. Root system is first affected in terms of poor branching, become thin and poorly developed due to Ca deficiency, Mg deficiency causes interveinal chlorosis and necrosis of older leaves on young plants. The tips of older leaves turn downwards and younger leaves upwards. Pale, yellow-green upper leaves where veins are lighter than rest of the leaf blade in case of sulphur deficiency and older leaves show interveinal chlorosis. Iron deficient plants show first interveinal yellowing of the newer leaves and then green colour is completely lost. Axillary shoots are also small and almost white in colour. Boron deficiency results in abortion of the terminal growing points including terminal spikes which results into growth of axillary shoots in the lower part of the stem. Over application of calcium accelerates the boron deficiency symptom. Hood *et al.* (1993) when comparing the four cultivars for their nutrient uptake at three developmental stages of growth, *i.e.* (i) seedling to flower initiation, (ii) initiation to visible flower bud, and (iii) visible bud to harvest, and reported that up to the harvest the nutritional levels should remain high without any reduction at the final developmental stage, which is contrary to Haney (1961), Rogers (1959) and Sanderson (1975) as they suggested that fertrility levels should be reduced during summer and eliminated during the final stage of spike development. Arthur and Hedley (1976) compared numerous hybrid cultivars for their nitrogen uptake and reported varying levels of nitrogen uptake by various hybrids. Laurie and Wagner (1940) described nutritional deficiency symptoms in snapdragon plants. Oertli (1970a-h) through a series of classic papers described optimal levels and deficiency symptoms of N, P, K, Ca, Mg, S, B and Fe using a standard Hoagland's solution and relating these to tissue nutrient levels. Howland (1946) tested various N, K and pH levels but did not record any major growth difference.

Rogers (1951) reported that due to various P and K level applications there was little variation in the growth of snapdragon plants but there was significant growth increasse when nitrate levels were increased from 5 ppm to 35 or 50 ppm. As compared to certain other major crops, snapdragons have little nutrient quirement. Flint and Asen (1953) provided 160 ppm N, 32 ppm P, 240 ppm K, 176 ppm Ca and 65 ppm Mg along with minor elements to snapdragon plants and recorded little variation between full rate and one-half rate or one-quarter rate, however, twice or four-times dose caused injury to the plants. Sanderson (1975) advocated 100 ppm N and K with each watering as optimum to snapdragon plants. Nitrogen is preferred in the form of nitrate than ammonia, and high nitrogen rates cause excessive vegetative growth. The uptake of nitrogen by the plants is more when it is in nitrate form than in ammonical (Hood *et al.*, 1993). Through the use of ammonium fertilizer in field-grown snapdragon, Kasten *et al.* (1990) recorded 60 per cent reduction in nitrate content of the soil while placement of straw fertilizer together resulted into 80 per cent reduction, and there was less fertilizer leaching in the groundwater. Snapdragons are sensitive even to low levels of boron so one should be certain that adequate levels are present (Mastalerz, 1957; Furuta, 1960). Carmichael (1968) stated that B becomes deficient when Ca levels are too high.

Antirrhinum being grown as an annual crop becomes very difficult to keep it weed-free as manual **weeding** is cost-prohibitive, hence the option is left only with chemical weeding. Small plantings can be kept weed-free manually but large or commercial plantings are not cost-effective so for this reason Duczmal (1989) with cv. 'Rose Mogenlicht' obtained maximum seed yield (1.1-12.1 kg/ha) when used 3.0-4.0 kg Trifluralin (Treflan) before transplanting. Lemont (1986) in the plantings of snapdragon found Alachlor, Napropamide, Oryzalin and chlorthal dimethyl quite effective in controlling 80 per cent of the grasses, Alachlor giving the best results in case of broad-leaved weeds (80 per cent control compared with 30-75 per cent of other herbicides), though Chloroxuron was not found effective. Tenoran 7 kg/ha, Ramrod (propachlor 65 per cent) 7 kg, Amiben-Granulat (chloramben), Kerb 50W 3 kg, Legurame flussig [(carbetamide (70 per cent)] 10 litres and Ronstar (oxidiazon [75 per cent]) 8 litres were tested for their selectivity in *Antirrhinum majus* where Kerb 50W and Legurame gave satisfactory weed control; Tenoran though gave best results but this and Ronstar caused some severe injury/ necrosis; Ramrod lacked adequate herbicidal activity; though Amiben-Granulat, Kerb 50W and Legurame flussig were well tolerated (Anon., 1974).

Snapdragons do not like a wet foot. Therefore overwatering should be avoided. In winter crop the intervals between irrigation may be long due to low production temperatures. The irrigation is highly dependent on the soil and weather conditions. Overwatering should also be avoided to avoid infection of *Pythium* root rot especially at harvesting time. Barring *Pythium*, these plants are tolerant of both high and low moisture levels (Hanan and Langhans, 1963; Hanan, 1965). Moisture content above 34 per cent or below 24 per cent is not good for the growth of snapdragons.

Growth, Development and Flowering

Temperature influences the rate of flower initiation and not the development once the plants are mature enough to perceive LD as has been advocated by Maginnes and Langhans (1961), and Edwards and Goldberg (1976) that temperature between 4 to 21 °C had little effect on days to flower once flower initiation occurred. Warm temperatures at early growth stage and low temperatures at later growth stage encourage bumper growth (Miller, 1962) and at low (cold) temperatures the growth is slower than at warm temperatures. However, the very low night temperatures, *i.e.* 10 °C if is followed by quite high day temperatures, *i.e.* 21 °C, the ability of the plants to absorb water is substantially inhibited which may cause moisture-stress. This fact has been explained by Rutland and Pallas (1972) when they observed wilting of snapdragon plants on bright sunny winter mornings when the soil temperatures were still cool. Those cultivars which did not wilt under such conditions, McDaniel and Miller (1976) stated that such cultivars, to sustain their life, had developed the ability to close their stomates under such circumstances so that water stress is reduced, and thus these complete their reproductive phase in normal fashion by producing quality blooms. However, to harvest the cut flowers from various production blocks at deferred dates, the crops may be grown at various temperature regimes, *i.e.* warm-cool, cool-warm, and through increase or decrease of the warm and cool temperature durations. Studies on the effects of differences (DIF) between day and night temperatures, and constant temperature regimes (15, 20 or 25 °C) on the elongation and development of *Antirrhinum majus* seedlings revealed that 25/15 °C (day/night, *i.e.* positive DIF) compared with constant 20 °C (zero DIF) regime treatment promoted plant height, internode and petiole lengths, whereas 15/25 °C (negative DIF) suppressed growth compared with zero DIF, however, leaf length and leaf unfolding rate responded significantly different to different constant temperatures, while not being significantly affected by DIF treatments, therefore, it is inferred that DIF effects may be advantageous to facilitate production of compact seedlings but that short-term control of elongation may be more easily achieved by altering average temperature than by changing DIF (Ito *et al.,* 1997).

Low light conditions at early stage of plant growing delay the floral induction and its further development. Long photoperiods hasten flowering, and supplementary lighting is beneficial to young plants when plant density is high (Flint, 1960; Peterson, 1955) though in older plants having wider spacing supplementary light is not economical. The time between visible flower bud and anthesis is not influenced by light duration or intensity (Rabinowitch *et al.,* 1976; Cockshull, 1985). Photoperiodic and light intensity response also differs from variety to variety. Photoperiod does not influence flowering over the entire period of growth from germination to flower bud initiation but only for a definite period that too before flower bud initiation, *i.e.* during juvenile phase which may occur some time at 40-65 days after germination when plants have attained 5-10 pairs of leaves. Normally long days hasten the flowering in most of the varieties while short days retard it but do not completely prevent the flowering though there had been exceptions. Laurie and Poesch (1932), Flint (1960), Sanderson and Link (1967), Hedley (1974), Hedley and Harvey (1975) and Cockshull (1985) stated that snapdragon is a facultative LD plant where flowering occurs faster under long photoperiods but earlier flowering can occur under SD also if right response group cultivars are chosen. Hedley (1974) stated that day length treatments exert quantitative effects and he classified snapdragons as quantitative long day plants. Rogers (1958, 1959, 1960, 1961) reported that just transplanted seedlings at a growing temperature of 10 °C when subjected to three weeks of supplemental lighting provided similar results to those grown at 16 °C temperature without supplemental lighting. During winter maximum light intensity is a cultural necessity. Use of high pressure sodium lamps supplemental lighting during winter increases flower quality and decreases flowering time (Stefanis and Langhans, 1982). Denesereau *et al.* (1998) in Canada recorded significantly better growth and flowering in snapdragon under single-layer glass than in polyethylene houses, *vis-a-vis* supplementary lighting of 60 $\mu mol\ m^{-2}s^{-1}$ accelerating flowering by 20-25 days, improving the flower quality, and eliminating the differences in plant growth and quality between covering treatments. Carpenter (1964) reported that even baffles of reflective aluminium foil kept to the north side of the benches in the greenhouse increased growth, and flowering occurred 5-7 days earlier. Snapdragon cultivars based on their response to light levels and photoperiod, are divided into response groups. The winter flowering cultivars have short juvenile phase so their flowering can not be delayed by short days or low irradiance, summer flowering cultivars have long juvenile phase, *vis-à-vis* more leaves under high irradiance and long days (Hedley, 1974; Maginnes and Langhans, 1961; Raulston, 1970; Singh and Jain, 1979) and the use of incandescent photoperiodic lighting extends the natural day (Maginnes and Langhans, 1961) while the night breaks from 2200 hours to 0200 hours from cyclic (10 or more seconds per minute) or continuous lighting were found equally effective (Maginnes and Langhans, 1967).

The varieties falling under Group I, II and to a lesser extent Group IV, are highly responsive to CO_2 during winter. The optimum CO_2 levels may range from 750 to 1500 ppm (Koths, 1964; Shaw and Rogers, 1964; Lindstrom, 1966; Duffett, 1968; Nelson and Larson, 1969). Shaw and Rogers (1964) recorded accelerated flowering by about 10 days though there had been little difference in stem length, stem weight or spike length due to CO_2 enrichment either at 10 or 15°C night temperatures or at higher day temperatures. Koths (1964) and Nelson and Larson (1969) recorded little benefit through CO_2 enrichment at 400 to 500 ppm in the former study and at 750 or 1500 ppm in the latter study, however, Lindstrom (1966) when grew cultivars belonging to different groups [(i) late spring and fall group, and (ii) summer group] at CO_2 level of 1500 and 4000 ppm at about 15°C night temperatures, recorded earlier and quality flowering though difference was negligible at both the levels, *i.e.* 1500 or 4000 ppm. Duffett (1968) in northern Ohio recommended 800 ppm CO_2 enrichmenat at 15°C night temperature. Rogers (1980) suggested 800 ppm CO_2 at 16°C night temperature as optimum during the low light periods. In fact, the cultivars of the later response groups respond more favourably. The investigations on the response of snapdragon to various light intensities and the growth stages are to go a long way still.

Certain growth substances such as IAA and GA_3 have been tried to regulate growth and flowering in snapdragons. Srivastava and Singh (1972; cited from Arora, 1989) when used 20 ppm and 60 ppm IAA at 17 days of transplanting, recorded improved vegetative growth with delayed flowering in former case whereas depressed vegetative growth with enhanced flowering in latter treatment. GA_3 at 5-100 ppm made remarkable increase in plant growth (Nijs and Boesman, 1981; cited from Arora, 1989) whereas at 50-200 ppm application there had been considerable increase in plant height and inflorescence length (Sarhan and Sayed, 1983; cited from Arora, 1989). Spraying with 50-1200 ppm CCC to the plants of snapdragon, Sarhan and Sayed (1983; cited from Arora, 1989) recorded largest number of flowers per inflorescence. Seeds of *Antirrhinum majus* cv. 'Hyacinthiflorum' were soaked for 24 h in solutions containing 50-200 mg GA_3/l, 50-200 mg GA_{4+7}/l, 5-20 mg NAA/l, 50-200 mg PP_{333} (paclobutrazol)/l, 250-100 mg ethephon/l, 25-400 mg RSW 0411 (triapenthenol)/l or 100-400 mg Sumi-7/l but the germination percentages were found negligible except for seeds soaked in 100 mg GA4+7/l or unsoaked control seeds (Grzesik, 1989).

Height Control

Application of granular ancymidol in the soil also influenced the plant growth (Murray *et al.,* 1986; cited from Arora, 1989). *Ethephon* at 1000 ppm caused stunting of plants, its higher doses delayed flowering and even proved toxic (Munshi *et al.,* 1980; cited from Arora, 1989). Application of SADH at 250-2000 ppm four weeks and again a month later after transplanting stunted the plants and delayed the flowering though there had been increase in the number of inflorescence as well as florets per inflorescence (Shedeed *et al.,* 1986; cited from Arora, 1989). Dole and Wilkins (1999) stated that Bonzi (paclobutrazol) drenches at 1.25 or 1.75 mg a.i. per 9-cm pot controls height of potted snapdragons though age of the plants is a factor which counts because drenching at earlier stage of growth is more effective. Early application also delays flowering and reduces the number of floral stems in case where plants had been pinched though there had been seasonal and cultivar differences. CCC (chlormequat) is also effective in controlling the height (Anon., 1970). A-Rest (ancymidol) and Sumagic (uniconazole) has also been reported effective in controlling snapdragon height (Adriansen, 1985). Wainright and Irwin (1987) advocated the use of Bonzi along with pinching to obtain acceptable size and shape of the plants at an affordable expense. The plug-grown *Antirrhinum majus* 'Tahiti Mix' were transplanted into jumbo six packs and when the seedlings had attained 5 cm in height or width, Kuehny *et al.* (2001) sprayed them with 1000/800, 1250/1250 or 1500/5000 mg l^{-1} of chlormequat/daminozide solutions which reduced the plant height significantly.

Postharvest

The commercial stage of flower harvesting in case of antirrhinum is when one-third of the florets are open and such spikes may last for about a week in ordinary tap or distilled water though through the use of preservative solutions the life may be prolonged up to three weeks. However, in case bud opening solutions are to be used, the spikes may be harvested when a few florets start showing colour or when the first florets are open. The spikes are graded according to the length of spike, the rigidity of stems, colour of petals and leaves, flower look (whether any defect is there) and the weight and number of open florets per spike. The standard grades established by the Society of American Florists are most commonly used by the growers in USA. U.S. system of classification defines various grades as 'blue', 'red', 'green' and 'yellow' which is roughly equivalent to European range from 'extra' through 'first' and 'second' class as is furnished in a Table under Classification, Species and Cultivars. According to this system the spikes should have straight stems, clean and uniform foliage and free of diseases and pests. Foliage of one-third of the lower part of the stem should be removed.

While handling the flowers of snapdragon, one may face two problems, one is negative geotropic curvature of spikes for which the stems should be kept straight, and the second is the shattering (abscission) of the flowers due to build up of endogenous ethylene (Wang *et al.*, 1977), therefore, the spike after cutting should be pulsed with STS at 21 °C ambient temperature for one hour. The life of freshly cut spikes kept under simulated shipping conditions is found almost doubled when treated with STS + sucrose for 20 hours. Nowak and Rudnicki (1975) with snapdragon cv. 'Super Tetra Crimson Giant' when used Proflovit-72 (0.3 g 8-HQS+0.05 g CCC+50 g sucrose/l) in comparison to 20-50 ppm $AgNO_3$, 4 ppm $KMnO_4$, 5 per cent sucrose or 50 ppm CCC, found Proflovit-72 giving good vase life with good quality flowers. Gast and Negm (1987) suggested treating the cut stems with auxins, *viz.* β-NAA, ñ-chlorophenoxyacetic acid or 2, 4, 5-trichlorophenoxyacetic acid as a dip or via stem uptake. Also against the curvature, the spikes immediately after cutting should be treated with 30 ppm N-1 napthylphthalamic acid. In case when the cut stems are to be shipped horizontally, these should be treated with B-Nine to reduce geotropic bending.

Cut snaps can be stored dry in dark at -0.6 to 1.7°C, preferably at 0-2°C for 3-4 days, and wet in preservative solution for 1-2 weeks whereas bedding plants can effectively be stored at 10 to 13°C. Halaba *et al.* (1983) suggested storage in preservative solution such as Proflovit-Universal, Everbloom, Oasis, etc. after treating the cut ends with a fungicide to protect against grey mould and this way these can be stored for 4-8 weeks. Even for shipping, the cut stems may be conditioned with silver thiosulphate or any other preservative containing STS such as Chrysal AVB, Oasis CT-2000, Silflor-50 or argylene. For long storage, the harvesting should be done when 2-3 florets are showing colour. After storage, the 2-cm cut end should be removed with a sharp knife and then the spikes are kept in preservative solution for bud opening at > 75 per cent RH, 20-30°C temperature, 2,000 lux light intensity and for 16 hours day length for successful bud opening (Vinod Kumar and Bhattacharjee, 2006). At 40 mm Hg pressure the flowers can be stored successfully for up to 42 days than 21 days maximum observed already with the best method used.

At the most, the cut flower life can be increased from 1-3 weeks together with increase in stem length and turgidity, floral pigmentation and opening of the number of florets. $AgNO_3$ 1000 ppm dip for 40 minutes and $Al_2(SO_4)$ at 0.08 per cent + KCl at 0.03 per cent + NaCl at 0.02 per cent @ 20 ml/l prolong the vase life of cut snaps. Larsen and Scholes (1966), Raulston and Marousky (1971) and Johnson (1972) stated 8-HQC 300 ppm + sucrose 0.5-1.5 per cent with or without B-Nine as optimum solution for snapdragon vase life. Larsen and Scholes (1966) suggested further addition of SADH. Halaba *et al.* (1983) suggested 300 ppm 8-HQS + 50 ppm CCC and 5 per cent sucrose. Marousky and Raulston (1970) stated that a floral preservative in presence of light is responsible for good pigmentation in the flowers. *Antirrhinum majus* being an ethylene sensitive cut flower so pretreatment with low doses of 1-MCP gives results similar to STS in negating ethylene effects and prolongs the vase life in an ethylene-free environment and in presence of ethylene at 1 μl/l, the vase life of 1-MCP-treated flowers increases up to four times than control (Serek *et al.*, 1995). Dole *et al.* (2005) found that neither of the chemicals (STS and 1-MCP) increased the vase life in *Antirrhinum* cvs 'Maryland White' and 'Yoshimite Pink'. Macnish *et al.* (2008) when added aqueous chlorine dioxide (ClO_2) at 2 or 10 μl l^{-1} in clean deionized water, the build up of bacteria in vase solution was prevented and the vase life was found extended in *Antirrhinum majus* cv. 'Potomic Pink', however, the vase life in the cut flowers was found increased even when ClO_2 was added in water already contaminated with 1011 CFUL-1 aerobic bacterium.

Insect-Pests, Diseases and Physiological Disorders

Parrini and Rumine (1989) recorded *Epichoristodes acerbella*, *Lipaphis erysimi*, *Myzus persicae*, *Nezara viridula* and *Tetranychus urticae* insect-pests feeding on snapdragons. Aphids (*Aphis gossypii*) feed on the underside of the leaves and on the tender parts in gardens and in greenhouses, and the peach aphids (*Myzus persicae*) discolour and distort the leaves similar to virus infection. These are controlled through 0.1-0.2 per cent spraying of malathion, lindane, parathion, endosulphan or oxydemeton methyl. Lacewing (*Chrysopa carnea*) effectively predates on the aphids if introduced into the greenhouses. Red spider mite (*Tetranychus telarius*) is a cumbersome pest during dry weathers. Cyclamen mite (*Steneotarsonemus pallidus*) and a 'red and black stink bug' also attack this crop. These are controlled through the use of aramite, dimite, kelthane or ovex. Looper larvae are controlled through foliar spraying of trichlorfon or methyl parathion. *Helicoverpa armigera* caterpillars are serious pests of antirrhinum which feed on whole of the plants including the seed capsules. Its control can be achieved by spraying the plants with 0.3 per cent thiodan or 0.45 per cent ripcord. Root-knot nematode (*Meloidogyne incognita*) is sometimes found attacking its roots which cause stunting of the plants. Through carbofuran the nematodes may be kept under check.

Rust (*Puccinia antirrhini*) is the most serious pest of snapdragons (Horst, 1990) which affects the crop in field as well as in greenhouses. In its infection brown pustules are observed on stems, leaves and sepals. The severity of the disease increases during dry weathers. Thiram or a mixture of sulphur and zineb at 0.2 per cent sprayings frequently will control this problem. There are now many varieties available which are genetically resistant to this disease. Parrini and Rumine (1989) in Italy described the diseases such as *Botrytis cinerea*, *Peronospora antirrhini*, *Puccinia antirrhini*, *Pythium*, *Rhizoctonia solani*, *Sclerotinia sclerotiorum* which were recorded infecting snapdragon. From Italy, Pasini *et al.* (1996) isolated *Rhizoctonia solani* from antirrhinum causing collar or foot rot. Damping off is a serious problem at seedling stage. This is caused either by ammonical fertilizer application or due to infection with *Pythium* (Horst, 1990), *Rhizoctonia solani* and *Pellicularia praticola* which cause a havoc when the temperature is high coupled with high soil moisture. To some extent these problems can be avoided through proper sanitation, thin sowing or distant planting and by regulating the water supply. Application of brassicol at 0.3 per cent or sterilized soil can also check these diseases. Antirrhinum wilt is caused due to infection with *Fusarium solani* and *F. solani* var. *merittii* which are favoured by high temperature (29-35°C) and high soil moisture. Seed dressing with 0.2 per cent captan is very effective in keeping this problem under check. Stem rot is also caused by *Sclerotinia sclerotiorum*. *Alternaria* and *Helminthosporium* also attack this crop. These all the pathogens can be kept under check by regulating the watering and through regular spraying of the crop fortnightly with bavistin 0.1 per cent alternate with captan 0.2 per cent. Snapdragon blight is caused by *Phyllosticta antirrhini* which produces brownish spots on the foliage and ashy-grey spots on stems, and the attack of this pathogen becomes severe during high temperatures. Spraying the crop with 0.2 per cent dithane M-45 or dithane Z-78 controls this pathogen. *Botrytis* (Horst, 1990) also infects this crop during windy, chilly and humid conditions. Its control is the same as to that of snapdragon blight. Powdery mildew [*Oidium* sp. (Horst, 1990), *Podosphaera leucotricha*)] is symptomized by the formation of a white powder coating on young stems, both the leaf surfaces and on the sepals, and its severity is more under high humidity conditions and dry soil. Through spraying with wettable sulphur or karathane at 0.2 per cent, or bavistin 0.1 per cent controls these pathogens. Downy mildew [*Peronospora antirrhini* (Horst, 1990)] forms mealy white patches on the lower side of the leaves, and in severe infection the plants become crippled. Seedlings are badly affected with this disease. Bordeaux mixture at 0.1 per cent or dithane M-45 at 0.2 per cent sprayings at fortnightly intervals will keep this pathogen under check. Metalaxyl and manzate are also quite effective. Snapdragon anthracnose is a common disease of greenhouse snapdragons caused by *Colletotrichum antirrhini* and *C. fuscum*. It is symptomized by pale-yellowish-green to grey spots with a narrow border on the stems and leaves, being sunken and oblong on the old stems. It is controlled by destroying the infected old debris and by spraying with dithane M-45, dithane Z-78 or captan 0.2 per cent or bavistin 0.1 per cent. *Septoria antirrhini* leaf spot causes small circular spots with a pale centre bordered with purplish-brown margins. *Cercospora antirrhini* also causes leaf spot. Dried infected leaves should be collected and destroyed. The control measures adopted in case of anthracnose will control these pathogens also. A strain of CMV which in nursery is transmitted through the aphids causes plant stunting and little leaf. Tomato spotted wilt virus and impatiens necrotic spot virus cause brown or black stem lesions which often do not appear until just before flowering (Laughner and Corr, 1996). Such plants should be rogued out and the aphids should be controlled regularly. Several researchers (Dimock, 1958; Forsberg, 1958; Nelsom, 1962; Williamson, 1962; Porter and Aycock, 1967; Engelhard, 1971) have reviewed the control of antirrhinum diseases.

Since antirrhinum plants continue absorbing high levels of nutrients even up to the time of harvest, Hood *et al.* (1993) objected the practice of reducing the mineral nutrition during the final stages of production prior to harvest as this causes tip-breaking problem. Excessive grassy growth can occur due to improper selection of cultivars and excessive use of nitrogen.

References

Anonymous, 1970. Pot plants – inexpensive winter production. *Rep. Efford exp. Hort. Stn 1969*, pp. 38-40.

Anonymous, 1974. *Ann. Report German Pl. Prot. Serv.*, No. 21, pp. 221-223.

Adriansen, E. 1985. Kemisk vaekstregulering. In: *Potteplanter I – Produktion, Metoder, Midler* (eds Christensen, O.V., A. Klougart, I.S. Pedersen and K. Wikesjö), pp. 142-162. GartnerINFO, Kbenhaven, Denmark.

Arora, J.S. 1989. Antirrhinum. In: Commercial Flowers (eds Bose, T.K. and L.P. Yadav), pp. 659-680. Naya Prokash, Kolkata.

Arthur, A.E. and C.L. Hedley, 1976. The effects of nitrogen on five varieties of *Antirrhinum majus*. *Ann. Bot.*, **41**: 627-636.

Bailey, L.H. and E.Z. Bailey, 1976. *Antirrhinum* L. In: *Hortus Third – A Concise Dictionary of Plants Cultivated in the*

United States and Canada, pp. 86-87. Macmillan Pub. Co., New York, USA.

Ball, V. 1991. Plugs – the way of the 1990s, pp. 137-153. In: *Ball RedBook* (15th edition, ed. Ball, V.). Geo. J. Ball Publishing, West Chicago, Illinois.

Boodley, J.1962. Fertilization. In: *Snapdragons: A Manual of Culture, Insect and Diseases and Economics of Snapdragons* (ed. Langhans, R.E.), pp. 28-34. Snapdragon School, New York State Extension Service and New York State Flower Growers Association, Ithaca, New York.

Carmichael, O.E. 1968. Boron toxicity of flowering plants. Master's Thesis, University of Missouri, Columbia, Missouri.

Carpenter, W.J. 1964. Reponse of snapdragons and chrysanthemums to supplemental reflective light. *Proc. Amer. Soc. hort. Sci.*, **84**: 624-629.

Cockshull, K.E. 1985. *Antirrhinum majus*. In: *Handbook of Flowering*, vol. I (ed. Halevy, A.H.), pp. 476-481. CRC Press, Boca Raton, Florida.

Dansereau, B., Y. Zhang, S. Gagnon and H.L. Xu, 1998. Stock and snapdragon as influenced by greenhouse covering materials and supplemental light. *HortSci.*, **33**(4): 668-671.

Dimock, A.W. 1958. Snapdragon diseases common in New York. *New York Fl. Grs Bull.*, No. 145, pp. 2-3.

Dole, J.M. and H.F. Wilkins, 1999. Antirrhinum. In: *Floriculture Principles and Species*, pp. 192-200. Prentice Hall, New Jersey, USA.

Dole, J.M., W.C. Fonteno and S.L. Blankenship, 2005. Comparison of silver thiosulfate with 1-methylcyclopropene on 19 cut flower taxa. *Acta Hort.* (*Proceedings of the Fifth International Postharvest Symposium*, eds Mencarelli, F. and P. Tonutti), held at Verona, Italy, on 6-11 June, 2004; No. 682 (vol. 2), pp. 949-956.

Duczmal, K.W. 1989. Weed control in ornamental plants grown for seed (Polish). *Biuletyn Instytutu Hodowli I Aklimatyzacji Roslin*, No. 169, pp. 59-76.

Duffett, W.E. 1968. Culture of greenhouse snapdragons. *Ohio Flor. Assocn Bull.*, No. 466, pp. 5-7.

Edwards, K.J.R. and J.B. Goldberg, 1976. A termperature effect on the expression of genotypic differences in flowering inductions in *Antirrhinum majus. Ann. Bot.*, **40**: 1277-1283.

Engelhard, A.W. 1971. *Botrytis*-like diseases of rose, chrysanthemum, carnation, snapdragon and king aster caused by *Alternaria* and *Helminthosporium*. *Proc. Florida St. hort. Soc.*, **83**: 455-457.

Flint, H.L. 1960. Reflective effects of light duration and intensity on growth and flowering of winter snapdragon (*Antirrhinum majus* L.). *Proc. Amer. Soc. hort. Sci.*, **75**: 769-773.

Flint, H.L. and S. Asen, 1953. The effects of various nutrient intensities on growth and development of snapdragon (*Antirrhinum majus* L.). *Proc. Amer. Soc. hort. Sci.*, **62**: 481-486.

Forsberg, J.L. 1958. Snapdragon diseases. *Illinois St. Florists Assocn Bull.*, No. 186, pp. 5-8.

Furuta, T. 1960. Test boron deficiency in snapdragon at Auburn. *Florist's Rev.*,**126**(3244): 25.

Gast, R.T. and F.B Negm, 1987. Analysis of geotropic curvature of snapdragon following chemical treatment. *HortSci.*, **22**: 182 (abstr. No. 867).

Grzesik, M. 1989. Effect of growth regulators on the seedling-growth of *Lathyrus odoratus*, *Zinnia elegans*, *Matthiola incana* and *Antirrhinum majus*. *Acta Hort.* (Proc. Symp. on Growth Regulators in Ornam. Hort., held at Skierniewice, Poland on September 5-10, 1988), No. 251, pp. 71-74.

Halaba, J., M. Kwiatkowski and R.M Rudnicki, 1983. Storage of cut snapdragon flowers. *Rosliny Ozdobne*, Ser. B (Skierniewice, Poland), **8**: 185-190.

Hanan, J.J. 1965. Efficiency and effect of irrigation regimes on growth and flowering of snapdragon. *Proc. Soc. hort. Sci.*, **86**: 681-692.

Hanan, J.J. and R.W. Langhans, 1963. Soil aeration and moisture control snapdragon quality. *New York St. Fl. Grs Bull.*, No. 210, pp. 3-6.

Hanan, J.J., R.W. Langhans and A.W. Dimock, 1963. *Pythium* and soil aeration. *Proc. Amer. Soc. Hort. Sci.*, **82**: 574-582.

Haney, W.J. 1961. Snapdragon culture. *Michigan Florist*, No. 366, pp. 25-26, 29.

Hedley, C.L. 1974. Response to light intensity and daylength of two contrasting flower varieties of *Antirrhinum majus*. *J. hort. Sci.*, **49**: 105-112.

Hedley, C.L. and D.M. Harvey, 1975. Variation in the photoperiod control of flowering of two cultivars of *Antirrhinum majus* L. *Ann. Bot.*, **39**: 257-263.

Hedley, C.L., A.E. Arthur and H.D. Rabinowitch, 1977. The effects of nitrogen level on the performance of *Antirrhinum* cultivars grown in greenhouse conditions. *Euphyt.*, **26**: 755-760.

Hood, T.M., H.A. Mills and P.A. Thomas. 1993. Develomental state affects nutrient uptake by four snapdragon cultivars. *HortSci.*, **28**: 1008-1010.

Horst, R.K. 1990. Snapdragon. In: *Westcott's Plant Disease Handbook* (5th ed.), p. 816. Van Nostrand Reinhold, New York.

Howland, J.E. 1946. Foliar dieback of the greenhouse snapdragon, *Antirrhinum majus* and a study of the influence of certain environmental factors upon flower production and quality. *Proc. Amer. Soc. hort. Sci.*, **47**: 95-97.

Ito, A., T. Hisamatsu, N. Soichi, M. Nonaka, M. Amano and M. Koshioka, 1997. Effect of altering diurnal fluctuations of day and night temperatures at the seedling stage on the subsequent growth of flowering annual. *J. Japanese Soc. hort. Sci.*, **65**(4): 817-823.

Jain, H.K., R.S. Rana, S. Jana, M.P. Alexander and J.N. Sharma, 1962. A study of fertility and plenotype in induced tetraploids of *Antirrhinum*. *Indian J. Genet.*, **22**: 154-159.

Johnson, C.R. 1972. Effectiveness of floral preservative on increasing the vase life of snapdragons. *Flor. Rev.*, **149**(3868): 47, 95-97.

Kasten, P., K. Sommer and A. Papenhagen, 1990. Ammonium placement fertilizing of cut-flowers. (German). *Deutscher Gartenbau*, **44**(26): 1710-1713.

Koths, J.S. 1964. The effects of CO_2 enriched greenhouse atmosphere on growth of snapdragons. *Michigan Florists*, No. 399, p. 15.

Kuehny, J.S., A. Painter and P.C. Branch, 2001. Plug source and growth retardants affect finish size of bedding plants. *HortSci.*, **35**(2): 321-323.

Lamont, G. 1986. Herbicide evaluation trials on cut flowers. *Australian Hort.*, **84**(7): 89-91.

Larsen, F.E. and J.F. Scholes, 1966. Effects of 8-hydroxyquinoline citrate, N-dimethyl amino succinamic acid and sucrose on vase life and spike characteristics of cut snapdragons. *Proc. Amer. Soc. hort. Sci.*, **89**: 694-701.

Larson, R.A., C.B. Thorne and R.R. Milks, 1985. Growth of geranium and snapdragon 'plugs' fertilized with controlled released micro-fertilizer to root-zone heating. *New York Fl. Grs Bull.*, **29**(3): 12-16.

Laughner, L. and B. Corr, 1996. Snapdragons: Formula for success. GrowerTalks, **60**(6): 57, 62.

Laurie, A. and G.H. Poesch, 1932. Photoperiodism – the value of supplementary illumination and reduction of light on flowering plants in the greenhouse. *Ohio agric. Exp.Stn Res. Bull.*, No. 512, pp. 8, 31, 35.

Laurie, A. and A. Wagner, 1940. Deficiency symptoms of greenhouse flowering crops. *Ohio agric. Exp. Stn Res. Bull.*, No. 61, pp. 19-21.

Lee, C.W., K.L. Goldsberry and J.J. Hanan, 1986. Production of cut snapdragons in rockwool. *Colorado Greenhouse Grs Assocn Bull.*, No. 438, pp. 1-2.

Lindstrom, R.L. 1966. Snapdragons – 60°F and CO_2. *Flor. Rev.*, **139**(3591): 18-19, 51-54.

Loeser, H. 1991. *Matthiola*, *Eustoma*, *Trachelium* and *Callistephus* in a closed system (German). *Gartenbau Mag.*, **38**(10): 51-53.

Macnish, A.J., R.T. Leonard and T.A. Nell, 2008. Treatment with chlorine dioxide extends the vase life of selected cut flowers. *Postharv. Biol. & Tech.*, **50**(2/3): 197-207.

Maginnes, E.A. and R.W. Langhans, 1961. The effect of photoperiod and temperature on initiation and flowering of snapdragon (*Antirrhinum majus* var. Jack Pot). *Proc. Amer. Soc. hort. Sci.*, **77**: 600-607.

Maginnes, E.A. and R.W. Langhans, 1967. Flashing light affects the flowering of snapdragons. *New York St. Fl. Grs Bull.*, No. 261, pp. 1-3.

Mahal, C. 1972. Cytogenetics of *Antirrhinum* and *Lantana*. Ph.D. Thesis, Kanpur University, Kanpur.

Mahal, C., M. Pal and T.N. Khoshoo, 1968. Artificial and natural polyploids in *Antirrhinum*. *Curr. Sci.*, **37**: 5-7.

Malik, R.S. 1979. Study on heterosis in antirrhinum (*Antirrhinum majus* L.). Ph.D. Thesis, Agra University, Agra.

Marousky, F.J. and J.C. Raulston, 1970. Enhancement of snapdragon floret color with light and floral preservatives. *HortSci.*, **5**: 355 (abstr.).

Mastalerz, J.W. 1957. Boron deficiency of snapdragon. *Pennsylvania Fl.Grs Bull.*, No. 75, pp. 3-6.

McDaniel, G.L. and M.G. Miller, 1976. Transpiration of snapdragon under southern summer greenhouse conditions. *HortScience*, **11**: 366-368.

Miller, R.O. 1962. Variations in optimum temperature of snapdragon depending on plant size. *Proc. Amer. Soc. hort. Sci.*, **81**: 535-543.

Nelson, P. 1962. Disease. In: *Snapdragons: A Manual of the Culture, Insects and Diseases and Economics of Snapdragons* (ed. Langhans, R.W.), pp. 70-80. Snapdragon School, New York State Extension Service and New York State Flower Growers Association, Ithaca, New York.

Nelson, P.V. and R.A. Larson, 1969. The effects of increased CO_2 concentrations on chrysanthemum (*C. morifolium*) and snapdragon (*Antirrhinum majus*). *North Carolina agric. Exp. Stn Tech Bull.*, No. 194, pp. 1-15.

Nowak, J. and R.M. Rudnicki, 1975. The effect of "Proflovit-72" on the extension of vase-life of cut flowers. *Prace Instytutu Sadownictea w Skierniewicach, B*, **1**: 173-179.

Oertli, J.J. 1970a. Nutrtient disorders in snapdragon. *Florist Rev.*, **146**(3773): 20-21.

Oertli, J.J. 1970b. Phosphorus deficiency in snapdragon. *Florist Rev.*, **146**(3774): 28-29.

Oertli, J.J. 1970c. Potassium deficiency in snapdragon. *Florist Rev.*, **146**(3775): 29.

Oertli, J.J. 1970d. Calcium deficiency in snapdragon. *Florist Rev.*, **146**(3776): 51.

Oertli, J.J. 1970e. Magnesium deficiency in snapdragon. *Florist Rev.*, **146**(3777): 23.

Oertli, J.J. 1970f. Sulfur deficiency in snapdragon. *Florist Rev.*, **146**(3778): 28.

Oertli, J.J. 1970g. Boron deficiency in snapdragon. *Florist Rev.*, **146**(3779): 65.

Oertli, J.J. 1970h. Iron deficiency in snapdragon. *Florist Rev.*, **146**(3780): 24.

Parrini, C. and P. Rumine, 1989. Diseases and pests of greenhouse-grown snapdragon and stock (Italian). *Colt. Prot.*, **18**(12): 29-38.

Pasini, C., T. Berio, P. Curir and F. D'Aquila, 1996. Further characterization of *Rhizoctonia solani* isolated from carnation and other ornamental plants. *Informatore Fitopatologico*, **46**(6): 33-36.

Peterson, H. 1955. Artificial light for seedlings and cuttings. *New York St. Fl. Grs Bull.*, No. 122, pp. 2-3.

Pfister, J.M. and J.M. Widholm, 1984. Plant regeneration from snapdragon tissue cultures. *HortScience*, **19**: 852-854.

Porter, D.M. and R. Aycock, 1967. Snapdragon leaf spot caused by *Cercospora antirrhini. North Carolina agric. Exp. Stn Tech. Bull.*, No. 179, pp. 1-31.

Rabinowitch, J.D., C.L. Hedley and A.E. Arthur, 1976. Variation in budding and flowering time of commercial cultivars of *Antirrhinum majus. Scientia Hort.*, **5**:287-291.

Raulston, J.C. 1970. Relationships of snapdragon response groups to cultivar performance in Florida field production. *Proc. Fla St. hort. Soc.*, **83**: 449-454.

Raulston, J.C. and F.J. Marousky, 1971. Effects of 8-10 day 5°C storage and floral preservatives on snapdragon cut flowers. *Florida Fl. Grs*, **8**(2): 4-10.

Rogers, M.N. 1951. Greenhouse Soil Fertility Analysis and Interpretation. Master's Thesis, University of Missouri, Columbia, Missouri.

Rogers, M.N. 1958. Year-round snapdragon culture. 1. Effects of lighting and shading snaps seeded during the summer months. *Missouri St. Florists News*, **18**(3): 3-7.

Rogers, M.N. 1959. Year-round snapdragon culture. 2. Summer snapdragons. *Missouri St. Florists News*, **20**(6): 3-5.

Rogers, M.N. 1960. Direct benching vs. potting snapdragon seedlings. *Pennsylvania Fl. Grs Bull.*, No. 118, pp. 3-5.

Rogers, M.N. 1961. The reactions of varieties of different response groups grown during the winter months at night temperaturesw of 60°F, 55°F, and 50°F with supplementary lighting. *Natl Snapdragon Bull.*, No. 13, pp. 1-6.

Rogers, M.N. 1980. Snapdragons. In: *Introduction to Floriculture* (ed. Larson, R.A.), pp. 107-131. Academic Press, San Diego, California.

Rutland, R.B. and J.E. Pallas, 1972. Transpiration of *Antirrhinum majus* L. in relation to radiant energy in the greenhouse. *J. Amer. Soc. hort. Sci.*, **97**: 34-37.

Sanderson, K.C. 1975. Fertilization, watering, temperature, light and photopertiod. *Flor. Rev.*,**156**(4038): 17, 59-61.

Sanderson, K.C. and C.B. Link, 1967. The influence of temperature and photoperiod on the growth and quality of winter and summer cultivar of snapdragon, *Antirrhinum majus* L. *Proc. Amer. Soc. hort. Sci.*, **91**: 598-611.

Sangwan, R.S. and H. Harada, 1975. Chemical regulation of callus growth, organogenesis, plant regulation, and somatic embryogenesis in *Antirrhinum majus* tissue and cell cultures. *J. Exp. Bot.*, **26**: 868-881.

Serek, M., E.C. Sisler and M.S. Reid, 1995. Effects of 1-MCP on the vase life and ethylene response of cut flowers. *Pl. Gr. Reg.*, **16**(1): 93-97.

Shaw, R.J. and M.N. Rogers, 1964. Interaction between elevated carbon dioxide levels and greenhouse temperatures on the growth of roses, chrysanthemums, carnations, geraniums, snapdragons and African violets. *Flor. Rev.*, **135**(3486): 23-24, 88-89.

Singh, V. and H.K. Jain, 1979. Floral organogenesis in *Antirrhinum majus* (Scrophulariaceae). *Proc. Indian Acad. Sci. Bull.*, **88**: 183-188.

Stefanis, J.P. and R.W. Langhans, 1982. Snapdragon production with supplemental irradiation from high pressure sodium lamps. *HortScience*, **17**: 601-603.

Vinod Kumar and S.K. Bhattacharjee, 2006. Antirrhinum. In: *Advances in Ornamental Horticulture* (vol. I; ed. Bhattacharjee, S.K.), pp. 234-246.

Wainright, H. and H.L. Irwin, 1987. The effect of paclobutrazol and pinching on *Antirrhinum* flowering pot plants. *J. hort. Sci.*, **62**: 401-404.

Wang, C.Y., J.E. Baker, R.E. Hardenburg and M. Liberman, 1977. Effects of two analogues of rhizobitoxine and sodium benzoate on senescence of snapdragon. *J. Amer. Soc. hort. Sci.*, **102**: 517-520.

Williamson, C.E. 1962. Root diseases and soil sterilization. In: *Snapdragons: A Manual of the Culture, Insects and Diseases and Economics of snapdragons* (ed. Langhans, R.W.), pp. 62-69. Snapdragon School, New York State Extension Service and New York State Flower Growers Association, Ithaca, New York.

7

Callistephus chinensis (Family: Asteraceae)

Sanyat Misra, R.L. Misra, Y.C. Gupta, Ajit Kumar, J.K. Ranjan and Pragya Ranjan

Introduction and Origin

Callistephus (China aster) is derived from the Greek *kallos* meaning beauty and *stephanos* meaning a crown, alluding to the large, colourful flower heads. First it was named as *Aster chinensis* by Linnaeus but subsequently Nees transferred this to *Callistephus chinensis*. This genus has only one species. It is native to China and Japan. *Callistephus chinensis* was introduced for the first time in Europe fron China in 1731 by a Jesuit Missionary. First were the German breeders who developed quilled type of varieties from *Callisphus chinensis* which became very popular in USA but only during 19th century. Initially, the seeds were being imported to various countries mainly from Germany, and then the China asters were being known as German asters. American breeders also developed several varieties and with the introduction of new branching type by James Vick in 1893, the USA became the main centre for developing new varieties and for production and supply of its seeds. The contribution of French florists is immense in developing double peony-flowered types, and in 1886, they developed cut flower types with long stems and spreading habits. In England, during 1752 double types were introduced and subsequently in 1807, variegated blue and white types were introduced. Dwarf types were developed in 1851, and in 1886, comet types having dwarfness and compact flat rays were developed which replaced the quilled types.

China aster is an half-hardy herbaceous annual with erect and branched stem growing up to 75 cm tall. Leaves alternate, triangularly and broadly ovate and toothed, basal stalked but upper ones sessile. The flower head solitary or a few together and surrounded by a tuft of leafy bracts, flowers are daisy-like, disc flowers which are perfect and fertile, being in the centre of the individual inflorescence are surrounded by the outer ray flowers, the ray florets (only with female organs) are white, pink or deep purple but never yellow though disc is yellow, and the flower colours may thus vary from white to yellow and purple via pink, rose, lavender, violet and blue. In the composite flower of China asters, seeds mature from outside of the ray florets to the inside of the disc florets taking many days from maturing of the first floret to the last floret. The fruit is a compressed achene. Though it is most suitable for bedding from late summer to autumn (by the onset of first frosts) and for cut flowers but with the development of the dwarf types, now it is also grown as popular pot plant. It is widely grown in Karnataka, Tamil Nadu, West Bengal and Maharashtra.

Genetics and Breeding

Its basic chromosome number is 9 and apart from usual diploid (2n = 18), it has tetraploid and other polyploid forms. Dua (1989, c.f. Guzenwski, 1976 and Petrenko, 1975) through the F_1 hybrids, the backcrosses and F_2 progenies of the eight cultivars and in the flowers of 13 different colours, reported four anthocyanins, *viz.* delphinidin, peonidin, cyanidin and pelargonidin.

Negi *et al.* (1983) worked out variability and correlations in 19 varieties of China aster for seven characters. Flower weight and stalk length showed highly genotypic coefficient of variation, heritability and genetic advance as percentage of mean due to additive gene action. Stalk length and flower weight had positive significant correlation among themself and with flower size, plant height and days to flower, flower size with plant height and days to flower, and number of flowers and days to flower with number of branches per plant. Rao and Negi (1992) studied 38 genotypes for 12 biometrical characters and found medium differences between genotypic and phenotypic coefficients of variation for number of laterals per plant and narrow for other characters. They recorded high heritability with high genetic gain for flower weight and number of ray florets, medium heritability with high genetic gain for number of laterals due to additive gene effects, and high heritability with medium genetic gain for plant height, plant spread, stalk length and stem girth and concluded that for further improvement of this crop selection should be based on flower weight, number of ray florets per floral head and number of laterals per plant. Aswath and Parthasarthy (1993) studied 12 cultivars and found that all the cultivars had high heritability as well as co-heritability estimates. Plant height was found positively correlated with flower weight, number of flowers and vase life, and number of laterals as well as flower weight with number of flowers per plant. Number of flowers showed direct as well as indirect influence on vase life.

In India, the work on China aster breeding is going on at IIHR, Bengaluru and at MPKV Regional Fruit Research Station, Ganeshkhind, Pune. Collection was made from various indigenous sources as well as from France and USA. These centres have evolved four varieties each, 'Kamini', 'Poornima', 'Shashank' and 'Violet Cushion' from IIHR and 'Phule Ganesh White', 'Phule Ganesh Pink', 'Phule Ganesh Violet' and 'Phule Ganesh Purple' from Ganeshkhind. Through systematic breeding efforts, IIHR has developed as many as 55 pure breeding lines through single plant selections from heterozygous lines (Janakiram, 2006). Dua (1989, c.f. Fleming, 1937) stated that purple colour is dominant to red and deep pink, red to white, while white is always recessive. Negi and Raghava (1990) stated that violet is dominant over other colours, blue dominating the creamy white, deep pink dominating the pink and white, deep pink being incompletely dominant over white-pink, and pink being dominant over white. Raghava and Negi (1992) stated droopyness incompletely dominant over erectness; and described that number of main branches per plant, laterals per plant, flowers per plant, days to flower, plant height, flower stalk length, duration of flowering, flower weight, and cut flower life are governed due to non-additive gene action while number of ray florets per flower head and flower size are governed due to additive gene action. Raghava and Negi (2001) stated that doubleness was monogenic dominant over singleness and pompon (one row ray florets) over chrysanthemum and fluffy types.

Through continuous pure line selections, Janakiram (2006) stated that wilt resistant lines in all the types (exceptions being the pompons) were developed long back by Jones and Riker. These cultivars include 'American Branching' (tall), 'Crego', 'Feather', 'Kirkwell', 'Laplator', 'Ostrich', 'Powderpuff' (bouquet type), 'Roment' and 'Stardust'. Kratika and Duskova (1991) through varietal evaluation against *Fusarium oxysporum* f. sp. *callistephi* reported that cvs 'Srdce Francie' and 'Bukett Bila' were resistant while 'Denisa' and 'Gaja' were susceptible. Janakiram (2006) on the basis of the work done at IIHR reported that against root-knot nematode, *Meloidogyne incognita*, the pure line IIHR-2 was highly resistant, AST-5, IIHR-17, IIHR-21 and cv. 'Shashank' were found resistant, cv. 'Poornima' moderately resistant while cvs 'Kamini', 'Violet Cushion' and AST-16 were found susceptible.

Janakiram (1995) when irradiated the seeds of China aster cvs 'Local Pink', 'Poornima' and 'Violet Cushion' with 10, 20 and 30 krad of ^{60}Co gamma rays, recorded that lower doses showed stimulatory effect while higher the inhibitory effect. Soaking of swollen achenes in 0.01, 0.03, 0.05, 0.1 and 0.3 per cent colchicines for 6, 12 or 18 hours, or through dropping of 0.1, 0.3, 0.5, 1.0 and 1.5 per cent colchicines between the cotyledon leaves after their opening at 7-8 a.m. for 2, 3 and 5 days, Hanzelka and Kobza (2001) found that low concentration as well as shorter time suppressed seed germination, and low concentration and longer treatment also suppressed emergence but not that much as higher concentration and shorter duration. Higher concentration with short duration of treatment resulted into deformed seedlings with poor root growth. Treatment of seeds with 0.3 per cent for 6 and 18 hours, and the treatment on seedling apex with 1.5 per cent for 5 and 3 days produced highest number of polyploids including mixoploids. High survival rate with minimal root damage was noted through colchicine application on the apex. Variation in the polyploid plants was in the form of thick, rounded and entire leaves.

Classification and Varieties

In China aster various cultivars are available to choose for cut flower production either in open field or in the greenhouses. However, dwarf cultivars can be used as flowering pot plant or for bedding in lawn or in the garden. The cultivars have varying growth habits, shape, size and appearance, flower types (double or single) and colours. These cultivars are classified according to flower-head shape and the height of the plant, as mentioned below:

- ☆ **Ball**: Height up to 75 cm, wilt resistant, neat and long stems, rounded heads quilled, fully double, incurved flowers, ideal for cutting. **Bouquet Powder Puffs**: It is also a ball type with height from 45-65 cm, erect and compact, form symmetrical, wilt and weather resistant, flowers large &.5-8.0 cm, up to 20 and fully double with quilled centres and having a good range of colours. 'Miss Europe' is its clear pink form.
- ☆ **Chrysanthemum-flowered**: Height 22.5- 25.0 cm, and the plant is highly compact with tight-petalled fully double chrysanthemum-type strap-shaped flowers, ideal for formal bedding. Its good rose-pink form is 'Chaterf's Erfurt'. **Duchess strain**: It falls under chrysanthemum- or peony-flowered type growing to a height of 60 cm, is wilt resistant with thick incurving petals, requires some support as flowers are quite large and double. Its form 'Fire Devil' bears bright red flowers.
- ☆ **Giants of California**: It grows to 75 cm height, growth is spreading type and bears large ruffled flowers. **Liliput strain**: Falling under 'Giants of California' type but height is only 38 cm, weather resistant, plant

shape is pyramidal with a mass of quilled and button-like flowers, and is available in a wide range of colours ideal for cutting and pot culture.

- ☆ **Ostrich Plume**: Height up to 38-65 cm, erect and much-branched, wilt resistant, a best known double type with many large ostrich-feather flower-heads having long wavy flowers with recurved petals and is a popular bedding variety in a full range of colours.
- ☆ **Peony-flowered strain**: Height up to 60 cm, the flowers are shell-pink in compact rounded heads with broad and incurving petals (florets).
- ☆ **Pompon strain**: Height up to 38-65 cm, weather resistant, and bears many neat button-like dense and compact flowers in many colours having rounded heads with contrasting centres. 'Pirette' is cerise-scarlet with white centre. Ideal for cutting.
- ☆ **Queen of the Market strain**: Height up to 38 cm with compact growth bearing large fully double flowers. Ideal for cutting and for pot culture. **Single Sinensis strain**: This falls under the 'Queen of the Market' strain type. This strain attains 60 cm plant height. It is wilt resistant and bears single flowers having clear yellow central discs with long and beautiful petals.
- ☆ **Unicum Mixed** or **Spider aster**: This grows erect to 45-60 cm tall and bears many large double flowers in many colours with long, slender and quilled or almost needle-shaped petals. Its form 'Waldersee Mixed' which grows to 30 cm is an upright bushy plant bearing profusion of small star-shaped (needle-shaped) flowers in many colours.

  Certain forms not fitting above but worthy of special mention are **Balsamina** bearing camellia-like flowers and **Charterhouse of Parma** bearing comet-type violet double flowers with tubular petals.

  Double flowered China asters are also classified on the basis of their height, *viz.* tall (75-90 cm), intermediate (medium-tall, 40-74 cm) and dwarf (below 40 cm) as under:

- ☆ **Tall**: Height 60 to 90 cm with various types such as: i) **American Branching** having long stems with different colours of flowers in different cultivars; ii) **Bouquet Powerpuffs** with stout stems having medium size of flowers in different colours; iii) **Princess** and **Giant Princess** with long woody, bushy and tall (75-90 cm) plants bearing many large extra double flowers with narrow petals on erect stems, in various colours, stems much branched and ideal for cut flowers; iv) **Peony-flowered** with upright growth bearing flowers in various attractive colours suitable for cut flowers; v) **Giant of California** having large flowers with good colour range but appearing late; vi) **Totem Pole** bearing large (10-12 cm across) bouquet type flowers in white, rose-pink, scarlet-cerise and blue; vii) **Chikuma** strain bearing pompon type ball-shaped fully double flowers with many cultivars in white, yellow, salmon-rose, rose, red with golden yellow centre, purple and red; viii) **Giant Massagno** has bushy plant bearing large flowers (10 cm across) with quilled petals in golden yellow, scarlet, rose, light blue and silvery blue colours in different cultivars; ix) **Giant Perfection** grows to 75 cm tall, vigorous, much-brnached and bushy, flowers have wide colour range, quite large, double with incurved petals and excellent for cutting; and x) **Madelina** growing to 75 cm, vigorous, erect and much-branched, and bearing a wide range of many single elegant flowers with long stalks, being perhaps most graceful of all the China aster varieties, and magnificent for cutting.
- ☆ **Intermediate (Medium tall)**: Height of the plant from 40 to 59 cm and has many types such as: i) **Ostrich Feather** bearing medium flowers with curved petals in various colours; ii) **Giant Comet** has hardy and much branched plants bearing flowers with curled petals; iii) **Giant Grego** having vigorous but bushy growth and flowers with curled petals in white, shell-pink, salmon, scarlet, deep rose, light and deep blue colours; iv) **Early Burpeana** which flowers early having 7-cm across chrysanthemum-type flowers with semi-incurved petals in white, scarlet, rose and blue shades; v) **Pompon** having globular flowers with quilled petals in shades of white, yellow, rose, scarlet, lilac, light blue, dark blue and with white centre; vi) **Rubens**, a cut flower type branched at base and flowering early with pompon type flowers (7-cm diameter); vii) **Liliput** having erect habit with fully double small flowers; and **Unicum** bearing large flowers with quilled petals.
- ☆ **Dwarf Double**: Under this category plant height ranges from 20 to 39 cm and has three types, *viz.* i) **Pinnochio** which has dwarf and compact growth where plants are most suitable for bedding, edging and window boxes and which bears numerous star-shaped flowers; ii) **Colour Carpet** having dwarf globular plants with chrysanthemum-type flowers; and iii) **Dwarf Chrysanthemum** type which has bushy and compact growth with a wide range of flower colours.
- ☆ **Asterix**: Single-flower cultivars good for cut flowers and those which are dwarf are useful for cut flowers and bedding.

In India wherever China aster is being grown commercially, they normally grow three colours, *viz.* pink, violet and white. In fact, in the subtropical climate of the country such as in Delhi, growers face the problem of seed production because if grown during its proper season, there is no seed setting or a very little seed set. This may be due to unfavourable climatic conditions, especially the temperature which plays an important role in China aster seed production. This problem is neither encountered in the temperate or sub-temperate regions nor in the tropical regions. Therefore, every season the growers have to depend for its seeds on either the seed nurseries which import the seeds or the growers directly import the seeds, and the continuous import costs the country dearly. The varieties growing 61-90 cm are grouped as tall, those growing 41-59 cm are grouped as medium tall and those growing 20-40 cm are grouped as dwarf or small. IIHR, Bengaluru has developed more than 55 pure lines of China aster. In the country, at MPKV, RFRS, Ganeshkhind, Pune; and at IIHR, Bengaluru the work on its breeding started and each of the centres have evolved four varieties of China aster, and these varieties are:

- 'Phule Ganesh White' (Pune): A medium tall and erect growing variety flowering late with snow-white double chrysanthemum-type flowers. Its stalk length is 45-50 cm and flower production is 45-50 lakh stems per hectare.
- 'Phule Ganesh Pink' (Pune): A semi-erect, early blooming excellent cultivar bearing large double chrysanthemum-type flowers of rosy-pink colour. Yield is 40-45 lakh flowers per hectare.
- 'Phule Ganesh Violet' (Pune): An excellent, early and semi-erect variety producing medium size double chrysanthemum-type flowers of violet colours. Its production is approximately 60 lakh flowers per hectare.
- 'Phule Ganesh Purple' (Pune): It is a semi-erect, medium tall and a very attractive variety with bright purple flowers. The yield is 45-50 lakh flowers per hectare.
- 'Kamini' (IIHR): It is a late flowering (138 days in Bengaluru) variety developed through pedigree method involving AST-6 and AST-36, grows to about 60 cm, stalk length 30 cm, flower colour deep pink, flower diameter 6 cm, flower weight 2 g each, vase life 8 days and yield per plant 50 flowers.
- 'Poornima' (IIHR): This early bloomer (105 days in Bengaluru) variety developed through pedigree method involving AST-29 and AST-3, grows to about 50 cm, stalk length 25 cm, flowers pure white powder puff type with tubular disc having 6 cm diameter with 5-6 rows of ray florets, flower weight 3.5 g, vase life 7 days and yield per plant twice of the local white.
- 'Shashank' (IIHR): A mid-season bloomer (124 days in Bengaluru) variety developed by crossing a Local Pink with AST-2 (a pure line) and through pedigree pure line selection as IIHR-42, grows to about 55 cm, stalk length 25 cm, flower colour creamy white powder puff type, flower diameter 6 cm, flower weight 2.5 g each, vase life 9 days and yield per plant 45 flowers (more than double the yield of the local white).
- 'Violet Cushion' (IIHR): A late season (130 days in Bengaluru) variety developed through pedigree selection by crossing a Local Pink and AST-2 which produces violet flowers of pompon type with 4-5 rows of ray florets and tubular disc florets, grows to 45 cm, stalk length 20 cm, flower diameter 4.5 cm, flower weight 2 g each, vase life 8 days and is highly floriferous yielding 70 flowers per plant.

With the nurseries, sometimes some named exotic varieties are available such as 'American Beauty', 'American Branching', 'Bouquet Powder Puff', 'Dwarf Queen', 'Giant of California', 'Pinnachio', 'Stardust' and 'Super Princess'.

Propagation

China aster is propagated through seeds which germinate within 4-7 days from 12 to 21°C temperature provided the seeds are healthy, however, those infected with *Botrytis cinerea* germinate slowly with poor quality of plants. The seeds can be germinated at a temperature range of 10-35°C, but optimum germination occurs at 21± 2°C (Thompson and Cox, 1979). The germination of fresh seeds occurs best at 5-15°C but favourable temperature range for germination is broadened when storage duration is increased up to six months (Grzesik *et al.*, 1997). For planting one hectare of land some 2.5-3.0 kg seeds are required. During seed germination, it is only temperature which affects germination as the light is of little significance. When sowing, the seeds are covered with 8 mm thick layer of soil or medium. After germination, the seedlings prefer 18°C further growing temperature (Honeywell, 1941).

Gujar *et al.* (2001) recommended drying of mature flower heads in shade for one week up to 6.8 per cent moisture level, then extracting the seeds and storing these at 2°C in cold storage or in refrigerators in sealed polyethylene bags which even after one year of storage remained viable even up to 78.95 per cent.

Cultural Practices, Nutritional Requirements and Watering

China asters grow best at a sunny situation in a loamy or sandy loam soil which is well-drained and has a **pH** of 5.5 to 7.5. Heavy soils with high Ca levels is not suitable for its cultivation. If the crop is being grown under protection, 600-900 ppm **CO_2** will encourage leaf growth and will advance the flowering (Dole and Wilkins, 1999). Growing at lower temperature (10°C) than at usual 18°C prolongs the time for floral initiation though after initiation flower development under SD is more rapid at 18°C than at 10°C (Lin and Watson, 1950), however, irrespective of the photoperiod the growing temperature is never recommended to be below 13°C (Post, 1934; Biebel, 1936; Lin and Watson, 1950). Its growth and further development is better when these receive 15-20°C night and 20-30°C day temperatures along with 50-60 per cent relative humidity. This can be grown throughout the year in the temperate regions, for summer seed sowing in February-March and transplanting after one month for flowering from June-July to October, but in winter seed sowing in September-October and transplanting after one month for flowering from May to September. In the plains it can be grown only during winter months because summer becomes too hot, however, in controlled environment it can be grown all the year round. From the date of seed sowing, it takes around four months for flowering. Connors (1933) stated that the China aster crop in New Jersey required 14 to 18 weeks from seed to flower, depending on season and the cultivars grown, which would allow three crops to be taken up in one year. In areas where day length is short, the plants should be supplied artificial lighting.

The **nursery raising** can be done in pots or seed pans when the plants are required only in limited quantity for garden planting but for large scale cultivation the seedlings should be raised in raised nursery beds. While preparing the nursery beds, the nursery soil should be mixed with compost or dung manure at the rate of 3 kg/m^2 + coarse sand, the soil is made to a fine tilth and raised up to 10 cm and then whole bed is fully drenched at least 5 cm deep with 0.25 per cent captan or thiride to kill any pathogen present in the soil. Seeds are sown thinly 10 cm apart and 8 mm deep. In the rows the seeds are sown at a spacing of 2-3 cm to avoid plants becoming lanky. After sowing the seeds, the bed is covered with straw and then watered with rose can gently. This requires watering daily by the evening just to wet the surface. When the seeds start throwing out the sprouts, immediately the straw is removed otherwise the seedlings will appear yellow, zigzag and lanky. The watering should be continued daily by the evenings but superficially.

When the seedlings have attained 3-4 leaf stage, which these attain at 30-45 days of seed sowing, these are **transplanted** in the prepared field beds. While preparing the field, the soil should be incorporated with 150-200 quintals of wellrotten farmyard manure (some 3 kg/m^2) and then it should be dug up some 20 cm deep with cultivators 2-3 times followed by planking after every digging, all the perennial weeds should be taken out through fork or manually and then beds of the convenient sizes are prepared, preferably 1.6 metre wide and 4.0 metres long. At the time of planting the beds should be quite levelled so that while irrigating the water may not stand anywhere. To both side of the beds there should be sunken clearance some 15-20 cm deep and 60 cm in width to facilitate tillage operations and also for irrigation so that the beds may suck the required water when water is discharged in these paths cum channels. This way the problem of water stagnation will not arise. Planting is done at 30 x 30 or 30 x 20 cm as per requirement, depending especially on the plant spread of a particular variety. In Pune, they recommend 30 x 30 cm, in PAU, Ludhiana 20 x 20 cm, and in Kalyani 20 x 30 cm **spacings** for optimum flower yield though IIHR, Bengaluru recommended 20 x 10 cm specing for higher flower and seed yield without any ill-effect on shelf or vase life (Janakiram *et al.*, 2001). For cut flower use, Honeywell (1941) suggested the spacing of 25 x 25 cm in the greenhouse China aster and 10 x 10 to 30 x 46 cm in the field in USA. Immediately after planting whether in the pots or in the field, it should be followed with watering to settle the roots and to avoid desiccation of the roots, *vis-à-vis* plants.

Patil *et al.* (1987) when planted the four week old seedlings of China aster cv. 'Ostrich Plume Blue' to three spacings from September to December at Pune, recorded 20 x 30 cm (highest density) spacing and November planting giving best flower and seed yield, followed by October planting. Dhomre *et al.* (1997) when planted the cv. 'White Powder Puff' at six plant densities during summer, monsoon and winter seasons in Pune, found 15 x 15 cm spacings and winter planting giving the best results.

Earthing-up is practiced at least twice, one when plants have started branching out and the other one month after the first earthing-up. This loosens the soil and gives proper support to the plants to avoid lodging.

China aster is heavy feeder which responds well to various **nutrition** applications. Nitrogen deficiency causes dwarfing, while phosphorus deficiency delays flowering. For getting higher flower yield, Pune centre recommended 400 kg N, 100 kg P_2O_5 and 200 kg K_2O, BCKV Kalyani 400 kg N, and 200 kg each of P_2O_5 and K_2O, PAU Ludhiana 200 kg N, 300 kg P_2O_5 and 400 kg K_2O, IIHR 90 kg N, 60 kg P_2O_5 and 60 kg K_2O as basal and 90 kg N/ha top-dressing 40 days after transplanting, and UAS Bengaluru recommended 5 tonnes/ha vermicompost + 180 kg N +120 kg P_2O_5 + 100 kg K_2O, each per hectare. UAS Dharwar recommended 180 kg N, 120 kg P_2O_5 and 100 kg K_2O for higher seed yield in China aster (Janakiram, 2006). Singh and Sangama (2000) with cv. 'Kamini' obtained highest flower yield with the application of 300 kg N/ha. Dole and Wilkins (1999) stated that when salt levels are high or when boron levels are over 2.5 ppm, plant size as well as flower size decrease. Kohl *et al.* (1957) through foliar nutrient levels recommended various fertilizers. In fact, plant's nutrient requirement is high when they are in active growth, as compared when they are not active. Conover (1969) reported that increased nitrogen levels decreases keeping quality but increases *Alternaria* disease problems. Gruis and Joiner (1960) found 195 g/m^2 of 8-0-8 fertilizer increasing the stem length and flower number in China aster.

Water stress in China aster is dangerous to the crop as this does not have deep root system. Continuity of the moisture should therefore be maintained throughout entire period of growth and development. Water requirement is affected due to various factors, *viz.* the type of the soil, the season of growing, weather conditions. Since it is a winter crop, hence it should be watered shallowly at 7-10 days interval. Conover (1969) recommended sub-irrigation so that foliage remains dry for disease prevention and to protect the protective chemicals not being washed away from the leaves.

Though China asters grow erect but their branches appear sub-horizontal and when the plants are fully grown, these create shady atmosphere which does not permit growing of the **weeds**, but before attaining full growth a lots of weeds come up which require to be weeded out regularly so that the nutrients available in the soil are not robbed by the growing weeds. Manual weeding is time consuming and expensive so can be practiced only in small plantings. In large plantings, however, only through chemical **weedicides** the growth of weeds may be controlled. The weedicides CDEC and chlorpropham at 6.75-10.0 kg/ha and dacthal at 9.0-13.5 kg/ha applied 4-6 weeks after transplanting are very effective in controlling the weeds. Tenoran 7 kg/ha, Ramrod (propachlor 65 per cent) 7 kg, Amiben-Granulat (chloramben), Kerb 50W 3 kg, Legurame flussig [(carbetamide (70 per cent)] 10 litres and Ronstar (oxidiazon [75 per cent]) 8 litres were tested for their selectivity in *Callistephus chinensis* where Kerb 50W and Legurame gave satisfactory weed control; Tenoran though gave best results but this and Ronstar caused some severe injury/necrosis; Ramrod lacked adequate herbicidal activity; though Amiben-Granulat, Kerb 50W and Legurame flussig were well tolerated (Anon., 1974). Lamont and O'Connell (1986) when in summer tested Alachlor (1.125 or 2.25 l), Oxadiazon (l, 2 or 4 kg), chlorthal dimethyl (7.5 or 15.0 kg), Oxyfluorfen (2 litres) and Napropamide (2.25 or 4.5 kg/ha) as pre-emergence in *Callistephus chinensis* plantings, they recorded only Oxadiazon and Oxyfluorfen giving 100 per cent control of all the weeds, and Napropamide was found quite safe in direct-sown crop giving 90 per cent control of grasses and <50 per cent control of broadleaved weeds, however, chlortham dimethyl was safe for the transplants controlling 70-90 per cent weed in summer and 50-70 per cent in winter but was injurious in direct-sown one. Pre-plant application with granular or liquid form of EPTC and vernolate at 8.7 kg/ha, trifluralin and dichlobenil at 2.25 kg/ha, or chlorpropham or alipur (cycluron + chlorbefam) each at 4.5 kg/ha one day before planting controls the weeds in China aster plantings effectively (Dhua, 1989). Duczmal (1989) obtained 55-58 per cent weed control in cvs 'Morelowe Igielkowe' and 'Duchese Cerine Rose' when applied Propachlor (Ramrod) at 7.0 kg/ha after transplanting, and recorded 1.1-12.1 kg/ha seed yield through use of 3.0-4.0 kg Trifluralin (Treflan) before transplanting. Basavaraju *et al.* (1992) recommended 0.75-1.25 kg/ha diuron as pre-emergence application for effective control of weeds except the sedges.

Pinching is an essential operation in China aster. One month after transplanting, the pinching of the main shoots encourages more number of shoots, more nodes, flowers per plant and ultimately the flower yield though with about one week delayed flowering as has been found with the cv. 'Ostrich Plume Purple' in UAS, Bengaluru. Disbudding is normally not carried out in this crop.

Glasshouse grown China aster plants will not require **support** but field-grown ones may require one or two layers of support through either individual stakes where required or through stretching of the strings. The plants grown for cut flower production or the genetically dwarf plants do not require **height control**. However, the plants in certain varieties may grow uncontrollably taller, the growth of such plants may be controlled through the use of ancymidol, chlormequat and daminozide (Adriansen, 1985).

The **yield** of the flowers is somewhere 10,000-15,000 kg per hectare, though through improved package of practices this can be increased up to 20,000 kg. Seed heads are harvested when they become dry and fuzzy. **Seeds** are shade-dried for about a week to the moisture level of 6.8 per cent, cleaned and stored in air-tight containers in a cool dry place, preferably at 2°C. Seeds remain viable for one year and one gramme may contain some 450-600 seeds.

Growth, Development and Flowering

China aster is classified as a facultative **LD** plant for rapid floral initiation at **temperatures** of 13° to 20°C but at or below 13°C temperature it is a obligate LD plant as this forms only rosette under SD, and after the initiation of the flowers it develops independent of temperature (Lin and Watson, 1950), though floral development is accelerated under SD (Laurie and Poesch, 1932; Post, 1934; Lin and Watson, 1950; Doorenbos, 1959; Hughes and Cockshull, 1965) or through GA treatment (Doorenbos, 1959). LD of 15 h (natural daylength + night break) for 7 to 13 weeks, depending upon the location where China asters are to be planted, can commence only when the plants have established (Kofranek, 1954a, b). In southern California, the China aster crop is given LD from August 20 to April 20 while in Florida, Conover (1969) used a 2-3 h incandescent night break from September 10 to April 20, though Gruis and Joiner (1960) advocated 10 weeks of LD optimum for Florida. Biebel (1936) stated that LD should commence 15 to 45 days after sprouting. Cockshull (1966) suggested only 1-h night break in winter which caused enhanced leaf growth and in turn caused rapid floral induction in New York. At 10°C production temperatures than at 18°C, LD prolongs the days for floral initiation, and once when flower initiation has taken place, flower development under SD becomes more rapid at 18°C than at 10°C, though growing temperatures should never be allowed to fall below 14°C. Doorenbos (1959) stated that floral induction requires six weeks under LD from the date of germination though is dependent on cultivar and the season. Doorenbos (1959) stated that when 3-week old plants did not flower after being shifted from SD or LD though 4- or 5-week old such plants flowered, seems that juvenility is present in China aster plants, and when the plants are once matured enough to perceive the LD signal, only 7 LDs are required for flower initiation, however, a critical number of leaves, *i.e.* four or more should have been formed under SD before these are shifted to LD. Kofranek (1954b) states that increase in number of SD before start of LD increases the number of leaves and, which in turn, increases the quality of flowers and the flower number.

In China aster, **growth regulators** have been found affecting growth, development and flowering. Janakiram (2006) states that GA_3 at 200 or 300 ppm spraying increases the number of flowers and flowering duration and NAA has little effect though MH delays the flowering. Through a study at UAS Bangalore, it was found that GA_3 spraying though did not affect flower yield but the quality was found increased with improved vase life at room temperature, its 250 ppm enhanced the anthesis by 52 days and 500 ppm produced more shoots, cycocel spray coupled with pinching 30 days after transplanting reduced the plant height, main stem nodes, days for flowering, branch numbers and flower size though only cycocel treated plants recorded highest number of flowers (Janakiram, 2006). GA at 300, 200 and 100 ppm and IAA at 100 and 50 ppm gave higher seed weight in China aster (Janakiram, 2006).

Postharvest

Flowers are harvested by evenings or by mornings when 60 per cent flowers are fully open and are stored at 0-4°C. Flower harvesting is done either of the individual stems or of the whole plant, lower leaves are removed and the cut ends are kept in distilled water immediately after cutting. When brought inside the packing room, these are graded according to stalk length, floral appeal, and flower colour, shape and size, and then bunches are made. Whole plant bunches may contain some 3-5 plants and 25-30 such bunches are bundled together. Some 400-500 such bundles are harvested from one hectare of land. The flowers should be wrapped in clear plastic films and then are securely packed in the cardboard boxes. High rates of nitrogenous fertilizers are not conducive in increasing the vase life, instead reduce the keeping quality and increase susceptibility to *Botrytis* and *Alternaria* (Conover, 1969). Kofranek *et al.* (1978) used silver nitrate (dipping for 10 minutes in 1000 ppm solution) or Physan-20 (200 ppm) a microorganism inhibitor increased the vase life by 2-3 times and addition of 5 per cent sugar and 75 ppm citric acid further increased the vase life. Janakiram (2006) stated that aluminium sulphate or sucrose at 0.2 per cent increases vase life.

Pulsing with 0.4mM STS + 7 per cent sucrose or 0.4 mM STS + 50 ppm GA_3 for 2 hours and then in the holding solutions containing 150 ppm 8-HQC + 1 per cent sucrose improved the vase life by 5-11 days in the cvs 'Songbon Red' and 'Songbon Clear Scarlet' and improved quality of the cut flowers while vase solution containing 70 per cent distilled water + 30 per cent carbonated water + 40 ppm NaOCl though did not extend the vase life but cut flower quality was found improved (Bang *et al.*, 1997).

Insect-Pests and Diseases

Aster beetle or black blister beetle (*Epicauta pennsylvanica*) feeds on foliage and flowers and destroys the whole plant. Methoxychlor weekly spraying during the infestation period will control this pest. **Asiatic beetle** (*Autoserica castanea*) feeds on the foliage during the night and hides itself in the soil at the base of the plant or in the vicinity. Chlordane or dieldrin soil application will eliminate the pest at its larval stage. Lead arsenate spraying on the plants once or twice during night will kill the feeding adults as well as larvae. **Chrysomelid beetle** (*Aulacophora foveicollis*) attacks the crop at initial stage by cutting holes in the leaves and cutting the tender shoots and its grubs feed on the underground parts including the roots which results into plant death. Carbofuran or phorate at 1kg/ha soil application will keep this pest under check. **Leaf and flower eating caterpillars** (*Helicoverpa armigera*) attack the crop at the onset of winter and feed on leaves and flowers while ***Phycita* green caterpillars** with brown heads feed on ovaries and stamens of the buds and flowers by grooving through the central receptacle. Spraying with 0.2 per cent Metasystox or methyl parathion will kill these larvae. Female moth of the **stem borer** (*Platyptilia molopias*) lays its eggs on the tips of the plants and the black-headed creamy larvae coming out after hatching of these eggs feed inside by boring the shoots, stems and the laterals thus affected shoots become hollow and wilt. Carbofuran or phorate at 1 kg/ha soil application will control this pest. **Semilooper**, a caterpillar (*Clenoplusia albostriata*) which is blue-green in colour feeds on the leaves and also on the flowers. Also sometimes the **cutworms, corn earworms** and **cabbage looper** attack this crop (Ranjith, 2013). Spraying with chlorpyriphos or quinalphos at 0.05 per cent or carbaryl at 0.1 per cent will control this pest. Numerous **root aphids** (*Macrosiphum* spp., Ranjith, 2013; *Anuraphis maidi-radicis*) are found feeding on the roots which results in yellowing and death of the plants. Chlordane soil application before planting or Lindane spraying on the crop controls this pest. **Aphids** (*Aphis* spp.) and **potato aphid** (*Macrosiphum salanifolii*) feed on the aerial part of the plants by sucking the sap which may be controlled through Malathion or Lindane sprayings. **Tarnished plant bug** (*Lygus lineolaris*) is light-brown and variously spotted which feeds on the plants by puncturing the terminal shoots below the flower buds causing the buds or flowers to droop. Spraying with 0.2 per cent Chlorpyriphos or methyl parathion will kill this pest. **Leaf hoppers** (*Macrosteles fascifrons*) transmit viral diseases from one plant to the other which may be controlled through 0.2 per cent spraying with methyl parathion. **Leaf miners** (*Liriomyza compositell*) feed inside the leaves and calyx through tunnelling and in its serious infestation leaves and calyx turn brown and die. Adult fly lays its eggs inside the leaves through puncturing, and after hatching its yellow maggots start feeding inside the leaves by making the mines. Spraying of monocrotophos, triazophos or methyl-o-demeton at 0.05 per cent controls this pest. Adult **thrips** (branded greenhouse thrips and western flower thrips) and its nymphs feed on the tender parts of the plant through piercing and in its severe infestation plants deform. These can be controlled by spraying with 0.2 per cent Metasystox. **Cottony cushion scale** (*Iceria purchasei*) may become a problem in temperate regions (Ranjith, 2013). Sometimes **soft scales** also attack this crop. Various species of **whiteflies** (greenhouse whitefly, Iris whitefly, silver leaf whitefly and sweet potato whitefly) have been observed attacking this crop (Ranjith, 2013). The infestation of **spider mite** (*Tetranychus telarius*) makes the foliage webby, discoloured and deformed which may be controlled through spraying with Kelthane at 0.1 per cent. **Root-knot nematode** (*Meloidogyne incognita*) infests the plants through galling of the roots and **foliage nematode** (*Aphelenchoides ritzema-bosi*) sucking the juice of the plants causing the plants arresting its growth, cupping, yellowing and malformation of the leaves and plant death in its serious infestation. Foliar nematodes cause leaf blight and leaf fall and its serious infestation may kill the plants. Nemacur soil application and demeton foliar spray will control both types of nematodes.

Diseases which are found attacking China aster are damping off (*Pythium splendens*), wilt (*Fusarium oxysporum* f. sp. *callistephi*, *F. conglutinans* var. *callistephi*, *Sclerotium rolfsii*, *Verticillium albo-atrum*, *Acrostalagmus vilmorinii*, *Phialophora fastigiata*), collar and root rot (*Phytophthora cryptogea*), stem rot (*Pellicularia filamentosa*), grey mould (*Botrytis cinerea*), rust (*Coleosporium solidaginis*), canker (*Phomopsis callistephi*), leaf spot (*Alternaria*, *Stemphylium callistephi*, *Septoria callistephi*, *Ascochyta asteris*) and viruses. From Italy, Pasini *et al.* (1996) isolated *Rhizoctonia solani* from China aster causing collar or foot rot. *Fusarium* infection affects the vascular system of the plants, expressed first as yellowing of the plants and then death. It has been found occurring along with *Acrostalagmus* and *Verticillium* causing wilt or with *Botrytis cinerea* or *Rhizoctonia* causing wilt. To control *Pythium splendens* on China aster seedlings, when Niebisch and Kelling (1986) used Previcur N (propamocarb) and Ridomil (metalaxyl) + zineb, they recorded Ridomil superior over Previcur N, and application of Captan 80 or Thiram FW on seeds and seedlings, they recorded increased yield of cut flowers. *Phialophora* also causes vascular wilt, first recorded from New Zealand. Against all these pathogens, soil sterilization with formalin or chloropicrin is a must before taking the next crop of China aster from the same field, however, resistant cultivars such as 'Bouquet Scharbach', 'Balls Blue', etc. are also available. Before sowing, the seeds should also be soaked in a 0.1 per cent solution of mercuric chloride for 30 minutes. Once disease occurs on the plants, there is no control so in disease prone areas, benomyl 0.1 per cent alternate with thiride 0.2 per cent should be sprayed to the standing crop fortnightly. Collar and root rot becomes serious under moist conditions. Its infection is symptomized by water-soaked areas at the stem base, rotting of the stem at base, root rot and wilting of the plants. Captan 0.2 per cent spraying is very effective against the infection of this pathogen. Restricted irrigation and use of resistant cultivars are suggested. Stem rot can be controlled through spraying with 0.2 per cent captan or thiride. Grey mould spreads more rapidly during cool and humid weathers. This disease generally occurs 3-4 days after sowing or 10-12 days after pricking out, and its serious infection causes plant blight. Spraying of such plants with dithane M-45 or dithane Z-78 at every 5-7 days during inclement weathers will keep this disease under check. Attack of rust to China aster plants causes bright yellowish-orange spots on underside of leaves, more serious being on the young plants. It can be controlled by spraying with wettable sulphur to the standing crop. Canker affects the lower part of the stem and distal portion completely collapses, though roots are not affected. Spraying with 0.1 per cent bavistin alternate with 0.2 per cent captan will control this problem. Leaf spots are first yellowish, turning brown and then black and increasing in size, lower leaves being infected first. Through maneb or zineb sprayings at 0.2 per cent will keep these diseases under check.

'Aster yellows virus' causes stunting of the plants, formation of numerous adventitious shoots, pale-yellowish tinge on the leaves, the blooms become off-colour and the rays usually yellowish-green, and the virus is transmitted through *Macrosteles fascifrons* leaf hopper. Its Californian strain is also reported. Control of the leaf hopper and destroying the infected plants will check further spread of this virus. A seed-borne viral disease 'chrysanthemum mosaic virus' causes mild dwarfing and serious bloom distortion. 'Spotted wilt virus' and 'curly top virus' also infect this crop so the affected plants should be uprooted and burnt.

References

Adriansen, E. 1985. Kemisk vaekstregulering. In: *Polleplanter I - Produktion, Metoder, Midler*(eds Christensen, O.V., A. Klougart, I.S. Pedersen and K. Wikesjö), pp. 142-162. Gartner INFO, København, Denmark.

Anonymous, 1974. *Ann. Report German Pl. Prot. Serv.*, No. 21, pp. 221-223.

Aswath, C. and V.A. Parthasarathy, 1993. Heritability and correlation studies in China aster (*Callistephus chinensis* Nees). *Indian J. Hort.*, **50**(1): 89-92.

Bang, C.S., C.Y. Song, K.Y. Huh, J.S. Song and B.H. Kim, 1997. Effects of pulsing and holding solutions on vase life and quality of cut flower (*Callistephus chinensis*). *RDA J. Hort. Sci.*, **39**(1): 77-81.

Basavaraju, C., J.V.N. Gowda and T.V. Muniyappa, 1992. Effect of pre-emergence herbicides on yield in China aster [*Callistephus chinensis* (L.) Nees]. *Curr. Res.*, UAS, Bangalore, **21**(3): 50-51.

Biebel, J. 1936. Temperature, photoperiod, flowering and morphology in cosmos and China aster. *Proc. Amer. Soc. hort. Sci.*, **34**: 635-643.

Cockshull, K.E. 1966. Effects of night-break treatment on leaf area and leaf dry weight in *Callistephus chinensis. Ann. Bot.*, **30**: 791-806.

Conover, C.A. 1969. Commercial aster production in Florida. *Florida Fl.Gr.*, **6**(9): 1-8.

Dhomre, J.K., N.S. Shirsath and A.S. Naphade, 1997. Effects of different plant densities and seasons on flowering of China aster. *J. Soils and Crops*, **7**(2): 136-138.

Dole, J.M. and H.F. Wilkins, 1999. *Callistephus.* In: *Floriculture Principles and Species* (eds Dole, J.M. and H. F. Wilkins), pp. 251-255. Prentice Hall, New Jersey.

Doorenbos, J. 1959. Response of China aster to daylength and gibberellic acid. *Euphytica*, **8**: 69-75.

Dua, R.S. 1989. China aster. In: *Commercial Flowers* (eds Bose, T.K. and L.P. Yadav), pp.681-696. Naya Prokash, Kolkata.

Duczmal, K.W. 1989. Weed control in ornamental plants grown for seed (Polish).*Biuletyn Instytutu Hodowli I Aklimatyzacji Roslin*, No. 169, pp. 59-76.

Gruis, J.T. and J.N. Joiner, 1960. The effects of periods of long days and levels of fertilization on China aster, *Callistephus chinensis*, 'All Saints'. *Proc. Fla St. hort. Soc.*, **73**: 378-381.

Grzesik, M., K. Gornik, M.G. Chojnowski, R.H. Ellis (ed.), M. Black (ed.), A.J. Murdoch (ed.) and T.D. Hong, 1997. Maternal effects on *Callistephus chinensis* seed yield and quality. Basic and applied aspects of seed biology. *Proc. of the Fifth intern. Workshop on Seeds* (held at Reading, U.K. on Sept. 10-15, 1995). *Current Pl. Sci. and Biotech. in Agric.*, **30**: 129-135.

Hanzelka, P. and F. Kobza, 2001. Genome-induced mutation in *Callistephus chinensis* Nees. I. Effect of colchicines application on the early plant development. *Zahradnictvi hort. Sci.*, **28**(1): 15-20.

Honeywell, E.R. 1941. Aster culture. *Purdue Univ. agric. Exp. St., Circ.*, No. 200, Lafayette, Indiana.

Hughes, A.P. and K.E. Cockshull, 1965. Interrelations of flowering and vegetative growth in *Callistephus chinensis* (variety 'Queen of the Market'). *Ann. Bot.*, **29**: 131-151.

Janakiram, T. and T.M. Rao, 1995. Radiation sensitivity analysis in China aster. *Indian J. Nemat.*, **52**(2): 149-151.

Janakiram, T., M.T. Patil and S.K. Bhattacharjee, 2001. China aster (*Callistephus chinensis*). *AICRP (ICAR) on Floriculture Tech. Bull.* No. 13.

Janakiram, T. 2006. China aster. In: *Advances in Ornamental Horticulture. 1. Flowering Shrubs and Seasonal Ornamentals* (ed. Bhattacharjee, S.K.), pp. 247-266. Pointer Publishers, Jaipur.

Kofranek, A.M. 1954a. Fall production of quality king asters grown out of doors. *Southern Florist and Nurseryman*, **67**(12): 76, 78-79.

Kofranek, A.M. 1954b. Artificial light promotes early flowering of asters. *Southern Florist and Nurseryman*, **67**(43): 18.

Kofranek, A.M., E. Evans, J. Kubota and D.S. Farnham, 1978. Chemical pretreatments for China asters to increase flower longevity. *Florists Review*, **162**(4206): 26, 70-72.

Kohl, H.C., Jr., A.M. Kofranek and O.R. Lunt, 1957. Photoperiodism – the value of supplementary illumination and reduction of light on flowering of plants in the greenhouse. *Ohio agric. Exp. St. Bull.*, No. 512, pp. 1-42.

Kratika, J. and E. Duskova, 1991. Evaluation of resistance of China aster (*Callistephus chinensis*) cultivars to *Fusarium oxysporum* f. sp. *callistephi*. *Sbornik UVTIZ, Ochrana Rostlin*,**27**(2): 127-135.

Lamont, G.P. and M.A. O'Connell, 1986. An evaluation of pre-emergent herbicides in field-grown cut flowers. *Plant Prot. Quart.*, **1**(3): 95-100.

Laurie, A. and G.N. Poesch, 1932. Photoperiodism – the value of supplementary illumination and reduction of light on flowering of plants in the greenhouse. *Ohio agric. Exp. St.Bull.*, No. 512, pp. 1-42.

Lin, L.C. and D.P. Watson, 1950. The influence of daylength and temperature on the growth and flowering of *Callistephus chinensis* Nees. *Proc. Amer. Soc. hort. Sci.*, **55**: 441-446.

Negi, S.S., S.P.S. Raghava, T.V.R.S. Sharma and V.R. Srinivasan, 1983. Studies on variability and correlation in China aster (*Callistephus chinensis* L.). *Indian J. Hort.*, **40**(1-2): 102-106.

Negi, S.S. and S.P.S. Raghava, 1990. Genetics of flower colour in China aster [*Callistephus chinensis* (l.) Nees]. *Euphytica*, **48**(2): 117-122.

Niebisch, R.M. and K. Kelling, 1986. Results of chemical control of fungal diseases in ornamental plant production (German). *Gartenbau*, **33**(7): 215-218.

Pasini, C., T. Berio, P. Curir and F. D'Aquila, 1996. Further characterization of *Rhizoctonia solani* isolated from carnation and other ornamental plants. *Informatore Fitopatologico*, **46**(6): 33-36.

Patil, J.D., N.R. Bhat, B.B. Chougule and B.A. Patil, 1987. Flower and seed yield in aster (*Callistephus chinensis* L.) cv. Ostich Plume Blue as influenced by planting date and planting density. *Curr. Res. Reptr*, **3**(2): 46-48.

Post, K. 1934. The effects of daylength and light intensity on vegetative growth and flowering of the China aster (*Callistephus chinensis*). *Proc. Amer. Soc. hort. Sci.*, **32**: 626-630.

Raghava, S.P.S. and S.S. Negi, 1992. Inheritance of growth habit in China aster [*Callistephus chinensis* (l.) Nees]. *Indian J. Hort.*, **49**(3): 281-283.

Raghava, S.P.S. and S.S. Negi, 2001. Inheritance of flower type and doubleness in China aster. *J. Ornam. Hort.* (n.s.), **4**(1): 7-12.

Ranjith, A.M. 2013. China aster. In: *Identification & Management of Horticultural Pests*, pp. 131- 132. New India Publishing Agency, Pitam Pura, New Delhi-110 088.

Rao, T.M. and S.S. Negi, 1992. Heritable components of biometric characters in China aster. In: *Floriculture, Technology, Trades and Trends* (eds Prakash, J. and K.R. Bhandary), pp. 318- 321. Oxford and IBH Pub. Co., New Delhi.

Singh, K.P. and Sangama, 2000. Effect of graded levels of N and P on China aster (*Callistephus chinensis*) cultivar Kamini. *Indian J. Hort.*, **57**(1): 87-89.

Thompson, P.A. and S.A. Cox, 1979. Germination responses of half-hardy annuals. 2. China aster (*Callistephus chinensis*). *Seed Sci. Techn.*, **7**: 201-207.

8

Campanula (Family: Campanulaceae)

Ritu Jain, Varun M. Hiremath and Sanyat Misra

[**Common names**: Bearded bellflower (*Campanula barbata*), Bellflower, Canterbury bell/Canterbury bells of English gardens/Coventry bells (*C. medium*), Carpathian bellflower (*C. carpatica*), Chimney bellflower (*C. pyramidalis*), Giant bellflower (*C. latifolia* syn. *C. macrantha, C. rotundifolia*), Harebell/Bluebell in Scotland (*C. rotundifolia*), Italian bellflower/Falling stars/Star of Bethlehem (*C. isophylla*), Peach-leaved bellflower (*C. persicifolia*), Rampion (*C. rapanculus*), Throatwort (*C. trachelium*), *etc.*]

Introduction

Campanula, is well known for its peculiar bell-shaped flowers as in Latin, the generic name is diminutive form of *campana* meaning 'a bell'. Though this is the very old name but came into official records only during second half of 16th century as it has a mention in the work of Rembert Dodoens, botanist and physician to Emperor Maximilian II. The genus originates in the northern temperate latitudes and into subtropical and tropical zones as a mountain plant (temperate regions of the Northern Hemisphere, and in particular the mountains of Europe, Asia Minor and in the Mediterranean regions). It is a large multifarious genus consisting of about 300 hardy and half-hardy annual, biennial (*C. medium, C. pyramidalis*), perennial and sub-shrubby species, mostly being herbaceous perennials. According to John Gerard, *Campanula* made its entry into European gardens towards the end of the 16th century. During 1596-97, Gerard collected specimens of *Campanula persicifolia, C. pyramidalis* and *C. medium* in England. Since none of those species were mentioned in William Hooker's ***Flora of the British Isles*** (1830-1842), therefore they must have been introduced there from elsewhere. Some of the species have been introduced into French gardens during late 17th century and these are *C. carpatica* in 1774; *C. alpina* and *C. pulla* in 1779; *C. siberica* in 1783; *C. versicolor* and *C. mollis* in 1788 (Pizzetti and Cocker,1968).

In Europe, the herbaceous campanulas such as *C. rapanculus, C. rapunculoides* and to a lesser extent *C. persicifolia* are cultivated as vegetables rather than ornamental plants. Botanists have reported that the fleshy roots have an agreeable sweetish taste. The solutions prepared from the aerial parts of the plants have been used against mouth and throat inflammations through the actions of their resinous substances. Apart from their medicinal value, camapanulas are commonly used as bedding plants, potted flowering plants (*C. pyramidalis*), basket plants (*C. fragilis, C. isophylla*) and as cut flowers. *Campanula* spp. are important ornamentals in Northern Europe, especially in Scandinavia where they are grown for spring sales in pots or in hanging baskets (Adams, 1991; Armitage, 1987; Whitman *et al.*, 1995).

Botany

Campanula flowers are predominantly bell-shaped or campanulate, but can be cup-, bowl-, tubular- or star-shaped. The five floral lobes may be shallow to deeply cut, thus enhancing the cupped, bell, or starry appearance. Flowers are borne singly or in many-flowered inflorescences, and may be upward facing or nodding in habit. Flower colour may be violet, blue, purple, pink or white. Bellflowers have characteristically different basal and stem leaves. Basal leaves may be oval, round, lanceolate, or heart-shaped and are usually larger, long petioled, and often toothed. Basal leaves of some species such as *C. carpatica, C. glomerata* and *C. rotundifolia* wither during the flowering period. Stem leaves are usually simple, smaller in size and either short-stalked or sessile. Foliar colour ranges from light to dark green with several yellow-leaved forms available. Bellflowers offer a rich diversity of plant habits too—from small, tufted alpine species (8 cm high, high spreader, *Campanula betulifolia*) to border plants (1.5 m high *C. laxiflora*) and from erect clump-formers (*C. trachelium*) to rambling-spreaders (*C. rotundifolia*). In the flowering of *Campanula* species, at an early stage the anthers release pollen to be deposited on hairs of the style, and for the time being for a shorter duration these hairs make available the pollen to the visiting insects including bees but when after sometime the flowers become old *vis-à-vis* stamens withered, the several stigma lobes separate and curl back into the stylar hairs, bringing about self-pollination as an insurance against earlier failure of cross-pollination by insects. It is a characteristic feature of the genus that the pollen is frequently shed while the flower is in bud (Everard and Morley, 1970). The fruit is a capsule containing numerous small seeds (Pizzetti and Cocker, 1968).

Propagation

Companula is commonly propagated by seed, vegetative cuttings, or divisions depending on the species. The most important species, *Campanula isophylla* is propagated from cuttings (Dole and Wilkins,1999). **Seeding** works well in plugs. Seeds are sown in October or in March-April in pots, pans of seed compost or plugs and placed in a greenhouse or cold frame. These germinate within 5-10 days under 18-20 °C but more constantly at 20 °C for four days (Dole *et al.*, 2001). When large enough to handle the seedlings are pricked off into pans or boxes, smaller species in 5cm pots filled with JIC I where these remain till transplanting in March or April, or September or October based on time of sowing. However, taller herbaceous species and the biennials should be there in the nursery until these are ready for planting in October. Care should be taken after transplanting that these should not be pot-bound and over-grown because it may lead to stunting due to shortened stems. Seeds of perennial species can be sown in the greenhouse at 18-21 °C in the early spring and moved outdoors after the frost. Biennial seeds can be sown in the summer, and then overwintered outside for flowering in the following year.

Certain species such as *C. persicifolia*, its hybrids which do not breed true to type from seeds and many others which produce more than single crowns are **divided** and replanted in October, March or April.

Excepting biennial species, the campanulas can also be multiplied from 2.5-5.0 cm **cuttings** taken from the vegetative (non-flowering) basal shoots in April or May. For growing these, peat and sand in equal parts (by volume) should be taken in the pots and the pots placed in a cold frame or greenhouse. After rooting, individual cuttings are placed in 7.5 cm pots containing JIC I where these are treated just as seedlings. Moe (1977) stated that 900 ppm CO_2 level in the greenhouse enhances the quality of cuttings, their production, their rooting and subsequent growth.

Classification, Species and Varieties

The most important commercial species in genus *Campanula* is *C. isophylla, C. carpatica, C. elatines, C. fragilis, C. medium, C. poscharskyana* and *C.pyramidalis*. The most commonly cultivated species are described here under (Bailey, 1942; Pizzetti and Cocker.,1968; Hay and Beckett, 1971; Beckett, 1985, 1987; Brickell, 1994).

Campanula alliariaefolia (*C. lamiifolia, C. macrophylla*): A native to Caucasus and Asia Minor, this mound-forming herbaceous perennial grows up to 60 cm in height and 50 cm in spread with woolly, striate and erect stems which are branched only at the top. Grey-green large basal leaves are cordate, tomentose and crenate while stem leaves have petioles being gradually shortened upward so uppermost ones become sessile. Floral stalks appear from the axils of the floral leaves which are larger and wide, and bear solitary, bell-shaped, creamy-white and drooping flowers some 5.0 cm long on short stalks during June and July. The flowers in var. 'Ivory Bells' are little larger than the type species.

Campanula allionii: A native to Piedmont, French Alps, this fully hardy herbaceous perennial grows up to 7.5 cm tall with the spread of 15 cm though not easy to cultivate at ordinary conditions as this survives only in the screes and alpine houses. It grows with underground creeping rootstocks (stolons) that send out stems at intervals of 1.25-2.5 cm. Its leaves are mid-green, linear-lanceolate, some 2.5-5.0 cm long, little hairy, sessile, entire and normally 7 on a stem and 5 whorled ones at the base. Purple solitary nodding flower some 3.75 cm long and across appears on a stem from June to July, which is larger to the size of the plant.

Campanula arvatica: A native to N. Spain, this dwarf herbaceous perennial species grows to a height of 5.0 cm and spread of 30 cm. it is a prostrate growing species which spreads through underground stolons. Its mid-green leaves are rounded cordate at the base, and ovate and shallowly toothed at the trailing stems. Its star-like blue flowers some 2.5 cm across appear in July. Its var. 'Alba' bears white flowers.

Campanula aucheri: A native of E. Caucasus, this tufted herbaceous perennial grows to 5.0 cm high with the spread of 10 cm. Its leaves are mid-green, oblong to ovate and crenately toothed. Flowers are large some 2.5 cm long, bell-shaped and deep purple which appear in July.

Campanula barbata: A native of European Alps, Norway, this evergreen perennial herbaceous plant is rosette-forming in clusters, each cluster producing one stem which bears a few mid-green and lanceolate leaves. It may assume a height from 7.5 to 45 cm and the spread of 15-30 cm. Panicles are sparsely flowered. Flowers are nodding, bell-shaped, 2.5-3.75 cm long, being borne only to one side of the panicle in June, and from the mouth long woolly and blue-purple to white hairs protrude out.

Campanula betulifolia: A native of Armenia, this tufted, prostrate and slender-stemmed fully hardy perennial attains the height of 2.0 cm but spreads up to 30 cm. Its leaves are ovate-cordate to wedge-shaped. Long branching floral stems some 15 cm or more appear in summer, each branch carrying a cluster of single open bell-shaped white to pink flowers, iinterior lighter and exterior deeper.

***Campanula* × *burghaltii*:** This perennial border hybrid forms clumps and grows to a height of 45 to 60 cm with erect but wiry stems which are set with tufts of ovate mid-green leaves. Drooping, bell-shaped and blue-grey flowers some 5.0-7.5 cm long appear during early summer. Var. 'Ivory Bells' bears creamy-white bell-shaped flowers on arching stems from early to late summer.

***Campanula carpatica*:** A native to Eastern Europe, it is a small hardy herbaceous perennial with compact habit which attains the height of 30 cm when in bloom. Floral stalks are erect, slender and spreading, bearing oval leaves though base rounded and the edge crenated. Flower stems are slender, erect and spreading. Flowers are blue, terminal, solitary, largest among the small campanulas and bell-shaped with wide mouth. For its successful cultivation, it requires fertile, loose and moderately moist soil at semi-shady location. Its most attractive varieties are 'Isabel' (violet), 'White Star' (white), 'Turbinata' (deep blue), *etc*.

Campanula cochleariifolia (*C. pusilla*)**:** A native of the Alps, this mat-forming alpine herbaceous perennial grows only up to 10 cm high with shining green leaves. Pale-blue masses of the bell-shaped pendulous flowers appear in late spring and continue giving enchantment at least for one month. It is excellent for planting in the wall pockets and in the rock gardens. It does well in partial shade with fertile and moderately moist soils.

***Campanula excisa*:** A native to Switzerland, this is a dwarf (some 10 cm high) herbaceous tufted and mat-forming perennial spreading up to 30 cm through underground stolons. This species is difficult to grow, best grown on a granite scree or in a small trough at a cool place with humid air but poor exposure to sun. Its stems are thin and wiry, leaves are mid-green and linear, basal leaves spatulate, and the bell-shaped and blue flowers some 2.5 cm long appear in June. The flowers have a circular hole at the junction of each petal lobe.

***Campanula fragilis*:** This frost-tender herbaceous perennial species is recorded growing in the rock crevices and old walls in alkaline soil of southern Italy, particularly on seasides. It is a prostrate-growing to a height of up to 13 cm and is ideal for pot-growing. Its basal leaves are cordate, stem leaves are smaller and oblanceolate, sky blue flowers are bowl-shaped and appear in terminal racemes. It likes sunny position with hot and dry atmosphere.

***Campanula garganica*:** A native of Italy (Gargano region), this herbaceous mat-forming prostrate perennial species is very attractive and makes a nice cushion. It grows up to 10 cm high, bearing glabrous stalks and bean-shaped small leaves with crenate edges. Pale-blue starry bell-shaped flowers appear from late spring to early summer and continue for about two months.

***Campanula glomerata*:** A native of Europe, this moderately hardy herbaceous perennial is most suitable for herbaceous border at a semi-shady and moist location, growing up to 45 cm in height. Oblanceolate leaves are alternate, hairy, serrated and folding around the base of the stalk. Terminal masses of funnel-shaped violet-blue flowers, individually guarded by bracts, appear with one central as terminal and others as lateral ones on stong stems. It has various colour forms in purple and white. During summer the soil, irrespective of its type, should remain moist.

***Campanula* × *haylodgensis*:** A garden hybrid probably between *C. carpatica* and *C. caespitosa*, it is a spreading fully hardy perennial growing to a height of 15-22.5 cm and spread of 20 cm. Its basal small leaves are tufted, roundish-cordate and slightly dentate though stem leaves toothed, ovate-cordate and light green and the pompon-like double flowers are deep lavender-blue appearing at the ends of the stems in June. This plant is most suitable for planting on a wall or in the rock gardens.

***Campanula isophylla*:** A native to NW Italy, it is a non-hardy tufted herbaceous perennial with about 20-30 cm long prostrate to trailing stems. It makes a good indoor pot plant than bedding and can effectively be used in hanging baskets, on walls and in the rock gardens at the warm and sunny situations. Moderately fertile porous and alkaline soils are ideal for its growing. Its small leaves are bright green, brittle, cordate, rounded and tooth-edged. Basal leaves are much smaller than stem leaves which fall off soon. If broken, its stems and leafstalks exude a milky white sap with a distinctive but not unpleasant odour (Catkins, 1979). It is free-flowering and the lilac-blue or white (mostly pale-blue) flowers which are produced from the leaf axils, are starry and funnel-shaped, and 2.5-3.75 cm across. It blooms from spring till late summer in so profusion that foliage is completely hidden. Its var. 'Variegata' has variegated leaves, 'Mayi' is softly hairy and 'Alba' bears white flowers.

***Campanula lactiflora*:** A native to Caucasus Mountains, this hardy herbaceous perennial grows from 90-180 cm tall and is quite good for cutting. It requires very fertile, deep, porous and moderately moist soil at a semi-shaded position. Its root is fleshy and tuberous. It is quite free-flowering and the white flowers are slightly tinted pale-blue and roughly 2.5 cm long which are arranged in loose panicles. Flowering plants can directly be raised through seeds sown at a warmer place during spring. Var. 'Loddon Anna' bears pale-pink flowers.

Campanula latifolia (*C. macrantha*)**:** Its native haunts extend from northern India to Europe. It is a hardy herbaceous perennial having strong, erect and smooth stems growing up to 1.5 m tall, bearing rough-textured, alternate, lanceolate, dentate and pointed leaves. Violet-blue flowers some 5.0-7.5 cm long appear on erect spikes during late spring and continue till mid-summer. It is an effective plant for beds and borders provided the soil is deep, fertile and moist at a cool and semi-shaded position.

Campanula latiloba (*C. grandis*)**:** A native of Mt. Olympus, Siberia, this fully hardy glabrous herbaceous perennial grows to 1 m in height with spread of up to 45 cm, bearing mid-green, long (7.5-12.5 cm), lanceolate but narrowing at both the ends and serrated leaves. Bell-shaped violet-blue, sessile and solitary or lightly clustered flowers some 5.0 cm across appear in June.

***Campanula medium*:** A native to Europe, this hardy herbaceous perennial is generally cultivated as a biennial plant though its one of the strains is also grown as an annual. Seed sowing in July in the greenhouse and transplanting during autumn will start producing flowers in the following summer while those sown directly in cold frame or greenhouse in March will produce flowers the same year. It grows to 45 cm tall and in England it is being cultivated since 1597. The basal leaves are rosette though stem leaves are alternate, oblanceolate, rough-textured and acuminate. Inflorescence is a tall, erect, rigid spike bearing dense masses of large bell-shaped flowers. It is the largest flower of all the campanulas and its hybrids have a range of colours including white, pink, and the various shades of blue. Its cultivation is easy but the flowers are not long-lived. It has double, semi-double and single flower forms.

***Campanula mollis*:** A native to Syria, Spain and Crete, it is a non-hardy herbaceous perennial whose beauty lies in its graceful prostrate habit and grey-green velvety appearance but its cultivation is cumbersome. It grows to a height of up to 20 cm. Basal leaves are obovate to spathulate and the stem leaves rounded to oval. Its inflorescence is a spike where bell-shaped deep purple-blue or lavender blooms having white lips, are spaced widely. In mild clime it is an excellent rock garden plant. It is easily propagated through seeds.

***Campanula morettiana*:** A native of S. Tyrol, Dalmatia, this half-hardy dwarf (up to 7.5 cm tall) species with spread of up to 15 cm is suitable for growing in the alpine house. It is a tufted plant producing mid-green, cordate (ivy-shaped) leaves with fine hairs. Floral stalk is arching, each carrying a solitary, 2.5 cm long, erect, bell-shaped and violet-blue flowers that appear in June.

***Campanula persicifolia*:** A native to Europe and Siberia, this evergreen, graceful and elegant hardy herbaceous perennial has produced many of the beautiful hybrids with single or double white or blue flowers excellent for cutting. This grows from 40-90 cm in height with slender stems and normally flowers twice, once during start of the summer and second during autumn. Its mid-green, leathery and linear-lanceolate leaves are alternate, dentate and in the form of peach tree leaves, and these form a basal rosette. Bell-shaped quite open (rather saucer-shaped), white, blue or purple-blue clustered flowers some 2.5-3.75 cm across with oval petals are borne on long slender racemose stems. Its cultivation is easy and prefers any deep, fertile and moderately moist soil. It can tolerate quite sunny situation compared to any other campanula. Its var. 'Planiflora' (*C. nitida*) grows only up to 30 cm high, 'Planiflora Alba' bears white flowers and 'Telham Beauty' graceful large rich blue flowers.

Campanula portenschlagiana (*C. muralis*)**:** A native of Dalmatia and other parts of South Europe, this hardy herbaceous perennial grows up to 20 cm high with spread of 45-60 cm and is long-lived invasive species. For its growing, it prefers lightly shaded position in the crevices of walls and paths where soil is porous, fertile and moist. Being prostrate in habit, it forms the carpet-like mass of vegetation with small, mid-green, rounded-cordate and sharply toothed leaves. Bell-shaped, graceful and deep blue-purple flowers some 2.0 cm long start appearing from May and continue until September. It is propagated through division throughout the year, through seeds sown during spring or through direct seeding.

***Campanula porscharskyana*:** A native of Dalmatia, this rampant growing spreading herbaceous perennial grows 15-30 cm high with 60-90 cm spread, bearing mid-green, round and sharply serrated leaves and long sprays of bell-shaped star-like violet flowers some 2.5 cm across on leafy stems from June to November. Since it spreads through runners, it is most suitable for growing along the bank or in the wild gardens.

***Campanula pulla*:** A native to Mountains of Austria at altitudes ranging from 1,200 to 1,800 m, this attractive short-lived herbaceous perennial is tufted to clump-forming, producing usually 1-flowered stem. Its height is 7.5-10.0 cm, spread 22-30 cm, and spreads by underground runners. Leaves are glabrous and crenulate-dentate, lower ones short-petioled and ovate-rotund though upper sessile and ovate-acute. Flowers are nodding, bell-shaped, some 2.0 cm long and rich purple which appear during June. It prefers partial shade and is good for scree or rock garden.

***Campanula × pulloides*:** A perennial hybrid of garden origin probably between *C. pulla* and *C. turbinata* but quite similar to *C. pulla* though more vigorous, this grows up to 15 cm high with spread of about 22 cm, bearing hairy leaves and glistening purple-blue flowers.

***Campanula pyramidalis*:** A native to SE Europe, it is a half-hardy clump-forming herbaceous perennial which when in blooms attains up to 2.0 m height and is most suitable for border planting. It is grown as a biennial as its deterioration starts in the second year of flowering. Its roots are fleshy and carrot-shaped. Leaves are rich green, obovate to oblanceolate and sub-cordate. Basal leaves are oval-oblong, stem leaves alternate, oblanceolate and both the types are serrated. White or pale-blue bell-shaped starry flowers some 3.0 cm across and 2.5-3.75 cm long are produced in large pyramidal panicles along the entire length of the inflorescence in July. It is propagated through offsets produced in the basal rosettes with pale-green leaves in the side of the plant. It requires deep and medium soil at warm and sunny position. Var. 'Alba' produces white flowers.

***Campanula rainerii*:** A native to Switzerland and N. Italy Mountains, this perennial herbaceous hardy species grows hardly up to 10 cm high but spreading up to 15 cm wide with branching and sub-erect stems spreading through underground runners, each branch with 1-3 flowers. It is a tufted plant with grey-green, subsessile, ovate and sparsely serrate leaves where lower ones are ovate to obovate and smaller. Flowers are large, solitary, bell-shaped, erect and deep purplish-blue some 2.5-3.75 cm across and are produced in June. It is an ideal alpine and rock plant at sunny to semi-shaded position. Waterlogged soils are not conducive for its growing.

Campanula rapunculus (*Rapunculus verus*)**:** A native to Italy, other parts of Europe, North Asia and North Africa, this hardy herbaceous perennial, though mostly cultivated as a biennial, has white, swollen, fleshy and some 1.25 cm thick long-tapered radish-like roots. The plant grows up to 60 cm height. The leaves and roots are eaten as salad. Basal leaves are short-petioled, obovate and little crenate though stem leaves entire and linear-lanceolate. Numerous small blue flowers with lilac calyx lobes, and funnel-shaped corollas appear in long narrow racemes, flowering for a long period from May to September.

***Campanula rotundifolia*:** A native of Europe, Siberia and North America, this dwarf hardy herbaceous perennial spreads rapidly and grows to a height of up to 30 cm. Basal leaves are petiolate and orbicular to cordate and dentate though stem leaves are smaller, entire and linear to lanceolate. Stalkless and pendulous sky blue flowers appear profusely which become darker blue if grown under shaded locations. It is ideal for growing in cool locations of rock gardens and for wall plantings provided the soil is light and remains moist.

***Campanula rupestris*:** It is a native to the Mediterranean region, especially of the rocks and cliffs of Greece and the Aegean Islands which was introduced into England in 1788 where since then it is being grown on the walls and in rock gardens due to its prostrate growing habit. It is a monocarpic (flowering and fruiting only once in its life and then dying). It forms a rosette of soft-grey leaves and handsome, bell-shaped lilac-blue flowers which at bud stage are yellow.

***Campanula thyrsoides*:** A native to the central mountainous regions of Europe at altitudes ranging from 1,520-2,630 m, this variable herbaceous biennial grows from 30 to 45 cm tall with grooved stems. All the leaves are covered with long hairs on their margins, upper narrower and acute though basal leaves in thick rosette are sessile, 6.25 cm long, 2.0 cm wide, spatulate or obtusely lanceolate and lie on the ground. Its dense clubs of greenish, sulphur or creamy-yellow flowers (15 cm long and 6.25 cm wide) and some 40-50 per thyrse-like spike are produced only once in its entire life and then it sets the seeds and dies.

***Campanula trachelium*:** A native of Europe, Caucasus, Siberia, Japan and North America, this hardy herbaceous perennial grows up to 1 m high, producing angular stems which are little bristly as in case of the flowers. In fact, entire plant is rough surfaced and hirsute. Basal leaves are short-stalked, ovate and cordate while stem leaves are rough, coarsely crenate-dentate and acuminate. Inflorescence is loosely racemose and 30-45 cm long, 1 3 flowered, flowers some 4.5 cm long, drooping-type, bell-shaped, short-stalked and blue or white. It is most suitable as border planting and for naturalizing. It prefers moderately moist soil and tolerates sun more than any other campanula.

Campanula* × *van houttei (*C. punctata, C. nobilis*)**:** It is a hardy (except coldest zones) herbaceous perennial of hybrid origin growing up to 60 cm high with compact habit but having long and loose hairs. Upper leaves are almost sessile, oval-lanceolate, irregularly bi-dentate, up to 10 cm long, strongly nerved and more sharply toothed than the lower which are long petioled, roundish-cordate and more or less lobed. Inflorescence is compact with masses of dark blue flowers some 5.0 cm long appearing in June. It is a superb hybrid with magnificent appearance.

Campanula vidalii (*Azorina vidalii*)**:** An endemic species to certain islands in the Azores where it was discovered in 1842 by Captain R.N. Vidal, this evergreen herbaceous sub-shrub is quite distinct than others. Its stems are thick, spotted with scars that mark the positions of the fallen leaves and exudes a milky juice. It grows from 30 to 60 cm tall having branches from the base where some branches are short and sterile while others glossy, grooved, long and floriferous. Leaves are glossy dark green, oblong-spatulate, 7.5-10.0 cm long, viscid, fleshy, coarsely serrate and the upper ones turning to bracts. Inflorescence is racemose in which 9 nodding flowers some 5 cm long appear in loose terminal raceme. Flowers are tubularly bell-shaped and white with yellow base or pink.

***Campanula zoysii*:** A native to Austrian Alps at 1,800-2,400 m altitudes, this rare and abnormal species is a frost-hardy tuft-forming perennial growing up to 10 cm high. The plants are tufted and glabrous, basal leaves are crowded, petiolate, ovate-obovate, entire and obtuse but stem leaves are linear obovate-lanceolate and both the types are glossy green, rounded and tiny. Stem is few-flowered, peduncles 1-flowered, mostly terminal but sometimes even axillary. Flowers blue, corolla long cylindrical with constriction at the apex though wider at base, and larger for the plant.

Cultural Practices

Campanulas are easy to grow, as mostly they do not have preference for soil types and there are species for growing at open sunny, semi-shaded and shaded positions. **Soil** should be well-drained with sufficient organic matter and pH range of 6.0 to 7.0. Most of the campanula species are sensitive to water-logging as it will increase disease problems particularly on open flowers. *C. poscharskyana* requires growing in dry (not arid) soils (Madson and Madson, 1986). Seed germination is good at 18-21 °C but best quality flowers and uniform plants are obtained at 14-17 °C. Plant **density** of 32/m^2 should be followed for production of campanula (Madson and Madson, 1986). For cut flower production spacing ranges from 15cm × 15cm to 25cm × 25cm.

For **pot cultivation**, seed should be sown in pots in summer or early autumn and care should be taken that pot does not dry out during hot weather since it has smaller seeds. Soil should be kept moist with the help of paddy straw or coconut husks until germination begins. The seed pans or pots should be placed under shade. *Campanula carpatica* and *C. garganica* can be grown in pots with diameter of 15-18 cm, while *C. pyramidalis* requires a larger size (30 cm diameter) while *C. glomerata* needs an intermediate size.

For single stem production, the plants should be spaced 15 cm and 25-30 cm apart for **pinched**- production. The longest stems are obtained in single stem-production system preferably in the greenhouse-growing (Dole *et al.*, 2001). *Campanula* species are usually cut back after the cold treatment and prior to forcing. In species such as *C. isophylla* and *C. medium*, no pinching should be effected. Chemical growth retardants are also used for **height control** of campanula plants. B-Nine (daminozide) or A-rest (ancymidol) can be used at visible bud stage (Adams,1991; Madson and Madson, 1986). B-Nine is used to reduce intermodal length of *C. isophylla* and *C. poscharskyana,* but flowering is delayed by one week. Cycocel has been used on *C. carpatica* (Whitman *et al.,* 1995).

Since campanulas have generally weak stems at the base, netting should be provided particularly during periods of low light intensity for **supporting** the plants. Netting is highly recommended for field-grown campanulas as their bell-shaped flowers catch hold of water and fall over (Dole *et al.,* 2001).

As mentioned with the species, there are a few species which can be planted under wet soil conditions and along the banks of the water sources but most of the *Campanula* species are sensitive to excess **watering**, as it encourages root rot diseases. At no time waterlogged conditions should prevail whether the crop is field-grown or pot-grown otherwise root rot problems may occur (Dole and Wilkins,1999). However, these like moist roots and when in active growth and to lengthen the flowering period, they should be watered plentifully but as much as necessary. Even during the winter rest period these plants are given some water enough to moisten the potting mixture fortnightly. Every fortnight when the pot is sufficiently root-filled (about $1^1/_2$ months after potting), they are given with standard liquid **fertilizer** till flowering is over (Catkins, 1979). It is recommended that nutrient levels should be at 200 ppm N from a complete fertilizer (Adams, 1991).

Soil-based **potting mixture** is advantageous that even after sterilization it contains microorganisms that break down organic matter into essential minerals which controls frequent manual feeding. Soil-based mixtures should have 1 part sterilized fibrous soil, 1 part medium grade peat-moss, ground tree bark or leaf mould, 1 part fine perlite or coarse sand, and 1 part dehydrated farmyard/cattle manure or balanced granular to powdered fertilizer (all by volume). This mixture will be quite advantageous for campanulas. Three to four rooted cuttings per 7.5 cm pot should be planted in early spring and after appearance of the roots on the surface of the mixture which takes 2-3 months, these are moved into 10.0-12.5 cm pots. For hanging baskets, several plants are required, and when blue and white are mixed together, these complement the effects of each other.

Growth and Flowering

Reproductive growth in *C. isophylla, C. carpatica* and *C. poscharskyana* occurs when the **light** period is longer than 14 h (Madson and Madson, 1986; Heide,1965; Whitman *et al.,* 1995). For flowering, 16 h daylength is recommended for *C. carpatica* though cold does not have any effect (Runkle *et al.,*1996a). Long day treatments must be maintained even after visible bud stage otherwise shoots may revert back to vegetative phase if shifted to short days. Critical daylength between 12-14 h with a 4-h night interruption than the 2-, 1-, or 0.5-h is most effective for encouraging uniform and rapid flowering when used in middle of 16-h darker period (Runkle *et al.,* 1996b). *C. fragilis* which is seed-propagated hanging basket plant is also a long day plant where too 4-h night interruption in middle of a 16-h dark period has been observed most effective (Zimmer, 1985a) though in case of *C. pyramidalis* where inflorescence size goes up to 1.52 m with 100-200 flowers and which forms rosette of offsets, it can be kept vegetative for up to five years by growing at 20 °C temperature (Zimmer, 1985b). Bolting in *C. pyramidalis* is regulated by low temperature (8 °C) treatment for 6-8 weeks and 14 to 15 h long days but the seedlings should be seven months old before these respond to LD and precooling (Zimmer, 1985b). The warmer the pre-induction temperature, the longer the low temperature treatment is required. In case of *C. medium,* only older seedlings respond to cold and SD signals as flowers initiate only after vernalization under SD, followed by LD, and stem elongation has been found to occur without flowering when treated with GA_3 under SD, and when these elongated plants are cold-treated these flower under LD matching with the time of vernalized ones grown under SD and then placed under LD (Wellensiek, 1985). Reduction of light levels at the end of crop cycle enhances flower colour (Herrick and Perry, 1997).

During growth period, various temperature regimes are required as stock plants are grown at 16-18 °C but medium temperature is maintained at 21 °C and of the atmosphere 18 °C for production of cuttings, for seed-propagation the germination occurs within 14-20 days at 18-21 °C, during production the growth is accelerated when subjected to 18 °C but for synchrony, floriferousness and quality it is 14-17 °C (Dole and Wilkins, 1999).

Postharvest

Campanula flowers are harvested when more than half of the buds have opened. Cut flowers are harvested when one or two flowers have opened. Though flowers are less sensitive to ethylene, but wilting is encouraged upon exposure to ethylene (Nowak and Rudniki, 1990). *C. carpatica* 'Clips' stored at 4 °C in the dark for 6 days had longer life than unstored control where flowers had turned darker blue. Plants should be stored at 2 to 6 °C for optimum vase life. Moe and Fjeld (1987) recommended spraying of STS at 0.2-0.5 mM to negate the effect of ethylene in potted plants and promote the vase life.

Pests and Diseases

Campanulas are susceptible to sucking pests such as **aphids** (*Dysaphis sorbi*), **spidermites** (*Tetranychus urticae*) and **fungus gnats** (Adams, 1991; Armitage, 1987; Madson and Madson, 1986). To avoid fungus gnats, make sure that containers have good drainage and plants are not too wet. Let the top 2.5-7.5 cm of soil dry before adding more water. Keeping the top of the soil dry will help stop fungus gnats from laying their eggs in the soil. Use of 'yellow sticky traps' placed horizontally at the soil surface will capture large number of egg laying adults. The gnats are attracted to yellow colour and are easily removed through the trap before they can lay more eggs.

Diseases such as ***Botrytis*** (Adams, 1991; Armitage, 1987; Madson and Madson,`986), root and stem rot (***Fusarium*** sp., ***Rhizoctinia solanii***, and ***Sclerotinia sclerotiorum***), **aster yellows**, and **powdery mildew** (*Erysiphe cichoracearum*) (Horst, 1990) are reported on campanulas. When plants are grown outdoors, there is possibility of infection of leaf spot diseases. In case of aster yellows the plants are uprooted and burnt. For other diseases, regular use of thiram alternate with bavistin fortnightly will keep all the diseases under check.

References

Adams, R. 1991. Pot culture: *Campanula*. *Grower Talks*, pp.50-51. Geo J. Ball Publishing, Geneva, Illinois, USA.

Armitage, A. 1987. Tips of the trade new crops: the present and the future. *Greenhouse Grower*, **5**(6): 122.

Bailey, L.H. 1942. *The Standard Cyclopedia of Horticulture* (vol.I), pp. 642-650. The Macmillan Co., New York, USA.

Beckett, K.A. 1985. *The Concise Encyclopedia of Garden Plants*, pp. 59-61. Orbis Publishing Ltd., London.

Beckett, K.A. 1987. *The RHS Encyclopaedia of House Plants Including Greenhouse Plants*, pp. 130-131. Salem House Publishers, Massachusetts, USA.

Brickell, C. 1994. *The Royal Horticultural Society Gardener's Encyclopedia of Plants and Flowers*, pp. 453-454. Dorling Kindersley Ltd., London.

Catkins, C.C. (ed.), 1979. *Reader's Digest Success with House Plants*, p. 128. The Reader's Digest Assocn, Inc., Pleasantville, New York/Montreal.

Hay, R. and K.A. Beckett, 1971. *Reader's Digest Encyclopaedia of Garden Plants and Flowers*, pp. 109-113. The Reader's Digest Association Ltd., London.

Heide, O.M. 1965. *Campanula isophylla* somlangdagsplante (Norwegian with English summary). *Gartneryket* (Oslo), **55**: 210-212.

Herrick, T.A. and L.P. Perry, 1997. Influence of freeze acclimation procedure on survival and regrowth of container-grown *Campanula takesimana* Nakai. *HortTechn.*, **7**: 43-46.

Horst, R.K. 1990. Campanula. In: *Westcott's Plant Disease Handbook* (5th ed.), p. 577. Van Nostrand Reinhold, New York.

Dole, J., T. Cavins and T. Bosma, 2001. Success with campanulas. *Greenhouse Product News*, pp. 28-35.

Dole, M.J. and F.H. Wilkins, 1999. *Floriculture Principles and Species*, pp. 255-260. Prentice Hall, New Jersey,USA.

Everard, B. and B.D. Morley, 1970. *Wild Flowers of the World*, plate 15. Peerage Books, London.

Madson, P. and K. Madson, 1986. *Campanula* - A bright star in the future. *Bedding Plants International News*, **16**: 7.

Moe, R. and T. Fjeld, 1987. Keeping quality of potted plants as influenced by ethylene (Danish). *Gartner Tidende*, **101**: 1580-1583.

Nowak, J. and R. Rudnicki, 1990. *Postharvest Handling and Storage of Cut Flowers, Florist Greens and Potted Plants*. Timber Press, Portland, Oregon, USA.

Pizzetti, I. and H. Cocker. 1968. *Flowers - A Guide for Your Garden*, pp. 174-182. Harry N. Abrams, Inc., Publishers, New York.

Runkle, E.S., R.D. Heins, A.C. Cameron and W.H. Carlson, 1996a. Effects of cold and photoperiod on flowering of several herbaceous perennial species (abstr.). *HortSci.*, **31**: 581.

Runkle, E.S., R.D. Heins, A.C. Cameron and W.H. Calson, 1996b. Effects of night interruption duration and cyclic lighting on flowering of obligate long-day herbaceous perennial plants (abstr.). *HortSci.*, **31**: 681.

Wellensiek, S.J. 1985. *Campanula medium*. In: *Handbook of Flowering* (vol. II, ed. Halevy, A.H.), 123-126. CRC Press, Boca Raton, Florida, USA.

Whitman, C., R. Heins, A.Cameron, and W. Carlson, 1995. Production guide for *Campanula carpatica* as a flowering plant. *Professional Plant Growers Association News*, **26**(4): 2-3.

Zimmer, K. 1985a. *Campanula fragilis*. In: *Handbook of Flowering* (vol. II, ed. Halevy, A.H.), pp. 117-118. CRC Press, Boca raton, Florida, USA.

Zimmer, K. 1985b. *Campanula pyramidalis*. In: *Handbook of flowering* (vol. II, ed. Halevy, A.H.), pp. 127-130. CRC Press, Boca Raton, Florida, USA.

9

Centaurea (Family: Asteraceae)

Ajit Kumar, Khushboo Kathayat, Jyoti Bajeli, Sanyat Misra and R.L. Misra

[**Common names:** Bachelor's buttons/Blue bottle/Bluet/Cornflower/French pink/Hurtsickle/Ragged sailor (*C. cyanus/C. arvensis, C. montana*), Basket flower (*C. americana/Plectocephalus americanus*), Blackmoor's beauty/Sweet sultan (*C. moschata/C. amberboii/C. imperialis/C. odorata/C. suaveolens/Amberboa moschata*), Brown knapweed/Brown ray knapweed (*C. jacea*), Diffuse knapseed (*C. diffusa*), Dusty miller (*C. argentea/C. cineraria/C. candidissima/C. gymnocarpa*), Giant knapweed (*C. macrocephala*), Great blue bottle/Mountain cornflower/Perennial cornflower (*C. montana*), Greater knapweed (*C. scabiosa*), Hardheads/Knapweed/Black knapweed/Common knapweed/Lesser knapweed (*C. nigra*), Mountain bluet (*C. montana*), Persian cornflower/Whitewash cornflower (*C. dealbata*), Spotted knapseed (*C. maculosa*), Wig knapseed (*C. phrygia*), Yellow knapweed (*C. salonitana*), *etc.*].

Introduction and Origin

Centaurea has origin from the Greek *centaur* meaning 'half-man and half-horse'. In Greek mythology, *kentauros* (in Latin, *centaurus*) is a wild creature with the head, arms and torso of a man joined to the body of a horse at its neck. Moreover, in Greek mythology, *centaur* = 'Chiron', who is known for his great wisdom, was a tutor of the Greek heroes such as Hercules, Achilles and Jason and who taught the mankind the healing virtues of herbs. The genus comprises 600 annual, biennial, perennial, sub-shrubs and shrubs from temperate and tropical areas of the Old World, Europe, N. Africa, India and China with a few in North and South America while one in Australia. Though the flowers in all the species are similar otherwise they are highly variable in growth form and leaf shape. Floral heads are composed of many tubular florets each opening to a bell-shaped 5-lobed mouth. The lobes may be either similar in shape or outer ring producing larger lobes similar to typical cornflower-like appearance. They are excellent for bedding, borders and edgings, for pot and basket planting and for conservatory, and when flowering starts, these can be taken indoors for enchantment. Though annual types and many others can be propagated through seeds but shrubby species through cuttings.

There is a story that when Queen Louise of Prussia with her children was fleeing from Napoleon's forces, she hid her children in the field of cornflowers. In Prussia, the military uniform colour is cornflower-blue now called as Prussian blue. Cornflowers were used together with other plant leaves and berries in an incredibly ornate natural burial wreath for Pharaoh Tutankhamun. During Victorian England, cornflowers were named as 'bachelor's button' as the young women wore them to signify that they were single and while courting, the men wore them in the buttonhole on their shirts as these flowers remain fresh for quite sometime. As a weed, especially in the corn fields of Europe and the USA, since these unsharpen the farmer's sickle, hence it was named as 'hurtsickle'. It was also the favourite flower of John F. Kennedy (www.landscapeofus.com).

Botany, Cytology and Breeding

The leaves in annuals and hardy & half-hardy perennials (*Centaurea gymnocarpa* and *C. pulchra* are not hardy) are alternate, and in some species the foliage is attractive silvery as in *C. cineraria*. Its 'flower' is in fact an inflorescence, a compact head of small flowers, surrounded underneath by whorls of bracts called an 'involucre' which is ovoid or globose, stiff and sometimes prickly which have the function of a calyx. Its receptacle is bristly and the marginal florets are elongated and usually sterile. Flowerheads have a globular base from where numerous slender and tubular flowers (disc florets) arise with stamens and ovaries, and the centre is thistle-like surrounded by a ring of slender ray petals. Anthers are fused into a tube around the style and the style having a ring of the hairs or brush-like. Flowers show protandrous condition and develop centripetally within the inflorescences (Rendle, 1975). Self-pollination is prevented by the two stigmatic surfaces being held together, face to face, at the tip of the style. The fruits are achenes which are yellowish, elliptic, flattish, fine-haired, 3.5-4.0 mm long with tip of the cypsela with short bristles. Seed is oblong with a notched large scar about 1/3rd the length of the achene.

In nature, *Centaurea* is diploid, tetraploid and hexaploid. Diploid species are *Centaurea diffusa* with 2n = 2x = 18 (Moore and Frankton, 1954), *C. gabrielis-blancae* with 2n = 20 and *C. cyanus* with 2n = 24 (Goralski *et al.*, 2014); tetraploids are *C. moschata* with 2n = 28 and 32 (Dey and Sharma, 1967), *C. maculosa* with 2n = 4x = 36 (Moore and Frankton, 1954), *C. goeksunensis* with 2n = 60, *C. ornata* and *C. haenseleri* each with 2n = 40, & *C. aspera* and *C. seridis* each with 2n = 44; and the hexaploids are *C. carystea* with 2n = 54, & *C. argecillensis* and *C. saxicola* each with 2n = 60; whereas the species *C. toletana* has diploid (2n = 20), tetraploid (2n = 40) as well as hexaploid (2n = 60) forms. Pink flowers of conflower yields a new anthocyanin acylated with succinic acid which is pelargonidin 3-(6"-succinylglucoside)-5-glucoside, apart from various secondary metabolites such as anthocyanin, flavonoids and their glycosides, phenol carboxylic acid, sesquiterpenes, coumarins, *etc.*

In general, the *Cyanus* species are entomophilous so are visited by a large number of insects for pollen and/or nectar. The disc florets containing functional gynoecium and androecium are actinomorphic and gamopetalous so when an insect lands on the capitulum during anthesis, it pushes its proboscis into florets with open corollas (Roche and Susanna, 2010). In plant population of various species, autogamy occurs in the genus (Koutecky *et al.*, 2012). Goodwillie *et al.* (2005) reported that presence of pseudo-self-compatible individuals indicates that a population is in the process of transitioning to self-compatibility. Bellanger *et al.* (2014) reported presence of genetic variation which is necessary for transition to selfing in self-incompatible plants of *C. cyanus*. They recorded that majority of the plants were self-incompatible and about 12 per cent were pseudo-self-compatible while one was self-compatible. Chalesworth and Charlesworth (1979) reported pseudo-self-compatible and self-compatible plants susceptible to inbreeding which may lead to inbreeding depression. It is a fact that selfing increases homozygosity, exposing deleterious alleles to selection which results in increased inbreeding depression and so it prevents the spread of self-compatible plants. They stated that progeny originating from pseudo-self-compatible plants had the fewest capitula per plant though did not differ from others for other traits. Bellanger (2011) reported that inbreeding depression causes a critical reduction in growth and reproduction of cornflower.

Propagation

All the annual species such as cornflower, sweet sultan and their varieties and many of the perennial species are propagated through **seeds**. Seed propagation is the easiest method of propagation. Seeds easily germinate at 21-24 °C temperature within 10 days. Seeds are sown from September to November whether in the plains, in sub-tropical regions or in temperate areas, however, in temperate areas these can also be sown in March-April but the seeds sown during

September are more vigorous, therefore, require more spacing. Transplanting is done after four weeks of germination, *i.e.* when these attain 8-12 cm height with 3-4 leaves on the seedlings, in the prepared field where these start flowering from January in the sub-tropical regions of Delhi but from April in the temperate regions while those sown in March-April start flowering from July. Ebadi *et al.* (2013) while studying effect of salinity levels on seed germination and further growth of the three cornflower varieties, recorded even 50 mM level injurious. Seeds of perennial species, except *C. gymnocarpa* are sown in temperate regions during March-April only in seed compost. Seedlings of *C. dealbata, C. hypoleuca* and *C. pulchra* are potted singly in soil mixture outdoors, in October these are taken to the flowering site and the pot-grown species are overwintered in a cold frame and then again planted out in the following April. In case of *C. gymnocarpa*, 7.5-10.0 cm long lateral **cuttings** are taken out during August-September, inserted in a mixture containing equal parts (by volume) of peat and sand and kept in a propagating chamber around 15 °C and then after rooting are potted singly in 7.5-8.0 cm pots containing potting compost and then overwintered at 5-7 °C and then finally potted on in 10.0-12.5 cm pots in March and hardened off in a cold frame before planting out in late May in the temperate regions.

Classification, Species and Varieties

There is no any special class under *Cyanus*, however, these are of two types, *viz.* annuals and perennials. Though many of its species are found as weeds in the field but there are certain species whose flowers are very attractive and therefore are used as cut flowers. Such species are being described here under.

Centaurea americana (*Plectocephalus americanus*) from USA to Mexico is almost smooth and hardy annual with stout and simple or sometimes little branched stems, growing up to 1.5 m tall. Leaves are usually entire, oblong-lanceolate and mucronate. Involucres are 1.25-3.75 cm in diameter and all its bracts are fringed with scarious appendages. Flowers are rose to purplish, disc diameter 2.5-7.5 cm, ray floret lobes are narrow and often 2.5 cm long. It is a very attractive species and excellent for cut flower production.

C. cineraria (*C. argentea, C. candidissima, C. gymnocarpa*) a perennial from Italy and Sicily (SE Europe) is a 90 cm tall sub-shrub grown for its grey-white-green and felty foliage on both the surfaces. Leaves are some 30 cm long, fern-like, arching and pinnati- to bipinnatisect with obtuse and linear-lanceolate lobes, lower ones petioled. Annually propagated crops rarely flower though flowers are rose-purple with clustered heads and head some 2-3 cm across. Involucre is ovate and bract-scales appressed with membranous black margin, long ciliate and bristled. It is most suitable for bedding.

C. cyanus (*Cyanus arvensis*) from whole Europe, especially Great Britain is an annual growing up to 90 cm in height with a spread of about 35 cm. It is mainly distributed in Europe, the Caucasus, WE Siberia, the Far East, Central Asia, western part of Asia Minor, Iran, NW India, North America, Australia and North Africa. It is commercially grown in Canada for cut flower industry. It was a popular plant in Tudor gardens and its blue petals yield a pigment once used by artists. It is the national flower of Estonia and Germany. It is an erect and branched annual with grey-green, alternate and narrowly lanceolate leaves some 1-4 cm long where basal leaves are stalked and sparingly toothed and sometimes pinnately lobed, and the stem leaves stalkless, lanceolate-linear with entire margins, and covered with white cobwebby which gives these dull grey appearance (Chiru *et al.*, 2013). It produces the branched sprays of flowers some 2.5-5.0 cm across from January to April in the plains of India and from June to September in the temperate regions with a ring of a few large and spreading ray florets surrounding a central cluster of disc florets. Ray florets are tubular and obliquely funnel-shaped with lobed tips, stamens 5, a pistil of two fused carpels and the lime green involucral bracts are overlapping in many rows, barrel-shaped, triangular, shallowly bilobed and the tips blackish-brown. Its varieties have flower shades in deep maroon (almost black), ligh pink, lavender and rose apart from colours already present in the type species, *i.e.* pink, red, purple, blue or white and in single and double forms. The flowers which have sweet to spicy (often clove-like) fragrance can be eaten raw or cooked, its shoots are also edible, and is also used as one of the ingredients in tea. Its honey which is reported to have antibacterial activity, and cornflower oil are prepared from this. It produces up to 95 kg/ha of pollen and 450 kg/ha of nectar (Ekbom, 2000), *vis-a-vis* extrafloral sepal nectarines before and after flowering (Geneau *et al.*, 2012). It contains tannin and potassium salts. It is effective in washing out wounds, its infusion for treatment of mouth ulcers, bleeding gums, dropsy and constipation. It also helps in treating eye infections as its petal-distilled water is effective against conjunctivitis, weak eyes and for making eye products, *vis-à-vis* skin care beauty products. Blue dye extracted from the flowers is used for colouring sugar and confectioneries. It helps in bringing resistance in the body, regulation of menstrual disorders, is antipyretic, astringent, mildly purgative, weakly diuretic, antitussive, tonic, and is used against indigestion, regulation of kidney, gall bladder and liver. Its seeds contain epoxylignans and berchemol. There are many garden varieties which are divided into two types as per their heights, *i.e.* tall and small. Tall varieties grow from 60-90 cm in height, include the **Ball** strain and are available in mixture and in separate colours. Its var. 'Blue Diadem' bears extra large flowers. Dwarf ones grow up to 30 cm in height, are quite compact and best for pot-growing and bedding, and its 'Polka Dot strain' is available in a range of colours. Other outstanding varieties are 'Dwarf Rose Gem' (rose-red) and 'Jubilee Gem' (dwarf, blue).

C. dealbata is a perennial plant from Asia Minor, Caucasus Mountains and Iran which grows sub-erect some 60 cm high. Its leaves are pinnate with deep lobes, glabrous grey-green above and white-silvery beneath, lower ones petioled, 30-45 cm long with obovate lobes auricled at the base or coarsely cut-toothed, and stem leaves sessile with oblanceolate lobes. Flowerhead solitary and just above the uppermost leaf, ray florets red and those of the disc rosy or white, outer scales of the involucre with lanceolate tips and the middle rounded, deeply fringed and ciliate. Its varieties are 'John Coutts' (glowing rose-pink with pale-yellow centre), 'Sternbergii' (deep pink), *etc.*

*C. **diffusa*** from Eurasia is a biennial to triennial pubescent plant growing up to 80 cm tall with heavily branched erect stems, bearing alternate leaves where basal ones are whorled, 20 cm long and 5 cm wide and much divided along with the lower stem leaves though upper stem leaves are bract-like and entire. Flowerheads some 1.4-1.6 cm high and 3 mm in diameter, numerous and radiating. Bracts are pale-yellowish-green and the upper part is narrowed into a stiff spine. Flowers are usually white, sometimes pink to purple, light brown to black achenes 2-3 mm long and pappus absent or more fringed and some 1 mm long. It is naturally self-compatible.

*C. **hypoleuca*** from Asia Minor, Armenia and Caucasia is a compact and vigorous perennial where leaves are grey-green and deeply dissected with broad lobes. It grows up to 45 cm in height, bears pink flowers some 5 cm across which appear in succession from May to August in the temperate regions. Its dwarf form *C. h. simplicicaulis* grows up to 25 cm tall bearing lilac-pink flowers.

*C. **isaurica*** is a perennial but endangered species found growing in the stony and rocky slopes, being endemic to Middle Taurus mountains in Turkey, having erect stem some 40-50 cm tall with few capitula. Leaves are grey-tomentose on both the sides, lanceolate and adpressed, with lower stem leaves being lyrate. Its flowers are yellow.

*C. **jacea*** from dry meadows and open woodlands of Europe is a perennial herbaceous species, growing from 10-80 cm tall. In this species the flowers are hermaphrodite hence there is self-fertilization and the pollination is through bees and flies.

*C. **macrocephala*** from Armenia and Caucasus is a robust clump-forming perennial species with stout stem which grows 0.9-1.5 m tall, bearing light green, ovate-lanceolate, scabrous, slightly decurrent, acute, somewhat serrate and rough leaves, leafy and subglobose flowerheads quite large (larger than an hen's egg) some 7.5-10.0 cm across with yellow flowers appearing from the papery silvery-brown bracts. Involucre comprises of some 8-12 rows of appressed, rusty, scarious-margined and fringed scales. Its flowers when cut are long-lasting and is most suitable for planting as border plant. It is usually propagated through seeds.

*C. **maculosa*** from Europe is a short-lived perennial growing up to 1 m tall, producing stout taproot and the seedlings usually remain in the rosette stage during winter but start growing in early April. Its leaves are pale-green, alternate, 2.5-7.5 cm long, rosette leaves deeply divided into lobes but stem leaves have fewer lobes and become smaller towards the tips of the branches. The plant has a shorter centre spine on the bracts and usually a black spot near the top. In temperate regions it bolts during May and flowers from July onward, individual flowerheads which appear at the top of main stem and upper branches bloom for 2-6 days in pink to light purple or sometimes in white shade before the bracts close but about 20 days later the bracts reopen and seeds are dispersed. It is entomophilous and self-compatible in nature.

*C. **montana*** from central and southern European meadows, especially Pyrenees, Alps and Carpathians is a perennial herb growing up to 80 cm in height and is mainly propagated through seeds and its rhizomes. It usually has lax growth with unbranched, winged and sparsely white-hairy stem and the leaves are mid- to deep green, stalkless, alternate, decurrent, white-hairy and obliquely oblong-lanceolate. Blue flowers some 6.25-8.0 cm across are produced in profusion during summer. Single flower-like capitula are surrounded by involucral bracts. Involucre is lime-green, oval-round, triangular-ovate and pinnate or ragged-toothed bracts overlap in many rows with dark brown tips. Ray florets are obliquely funnel-shaped and dark blue where tip is lobed, disc floret is tubular and purple, stamens are 5 and the gynoecium is composed of two fused carpels. Several capitula individually terminating each stem. Fruits are flattish cypsela with quite short bristles at tips. Its 'Gold Bullion' is a golden-leaved. Varieties 'Alba' (white), 'Rosea' (pink) and 'Violetta' (deep blue-purple) are highly promising.

*C. **moschata*** (*C. amberboii, C. odorata, C. suaveolens, Amberboa moschata*) from SW Asia and Eastern Mediterranean region is a fast and upright-growing and slender-stemmed annual to biennial growing erect up to 60 cm tall and with 20 cm spread with basal branching thin stems, bearing greyish-green, narrowly lanceolate, pinnatifid leaves with dentate lobes. Entire plant is smooth and bright green. It is commonly known as sweet sultan as the flowers are fragrant and under the rule of Ottoman Sultan in 17th century it was introduced from Near East. It is tolerant to drought. Highly fragrant large flowers in the shades of red, mauve, purple, reddish-violet, pink, yellow or white appear in long-petioled solitary heads some 4.0-7.5 cm across. Involucres are round or ovate and smooth but innermost of the scales are with scarious margins. *C. imperialis* Hort. (*C. moschata* × *C. margaritae* Hort.) is a tall growing hybrid with quite large flowers in shades of white, rose, lilac and purple which is highly fragrant and as cut flower last for up to 10 days. Its other important varieties are 'Alba syn. *C. margaritae*' (highly fragrant and white), 'Bride' (white), 'Flavo' (yellow), 'Rosea' (pink), 'Rubra' (red with long-stalked heads), *etc.*

*C. **nigra*** from Europe is a perennial growing up to 1 m in height with branched, rough and pubescent stems, having lanceolate and entire leaves, lower ones stalked, sparingly toothesd or lobed but not pinnatifid, upper ones stalkless, some 25 cm long and usually deeply lobed and hairy. Inflorescence with a few flowerheads, each in the form of hemisphere and with bristly, black or brown phyllaries, each head with many small bright purple flowers of similar shape and size whether marginal or disc and these appear in late summer. Involucral bracts are with pectinate-ciliate-fringed black appendages. Fruit is a tan and hairy achene 2-3 mm long, sometimes with a tiny dark pappus. Its variety 'Variegata' is tufted and a very striking border plant in dry and open places, bearing leaves edged with creamy-white.

C. phrygia from Europe is a hardy perennial species with pink flowers, flowering during mid-summer in temperate areas though from January to April in the sub-tropical regions.

C. pulcherrima (*Aethiopappus pulcherrimus*) from Caucasus is an upright-growing perennial (75 cm plant height and some 60 cm plant spread) with deeply cut silvery leaves. Solitary flowerheads some 6.25 cm across appear during summer on slender stems, bearing rose-pink flowers with thistle-like paler centres and star-shaped ray petals.

C. pulchra from Kashmir is a half-hardy perennial species growing up to 60 cm tall with about 30 cm spread. The leaves are smooth, silvery and pinnate and flowers rose-pink some 5 cm across which appear in late spring in the sub-tropical regions while in August in temperate regions. *C. p. major* is more widely grown than the type species.

C. ruthenica from East Europe and Siberia is a smooth-stemmed and branching perennial species growing erect up to 1.2 m in height with about 50 cm spread. Its deeply cut leaves with linear-toothed lobes sharply narrowing at both the ends, are glossy dark green. Flowerheads are usually solitary, rays sulphur-yellow, some 2.0 cm long and thistle-like, heads some 3.75 cm across, and appear on slender stems in February-March in sub-tropical regions while during July-August in temperate regions.

C. salonitana from Balkans and Aegean Islands has spine-tipped involucral bracts which vary in length of the spine.

C. scabiosa from dry grasslands, hedgerows and cliffs in lime-rich soil of Europe is a perennial species with upright growing branched stems which terminate into a single thistle-like flowerhead which bears purple flowers. Flowerheads have an outer ring of extended purple-pink ragged bracts. The leaves are deeply dissected and at the base these form a clump.

Centaurea cyanus, C. americana, C. montana, C. moschata and *C. macrocephala* are most desired commercial species for their deep blue to purple and pink to rose flowers. However, the pure white flowers of *C. moschata* are also admired most by the people. There are many single or double flowering cultivars, *e.g.* 'Black Boy', 'Black Gem', 'Black Magic' (dark purple), 'Blue Ball', 'Blue Boy' (double, blue), 'Blue Carpet', 'Blue Diadem', 'Blue Midget', 'Double Blue', 'Dwarf Blue Midget' (bright blue), 'Emperor William' (single, blue), 'Florence Mixed', 'Frosted Queen', 'Frosted Queen Mixed', 'Frosty Mixed', 'Garnet' (dark burgundy), 'Jubilee Gem', 'Little Boy', 'Little Lady', 'Midnight', 'Mystic Blue', 'Persian', 'Pinky', 'Polar Bear', 'Polka Dot' (various colours), 'Red Boy', 'Snowman' (double, pure white), 'Southern Garden', 'Tall Mixed', *etc.*

Cultural Practices

By and large, *Centaurea* species require long **photoperiods** for floral induction. Keeping seedlings under SD will increase basal branching, *vis-à-vis* number of stems. Once floral induction initiates, LD treatment accelerates floral development (Cox, 1985) and it takes three weeks for completion of induction. During winter in Israel, Kadman-Zahavi and Yahel (1985) found that far-red light treatments or GA_3 sprays were effective only for shoot elongation but not for floral induction in *C. cyanus*. Use of 55 per cent shade cloth in case of *C. americana* reduced flower number less than half though quality of flower was found excellent when compared to full sun exposure but here quality can not compensate such a poor yield (Armitage, 1991). In fact, no cold treatment is required in case of annual species, however, yet no satisfactory work is available on perennial species. Annual species such as *C. cyanus* has been recommended doing well at day/night **temperature** of 10-13 °C (Nau, 1993) though Cox (1987) recommends 21/18 °C day/night temperatures. Nau (1993) for *C. montana* recommended 13-16 °C optimum temperature range.

Cyanus species do well in various **soils**, but the performance is best when it is well-drained clay- to sandy-loam soil having sufficient moisture and a pH of 6.6-7.8, *i.e.* neutral to slightly alkaline. There are species and varieties which even endure arid and low fertility soils. Cho *et al.* (2012) studied cornflower in two seasons in different soils and recorded clay-loam soil giving best results, and the spring seedlings outperforming the autumn ones. For raising seedlings, the **nursery** can be raised in any container, preferably trays or pans and at small scale even in earthen pots. In large scale, the nursery can be raised directly in the 10-15 cm raised nursery beds having proper drainage and aeration so that seedlings are healthy. For preparing nursery beds, the garden soil should be mixed with coarse sand and well rotted farmyard manure or compost, all by volume in equal proportion. In the hills nursery beds should be prepared during August or March but in tropical and sub-tropical regions only during August and after the beds are ready the soil should be treated with thiram. Seeds should be sown thinly in the rows some 5-6 cm apart by drawing the furrows through fingers, and then covered thinly with the same soil some 5 mm deep and then covered with straw over which watering through fine nozzle should be given daily by evening so that soil remains moist. Daily it should be checked if seeds have started germinating and if so the straw should immediately be removed otherwise the seedlings will become crooked and lanky. Seedlings become ready for transplanting with four weeks of seed germination. However, there is an alternative to seedling raising in the nursery, and this way the seeds can be **sown directly** in the permanent field. After germination, the seedlings are thinned out to maintain proper distance between rows and the plants. In this method there is no risk of root damage but requirement of seeds become manifold.

For **transplanting** the field should be prepared properly by mixing some 200 quintals of organic manure with the soil. The field should be cultivated thrice to a depth of at least 20 cm, followed by planking each time and then beds of convenient sizes should be prepared after levelling after taking out all the rootstocks of weeds and other foreign material. At the third ploughing there should be sufficient moisture in the soil. There should be ample space left for irrigation channels and for bunds. The width of the beds is normally kept 1.6 m though length some 4.5 m. Planting is normally done 25 cm apart in every case, in the month of September and since September-planted crop has vigorous growth so here spacing should be 30 cm apart but April-planted crop as in the temperate regions, the distance should be only 25 cm apart. After transplanting, immediately it should be followed by irrigation to settle the roots.

When plants have attained some 15 cm of length, the terminal growing point of the plant should be **pinched out** to induce more lateral shoots. Pinching may be repeated every fortnight when the plant is in vegetative phase so that plants become compact with so many secondary and tertiary shoot growths. **Disbudding** of the just emerging buds may also be carried out to harvest larger flowers with long stems. In case of the taller plants, the plants should be given support of **stakes** so that wind or lodging may not break or uproot them. Support may also be given through strings stretched lengthwise by putting tall and strong pegs at all the four corners of the beds. However, plants can also be dwarfed through use of **growth regulators** such as Dazide or B-Nine at 2,500-5,000 ppm and Abide or A- Rest at 7-26 ppm which controlled growth in cornflower (Whipker, 2013). **Weeds** compete with the main crop for nutrients, water, light and space, *vis-à-vis* spread diseases and harbour insect-pests. Therefore these should be removed continuously. Faded flowers are immediately **deadheaded** so that energy is saved for producing subsequent succession of flowers.

Its **water requirement** is normal. This benefits from regular watering though over-watering is dangerous, however, to some extent these can tolerate drought or dry conditions. Water-logged soils should also be avoided. During winter, these require irrigation once in 10-12 days but in summer season at every 5-7 days. Water requirement also depends on the soil conditions. If it is sandy loam soil, water requirement is so frequent but in clayey soils it is quite shallow and at long intervals. If the soil has been fortified with FYM or any other organic manure before planting, further there is no requirement of any **nutrient** but if the crop is showing sign of any deficiency symptom, the required nutrient is immediately applied. Application of N and K at 10-15 g/m^2 improves growth and flowering (Arora, 1993). Cox (1985) advocated 200 ppm N (20N-10P-20K) in greenhouse production of *C. montana* and Armitage (1993) stated that in no case more than 200 ppm N should be applied, however, it would be better if only 100-150 ppm N from a complete fertilizer is applied. Karoly *et al.* (1999) reported higher concentrations of macro-nutrient uptake at vegetative phase though micro-nutrient at reproductive phase in cornflower, and they further revealed that Mg, Fe, Mn, Zn and Cu contents of leaves were high at anthesis.

Postharvest

In cornflower, flowering starts after 90-100 days of transplanting and they are harvested before they are fully open when the centres are still curving inward, *i.e.* when terminal flower is 3/4th to fully open. Cornflower yields 10-15 flowers per plant (Anderson and Brown, 2015). Harvesting at fully open stage is not preferred as such flowers are subject to damage during transport and moreover their life is also short. The flowers should be harvested during cool hours preferably by the evening when there is no dew on the plants as this will entice *Botrytis cinerea* infection during transit. After harvesting, the cut ends should be just immersed in a bucket of water and taken to cold storage for removing field heat and where according to the required standard these should be graded and bunched. They are packed in bunches of 6-10 stems. The flowers are cold-stored at 4.5 °C and at 90-95 per cent relative humidity (McGregor, 1989). Before using, these are hydrated in a warm water solution along with some floral food at least for two hours before use.

Generally the flowers last for 6-10 days under ambient storage conditions provided a little portion of their cut ends are removed every day or two, by changing the water daily and by removing all the leaves which become submerged in the water otherwise bacterial decay will occur and block the vascular system. *Centaurea* has hollow stems so sometimes these require to be strengthened by inserting a fine wire up the hollow stem and into the flower base. *Cyanus* flowers **dry** well and retain their colour well after drying so for this purpose 3-4 stems are bundled with rubber band and hanged upside down at a warm, well ventilated and dry place for about a week. *Centaurea* is also used for **value addition** apart from vase arrangements such as hand-tied bouquets, corsages, buttonnieres and hair adornments. Various value added products are obtained from the flowers such as blue dye, petal-distilled water, cornflower honey and cornflower oil.

Insect-Pests and Diseases

Insect-pests are generally not so problematic. However, aphids (*Brachycaudus helichrysi*), leaf hoppers (*Macrosteles fascifrons*), leaf rollers (*Plusia* species), stalk borers (*Papaipema nebris*) and spider mites (*Tetranychus telarius*) may sometimes pose a problem. Aphid population increases at an alarming stage and the nymphs and adults cluster around the tender parts where these suck the plant sap, excrete honeydew which attracts ants and produces sooty mould. These should be controlled through jet sprays or through spraying of tobacco decoction. Leaf hoppers (nymphs and adults both) also damage the tender parts and their feeding points become discoloured. These also spread viruses. Leaf roller Lepidopterous larvae roll the leaves and keep on feeding inside during daytime but in nights these come out to feed on other leaves. Stalk borer larvae while feeding on the tips bore the stalk and damage whole plant. Nymphs and adults of spider mites stick to the plant stems and branches and continue sucking the cell sap. These build up their population alarmingly very rapidly. These all the pests can be controlled by using contact insecticide. Methyl parathion spraying is effective in controlling all the insects infesting on this crop.

Diseases which infect this crop are rusts (*Puccinia cyani, P. irrequiseta*), downy mildew (*Bremia lactucae, Plasmopara halstedii*), powdery mildew (*Erysiphe cichoracearum*), wilt (*Verticillium albo-atrum*), stem rots (*Sclerotinia sclerotiorum, Phytophthora cactorum, Fusarium oxysporum*), grey mould or *Botrytis* blight (*Botrytis cinerea*), southern blight (*Sclerotium rolfsii*), root rot (*Phymatotrichum omnivorum, Pythium* spp., *Rhizoctonia solani*), and aster yellows virus which spreads through leaf hoppers (Post, 1949; Horst, 1990; Armitage, 1993).

Attack of **rust** makes covering of brown pustules on the stems and leaves which causes spotting on the plant, defoliation and reduced growth of the plant. This disease is favoured by warm and wet conditions. Overhead irrigation should be avoided, there should be proper spacing among the plants, field should be cleaned of the debris, infected part or plant should be removed or uprooted and destroyed and disease-free planting material should be used. It can be controlled through use of wettable sulphur on infected plants. **Downy mildew** infection produces irregular pale-greenish or reddish spots on upper leaf surface while soft mouldy growth on lower surface, especially on young plants which ultimately causes collapse of the affected parts. Affected parts should be removed and badly affected plants should be uprooted and destroyed. Field should be kept clean and proper spacing should be maintained for good aeration. Application of Bordeaux mixture or spraying of the crop with Blitox (3g/l of 50 per cent copper oxychloride) controls this problem when infection is initiating. **Powdery mildew** is a serious problem especially when days are warm and nights are cool. In its infection, young growing shoots and leaves are covered with white powdery growth which ultimately causes the infected leaves to become purplish and falling off and the floral buds failing to open. The spread of the disease is favoured by dampness or high humidity in the atmosphere, poor air circulation and high plant density. At the time of pinching all such leaves should be removed, field should be kept clean, overhead irrigation should be avoided and the plants should be thinned out to facilitate proper aeration. Infected plants should also be dusted with 80 per cent sulphur or sprayed with 0.1 per cent Karathane at fortnightly intervals. **Wilt** caused by *Verticillium albo-atrum* is a very serious disease where plants show wilt symptoms. In its infection, 0.2 per cent Thiram drenching will prove quite effective. **Stem rots** are also a type of wilt disease, usually caused by *Sclerotinia sclerotiorum, Phytophthora cactorum* and *Fusarium oxysporum* and the symtoms are also quite similar to wilt. Infected seedlings wilt suddenly though aged plants gradually turn yellow, starting from the lower leaves, followed by wilting. In case of *Fusarium* infection the plants show one-sided development as to the other side black streaks on the stems are formed due to formation of dark brown vascular ring just below the bark and the stem base is often blackened with masses of pink-coloured *Fusarium* spores. This stunts the plants, roots rot, vascular system is blocked and ultimately the plants die. In such cases only resistant cultivars should be planted, crop roation should be followed, overwatering and deep planting should be avoided, nitrogenous or phosphatic fertilizers should be given only when soil is quite exhausted and in no case with any type of tillage operation the plants should be injured. Starting from nursery stage to flowering the plants should be sprayed fortnightly with 0.2 per cent Captafol alternate with 0.1 per cent Bavistin. **Root rot** also occurs due to infection of *Phymatotrichum omnivorum, Pythium* spp., *Rhizoctonia solani*. The measures applied in case of wilt and stem rot will control this problem also. ***Botrytis* blight** is a highly problematic disease whose infection causes, grey, off-white or grey-brown fuzzy fungal growth on infected areas. It lives on most living and dead plant material so its spores are always present in the air. It enters the plants through weakened or wounded parts and affects every above-ground part of the plant making a complete loss and its spread is favoured by humid, windy and chilly weathers as its spores readily travel through rain or water splash. The fungus may carry over from year to year as sclerotia in the soil or infected plant debris. All the injured and dead plant parts should be removed before spread of the disease, field should be cleaned of all plant debris, overhead watering and field dampness should be avoided and carbendazim should be sprayed immediately and weekly. **Southern blight** (*Sclerotium rolfsii*) infection makes the entire plant including the stem discoloured as large black sclerotia (fungal resting bodies) are embedded in the dense, white and fluffy (cotton wool-like) fungal growth. Its infection and spread is favoured by cool and damp conditions. Sclerotia fall in the soil where these remain dormant until spring and then produce apothecia (spore-producing cup-shaped fungal growths) to infect the next crop. Infected parts and plant debris should be promptly removed and burnt and a 4-year crop rotation should be followed. **Aster yellows** spread due to infestation by leaf hoppers so prompt uprooting and destroying such plants and controlling the leaf hoppers will control this disease also. In its infection plants become straw-yellow and distorted in various forms.

References

Anderson, R.G. and J. Brown, 2015. Greenhouse production of Bachelor Buttton. Retrieved Feb. 20, 2015 from http://www.uky.edu/Ag/CDBREC/bachbut.htm.

Armitage, A.M. 1991. Shade affects yield and stem length of field-grown cut-flower species. *HortSci.*, **26**: 1174-1176.

Armitage, A.M. 1993.*Centaurea*. In: *Specialty Cut Flowers*, pp. 67-72, 183-195. Varsity Press/Timber Press, Portland, Oregon.

Arora, J.S. 1993. Corn flower. In: *Commercial Flowers* (eds Bose, T.K. and L.P. Yadav), p. 859. Naya Prokash, Calcutta.

Bellanger, S. 2011. Etude de la biologie d'une messicole en regression: le bluet (*Centaurea cyanus* L.). Dissertation, University of Burgundy.

Bellanger, S., J.P. Guillemin and H. Darmency, 2014. Pseudo-self-compatibility in *Centaurea cyanus* L. *Flora*, **209**: 325-331.

Chalesworth, D. and B. Charlesworth, 1979. The evolution and breakdown of S-allele systems. *Heredity*, **43**: 41-55.

Chiru, T., C. Tatiana and A. Nistreanu, 2013. Morphological and anatomical studies of *Cyani herba*. *Modern Phytomorph.*, **4**: 65-68.

Cho, H., K.Y. Seong, T.S. Park, M.C. Seo, W.T. Jeon, H.W. Kang and H.J. Lee, 2012. Yield of green manure and nitrogen of cornflower (*Centaurea cyanus* L.) in different upland soil textures. *Korean J. Soil Sci. Fert.*, **45**(4): 664-670.

Cox, D.A. 1985. Production of perennial bachelor's button (*Centaurea montana*) in containers. In: *Processing Alabama Greenhouse Nursery Seminar*, pp. 98-99, held at Auburn, Alabama on June 14-15, 1985.

Cox, D.A. 1987. Gibberellic acid induced flowering of containerized *Centaurea montana* L. *Acta Hort.*, No. 205, pp. 233-236.

Dey. D. and A.K. Sharma, 1967. Chromosome studies in the genus *Centaurea*. *Fol. Biol. Polsk. Akad. Nauk.*, **15**: 191-207.

Ebadi, M.T., G.H. Homayouni and A. Farzaneh, 2013. Do different cultivars of cornflower (*Centaurea cyanus*) have different responses to salinity stress? *Int. J. Agron and Pl. Prod.*, **4**(8): 1898-1902.

Ekbom, B. 2000. *Interchanges of Insects Between Agriculture and Surrounding Landscape.*, 134 pp. Kluwer academic Publishers, Dordrecht, Holland.

Geneau, C.E., F.L. Wackers, H. Luka, C. Daniel and O. Balmer, 2012. Selective flowers to enhance biological control of cabbage pests by parasitoids. *Basic and Appl. Ecol.*, **13**: 85-93.

Goodwillie, C., S. Kalisz and C.G.Eckert, 2005. The evolutionary enigma of mixed mating systems in plants: occurrence, theoretical explanations, and empirical evidence. *Annual Rev. Ecol. Evol. Syst.*, **36**: 47-79.

Goralski, G., A. Judasz, P. Gacek, A. Grabowska-Joachimiak and A.J.Joachimiak, 2014. Polyploidy, alien species and invasiveness in Polish angiosperms. *Plant Syst. Evol.*, **300**: 225-238.

Horst, R.K. 1990. *Centaurea*. In: *Westcott's Plant Disease Handbook* (5th ed.), p. 587. Van Nostrand Reinhold, New York.

Kadman-Zahavi, A. and H. Yahel, 1985. *Centaurea cyanus*. In: *Handbook of Flowering*, vol. II (ed. Halevy, A.H.). CRC Press, Boca Raton, Florida.

Karoly, H., G. Kazinazi and D. Lukaes, 1999. Data on the biology of cornflower (*Centaurea cyanus* L.). II. Nutrient uptake: allelopathy and seed production. *Novenyvedalam*, **35**(9): 425-430.

Koutecky, P., J. Stepanek and T. Badurova, 2012. Differentiation between diploid and tetraploid *Centaurea phrygia*: mating barriers, morphology and geographic distribution. *Preslia*, **84**: 1-32.

McGregor, B.M. 1989. *Tropical Products Transport Handbook*. USDA Office of Transportation, Agricultural Handbook p. 668.

Moore, R.J. and C. Frankton, 1954. Cytotaxonomy of three species of *Centaurea* adventives in Canada. *Canadian J. Bot.*, **32**: 182-186.

Nau, J. 1993. *Centaurea*. In: *Ball Culture Guide: The Encyclopedia of Seed Germination* (2nd ed.), pp. 70, 88. Ball Publishing, Batavia, Illinois.

Post, K. 1949. *Centaurea*. In: *Florist Crop Production and Marketing*, pp. 378-381. Orange Judd Publishing, New York.

Rendle, A.B. 1975. *The Classification of Flowering Plants*. II. Cambridge University Press, Cambridge, U.K.

Roche, C.T. and A. Susanna, 2010. New habitat, new menaces: *Centaurea × kleinii* (*C. moncktonii × C. solstitalis*) a new hybrid species between 2 alien weeds. *Collectanea Botanica*, **29**: 17-23.

Whipker, B.E. 2013. Plant growth regulator guide. *Grower Talks*, p. 17.

10

Cut Foliage/Cut Greens and Other Fillers

D.V.S. Raju, Sanyat Misra and R.L. Misra

There are not only the cut flowers of various flower crop species which are used for enchantment but various other foliage materials are also used together with to fill in the gaps, as a background for various cut flowers, for making bouquets or in arrangements, for providing mass effects and over all to enhance the beauty of the arranged true cut flowers, therefore these are known as **cut foliage, cut greens** and **other fillers**. Not only for these purposes, but these are also used along with the cut flowers as a welcome gesture to the guests, along with those used for button holes, for making doles and various other purposes. The main characteristic of such foliage plants is that these should match the company of the cut flowers used and should endure their longevity similar to those of companion cut flowers. Though, in number, these are many but only a few very popular ones are being described here as defined by various authorities (Bailey, 1960; Hay and Beckett, 1971; Beckett, 1983, 1987; Nowak and Rudnicki, 1990; Salunkhe *et al.,* 1990).

Abies (Pinaceae): Commonly it is known as 'fir'and has some 50 evergreen tree species from northern hemisphere so can be cultivated only in temperate areas. These have central erect trunks and whorls of branches in tiers, forming narrowly conical or columnar trees. The leaves are lanceolate or oblanceolate, entire, sessile, shining dark green, needle-like, small and blunt, and the stalk of each one forms a small disc of attachement and this structure distinguishes it from spruce (*Picea*). The leaves persist for many years and are flattened on young plants and lower sterile branches, whereas on upper fertile branches these are more or less erect, crowded and often incurved or falcate, thickened or quadrangular and obtuse or acute. The flower spikes are known as strobili, the males catkin-like and the females like soft miniature cones. The cones in the ovoid or cigar-shape stand erect on the branches and fall down when ripe. They require moist but not waterlogged soil which is slightly acidic to neutral and semi-shady to sunny situations. They are propagated through seeds. *A. balsamea* and *A. procera* are most important with respect to use of their leafy twigs in the flower arrangement. They are stored at <5 °C, and their twiggy-needles last for 2-3 weeks in vases.

Acalypha (Euphorbiaceae): It is one of the largest genus in the family having approximately 450 shrubby species. They commonly grow in tropical and subtropical regions as they have originated so. The leaves are alternately arranged, stipulated and toothed. *A. wilkesiana* (copper leaf and Jacob's coat) is grown for the coloured foliage and has many cultivars. It grows well in full sun. The plant branches less in partial shade. They grow in a wide variety of soils and are drought tolerant. Some of the common cultivars are 'Godseffiana' (leaves green with creamy-white margins); 'Macafeeana' (leaves red, marked crimson and bronze); 'Macrophylla' (leaves brown); 'Marginata' (leaves margined with crimson or differently coloured); 'Miltoniana' (leaves oblong, somewhat drooping with irregularly cut, white margins); 'Musaica' (leaves green with orange and red markings); 'Obovata' (leaves obovate, bronzy-green with rosy-pink margins); *etc.* Their stemmed leaves are used as filler in floral arrangements. They can be wet-stored at 2-5 °C for up to two days. Their vase life is up to one week.

Aglaonema (Araceae): A native to tropical and sub-tropical SE Asia, it is a shade-loving herbaceous tropical genus with about 50 evergreen species, suitable as cut foliage in floral arrangement. They thrive in warm, moist and shaded conditions. The vase life of its foliage ranges from 12-15 days. These are stored and transported at 12-13 °C temperature. ***A.commutatum pseudobracteatum*** leaves are narrower and longer with green and yellow mottling with distinct white veins. ***A. oblongifolium*** grows up to more than 1 m in height producing 23 cm long dark green leaves with silvery markings and green-white spathes some 5-10 cm long. *A.o. curtisii* is the form with largest leaves.

Alocasia (Araceae): An evergreen perennial genus with 70 species from tropical Asia valued for its elephant ear-like leaves. *Alocasia* derives from the Greek *a* for 'without' and *Colocasia*, a closely related genus from which the species were separated to form *Alocasia*. This has thick and fleshy rhizomes which are erect and stem-like or have tuberous rootstocks. Its usually long-stalked leaves are peltate, sagittate to cordate and occasionally deeply lobed. Undecorative flowers appear in old specimens which are petalless and are borne on a spadix within an arum-like spathe. These are clump-forming plants most suitable for pot planting indoors or in conservatories. Its propagation is through suckers or rhizome cuttings during spring. Various cultivated species are *A.* × *amazonica* (*A. sanderiana* × *A. lowii*) bearing highly attractive very dark glossy green sagittate-peltate leaves some 30-40 cm long with a wavy margin. The leaf margin is narrow-white while mid and axillary veins are wide white which contrast with the leaf colour. *A.* × *argyraea* (*A. longiloba* × *A. pucciana*) bears peltate leaves which are leathery and ovate with cordate base, 30-50 cm long and with a silvery patina. *A.* × *chantrierana* (*A.* × *chantrieri*) is a tuberous plant bearing cordate-oval, 30 cm or more long leaves where margins are undulating, the basal lobes are smaller, major veins are curved, midrib and lobe-ribs white and the side veins grayish and in between the veins the leaves are quilted. *A.* × *chelsonii* bears metallic green with purple beneath, oval and pointed leaves with no lobes, length up to 37.5 cm, width up to 17.5 cm, midribs greenish-white and side veins silvery. *A. cucullata* from Bengal and Myanmar, bearing glossy rich green leaves with conspicuous veins. Leaves are cordate, pointed, up to 30 cm long and the margins are slightly upturned. *A. cuprea* from Borneo bears leaves dark metallic green above, purplish beneath, peltate, ovate-cordate and about 30 cm long with depressed veins. *A. guttata* var. *imperialis* produces purple-spotted leaf stalks,and broadly cordate and deeply lobed 25 cm long convex leaves. Leaves are thick, leathery, blue-gray but blue- black near the veins. Midrib and main side veins deeply imbedded. Reverse light to dark purple. *A. korthalsii* from Malaya, bears leaves with deep olive-green at upper surface with grey flushing, purple lower, ovate-cordate and about 60 cm long. *A. longiloba* from Borneo, Java and Malaysia, bearing lustrous green leaves at upper surface, purple lower, margins and main vein grey, 30-65 cm long and sagittate but narrowly triangular. Two lower lobes are more than half longer than the main lobe. The leaves in *A.l.* 'Magnifica' is longer. *A. lowii* from Borneo bears deep green leaves with a metallic sheen and silvery veins above, purple below, broadly sagittate, and about 50 cm long. Two basal lobes join the main lobe with about 1/4th of their length. *A. macrorrhiza* (giant elephant's ear) from Sri Lanka and Malaya bears some 4.5 m tall trunk-like rhizome. Leaves are broadly sagittate, up to or more than 60 cm long and with 9-12 pairs of veins. Yellow-green spathe is some 20 cm long. *A.m.* 'Variegata' bears cream and grey-green blotched leaves. *A. micholitziana* from Philippines is a very handsome species bearing velvety-green leaves contrasting with white midrib and main veins, narrowly sagittate and some 25 cm long. *A. odora* from tropical Asia though is confused with *A. macrorrhiza* but the leaves are peltate with 6-10 pairs of veins on the main lobe. *A. plumbea* (*A, macrorrhiza rubra*) from Java looks as dwarf form of *A. macrorrhiza* but is purple flushed and the leaves are up to 75 cm long. *A. portei* from Philippines is suitable only for the largest conservatory, bears deep green and up to 2 m long ovate-sagittate leaves with metallic luster, which in the juveniles are only deeply sinuous while in mature plants deeply pinnatisect. Leaf stalk matches to the length of leaves, is green marbled with red-purple and quite robust. *A. sanderiana* (Kris plant) from Philippines bears metallic silver-green leaves having white margin, veins patterned with grey, narrow-sagittate, lobed and 30-40 cm long. *A.* × *sedenii* (*A. lowii* × *A. cuprea*) is though quite similar to its first parent but basal lobes are joined to 2/3rd of their length. *A. veitchii* (*A. lowii veitchii*) from Malaya bears deep green leaves above, purple below, veins and margins light grey, and these are sagittate but narrowly triangular and some 45 cm long. *A. watsoniana* from Sumatra bears ovate-sagittate leaves where upper surface is glaucous tinted, veins silvery, some 90 cm long and basal lobes joined for about half their length. *A. putzeyi* appears as small version of the above. *A. wentii* (*A. whinckii*) from New Guinea bears peltate, ovate or sagittate deep glossy green leaves where margins are slightly undulated and tips pointed. Its basal lobes are rounded and joined for ½ to $^{2}/_{3}{}^{rd}$ of their length. *A. zebrina* from Philippines bears up to 35 cm long, green, sagittate and leathery-textured leaves. The species is known so because of the brownish-black zebra stripes on 50-90 cm long leaf-stalks.

Alpinia (Zingiberaceae): A native to tropical and sub-tropical Asia and Polynesia, it is a genus of about 250 species, mainly of evergreen rhizomatous perennials. Their stems are erect clad with simple ovate to lanceolate, green, purple or variegated leaves, and terminate into a racemose or panicled inflorescence bearing beautiful flowers, mostly enclosed with colourful bracts. Leaves are useful in floral arrangements with a vase life of 1-2 weeks. These can be stored and transported at 12-15 °C temperature.

Anthurium (Araceae): A genus of evergreen perennials comprising of 550 species from humid rain forests of tropical America and the West Indies. A number of species are grown for their decorative foliage and flowers, some are tufted while some others climbers as have aerial roots. The leaves are lustrous dark green, leathery and sometimes beautifully patterned. Flowers are small and closely arranged in spadix, and in some species are quite bright and attractive with a large spathe. They are propagated through division or seeds. Important species which can be used as cut greens are *A. andraeanum, A. clavigerum, A. cordifolium, A. crassinervium, A. crystallinum, A. digitatum, A. grande, A. hoffmannii, A. magnificum, A. palmatum, A. pentaphyllum, A. podophyllum, A. regale, A. scherzerianum, A. veitchii, A. warocqueanum, A. watermaliense, etc.*Their mature leaves are harvested, wet-stored at 13 °C or more and the vase life is more than 10 days.

Aralia (Araliaceae): A genus of 68 species of deciduous or evergreen trees, shrubs, and rhizomatous herbaceous perennials, native to Asia and the Americas, with most species occurring in mountain woodlands. *Aralia* plants vary in size, with some herbaceous species only reaching 50 cm tall, while some are trees growing to 20 metres. They have large bipinnate (doubly compound) leaves clustered at the top of their stems or branches, and only in some species the leaves are covered with bristles. They like filtered sun and moderately rich soil which is not water-logged but is moisture-retentive. They are propagated by seed, root cuttings and suckers. Their mature leaves are harvested, wet-stored at 3-7 °C and the vase life is about one week.

Araucaria (Araucariaceae): An evergreen coniferous genus of 19 tree species native to New Caledonia, Norfolk Island, eastern Australia, New Guinea, Chile, Argentina and southern Brazil. It was discovered in about 1780 by a Spanish explorer and introduced into England by Archibald Menzies in 1795. The horizontal, spreading branches grow in whorls and are covered with leathery needle-like leaves. The trees are mostly dioecious, with male and female cones found on separate trees. The female cone is globular and up to 20 cm in diameter; the male cone is cylindrical and up to 15 cm long. The optimum day/night temperature for araucaria is 15-22°C/10-18°C. They require well-drained and moderately fertile soil, and tolerate both, acidic or salty soils. Though these prefer filtered light but full sun is not detrimental if it is not very hot summer. They are propagated by seed and terminal cuttings. The cuttings taken from horizontal branches exhibit topophysis, as a result only terminal cuttings are used for propagation. Since the plants are very expensive so their foliage should be taken when it has grown into a full tree. The length of the branches should be removed as per requirement for the arrangement. Leaves with stems are harvested at any stage, stored dry at 5-7 °C for about 2 weeks and their vase life is over 3 weeks.

Ardisia (Coralberry; Myrsinaceae): It is a native to the tropics and sub-tropics, mostly from Asia and America, and some in Australia. It is a genus of 400 evergreen shrubs and trees. Usually they have alternate elliptic to lanceolate leaves and small flowers in terminal or axillary clusters. ***A. crenata*** (*A. crenulata, A. crispa*) from either West Indies or China, is a clear growing compact shrub with alternate, lanceolate-oblong, up to 10 cm long and wavy-margined dark green leaves. Its panicled sweet-scented flowers are starry, 5-petalled and white but sometimes flushed red. Its fruits are bright red, round, 6-8 mm in diameter and persisting for long. Its long-lasting leaves and drooping fruits with twigs are used in floral arrangements. It can be stored for more than one fortnight at 7 °C and transported.

Asparagus (Liliaceae): *Asparagus* is either evergreen or deciduous sub-shrubby or climbing perennial with tuberous roots. Stems bear both inconspicuous scale-like leaves and larger, leaf-like phylloclades, with small white or pinkish flowers followed by red berries. Asparagus is noted for its dense fern-like foliage that forms an arching mound. Plants usually develop weak prickles. Important *Asparagus* species are *A. asparagoides, A. densiflorus* (syn. *A. sprengeri*), *A. pyramidalis* and *A. setaceus* (syn. *A. plumosus*). It tolerates a wide range of temperatures (20 °C or above day and 10-15 °C night) but becomes uncomfortable at 10 °C or lower temperatures, and does not require high humidity. It thrives in bright indirect light because direct bright sun yellows its leaves. The plant resumes the growth as long as the underground tubers survive. Indoors, it requires artificial light of at least 400 fc. It is best grown in a well-drained, peaty potting mixture. Asparagus ferns tolerate clay, sandy and salty soils but not the waterlogged conditions, however, pH for optimum growth should be 6.5 to 7.5. Amending the soil with lots of compost or peat moss ensures healthy growth. ***A. densiflorus*** 'Sprengeri' is the most common form. It has long been favoured as a foliage compliment in outdoor containers. As the hardiest of the lot, it can survive temperatures well below freezing and can last well into the winter, sometimes adorning itself with showy, bright red (but poisonous) berries. Because of cascading habit, it is suitable for pots or baskets. ***A. densiflorus*** 'Meyersii' (foxtail fern) from South Africa is a slow-growing dramatic form, growing up to 60 cm in height and some 90 cm in spread and produces spire-like fronds which radiate reliably from a central core. ***Asparagus setaceus*** (plumosa fern), a climbing form is though delicate and daintier but a tough camper which is invasive once becoming pot-bound. ***A. retrofractus***, syn. ***A. macowanii*** (Ming fern), a semi-cascading form is rarely available, and is identified by its clusters of soft green tufts randomly spaced along its barbed branches. Its new growth is bright green which makes it quite informal bushy plant with unique appeal. It is highly suitable as a house plant, bonsai, or container accent, and by the florists for it's desirable texture. Cut stems in water can last for weeks. Stamps *et al.* (2005) evaluated vase life of 10 species and cultivars of asparagus. Average overall vase life duration ranged from 24.4 days for *A. densiflorus* 'Myers' to 6.2 days for *A. pseudoscaber*. Ewa Skutnik *et al.* (2006) reported an average vase life of 10 days in *A. densiflorus* 'Meyerii', and about 20 days in *A. densiflorus* 'Myriocladus' and *A. setaceus*. Chrysal RVB® doubled vase life in 'Meyerii', had no effect in 'Myriocladus', and actually shortened vase life in *A. setaceus*. Chrysal SVB® tripled vase life in 'Meyerii', and doubled vase life in 'Myriocladus', while shortened in *A. setaceus*. Gibberellic acid prolonged vase life in 'Meyerii' and 'Myriocladus', but had no effect in *A. setaceus*. BA prolonged vase life in all three taxa tested, even in *A. setaceus*. The 8-HQC and sucrose solution doubled vase life in 'Meyerii', had no effect in 'Myriocladus', and shortened vase life in *A. setaceus*.

Higher planting density (12 plants/m^2) increases the number of stems and total fresh weight in *A. plumosus* and *A. densiflorus* (Marino *et al.*, 2003). Pinching back is done of the stem tips as per requirement to maintain plant form and to promote dense foliage growth. If plant loses its attractive shape, stems may be cut back close to the soil level to regenerate. Chlorosis in asparagus is mostly associated with Mn deficiency (Stamps and Rock, 1999). Micronutrients, except Fe, do not affect average stem yields. N in Florida is applied at the rate of 2928 kg/ha. Weekly watering is sufficient from spring until autumn, however, mulching will drastically reduce the water requirement. Though not very particular about repotting but when these become quite pot-bound, these require to be divided and repotted. Its mature stems are harvested, wet-stored but for transportation dry packing in plastic bags or moisture-retentive boxes is done only for 1-2 days at 4-5 °C. It requires no preservative and its vase life is 1-4 weeks.

Aspidistra (Liliaceae): Its common name is cast-iron plant. Some eight species of this plants native to East Asia, especially China exist. *A. elatior* is a slow-growing, rhizomatous, perennial and evergreen interiorscape plant, and has upright, glossy dark green, leathery, elliptic, lanceolate and durable leaves that are used fresh and dried, as filler in large floral arrangements. Apart from this, it is also used as ground cover in shade gardens, in dry *cum* shady areas, under tree shades, as accents, in edgings and as container plants. *Aspidstra* leaves are reportedly non-toxic and quite long-lasting. *A. caespitosa* 'Jade Ribbons' leaves can be used as linear elements. In addition, *Aspidistra* leaves can be rolled, twisted, tied, and pinned into all kinds of shapes. *A. elatior* 'Okame' (syn. *A. elatior* 'Variegata') leaves are of similar size to *A. elatior* but are irregularly marked with light green and white streaks. *A. lurida* 'Ginga' (syn. 'Starry Night') bears shiny green leaves covered with cream- to white-streaks and spots. 'Ginga' leaves can grow as long as those of *A. elatior* but are narrower. The leaves of the cultivar 'Milky Way' ('Minor') are also covered with ivory- to white-dots and dashes like 'Ginga', however, the blades are shorter, narrower and dull. *A. caespitosa* cv. 'Jade Ribbons' has long but narrow leaves. These can be grown under conditions where temperature does not fall below -5 °C. Though seed formation is rare but if formed it can be propagated through them, otherwise it is propagated only by division of the rhizomes. Plants can be started from each underground node. They are tolerant to a wide range of nutrient conditions. In an experiment conducted at the mid-Florida Research and Education Center (MREC) in Apopka (Florida), yield (leaf number, and fresh and average weights) of *A. elatior* leaves from plants growing in ground beds of fine sand soil increased with increasing nitrogen application rate up to about 224 kg/ha/year. The plants of variegated aspidistra were grown in 9.5-liter pots containing a Florida sedge peat:pine bark:builders' sand (6:3:1) potting medium. A 19N-2.6P-10K controlled-release fertilizer (Osmocote® 19N-6P_2O_5-12K_2O from Scotts Company was applied at four nitrogen application rates with no effect on yield (leaf numbers, fresh and average weights) of variegated leaves was recorded but after 3¾ years, 45 per cent of the plants had produced harvestable green leaves under the dose of 792 kg/ha/year. *Aspidistra* is somewhat fluoride sensitive. In addition, care must be taken when applying granular fertilizers, *vis-a-vis* other granular materials so that they do not get trapped inside the tubes formed by the rolled-up emerging leaves. Once established, *Aspidistra* is tolerant to a wide range of well-drained soils and potting media, and droughty conditions though grow better in media containing organic matter with high cation exchange and good water-holding capacities under evenly moist conditions. They are also classified as moderately salt tolerant. Overhead irrigation may help reduce mite problems but can increase foliar diseases.

Aspidistra cv. 'Milky Way' establishes well under both 50 and 80 per cent shade under Florida conditions. These shade levels are equivalent to maximum photosynthetically active radiation levels of 1,194 and 478 $u$mol•m^{-2}•s^{-1}, respectively. Under 30 per cent shade, the plants barely survived, and in full sun the plants essentially died (Stamps *et al.*, 1995). Leaf lengths and fresh weights, as well as vase life increased with increasing shade level and peaked under 80 per cent shade. While vase life was excellent under 30 per cent shade (43 days), it was 46 per cent longer for leaves grown under 80 per cent shade. Numbers and fresh weights of *A. elatior* 'Okame' ('Variegata') leaves were 58 per cent and 66 per cent greater under 63 per cent shade compared to 73 per cent shade though it had no effect on the production of green leaves. Fresh weight of harvested leaves of *Aspidistra elatior* 'Variegata' was higher under black netting than under blue, gray or red, and total number of harvestable leaves was higher from black than the blue or red netting (Stamps and Chandler, 2008). Leaf variegation and the percentage of all green leaves produced were the same under black, blue, gray and red netting (Stamps and Chandler, 2008). Netting colour had no significant effect on leaf vase life, a critical factor for any cut foliage. For cut foliage use, *Aspidistra* leaves are harvested with clippers and are frequently bundled ten per bunch using rubber bands. Leaf lengths vary with cultivar and markets; however, larger and variegated leaves generally command higher prices than smaller and/or all green leaves. Leaves of cut and potted *Aspidistra* should be protected from being mechanically damaged during handling. Wounding the leaves can lead to fungal infection and damage. Harvested leaves should be stored and shipped at temperatures around 4°C. Stems will hold for weeks in waxed corrugated fibreboard containers. Acclimatized containerized *Aspidistra elatior* plants can be transported for two weeks if held at 10-13 °C. Typically the stems will last in arrangements for a month or longer.

Buxus (Buxaceae): A genus of 70 evergreen shrubby or tree species from Europe, Africa, America (north and south) and Asia. They have opposite or alternate, small, oval to oblong leaves and clustered, petalless, inconspicuous and monoecious flowers where males have four stamens. For its growing, it requires well-drained soil in sun or shade. Propagation through cuttings or seeds. ***B. sempervirens*** is the most popular species. Its mature stems for fillers are removed, stored wet in water at 2-5 °C for 1-2 weeks or in water-retentive boxes. Its vase life is up to 2 weeks.

Caladium (Araceae): A native to tropical America and West Indies, it is a genus of 15 perennial species with underground tubers. The leaves are long-stalked, heart-shaped to lanceolate patterned and blotched in various colours. It likes shaded and warm position. Its cut foliages are grown throughout the Europe for use as cut greens. Its vase life is from 1 to 2 weeks. Its foliage can be stored or transported at 10-13 °C.

Calathea (Marantaceae): They are commonly known as peacock plants due to leaf variegation. A perennial genus of 150 evergreen, tufted or clump-forming species from tropical America and West Indies. Their leaves are attractively patterned and ovate to lanceolate. They are excellent house and conservatory plants propagated by division. Its asymmetrical flowers may appear in clusters. Its most important species for the purpose of fillers are *C. argyraea, C. bella, C. carolina, C. crocata* (much preferred), *C. cylindrica* (quite long leaves up to 90 cm), *C. lancifolia* (syn. *C. insignis*), *C. leopardina, C. lietzei, C. lindeniana, C. makoyana* (syn. *Maranta makoyana*), *C. medio-picta, C. metallica, C. ornata, C. picturata, C. roseopicta, C. rotundifolia, C. rufibarba, C. veitchiana, C. vittata, C. warscewiczii, C. zebrina, etc.* Only mature leaves are harvested and placed immediately at an illuminated place in water immediately after cutting. They are stored wet at 7-10 °C and their vase life is from 1-4 weeks.

Camellia (Theaceae): An evergreen genus of some 80 species of trees and shrubs from India to Indonesia, China and Japan. They have beautiful fragrant flowers with some of the species even with handsome foliage used as fillers. For their cultivation these require neutral to acid medium. They are propagated through layering and stem or leaf-bud cuttings. The important species used as florist green is *C. japonica*. The leaves are available in spring. Stems are cut with mature leaves, wet-stored at 2-5 °C and their vase life is 1 week.

Cedrus (Pinaceae): A temperate genus of 3-4 large tree species of evergreen conifers from North Africa, east Mediterranean and Himalaya. These bear imposing needle-like leaves, They bear two types of branches, one so long that extending the size of the tree each year bearing spirally arranged leaves and short spurs which grow quite slow and bear terminal rosette of leaves, and the others grow into long shoots and produce strobili (fruits), males being catkin-like and shed abundant pale-yellow pollen when ripe, and the females are small, erect and barrel-shaped cones which may be green or purplish and mature in two years. These require well-drained moisture-retentive soil at a sunny site. They are propagated through seeds. Important filler species are *C. atlantica, C. brevifolia, C. deodara* and *C. libani*. Its flat, medium green and lacy foliage with stems are harvested, stored at 2-10 °C or even at room temperature with water misting for up to 2 weeks, and the vase life is 2 weeks or more.

Chlorophytum comosum (Liliaceae): Commonly it is called the spider plant, airplane plant or hen-and-chickens. It is a perennial herb native to tropical and southern Africa, but has become naturalized in other parts of the world, including western Australia and San Francisco (California). Variegated forms in particular are used as house plants. *Chlorophytum comosum* grows to about 60 cm high. It has fleshy and tuberous roots about 5–10 cm long. The long narrow leaves reach a length of 20–45 cm and a width of 6–25 mm. The inflorescences carry plantlets at the tips of their branches, which eventually droop and touch the soil, developing adventitious roots. The stems (scapes) of the inflorescence are sometimes called as 'stolons' though the term is incorrect. Spider plants are easy to grow, being able to thrive in a wide range of conditions. They will tolerate temperatures down to 2 °C, but grow best at 18-32 °C. Plants can be damaged by high fluoride or boron levels. Spider plants have also been shown to reduce indoor formaldehyde air pollution, and approximately 15 plants would neutralize formaldehyde production in a representative energy-efficient house. The mature stems should be harvested, stored in moisture-retentive boxes at 2-7 °C and their vase life is about one week.

Codiaeum (Euphorbiaceae): A genus of 15 evergreen shrubs from Malaysia to the Pacific Islands. Only one species, *C. variegatum* is in cultivation. It is an excellent house plant, and those varieties having long and pendulous leaves are quite suitable as cut greens. It is propagated through tip cuttings. For use as fillers, its mature leaves are harvested, stored in moisture-retentive boxes at 2-5 °C and its vase life is 1-2 weeks if kept in water.

Coleus (Lamiaceae): It is commonly known as coleus or painted nettle. An annual to perennial genus of 150 herbaceous species native to tropical Africa and SE Asia. Its important cultivated species are *C. blumei* (*Solenostemon scutellarioides*), *C. frederichii, C.pumilus* and *C. thyrsoideus*. Coleus is used for its brightly coloured leaves that come usually in multicolours. In addition, leaf margins may be entire, serrated, or lobed and fimbriated. It is a tender perennial and is highly susceptible to frost. Different coleus cultivars are adapted to different amounts of light from shade to full sun. Coleus is propagated by herbaceous cuttings. Coleus thrives well at temperatures above 21°C. It is susceptible to temperatures below 10°C. Pinching is an essential practice to get bushy growth. Its stems with mature leaves are harvested and kept immediately in water. These are wet-stored at 10-15 °C or in moisture-retentive boxes. Their vase life is about one week.

Cordyline (Agavaceae or Liliaceae): Cordyline is a genus of woody monocotyledonous plants having 15 species. *Cordyline terminalis*, a native of East Asia is the most popular species for indoor potted plants. Some selections of this plant are also used extensively as cut florist greens. Light levels can affect appearance of multi-coloured cultivars. The optimum light level is about 3,000 to 3,500 fc for producing plants with good coloration, and optimum temperatures for best growth are 18-35 °C. It can tolerate lower and higher temperatures, but growth rate will be reduced. This is propagated through cuttings and micro-cuttings or plugs from tissue culture laboratories. Its fully mature leaves are harvested, placed in water immediately, stored in moisture-retentive boxes at 2-4 °C and 98 per cent relative humidity where these can be stored up to 2-3 weeks.

Dracaena comprises of 150 species of trees and shrubs from tropical and sub-tropical Africa and Asia. Leaves are lanceolate and are forms either in tufts on the tips or along the length of the stem and make them attractive foliage plants most suitable as indoor plants. They are propagated through basal, stem or tip cuttings. Important cultivated species are *D. angustifolia, D. deremensis, D. draco, D. goldieana, D. hookeriana, D. marginata, D. phrynioides, D. sanderiana, D. surculosa, D. thalioides* and *D. umbraculifera*. Their fully mature leaves as filler are harvested and stored dry in moisture-retentive boxes at 2-4 °C.

Cycas (Cycadaceae): A genus of 20 species of primitive seed-bearing palm-like plants from Malagasy, E&SE Asia, Indo-Malaysia, Australasia and Polynesia. Their trunk is stout and woody and suckers at the base. Leaves are leathery, hard-textured, pinnate and in terminal rosette. Male flowers are cone-like and the females appear as much reduced woolly leaves bearing scattered globules. These are propagated through seeds, and well developed suckers. Important cultivated species are *C. circinalis* and *C. revoluta*. As fillers their mature leaves are harvested and stored at 5-10 °C, and these last for a few days even without water but with water it lasts for over 3 weeks.

Cyperus (Cyperaceae): An evergreen genus of mostly perennial but sometimes annual 550 species native to tropical, sub-tropical and warm temperate regions throughout the world. All cultivated ones are perennials, mostly rhizomatous or stoloniferous, and some with tubers. They all have grass-like tapering leaves. Flowers are umbellate, petalless, each composed of a single ovary and 1-3 stamens arranged in flattened and grass-like spikelets. Though these are excellent aquatic plants but leaves along with the portion of stems are used as fillers in floral arrangements. These are propagated through division and seeds. Important cultivated species are *C. alternifolius, C. diffusus, C. esculentus* and *C. papyrus*, and out of these, *C. alternifolius* and *C. papyrus* are most suitable cut greens. The stems up to 1 m long with beautiful radiating leaves are harvested and kept in water immediately. They are stored at 2-5 °C and these last for 2-4 weeks.

Cytisus (Leguminosae): A deciduous or evergreen genus of 30 species of shrubs from Europe, Mediterranean region, and North Atlantic Islands. The leaves are alternate, simple or trifoliate and short lived. They are excellent for conservatory and windowsill planting where these produce typical pea-like attractive flowers. Important cultivated species are *C. canariensis, C. fragrans* and *C. scoparius*. The needle-like leaves of *C. scoparius* grow some 60 cm long and are available during winter months in the temperate areas. These are severed along with a portion of stem and used as cut greens. They are good for curving, cut and stored for up to 2 weeks in water at 2-5 °C and their vase life is up to 3 weeks.

Dieffenbachia (Araceae): A perennial genus of 30 species from tropical America and West Indies. Their stems are erect, robust, mostly fleshy and carry a terminal tufts of large oblong or ovate leaves. Flowers are similar to arum but insignificant and these are valued only for their beautiful patterned foliage. These are propagated through cuttings of stem sections or tip cuttings. Important cultivated species are *D. amoena, D. bowmannii, D. imperialis, D. leopoldii, D. longispatha, D. maculata, D. oerstedii* and *D. seguine*. As filler, their mature leaves are harvested and stored dry in moisture-retentive boxes at 13 °C.

Duranta [Verbenaceae (Durantaceae)]:The genus *Duranta* contains around 36 species of tropical trees and shrubs native to Mexico, C&S America, Florida and West Indies. The species most cultivated is *Duranta erecta* (syn. *D. plunieri, D. repens*), commonly called as golden dewdrop or Brazilian sky flower. Duranta grows as a woody shrub 1-4 m in height. Its leaves are green, golden or variegated, depending upon the variety. The flowers are mauve, blue or white and are grouped together. They develop into a round orange berry, about 1cm across, which hang in loose drooping bunches from the ends of the branches. Durantas have the ability to tolerate sun or shade but bloom best in full sun. The variegated ones such as 'Gold Edge' actually look best in partial shade. The leaves are removed along with a part of the stem and immediately kept in water. These are packed in water-retentive boxes and stored or transported at 3-5 °C. Vase life of leaves is 1 week.

Epipremnum (Araceae): See *Scindapsus*.

Eucalyptus (Myrtaceae): An evergreen genus of about 500 species of trees and shrubs from Australia and Indo-Malaysia. Many of these are known as gums. They have aromatic, waxy and leathery leaves where shape of the juveniles (whether in young plants or leaves emerging after annual cutting back) may differ and appear in pairs in most species to those of senile or adult plants where the leaves are often willow-like. *E. globulus* (blue gum) has brightly hued leaves, the juvenile leaves are ovate-cordate and blue-white while the adult ones are lanceolate and more green than grey. To retain the juvenile leaves, they are required to be cut back annually or every two years at the end of harvest season or in spring when new growth is to begin. Plants grown for cut foliage are maintained as a coppiced or pollarded crop. Plants kept juvenile usually have more attractive leaf shapes and colours. It is the juvenile foliage which is demanded by the trade.Those having attractive green, blue-grey and grey leaves are used as fillers in mixed floral bouquets and other arrangements. The flowers are petalless and multi-stamened which may be white, cream, yellow, pink or red. The species grown for cut foliage are *Eucalyptus gunnii, E. parvula, E. perriniana, E. pulverulenta, E. mourei* and *E. rubida*. Eucalyptus can grow under a wide variety of climatic and soil conditions. They prefer well-drained loams. They can be planted near water sources as for their growth these require a lot of water. These are propagated by seeds. For cut greens they can be planted at a distance of 2 × 2 m. Initially the plants can be supported by bamboo poles of 2 m. *E. cinerea* 'Silver Dollar' and *E. gunnii* bear rounded twin leaves, *E.populus* and *E. stuartina* oval leaves, while *E. parvifolia* and *E. nicolli* bear thin and longer leaves. They are harvested with almost mature stems, stored in moisture-retentive boxes at 2-5 °C for 2-3 weeks, and floral preservative Chrysal and Roselife are effective. These are transported in water.

Euonymus (Celastraceae): A genus of about 170 deciduous or evergreen species of green shrubs, trees and climbers of cosmopolitan distribution, mainly Asia. They have simple leaves in opposite pairs and insignificant greenish, whitish or purplish flowers in small axillary cymes. They are propagated through seeds and cuttings. The species most commonly used as fillers are *E. fortunei and E. japonicus*. Other species of interest is *E. lucidus* (*E. fimbriatus*). For cut greens, they are harvested with 50-60 cm stem length and stored in water at 2-5 °C. While arranging, the recutting of 2.5 cm stem ends in water is recommended. The cut greens last over 2 weeks.

Fatsia japonica and ×***Fatshedera*** (Araliaceae): A native to subtropical regions of Japan, Korea and Taiwan, this genus consists of 1-2 species of evergreen shrubs with dark green, alternate and palmate leaves having clear white veins. ***F. japonica*** (*Aralia japonica, A. sieboldii*; false castor oil plant) from South Korea and Japan is though a small tree in the wild but in pots grows some 2.0 m high with stems showing conspicuous leaf scars. Its leaves are 20-40 cm wide having 5-9 lobes with wavy margins. It produces milky-white flowers some 5 mm across, followed by rounded glossy black berries. The leaves of *F.j.* 'Moseri' are larger on compact plant and of *F.j.* 'Variegata' are cream-margined. ×*Fatshedera lizei* (*Fatsia japonica* 'Moseri'× *Hedera helix hibernica*; tree ivies) was developed by the nursery firm Lizé Frères of Nantes, France in 1910, in nature this has acquired the partially climbing habit from *Hedera*, *vis-a-vis* is heat, cold or poor light tolerant, and produces wavy-margined palmately-lobed (5-9) leaves some 20-40 cm wide and pale-green flowers in rounded umbels, each flower about 5 mm across. Var. 'Variegata' leaves are cream-margined. Their large, glossy and deeply lobed leaves make effective background in large floral arrangement. Their leaves can be stored dry and transported at 2-5 °C in moisture-retentive boxes. Their vase life is 2-3 weeks.

Ferns: Almost all the fern species can be used as cut greens in various arrangements. Those ferns and allies commonly available and used are described here. ***Adiantum*** (maidenhair fern) is a delicate fern with soft green pinnules. Many of its 200 known species have attractive purplish-black stripes on leaf stalks. Though these are popular cut foliage plants but have only 5-7 days vase life. Its mature leaves are severed. wet-stored at 0-4 °C and transported in moisture-retentive boxes or sleeved in plastic foil. Certain floral preservative such as 25 mg/l $AgNO_3$ or Florisant 100 increases the vase life. ***Cyrtonium falcatum*** (holly fern) is a slow-growing fern producing dark green pinnae on fronds up to 75 cm long. Its fronds have short vase life. ***Davallia solida*** is a tropical fast-growing fern, growing well in high shade, where infertile fronds are popular as cut foliage. In case of ***Dryopteris erythrosora*** and ***Rumohra adiantiformis***, the fronds measuring 15-60 cm long are severed and stored at 2-5 °C with 95 per cent humidity. In vases these require frequent water changing and stem end removing. The vase life is about 6 weeks from the date of cutting. The cut greens in case of ***Nephrolepis exaltata bostoniensis*** having short light green leaves with some 30 cm long stems, and ***N. exaltata cordifolia*** with darker green and longer foliage are removed with fully mature leaves and stored wet at 2-5 °C up to 2 weeks. Their vase life is 5-10 days. The life can be increased with the use of preservative 8-HQC at 300 mg/l + sucrose at 22.5 g/l. The cut foliage of ***Polystichum munitum*** are removed with stem length of 50-60 cm when leaves are fully mature, and immediately placed in water and the foliage is misted regularly. For transportation or storage these are wrapped dry in wet boxes where these last up to 3 weeks at 2-5 °C. Their vase life is 10-14 days. The life can be increased with the use of floral preservative silver nitrate at 25 mg/l or citric acid at 75 mg/l. The mature leaves of ***Pteris*** species are harvested and placed immediately in water at 4 °C for 10-20 hours then transported dry in moisture-retentive boxes for 2 days. The fronds with fork-like branches, bearing pine-like needles and measuring some 50 cm of a fern ally, ***Lycopodium taxifolium*** are harvested and stored at 5-10 °C in water.

Ficus (Moraceae): A native to tropical and sub-tropical regions of the world, most common being in India and SE Asia, there are some 800 species under the genus with alternate leaves. The leathery leaves in many size, shape and colours of *Ficus altissima* (council tree), *F. elastica* (rubber plant) and *F. lyrata* (fiddleleaf fig) are suitable for floral arrangements with the vase life of 2-4 weeks. These can be stored and transported at 10-13 °C.

Galax (Diapensiaceae): An evergreen and perennial genus with only one rhizomatous species, *G. aphylla* from E. USA. It forms dense clumps or colonies, and bears lustrous deep green to reddish-green, basal, long-stalked, some 3-16 cm wide (mostly around 10 cm), leathery, rounded, cordate and crenate-toothed leaves, and small, 5-petalled, white and dense flowers in more than 60 cm long spike-like racemes. For its growing, it requires humus-rich soil having 6.0-7.5 pH reaction and semi-shaded condition. It is propagated through seeds and division of clumps. Its leaves are durable and most suitable as fillers but suitable only for local sale and not for distant markets. Immediately after cutting the leaves are placed in water. These last up to 10 days.

Gaultheria (Ericaceae): A low-growing genus of evergreen shrubs, mostly rhizomatous, comprising of some 200 species from countries around the Pacific, the Himalayas, EN America, and Brazil. To some extent this is poisonous plant. The leaves are alternate, lanceolate to oval to obovate, nodding and bell- or urn-shaped. The flowers are white and appear in racemes. Propagation through seeds. The cultivated species are *G. adenothrix, G. cuneata, G. hispida, G. itoana, G. miqueliana, G. nummularoides, G. shallon* and *G. trichophylla*. The length of leaves in all the species range from 1.0 to 6.0 cm. From the angle of cut green production, *G. shallon* is most important. Long branches with large and mature leaves which are available only during winter months and that too in temperate areas, are removed, stored dry at 0-5 °C in moisture-retentive packages for 1-2 weeks, and their vase life is up to 3 weeks.

Grevillea (Myrtaceae):There are about 270 species of evergreen trees and shrubs with alternate leaves, native to Australia, and New Caledonia to Malaysia. Their leaves are pinnately divided. With excecption of *G. robusta* which is a tree, they are shrubs grown for their racemes of conspicuous petalless flowers, consisting of a tubular calyx and a prominent curved style. They prefer light shade to sunny and dry situations, well-drained soils and are sensitive to phosphorus fertilization.The colour and structure of the flowers and leaf size and shape are quite different in the many species. Regular pruning after flowering is a regular practice for rejuvenation of plants and enhancement of flowering. Some species are frost-hardy and tolerate even up to 6 °C temperature. Though *G. robusta* is a tree and foliage species but others are flowering species such as *G. banksii* (red silky oak), *G. filoloba* (spidernet grevillea), *G. lonigera* (*G. woolly grevillea*), *G. rosemarinifolia, G. sulphurea* (*G. juniperina*), and *G. victoriae* (Victorian grevillea). The leaves of grevillea are most commonly used as fillers. These are harvested with a portion of stem and dry-stored at 2-5 °C. Its stem-ends require regular cutting in water. Its foliage lasts from 7 to 10 days.

Gypsophila (Caryophyllaceae): Commonly known as 'baby's breath' or 'soapwort', is a genus of about 125 species from Mediterranean region, Europe, Asia and North Africa. Since its several species have been found growing among gypsum rocks or limestones hence botanically it is known as *Gypsophila*, Greek *gypsos* means 'gypsum' and *philos* meaning 'loving'. They are herbaceous perennials or annuals, stems 10-120 cm long, slender, erect to spreading and swollen at the nodes, leaves in opposite pairs, linear to narrow triangular, often falcate, 1-7 cm long and 2-8 mm broad, and white to rose-pink flowers are produced usually in dichasial or panicled cymes on large inflorescences which are profusely branched. Each flower is small, 3-10 mm in diameter, with 5 white to pink petals. The basic chromosome number of gypsophila is x = 17 and most of the species are diploid with 2n=34. Though in itself it is an ornamental but is even more valuable as filler in flower arrangement. The young little girls when attending wedding ceremonies in U.S. decorate their hairs with its flowers. *Gypsophila paniculata* has become an invasive species in parts of North America. *Gypsophila rokejeka* is used to provide saponins. The root of the *Gypsophila repens* is used to make the whip cream topping of the Turkish dessert Kerebic. The most commonly encountered species in the garden are *G. elegans*, *G. muralis* and *G. paniculata*, former two being annuals and the latter one perennial. Other species are *G. acutifolia, G. altissima, G. aretioides, G. arrostii, G. bicolor, G. bungeana, G. capituliflora, G. cephalotes, G. cerastioides, G. davurica, G. desertorum, G. fastigiata, G. huashanensis, G. lasericea, G. licentiana, G. nana, G. oldhamiana, G. pacifica, G. patrinii, G. perfoliata, G. petraea, G. pilosa, G. repens* (syn. *G. dubia, G. fratensis, G. prostrata fratensis*), *G. rokejeka, G. scorzonerifolia, G. spinosa, G. tenuifolia, G. tschiliensis, etc.* Commercially only *G. elegans* and *G. paniculata* 'Bristol Fairy' are cultivated, though the perennial type has many other important cultivars such as 'Bridal Veil', 'Danninger', 'Flamingo', 'Rosy Veil' (syn. 'Rosenschleier'), etc. Annual type cultivars ('Paris Market', 'Covent Garden', 'Grandiflora Alba', etc.) grow 40-50 cm in height with-large or small, white, carmine, red, rose and deep rose flowers, but perennial types are taller up to 90 cm with single or double flower in white and pinks. *Gypsophila repens* is a trailing species with rose or white flowers useful for rockeries. They are easily propagated from seed, cuttings, root division and tissue culture. *G. elegans* is exclusively multiplied through seeds though *G. paniculata* through sexual and asexual both except the double flowered ones which are propagated only asexually. Seeds germinate within 10-15 days. When the plants are only in vegetative phase and have attained 6-8 leaf stage, the cuttings are taken, treated with IBA powder and inserted in a well-drained medium in a mist chamber. Apical shoot cuttings of *G. paniculata* cv. 'Bristol Fairy' rooted and established well under intermittent mist when a 5- or 30-second dip was given in IBA at 3,000-10,000 mg/l (Kusey and Weiler, 1980) and planted in peat cylinders, in plastic foam or in cellulose fibre blocks where it took 3-4 weeks to strike roots. Even otherwise when grown in polyhouse or in the open, being a winter crop it should be grown at an well aerated place with full sunlight and around 75 per cent RH but should have protection from untimely rains to avoid flowers becoming brown. Though it can be grown in a variety of **soils** with a pH of 5.5-7.5, preferably calcareous for *G. paniculata,* which is well-drained, but sandy-loam soils rich in organic matter is best being highly porous and with good water-retentive capacity. The salinity level should not be more than 1ms/cm. In water- logged soils it suffers from *Phytophthora* root rot so it would be wise to grow them on raised beds. It would be better if media in greenhouses are sterilized either through formalin or by solarization. Long days and high night temperature enhance flower initiation (lowest number of leaves below apical buds), flower bud differentiation and development. The optimum **temperature** for gypsophila is 26-27°C. Increased night temperature reduces plant height and number of flower stems per plant. Low temperature treatment (12 °C) of young vegetative plants for some weeks enhances lateral branching and increases the yield of flowers. In greenhouses, a **light intensity** of 55,000 to 75,000 lux is desirable. Gypsophila is a long day plant and depending on the varieties, for better flowering it requires 12-18 hours **day length** through incandescent lights, especially during elongation phase after pinching so that apart from profuse flowering the shoots may also be longer (de Graaf-van der Zande *et al.*, 1997). Long days are required both for flower initiation and flower development. The long photoperiod was effective only at relatively high temperatures (27/22 °C or 22/17 °C, day/night). At night temperatures below 12 °C, the plants remained vegetative even in long days. *Gypsophila* generally is grown as a protected crop, with night lighting supplied throughout the growth cycle during the winter to provide the long day (LD) conditions necessary for flowering (Shillo and Halevy, 1982). Spraying **GA_3** at 150 ppm thrice in a crop cycle in commercial growing reduces low-temperature growth-inhibition, and thereby increases flower yield. In greenhouse conditions it gives three flushes and under open field conditions only two, per cycle (flush) number of flowers being 8-12.

The beds of any length but of one metre width may be **prepared**, and may be raised up to 30 cm. Sand, rice husk, coco-peat and organic matter may be added to improve drainage. In case of very heavy soils, gravel may be placed below beds to facilitate drainage. Organic matter should be 30 per cent of the bed composition. Plants are **spaced** in double rows, with 50 cm between rows and 30 cm within the row. One additional plant is to be planted for every square metre area. The plugs can be placed in the holes already made in the bed. The best plant density is approximately 6-8 plants per square metre. The specific **water** requirement depends on soil and climatic conditions. Initially after planting it can be watered with a shower can or misting system. After one week, it can be irrigated with a drip system. Organic matter should be regularly added in such a way that they form 30 per cent of the bed composition. In vegetative stage the ratio of N, P and K application should be 2:2:1, whereas in reproductive stage it should be 1:1:1. The **nutrients** are applied through fertigation. The nutrient application should be based on testing of nutrient status in soil and plant. Gypsophilas for their erectness need **support**, preferably with net at a height of 25 cm, which can be erected just after planting. Wires on both sides of the bed provide additional side support at heights of 60 cm, and if necessary again at 100 cm. Shoots that grow out into the aisles should be regularly trained back inside the side wires. First year stock of gypsophila requires sole **pinching** 5-6 weeks after transplanting, leaving only 8-10 pairs of leaves on the plants, so that apical dominance is broken and as a result numerous lateral shoots are encouraged to appear, and from this first crop is harvested. **Pruning** is performed after each cycle of production to regulate plant height, *vis-a-vis* sanitation, followed by fungicide spraying to avoid any further infection.

Gypsophila is harvested when 50 per cent of the flowers in a stem are open, and then these are bunched and sleeved five stems each as per the international specifications. It is highly sensitive to endogenous and environmental ethylene, causing early senescence (drying), flower drop and drooping. However, longevity of over two weeks can be achieved when flowers are pretreated with STS at 4 mmoles per litre for 30 minutes, and then held in a solution containing 200 ppm Physan (a proprietary quaternary ammonium compound which acts as a bactericide) and 15 g/l sucrose (Newman *et al.*, 1998). Postharvest treatment of *Gypsophila paniculata* in a sugar solution and a bactericide, or in STS improves the quality of the flowers and their vase life with better flower opening, whiter colour and the keeping quality can also be doubled to about 10 days (Barendse, 1986).

The grey-brown **caterpillars** of *Spodoptera exigua* having triangular black spots on upper surface feed on the leaves and stem of the plant. Very high population of **leaf miner** (*Liriomyza* spp.) may result into dropping of leaves. **Thrips** (*Frankliniella* spp.) cause silvering, yellowing and browning of flowers. The control measures are spraying of Metasystox and Detamethrin against caterpillars, Triazophos against leaf miners, Monocrotophos and Metasystox against thrips. **Red spider mites** (*Tetranychus urticae*) feed on the plants with their needle-like mouth parts and web leaf undersurfaces. These are controlled by Dicofol and Abamectin. ***Alternaria* flower blight** (*Alternaria* spp.) causes blackening of flowers which can be managed through mancozeb. ***Botrytis* blight** (*Botrytis cinerea*) causes graying of bud scales and stems and ultimate death of the plants. No overhead or evening irrigation, proper plant spacing and ventilation, proper sanitation and application of mancozeb will keep this disease under check. ***Phytophthora* crown rot** (*Phytophthora parasitica*) starts at the soil level as wet rot. ***Pythium* root and stem rot** (*Pythium* spp.) starts as a wet rot at the base of young plants. ***Rhizoctonia* stem rot** (*Rhizoctonia* spp.) starts at the soil level. The stem appears dry and when split longitudinally shows a dry and shredded appearance. In case of all these three diseases the soil should be disinfested and the plants should be sprayed or drenched with thiophanate methyl and PCNB.

Hedera (Araliaceae): A genus of about 15 evergreen climbing species from Europe to Caucasus, Himalaya and Japan, Canary Islands and Madeira. Juvenile stems (runner growth) have short clinging roots which help these plants to climb.They have usually palmately-lobed ovate leaves though mature plants (arborescent growth) have unlobed leaves but with wavy margins. The arborescent growth is produced from the summit of the runner growth when it reaches the top of its support. Small greenish-yellow and 5-petalled flowers are produced in terminal umbels. They make an ideal pot and hanging basket plants. They are propagated through cuttings, however, those taken from non-climbing stems produce bushy, i.e so-called 'tree ivies'. Important cultivated species are *H. canariensis*, *H. colchica* and *H. helix*. Their fully mature stems are harvested and stored dry in moisture-retentive boxes for 2-3 weeks at 2-5 °C.

Ilex (Aquifoliaceae): A genus of 400 species of tender and hardy deciduous and evergreen trees and shrubs of world-wide distribution, mainly North and South America. They have alternate, rounded to narrowly oblong leaves, and axillary clusters of small, 4-6 petalled, white or greenish, mostly unisexual flowers, and afterwards with red berries. They are grown in well-drained water-retentive soil at light shade to sunny sides. They are propagated through heel cuttings or from seeds which may take long time to germinate. Important cultivated species are *I. aquifolium*, *I. cornuta*, *I. perado* and *I. pernyi*. The fully mature spiky leaves, available during November and December only in the temperate areas, are harvested and stored dry in moisture-retentive boxes at 0-2.5 °C for 2-3 weeks.

Juniperus (Cupressaceae): A genus of 60-70 evergreen coniferous tree or shrub species from northern hemisphere. They have medium to light green and needle- or awl-like leaves in pairs or whorls of 3 when young, but on maturity some species show closely overlapping pairs of scale leaves. Male flowers are small with ovoid to columnar catkins while the females similar to quite small cones where scales become fleshy and fuse together to form a berry-like fruit. They are grown in well-drained soil at sunny site in temperate regions. They are propagated through heel cuttings of young shoots or by grafting on established stock of their type. They can also be propagated from seeds but seeds may take up to 18 months for germination. Important cultivated species are *J. chinensis*, *J. communis*,*J. conferta*, *J. coxii* (*J. recurva*), *J. horizontalis* (*J. prostrata*, *J. sabina prostrata*), *J. procumbens* (*J. chinensis procumbens*), *J. sabina*, *J. sibirica* (*J. communis nana*), *J. squamata* (*J. recurva densa*, *J. recurva squamata*) and *J. virginiana*. The needles from fully mature stems should be removed and stored dry at 0 °C in moisture-retentive boxes for 1-2 months.

Magnolia (Magnoliaceae): a genus of some 80 species of evergreen and deciduous trees and shrubs from Himalaya to Japan, Java, Borneo, EN&C America and Venezuela. They have large, solitary, terminal, bowl- to globlet-shaped, frost-tender and mostly fragrant flowers with 6-15 petals. The leaves are large, blue-green, alternate, leathery and ovate to oblanceolate. They grow best from partial shade to fully sunny site having fertile, well-drained but water-retentive soil with 6.5-7.5 pH. If plant in cooler regions, these will tolerate some lime. Either these are propagated through layering, by cuttings or through seeds though seeds can take up to 18 months for germination. Important cultivated species are *M. campbellii*, *M. conspicua* (*M. denudata*, *M. heptapeta*, *M. yulan*), *M. delavayi*, *M. discolor* (*M. liliflora*, *M. quinquepeta*), *M. grandiflora*, *M. halliana* (*M. stellata*), *M. hypoleuca* (*M. obovata*), *M. kobus*, *M. mollicomata*

(*M. campbellii mollicomata*), *M. nicholsoniana* (*M. taliensis, M. wilsoniana*), *M. parviflora* (*M. sieboldii*), *M. salicifolia, M. sinensis, M. × soulangiana* and many interspecific hybrids and their varieties. Their oval and leathery leaves some 15 cm long are removed and immediately placed in water. These are stored at 2-5 °C, and lasts for 5-7 days. They are quite popular in funeral floristry.

Maranta (Prayer plant; Marantaceae): These clump-forming herbs are indigenous to tropical Americas, primarily South America. In 1975, Maranta was estimated to represent 3 percent of the foliage plant product mix value in Florida. With the large number of new plants introduced from other genera since 1975, it is estimated that the marantas constitute about 1 to 2 percent of the present product mix volume. Most marantas grown in Florida are produced in central Florida greenhouses. Marantas are versatile plants indoors because they can be used as small specimen plants, in hanging plants which cascade, as ground covers in interiorscapes and in dish gardens and other combination planters. Marantas are usually started from cuttings selected from stock beds. Cuttings are commonly provided by tropical American producers or a few local nurseries that maintain stock. Some growers have attempted to grow their maranta stocks under the benches used for finished potted plants, with poor results. Growers are not able to maintain adequate and uniform levels of light, water and fertilizer in the under-bench areas where management of diseases and insect-pests is very difficult. Stems cut from stock are usually cut down to single-node and double-node cuttings that are generally stuck in hanging baskets. Two cuttings are usually used in a 7.5 to 8.0 cm pots and 3 cuttings in a 10 cm pot. Marantas are best grown under a light level range of 1,000 to 2,500 fc in greenhouses where moisture and temperature can be controlled. Temperatures of 21-27 °C are ideal for maranta rooting and growth. Good growth occurs up to 32.2 °C temperature but beyond this the growth is hampered. Maranta grow well in a potting medium with good aeration, high water holding capacity and a pH of 5.5 to 6.0. Peat-based mixes generally require addition of dolomite to raise the initial pH and addition of a micronutrient blend product, such as MicroMaxR (500 g/m^3) is recommended. Fertilization of maranta at the rate of 900-1,300 g of nitrogen from a 3-1-2 or similar ratio fertilizer per 90 m^2 per month will provide adequate nitrogen, phosphorus and potassium. They are stored wet at 7-10 °C and their vase life is from 1-4 weeks.

Monstera (Araceae): A herbaceous genus of about 60 species of evergreen root climbers, native to tropical America, many of the species being epiphytic in nature. They have leathery, dark green and very large, often with holes in the leaf blade, alternate and long-stalked leaves. The flowers are borne on a specialised inflorescence called a spadix, 5–45 centimetres long; the fruit is a cluster of white berries, edible in some species, especially *M. deliciosa* which taste like a combination of banana and pineapple. They are commonly grown indoors as house plants. It is propagated by seed, cuttings or layering. Its large leaves are used as cut greens with large floral arrangements or as background to the arrangement. The leaves when mature are removed and dry stored in moisture-retentive boxes at 13 °C for 1-2 weeks. Its vase life is up to 10 days.

Murraya exotica (*M. paniculata*; Rutaceae): A native to SE Asia, Indo-Malaysia and Pacific Islands, the genus comprises some 4-12 species of evergreen shrubs and trees related to *Citrus*. *M. paniculata* is a large shrub with well-branched stems, bearing dark glossy green and pinnatifid leaves which are composed of 3-9 obovate and 5-7 cm long leaflets. White fragrant corymbose flowers with pointed petals, 1.2-2.0 cm long continue appearing if weather remains warmer throughout the year. Its twigs containing the leaves are detached and stored dry in moisture retentive boxes for 10-12 days at 7-10 °C.

Myrtus (Myrtaceae): An evergreen genus of 100 shrub or tree species from the tropics and sub-tropics. They have opposite pairs of entire, and obovate to lanceolate leaves. Flowers are 4-5 petalled, either solitary or in small cymes, which are followed by fleshy and often edible berries. These grow well at sunny location having well-drained but moisture-retentive soil. They are propagated through seeds at 15-18 °C temperature or with heel cuttings at 18 °C bottom heat. The important cultivated species are *M. apiculata* (*M. luma*), *M. bullata, M. chequen, M. communis, M. lechlerana, M. nummularia, M. obcordata,* and *M. ugni*, most popular one being *M. communis*. There are two types, one up to 120 cm long stems with large glossy-green leaves, and the other some 30 cm long with smaller leaves. These are cut with mature stems and stored dry at 2-5 °C in moisture-retentive boxes or in water up to 5 days. This lasts for 10 days.

Ophiopogon (Lily turf or Mondo grass; Liliaceae): A genus of more than 10 species of evergreen perennials from Himalayas to Japan and Philippines. They are tufted, rhizomatous or stoloniferous, bearing leathery, grassy leaves and racemes of small. 6-tepalled, bell-shaped, nodding and normally white flowers. Though it may often be confused with *Liriope* but the latter bears individual flowers facing outwards or upwards on straight stalks. They can be grown in humus-rich well-drained soil in shade. They are propagated through plant division and through seeds. Important cultivated species are *O.arabicus* (*O. planiscapus*), *O. jaburan* (variegated), *O. japonicus,*and *O. nigrescens*. Its mature leaves are removed and packed in moisture-retentive boxes at 2-5 °C. Leaves last for one week.

Palms: There are many genera and species that can be used as cut greens. ***Adonidia*** (Christmas palm; Arecaceae) is a monotypic genus of flowering plants. It consists of one species, *Adonidia merrillii* (syn. *Normanbya merrillii,. Veitchia merrillii*; Christmas palm, Manila palm). This palm is moderately salt-tolerant and usually not affected by salt spray, cold-susceptible, typically fairly small and slender, slow-growing and normally attaining 3.6–7.5 m in height but has capacity to grow up to 10.8 m. Performing best in full sun and tolerates partial shade (when young essentially these require some shade) though deep shade is dangerous. Prefers variety of soils including limestone, which are well-drained. The substitute of *Adonidia merrillii* is the (*Ptychosperma macarthurii*; Carpentaria palm) which is resistant to lethal yellowing disease. It is propagated by seeds. ***Chamaedorea*** is a genus of small bamboo-like palms of about 100 species from Mexico to northern South America. A few of the species are solitary but many are of suckering habit with pinnate and rarely entire leaves. They are propagated by seed or by division of suckers. Important filler species are *C. cataractarum* (cat palm), *C. elegans* (syn. *Neanthe elegans*), *C. erumpens* (bamboo/jade palm) and *C. seifrizii* (reed palm). Leaves are pinnate, dark green, leathery, 60-120 cm long with broadly lanceolate leaflets. As filler, fully mature leaves are severed and stored in moisture-retentive boxes at 2-7 °C for 2-4 weeks. ***Chrysalidocarpus lutescens*** (areca palm) is another palm used as florist's green for background arrangements but these should be grown under shade. ***Sabal palmetto*** has some 50 cm long palmate leaves which can be stored at 5-15 °C and even at room temperature in water though stalk is short but with water misting or in moisture-retentive boxes for up to 10 days and their vase life is 1-2 weeks. ***Chrysalidocarpus lutescens*** (*Areca lutescens*; common names: areca palm, bamboo palm, butterfly palm, golden cane palm; Arecaceae;): A genus of 20 species from Madagascar, Comoro and Pemba Islands, areca palm grows to 6-12 m tall with branched basal stems and prefers bright filtered sun as direct bright light burns the fronds. Its leaves are 2–3 m long, arched and pinnate, with 40-60 pairs of leaflets. It bears panicles of yellow flowers in summer. Arecas are propagated from seeds by soaking them for 10 minutes in a solution of hot sulphuric acid, which may germinate in about 6 weeks. Fresh yellow to ripe seeds should be planted with the top of the seed barely visible and germination temperature maintained between 26.6-29.4 °C. Lower temperatures may double the germination time. Seed storage at low humidity and low temperature is detrimental to germination. Cleaning seed is not essential if they are planted immediately. If seeds are to be stored, clean the yellow to fully-ripened red seeds, air-dry them properly, treat them with a seed protectant, and store at 24 °C. As cut green, its mature leaves of proper colour should be harvested, and stored dry in water-retentive boxes at 2-5 °C for 2-3 weeks. Its vase life is 2-3 weeks but cut ends require frequent cuttings in vase water. ***Livistonia*** (Chinese fan palm, fountain palm; Palmae): A genus of about 30 species mainly of large fan palms from Indo Malaysia and Australia. They have single straight trunks and terminal cluster of leaves that are halfway between pinnate and palmate (fan-shaped) and grow up to 1.8 m long including sharply spined petioles. *Livistonia chinensis* can attain height of about 9.1 to 15 m and a spread of 3.7 m, bearing large leaves which have drooping tips. The divided leaves have long, tapering, ribbon-like segments which gracefully sway beneath the leaves, creating an overall fountain-like effect. The inconspicuous flowers are hidden among the leaves and are followed by small, blue-black and olive-like fruits. Although Chinese fan palm has long been used as a container plant, its neat leaf habit and interesting form make it ideal for further landscape uses, such as for staggered groupings, as free standing specimens or as street trees. They form a closed canopy when planted about 3 m apart along a walk or street. They grow well in confined soil spaces. The palm is self-cleaning of old leaves and will require little or no pruning. Tolerant of full sun, young specimens should be partially shaded. Any reasonably fertile and well-drained soil including the alkaline ones is suitable for its cultivation. Chinese fan palm will give good results if fertilized 2-3 times during the year. Plants should be watered during dry spells and will benefit from an organic mulch. Its mature leaves along with petioles are harvested and stored dry in water-retentive boxes at 2-5 °C for 2-3 weeks. Its vase life is about 2-3 weeks. Its cut ends require frequent cuttings in water.

Philodendron (Araceae): An evergreen genus of 200 or more species of epiphytic climbers or shrubs from tropical Central and South America. The leaves in many of the species at juvenile stage are different to those of senile stage. They have almost dark glossy-green, alternate in climbing species though rosette or crowded on erect growing ones, arrow to cordate in shape and often leathery leaves. Small flowers are borne on small spadices, partly surrounded by white, green or red spathes. They are shade-loving plants thriving in moist soils rich in organic matter. These grow well at the day/night temperatures of 24.0-29.5/18.3-21.0 °C. They are propagated through tip cuttings, stem sections or leaf buds. Important cultivated species are *P. andreanum* (*P. melanochrysum*), *P. auriculatum*, *P. bipennifolium* (*P. panduriforme*), *P. bipinnatifidum*, *P. cannifolium* (*P. crassum*, *P. martianum*), *P. cordatum*, *P. crassinervium*, *P. domesticum* (*P. hastatum*), *P. eichleri*, *P. erubescens*, *P. fenzlii*, *P. glaziovii*, *P. gloriosum*, *P. ilsemannii*, *P. imbe*, *P. lacerum*, *P. laciniatum* (*P. pedatum*), *P. mamei*, *P. micans* (*P. oxycardium*, *P. scandens*), *P. pinnatifidum*, *P. quercifolium*, *P. sagittifolium* (*P. sagittatum*), *P. selloum*, *P. sellowianum*, *P. squamiferum*, *P. trisectum* (*P. tripartitum*), *P. verrucosum*, *P. warscewiczii* and *P. wendlandii*. Mature leaves are removed and placed immediately in water. They can be stored dry at 13-16 °C at 80-90 per cent RH. Their vase life is 7-10 days.

Pinus (Pinaceae):An evergreen and coniferous genus of 80 tree species from northern hemisphere which grow only in temperate regions. They have erect trunk, when young conical in shape and regular whorls of smaller branches, but when quite grown they become flat-topped with long and short lateral shoots having quite irregular shape. Longer shoots extend the height and spread of the tree though shorter ones grow only as small leaf-bearing pegs. The needle (fascicles)-like leaves are grey- to yellow-green, drooping and come out in pairs, 3s or 5s on each short shoot when plants are quite grown though nursery plant show only dense spiral of solitary leaves. Strobili are catkin-like, monoecious, males cylindrical while females smaller and barrel-shaped.They are grown in well-drained soils which are moderately fertile and under sky-open sites. They are propagated through seeds. *Pinus patula* is a coniferous tree which grows to a height of 30 m. Popularly grown species are *P. aristata*, *P. armandii*, *P. austriaca* (*P. nigra*), *P. ayacahuite*, *P. banksiana*, *P. cembra*, *P. contoria*, *P. densiflora*, *P. excela* (*P. griffithii*, *P. wallichiana*), *P. halepensis*, *P.insignis* (*P. radiata*), *P.jeffreyi*, *P. lambertiana*, *P. laricio* (*P. nigra maritima*), *P. longifolia*, *P. montezumae*, *P. mugo* (*P. montana*, *P. mughus*), *P. muricata*, *P. nigra*, *P. parviflora*, *P. patula*, *P. pinaster*, *P. pinea*, *P. ponderosa*, *P. rigida*, *P. strobes*, *P. sylvestris*, and *P. thunbergii* (*P. thunbergiana*). Its mature stems with needles are harvested and stored at 2-5 °C in moisture-retentive boxes. When arranged, these require regular misting at room temperature. Vase life of needles is above 10 days.

Pittosporum (Pittosporaceae): An evergreen genus of trees and shrubs with 150 species from E. Asia, Australasia, Pacific Islands and Africa. Leaves are alternate, entire, and normally ovate to oblong or linear to obovate. Flowers are solitary or clustered, 5-petalled and often with shortly tubular bases. They are excellent tub plants and quite amenable to geometrical forms. They prefer moderately fertile soil which is well-drained and at sheltered sunny sites. They are propagated through seeds at 13-15 °C temperatures or through cuttings given with bottom heat of 18 °C. Important cultivated species are *P. crassifolium, P. dallii, P. eugenioides, P. tenuifolium* (*P. mayi*), *P. tobira* and *P. undulatum*. Most of these species emit amicable fragrance. Among all these, *P. tobira* is quite popular. Clustered leaves with thick woody stems which are mature are harvested and stored at 2-5 °C in water as they do not like dry storage. Vase life is 1-2 weeks.

Podocarpus (Podocarpaceae): An evergreen coniferous genus of about 100 species of trees and shrubs from the southern hemisphere, north to tropical and E. Asia. They are hardy dioecious plants with linear, spirally arranged leaves and catkin-like male inflorescences though females are small with a few scales and ovules, out of which only one develops into an exposed and ovoid seed. They are grown in moisture-retentive well-drained fertile soil at exposed position. They are propagated through seeds and by cuttings. The important cultivated species are *P. alpinus, P. andinus, P. macrophyllus* and *P. nivalis*, out of which *P. macrophyllus* is comparatively more popular. Its long, slender and flat needles on tall branches are available during whole winter. These are harvested and stored at 5-10 °C in water for 1-2 weeks. Their vase life is about 10 days.

Rhododendron (Ericaceae): A genus of about 800 species of evergreen or deciduous shrubs and trees or prostrate mat-forming plants from temperate regions of the northern hemisphere, particularly the Himalayas to China, extending to the mountains of SE Asia, from Malaysia to New Guinea, and one species in NE Australia. They bear often leathery leaves which are linear to ovate, entire and mainly aggregated towards the tips of the stems. They produce profusion of terminal blooms in umbel-like clusters (sometimes solitary) which are tubular, funnel- or bell-shaped and 5-8 lobed. They prefer humus-rich, acidic, well-drained and water-retentive soil at filtered or sunny site and moist atmosphere. They are propagated through seeds at 13-16 °C temperature, and through layering and cuttings. Important cultivated species are *R. bullatum* (*R. edgeworthii*), *R. ciliatum, R. cubittii* (*R. veitchianum*), *R. javanicum, R. lochae, R. macgregoriae* and *R. simsii* (*Azalea indica, R. indicum, R. indicum simsii*). They are harvested with fully mature leaves and stored dry at 0 °C in moisture-retentive boxes for 2-4 weeks. Their life is about 7-10 days.

Ruscus [Ruscaceae (Liliaceae)]: An evergreen, tufted to clump-forming genus of 3-7 shrubby species of rhizomatous rootstock from Atlantic Islands, through Mediterranean Europe to the Caucasus and Iran. They have green stems which broaden to form leaf-like cladodes, and from the axils of which tiny 6-tepalled starry flowers emerge. The plants are dioecious. They grow in humus-rich soil in shade to partial shade and are propagated through seeds and division of clumps. Important cultivated species are *R. aculeatus* with dark green, erect, stiff and branched stems, bearing dark green, rigid, 2-4 cm long, ovate and spine-tipped cladodes; and *R. hypoglossum* with flexible, arching and branched stems, bearing 5-11 cm long, glossy and elliptic to oblanceolate cladodes. The mature stems are harvested with 40-90 cm length bearing small oval leaves and stored at 2-5 °C with high (95 per cent) R.H. where these will last dry for about 10 days and with water additional 2-3 weeks. Their vase life is about 2 weeks. Preservative 8-HQC at 300 mg/l + sucrose at 22.5 g/l is useful.

Sansevieria [(mother-in-law's tongue, snake plant; (Agavaceae, formerly Liliaceae): The genus originated in Africa and southern Asia. It is a genus of about 60 succulent evergreen rosette or clump-forming species, some epiphytic and others terrestrial, where some have sword-shaped leaves while others cylindrical. The leaves are very thick and mostly variegated. These last for more than a month in vase solution and these can be stored dry at 7-10 °C. Important cultivated species suitable for cut foliage are *S. aethiopica, S. cylindrica, S. grandicuspis, S. hyacinthoides, S. kirkii, S. trifasciata,* and *S. zeylanica*.

Scindapsus (Araceae): A genus of 20-40 species of evergreen climbers native to SE Asia from China to Malaysia, and Pacific Islands. They climb with the help of their aerial roots. Their leaves are entire and leaf-stalk bases ensheath the stem. Flowers are insignificant and arum-like. *Scindapsus* is though similar to *Epipremnum* but the former has only one ovule in each ovary whereas the latter a few. They are indoor plants and take root even when severed and placed in water indoors. They are propagated by cuttings. They can be stored dry at 13-16 °C at 80-90 per cent RH and their vase life is 7-10 days.

Solidago (Asteraceae): A clump-forming genus of 100 to 120 species of flowering plants mainly from North, America, with a few in South America, Europe and Asia. Most are rhizomatous herbaceous perennials found in the meadows, pastures, and waste areas of North America. Its branched central hairy stem (panicle) grows to 60-180 cm tall with masses of tiny yellow flowers in compact heads, and each head less than 6 mm across. Because of the wide distribution and the existence of several varieties, there is significant variability in the characteristics of local ecotypes. The leaves are alternate, lanceolate to broadly linear, shortly toothed along the margins, underside being pubescent, about 10-15 cm long & 2.5 cm wide and become gradually smaller towards the apex. They like moderately rich soil, moist but not water-logged and sunny to filtered situation, and can be grown from tropical to temperate regions, however, in tropical region of India it continues flowering throughout the year, in temperate regions twise (spring and autumn) and under sub-tropical conditions only once during autumn. They are propagated through division of clumps and seeds. Important species are

S. canadensis and S. virgaurea (*S. brachystachys*). Its flowers as well as non-flowered mature stems are used as fillers. These are harvested for flowers when a few of the tiny flowers have started to open though overall colour is still green, however, for foliage when the leaves are mature on the stem but have yet not initiated flowering. After harvesting, these can be held dry for up to 5 days at 2-5 °C. Once these have been placed in water, they will start opening or greening. The vase life of cut flower is up to 10 days though of the foliage 10-14 if regular cutting of stem ends in the water is being carried out.

Spathiphyllum (peace lily; Araceae): A native of tropical areas of CS America and the islands of SE Asia, it is a genus of about 35 dense clump-forming evergreen perennial species, bearing long-stalked ovate to lanceolate, slender and glossy green leaves. In Greek, *spathe* is for the 'bract' and *phyllon* is for the 'leaf', referring to the spathe being leaf-like. These are good container plants for warm conservatory or indoor having humid atmosphere. Its leaves are very suitable in arrangement of flowers and in vases these last for 3-4 weeks and can be stored or transported dry in moisture retentive boxes at temperatures of 10-13 °C. Its suitable species are *S. cannifolium*, *S.* × 'Clevalandii' (*S. kochii*), *S. cochlearispathum*, *S.* 'Mauna Loa' and *S. wallisii*.

Strelitzia [Musaceae (Strelitziaceae)]: An evergreen perennial genus of five tender species from sub-tropical to South Africa. Some species grow to 1 m while certain others even to tree-forms growing up to 8 m tall, but their growth is always palm-like. Their leaves are long, undivided, leathery and in two ranks. They like humus-rich moist but not water-logged soil at filtered situation with high atmospheric humidity and at 16-27 °C temperature. They are propagated through seeds and division. *S. alba* (*S. augusta*) grows up to 5 m in height and bears smaller leaves and white flowers. *S. nicolai* branches at ground level and has woody stems growing up to 8 m in height. It is palm-like and erect, bearing glossy rich green oblong-ovate leaves 90-120 cm or even more in length with stalks as long. Flowers are pale-blue to light mauve or even white, and emerge from some 30 cm long boat-shaped baract which are reddish-brown. Leaves in *S. reginae* form a large oblong-lanceolate fan, are blue-green, 25-45 cm long on some 45 cm long stalks, glaucous beneath and leathery in texture. Flowers are with three orange or yellow sepals, one small orange petal and the joined pair blue, and these emerge from a 15-20 cm long, narrowly red-edged and boat-shaped bracts. *S. reginae juncea* bears rush-like robust leaf stalks usually without terminal blades (sometimes very small), *S. reginae humilis* is quite dwarf, and *S. reginae intermedia* is halfway to *S.r. juncea* and *S. reginae* with quite reduced leaf blades. Mature leaves 50 cm or more long with stalks are harvested. The wide and oval ones on long stalks are suitable for tall arrangements. They are stored at 2-5 °C in water. Its vase life is 2 weeks.

Syngonium (Araceae): A native to the tropical areas of Americas and the West Indies, it is a genus of 20 evergreen climbers which in the wild are epiphytic. In Greek, *syn* is for 'together' and *gone* for a 'womb', owing to its united ovaries. Species are *S. auritum* (*Philodendron auritum*; five fingers) from Jamaica, *S. hoffmannii* (*Nephthytis hoffmannii*) from Costa Rica, *S. podophyllum* (arrowhead vine, goosefoot plant) from Mexico to Panama with various varieties, *S. wendlandii* (*Nephthytis wendlandii*) from Costa Rica, *etc.* In *Syngonium*, the leaves vary in shape and size from juvenile to adult form, initially being entire or arrow-shaped but afterwards becoming trifoliate and pedate. Sometimes its leaves are used as cut foliage and their vase life is about 2-3 weeks.

Thuja (Cupressaceae): A hardy juniperous (coniferous) genus comprising of five evergreen species from E. Asia (China, Formosa and Japan) and N. America. They are quite similar to *Chamaecyparis* but in *Thuja* the flat scales (leaves) are somewhat larger and densely clothe the shoots. The scale leaves are arranged in alternating decussate pairs in four rows along the twigs. The leaves are arranged in flattened fan shaped groupings with resin-glands, and oppositely grouped in 4-ranks. The mature leaves are different from younger leaves, with those on larger branchlets having sharp, erect and free apices. The leaves on flattened lateral branchlets are crowded into appressed groups and scale-like and the lateral pairs are keeled.The cones are upright, urn-shaped or ovoid to cylindrical and less than 2.5 cm in length. The solitary flowers are produced terminally. The male cones are small, inconspicuous, and are located at the tips of the twigs. The female cones emerge out similarly inconspicuous, but grow to about 1–2 cm long at maturity when 6–8 months old. They have 6-12 overlapping, thin and leathery scales, each scale bearing 1–2 small seeds with a pair of narrow lateral wings. Seed cones are ellipsoid, typically 9-14 mm long and they mature and open the first year. The thin woody cone scales numbering 4-6 pairs are persistent and overlapping, with an oblong shape and are basifixed. The crowns are densely conic or irregularly rounded but the branches are erect with flat shoots. They withstand heavy pruning and shearing because new branches develop from concealed buds in the branch crotches. Important species are *T. occidentalis, T. orientalis* and *T. plicata* (*T. lobbii*). They prefer sunny to filtered situation, and humus-rich soil which is not water-logged but has good water-retention capacity. They are propagated by seed or heel cuttings. Their twigs with mature scale leaves are removed and stored wet in moisture-retentive boxes at 2-5 °C for 2-3 weeks. Their vase life is 2-3 weeks. However, in vase the cut ends require frequent recutting.

Ulex (Leguminosae): A genus of some 20 species of densely spiny shrubs fromW. Europe and NW Africa, but only *U. europaeus* is grown in the gardens with the common names of 'Furze', 'gorse' and 'whin', native of W. Europe to SW Scandinavia. This grows up to 2.0 m tall, is bushy where all the stems and shoots are spine-tipped. Seedlings have soft and trifoliate leaves while adult plants have leaves reduced to scales or spines. Flowers are pea-shaped, bright yellow, fragrant, about 1.5 cm long and normally crowded on the spine-tipped lateral shoots at stem ends. It likes moist but not water-logged, moderately rich sandy-loam or chalky soils in full sun and is propagated through seeds. *U. europaeus* stems with thin and pine-like leaves are available mainly during winters. Immediately after harvest, these are placed in water and then stored at 2-5 °C. Recutting of stems under water is recommended. Their vase life is about 1 week.

Vaccinium (Blueberry, bilberry; Ericaceae): A genus of 400 species of tender, half-hardy or hardy evergreen or deciduous shrubs (or rarely small trees) from temperate regions in the world. They have alternate, obovate or ovate to linear leaves, and bell- to urn-shaped, 4-5 lobed and nodding flowers appear singly or in small clusters or racemes, followed by globular fleshy berry which are edible in all the species. These prefer moisture-retentive, peaty, or lime-free soil in an open position. These require spring-pruning every year for maintaining bushy growth. They can be propagated through layering by pegging the branches down in the soil, by taking 5-8 cm long half-ripened cuttings of shoots, division of the plants or by seeds. Important species are *V. angustifolium, V. corymbosum, V. delavayi, V. floribundum* (*V. mortinia*), *V. glaucoalbum, V. moupinense, V. myrtillus, V. nummularia, V. ovatum, V. oxycoccus* (*Oxycoccus palustris, Vaccinium macrocarpum*) and *V. vitis-idaea*. The small, elliptic and dark green leaves in 'tips' and brownish red variety 'Red Huck', both commonly known as 'florist greens', which bear up to 75 cm long stems. These are available during winter months in the temperate regions, from which only tender upper parts are cut, dry stored or wrapped in wet paper for up to 2 weeks at 2-5 °C and 95-98 per cent relative humidity. Their vase life is 2 weeks.

Xerophyllum (Liliaceae): An woody rhizomatous genus of three grassy species from North America. Stems are simple, tall and erect and bear basally rosetted leaves which are long, linear and stiff with scabrous margins, and the cauline leaves are much smaller and sparse. Numerous white flowers appear in terminal raceme having oblong or lanceolate and persistent perianth. *X. asphodeloides*, a hardy perennial resembles asphodel and grows from 30-120 cm with 7.5-15.0 cm long racemes, and *X. tenax* grows 60-150 cm in height with 30-60 cm long racemes. They are grown in moderate fertile and moist soil which is not water-logged. They are propagated through seeds and division of rootstocks.Grass stems some 1 m in length are severed from the plant with mature leaves having 1.25 cm width. The leafy stems last over 2 weeks.

Apart from these, there are a number of other plants whose foliage is also used as fillers. Some of such plants are *Alpinia, Aucuba japonica, Brassaia, Callistemon citrinus, Camellia japonica, Canna, Dizygotheca elegantissima, Ficus benjamina, F. elastica, Limonium* (spike), *Molucella laevis* (spike), *Murraya exotica, Putranjiva roxburghii, Sansevieria, Saponaria* (spike), and so on.

References

Bailey, L.H. 1960. *The Standard Cyclopedia of Horticulture* (3 vols.).The Macmillan Co.,NY, USA.

Barendse, L.V.J. 1986. Postharvest treatment of *Gypsophila paniculata. Acta Hort.*, No. 181, p. 338.

Beckett, K.A. 1983. *A Concise Enclopedia of Garden plants.* Orbis Pub. Ltd., London, England.

Beckett, K.A. 1987. *The RHS Encylopedia of House Plant Including Greenhous Plants.* Salem House Publishers, Massachusetts, USA.

de Graaf-van der Zande, M. Th. And T. Blacquiére, 1997. Alternative sources for pH to periodic lighting of gypsophila. *Acta Hort.*, No. 418, pp. 119-126.

Ewa Skutnik, Julita Rabiza-Świder and Aleksandra J.Łukaszewska, 2006. Evaluation of several chemical agents for prolonging vase life in cut asparagus greens. *J. Fruit & Ornam. Pl. Res.*, **14**: 233-240.

Hay, R. and K.A. Beckett, 1971. *Reader's Digest Encyclopaedia of Garden Plants and Flowers*. The reader's Digest association Ltd., London. Great Britain.

Kusey, W. E. and T.C. Weiler, 1980. Propagation of *Gypsophila paniculata* through cuttings. *HortSci.*, **15** (1): 85-86.

Marino, A., A. Ferrante, M. Maletta and A. Mensuali-Sodi, 2003. Production and postharvest evaluations of ornamental *Asparagus* spp. *Adv. Hort. Sci.*, **17**(2): 88-92.

Newman, J.P., L.L. Dodge and M.S. Reid, 1998. Evaluation of inhibitors for postharvest treatment of *Gypsophila paniculata* L. *HortTech.*, **8**: 58-63.

Nowak, J. and R.M. Rudnicki, 1990. Guidelines for Particular Plant Species. In:*Postharvest Handling and Storage of Cut Flowers, Florsit Greens, and Potted Plants*, pp. 129-199. Timber Press, Portland, Oregon, USA.

Salunkhe, D.K., N.R. Bhatt and B.B. Desai, 1990. Foligae Plants. In: *Postharvest Biotechnology of Flowers and Ornamental Plants*, pp. 333-366. Naya Prokash, Calcutta,

Shillo, R. and A.H. Halevy, 1982. Interaction of photoperiod and temperature in flowering-control of *Gypsophila paniculta* L. *Scientia Hort.*, **16**: 385-393.

Stamps, R.H. and A.L. Chandler, 2008. Differential effects of colored shadenets on three cut foliage crops. *Acta Hort.*, No.770, pp.169–176.

Stamps, R. H. and D. K. Rock, 1999. Effect of four elements on color, yield and vase life of tree "fern"(*Asparagus virgatus*). *Proc. Fla. St. hort. Soc.*, **112**: 282-285.

Stamps, R. H., F. P. Beall, B. W. Beall, J. P. Carris, M. R. Jackson, S. M. Motley, P. S. Smith and N. L. Freeman, 1995. Effects of shade level and radiation freezes on survival and growth of ground cover plants. *Proc. Fla St. hort. Soc.*, **107**: 432-436.

Stamps, R. H., K. Diane, D.K. Rock and A. L. Chandler, 2005. Vase life comparison of ornamental asparagus species and cultivars, *Proc. Fla. St. hort. Soc.*, **118**: 365-367.

11

Dahlia (Family: Asteraceae)

R.L. Misra and Sanyat Misra

[**Common name**: Cactus dahlia (*Dahlia juarezii*), Collarette dahlia, Dahlia, Decorative dahlia, Old-fashioned dahlia, Fancy dahlia/Pompon dahlia/Show dahlia/Single dahlia (*D. rosea* syn. *D. barkeriae/D. coronata/D. crocata/D. nana/D. purpurea/D. royleana/D. superflua/D. variabilis*), Paeony-flowered dahlia, Tom Thumb dahlia, Tree dahlia (*D. imperialis, D. excelsa* syn. *D. arborea*), etc.]

Introduction and Origin

The genus *Dahlia* was erected after Dr. Andreas Dahl (1751-1789), an 18th century Swedish Professor of botany who had been author of '**Observationes Botanicae**' and pupil of Linnaeus. *Dahlia* species are half-hardy tuberous perennial herbs or sub-shrubs, sub-trees and a few epiphytes originating from Central and South America between higher parts of Mexico to Guatemala and Colombia but the great majority of plants grown today are variations of the true species and hybrids raised from variants that have arisen in cultivation. Lawrence (1942) stated that garden dahlia (*Dahlia variabilis*) already domesticated in Mexico before 1575, its tuber was first introduced in 1789 into Europe. Rünger and Cockshull (1985) and De Hertogh and Le Nard (1993) stated that migration and hybridization of species from Mexico and South America probably occurred in the early 17th century. Dahlia is a prolific and long-flowering plant with wide ranging forms and colours which make them outstanding plants for garden decoration and as cut flowers. Its flowers have a sharp typical fragrance. It is a very popular show flower, as a specimen plant, used in perennial borders, as bedding plant, for cultivation in pots and tubs and for cut flowers. The dwarf plants that require no support are used in massed bedding. Dahlia has almost all the colours, *viz.*, white, cream, yellow, orange, pink, crimson, rose, red, maroon, mauve, purple, brown, bicoloured and multicoloured but lacks true blue, green and black colours.

Botany

It is a tuberous-rooted (root tuber) plant, the tuber having growing points only in the crown where it is attached with the aerial stem, and when during transit or otherwise the crown is damaged the tubers do not sprout, stems hollow and mostly erect and branched, glabrous or scabrous, leaves opposite, pinnatifid, heads long-peduncled consisting of several hundred individual florets in a cyme, small to large (10 to 25 cm in diameter), open to ball shape, disc yellow and fertile, ray florets neutral or pistillate and spreading, and fruits oblong or obovate with many flat seeds.

Dahlia tubers contain significant amounts of inulin and fructose and small quantities of medicinally active compounds such as phytin and benzoic acid (Whitley, 1985). Nordström and Swain (1953) extracted flavonoid **glycosides** such as cyanidin, apigcnin and lutcolin from *D. variabilis* cv. 'Dandy'. Harborne *et al.* (1990) extracted malonylated chalcone glycosides (the yellow pigments, *viz.* 4'-malonylsophoroside and 4'-malonylglucoside of butein) in the petals of *D. variabilis* (*D. pinnata*) cv. 'Cocquencourt' and *D. coccinea*. Forkmann and Slotz (1984) stated that flavanone 3-hydroxylase, which catalyses the conversion of flavanones to dehydroflavonols, was detected in flower extracts of cyanic strains of dahlia, but not in the white-flowering mutants. Fischer *et al.* (1988) reported that flowers of *D. variabilis* (*D. pinnata*) cv. 'Scarlet Star' at early developmental stages contained relatively high activity of

(+) – dihydroflavonol (DHF) 4-reductase containing only one polypeptide chain with a MW of about 41000, and the reductase required NADPH as co-factor and catalysed the transfer of the pro-S hydrogen of NADPH to the substrate. Flavanones and dihydroflavonols (3-hydroxyflavanones) were substrates for DHF reductase with pH optima of about 6.0 for flavanones and of about 6.8 for dihydroflavonols. Flavanones were reduced to the corresponding flavan-4-ols, and (+)-dihydroflavonols to flavan-3, 4-*cis*-oliols. *D. pinnata* polyacetylenes contained the sufficient amount of an ene-diynediene chromosphore which was extracted only in traces in *D. coccinea* (Bendixen *et al.*, 1969), *D. imperialis* roots contained 10 acetylenic compounds and *D. tenuicaulis* 9 (Lam, 1971), and 26 polyacetylenes of known structures together with cosmene and eugenol were isolated from roots, leaves and flowers of *D. australis*, *D. coccinea* and *D. sherffii* (Lam *et al.*, 1991). Kannangara and Booth (1979) reported that transfer-RNA extracts from radicles of *D. variabilis* showed cytokinin activity upon hydrolysis, and the properties of the active component were similar to those of 6-Δ^2-isopentenyladenosine.

They have haploid **chromosome** numbers of 16, 17, 18 and 32 with most of the garden cultivars including *Dahlia variabilis* being tetraploids having 2n = 64. Lawrence (1942) stated that since Humboldt sent home the seeds of dahlia in 1804, it took 12 years for evolution of every colour and form what we find today. The reasons for the profusion of form and colour are: i) that dahlias being self-incompatible can never breed true, ii) the factors producing flower colour are in duplicate, 2 for flavone (ivory and yellow) and 2 for anthocyanin (pale and deep pigmentation), and iii) each of these factors is represented 4 times. It was therefore evidently suggested that the plant arose as a hybrid between polyploid members of 2 colour groups and that doubling of the chromosome number accompanied this hybridization.

Suman *et al.* (1984) while studying the genetic divergence in 39 dahlia cultivars for seven quantitative traits, grouped them into nine clusters and through principal component analysis major portion of variability in the populations was explained by the first two canonical roots. Choudhary (1989) through path analysis reported that branches per plant exerted greatest direct effect on tuber production though maximum indirect effect on tuber number per plant was exerted by branches per plant via days to first flowering. Genotypic and phenotypic correlations between branches and tubers per plant were high (r = 0.48 and 0.43, respectively).

Classification, Species and Varieties

Dahlia is a genus of some 27 species where cultivars and hybrids run in thousands (more than 20,000). Modern garden dahlias are derived from three Mexican species, *Dahlia pinnata* (low growing up to 90 cm height, leaves roundish, leaflets up to 25 cm long, flowers large and double of purple colour or red with bluish tinge), *D. coccinea* (*syn. D. bidentifolia, D. cervantesii, D. crocea*) has pinnate or bipinnate leaves, flowers on the stem appear at 90 cm height, as single red or scarlet, sometimes with orange or yellow centres, and is summer flowering, and *D. rosea* (leaves typically one pinnate, sometimes bipinnate, and the flowers are single rose, and is the parent of old fashioned dahlias, *viz.*, Single, Pompon, Show and Fancy types, sometimes double forms are also available) which may themselves all originate from one species, *D. variabilis* (sometimes semi-double type of *D. rosea* is called as *D variabilis*). *D. imperialis*, an erect species up to 8-9 meters tall with 2-3-pinnately parted leaves, flowers 10-18 cm across, white or purple during summer and autumn, grows as escape on Indian hills throughout the country and is known as tree dahlia. *D. scapigera* is a dwarf species, some 50 cm high, leaves pinnate, segments 3-7, capitula few, usually solitary, slightly drooping, ray florets white to pale purple, base sometimes yellow, disc florets yellow, apex purple, and flowering during summer and autumn. *D. excelsa* (syn. *D. arborea*) grows up to 6 metres or more, stems become woody in mild climates, leaves up to 75 cm long and 60 cm wide with as many as 25 leaflets, flowers purple or crimson-pink and up to 12 cm across, *D. juarezii* grows to 90-150 cm, is cactus type with florets of irregular length and overlapping, flowers bright scarlet, and flowering in mid- to late-summer, and *D. merckii* (syn. *D. glabrata*) has bipinnate leaves, flowers lilac, rays pistillate and is much valued for its more attractive foliage.

Broadly, dahlias are divided in two groups: border dahlias most suitable for mixed borders, and bedding dahlias which are grown as annuals from seeds.

Border Dahlias

Border dahlias flower freely for 4-5 months, on the hills from July onward until cut down by the frost though in the plains from December to April and sometimes even up to May when the flowers dry up due to scorching sun. Border dahlias are classified into 10 groups.

Single-Flowered

Height 45-75 cm, planting distance 45-60 cm, flower diameter up to 10 cm, each flower has only one outer ring of florets which may overlap and a central disc, and flowering by July end on the hills and in January in the plains. This can further be divided into *show singles* (flower heads not more than 7.5 cm across, only 8 rays, smooth, slightly recurved at tips and overlapping in a perfect round form), *singles* (rays not so overlapping as in *show singles* and tips separated) and *mignon* dahlias (similar to *singles* but plants not more than 45 cm high). It is ideal for pot culture or patio plantings. The varieties are 'Bambino' (white), 'Chessy' (bronze-yellow), 'Coltness Gem' (a dwarf bedder, seed-sown various colours), 'G.F. Hemerik' (soft orange), 'Inflammation' ('Red Liliput', red), 'Irene van der Zwet' (soft yellow), 'Murillo' (lilac-pink with deeper centre), 'Nellie Geerlings' (scarlet), 'Orangeade' (flame), 'Princess Marie Jose' (lilac-pink), 'Reddy' (red), 'Sion' (bronze), etc.

Anemone-flowered

Height 60-105 cm, planting distance 60 cm, flower diameter up to 10 cm, double flowers having flat outer florets surrounding a densely packed group of shorter tubular florets, and flowering from July end on the hills and in January in the plains. Ideal for cutting and for floral arrangements. Varieties are 'Asahi Chohje' (red and white outer petals and white centre), 'Bridesmaid' (white outer petals, lemon centre), 'Comet' (maroon), 'Evita' (lilac with orange centre), 'Floorimoor' (pink-ornage), 'Guinea' (yellow), 'Honey' (apricot-pink), 'Lucy' (outer petals purple, centre yellow), 'Purple Puff' (purple), 'Scarlet Comet' (a lighter red sport of Comet), 'Siemen Doorenbosch' (light magenta), 'Slow Flox' (pink-yellow), etc.

Collerette

Height 75-120 cm, planting distance 60-75 cm, flower diameter 10 cm, flower heads with one or more series of flat ray florets as in group I, ring of florets (the collerette) only half the length of rays and usually of contrasting colour to the main rays. This may further be divided into (a) *Collerette Singles* (heads with single series of true rays and one collerette with yellow disc), (b) *Collerette Paeony-flowered* (flower heads with 2-3 series of rays and collars with yellow disc), and (c) *Collerette Decorative* [as to that of b) but fully double]. These start flowering from late July. The varieties are 'Can Can' (pink petals, yellow collar), 'Claire de Lune' (yellow, paler-yellow collar), 'La Cierva' (purple, tipped white on the outer petals, collar white), 'La Giaconda' (scarlet with golden yellow collar), 'Mrs. H. Brown' (orange and flame petals, yellow collar), 'Nonsense' (cream white petals, orange collar), 'Ruwenzori' (scarlet petals, yellow collar), etc.

Waterlily or *Nymphaea*-Flowered

Flowers fully double with broad and generally sparse flat or slightly with recurved or incurved margins of petals, giving overall effect of a flat or shallow bloom, height 90-120 cm, flower diameter 10-12 cm and blooming in early August on the hills. The varieties are 'Peace Pact' (white), 'Pearl of Heemstede' (dark pink), etc.

Decorative

Blooms fully double, showing no central disc, with broad, flat or slightly involute ray florets, seldom twisted and usually bluntly pointed. This may further be classed into two: *Formal Decorative* (fully double with no visible disc and arrangement of florets regular with slightly incurved margins, tips flattened and pointed or rounded) and *Informal Decorative* (as *Formal Decorative* but rays irregularly arranged). As per size of the blooms, these are further divided in **Giant** with flower size >25 cm [varieties include 'Burgess Ray' (yellow), 'Crossfield Festival' (bicoloured, light red and white), 'Hamari Girl' (pink), 'Holland Festival' (rich orange with white-tipped petals), 'Jocondo' (purple), 'Lavender Perfection' (lavender-pink), 'Liberator' (crimson), etc.], **Large** with flower size 20-25 cm [varieties include 'Bunratty' (purple), 'Dutch Triumph' (pink, yellow centre), 'Enfield Salmon' (pink), 'Robert Damp' (yellow), etc.], **Medium** with flower diameter of 15-20 cm [varieties include 'Betty Russell', syn. 'Yellow Terpo' (yellow), 'Breckland Joy' (bronze), 'Deuil du Roi Albert' (violet-purple tipped white), 'Edinburgh' (maroon and white), 'First Lady' (yellow), 'Golden Turban' (shades of yellow, orange), 'Majuba' (blood-red), 'Peter' (purple-pink), 'Peter's Glory' (lilac-rose, white centre), 'Red & White' (red & white), 'Rosella' (lilac-pink), 'Snow Country' (white), 'Sterling Silver' (white), 'Terpo' (crimson), etc.], **Small** with flower diameter of 10-15 cm [varieties include 'Arabian Night' (maroon, appearing quite black), 'Chinese Lantern' (yellow splashed orange-red), 'Dedham' (shades of pink and mauve), 'Gerrie Hoek' (deep pink), 'Glorie Van Heemstede' (yellow), 'House of Orange' (amber-orange), 'Millbank Inferno' (shades of red orange), 'Snow Queen' (white), 'That's It' (flame), 'Towneley Class' (white, flushed lavender), etc.], and **Miniature** with up to 10 cm flower diameter [varieties include 'David Howard' (orange), 'Doris Duke' (pink), 'Jo's Choice' (scarlet), 'Lilianne Ballego' (bronze), 'Newby' (peach), etc.].

Ball

Fully double ball-shaped blooms, seldom flattened at top. The sides of the ray florets curve inwards for more than half its length, tips blunt or round and spirally arranged. This is divided into **Ball** with 90-120 cm height, planting distance 75 cm and flower diameter 10-15 cm and the varieties are 'Bernard Colwyn Hayes' (deep pink), 'Doreen Hayes' (scarlet), 'Esmonde' (yellow), 'Gloire de Lyon' (white), etc., and **Miniature Ball** with 90-120 cm height, planting distance 75 cm and flower diameter up to 10 cm, and the varieties are 'Attlami Cherry' (red), 'Blyton Scfter Glean' (orange-purple), 'Camano Candy' (pink) 'Cornel' (red), 'Crusader' (red), 'Dr. John Grainge' (orange), 'Florence Vernon' (lilac), 'Nelly Birch' (scarlet), 'Snowfall' (white), 'Risca Miner' (purple), 'Sulphurea' (yellow), 'Swiss Miss' (shades of crimson-purple and dark pink), 'White Polventon' (white), etc.

Pompon

Height 90-120 cm, planting distance 75 cm, flower diameter up to 5 cm, similar to ball dahlias but much smaller and more globular, florets curving inwards along their whole length and flowering late July. The varieties are 'Albino' (white), 'Andrew Lockwood' (lilac), 'Deepest Yellow' (deep yellow), 'Little Conn' (dark red), 'Lydia' (red), 'Moor Place' (purple), 'Nero' (dark red with purple sheen), 'Pom of Poms' (scarlet), 'Potgieter' (primrose yellow), 'Rhondda' (lilac, centre white), 'Stolze von Berlin' (pink), 'Zonnegoud' (canary yellow), etc.

Cactus

Cactus and semi-cactus groups are largely derived by crossing *Dahlia hortensis* with other species, both groups are further divided on the basis of flower size. The cactus types have fully double blooms showing no disc and with long pointed ray florets, finely quilled (margins strongly revolute) for more than half its length. They are further divided into **Giant** with 120-150 cm plant height, 120 cm planting distance, flower diameter >25 cm and flowering by last August on the hills [varieties are 'Danny' (pink), 'Gladys M. Reynolds' (bronze), 'Polar Sight' (cream white), etc.], **Large** with 120-150 cm height, planting distance 120 cm, flower diameter 20-25 cm, and flowering in August [varieties are 'Cum Laude' (deep pink), 'Drakenburg' (purple), 'Irish Visit' (crimson), 'Royal Highness' (pink), 'Soest Vooruit' (orange), etc.), **Medium** with105-135 cm plant height, planting distance 90 cm, flower diameter 20-25 cm and flowering in early August [varieties are 'Authority' (shades of bronze and orange), 'Priscilla' (flame),

'Sure Thing' (red), 'Topaz' (yellow), 'Tornado' (orange-red), etc.), **Small** with 105-120 cm height, planting distance, 75 cm, flower diameter 10-15 cm and flowering in late July [varieties are 'Doris Day' (crimson), 'Klankstad Kerkrade' (yellow), 'Paul Chester' (orange with yellow centre), 'Worton Sally Ann' (lilac), etc.], and **Miniature** with 90-120 plant height, planting distance 75 cm, flower diameter up to 10 cm and flowering by late July [varieties are 'Charmer' (purple), 'Chasamay' (pink), 'Hazel' (lilac), 'Pirouette' (yellow), 'Poppet' (blended salmon and yellow), etc.].

Semi-cactus

Blooms are fully double with pointed ray florets, broader and quilled for less or half their length. Similar to the **CACTUS** group, this is also divided in five sub-groups with the same characters as to their counterparts in **CACTUS** group. However the varieties are 'Cocorico' (scarlet), 'Frontispiece' (cream white), 'Hamari Boy' (yellow), 'Respectable' (yellow), etc. in **Giant** sub-group, 'General Mourer' (purple), 'Nantenan' (yellow), 'Royal Sceptre' (orange to deep yellow), 'Royal Wedding' (flame with yellow centre), etc. under **Large** sub-group, 'Apache' (bright red), 'Autumn Fire' (apricot -yellow tinged flame at tip), 'Countess Dunraven' (pink), 'Hamari Bride' (white), 'Rotterdam' (crimson), 'Sweet Secret' (mixed deep pink and yellows), etc. under **Medium** sub-group, 'Hoek's Yellow' (yellow), 'Marilyn' (pink), 'Rothesay Red' (crimson), 'White Swallow' (white), etc. under **Small** sub-group and 'Happy Mood' (yellow), 'Little Ann' (pink), 'Lynne Bartholomew' (cherry-purple), 'Snip' (bronze tinged orange), 'Yellow Mood' (yellow), etc. under **Miniature** sub-group.

Miscellaneous

This group contains small, dissimilar classes *viz.* **Orchid** cultivars which are similar in shape to the Single dahlias except that their ray florets are revolute for at least two-thirds of their length as var. 'Giraffe', banded yellow and bronze; the **Star** dahlias with small incurving flowers, consisting of 2-3 rows of scarcely overlapping pointed rays encircling a central disc; the **Chrysanthemum** types looking incurved exhibition chrysanthemum as var. 'Akita'; and the tiny **Liliput** series with 2.5 cm blooms. Others which are included in this group are **Paeony-flowered** (originally group 4 of the British National Dahlia Society classification) which grows 90 cm, planting distance 60 cm, flower diameter 10 cm, the flowers consist of two or more rings of flat ray florets and a central disc, flowering from late July and the varieties are 'Bishop of Llandoff' (scarlet), 'Fascination' (purple), 'Grenadier' (scarlet), 'Oranje Flora' ('Orange Flora', orange), etc.

Senaratna (1947) developed two new paeony-flowered dahlia hybrids which he named 'Fr. Le Goc' and 'Ceylon Sunrise'. Protich (1988) from Bulgaria also described 5 new decorative-flowered cultivars, *viz.* 'Topajo', 'Sourie de Cozon', 'Esmeralda', 'Faluna' and 'Rosela', and 2 cactus-flowered, *viz.* 'Véritable' and 'Top Choice', all flowering for two months during July, August and September and were suitable for cut flower production. Broertjes and Ballego (1968) developed five cultivars through irradiation of the cv. 'Arthur Godfrey'. Singh (1970) isolated one spontaneous mutant of rosy-magenta flowers with more attractive shape than its parent cultivar 'White Pearl', and suggested that an inhibitor gene was responsible for the new type. Dohare *et al.* (1974) released one cv. 'Manali' of pure white colour obtained through spontaneous mutation. Das *et al.* (1978) through ^{60}Co gamma irradiation of 14 dahlia cultivars developed 19 flower colour mutants under 2-3 krad, out of which 11 have been released as varieties. In dahlia, there is progressive degeneration in terms of branching, the flower numbers and head structures, and tuber size the cause for which was attributed by Lavrichenko (1975) as the ontogenic character associated with age changes in the shoot and root structure whether grown from cuttings, tubers or seeds. In a varietal trial at Lucknow with 22 dahlia cultivars, Sharga (1975) recommended 'Axford', 'Triumph', 'Black-Out', 'Black-Out Sport', 'Cheroky Beauty', 'Ginga', 'Kelvin Rose', 'Mrs. John Ralph' and 'Nirmal Chandra' as most promising. Through a trial of new and old cultivars of dahlia in Germany, Mehlis (1976, 1978a, b) found 'Bodenseeperle', 'Kurdirektor Diekmann' and 'Rote Spinne' as outstanding in the first study while out of 27 and 25 cultivars in the second and third studies, respectively, 'Hanna Hofbauer' (yellow, semi-cactus), 'Pipers Pink' (semi-cactus), 'Stadt Wiehl' (wine-red, cactus) and 'Wolstad' (lilac, decorative) were found outstanding. Mishra *et al.* (1990) with seven cultivars reported that cvs 'Kenya' and 'Vigour' were outstanding.

Bedding dahlias are grown from seeds. These grow 30-50 cm in height, planted at 30-50 cm across, flowers of almost all the hues and colours and may be single, semi-double or double. Flowers measure 5.0-7.5 cm across and appear from June end and continue until the occurrence of first frost on the hills, but in the plains these continue blooming from December to April or May when flowers burn due to scorching sun. This group has many seed strains and mixtures out of which 'Coltness Hybrids' is the best known group, seeds with mixtures and separate colours being available in this group. 'Pink Radiance' has single rose-pink flowers with a bold crimson zone. 'Unwin's Dwarf Hybrids' produces semi-double flowers in a wide range of bright colours having yellow, orange, pink and crimson. 'Early Bird' is a selected dwarf form (growing hardly to 30 cm high) with flowers in the similar colour range.

'Fascination' is a pink or purple, and paeony-flowered dwarf variety. 'Princess Mary Jose' produces single lilac-pink flowers. 'Redskin' (mixed colours, double and semi-double) and 'Rigoletto' (mixed colours: white, yellow, orange, pink and red, with double flowers) are two other outstanding varieties being grown from seeds.

Propagation

Dahlias are propagated through **seeds**, by tuberous rooted divisions, stem cuttings and by *in vitro* methods. Small flowered dahlias are propagated through seeds. Safronova (1960) obtained fully viable seeds from open pollination of double dahlias and obtained best germination by sowing the seeds directly in the open ground, and the plants were healthier, flowers were better and the tubers were sturdier than their parents. For seedling sale these are raised in plugs and for this the seeds are sown in the plugs, just covered for

3-4 days and kept at 26-27° C, then they are provided with 20-21° C temperature for 7 days and fed with 50 ppm N per week, then for 7 days at 18-21° C with 100 ppm N once or twice per week, and finally they are kept at 15-17° C for 7 days and fed with liquid fertilizer as per requirement (Ball, 1991). This way the plugs are ready for sale within 3-4 weeks. For usual seed growing, the bedding dahlias are grown through seeds. Seeds are sown 3 cm apart in boxes or pans filled with seed compost. For early growing these may be started inside the glasshouse in February otherwise in the open in March. After sowing they are covered with a very thin layer ($^1/_3$ cm) of soil and watered with a fine rose can where within 15 days all the seeds will germinate at 13-16° C temperature. As soon as these grow enough for handling, these are planted outside individually in the standard potting mix at around 10-13° C night temperature for further growth and hardening. These require full sun but not piercing one at seedling stage. For field planting, these can be used when these have attained 5 cm height. Now these are treated as usual as other plants where these will flower the same season like to those grown through tubers. Aoba *et al.* (1960) when planted dahlia seedlings in first week of June in Japan and at every 10 days until November a batch was being dug out to observe their root development. It was found that adventitious roots started appearing from early June and continued till early August and these roots developed into tubers by October. Early foliage removal (not late), and shade treatment reduced root diameter but SD (8 h for 20 or 40 days) caused reduction in the number of roots though increased root diameters. In dahlia one gramme may contain some 98 seeds.

Germinating dahlia seeds on moist filter paper and when their roots are 0.2 to 0.5 mm long the halving of young plants longitudinally and then the two halves are induced to grow independently into normal plants (Rudionenko and Zaar, 11951), and this way one can get two **identical genetic plants** from one seedling.

Propagation through **stem cuttings** is the most commercial method. For this healthy mother tubers are selected and planted in March (temperate areas) or in October (plains) under full sunlight with their collars just peeping out and the night temperature should not be less than 13° C. The soil at planting should have sufficient moisture so that up to sprouting there may not be any need for watering. After a week young shoots may emerge. These are allowed to develop until these attain 8 15 cm height. Now these are removed with a sharp knife, cutting close to the tuber but without removing any part of it. Preferably the cuttings should be 7.5 cm long. This way each mother tuber shall be producing 20-30 cuttings, depending on the cultivar. After dipping them in rooting hormone, these are planted by inserting its half lower part by removing the leaves if any, in pure white sand in flats or pots at 18-24°C temperature where it may take 10-18 days for rooting, depending on the temperature range provided. After planting, the pots should be watered and covered with a plastic tent quite above the seedlings with the help of some frames. After 3-4 days these may be opened through side flaps for ventilation but after 8-10 days these tents are completely removed. After rooting, these may be potted in individual pots containing 7 parts top soil, 3 parts peat and 2 parts sharp sand and then taken to a frame until planting outdoors in beds in May on the hills or in November-December in the plains. In the frame, a night temperature of 13° C is provided. Within 8 weeks, the cuttings will produce a tuber having 2.5-4.0 cm in diameter, along with three or more tubers. It takes some six weeks from the time they are first taken until of planting size. The most suitable soils for its growing are sandy loam at a sunny situation having pH from 5.5 to 7.5. Marston and Evans (1972) took nodal cuttings of dahlia cvs 'Autumn Fire' and 'Hamari Bride', inserted in a peat or perlite compost under mist without bottom heat in last week of June, and after rooting these were potted in JIP compost No. 1 in 9-cm plastic pots on a capillary bench in a cool greenhouse. Fibrous roots grew through the pot holes into the sand and anchored the pots, and without feeding these plants were allowed to flower. After flowering, in mid-October the pots were removed and the plants dried off gradually before the tops were cut to leave about 7.5 cm of stem, and then average fresh weight per pot-tuber was recorded as 48 g for 'Autumn Fire' and 50.5 g for 'Hamari Bride'. These tubers stored well and were later used to produce cuttings.

After planting the mother **tuber**, it sends out one or many aerial shoots which afterwards form many enlarged tuberous roots as the primary storage organ. The development of these roots is facilitated by the short days. When plants have matured, the aerial part is cut down and the bunch of these tuberous roots are gently pulled out from the soil with the help of fork or shovel, cleaned and stored in a well ventilated room in saw dust or sand from June to September in the plains and from December to March on the hills. At planting time they are separated gently with the help of a sharp knife or just by pulling them gently from one another in a manner that crown (where it is attached to the aerial shoot) is intact on the tuberous root, each crown having at least one eye (adventitious bud). Sometimes there may be only one-eyed two-tuberous roots jointed together, and in such cases both the tubers should be taken as one entity while separating from the aerial shoot. On planting, these root tubers will produce new plants which in turn will produce some 6-15 root tubers per plant by the time they are lifted. If minimum size has been attained there is insiginificant difference between various sizes of root tubers, and the performance of tubers weighing 80 grammes goes at par to those tubers which weigh 120 grammes. Reduction in number and/or size of the flowers increases the tuber weight (Wasscher, 1955; Newton, 1960). Newton and Tullis (1962) stated that flower bud removal, especially in late flowering cultivars, increases tuber weight, and the presence of flowers even for a brief period reduces the weight of tubers. Dahlia cuttings taken in April, May or June, grown in peat/perlite compost in pots containing JIP-2 or peat/perlite in greenhouse under natural day length and subjected to 10-h SD from mid-June formed good tubers by mid-October (Lloyd, 1968). Read (1968) through foliar sprayings of Alar at 2500 ppm and CCC at 1500 ppm found increased size (up to 200 per cent) and number of tuberous roots in dahlia grown under short day conditions conducive to tuberous root formation, and the development of tuberous roots on cuttings grown under LD conditions was also enhanced by these treatments whereas control produced only fibrous roots. Moser and Hess (1968) found critical day length of 11-12 hours and 15.6 or 21.0°C night temperature for maximum tuberization in cv. 'Sneezy' though 10.0 or 26.7°C night temperatures inhibited it, and temperature could not alter the SD requirement for tuberization. Spraying with ABA or ethephon promotes tuberization in dahlia plants in long days though less effective than at short days, SADH and ethephon treatments of budless leaf-cuttings inhibit tuberization whereas ABA slightly enhances it, and after the start of SD treatments endogenous ethylene evolution reaches a peak between second and third week, *i.e.* one week before the onset of tuberization and then decreases to the low level under LDs (Biran *et al.*, 1972). Read *et al.* (1972) when sprayed SADH at 2500 ppm and chlormequat at 1500 ppm found increased number and size (sometimes even three times) of tuberous roots grown under SD in plants of *Dahlia pinnata* cultivars irrespective of whether the cultivars were quick tuberous-root forming or those forming poor quality roots, and these plants were also excellent in growth. Even under LD conditions these treatments caused formation of tuberous roots but untreated cuttings produced only fibrous roots. Èaplygin and Šahova (1961) suggested supply of cheaper planting material by taking of **cuttings** in June instead of usual March or April when heating is also necessary and these cuttings root to a tune of 87 per cent within 10 days. They also advocated that keeping of rooted cuttings in soil in pots at 2-4°C overwinter better than those stored in the usual way in sand. Contrary to this, Mustafajeva (1985) advocated taking of cuttings in March-April in a perlite + sand mixture for up to 100 per cent rooting, and then planting 50 cm apart both ways. The cuttings trimmed to a node were when inserted in the soil or medium though produced more but smaller tubers than untrimmed cuttings which produced slightly less but heavier and longer tubers and these tubers produced more cuttings (Newton, 1962). Biran and Halevy (1973) found an inverse relationship between rooting percentage of the cuttings and growth rate of the buds in the cuttings which sprouted during the rooting period, reproductive buds suppressing rooting more than the vegetative ones, and removing the growing point together with a few terminal leaves there was an increase in the rooting percentage, particularly in largest cuttings which root poorly (33 per cent) than the smaller ones (85 per cent). Maiko and Yashchenko (1980) found early tuber formation with improved subsequent tuber growth through pre-plant soaking of dahlia cuttings in Alar at 20 mg/litre though foliar spraying with ethrel at 100 mg/litre or Alar at 200 mg/litre was even more effective than dipping, and the flowering in these treatments was also stimulated but only in dwarf cultivars. Cayeux (1936) and Safranova (1972) advocated dahlia propagation through heel cuttings (single leaf cutting with a small portion of the main stem attached) by taking two cuttings from each node (leaves are opposite on the stem) in April-May and the new growth strikes from the axil of the leaf, irrespective of whether at the time of taking the cutting the buds were visible or not, if planted in turf soil and sand in a 1:2 ratio in a cold frame. These cuttings should be planted out at 50 x 30 cm spacings, followed by saw dust mulching immediately.

This may be cultured *in vitro* through **meristem tip culture** but it is not a commercial practice (Morel and Martin, 1952). Through this practice one can get rid of 'dahlia mosaic viruses' (Mullin and Schlegel, 1978) and 'tomato spotted wilt viruses'. Gavinlertvatana *et al.* (1979, 1982) when grew *in vitro* the leaf segments of stock plants grown under 8 h SD or 16 h LD, on a modified MS agar medium containing various levels of NAA and/or kinetin, larger callus and higher levels of ethylene and CO_2 were detected when stock plants grown under SD were further incubated under SD (SD to SD) than SD to LD, and tissues under LD to LD or LD to SD formed similar amounts of callus to SD-LD cultures but with higher ethylene and CO_2 levels. Daminozide spray on SD stock plants one day before taking the explants promoted callus formation, *vis-a-vis* ethylene and CO_2 production in tissues incubated under either SD or LD though spray on LD stock plants had no effect on callus or ethylene but CO_2 level was found increased under SD or LD incubation. They further stated that the ethylene accumulation in the flask atmospheres was dependent mainly on NAA and to a lesser extent on kinetin levels. The usual propagation methods in dahlia provide ample opportunity in multiplying the elite germplasm lines (varieties) rapidly therefore tissue culture remains only of academic interest and not a commercial proposition.

Soils, Preparation of Land, Planting, Pot Culture and Weed Control

Dahlia requires an open sunny location where temperature ranges from 13-25° C. Temporary rise or fall of temperature is immaterial until it persists for long. Though it can be grown in a wide range of **soils** but a porous and rich soil which has good water-holding capacity but not water-logged, is most suitable for its growing provided its pH is in the range of 5.5-7.5, preferably 6.5-7.0. Heavy soils may also be improved with addition of organic matter and sand. Dahlia responds well to nutrients, hence for its cultivation soil should have sufficient humus, more than 3 per cent of organic matter.

The land where dahlia is to be grown should be quite deep, *i.e.* more than 30 cm. Three deep ploughings before planting are required, followed by planking each time. Land preparation should start two months prior to planting and a green manure crop may also be taken up which may be knocked down just after four weeks of sowing after filling the water in the field so that it decomposes thoroughly. With second ploughing, farmyard manure or compost at the rate of 50 tonnes per hectare should be mixed in the soil so that any weed seed present in it may germinate by the third or final ploughing. All the rootstocks of perennial weeds should also be taken out at the time of **bed preparation**. At the time of third ploughing there should be sufficient moisture in the soil so that after planting it may require water only after sprouting. The beds after third ploughing are properly levelled so that unnecessary rain or otherwise water may not stand anywhere in the field. Proper chanelling and bunding are laid out keeping width of the beds two metres and length to any convenient size. **Planting** is carried out normally in March-April on the hills and in October in the plains, giving

75 x 75 cm **distance** for taller varieties and 45 x 45 cm for dwarf, pompons or bedding types. De Hertogh (1996) stated that in the Northern Hemisphere the tubers are planted from January to April for greenhouse forcing. Quagliotti and Gullino (1967) out of the four spcings (35 x 50, 45 x 65, 55 x 80 and 65 x 95 cm) tried, obtained increased flower heads per plant with increased spacing though per square metre yield was found reduced while at close spacing the yield was more. Armitage (1993) recommends 90 x 90 cm spacing for ease of harvest and also for prevention of diseases, but this may not be economical. While planting, the crown should be covered only with one cm of soil and not more. Root tubers require 8-12 weeks to flower from planting, and 10-12 weeks to flower from seeds, provided seeds are already grown and shifted to 10-cm pot which requires 1-3 weeks further. Bose and Mukhopadhaya (1967) recorded that in rooted cuttings of dahlia tuber formation starts within 30 days of planting, being most active during the period of maximum vegetative growth and flower initiation, contents of total and non-reducing sugars in the stems increases markedly during flowering, followed by an abrupt decline at the peak flowering, leaf contents of sugars decline sharply when the rooted cuttings are planted and remains low until the final sampling when flowers start fading, the contents of soluble N in leaves and stems decline appreciably when the buds become visible.

In **pots**, the dwarf ones and bedding types are grown. The medium in pots should be well-drained and should contain sandy garden loam, sharp sand and well rotten farmyard manure, all in equal volume with a little bone meal (15-20 g per pot). Winter (1977) used four types of media with or without a basic slow-acting inorganic fertilizer, and drenching with four rates (0, 2, 4 or 8 ml/15-cm pot) of ancymidol (Reducymol) to dahlia cv. 'Miramar', and recorded best growth and shoot production with advanced flowering in compost medium with 6 kg/m^3 of 14:14:14 Osmocote but 8 ml/pot of ancymidol delayed flowering only by 2-3 days though the same strength was quite effective in a mixture of black soil, sphagnum moss and FYM to overecome strong vegetative growth. He found 4 ml adequate for a 1:1:1:1 mixture of black soil, sphagnum peat, clay and perlite mixture or a similar mixture + sand, whereas only 2 ml were required for a sand medium.

Barrett and Hertogh (1978) stated that **pinched** plants produce more flowers but with smaller and delayed flowering, pinching the terminal at node four being the optimum, and the pinching at the more distal side encourages formation of more laterals. Pinching is practiced when as per the requirement of the variety the shoots have not appeared or only one shoot is developing per tuberous root though more shoots are required, when more profusion of blooms are required or when delayed flowering is desired. Vigorous dominant shoots should also be pinched on multistem plants. **Disbudding** the terminal flower should only be done when flowering is ahead of schedule for shows or for marketing, and this practice will force the axillary flowers to develop, though for cut flower growing opposite is done, axillary flower buds are removed to increase size of the terminal flower (Dole and Wilkins, 1999).

In dahlia plantings, **mulching** is carried out with grass clippings, old hay, saw dust, straw, dried leaves, compost, etc. which helps in conserving moisture, maintains an even soil temperature, keeps down the weeds and in the end decomposes and provides organic matter to the field. Black polythene does not allow weeds to grow, conserves the moisture and increases soil temperature. Simazine, Lenacil and Linuron are effective **weedicides** for controlling weeds in dahlia plantings. Clay and Ivens (1964) recommended simazine application under moist soil conditions in sandy loam soil at 1.15 kg/ha 11-14 days after planting and chlorpropham at 2.25 kg/ha + diuron at 0.9 kg/ha as quite safe though there may be varietal differences. Clay and Davison (1968) recorded lenacil at 2.25 kg and 4.5 kg/ha spraying as the most promising in dahlia field as compared to chloramben, linuron, prometryne, simazine and trifluralin.

Irrigation and Nutrient Management

Dahlias require copious **watering** throughout the growing period. In the plains they are grown during winter season where these require watering at 10-days interval from November to February and at 5-7 days interval in October, March and April but on the hills they are watered every week as it is grown during summer. After every application of fertilizer also these are immediately watered. If there had been rains, the irrigation is adjusted accordingly. These are watered only when these need it. The plants planted in pots are watered every fourth day. Cavalchini (1968) recommended the soil moisture tension in a plastic house to dahlia plants at 0.25 and 0.1 atm during summer and 0.25 atm during late summer and early autumn.

For successful dahlia cultivation, balanced **nutrition** levels are required at right time. Nitrogen at 100 kg, phosphorus at 125 kg and potassium at 100 kg per hectare of land should be applied for getting flower yield with good quality. Though phosphorus and potassium are applied as basal fertilizers but nitrogen can be given in three split doses, one-third as basal dose, and other one-third at the time when these are at the youth of their vegetative growth and the remaining $^1/_3$rd when the plants have started producing reproductive buds. Potassium fertilizers may be applied only when soil is lacking in this element. However, the use of fertilizers should be based on soil and leaf analysis. Sang (1974) applied $(NH_4)_2SO_4$, NH_4NO_3 and urea to dahlia plants grown in sandy-loam soil, in sandy-clay-loam soil and in a loam soil in Korea and found ammonium sulphate promoting most growth, followed by ammonium nitrate but urea least effective, however, tuber yields were found highest with ammonium nitrate and lowest with urea where flowering was also poor. Regarding the media, ammonium nitrate was more effective as the clay content increased. Through four applications (mid-July to mid-October) of ammonium nitrate at 20, 40 or 60 g N/ha in Italy, only flower numbers were found significantly increased with the lowest rate (Sasso, 1961). El-Gamassy and Moustafa (1963a, b) had grown dahlias in Cairo with high, medium and low levels of N, P, K, Ca and Mg solutions in various combinations and recorded best growth, flowering and tuber yield from highest dose of complete treatments, N and P being the most important elements but where K and Ca were omitted the result was slightly poor. They stated that though with age N, P and K contents in plant parts decrease in dry weight basis but total amounts were found increased though Ca increased with age, and the omission of N reduced the N, P, K and Ca contents of the plants and omission of P, Ca or K also reduced N content of the plants, however, absence of K resulted in a slight increase in the percentage of Ca. El-Gamassy and Moustafa (1964a, b) applied various amounts of calcium nitrate, calcium superphosphate and potassium sulphate to dahlias in pot and field trials and recorded best growth, flowering and tuber yield with a fertilizer mixture of 46:60:10 g/plant of calcium nitrate : calcium superphosphate : potassium sulphate, followed by one of 40:40:20, those receiving P and K only were poor than those also receiving N. No superphosphate or only 20 g/plant reduced the number of flowers and increase in the K levels in presence of N and P increased the number of flowers but high rates of P or 30 g of potassium sulphate reduced the flower length and weights. Calcium nitrate 40 g, calcium superphosphate 60 g and potassium sulphate 20 g to the potted dahlias with nitrate being added as 3-monthly dressings and other compounds either at or after planting, they obtained best results in terms of growth, flowering and tuber yield with P and K applied in 3-monthly doses starting one month after planting. Rahman and Mitra (1974) applied N at 0, 28, 56 or 84 kg/ha and P and K at 28 kg/ha each, and obtained maximum flower size and number at 56 kg N though 84 kg N yielded highest tuber yield and earliest flowering, though plants receiving no N were poor in growth. Boertje (1979) in Holland found 1.5-2.0 g/l of 17:6:18 NPK fertilizer when applied four weeks after pricking out in 160 ml potting pots filled with standard potting compost. Bhattacharjee and Mukherjee (1983) when applied P_2O_5 at 30, 40 or 50 kg/ha along with a basal dose of 40 kg/ha K_2O, and first half of the 20, 30 or 40 kg N/ha at planting and the other half 30 days after planting to dahlia cv. 'Blackout', they obtained best growth and flower production with the highest NP rates.

Potted dahlias may be given with slurry every fortnight. Slurry is prepared by diluting the fresh cow dung to its four times and then kept in an earthen pot under sun with covered mouth for decomposition for 10-15 days by stirring daily and then again diluting it to its four times, so one kg of cow dung will become 16 litre in which 320 g of urea is added at the final stage, and this is the slurry which is very effective in case of potted as well as to those plants growing in the field.

Growth, Development, Flowering and Dormancy

Freshly harvested root tubers of dahlia do not sprout immediately because of **dormancy**. These sprout only when these have attained physiological maturity. At 15° C temperature and at 13-14 h photoperiods, many dahlia cultivars will grow continuously. Konishi and Inaba (1967a, b) obtained reducing percentage of sprouting in dahlia tubers planted between the end of October and early December and those sprouted also the shoots did not elongate normally but those planted from mid-December onwards the sprouting as well as those planted at the end of December even shoots also grew normally, and digging of the tubers after November 20 and stored for 40 days at 0°C sprouted well with normal growth of shoots though there may be varietal differences. Adventitious or axillary shoots from newly planted tubers do not elongate if photoperiod is less than 12 hours (Konishi and Inaba, 1967b). They further stated that with tubers from plants propagated from cuttings taken during or after June the dormancy was weak whereas the dormancy of tubers from cuttings taken in or before April was similar to that of the mother plants. A **photoperiod** of 13 h and 10° C night temperature enhances the flower quality as compared to 5 or 15° C. Photoperiods also determine number of total florets in the inflorescence, 14 h photoperiod producing more number of flowers though 8 h photoperiod drastically reducing it. Short days, apart from reducing the number of florets, increase the proportion of disc florets as well as induce abortions. Low light intensities also behave like photoperiods, these delay flowering and reduce the percentage of flowering plants. Tuberization is induced by short days (11-12 h) for 5 days at 16-21° C temperature where maximum fresh weight of the tuberous root is obtained, though the root tuber growth is inhibited at 10 and 27° C. Potted dahlia plants can be started in March on the hills if the tubers are stored at 7-9° C for 2-3 months or in October in the plains when stored for 4 months. For glasshouse forcing, in the Northern Hemisphere the tubers are planted from January to April (De Hertogh, 1996). Photoperiods less than 12 h promote tuberous root formation and check shoot growth. Brøndum and Heins (1993) reported optimum production conditions for dahlia plants of cv. 'Royal Dahliette Yellow' grown in 10-cm pot as 12-14 h photoperiod and 20°C temperature; with decrease in photoperiod from 16-h to 10-h, the number and length of axillary shoots decreased; and photoperiods shorter than 12-h inhibited shoot growth though promoted tuberous root formation. Maatsch and Rünger (1955) reported that in dahlia lateral shoots are formed only in the upper leaf axils in 10- to 12-hour days and with increase in the day-length up to 12 h tuber growth also increases but tuber growth under long day conditions in cuttings is inhibited though all the leaf axils produce shoots under long day conditions (14 h days), however, a short day treatment from beginning of August for 20-30 days promotes tuber formation. Short days treatment of 20-30 days, especially for 4-weeks, is adequate between July end to beginning of September or from latter half of July to early August to increase tuber size 3-5 times in varieties which are slow in tuber formation (Rünger, 1955; Wasscher, 1955) though flower number becomes considerably less while long days stimulate aerial growth (Rünger, 1955; Yasuda and Yokoyama, 1959; Konishi and Inaba, 1964). Yasuda and Yokoyama (1959) obtained 446 flowers under 13-h photoperiod, 235 under 10-h and only 158 under 7-h photoperiod. Increase in day length (up to 13 or in some varieties 14 hours) with supplementary lighting with 20-36 lux increased the total number of florets and the number of ray florets but decreased the number of disc florets. Konishi and Inaba (1966a) obtained floral primordia initiation under 8 to 16 h day lengths but their further development was found inhibited by LDs, optimum day length for flower initiation being 10 hours or less but later stages of flower development required 12-13 hours of LD. They recorded futher that growing the plants for 40 days at optimum day length after pinching and then shifting to less than 12-h day lengths many buds become blind. Exposure of dahlia cuttings to 12-h SD and 14-h LD with and without interruptions of SD and LD treatment during early period of lateral growth, floral primordia were formed under both day lengths but the longer the SD treatment the earlier was the flowering with reduced shoot length and weight, floret number, *vis-à-vis* ray florets and double flowers but increased the disc floret number when applied only for five days, though LD for less than 20 days did not affect the time of flowering but beyond this the percentage of ray florets and double flowers increased. Konishi and Inaba (1966b) recorded a day length of 13-h with a minimum night temperature of 10°C optimum for flowering, a high night temperature of 15°C extending the flowering period while 5°C night temperature retarded the flowering and shortened the flowering duration. Reduced light intensity condition was compensated by increasing the critical day length for flowering with same amount of blooms but with delayed flowering. Konishi and Inaba (1966c) reported that photoperiodic conditions prior to stopping of shoot had no effect when after stopping the plants were grown in 13-h days but when plants were grown at SD below the critical day length of 12-h prior to stopping the flowering was considerably retarded. Early photoperiod treatment did not influence the number of florets developed. Mastalerz (1976) found that in some of the dahlia cultivars the flower initiation and development could not be prevented by continuous LDs, though in certain other cultivars plants remained vegetative in 16-h or longer days and initiated flowers in days shorter than 14-h and when these second type of plants were given 9-h days until the buds opened it became too late as stems were very short with reduced floret number but this problem was overcome by subjecting the vegetative plants to 14 SDs of 9-h followed by 18- to 24-h long days until the buds opened. Post (1949) states that a great profusion of blooms in dahlia is usually in September and October as plants respond well to short days for bud formation, and high quality blooms are produced in regions of cool nights but shorter nights during autumn promote tuber formation though frosty nights during autumn terminate the growth cycle. **Temperature** affects storage of root tubers, after interacting with photoperiod it controls tuberization of the roots, affects the growth rate of the vegetative shoot, after interacting with photoperiod it affects flowering of the shoot, and it also affects termination of shoot growth. For vegetative shoot development the optimal temperature is 13-25° C, and for flower initiation and development it is 10-15° C which is optimum. After tuber harvesting, they exhibit dormancy to save them from climatic adversities because in the plains it is very hot from May to August and on the hills from November to March it is very cold while their lifting time in the plains and on the hills is May and October, respectively. There is no significant difference in flower production due to size of the root tubers planted if this has attained minimum required size and the tubers having 80 g weight will behave similar to those having 120 g weight. After lifting, if the root tubers are stored at 0° C for 40 days, the dormancy is released and the tubers sprout but then the climatic conditions are not congenial for its growing until in the glasshouses the required temperatures and photoperiods are provided artificially. If such tubers are exchanged from the plains to the hills and *vice versa,* there too the crop fails as the tubers have not attained physiological maturity though dahlias grown during winter season (as in the plains) have little dormancy than those grown during summers (as on the hills) which have maximum dormancy. Okada and Harada (1955) found small to very large disc to ray floret ratios due to seasonal variations (some due to temperature, some due to photoperiod while others due to both the factors) in various dahlia varieties, and reported that total number of florets per head was affected due to temperature and not the photoperiod. Tubers of five dahlia cultivars planted in late February at a minimum temperature of 13°C with half the plants being exposed to 2-h night break through fluorescent lighting until mid-May, started flowering between 5 to 19 April, and their floret count revealed that night break influenced the plants favourably (Hall, 1969). He noted semi-daisy-eyed blooms in cv. 'David Howard' until April 26 in lighted plants and until May 31 in unlighted plants; in 'Horn of Plenty' the normal blooms were produced five weeks ahead in lighted plants than the unlighted; in 'Larkford' the blooms became normal by April 26, *i.e.* three weeks ahead of unlighted plants; in cv. 'Minley Boy' such blooms appeared until May 3 in lighted plants though further up to two weeks after in unlighted plants; and in 'Brandaris' such blooms appeared up to April 26 under lighted plants and the blooms were normal four weeks later on the unlighted plants.

When root tubers of dahlia are planted, it is not so that immediately these start **flower initiation**. In fact, in glasshouses, flower initiation occurs after the formation of five leaf-pairs while in the field after seven leaf-pairs. At 17° C night temperature in the greenhouse, tubers start floral initiation after 15-25 days of planting. Various **growth regulators** have been found altering growth and flowering in dahlia. Moser and Hess (1968) stated that GA inhibits tuberization either under long or under short days but Alar promotes it under LDs though simultaneous spray of GA again suppresses the effect of Alar. Pappiah and Muthuswamy (1974) found increased number of laterals (nearly double of control) through ethrel at 1,000 ppm but its 500 ppm followed by 1,000 ppm shortened the time for flowering, with 90 per cent increase in flower yield under 1,000 ppm, and where one concentration was effective other was also found effective but slightly less. Biran *et al.* (1974) tried ABA and GA to budless dahlia leaf cuttings under LD, and found GA inhibiting the growth of tubers and roots and enhancing thickening of the petiole base though ABA promoted growth of the tuberous roots. SADH and ethephon promote tuberization but not in budless cuttings, ABA also enhances tuberization but GA inhibits it, and SD conditions which promote tuberization increases the endogenous level of ABA (Halevy and Biran, 1975). Mittal (1967) found height increase with fewer internodes and more lower branches, *vis-à-vis* advanced flowering when dahlia plants were sprayed twice with GA, especially with 100 to 500 ppm. Mather and Sharma (1969) stated that dahlias behave as short day plants with a critical day length of 16 hours, with SDs flowering occurs significantly earlier but with LDs or in continuous light no flowering occurs, and such plants flower immediately on being transferred from non-inductive to inductive day lengths. They stated that GA application accelerates flowering under inductive day lengths but the critical day length is not affected. Bhattacharjee *et al.* (1976, 1978, 1984) when used IAA and GA both at 100 and 1000 ppm, found marked promotion of shoot growth of dahlia with increased number of flowers under GA treatment, and IAA at higher concentration and GA at both the dilutions counteracted the retarding effect of B9 while IAA at higher rate and GA at lower rate suppressed the effect of floral initiation induced due to B9 or CCC treatments. Suma (1993) tried Alar and CCC on dahlia propagated through cuttings and recorded retarded plant height, increased tuber weight, increased number of branches and leaves, delayed flowering by 3.9 to 5.9 days but with size increase of flowers, number of florets and lasting quality of flowers on stalk and in vase, *vis-à-vis* maximum tuber production especially with 4,000 ppm concentration through Alar though plants treated with CCC had increased amount of total chlorophyll content, length of internodes, thickening of the internodes and the nodes, and earlier flowering by 8-10 days. Durso and Hertogh (1977) forced dahlia cvs 'Kolchelsee' and 'Park Princess' from January to June where plant heights were controlled by applying 0.5 mg/pot ancymidol two weeks after planting, and obtained best height control at 25 °C day and 16 °C night temperatures, accelerated flowering at 28/17 °C and 29/20 °C day/night temperatures but with poor quality of flowers, and delayed flowering at 12/12 °C day/night temperatures. They found natural spring photoperiod increase from 10 to 14 h as optimal but with high light intensities for forcing, long days given either as a 16-h photoperiod or as a 4-h night break delayed the flowering slightly, and those grown at 8-h photoperiod though flowered earliest but only scarce, and under 50 per cent shading the plants were too tall to control even with ancymidol so were unfit for pot growing.Kumar *et al.*(1978) had grown the dahlia plants in 8-h SD or 24-h LD and applied 10 or 100 mg/litre of GA to the plant apices on alternate days, which resulted into much taller plants especially under LD with more leaves and bud initiation in several plants after 62 and 85 days at the two concentrations, respectively, though with SD all the plants flowered and with GA number of days for floral initiation was found reduced from 56 to 40 days.

Height Control, Pinching & Disbudding, and Staking

Height control is required only for potted plants and not for cut flower production. Internode elongation is controlled with ancymidol (A-Rest) drench at the rate of 0.5-2.0 mg a.i. per 15-cm pot in moist medium and at the time when the shoots are 0.6 cm long or after two weeks of planting. Ancymidol is consistently effective in reducing total plant height at flowering without affecting the flowering date, flower size or number of shoots per tuber. Paclobutrazol (Bonzi) at 2-4 mg and uniconazole (Sumagic) at 0.25-0.5 a.i. are applied per 15-cm pot (Whipker and Hammer, 1997). De Hertogh (1996) advocates application of drenches to moist media and when the shoots are at least 0.5 cm long. In rooted cuttings, B-Nine (daminozide) also reduces the total plant height. Daminozide or chlormequat increases the number and size of the tuberous roots produced under short days. Pappiah and Muthuswamy (1974) through twice spraying at fortnightly intervals of MH (1000, 2000 ppm), chlormequat (1000, 2000 ppm) and ethrel (ethephon 39.56 per cent at 500 and 1000 ppm) to single-flowering dahlia plants 45 days after planting and recorded highly reduced height under MH but only little through chlormequat though its both the concentrations greatly increased the number of tubers per plant. To the cuttings of 10 dahlia varieties when B9 was applied at 5000 and 10000 ppm as a spray, Bhattacharjee *et al.* (1971) obtained height reduction by 27 per cent and 49 per cent though CCC as dreanch or spray did not inhibit growth and in some cases it promoted the growth. Bhattacharjee *et al.* (1974) obtained retardation of 17.4-64.7 per cent plant height in 15 cultivars at 10000 ppm B-Nine application though flower size was larger; when pinched and non-pinched plants of cv. 'Avalanche' were treated with B-Nine, its 5000 ppm caused more retardation in the pinched plants; and chlormequat as drench or foliar spray at 1000 or 2000 ppm promoted plant height by 0.6-12.0 per cent but its 4000 and 8000 ppm concentrations retarded the growth by 1.5-12.6 per cent. De Hertogh *et al.* (1976) used ancymidol, chlormequat and daminozide at various rates as soil drenches or foliar applications to potted plants of various cultivars 2 or 4 weeks after planting and then the plants were forced under glass in natural days at 16/17° night and 18/20°C day temperatures, and found that only ancymidol consistently reduced plant height at flowering, its 2 mg/plant soil drench being more effective than its 0.5 mg or the spraying, though noted that for early planting on February 8 the time of drenching was not important but for late planting on March 8 the treatment two weeks after planting was best. Durso and Hertogh (1977) controlled the plant heights in dahlias through ancymidol application at 0.5 mg/pot two weeks after planting at 25°C/16°C day/night temperatures and obtained best height control. Bhattacharjee *et al.* (1978) found highly suppressed plant height in dahlias through B-nine 500 ppm application and in *Dahlia conzattii* by CCC at 500 ppm. Bhattacharjee (1984) recorded markedly reduced plant height through 2000 ppm TIBA or 500 ppm MH when applied at 15-18 cm tall plants of dahlia cv. 'Kelvin Rose'.

Pinching is done to encourage more shoots or flower bearing branches. It is the removal of apical growing vegetative bud to encourage more axillary buds so that per plant number of flowers becomes manifold. Pinching in dahlia is applied when only one shoot develops per tuberous root or if more number of flowers is required per pot for good display. The proper time for pinching is when 3-4 leaf pairs are fully developed, then soft pruning of the terminal bud is done leaving 3-4 nodes from where new axillary shoots will develop. In the multistemmed plants also the vigorous dominant shoot (the central one from each stem or branch) is nipped for prolific flowering. Pinching of terminal bud is also done if flowering is too advanced for date of planned flowering for shows and exhibitions or for marketing and this, in fact, is known as stopping. The practice is exactly the same as pinching but the purpose is slightly different. However, **disbudding** is the removal of the axillary buds so that development of apical bud or sometimes two more side buds are encouraged. Normally all the buds are removed leaving the central one so that the energy saved from disbudded ones could be diverted towards development of the central bud with more strength and vigour. Disbudding is normally carried out for producing single exhibition bloom per plant. If the plant has become messy, the weak shoots are removed so that main shoots get more energy and light for their proper development and this practice is known as thinning.

Pot-grown dahlias are already dwarf so these do not require **staking** for their support. Field grown dahlias sometimes go very tall which require strong support. Support of bamboo sticks or better of *Salix* twigs should be used. One stick per plant will not work but in dahlia, 2-4 sticks per plant should be inserted in the plant periphery at equal distance, number depending on the growth of the plant or the variety used. These sticks are tied around the plant in 3-4-tier system with jute strings to keep the plants securely on the place. After the crop is over, these sticks are removed to be used in the next crop.

Postharvest

The dahlia flowers are cut when they are half-open. After cutting, they are immediately placed in water which is 50° C hot, to prevent wilting and to encourage water uptake (Post, 1949; Nowak and Rudnicki, 1990) though Arnosky and Arnosky (1997) recommend 71°C water and a floral preservative. Holding solutions should be adjusted at a pH of 3.5 to 4.0. They are cut with as much stem as possible. Too immature buds if cut will fail to open. Transportation or storage for up to 5 days at 4-5° C is possible (Nowak and Rudnicki, 1990). Bar-Sella (1959) stated that cut dahlias have short vase life as two days after picking, uptake of water is impeded by plugs originating in the xylem vessels and antibiotics increase the storage life only by about one day, therefore, reduction of the respiration rate is important and dipping the stems in MH at 50 ppm may solve this problem to some extent. Staden (1976) recommended the product Anger (Carnation) Chrysal as better than Tulpen (Tulip) Chrysal by which a vase life of at least one week was achieved. Lukaszewska (1983a, b) tried semi-open dahlia cv. 'Purple Gem' cut blooms with 30 cm leafless stalks by placing them for 24 h in glucose at 10 per cent + $AgNO_3$ 0.2 mM or with glucose 10 per cent + $AgNO_3$ at 0.2 per cent + 8-HQS at 200 mg/litre, and obtained vase life of 8.7 days in the former case while 8.2 days in the latter though cut flowers lasted only for 4.3 days in control and the life in sugar was also poor. Further, he sampled the flowers at harvest, 24 h later and again at wilting and found that when flowers were kept continuously in glucose or sucrose solutions resulted into accumulation of reducing sugars and sucrose in the flowers, the amount of endogenous carbohydrates doubled in the inner rays and receptacles after 24 h treatment, at wilting this amount was higher in flowers held in water but lower than in flowers held continuously in sugar solutions, and the presence of sugars in the holding solution prevented an undesirable accumulation of free amino acids in the flowers. Further he reported that changes in the level of free amino acids in the outer rows were negligible but in the inner rows a significant increase occurred at wilting in all variants but least of all in flowers held continuously in the sugar solutions. Swart (1989) advocated picking of dahlia flowers at a relatively immature stage of flower opening and their immediate placing in Chrysal AKC solution at 24 g/litre or Anjer-Chrysal at 13 g/litre for improving the vase life, and low temperature storage at 2°C for two days also improves subsequent vase life at 20°C. Dahlias respond well to glucose, $AgNO_3$ and 8-hydroxyquinoline sulphate provided the water pH is adjusted to 3.5-4.0. Glucose 10 per cent + $AgNO_3$ 0.2 mM or 10 per cent glucose + 0.2mM $AgNO_3$ + 200 mg 8-HQS/litre are most favourable for prolonging the vase life in dahlia cut flowers. The decreasing effect of Ethephon 50 ppm is negated when glucose 10 per cent + $AgNO_3$ 50 ppm treatment is given to dahlia cut flowers, and this treatment also increases the vase life.

The pot dahlias are ready for marketing when the first flowers begin to open. Pot dahlias when in flower are not recommended for cold storage. If flowering is too early, the flowers are disbudded and the laterals are allowed to develop which may take further 5-7 days. In case flowering is ahead of the schedule, the greenhouse temperature may be reduced to 10-13° C which will slow the development of the plants. For post-greenhouse handling, a plastic sleeve is used.

Lifting of Tubers and Storage

On the hills, by the first frost the plants burn when these should be lifted, *i.e.* by late October or early November. Leaving some 15 cm stems from the ground level the plants are cut and the tubers are lifted gently with the help of fork, shovel, spade or *khurpi* just by loosening the soil and pulling the whole clump by catching the left out portion of the aerial stem. A frost of -2°C before lifting in October kills all the dahlia leaves but only partially injures the flowers and the osmotic values of frozen cells are considerably higher than normal cells (Pardatscher, 1951). Dogonadze (1957) also observed that the seedlings of a dahlia grafted on a yellow chrysanthemum acquired an improved resistance to light autumn frosts as its resistance became similar to that of chrysanthemum. Instead of manual removal of the top growth at lifting which may disseminate viral and bacterial infection, Berthier (1987) found dinoseb at 121 kg/ha quite effective when applied in last week of September at 17-20°C temperature and 80 per cent RH, then tubers were lifted in first week of November and stored in moist peat at 12-15°C in mid-February, these all the tubers sprouted cent per cent. He recommended diquat and dinoseb quite promising and safe but suggested to apply dinoseb at a minimum temperature of 15°C. Root tubers after lifting and cleaning are shade-dried for 5-7 days, then stored at 5-10° C in peat with 50 per cent water content to prevent moisture loss, as in storage, the tubers should not be allowed to dry out, alternatively the root tubers can also be dipped in paraffin but in any case the freezing or lethal temperature for tubers is -2° C which should be avoided (Allen, 1937; De Hertogh and Le Nard, 1993). On the hills, either these are not lifted from the field until planting time or the clumps of root tubers intact with a portion of aerial shoots are stored in a room with open windows for ventilation and these start throwing out the sprouts in March-April and that is the proper planting time on the hills. If the lifted clumps as such are buried elsewhere in the field, these rot, and the situation is aggravated when divided tubers are buried because of the injury at cut portion from where cold or infection enters the tubers. Room storage may not pose this problem. In the plains, in May when flowers start burning and drying, these are lifted the same way as on the hills, cleaned and stored at 7-9° C temperature with peat having 50 per cent moisture. These are taken out from cold storage in September for planting in October. Steffen (1958) advocated the storage of tubers at 10°C in boxes containing dry sand/peat mixture or in plastic bags after drying the tubers. Ailincai (1972) stated that dahlia tubers stored at 6 or 9°C had lower respiration rates, *vis-à-vis* lower peroxidase, catalase and ascorbinoxidase activities and a higher polyphenoloxidase activity than those held at 20°C.

Insect-Pests, Diseases and Physiological Disorders

Aphids (*Aphis fabae, Brachycaudus helichrysi* and *Myzus persicae*), thrips (*Heliothrips haemorrhoidalis, Frankliniella tritici* and *Thrips tabaci*), red spider and other mites (*Tetranychus urticae*), whiteflies, European corn borer (*Pyrausta nubilalis*), capsid bugs, tarnished plant bugs, Japanese beetles (*Popillia japonica*), the carrot beetle (*Ligyrus gibbosus*), mordellid beetle (*Mordellistena ghanii*), the flea beetle (*Nodonta clypealis*), flea hoppers (*Halticus bractatus*), leaf hoppers (*Empoasca fabae,* and *E. solana*), grasshoppers, earwigs and wireworms are the insect-pests which infest on dahlia.

Aphids damage the plants by sucking the sap and transmitting various viruses. This pest can be controlled through spraying with 0.2 per cent Metacid-50. Thrips rasp the surface of the petals and feed on the exuding juice and the lower surface of the petals turns whitish and petals wither. Olisevič (1959) found good control of aphids (*Myzodes dianthi* and *Neomyzus circumflexus*) through spraying with 0.1-0.2 per cent Mercaptophos and octamethyl (schradan) at 15-20 day intervals, first remaining toxic to aphids up to or beyond three months while the second for two months. While controlling the aphid, this will control the thrips also. Both adult and larval stages of broad mite (*Hemitarsonemus latus*) causing curling and distortion of foliage can be controlled by dusting with finely ground sulphur or spraying with water-soluble sulphur at 6 g/litre of water (Anon., 1951) and red spider mites (*Tetranychus urticae*) feed on the under side of the leaves and cause white specklings and make the webs, and transmit dahlia mosaic virus (Korneeva *et al.,* 1970) which are controlled with Plictran (cyhexatin) within 11 days and through anise oil giving acceptable control (Wickizer and Meisch, 1977). Fluvanilate tuber dip or foliar spray will also control the mites. Harris (1971) because of the resistance being developed of organophophorus compounds to red spider mite, advocated introduction of the predacious mite (*Phytoseiulus persimilis*) which gave effective control even in the absence of chemical sprays. European corn borer larvae feed on the bud ends, floral parts and leaves and cause distortion. The larvae also bore the stems and cause extensive damage. Measures applied for aphids will control this pest also. Flea beetle adults and larvae feed on leaves, flowers and seed pods. Mordellid beetles bore into the stems and cause extensive damage. Adult Japanese beetle feeds on the leaves, flowers and stems but grubs damage the roots and tubers. Commercial preparations of *Bacillus popiliae* (milky spore disease) in soil infects the grubs and kill them. The carrot beetle larvae feed on the root and tuberous roots of dahlia. Soil application of Temik 10-G in the soil and spraying of Nuvacron at 0.2 per cent on the aerial parts will control all sorts of beetles. Flea hoppers, leaf hoppers and grasshoppers feed on the aerial parts of the plants and wireworms on roots and tubers so while controlling beetles these insects will automatically be controlled. Neiswander (1955) found 0.065 g of 25 per cent malathion powder + 0.048 g of 50 per cent DDT powder per litre of water when sprayed at every 10 days starting when plants were 30 cm, found excellent control of potato leafhopper, tarnished plant bug, onion thrips, green peach aphid and common red spider mite (two-spotted spider mite). Widmer (1965) found effective insect control when applied 10 per cent granular Thimet (phorate) in the well prepared soil at 21 g and 35 g/plant and 10 per cent granular Di-Syston (disulfoton) at the same rates as Thimet and also at 112 g/plant, followed by immediate irrigation, and found effective control with both the chemicals at 35 g per plant (12 sq. ft.). At 35 g of Dimetilan and Thimet, with Meta-Systox-R for an area of 2 x 6 sq. ft. containing 2 plants were also recorded very effective.

Through a survey in the Riyadh region of Saudi Arabia, Ibrahim and Al-Yahya (2002) recorded *Meloidogyne arenaria* from galled roots of *Dahlia variabilis* (*D. pinnata*). Nematodes like *Meloidogyne hapla* causes root galling, *Trichordoridorus pachydermus* found in the soil has been recorded from Michigan causing poor development of root system with only few feeder roots and leaf notching in *Dahlia pinnata* (Bird and Knobloch, 1976), *Aphelenchoides ritzemabosi* attacks leaves and floral buds, and *Rotylenchus robustus* infests the aerial parts, especially the leaves. All these nematodes can be controlled by the use of Furadan @ 4-5 kg when mixed in the soil at the time of preparation of the soil and by spraying the infested crop with Dimecron or Nuvacron @ 0.1 per cent. Pearman (1959) suggested crop rotation and burning of all the infected plants, *vis-à-vis* soil fumigation with ethylene dibromide, methyl bromide, DD or Nemagon. Gill (1981) recommended HWT of infected cuttings or shoots at 50°C for 5 minutes or at 44.4°C for 30 minutes, and phorate or aldicarb at 1.5 kg/ha applied as basal dressing at the time of transplanting.

Beaumont (1950) stated that in dahlia damping off is caused by *Phytophthora cryptogea, Rhizoctonia solani* and *Botrytis cinerea*; wilting by *Verticillium dahliae* and *Sclerotinia sclerotiorum*; leaf spots by *Entyloma dahliae* and *Ascochyta dahliaecola*; crown gall by *Bacterium tumefaciens*; leafy gall by *Corynebacterium dasciens*; and dahlia mosaic such as tomato-spotted wilt and cucumber mosaic. Jain and Nema (1952) recorded blossom blight of dahlia caused by *Choanephora infundibulifera. Botrytis* blight/grey mould (*Botrytis cinerea*) attacks the crop when there is cloudy weather, excess humidity, and poor ventilation. This grey mould infects the young shoots and the flower buds, afterwards death of the whole plant by causing soft rot and then covering with white mould. Zineb sprayings 0.2 per cent at weekly intervals will keep this disease under check. Pearman (1959) recommended thiram and captan for controlling grey mould. Dahlia smut (*Entyloma dahliae*) though is not a serious disease of dahlia but it causes black spots and shot holes on the foliage. Dithane M-45 or Dithane Z-78 at 0.2 per cent sprayings will control this disease. Pearman (1959) recommended thiram and captan for its control. Copper spraying in summer also controls this disease (Pape, 1953). Grouet (1962) recommended spraying of the crop with 0.25 per cent captan. Joshi (1967) when sprayed infected dahlia plants with Bordeaux mixture at 4 gal/100 plants thrice, *i.e.* July, August and September and found good control of smut. Misra (1978) sprayed several fungicides to the smut infected dahlia plants at 15-days intervals from June 20 to October 20 but obtained good control only with dithane Z-78 (zineb) and dithane M-45 (mancozeb), each at 0.2 per cent. Upadhyay and Bhandari (1985) found good control of dahlia smut through dithane Z-78, blitox (copper oxychloride) and benomyl. Stem rot (*Sclerotinia sclerotiorum*) in the form of white mould encircles the main stem and sister branches near the base of the plant which causes them to wilt and die. To avoid this disease, the soil should be quite porous with good drainage facilities, and the spacing among the plants should be sufficient to permit proper aeration. Soil sterilization with chemicals or steam will prevent this problem. Pearman (1959) suggested burning of all the infected plants and crop rotation for *S. sclerotiorum* as well as *Sclerotium rolfsii*. For dahlia tubers infected with white rot, Šumilenko and Šaranova (1963) treated the tubers with 50 per cent TMTD at 8 g per kg of tubers four months before planting which produced vigorous plants with large number of flowers and when tubers were lifted there appeared no sign of white rot. Root and stem rot occurs due to attack of *Verticillium alboatrum, Fusarium, Rhizoctonia solani, Pythium* and *Phytophthora* fungi and some of these fungi attack the tubers even in storage and rot them. Only healthy tubers should be used for planting after treating them with Captan 0.2 per cent. Regular fortnightly spraying of benomyl 0.2 per cent alternate with Captan 0.2 per cent throughout the growing period will save the crop from all the fungal diseases. Powdery mildew (*Erysiphe polygoni* and *E. cichoracearum*) causes white to grey powdery mass on the upper surface of the leaves which can be controlled with spraying of 0.1 per cent Bavistin or Karathane. Pearman (1959) recommended sulphur preparations for control of powdery mildew (*Oidium* sp.) and Roy (1965) against the infection of *O. erysiphoides* recommended spraying the crop with Thiovit, barium polysulphide or lime-sulphur at the first appearance after removal of infected leaves and again a fortnight later. Crown gall disease (*Agrobacterium tumefaciens*) causes large tumours of abnormal tissues at the base of the plants and on roots. Infected plants become stunted and deformed. The disease can be prevented by dipping the roots in streptomycin solution before planting and after lifting. Before planting, soil should also be disinfected. Clark and Paton (1956) in Scotland observed a wet and black rot in dahlia, on young cuttings in propagation beds and sometimes even of mature plants due to infection of *Pseudomonas marginalis*, which to some extent could be controlled by pre-planting dip or spraying the plants in field after planting with streptomycin. *Pearman* (1959) suggested burning of all the infected plants and crop rotation against this bactierium. Plants affected with *Pseudomonas solanacearum* droop and wilt suddenly from the base. Control measures are the same as described under crown gall disease. *Erwinia carotovora, E. chrysanthemi* and *E. cytolytice* cause browning of stems and soft rot of stems and root tubers. The infection is aggravated during cold weathers. These can also be controlled as described under crown gall disease. Saaltink (1969) suggested careful selection of planting material which can eliminate the bacterial wilt of dahlia (*E. chrysanthemi*) almost completely. Kamerman and Saaltink (1968) recommended formalin or steaming treatment of infected soil, use of clean water for irrigation and avoidance of any dubious planting material. Shrotri *et al.* (1984) stated that out of 21 fungi associated with dahlia seeds, *Alternaria alternata, Curvularia pollescens, Phome putaminum, Aspergillus flavus, A. niger, A. repens, Drechslera (Cochliobolus) hawaiiensis, Cochliobolus spicifer* and *Fusarium* spp. caused varying reductions in seed germination in pathogenicity tests, and for complete control of seed-borne fungi they recommended Ceresan, Dithane M-45 (mancozeb) and Aureofungin.

Brierley (1933a, b) stated that dahlias are infected by mosaic (stunt as dwarf, rugose mosaic and veinal mosaic) virus transmitted by aphid *Myzus persicae*, ring-spot, yellow ring-spot and oakleaf viruses. *Dahlia* plants are also attacked by cucumber mosaic virus, dahlia mosaic virus, tobacco stripe virus, tomato fleck virus and tomato spotted wilt virus (bronzing). Marini (1951) from Italy reported dahlia virus 1, the spotted wilt lycopersicon virus 3 and the cucumber virus 1. Lawson (1966) isolated from dahlia cv. 'Helen Anita' plants a strain of CMV producing oakleaf patterns. Brunt (1968) isolated tobacco streak virus for the first time from dahlia plants. Sansdrap (1969) described the symptoms of dahlia mosaic, bronzing (tomato spotted wilt virus), potato virus Y and cucumber virus 1. Holmes (1955) suggested elimination of spotted wilt virus by growing them through their small stem-tip cuttings as this virus does not move freely towards the growing tips of dahlia plants. Brierley (1933b) stated that mosaic symptoms appear as yellowish or pale-green banding along the midribs and larger branch veins of affected leaves, and the dwarfing is a second type of symptoms evident in mosaic plants of all but the most tolerant varieties and includes shortening of the roots. For its control he suggested fumigation of glasshouses with nicotine once a week, periodic rogueing and control of *Myzus persicae* with insecticides. Limasset (1946) reported from France that tomato spotted wilt virus may be transmitted by *Thrips* spp. Martin (1954) stated that biochemical activity of the enzymes cytochrome-oxidase, peroxidase and tyrosinase were greater in dahlia mosaic, cucumber mosaic and spotted wilt infected than in healthy dahlia plants. Martin (1959) also stated that mosaic-affected leaves of dahlia had higher contents of phenols. Korneeva *et al.* (1970) suggested destruction of mosaic virus infected plants, control of aphids and mites and of weeds harbouring the mites, taking the cuttings only from the healthy plants and sanitation measures. Korneeva (1971) stated that through healthy shoot grafting of dahlia on mosaic infected plants, symptoms appeared after three weeks and grafting of mosaic infected plants on healthy plants the symptoms were found developing only in the third year. Breman (1990) stated that though dahlia mosaic caulimovirus (DMV) is present in USA since 1927 and as there is no any effective control so the plants should be started from seeds as it does not transmit through seeds. Pearman (1959) stated that thrips being the vector for spotted wilt and aphids for stunt which should be controlled by spraying the plants with 0.05 per cent malathion every 10 days. Asjes (1968) suggested placing of 12 mm wide strips of reflective aluminium foil horizontally just above dahlia plants in the field. In any case the infected plants should be rogued out and burnt immediately. Aphids, hoppers, mites and thrips which spread the disease should also be controlled effectively by use of insecticides. The field should also be kept clean of debris.

Due to imbalanced temperature and/or photoperiod, flower bud abortion and terminal shoot blindness are caused while forcing the potted dahlias. Those cultivars which are amenable to the standard treatment of temperature and photoperiod should only be forced while others may be forced only when proved through experiments. Blindness usually occurs due to improper propagation practices through stem cuttings when a piece of crown was not taken while removing the cuttings from the crown of the root tuber.

References

Ailincai, N. 1972. Cîteva procese biochimice caracteristice rdcinilor tuberizate de dahlia, depozitate în conditii diferite de temperatur. *Lucrri Stiintifice, Institutul agronomic* "Ion Ionescu de la Brad", **I**: 325-331.

Allen, R.C. 1937. Temperature and humidity requirements for the storage of dahlia roots. *Proc. Amer. Soc. hort. Sci.*, **35**:770-773.

Anon. 1951. The broad mite (*Hemitarsonemus latus*) as a pest of dahlias. *Agric. Gaz. N.S.W.*, **62**: 252-253.

Aoba, T., S. Watanabe and C. Saito, 1960. Studies on tuberous root formation in dahlia. I. Periods of tuberous root formation in dahlia (Japanese). *J. hort. Ass. Japan*, **29**: 247-252.

Armitage, A.M. 1993. *Dahlia* hybrids. In: *Specialty Cut Flowers*, pp. 268-271. Varsity Press/Timber Press, Portland, Oregon.

Arnosky, P. and F. Arnosky, 1997. Dahlias are difficult but worth the effort. *Growing for Market*, **6**(10): 11-13.

Asjes, C.J. 1968. Effect of aluminium strips on the spreading of two aphiborne viruses in dahlias. *Meded. Rijksfac. Landb Wetensch., Gent*, **13**: 1171-1177.

Ball, V. 1991. Plugs – The way of the 1990s. In: *Ball Redbook* (15th edition), pp. 137-153. Geo. J. Ball Publishing, W. Chicago, Illinois.

Barrett, J.E. and A.A.De Hertogh, 1978. Pinching forced tuberous-rooted dahlias. *J. Amer. Soc. hort. Sci.*, **103**(6): 775-778.

Bar-Sella, P. 1959. Causes of rapid wilting of cut dahlias and means to improve their keeping qualities. *Bull. Res. Counc. Israel, Sect. D*, **7D**: 35-38.

Beaumont, A. 1950. Diseases of dahlias. *Gdnrs' Chron.*, **128**: 247.

Bendixen, O., J. Lam and F. Kaufmann, 1969. Polyacetylenes of *Dahlia pinnata. Phytochem.*, **8**:1021-1024.

Berthier, G. 1987. Etude de défoliants appliqués sur dahlia. *P.H.M.-Revue Horticole*, No. 278, pp. 39-42.

Bhattacharjee, S.K. 1984. Effect of growth regulating chemicals on growth, flowering and tuberous-root formation of *Dahlia variabilis* Desf. *Punjab hort. J.*, **24**(1/4): 138-144.

Bhattacharjee, S.K. and T. Mukherjee, 1983. Influence of nitrogen and phosphorus on growth and flowering in dahlia. *Pb. hort. J.*, **23**(1/2): 111-115.

Bhattacharjee, S.K., K.D. Gupta and T.K. Bose, 1971. B-Nine, an effective growth retardant on *Dahlia;* in effectiveness of CCC. *Phyton, Argentina*, **28**(1): 61-65.

Bhattacharjee, S.K., T. Mukhopadhyay and T.K. Bose, 1974. Experiments with growth retardants on *Dahlia. Indian J. Hort.*, **31**(1): 86-90.

Bhattacharjee, S.K., T. Mukhopdhyay and T.K. Bose, 1976. Interaction of auxin and gibberellin with growth retardants on growth and flowering of *Dahlia variablils* and *Malvaviscus conzattii. Indian Agricult.*, **20**(3): 193-199.

Bhattacharjee, S.K., T. Mukhopadhyay and T.K. Bose, 1978. Changes in nitrogen and carbohydrate fraction in *Dahlia variabilis* cv. "Avalanche" and *Malvaviscus conzattii* treated with CCC, B-nine and GA. *Indian Agricult.*, **22**(1): 41-48.

Biran, I., I. Gur and A.H. Halevy, 1972. The relationship between exogenous growth inhibitors and endogenous levels of ethylene, and tuberization of dahlias. *Physiologia Plantarum*, **27**(2): 226-230.

Biran, I. and A.H. Halevy, 1973. The relationship between rooting of dahlia cuttings and the presence and type of bud. *Physiologia Plant.*, **28**(2): 244-247.

Biran, I., B. Leshem, I. Gur and A.H. Halevy, 1974. Further studies on the relationship between growth regulators and tuberization of dahlias. *Physiologia Plantarum*, **31**(1): 23-28.

Bird, G.W. and N. Knobloch, 1976. Occurrence of *Trichodorus pachydermus* in Michigan. *Pl. Dis. Reptr*, **60**(1): 76-77.

Boertje, G.A. 1979. Bijmesten perkplanten in setjes : hoe vaak en hoeveel? *Vakblad Bloemist.*, **34**(12): 26-29.

Bose, T.K. and A. Mukhopadhyay, 1967. Changes in the carbohydrate and nitrogen during growth, flowering and tuber formation in dahlia. *Indian J. Hort.*, **24**: 191-196.

Breman, L.L. 1990. Dahlia mosaic virus. *Plant Pathology Circ.* (Gainesville), No. 331, 2pp.

Brierley, P. 1933a. Studies on mosaic and related diseases of dahlia. *Contrib. Boyce Thompson Inst.*, **5**: 235-288.

Brierley, P. 133b. Dahlia mosaic and its relation to stunt. *Boyce Thompson Inst.*, **1**(25) 240-246.

Broertjes, C. and J.M. Ballego, 1968. Addendum to : Mutation breeding of *Dahlia variabilis. Euphytica*, **1**: 507.

Brøndum, J.J. and R.D. Heins, 1993. Modeling temperature and photoperiod effects on growth and development of dahlia. *J. Amer. Soc. hort. Sci.*, **118**: 36-42.

Brunt, A.A. 1968. Tobacco streak virus in dahlias. *Plant Path.*, **17**: 119-122.

Čaplygin, B.K. and G.I. Šahova, 1961. Accelerated propagation of dahlias (Russian). *Priroda*, **50**(9): 116-117.

Cavalchini, A. 1968. studies on dahlia irrigation (Italian). *Ann. Fac. Sci. agrar. Torino*, **4**: 345-358.

Cayeux, H. 1936. Un mode nouveau de bouturage du dahlia. *Rev. hort. Suisse*,**9**: 9-12.

Choudhary, M.L. 1989. Correlation and path analysis in dahlia (*Dahlia variabilis* L.). *Annals agric. Res.*, **10**(1): 21-24.

Clark, M.R.M. and A.M. Paton, 1956. A new bacterial disease of dahlias caused by *Pseudomonas marginalis. Plant Path.*, **5**: 32-35.

Clay, D.V. and J.G. Davison, 1968. The response of dahlias to soil-applied herbicides. *Proc. 9th Brit. Weed Control Conf.*, pp. 965-970.

Clay, D.V. and G.W. Ivens, 1964. Herbicide trials on dahlias. *Proc. 7th Brit. Weed Control Conf.*, pp. 241-247.

Das, P.K., S. Dubey and A.K. Dey, 1978. New dahlias through irradiation. *Indian Hort.*, **23**(1): 19, 21.

De Hertogh, A.A. 1996. Dahlia – Potted Plants. In: *Holland Bulb Forcer's Guide* (5th edition). International Flower Bulb Centre, Hillegom, The Netherlands.

De Hertogh, A.A., N. Blakely and W. Szlachetka, 1976. The influence of ancymidol, chlormequat and daminozide on the growth and development of forced *Dahlia variabilis* Willd. *Scientia Hort.*, **4**(2): 123-130.

De Hertogh, A.A. and M. Le Nard, 1993. *Dahlia*. In: *The Physiology of Flower Bulbs*, pp. 273- 283. Elsevier, Amsterdam.

Dogonadze, D.S. 1957. Vegetative hybrids of dahlia and chrysanthemum (Russian). *Agrobiologija*, No. 1, pp. 127-128.

Dohare, S.R., H.S. Gill and R.S. Arora, 1974. 'Manali' a new dahlia. *Indian Hort.*, **19**(2): 21.

Dole, J.M. and H.F. Wilkins, 1999. Dahlia. In: *Floriculture Principles and Species*, pp. 287-291. Prentice Hall, New Jersey.

Durso, M. and A.A. De Hertogh, 1977. The influence of greenhouse environmental factors on forcing *Dahlia* Willd. *J. Amer. Soc. hort. Sci.*, **102**(3): 314-317.

El-Gamassy, A. and M.B. Moustafa, 1963a. Nutritional requirements of *Dahlia pinnata*. 1. Effect of nutrient solutions on the growth, flower quality and yield of tuberous roots of dahlia plants in sand culture. *Ann. Agric. Sci., Cairo*, **8**(2): 357-381.

El-Gamassy, A. and M.B. Moustafa, 1963b. Nutritional requirements of *Dahlia pinnata*. 2. Effect of nutrient solutions on the chemical composition of dahlia plant parts in sand culture. *Ann. agric.Sci., Cairo*, **8**(2): 383-416.

El-Gamassy, A. and M.B. Moustafa, 1964a. Nutritional requirements of dahlia plants. 3. Effect of fertilizer rates on the growth rate, flower quality and yield of tuberous roots of dahlia plants grown in pot and field. *Ann. agric. Sci., Cairo*, **9**(1): 395-421.

El-Gamassy, A. and M.B. Moustafa, 1964b. Nutritional requirements of *Dahlia pinnata*. 4. Effect of appliction time of phosphatic and potassic fertilizers on the growth, flower quality and yield of tuberous roots of dahlia plants. *Ann. agric.Sci., Cairo*, **9**(1): 423-431.

Fischer, D., K. Stitch, L. Britsch and H. Griesbach, 1988. Purification and characterization of (+) dihydroflavonol (3-hydroxyflavanone) 4-reductase from flowers of *Dahlia variabilis. Archives Biochem. Biophysics*, **264**(1): 40-47.

Forkmann, G. and G. Stotz, 1984. Selection and characterization of flavanone 3-hydroxylase mutants of *Dahlia*, *Streptocarpus*, *Verbena* and *Zinnia*. *Planta*, **161**(3): 261-265.

Gavinlertvatana, P., P.E. Read, H.F. Wilkins and R. Heins, 1979. Influence of photoperiod and daminozide stock plant pretreatments on ethylene and CO_2 levels and callus formation from dahlia leaf segment cultures. *J. Amer. Soc. hort. Sci.*, **104**(6): 849-852.

Gavinlertvatana, P., P.E. Read, H.F. Wilkins and R. Heins, 1982. Ethylene levels in flask atmospheres of *Dahlia pinnata* Cav. Leaf segments and callus cultured *in vitro. J. Amer. Soc. hort. Sci.*, **107**(1): 3-6.

Gill, J.S. 1981. Fighting the foliar nematode of chrysanthemum. *Indian Hort.*, **25**(4): 21-22.

Grouet, D. 1962. A contribution to the study of dahlia *Entyloma* (French). *Ann. Épiphyt.*, **13**: 5-22.

Halevy, A.H. and I. Biran, 1975. Hormonal regulation of tuberization in dahlia. *Acta Hort.*, No. 47, pp. 319-329.

Hall, O.G. 1969. Dahlias for early cut flower production. *Shinfield Progr.*, **14**: 36-37.

Harborne, J.B., J. Greenham and J. Eagles, 1990. Malonylated chalcone glycosides in *Dahlia. Phytochem.*, **29**(9): 2899-2900.

Harris, K.M. 1971. A new approach to pest control in the dahlia trial at Wisley. *J. Roy. Hort. Soc.*, **96**: 200-206.

Holmes, F.O. 1955. Elimination of spotted wilt from dahlias by propagation of tip cuttings. *Phytopath.*, **45**: 224-226.

Ibrahim, A.A.M. and F.A. Al-Yahya, 2002. Phytoparasitic nematodes associated with ornamental plants in Riyadh Region, Central Saudi Arabia. *Alexandria J. agric. Res.*, **47**(3): 157-167.

Jain, A.C. abd K.G. Nema, 1952. Blossom blight of dahlia. *Curr. Sci.*, **21**: 71-72.

Joshi, N.C. 1967. Varietal resistance of dahlia to smut disease (*Entyloma dahliae*) and efficacy of bordeaux mixture against it. *Sci. Cult.*, **33**: 452-453.

Kamerman, W. and G.J. Saaltink, 1968. Bacterial wilt in dahlias. *Praktijmeded. Lab. BloembollOnderz., Lisse*, No. 28, pp. 10.

Kannangara, T. and A. Booth, 1979. Cytokinin activity in transfer-RNA from seedling roots of *Dahlia variabilis. Zeitschrift für Pflanzenphysiologie*, **92**(2): 187-190.

Konishi, K. and K. Inaba, 1964. Studies on flowering control of dahlia. I. On optimum day length (Japanese). *J. Jap. Soc. hort. Sci.*, **33**: 171-180.

Konishi, K. and K. Inaba, 1967a. Studies on flowering control in dahlias (Japanese, summary English). VII. On dormancy of crown-tubers. *J. Jap. Soc. hort. Sci.*, **36**: 131-140.

Konishi, K. and K. Inaba, 1967b. Studies on flowering control in dahlias. VIII. Effect of day-length on dormancy in axillary bud (Japanese, summary English). *J. Jap. Soc. hort. Sci.*, **36**: 243-249.

Konishi, K. and K. Inaba, 1966a. Studies on flowering control of dahlias. III. The effect of day length on the initiation and development of flower buds (Japanese). *J. Jap. Soc. hort. Sci.*, **35**: 73-79.

Konishi, K. and K. Inaba, 1966b. Studies on flowering control of dahlias. IV. The effect of day length at the early stage of shoot growth upon the flowering date and the quality of cut flowers (Japanese). *J. Jap. Soc. hort. Sci.*, **35**: 195-202.

Konishi, K. and K. Inaba, 1966c. Studies on flowering control of dahlias. V. The effects of night temperature and light intensity and duration on flowering (Japanese). *J. Jap. Soc. hort. Sci.*, **35**: 317-324.

Korneeva, I. T. 1971. The development of mosaic virus symptoms in dahlias. *Byulleten' Glavnogo Botanicheskogo Sada*, No. 82, pp. 120-123.

Korneeva, I.T., L.A. Zinov'eva and N.L. Rostova, 1970. The role of the mite *Tetranychus urticae* in the spread of mosaic virus on dahlias (Russian). *Bjull. Glav. Bot. Sada*, No. 76, pp. 88-92.

Kumar, S., R. Sharma, K.S. Datta and K.K. Nanda, 1978. Gibberellic acid causes induction of flowering in *Dahlia palmata* under non-inductive photoperiods. *Indian J. Pl. Physiol.*, **21**(3): 261-264.

Lam, J. 1971. Polyacetylenes in *Dahlia imperialis* and *Dahlia tenuicaulis. Phytochem.*, **10**(9): 2227-2228.

Lam J., L.P. Christensen and T. Thomsen, 1991. Polyacetylenes from *Dahlia* species. *Phytochem.*, **30**(2): 515-518.

Lavrichenko, E.V. 1975. Morphogenesis of the vegetative parts and causes of degeneration of *Dahlia cultorum. Botanicheski Zhurnal*, **60**(3): 412-424.

Lawrence, W.J.C. 1942. The origin of the garden dahlia. *Publ. John Innes hort. Inst.*, 8 pp.

Lawson, R.H. 1966. Oakleaf chlorosis symptomatic of cucumber mosaic virus infection in dahlia. *Phytopath.*, **56**: 343-344.

Limasset, P. 1946. Les maladies à virus du dahlia. *Rev. hort. Paris*, **118**: 11-14.

Lloyd, F.G. 1968. Dahlias – production of healthy tubers under glass. *9th Progr. Rep. exp. Husb. Fms exp. Hort. Stats*, pp. 66-67.

Lukaszewska, A.J. 1983a. The effect of continuous and 24 hour sugar feeding on the keeping quality of cut dahlias. *Prace Instytutu Sadownictwa i Kwiaciarstwa w Skierniewicach, B (Roœliny Ozdobne)*, **8**: 199-205.

Lukaszewska, A.J. 1983b. The effect of continuous and 24 hour sugar feeding on carbohydrates free aminoacids in the inflorescences of cut dahlias. *Prace Instytutu Sadownictwa i Kwiaciarstwa w Skierniewicach, B (Roœliny Ozdobne)*, **8**: 207-214.

Maatsch, R. and W. Rünger, 1955. Über die photoperiodische Reaktion einiger Sorten von *Dahlia variabilis* Desf. *Gartenbauwiss.* (n.s.), **1**: 366-390.

Maiko, T.K. and N.P. Yashchenko, 1980. Effect of retardants on tuber formation and flowering in dahlias. *Byulleten' Glavnogo Botanicheskogo Sada*, No. 118, pp. 61-64.

Marini, E. 1951. Segnalazione di virosi della dalia nelle culture di produzione in Italia. *Not. Mal. Piante*, No. 17, pp. 13-16.

Marston, M.E. and G. Evans, 1972. Capillary bench technique could prove invaluable in commercial production of dahlia pot tubers. *Grower*, **77**(17) 950.

Mastalerz, J.W. 1976. Dahlias have potential as new cut flower crop. *Science in Agric.*, **24**(1): 11.

Martin, C. 1954. Recherches sur les maladies à virus du dahlia. *Ann. Epiphyt.*, **5**: 63-78.

Martin, C. 1959. Study of some metabolic abnormalities in plants affected by virus diseases (French). *Ann. Physiol. Vég.*, **1**: 59-82.

Mather, S.N. and R.A. Sharma, 1969. Influence of gibberellic acid on flowering and critical daylength of dahlia. *Abstracts 11th int. bot. Congr., Seattle*, p. 142.

Mehlis, E. 1976. Alte und neue Sorten gezeigt. *Deutsche Gärtnerbörse*, **76**(49): 1070-1071.

Mehlis, E. 1978a. Dahlien – Neuheiten 1978 (2. Prüfung). *Gb + Gw*, **78**(44): 1055.

Mehlis, E. 1978b. Dahlien – Neuheitenprüfung 1978 (zweite Prüfung). *Deutscher Gartenbau*, **32**(44): 1833.

Mishra, H.P., K.P. Singh, G.M. Mishra and B. Prasad, 1990. Performance of some dahlia (*Dahlia variabilis*) varieties under late planted condition in calcareous soil of plains. *Haryana J. hort. Sci.*, **19**(3-4): 284-290.

Misra, R.L. 1978. Combating dahlia smut. *Progressive Hort.*, **10**(1): 43-46.

Mittal, S.P. 1967. studies on the effect of gibberellin on growth and flowering of dahlia. *Madras agric. J.*, **54**: 103-107.

Moser, B.C. and C.E. Hess, 1968. The physiology of tuberous root development in dahlia. *Proc. Amer. Soc. hort. Sci.*, **93**: 595-603.

Morel, G. and C. Martin, 1952. Guérrison de dahlias atteints d'une maladie á virus. *Comptes Rendus Academie des Sciences*, **235**: 1324-1325.

Mullin, R.H. and D.E. Schlegel, 1978. Meristem-tip culture of dahlia infected with dahlia mosaic virus. *Plant Dis. Reptr*, **62**(7): 565-567.

Mustafajeva, A.K. 1985. Propagation of dahlias by cuttings in Apsheron (Russian). In: *Introduktsiya i Akklimatizatsiya Rastenii, Baku, Azerbaidzhan SSR*, pp. 89-92.

Neiswander, R.B. 1955. Control of insects results in more beautiful dahlias. *Ohio Fm Home Res.*, **40**: 80-81.

Newton, P. 1960. Dahlia researches. *Gdnrs' Chron.*, **148**: 70.

Newton, P. 1962. Investigating tuber production by dahlias. *Gdnrs' Chron.*, **152**: 140-141.

Newton, P. and D. Tullis, 1962. Investigating tuber production by dahlias. *Gdnrs. Chron.*, **152**: 176-177.

Nordström, C.G. and T. Swain, 1953. The flavonoid glycosides of *Dahlia variabilis*. Part I. General introduction. Cyanidin, apigenin, and luteolin glycosides from the variety "Dandy". *J. Chem. Soc.London*, pp. 2764-2773.

Nowak, J. and R.M. Rudnicki, 1990. *Postharvest Handling and Storage of Cut Flowers, Florist Greens, and Potted Plants.* Timber Press, Portland, Oregon.

Okada, M. and H. Harada, 1955. Effects of daylength and temperature on the ratio of ray-flowers to disk-flowers in dahlia flower heads (Japanese). *J. hort. Ass. Japan*, **23**: 259-263.

Oliseviè, G.P. 1959. The use of chemicals with systemic activity for the control of aphids under glass. *Bjull. Glav. Bot. sada*, No. 35, pp. 91-95.

Pape, H. 1953. Eine häufige Blattfleckenkrankheit der Dahlie. *Gartenwelt*, **53**: 255-256.

Pearman, J.A. 1959. Diseases of dahlias. *Agric. Gaz. N.S.W.*, **70**: 369-375.

Pappiah, C.M. and S. Muthuswamy, 1974. Effect of some plant growth regulants on growth and flowering of dahlia (*Dahlia variabilis*). *South Indian Hort.*, **22**(3/4): 77-80.

Pardatscher, G. 1951. Protoplasmatische Studien an Blütenzellen von Dahlia. *Portugal Acta biol., Ser. A*, **3**: 171-186.

Post, K. 1949. *Dahlia pinnata*. In: *Florist Crop Production and Marketing*, pp. 439-444. Orange Judd Publishing, New York.

Protich, N. 1988. New dahlia cultivars. *Rasteniev"dni Nauki*, **225**(4): 52-55.

Quagliotti, L. and G. Gullino, 1967. The effect of planting density on dahlia growth and flowering (Italian). *Att. Conv. Naz. Floric., Trieste*, pp. 18.

Rahman, M.M. and S.N. Mitra, 1974. Effects of nitrogen on growth and flowering of *Dahlia variabilis* Willd. *Bangladesh Hort.*, **2**(2): 15-18.

Read, P.E. 1968. Effects of succinic acid 2, 2-dimethyl hydrazide and (2-chloroethyl) trimethylammonium chloride on tomatoes (*Lycopersicon esculentum* Mill.) and on tuberous root formation in *Dahlia pinnata* Cav. *Diss. Abstr., sect. B*, **29**: 838.

Read, P.E., C.W. Dunham and D.J. Fieldhouse, 1972. Increasing tuberous root production in *Dahlia pinnata* Cav. with SADH and chlormequat. *HortSci.*, **7**(1): 62-63.

Roy, A.K. 1965. Occurrence of powdery mildew caused by *Oidium erysiphoides. FAO Plant Prot. Bull.*, **13**: 42.

Rudionenko, G.I. and E.I. Zaar, 1951. Obtaining dahlia twin plants (Russian). *Priroda*, **40**(5): 62-63.

Rünger, W. 1955. Die photoperiodische reaction von dahlien. *Gartenwelt*, **55**: 223-224.

Rünger, W. and K.E. Cockshull, 1985. *Dahlia*. In: *Handbook of Flowering* (ed. Halevy, A.H.), vol. II, pp. 414-418. CRC Press, Boca Raton, Florida.

Saaltink, G.J. 1969. Bacterial infections of flower bulbs (Dutch). *Tuinbouwmededelingen*, **32**: 274-285.

Safronova, A.I. 1969. Seed propagation and selection of double dahlias in Crimean conditions. *Bjull. gos. nikitsk. Bot. Sada*, No. 1(8), pp. 14-17.

Safranova, A.I. 1972. Characteristics of dahlia (*Dahlia x cultorum*) cultivation in the Crimea (Russian). *Trudy Gosudarstvennogo Nikitskogo Bonicheskogo Sada*, **59**: 27-43.

Sang, C.G. 1974. Studies on propagating dahlias from cuttings. II. The effect of nitrogen forms on plant growth, flowering and tuberous root formation (Korean).*J. Korean Soc. hort. Sci.*, **15**(2): 178-186.

Sansdrap, A. 1969. Virus diseases of dahlias (French). *Pépiniéristes Horticulteurs, Maraîchers*, No. 96, pp. 5657-5663.

Sasso, G. 1961. An experiment on mineral nitrogen fertilizing of dahlias grown for cut flowers (Italian). *Riv. Ortoflorofruttic. ital.*, **45**: 119-126.

Senaratna, J.E. 1947. Two new dahlias bred in Ceylon. *Ceylon J. Sci.*, **12**: 217-219.

Sharga, A.N. 1975. Horticultural description and performance studies of some dahlia cultivars. *Progressive Hort.*, **7**(3):

29-38.

Shrotri, S.C., J.S. Gupta and R.N. Srivastava, 1984. Mycoflora associated with dahlia seeds, its pathogenicity and control. *Indian J. Mycol. Pl. Path.*, **14**(3): 266-269.

Singh, J.P. 1970. A spontaneous mutant for flower colour and shape in a white flowering dahlia. *Euphytica*, **19**: 261-262.

Staden, O.L. 1976. Houdbaarheidsmiddelen bijzonder waardevol. *Vakblad Bloemist.*,**31**(34):44-45.

Steffen, L. 1958. Winter storage of dahlias. *Gartenwelt*, **58**: 398.

Suma, B. 1993. Effect of Growth Retardants on Growth, Flowering, Vase-life and Tuber Formation of Dahlia (*Dahlia variabilis* Desf.) Propagated through Cuttings (M.Sc. Thesis). College of Agriculture (K.A.U.), Vellayani, Kerala.

Suman, C.L., S.D. Wahi and S.K. Bhattacharjee, 1984. Genetic divergence in *Dahlia*. *Indian J. Hort.*, **41**(3/4): 299-303.

Šumilenko, E.P. and A.J. Šaranova, 1963. The distinction of planting material of flower crops. *Zašè. Rast. Vred. Bolez.*, **8**(4): 38.

Swart, A. 1989. Dahlia als snijbloem. Korte houdbaarheid blijft problem voor afzet. *Bloembollencultuur*, **100**(5): 24-25.

Upadhyay, J. and T.P.S. Bhandari, 1985. Fungicidal control of leaf smut of dahlia caused by *Entyloma dahliae*. *Indian Phytopath.*, **38**(2): 338-339.

Wasscher, J. 1955. De invloed van een kortedagbehandeling op de knolvorming bij dahlia's, knolbegonia's en gesneria's. *Meded. Dir. Tuinb.*, **18**: 342-352.

Whipker, B.E. and P.A. Hammer, 1997. Efficacy of ancymidol, paclobutrazol, and uniconazole on growth of tuberous-rooted dahlias. *HortTechn.*, **7**: 269-273.

Whitley, G.R. 1985. The medicinal and nutritional properties of *Dahlia* spp. *J. Ethnopharmacology*, **14**(1): 75-82.

Wickizer, S.L. and M.V. Meisch, 1977. Effects of non-commercial miticides against mites on dahlias and soybeans. *Arknsas Farm Res.*, **26**(6): 13.

Widmer, R.E. 1965. Systemic insect control for dahlias. *Minn. Hort.*, **93**: 26-27.

Winter, J.A.T. De, 1977. De teelt van dahlia's als pot plant. *Bloembollencultuur*, **87**(30): 602-603.

Yasuda, I. and N. Yokoyama, 1959. Effect of day length on growth and root formation of dahlia. I. Short-day treatment in summer (Japanese). *Sci. Reps Fac. Agric. Okayama Univ.*, No. 13, pp. 57-62.

12

Delphinium (Family: Ranunculaceae)

Sanyat Misra and R.L. Misra

[**Common names**: Alabama larkspur (*Delphinium alabanicum*), Anderson's larkspur (*D. andersonii*), Baker's larkspur (*D. bakeri*), Bee larkspur (*D. elatum*), Chiricahua mountain larkspur (*D. andesicola*), Colorado larkspur (*D. alpestre*), Knight's spur, Forking larkspur (*D. consolida*), Lark's heel (a name given by Shakespeare), Lark's paw, Larkspur/Common larkspur/Candelabrum larkspur (*D. regale*, syn. *D. consolida*), Musk larkspur (*D. brunonianum*), Rocket larkspur (*D. ajacis*, syn. *D. ambiguum*), Subalpine larkspur (*D. barbeyi*), Tracy's larkspur (*D. antonium*), Yellow larkspur (*D. luteum*)]

Introduction and Origin

Delphinium takes its name from the Greek *delphinion* which has come from *dolphin*, alluding to the shape of the flowers of certain annual species. The Greek name for the plant means literally 'wild raisin'. Parrett (1961) mentions that at least 15 species of *Delphinium* abound in Greece, the Aegean Islands, and the Asia Minor alone, and one of them is *D. peregrinum*, the unopened buds of which faintly resemble a dolphin in miniature, thus inspiring the Latinised name *Delphinium*, from the Greek 'a little dolphin'. Alexander the Great in the spring of 328 B.C. when crossed Hindu Kush to conquer Afghanistan and NW India, found natural haunts of the golden-yellow *D. zalil* both in the high lands of NE Persia as well as NW Afghanistan as this plant for centuries provided a dye for dyeing silk in these parts, as they do even today (Parrett, 1961). The genus comprises some 300 species of annuals and herbaceous perennials from northern temperate zone and mountains farther south. The annual larkspur is native to the Mediterranean, North Africa, southern Europe and western Asia; and the original perennial species commonly known as delphiniums are from the cold areas such as Siberia, China and Europe. The common flower colours available in larkspur are white, yellow, orange, pink, rose, scarlet, red and light to dark blue. These have erect and branching herbaceous stems; alternate, palmately or digitately lobed leaves, typical of Ranunculaceae family; the flowering stalk is erect and varies greatly in size among the species, from 10 cm in some alpine species to 2 metres tall in larger meadowland species, topped with a raceme of many flowers; terminal racemes of spurred flowers with 5 petaloid sepals where posterior one is prolonged as spur, and 2-4 much smaller (the posterior ones spurred), sometimes contrastingly coloured petals, 1 or 2 of them (1 petaloid sepal and 1 petal) with nectiferous spurs inserted into the large sepal spur; pollination occurring through butterflies and bumble bees; the fruit consists of erect clusters of 1-3 or sometimes 5, many-seeded cylindrical follicles, and the seeds are small and often shining black. Most species are toxic due to presence of delphinine alkaloid in all the parts of the plant including seeds and this alkaloid is known to be cardiotoxic and affects neuro-vascular systems, even though it is attacked by Lepidopterous larvae. The juice of flowers mixed with alum provides a blue ink.

They are largely represented in the gardens by the annual larkspurs and the race of perennial hybrids derived from *D. elatum* in the nineteenth century in France by Victor Lemoine. The common groups include Pacific Giant, Dwarf Pacific, New Millenium and Magic Fountain. This species is native of Alps. The hybrids are popular as garden plants for their stately showy spires of true blue, purple, white and the flowers of other colours used for cutting, as potted perennial and for herbaceous borders. These hybrids are of various heights to suit every occasion. The tall and stately hybrids have been developed by crossing perennial larkspur, *Delphinium elatum* (a taller, growing up to 1.6 metres) with other species, and this hybrid group as a whole is known as *D.* x *cultorum*. *D.* x *belladonna*, also a perennial, was developed by crossing *D.*

elatum and *D. grandiflorum* (which has dwarf forms also, and is a more informal and shorter than *D. elatum* as well as *D.* x *belladonna* hybrids), but these hybrids are shorter and more heavily branched than *D.* x *cultorum* with flower spurs up to 2.5 cm long (Hay, 1971; Bailey and Bailey, 1976; Armitage, 1993; Dole and Wilkins, 1999).

Classification, Species and Cultivars

There are some 300 species under this genus. Some 60 species are native of the north temperate zone, four of which are of much greater popularity, *viz. D. ajacis* (annual species), *D. grandiflorum, D. hybridum* and *D. formosum* (Bailey, 1942).

The important ones under cultivation are (i) the annual ones such as *D. ajacis* (syn. *D. ambiguum*, Mediterranean), a hardy annual growing upright from 30 to 90 cm with sparsely branched stems, racemes loose and some 60 cm long, and flowers blue or violet, and this species has given tall as well as dwarf garden hybrids of 'hyacinth-flowered' group with double flowers, excellent for cutting and for growing in pots; and forking larkspur, *D. consolida* (Europe including Great Britain), a hardy annual preferring chalky loam for its cultivation, usually well branched from the base, growing to 120 cm, spread 30-45 cm, racemes dense and 23-38 cm, bearing white, pink, red, purple and blue flowers, and its garden strains are 'Giant Imperial strain' growing some 120 cm high with varieties such as 'Blue Spire' and 'Dazzler' (carmine scarlet), and 'Stock-flowered strain' growing up to 90 cm where flowering is earlier than 'Giant Imperial strain', and its var. 'Rosamund' is bright rose; and (ii) the perennial species being numerous such as *D. altissimum* (Himalayas), plants tall, branched and slender, and flowers blue or purple in long branching racemes; *D. bicolor* (Colorado and Alaska), roots fascicled, growing to about 30 cm with erect and stout stem, raceme with only a few flowers where spur and sepals are blue and upper petals pale-yellow or white but blue-veined and lower petals blue; *D. brunonianum* (Afghanistan to W. China), stems erect, 15-45 cm long, segments deeply cut and musk-scented, flowers large, light blue with purple margins but centre black, sepals 2.5 cm long and spur short; *D. cardinale* (California, Mexico), hardy, tufted perennial, roots somewhat fleshy, stems erect, height 60-105 cm or more, racemes loose and 22-30 cm or more long having many bright red cup-shaped 2-3 cm wide flowers shaded yellow in the centre; *D. carolinianum* syn. *D. azurem* and *D. virescens* (North Carolina to Illinois), stems 45-75 cm high, less branched, racemes spicate and usually many-flowered, flowers azure-blue but varying to white, and sepals normally with a brownish spot, its var. *albidum* is taller than the parent with creamy-white flowers, and var. *vimineum* is sometimes branched bearing violet or white flowers; *D. cashmerianum* (Himalayas), stems erect and 25-45 cm long, inflorescence corymbose, flowers 5 cm long, deep azure-blue, and upper petals almost black and 2-lobed while lateral ones greenish, and its var. *walkeri* quite short and many-flowered, flowers quite large, and light-blue with yellow petals; *D. cheilanthum* (syn. *D. magnificum*, Siberia), stems 60-90 cm, erect, simple or branched, flowers dark blue, but upper petals sometimes pale-yellow; *D.* × *cultorum* (syn. *D. hybridum* Hort and not Steph., hybrid form of single, semi-double and double forms), of many statures and colours which comprise of many choicest garden and border delphiniums; *D. decorum* (California), stems weak and 15-45 cm long, flower in a loose raceme, sepals 1.25 cm long, blue and almost equal to spur and upper petals tinged yellow; *D. elatum* (syn. *D. pyramidale, D. alpinum* or *D. intermedium*, Pyrenees to Siberia and Mongolia), the true species is hardly in cultivataion and is represented by a race of hybrids developed by crossing through *D. grandiflorum* and other species, hardy, very close to *D. exaltatum*, a species which through hybridization has resulted in many tall and intermediate hybrids and 1000s of cultivars are available with single or double blooms in a wide range of colours, *viz.* blue, purple, pink, cream and white, and more recently the red species *D. cardinale* and *D. nudicaule* have been used and University Hybrid group has been evolved with a few hybrids having brilliant red flowers, especially *D. cultorum, D. belladonna* and many others by crossing with *D. grandiflorum, D. exaltatum, D. formosum*, and other species, grows to 100-200 cm tall, with about 30 cm of raceme, raceme more towards a spike, flowers 2.5-4.0 cm wide, petals blue with dark violet to yellow or blackish-brown and sepals almost equal to spurs, and its hybrids are classified into two major groups: large-flowered (Elatum varieties) and Belladonna varieties. Elatum varieties are erect with large and flat florets which are often semi-double or double, growing up to 240 cm with a spread of up to 90 cm, and may have tall (135-180 cm) which includes 'Betty Hay' (pale-blue, eye white), 'Butterball' (cream, eye yellow), 'Cressida' (pale-sky-blue, eye white), 'Daily Express' (bright sky-blue, eye black), 'Mullion' (cobalt-blue, eye black-brown), 'Purple Ruffles' (double, deep purple, shaded deep blue), 'Purple Triumph' (violet-blue, eye balck and gold), 'Royal Marine' (deep purple), 'Silver Moon' (silver-mauve, eye white), 'Swanlake' (white, eye brown), 'Xenia Field' (palest lavender-white, eye white), also this group has one mainly seed-grown tall varieties known as Pacific Hybrids growing up to 120-180 cm, bearing semi-double or double flowers in colour range of white, pink, purple, blue, and some single-colour strains are 'Astolat' (lilac and pink), 'Black Night' (deep violet), 'Galahad' (white), 'Summer Skies' (blue), etc. and the dwarf (90-135 cm) varieties include 'Baby Doll' (pale-mauve, eye yellow-white), 'Blue Tit' (indigo-blue, eye black-brown), 'Blue Jade' (sky-blue, eye brown), 'Cinderella' (mauve), 'Mighty Atom' (deep lavender, flowers large), 'Page Boy' (blue, eye white), and the Belladonna varieties (height 105-135 cm) are smaller with much branched loose but wiry spikes bearing cupped florets, and the varieties under this group are 'Blue Bees' (pale-blue), 'Bonita' (blue), 'Loddon Blue' (royal blue), 'Lamartine' (violet-blue), 'Moerheimii' (white), 'Pink Sensation' (pink), 'Wendy' (blue, flecked purple), etc.

D. exaltatum (Alaska to Minnesota), stems stout and 60-120 cm, flowers are crowded on pyramidal racemes, colour blue but yellow on upper petals and sepals almost equal to the length of spurs; *D. formosum* (Asia Minor), hardy, good for naturalizing, stems stout and 60-90 cm long, racemes bear many flowers of blue colour with indigo margins, spur long, bifid at the tip and violet in colour, and its var. *coelestinum* bears light blue flowers, and it has produced certain good hybrids by crossing with certain other species such as *D.*

elatum; *D. grandiflorum* (syn. *D. sinense*, Siberia, China) is a tufted hardy perennial with 30-90 cm tall stems which are branched and 23-30 cm long racemes, widely funnel-shaped flowers are blue to violet and varying to white and large (up to 4 cm wide), spurs long and lower petals often violet while upper petals often yellow, its var. *album* is white, var. *albo-pleno* bears double white flowers, var. *flore-pleno* (var. *hybridum flore-pleno*) bears graceful double blue flowers, and var. *sinense* (*chinense*) has less-branched stems with numerous flowers and is a favourite garden form, form *breckii* bears double blue flowers, and the var. 'Blue Butterfly' which rarely grows 30 cm in height bears bright blue flowers and this can be grown as an annual by sowing in the greenhouse; *D. hybridum* Steph. (Mountains of Asia), its roots are somewhat bulbous, stems 90-120 cm, racemes dense, flowers blue, lower limbs white and bearded, and spurs straight and longer than sepals; *D. maackianum* (Siberia), growing to 90 cm and erect, panicles loose, sepals blue, spurs small and petals dark violet; *D. menziesii* (California, Alaska), a tuberous-rooted, stems simple and 15-45 cm high, inflorescence a simple and conical raceme bearing up to 10 flowers about 2 cm wide, sepals rich deep blue and equal to spurs in length, lower petals sometimes white-lined and upper petals yellowish-blue; *D. mesoleucum* (origin not certain), stems up to 90 cm, and flowers blue with pale-yellow or whitish petals; *D. muscosum* (Bhutan), a tufted perennial, stems 10-15 cm, and flowers deep-violet on long stalks and some 3 cm wide; *D. nudicaule* (California, Oregon), a tufted perennial, stems branched and 25-50 cm long, flowers cupped, borne in loose panicles, 2.5-3.5 cm long including spur as spurs are longer than sepals, and panicled (long stalked), sepals bright red or orange-red, and petals almost equal to sepals and yellow; *D. nuttallii* (syn. *D. columbianum*, Columbia riverside), stems simple, leafy, erect and 45-75 cm long, racemes are many-flowered and long, sepals deep blue, ovate and shorter than spur, and petals blue or upper ones yellow while lower ones white and bearded; *D. pauciflorum* (Colorado to Washington and California), stems 60-150 cm tall with simple and densely flowered racemes bearing many flowers of blue, purple or rarely of white colour, upper petals often yellow, and spurs equal to sepals and some 1.25 cm long; *D. przewalskii* (syn. *D. prsewalskianum*, Asia), varying in height, erect but branched at the base, flowers yellow but sometimes tipped blue and spurs equal to sepals; *D. scopulorum* (west of Rockies at moist locations), stems 60-150 cm bearing simple, dense and many-flowered raceme, flowers blue or purple but rarely white, upper petals mostly yellow and spurs 1.25 cm long and equal to sepals, and its var. *subalpinum* syn. *D. occidentale* is smaller plant with shorter racemes but flowers are large and deeper than the parent; *D. simplex* (Idaho and Oregon mountains), stems 60-90 cm and almost simple, racemes little branched and dense, flowers pale-blue, sepals and spur equal in length, upper petals yellow and lower white and bearded; *D. tricorne* (North USA), stems succulent and about 30 cm in height, flower large, blue but sometimes whitish, upper petals occasionally yellow with blue veins, lower ones white and bearded and spurs equal to sepals in length; *D. tatsienense* (China), very similar to *D. grandiflorum* with the same size of the spurs, though as long as double of the sepals in original species, height 23-38 cm and most suitable for rock gardens, panicles branched and about 15 cm long, flowers large and funnel-shaped, deep purple-blue and long spurred; *D. trollifolium* (moist location of Columbia river), height 60-150 cm, racemes long (30-60 cm) but quite loose, flowers blue, upper petals white and sepals and spurs some 2 cm long; *D. zalil* (syn. *D. sulphureum* or *D. hybridum* var. *sulphureum*, Persia), roots tuberous, stems 30-60 cm high, erect and almost simple, racemes long, and flowers are large and light yellow; etc. Baker's larkspur (*D. bakeri*) and yellow larkspur (*D. luteum*), both native to very restricted areas of California are highly endangered species.

Propagation

The Belladonna and Elatum groups, and most of other perennials can best be increased through 7-10 cm basal **cuttings** taken close to the rootstock in April in the temperate areas of the country. The cuttings are inserted in the medium containing equal part of peat and sand in a cold frame to protect from severe frosts, and after rooting these cuttings can be lined in the nursery beds or in the 10-cm pots up to August. In September these are transplanted to their permanent positions (Hay and Beckett, 1971). Cuttings are taken when the young leaves start to expand, severed at the ground level quite close to the crown, however, those with hollow centre are discarded. Hybrids and other perennials can also be raised through **root-division**, though it is not the commercial practice, each division having axillary buds, by sowing in March or April or also during autumn, and these will flower the following autumn but this practice may be adopted only every 2-3 years, however, established perennial species and cultivars can last for a decade or more in its favourable cool climatic conditions as prevail in the temperate regions of India though under warmer climates even the perennials are treated as annuals (Armitage, 1993).

One gramme of *Delphinium* hybrid may contain some 350 seeds. **Seeds** are short-lived so these should be sown in autumn either just after harvesting or in the following spring. If this is not possible, seeds should be stored at 2-5°C temperature until sowing. Long term storage of *D.* x *cultorum* seeds at 5°C and at 30-50 per cent relative humidity earlier, germinate uniformly and with higher percentage than nonstored ones (Dole and Wilkins, 1999), however, prior to germination these seeds should be shifted at 2°C for three weeks and then should be hydrated and held constantly at 15 or 20°C in either light or in dark though there had been cultivar differences (Carpenter and Boucher, 1992). However, the annual larkspur seeds germinate rapidly under LD after six weeks of cold stratification at 2°C and then sowing at 13-18°C (Schwabe, 1971). All the species whether annual or perennials and the varieties can be started through seeds in the greenhouse or in the hotbeds during autumn or in February-March and transplanted about one month later. The seeds of only annual species and varieties are sown in the plains during September and transplanted at their permanent positions when these attain 7-10 cm of height which takes 30-45 days. Seeds of *Delhpinium grandiflorum* and *D. tatsiense* may also be sown in pots or pans filled with seed compost, in February around 13°C temperature, pricked off in boxes and hardened off in the cold frame, and then in April it should be planted out in permanent positions for flowering the same year (Hay and Beckett, 1971). Seeds may also be grown through plug system. Plug plants do best when transferred either into 13-cm containers for potted plant production or directly into the bed for cut flower production. Also, these can be propagated through *in vitro* (Garner *et al.*, 1997).

Cultural Practices

Delphiniums require sunny situation but protection from wind, and a sandy-loam **soil** having pH from 5.5 to 7.5, most suitable being when it is around 7. Soil should have sufficient organic matter to support the plants throughout as perennial plants are left as such for many years at the same site. Though young plants are not very demanding but grown up ones have good demand. It would be better if soil is incorporated with 250-300 quintals of well-rotten dung manure or farmyard manure at the time of **soil preparation.** Three deep cultivations are required to make the soil to a fine tilth. Beds of the convenient sizes are made with a proper network of bunds and channels for cultural operations and for movements. **Planting** is done in the month of October irrespective of whether it is tropical or temperate region, though in the plains survival of the plants becomes difficult, especially for the perennial ones in summer until these are grown under protected environment. Distances apart are kept 15 to 20 x 20 for dwarf ones, 30 x 30 for those having normal spraed and 40 x 30 for Belladonna types and for *D. elatum*. Though it does not like wet feet but the soil should be moist and not wet, throughout the growing season. The planting should be followed by light watering. For supporting the tall growing plants, these should be **staked** in April-May. Tall varieties with large flower spikes need stout stakes or canes and the dwarf or annual larkspurs require twiggy sticks and if grown under sheltered gardens no stakes are required. After flowering, the stems are cut back to the nearest healthy leaf below the raceme so that in favourable summers a second crop may also appear. In autumn end, all the stems should be cut back to the ground level. Before setting of the winter, now the beds are **mulched** with about 5-cm thick layer of farmyard manure for protection of the underground buds against frosty winter and will greatly enrich the soil. This manure is spaded up in the soil during spring. During mid-summer, a top-dressing of manure near the plants is used to force the plants for the second crop of flowers during autumn. This dressing also conserves the soil moisture, prevents weeds, increases fertility of the soil and helps us getting more adventitious buds in the spring for division. Garner *et al.* (1997) suggests **pinching** of *Delphinium* seedlings for delaying the harvest and for increasing the stem lengths and yields. Whenever, the **weeds** appear (in the plains whole of the winter and in the hills from March to May), these are taken out so that soil nutrients are fully utilized by the main crop and there may not be any competition for sunlight. After sometime when there is full growth of the plants of the main crop, the weeds will not get any chance to appear due to thick canopy of the main crop. Since it a winter crop in the plains hence it may require one light **watering** every 7-10 days, whereas in the temperate regions as it becomes the summer crop so there it should be watered every 5-7 days if the weather is sunny and there is no rain as watering is highly dependent on weather conditions. Regarding **fertilizer** applications, if sufficient farmyard manure has been added in the soil at the time of soil preparation, during midsummer to individual plants and again by the end of autumn if mulched again with farmyard manure, it may not require any further nutrients but in case if plants are in poor growth, 75 kg/ha of nitrogen may be applied to the plants about a fortnight before flowering. **Harvesting** of the spikes is done when half of the florets have opened.

Growth, Flowering and Height Control

After stratification for six weeks at 2°C, the seeds of annual larkspur germinate rapidly at 13-18°C (Schwabe, 1971) and since plants at SD rosette at 13°C, long days are also required. Incandescent night interruption and 11°C temperature cause elongation but when very young annual seedlings are subjected to high temperatures and LD, there occurs premature flower induction, hence, quality becomes poor, though at low temperature of 10°C and natural short winter days, annual larkspurs fail to flower as occurs in Michigan (Dole and Wilkins, 1999). Lindstrom *et al.* (1957) stated that at 15°C, it took 187 days to appearance of floral buds in annual larkspur but when GA was applied it took 203 days at 10°C and 178 days at 15°C.

In perennials, the axillary or dormant buds are still vegetative in the autumn and during spring these buds elongate with complete leaf initiation and initiate and develop inflorescences (Dole and Wilkins, 1999). Wilkins (1986) reported that for shoot elongation and flower initiation and development, annual or perennial larkspurs require cold environment, and Post (1952) advocated a cold **temperature** of 10-13°C until February for perennial delphiniums, though Holcomb and Beattie (1988, 1990) opined that after germination the plants should be kept at 13/16°C during night and up to 24°C during daytime where successful flowering occurs under high irradiance. Visible bud and flowering in *D. grandiflorum* 'Blue Mirror' were found delayed up to 27°C temperatures (Wang *et al.*, 1997). During winter, high intensity discharge (HID) **light** used as supplementary light is more important than photoperiod (incandescent night breaks) since under HID light from 5 P.M. to 8 A.M. (15 hours) larkspur plants flowered within 82 days whereas those under a 10 P.M. to 2 A.M. (4 hours) incandescent night break required 152 days (Dole and Wilkins, 1999) while contrarily, Garner *et al.* (1997) stated that from 10 P.M. to 2 A.M. incandescent light night breaks reduced the number of days required for flowering and increased the yield.

Keeping day and night temperatures constant, the **height** of potted plant will not be that much problematic (Hamaker *et al.*, 1996). A-Rest (ancymidol) spray at 100 ppm and Sumagic (uniconazole) at 5 ppm drench are effective in controlling the height of potted larkspur (Dole and Wilkins, 1999). Holcomb and Beattie (1988) reported that effect of either A-Rest or Sumagic on the plants is nullified through incandescent night breaks.

Postharvest

When half the flowers on the spikes are open, it is proper time to harvest the flowers. Pulsing with STS reduces ethylene responses as the flowers are highly sensitive to ethylene, and increases longevity. Holding solution containing a floral preservative in warm water at pH 3.5 to 4.0 is effective to larkspur flowers (Dole and Wilkins, 1999). The flowers can be stored and transported at 2 to 5°C effectively (Dole and Wilkins, 1999). Apical flower portion of the cut blooms have tendency to bend upward during transit when packed horizontally, hence, the flowers should be transported in upright position (Nowak and Rudnicki, 1990; Sacalis, 1993). *Delphinium* cut flowers were when held immediately after cutting in water or in various conditioning solutions before marketing, best results were obtained with Chrysal-avb which extended the vase life by 4.3 days (Kalkman, 1983). The effects of STS and 1-MCP were studied by Dole *et al.* (2005) on cut *Delphinium* cv. 'Light Blue' by treating the unpacked cut stems and placing them either in deionized water (DI) and subjected to 1-MCP (740 nl l^{-1}) or ambient air for 4 h or DI + STS at 0.2 mM for 4 h. After treatment, stems were removed, placed in polyethylene sleeves and stored either wet in DI water or dry in plastic-lined floral boxes at 5 °C in the dark for 4 days. After storage, bunches were placed in DI water under 12 h light (76 to 100 μmol $m^{-2}s^{-1}$) light per day where it was found that STS increased the vase life though 1-MCP did not.

Insect-Pests and Diseases

Delphiniums are very hardy plants. Of garden pests, only slugs pose a serious threat though there are certain other insect-pests which may attack the crop. Apart from slugs and snails, they are infested by cyclamen mites, red spider mites, aphids, thrips, caterpillars and nematodes. Slugs check early growth of the plants by feeding on the young shoots and by making holes in the leaves of older plants. They are of four types, *viz.* Grey Field, Black Field, Large Black and Keeled, lattest being more serious during winters which damages the underground plant parts. Contact killers containing alum and the poison metaldehyde baits can control these pests. Cyclamen mite infestation symptoms appear as the symptoms of the disease since this causes stunting, malformations and blackening of the buds. The larvae of the moth, *Tortrix cnephasia* feed on young buds and this can be controlled by spraying the plants with 0.2 per cent Metasystox. This will also control aphids and thrips together. Spraying with 0.1 per cent Keltahne will control the mites. The nematode *Pratylenchus pratensis* infests on the larkspur roots and stools which can be worsened by millipedes and by stagnant water. To get rid of this nematode the soil should be disinfected and resistant varieties should be planted.

Bazzi *et al.* (1987) reported *Pseudomonas syringae* pv. *delphinii* infecting *Delphinium* in Italy.

Diseases that attack the crops are powdery mildew (*Erysiphe polygoni*), crown rot (*Sclerotium rolfsii*), root and stem rot (*Pythium* sp., *Rhizoctonia solani* and *Sclerotinia sclerotiorum*), stem canker (*Fusarium* sp.), bacterial foot rot (*Erwinia carotovora* pv. *atroseptica*), bacterial leaf spot (*Pseudomonas syringae* pv. *delphini*) and aster yellows phytoplasma (Post, 1952; Farnham and McCain, 1982; Bazzi *et al.*, 1987; *Horst*, 1990), *vis-a-vis* fasciation. Cucumber mosaic and tomato spotted wilt cause stunting of plants and leaf distortion. Brilliant yellow rings and line patterns in concentric rings or variable areas of mottling also occur. Powdery mildew makes a white powdery coating on leaves, stems and flowers in July and continue throughout the whole winter. This can be controlled by spraying with lime-sulphur when its appearance is seen. Resistant strains through breeding is the best solution and many such varieties are already in existence. Crown rot and root rot are also caused by many fungi including the black root rot fungus and here root and crown tissues become black and rotten and the top growth dies down. A similar disease also shows where inner tissues of the crown become black and rot away, and on cutting a hollow centre is seen. Stem rot is caused by many fungi including grey mould which appears as grey velvety fungal growth at the base of the stem. Through proper sanitation practices, good drainage and sensible feeding (use of too less nitrogenous fertilizers) are the common practices to minimize these problems. Spraying with benomyl (0.1 per cent) alternate with Captan (0.2 per cent) fortnightly will keep this problem under check and will control even other fungal diseases including canker. Bacterial diseases can be prevented by dipping the roots in streptomycin solution before planting and through spraying, *vis-a-vis* disinfection of the soil. If all the insectpests are controlled, there is little chance of appearance of viral problems, however, infected plants should immediately be rogued out. Fasciation causes flattening of the stems and russetting of the leaves due to environmental fluctuations and may also be due to viral infection in perennial larkspur. Such plants should also be rogued out.

References

Armitage, A.A. 1993. Delphinium. In: *Specialty Cut Flowers*, pp. 190-194. Varsity Press/Timber Press, Portland, Oregon.

Bailey, L.H. 1942. *The Standard Cyclopedia of Horticulture*, pp. 975-978. The Macmillan Co., New York.

Bailey, L.H. and E.Z. Bailey, 1976. *Delphinium* L. In: *Hortus Third : A Concise Dictionary of Plants Cultivated in the United States and Canada*. Macmillan, New York.

Bazzi, C., P. Minardi and U. Mazzucchi, 1987. Bacterial diseases of flower and ornamental plants in Italy (Italian). *Informatore Fitopatologico*, **37**(6): 15-24.

Carpenter, W.J. and J.F. Boucher, 1992. Temperature requirements for storage and germination of *Delphinium* x *cultorum* seed. *HortSci.*, **27**: 989-992.

Dole, J.M. and H.F. Wilkins, 1999. Delphinium. In: *Floriculture Principles and Species*, pp. 566- 568. Prentice Hall, New Jersey.

Dole, J.M., W.C. Fonteno and S.L. Blankenship, 2005. Comparison of silver thiosulfate with 1- methylcyclopropene on 19 cut flower taxa. *Acta Hort.* (*Proceedings of the Fifth*

International Postharvest Symposium, eds Mencarelli, F. and P. Tonutti), held at Verona, Italy, on 6-11 June, 2004; No. 682 (vol. 2), pp. 949-956.

Farnham, D.S. and A.H. McCain, 1982. Delphinium in the greenhouse, growing larkspur as a cut flower. *Florists' Rev.*, **169**(4390): 16-17.

Garner, J.M., S.M. Jones and A.M. Armitage, 1997. Pinch treatment and photoperiod influence flowering of *Delphinium* cultivars. *HortSci.*, **32**: 61-63.

Hamaker, C., W.H. Carlson, R.D. Heins and A.C. Cameron, 1996. Diurnal temperature alterations influence final height of herbaceous perennials (abstr.). *HortSci.*, **31**: 678.

Hay, H. (ed.) and K.A. Beckett (Tech. Advisor), 1971. *Reader's Digest Encyclopaedia of Garden Plants and Flowers*, pp. 216-218. The Reader's Digest Association Ltd, London.

Holcomb, E.J. and D.J. Beattie, 1988. Effect of growth retardants on perennials. *Res. Rpt, Bedding Plant Foundation*, **032**: 1-4.

Holcomb, E.J. and D.J. Beattie, 1990. Potted Delphiniums. *GrowerTalks*, **8**: 14-15.

Horst, R.K. 1990. Delphinium. In: *Westcott's Plant Disease Handbook* (5th ed.), pp. 620-621.

Kalkman, E.C. 1983. Pre-treatment improves the quality of summer cut flowers (Dutch). *Vakblad voor de Bloemist.*, **38**(50): 26-29.

Lindstrom, R.S., S.H. Wittwer, and M.J. Bukovac, 1957. Gibberellin and higher plants, IV. Flowering responses of some flower crops. *Quart. Bull. Mich. Agric. Exp. Stn*, **39**: 673-681.

Nowak, J. and R. Rudnicki, 1990. Posthaarvest Handling and Storage of Cut Flowers, Florist Greens and Potted Plants. Timber Press, Portland, Oregon.

Parrett, R.C. 1961. Delphiniums. Penguin Books Ltd., Middlesex, Great Britain.

Post, K. 1952. Delphinium. In: *Flower Crops Production and Marketing*, pp. 446-451. Orange Judd Publishing, New York.

Sacalis, J.N. 1993. Delphinium. In: *Cut Flowers, Prolonging Freshness* (ed. Seals, J.L.; 2nd edition), pp. 45-46. Ball Publishing, Batavia, Illinois.

Schwabe, W.W. 1971. Physiology of reproduction. In: *Plant Physiology*, vol. VI-A (ed. Steward, F.C.), pp. 258-259. Academic Press, New York.

Wang, S.Y., R.D. Heins, W.H. Carlson aand A.C. Cameron, 1997. Effect of forcing temperature on flowering of four herbaceous perennial species (abstr.). *HortSci.*, **32**: 501.

Wilkins, H. 1986. Delphinium. In: *Handbook of Flowering*, vol. V (ed. Halevy, A.H.), pp. 89-91. CRC Press, Boca Raton, Florida.

13

Dendranthema (Family: Asteraceae)

R.L. Misra and Sanyat Misra

[**Common names**: Annual chrysanthemum (*Chrysanthemum carinatum, C. coronarium, C. segetum*), Chrysanthemum, Corn marigold (*C. segetum*), Costmary/Mint geranium (*C. balsamita*), Dalmatian pyrethrum (*C. cinerariifolium*), Dusty miller (*C. ptarmiciflorum*), Feverfew (*C. parthenium*), Florist's chrysanthemum (*C.* x *morifolium*), Garland chrysanthemum/Rainbow daisy (*C. coronarium*), Giant daisy (*C. uliginosum*), Glaucous marguerite (*C. anethifolium* syn. *C. foeniculaceum* or *C. f. bipinnatifidum*), High daisy (*C. serotinum*), Marguerite daisy/Paris daisy (*C. frutescens*), Moon daisy/Moonpenny daisy (*C. maximum, C. uliginosum*), Ox-eye daisy/Whiteweed (*C. leucanthemum*), Painted daisy (*C. frutescens*), Pyrethrum (*C. cinerarifolium*), Queen of the East, Shasta daisy (*C. maximum*), Tansy (*C. vulgare*), Tansy chrysanthemum (*C. macrophyllum*), Tricolour chrysanthemum (*C. carinatum*), Turfing daisy (*C. tchihatchewii*), etc.].

Anderson (1987) mentioned that scientific names of several species formerly in the genus *Chrysanthemum* have been changed, *viz.* florist's chrysanthemum (from *C.* × *morifolium* to *Dendranthema* × *grandiflorum* Kitam.), costmary (from *Chrysanthemum balsamita* to *Tanacetum balsamita* L.), dusty miller [from *C. ptarmiciflorum* to *T. ptarmiciflorum* (Webb and Berlin.) Schultz-Bip], feverfew [from *C. parthenium* to *T. parthenium* (L.) Schultz-Bip], pyrethrum [from *C. cinerarifolium* to *T. cinerarifolium* (Trev.) Schultz Bip], tansy (from *C. vulgare* to *T. vulgare* L.), tansy chrysanthemum [from *C. macrophyllum* to *T. macrophyllum* (Wldst. and Kit.) Schultz-Bip], high daisy [from *C. serotinum* to *Leucanthemella serotinum* (L.) Tzveleu.], ox-eye daisy (from *C. leucanthemum* to *Leucanthemum vulgare* Lam.), Shasta daisy [from *C.* × *superbum* to *Leucanthemum superbum* (J. Ingram) Bergmans ex Kert], and marguerite daisy [from *C. frutescens* to *Argyranthemum frutescens* (L.) Schultz-Bip].

Introduction and Origin

[*Dendranthema* × *morifolium* (Florist's chrysanthemum)]

Chrysanthemum takes its name from the Greek *chryos* meaning gold and *anthos* meaning a flower, as many species have yellow flower-heads. The genus comprises some 200 species of annuals, perennials and shrubs from northern temperate zone (chiefly Europe and Asia) whereas some claim it to have originated only in China. Though botanists have retained only the annual species under the genus *Chrysanthemum* (Beckett, 1983; Dole and Wilkins, 1999) but growers and florists are finding very hard to accept it because the florist's chrysanthemum has centuries-old popularity, and also as because of this name the florist's chrysanthemum has become synonymous to chrysanthemum which is also the national flower of Japan, though now botanically the florist's chrysanthemum has become *Dendranthema* × *morifolium* (syn. *C. hortorum*; Bailey, 1942). Florist's chrysanthemum is of complex hybrid origin and has derived from a number of species such as *C. boreale* [(Kher, 1989; Arumugam *et al.*, 2006), *C. indicum* (L.) Desmoul. (Thistlethwaite, 1960; Beckett, 1983; Anderson, 1987)] bearing single yellow flowers, *C. japonicum* (Mak.) Kitam (Anderson, 1987), *C. ornatum* (Kher, 1989), *C. sinense* bearing single white ray flowers (Thistlethwaite, 1960), *C. satsumense* (Kher, 1989) and *C. vestitum* syn. *C. morifolium* (Beckett, 1983), and certain other species native of eastern Asia, *viz.* China and Japan (Hay and Beckett, 1971). So-exquisite modern cultivars have developed from small, single and not so attractive types, not only due to natural crossing and extensive manual cross pollination but through spontaneous and artificially induced mutations, chromosomal differentiation, *vis-a-vis* repatterning and polyploidy (Nazeer and Khoshoo, 1982). In the international cut flower trade rose is first and chrysanthemum ranks next to it. In India also, it is highly prized crop and has occupied a place of highly esteemed commercial flower with the name of 'guldoudi', derived from 'gul-e-doud' meaning the 'flower of Doud', and a cultivar from India named 'Gul-doodi' is reported to be growing in Holland since 1,690. Commercially it is being cultivated in Maharashtra, Tamil Nadu, Andhra Pradesh, West Bengal, Rajasthan, Madhya Pradesh and Bihar. For local consumption it is also being grown in all the temple cities of the country. Its cultivation history is very old as evidently China had been cultivating it since 500 B.C. and Confucius, the Chinese philosopher has also described its yellow glory as early as 550 B.C., and the cultivated chrysanthemums after over 1,000 years of development were exported from China to Japan about A.D. 750 where Japanese further developed the Chinese cultivated varieties through cross-fertilization with wild Japanese varieties (Thistlewaite, 1960). The introduction of cultivated chrysanthemum varieties from Far East to Holland in 1,688 was of no meaning as none survived. From China, the varieties reached France in 1,789 and to England 1,795, and when in 1,846 the first shows of National Horticultural Society of England was held all the chrysanthemum varieties being grown there were of Chinese origin. Mr. Robert Fortune on behest of The Horticultural society of London went to Japan in 1,861and sent home a large collection of Japanese varieties which were developed by skilled Japanese gardeners for over 1,000 years of their concerted efforts and devotion, and this introduction to Western civilization is considered a great event in the long history of chrysanthemum growing there. Then the Chinese types had only tightly incurved forms with a limited colour range, *i.e.* white, yellow and pale-mauve but Japanese varieties had great size and many new colours and forms, *viz.* loose incurving and reflexed types with even those having twisted florets in all the directions. Chinese varieties were when crossed with new forms and colours of the Japanese varieties by Mr. John Salter of Hammersmith, within 20 years he got developed new colours with better forms of flowers. This time also matched with that of France where Mr. Simon Delaux developed the first of the early-flowering outdoor dwarf spray type varieties by using new Japanese types. These vartieties were distributed between 1,880 and 1,890. The known history of chrysanthemum introduction into USA goes back to 1,798 when a scientist, Mr. John Stevens introduced 'Dark Purple' as the first variety which later on was followed by many more introductions from China.

In red, chrysanthemum says 'I love you' and in white it symbolizes 'truth'. Chrysanthemums bloom during autumn-winter period in India, and scientists at NBRI, Lucknow have developed varieties which flower one by one throughout the year. The erect and tall cultivars are suitable for cut flowers and for border plantings in the background whereas the dwarf ones are suitable for planting in the front of the border and in pots for various landscape uses including the cascading effect. Small blooming dwarfs are suitable for loose flower use, especially for garland making and hair decoration whereas the taller ones as cut flowers. Small plastic potted plants with abundance of blooms per pot are known as 'potmums' which are suitable for home use and for export. Flower colour range in chrysanthemum is wide as it has white, yellow, orange, pink, scarlet, salmon, red, mauve and violet, purple, bicolours and mixed ones in various shades and hues. Pyrethrum insect powder is prepared from the dried flowers of *C. cinerariaefolium* (Dalmatia mountains of Austria) being cultivated largely in France and *C. coccineum* is commercially cultivated in California for extraction of 'buhach' insect killer.

Botany, Genetics and Breeding

Florist chrysanthemum, *Dendranthema* × *morifolium*, a perennial group of plants, are half-hardy border perennials and hardy greenhouse plants which are usually heavily scented. They have alternate, lanceolate to obovate leaves, from nearly or quite entire, often lobed or more deeply dissected and terminal. Flower head is known as capitulum, many-flowered and is formed of two types of flowers. In a single chrysanthemum the outer ring of florets have obvious petals and these are called as ray florets which are unisexual with only female parts, whereas in the centre the appearance of the yellow flowers, is in fact, stamens and stigmas which are collectively called as disc florets which contain both male and female parts. Various classes of chrysanthemum are based on the forms of the flowers, angles of the petals of both types of florets, and the size of the capitulum they possess when in flower.

Nazeer and Khoshoo (1982) stated that though basic chromosome number for the genus is n = 9, a wide range of ploidy level (hexploidy, aneuploidy, polyaneuploidy, etc., *i.e.* 2n = 36, 45, 47, 51-75) is found in different cultivars. Self-incompatibility found in chrysanthemum is either protandry or sporophytic and involves more than one locus (Fryxwell, 1957; Drewlow *et al.*, 1973). Drelow *et al.* (1973) also reported that nine sibling clones selected from a highly compatible cross were all self-incompatible. Zagarski *et al.* (1983) when crossed 11 sibling lines obtained through 2-generation single plant self-pollination, in a complete diallel selfing and crossing with completely unrelated individuals, found that at least three genes were involved which governed self-incompatibility in chrysanthemum. While studying the somatic-genetic analysis of the apical layers of 16 chimeral sports of cv. 'Indianapolis', Steward and Derman (1970) reported that a cultivar can be genetically of one colour in L_1 and of another colour in L_2, and as sex cells arise from L_2, a L_1 white but L_2 pink cultivar which appears of white flower colour will breed pink and *vice versa* on crossing. It is also recorded that for flower colour inheritance, the presence of one gene *A* ensures anthocyanin production, gene *I* inhibits carotene production, absence of *A* and *I* both give yellow colouration, absence of only *A* expresses white flower colours, *A* without *I* results into bronze and brownish-red flowers, while combination of *A* and *I* results into pink, carmine and bluish-red flowers (Kher, 1989). Singleness is partially dominant over doubleness, in which semi-doubles are comparatively much more in the progeny, followed by singles including anemone types, and then doubles (Kher, 1989).

Chrysanthemums are required to be bred for flowering during low temperature period, for having rapid growth following a short period of long days, for long (up to 80 cm) stems as well as compact and short growth habits, for good sucker production, for less and smaller leaves to save planting space, and for pollen-less cultivars. Kher (1986) states that a large number of outstanding spray and loose type chrysanthemum cultivars in India have been evolved through selection such as 'Apsara', 'Birbal Sahni', 'Co_1', 'Co_2', 'Jaya', 'Jayanthi', 'Sharad Singar', etc.

In chrysanthemums, sports are of frequent occurrence so the better ones are isolated and multiplied. Artificially also, through irradiation with X-rays or ^{60}Co gamma rays the mutations in chrysanthemums are induced. Broertjes (1966) reported optimum X-ray dose as 1500 R for producing chrysanthemum mutants though mutation frequency was low. Gupta (1971) reported three groups of cultivars with respect to optimum dose of gamma rays for induction of mutations in chrysanthemum, *viz.* those highly sensitive to gamma rays where LD-50 appears below 1.5 KR dose, those cultivars where LD-50 appears between 1.5 and 2.5 KR, and those cultivars where LD-50 appears above 2.5 KR dose. Induction of significant increase in the ray foret length through gamma irradiation has also been possible. Pink cultivars have been recorded producing more mutants in the direction of obtaining bronze, yellow and whites, though the mutants of reverse colours are difficult to obtain. Likewise, whites generally produce yellow, and bronze red-yellow mutants. Broertjes *et al.* (1976) produced non-chimeral solid mutants by growing *in vitro*, the irradiated single cells of the epidermis, stating that mutation of L_1 layer produces flower colour mutations while L_2 layer mutation causes effect on plant height, vigour and flower size.

Classification and Varieties

As a result of interbreeding of various species (*C. boreale, C. indicum, C. japonicum, C. ornatum, C. satsumense, C. sinense, C. vestitum*, etc.) from eastern Asia (China and Japan) before 500 B.C., and further developed by the Japanese from about A.D. 800 onwards, a range of different flower types has been developed. Chinese cultivars reached England in 1,795 via France while Japanese cultivars reached there directly in 1,861, and then several cultivars have been bred in the western world with various size, shape and disposition of the florets in the flower. Except for the single or semi-double cultivars, the florets of all florist's chrysanthemums are strap-shaped, variously rolled and twisted. Therefore, based on the floral characters, the English National Chrysanthemum Society put these cultivars into 7 main groups such as:

1. **Incurved:** Flowers perfectly globular with fully double flowerheads having firm and incurving petals closing tightly over the crown, *e.g.* var. 'Alison Kirk'. This group includes *Exhibition Incurved* (November flowering) and *Early-flowering Incurved.*
2. **Reflexed:** Blooms fully double rounded with petals falling (curving) outwards and downwards to touch the stem, often with curl or sideways twist and overlapping like feathers on a bird, *e.g.* 'Primrose West Bromwich' in case of *Fully Reflexed*, while those with partly reflexed petals and with a spiky outline known as *Reflexed, e.g.* 'Rose Yvonne Arnaud'. In this group the blooms may have straight or drooping florets, and also those blooms with outer florets reflexing and inner florets incurving. This group includes *Large Exhibition* and *Reflexed Decoratives* (both November flowering), October-flowering *Reflexed* and Early-flowering *Reflexed.*
3. **Intermediate** (formerly **Incurving** group)**:** Fully double and globular flowerheads but being halfway between the previous groups; and where petals are loosely incurving with regular shape is *Formal Intermediate, e.g.* 'Skater's Waltz', whereas those with irregularly incurving petals are *Informal Intermediate, e.g.* 'Green Satin'. This includes November flowering *Large Exhibition Incurving* and *Incurving Decoratives, vis-à-vis* October-flowering *Incurving* and Early-flowering *Incurving.*
4. **Single:** Flowerheads daisylike with prominent central discs and 5 or less radiating rows of flat-petalled ray florets (other than anemones), *e.g.* 'Roytorch'.
5. **Anemone-centred:** Similar to group 4 in singleness of flowerheads but the central discs consist of tubular, flat or occasionally spoon-shaped florets forming a raised cushion, *e.g.* 'Pennine Oriel'. These normally flower in November.
6. **Pompon:** Fully double small cushion-shaped to globular or button-shaped blooms formed of numerous tightly packed short, broad, and tubular ray florets with flat and rounded petals, *e.g.* 'Maria'.
7. **Miscellaneous:** It is general category for the several small groups that do not fit into the previous 6 groups, and these are: **Spider** with long thread-petalled, quilled or spidery blooms with long, tubular or spoon-shaped petals, *e.g.* 'White Spider'. **Spoons** are with single flowerheads having tubular florets but the petals open at the tips to form the shape of a spoon, *e.g.* 'Pennine Jewel'. This also includes **Korean** group having single or semi-double and flat florets on the flowerheads with clear open discs, *e.g.* 'Salmon Pye'. Also included are **Stellate** group where flowerheads are similar to Korean but the ray florets are twisted to give the bloom a star-like appearance and **Charm** or **Cineraria** group where the flowerheads are single or semi-double attaining the maximum size of 3 cm across. **Decorative** under small group is also similar to class 7 except that the flowerheads are small. **Anemone**, **Quill**, **Pompon**, **Button**, **Spray** and **Cascade** also bear small flowerheads produced in profusion.

There is a further division within these seven groups of chrysanthemum varieties determined by bloom-size and the normal time of flowering, *e.g.* Early-flowering, October-flowering, and Late-flowering. Early-flowering chrysanthemums are those that bloom in the normal season (before October, *i.e.* from August to late September) in the open ground without any weather-structures, Late-flowering (from October to December) chrysanthemums blooming in the greenhouse and are grown in pots for its entire life, and Late-autumn or October-flowering chrysanthemums, bridging the gap, either flowering outside or under glass, according to season and location. The plants which carry such diverse blooms are less variable, are clump-forming hardy or so, perennials, with erect and woody stems 60-150 cm tall, and the leaves variously lobed, somewhat oak-like, rough, ovate, glossy rich green or grayish. The flowerheads in shades of yellow, pink, red, bronze, purple and white are carried in branched terminal clusters as sprays.

For show purposes the Floral Committee of the National Chrysanthemum Society of America has sub-divided the aforementioned seven groups as follows:

Division A

Ray florets flattened to concave or convex. Visible portion never tubular.

Section I

Disc prominent and circular in outline, and composed of many disc florets.

- ☆ **Class 1**. Single: Ray florets in a single row while disc florets flat to slightly rounded.
- ☆ **Class 2**. Semi-Double: Disc as in Class 1 while ray florets in more than one row and may incurve downward at the tips.
- ☆ **Class 3**. Anemone: Ray florets variable or equal in length and shape, *i.e.* flat, broad or reflexing. Disc florets numerous, tube-like and elongated to form a clear flat or hemispherical shape.

Section II

Disc not apparent as disc florets may be entirely absent, and if present either in the centre or scattered over the receptacle, should be concealed.

- ☆ **Class 4**. Pompon: Blooms globular or small button (2-3 cm in diameter) type, initially may be flat. Ray florets broad, incurved, smooth and firm with a good substance.
- ☆ **Class 5**. Incurved: Flowerheads globular with equal breadth and depth. Ray florets narrow to broad and smooth to incurved (incurving regular or irregular) with no open centre, *e.g.* 'Audrey', 'Maylen', 'Mountaineer', 'Nob Hill', 'Snow Ball', etc.
- ☆ **Class 6**. Reflexing Incurved: The depth and breadth of the bloom almost equal to form either a perfect globe or slightly flattened but less compact than incurved, and florets neither completely incuving nor completely reflexed though sometimes lower florets may reflex giving a skirted effect, and the ray florets usually broad and smooth, *e.g.* 'Dream castle', 'Indianapolis', etc.
- ☆ **Class 7**. Decorative: Blooms more flattened than globular, and ray florets broad to narrow, and short to long and/or pointed, normally reflexing though upper florets may incurve, *e.g.* 'Firecracker', 'Loveliness', 'Otome Pink', 'Princess Anne', etc.
- ☆ **Class 8**. Reflexed: Length and breadth of the blooms globular or slightly flattened with full centre, and ray florets narrow to broad, reflexed and overlapping regularly or irregularly, *e.g.* 'Coronation Pink', 'Pinksmoor', 'Tracey Waller', etc.

Division B

Ray florets tubular, coiled or straight, thready to coarse, closed, spatulate, flattened, or hooked at the distal end.

- ☆ **Class 9**. Spoon: These are with single flowerheads, *i.e.* open disc, having usually straight, regular and tubular ray florets but the distal end open at the tips to form the shape of a spoon.
- ☆ **Class 10**. Quilled: Disc not visible, blooms fully double, ray florets straight and tubular and either with closed and pointed tips or open and spatulate.
- ☆ **Class 11**. Spider: Disc unclear, ray florets long, tubular, fine to coarse and spanning to all the directions, distal end closed or open but spatulate tips typically hooked or coiled. Further, on the basis of tube fineness, divided into four sub-classes, *viz.* thread, fine, medium and coarse-tubed.

Division C

Disc inconspicuous or prominent, and the ray florets flattened or tubular.

- ☆ **Class 12**. Lacinated: Flowreheads of any of the shapes as in classes 1 to 11, and the ray florets lacinated or feathered at tips, *e.g.* 'Jack Straw'.
- ☆ **Class 13**. Brush or Thistle-like: Flowerheads with almost hidden disc and the ray florets are fine tubes in the form of a brush or thistle-shape, *e.g.* 'Saga', 'White Isle', etc.

Apart from the classes described above, there is a class known as 'ball' or 'rayonante' in which the broad ray florets from centre radiate in all the directions like sun rays, *e.g.* 'Pride of Madford'.

There may be more than 30,000 chrysanthemum cultivars worldwide today, some two-third listed in Japan alone. The widely grown cultivars worldwide today are given below:

I. Early-flowering cultivars suitable for cutting and for garden display (all the spray and pompon cultivars are suitable for garden display)

- ☆ **White:** American Beauty, Apsara, Arctic, Baggi, Bonnie Jean, Birbal Sahni, Brumas, Cloudbank, Divinity, Dorothy Else, Elegance, Favourite, Fred Shoesmith, Globemaster, Himani, Horizon, Hurricane, Illini Cascade, Japanerin, Jacqueline, Jyotsna, Mary Morin, Memento, Niharika, Nimbo, Pennine Snow, Polaris, Rita, Schnesstern, Sharad Shobha, Sharad Singar, Snowdance, Snowdon, Super White, White Illini Springtime, White Loveliness, White Marble, White Sands, White Spider, White Taffeta, etc.

- ☆ **Yellow:** Ada Stevens, Agneta, Archana, Arun Singar, Basanti, Celebrate, Delightful, Freedom, Friendly Rival, George McLeod, Golden Crystal, Golden Delightful, Golden Hurricane, Golden Polaris, Golden Sands, Golden Seal, Golden Vedova, Golden Winner, Henry Shoesmith, Indira, Jane Ingamells, Jayanti, Jubilee, Kundan, Mayford Supreme, Nanako, Sharad Kranti, Sonali Tara, Souvenir, Starbrite, Sujata, Sunavon, Sunbeam, Super Yellow, Yellow Agneta, Yellow Bonnie Jean, Yellow Divinity, Yellow Fred Shoesmith, Yellow Galaxy, Yellow Horim, Yellow Illini Springtime, Yellow Marble, Yellow Nimbo, Yellow Snowstar, Yellow Spider, Yellow Tuneful, Yellow Verinete, etc.
- ☆ **Pink:** Actress, Amy Shoesmith, Belair, Bluechip, Blue Marble, Blue Winner, Brenda Talbot, Dolly, Enid Walters, Eugene, Exmouth Pink, Favourite Supreme, Florence Horwood, Illini Springtime, Gaiety, John Woolman, Lilac Loveliness, Lilac Nymph, Loveliness, May Wallace, Mason's Orange, New Princess, Nilima, Patricia Hurst, Pink Fred Shoesmith, Pink Marble, Pink Pride, Pollyanne, Princess Anne, Riviera Spider, Rose Harrison, Sarath Riley, Sheila Rose, Snapper, Sylvia Riley, Taffeta, Una, Vedova, Worthing Success, etc.
- ☆ **Red:** Appeal, Blaze, Chesswood Beauty, Cotswold Flame, Crackerjack, Dainty Maid, Escort, Flirt, Gem, Jaya, Mayford Crimson, Portrait, Rakhee, Red Belcombe Perfection, Red Flare, Red Frandango, Red Galaxy, Red Leader, Red Nero, Red Tuneful, Shirley Late Red, Starfire, Teddy Dog, Working Perfection, etc.
- ☆ **Bronze** and **Light Bronze:** Ajay, Apricot Marble, Apricot Sylvia Riley, Apricot Winner, Balcombe Perfection, Belreef, Bronze Nero, Bronze Rosado, Bronze Rose, Covent Graden, Dramatic, Finale, Flame Belair, Flaming Sunset, Galaxy, Invicta, J.R. Johnson, Mason's Bronze, Orange Aglow, Orange Morn, Topper, Tuneful, William Dodd, etc.
- ☆ **Purple** and **Mauve:** Alison, Apsara, Fandango, Fantasy, Flamenco, Gaiety, Hemant Sagar, Mohini, Monarch, Nilima, Regalia, Sharad Prabha, Silver Dollar, Suhag Singar, Wyvern, etc.
- ☆ **Salmon:** Aileen, Coralie, Coral Marble, Daydream, Peach Blossom, Percy Salter, Salmon Loveliness, Salmon May Wallace, etc.

II. Standard Cultivars for Exhibition

- ☆ **Green**: Green Goddess, Madam E. Rogers, Woolman's Century, etc.
- ☆ **White:** Ajina White, Beauty, Casa Grande, Chandrama, Dorothea, Dorridge Queen, Ermine, Evelyn Bush, Fantastic, Gazelle, General Petain, Giant Indianapolis White, Green Goddess, Improved Mefo, Improved Snowball, Innocence, Kasturba Gandhi, Maylen, May Shoesmith, North Light, Premier, Premier Snowball, Rex, Snowball, Triumphant, Valiant, White Ball, White John Woolman, White Reflex, White Wings, William Turner, etc.
- ☆ **Yellow:** Balcombe Yellow, Bill Gillibrand, Bright Golden Anne, Brighton Yellow, Bright Yellow May Shoesmith, Charles Horwood, Coronation Gold, Duke of Kent, Evening Star, Garden State, George McLeod, Golden Harvest, Golden Peter Shoesmith, Golden Rule, Gold Plate, Greenwich Pride, Harold Park, Henry Shoesmith, Imperial Yellow, J.S. Lloyd, Kikubiori, Lady Frank Clarke, Martin Riley, Melody Lane, Moonbeam, Mountaineer, Orange Fair Lady, Primrose Evelyn Bush, Primrose Kathleen Doward, Rivalry, Solorama, Sonar Bangla, Spectrum, Springtime, Sunblest, Super Giant, Swarnima, Triumphant, Yellow Delight, Yellow Fred Shoesmith, Yellow Triumph, Yellow Wings, etc.
- ☆ **Pink:** Amy Shoesmith, Anemone Pink, Audrey Shoesmith, Cassandra, Classic Beauty, Deep Champagne, Denebola, Fantastic, John Woolman, Kathleen Doward, Mermaid, Pink Casa Grande, Pink Champagne, Pink Cloud, Pink Pride, Pink Turner, Promenade, Regal Anne, Rose Harrison, Sarah Riley, Sheila Rose, Sweet Seventeen, Una, etc.
- ☆ **Red:** Alfred Wilson, Claret Glow, Cossack, Crimson Anne, Chrimson Tide, Diamond Jubilee, Distinction, Escort, Firecracker, Red Anne, Red Resilient, Sultan, Working Scarlet, etc.
- ☆ **Bronze and Light Bronze:** Achievement, After Glow, Alert, Alfred Simpson, Alfred Wilson, Appart, Autumn Blaze, Autumn Delight, Bravo, Bronze Pride, Bronze Princess Anne, Charles Shoesmith, Cheyenne, Distinction, Florence Shoesmith, Gay Anne, Harry James, Mrs. W.A. Reid, Peter Shoesmith, Pinksmoor, R.C. Pinch, Resilient, Rustic, Shirley Monarch, S.L. Andre Reffaurd, The Dragon, Tracey Walker, Westfield Bronze, etc.
- ☆ **Purple and Mauve:** Ajina Purple, Auto Me Jakura, Classic Beauty, Cover Girl, Fish Tail, Kenroku Kangiku, Loveliness, Mahatma Gandhi, Marion Stacey, Peacock, Peter May, Pink Cloud, Pink Giant, Potomac, Preference, Purple Anne, Raja, Royal Purple, Shirley Perfection, Silk Brocade, etc.

III. Suitable for Pot Culture

- ☆ **White:** Altis, Birbal Sahni, Bonnie Jean, Carol, Honeycomb, Illini Cascade, Jyotsna, Lilith, Mercury, Mountain Snow, Neptune, Perfecta, Sharad Mala, Sharad Shobha, Snow Crystal, Topaz, White Ann, White Popsie, Windsong, etc.
- ☆ **Yellow:** Archana, Aparajita, Armelle, Basanti, Exquisite, Freedom, Golden Crystal, Hosur Yellow, Kundan, Liliput, Mountain Peak, Pride, Prof. Harris, Reaper, Sharad Singar, Spic, Stargold, Topaz, Yellow Bonnie Jean, Yellow Delaware, Yellow Hector, Yellow Illini Spinwheel, Yellow Mandalay, Yellow Paragon, Yellow Popsie, Yellow Tuneful, etc.

- **Pink:** Always Pink, Charm, Dark Maritime, Deep Louise, Deep Popsie, Distinctive, Illini Trophy, Judith, Maritime, Megami, Princess Anne Superb, Proud Princess Anne, Regal Anne, Rose Hostess, Royal Trophy, Wedgewood, etc.
- **Red:** Crimson Anne, Crimson Torch, Red Torch, Rory, Rufus, Woking Scarlet, etc.
- **Bronze and Light Bronze:** Bronze Princess Anne, Bronze Popsie, Copper Hostess, Dramatic, Flirt, Garnet, Gay Anne, Gay Louise, Gem, Glowing Mandale, Jean, Mandalay, Orange Aglow, Orange Bowl, Rascal, Red Anne, Sparkling Mandale, Tuneful, Winifred, etc.
- **Purple and Mauve:** Alison, Cerise Anne, Charm, Megami, Modella, Purple Magnum, Royal purple, Singar, etc.

IV. For Loose Flowers and Garland Making

- **White:** Apsara, Baggi, Birbal Sahni, Carol, Himani, Illini Cascade, Lilith, Sharad Shobha, etc.
- **Yellow:** Basanti, Freedom, Hosur Yellow, Kundan, Meghdoot, Prof. Harris, Sonali Tara, etc.
- **Mauve:** Sharada, etc.

V. Indian Bred Varieties

- **IIHR Bangalore:** Arka Ganga, Arka Ravi, Arka Swarna, Chandrakant, Chandrika, Indira, Kirti, Nilima, Pankaj, Rakhee, Ravi Kiran, Red Gold, Usha Kiran, Yellow Gold, Yellow Star, etc.
- **PAU Ludhiana:** Baggi, Basanti, Gul-E-Sahir, Punjab Gold, Santi, etc.
- **TNAU Coimbatore:** Co. 1, Co. 2, MDU 1, etc.
- **NBRI Lucknow:** Bindiya, Birbal Sahani, Diana, Jayanti, Jubilee, Kargil 99, Kundan, Mother Teresa, Purnima, Nanako, Sadbhavana, etc.

1. **No pinch no stake type:** Appu, Arun Kumar, Arun Singar, Apurva, Guldasta, Haldi Ghati, Hemant Kumar, Sharad Singar, Suhag Singar.
2. **Off-time bloomers:**

   January-February: Maghi

   April-May and **October**: Himanshu, Usha

   May-June and **October-November**: Jwala, May Day

   June-July and **November**: Jyoti, Tushar

   July-August: Meghdoot, Phuhar

   September-October: Ajay, Sharda

   October: Sharad Mala

   October-November: Sharad Singar

   December-January: Jaya, Vasantika
3. **Pompons**: Apsara, Birbal Sahni, Jayanti, Jubilee, Kundan, Maghi

Propagation

For taking cuttings, only healthy and vigorous **stock** (mother) plants should be selected. Such plants should be tagged at flowering so that only true to type plants are selected which bear quality blooms taking into account size of the blooms as well as intensity of colour. During winter when flowering is going on or is over, the selected plants are cut down to about 20-25 cm height from the ground in the open and all the new and sappy growths, *vis-à-vis* large leaves are removed.

Cuttings in 'Early-flowering' chrysanthemums are normally taken four to six weeks after new shoots have appeared and then parent plants may be discarded. Only close jointed shoots are selected near the base of the stem with soft but firm growth, and are removed either with a sharp knife or these are snapped off and trimmed to about 4 cm cutting just under a node, bases dipped in some rooting hormone, then inserted in standard cutting compost and are then placed in a chamber at 10 °C where these root within 2-3 weeks. Once rooted, these cuttings are moved to slightly cooler conditions for a week and then are planted in batches of 6-8 in a pot filled with a loam based potting compost. After about a month these plants are transferred to a cold frame for hardening. In late-flowering cultivars, after flowering the plants are cut back leaving about 22 cm, stools may remain in their positions or lifted but the pots are kept in a well ventilated cool greenhouse at about 5 °C for a month and then to 7 °C to produce shoots for cuttings about 4-6 weeks later in late winter. Cuttings of the 'late-flowering' cultivars are also treated similar to those of early-flowering cultivars but the cuttings of the late-flowering spray cultivars do not require to be rooted until early summer, and once rooted these are treated similar to early-flowering ones. In late spring when the pots are full with roots, these are planted in their final pots of 24 cm size.

If cool greenhouses are available the **stools** of the open-grown plants should be dug up immediately after flowering and planted in the beds inside or kept in boxes inside covered with a little soil and lightly watered to avoid them from shrinking, though those flowered in pots inside do not require to be disturbed but inside environment should be well aerated and illuminated. The large fleshy basal growths appearing should be sheared off by mid-January. These stools or the rooted cuttings taken from the mother plants, if provision is for temperature control, should be rested for 3-4 weeks at 4.5 °C for successful growing in the following year, and if this temperature is around 10 °C for whole of the winter and spring, following year the growth will become messy without any main shoot. Full precautions will have to be taken that stock plants remain free from leaf miners and aphids just by spraying the plants fortnightly with nicotine insecticide. Aphids are vectors of aspermy virus and this period is crucial for the infection. Cuttings are by far the best way of getting vigorous growing and healthy plants than division of the old stools or **suckers**. Though initially rooted cuttings seem to be taking up poorer than the plants started from divisions but with the advancement of the season the plants from cuttings overtake the plants started through divisions. Cuttings may be taken just after flowering from November to January in

Large Exhibition varieties, in January and early February of the *Exhibition Incurved*, late Jnauary and February for *Decoratives, Late-flowering Singles* and *Pompons*, in mid-February and March for *Early-flowering* outdoor varieties, and in April and early May for *Decoratives* to be grown as dwarf pot plants (Thistlethwaite, 1960). Taking of cuttings in March and April from the open-grown varieties, in January and February of the large-flowered greenhouse varieties, in April and May of early-flowering greenhouse types, and in June of the late-flowering greenhouse types have been found giving excellent results (Hay and Beckett, 1971).

Before taking any cutting the stools must be started in active growth as cuttings taken from the dormant stock plants will not root well so while taking the cuttings, somehow, the temperature of the greenhouse or frame should be raised to 10 °C for about 7-10 days somewhere from mid-February to early January. When the growth has started, the preference should be given for basal cuttings over the stem cuttings, however, **terminal stem cuttings** are the best means for commercial propagation. Stock plants, for obtaining healthy stock, are commonly grown through terminal stem cuttings. One or two days prior to taking of cuttings, the stock plants should be thoroughly watered, and when taking the cuttings the ones which are hollow, pithy and limp are discarded. The shoots some 6.5 to 8.0 cm long (Thistlethwaite, 1960; Machin and Scopes, 1978) of proper thickness which have appeared 4-6 weeks before, and close-jointed are severed just below a node with a sharp knife, having lower leaves (up to 3-4 cm length of shoot) removed, treated with Seradix powder and then inserted 2.5 cm deep in 5 cm deep seed trays or in 8-10 cm pots 5 cm apart, in the trays or pots filled with compost mixture (2 parts loose bulk of medium loam, 1 part peat, 1 part sand, and to 1 litre volume of this mixture some 1 g superphosphate of lime + ½ g finely ground chalk) which requires light watering twice daily. Medium with low pH causes manganese toxicity, however, for cuttings the medium should have a pH reaction of 7 to 8. At 15.5-18.0 °C, these cuttings root well in 7-20 days (Thistlethwaite, 1960) under mist, fog or plastic laid directly over the cuttings. Sidhu and Singh (2002) tried NAA and IBA at 250, 500 and 1000 ppm as quick dip treatment of chrysanthemum cuttings before inserting into river sand for rooting and found 250 ppm in both the cases as optimum for inducing rooting. During summer the plastic should be white while during winter completely transparent. Direct sun is injurious to the cuttings, so these should be kept only at filtered position. At 1-3 °C temperatures, the rooted cuttings of some cultivars can be stored for 4-6 weeks whereas certain others only for one week (Rajapakse *et al.*, 1996) at 67 fc (10 µmol. $s^{-1}m^{-2}$) light intensity from cool-white fluorescent lamps, for improving the cutting quality (Dole and Wilkins, 1999). When young plants have filled the pots with roots after about 4-5 weeks, these are potted on into 12.5 cm pots and finally into 20-25 cm pots. Cuttings taken during April and May can be planted directly to their final beds.

Mukhopadhyay and Das (1976) stated that suckers produce taller plants, and while growing through suckers, these should be separated when have attained 5-6 leaves. However, cuttings taken from the previous plants for making stock plants is a sound practice. Such stock plants are pinched and repinched to induce more lateral shoots for taking more cuttings (Furuta, 1954; Machin and Scopes, 1978). Cathey (1955) reported that a large number of cuttings are obtained from stock plants at 21-27 °C than at 16 °C, which root optimally at 18 °C night air temperature, and Machin and Scopes (1978) stated that temperature between 18 °C and 21 °C promotes rooting while below 16 °C inhibits it. Cuttings taken from young stock plants produce better rooting. Growth retardants, especially IBA and SADH promote rooting though CCC inhibits it (Arumugam *et al.*, 2006). Pinching, feeding especially phosphorus, growth retardants and supplementary lighting increase sturdiness in cuttings and encourage rooting. Misting the cuttings and CO_2 concentration at 10,000 to 20,000 ppm for 12 h per day also encourage rooting.

Shoot apex, leaves and peduncle and other floral part explants can be used for **micropropagation**. Terminal cuttings 2-3 cm long containing at least one node, washed, and sterilized with 70 per cent ethanol and then with 4 per cent sodium hypochlorite and then placed on agar substrate containing MS medium amended with 0.5-1.0 mg/l BA and 0.1 mg/l IBA for initial proliferation and then keeping at 23.2°C for 16h/day under 4000 lux, produce about 8 mini-cuttings at every 45 days which could be proliferated many times, and these mini-cuttings are rooted in substrate lacking BA but containing IBA and/or NAA at 1 mg/l, and which after forming the roots should be potted in peat and vermiculite and kept at saturated atmosphere for 10-15 days at 20-25°C before planting out (Kher, 1989).

Certain cultivars which have conspicuous discs generally produce **seeds**. Spray chrysanthemums set copious seeds. These seeds are collected after complete drying of the capitulum and sown in January or February directly in the prepared beds or in seed pans or pots containing proper soil medium, *i.e.* peat, compost and sand where temperature is 16 °C to 21 °C (Nau, 1993). Hay and Beckett (1971) have, however, suggested 13-16 °C germination temperature. After spreading the seeds on the substrate of the containers, a fine layer of compost is used to cover these and then it is watered with fine rose can gently so that seeds may not expose. Within a few days these germinate, and when these attain 6-8 cm height are transplanted in individual pots of the size of 7.5 cm.

Cultural Practices

Light and temperature are the two important environmental factors influencing the growth and flowering in chrysanthemums, light (photoperiod and intensity both) dominating the autumn-flowering and temperature dominating the summer-flowering cultivars. Chrysanthemums flower when **daylength** becomes short that is why these are categorized as short-day plants. During their vegetative growth period, however, these require high **light intensity** otherwise the plants will become lanky with thin stems and large leaves. There is also a critical level of **temperature**, above which only floral buds develop but below that only vegetative growth occurs, and that is why most cultivars require warm nights at the time of floral bud formation. At the start of short days, these require a minimum night temperature of 15.5 °C and when the flower buds are clearly visible it requires lowering of 1 °C every two-nights up to 12.7 °C for better quality of blooms, especially with 12- to 15-week cultivars (Laurie *et al.*, 1968). High temperatures delay the development of floral buds, though this varies with the variety. Though 16 °C night temperature is reported critical (minimum) for American and British cultivars for bud initiation but for most rapid and uniform development, 10 °C to 12 °C is most desirable after the buds are visible (Kher, 1989). Salinger (1985) mentions that (a) those flowering in response to warm temperatures (15 °C and above) independent of daylength are considered **thermopositive**, and are called early-flowering or summer types; and (b) those responding to photoperiod, *i.e.* short days, *vis-à-vis* temperatures but to a lesser extent, are called 'all year round' as these can be flowered any time of the year by manipulating the daylength. These **All Year Round** chrysanthemums are subdivided further into response groups as per their time requirement (number of weeks from 8 to 14) from flower bud initiation to actual flowering. The majority of commercial cut flowers are obtained from 10-12 week cultivars where required minimum temperature for sufficient flower bud growth in 8-12 week cultivars is also 15 °C (mean of day and night temperatures with day temperature not going above 25 °C and night temperature above 10 °C), and temperatures above 27 °C slow down bud development, and so these are known as **thermozero** types, however, Cockshull and Kofranek (1992) when studied eight cultivars, they found these to be in the 6- to-8-week response groups from the start of short days, all forming flower buds even under long days; and those requiring 13-14 weeks and can tolerate cooler growing conditions, these are known as **thermonegative**, such as 'Cavell', 'Elegance' and 'Kitchener' (Salinger, 1985). Cathey (1955) categorized chrysanthemum cultivars into three response groups based on temperature requirement: (i) **thermozero** where cultivars flower at any temperature between 10 °C and 27 °C, more consistently at 16 °C night temperature, *e.g.* 'Shasta'; (ii) **thermopositive** where bud initiation is delayed or inhibited at temperatures between 10 °C and 13 °C, minimum for bud initiation is 16 °C, but rapid at 27 °C though at this temperature flowering is delayed, and all these being independent of daylength, *e.g.* 'Cameo', and (iii) **thermonegative** where bud initiation occurs at any temperature between 10 °C and 27 °C but continuous high temperature delays bud development. Arumugam *et al.* (2006) stated that chrysanthemums have two critical day lengths (photoperiods), one for floral initiation and the other for flower development, and these vary due to temperature and variety. Hicklenton (1984) reported that supplementary illumination with high pressure sodium lamps increased the number of flowering stems, stem break length and dry weight. The hypothesis that chrysanthemums are short day plants has been proved otherwise as flower buds can be initiated even in the absence of short days and through an experiment where Cockshull and Kofranek (1985) tried 38 cultivars in continuous long days and found only two remaining vegetative after 19 weeks. Only after certain growth the plants may become responsive to photoperiod though there may be cultivar variation for this period of growth (Stefanis and Langhans, 1983). A larger plant responds to short days more rapidly than a smaller plant.

Quality of light and light intensity during the day influence the response to short days, high intensity inducing early flowering under short day conditions. Post and Kamemoto (1950) reported a minimum of 9.0-9.5 hours of continuous dark period every day necessary for flower bud initiation. Post (1948) reported that flower bud initiation may take place even during long days of 14.0-14.5 hours but in most cultivars the development of initiated buds require a longer and continuous dark period, *i.e.* 10.5- 11.0 hours. For chrysanthemum growing environments, 70-90 per cent relative humidity is optimum (Kher, 1989).

'Early flowering' or 'outdoor' chrysanthemums, which are normally grown in open situations, require a fairly sunny and sheltered site either in the beds or in pots, though 'late' or 'indoor chrysanthemums' are grown in pots under net-houses or protected structures. Chrysanthemums prefer a well-drained slightly acid **soil** with a pH value of 6.5 (Brickell, 1992). Dole and Wilkins (1999) suggests a soil pH ranging from 5.7 to 6.2, and Kher (1989) from 6.2 to 6.7. Chrysanthemums have shallow and fibrous root system, so if there is no properly aerated soil, the plants may be affected with root rot and wilt as its root system is highly sensitive to waterlogging. Clay and clayey-loam soils are not suitable for its growing as these retain too much of moisture which encourages rot diseases and when dry become too compact to permit the roots to grow properly, therefore it is only sandy-loam soil which is well aerated and most suitable for its growing though it permits free leaching of most of the nutrients supplied to the plants. An ideal soil is that which is light (porous to permit free aeration), having sufficient humus and minerals, contains useful microorganisms sufficiently, and has adequate water holding capacity. An ideal soil should have nitrogen 10-50 ppm, phosphorus 5-10 ppm, potassium 30-50 ppm, calcium 100-150 ppm, manganese 3-4 ppm, zinc 6-8 ppm, boron 20 ppm and copper 5 ppm (Kher, 1989). However, even heavy soils may be improved through addition of coarse sand and organic matter. Therefore, at the time of field preparation, some 250-300 quintals of organic matter in the form of manures (cow-dung, farmyard, compost, poultry, pig or stable) should be incorporated in the soil and mixed thoroughly so that it is uniformly mixed with 15-20 cm of soil surface.

Field preparation starts in late autumn or early winter, especially the heavy ones or in early spring especially the light soils. Field preparation is required only for the early-flowering cultivars as late-flowerring cultivars are grown in pots for growing in the greenhouses. Cultivation of soil is done to a depth of more than 20 cm, and thus it requires three ploughings one by one, followed by levelling each time after taking out the rootstocks of perennial weeds, brick and stone pieces and any other material which may hinder the cultivation of chrysanthemum. Finally the soil should reach to a fine tilth. After thorough levelling, the beds are prepared of the convenient sizes keeping 60 cm clearance either way for irrigation and cultural operations.

In the beds, the early-flowering florist's chrysanthemums are **planted** in late spring shallowely just to cover the roots well extended, *i.e.* not more than 1.2 cm, however, late-flowering ones should preferably be planted in the pots in May. It should be ensured that all the cuttings which are being planted together are almost of the same size so that there may not be improportionate competition for light. For garden use, planting may be carried out until September for obtaining the blooms throughout the year. Looking into the spread of the plants, the distances per plant have been recommended as 12.5 x 12.5 cm in summer and 15.0 x 15.0 cm during winters (Machin and Scopes, 1978). Yadav (1983) for a small flowered cultivar found 30 x 20 cm spacing optimum for flower production under Nadia (W.B.) conditions as compared to 30 x 30 cm and 40 x 30 cm spacings, though South Indian conditions permit 30 x 30 cm spacing for optimum yield. In Holland, they recommend 32-40 plants per square metre. Spacing requirement varies with the variety used, the time of planting, the place and the purpose whether for spray or for standard cut flowers. Immediately after planting, the beds or pots are watered thoroughly.

It is important that early-flowering cultivars are **staked** and tied as soon as possible after planting in order to guard against wind damage. Split bamboo canes 1.2 metre high after planting are inserted 45 cm apart in the field. Those growing for exhibition often use three or four canes per plant, one to each break. First tie with stake is given when the plants are 30-45 cm tall. Second tie is given later in the season to support the lateral growths forming the first break. A single layer of wire and string, fish net or nylon net with a mesh equivalent to the plant spacing to be used for greatest length of time, will support the plants in a more effective way when grown inside the greenhouse or in the beds outside and such nets will be economical to be used year after year. In this way the nets will be lowered to the ground to act as marker for planting and raised by means of supporting rods at each end of the bed. During summer on *Dendranthema*, all the herbicides (Alachlor, Napropamide, Oxadiazon, Oxyfluorfen and chlorthal dimethyl) tested were effective against controlling >90 per cent of grasses and 67-100 per cent of broad-leaved **weeds**.

Chrysanthemum flowers are in demand throughout the year but quality blooms can not be obtained outdoors, hence, these are grown under **protected structures** where through control of temperature, light, humidity and CO_2 as per requirements the flowers can be obtained year round. Generally late-flowering cultivars are grown under protected cultivation whether planted directly in the beds, in pots or on the benches. Erection of any structure for growing chrysanthemums should be at a location where full winter sunlight is available. Aluminium, or some cheaper substitutes such as wood logs and bamboos are used for making the frames and polyethylene for covering the structure, equipped with vents on the sides of the tops and below cooling system with cladding material at one side and exhaust fans to the other. Fertigation is normally given through drip system. Vents and 'pad and fan' system work during summers when temperature rises whereas during winters, if need be, heaters are switched on. However, the low height poly-structures or tunnels may be used against winter protection but quite high structures with vents and cooling structures are more advantageous. Incandescent lamps or low pressure sodium lamps are used for cyclic lighting otherwise 60-watt lamps, fluorescent lamps, high pressure mercury fluorescent lamps and fluorescent tubes are used to provide supplementary lighting. For providing shade there is provision of inverted 'U' frames for black polythene film of 150 gauge. CO_2 at 0.1 per cent from usual 0.03 per cent can be provided at a constant rate of 56 kg/ha/hour at a temperature not less than 15.6 °C. CO_2 is given during vegetative phase. CO_2 at 800 to 1000 ppm has been found giving more dry weight and earlier flowering. Co_2 is used only during daylight with or without supplementary high density discharge (HID) lighting, however, the HID lighting during night has insufficient intensity to make supplemental CO_2 beneficial (Hughes and Cockshull, 1972; Rex *et al.*, 1983b). Rex *et al.* (1983a) reported improved vegetative growth due to CO_2 enrichment of the environment. Mortensen (1986) stated that supplementing CO_2 at 900 ppm enhanced flower number. Commercially, the greenhouses are at present using 1000 ppm CO_2 starting one hour before sunrise to one hour before darkness and are harvesting about one week earlier and large flowers being born on strong stems (Dole and Wilkins, 1999). Gaikwad and Dumbre-Patil (2001) obtained 35-40 per cent more flower yield along with two times more vase life by growing chrysanthemums under protected conditions.

For growing **pot planted** chrysanthemums and also for raising **potmums**, the media used should be loose, coarse, moisture retentive together with adequate drainage capacity, and should have proper anchorage. A variety of organic and inorganic materials are used for potting. Loam, garbage compost, peat or bark humus, sand, perlite, vermiculite, leaf mould, pine bark, peanut hulls, saw dust, polysterene peat mixtures and plastic foam are used for this purpose. Peat, vermiculite and sand mixture is most commonly used. Peanut hulls 20 per cent +40 per cent peat + 40 per cent sandy loam soil; garbage compost in combination with vermiculite, pine bark, peat and sand; peat + sand; bark humus; polysterene peat mixtures, John Innes Compost Nos. 2, 3 and 4 are the best media for chrysanthemum growing, however, these substrates should be fortified with N, P or K fertilizers in slow or quick release forms. However, the pH of the medium should be adjusted to 5.7-6.2 by using dolomitic limestone. The soil of the bench should also have plenty of organic matter in any of the substrates in coarse form as defined earalier (Kher, 1989). Dutt *et al.* (2002) used cocopeat, soilrite, compost, garden soil, cocopeat+rice husk, soilrite+rice husk, compost+rice husk, cocopeat+soilrite, soilrite+compost, cocopeat+compost, cocopeat+soilrite+rice husk, soilrite+compost+rice husk and cocopeat+compost+rice husk, all in equal volume and recorded that compost+soilrite, followed by compost+cocopeat gave highest values with regard to fresh root, *vis-à-vis* dry weight of roots. Single unrooted cuttings are used for a 8-10 cm pot, 2-3 cuttings in a 13 cm pot, and 3-5 cuttings are used in case of 15-18 cm pots and except the 8 cm pot size, the plants in other pots require pinching. Initially the pots should be of 7.5

cm size, at second potting the size should be 11.0-12.5 cm and finally 20.0 cm, 22.5 cm or 25.0 cm as per type and vigour of the variety. However, the most preferable size is 15 to 16 cm, followed by 10 cm. Pot plants are fed generally after one month of final potting when final pots are full of roots.

Blooms grown outdoors for **exhibition** require protection from adversities. Yellow and white cultivar blooms are covered with special grease-proof bags and in case there are some insects these may be sprayed with some effective insecticide before covering these. These plants should also be supported with sticks, preferably rounded ones so that winds may not break them. Other cultivars having different other colours, large reflexes, sprays and many others are grown under protected structures. Looking into the dates of the 'Flower Shows or Exhibitions', it is possible to advance the flowering by providing them short days and delay the same through additional illumination. Bud-forming, which normally requires 10 hours of continuous darkness, is advanced by excluding day light, or delayed by extending the period of day light through artificial lighting. Late-flowering spray cultivars may be covered with opaque material to provide at least 14 hours of total darkness in the night for a period of three weeks which will encourage bud initiation on the required date. For exhibition only fully open blooms are cut and the cut stems are immediately placed in salt-free water for 24 hours

Soon after planting when plants have attained some 1.5 cm new growth which normally occurs after 10 days of planting out, the plants need to be **stopped (pinched)** in which growing tips some 1.0-1.2 cm are removed leaving some 6-8 nodes on the stem-top to encourage lateral shoots which will bear the flowers. This is the only stopping required for early-flowering varieties as these produce their best flowers on first crown buds developing at the end of a lateral growth from the main stem of the plant. Too early pinching causes peduncles of the sprays becoming too long and too late pinching makes the spray clubby (Post, 1950). After appearance of the 1st crown bud, all other side growths developing in the leaf axils on the laterals are removed, and so not more than six blooms per plant are retained. Generally, when the laterals are about eight cm long, these are reduced to only from 4 to 5 each main stem in spray cultivars for general display whereas 4 to 6 may be retained for disbudded plants and those meant for exhibition. In potted late-flowering chrysanthemums, the plants are stopped in mid-summer to produce their flower buds 10-12 weeks later. Disbudding and side shoots are also removed in the same way as for ealy-flowering chrysanthemums. A month after stopping, a light dressing of balanced fertilizer some 60 g/m^2 is supplied and watered which is further repeated after about one month to encourage the strong growths. During this period no any other side shoots should be allowed to appear so that already growing laterals appear strong. Beach and Leopold (1953) found maleic hydrazide and certain alkyl esters suppressing the apical dominance and promoting branching in chrysanthemum. Cathey (1976) recommended spraying of UNIP. 293 at 0.5 per cent on 18th SD and 1.0 per cent on 21st SD quite effective as chemical pinching by stopping cell division on the top eight nodes. Chemical pinching plus one manual disbudding provides good results in many cultivars. Kofranek (1980) obtained variable results with application of methylesters of fatty acids such as Emgard 2077 and Off-Shoot-0. The removal of all lateral flower buds from each main stem of the potted chrysanthemum to divert the resources to develop one large and high quality terminal flower is, in fact, known as **disbudding**. The practice of removing undesirable immature flower buds to provide either a small number of large flowers or a large number of small flowers is called disbudding. This practice is most common in potted and cut chrysanthemum cultivars. Decorative, Incurved, Fuji, and Spider flower types are most commonly disbudded. To obtain desired size and for equal distribution of flowers, the laterals are trimmed 7 to 8 weeks after stopping. In case when only one flower per stem is required, all the side shoots from each lateral are disbudded so that the main apical bud develops to its perfection otherwise the laterals are trimmed so as to have one main apical bud surrounded by side shoots. However, to obtain the spray full of innumerable small- and even-sized flowerheads with even distribution on whole plant, only the apical buds on main laterals are pinched. Partly initiated buds can be killed through chemicals (Cathey *et al.,* 1966) when the desired ones to be retained are fully developed. Stopping (pinching) and disbudding make the plants compact and full of sprays all around, reduce the plant height encouraging the horizontal growth of the axillary branches, delay flowering and avoid russetting of flowerheads.

Weed control is major problem in outdoor culture. Nitrofen at 2-3 kg a.i./ha should be applied with sufficient water to wet the weeds. Chloroxuron 5 kg a.i./ha should be applied to weed-free soil followed by immediate washing off the plants. These will control all the annual weeds (Salinger, 1985). Tenoran (chloroxuron 50 per cent) at 7 kg/ha, applied 14 days after planting chrysanthemums on a friable, moist soil was well tolerated by the cvs 'Twinkle', 'Orchid', 'Helen', 'Amber Glory' and 'Denise', though 'Weisses Bukett' showed light chlorosis of the leaves (Anon., 1974).

Kher (1989) described that average yield of loose flowers suitable for making garland and *veni* may range from 7.5 to 15 tonnes or about 100,000 good quality sprays while the small-flowered varieties may produce 28.6-41.4 lakhs of cut flowers per hectare, though this depends much on the variety used, the plant density, growing environment, pinching and disbudding and the purpose of cultivation, *viz.* exhibition, cut flowers, loose flowers, or potmums.

The ideal potmums are those which grow only two to two-and-half times the pot size (pot length). However, the height of chrysanthemum plants is controlled to some extent under long days from the stage of pinching to beginning of the short day's treatment. B-Nine (SADH, Alar or daminozide) at 0.0625 to 0.125 per cent was being used and is still being widely used as one of the **growth retardants** in chrysanthemum. Its concentration and frequency depends upon the cultivar and time of the year. When lateral shoots from the pinch are some 3.5 to 5.0 cm long, it is applied at the concentration of 2500 to 5000 ppm, and if need be, the second or even the third application may also be repeated. On spray cultivars, it is used two weeks after the start of short days. Paclobutrazol at 50 ppm is helpful in regulating year round production of chrysanthemum cv. 'Co. 1'. Jung *et al.* (2000) when used uniconazole at 0.05, 0.10 and 0.15 mg a.i. per pot at 14 days after planting of rooted chrysanthemum cuttings coupled with SD treatment and pinching, recorded reduced plant height and reduction in number of stomata, length of vascular tissues and size of xylem vessels.

Chrysanthemum Growing Calendar Throughout the Year

A. For Exhibition Type

Suckers are taken and planted in earthen pans in January-February, again the suckers are planted in beds in March for taking cuttings in July and the already January-February planted suckers in pans are transplanted in 10 cm pots in March and maintained properly in April for checking its growth through restricted watering, in May these are repotted in 15 cm pots and the plants are protected from hot winds in May and June, in July these are repotted in 20 cm pots and the softwood cuttings from the bed-planted ones are to be taken for rooting, and the pot-planted crop is to be fed with liquid manure for encouraging vegetative growth. In August, rooted cuttings are planted in 25 cm pots but excessive growth is restricted through pinching to avoid growing taller and the plants are fed with liquid manure to encourage vegetative growth. In September, the plants are finally repotted in 25 cm pots, fed with liquid manure but lateral shoots are removed, and the stakes are inserted and tied for proper support. In October, the deshooting should continue for a balanced growth and the plants should be fed with superphosphate and top-dressed with oil cake. In November, further top-dressing with oil cake and potash and further disbudding to encourage vigorous growth of desired buds, however, the faded leaves are removed and the pots are coloured for maintaining its glow. In December, the faded flowers are removed

B. For Small-Flowered Type

In January-February the suckers are taken and planted in pans, in March these are transplanted in prepared beds, in April these plants are heavily pinched to encourage branching, in May again these are given second pinching of apical growth and the plants are protected from hot winds, and in early June these are pinched for the third time and the plants are further protected from heat. In July these plants are taken out from the beds and transplanted in 25 cm pots, and applied with liquid feeding. In August, these are pinched for the fourth time in succession and fed with liquid manure, however, faded and drying leaves are removed. In September, these are finally pinched for the fifth time, fed with liquid manure and top-dressed with oil cake for encouraging vegetative growth. In October, again the plants are top-dressed with superphosphate and oil cake, plants are staked and faded leaves are removed. In November, again the plants are top-dressed with oil cake and potash, dead leaves are removed and the pots are coloured for maintaining the glow. In December, the dead flowers and leaves are removed.

Irrigation, Manures and Fertilizers

Chrysanthemums are heavy feeder and have large requirement for nitrogen, phosphorus and potassium. Composts provide these elements to some extent and improve the texture and soil structure. The foundation of all chrysanthemum composts is fibrous loam which is prepared through decomposition of organic garbage, kitchen waste, plant debris and weeds, coconut fibres, grasses and their rootstocks, waste animal feed, sawdust, bagasse, ash, dry leaves, rice husks, peat moss, etc. Compost or manure (farmyard, dung) at 250 quintals per hectare should be mixed in the soil thoroughly at the time of preparing the field. As soon as the cuttings are rooted, **fertilization** commences and continues until the flowers are in colour, however, afterwards it is stopped for optimum postharvest life. After the vegetative phase is over, the fertilizer requirement also lessens. Nitrogen deficiency causes reduction in plant vigour, flower and leaf size, yellowing of foliage, delayed flowering and in severe cases the lower leaves become chlorotic with reddish veins and margins; phosphorus deficiency causes deep purple colouration on lower portion of stems, reduced new leaf's size and lower leaves turning reddish, then yellow and finally brown starting from the leaf apex; potassium deficiency causes lack of plant vigour, weak stems and small leaves and in severe cases leaves develop interveinal and marginal chlorosis; calcium deficiency causes small, curled and thickened leaves around growing points, breaking of peduncles at the time of colour breaking stage, poor shelf or vase life, and stubby and brown roots, and in severe cases leaves rosette and growing points die; magnesium deficiency causes interveinal chlorosis but with green veins and downward curling of older leaves, while in severe cases reddish spots appearing along leaf margins interveinally, gradually moving to upper leaves; iron deficiency causes interveinal chlorosis of young leaves; manganese deficiency causes pale-greening of plants with mild interveinal chlorosis of young leaves, and in severe cases small necrostic spots on middle leaves while interveinal areas are first white or grey and finally tan; zinc deficiency causes though very rare but small chlorotic spots anywhere on middle or upper leaves which afterwards may turn necrotic in the centre; copper deficiency causes dull green leaves, veinal chlorosis but with green margins which turn upward and such leaves wilt during day time, and the flowers become small, reflex and soft; and boron deficiency causes red pigmentation on veins with interveinal chlorosis, veins corky, petioles with brittle sides, leaves curving downward, larger flowers open partially and become incuved, terminal bud may die or secondary buds fail to develop, and roots brown and stubby (Waters, 1964; Raulston *et al.*, 1972; Eysinga and Smilde, 1980; Harshey and Paul, 1981; Woodson and Boodley, 1983; Hammer, 1989; Huang *et al.*, 1997; Dole and Wilkins, 1999).

Requirement of nitrogen is at early stage of growth, especially up to two months of growth (Salinger, 1985), and the deficiency during this period can not be compensated afterwards though to some extent its deficiency is corrected through application of ammonium sulphate at 1.5 g/l, but when buds start appearing the great emphasis should be given on potassium. Nitrogen at 200 kg, and potassium and phosphorus each at 100 kg/ha is adequate for good chrysanthemum growth. Belgaonkar *et al.* (1997) stated that 100 kg N and 200 kg P_2O_5 increased plant height, primary and secondary branchings, yield and number of flowers per plant and improved the vase life in chrysanthemum. Dole and Wilkins (1999) mentions that frequent use of 200 ppm constant liquid fertilizer to field crops and from 250 to 400 ppm to the plants grown in pots in soilless medium with almost each watering produce quality blooms. Potassium deficiency can be corrected by applying K_2SO_4 at 1.25 g/l five times. Excess of potassium results in Mg or Ca deficiency. Phosphorus is applied as basal dressing because it is available to the plants slowly and its requirement to the plants is throughout. Its deficiency can be corrected through application of mono-ammonium phosphate at 0.3 to 0.4 g/l of water five times or 30 to 60 g superphosphate per square metre. Excess of P results into deficiency of iron, manganese, copper and zinc. Chrysanthemum have large requirement of calcium, magnesium and sulphur. Calcium deficiency can be corrected by spraying calcium nitrate at 3 g/l of water. Its excess may cause boron, iron, manganese and copper deficiency and reduces availability of phosphorus to the plants. Magnesium deficiency is corrected through foliar spray of epsum salt at 10 g/l fortnightly. Its excess induces potassium and Ca deficiencies. However, the requirement of trace elements such as iron, manganese, boron, copper and zinc is very little. Iron deficient plants are more vulnerable to fungal rot. Its deficiency is corrected through application of iron sequestrene every 7-10 days. Deficiency of manganese is corrected by applying manganous sulphate at 0.6 g/l. Boron deficiency can be corrected by applying borax at 0.02 g/l for 3-4 weeks alternate with irrigation. Copper deficiency is corrected by spraying copper sulphate at 0.35 g/l. A month after stopping (pinching), a light dressing of balanced fertilizer some 60 g/m^2 is supplied and watered which is further repeated after about one month to encourage the strong growths. The beds and pots are watered immediately after planting.

Overhead **irrigation** is preferable in case of chrysanthemum but when flowers start opening frequency of watering is substantially reduced. Existing climatic conditions and the stage of growth are the major deciding factors for frequency of watering. During active growth when new leaves are being formed, these require ample watering but after the formation of flower buds when formation of the new leaves stops, these require less watering. Post (1949) stated that those plants which were grown constantly in moist soil from planting to their establishment, the plants grew best. Soil moisture content also influences the shelf or vase life of flowers. In pots the media should remain always moist but not wet. However, the roots being highly sensitive to waterlogging, this condition should never be allowed to occur otherwise plants may collapse. Both rotary head and perforated pipe irrigation are suitable, the latter being placed on posts above the plants at a low level height. Overhead mist spray lines and other types of sprinkler and self-travelling sprayers are quite good but these wet the flowers also with every application so trickle irrigations may be more preferable as this saves up to 90 per cent of water (Kher, 1989). Small growers can water the individual pots manually. Many growers use capillary mat and sub-irrigation systems.

Control of Growth and Flowering

Chrysanthemums can be scheduled to flower for any day through lighting to keep them vegetative and through covering them with black cloth to induce flowering. From late rainy season to autumn when natural shortening of daylength occurs, it becomes conducive for initiation of flowering from early autumn to winter (Post, 1949). In the greenhouse this type of environment can be created for getting blooms all the year round. Long days (14 hours or more) encourage vegetative growth and short days (12 hours or less) encourage flowering (Cockshull, 1985). Under low irradiance levels, both the vegetative growth and floral initiation *cum* development remain slower in SD though under high light conditions these processes are accelerated (Mastalerz, 1968; Cockshull, 1972; Karlssen *et al.*, 1989a; Schoellhorn *et al.*, 1996). Flower induction may occur at 14 hours or below but there is no usual development of the flowers or under LD flowers may occur but development is inhibited. Such cultivars are in longer response groups and include late-flowering potted ones and cut flower cultivars. In late flowering cultivars, in fact, initiation and development are obligate SD responses where flower initiation may occur under LD but flowers may never develop. Cuttings taken from physiologically matured plants (aged meristem) flower earlier than those taken from younger stocks which shows that juvenility exists in chrysanthemum. Stock plants under long days form terminal crown bud, especially the old shoots with a large number of leaves, which rarely reaches anthesis, it is therefore, advised that stock plants should be renewed up to four times each year though it is to some extent dependent upon the cultivar (Sealey and Weise, 1965; Cathey, 1969; Cathey and Borthwick, 1971; Cockshull and Kofranek, 1985; Dole and Wilkins, 1999).

With 'all the year round' cultivars, the critical day length for flower bud initiation is 14.5 hours, so when the days are longer than this the plants remain vegetative and when shorter than this the plants go to reproductive phase (Salinger, 1985). To obtain **long days** from December in a country like India, the lights of 200 lux intensity above the plants should be switched on during nights in a manner that no night period exceeds six hours. Dole and Wilkins (1999) state that night break in the winter varies with the latitude, a 5 h for locations at 40 to 50° latitude, 4 h at 24 to 40° latitude and 3 h at 25° latitude to the equator from August 1 to May 30. For vegetative growth the lighting is necessary year round from 15° latitude south or north to the equator (Dole and Wilkins, 1999). In northern latitudes, high intensity discharge (HID) lighting at 175 to 300 usually supplements natural winter days during the LD vegetative period (Dole and Wilkins, 1999) as this is required for producing high quality plants there. During summer, shading of plants is done to reduce only greenhouse temperatures. Carpenter (1976) states that though the time span from planting the unrooted cuttings to shifting of plants into SD is only 4-6 weeks but HID lighting greatly increases quality, however, it is rarely used during SD as being of little importance (Hicklenton, 1984, 1985), and also as it is generally economical only to potted flowering plants as these are kept pot-to-pot (very close) before being shifted to SD where these are spaced and pinched (Hughes and Cockshull, 1972) and not even to cut flowers. Normally for the 4 h night break, the cyclic lighting is provided with 6 minutes lights on after every 24 minutes off (Dole and Wilkins, 1999). Generally long days are created through night break by using incandescent lamps commercially (cool-white fluorescent lamps are more effective for inhibiting reproductive growth, *vis-à-vis* less electric consumption; Cathey and Borthwick, 1961; Accati-Garibaldi *et al.*, 1977) and with increase of the light duration or the intensity there is increase of the reproductive growth inhibition. Mastalerz (1977) states that light intensity for cyclic night-break should be at least 10 fc (2.2 $\mu mol \cdot s^{-1} m^{-2}$), though for continuous lighting 7 to 10 fc is sufficient (Sachs and Kofranek, 1979), at the shoot apex. Stefanis and Langhans (1983) state that HID lighting can be used for maintaining vegetative growth under long days. The period from beginning of SD to flower initiation is normally uniform for all the cultivars but it is the time from initiation to development of the inflorescence which varies among the cultivars. Those flowering within 6-8 weeks after start of SD are generally garden types, cultivars flowering from 8-11 weeks after SD are normally flowering potted plants, and those requiring 8-15 weeks after SD are cut flower chrysanthemums (Dole and Wilkins, 1999).

Irrespective of the response groups of the cultivars, under absolute SD the floral initiation and development to anthesis are more rapid, therefore, when the plants have made sufficient growth to colour showing stage, total **blackout** from mid-March (or even sooner) is provided until colour showing stage, *i.e.* up to mid-September from 5 P.M. to 7 A.M. (14 hour), especially in greenhouse-grown those cultivars which are light-responsive, grown in summer and are meant for 'all year round' flower production (Salinger, 1985; Dole and Wilkins, 1999). Even a slightest leakage of light may jeopardize the cause. Under the blackout duration, the environment should be kept cool and properly ventilated.

Incongenial **temperatures** provided during SD delay floral initiation and its further development as these are temperature dependent, and its influence is great during the first week of SD though before SD treatments its (5 to 27 °C) influence is negligible because there is almost similar floral initiation and development when before SD treatments the plants are subjected either to 21/5 °C or 27/13 °C or 21/18 °C (day/night) temperatures, and when Wilkins *et al.* (1990) grew the plants at 17/13 °C or 20/13 °C (day/night) temperatures at the visible bud stage they found no difference in the date of harvest thouigh the plants were grown at 12/12 h (day/night) photoperiods for vegetative growth and 8/16 h day/night photoperiods for flowering. Cathey (1954a, b) and Vince (1960) both recommended 16 °C night temperature from floral initiation to visible bud stage and afterwards, Vince and Mason (1959) recommended 10-13 °C until harvest. Karlsson *et al.* (1989a, b) reported that leaf unfolding increases linearly from 0.2 to 0.5 per day with the increase of temperature from 10 to 30 °C, most rapid being at 20/16 °C (day/night) or to a constant 20 °C from planting to harvest. He advocated optimum temperatures from start of the SD to visible bud stage as 21.3 °C, from visible bud to disbud stage as 20.3 °C, from disbud to flower colour stage as 23.1 °C and from flower colour to harvesting stage as 19.1 °C. Whealey *et al.* (1987) stated that when during early period of SD the temperature is above 29 °C, the delay of floral induction and early development occur and this problem is known as **heat delay** which is corrected by covering the plants with black cloth late in the evening when temperatures drop and removed at mid-night to release the heat generated inside the cover or after more than 12 hours in the morning when mornings are very cold. Will *et al.* (1997) while working with 15 garden cultivars reported that at 18 °C it required 8-12 hours day length producing good flowering but when the temperature was increased from 18 °C to 24 °C, it required only 8-10 hours day length.

The **inflorescence forms** can be influenced through adjustment of pinching date relative to SD, disbudding, and by interrupting the SD flower induction period by LD, the last one being used for cut flower production in the spray cultivars (Dole and Wilkins, 1999). Most garden chrysanthemums are in the 6-7 week response group, most potted flowering chrysanthemums in 8 week response group and most cut flower cultivars in 11 week group, and irrespective of the response groups most of the pot mum cultivars are given two weeks of LD whereas all the cut flower cultivars are subjected to 3-4 weeks of LD so that floral stem may attain sufficient length. Some globe shaped standard or disbudded cultivars during winters develop small and flat inflorescences which can be prevented by subjecting the plants to SD for 35 to 42 days then coming back to LD when buds have started showing colour. Formation of clubby inflorescences or short peduncles can be prevented by providing 12SD-10LD-SD sequence until flowering (Dole and Wilkins, 1999). It has also been reported that ray florets develop under long days and disc florets under short days, and short days followed by 8-15 long days and again short days until maturity produces more ray florets than short days alone (Misra *et al.*, 2002).

Many **growth regulators** have given good response with chrysanthemums such as triacontanol at 100 µg/l, β-sitosterol and lanosterol each at 100 ppm, glycocholic acid at 500 ppm, ethephon, morphactin at 20-100 ppm, BA at 40 ppm, PBA at 200 ppm, CEPA at 100 or 1000 ppm, phthalimides at 250-1000 mg/l and GA are reported affecting various growth parameters in chrysanthemum (Kher, 1989). Alar at 500, 1000 and 1500 ppm induced earliness in flowering whereas ethrel at 500, 750 and 1000 ppm delayed it though both have been found increasing number of flowers per plant, in 'Punjab Gold' the flower numbers being maximum while 'Ratlam Selection' exerted maximum duration of flowering, and var. 'Baggi' exhibited maximum plant height (Namika *et al.*, 2002). GA_3 treatment during third week of SD causes stem elongation, application in the fourth week of SD causes pedicel lengthening and application in the seventh week of SD treatment causes floral development (Kher, 1989). It is also applied to supplement low temperature requirement. Dahiya and Rana (2001) applied 100 ppm GA_3 at 45 days of planting under 50 per cent shade and found increased duration of flowering, higher flower yield and increased stem length in the cv. 'Vasantika'.

Postharvest

Chrysanthemum flowers are harvested when outer petals have fully expanded though inner ones are still enlarging. Such cases when appear in 3-4 upper flowerheads and when pollen shedding has not started, the sprays are harvested from the base of the stem and kept in bucket having water and when the buckets are full these are carried to the packing site where the cut ends are immersed in boiling water for 30 seconds and then shifted in buckets containing more deep water. In standards the proper harvesting stage is when outer petals are fully elongated. These both the types can be held in holding solutions for opening. Dole and Wilkins (1999) state that longevity is prolonged if there is high light (5000 to 6000 fc, 1000 to 1200 $\mu mol s^{-1} m^{-2}$) during the last 3-4 weeks of production and lowering of temperature at 13-16 °C during the last 2-4 weeks will enhance the flower colours of many cultivars, though white cultivars may develop a pink cast, however, life of the leaves is enhanced by keeping the flowers under 100 fc (20 $\mu mol s^{-1} m^{-2}$) light or the senescence of the leaves can be reduced by using 10-100 ppm BA (Dole and Wilkins, 1999). Roude (1991a, b) reported that nutrition regimes with 6-70 per cent of the nitrogen should be in the nitrate form as ammonium level above this level reduces floral longevity. Nell *et al.* (1989) and Nell (1993) reported enhanced longevity without reduction in the quality when no fertilization was given at disbud stage 3-4 weeks prior to marketing.

Cut stems should be pulsed with 2-3 per cent sugar and 150-200 ppm 8-HQC at 21°C temperature and day-and-night lit area with fluorescent lights (Salinger, 1985; Dole and Wilkins, 1999). Sacalis (1993) suggested cutting of the stems every time and placing them in 38 °C hot water when these are taken out of the bucket during marketing process. Pulsing solution should be adjusted at a pH of 3.5 through the use of citric acid and this will check the growth of microorganisms, citic acid in the form of 8-HQC at 200 ppm + sucrose less than 3 per cent (preferably 1.5 per cent) being most appropriate. Excess sugar causes the leaves to become hard and brittle. Alternatively, 25 ppm silver nitrate + 75 ppm citric acid is also equally effective. Cut flowers of chrysanthemum cvs 'Co. 1' and 'Co. 2' at three-fourth developed stage when treated in Coimbatore in STS at 4 per cent + sucrose at 2 per cent + aluminium sulphate at 0.1 per cent, recorded highest percentage of fresh weight (Arumugam *et al.*, 2006). The vase life in *Dendranthema grandiflora* cv. 'Albatron' was found increased even when ClO_2 was added in water already contaminated with 1011 CFUL-1 aerobic bacterium (Macnish *et al.*, 2008).

The hydrated cut stems can be cool-stored at 0-4 °C (Dole and Wilkins, 1999), or at 2-3 °C (Salinger, 1985) for up to two or at the most three weeks. Standard large-flowered (13 cm diameter) cultivars can be dry stored in wax paper lined boxes for up to three weeks while smaller ones (7-<13 cm) for only two weeks. After taking out from the cold storage, these are hydrated the same way as elaborated for pulsing. For packing, the cut flowers are graded depending on stem length and strength, colour and diameter of flowers. Pompons are graded into 250-340 g bunches having several stems while standards of equal size are generally graded in groups of 10 or 12. In USA, the grading is done according to the Society of American Florists (SAF) while in Europe as per EEC. SAF (as quoted by Laurie *et al.*, 1968) for fully open standards has suggested following grades (grade 'Yellow' is being omitted as it is almost the grade 'Green' with no specification for floral diameter):

Grade Name	Fancy	Standard	Short
Label colour	Blue	Red	Green
Minimum diameter	14 cm	12 cm	10 cm
Minimum length (flower + stem)	76 cm	76 cm	61 cm

Machin and Scopes (1978) have quoted a metric grade specification for sprays for bulk packing which is given below (stem lengths should not be less than 66 cm for the British markets and those less than 51 cm should be marked short):

Grade	Stems/Sleeve	Specification
Gold	(10)	6 flowers or more out and some to come
Silver	(15)	4 or more flowers out and some to come
Bronze	(20)	3 flowers out and some to come
Make-up	(- -)	All stems not covered above, filling sleeves to same extent as other grades

Standard chrysanthemums are placed in sleeves and packed as per their grades in cardboard boxes measuring 91 x 43 x 15 cm or sprays in 80 x 50 x 23 cm boxes. For large and loose blooms, below the neck cotton or paper cuttings are dressed for cushioning them so that during transit these are not damaged. For bulk packing of sprays 10, 15 or 20 stems are placed in sleeves according to the grades, putting six sleeves, three at each end in a box. Potmums are ready for business when at least one-third of the buds have opened and have developed the pigments characteristic to the cultivar. Such plants can be transported or stored at 2 to 4 °C. Such plants are kept at a minimum of 50 fc lighting. These are packed by placing in paper or in polythene sleeves and then in cardboard boxes in groups of six.

Insect-Pests and Diseases

A number of insect-pests are found attacking chrysanthemums. Aphids, thrips, leaf miners, whiteflies, gall midge, caterpillars, hairy caterpillars, root cutting grub, mites and nematodes are the most common pests of chrysanthemum as reported by Waters and Conover (1969), Hay and Beckett (1971), Scopes (1979), Salinger (1985), Lindquist (1989) and Daughtrey and Chase (1992). Heinz and Thompson (1997) also reported that cultivars vary in their resistance to aphids, leaf miners and thrips. **Aphids** (*Myzus persicae, Aphis gossypii, A. fabae, Aulocorthum circumflexum, Brachicaudus helichrysi, Macrosiphoniella sanborni*) suck the sap from growing points and from the under surface of leaves and also act as vectors of certain viruses. *Aphidius matricariae* wasps predate on the aphids. Smoking with 98 per cent deltamethrine at 1 g/m^3 for one hour and phosphine fumigation for another five hours have been found controlling aphids completely. Apart from these, Malathion, Metasystox and Lindane sprayings are also effective. Introduction of the fungus *Verticillium lecani* in the aphid infested polyhouses will also control this pest. **Thrips** [*Microphalothrips abdominalis* (Ranjith, 2013), *Frankliniella tritici, Hercinothrips* sp., *Thrips tabaci, T. nigrophilosus*] also feed on growing points causing mottling and distortion of leaves. Leaves due to their attack become silvery. These also attack the flowers. Machin and Scopes (1978) recommended 200 g of BHC dissolved in 100 litres of water or 40 g per 100 litres of water diazinon drenching as the insect pupates in the soil. **Leaf miners** [*Phytomyza syngenesiae; Liriomyza trifoli; Chromatomyia horticola* (Ranjith, 2013)] tunnel into the leaves, lay eggs, after hatching the larvae develop and feed inside and pupate. These can be controlled through use of pyrethroids and organophosphate insecticides. Cypermethrin and Permethrin are quite effective in controlling the leaf miners. The best way is to trap the adults on yellow sticky boards and through using the fungicide pyrazophos at 0.5 ml/l. **Whiteflies** though have become quite resistant to many of the insecticides but pyrethroids have proved useful. Spreading of bright yellow greasy boards in the beds also attract the whiteflies, where these stick and die. **Gall midge** (*Diarthronomyia chrysanthemi*) infests on the stems and causes raised areas while distorts the floral buds. Diazinon is quite effective in controlling this pest. **Caterpillars** (*Helicoverpa armigera*) feed on the flower buds voraciously at bud formation and petal expanding stages which can be controlled by applying bacterial spray of *Bacillus thuringiensis* at flower bud stage. **Hairy caterpillars** [*Diacrisia* (*Spilosoma*) *obliqua*] which have hairs on their whole body attack the chrysanthemum plants during rainy season and continue feeding up to winter. These eat up the leaves leaving only papery skeleton which dries up later. Brown headed green larvae of the **soybean leaf roller** (*Hydylepta indicata*) eats, the buds and flowers by folding the leaves, and the green matter of the leaves becomes completely skeletonized, whereas **semilooper** (*Autographa nigrisegna*) feeds on flowers (Ranjith, 2013). These can be controlled through trapping at earlier stage of their infestation and through 0.2 per cent spraying with Metasystox. **Root cutting white grubs** (*Holotrichia* sp.) are not easy to control, however, at the time of field preparation, soil application of BHC, Thimet 10-G or Solvirex granules will control this pest effectively. **Flower** or **blister beetles** (*Mylabris macilemta, M. phalerata, M. pustulata*) are voracious feeders of petals making the plants completely denuded of flowers, **leaf weevil** (*Alcidodes fabricii*) adults feed on the foliage, whereas adults of the **flea beetle** though also feed on the leaves making small shot holes (Ranjith, 2013). Since these are nocturnal in habit so its control is a bit difficult, however, trappers in the night may control these pests. The green nymphs and adults of **grasshopper** (*Attractomorpha crenulata*) feed on the edges of the leaves and on petals (Ranjith, 2013). These are killed through use of methyl parathion. **Termites** (*Microtermes obesi*) are a serious problem during dry seasons. These feed on the roots causing plant wilting. Chlorpyriphos drenching will control this pest but it is not easy to eradicate its colonies which are many feet below the ground. **Capsid bugs** and **frog-hoppers** also infest chrysanthemums by puncturing the tender stalk below the flower bud as well as the buds and suck the sap, both giving similar symptoms of their attack. BHC or DDT application will control these pests. **Earwigs, woodlice, wireworms** and **leather jackets** also attack this crop but are of little significance. **Red spider mites** (*Tetranychus urticae* and *Bryobis* sp.) are red dot-like bodies on the under-surface of leaves, occurring during summer and damage the leaves and buds which become pale afterwards due to its attack. Metasystox, Kelthane, Dicofol, Tetradifon, Aldicarb, demeton S-methyl (22 g/100 l), diazinon (16 g/100 l), quinomethionate (125 g/100 l), Cyhexatin (25 g/100 l), etc. control this pest. In polyhouses, introduction of the predator *Phytoseiulus persimilis* has been found very successful in controlling red spider mite.

Foliar (*Aphelenchoides ritzemabosi, Belonolaimus longicaudatus* and *Pratylenchus penetrans*) **nematodes** cause serious damage during the wet season. Field-grown early-flowering plants are normally more seriously affected than the pot-grown greenhouse plants. Foliar nematodes cause triangular browning and black patches in the interveinal regions of the leaves, starting from the lower leaves, which later on turn black and die. For its dispersal, it depends solely on water and enter the leaves through the breathing pores on the underside. **Root-knot** (*Meloidogyne incognita*) **nematodes** cause root galling, cupping of the leaves and poor growth. The infestation by **lesion nematode** (*Pratylenchus coffeae*) causes heavy root damage and root lesions which ultimately results into plant stunting, premature yellowing, drying of leaves and reduced flower size (Ranjith, 2013). Aldicarb application before planting in the soil has been quite successful venture. Parathion and Mocap (O-ethyl S, S-dipropyl phosphorodithioate) are also reported as quite effective against the nematodes.

Slugs and **snails** feed on the young shoots on chrysanthemum stools. Before planting of the stools, these should be washed thoroughly to free them of their eggs. A proprietary slug bait containing metaldehyde should be placed around the stools in small heaps which will control the slugs within three nights.

Many **fungal diseases** attack chrysanthemums. ***Phytophthora*** spp. and many species of *Pythium* cause stem rot of cuttings in rooting bed and root rot, followed by sudden wilting (a typical symptom of *Pythium* infection) in chrysanthemum under warm moist conditions. The fungus may also enter the plants through the wounds caused through pinching. Quick removal of infected cuttings, good drainage and sterilization of medium are some of the measures adopted to prevent this malady. Drenching of soil with Thiram or Captan at 2.5 g/m^2 is quite useful in preventing its infection. *Phoma* root rot is symptomized by chlorotic spots first appearing on lower leaves and then spreading to upper leaves together with limping, and is favoured by inadequate supply of nitrogen and phosphorus. Nabam drenching at 140 g/100 litres of water two weeks before planting has been found quite effective. From Italy, Pasini *et al.* (1996) isolated *Rhizoctonia solani* from chrysanthemum causing collar or foot rot though from certain other countries it was reported even earlier.

Rhizoctonia solani is serious during warm moist conditions and causes mushy brown rot of stem and leaves. A saturated soil condition reduces its infection. Thiram soil drenching at 150 g/100 litres of water or dusting with Quintozene at 13 g/m^2 or spraying with benomyl at 240 g/100 litres of water control this malady. *Sclerotinia sclerotiorum* stem rot or wilt may affect the plants in warm humid conditions. In its infection, brown lesions on stems are produced over which a fluffy fungal growth develops. It is controlled through spraying benzamidazole, thiram and benomyl. *Ascochyta chrysanthemi* (*Mycosphaerella ligulicola, Didymella ligulicola*) infection causes blackish lesions on stems and lower leaves and browning of petals on one side. It also affects the cuttings. Proper sanitation, rogueing, dipping of the cuttings in benomyl at 50 g/100 litres of water and spraying with mancozeb at 80 g/100 litres of water can keep this disease to the minimum. *Verticillium dahliae* and *V. albo-atrum* infection to chrysanthemum plants is symptomized as immediate wilting of the plants after appearance of the floral buds. In its infection, the vascular tissues are stained brown. Removal of infected plants and drenching with carbendazim at 25.3 g/100 litres of water controls the disease successfully. *Fusarium oxysporum* infection causes dark brown colouration on the stem at soil level and ultimate drying. Lower leaves turn yellow and plants show wilting during day time and recover by night for a few days, and then finally collapse. Symtoms are apparently seen after formation of the bud or at colour splitting stage. Thiride or Captan application in the soil checks the infection of this pathogen. Thiram soil drench at 150 g/100 litres of water or spraying with benomyl at 240 g/100 litres of water has been reported controlling this disease when plants are young. *Botrytis cinerea* is a very cumbersome pathogen which attacks the crop at all the stages, starting from rooting bed to flowers in transit. It causes grey mould through rotting of flowers, leaves, buds and stems, characterized by a grey fluffy fungal growth on the affected tissues. Moist and cool atmosphere favours the spread of this disease. Good aeration even through density adjustments, avoiding moist conditions and spraying with Thiram at 0.3 per cent or Dithane M-45 at 0.3 per cent during inclement weathers controls this malady. Petal blight or flower scorch is caused by *Itersonilia perplexans,* first appearing on the outer parts of the petals as brown spots, afterwards enlarging to oval in shape and complete rotting of the flowers. This disease is serious in damp weather in late summer. Chlorothalonil application checks further spread of this disease. *Erysiphe cichoraceum* causes white to grey powdery growth on the leaves. Benzomidazole and dinocap are potent fungicides to control this pathogen. *Oidium chrysanthemi* causes powdery mildew in which white mealy covering on upper side of the leaves are formed. High temperatures favour its infection. Dry environment and use of triforine at 30 g/100 litres of water or sulphur formulations will control this disease. *Stemphylium floridanum* and *Alternaria chrysanthemi* cause dusky specks on ray florets starting from the centre of the flowerhead. A black discoloration extends to the base of the petals. It also infects the cuttings killing the growing point. Overhead watering at flowering stage should be avoided. Spraying the crops with maneb or zineb will check its further spread. *Septoria chrysanthemella* and *S. obesa* are the serious pathogens of chrysanthemums causing round, brown spots on both sides of leaves which afterwards enlarge and coalesce, leaves yellow and die. Leaf wetting facilitates its infection. Dry air environment, destroying of infected leaves and spraying the infected plants with 0.1 per cent Bavistin, benomyl, mancozeb, chlorothalonil or triforine, etc. will keep this disease under check. *Puccinia horiana* causes white rust, and *P. tanaceti* brown rust by forming whitish in former case and brownish pustules in latter case on underside of leaves. *P. chrysanthemi* produces orange and rust-coloured round sporing bodies on the underside of the leaves late in the season. Disinfection of soil and pots, keeping the humidity below 75 per cent and raising the temperature to 35 °C for 20 hours will control this malady. Mancozeb or triforine application will also control the diseases successfully.

Bacterial pathogens such as *Erwinia chrysanthemi* causes bacterial blight and the symptoms start with wilting of one or more branches during a sunny day but recovering at night, afterwards stem tips turning brown and collapsing. Stems become hollow and streaked brown up to the base. Disease is favoured by high temperature and humidity. *Pseudomonas cichorii* causes leaf spot and floral blight. Circular to elliptical leaf spots appear which enlarge and coalesce to make large chlorotic areas. Disease goes upward infecting even the flower buds which become dark and die prematurely. Bazzi *et al.* (1987) reported bacteria *Agrobacterium tumefaciens* and *Erwinia carotovora* infecting chrysanthemum in Italy. Soil sterilization, use of disease-free cuttings, avoiding contamination during pinching and spraying the plants with tribasic copper sulphate at 1.8 kg/100 gallons of water at weekly intervals during rainy season will control these diseases. Streptomycin is also very effective when used in propagation benches. Leafy galls are also formed of numerous short, thickened or distorted buds at the base of each stem or dense clusters of twisted and malformed shoots on the stool near the base of the old stem due to a bacterial infection. Since there is no cure for this disease so the affected plants or stools should be burnt.

Certain **viruses** are of high significance in chrysanthemum growing. Machin and Scopes (1978) describe some 20 viruses infecting chrysanthemums, out of which six are of more significance. 'Chrysanthemum stunt virus' is transmitted through pinching. It is characterized by overall reduction of plant size, paling of foliage with no enlargement of leaf margins, and red and bronze flowers often bleached and open prematurely. To get rid of this problem, the cuttings should be taken from virus-free stocks and the hands or tools should be washed with spirit after pinching of one plant. 'Chrysanthemum flower distortion' is transmitted through grafting. It is characterized through plant dwarfing and flower distortion, mostly with unopened buds and the ray florets become narrow, short and incurved. To prevent this virus the cuttings should be taken from virus-free stocks. 'Chrysanthemum mosaic' is transmitted through aphids. In its attack the plants are weak and small and flowers show breaking of colours though leaf symptoms inconspicuous. Control of aphid will control this virus. 'Chrysanthemum rosette' is transmitted through grafting. It is characterized through yellow mottling, sometimes with mosaic patterning, veins enlarged, wrinkled and rosette formation on terminal part. Cuttings should be taken from virus-free plants to prevent the appearance of this virus. 'Aspermy (flower distortion) virus' is transmitted through aphids, especially at rooted cutting stage. It causes severe twisting of the ray florets and ragged appearance coupled with paling of ray floret in some varieties. To eliminate this virus, affected plants should be subjected to 36.1 °C for 3-4 weeks (Thistlethwaite, 1960). From Markazi (Iran), Farzadfar *et al.* (2005), for the first time in 2003, recorded TuMV causing mottling on *Chrysanthemum* sp. 'Tomato spotted wilt virus' is transmitted through thrips and many weeds are its host plants. Infected plants show reduced growth with one-sided development, leaves may show ring and line pattern and necrotic streaks on stems and leaves. Thrips and weeds should be controlled regularly. English stunt, American stunt and vein mottle viruses are also of some significance. For getting rid of viruses the infected plants should be destroyed and insects will have to be controlled regularly.

References

Accati-Garibaldi, E., A.M. Kofranek and R.M Sachs, 1977. Relative efficiency of fluorescent and incandescent lamps in inhibiting flower induction in *Chrysanthemum morifoilum* 'Albatross'. *Acta Hort.*, No. 68, pp. 51-58.

Anderson, N.O. 1987. Reclassification of the genus *Chrysanthemum* L. *HortSci.*, **22**: 313.

Anon., 1974. *Annual report of the German Plant Protection Service* (German), 21 Jg, p. 266.

Arumugam, T., M. Jawaharlal and M. Vijayakumar, 2006. Chrysanthemum. In: *Advances in Ornamental Horticulture*. Vol. 2. *Herbaceous Perennials and Shade Loving Foliage Plants* (ed. Bhattacharjee, S.K.), pp. 81-108.

Bailey. L.H. 1942. *The Standard Cyclopedia of Horticulture* (vol. I), pp. 753 766. The Macmillan Co., New York.

Bazzi, C., P. Minardi and U. Mazzucchi, 1987. Bacterial diseases of flower and ornamental plants in Italy (Italian). *Informatore Fitopatologico*, **37**(6): 15-24.

Beach, R.G. and A.C. Leopold, 1953. The use of maleic hydrazide to break apical dominance of *Chrysanthemum morifolium*. *Proc. Amer. Soc. hort. Sci.*, **61**: 543-547.

Beckett, K.A. 1983. *The Concise Encyclopaedia of Garden plants*, pp. 80-82. Orbis Pub. Ltd., Great Britain.

Belgaonkar, D.V., M.A. Bist and M.B. Wakde, 1997. Influence of nitrogen, phosphorus and different spacings on flower quality of annual chrysanthemum. *J. Soils and Crops*, 7(1): 90-92.

Brickell, C. (Ed.-in-Chief), 1992. The Royal Horticultural Society Encyclopedia of Gardening, pp. 152-154. Dorling Kindersley, London.

Broertjes, C. 1966. Mutation breeding of chrysanthemum. *Euphytica*, **15**: 156-162.

Broertjes, C., S. Roest and G.S. Bokelmann, 1976. Mutation breeding of *Chrysanthemum morifolium* using *in vivo* and *in vitro* adventitious bud techniques. *Euphytica*, **25**: 11-19.

Carpenter, W.J. 1976. Photosynthetic supplementary lighting of spray pompon *Chrysanthemum morifolium* Ramat. *J. Amer. Soc. hort. Sci.*, **101**: 155-158.

Cathey, H.M. 1954a. Chrysanthemum temperature study. B. Therma modifications of photoperiods previous to and after flower bud initiation. *Proc. Amer. Soc. hort. Sci.*, **64**: 483-489.

Cathey, H.M. 1954b. Chrysanthemum temperature study. C. The effect of night, day, and mean temperature upon the flowering of *Chrysanthemum morifolium*. *Proc. Amer. Soc. hort. Sci.*, **64**: 499-502.

Cathey, H,.M. 1955. Chrysanthemum temperature study. Effect of temperature shifts upon the spray formation and flowering of *Chrysanthemum morifolium*. *Proc. Amer. hort. Sci.*, **66**: 386.

Cathey, H.M. 1969. *Chrysanthemum morifolium* (Ramat.) Hemsl. In: *The Induction of Flowering* (ed. Evans, L.T.), pp. 268-290.

Cathey, H.M. 1976. Influence of a substituted oxathiin, a localized growth inhibitor, on the stem elongation, branching and flowering of *Chrysanthemum morifolium* Ramat. *J. Amer. Soc. hort. Sci.*, **101**(5): 599-604.

Cathey, H.M. and H.A. Borthwick, 1961. Cyclic lighting for controlling flowering of chrysanthemums. *Proc. Amer. Soc. hort. Sci.*,**78**: 545-552.

Cathey, H.M. and H.A. Borthwick, 1971. Phytochrome control of garden and greenhouse chrysanthemums. *Proc. Amer. Soc. hort. Sci.*, **87**: 464-471.

Cathey, H.M., A.H. Yeoman and F.F. Smith, 1966. Abortion of flower buds in chrysanthemum after application of a selected petroleum fraction of high aromatic content. *HortSci.*, **1**: 61-62.

Cockshull, K.E. 1972. Photoperiodic control of flowering in the chrysanthemum. In: *Crop Processes in Controlled Environments* (eds Rees, A.R., K.E. Cockshull, D.W. Hard and R.G. Hard), pp. 235-250. Academic Press, London.

Cockshull, K.E. 1985. *Chrysanthemum morifolium*. In: *CRC Handbook of Flowerting* (vol II, ed. Halevy, A.H.), pp. 238-257. CRC Press, Boca Raton, Florida.

Cockshull, K.E. and A.M. Kofranek, 1985. Long day flower initiation by chrysanthemum. *HortSci.*, **20**: 296-298.

Cockshull, K.E. and A.M. Kofranek, 1992. Responses of garden chrysanthemum to day length. *HortSci.*, **27**: 113-115.

Daughtrey, M. and A.R. Chase, 1992. Chrysanthemum. In: *Ball Field Guide to Diseases of Greenhouse Ornamental*, pp. 30-39. Ball Publishing, Geneva, Illinois.

Dahiya, D.S. and G.S. Rana, 2001. Regulation of flowering in chrysanthemum as influenced by GA and shade houses of different intensisties. *South Indian Hort.*, **49** (Special issue): 313-314.

Dole, J.M. and H.F. Wilkins, 1999. *Floriculture Principles and Species*, pp. 292-304. Prentice Hall, New Jersey.

Drewlow, L.W., P.D. Asacher and R.E. Widmer, 1973. Genetic studies of self incompatibility in the garden chrysanthemum *C. morifolium* Ramat. *Theor. Appl. Genet.*, **43**: 1.

Dutt, M., M.T. Patil and P.C. Sonawane, 2002. Effect of substrates on root growth in *Chrysanthemum*. In: *Floriculture Research Trend in India* (eds Misra, R.L. and Sanyat Misra), pp. 287-289. Indian Society of Ornamental Horticulture, Division of Floriculture and Landscaping, I.A.R.I., New Delhi.

Eysinga, van J.P.N.L.R. and K.W. Smilde, 1980. *Nutritional Disorders of Chrysanthemums*. Centre for Agricultural Publishing and Documentation, Wageningen, The Netherlands.

Farzadfar, S., K. Ohshima, R. Pourrahim, A.R. Golnaraghi, S. Jalali and A. Ahoonmanesh, 2005. Occurrence of turnip mosaic virus on ornamental crops in Iran. *Pl. Path.*, **54**(2): 261.

Fryxwell, P.A. 1957. Mode of reproduction in higher plants. *Bot. Rev.*, **23**:135-233.

Furuta, T. 1954. Photoperiod and flowering of *Chrysanthemum morifolium*. *Proc. Amer. Soc. hort. Sci.*, **63**: 457.

Gaikwad, A.M and S.S. Dumbre-Patil, 2001. Evaluation of chrysanthemum varieties under open and polyhouse conditions. *J. Ornam. Hort. (New Series)*, **4**(2): 95-97.

Gupta, M.N. 1971. Mutation breeding of chrysanthemum. I. Production of new cultivars by gamma ray induced somatic mutations in vM_1. *Proc. Intern. Symp. on Use of Isotopes and Radiation in Agriculture*. New Delhi.

Hammer, P.A. 1989. Nutrition. In: *Tips on Growing Potted Chrysanthemums* (eds Tayama, H.K. and T.J. Roll). Ohio Coop. Ext. Serv., The Ohio St. University, Columbus, Ohio, U.S.A.

Harshey, D.R. and J.L. Paul, 1981. Critical foliar levels of potassium on pot chrysanthemum. *HortSci.*, **16**: 220-222.

Hay, R. and K.A. Beckett, 1971. *Readrer's Digest Encyclopaedia of Garden Plants and Flowers*, pp. 141-149. The Reader's Digest Association Ltd., London.

Heinz, K.M. and S. Thompson, 1997. Use resistant varieties for chrysanthemum pest management. *GrowerTalks*, **60**(13): 82, 84, 86.

Hicklenton, P.R. 1984. Response of pot chrysanthemum to supplemental irradiation during rooting, long day and short day production stages. *J. Amer. Soc. hort. Sci.*,**109**: 468-472.

Hicklenton, P.R. 1985. Influence of different levels of timing of supplemental irradiation on pot chrysanthemum production. *HortSci.*, **20**: 374-376.

Huang, L., E.T. Paparozzi and C. Gotway, 1997. The effect of altering nitrogen and sulfur supply on the growth of cut chrysanthemums. *J. Amer. Soc. hort. Sci.*, **122**: 559-564.

Hughes, A.P. and K.E. Cockshull, 1972. Further effects of light intensity, carbon dioxide concentration, and day temperature on the growth of *Chrysanthemum morifolium* cv. Bright Golden Anne. *Ann. Bot.*, **36**: 533-550.

Jung, S.S., H.W. Jeong and K.S. Kim, 2000. Effects of uniconazole treatment on the growth and flowering of *Chrysanthemum indicum* L. *Korean J. hort. Sci. Tech.*, **18**(1): 28-32.

Karlssen, M.G., R.D. Heins, J.E. Erwin, R.D. Verghage, W.H. Carlson and J.A. Biernbaum, 1989a. Temperature and photosynthetic photon flux influence chrysanthemum shoot development and flower initiation under short day conditions. *J. Amer. Soc. hort. Sci.*, **114**: 158-163.

Karlsson, M.G., R.D. Heins, J.E. Erwin and R.D. Berghage, 1989b. Development rate during four phases of chrysanthemum growth as determined by preceding and prevailing temperatures. *J. Amer. Soc. hort. Sci.*, **114**: 234-240.

Kher, M.A. 1986. Chrysanthemum. In: *Ornamental Horticulture in India* (eds Chadha, K.L. and B. Choudhury). ICAR, New Delhi.

Kher, M.A. 1989. Chrysanthemum. In: *Commercial Flowers*, pp. 53-86. Naya Prokash, Kolkata.

Kofranek, A.M. 1980. Cut chrysanthemum. In: *Introduction to Floriculture* (ed. Larson, R.A.). Academic Press, Langton, U.S.A.

Lamont, G. 1986. Herbicide evaluation trials on cut flowers. *Australian Hort.*, **84**(7): 89-91.

Laurie, A., D.C. Kiplinger and K.C. Nelson, 1968. *Commercial Flower Forcing*, McGraw Hill, New York.

Lindquist, R.K. 1989. Insects and related pests. In: *Tips on Growing Potted Chrysanthemums* (eds Tayama, H.K. and T.J. Roll), pp. 40-49. Ohio Coop. Ext. Serv., Columbus, Ohio.

Machin, B.J. and N. Scopes, 1978. *Chrysanthemums-Year Round Growing*. Blandford Press, U.K.

Macnish, A.J., R.T. Leonard and T.A. Nell, 2008. Treatment with chlorine dioxide extends the vase life of selected cut flowers. *Postharv. Biol. & Tech.*, **50**(2/3): 197-207.

Mastalerz, J.W. 1968. Too much shade delays pot mums. *Pennsylvania Fl.* Grs, **208**: 3.

Mastalerz, J.W. 1977. Duration of irradiation. In: *The Greenhouse Environment*, pp. 221-270. John Wiley and Sons, New York.

Misra, R.L., Vinod Kumar and Sanyat Misra, 2002. Greenhouse Management of Ornamental Plants – an Overview. In: *Floriculture Research Trend in India* (eds Misra, R.L. and Sanyat Misra), pp. 13-18. ISOH, Division of Floriculture and Landscaping, IARI, New Delhi-110012.

Mortensen, L.M. 1986. Effect of intermittent as compared to continuous CO_2 enrichment on growth and flowering of *Chrysanthemum* x *morifolium* Ramat. and *Sainpaulia ionantha* H. Wndl. *Scientia Hort.*, **29**: 283-289.

Mukhopadhyay, A. and P. Das, 1976. Better chrysanthemums through air layering. *Indian Hort.*, **21**: 13, 15-16.

Namika, J.S. Arora, Kushal Singh and G.S. Sidhu, 2002. Effect of ethrel and alar on *Chrysanthemum*. In: *Floriculture Research Trend in India* (eds Misra, R.L. and Sanyat Misra), pp. 139-142. Indian Society of Ornamental Horticulture, Division of Floriculture and Landscaping, I.A.R.I., New Delhi.

Nau, J. 1993. Chrysanthemum. In: *Ball Culture Guide, The Encyclopedia of Seed Germination* (2nd ed.), p. 88. Ball Publishing, Batvia, Illinois.

Nazeer, M.A. and T.N. Khoshoo, 1982. Cytological evaluation of garden chrysanthemum. *Curr. Sci.*, **51**: 583-585.

Nell, T.A. 1993. *Dendranthema grandiflora*. In: *Flowering Potted Plants, Prolonging Shelf Performance*, pp. 35-38. Ball Publishing, Batavia, Illinois.

Nell, T.A., J.E. Barrett and R.T. Leonard, 1989. Fertilization termination influences postharvest performance of pot chrysanthemums. *HortSci.*, **24**: 996-998.

Pasini, C., T. Berio, P. Curir and F. D'Aquila, 1996. Further characterization of *Rhizoctonia solani* isolated from carnation and other ornamental plants. *Informatore Fitopatologico*, **46**(6): 33-36.

Post, K. 1949. *Chrysanthemum morifolium* (Chrysanthemum). In: *Florist Crop Production and Marketing*, pp. 385-417. Orange Judd Publishing, New York.

Post, K. 1950. Controlled photoperiod and spray formation of chrysanthemum. *Proc.Amer. Soc. hort. Sci.*, **55**: 467-472.

Post, K. and H. Kamemoto, 1950. A study on the number of short photoperiods required for flower bud initiation and effect of interrupted treatment on flower spray formation in two commercial varieties of chrysanthemum. *Proc. Amer. Soc. hort. Sci.*, **55**: 477.

Rajapakse, N.C., W.B. Miller and J.W. Kelly, 1996. Low temperature storage of rooted chrysanthemum cuttings: Relationship to carbohydrate status of cultivars. *J. Amer. Soc. hort. Sci.*, **121**: 740-745.

Ranjith, A.M. 2013. Chrysanthemum. In: *Identification & Management of Horticultural Pests*, pp. 133-136. New India Publishing Agency, Pitam Pura, New Delhi-110 088.

Raulston, J.C., W.E. Waters, S.S. Woltz and C.M. Geraldson, 1972. Summary of chrysanthemum fertilization programs for field production in Florida. *Fla Fl. Gr.*, **9**(10): 9.

Rex Y.N. Eng, M.J. Tsujita and B. Grodzinski, 1983a. The effects of supplementary HPS lighting and carbon dioxide enrichment on the vegetative growth, nutritional status and flowering characteristics of *Chrysanthemum morifolium* Ramat. *J. Hort. Sci.*, **60**: 389-395.

Rex, Y.N. Eng, M.J.Tsujita, B. Grodzinski and R.G. Dutton, 1983b. Production of chrysanthemum cuttings under supplementary lighting and CO_2 enrichment. *HortSci.*, **18**: 878-879.

Roude, N., T.A. Nell and J.E. Barrett, 1991a. Nitrogen source and concentration, growing medium, and cultivar affect longevity of potted chrysanthemums. *HortSci.*, **26**: 49-52.

Roude, N., T.A. Nell and J.E. Barrett, 1991b. Longevity of potted chrysanthemums at various nitrogen and potassium concentrations and NH_4:NO_3 ratios. *HortSci.*, **26**: 163-165.

Sacalis, J.N. 1993. *Dendranthema grandiflora*. In: *Cut Flowers, Prolonging Freshness* (ed. Seals, J.L, 2nd ed.), pp. 47-49. Ball Publishing, Batavia, Illinois.

Sachs, R.M. and A.M. Kofranek, 1979. Radiant energy required for the night break inhibition of floral initiation as a function of day time light input in *Chrysanthemum* x *morifolium* Ramat. *HortSci.*, **25**: 609-610.

Salinger, J.P. 1985. *Commercial Flower Growing*, pp. 163-179. Butterworths Horticultural Books, Wellington, New Zealand.

Schoellhorn, R.K., J.E. Barrett and T.A. Nell, 1996. Branching of chrysanthemum cultivars varies with season, temperature, and photosynthetic photon flux. *HortSci.*, **31**: 74-78.

Scopes, N.E.A. 1979. *Pests, Diseases and Nutritional Disorders of Chrysanthemums*. National Chrysanthemum society, London.

Seeley, J.G. and A.H. Weise, 1965. Photoperiod response of garden and greenhouse chrysanthemum. *Proc. Amer. Soc,. hort. Sci.*, **87**: 464-471.

Sidhu, G.S. and P. Singh, 2002. Effect of auxins on propagation in *Chrysanthemum morifolium*. In: *Floriculture Research Trend in India* (eds Misra, R.L. and Sanyat Misra), pp. 285-286. Indian Society of Ornamental Horticulture, Division of Floriculture and Landscaping, I.A.R.I., New Delhi.

Stefanis, J.P. and R.W. Langhans, 1983. Photoperiod and continuous irradiation studies of chrysanthemum with high pressure sodium lamps. *HortSci.*, **18**: 202-204.

Stewart, R.N. and H. Derman, 1970. Somatic genetic analysis of the apical layers of chimeral sports in chrysanthemum by experimental production of adventitious shoots. *Amer. J. Bot.*, **57**:1061.

Thistlethwaite, E.T. 1960. *Chrysanthemum*. Penguin Books, Middlesex, England.

Vince, D. 1960. Low temperature effects on flowering of *Chrysanthemum morifolium* Ramat. *J. Hort. Sci.*, **35**: 161-175.

Vince, D. and D.T. Mason, 1959. Low temperature effects on internode extension in *Chrysanthemum morifolium. J. Hort. Sci.*, **34**: 199-209.

Waters, W.E. 1964. The effects of soil mixture and phosphorus on growth responses and phosphorus content of

Chrysanthemum morifolium. *Proc. Amer. Soc. hort. Sci.*, **84**: 588-593.

Waters, W.E. and C.A. Conover, 1969. *Chrysanthemum Production in Florida* (Bulletin 730). Agric. Exp. Stat., Instt. of Food and Agric. Sci., Univ. of Florida, Gainesville, Florida.

Whealey, C.A., T.A. Nell, J.E. Barrett and R.A. Larson, 1987. High temperature effects on growth and development of chrysanthemum. *J. Amer. Soc. hort. Sci.*, **112**: 464-468.

Wilkins, H.F., W.E. Healy and K.L. Grueber, 1990. Temperature regime at various stages of production influences growth and flowering of *Dendranthema* x *grandiflorum*. *J. Amer. Soc. hort. Sci.*, **115**: 732-736.

Will, E., T.W. Starman, J.E. Faust and S. Abbitt, 1997. Photoperiodic responses of garden chrysanthemums (abstr.). *HortSci.*, **32**: 502.

Woodson, W.R. and J.W. Boodley, 1983. Accumulation and partitioning of nitrogen and dry matter during the growth of chrysanthemum. *HortSci.*, **18**: 196-197.

Yadav, L.P. 1983. Ph. D. Thesis. Bidhan Chandra Krishi Visweavidyalaya, Nadia (W.B.).

Zagarski, J.S., P.D. Ascher and R.E. Widmer, 1983. Multigenic self incompatibility in hexaploid chrysanthemum. *Euphytica*, **32**: 1-7.

14

Dianthus caryophyllus (Family: Caryophyllacae)

Y.C. Gupta and Kalkame Ch. Momin

[**Common names for carnation:** Carnation/Clove pink/Cut carnation/Divine flower/Florist carnation/Gillyflower/Grenadine/Picotee (*Dianthus caryophyllus*), Miniature carnation/Dwarf carnation/Pot carnation (*D. carthusianorum*, syn. *D. hybriden* though latter name does not have any botanical standing), *giroflées* (French), *Garofolo* (Italian), etc.]

Introduction

Carnation is one of the important cut flower crops grown throughout the world and occupies a prime position because of variety of colours, forms and a good vase life. Due to the excellent keeping quality, ability to withstand long distant transportation and remarkable ability to rehydrate after continuous shipping, carnation is preferred by growers. Carnations are excellent for cut flowers, for bedding, for pots, in borders, as edging and in rock gardens. Carnation is preferred by growers over rose and chrysanthemums in several flower-exporting countries and it stands number two in importance in the world cut flower trade. Early exporting countries were Republic of South Africa, Colombia and Kenya but later on Morocco, Turkey, northern Europe (Poland, the Netherlands, France, Italy, Spain), USA and SE & SW Asia (Japan and Israel) also joined for exporting its cut flowers to North America and Europe. Though carnations are grown throughout the year, it is in great demand especially during Valentine's Day, Mother's Day, Easter and Christmas in the Western countries. While standard carnations (*Dianthus caryophyllus*) are in great demand, the miniature types (*Dianthus carthusianorum*) are gaining popularity for their potential role in floral arrangement and also as a cut flower

Various Carnation Cultivars.

at comparatively low price. The preference for flower colour depends upon time, season, occasion and personal preferences, as during Mother's Day white carnations are in great demand, and even otherwise the white and pink standard carnations are in great demand followed by red, yellow and bicolours.

For the most part, carnations express love, fascination and distinction, though there are many variations dependent on colour. Carnation is considered as the birth flower for those who are born in January, light red for admiration, dark red for deep love and affection, pinks the symbol of mother's undying love as there is a legend that when Jesus carried the cross the Virgin Mary shed tears at Jesus plight and the earth where these tears fell there carnations sprang up, whites for pure love and good luck, stripes for regretting that love can not be shared, purple for capriciousness as in France it is a traditional funeral flower and even otherwise in France and in Francophone culture it symbolises bad luck and misfortune,

Origin, History and Distribution

Carnations were mentioned in Greek literature 2,000 years ago. *Dianthus* was coined by Greek botanist Theophrastus, and is derived from the Greek words *dios* for divine and *anthos* for flower. Some scholars believe that the name 'carnation' comes from 'coronation' or 'corone' (flower garlands) as it was one of the flowers used in Greek ceremonial crowns. Others think the name stems from the Latin *caro* (genitive carnis) meaning flesh, which refers to the original colour of the flower, or *incarnatio* (incarnation), which refers to the incarnation of God. Linnaeus chose the species name *caryophyllus* after the genus of clove as the fragrance from carnation is reminiscent of clove. The name carnation probably comes from its use by the ancient Greeks as a coronation flower.

Carnation is commercially cultivated throughout the world. The *Dianthus* species are adapted to the cooler Alpine regions of Europe and Asia and are also found in Mediterranean coastal regions. Natural climate for carnation occurs near 30° N or S latitude and on the western edges of continents. Today, the largest growing areas of carnations are in Bogota (Colombia). They are also grown on a large scale in Italy, Spain, Kenya, Sri Lanka, Canary Islands, France, Holland, Germany, Israel and USA. Carnations are also being cultivated commercially in India but for domestic consumption. The major production centres are located around Pune, Bangalore, Himachal Pradesh, Punjab, West Bengal and Kashmir.

Botany, Genetics and Breeding

D. caryophyllus, a native of C. Mediterranean is a perennial of loosely tufted habit, growing up to 90 cm with prominent joints, and the branched stems become woody with age. Leaves are linear, smooth, glaucous-grey (silvery), keeled, 5-nerved and up to 8 cm long. Flowers highly scented, mostly single, up to 4 cm across, colours mainly white, pink and red, petals dentate, and calyx-bracts 4, broad and pointed. *Dianthus* has a basic chromosome number x=15 and most of the species are diploid (2n=30) though *Dianthus chinensis* is a tetraploid (2n=60) and *Dianthus gratianopolitanus* is found in the forms of tetraploid (2n=60) and hexaploid (2n=90). Reciprocal diploid-tetraploid crosses can be used for production of triploids in carnation. Most of the F_1 hybrids are triploid and sterile due to production of unreduced gametes in *D. caryophyllus*.

Carnation is grouped into three categories, *viz.* single (5 petals), semi-double (30-60 petals) and double or super double (100-350 petals). Singleness in carnation is a monogenic character and incompletely recessive to doubleness, while heterozygotes yield semi-double flowers (Bhatt, 1989). Based on the flower colour, carnation is divided in three main groups. The first acyanic group contains yellow and white, the second group is cyanic containing salmon, lavender, pink and crimson, and the third group is a transition containing salmon-yellow, orange and maroon. The expression of self colour in carnation depends on six factors, *viz.* 'Y' controlling the production of yellow anthoxanthin, 'I' controlling the production of ivory anthoxanthin and is epistatic to 'Y'. 'A' is the basic anthocyanin factor and its recessive allele 'S' dilutes red or salmon. 'R' is responsible for crimson-red anthocyanin and modification of red series to series of magenta is controlled by 'M' (Misra *et al.*, 2006). Lutescent seedlings (greenish-yellow with weak growth), club neck (swelling of upper portion of roots), virescent (deficiency of chlorophyll as seen on leaves and stem as longitudinal stripes), lazy virescent (chlorophyll towards leaf edges), and dwarfness and thickness (absence of branches at the base) are all monogenic and differ from normal plants in a single recessive factor (Bhatt, 1989).

Still there is much scope to obtain a qualitatively better carnation with high yield and prolonged vase life, more continuous and rapid flowering and for generating more varieties with better colour variations and combinations. The breeding procedure typically consists of hybridization, self-pollination and selection (Holley & Baker, 1992). If the desired trait is recessive, it may not be expressed in F_1 progeny. By self-pollinating F_1's and growing of large population of F_2's, selection of one or more individuals with desirable traits will be possible. The process of inbreeding (self-pollination) may, however, hinder the breeding objectives by generating recessive homozygotes expressing undesirable traits. While inbred parental lines are necessary to breed homogeneous F_1 hybrid varieties, inbreeding detrimentally affects the inbred plants (Galbally and Galbally, 1997). Inbreeding depression appears in the third selfed generation (S_3) and, therefore, it is almost impossible to produce S_4 seeds (Sato *et al.*, 2000).

Hybrids between carnation and other *Dianthus* species can provide useful sources of genetic traits to achieve the above objectives (Segers, 1987). For instance, interspecific hybrids have been obtained through crossing *D. caryophyllus* and *D. capitatus*. These hybrids are highly resistant to bacterial wilt caused by *Pseudomonas caryophylli*. However, the flower quality is adversely affected and further improvement through backcrossing is necessary before commercial production (Onozaki *et al.*, 1998). Interestingly, traditional techniques for carnation breeding were developed before the discovery of inheritance by Mendel. As a result, these traditional techniques do not include a self-pollination step. In the absence of self-pollination, continuous hybridization has inadvertently resulted in highly heterozygous carnation varieties. The high level of heterozygosity promotes recombination mutations resulting in further new varieties (Holley and Baker, 1992). Mutation breeding has also been employed to create new colour mutants. More recently, the development of doubled haploidy techniques has also permitted breeders to accelerate breeding and selection (Holley and Baker, 1992). Dwarf carnations have been commercialised as alternatives to potted chrysanthemums.

Carnations have been genetically modified for prolonged vase life or to produce violet, mauve or purple flowers. These have been commercially released in Europe. Several GM carnations with modified vase life, flower colour or resistance to fungal pathogens have also been approved for limited and controlled release in Australia. In 1995, a GM carnation with improved vase life and another GM carnation with modified flower colour were approved for commercial release in Australia under the former voluntary system for regulation of gene technology overseen by 'Genetic Manipulation Advisory Committee'. The GM carnation with modified flower colour was subsequently re-licensed by the Gene Technology Regulator, under the current regulatory system in 2003.

Species, Varieties and Classification

Dianthus is a genus of about 250-300 species out of which *Dianthus caryophyllus*, *Dianthus barbatus* and *Dianthus chinensis* are the most commonly cultivated ones. *Dianthus chinensis* gave rise to garden pinks (var. 'Bedde Wigii') and fringed pinks (var. 'Laciniatus'). *D. caryophyllus* is the ancestor of border carnations. *D. caryophyllus* has been extensively used by the breeders for centuries and as a result many cultivated varieties and hybrids exist today, each with a name that usually describes its feature (Galbally and Galbally, 1997). Other important *Dianthus* species are *D. alpinus, D. arenarius, D. attenuatus, D. caesius, D. callizonus, D. capitatus, D. carthusianorum, D. cinnabarinus, D. cruentus, D. deltoides, D. diadematus, D. fimbriatus, D. glacialis, D. grandiflorus, D. imperialis, D. knappii, D. laciniatus, D. latifolius, D. monspessulanus, D. nobilis, D. pancicii, D. petraeus, D. plumarius, D. squarrosus, D. superbus, D. sylvestris, D. versicolor, D. viscidus* and *D. winteri.* Since only carnation is being dealt with here, a brief description only of *Dianthus caryophyllus* is being furnished. *D. caryophyllus*, is a perennial branched herb of loosely tufted habit with swollen joints and growing up to 90 cm. Old stems become woody and bear smooth, silvery, linear and 5-nerved leaves. Flowers highly clove-scented, single, 3-4 cm across and in white, pink or red. Petals dentate and calyx-bracts 4, wide and pointed. Popular ***standard*** carnation varieties under whites are 'Baltico', 'Candy White', 'Calypso', 'Dover', 'Flannel', 'Fragrant Ann', 'Icecap', 'Madonna', 'Niva', 'Nordika', 'Rivera', 'Roma', 'Snow Clove', 'Sonsara', 'Solo', 'White Giant', 'White Perfection', 'White Sim', 'White Wedding', etc.; under yellows are 'Calibra', 'Esty', 'Exotica', 'Golden Rain', 'Helios', 'Liberty', 'Murica', 'Pallas', 'Pinto', 'Raggio di Sole', 'Sunrise', 'Tahiti', 'Tobago', 'Yellow Dot Com', 'Yellow Sim', 'Yellow Star', etc.; under oranges are 'Amstel', 'Orange Isae', 'Orange Pinto', 'Salva', 'Solar', 'Solar Chiaro', etc.; under pinks are 'Bizet', 'Candy', 'Calypso', 'Castellaro', 'Charment', 'Cipro', 'Crowley's Sim', 'Dark Rendezvous', 'Fiery Cross', 'Lena', 'Linda', 'Nora', 'Novona', 'Oriana', 'Paola', 'Pink Diamente', 'Pink Sim', 'Pirandello', 'Rubesco', 'Sensi', 'Sharina', 'Shocking Pink', 'Tabor', 'Tasman', 'Varna;, etc.; under reds are 'Desio', 'Diplomat', 'Domingo', 'Firato', 'Hildelgo', 'Kazhuca', 'Master', 'Norman', 'Perfect Clove', 'Prasio', 'Sangria', 'Scania', 'Tanya', 'William Sim', etc.; under reds are 'Gaudina', 'Granada', 'Grand Slam', 'Killer', 'Master', 'Red William', 'Scania', 'Tanga', etc.; under purples are 'Exquisite', 'Imperial Clove', 'La Royale', 'Lavender Lace', 'Margaret', 'Sparta', 'Storm', 'Vanessa', etc.; and under bronzes are 'Consul', 'Exquisite', 'Galatec', 'Harvest Moon', 'Mandarin Sim', 'Tangerine Sim', etc. whereas **spray type** carnation varieties under whites are 'Amalia', 'Close Up', 'Excel', 'Hermon', 'Iceland', 'Milkway', 'Opal', 'Tibet', 'Virgo', 'West Crystal', 'White Prestige', 'White Royalette', etc.; under yellows are 'Alicetta', 'Annelies', 'Cartouche', 'Castillo', 'Eureka', 'Koreno', 'Limon', 'Lior', 'Yellow Odeon', etc.; under oranges are 'Amaya', 'Macarena', 'Marisol', 'Niky', 'Target', 'Tip Top, etc.; under pinks are 'Annelies', 'Barbara', 'Donora', 'Fantasia', 'Karina', 'Krystie', 'Madea', 'Medley', 'Nathalie', 'Noelia', 'Nuria', 'Opale', 'Rossini', 'Silvery Pink', etc.; under reds are 'Enzo', 'Etna', 'Karma', 'Rony', etc.; and under purples are 'Boreal', 'Duraongo', 'Purple Chopin', 'Roxette', etc. At present, the top carnations in cut flower trade are: (i) **Standards** ~ 'Brecas', 'Castellaro', 'Decio', 'Delphi', 'Donna', 'Fransesco', 'Ivonne', 'Magic Brepi', 'Manon', 'Miledy', 'Nelson', 'Pallas', and 'Raggio di Sole'; (ii) **Sprays** ~ 'Albatross', 'Bagatella', 'Barbara', 'Carosel', 'Elsy', 'Kandinskij', 'Karina', 'Kortina', 'Lior', 'Medea', 'Nargiz', 'Natalia', 'Princess', 'Rondine', 'Rony', 'Scarlette', and 'Stephanie'.

Border and picotee carnations are the earliest types and have symmetrical flowers that are frilled with open centres and broad & smooth-edged petals. There are dwarf and hardy perennials branching mainly at the base and produce single stem during first year and become bushy in subsequent years. The flower colour varies from a single blended colour and makes them more attractive. Broadly it is further divided into four clasess, *viz.* ***Flakes*** having clear grounded flakes with one colour, *viz.* scarlet, purple or rose; ***Selfs*** having only single colour in the petals without any stripe or spot; ***Bizarres*** having pure ground as in flakes but are marked with 2-3 colours as crimson, scarlet, pink or purple; ***Fancies*** which do not fit in any of the above groups, and the petals have various colours arranged in bands on light grounds; and ***Picotees*** having pure ground of white or yellow, bordered with a band of colour at the margin of each petal. The name **perpetual carnation** implies to their habit of flowering throughout the year. These are hybrids involving many *Dianthus* species. Flowers are generally self-coloured with long stems and have a great ability to withstand long distance transportation that make it best suited for cut flowers. Perpetual carnations are further divided into two major groups, *viz.* standard and spray. The standard types produce larger single flowers on a sturdy long stem, while spray types produce many flowers of small size on a short stem providing a tight cluster, or longer stems originating lower down the stem giving a more open spray. **Chabaud or marguerite carnations** are annual carnations and have clove-scented single or double flowers that reproduce from seeds. The blooms are large with fringed petals, which do not last very long after harvest as compared to the perpetual flowering types, but are produced more freely over a long period and are comparatively easier to cultivate than perpetual types. **Malmaison carnations** are stiffer and broad-leaved, giving them a massive appearance. The flowers are large double with well-filled centres, fragrant and generally have a pink shade. The buds are round and chubby unlike the perpetual carnations which have long and pointed buds.

Propagation

Seeds in the temperate regions are sown from April to June whereas in the plains in September as for the species and the garden pinks. Seeds are sown thinly in pans or boxes and kept at 15 °C temperature until germination. Nau (1993) described that seeds of garden cultivars of carnations are successfully germinated at 18-21 °C. The medium for seed sowing should consist of loam 2 parts + loose peat 1 part + coarse sand 1 part, and to one litre of this mixture some 1.2 g of lime superphosphate + 0.6 g of finely ground chalk or limestone, and this mixture is sieved through 1.12 mm mesh. After 10-30 days of germination, these are pricked out in boxes at 4 × 4 cm, or 5 × 5 cm (in case of delayed planting) apart in John Innes Potting Compost (JIPC) No. I, *i.e.* loose loam 7 parts, peat 3 parts, coarse sand 2 parts, and to each litre of this mixture some 3 grammes of a mixture consisting of bone meal 2 parts + lime superphosphate 2 parts + potassium sulphate 1 part, and some 0.6 g of finely ground chalk are added. For about a month of pricking, the plants are kept at

Carnation Mother Stock

Selection of cuttings (terminal 10-15 cm)

Placing cuttings in fresh water

Lower leaf removal and preparation of cuttings

Placing of cuttings on medium

Misting of carnation cuttings

Rooting of cuttings after 4 weeks

Cuttings ready for packaging and transportation

13-15 °C, then at 10 °C and then hardened off in a cold frame and when seedlings attain 3 cm of height these should be planted outdoors at their permanent position in October in the plains or in March-May in temperate regions with a fair ball of soil to their roots. During germination there should be sufficient moisture in the medium, but after germination these are moderately watered, however, during frosty weathers there should be no watering. The seeds of perpetual-flowering carnations should be sown in February in the temperate regions in the greenhouse as for Chabauds, to get the blooms in majority of the cases by September, and by all means by October. Similar to the perpetual-flowering carnations, the seeds of perpetual border carnations are also grown. In both these cases the temperature requirement for seed germination is 15.5 °C. Seedlings appear normally within 10 days, and then any time after 10 days of germination these are pricked out and planted either directly in 5.0 to 7.5 cm pots or in other boxes some 5 cm apart containing JIPC No. I. No pinching is recommended in case of perpetual-flowering and perpetual border carnations if raised through seeds so that immediately after flowering the singles and inferior ones are discarded and good ones retained.

Commercial production of carnation plants is started from unrooted top **cuttings**. The cuttings are harvested from selected disease-free mother plants that have been procured from a well known source. A typical carnation cutting is a sturdy vegetative stem tip, 10-15 cm long with 4 to 5 visible leaf pairs and weighs about 10 grammes. Cuttings should be broken from the stock plant to avoid spreading of the disease through wounded ends (as with a knife). Bharathy *et al.* (2004) reported that the cuttings set in vermicompost root earliest (9.78 days), followed by perlite+cocopeat and sand+vermicompost. The percentage of rooted cuttings, root number, mean root length and fresh dry weight of roots was also highest in vermicompost, followed by perlite+cocopeat. However, Thomas *et al.* (2003) observed that the medium containing 3 parts sand and 1 part soil proved beneficial to increase the percentage of rooted cuttings and number of roots per cutting in carnation cv. 'Mixed Super Chabaud'. Spraying ethrel at 15 ppm after 3-4 days of harvesting of cuttings increased the number of cuttings in the next lot (Gupta and Bhattacharjee, 2002). These nucleus blocks need removal after every one or two years. Unrooted cuttings should be treated with a combination of Dithane M-45 (0.25 per cent)+Bavistin (0.1 per cent) for 30 minutes to avoid the incidence of fungal diseases during the rooting. Quick dip treatment of cuttings in 50 ppm NAA for 5 seconds encourages better rooting. Study on the use of growth regulator on rooting of cuttings at Solan (H.P.) showed that maximum percentage of rooting and best quality roots are recorded with NAA at 500 ppm. Rooting occurs in 3-4 weeks and rooted cuttings are then transferred to hardening chambers before transplanting in the main field. Sanitation is important in propagation. The rooting medium should be steam- pasteurized for each successive group of cuttings. Dipping the cuttings in fungicidal solutions has been generally avoided, since bacterial wilt disease can spread with the dips. Fungicide drenches can be applied over the cuttings in the bench, however.

Micropropagation is another commercial method for propagating carnations on a large scale. *In vitro* culture makes it possible for a grower to obtain disease-free planting material in bulk. Wankhede *et al.* (2006) observed axillary bud explants responding well compared to the shoot tip and stem segment explants. Raja Ram and Zaidi (1999) cultured meristem-tips to obtain virus-free plants.

Cultural Practices

Selection of **site** plays an important role in commercial production of carnations. A well-drained sandy or sandy loam **soil** is most suitable for successful cultivation. The site chosen for cultivation should be free from weeds, nematodes and soil-borne pathogens. Pre-plant soil sterilization with steam solarisation or 5 per cent formalin is beneficial in eradicating

Carnation at the Farmer's Field in Solan.

Carnation at Blooming Stage.

pathogens and other pests. Ideal soil pH is between 6.0 and 7.0. Bhatia *et al.* (2004) reported that the **growing medium** consisting of soil+FYM+cocopeat (2:1:1) is the best for important parameters such as plant height, flower bud diameter, stem length and vase life when inoculated with biofertlizers and fertigated with water-soluble fertilizers. Carnation cv. 'Raggio-de-Sole' when grown in sand+soil+vermicompost (1:1:1 v/v)+inorganic fertilizers+biofertilizers at 2g/plant (*Azospirillum* and phosphate soubilizing microorganisms) produced maxium plant height, number of flowers, length of flower stem, flower size, earliness in flowering, maximum percentage of A grade flowers and vase life (Bhalla *et al.*, 2007).

Most of the perpetual carnations are grown commercially under protection with sufficient **light** and proper ventilation obtained through perfect designing and orientation of greenhouses. Carnation produces flowers in flushes and a heavy flush of flowers during a short period is not desirable. Time and duration of flushing and subsequent production are regulated through planting time, planting density and method of pinching. In India, carnation is **planted** during October-November in plains, October to February in foot hills where there is no snowfall and during February to April in temperate areas having seasonal snowfall (Misra *et al.*, 2006). Places in mid-hills of Himachal Pradesh are practically suitable for year round production. Kumar *et al.* (2002) noted that early planting (30 September) of carnation cv. 'Red Corso' produced greatest plant height, leaf length, flower number, and weight, length & longevity of individual flowers, number of branches and stalk length.

The **depth** that the rooted carnation cuttings are planted can be a critical factor for success or failure. Carnations are sensitive for too deep planting, hence, should be planted at a shallower depth to encourage rapid root growth. Deeper planting results in root and stem rot, therefore, rooted cuttings should be planted at same depth as they were in the rooting medium (Besemer, 1980). **Spacing** depends on the cultivar, light level, timing and pinching programmes. Standard carnations are spaced at 20 × 20 cm and spray types at 30 × 30 cm, and with the spacing of 20 × 20 cm, 25 plants are accommodated per square metre area. High density is used for one-year-old crop and more generous spacing for the 2-year old crops to maintain high yields. Increased density intensifies the first flush but reduces the total yield in the first year, whereas wider spacing spreads the yield and helps in producing quality flowers. Therefore, closer spacing suits the single cropping system but for two years or more cropping systems, wider spacing is recommended (Besemer,1980). Langhans (1961) recommended 10 × 15 cm for one year production and 15 × 15 cm and 15 × 20 cm for 2-year production.

For successful production of top quality carnations, **pinching** is a must. During pinching, the tip of the stem is removed to encourage growth of laterals. Pinching is very necessary for standard (Sim) carnations. Pinching is done by hand to reduce the possibility of transmitting pathogens by using knives or clippers. The '**single pinch method**' results in the most rapid flowering but all stems flower about the same time. After the plants have been established, about three to four weeks after planting the shoot apex is pinched leaving 4-5 lateral shoots to develop. Pal and Biswas (2004) recorded single pinched plants of carnation cvs 'Desio' and 'Supermix' resulting into greatest plant height and longest flower stem. Verma *et al.* (2003) reported single pinching producing good quality flowers in carnation cv. 'Impala'. Single pinching of carnation cvs 'Angelica', 'Golden Boy' and 'Domingo' at 20 days after planting was most promising resulting in earliest flower bud appearance, maximum flower diameter and maximum number of flowers (Rao *et al.,* 2008). Carnations can also be pinched using a method called '**pinch and a half**'. After the initial pinch, one-half of the breaks from the first pinch are pinched second time after they are 5 cm long or have 5-6 pairs of leaves. Although this method results in lower yields on the first harvest compared to the single pinch method, it permits more evenness of production over a span of time. Pathania *et al.* (2000) found pinch and a half method (with 6 laterals) as the best treatment which staggered the flower production in two flushes (3rd week of May to July end) and yielded 66.66 quality flowers/m^2 in one cropping season. Chavan *et al.* (2004) reported pinch and a half method as most effective as it staggered flower production and yielded a moderate number of flowers with intermediate quality. In the '**double pinch method**', the main shoot is pinched once, followed by pinching of all the lateral shoots arising from the first pinch when they are of desired length. Pal and Biswas (2004) reported double pinching delaying the flower production significantly compared to other methods of pinching. However, they found that double pinching results in the highest number of flowers per plant followed by the pinch and a half method. Double pinching of carnation cv. 'Tasman' resulted in maximum plant spread and maximum branch number (Singh *et al.,* 2005). Studies conducted by Dalal *et al.* (2006) also showed that double pinching of carnation cv. 'Yellow Solar' resulting in highest number of shoots per plant, *vis-à-vis* flower yield per plant and per square metre. A fourth method, '**single pinch plus pull pinch**' is similar to double pinching. It is not a very common method of pinching carnation. This method involves a single pinch of the main shoot followed by pulling of growing tips when the lateral shoots are long enough. Pull pinching of shoots is followed for about two months to provide constant flower production over a longer period of time. It has been found that this pinch should be practiced only in high density planting.

Disbudding of standard carnations is done as soon as the lateral buds are easily removable and should be done on a continuous basis. In standard carnations, lateral buds below the sixth node from the terminal flower bud are removed to encourage production of best quality terminal flowers. The best time for disbudding is when the apical flower bud is about 15 mm in diameter. Unlike the standard type, spray type carnations need a different kind of disbudding where terminal flower bud is removed from main axis to encourage lateral flower buds to develop.

Carnation plants require **support** for their growth as the stems are very weak, and without support it may result in bent stems decreasing the stem length and quality. Poorly supported branches easily break close to the stem making them susceptible to diseases. Individual plant staking is done with the wooden or iron rods. Wire mesh system (15 × 15 cm mesh) is laid over the plants supported by stakes of metal or fixed at the corner of the beds. The mesh in 2-3 layers is to be laid together on the soil surface (15 cm above the soil) and the above layers are separated by 20 cm apart. As the plants grow, the stems must be constantly stalled or caged within the respective mesh opening to maintain straight stems.

During summer season on *Dianthus caryophyllus*, all the **herbicides** (Alachlor, Napropamide, Oxadiazon, Oxyfluorfen and chlorthal dimethyl) tested were found effective against >90 per cent of grasses and 67-100 per cent of broad-leaved weeds.

Feeding and Irrigation

Proper **nutrition** in carnation is very important to get better yield with high quality flowers. Always nutrients should be given only after analysis of the soil and leaf tissues. The application of nutrients in small doses but more frequently favours better growth and flower production. At the time of land preparation sufficient quanity of organic manure must be added. Generally farmyard manure at 5-10 kg/m^2, NPK at 30:20:10 g/m^2 for standard carnations, while for spray types it is almost the same except that N is 40 g/m^2. Nitrogen is given in two splits ~ one before planting and the remaining half at one month after planting. The optimum nitrate level in the medium should be between 25 and 40 ppm. During low light conditions, nitrate rather than ammoniacal nitrogen is applied for toning and stronger stems. However, excess nitrogen causes weak stems in carnation. Low nitrogen levels inhibit flower bud initiation. Verma (2003) reported nitrogen at 1,000 ppm recording greatest plant height and highest number of flowers per plant in carnation cvs 'White Candy' and 'Red Corso'. The split application of N and P (200 ppm) to foliage twice a week has been recorded producing longest and widest leaves, *vis-à-vis* largest flowers, whereas N and P applied to soil once in fortnight gave the earliest flowering and longest field life (Sarkar and Roychowdhury, 2003). The optimum potassium level in the medium for carnation is between 25 and 40 ppm. Low potassium levels reduce yield, grade, stem length and longevity. Calyx tip dieback is associated with potassium deficiency. Optimal calcium levels are 150 to 200 ppm. Calcium deficiency appears as crescent-shaped necrotic lesions 2.5-5.0 cm from the leaf tip and calyx scorching. Phosphorous deficiency reduces growth and causes stunted plants. The optimal phosphorous level in the medium is 5 to 10 ppm which should be attained with addition of superphosphate. Boron deficiency increases calyx splitting and induces bud abortion. Less than optimal foliar boron levels (below 20-25 ppm) cause shortened internodes, clubbiness, distorted flower buds and 'witches broom' symptoms. The best results of vegetative growth and flowering characteristics were obtained by foliar application of carnation plants with orthophosphoric acid at 200 mg per litre and boron at 50 mg per litre (Sharaf and El-Naggar, 2003). Kumar *et al.* (2003) reported highest plant height, number of branches per plant, number of flowers per plant, flower diameter, flowering duration and earliness with 800 ppm Zn.

Carnation plants require ample quantity of **water** for its proper growth and development. Watering should be done immediately after planting. Therefore, drip irrigation is adopted for regular irrigation. Sprinklers are not used as they encourage fungal diseases. After establishment of plants, these should be regularly watered during warm sunny days. The crop requires 4-5 litres of water/m^2/day. Stagnation of water should be avoided to minimize the incidence of diseases. Frequency of irrigation depends on the weather conditions and soil type. Low soil moisture reduces flower quality though it is also dependent on soil N and electrical conductivity.

Yield of flowers depends mainly on variety, pinching programme and on the period the crop is harvested. In general, sprays produce more flowers than standards. Generally, 200 flowers/m^2 can be obtained from standard types while 250 flowers/m^2 from sprays.

Growth and Flowering

The character of any plant is basically governed by its genotype, but plants of the same genotype can give different responses according to the environment in which they grow. **Temperature** is the major factor that influences the growth and flowering of carnation. Best quality carnations are produced in areas having high light intensity during winter and at the same time the temperatures during summer months are mild. Carnation is a cool temperature crop and optimum temperature for quality flower production in carnation is 10-12 °C (night). A 16-18 °C results in production of low quality flower with decrease in flower diameter (Misra *et al.*, 2006). Ideal day and night temperature is 28 °C and 16 to 18 °C, respectively. Temperature at night is very important for quality. The high difference between day and night temperatures but under the prescribed range only provide a good carnation crop. Higher day and night temperatures especially during flowering results in abnormal flower opening and calyx splitting. Bunt (1973) studied the effect of temperature on development of floral buds and found that the time taken from visible bud to anthesis was 100 days at 10 °C while it was 33 days at 20 °C. However, low temperatures along with photoperiods were found advancing floral initiation (Bunt and Cockshull, 1985). The rate of development of floral buds is entirely related to temperature and is rapid at higher temperatures.

The carnation is a facultative long day plant, which means that they form the flowers faster during long days than in short days. Carnations require high levels to produce high quality flowers. The **photoperiod** is more important factor than **light intensity** in flowering, which influences the lateral shoot development and flowering in carnation. The initial vegetative growth is directly proportional to the light intensity. Time taken to produce flower buds directly depends on both photoperiod and intensity, *i.e.* more flowers are produced during long days with high light levels and low temperature. The rate of flower initiation in carnation is also affected by irradiance (Bunt *et al.,* 1981). The amount of light has had a tremendous effect on determining the world's carnation production centres. Carnations require a minimum of 21.5 klux to produce sufficient photosyntates for growth and flowering. Shoots become receptive to photoperiod induction when 6-8 leaf pairs are formed (Kumar *et al.,* 1999). Under normal environmental conditions, 15-18 leaf pairs are formed prior to floral initiation. Fewer leaves subtend to the flower under long days while there will be more leaves under short days, hence more flowers are produced in long days with their high light levels. Long days from night interruption lighting are more effective than day extension. Cyclic lighting is as effective as continuous night break lighting (Bunt and Cockshull, 1985; van der Hoeven, 1987; Lolapori and Arora, 1995). Long day treatments inhibit lateral branching, so stock plants are produced under short days to increase lateral shoot production and cutting quality, and to ensure the production of vegetative cuttings. Hlatshwayo and Wahome (2010) observed that the highest plants resulted from carnation plants provided with 70 per cent shading. The highest number of leaves per plant, leaf area and number of lateral shoots per plant were observed in plants provided with 20 per cent. Kumar *et al.* (2002) reported that the additional light given to carnation plants from 2200 to 0200 h for six weeks after three weeks of planting produced greatest plant height, leaf length, number, weight, length and longevity of individual flowers, number or branches and stalk length.

Carbon dioxide levels affect both the growth and quality of carnation plants. Low levels of carbon dioxide, *i.e.* 100 to 150 ppm in closed greenhouses during the day inhibit growth as the rate of photosynthesis becomes equal to the respiration rate. Thus, ventiliating or adding supplemental carbon dioxide is necessary to maintain adequate growth and enhancing the quality. The greenhouse carbon dioxide level should be maintained at 300 to 500 ppm on cloudy days and at 750 to 1,000 ppm on sunny days. Carbon dioxide injection (1,000 ppm) can increase 30 to 35 per cent yield, may reduce the harvesting time and increases per cent dry weight. Increased yield and reduced time between harvesting adds up to increased and more uniform production. An increase in per cent dry weight is manifested in enhanced stem length and flower longevity.

Growth regulators play an important role in regulation and production of carnation flowers. Spraying of chlormequat at 10-100 ppm or 0.25 per cent and SADH at 0.4 per cent advanced flowering and increased production while MH delayed flowering (Bhatt, 1989). NAA when applied alone at 550 ppm showed best results for earliness in root formation, rooting percentage, root number, root length and weight (Kumar *et al.*, 2006). Studies conducted by Singh and Singh (2003) revealed that gibberellic acid treatment significantly affected the vegetative and flowering characters. Cycocel according to its mode of action evidently retarded certain growth characters, but at 1,000 ppm, this increased the number of branches and flowers. GA_3 at 100 ppm was found effective in increasing the plant height, stem length and flower yield (Verma *et al.*, 2000). Two sprays of GA_3 at 100 ppm, one at first pinch and second when 8-10 cm of length has been attained by axillary shoots, produced early flowering and long stems (Gupta and Bhattacharjee, 2002).

Postharvest

Carnation flowers **harvested** at tight bud stage or partially open stages store for longer periods of time and have greater longevity than flowers harvested when fully open (Reid *et al.*, 1983). Standard carnations are harvested at tight bud or at cross-bud (visible petal) stage when the outer petals are unfolded nearly perpendicular to the stem. In case of spray carnations, harvesting should be effected when at least two flowers have opened and the other buds are showing colour. The best time to harvest carnations is during the morning hours when they are turgid but after dew is evaporated. Flowers are harvested with sharp knives or shears. On standard carnations, 2-3 nodes and on spray carnations 3-4 nodes are left on the shoots for the next flowering.

Grading is essential for selection of flowers according to quality parameters which brings better price than flowers of equal quality but of poor grade. Carnations are graded on measurable characteristics such as stem length, stem strength and flower diameter. The Society of American Florists (SAF) suggested the following inspection standards for cut carnations (Staby *et al.*, 1978).

1. Bright, clean, firm flowers and leaves
2. Fairly tight petals near the centre of unopened flowers
3. Symmetrical flower shape and size characteristics of the cultivar
4. No split or mended calyx
5. No lateral buds or suckers
6. No decay or damage
7. Straight stem and normal growth

SAF has suggested the following grades:

Minimum Flower Diameter (mm)	*Grades*		
	Blue (Fancy)	*Red (Standard)*	*Green (Short)*
Tight	50	44	None
Fairly tight	62	56	None
Open	75	60	None
Minimum stem length (cm)	55	43	30

The following are the grades being followed locally:

Grade	*Stem Length (cm)*
A	> 45
B	<45 >30
C	> 30

Flowers harvested at paint brush stage can be **stored** dry in moisture retentive boxes at -0.5 to 0 °C for 3-4 weeks. Fully open flowers are better stored in a floral preservative solution at 3-4 °C because lower temperatures may cause injuries and discolouration of petals in some cultivars, especially when the storage period is longer than two weeks. Prior to prolonged storage, either dry or wet, flowers should be treated with STS. During wet storage, high RH of 90-95 per cent should be maintained and flowers must be shielded against direct streams of cold air from cooler. Flowers destined for prolong storage should be protected against grey mould by dipping whole flowers in a fungicide solution for a few seconds just after harvest. Buds should then be conditioned in a solution containing STS and 70-100 g/l of sucrose. Conditioning may be carried out in cold storage at 0-1 °C using a solution warmed to 40 °C in which the flower stems are submerged at 3-5 cm. Buds should remain in the conditioning solution for 20 to 24 h. The buds should be packed at low temperature close to 0°C, wrapped in paper and then placed in foil bags and tightly sealed inside. The optimal temperature for storing carnations is 0 to 0.5 °C. Singh *et al.* (2002) observed that carnation flowers can be wet-stored for up to 9 days and dry stored for upto 9-12 days at 4(±0.5) °C and 80-85 per cent RH without any loss in vase life. In another experiment conducted by Bhatia *et al.* (2002), it was found that the cut flowers when stored dry at 2 °C and 70 per cent relative humidity for 24 hours gave longest vase life in carnation cvs 'Impala' (13.33 days) and 'Purple Choppin' (9.83 days).

Luo Hongyi *et al.* (2003) reported that the cut carnations when **pulsed** with 50 mg $AgNO_3$+40 mg benzoic acid+200 mg 8-HQ/litre+3 per cent sucrose resulted in delayed petal withering (caused by water loss) and greater longevity of cut flowers. Maximum vase life (15.06 days) was observed from carnation flowers treated with 200 ppm 8-HQC+0.5 mM STS+2 per cent sucrose (Verma *et al.*, 2002). Chandrashekar and Gopinath (2001) stated that carnation cvs 'Acapalca' and 'Pink Dona' when held at 1mM STS+3 per cent sucrose resulted in higher cumulative water uptake, cumulative fresh weight and vase life of cut flowers. Pulsing solution of 1mM STS+200 ppm of 8-HQC supplemented with 10 per cent sucrose was found to be the best for carnation cvs 'Impala' and 'Purple Choppin' (Bhatia *et al.*, 2002). The cut flowers of cv. 'Pentara' pulse-treated with 0.4 mM STS+6 per cent sucrose+400 ppm 8-HQ resulted in maximum vase life (10.80 days). The maximum vase life (15.8 days) of 'Sunrise' was obtained from 100 ppm $AgNO_3$+4 per cent sucrose+200 ppm 8-HQ (Chikkasubanna and Sharada, 2002). Singh *et al.* (2007) found that cut flowers pre-treated with solution containing sucrose

(10 per cent)+$Al_2(SO_4)_3$ at 200 ppm+STS (0.2 mM) recorded maximum vase life (11.89 days).

Flowers are kept in **holding solutions** by the florists or consumers. Cut flowers of nine carnation cultivars were when subjected to 'Proflovit-72' (0.3 g 8-HQS + 0.05 g CCC + 50 g sucrose/l) and compared with $AgNO_3$ (20-50 ppm), $KMnO_4$ (4 ppm), streptomycin (200 ppm), or 8-HQS (200 ppm) and in some cases, 5 per cent sucrose and/or CCC at 50 ppm, 'Proflovit-72' was found to be the best for flower quality and vase life (Nowak and Rudnicki, 1975) as this provided 100 per cent flower opening even when at small bud stage the flowers were harvested and placed in 'Proflovit-72' though in the control it was utterly poor. Singh *et al.* (2007) reported that vase life of cut flowers was maximum in solution containing sucrose 5 per cent +aluminium sulphate (200 ppm). Holding solution containing 7.5 per cent sucrose+50 ppm $AgNO_3$+300 ppm 8-HQS or 5 per cent sucrose+50 ppm $AgNO_3$+300 ppm 8-HQS resulted in longest vase life of carnation cvs 'Kristina', 'Aleda', 'Master' and 'Vienna' (Krishnappa *et al.*, 2000). Vase life of carnation cvs 'Red Corso' and 'Cabaret' was improved by placing in solutions containing 8-HQC, 200 ppm $AgSO_4$ and 100 ppm $ZnSO_4$ (Kumar *et al.*, 1999). Studies conducted by Khaligi and Shafie (2000) revealed that maximum vase life of cut carnations was recorded with 200 ppm 8-HQC+STS+2 per cent sucrose. Six hours pretreatment with low doses of 1-MCP (a cyclic ethylene analogue) acted similar to STS in negating ethylene effects and inhibited the normal wilting response of cut carnations exposed continuously to ethylene at 0.4 μl/l, *vis-à-vis* prolonged vase life in an ethylene-free environment (Serek *et al.*, 1995). Macnish *et al.* (2008) when added aqueous chlorine dioxide (ClO_2) at 2 or 10 $\mu l\ l^{-1}$ in clean deionized water, the build up of bacteria in vase solution was prevented and the vase life was found extended in *Dianthus caryophyllus* cv. 'Pasha'. The effects of STS and 1-MCP were studied by Dole *et al.* (2005) on cut carnation cvs 'Icardia', 'Silk Road' and 'Tasman' by treating the unpacked cut stems and placing them either in deionized water (DI) and subjected to 1-MCP (740 $nl\ l^{-1}$) or ambient air for 4 h or DI + STS at 0.2 mM for 4 h. After treatment, stems were removed, placed in polyethylene sleeves and stored either wet in DI water or dry in plastic-lined floral boxes at 5 °C in the dark for 4 days. After storage, bunches were placed in DI water under 12 h light (76 to 100 $\mu mol\ m^{-2}s^{-1}$) per day where it was found that both STS and 1-MCP increased the vase life, more being under STS.

Carnations are **packed** in corrugated cardboard containers. About 800 carnations are packed in a standard size carton of 30 cm height, 50 cm width and 122 cm length. The boxes or containers should be well insulated and should have a polythene lining as vapour barrier to help maintain high relative humidity inside the packages. Bunches of 25 flowers are then placed in these boxes with half of the total number of bunches oriented on each of the container. Close packing of flower helps to minimize the distance for heat exchange between flower layers. In order to maintain high humidity, insulating paper is put across the box to cover the flowers completely. Moreover, shipping cartons should have holes to ensure that the cold air can flow freely through the cartons to avoid ethylene build up in the carton. The water balance in the cut carnation should be at optimal level for maximum longevity. Moisture stress and low humidity can lead to reduced longevity and petal burn. Transportation should be done in a refrigerated van at 2-4 °C temperature.

Carnation Flower Packaging.

Insect-Pests, Diseases and Physiological Disorders

Misra *et al.* (2006) have given the detailed account of insect-pests and diseases of carnation. **Red Spider Mite** (*Tetranychus urticae*) is the most serious pest on carnations. The mites are minute red insects which feed on the undersides of the leaves, suck the sap and eventually the leaves turn pale to bronze, wither and show severe webbing. In its attack, plant growth, crop quality, yield and vase life of carnation flowers are drastically reduced. In greenhouse, this pest is best controlled with miticide aerosols. Spraying schedule with Profenophos (0.05 per cent) when mite population reaches 2-5 adults per plant, followed by Ethion (0.05 per cent) after 10 days will control this pest. **Aphids** (*Myzus persicae*) suck the sap from the leaves and disfigure the young growth. In severe attack, they leave sticky deposits on the leaves and flower buds. Aphids can be responsible for the transmission of viruses. These can easily be controlled with Malathion or Sevin sprays (0.2 per cent). **Thrips** (*Thrips tabaci*) also suck the sap from the leaves, causing them to turn yellow and patchy, often with black specks and slight wrinkling. They also cause streaks in the flowers making them unmarketable. Severe thrips attack in greenhouse is often associated with poor growing conditions resulting from under-watering and overheating. Regular watering and maintenance of a cooler, and humid atmosphere can therefore help preventing its infestation. Most thrips are easily controlled by spraying Endosulfan (0.1 per cent) or Monocrotophos (0.1 per cent). **Bud Borer** (*Helicoverpa armigera* and certain others) caterpillars are mostly a problem of the carnation bud. The eggs are laid in the buds and the larvae eat into the bud and completely damage it. These also feed on other plant parts.

These are easily controlled by early application of *Bacillus thuringiensis*. Regular sprays of Endosulfan (0.05 per cent), Sevin or Polyfrin (0.1 per cent) at forthnightly intervals is quite effective for controlling caterpillars.

Fusarium wilt (*Fusarium oxysporum* f.sp. *dianthi*) is one of the most serious diseases of carnation. Symptoms include wilting of foliage, often only on a few branches followed by death of whole plant. Stem rot below ground level is common due to its infection by which there becomes internal brown streaking. If pulled, the plant breaks off easily while the firm roots remain in the soil. Infected cuttings wilt and die rapidly. The best control measures are soil sterilization or chemical fumigation of the soil, use of pathogen-free plants and general sanitation in the greenhouse. Spraying of plants with Bavistin (0.1 per cent)+Dithane M-45 (0.25 per cent) at forthnightly intervals is quite effective. **Fusarium stem rot** (*Fusarium roseum* f. sp. *cerealis*), a soil-borne disease initially starts by rotting of tissues at the stem surface but afterwards inner tissues also get infected. Incidence increases with high soil moisture content and when temperature is above 25 °C. Copper oxychloride (0.4 per cent) is known to be effective against this pathogen. Spraying of benomyl or Bavistin (0.1 per cent) also controls this disease. **Phlalophora wilt** (*Phlalophora cinerescens*), a soil-borne disease is characterized as uniform dull foliage colouration, light colouring of woody plant parts and loss of cell turgidity which causes plant to die finally. Soil fumigation before planting is must against this problem. Chemical control measures are the same as for *Fusarium* wilt. **Rhizoctonia stem rot** (*Rhizoctonia solani*) is a soil-borne disease. From Italy, Pasini *et al.* (1996) also isolated *Rhizoctonia solani* from carnation causing collar or foot rot. It affects the plant at soil level and wilting and yellowing of foliage is seen, followed by death of entire plant. Stems sometimes show a brown discolouration and cracking just below soil level. Fluffy and light brown fungal hyphae can sometimes be observed on the surface of the rotting tissues. Relative resistance of carnation to *R. solani* is increased by good air circulation, good drainage, shallow planting of cuttings and a low or medium fertility level in the soil. Incidence of disease is reduced by drenching with fungicide (Bavistin or Benomyl @ 2g/l) before planting. **Grey mould** (*Botrytis cinerea*) is the postharvest disease of carnation. Initially a wet and tan-coloured blotch develops on petal tips and spreads rapidly through the petals to produce a fluffy grey mould. This disease can develop on cut flowers while in transit. The disease is favoured by high humidity. It is essential to reduce humidity, and maintain good ventilation and hygiene practices in order to prevent the disease. Rovral @ 0.5g/l or Benlate @ 1.5g/l is registered for control of *Botrytis* on carnations. **Rust** (*Uromyces caryophyllinus* or *U. dianthi*) is observed under high moisture conditions as the long and narrow brown patches on leaves, stems and floral-buds where later on leaves curl and plants stunt. To get rid of this problem, the polyhouse atmosphere will have to be kept dry by not irrigating through overhead system. Bavistin at 0.1 per cent or Dithane M-45 at 0.2 per cent spraying will also control this problem. **Fairy ring spot** (*Cladosporium echinulatum*) is characterised as fairy ring-like spots which grow and merge together resulting in the death of the leaves. Whitish-tan spots up to 5 mm in diameter are surrounded by a narrow red-purple margin. Black pin-head spore masses occur in concentric rings on the surface of the spots. The disease occurs on the leaves, stems and calyces. Spraying of Dithane M-45 (0.2 per cent) or Bavistin (0.1 per cent) regularly at 10 days interval reduces disease incidence. **Alternaria leaf spot** (*Alternaria dianthi*) is characterized by small purple spots which appear on the leaves, stems and occasionally on the flowers. These develop into spots upto 5 mm in diameter with brown centre surrounded by broad purple margin. Spores resembling black specks develop randomly in the centre of spots. Heavily infected leaves may die. Infected branches may be girdled, particularly at the nodes. Temperatures beyond 23.8 °C are known to aggravate the situation. Spraying Dithane M-45 (0.2 per cent) or Bavistin (0.1 per cent) controls this disease. **Bacterial wilt** (*Pseudomonas caryophylli*) is a soil-borne malady entering the plants, especially the old ones, through the wounds, and occurs when night temperatures are 23.8 °C or above. Bazzi *et al.* (1987) reported bacteria *Erwinia chrysanthemi* pv. *dianthicola* and *Pseudomonas woodsii* attacking carnation in Italy. The symptoms of bacterial infection are similar to *Fusarium* wilt and is controlled through 0.01 per cent streptocycline. When the seeds or rooted cuttings of carnation were inoculated with *Pseudomonas fluorescens* strain E6, Yuen and Schroth (1986) recorded increased growth and fresh weight (18-41 per cent greater) after 3-4 weeks of treatment.

Carnation mottle virus infects the carnation plantings. Usually infected plants are symptomless, however, some cultivars can show a mottling pattern on the leaves. It reduces flower quality and production and may cause inconsistent colour integrity in the petals. Other viruses which attack this crop are mosaic, vein mottle, streak, ring spot and etched ring. In China, Duan YongJia and Cai Hong (1998) observed brilliantly variegated carnation flowers infected with tulip breaking potyvirus. Many of these viruses are transmitted through insect vectors and cutting tools. Such plants should be rogued out immediately and there should be full precaution to control any insect found in the polyhouse. Either the cuttings should be taken from the indexed plants or the plants should be raised through micropropagation by using shoot tips. While cutting flowers, sterilized tools should be used.

Calyx splitting is a major problem in standard carnations as flowers are asymmetrical and value is reduced. As the flower bud opens and petals reach to their full size, the calyx may split down either half or completely. Cultivars vary in susceptibility to splitting. Those with short and broad calyces are less likely to split than those with long and narrow calyces. Splitting is attributable to conditions that favour proliferation of petals and accessory flowers or conditions that prevent normal calyx expansion. Low temperature causes an increase in the production of extra growth centres inside the calyx, but the calyx is not able to contain these extra petals or petaloids and split. Rapid temperature drop at night enhances growth centre proliferation. A gradual reduction in night temperature will not cause splitting. Carnation buds are the most sensitive to this rapid temperature drop, approximately 3 to 5 weeks before harvest or when buds are beginning to open. Moreover, wide variations in day and night temperatures also favour splitting. Both the boron and nitrogen levels in carnations are important factors in calyx splitting. Low nitrogen, high ammoniacal nitrogen or low boron levels enhance splitting by attributing to weak calyces. Higher nitrate to ammoniacal nitrogen ratios during low light periods are recommended to reduce splitting. Small rubber bands and clear tape have been used to maintain calyx integrity. Calyces can be banded when the bud shows a small opening. Additionally, chosing cultivars like 'Espana', 'Cabaret', 'Red Corso', 'Pamir' and 'Raggio-de-Sole' that are less prone to splitting will reduce this problem. **Slabside** disorder may arise during cooler periods and the buds do not open evenly so that the petals protrude on one side only. This can be overcome by gradual increase of temperature to the optimum.

References

Bazzi, C., P. Minardi and U. Mazzucchi, 1987. Bacterial diseases of flower and ornamental plants in Italy (Italian). *Informatore Fitopatologico*, **37**(6): 15-24.

Besemer, S T. 1980. Carnation. In: *Introduction to Floriculture* (ed. Larson, R.A.), pp. 49-79. Academic Press Inc., New York.

Bhalla, R, M.H.S. Kumar and R. Jain, 2007. Effect of organic manures and biofertilizers on growth and flowering in standard carnations (*Dianthus caryophyllus* Linn.). *J. Ornam. Hort.*(New Series), **10**(4): 229-234.

Bharathy, P.V., P.C. Sonawane and A. Sasnu, 2004. Effect of different planting media on rooting of cuttings in carnation (*Dianthus caryophyllus* L.). *J. Maharashtra agric. Univ.*, **28** (3): 343-344.

Bhatia, S, Y.C. Gupta, S.R. Dhiman and K.S. Thakur, 2002. Studies on the pulsing and storage of carnation flowers. *J. Ornam. Hort.* (New Series), **5**(2): 24-26.

Bhatia, S, Y.C. Gupta and S.R. Dhiman, 2004. Effect of growing media and fertilizers on growth and flowering of carnation under protected condition. *J.Ornam. Hort.* (New Series), **7**(2): 174-178

Bhatt, R.N. 1989. Carnation. In: *Commercial Flowers* (eds Bose, T.K. and L.P. Yadav), pp. 355-416. Naya Prokash, Calcutta.

Bunt, A. 1973. Effect of season on carnation (*Dianthus caryophyllus* L.). II. Flower Production. *J.Plant Sci.*, **48**:315-323.

Bunt, A C and K.E. Cockshull, 1985. *Dianthus caryophyllus*. In: *Handbook of Flowering*, vol. II (ed.Halevy, A.H., pp. 433-440. CRC Press, Boca Raton, Florida (USA).

Bunt, A.C., M.C. Powell and D.O. Chanter, 1981. Effects of shoot size, number of continuous life cycles and solar radiation on flower initiation in carnations. *Scientia Horticulturae*, **15**: 267-276.

Chandrashekar, S.Y. and G. Gopinath, 2001. Influence of chemicals on the postharvest quality of carnation cut flowers. *Karnataka J.agric. Sci.*, **14**(3): 731-735.

Chavan, S.K., K.B. Jagtap and K.B. Chavan, 2004. Effect of pinching methods on growth and flowering in carnation (*Dianthus caryophyllus*) cv. 'Gaudina'. *J.Maharashtra agric. Univ.*, **29**(3): 350-351.

Chikkasubbanna, V. and R. Sharada, 2002. Effect of floral preservatives on the longevity of cut carnations. *Crop Research*, **23**(2): 357-361.

Dalal, S.R., D.R. Nandre, S.G. Bharad, U. Swarupa and R.D. Shinde, 2006. Effect of pinching on carnation cv. Yellow Solar under polyhouse conditions. *International J. agric. Sci.*, **2** (2): 356-357.

Dole, J.M., W.C. Fonteno and S.L. Blankenship, 2005. Comparison of silver thiosulfate with 1- methylcyclopropene on 19 cut flower taxa. *Acta Hort.* (*Proceedings of the Fifth International Postharvest Symposium*, eds Mencarelli, F. and P. Tonutti), held at Verona, Italy, on 6-11 June, 2004; No. 682 (vol. 2), pp. 949-956.

Duan YongJia and Cai Hong, 1998. Studies on the variegated flowers of flowering plants (Chinese). *Acta Phytopath. Sinica*, **28**(1): 85-89.

Galbally, J. and E. Galbally, 1997. *Carnations and Pinks for Garden and Greenhouse*, 310 p. Timber Press, Portland (Oregon, USA).

Gupta, Y.C. and S.K. Bhattacharjee, 2002. *Carnation* (Tech. Bullettin No.15). AICRP on Floriculture, ICAR, New Delhi.

Hlatshwayo, M.S. and P.K. Wahome, 2010. Effect of shading on growth, flowering and cut flower quality in carnation (*Dianthus caryophyllus*). *J. agric. Social Sci.*, **6**(2): 34-38.

Holley, W.D. and R. Baker, 1992. Breeding for better varieties. In: *Carnation production* (eds Holly, W.D. and R. Baker), pp. 21-30. Colorado State University, USA.

Khaligi, A. and M.R. Shafie, 2000. Effects of chemical and temperature treatments and harvesting stages on cut flower longevity and some other characteristics of carnation (*Dianthus caryophyllus* L.). *Iranian J. agric. Sci.*, **31**(1): 119-125.

Krishnappa, K.S., N. Shivareddy and M. Anjanappa, 2000. Effect of floral preservatives on the vase life of carnation cut cultivars. *Karnataka J. agric. Sc.*, **13**(2): 395-400.

Kumar, A., M.S. Verma, S.K. Lodhi and S.K. Tripathi, 2006. Effect of growth chemicals, type of cutting and season on root formation of carnation (*Dianthus caryophyllus* L.) cutting. *International J. agric. Sci.*, **2**(2): 596-598.

Kumar, J., Mira Amin and P.V. Singh, 2003. Effect of Mn and Zn sprays on carnation. *J. Ornam. Hort.* (New Series), **6**(1): 83.

Kumar, P.N., B. Singh and S.R. Voleti, 1999. Effect of chemicals on the vase life of standard carnations. *J.Ornam. Hort.* (New Series), **2**(2): 139-140.

Kumar, R., K. Singh and B.S. Reddy, 2002. Effect of planting time, photoperiod, GA_3 and pinching on carnation. *J. Ornam. Hort.* (New Series), **5**(2): 20-23.

Lamont, G. 1986. Herbicide evaluation trials on cut flowers. *Australian Hort.*, **84**(7): 89-91.

Langhans, R.W. 1961. *Carnation - A Manual of Culture, Insect and Diseases and Economics of Carnation.* New York State and County Agricultural Extension Services, Ithaca, New York.

Lolapori, N. and J.S. Arora, 1995. Response of Sim carnations (*Dianthus caryophyllus* L.) to photoperiodsm (exetended day length) and gibberellic acid. *J. Ornam. Hort.*, **3**(1-2): 14-22.

Luo Hongyi, Jing Hong Juan, Li Ju Rong and Luo Sheng Rong, 2003. Effect of different preservatives on freshness of cut carnation flowers. *Plant Physiology Communications*, **39**(1): 27-28.

Macnish, A.J., R.T. Leonard and T.A. Nell, 2008. Treatment with chlorine dioxide extends the vase life of selected cut flowers. *Postharv. Biol. & Tech.*, **50**(2/3): 197-207.

Misra, Sanyat; Pragya; and R.L. Misra, 2006. Carnation. In: *Advances in Ornamental Horticulture* (eds Bhattacharjee, S.K.), pp. 66-80. Pointer Publishers, Jaipur.

Nau, J. 1993. *Ball Culture Guide, The Encyclopedia of Seed Germination* (2nd ed.). Ball Publishing, Batavia, Illinois.

Nowak, J. and R.M. Rudnicki, 1975. The effect of "Proflovit-72" on the extension of vase-life of cut flowers. *Prace Instytutu Sadownictea w Skierniewicach, B,* **1**: 173-179.

Onozaki, T., H. Ikeda, T. Yamaguchi, and M. Himeno, 1998. Introduction of bacterial wilt (*Pseudomonas caryophylli*) resistance in dianthus wild species to carnation. In:: *Proc. Third intern. Symp.on New Floricultural Crops* (ed. Considine, J.A.), held in Perth (Australia), Oct. 1-4, 1996. *Acta Horticulturae* (vol. III):*New Floricultural Crops*, No. 454, pp. 127- 132.

Pal, A.K. and B. Biswas, 2004. Effect of pinching methods and fertilization on growth and flowering of standard and spray carnations under open field condition. *Hort. J.*, **17** (2): 175-180.

Pasini, C., T. Berio, P. Curir and F. D'Aquila, 1996. Further characterization of *Rhizoctonia solani* isolated from carnation and other ornamental plants. *Informatore Fitopatologico*, **46**(6): 33-36.

Pathania, N.S., O.P. Sehgal and Y.C. Gupta, 2000. Pinching for flower regulation in Sim carnations. *J. Ornam. Hort.* (New Series), **3**(2): 114-117.

Raja Ram and A.A. Zaidi, 1999. Meristem tip culture and carnation mottle virus tested on perpetual flowering carnation. *Indian J.Virol.*, **15**(1): 43-46.

Rao, K.U.M., R.C. Sekhar, J.D. Babu and M.R. Kumar, 2008. Effect of pinching at different days after planting on flowering behaviour in three cultivars of carnation (*Dianthus caryophyllus* Linn.) *J.Res.*, ANGRAU, **36**(1): 30-35.

Reid, M.S., A.M. Kofranek and S.T. Besemer, 1983. Postharvest handling of carnations. *Acta Hort.*, No. 141, pp. 235-238.

Sato, S., N. Katoh, H. Yoshida, S. Iwai and M. Hagimori, 2000. Production of double haploid plants of carnation (*Dianthus caryophyllus* L.) by pseudo-fertilized ovule culture. *Scientia Hort.* **83**: 301-310.

Sarkar, I. and N. Roychowdhury, 2003. Effect of nitrogen and phosphorous on the growth and flowering of carnation cv. Chabaud Mixed under open conditions. *Environment and Ecology*, **21**(2): 696-698.

Segers, A. 1987. The development of interspecific carnation hybrids. *Acta Hort.*, No. 216, pp. 373-375.

Serek, M., E.C. Sisler and M.S. Reid, 1995. Effects of 1-MCP on the vase life and ethylene response of cut flowers. *Pl. Gr. Reg.*, **16**(1): 93-97.

Sharaf, A.I. and A.H. El- Naggar, 2003. Response of carnation plant to phosphorous and boron foliar fertilization under greenhouse conditions. *Alexandria J. agric. Res.*, **48**(1): 147-158.

Singh, B.K. and J.N. Singh, 2003. Effect of gibberellic acid and cycocel on growth and flowering of carnation. *New Agriculturist*, **14**(1/2) : 53-55.

Singh, K., P.J. Singh, J.S. Arora and R.P.S. Mann, 2002. Studies on refrigerated storage of carnation flowers. *Proc. Nat. Symp.Indian Flor. in the New Millenium*, Lal Bagh, Bangalore, February 25-27, pp. 303-304.

Singh, K., P. Singh and M. Kapoor, 2007. Effect of vase and pulsing solutions on keeping quality of standard carnation (*Dianthus caryophyllus* Linn.) cut flowers. *J. Ornam. Hort.* (New Series), **10**(1): 20-24.

Singh, R., R. Kumar and K. Singh, 2005. Effect of pinching and nitrogen application on growth and flower production in carnation (*Dianthus caryophyllus* Linn.) cv. Tasman. *J. Ornam. Hort.* (New Series), **8**(3): 239-240.

Staby, G.L., J.L. Robertson, D.C. Kipliger and C.A. Conover, 1978. *Chain of Life*. Ohio Florists Association, Ohio State Univ., Columbus.

Thomas, L.M., S. Gonsalves, T. Mondal and N. Roychowdhury, 2003. Effect of different media on rooting of cuttings of carnations cv. Mixed Super Chabaud. *J.Interacademicia*, **7**(3): 262-264.

van der Hoeven, A P. 1987. The influence of daylength on flowering of carnation. *Acta Hort.*, No. 216, pp. 315-323

Verma, V. K. 2003. Response of foliar application of nitrogen and gibberellic acid on the growth and flowering of carnation (*Dianthus caryophyllus* L). *Himachal J. agric. Res.*, **29** (1/2): 59-64.

Verma, V. K., O.P. Sehgal and Y.C. Gupta, 2000. Effect of nitrogen and GA_3 on carnation. *J. Ornam. Hort.* (New Series), **3**(1): 64.

Verma, V.K., Y.D. Sharma and Y.C. Gupta, 2002. Effect of holding solutions under different planting dates and nitrogen levels on vase life of cut carnations. *Proc.Nat. Symp. Indian Flor. in the New Millenium,* Lal Bagh, Bangalore, February 25-27, pp. 169-172.

Verma, V.K., Y.D. Sharma and Y.C. Gupta, 2003. Response of carnation to foliar application of nitrogen. *J. Ornam. Hort.* (New Series), **6**(2): 89-94.

Wankhede, M., S. Patil and L. Komma, 2006. *In vitro* propagation of carnation cv. Supergreen. *J. Soils and Crops*, **16**(1): 165-169.

Yuen, G.Y. and M.N. Schroth, 1986. Interactions of *Pseudomonas fluorescens* strain E6 with ornamental plants and its effect on the composition of root-colonizing microflora. *Phytopath.*, **76**(2): 176-180.

15

Digitalis [Family: Plantaginaceae (Olmstead *et al.*, 2001), formerly: Scrophulariaceae]

Namita, Sapna Panwar, Meenu Kumari, Shisa Ullas P. and Sanyat Misra

[**Common names**: Foxglobe/Purple foxglobe (*Digitalis purpurea*), Common foxglobe/Fairy globe/Finger flower (*D. tomentosa*), Yellow foxglobe (*D. ambigua* syn. *D. grandiflora*), Hindi- Tilpushpi].

Introduction

Digitalis has derived from the Latin word *digitale* meaning 'finger of a glove' which refers to the ease with which a flower can be fixed over a human finger tip, first referred by the German writer Leonhart Fuchs (Fuchsius) during 16th century in his ***Historia Plantarum***, hence, its specific name becomes *D. fuchsii*. The genus is native to W&SW Europe, W&C Asia, Australasia and NW Africa. Geographically, it is mainly distributed throughout the Iberian Peninsula, North-Western Africa, Macaronesia and Balkan Peninsula, and Asia Minor. Some 25 species are found in the genus *Digitalis*. Three species *Digitalis purpurea, D. grandiflora* and *D. lutea* were found in Germany whereas *D. nervosa, D. ciliata* and *D. ferruginea* ssp. *schischkinii* in the Caucasian Mountains. The geographical distribution of the species coincides with their taxonomic relationship. The crop is being cultivated in France, Germany, United Kingdom, Hungary, USA and other Asian countries for medicinal uses. It was introduced as an ornamental plant in North America, Canada, Mexico, Central America and Asia. The Northern states and southern hilly areas of India have vast potential to produce sufficient quantity of seeds and leaves to meet the demand of pharmaceutical industries (Farooqi and Sreeramu, 2004).

The species *Digitalis purpurea* and *D. lanata* are used in life-saving medicines due to its glucoside content. The leaves of *Digitalis purpurea* are rich source of digitoxin, digitalin and gitalin. *D. lanata* is the source of acetyldigoxin, deslanoside, digoxin - an active cardiac glycoside and lanatosides A, B & C (Launert, 1981; Grive, 1984). *Digitalis* is used mainly for its effect on cardiovascular system. It is used as myocardial stimulant in congested heart failure, auricular flutter and rapid auricular fibrillation. It is also more selective in preserving the cells severely injured by heat.

Many varieties of *Digitalis purpurea, D. lanata, D. ferruginea, D. grandiflora, D. lutea* and *D. thapsi* are used as ornamental and cultivated worldwide as garden herbs due to their attractive and appreciated inflorescences (Duke, 2002). *Digitalis purpurea* is cultivated in several countries of Europe, Asia and America whereas *D. lanata* in Europe, India, Nepal and Brazil.

Botany

All *Digitalis* sp. are biennial or perennial herbs with erect stems that grow 91-183 cm tall. It is rarely small shrubs with simple alternate leaves crowded in basal rosettes. Basal leaves are up to 30 cm long and 5 cm wide. They have toothed margins and are soft-hairy above. Foxglove grows as a rosette during the first year and in 2nd year produces leafy stock with bell-shaped flowers on a spike. The inflorescence in all the species is racemose but the pedicels are very short in *D. lanata*. Flowers are zygomorphic, pentamerous, hypogynous, bisexual with 4-5 partite bracts and arranged in terminal, bracteate racemes and vary from purple to pink, white and yellow. Flowers are 4 to 6 cm long with dark spots on their lower interior surfaces. The calyx is of five equal lobes, shorter than corolla tube and the sepals gamosepalous. The corolla with a cylindrical-tubular to globose tube is constricted at the base and the limb is more or less two-lipped. The upper lip is shorter than the lower which is spotted or veined inside. There are four petals which are gamopetalous, 2-lipped, sometimes spurred and with imbricate aestivation. There are 2-4 stamens. The anthers are bithecate, hairy, dehiscing by longitudinal slits and epipetalous. The ovary is bicarpellary, syncarpous, superior with many ovules and single and bilobed style. In *D. lutea* and *D. lanata,* the long stamens are within 0.5 mm of the corolla rim and may even protrude slightly at the time of dehiscence. In both the species, the anthers become soaked if it rains. The anthers and stigmas of *D. purpurea, D. amandiana* and *D. ambigua* are contained within the corolla bell and remain dry during rain. Fruits are ovoid and approximately 13 mm long with many minute seeds (Harris, 2000; Whitson *et al.*, 2000).

Cytological studies revealed that *Digitalis* has 2n=56 chromosome numbers. Its petal colours are known to be determined by at least three genes that interact with each other. The M gene determines the production of a purple pigment, a type of anthocyanin though not the m gene. The D gene is an enhancer of the M gene, and leads it to produce a big amount of the pigment though d gene does not enhance the M gene, and only a small amount of pigment is produced. Lastly, the W gene makes the pigment deposited only in some spots, while the w gene allows the pigment to be spread all over the flower.

Four *Digitalis* Genotypes.

By applying biotechnological tools, Lehmann *et al.* (1995) developed transgenic plants in *D. lanata* and *D. minor*; haploid cultures were standardised in *D. lanata* (Diettrich *et al.*, 2000), *D. obscura* (Perez-Bermudez *et al.*, 1985a,b) and *D. purpurea* (Corduan and Spix, 1975); and the protocols were developed in cryopreservation of *D. lanata* (Diettrich *et al.*, 1986a,b; 1987), *D. obscura* (Sales *et al.*, 2001) and *D. thapsi* (Moran *et al.*, 1999).

Propagation

Purple foxglove reproduces by **seeds** that have a viability of two years of harvesting, however, can remain viable in the soil for at least five years (Harris, 2000). Seed germination will be high at 21-26 °C temperatures and does not require cold scarification (Anon., 2002) for germination. About 1,500 seeds weigh one gramme. The seeds should be presprouted to hasten germination by soaking them in water and incubating them at about 30 °C for 2-3 days. About 5-8 kg of seeds is required for direct seeding one hectare area, while for nursery raising just 1-2 kg of seeds are sufficient. Outside the seeds are sown from April to June in seed beds and when germinated and have attained substantial height the seedlings are pricked out into nursery beds and planted in their permanent position during October-November or in following spring some 60 cm apart. Seeds can also be sown directly into their permanent position in spring and to space them properly, these are thinned out afterwards. These also naturalize easily if after flowering the plants are cut back, many basal rosettes appear that will produce the flowers in the following season. Various organs of various species of *Digitalis* were raised successfully through micropropagation such as culture of hairy roots in *D. lanata* (Pradel *et al.*, 1997) and *D. purpurea* (Saito *et al.*, 1990); shoot culture in *D. lanata* (Schoner and Reinhard, 1982; Diettrich *et al.*, 1990), *D. obscura, D. purpurea* (Hagimori *et al.*, 1982b) and *D. thapsi* (Herrera *et al.*, 1990); shoot organogenesis in *D. lanata* (Luckner and Diettrich, 1988; Rucker, 1988), *D. obscura* (Perez-Bermudez *et al.*, 1985a,b), *D. purpurea* (Rucker *et al.*, 1981; Hagimori *et al.*, 1982a,b) and *D. trojana* (Corduk and Aki, 2010); cell culture in *D. lanata* (Nickel and Staba, 1997) and *D. purpurea* (Pilgrim, 1977); somatic embryogenesis in *D. lanata* (Kuberski *et al.*, 1984; Reinborthe *et al.*, 1990) and *D. obscura* (Arrillaga *et al.*, 1987); and protoplast-derived plants in *D. lanata* (Diettrich *et al.*, 1982), *D. obscura* (Brisa and Segura, 1987) and *D. purpurea* (Diettrich *et al.*, 1980). Explants such as shoot tips and leaves in *D. minor* (Sales *et al.*, 2002), shoot tips, leaves, hypocotyls and roots in *D. thapsi* (Cacho *et al.*, 1991) and only leaves in *D. trojana* (Corduk and Aki, 2010) were successfully cultured *in vitro* and micropropagated plants were obtained.

Classification, Species and Varieties

Werner (1961, 1965) divided the genus into five sections on the basis of morphological characterization. These sections are Frutescentes, Digitalis, Grandiflorae, Tubiflorae and Globiflorae. The genus *Digitalis* comprises approximatey 23-25 species including four species of the former genus *Isoplexis* (Herl *et al.*, 2008) as mentioned here below.

Werner (1965)	*Brauchler et al. (2004)*
Digitalis atlantica Pomel	*Digitalis atlantica*
D. cariensis Boiss	*D. cariensis*
D. ciliata Trautv.	*D. ciliata*
D. davisiana Heyw	*D. davisiana*
D. ferruginea L.	*D. ferruginea*
D. grandiflora Mill.	*D. grandiflora*
D. heywoodii P. et. M. Silva	*D. heywoodii*
D. laevigata Waldst	*D. laevigata*
D. lanata Ehrh	*D. lanata*
D. lutea L.	*D. lutea*
D. mariana Boiss	*D. mariana*
D. dubia Rodr.	*D. minor*
D. nervosa Steud	*D. nervosa*

Werner (1965)	Brauchler et al. (2004)
D. obscura.L	*D. obscura*.
D. parviflora Jacq.	*D. parviflora*
D. purpurea L.	*D. purpurea*
D. subalpina Br. - Bl.	*D. subalpina*
D. thapsi L.	*D. thapsi*
D. viridiflora Lindl.	*D. viridiflora*
Isoplexis isabelliana (Webb & Berthel.) Morris	*D. isabelliana* (Webb) Linding
I. canariensis (L.) Loudon	*D. canariensis* L.
D. sceptrum (L.) Loudon	*D. sceptrum*
I. chalcantha Svent. & O'Shan.	*D. chalcantha* (Svent. & O'Shan.) Albach, Bruchler & Heubl

The details regarding some of the most important *Digitalis* species are furnished below (Bailey, 1942; Everard and Morley, 1970; Hay and Beckett, 1971; Pizzetti and Cocker, 1975; Beckett, 1983; Brickell, 1994).

Digitalis ambigua (*D. grandiflora, D. ochroleuca*)**.** A native to generally dry meadowland of Central Europe, France, Belgium, the Balkans, C&S Russia and W. Siberia, it is a fully hardy, clump-forming, evergreen herbaceous biennial to perennial with slight hairs, growing up to 100 cm and producing more than one spike per plant. Finely dentate, ovate-lanceolate and lustrous mid-green leaves are hairy at the margins and along the veins, lower leaves are highly short-stalked while upper ones sessile with base overlapping the stem. Pendulous sulphur-yellow blooms netted with brown and some 4.5 cm long appear to one-side of 50 cm long spike during June to August and the flowers are marked and veined brown internally.

Digitalis dubia. A native of Balearic Islands, this half-hardy herbaceous perennial species appears quite similar to *D. purpurea* though quite smaller with only 15-25 cm tall or rarely even up to 50 cm tall stems. Its leaves are mostly basal, oblong-lanceolate and white-haired beneath. Its spikes are loose and few-flowered, flowers pale purple-rose and 3-4 cm long, spotted dark internally

Digitalis ferruginea. A native of S. Europe, especially Italy eastwards, this is a perennial species but best treated as a biennial species growing up to 1.5 m tall bearing leafy spikes that emerge from among the rosette of mid-green and lanceolate leaves that may either be smooth or fringed with hairs. Many yellowish or reddish-brown funnel-shaped flowers 2.5-3.5 cm long with a network of darker veins on some 60 cm long spikes appear during July.

Digitalis lanata. A native along the Danube River and Greece, this fine biennial to perennial species grows up to 90 cm tall, bearing ciliate, oblong or lanceolate leaves. Floral racemes are dense and many-flowered, and the flowers some 2.5-3.75 cm long, creamish-yellow to greyish or rarely whitish or purplish and downy appear during July and August.

Digitalis lutea. A native of SW Europe, NW Africa, this fully hardy herbaceous perennial species grows up to 90 cm tall bearing erect tapering stem and glossy mid-green, lanceolate and finely serrated leaves and some 2 cm long, narrowly tubular, and downward-pointing creamy-yellow flowers with a pointed tongue-like lower lobe.

Digitalis* × *mertonensis. A fully hardy, clump-forming, true breeding and tetraploid perennial hybrid between *D. grandiflora* and *D. purpurea* developed at John Innes Institute, this grows to a height of up to 90 cm bearing mid-green lanceolate leaves and about 2.5-5.0 cm long, downward-pointing, tubular and rose-mauve to pinkish flowers from June to September. It is much like to *D. purpurea* but is more compact.

Digitalis purpurea (*D. tomentosa*)**.** A native to the W&C Europe as far north as Scandinavia, it is a hardy herbaceous biennial or short-lived perennial growing up to 1.8 m high, where entire plant is covered with white hairs. It is most suitable for cutting. Leaves are rich green, oblong, rough and wrinkled, several basally in the rosette that are ovate-lanceolate and long-stalked while the stem leaves have shorter stalks. Flowering occurs from May to August on erect, dense and tall stem some 90 cm long only to one side with downward inclined bell-shaped spotted flowers from pink to pinkish-purple, maroon or rarely white some 6.25 cm long in the true species. Its var. 'Alba' is white, 'Maculata Superba' with bold corolla markings, 'Gloxiniaeflora' with greater range of tints such as white, yellow, pink, salmon and carmine, *vis-à-vis* internally brown striped, 'Monstrosa' ('Campanulata') with large and double flowers, *etc.* Its race 'Excelsior Hybrids' has distribution of flowers all around the compact stem in perpendicular form. 'Foxy' strain is similar to 'Excelsior' but not as tall as it is only up to 75 cm and produces up to nine shoots per plant. 'Shirley Hybrids' are similar to the type species but more taller and the flowers range from various shades of white to red and purple.

Digitalis sibirica. A native to Siberia, this perennial species in habit is similar to *D. ambigua* but the flowers are similar to those of *D. lanata*. Its leaves are downy, ovate-lanceolate, upper entire though lower serrate. Flowers are yellowish, villose and ventricose. This species is rarely cultivated.

Digitalis thapsii. A native to Spain and Portugal, this perennial species is much like *D. purpurea* but is covered with yellowish glandular hairs. Height is up to 1.2 m, leaves are ovate-lanceolate, or oblong, rugose and decurrent. It produces purple flowers from June to September with paler throat and internally with red spots.

The strain EC 115996, D-76, D-21, DPF and DYF have been reported to give higher foliage yield and high glucoside content. These strains are utilized in many hybridization and selection programmes for further identification of promising lines.

Cultural Practices

Digitalis purpurea is adapted to fine and medium-textured **soils** with a pH range of 5.5-7.0. It prefers semi-shady position with well-drained, acidic, light, sandy loam soil rich in organic matter but hates lime. Water-logging and too-drained soil conditions and clay and sandy soils are detrimental to crop.

Digitalis purpurea requires acidic soils for better growth whereas *D. lanata* needs lime rich soils for proper growth and glucoside content in the leaves. *Digitalis* prefers warmer and humid **climate** with a range of temperature from 20-30 °C. *Digitalis purpurea* and *D. lanata* are commercially cultivated in India. It requires 190 frost-free days for successful growth and reproduction. Purple foxglove can withstand temperatures as low as -25 °C. It is shade-intolerant (Anon., 2002). Seed germination, growth and flowering and content of active principles are maximum at 20-30 °C temperature range. Well distributed rainfall and sunshine during winter is ideal for luxuriant growth.

Field preparation should be done by ploughing followed by planking thoroughly 4-5 times using a disk and harrow. Weed's rootstocks, plastic pieces, brick and stones and any other undecompsing material should be removed at the time of preparation of the field and the soil should be brought to a fine tilth. At the time of second ploughing farmyard manure at the rate of 40-60 tonnes per hectare of land should be incorporated and thoroughly mixed in the soil. When the soil is fully prepared and there is sufficient moisture, the beds of convenient sizes should be made, preferably with a 1.6 m width though length as per convenience and depending upon the soil level. The best **time of planting** is mid-April in cooler regions and July-August in hilly areas of southern states. Seeds can directly be sown in the main field at a depth of not more than 2 cm to avoid failure of seed germination. A **spacing** of 45 cm between rows and 30 cm between plants in a row is maintained.

Raised **nursery beds** of 3 m long, 1 m wide and 15 cm high are prepared in a soil enriched with organic matter To each bed, 3-4 baskets of well-decomposed farmyard manure is incorporated and land is brought to a fine tilth. The seeds are sown in lines 4-6 cm apart. The surface of bed is covered with a 1 cm layer of sand. The beds are irrigated lightly every evening with rose-fitted cans. The seeds germinate in 15-20 days and seedlings will be ready for transplanting in 35-45 days after sowing when they have attained a height of 8-10 cm. The field is kept ready by preparing ridges and furrows at a 45-60 cm distance.

Nitrogen 100-150 kg, **P_2O_5** 50 kg and **K_2O** 25-30 kg per hectare is optimum for better growth of the crop. One fourth of N along with full dose of P and K should be applied at the time of planting. The remaining N is applied in 3-split doses, *i.e.* first top-dressing at 60 days after transplanting, and 2nd and 3rd doses at 1.5 to 2 month intervals thereafter. Immediately after transplanting the crop is thoroughly **watered** to settle the roots. During summer season, *i.e.* April to July if there is no rain, the crop is watered at an interval of 7-10 days, then 2-3 irrigations will be sufficient each month until the dormant period sets in. **Weeds** compete with crop for light, moisture and nutrients and also act as host for many diseases and pests, causing drastic losses to the crop. During rainy season, hand weeding is preferred. **Hoeing** should be done during summer and after each harvest. At least two hoeings before the first harvest and another afterwards are beneficial to the crop.

Growth and Flowering

Once flower induction and initiation have occurred, successful flower development must follow for proper flowering. This can require specific environmental conditions. Meeting the photoperiodic requirement is must for successful flower development (Anderson, 2007). In addition to photoperiod, irradiance can affect earliness of flowering of many species (Armitage and Tsujita, 1979; Armitage *et al.,* 1981; Zhang *et al.,* 1996; Erwin *et al.,* 1997). In general, increased irradiance reduces the length of the juvenile period with many plant species. In contrast, low irradiance can extend the length of the juvenile period with some species. Fausey *et al.* (2001) noted that 100 per cent flowering in *Digitalis purpurea* 'Foxy' was only reached with DLI greater than 11 µmol m-2 d-1.

Postharvest

In Digitalis, the flowering period lasts for about six weeks. The best time for spike harvesting is the coolest part of the day and when there is no surface water from dew or rain on the plants. Optimal stage of development for harvesting the fresh cut flowers of *Digitalis purpurea* is when one-half florets open. Pollinated flowers produce ethylene gas more rapidly so flowers die at faster rate. Foxgloves are sensitive to ethylene and require postharvest treatment to prevent damage. After cutting the *Digitalis* stems, the foliage touching the vase water or below the water line are sheared off. Thereafter, dip or place the stems into a hydration solution. After this treatment, stems are placed into a clean container. Flowers should be stored upright at temperatures of 1-2 °C. These tall linear flowers are geotropic, *i.e.* respond to gravity and will begin to curve upwards if stored diagonally or horizontally. Vase life varies with cultivar and generally, they will last 5- 10 days with optimum care and handling. *Digitalis* absorbs more water, therefore, water levels should be checked frequently.

Insect-Pests, Diseases and Physiological Disorders

Anthracnose (*Colletotrichum fuscum*) appears on leaves as light or purplish brown, circular or somewhat angular spots bordered with purplish margins. Initially the spots are smaller but later on become larger in diameter with numerous small black fungal bodies visible at the centre of the spot. In its severe attack, the seedlings damp off. **Leaf spot** (*Ramularia variabillis*) occurs on lower leaves more frequently than upper leaves and causes formation of brown irregular spots with reddish margins of 1.5 cm diameter. Heavy infection causes drying and death of entire leaf. The development of conidia/ spores on leaves gives a white mouldy appearance. **Root and stem rot** (*Fusarium* sp., *Pellicularia filamentosa, P. rolfsii, Sclerotinia sclerotiorum*) and **wilt** (*Verticillium alboatrum*) have also been sometimes observed infecting foxgloves. **Tobacco mosaic virus** occurs occasionally on foxglove, especially on the plants of *Digitalis lanata*. **Beet curly top virus** causes leaf curling, *vis-a-vis* stunting which is transmitted through beet leafhopper (*Circulifer tenellus*). Against viruses, infected plants should be rogued out and burnt and vectors controlled by spraying malathion 0.2 per cent. Leaf spot diseases should be controlled by spraying 0.2 per cent dithane Z-78. Spraying and/ or drenching of 0.2 per cent thiram will control rots.

Aphids (*Acyrthosiphon solani* and *Neomyzus circumflexus*) infest the foxgloves which can be controlled by spraying 0.2 per cent malathion. **Beetles** (Asiatic garden, Japanese beetles and rose chafer) feed on foxglove leaves and flowers mostly after sunset so spraying malathion during night will kill these. **Mealybug** (*Pseudococcus fragilis*) suck cell sap from stems and the leaves. During summer when temperatures are high, introduction of natural ladybird predator (*Cryptolaemus montrouzieri*) will reduce its population. Spraying with insecticidal soap or malathion is quite effective. **Stem and bulb nematode** (*Ditylenchus dipsaci*) infestation causes angular leaf spots. In moist conditions, nematodes may enter the stomata and cause leaf spots similar to those caused by leaf nema, *Aphelenchoides ritzima-bosi*. Such leaves should be picked up and destroyed.

References

Anderson, Neil O. 2007. *Flower Breeding and Genetics - Issues, Challenges and Opportunities for the 21st Century*, 824 p. University of Minnesota, St. Paul, Minnesota, U.S.A. Published by Springer, P.O. Box 17, 3300 AA Dordrecht, The Netherlands.

Anonymous, 2002. The PLANTS Database, Version 3.5. USDA, Natural Resource Conservation Service (http://plants.usda.gov). National Plant Data Center, Baton Rouge, LA 70874-4490 USA.

Arillaga, I., M.C. Brisa and J. Segura, 1987. Somatic embryogenesis from hypocotyls callus cultures of *Digilalis obscura* L. *Plant Cell Rep.*, **6**: 223-226.

Armitage, A.M. and M.J. Tsujita, 1979. The effect of supplemental light source, illumination and quantum flux density on the flowering of seed-propagated geraniums. *J. hort. Sci.*, **54**(3): 195-198.

Armitage, A.M., W.H. Carlson and J.A. Flore, 1981. The effect of temperature and quantum flux density on the morphology, physiology, and flowering of hybrid geraniums. *J. Amer. Soc. hort. Sci.*, **106**(5): 643-647.

Brisa, M.C. and J. Segura, 1987. Isolation, culture and plant regeneration from mesophyll protoplasts of *Digitalis obscura*. *Physiol. Plant.*, **69**:680-686

Cacho, M., M. Moran, M.T. Herrera, J. Fernandez-Tarrago and P. Corchete, 1991. Morphogenesis in leaf, hypocotyls and roots explants of *Digitalis thapsi* L cultured *in vitro*. *Plant Cell Tiss. Org. Cult.*, **25**: 117-123.

Corduan, G. and C. Spix, 1975. Haploid callus and regeneration of plants from anthers of *Digitalis purpurea* L. *Planta*, **124**: 1-11.

Corduk, N. and C. Aki, 2010. Direct shoot organogenesis of *Digitalis trojana* Ivan., an endemic medicinal herb of Turkey. *Afr. J. Biotechnol.*, **9**: 1587-1591.

Diettrich, B., S. Ernst and M. Luckner, 2000. Haploid plants regenerated from androgenic cell cultures of *Digitalis lanata*. *Planta Med.*, **66**: 237-240.

Diettrich, B., U. Haack and M. Luckner, 1986b. Cryopreservation of *Digitalis lanata* cells grown *in vitro*. Pre-cultivation and recultivation. *J. Plant Physiol.*, **126**: 63-73.

Diettrich, B., H. Mertinat and M. Luckner, 1990. Formation of *Digitalis lanata* clone lines by shoot tip culture. *Planta Med.*, **56**: 53-58.

Diettrich, B., D. Neuman and M. Luckner, 1980. Protoplast derived clones from cell cultures of *Digitalis purpurea*. *Planta Med.*, **38**: 375-382.

Diettrich, B., D. Neuman and M. Luckner, 1982. Clonation of protoplast-derived clones from cell- derived cells of *Digitalis lanata* suspension cultures. *Biochem. Physiol. Pflanzen.*, **177**: 176-183.

Diettrich, B., C. Steup, D. Neumann, H. Scheibner, C. Reinbothe and M. Luckner, 1986a. Morphogenetic capacity of cell strains derived from filament, leaf and root explants of *Digitalis lanata*. *J. Plant Physiol.*, **124**: 441-453.

Diettrich, B., T. Wolf, A. Borman, A.S. Popov, R.G. Butenko and M. Luckner, 1987. Cryopreservation of *Digitalis lanata* shoot tips. *Planta Med.*, **53**: 359-363.

Duke, J.A. 2002. *Handbook of Medicinal Herbs* (2nd ed.). CRC Press, Boca Raton, FL, USA.

Erwin, J.E., R. Warner, T. Smith and R. Wagner, 1997. Photoperiod and temperature interact to affect *Petunia x hybrida* Vilm. development. *HortSci.*, **32**: 501.

Farooqi, A.A. and B.S. Sreeramu, 2004. *Cultivation of Medicinal and Aromatic Crops*, p. 506. Universities Press (India) Pvt. Ltd., Hyderabad.

Fausey, B.A., A.C. Cameron and R.D. Heins, 2001. Daily light integral, photoperiod, and vernalization affect flowering of *Digitalis purpurea* L. 'Foxy'. *HortSci.*, **36**(3): 565.

Grive, M, 1984. *A Modern Herbal*. Penguin Books,London,UK.

Hagimori, M., T. Matsumoto and Y. Obi, 1982a. Studies on the production of *Digitalis* cardenolides by plant tissue culture. 2.Effect of light and plant growth substances on digitoxin formation by undifferenciated cells and shoot forming cultures of *Digitalis purpurea* L. grown in liquid media.*Plant Cell Physiol.*, **23**: 1205-1211.

Hagimori, M., T. Matsumoto and Y. Obi, 1982b. Studies on the production of *Digitalis* carotenoids by plant tissue culture. 3.Effects of nutrients on digitoxin formation by shoot forming cultures of *Digitalis purpurea* L.grown in liquid media. *Plant Cell Physiol.*, **23**: 1205-1211.

Harris, S.A. 2000. *Digitalis purpurea* L. In: *Invasive Plants of California's Wildlands* (eds Bossard, C.C., J.M. Randall and M.C. Hoshovsky), pp. 158-161. University of California Press, USA.

Herl, V., D.C. Albach, F. Muller-Uri, C. Brauchler, G. Heubl and W. Kries, 2008. Using progesterone 5β-reductase, a gene encoding a key enzyme in the cardenolide biosynthesis, to infer the phylogeny of the genus *Digitalis*. *Plant Syst. Evol.*, **271**: 65-78.

Herrera, M.T., M. Cacho, M.P. Corchete and J. Fernandez-Tarrago, 1990. One step shoot tip multiplication and rooting of *Digitalis thapsi* L. *Plant Cell Tiss. Org. Cult.*, **22**: 179-182.

Kuberski, C., H. Scheibner, D. Steub, B. Diettrich and M. Luckner, 1984. Embryogenesis and carotenoid formation in tissue cultures of *Digitalis lanata*. *Phytochem.*, **23**: 1407-1412.

Launert, E. 1981. Edible and medicinal plants. Hamlyn, London, UK.

Lehmann, U., D. Moldenhauer, S. Thomar, B. Diettrich and M. Luckner, 1995. Regeneration of plants from *Digitalis lanata* cells transformed with *Agrobacterium tumifaciens* carrying bacterial genes encoding neomycin phosphotransferase II and β-glucuronidase. *J. Plant Physiol.*, **147**: 53-57.

Luckner, M. and B. Diettrich, 1988. *Carotenoids ~ Phytochemicals in Plant Cell Cultures*. In: *Cell Culture and Somatic Cell Genetics of Plants* (vol 5, eds Constabel, F. and K. Vasik), pp. 193- 212. Academic Press, San Diego, CA, USA.

Moran, M., M. Cacho and J. Fernandez-Tarrago, 1999. A protocol for the cryopreservation of *Digitalis thapsi* L. *Cell cultures,CryoLett.*, **20**: 193-198.

Nickel, S.L. and E.J. Staba, 1997. *RIA-test of Digitalis plants and tissue cultures*.In: *Plant Tissue and its Biotechnological Application* (eds Barz, W. and M.H. Zenk), pp. 278-284. Springer, Berlin.

Olmstead, R.G., C.W. dePamphilis, A.D. Wolfe, N.D. Young, W.J. Elisons and P.A. Reeves, 2001. Disintegration of the Scrophulariaceae. *Amer. J. Bot.*, **88** (2): 348–361.

Perez-Bermudez, P., M.J. Cornejo and J. Segura, 1985a. A morphogenetic role for ethylene in hypocotyls cultures of *Digitalis obscura* L. *Plant Cell Rep.*, **4**: 188-190.

Perez-Bermudez, P., M.J. Cornejo, J. Segura, 1985b. Pollen plant formation from anther cultures of *Digitalis obscura* L. *Plant Cell Tiss. Org. Cult.*, **5**: 63-68.

Pilgrim, 1977. Ein beitrag zur suspensionskultur (batch) von *Digitalis purpurea -Geweben. Pharmazie*, **32**: 130-131.

Pradel, H., U. Lehmann, B. Diettrich and M. Luckner, 1997. Hairy root cultures of *Digitalis lanata* :secondary metabolism and plant regeneration. *J. Plant Physiol.*, **151**: 209-215.

Reinbothe, C., B. Diettrich and M. Luckner, 1990. Regeneration of plants from somatic embryos of *Digitalis lanata*. *J. Plant Physiol.*, **137**: 224-228.

Rucker, W. 1988. *Digitalis* spp: *in vitro* culture regeneration and the production of cardenolides and other secondary products. In: *Biotechnology in Agriculture and Forestry* (vol 4). *Medicinal and Aromatic Plants* (ed. Bajaj, Y.P.S.), pp. 388-418. Springer, Berlin, Germany.

Rucker, W., K. Jentsch and M. Wichtl, 1981. Organ differenzierungund glykosidbildung bei *in vitro* kultivierten Blattexplantaten von *Digitalis purpurea* L. *Z Pflanzenphysiol.*, **102**: 207-220.

Saito, K., M. Yamazaki, K. Shimonura, K. Yoshimatsu and I. Murakoshi, 1990. Genetic transformation of foxglove (*Digitalis purpurea*) by chimeric foreign genes and production of cardioactive glycosides. *Plant Cell Rep.*, **9**: 121-124.

Sales, E., S.G. Nebauer, M. Mus, I. Arrillaga and J. Segura, 2001. Cryopreservation of *Digitalis obscura* L. selected genotypes by encapsulation-dehydration. *Planta Med.*, **67**: 833-838.

Sales, E., S.G. Nebauer, M. Mus, I. Arrillaga and J. Segura, 2002. Plant hormones and *Agrobacterium tumefaciens* strain 82.139 induce efficient plant regeneration in the cardenolide producing plant *Digitalis minor*. *J. Planta Physiol.*, **159**: 9-16.

Sales, E., S.G. Nebauer, M. Mus, I. Arrillaga and J. Segura, 2003. *Agrobacterium tumifaciens*- mediated genetic transformation of the cardenolide producing plant *Digitalis minor* L. *Planta Med.*, **69**: 143-147.

Schoner, S. and E. Reinhard, 1982. Clonal multiplication of *Digitalis lanata* by meristem culture. *Planta Med.*, **45**: 155.

Whitson, T. D., L. C. Burrill, S. A. Dewey, D. W. Cudney, B. E. Nelson, R. D. Lee and R. Parker, 2000. Weeds of the West. The Western Society of Weed Science in cooperation with the Western United States Land Grant Universities, Cooperative Extension Services. University of Wyoming. Laramie, Wyoming, 630 p.

Zhang, D., A.M. Armitage, J.M. Affolter and M.A. Dirr, 1996. Environmental control of flowering and growth of *Achillea millefolium* L. 'Summer Pastels'. *HortSci.*, **31**(3): 364-365.

16

Eustoma (Family: Gentianaceae)

Sanyat Misra, R.L. Misra, M. Kannan, S. T. Bini Sundar and P. Ranchan

(**Common names**: Flor-de-Muerto, Lisianthus, Prairie gentian)

Introduction

Lisianthus means compound in Greek and *Eustoma* derives from the Greek *eu* meaning good and *stoma* meaning mouth (Anthony Huxley *et al.*, 1992), in more literal sense for a 'pretty face', *i.e.* lovely flowers. There are some 50 herbaceous, woody and shrub-like gentian species of the warm regions extending from Mexico and the West Indies into tropical South America (southern United States, Caribbean and northern South America). Commonly called as 'Flor de Muerto' as it is a favourite flower for decorating graves in southern Mexico (Everard and Morley, 1970). This name has become its common name though scientifically it is now *Eustoma*.

Eustoma grandiflorum (syn. *E. russellianum*, *Lisianthus russellianus*), a herbaceous species found from Colorado and Nebraska to Texas and northern Mexico (Beckett, 1987) is the only species of the commerce and is a newly introduced into India as the high value cut flower crop as well as indoor pot plant with attractive colours. It has potential to compete with tulip for its beauty. Flower is elegant in form and is easily mistaken for a rose. This is highly potential to be grown under protected conditions in India for export (Sheela, 2008). *Eustoma* varieties, both in singles and doubles resembling Canterbury bells, tulips or rose are available for cultivation and for trade as Americans introduced this into the cut flower trade in the early 1980s. Halevy and Kofranek (1984) evaluated the worth of this crop and advocated this as a potential crop for cut flower use.

Botany and Breeding

Eustoma grandiflorum is an erect perennial with simple stem or with a few opposite branches though grown as annual or biennial, 60-90 cm high bearing blue-green connate, ovate or oblong-ovate, 3-5 nerved, in simple or opposite pairs and 7-8 cm long foliage, and the panicled flowers which are borne more than 10 on a plant are large (5.0-10.0 cm across), ruffled, long-lasting (individual flower lasting for up to three weeks; Bailey, 1960) and showy, funnel- or bowl-shaped, 5-petalled blue, pink, lavender, rose, white and bicoloured (blue-violet, brown-black and yellow-green) in loose, terminal, corymbose clusters with delicate petals on long straight stalks in summer from the upper leaf axils, lobes obovate and spreading, stigma of 2 very large, green and velvety, pod oblong, and the seeds minute and pale-brown. Flowers in singles consist of five wide ruffled petals while in doubles many (10-20). Though original species is not easy to grow but continuous efforts of the breeders have brought about new compact seed strains 35-45 cm tall.

In plants the flavonoids are considered to be located predominantly in the vacuoles of epidermal cells and in the cuticular wax of the terrestrial plants, but Markham *et al.* (2000) through their studies on lisianthus stated that these may be present even elsewhere in the cells of green leaves, and in flower petals they stated that 30 per cent of the whole petal flavonol glycosides are located in the cell wall though these lack acylation when compared to vacuolar glycosides. Hashimoto *et al.* (2004) studied the inheritance of three major anthocyanidins ~ pelargonidin (Pg), cyanidin (Cy) and delphinidin (Dp) through self and cross pollinated progenies of lisianthus and recorded clear phenotypic segregation of progenies in which the B-ring hydroxylation during flavonoid synthesis was under regulation completely complimenting each other between Pg- and Dp-synthesis, and on the basis of 782 progenies studied it showed that flavonoid B-ring hydroxylation is controlled by multiple alleles. The PgDp-phenotype did not exist and may accompany Cy-synthesis. Segregation patterns could be explained by the presence of at least four alleles designated H^T, H^F, H^D and H^o in lisianthus flowers, controlling the 3'-; 5'-; 3',5'-; 3'- and 3',5'-hydroxylation, respectively. The level and pattern of pigmentation in purple-flowered lisianthus cultivars was altered by introduction of a lisianthus chalcone synthase [CHS (naringenin-chalcone synthase)] cDNA in an antisense orientation (Bradley *et al.*, 2000). Zaccai *et al.* (2001) tested the effect of *LFY* overexpression on floral development in lisianthus plants transformed with *LFY* cDNA under constitutive control of the 35S CaMV promoter and the F_1 progenies from the two transformed plants were observed flowering 1-2 weeks earlier for two consecutive years. They further stated that they are on the way in introducing fragrance in lisianthus by introducing the gene from scented *Clarkia breweri* ~ acetyl CoA benzyl alcohol acetyl transferase (BEAT) and S-linalool synthase (LIS).

The chromosome number of lisianthus is 2n=36. Neil (2007) stated that lisianthus is easy to breed compared to many floricultural crops. It is an out-crossing species that is self-compatible, but subject to inbreeding depression. The stigma is positioned well above the anthers in most cases. Pollen starts to dehisce as flowers open and one or two days after when flowers have opened, pollen is usually plentiful. The stigma generally does not become receptive until 3-5 days after flowers begin to open so emasculation is easy to accomplish without accidental selfing.

Species and Varieties

The shrub gentian of Mexican and Guatemalan origin is *L. nigriscens* growing some 1.5 m high producing 60-90 cm tall but diffuse inflorescence bearing 1-3 drooping flowers of blue-black in colour. The Jamaican shrubs are *L. capitatus, L. longifolius* and *L. umbellatus*, the latter growing some 2.4-3.5 m high with clustered leaves at the tips of its branches, and from these tips long inflorescences appear bearing densely and beautifully packed cluster of yellow and green flowers subtended by leaves at the base of the flowers but at the tip of the peduncle. Five such more *Lisianthus* species (shrubs) are recorded from Jamaica. Apart from shrubs there are also tree species.

Though most of them are quite charming but the one herbaceous species, *Eustoma grandiflorum* is the only commercial species of recent introduction to India. Hence, only this shall be dealt here with further. It is a valued flowering pot plant producing beautiful cut flowers. It is most suitable for growing in the conservatory (Armitage, 1993) but can be brought home when in bloom. Cultivars are divided into double and single flowering types. Single flowering ones are appealing in form, however, there appears to be a strong consumer preference for double flowering forms. Doubles have 10-20 flower petals per flower that show a rose-like appearance when fully open though single ones being similar to poppy. Although colour preference differs from person to person and season to season even though the most preferred ones are bicolours, white with blue-rim and white with pink rim. In addition, some of the pure white to ivory cultivars are also having higher demand in the market. Double types are 'Avilia series' which contains ivory, blue-rimmed, deep rose and purple which all were developed to flower under the low light levels and cool air temperatures during winter months with crop duration of 14 weeks from transplanting where supplemental lighting (cyclic or continuous) increases stem length and shortens crop duration, 'Bulboa series' which possesses tetra-branched 50 cm floral stalk with about 30 flowers per plant of blue, blue-blush and blue-rimmed colours which all have strong market appeal and these all flower under long day conditions and warm temperatures from February to May, and require 11-14 weeks from transplanting to harvesting, 'Candy series' has high uniformity in flowering and produces flowers under moderate light intensity and for short day conditions under winter, 'Catalina series' was developed to produce blue-blush and yellow flowers during long days and high heat of summers which can be harvested between 11 and 14 weeks of transplanting, 'Echo series' with colour range of various shades of blue and pink, and pure white, which is most commonly grown one among the double forms for long crop duration under lower light levels but with weak stalks and produces longest stalks in February though in summer stalks become too short, *viz.* 'Echo Blue', 'Echo Yellow', and 'Mariachi series' is most suitable for summer production with a range of flower colours in white, pink-blush, blue-blush and blue, flowers numbering 21 and being borne on the 48 cm long 4-stemmed plant, where floral petals are recurved. Single types are 'Flamenco series' that flowers in various shades of blue, pink and white on long and strong stalks during long day conditions and high summer temperatures, 'Heidi series' which flowers in wide range of colours being most suitable for growing in greenhouses for production of blooms during winter and spring months and grows well under moderate light intensity and day length, 'Languna series' is available in deep blue, blue-blush, deep rose and pink-rose, each plant producing three stalks, and flowering in summer on 48 cm long stalks with 25 flower buds per plant from 11-13 weeks of transplanting, 'Malibu series' produces flowers during autumn and spring but not in summer, the flower colours being in lilac, deep blue, blue-rimmed, rose and white which appear in 11-12 weeks after transplanting, and 'Yodel series' produces flowers in blue, deep blue, lilac, pink, rose and white on 45-50 cm long stems. Worldwide, around 100 varieties are available with varied colours and petal shapes, the leading ones being the F_1 series; (**A**) **Spring/autumn growing**: A fast grower, performing well under low light conditions (high light conditions shorten the stem) - (i) **Fuji** (it is improved Yodel series, growing to 70 cm, flowers more upright and appearing on top, less branched and outstanding as cut flower), *viz.* 'Apricot' (salmon apricot), 'Blue Vein', 'Cherry Blossom' (cherry rose), 'Deep Blue' (intense blue, highly grown variety), 'Lilac Improved' (light lilac-rose), 'Lilac Rose' (lilac-rose), 'Pearl' (white splashed light-blue), 'Picotee Blue Improved' (white edged blue), 'Picotee Pink' (only a few white, others white edged light pink), 'Pink Vein' (white, aging to soft rose), 'Pure White' (white), 'Rose' (bright rose), 'Silver Blue' (attractive soft blue), 'Yellow Improved' (unique soft

Catalina Series

Echo Series

Mariachi Series

Single Flowering Type

Flamenco Series

Heidi Series

Laguna Series

Malibu Series

yellow); (ii) **Misato** (an entirely new type of very free-flowering lisianthus growing up to 70 cm in height with many small flowers occurring on the top on more upright stalks, and the stems are less branched), *viz.* 'Picotee Rose' (white edged distinctly soft rose), 'Pure White' (white), 'Rose' (rose); (iii) **Yodel** (height 70 cm, stems long and strong and flowers large, up to 8 cm in diameter), *viz.* 'Blue' (blue), 'Deep Blue' (deep blue), 'Pink'(soft rose), 'Rose' (dark rose) 'White' (cream white); (iv) **Dream** (height 70 cm, flowers double, flowering time can be adjusted through cultural practices up to three weeks later than above-cited three groups but the stem is thin to bear the load of the flowers), *viz.* 'Blue Improved' (blue), 'Lilac Rose' (lilac-pink), 'Rose' (pink), 'Pearl' (white splashed soft blue), 'Picotee Blue Improved' (white edged blue, 15 per cent off-type), 'Picotee Pink' (white edged light pink, 15 per cent off-type), 'Pure White' (white); (**B**) **Summer growing:** Provides cut flowers during summer with long, thick and stout stems and with one fortnight extended flowering period than the group '**A**' – (i) **Kyoto** (plant height 80-90 cm, flowers very beautiful, stems less branched and flowering on top on strong and upright stems), *viz.* 'Deep Blue' (intense blue), 'Donker Blue' (blue), 'Picotee Blue' (white edged blue), 'Picotee Pink' (white edged light rose), 'Picotee Rose' (pink edged darker), 'Pure White' (white but flower shape a bit different), 'Purple' (self colour), 'Rose Improved' (intense rose), 'Rose Pink' (pink), 'Silver Blue'; and (ii) **Sapporo** (height and habit same as above but with smaller flowers and has only one colour), *viz.* 'Picotee Rose' (white but edged distinctly rose); (iii) **Double-flowering**: 'Queen of Rose series' (first improved double-flowering lisianthus growing 80-90 cm high with stout, upright and long stems); and **(C) autumn growing -'Charm series',**are the specially selected series to be planted in summer for autumn flowering bearing stout and long stems (90-100 cm), about 15-25 cm taller than 'Kyoto series' and flowering occurring up to two weeks later than the group **'B'** varieties, *viz.* 'Charm Pink' (rose), 'Picotee Blue' (white, edged blue), 'Pure White' (white), etc. Other outstanding varieties are 'Asuka-no-Asa', 'Asuka-no-Moegi', 'Asukanosakura', 'Asuka-no-Shizuku', 'Asuka-no-Shinsetsu', 'Asuka-no-Shirabe', 'Asukanoyasoo', 'Azuma-no-Ginga', 'Azumanosakura', 'Candy White', 'Clear Marine', 'Double Up Pink', 'Eculousa Green', 'Eculousa Pink', 'King Violet', 'Micky Pink', 'Micky Rose', 'Miyakomomo', 'Nagoya Dark Blue', 'Nagoya Pink Picotee', 'Nagoya Rosy Red', 'Nagoya Salmon Pink', 'Nail Light Pink', 'Nail Marine Neo', 'Nail Pink', 'Osaka Blue Picotee', 'Osaka Pastel Yellow', 'Osaka Red Picotee', 'Prince Purple Red', 'Pure White Limo', 'Rainy Orange', 'Rainy Pink', 'Royal Violet', 'Sentinel Porcelain', 'Seoul Lavender', 'Shimane', 'Tsukushinoyuki', 'Tyrol Blue', 'Tyrol Donker Rose', 'Tyrol Rose', 'Venture Blue', 'Winter Pink', etc. Out of these varieties, those with purple colours are in great demand, followed by pink or white and then of mixed colours, however, the colours which are poor in demand are blue, yellow and bicolours. The monthly demand of its cut flowers is highest from May to October, i.e summer and autumn (Anon., 1997). In USA, Harbaugh and Scott (2001, 2005a,b) developed F_1 hybrids 'Florida Silver', a semi-dwarf heat-tolerant lisianathus with blue overtone under certain light conditions and violet-blue eye-spot, being most suitable as pot plant and for bedding, through inbred lines UF99-16 and UF99-49 (Harbaugh and Scot, 1999, 2001); 'Maurine Dawn'- a heat-tolerant lisianthus (a little shorter) with pink-white bicoloured flowers & 'Maurine Blue' (a little taller than former), again a heat-tolerant developed through inbred lines UF03-596 and UF-3-573 (Harbaugh and Scot, 1998); and 'Florida Blue Frill' bearing blue-pink bicoloured flowers through inbred lines UF03-529 and UF03-524 & 'Florida Pink Frill' bearing pink and bicolourd flowers through inbred lines UF02-1381 and UF02-1377, these both being semi-dwarf and heat-tolerant. Other heat-tolerant varieties are 'Maurine Lilac', 'Maurine Pink', 'Maurine Pink Lilac', 'Maurine White' and 'Maurine White on Blue' (Harbaugh and Scott, 1998) whereas semi-dwarf heat-tolerant varieties are 'Florida Blue', 'Florida Light Blue' and 'Florida Pink' (Harbaugh and Scott, 1999, 2001). In Korea, Huh KunYang *et al.* (2003) reported about a newly developed F_1 cultivar 'Little Jewel Pink', developed in 2002 by crossing pure breeding lines of 'Wonkyo A6-1' and 'Wonkyo A-6-17'. This has narrow, ovate and wide leaf shape and blooms later with many wide funnel-shaped flowers of light pink colour having round red edged petals. There are natural dwarf varieties exclusively for **pot growing** without treatment of any growth retardant and these are 'Blue Lisa', 'Little Belle Blue' and 'Mermaid Blue', 'UF Double Joy series' having 5 colours and 10 or more petals per flower instead of usual 5, 'UF Savanna series' being heat-tolerant, intermediate in height having 8 colours, 'UF Savanna Blue Frost' and 'UF Savanna Pink Frost' both being the first pot types double-flowering and heat-tolerant varieties bearing vivid bicoloured flowers (Zhanao Deng and Harbaugh, 2006). The cultivars 'Bridal pink', 'Hallelujah Purple', 'Heidi Pure White' and 'Ventura Deep Blue' under blue-purple, pink and white colour group are less susceptible to *Fusarium* crown and stem rot whereas the cultivar 'Magic Champagne' is resistant and 'Echo Lavender' & 'Echo White' are less susceptible to grey mould.

Harbaugh *et al.* (2000) laid out an experiment of 47 lisianthus cultivars and recorded that at seedling stage cvs 'Malibu Purple', 'Catalina Blue Bush' and 'Alice Pink' were found best as these expressed only 5 per cent rosettes and with large leaves and vigorous root systems while they recorded 'Malibu Purple', 'Malibu Bush Blue', 'Alice Purple', 'Balboa Blue', 'Avila Blue Rim', 'Mellow Pink', 'Flamenco Wine Red', 'Flamenco Rose Rim', 'Alice Pink', 'Avila Rose', 'Echo Pink', 'Alice White' and 'Marachi White' as the best cultivars with regard to rosetting, plug performance, vegetative and floral attributes, flower size, number of petals and the postharvest characters.

Propagation

Lisianthus is commercially propagated only through seeds (Armitage, 1993). For convenience of sowing the seeds are pelleted. Seeds are sown in February-March in trays containing sterilized peat-perlite mixture. They germinate in 14-20 days. During germination, a temperature of 20- 22 °C with 95 per cent relative humidity and 13-14 hours of day length are ideal. The seeds of lisianthus have to be soaked in water for a few hours before sowing. Pre-sowing treatment at 3 °C for four weeks causes earlier germination and a higher percentage of bolted and flowered plants. Thus vernalization of lisianthus seeds could be a useful method to shorten commercial production cycle. Eustoma plugs may also be purchased to get earlier flowering. It normally takes 15-18 weeks from sowing to planting, and further some 8-9 weeks for flower production.

Seed propagation is the best method to start a crop although rooting of cuttings and tissue culture are also possible. Germination *in vitro* of lisianthus seeds allows reducing time to obtain plants for transplanting. *In vitro* germination percentage is 100 per cent on Brooks and Hough medium added with gibberillic acid and activated charcoal. Paek KeeYeup and Hahn EunJoo (2000) stated that in seed-propagated lisianthus, the shoot-tips are when subjected to cytokinin, auxin and activated charcoal (AC) the organogenesis and anatomical characteristics are influenced. BA and kinetin at 13.3-22.2 and 13.94-23.23 µM give good shoot formation but together high percentage of hyperhydric shoots are also observed, and culturing in MS medium with 4.44 µM BA and 1.47-4.92 µM IAA and IBA, 15 shoots per explant after four weeks of culturing were obtained, however, IAA and IBA favoured root formation though NAA affected it adversely, and AC suppressed both, the root and shoot formation. Ruffoni *et al.* (1990) used leaf fragments of *Lisianthus russellianus* (*Eustoma grandiflorum*) for direct shoot organogenesis and somatic embryogenesis and cultured these in liquid media containing zeatin, isopentenyladenosine (2iP) or BA and then recultured in the same medium, and obtained shoot regeneration after four weeks with zeatin or low concentration of 2iP. One cm long terminal root segments of *E. grandiflorum* were when cultured in MS medium with 0-1 mg NAA or IBA and 20 g sucrose/l, both NAA and IBA decreased growth of the original meristem but all except the lowest concentration of IBA increased production of the laterals. More lateral roots were obtained in those supplemented with NAA than IBA, and 1 mg NAA produced a callus-lke growth of multiple roots. The cultured roots transferred to solid medium with combinations of 0-0.1 mg each of BA and NAA, regenerated adventitious shoots. NAA decreased the shoot formation while BA increased, best being with 0.1 mg BA alone/l.

Cultural Practices

A full account of its culture has been given by Corr and Katz (1997). **Soil** pH was found to be another major cultural factor that could be a stumbling block to quality production of lisianthus. A relatively high soil pH is required to ensure a good root system. When the soil pH is < 6.2, micro-element toxicities, especially zinc limits the growth and causes undesirable leaf chlorosis and necrosis. Ideal pH for its growing is 6.3-7.0 in case if other things remain the same. Soil EC should never be more than 1.0 as afterwards, when fertilizers are applied or if water is heavy, the level rises further. The crop can never tolerate EC level of 1.5 or above. **Medium** for its successful cultivation should preferably be of good nutrient status and sufficiently porous with proper drainage facilities. Soils of other texture should be improved through addition of sand and decomposed organic matters. In its commercial production in other countries, it is usually grown in peat-perlite (1:1) mixture. This costly medium can effectively be substituted with a mixture of composted coir peat, sand and soil (2:2:1) under Indian conditions (Sivakumar *et al.*, 2002). Bailey (1960) recommended good loam, sand and well-rotted manure. In an ordinary polyhouse when Loeser (1991) grew *Eustoma* in two batches, the flowers produced were of good quality in a cellular potting system with 96 plants/m^2 or with 90 plants/m^2 with 34 litres substrate/m^2. Lisianthus grows well at 24 °C day and 18 °C night **temperatures**, however, a day temperature range of 20-25 °C and night temperature range of 15-18 °C will be quite encouraging. Though there are varieties quite insensitive to high temperature but most of the cultivars do not like a growing temperature lower than 15 °C or higher than 25 °C as this causes rosetting (Harbaugh *et al.*, 1992). Ohkawa *et al.* (1991) stated that when they raised *E. grandiflora* cv. 'Fukushihai' seedlings from seeds germinated at 33/28 °C day/night temperatures, these formed rosettes whereas seeds exposed to 23/18 °C their most of the seedlings bolted. The exposure of the germinating seedlings to high temperatures for as little as three days reduced elongation though extending this treatment for more than two weeks inhibited the elongation completely even when the plants were subsequently grown at lower temperatures. Temperatures below 18°C significantly slow down root and shoot growth but do not hinder subsequent flower development. A greenhouse temperature of 10-12 °C drastically slows down plant growth though 20-25 °C causes rapid plant development but flower quality is hampered when temperature rises, however, a temperature range of 17-18 °C with 70-80 per cent relative humidity is recommended as optimum for its proper development and high flower quality (Anon., 1997). Temperature has significant effect on flowering time. Plants grown at 18 °C night temperature bloom 11-23 days earlier than those grown at 13 °C. Under **long day conditions**, the bud formation in lisianthus is earlier while under short days though blossoming occurs but remains erractic and inhibited as it is a quantitative long day plant, therefore, under short days (shorter than 12 hours of natural illumination) artificial lighting to lengthen the day length for 16 hours using 15 W/m^2 bulbs will be quite effective (Halevy and Kofranek, 1984; Anon., 1997). For floricultural use, three types of lamps are available, *viz.* incandescent, fluorescent and high intensity discharge (HID) lighting, the first one being used more frequently for photoperiod regulation, the second one for propagation and the third one, *i.e.* HID lamps for supplementing sunlight in greenhouse production. HID lighting or the 150 W incandescent lights spaced 3 m apart for each 4.5 m wide greenhouse will be quite effective for harvesting a good quality lisianthus flowers with higher yield. Since India is bestowed with wide climatic conditions, lisianthus can successfully be grown in subtropical and temperate parts of our country. Seedlings attaining 4-6 leaf stage, *i.e.* two months old, are ready for **planting**. Planting is neither too deep nor too shallow. To reach this stage, it takes about 15-18 weeks from seed-sowing to transplanting, then again some two months to attain 5-6 pairs of leaves when these become ready for final planting. Beds of one metre width are prepared and some 64-80 plants are maintained per square metre area with pathways of 30-40 cm. These are **supported** with a single layer of mesh netting of 12.5 × 12.5 cm mesh size. **Pinching** is carried out about six weeks of planting at the stage when these are to attain active growth. Pinching is carried out at the third node but only in those cases where single flower stem is not required. This may be carried out twice for maximum flower production though this delays flowering by 2-3 weeks.

Use of plastic **mulch**, saw dust, dry leaves and other organic mulches, and use of non-selective herbicides are a few possibilities to control the weeds. An herbicide such as Glysophate will be translocated through entire plant to kill both above and below ground parts. Soil fumigant Basamid (a granular formulation) controls a variety of weed seeds, nematodes and soil pathogens.

Feeding and Irrigation

Fertilizer application should always be based on soil (or medium) and leaf analyses to derive any conclusive recommendation. Only in poorly nurtured crop, deficiency symptoms of major nutrients appear and if at planting the plants are healthy and even when the plants have taken growth sufficiently, even deficiency of major nutrients in the form of chlorosis does not appear. The chlorosis due to deficiency of major nutrients appear only when the plant growth is initially poor. Poor growth and weak stems, *vis-à-vis* interveinal chlorosis to bleached foliage appear even due to low pH of medium. Until and unless it is known that the soil is deficient in potassium, it should never be used because normally Indian soils are not deficient in potash. Though requirement of potassium by lisianthus is of quite significance and when buds start initiating the requirement also increases. Lisianthus produces good quality flowers with excellent postharvest life when **fed** with nitrogen and potassium in a 1:1.5- 1.7 ratio. Camargo *et al.* (2004) when studied cv. 'Echo' which was transplanted in October, 1999 (60 days after sowing), at the end of the cycle (120 days after nutrient application) when height and dry matter weight were maximum, the nutrient uptake per hectare of land was 238.8 kg N, 157.1 kg K, 33.9 kg S, 17.5 kg Mg, 14.9 kg P, 10.6 kg Ca, 1281.3 mg Fe, 294.4 mg B, 127.1 mg Mn, 121.1 mg Zn and 35.8 mg Cu. Soil mix of Chitosan (1 per cent w/w) at sowing time markedly enhances growth and 15 days earlier flowering with increased number and weight of cut flowers. Highly humid weathers restrict translocation of calcium even when the medium or soil is rich in Ca, and under such circumstances the only alternative is to administer foliar feeding. Calcium deficiency causes tip burn, bud abortion and thin growth. Zn toxicity also appears as interveinal chlorosis to bleached foliage. It is very sensitive to waterlogging. **Irrigation** frequency, *vis-à-vis* enormity should be gradually reduced to avoid high temperature and high humidity environment, otherwise the plants may be victim of many of the fungal pathogens. It should further be reduced during harvesting and poor light conditions. However, during whole of the cultivation period of the crop, the soil should be kept moist.

Growth and Flowering

Tanigawsa *et al.* (2001) subjected the plants of lisianthus cultivars 'Asukanoasa', 'Asukanosakura', 'Asukanoyosoo', 'Azumanosakura' and 'Miyakomomo' to high and low growing temperatures for different durations, first two cultivars being grown at 11 °C for three weeks after sowing and then at 30/25 °C (day/night), 27.5/17.5 °C and 22.5/12.5 °C under a 12 h photoperiod for 0-6 weeks where they obtained 100 per cent bolting in the seedlings at 22.5/12.5 °C temperature range though only 13-88 per cent at 30/25 °C. They grew'Asukanoyosoo' and 'Miyakomomo' cultivars after sowing at 30/25 °C for 0-8 weeks under natural photoperiod and then subjected to 11 °C for 4-6 weeks, and these all the plants in both the cultivars bolted after attaining 1.1-2.3 nodes per stem, respectively, however, those seedlings grown only for 0-2 weeks before chilling treatment exhibited rosetting. The cultivars 'Azumanosakura', 'Asukanosakura' and 'Asukanoasa' when raised at 35/25 °C for 0-2 weeks under natural photoperiod after sowing, then exposed to 11 °C for 1-5 weeks and then cultured at natural day/night temperatures in Japan, most of the seedlings bolted when exposed to chilling temperatures for 3-5 weeks irrespective of initial duration of treatment, and the exposure of 35/25 °C for 4-8 weeks after sowing and then at 11 °C for 3-6 weeks, all the seedlings were found bolting after five weeks of chilling. They concluded that in the cultivars hard to form rosette, the low temperature exposure is primarily responsible for bolting. Likewise, Li Jie *et al.* (2004) subjected the seedlings of 13 lisianthus cultivars to 25/15, 27.5/17.5, 30/20, 32.5/22.5 and 35/25 °C day/night temperatures, a 14-h photoperiod and 100 μmol $m^{-2}s^{-1}$ photosynthetic photon flux density (PPFD) and found decreased bolting rates with increase in temperatuare. The transition from the vegetatiuve rosette stage to the reproductive (bolting) growth stage, lisianthus essentially has vernalization requirement, a treatment which causes oxidative stress, but addition of glutathione (GSH) or its precursor cysteine instead of vernalization induces bolting which shows that vernalization stimulates GSH synthesis and the synthesized GSH specifically determines the bolting time of *Eustoma grandiflorum* (Yanagida *et al*, 2004). Hisamatsu *et al.* (2004) stated that heat-induced russetting is associated with a reduction of GA_1 content, and to estimate the *ent-kaurene* biosynthesis activity, the use of uniconazole on *ent-kaurene* oxidase inhibitor, showed that in stems of the cold-treated seedlings it was approximately 1.8 times that of the non-cold-treated seedlings, though no difference was found in the leaves. Also there was no difference in endogenous levels of GA_1 and GA_{20} in stems of the heat-induced rosetted plants during cold treatment, whereas their levels increased with stem elongation after transfer to warm conditions. Hisamatsu (2001) stated that GA_1 in the early-13-hydroxylation pathway was physiologically important for regulation of stem elongation, leaf expansion and flower bud development but not for flower bud initiation, and the GA-biosynthesis pathway prior to GA_{53} might be blocked in the plants with rosette formation. Shidahara and Ohta (2004) studied complementary lighting with fluorescent lamps (FL) of blue, red and white colours to two cvs 'Shimane Original Purple' and 'Tsukushinoyuki' and found remarkably enhanced seedling growth with blue FL when exposed to 12 h per day at 10 μmol m^{-2} s^{-1}. Gibberellic acid sprays (15- 25 ppm) are used for rapid stem development after rosette plants begin to bolt and GA_3 sprays (250 ppm) promote stem elongation of cut flowers. Lee JongWon *et al.* (2000) recorded increased floral stalk length and leaf number under the treatment of GA_3 (200 or 400 mg/l) and GA_3 (200 mg/l) + ethephon (200 mg/l) though marketability of flowers was best when only one application of GA_3 (200 ppm) was applied 80 days after sowing, however, they recorded delayed flowering of 2-5 days in GA_3 treatment and 10-12 days under ethephon. Oha *et al.* (2001) stated that growing vernalized lisianthus in a medium containing GA_3 or Ancymidol, vegetative rosetting occurred, more pronounced being in vernalized plants than the non-vernalized ones though bolting rates were monitored to some extent through use of different concentrations of GA_3. Hirai and Mori (1999) stated that when seedlings are exposed to high temperature, it induces rosetting, hence, suggested that during summer for seedling raising a cooling system (<20 °C) should be used. They further mentioned that plants should be grown above 10 °C temperature with an exposure to 16 h photoperiod to get rid of rosetting.

Daminozide foliar sprays applied once at 5,000 ppm or twice at 2,500 ppm, paclobutrazol drench with 2-4 ppm and/or uniconazol foliar sprays at 5 or 10 ppm controls the height in potted and bedding plants. Starman (1991) stated that uniconazol foliar spray at 10 mg/l two weeks after pinching and foliar spray of 5 mg/l at 2-3 weeks of pinching, or drenching at 1.6 mg/pot two weeks after planting are quite useful for conforming the plants to their pots. To keep the plants dwarf in the pots, these require high rates of retardant (Tjia and Sheehan, 1986; Whipker *et al.*, 1994). Tjia and Sheehan (1986) recommended ancymidol drench at 0.25 and 0.30 mg a.i./1.8 litre pot or 66 mg/l foliar spray, and daminozide at 2,500, 5,000 or 7,500 mg/l foliar spray for getting a good control of lisianthus plant height.

Postharvest

The flower stalks can be harvested in the morning when at least two basal flowers have opened and other buds develop characteristic colour. The first flower normally appears untimely which should be removed and subsequently when two flowers appear together then it becomes ready for distant marketing. For fresh market sales, one can wait for up to four flowers to open before harvesting. For cutting, the whole plant is pulled up and the root portion is removed. Prior to export, the cut flower stems are pre-cooled at 13 °C. Pulsing the flower stalks with sucrose solution (10 per cent) for 24 hours increases the vase life of lisianthus flower. Flowers of 10 stems are bunched in perforated sleeves to avoid *Botrytis* infection and then are packed in corrugated fabricated boxes preferably with ethylene absorbents such as potassium permanganate though not highly susceptible to ethylene, and then these are transported to their desired destinations. Since, like to those of gladioli, lisianthus flowers also have tendency to bend upward, therefore, the aqua-packed cut flowers are marketed in an upright position.

Pollination and stigma crushing or removing accelerates senescence though has not been observed through pistil removal, and in such cases 2mM STS extends the vase life of pollinated lisianthus flowers (Ichimura and Goto, 2000). Liao LiJen *et al.* (2001) treated the cut flowers of lisianthus cv. 'Hei Hou' having 40 cm stalk length with an opening solution of $Al_2(SO_4)_3$ at 0, 50, 100 or 150 mg/l with 70 per cent RH and 25 °C temperature and recorded 15 days of vase life with 100 per cent bud opening than usual 8 days. Shimizu and Ichimura (2005) studied lisianthus inflorescences with two open flowers and four buds by subjecting them to pulsing with 0.2 mM STS, 4 per cent sucrose and 0.2 mM STS combined with 4 per cent sucrose at 23 °C and 70 per cent RH in dark for 20 hours and recorded better condition of flowers in other treatments than in STS alone. Pulsing in 6 per cent sucrose for 24 or 48 hours and then placing the cut flowers in deionized (DI) water increases floret opening and vase life as compared to those kept only in DI water (control), and 3 per cent sucrose as vase solution also increases the life for more than 18 days (Cho MoonSoo *et al.*, 2001). Pulsing with BA at 50 mg/l followed by 4 per cent sucrose promotes floral longevity in lisianthus cut flowers (Huang KuangLiang and Chen WenShaw, 2002). Uddin *et al.* (2004) tried *in vitro* the buds of lisianthus cvs 'Asukanoasa', 'Micky Rose' and 'Royal Violet' in holding solutions containing different sugars with or without chitosan where chitosan with fructose recorded special vivid petal colouration in all the cultivars which shows that *in vitro* it helps accumulation of anthocyanin in petals. Dole *et al.* (2005) found that neither of the chemicals (STS and 1-MCP) increased the vase life in *Eustoma* cv. 'Echo White'. Akbudak *et al.* (2005) studied storage of lisianthus flowers grown under saline conditions. They studied two cvs 'Pure White' and 'Pink Picotee' through drip irrigation of NaCl at 0, 2, 4, and 8 dS m^{-1} at two stages, *viz.* vegetative and germinative phases, and the flowers after cutting were stored under normal atmosphere (NA) and modified atmosphere packaging conditions (MAP) with 5±0.5 °C and 80±5 per cent RH. They found drastic differences in both the varieties studied as cv. 'Pure White' showed best results under 4 dS m^{-1} in NA and under 2dS m^{-1} in MAP while cv. 'Pink Picotee' recorded just the reverse. This way the flowers of 'Pure White' could be stored for 2 weeks in NA and 5 weeks in MAP while the flowers of 'Pink Picotee' for 2 weeks in NA and 3-4 weeks in MAP. In fact, MAP is supplemental to refrigeration and helps in increasing the efficiency of refrigeration in extending the storage life of cut flowers.

Insect-Pests and Diseases

White flies (*Trialeurodes vaporariorum*) which suck the cell sap of the tender parts including flowers are controlled through parasitoides, *Encarsia formosa* & *Eretmoceus eremosces*, and fungal pathogen *Beauveria bassiana*. **Aphids** (*Aphis gossypii, Myzus persicae*) also the sucking insects where nymphs and adults both damage the leaves and flowers, and **beet armyworms** (*Spodoptera exigua*) which feed on leaves and flowers and even the emerging shoots are controlled through spraying with 0.2 per cent Metacid-50, Pirimor or Ambush. **Thrips** (*Thrips palmi, Frankliniella occidentalis* and many others) are controlled through predatory phytoseid mites, *Amblyseius cucumeria* and *Amblyseius degenerans*. Nymphs and adults, both feed on the tender parts of the plants including the flowers which can also be controlled while controlling the aphids. Ambush or Nuvan at 0.1-0.2 per cent are also very effective. **Leaf miners** which tunnel inside the leaves are controlled through Vertimec at 0.1 per cent spraying. Gill Stanton *et al.* (1998) have described many of the insect-pests and diseases which attack lisianthus at different stages of its growth along with their management.

Daughtrey *et al.* (1995) described various pathogens infecting lisianthus crop at different stages along with their management practices. ***Pythium*** root rot also casues rhizome rot with formation of the black necrotic spots and in severe cases even the death of the plant. The fungus is spread primarily through soil, fertilizer, watering and so on. This can be controlled by proper sanitation, air circulation, drainage and drenching with 'root shield' or 'soil guard'. **Downy mildew** (*Peronospora chlorae*) during warm humid conditions causes pale spots on the upper surface of leaves whereas dark-white to light-purple mouldy fluff on the underside and necrotic brown stripings on the stalks. To a greater extent, the infection can be minimized by reducing air humidity levels in the crop, by keeping the aerial parts dry and when the crop is already infected, it should be sprayed with 0.2 per cent zineb or Daconil, and together this will also control *Phytophthora* and *Pythium* infections. ***Fusarium*** crown and stem rot is caused by *Fusarium avenaceum* where stems become brown, wither and die (rot), and this can be prevented through seed treatment with Thiram and, in the field, by managing proper drainage. Truter and Wehner (2004) found lisianthus crown and root infection caused by *Fusarium solani* which covered whole of the crown with masses of fusoid, curved hyalophragmospores in its advanced stage of infection. In South Africa this was its first record. ***Rhizoctonia*** crown and stem rot may be controlled through fortnightly spraying with 0.1 per cent benomyl or Bavistin alternate with Captan or Thiram 0.2 per cent throughout the cropping period, and these treatments will also control other diseases, if any. Rozolex is quite effective against *Rhizoctonia* rot. **Gray mould** (*Botrytis* blight), ***Curvularia*** leaf blotch, and **viruses** such as Impatiens necrotic spot virus (INSV), broad bean wilt virus (BBWV), bean yellow mosaic (BYMV), tobacco mosaic virus (TMV) and cucumber mosaic virus (CMV) are the common diseases of this crop. Gera *et al.* (2002) found systemic necrosis, necrotic spots and rings on lisianthus leaves due to infection with iris yellow spot virus (IYSV) and Chen *et al.* (2002) recorded yellow spots and an irregular concentric ring pattern of dark green and pale-green tissues which ultimately develop into necrotic concentric rings, necrotic local lesions, mottling and sometimes malformation in *Eustoma russellianum* leaves that have been found to be caused by *Fabavirus*. Gera *et al.* (2006) recorded phytoplasma and/or spiroplasma disease on *Eustoma grandiflorum* in Israel. Mycoplasma and viruses can be controlled through roguing and through control of vectors. *Botrytis* and *Curvularia* infections can be controlled through benomyl, Bavistin, Dithane M-45 or Dithane Z-78 sprayings at

0.2 per cent whenever these diseases appear. Its 2-3 sprayings at weekly intervals will control these diseases.

Constant high RH or a few hours of high RH at night, *vis-à-vis* plants taking longer time to flower cause incidence of tip burn though differences in cultivars prevail, however, the cause of tip burn is not related to high calcium concentration (Islam *et al.*, 2004). Ca deficiency causes tip-burn of young foliage, bud abortion and poor stem length and the availability of Ca to the plants in Ca-rich soil also becomes poor if the air is very humid, therefore, in such cases it would be advisable to give foliar application. High RH and high soil temperature cause breaking of stems for which overhead irrigation at bud formation stage and then drip irrigation can solve this problem. Regulation of a temperature range of 20/22 °C during day time and 18 °C in the night is very congenial for the crop, however, high temperatures of more than 24-26 °C keep the plants vegetative and/or produce smaller flowers so in this case GA_3 at 10-15 ppm spraying will help otherwise one should go for high temperature-insensitive varieties such as 'Malivu' and 'Ventura'

References

Akbudak, B., A. Eris and O. Kucukahmetler, 2005. Normal and modified atmospheric packaging storage of lisianthus (*Lisianthus grandiflorum*) grown in saline conditions. *New Zealand J. Crop & hort. Sci.*, **33**(2): 185-191.

Anonymous, 1997. *Production Manual of Eustoma russellianum (Lisianthus)*, 10 pp. Agricultural and Processed Food Products Export Development Authority. August Kranti Marg, New Delhi.

Anthony Huxley, Mark Griffiths and Margot Levy, 1992. The New Royal Horticultural Society Dictionary of Gardening (A-C). The Macmillan Press Ltd., London.

Armitage, A.M. 1993. Specialty cut flowers. In: *The Production of Annuals, Perennials, Bulbs and Woody Plants for Fresh and Dried Cut Flowers.* Portland, Oregon, Timber Press.

Bailey, L.H. 1960. *The Standard Cyclopedia of Horticulture* (vol. II), pp. 1890-1891. The Macmillan Company, New York.

Beckett, K.A. 1987. The RHS Encyclopedia of House Plants Including Greenhous Plants, pp. 149. Salem House, Massachusetts, USA.

Bradley, J.M., S.R. Rains, J.L. Manson and K.M. Davies, 2000. Flower pattern stability in genetically modified lisianthus (*Eustoma grandiflorum*) under commercial growing conditions. *New Zealand J. Crops Hort. Sci.*, **28**(3): 175-184.

Camargo, M.S.De, L.K. Shimizu, M.A. Saito, C.H. Kameoka S. da C. Mello and Q.A. da C. Carmello, 2004. Growing and nutruient absorption by lisianthus (*Eustoma grtandiflorum*) cultivated in soil (Brazilean). *Horticultura Brasileira*, **22**(1): 143-146.

Chen, C.C., C.C. Hu, Y.K. Chen and H.T. Hsu, 2002. *Fabavirus* inducing ringspot disease in lisianthus. *Acta Hort.* (as *Proc. 10th intern. Symp. on Virus Diseases of Ornam. Plants*, held at Annapolis, Maryland, USA from May 27 to June 1, 2000; ed. Hammond, J.), No. 568, pp. 51-57.

Cho MoonSoo, F. Celikel, L. Dodge and M.S. Reid, 2001. Sucrose enhances the postharvest quality of cut flowers of *Eustoma grandiflorum* (Raf.) Shinn. In: *Proc. 7th Intern. Symp. on Postharvest Physiology of Ornamental Plants* (eds Nell, T.A. and D.G. Clark), held at Ft. Lauderdale, Florida (USA) on November 13-18, 1999. *Acta Hort.*, No. 543, pp. 305-310.

Corr, B. and P. Katz, 1997. *A Grower's Guide to Lisianthus Production. Floraculture International*, **7**(5): 16-19.

Daughtrey, M., R. Wick and J. Peterson, 1995. *Compendium of Flowering Potted Plant Diseases.* APS Press.

Dole, J.M., W.C. Fonteno and S.L. Blankenship, 2005. Comparison of silver thiosulfate with 1-methylcyclopropene on 19 cut flower taxa. *Acta Hort.* (*Proceedings of the Fifth International Postharvest Symposium*, eds Mencarelli, F. and P. Tonutti), held at Verona, Italy, on 6-11 June, 2004; No. 682 (vol. 2), pp. 949-956.

Everard, B. and B.D. Morley (eds Stearn, W.T. and P.S. Green), 1970. Wild Flowers of the World, pp. 172. Peerage Books, London.

Fukai, S., M. Goi, M. Tanaka and H. Furukawa, 1991. Multiple shoot regeneration from root cultures of prairie gentian (*Eustoma grandiflorum*). *Kagawa Daigaku Nogakubu Gakuzyutu Hokoku = Tech. Bull. Fac. Agric., Kagawa Univ.*, **43**(1): 31-34.

Gera, A., A. Kritzman, H. Berckelman, J. Cohen and B. Raccah, 2002. Detection of *Iris yellow spot virus* in lisianthus. *Acta Hort.* (as *Proc. 10th intern. Symp. on Virus Diseases of Ornam. Plants*, held at Annapolis, Maryland, USA from May 27 to June 1, 2000; ed. Hammond, J.), No. 568, pp. 43-49.

Gera, A., L. Maslenin, A. Rosner, M. Zeidan and P.G. Weintraub, 2006. Phytoplasma diseases in ornamental crops in Israel. *Acta Hort.* (*Proceedings of the XIth Int. Symp. on Virus Dis. of Ornam. Plants*, held at Taichung, Taiwan, on March 9-14, 2004; ed. Chang, C.A.), No. 722, pp. 155-161.

Gill Stanton, A., E. Dutky and D. Clement, 1998. *Destructive and Beneficial Insects, Mites and Diseases of Herbaceous Perennials, Ornamental Grasses and Aquatic Plants.* Ball Publishing, Batavia, Illinois.

Halvey, A.H. and A.M. Kotranek, 1984. Evaluation of lisianthus as a new cut flower crop. *HortScience*, **19**: 845-847.

Harbaugh, B.K., M.L. Bell and Liang RongNa, 2000. Evaluation of forty-seven cultivars of lisianthus as cut flowers. *HortTechn.*, **10**(4): 812-815.

Harbaugh, B.K., M.S. Roh and B. Pemberton, 1992. Rosetting of *Lisianthus* cultivars exposed to high temperature. *HortSci.*, **27**(8): 885-887.

Harbaugh, B.K. and J.W. Scott, 1998. Six heat-tolerant cultivars of *Lisianthus. HortSci.*, **33**(1): 164-165.

Harbaugh, B.K. and J.W. Scott, 1999. Florida Pink and Florida Light Blue ~ a semi-dwarf heat tolerant cultivars of *Lisianthus. HortSci.*, **34**(2): 364-365.

Harbaugh, B.K. and J. W. Scott, 2001. 'Florida Silver' – a semidwarf heat-tolerant lisianthus. *HortSci.*, **36**(5): 988-989.

Harbaugh, B.K. and J. W. Scott, 2005a. 'Maurine Dawn' – a heat-tolerant lisianthus with pink/white bicolored flowers. *HortSci.*, **40**(3): 858-860.

Harbaugh, B.K. and J. W. Scott, 2005b. 'Florida Blue Frill' and 'Florida Pink Frill' – a semi-dwarf heat-tolerant *Lisianthus* with bicoloured flowers. *HortSci.*, **40**(3): 862-863.

Hashimoto, F., A.F.M.J. Uddin, K. Shimizu and Y. Sakata, 2004. Multiple allelism in flavonoid hydroxylation in *Eustoma grandiflorum* (Raf.) Shinn. Flowers. *J. Japanese Soc. hort. Sci.*, **73**(3): 235-240.

Hirai, H. and G. Mori, 1999. The utilization of a spot air-conditioner in raising of seedlings for resetting prevention in *Eustoma* and *Delphinium* (Japanese). *Environm. Contr. Biol.*, **37**(3): 191-196.

Hisamatsu, T. 2001. Studies on the role of gibberellin biosynthesis in the control of growth and development in *Matthiola incana* (L.) R. Br. and *Eustoma grandiflorum* (Raf.) Shinn. (Japanese). *Bull. Nat. Res. Instt. Vegetables, Ornam. Pl. & Tea*, No. 16, pp. 79-133.

Hisamatsu, T., M. Koshioka and L.N. Mander, 2004. Regulation of gibberellin biosynthesis and stem elongation by low temperature in *Eustoma grandiflorum. J. hort. Sci. & Biotech.*, **79**(3): 354-359.

Huang KuangLiang and Chen WenShaw, 2002. Ba and sucrose increase vase life of cut *Eustoma* flowers. *HortSci.*, **37**(3): 547-549.

Huh KunYong, Bang ChangSeok, Song JeongSeob and Kim WanSoon, 2003. Breeding of a dwarf type *Eustoma* cultivar 'Little Jewel Pink' (Korean). *Korean J. hort. Sci. & Tech.*, **21**(2): 146-148.

Ichimura, K. and R. Gota, 2000. Acceleration of senescence by pollination of cut 'Asuka-no-nami' *Eustoma* flowers (Japanese). *J. Japanese Soc. hort. Sci.*, **69**(2): 166-170.

Islam, N., G.G. Patil, H. Torre and H.R. Gislerød, 2004. Effects of relative air humidity, light and calcium fertilization on tip burn and calcium content of the leaves of *Eustoma grandiflorum* (Raf.) Shinn. *European J. hort. Sci.*, **69**(1): 29-36.

Lee JongWon, Kim TaeJoung, Kim JuHyoung, Lee HeeDoo, Kim SiDong, Yun Tae and Paek KeeYoeup, 2000. Effects of plant growth regulators on the growth and cut flower quality of *Eustoma grandiflorum* cv. Plantina Violet (Korean). *J. Korean Soc. hort. Sci.*, **41**(4): 419-422.

Liao LiJen, Lin YuHan, Huang KuangLiang and Chen WenShaw, 2001. Vase life of *Eustoma grandiflorum as affected by aluminium sulfate. Bot. Bull. Academia Sinica,* **42**(1): 35-38.

Li Jie, H. Ohno and K. Ohkawa, 2004. Influences of temperature and photoperiod on rosetting characteristics of *Eustoma grandiflorum* (Raf.) Shinn. cultivars grown in growth chamber (Japanese). *Environ. Contr. in Biol.*, **42**(2): 131-136.

Loeser, H. 1991. *Matthiola, Eustoma, Trachelium* and *Callistephus* in a closed system (German). *Gartenbau Mag.*, **38**(10): 51-53.

Markham, K.R., K.G. Ryan, K.S. Gould and G.K. Richards, 2000. Cell wall sited flavonoids in lisianthus flower petals. *Phytochem.*, **54**(7): 681-687.

Neil, O. Anderson. 2007. *Flower Breeding and Genetics*, Vol.II, pp. 645-663. Springer, The Netherlands.

Ohkawa, K., A. Kano, K. Kanematsu and M. Korenaga, 1991. Effects of air temperature and time on rosette formation in seedlings of *Eustoma grandiflorum* (Raf.) Shinn. *Scientia Hort.*, **48**(1-2): 171-176.

Oka, M., Y. Tasaka, M. Iwabuchi and M. Mino, 2001. Elevated sensitivity to gibberellin by vernalization in the vegetative rosette plants of *Eustoma grandiflorum* and *Arabidopsis thaliana. Pl. Sci.*, **160**(6): 1237-1245.

Paek KeeYoeup and Hahn EunJoo, 2000. Cytokinins, auxins and activated charcoal affect organogenesis and anatomical characteristics of shoot-tip cultures of *Lisianthus* [*Eustoma grandiflorum* (Raf.) Shinn.]. *In vitro Cellular & Developmental Biol.-Plant*, **36**(2): 128-132.

Ruffoni, B., C. Damiano, F. Massabo and P. Esposito, 1990. Organogenesis and embryogenesis in *Lisianthus russellianus. Acta Hort.* (*Proc. first ISHS symp. on in vitro culture and hort. breeding*, Cesena, Italy, May 30-June3, 1989), No. 280, pp. 83-88.

Sheela, V.L. 2008. *Flowers for Trade*, pp. 279-281. New India Publishing Agency, New Delhi.

Shidahara, T. and K. Ohta, 2004. Reduction of seedling period of *Eustoma grandiflorum* (Raf.) Shinn. by complementary light treatments with colored fluorescent lamps (Japanese). *Environ. Control in Biol.*, **42**(1): 95-98.

Shimizu, H. and K. Ichimura, 2005. Effects of silver thiosulfate complex (STS), sucrose on their combination on the quality and vase life of cut *Eustoma* flowers. *J. Japanese Soc. hort. Sci.*, **74**(5): 381-385.

Sivakumar, K.C., Deepu Mathew and Sachin Shetty, 2002. Effect of media composition of growth and flower attributes of lisianthus [*Eustoma grandiflorum* (Raf.) Shinner]. *Res. On Crops*, **3**(1): 149-153.

Starman, T.W. 1991. Lisianthus growth and flowering responses to uniconazole. *HortSci.*, **26**: 333.

Tanigawa, T., Y. Kobayashi, H. Matsui and T. Kunitake, 2001. Effects of seedling age and high and low growth temperatures on bolting of *Eustoma grandiflorum* (Raf.) Shinn. cultivars (Japanese). *J. Japanese Soc. hort. Sci.*, **70**(4): 501-509.

Tjia, B. and T.J. Sheehan, 1986. Chemical height control of *Lisianthus russellianus. HortSci.*, **21**: 147-148.

Truter, M. and F.C. Wehner, 2004. Crown and root infection of lisianthus caused by *Fusarium solani* in South Africa. *Pl. Dis.*, **88**(5): 573.

Uddin, A.F.M.J., F. Hashimoto, K. Shimizu and Y. Sakata, 2004. Monosaccharides and chitosan sensing in bud growth and petal pigmentation in *Eustoma grandiflorum* (Raf.) Shinn. *Scientia Hort.*, **100**(1-4): 127-138.

Whipker, B.E., R.T. Eddy and P.A. Hammer, 1994. Chemical growth retardant application to *Lisianthus*. *HortSci.*, **29**: 1368.

Yanagida, M., M. Mino, M. Iwabuchi and K. Ogawa, 2004. Reduced glutathione is a novel regulator of vernalization-induced bolting in the rosette plant *Eustoma grandiflorum*. *Pland & Cell Physiol.*, **45**(2): 129-137.

Zaccai, M., E. Pichersky and E. Lewinsohn, 2001. Modifying *Lisianthus* traits by genetic engineering. *Acta Hort.* (*Strategies for new ornamental, Proc. 20th intern. EUCARPIA Symp.Section Ornamentals*; held on July 3-6, 2001 at Melle, Belgium; eds Huylenbroeck, J. van, E. van Bockstaele and P. Debergh), No. 552, pp. 137-142.

Zhanao Deng and Harbaugh, B.K. 2006. New *Caladium*, *Gerbera* and *Lisianthus* cultivars for Florida. *Proc. Fla St. hort. Soc.*, **119**: 409-412.

17

Gaillardia (Family: Asteraceae)

Atul Batra and B.K.Banerji

(**Common names**: Blanket flower, Indian blanket flower, Indian flower and Brown-eyed Susan)

Introduction and Botany

The genus *Gaillardia* was named so by a French botanist after the 18th century patron of botany as well as a French magistrate, M. Gaillard de Charentonneau (Armitage, 2001), and its epitaph in Latin means pretty. In 1786 when it was named so, *Gaillardia* included only 14 species though now it contains some 28 species of annuals and perennials, all native to North America, especially United States except *G. megapotamica* which is indigenous to South America (Anon., 2003). To European gardens, it was introduced towards the end of the 18th century. Three species that have played the greatest part in the evolution of modern garden forms are *G. amblyodon* (annual), *G. aristata* (perennial) and *G. pulchella* (annual).

Gaillardia is a low-growing, medium-sized, free-flowering plant, blooming for a longer period and is a state flower of Oklahoma which is an impressive and beautiful native flower found growing along roadsides, in fields and pastures, sometimes covering large areas. These are low-maintenance summer-growing plants suitable for herbaceous borders and as ground cover, especially in coastal dune areas or elsewhere in gardens and is used for garden decoration, as loose flowers as well as cut flowers. Plant grows erect, some 45-75 cm high with a spread of 30-60 cm, foliage hairy, leaves alternate, more or less toothed, flowers long-stalked (up to 20 cm), solitary and very showy rose-purple and the dense, frilly petals are yellow, orange, crimson or copper-scarlet, rays yellow, red and neutral while disc flowers purple and fertile. Bracts broad, hairy, in 2-3 series, ligules tri-cleft and making a fringed appearance.

Stoutamire (1955) studied 18 collections of seeds of *Gaillardia pulchella* from different parts of standard habitats, at the Indiana University Experimental Garden. Chromosome counts of these races were determined from microsporocyte and the root top preparations. All the collections had n=17. All coastal races of *G. pulchella* appeared to be cytologically identical. Stoutamire (1977) stated that the cytotypes in *G. pulchella* were in some cases recognizable. *G. gypsophila*, *G. powellii* and *G. pinnatifida* were found diploid with 2n=17, and *G.comosa* tetraploid with 2n=34 (Turner, 1972). *G. pulchella* was recorded diploid with 2n = 36 as well as tetraploid with 2n=72 (Pal, 1989).

Species and Varieties

Original species are rarely cultivated and these are only varieties or hybrids which we find flowering in the garden. However, the species ***Gaillardia aestivalis*** has a narrower range, with Kansas as its northwestern reach, south to Texas and Florida, and as far north as South Carolina on the east coast (Anon., 2004). ***G. amblyodon***, a free-flowering annual plant with leafy rigid stems of 60 cm high, bearing highly hairy rounded-oblong leaves and reddish-maroon flower some 5 cm in diameter, is native to Texas. ***G. aristata*** (syn. *G. grandiflora, G. maxima, G. perennis*), a vigorous, erect and robust perennial growing some 90 cm heigh with elongated and deeply cut leaves, bearing rich vivid yellow flowers of about 12.5 cm in diameter, most suitable for cutting, normally grown as annual or biennial, and is native to western United States (Prairies). It has given vast number of perennial garden forms, the main varieties being *G. grandiflora* and *G. maxima*. Its various other noteworthy varieties are 'Baby Cole' (plants 20 cm tall, flowers red, margined yellow), 'Burgundy' (wine-red), 'Copper Beauty' (orange-yellow), 'Dazzler' (rays bright orange-yellow with maroon-red centre), 'Firebrand' (orange and maroon), 'Goblin' (a dwarf form with yellow and red floral colour), 'Golden Goblin' (golden-yellow), 'Ipswich Beauty' (orange and brown-red), 'Kobold' (plants 20 cm, flowers red margined yellow), 'Mandarin' (flame-orange and red), 'Monarch Strain' (yellow to deep red or bicolours), 'Mrs. H. Longsten' (golden-yellow with yellow centre), 'Nana Nieske' (yellow and red), 'Red Plume' (bright red), 'Sundance Bicolor' (a double-flowered with a low trailing habit), 'Tangerine' (orange), 'The King' (crimson and yellow), 'The Prince' (yellow and crimson), 'The Sun' (golden-yellow), 'Tokajer' (yellow with red base, centre darker), 'Wirral Flame' (deep-red, petal tipped gold), etc. ***G. pulchella*** (syn. *G. bicolor, G. drummondii*), a free-flowering (from summer to autumn) annual species native to central United States with a wide growing range stretching from the northernmost to southernmost states (continental US, Arkansas, Virginia, Colorado, New Mexico, Florida and Hawaii) to Mexico, originally it is found growing in dry or gritty soils prone to summer drought, bearing leafy floral stems some 30-45 cm long, leaves spatulate, hairy and deeply cut, and attractive yellow flowers some 7.5-10.0 cm in diameter, enlarged ray florets with quilled central discs or ball-shaped flower heads, and external ray petals purple-pink at base, is a parent of many of the annual hybrids, many breeding true through seeds though are normally available in mixed forms. Its varieties are 'Lollipops' (weather resistant and free-flowering with red or red and yellow flowers), 'Picta' (brown-red and golden-yellow), 'Picta Alba Marginata' (reddish-maroon and white), 'Picta Aurea' (yellow), 'Picta Chameleon' (scarlet, white and yellow), 'Picta Indian Chief' (red), 'Picta Lorenziana' (yellow, bronze and red, a very showy hybrid with double and tubular petals), 'Picta Lorenziana Bicolor' (red and yellow), 'Picta Lorenziana Rubra' (red and white), 'Picta Lorenziana Sulphurea' (golden-yellow), 'The Bride' (cream-white), etc. ***G.* × *grandiflora*** is the derivative of *A. aristata* × *A. pulchella*, and covers all the known hybrids from which several popular cultivars are widely available including selections with single, double (having tubular-shaped petals that completely hide the central disc), and semi-double flower forms. Other popular species are *Gaillardia aestivalis, G. arizonica, G. coahuilensis, G. flava, G. multiceps, G. parryi, G. pinnatifida, G. spathulata, G. suavis, G. cabrerae, G. megapotamica, G. tontalensis*, etc.

Propagation

Many gaillardias are grown from seeds, especially the annual species and hybrids, and as perennial species may not appear true from seeds so the hybrids may be propagated through root cuttings (Anon., 2003), stem cuttings (Anon., 2005) or by division of the rootstock. **Seeds** have no dormancy and may remain viable up to 31 months if moisture content in seeds is not more than 9 per cent (Ramakrishnan *et al.*, 1970). These are sown in September-October for winter cropping, in February-March for summer cropping and in May-June for rainy season cropping, on raised beds to produce blooms in winter, summer and rainy seasons, respectively, but the seedling raising temperature should be around 15.6-18.3 °C at which these germinate within 10 days. Seedlings are transplanted in beds 25-30 cm apart in case of annuals while perennials 45 cm apart. The ideal time for transplanting is when seedlings attain 30-60 days of maturity. For a very late planting, seeds can be sown even directly in the prepared field. Through seeds, irrespective of their being annual or perennial, these flower the same year. Though not in case of annual types but the seed germination in perennial types (*G. pinnatifida* and *G. multiceps*), are affected by $CaSO_4$. Banker (1980) reported that seed germination in *G. pulchella* is accelerated by 1-2 days with marginal increase in the germination percentage when treated with GA_3 at 60 ppm. Gaillardias are also propagated through **stem cuttings** which should be inserted in a mixture consisting of 2 parts sand and 1 part peat. Likewise, it is also multiplied through **division** and **root cuttings.**

Bourque *et al.* (1989) studied young leaves and internodal stem segments of *G. pulchella* on MS medium supplemented with NAA at 2 mg/l and BA at 0.4 mg/l in dark and induced callusing, and when the cultures were transferred to light conditions for organogenesis, flowers were obtained when regenerated shoots were maintained on MS medium but for rooting plants required planting in soil medium. Pillai *et al.* (1992) took leaf epidermis segments from 45-days old *Gàillardia picta* seedlings and cultured on the semi-solid Gamborg's B5 culture medium with 0.5 per cent agar and 2 per cent sucrose and with 1 µM NAA and 1 µM BA where by four weeks the embryogenesis at 83.3 per cent of the cultures was obtained with 1 µM NAA and 1 µM BA medium.

Cultural Practices

Plant needs sunny place to perform well. This being a **long day plant** requires short days (10-12 h) for good vegetative growth while long days (12-16 h) for flower initiation. In the early stages of growth, vegetative growth takes place under short days and when sufficient aerial growth with good number of leaves have developed, this requires long days with high light intensity (20-30 klx) for flower development. Therefore, in winter additional light is required as poor light can cause poor growth. Warm cultivation temperatures require high light intensity in short days as well as in long days. It can withstand heat and drought conditions considerably. Gaillardias grow well at temperature range of 20 to 30 °C. The perennial forms are reasonably hardy and tolerate temperatures as low as -4 °C but persisting temperatures below 5 °C are dangerous. During persisting low temperatures the plants should be protected through thick layer of organic mulches with very little watering otherwise root rot will occur when soil becomes soggy due to excessive cold coupled with high moisture level. However, the crop is drought tolerant. A medium well-drained **loam-soil** is best for cultivation of gaillardias though by 20 per cent addition of coarse sand even the heavier soils can be improved. Soils for gaillardia cultivation should consist of 15-30 per cent clay, 0-15 per cent (bark, perlite, sand), 1-3 kg/m^3 complete balanced fertilizer, 0-3 kg/m^3 slow release fertilizer (3-9 months), iron-chelates and certain other micronutrients. It can tolerate as high as 8.5 pH but ideally the soil pH should be 5.5 to 7.8. Khimani and Patil (1994) stated that increase in salinity levels from 1.2 to 2.8 dS/m, dry weight and flower yield decrease significantly due to increase in the leaf proline. Though the crop is not more demanding but the soil should have medium level of **nutrients** and it would be better if the soil is incorporated with slow acting nutrients such as bone meal. However, at the time of soil preparation, the soil should be incorporated with some 150 quintals of farmyard manure per hectare of land. Since it is a crop coming up throughout the year, it can be **planted** in any soil and sown throughout the year in summers (February-March), rainy season (May-June) and in winters (September-October). *Gaillardia aristata* becomes perennial on the hills and a care-free annual in the plains that can be grown in any soil and sown throughout the year. **Planting distance** followed should be 30 × 30 cm in case of annual species and varieties whereas 45 × 45 cm in case of perennials. Singatkar *et al.* (1995) under Maharashtra conditions obtained best flowering span and floral yield in gaillardia through application of 173 kg N +103 kg P + 63 kg K/ha. Under Bihar conditions, Mishra (1998) recommended 30 g N/m^2 and 30 × 40 cm spacing per plant for quality and better yield in gaillardia, however, 20 g N/m^2 and 30 × 30 cm spacing was found as the second option. Hugar and Nalawadi (1999a) in Karnataka suggested 30 × 20 cm spacing and 75 kg N/ha (19.6 and 18.0 t/ha in *Rabi* and *Kharif*, respectively, with the seed yield of 7.0 and 6.8 quintals per hectare, respectively) for good flower and seed yield.

During warm weathers, gaillardias require plenty of **water** for successful crop though tolerate drought conditions considerably. However, under excessive cold conditions the wet soils will damage the crop due to root rot so only shallow watering should be given fortnightly. However, in warm weathers it should be watered shallowly but frequently at 4-5 days intervals.

Through pegs and strings, the plants should be **supported** from the border of the beds. **Deadheading** the spent blossoms in summer is recommended to promote active blossoming throughout the autumn as has been reported on *Gypsophila paniculata* by Farina *et al.* (1987). Every three years, gaillardia clumps are to keep new growth from out-competing the older plants. For good growth and flowering, gaillardias require regular **weeding**. Once the crop is fully grown, normally these do not require weeding if there is no gap in between the plants. From the field of gaillardia crop, Derr (1994) recorded excellent control of weeds such as *Euphorbia maculata, Oxalis stricta*, etc. through pre-emergence weedicides, dithiopyr, pendimethalin and prodiamine though little damage to gaillardia crop was obtained. Pellett and Heleba (1995) controlled the weeds effectively in gaillardia fields through mulching with chopped newspaper at 2.3 and 3.6 kg/m^2.

Gailardias take 3-4 months for flowering after seed sowing and continue flowering for a quite long time, however, fierce summers and the severe winters encounter poor flowering. Flower yield depends upon environment and the varieties.

Growth and Flowering

Through manipulation of **temperatures** and **photoperiods**, the peak flowering time may be adjusted. Increase in temperature advances the flowering (Yuan *et al.*, 1998). Trippi (1965) reported that continuous lighting stimulates flowering and accelerates ageing process. *G. pulchella* var. 'Picta' may flower in winter through extended day length. Evans and Lyons (1988) induced flowering in the cvs 'Dazzler' and 'Goblin' through longer days though reported that GA $_{4+7}$ substitutes the long days and induces flowering under short days.

Through treatment of gaillardia plants with chlormequat and daminozide (500 ppm), Khimani *et al.* (1994) obtained increased plant dry weight, *vis-à-vis* flower yield while through MH sprayings they recorded reduced dry weight, *vis-à-vis* flower yield. Hugar and Nalawadi (1999b) through spraying with 200 ppm TIBA and 50 ppm kinetin obtained incresead floral and seed yield in *G. pulchella* var. 'Picta'. Daminozide has also been reported by Mukherjee and Bose (1972) increasing flower number in gaillardia. Koning (1983, 1984) reported that GA activity alone is responsible for corolla growth and not the multiple hormone interactions though style elongation is inhibited by GA and ethylene whereas fusicoccin and acidic buffers stimulate elongation of isolated filaments. Evans and Lyons (1988) found GA increasing the scape length though reducing scape calliper, floralhead diameter and ray floret number.

Postharvest

In case of gaillardia, fully open flowers with long stems but without leaves for cut flowers and without stalks for loose flowers are harvested early in the morning. Cut flowers are kept in deep water with flowers above the water level. These remain fresh for up to a week in vases. For marketing, the graded cut flowers are bunched in a dozen or so. Khimani and Patil (1995) reported maximum weight loss in gaillardia cut flowers after 24 hours with 150 kg N/ha while for loose flowers, both 100 and 150 kg N/ha caused maximum weight loss, after 48 hours with 100 kg N/ha there was maximum weight loss in packed cut flowers while after 36 hours in loose flowers. They found phosphorus only in the later stages showing greatest weight loss with 100 kg P/ha; however, at the end of storage period minimum weight loss was observed with 0 kg N/ha and 100 kg P/ha for cut flowers, and 0 kg N/ha and 50 kg P/ha for loose flowers.

Harode *et al.* (1993) with cut flowers of *G. pulchella* var. 'Picta' found 1 per cent sucrose increasing the vase life and flower colour. Rajitha *et al.* (1999) with *G. pulchella* obtained best vase life with $AgNO_3$ alone, followed by sucrose and BA, respectively. Though $AgNO_3$ does not have any impact on the negative effect of fluoride but BA has been reported to have it.

Insect-Pests and Diseases

Perennial gaillardias are generally resistant to **pest** infestation. Occasionally slugs & snails, aphids, four-lined plant bugs (*Poecilocapsus lineatus*), Japanese beetles, leafhoppers (*Macrosteles fascifrons*), thrips, whiteflies and spider mites can be seen feeding on the leaves of the plant. Spraying of 0.2 per cent Rogor, Kelthane or Malathion will control all these pests, however, for slugs & snails these should be hand-picked in the night or can be trapped by spreading vegetable refuse nearby or through watering such plants with a liquid killer based on metaldehyde or through metaldehyde poison baits. Through a survey, Pallavi *et al.* (2002) recorded *Austroagallia* sp., *Empoasca* sp., *Empoascanra* sp., *Eusarcocoris montivagus*,

Helicoverpa armigera, Nesidiocornis tenuis, Plusia orichalsia, Porthesia scintilans and *Spodoptera litura* for the first time in Maharashtra as gaillardia pests.

Among fungal diseases, the most common ones are root rot or damping off and powdery wilt. **Root rot** pathogen is soil-borne and enters the plant through wounds in the roots or seed coats, causes leaf yellowing, stem wilting and eventual death of the plant. Seed-borne fungi such as *Alternaria alternata, Aspergillus, Botrytis cinerea, Curvularia pallescens, Drechslera* and *Fusarium* cause seed and seedling mortality while *Alternaria* and *Botrytis* cause foliar infection. Shotri *et al.* (1983) recommended control of these pathogens through seed treatment with Dithane M-45 (0.3 per cent) and Aureofungin (0.01 per cent) *in vitro*. **Root and stem rot** is caused by various pathogens (*Pythium, Sclerotinia, Thielaviopsis* and *Rhizoctonia*). Its initial infection causes colour change in stem and lower stems turn tan to dark brown. Sometimes white webs of fungal growth can be seen. Over-watering and mulch-mounding around the plant are avoided. **Powdery wilt** is caused by *Sphaerotheca* and *Erysiphe cichoraceorum* which appear as a fuzzy white coating over both the surfaces of the leaves. Application of myclobutanil prevents this disease. Potassium sulphide solution at 2 g/l spray or any colloidal copper fungicide will control this disease. Sulphur dusting is also effective. **White smut** is caused by *Entyloma* spp. Initially it invades the foliage as light green spots, some with tan centres and such spots form the white spores. These spots become dark brown at later stage of infection. Such leaves should be removed and destroyed. Application of myclobutanil will safeguard the healthy plants. **Septoria leaf spot** caused by *Septoria gaillardiae* is symptomised by brown flecks with reddish-purple borders on leaf surface, later on turning ash-grey. Afterwards shot holes in the leaves are formed and such leaves may wither and fall. Affected leaves should be collected and burnt and the plants should be dusted with copper fungicide or treated with Bordeaux mixture. Seed disinfection with $HgCl_2$ 0.1 per cent or field treatment with formalin 2 per cent before sowing should be carried out as a precautionary measure.

Bacterial leaf spot is caused by *Pseudomonas*. In its infection the colour change between veins of the foliage is quite obvious. In severe infection the foliage turns dark brown resulting into collapse of entire plant. Once the disease is established, it is very difficult to control. Removal and destroying of infected plants are the best measures as soon as symptoms appear. Sprinkler irrigation is strictly prohibited as it helps dissemination of pathogen.

Some of the common **viral diseases** of *Gaillardia pulchella* are 'impatient necrotic spot', 'tomato spotted wilt' and 'cucumber mosaic virus'. The symptoms appear on the leaves as distortion or patches of light and dark green colouration. Removal and destroying the infected plants are the best option. There should be timely control of thrips (vector of impatiens necrotic spot and tomato spotted wilt) and aphids (vector of cucumber mosaic) to stop the spread of the diseases. *Gaillardia* x *grandiflora* is susceptible to aster yellows (phytoplasma) that stunts the growth. Infected leaves turn yellow and flowers are distorted in shape and colour. Such plants should be destroyed. Leaf hoppers which spread this disease should be controlled in the field and in the vicinity. MLOs cause flower phyllody so such plants should also be uprooted and destroyed.

References

Anon., 2003. *Gaillardia pulchella. Floridata*, July 28, 2005.

Anon., 2004. The Plants Database, Version 3.5. *National Plant Data Center*, USDA, July 28, 2005.

Anon., 2005. *Propagation Guide. Ball Flora Plant*. Ball Horticultural Co., United States.

Armitage, A.M. 2001. *Armitage's Manual of Annuals, Biennials, and Half-hardy Perennials*. Timber Press, Portland, USA.

Banker, G.J. 1980. Effects of seed treatments with gibberellic acid on germination of winter season annuals. *South Indian Hort.*, **28**(2): 60-61.

Bourque, J.E., S. Tanner and T.J. Mabry, 1989. *In vitro* regeneration of *Gaillardia pulchella* Foug. *Plant Cell, Tissue, Organ* Culture, **16**(1): 67-72.

Derr, J.F. 1994. Weed control in container-grown herbaceous perennials. *HortSci.*, **29**(2): 95-97.

Evans, M.R. and R.E. Lyons, 1988. Photoperiodic and gibberellin induced growth and flowering responses of *Gaillardia grandiflora. HortSci.*, **23**(3): 584-586.

Farina, E., T. Paterniani, M.de Vita, G.V. Zizzo and G. Pergola, 1987. Yield comparisons between types and selections of *Gypsophila paniculata* in different environments. *Annali dell'Istituto Esperimentale per la Floricoltura*. **18**(1): 65-80.

Harode, S.M., V.G. Kotake and J.P. Zend, 1993. Effects of selected treatments on longevity and freshness of selected flowers in home decoration. *J. Maharashtra agric. Univ.*, **18**(3): 448-450.

Hugar, A.H. and U.G. Nalawadi, 1999a. Studies on influence of spacing and nitrogen levels on growth and yield parameters in gaillardia (*Gaillardia pulchella* var. Picta Fouger). *Karnataka J. agric. Sci.*, **12**(1-4): 137-141.

Hugar, A.H. and U.G. Nalawadi, 1999b. Effect of growth regulators on morphological characters, flower production and seed yield of *Gaillardia. Karnataka J. agric. Sci.*, **12**(1-4): 226-229.

Khimani, R.A. and A.A. Patil, 1994. Response of gaillardia (*Gaillardia pulchella* Foug.) to varying levels of naturally saline soil. *Recent Hort.*, **1**(1): 97-101.

Khimani, R.A. and A.A. Patil, 1995. A note on shelf life studies in gaillardia. *Gujarat Agric. Univ. Res. J.*, **20**(2): 190-191.

Khimani, R.A., A.A. Patil and M.B. Chetty, 1994. Influence of CCC, Alar and MH on yield and physiological growth parameters of gaillardia during *rabi* season. *Recent Hort.*, **1**(1): 91-96.

Koning, R.E. 1983. The roles of plant hormones in style and stigma growth in *Gaillardia grandiflora* (Asteraceae). *Amer. J. Bot.*, **70**: 978-986.

Koning, R.E.1984. The roles of plant hormones in the growth of the corolla of *Gaillardia grandiflora* (Asteraceae). *Amer. J. Bot.*, **71**: 1-8.

Mishra, H.P. 1998. Effect of nitrogen and planting density on growth and flowering of gaillardia. *J. Ornam. Hort.* (N.S.), **1**(2): 41-47.

Mukherjee, D. and T.K. Bose, 1972. Comparative effects of B-nine and CCC on some species of plants under Compositae. *Plant Sci.*, **4**: 95-100.

Pal, P. 1989. Gaillardia. In: *Commercial Flowers* (eds Bose, T.K. and L.P. Yadav), pp. 852-853. Naya Prokash, Calcutta.

Pallavi, T., A.E. Rane, S.S. Ali and P.S. Burange, 2002. Occurrence of insect-pests of gaillardia (*Gaillardia pulchella*). *J. Soil & Crops*, **12**(1): 101-104.

Pellett, N.E. and D.A. Heleba, 1995. Chopped newspaper for weed control in nursery crops. *J. Environ. Hort.*, **3**(2): 77-81.

Pillai, K.G., I. Usha Rao, I.V.R. Rao and H.Y. Mohan Ram, 1992. *Plant Cell Rep.*, **10**(12): 599-603.

Rajitha, B.S., S.G. Rao, Y.B.N. Rao and D. Singh, 1999. Postharvest effect of fluoride and other chemicals on the vase life of *Gaillardia pulchella* cut flowers. *J. phytol. Res.*, **12**(1-2) 107-109.

Ramakrishnan, V., W.M.A. Khan and A.M.M.A. Khan, 1970. Studies on the germination of seeds of a few ornamental flowering annuals. *South Indian Hort.*, **18**(3-4): 93-95.

Shotri, S.C., J.S. Gupta and R.N. Srivastava, 1983. Seed-borne mycoflora of *Gaillardia:* its pathogenic effects and control. *Acta Bot. Indica*, **1**(2): 253-255.

Singatkar, S.S., R.B. Sawant, S.A. Ranpise and K.N. Wavhal, 1995. Effects of different levels of N, P and K on growth and flower production of *Gaillardia. J. Maharashtra agric. Univ.*, **20**(3): 392-394.

Stoutamire, W.P. 1955. Cytological differentiation in *Gaillardia pulchellla. Amer. J. Bot.*, **40**(10): 912.

Stoutamire, W.P. 1977. Chromosome races of *Gaillardia pulchella* (Asteraceae). *Brittonia*, **29**(3): 297-309.

Trippi, V.S. 1965. Studies on ontogeny and senility in plants. XI. Leaf shape and longevity in relation to photoperiodism in *Gaillardia pulchella. Phyton Argentina*, **22**: 113-117.

Turner, B.L. 1972. Two new Gypsophilous species of *Gaillardia* (Compositae) from north Central Mexico. *The South Western Naturalist*, **17**(2):181-190.

Yuan, M., W.H. Calson, R.D. Heins and A.C. Cameron, 1998. Effect of forcing temperature on time to flower of *Coreopsis grandiflora, Gaillardia grandiflora, Leucanthemum superbum* and *Rudbeckia fulgida. HortSci.*, **33**(4): 663-667.

18

Gerbera (Family: Asteraceae)

Sanyat Misra and R.L. Misra

[**Common names**: African daisy/Barbertson daisy/Transvaal daisy (*Gerbera jamesonii*), Hilton daisy (*G. aurantiaca*) and Gerbera]

Introduction and Origin

The genus is named after Traugott Gerber, a German naturalist who travelled in Russia, and died in 1743. The true *Gerbera jamesonii* species (native to Natal, Transvaal and Swaziland) along with *G. viridifolia* (native to Cape and elsewhere from 5 °S of the equator and 35 °S latitude in Africa), when for the first time in 1880 this was introduced to the Cambridge Botanic Gardens (England) by Robert Jameson (a Scotsman) who while operating a gold mine near Barberton in the Transvaal (South Africa) collected it. R.I. Lynch in 1887 crossed both the species together and named the resultant hybrid as *G. cantabrigiensis,* which afterwards used to be called as *G. jamesonii* itself or sometimes even *G. hybrida* and this hybrid species with the name of the original species, *i.e. Gerbera jamesonii* is being cultivated presently worldwide with its many varieties in gardens and greenhouses. Its species, apart from Africa, are also found in Asian regions (temperate Himalayas from Kashmir to Nepal). Extensive breeding at the University of California at Davis during 1970s led to development of many gerbera varieties suitable for garden use. A German, Robert Diem established his famous 'Gerbera Nursery' in Bordighera (Italy) which is a very reputed nursery even today though he died in 1955. As a garden or a cut flower plant, the gerbera was well established in Riviera (France) by 1910, in North America by 1930s and it has occupied the most prestigious place as a cut flower plant in trade in Holland in the present time. The majority of cut gerberas come from Columbia and the surrounding countries in South America. In India, it is mainly cultivated in West Bengal, NEH States, Maharashtra, Andhra Pradesh, Karnataka, Tamil Nadu and Uttarakhand, and there are commercial projects around Pune, Karnataka and Tamil Nadu. They are suitable for planting in the beds, the borders, rockeries, dish gardens, pots, patio pots, mixed containers and for various floral arrangements. In vase water, these last longer. Gerbera is remarkable for the unique geometrical regularity in its flower forms, singles and doubles, and delicate colours (self and bicolours).

Botany, Genetics and Breeding

The plants are herbaceous perennials bearing petioled (15 cm), radical, lanceolate, deeply-lobed (sometimes entire) leaves (12-20 cm) long but narrower at base and wider at top, coarse in appearance, wrinkled on the upper surface and arranged in a large rosette at the base. Its inflorescence is capitulum. Its floral stems are long, erect, leafless but hairy throughout and bear solitary terminal heads with many flowers. The flowers are daisy-like and may be single, giant-flowered single and double, 7-15 cm across (larger in the hybrids), and with regard to colouration these have white, amber, cream, yellow, pink, terracotta, orange, scarlet, salmon, and red colours. Ray flowers in outer 1-2 rows which are functionally female where stamens are converted into staminods made up of thin sterile filaments. Three of the five petals in these ray florets are joined to form a broad ligula and the remaining two making a smaller lip, and the style is long with forked stigma. Disc flowers which remain in the centre are tubular, functionally male and may be of the same colour as the rays, mostly yellow or vividly contrasting black, and are many to almost nil (Tourjee *et al.,* 1994). Also there are transflorets, *i.e.* hermaphrodite or bisexuals in between the ray and disc florets where anthers liberate the pollen before stigma receptivity, a case of protandry, and since flowering proceeds centripetally, hence, whole inflorescence should be taken as protogynous, *i.e.* first flowers to mature are entire female ones in the outer row(s) and in the last the male ones in the central disc (Aswath, 1999). Achenes are beaked and with pappus.

The basic chromosome number in gerbera is n = 25, and *G. jamesonii* is diploid with 2n = 50. Polyploid forms are also reported with desirable characters. On the basis of fertility status in gerbera, the plants could be divided into three types, *viz.* (i) completely self-sterile plants, (ii) distinct differences in seed set in a flower head when pollination occurs here with inflorescences of the neighbouring plants, and when pollination occurs within an inflorescence itself, and (iii) when these differences are observed minimum. For the purpose of large quantity of seed production, third type of pollination is seldom preferred as mostly the seeds are produced through self-pollination within the same inflorescence though with less variation, and mostly with poor characteristics. Gerbera petals contain pelargonidin, cyanidin-3-malonyglucosides, cyanidin-3-glucosides, and the related flavonoid copigments apigenin, luteolin 4-glucosides, luteolin 7-glucosides, apigenin 7-malonylglucoside, kaempeferol, quercetin 3-glucosides, quercetin 4-glucosides and quercetin 3-malonylglucosides (Das and Singh Samanta, 1989).

Wernett *et al.* (1996) when evaluated a large number of *Gerbera* × *hybrida* cultivars, observed the range of heritability from 22 to 35 per cent and reported that additive gene action is postulated to control vase life. In fact, in gerbera, productivity and other major floral characters are determined by additive gene action. There is high heritability for inflorescence diameter and number of ray florets in selfed progeny but relatively low for disc diameter and stem length, though there had been differences in different clones with regard to general and specific combining ability. Anuradha and Gowda (1999a) reported high heritability combined with high genetic advance for weight of ray florets, disc diameter, length and fresh stalk weight, and fresh inflorescence weight, while Mahanta *et al.* (1998a, 1988b) recorded high heritability and high genetic advance for days to bud visibility, days to bud opening and days to full floral opening in 10 gerbera cultivars, and further suggested that leaf area, stalk girth and two flower bud opening could be considered for selection to improve number of flowers per clump.

Aswath *et al.* (1999) reported that flower colour is controlled by two pairs of genes, 'A' causing intense red colour, 'a' dull-red colour, 'H' for causing lighter colouring in case of 'A' and 'a', and 'B' causing ivory or yellow colouration, more pronounced being in absence of 'A'. Huang and Harding (1998) stated that quantitative genetic structure of gerbera flowers is controlled by a few independent components and that the principal component analysis is a useful tool to find out variation in this structure, and these composite traits are heritable. Aswath *et al.* (1998a) found positive correlation between stalk length and days to bud burst but negative correlation between flower diameter and days to bud burst. Anuradha and Gowda (2000) when studied 25 gerbera genotypes at Bengaluru, recorded cut flower yield exhibiting a high level of significantly positive correlation and positive direct effect with number of leaves per plant, *vis-à-vis* significant positive correlation with weight of ray florets and days to flower opening.

Species and Varieties

There are some 45 to 70 species of *Gerbera* scattered over Africa (13 species only from South Africa), Madagascar and Asia, especially Himalayan range from Kashmir to Nepal at altitudes ranging from 1,300 to 3,200 metres (Hansen, 1999), but the only species in cultivation is *Gerbera jamesonii*. *G. asplenifolia* has 10-15 cm long and narrow leaves mostly roundish-lobed, margins conclave, lobed deeply, leathery, glossy above, flowerheads revolute and purple and being borne on a hairy scape. *G. aurantiaca*, a native to Natal and Transvaal, bears 12-15 cm long leaves which are lanceolate to oblong, acute, entire or toothed and the flowerheads orange to red with bright yellow anthers. *G. hintonii* (previously known as *Chapatalia hintonii*) is endemic to Mexico, bearing flowers resembling Gerbera, hence this has been included under this genus (Katinas, 1998). *G. linnaei* bears lanceolate and pinnatifid to deeply pinnatisect leaves, ray florets white or white above and purple, red-brown or yellow below and the disc florets are usually dark. *G. maxima* has been rediscovered after a century from Pauri Garhwal by Silas and Ansari (1999). *G. kunzeana* is an Himalayan species whose flowers open seldom. *G. viridifolia*, a native of Cape (South Africa) is intolerant to any frost, height 45 cm, and bears leaves elliptical to oblong, obtuse, almost smooth, floral stalk small, and floral head small and off white in colour but base of petals yellow. *G. jamesonii*, a native to South Africa which is the only species in cultivation, not hardy, height about 75 cm, its leaves are long-stalked and deep green, undersurface hairy and woolly, lobes dentate, floral stalk long and stout but hollow and bears daisy-like solitary flowerhead some 5-8 cm in diameter having some 30 long but narrow petals of yellow, orange or red colour. It has numerous varieties in various shapes and colours.

Majority of gerbera cultivars are exotic. The export quality cut flower varieties should have thick stems terminating in the absolute circular, double and flat blooms facing outwards, and the blooms should have charming clear colours in their ray as well as disc florets. The petals in double forms should be firm with easy recognizable stage of harvest. Further, the cut flowers in vase should last longer with no bent neck. The variety should be easy grower with uniform growth and short gestation period, suitable to various agro-climatic conditions, flowering should be continuous for a long period, blooming even during winters, and should be insect- and wilt-resistant. Out of the four varieties of gerbera ('Eoliet', 'Presley', 'Pritty' and 'Sunbird') studied in Kerala, Suma (1993) recorded earliest flowering in 'Presley' with maximum number of blooms and latest in 'Eoliet' with minimum number of blooms. Choudhary *et al.* (1998) evaluated 10 gerbera cultivars at Jorhat and reported 'Popular', 'Evening Bells', 'Red Monarch' and 'General Kaiser' performing well there. On the basis of a trial at PAU, Ludhiana with five cultivars ('Blorosa', 'Goldspot', 'Lilabella', 'Lily' and 'Rosabella'), Deepak Kumar and Ramesh Kumar (2000) recorded 'Goldspot' as the best cultivar for quality and quantity flower production. At Yercaud, Jawaharlal *et al.* (1998) evaluated 49 accessions and reported 14 aaccessions producing 8.00 to 9.45 cm diameter of flowerheads, six accessions producing more than 80 flowers per clump, and the vase life ranged from 3.95 to 8.55 days. Under greenhouse conditions at IIHR, Bengaluru nine exotic varieties, *viz.* 'Diable', 'Lyonella', 'Ornella', 'Sunset', 'Tara', 'Thalassa', 'Tiramisu', 'Twiggy', and 'Whitesun' were tested where 'Lyonella', 'Ornella', 'Tiramisu' and 'Twiggy' were observed quite promising.

Cultivars 'Cidda', 'Jautriite', 'Mikus', 'Raunis', 'Ruusinjs' and 'Toms' flower in winter with good quality of blooms. 'Aime', 'Aira', 'Lina' and 'Zeltene' have all the good characteristics which a florist gerbera should have.

There are even seed-raised gerbera in various shades of yellow, apricot, orange, pink, scarlet, red and lavender. Such important series are California Giant, Pandara series, Parade series, Sunburst series and Tempo series (Brickell, 1996).

The cultivars commercially famous throughout the world are 'Aalsmeera' (orange), 'Adamant', 'Agnes', 'Aida' (orange), 'Alcatraz' (red), 'Alexis', 'Aligator' (red), 'Alp' (white), 'Amber', 'Anke', 'Ansofie' (off-white), 'Appelbloesem' (pink), 'Arabella', 'Arka Krishika', 'Aruba' (yellow), 'Aviva', 'Balance', 'Baron' (red), 'Basic' (red), 'Bilitis','Bloemfontein', 'Blush' (red), 'Bonaparte', 'Cabana', 'Cacharelle', 'Caprice', 'Carrera', 'Casava' (red), 'Clivia', 'Cornice' (orangish-yellow, disc orange, centre black), 'Corazon', 'Cream Clementine' (cream), 'Croliva', 'Crossroad' (yellow, centre red), 'Daikiri', 'Dalma', 'Davis Memory' (light mauve), 'Dazzling', 'Diablo', 'Diana Ellen', 'Delphi' (white), 'Denise', 'Doni', 'Dusty' (red), 'Easter Star', 'Elegance', 'Ellymay', 'Enzelt Orion' (golden, large double), 'Esmara', 'Essandre', 'Estelle', 'Evening Bells', 'Fabiola', 'Faith' (orange), 'Farida', 'Fermin', 'Feugo', 'Fogaro', 'Fiona' (light pink), 'Fiorenza', 'Flamingo' (pale-rose), 'Forza', 'Foske' (orangish-yellow, disc small and orange), 'Fredaisy' (pink), 'Fredeking' (yellow), 'Fredigor', 'Fredorella' (red), 'Fresaline', 'Fresinelle', 'Froschkonig', 'Fuego' (red), 'General Kaiser' (mauvish-orange), 'Golden Gate', 'Goldspot' (orangish-yellow), 'Goliath' (orange, centre yellow), 'Gomarchi', 'Gracia' (large, violet-red), 'Gucci' (pure yellow), 'Hildegard', 'Ibiza', 'Ice Queen', 'Ira', 'Italasa', 'Jaffa', 'Jankfrau', 'Joyce','Julia', Kabada', 'Kabandha-Mufeta', 'Labalgo' (lilac), 'Labazzo', 'Labignon', 'Lablinel', 'Lancaster', 'Laureen', 'Laurentius' (yellow), 'Lilabella' (purple-red), 'Lindessa', 'Lyonella', 'Magnum' (pink), 'Malibu', 'Mammut', 'Marasol', 'Maria', 'Marleen', 'Marmara', 'Maron Clementine' (orange), 'Michelle', 'Monaco', 'Nadja' (yellow), 'Nena', 'Nevada' (light yellow), 'Onyx' (lilac), 'Optima', 'Orange Elegant', 'Oranje Nassau', 'Ornella' (red), 'Paintball', 'Paradizo', 'Pascal', 'Peleton', 'Perdita', 'Pink Elegance' (deep pink), 'Pioneer', 'Pink Pampadour', 'Piton' (yellow), 'President' (red), 'Princess' (yellow), 'Red Monarch', 'Red Uranus', 'Roma', 'Romeo', 'Romilda', 'Rosabella' (pinkish-mauve), 'Rosalin' (soft mauve), 'Rosula', 'Rotelli', 'Rozamunde', 'Ruby Red', 'Rutina', 'Salina', 'Salmorosa', 'Salvadore', 'Sangria' (red), 'Saskia', 'Sazou' (light yellow), 'Silvester', 'Snow Flake', 'Snow Star', 'Stanford', 'Star Bright', 'Sundance' (yellow, margins intense), 'Sun Gold', 'Sunset' (red), 'Sun Way', 'Supernova', 'Suzan', 'Sylvia', 'Sympathie', 'Symphonie' (white), 'Talasa', 'Terracalypso', 'Terracorso', 'Terrafame', 'Terramaxima', 'Terramint', 'Terramor' (light red), 'Terranutans', 'Terraparva', 'Terra Parade', 'Terraqueen' (pink), 'Terrascarlet', 'Terraspirit', 'Terra Sun', 'Terravisa', 'Tropic Blend' (yellow), 'Uranus' (yellow), 'Valentine' (pink), 'Venturi' (red), 'Virginia', 'Veronica', 'Vesta' (red), 'Vino' (red), 'Whitsun' (white), 'Winter Queen', 'Xenia', 'Yanara', 'Yesnara', 'Zalmora', 'Zitronella' (yellow), etc. Some of the mini-gerberas are 'Ashley' (purplish), 'Laurie' (red), 'Lily' (orange), 'Nini' (yellow), 'Northstar' (red), 'Siby' (white), 'Sissy' (purple), etc. Korean are also actively engaged in gerbera breeding and about eight years back they developed cv. 'Honeymoon' (single-type, large and red with brown centre having vase life of 9.1 days) and 'Pink Angel' (semi-double, pink with brown centre having vase life of 11.5 days and producing some 47.5 flowers per plant per year).

Messers Terra Nigra B.V. of Holland has developed some of the excellent cultivars such as 'Easter Star', 'Estelle', 'Nena', 'Terra Parade', 'Terra Sun', etc. Florist de Kwaket of Holland also developed certain beautiful cultivars which are 'Ansofie', 'Aruba', 'Ashley' (mini), 'Fiona', 'Fuego', 'Goldspot', 'Laurentius', 'Laurie' (mini), 'Lily' (mini), 'Nevada', 'Nini' (mini), 'Northstar' (mini), 'Ornella', 'Princessa', 'Rosabella', 'Sangria', 'Siby' (mini), 'Sissy' (mini), 'Sundance', 'Sunset', 'Venturi', 'Vino' and 'Whitsun', and these varieties in India are being multiplied through tissue culture.

The cultivars which are being grown commercially in Maharashtra, Karnataka, Kashmir and elsewhere in India are 'Aida' (peach), 'Alberto', 'Alexis', 'Aalsmeera', 'Beverly' (yellow), 'Black Heart', 'Cabana' (yellow), 'California', 'Caressa' (pinkish-white), 'Carrona', 'Congo', 'Dana Ellen' (yellow), 'Detty', 'Diablow' (red), 'Dingo', 'Divas Memory', 'Ellymay' (red), 'Essence' (pink), 'Estoril' (orange), 'Evening Bell', 'Francella' (pink), 'Funda', 'Gloria' (white), 'Golden Gate', 'Goliath' (orange), 'Grizzle', 'Kalika', 'Lila Bella', 'Lyonella' (yellow), 'Magnum' (pink), 'Nevada', 'Optima', 'Ornella' (red), 'Pink Elegance' (pink), 'Pink Sparklet' (pink), 'Pink Star', 'Piton' (yellow), 'Racko', 'Red Explosion' (red), 'Regilla', 'Reward', 'Rosalin' (pink), 'Rosna', 'Ruby Red' (red), 'Salmon Spray', 'Sangria', 'Savannah' (red), 'Snowflake' (white), 'Sunanda', 'Sunset', 'Supernova' (yellow), 'Tavita', 'Testa Rassa', 'Thalassa' (yellow), 'Tiramisu', 'Twiggy', 'Vina', 'Xamena', 'Zankfrau', etc.

Some of the gerbera varieties which Kumar Florist is multiplying commercially for sale in India are 'Golden Gate' (double, yellow with black centre), 'Junkfrau' (double, white), 'Foske' (single, peach with black centre), 'Mystique' (single, orange with black centre), 'Pink Sparklet' (single, pink with black centre), 'Rubian' (single, red with black centre), 'Ruby Red' (single, red with black centre), 'Savannah' (single, red with black centre), 'Ringo' (single, lilac), 'Tonga' (single, magenta), 'Sangria' (semi-double, red with black centre), and the rest are single and double both which are being furnished colourwise such as **reds:** [Diablo (black centre), Testarossa], **yellow:** [Cabana (black centre), Piton (black centre), Tonneke (black centre), Thalassa (black centre)], **white:** [Dalma (black centre), Winter Queen (black centre in doubles)], **peach:** Aida (black centre in doubles), **pink:** Charmander, Fleurance, Pink Elegance, Primrose, **orange:** Careera, Gollath, Oriella, Rodis (black centre), Sunset (black centre), Sunway, **cream:** Tiramisu (black centre), Vital (black centre), **lilac:** Magnum (black centre), and **magenta:** Aquilla. This Company would have introduced even many other varieties by now.

Propagation

Commercially important clones are multiplied only through vegetative means. During monsoon in the plains and during March-April in the temperate regions, the plants are lifted, leaves removed, roots pruned and then clump is **divided** into separate units containing 2-4 prominent buds (crowns). These divided sub-plants are planted in ready-to-plant pots or in the greenhouses as per the requirement. New roots emerge with the commencement of plant growth. To accelerate its growth or for even better multiplication, these are multiplied through **cuttings**. Lifted crowns are removed of their entire foliage and excess roots, even by shortening, and are kept closely on a bench with damp peat surrounding these, in the greenhouse around 25 °C under high humidity where new leaves come up soon. These shoots with one pair of leaves are severed from the mother plants with a small piece of young tissue, base being dipped in 0.8 per cent IBA and are inserted in the rooting medium at 25-28 °C under high humidity, preferably in mist chamber where these root soon and then these are planted in the separate containers or directly in the field for further growth and flowering (Salinger, 1985). Through cuttings it takes 70-80 days for transplanting.

The quickest way for bulk multiplication of a gerbera clone is only through **micropropagation**. In India, for commercial cultivation gerbera is propagated through tissue culturing. There are many such laboratories in different regions of the country for its rapid multiplication to meet the demand of commercial growers. Pierik *et al.* (1973, 1982) succeeded with capitulum explants through activation of tissues with auxin and cytokinin though with slow shoot formation; and when such explants were cultured on MS medium, MS organic constituents, macro elements at half strength, Heller's micro elements, Na2Fe-EDTA at 21.4 mg/l, BA at 10 mg/l, 1 per cent sucrose and Bacto-agar at 0.8 per cent with adjustment of pH at 5.6 and then subjected to darkness for two weeks, followed by illumination at 800 lux for 16 h daily, Chu and Huang (1983) succeeded in getting >5 shoots per explant after 16 weeks of culturing. Laliberte *et al.* (1985) obtained excellent shoot multiplication through young capitulum explants on medium containing 0.1 mg/l IAA and 1-2 mg/l BA though half strength MS caused callus formation. Huang *et al.* (1987) got success on half strength MS medium with 0.03 mg NAA/l where plants flowered four months after transplanting. Capitulum explants of the cvs 'Applebloesem', 'Clementine', 'Marleen' and 'Pimpernel' were when cultured on modified MS medium, Arello *et al.* (1991) observed best results with addition of 2 mg BA and 0.5 mg IAA/l. Shoot tips were found as the best explants by Murashige *et al.* (1974), and Hedtrich (1979) observed shoot formation in gerbera cv. 'Vulkan' on modified MS medium supplemented with 1 mg/l benzylamino purine and 0.1 mg/l GA_3. Parthasarthy and Nagaraju (1999) studied shoot explants (1.0-2.0 cm long) of the cvs 'SWM' and 'Dilmaya' on MS medium with varying concentrations of benzyladenine (BAP) and observed best results with 1.0 mg/l (BAP) with regard to culture growth and callus weight though 0.5-1.0 mg/l BAP for shoot growth. Parthasarthy *et al.* (1997) recorded success even with leaf explants from fully expanded leaves on MS medium and stated that callus growth was higher with NAA, BA and IBA at 1.0 mg/l but adventitious roots were observed when MS medium was supplemented with 1 mg NAA + 0.75 mg BA or 0.75 mg BA/l, and Orlikowska *et al.* (1999) stated that young leaves are better than petiole bases of old leaves for adventitious shoot regeneration. Petru and Matoush (1984) got success with young floral buds cultured on modified MS medium supplemented with 7 mg kinetin and 0.2 mg IAA/l, though for rooting IAA level was increased to 8 mg/l and kinetin was dropped. Radice and Macaroni (1998) studied immature inflorescences of six gerbera cultivars in the dark or in the light on half-strength MS medium supplemented with adenine sulphate, IAA, BA and sucrose and obtained 70-100 per cent rooting with 0.5 mg IBA, and some 80 per cent of plants were successfully transferred after acclimatization. Floral peduncle explants of gerbera were successfully regenerated *in vitro*, rooted and after acclimatization were transferred to soil by Lê *et al.* (1999). Ruffoni and Massaba (1991) used leaf, petiole and apex as explants to establish tissue and callus cultures and reported numerous root formation in root explants while more shoot and callus formation in the apex explants, shoot development being better in light than in darkness. Mitika Kalra *et al.* (2008) taking young leaf and capitulum explants reported 70 per cent alcohol for 25-30 seconds best for explant disinfection in gerbera cv. 'Terraqueen', basal MS salts supplemented with 5 mg/l BAP and 0.1 mg/l IAA for maximum shoot induction, MS basal medium containing 5 mg/l BAP + 0.25 mg/IAA for callus induction in young leaf explants, and full strength MS basal salts supplemented with 0.2 mg/l IAA for minimum days to root induction and maximum rooting. Kaur *et al.* (1999) tried to replace certain expensive organic and inorganic chemicals of MS medium and got success without affecting multiplication rate even during subcultures and with 100 per cent survival rate after transfer to pots.

For **seed** propagation, single types are best suited as these set seeds freely and there is no much risk of losing the original form and colour as mostly these are not commercial varieties. Elite clones can be propagated through seeds when only new varieties are to be developed because commercially people, as per their mind-set, have preference for the same colour and form a variety possesses. Gerbera seeds are very expensive, delicate and sensitive to germinating conditions. Propagation from seed is not easy as it is not easy to obtain fresh seeds, and after three months the seeds lose their viability. Storing the seeds at -5 to 5 °C prolongs the viability up to 12 months (Sheela, 2006) provided the seeds are dried to a moisture level of 3.5 to 5.7 per cent and stored at 32 per cent relative humidity but before planting these should be subjected to 52 per cent moisture level for maximum germination (Carpenter *et al.*, 1995, Dole and Wilkins, 1999). Considering that the crop requires 14-18 weeks from seed to flower or 16 weeks from germination to flowering or after 11 to 16 weeks after transplanting *in vitro* propagated plugs (Dole and Wilkins, 1999), many of the commercial gerbera growers procure established plugs from specialist propagators. Once after opening the packet, whole lot should be sown at once as the seeds loose their viability rapidly on exposure to room conditions. Seeds may be sown vertically, *i.e.* the points downwards so that the little tuft of hairs at the tip remains at soil level (Pizzeti and Cocker, 1975) in the beds, pans, polythene bags, open flats or in plug trays, without covering the medium (Moe *et al.*, 1996). Germination by all means is complete within 15 days. Transplanting from beds, pans or flats is labour-intensive and delays establishment in the new containers or directly in the field, *vis-à-vis* encounters some casualty, hence also the plug planting is considered best. Sowing media is watered before sowing and after sowing the media are covered with a very thin layer of vermiculite and then sowing containers are subjected to high humidity conditions (up to 85 per cent) and 16-18 °C temperature (Hay and Beckett, 1971). High pressure sodium lamps supplementing natural light at 500 f.c. improves germination and seedling's dry weight, and decreases days from germination to visible buds (Erwin *et al.*, 1991). Seedlings come up between 7 to 14 days from sowing at an optimum temperature of 20-23 °C (Erwin *et al.*, 1991) and now these containers are taken to greenhouse. These seedlings are pricked off when large enough to handle, *i.e.* after four weeks, into boxes, and later transferring singly to 7.5-10.0-cm pots of the growing compost and when good root ball has formed these can be replanted into larger receptacles or planted out in flowering positions.

Cultural Practices

Its rootstock is compressed and rhizomatous. In seedlings, initially the vegetative shoot is formed terminating into a flower, followed by second flower, then third flower and so on the process goes on repeating. Before first flower appearance some 2-8 leaves are formed on the plant. Subsequently, every time for production of flowers new lateral vegetative shoots emerge with almost same number of leaves and these vegetative buds are known as crown and so in a course of time the plant becomes quite leafy due to primary rootstock, secondary and so on. A night **temperature** of 20 °C is preferable until establishment of the root system, and afterwards 18 °C. During day time when temperature is above 23 °C, there should be proper ventilation in the greenhouse and during night it should have full aeration otherwise *Botrytis* can be a problem during dark and cold periods under high humidity conditions. Relative **humidity** should be around 70 per cent but with optimum aeration. Warmer temperature encourages rapid growth though cold temperature slows it down. Gerbera plants are slightly photoperiodic but are highly responsive to high light intensity and its duration. Short days speed up flowering while long days delay it. Supplemental high-pressure sodium lighting of 500 foot candles can be used for good and regulated flowering. CO_2 concentration at 600-800 ppm in the greenhouse will be quite useful to the plants in presence of supplementary lighting.

A light and well-drained **soil**, *i.e.* sandy-loam which is rich in organic matter (at least 30 per cent with high nitrate content) and has a medium pH range, *i.e.* 5.5 to 6.0 or at the most 6.5, is most suitable for its growing. Soil reaction at pH range of 5.0 to 7.2 is not injurious to the crop if there is sufficient organic matter in the soil with good soil structure and where water table is medium, *vis-à-vis* no drainage problem (Das and Singh Samanta, 1989). EC of the soil should be around 0.5, however, around 1.0 is dangerous. In high water table areas, desired water table can be achieved by providing sub-surface drainage three metres apart at a depth of more than 50 cm (Mondal *et al.*, 2007). Moderately alkaline (calcareous) soil having the above characteristics is excellent for gerbera cultivation, however, the excess of lime or excess moisture at the roots causes chlorosis where leaves show yellowing. Peat soils are quite unsatisfactory, specially where water table is high. The soil should be sufficiently deep as roots of gerbera grow below up to 50-70 cm in porous soils. An ideal soil mixture for gerbera cultivation would be 1 part coarse sand + 1 part fibrous loam + ½ part peat + ½ part well-decomposed leafsoil + ¼ part well-rotten manure (Pizzetti and Cocker, 1975). Das and Singh Samanta (1989) stated that some of the uncertainties associated with gerbera cultivation may be removed by growing them in peat with pH regulation.

Gerberas adapt well to most commercial **potting** materials that are rich in organic matter, are quite porous for aeration, have good water retaining capacity and permit proper drainage. Potting mixture as peat should have to a maximum of 1 kg/m^3 of lime. Pine bark 9 parts:birch bark 1 part through frequent treatment with hot water for removing excess tannins and then composting of the same for about six months will make them suitable as gerbera potting mixture. Addition of urea at the rate of 3.0-5.5 kg/m^3 and by volume some 10-15 per cent of active sludge to this bark medium will be quite beneficial as this will substitute the nitrogen loss during composting. Composted pine bark and sphagnum peat moss in equal quantity by volume is also good. However, soils or soil composts are used as potting media, these should contain about 20 per cent of hydrogell or hydramull to improve their moisture retention capacity. Bik and Berg (1984) when tried gerbera cv. 'Fleur' in 25 or 75 per cent mixture of perlite + sphagnum peat or fibrous peat + sphagnum peat or in rock wool having 8.5 to 25.0 cm depth, highest number of flowers was recorded with plants in peat mixtures having depth of 17 and 25 cm. Vidalie *et al.* (1985) recorded yield of gerbera flowers better in Grodan than in Cultilene, though Charpentier *et al.* (1986) did not record any difference between the two types of rockwool. Lamanna *et al.* (1991) with gerbera cv. 'Valentine' recorded better plant growth under peat + compost as compared with peat or compost material alone such as grape stalks, peat, clay granules, perlite, hydrogel, bark, sewage, sludge and phenolic resins. 'Amber' and 'Joyce' varieties of gerbera when were grown for 15 months by adjusting the average pH between 5.6 to 6.7 through use of NH_4 to the nutrient solution or increased Mn supply even at the highest pH, chlorosis was found drastically reduced, and though Mn did not play any role in flower yield, but at the lowest pH flower yield was 26 and 16 per cent higher in both the respective varieties, as compared to higher pH (Sonneveld and Voogt, 1997).

It would be wise to **disinfect** the soil before planting to avoid the menace of *Phytophthora* root rot. Soil may be solarized by covering it with black polythene film for 6-8 weeks so that solar heat is generated and all the spores of harmful fungi are killed. Alternatively, formalin 7.5-10.0 litres/100 m^2 through 10 times dilution should be drenched and immediately covered for a week with polythene sheets. After taking out the plastic the soil should be flooded with clean water to drain out the traces and then only after two weeks planting should be carried out. Methyl bromide @ 25-30 g/m^2 or Basamid (dazomet) @ 30-40 g/m^2 is equally effective.

For planting soil should be made fully pulverized through deep and thorough ploughings, weeds and the rootstocks of the perennial weeds along with other foreign material taken out and then properly levelled. Usually gerberas are grown on 15 to 20 cm raised beds for better drainage. Bed width may be adjusted at 160 cm but to any length as per the convenience. Between two beds, there should be clearance of 30 cm to facilitate walking and for irrigation, *vis-à-vis* to attend to all the cultural operations in the field. While **planting**, the crown should be well above the surface of the soil. In case of annual cropping the spacing should be close while for perennial cropping it should be wide. Pots planted with gerbera should be sufficiently spaced so that the leaves from one pot should not overlap the other otherwise there will be competition for light. According to Dufault *et al.* (1990) and Rogers and Tjia (1990), the **spacing** for 10-cm pots should be 18.0-21.5 cm (4.2-9.0 plants/m^2), 13-cm pots 28 × 28 cm and 15-cm pots 36 × 36 cm, however, in the beds or in the large pots for the plants being grown for cut flower production the recommended spacing is 33 × 33 cm to 38 × 38 cm apart. In India, the proper **planting time** for most of the tropical and sub-tropical regions should be October while in Kerala it has been found as June (Suma, 1993). Steen (1975) grew gerberas in rows at 20-40 cm with plant density of 9.4-4.7 plants/m^2 and obtained more number of flowers in closest spacing, the results being more pronounced in the first three months (September-November) of flowering though afterwards there was little difference in yield. However, it is wise to have plant density of 7.5 plants/m^2. Buisman (1985) took 20 cm (9.37 plants/m^2), 25 cm (7.50 plants/m^2) and 30 cm (6.25 plants/m^2) spacing in the row and obtained better yield under closest spacing with flower quality similar to widest spacing, however, yield per plant was lowest but per square metre area it was more. Lisiecka (1991) also obtained similar findings with 30 × 15 cm (22.2 plants/m^2), 30 × 20 cm (16.6 plants/m^2) and 30 × 25 cm (13.3 plants/m^2) spacings. Planting time in India in the greenhouse is September-October. Planting on the hills can be carried out in February-March. In case of gerbera no pinching is required, however, the very first flower may be nipped to promote vegetative growth and uniformly in flowering. If planting has been done in sterilized soil, no weed problem may be encountered, however, in case of any weed coming up, immediately that should be taken out.

Feeding and Irrigation

Gerbera is heavy **feeder**. Frequent applications of nutrients at regular intervals are useful for optimum growth and flowering of gerbera. Das and Singh Samanta (1989) suggested application of 750 quintals of well-rooten stable manure per hectare in fairly light sandy soil at the time of preparation of land for planting of gerbera, and particularly in the second year they mention incorporation of specially prepared peat as a substitute for manure provided the pH of the soil is adjusted in between 6.0 to 7.5 through liming. Gutmane and Grivina (1975) suggested that largest number of gerbera flowers with better vegetative growth and increased NPK concentration in the leaves and roots can be produced by growing in peat with ammonium nitrate + superphosphate + potassium sulphate at 4 g/l than 2 g/l. To determine nutrient status in gerbera plants, leaf blade is considered more reliable than petiole, and the good flower yield is recorded when its leaves contain 2.7-3.13 per cent N, 0.19-0.35 per cent P, 3.06-3.64 per cent K, 1.66-2.18 per cent Ca and 0.3-0.48 per cent Mg, so it is always better to apply any of the chemical nutrients only after soil and leaf analysis. Through the use of ammonium fertilizer in greenhouse gerbera, Kasten *et al.* (1990) recorded 60 per cent reduction in nitrate content of the soil while placement of straw fertilizer together resulted into 80 per cent reduction, and there was less fertilizer leaching in the groundwater. Based on young leaf stage, *i.e.* less than 2.5 cm long, recently mature, *i.e.* 5.0 to 8.75 cm long or mature ones, *i.e.* greater than 10 cm in length, the following nutrient contents in potted gerbera leaves were reported (Dole and Wilkins, 1999):

Nutrient Element	*Concentration*		
	Young	*Recently Mature*	*Mature*
N per cent	2.5	2.7	3.0
P per cent	0.5	0.5	0.5
K per cent	3.2	3.2	3.8
Ca per cent	0.5	0.4	1.3
Mg per cent	0.2	0.4	0.6
Fe (ppm)	62	62	132
Mn (ppm)	17	30	62
Cu (ppm)	2	2	4
Zn (ppm)	19	19	24
B (ppm)	19	19	24

Excess of N reduces yield and vase life (Moulinier and Montarone, 1978) and its deficiency causes yellowing of old leaves as this translocates in the new leaves upward. Along the underside veins of old leaves, brown discolouration is found in case of phosphorus deficiency, potassium deficiency causes marginal necrosis of old leaves, and Ca deficiency causes intense yellowing of new leaves. Mg deficiency is characterized as thick and crispy interveinal chlorosis on older leaves. Magnesium and iron deficiencies are most common in gerbera field as the plants have high requirement for these minerals (Dole and Wilkins, 1999). Shoemaker and Carlson (1985) recommend monthly application of magnesium sulphate at 1.2 g/l and iron chelate constantly at 0.037 g/l. In fact, nitrogen to the plants in the form of top-dressing should be applied in the range of 5 to 25 g/m^2 and never more. Tjia and Rogers (1984) recommended that a 70:30 nitrate:ammonia enhances earliness and flower yield. Strojny *et al.* (1976) for the composted pine bark medium recorded optimum level of N as 250 mg/l. Mantrova *et al.* (1982) while using 3 part peat:1part soil:1 part sand as the growing medium, applied various fertilizer combinations [(i) N at 0.025 g/kg of soil applied at 15-17 days after planting, (ii) 0.1g N and 0.15 g K_2O/kg of medium at active vegetative growth stage, (iii) N 0.1 g and P_2O_5 0.1 g/kg of medium at bud formation stage, and (iv) 0.1 g N as well as P_2O_5 and 0.23 g K_2O/kg of medium at flowering stage] and observed better plant growth as well as best flower production. Kamel *et al.* (1977) found decreased number of flowers with increased doses of slow-release NPK fertilizer, and Blomme and Dambre (1979) also reported that increased doses of slow-release NPK fertilizer Plantoson 4D as top-dressing from 20 g to 60 g/m^2 decreased flower production in gerbera. In a further trial with gerbera, Blomme and Dambre (1984) in Belgium applied N and K_2O each at 150 mg/l with irrigation water and recorded higher number of flowers from January 5 to April 10, though from April 13 to May 31 it was slow-release fertilizer mixture of N + K_2O at 20 + 10 g/m^2 which provided better yield, *i.e.* 34 flowers/m^2. Lisiecka (1976) tried 25 NPK combinations and found best results with 2:1:3, *i.e.* 281-293 mg N, 127-133 mg P_2O_5 and 360-440 mg K_2O/l of peat substrate, while Kamel *et al.* (1975) reported 2g N + 4 g P_2O_5 + 1 g K_2O/l as the best for a 25-cm pot for increased number of flowers though for tallest floral stalks he suggested 1:2:0.5 g NPK/25-cm pot. Skalska (1979) found 1.5 N_2:0.8 P_2O_5:2.5 K_2O at 2 g/l of substrate giving best results. Skalska (1980, 1985) grew gerbera cv. 'Lada' in 1 part loam:2 parts peat substrate with four NPK treatments and recorded 40 flowers per year when N level in spring, summer, autumn and winter was 40, 20, 10 and 5 mg/100 g of substrate, respectively, and when 30 mg N/100 g of potting substrate was used the yield of the flower recorded was 37.77 per plant, *i.e.* 244.86 per square metre. Kacperska (1985) with gerbera cv. 'Peter' recorded quality and quantity flower production with 1.0 N:0.21 P_2O_5:1.21 K_2O at 1-2 g/l of peat. Valpi and Farina (1987) recorded best dose for gerbera annually as 60 kg N, 48 kg P_2O_5 and 90 kg K_2O/1000 m^2. Daufault *et al.* (1990) found increased number of marketable flowers by increasing N and K doses up to 110 kg/ha though increase of N to 220 kg/ha further increased cut flower production but increase in K level did not have any better result. Gerbera requires ample amount of phosphorus and potassium for profuse flowering. Bose and Jana (1978) reported better growth and flower yield in gerbera with NPK application. Minimum foliar level of Mn at 40 ppm prevents chlorosis. When Koch and Holcomb (1982) grew gerbera in the substrate of equal volume of sphagnum moss and vermiculite in the 15-cm pot and top-dressed with half and the recommended dose of Osmocote 14-14-14, Nutricote 14-14-14 or Lesco 14-14-14 NPK fertilizer, the flower quality in half dose of Nutricote was recorded poor.

Micronutrients play a very important role in gerbera cultivation. The plants showing interveinal chlorosis on young leaves is due to iron deficiency and in its serious deficiency it shows yellowish-white colouring. In case of chlorosis where one-half of leaf blade ceases to expand and develop while other half being normal, making the leaves in 'C' shape, is a symptom of zinc deficiency. In case of copper deficiency, the young leaves show chlorosis and flowers also develop poorly. Leaf edge chlorosis is due to molybdenum deficiency and in case of boron deficiency the base of leaves becomes black. Mn deficiency which also causes interveinal chlorosis can be corrected through spraying with 0.1-0.2 per cent $MnSO_4$.H_2O (Voogt, 1976). Iron chlorosis can also be corrected through application of 1 g chelate/m^2. Zn deficiency, apart from chlorosis, also causes rosetting and little leaf which can be corrected through spraying with 0.2-0.4 per cent $ZnSO_4$ (Prabhat Kumar, 2002). Lisiecka (1976) while studying the role of Cu and Mo on the quality of seedling and flowering in gerbera grown in a fertilizer enriched peat substrate having pH range of 4.8-5.5, deficiency of Cu was corrected with application of 60 g $CuSO_4$ and 2 kg Ca CO_3/m^3 of peat and then quality of commercial grade flowers was also found increased with 46.5 per cent more flower yield. However, Fischer and Forchthommer (1987) also advocated 60 mg Cu (5-6 g solution of $CuSO_4.5H_2O$) per plant at 6-8 weeks intervals for cvs 'Hanny' and 'Veronica'. Rodriguez-Navarro *et al.* (2000) when treated gerbera plants with 20 per cent vermicompost with or without chemical fertilizers, obtained optimum levels of macro and micronutrient content in plants, except K and Mn, while without treatment of vermicompost, the plants had lower macro and micronutrient contents except K and Cu. Jacob (1997) reported that though application of *Bacillus subtilis* and VAM (*Vomfungus glomus etunicatum*) to gerbera crop is cultivar dependent but, by and large, is beneficial to plant growth and yield.

Ten gerbera cvs 'Amulet', 'Dana Ellen', 'Doni', 'Goliath', 'Paco', 'Rosalin', 'Salvadore', 'Savannah', 'Sunway' and 'Winter Queen' were evaluated under high-tech greenhouse for their leaf tissue NPK and chlorophyll contents, and observed that though there was no significant difference in N and P content among the different varieties which was from 2.01 to 2.22 per cent for N and 0.19 to 0.30 per cent for P, but K content was recorded significantly higher in 'Dana Ellen' (3.96 per cent) and 'Rosalin' (3.85 per cent), however, the total chlorophyll content was significantly higher in 'Winter Queen' (1.11 mg/g) followed by 'Sunway' and 'Diana Ellen' (Soni *et al.*, 2010). Soni *et al.* (2010) through foliar application of borax, $FeSO_4$ and $MnSO_4$ at 0.1 and 0.3 per cent at fortnightly and monthly intervals on greenhouse gerbera cv. 'Winter Queen' recorded higher leaf N content (3.13 per cent), P content (0.24 per cent) and K content (3.53 per cent) under 0.3 per cent $FeSO_4$ when these were applied fortnightly and under this treatment even chlorophyll content (1.06 mg/g chlorophyll-a, 0.21 mg/g chlorophyll-b and 1.20 mg/g total chlorophyll) was found significantly improved. B content of leaf (38.25 ppm to 33.50 ppm) with borax application, Fe content (324.00 to 300.25 ppm) with $FeSO_4$ and Mn content (136.75 to 123.75 ppm) with $MnSO_4$ were recorded which were found maximum.

Gerberas never like over-watering at any stage of their growth and development. The plants should be allowed to dry slightly between two **irrigations** which will cause plants to retard naturally without any use of growth retardant on potted plants. When watering, these should be sufficiently watered, at least 5 cm flooding throughout so that soil becomes moderately moist to a depth of at least 10 cm. Water requirement of the full-grown gerbera plant is from 500-700 ml per day depending upon the prevailing weather conditions and the growth stage of the crop. Frequent sprinkling of water is avoided in gerbera. Slight dryness of the soils between two irrigations will ensure that there may not be any infection of *Phytophthora* root rot. Sub-surface drip irrigation is the best for maximizing flower production. In the beds under greenhouse conditions, it is only drip irrigation which should be followed and volume of water for each irrigation should be based on evaporation losses and the crop coefficient which is usually 0.32 to 0.52 for vegetative phase and 0.5 to 0.71 for flower production phase. However, irrigation should be based on the soil type, the weather conditions and the crop stage. First watering is given immediately after transplanting. Irrigation water should be of good quality having proper Ec and pH.

Growth and Flowering

If optimum temperature to the gerbera crop is given at a sunny position, it will remain evergreen and continue blooming throughout the year. Whether gerbera plants are raised vegetatively including micropropagation or through seeds, their time requirement for flowering is almost the same irrespective of their growing as potted plants or as cut flowers. The sale of potted plants in 10-cm pots starts 10-11 weeks of transplating when transplanted from April to September and 14-16 weeks when transplanted from September to April (Rogers and Tjia, 1990; Erwin *et al.*, 1991). The first visible flower buds are observed when some 10-14 primary leaves (in the primary leaf whorls) and 2-6 secondary leaves (in the lateral leaf whorls) have appeared, and thereafter 11 days are required for the next flower to become visible under SD conditions where primary flower requires 65 days to attain anthesis after visible bud or 18 days under LD conditions where secondary flower appears after 90 days (Dole and Wilkins, 1999). Roh and Lawson (1984) stated that slower development of secondary flower is due to apical dominance and not due to photoperiod, though there may be variation in photoperiodic response in different varieties (Lin and French, 1985), however, Erwin *et al.* (1991) reported only insignificant differences among various cultivars due to photoperiod and stated that flower initiation and development are mainly due to temperature (above 24 °C can cause no flowering) and light intensity which is rather more important during seedling growth, *i.e.* high light intensity, as flower initiation starts when 2-3 true leaves become visible. These are, in fact, evergreen plants but growing **temperatures** below 12 °C force these to die or becoming dormant. For seed germination, the optimum temperature requirement is 20 to 23 °C, day/night temperatures for pot plants from potting to visible bud are optimum at 21/17 °C and afterwards from visible bud to colour showing stage it is 24 °C though at higher temperatures, *i.e.* 25 °C leaf area increase is maximum, high difference between night and day temperatures lengthens the peduncle, day/night temperatures of 25/14 °C are optimum for maximum dry weight production of plants (Erwin *et al.*, 1991) while day temperatures of 21-24 °C and night temperatures of 14-19 °C are optimum for cut and pot-gerberas (Dole and Wilkins, 1999). A low temperature of 13 °C during growing period induces more vegetative axillary shoots, consequently causing more flower yield at the later stage (Dole and Wilkins, 1999). Lin and French (1985) and Rogers and Tjia (1990) stated that root zone heating at 21 °C can be used to maintain production. High total **light** integral is essential from germination onward to ensure rapid and uniform flowering (Dole and Wilkins, 1999). Gerbera plants are partially photoperiodic but dramatically respond to light intensity × duration (Rogers and Tjia, 1990). Gerberas like sunny sheltered position preferably at the foot of a wall. However, in greenhouses these can be overwintered at a temperature of 7 °C by keeping the plants just moist at all the time. The greenhouses should be ventilated freely during summer and the shade-screens should be used from April to September in the plains and from May to September in temperate regions. Plants should also be fed fortnightly with a weak liquid manure. *Botrytis* can be a serious problem with gerbera during winter, especially dark periods when greenhouse **humidity** is very high, higher than 70 per cent during day time and 85 per cent during nights and ventilation during day time *cum* air circulation in the nights is poor. High humidity also contributes to floral-stalk stretching. Over-watering should also be avoided. **CO_2** concentration at 600 to 800 ppm in the greenhouse facilitates good growth, increases stalk length and enhances yield provided supplementary light is used (Tsujita, 1983), therefore, this may not be beneficial to the potted plants as stalk length also increases.

Growth regulators play important role in regulating the flowering in gerbera crop. For uniform flowering and to harvest more number of flowers, gibberellic acid and dikegulac sodium have been used. El-Shafie and Hassan (1978) obtained early flowering in gerbera through 50 ppm GA_3, more number of flowers in 100 ppm GA_3 during first season and at 100-200 ppm during second season with slight increase in floral diameter though chlormequat promoted but delayed the flowering and its 750 ppm produced fewer but heavier flower with reduced peduncle length. Guda *et al.* (1985) sprayed 100 ppm GA_3 with 0.2 per cent DMSO and 0.2 per cent Tween-20 on gerbera cvs 'Cocktail' and 'Peter' and recorded increased number of flowers. Farina *et al.* (1989) through monthly spraying of GA_3 at 100 ppm with gerbera cv. 'Joyce' recorded enhanced flower production with increased floral diameter but peduncle length was found reduced. Suma (1993) recorded hastened flowering in gerbera through 50 and 100 ppm application of GA_3, with 50 ppm increasing the flower longevity and 100 ppm number of the blooms.

There are varieties which are desired for their plant diameter as this and floral scape length are controlled genetically so such varieties for pot culture should be selected. Difference between day and night temperatures controls the scape length, more the difference longer the scape (Dole and Wilkins, 1999). Basu and Bose (1980) applied 100-200 ppm ancymidol (A-Rest) to gerbera cv. 'Pink' at 2-3 leaf stage which delayed bud appearance by 7-18 days. Armitage *et al.* (1984) used two applications of ancymidol at 200 ppm, daminozide (B-Nine) at 4,000 ppm or chlormequat (CCC) at 1,500 ppm and found ancymidol more effective than daminozide when applied at 6-7 weeks after transplanting and reduced even the flower size though chlormequat was not effective at all. The retardant B-Nine is favourite among the gerbera growers, its 1,000-1,500 ppm may be applied in the plug stage when seedlings have 4-5 mature leaves, and its first application of 2,500 ppm is given 10-14 days after potting to the final container. It is applied from fall to spring 1-2 times to 11.0-12.5 cm pot size or only one application to 15 cm pot size, but during summers 2-3 applications to 11.0-12.5 pot size and 1-2 times to 15 cm pot size. CCC at 500 and 750 ppm application delayed the flowering in gerbera though 750 ppm increased number and field longevity of flowers (Sumo, 1993). Lee and Lee (1990) used drenching of ancymidol at 1.0 mg/pot or paclobutrazol at 0.25-0.5 mg/pot which caused reduction in foliage size, leaf area and peduncle length.

Postharvest

Mondal *et al.* (2007) suggest that first flowers are produced 50-70 days of planting, and when outer two whorls of disc florets are straight on the flowerhead showing pollen grains in case of singles but in doubles also the disc flowers should be fully developed, these are harvested during cool hours of the day by pulling (not cutting) near the base of stalk. Approximately 150-250 flowers are harvested per square metre annually though this greatly differs from variety to variety. It is 'Maron Clementine' which is high yielder and produces some 400 flowers per square metre annually. Though there is great difference in vase life of various cultivars but normally freshly harvested gerbera stems last 2-3 weeks.

Pulsing treatment to the cut stems is given with 1,000 ppm silver nitrate which is more effective than STS, sodium hypochlorite 600 ppm, sucrose 6 per cent and Tween-20 surfactant 0.1 per cent for 10 minutes for reducing bacterial stem blockage and to prolong vase life, but sucrose will encourage scape elongation at neck of the flower (Dole and Wilkins, 1999). Lupi *et al.* (1977) reported that immersing of gerbera cut ends for 24 hours in a solution containing 200 mg 8-HQS, 50 mg $AgNO_3$ and 5 per cent sucrose in one litre of water prolongs the life by 4-5 days though found that stem length was negatively correlated with storage life. However, its flowers are not damaged by chlorinated water, and Barendse (1978) reported that 0.5 mg of 10 per cent chlorine kills the bacteria and prolongs the vase life. Nowak (1981) recorded longest cut gerbera life after pretreatment with 25 mg/l $AgNO_3$ + 200 mg/l 8-HQC + 7 per cent sucrose, as $AgNO_3$ and sucrose in vase water increases the rigidity and mechanical stability of flowers. Tjia *et al.* (1987) when tried cut gerbera in fluorinated water and floral preservatives, observed petal necrosis, but when they used deionized water or water containing 8-HQC + sucrose there was no petal necrosis. Suma (1993) found significantly increased vase life when applied 100 ppm GA_3 and 750 ppm CCC to the plants, and with the use of 5 per cent sucrose + 20 ppm $AgNO_3$ in the vase water also extended longevity of flowers significantly. Meera Manjusha and Patil (2002) found 1.0 mM cobalt sulphate + 7 per cent sucrose providing vase life in gerbera cv. 'Sath Baba' of 15.70 days than 8.17 days in control. Manreet Sooch *et al.* (2002) found best results with 25 ppm bleach (containing 25 per cent chlorine) giving 8.87 days of vase life in gerbera cv. 'Glory', followed by 8.47 days in $CoCl_2$ as compared to 4.4 days in control. Mondal *et al.* (2007) mentions 200 mg 8-HQS, 50 mg $AgNO_3$ and 5 per cent sucrose or pulsing in $AgNO_3$ at 200 mg/l or 8-HQC at 200 mg/l with sucrose at 100 g/l for 24 hours at 20 °C reducing the number of bends or folded stalks and prolonged vase life both for gerberas kept in water or in preservative solution after transport. Macnish *et al.* (2008) when added aqueous chlorine dioxide (ClO_2) at 2 or 10 µl l^{-1} in clean deionized water, the build up of bacteria in vase solution was prevented and the vase life was found extended in *Gerbera jamesonii*' cv. 'Monarch', and the efficacy of 10 µl l^{-1} ClO_2 in vase water containing 0.2 g l^{-1} citric acid and 10 g l^{-1} sucrose to extend the display life of gerbera 'Lorca' and 'Vilassar' flowers was equal to or greater than other tested biocides, *i.e.* aluminum sulphate, dichloroisocyanuric acid, 8-HQS, Physan 20TM, sodium hypochlorite, so this can be used as an alternative antibacterial agent in vase solutions. Mondal *et al.* (2007) further reported various combinations such as 20-30 mg Ag NO_3 + 3-6 per cent sucrose, 20 mg Ag NO_3 + 2 per cent sucrose + 150 mg $NiCl_2$, 200 mg 8-HQC + 3 per cent sucrose, and 50 mg DICA + 2 per cent sucrose, each combination per litre of water where cut flower life may be prolonged for up to 4-5 days. Its flowers can be precooled at 5 °C up to three days, and these can be stored and transported at 2 °C (Sacalis, 1993) though pot plants can be transported at 12 °C. Tjia *et al.* (1987) stated that gerbera flowers are highly sensitive to even 1 ppm fluorides in water. Ethylene is injurious to cut blooms as well as the pot plants as this encourages senescence, therefore, when the flowers are to be stored for long duration (3-4 weeks), these should be only dry-stored at 4 °C (Nowak and Plich, 1982; Tjia and Rogers, 1984; Tjia *et al.*, 1987; Nowak and Rudnicki, 1990; Rogers and Tjia, 1990; Sacalis, 1993). Sodium hypochlorite can be substituted with 200 ppm 8-hydroxyquinoline citrate which also makes solution acidic due to its citric acid. Bruyan and Hulsama (1972) suggested that fluoride injury could be reduced or eliminated by addition of aluminium sulphate at 600-750 ppm or crystal at 15 g/l to the fluorinated water.

For transportation, insulated boxes are used for packing. Plastic coated metal grids of 50 × 70 cm with mesh size of 2 × 2 cm are used to pack gerbera flowers in Holland, the flower heads are supported by the grids (on the place of metal grids cardboard cups or minisleeves can effectively be used as is the practice in Holland) which are suspended above a plastic tray measuring 48 × 70 × 30 cm, at a height which can be adjusted as per the length of the floral stalk so that cut ends are immersed in water to 15 cm depth in the tray (Bhat and Singh Samanta, 1989). In fact, here each flower is hung from a perforated cardboard sheet, sheet being inserted in an outer carton. Individual flowers, are nowdays packed in specially prepared transparent alkathene bags but with cut ends inserted into water or vase-solution filled air-tight plastic tube. Locally these are packed in bunches of 10-12 with their cut ends placed in water or vase solution (Van Meeteren, 1978). The stem length of gerbera should not be less than 40 cm, should be straight and firm and the flower diameter not less than 7 cm.

After receiving the consignments the cut stems require to be rehydrated immediately with 43 °C warm water. Solution of 25 ppm silver nitrate after being adjusted with citric acid to pH 3.5 should be used as holding medium (Dole and Wilkins, 1999). Dipping flowerheads in 0.1 mM benzyladenine delays the senescence and maintains flower weight (Dole and Wilkins, 1999). Gerbera cut flowers respond very well to recutting of cut ends before placing in preservative solution or up to marketing. Shortage of water as a result of vessel blockage or microbial growth may cause flower drooping or wilting. Its stalk being hollow, sometimes due to filling of air inside water absorption is held up so pricking of the stalk 10 cm below the flowerhead will facilitate escaping of the air, *vis-a-vis* water absorption.

Insect-Pests, Diseases and Physiological Disorders

Crown rot/Foot rot/Root rot or Wilt is caused by *Phytophthora cryptogea* which is a very serious soil-borne pathogen occurring at soil level in poorly managed irrigation water. In its infection the stalk-foot and/or plant roots turn black-brown and rot, *vis-à-vis* causes crown rotting, which ultimately causes death of the whole plant. **Foot rot** of gerbera is also caused by *Fusarium oxzysporum* whose symtoms are almost akin to the infection of *Phytophthora cryptogea* and such infected plants if are cut the vascular system shows blackening. *Fusarium oxysporum, Penicillium* sp. and *Trichoderma* sp. colonizing the gerbera substrate (sphagnum peat and sphagnum peat + gravelite) limit *Phytophthora cryptogea* and *Pythium ultimum* growth (Kurzawiska and Pacyna, 2000), and Malvolta Jr.(1994) reported that application of Fe-EDDHA, ferrous sulphate, Zn-EDTA or zinc sulphate and calcium nitrate nutrition with the top-dressing medium suppresses the incidence of *Phytophthora* foot rot up to 4-month cultivation period of gerbera. Soil sterilization with Vapam at 100 ml/m^2 keeps this problem under check, and warming of the soil at 26 °C also reduces incidence of this disease, and soil sterilization together with regular application of fungicides such as copper oxychloride in the soil will control even the **root rot** caused by soil-borne pathogens, such as *Pythium irregulare, P. ultimum, Sclerotium rolfsii, S. sclerotiorum* and *Rhizoctonia solani*. From Italy, Pasini *et al.* (1996) also isolated *Rhizoctonia solani* from gerbera causing collar or foot rot. *Pythium* causes even growth stunting where root skin easily removes, and *Sclerotium* rotting of the aerial plant parts which becomes serious at high atmospheric relative humidity and persistently high temperature (30-34 °C). *Sclerotium* infection appears first as water-soaked lesions on collar and other parts touching the soil, and then afterwards plants turn yellow, brown to black and die down. Methyl bromide or Metam at half doses under plastic is very effective though its full doses cause phytotoxicity (Minuto *et al.*, 2000). Prothiocarb at 0.15 per cent on young infected plants or Furadaxyl at 2 g/l is quite effective in controlling these pathogens. Preplanting treatment with garlic homogenate at 5 g/l of substrate keeps the plants healthy up to 10-week cultivation period (Saniewska and Orlikowski, 1994). *Rhizoctonia solani* also causes plant stunting and ultimate death of the plant. Soil sterilization and drenching with copper oxychloride at 0.4 per cent or Dithane M-45 0.2 per cent controls this disease.

Powdery mildew is caused by *Erysiphe cichoracearum, E. sclerotiorum* and *Oidium erysiphoides* f. sp. *gerberae* as white powdery coating on gerbera foliage and floral stalks. *E. cichoracearum* can be controlled by spraying with Karathane or wettable sulphur, and *Oidium erysiphoides* f. sp. *gerberae* through reduction of relative humidity, proper ventilation, by removing the old leaves, and 0.1 per cent spraying of benomyl in the greenhouse. Spraying of dinitro capryl phenyl crotonate (Mildex) at 0.25 per cent + a wetting agent controls the disease without any plant injury. **Downy mildew** is caused by *Albugo tragopogonis* whose infection on gerbera leaves appears as white lesions on the underside, hence, sometimes it is also incorrectly called as white rust, and *Bremia lactucae* causes white lesions on the underside of leaves. Downy mildews can be controlled effectively through spraying with 0.3 per cent Dithane M-45, Metalaxyl or 0.3 per cent Captafol 7-8 times at 10 day's interval.

Gerbera **blight** (gray mould) is caused by *Botrytis cinerea*. Its symptoms first appear as small black spots on ray florets, especially due to deep planting, poor ventilation and aeration, and poor drainage in the greenhouses. When relative humidity of the air is more than 92 per cent for two hours in the morning, this develops gray spots on the flower petals. It also kills the young growing tissues, and its infection also causes damping off in seedlings. Benlate at 0.1 per cent, Dithane M-45 at 0.2 per cent, Dithane Z-78 at 0.2 per cent or Thiram at 0.1 per cent weekly sprayings may control this problem. Sprinkler or overhead irrigation should be avoided and watering should be carried out only to the required quantity and whenever the real need be. **Leaf spot** of gerbera is caused by *Alternaria brassicola, Phyllosticta gerbericola* and *Cercospora gerberae*. These pathogens cause various kinds of leaf spot symptoms on gerbera leaves. Such problems may be controlled through spraying with 0.2 per cent Dithane M-45 or Dithane Z-78. Sometimes, *Verticillium dahliae* causes severe stunting and slow wilting of plants, and this should be controlled through soil sterilization.

In 1985 in São Paulo (Brazil), a **bacterial** disease causing leaf blight was observed whose pathogenicity confirmed it to be *Pseudomonas cichorii* (Malavolta, 1994). This can be controlled through use of streptocycline 0.01 per cent sprayings.

Viruses so far recorded on gerbera crop are 'cucumber mosaic virus' (CMV), 'gerbera ilar virus' (GIV), 'gerbera mosaic virus' (GMV), 'tobacco rattle virus' (TRV), 'tomato spotted wilt virus' (TSWP), which spread through aphids, thrips and nematodes. CMV causes distortion of colour with broken streaks on flowers, stalks and leaves, *vis-à-vis* reduced and mottled leaves. GIV causes growth retardation together with chlorosis (diffusion, mottlings, rings, and line patterns). GMV is transmitted through grafting and causes mottling in the plants. TRV expresses yellow or black annulated ring spots on the leaves which is transmitted through *Trichodorus* nematodes which can be controlled by destroying these nematodes through soil steaming before crop planting. TSWV infection of gerbera plants checks the vigour of the plants and causes poor quality flowers. One can get rid of viruses through sanitation, immediate destroying of the infected plants, use of virus-free material, soil steaming to kill nematode vectors and management of virus vectors.

Bent-neck (insufficient floral stem tissue hardening or maturation below harvested flower resulting into stem collapse) is the problem where flowerheads at neck bent from where it may break easily. Bentneck occurs due to poor winter growing conditions. Harvesting of flowers at lower temperature or when their stems are immature should be avoided. Sometimes **leaf yellowing** similar to mineral deficiency occurs due to poor drainage or because of root infection with some insect-pests or diseases which should be rectified after ascertaining the cause.

Root-knot **nematode** (*Meloidogyne incognita*) by making galls in the roots restricts the root growth and causes foliage yellowing. This can be controlled through application of organophosphate chemicals in the soil. *Neem* cake 30-50 g/ plant, Carbofuran (Furadan) 10 g/plant or Diazinon (Suzon) 0.015 per cent drenching will kill the nematodes to a great extent.

The crop gerbera is infested with greenhouse **whiteflies** (*Trialeurodes vaporariorum*) during hot and dry weathers in the greenhouses and has been recorded infesting some 80 families of plants (Mound and Halsey, 1978). These are tiny (1.5 mm long) white insect laying their eggs on the underside of the new leaves in circle or in crescent-shape. Eggs hatch in 5-7 days and the new crawlers move from leaf to leaf for a day or two and then start sucking the sap from the young leaves from where these do not move until their maturity, then moulting into nymphs of yellow colour and red eyes and then into pupae, and finally becoming adults of whitish-yellow colour. Its infection causes chlorosis, yellowing of leaves, premature leaf defoliation *cum* bud shedding, and gradual drying of the plants, and over all infested plants become aesthetically unappealing. The insecticides such as Confidor (Imidacloprid) at 0.005 per cent ; Astra, Lanate (Methomyl) or Pride (Acetamiprid) at 0.004 per cent ; and Neemazol, Rogor (Dimethoate) or Malathion at 0.02 per cent sprayings continuously thrice at 5 7 days should control all the stages of this pest. The yellowish brown larvae of **leaf miners** (*Liriomyza trifolii, L. huidobrensis*) feed on the plants by tunnelling young leaves in the form of serpentine mines and their severe infestation causes drying and dropping of the leaves. *Neem* cake extract is highly effective in this case which may be used regularly. Two sprayings of Imidacloprid, pyrazophos or Esfenvalerate at fortnightly intervals will control this pest (Paradikovi, 1998). Vertimec (Abamectin) at 0.005 per cent ; Chlorpyriphos or Dichlorovos at 0.01 per cent ; Acephate or Methomyl at 0.015 per cent ; or Dimethoate at 0.1 per cent spraying will control this pest. Stolz (1996) suggested integrated control measures through use of bracanoid (*Dacnusa gibirica*) and eulophid (*Diglyphus isaea*). The nymphs and adults of the **aphids** (bean aphid, *Aphis fabae*; green peach aphid, *Myzus persicae*) suck the sap of tender leaves and buds and distort them. The clustering of the insects on newly emerged leaves and the buds is quite obvious on such plants during warm weathers. These can be controlled through fumigation with Sulfotep smoke cartridges at 200 m^3 or Primicarb smoke candles at 700 m^3, and chemical control through spraying with 0.01 per cent Ambush; 0.02 per cent Malathion; 0.1 per cent Dichlorvos, Methomyl or Oxamyl; 0.04 per cent Decamethrin or Cypermethrin; or 0.05 per cent Oxamyl. Spraying water with jet nozzle or spraying of the tobacco decoction will also control these pests. **Mites** that infest gerbera are *Hemitarsonemus latus*, bulb scale mite (*Steneotarsonemus pallidus*), spider mites (*Tetranychus urticae*), *Polyphagotarsonemus latus,* and cyclamen mites (*Phytonemus pallidus*). *Hemitarsonemus latus* and *Steneotarsonemus pallidus* mites cause crooked and deformed flowers by spinning webs and feeding on the leaves and flowers. Sulphur dusting of plants before flowering and Mitac, Torque and Rospin application to the flowering plants will control these mites. Spider mites feed on the leaves by piercing the epidermis and while sucking the sap the mesophyll tissue in the pierced region collapses and form the chlorotic spots. The adults in winter hibernate in the soil or on weed hosts but during summer these become very active. At early stage of their infestation, daily spraying with a garden hose will be very effective. Vertimec 0.004 per cent spraying twice at 4-day's intervals will control this pest. *Phytoseiulus persimilis* parasitizes this mite. *Polyphagotarsonemus latus* can be controlled effectively by applying 20 kg/ha elemental sulphur before flowering but flowering plants should be treated with Hostathion (Triazophos) or Vydate (Oxamyl). *Phytonemus pallidus* infestation makes the gerbera flowers discoloured, shrivelled, puckered, crinkled and brittle, *vis-à-vis* curling of leaves. This mite remains at darker place and prefers high humidity conditions. They are sensitive to acaricides. Dipping the affected plant parts in 43 °C hot water for 15 minutes will kill the mites though plants will not be damaged. **Root mealy bug** (*Rhizoecus falcifer*) feeds on gerbera roots which may be killed through smoke generator formulations though not very effective as they are found underground. Western flower **thrips** (*Frankliniella occidentalis*) infestation on gerbera causes white specks or small stripes on the ray florets and in its severe infestation the flowerheads are deformed. Leaves also show grayish spots. Seedox 80W at 0.1 per cent spraying at 4-5 days interval will control this pest. **Vegetable bugs** (*Nezara viridulae*) chew the foliage and their juvenile stages normally harbour around the leaf bases. **Caterpillars** (*Helicoverpa armigera*) are polyphagus in nature and feed on every plant part. *Bacillus thuringiensis* based parasites are very effective in keeping such insects under check. **Fungus gnat** maggots are often present in potting mixture containing peat moss. These can be controlled through *Bacillus thuringiensis israelensis.* These are also highly sensitive to insecticides ao any insecticide in the mixture will kill these. Other insect-pests which infest gerbera but are of little importance are *Mamestra brassicae, Spodoptera littoralis*, leaf roller and loopers. While controlling any of the insect-pests, these will automatically be killed. In any case, there should be proper sanitation of the field, no waterlogging and continuous inspection of the crop will mostly prevent the use of chemicals.

Slugs and **snails** are the problems in high humid areas having diffused or no light and where temperature is below 21 °C. In the day time these hide themselves below the debris but become active during night hours when these feed on foliage, new shoots and flowers. Proper cleanliness will prevent their multiplication and will also destroy their hiding places. These can be trapped by spreading vegetable refuse. Finely ground lime in dry soil and during dry period and copper sulphate barriers will kill these pests if these will try to cross the line.

References

Anuradha, S. and J.V.N. Gowda, 1999a. Quantitative genetic studies in gerbera. *Mysore J. Agric. Sci.*, **33**(2): 224-227.

Anuradha, S. and J.V.N. Gowda, 1999b. Genotypic differences in the postharvest life of gerbera cultivars. *Mysore J.agric. Sci.*, **33**(4): 312-316.

Anuradha, S. and J.V.N. Gowda, 2000. Association of cut flower yield with growth and floral characters in gerbera. *Crop Res.* (Hisar), **19**(1): 63-66.

Arello, E.F., M. Pasqual, J.E.B.P. Pinto and M.H.P. Barbosa, 1991. *Pesquisa Agropecuaria Brasileria,* **26**(2): 269-273.

Aswath, C. 1999. Gerbera cultivation and improvement. *Proc. Short Course on Hitech Prod. of Ornam. Crops*, pp. 53-54, IIHR, Bengaluru.

Aswath, C., V.A. Parthasarthy and G. Bhowmik, 1998a. Genetic studies in economic characters of gerbera. *J. Maharashtra agric. Univ.*, **23**(2) 140-142.

Aswath, C., V.A. Parthasarthy and G. Bhowmik, 1998b. Dry storage as an aid in selection for longevity in gerbera. *J. Ornam. Hort.* (N.S.), **1**(2): 55-60.

Aswath, C. V.A. Parthasarthy and G. Bhowmik, 1999. Role of biochemical component in vase life of gerbera. *Ann. Pl. Physiol.*, **12**(1): 32-37.

Barendse, L.B.J. 1978. The addition of chlorine to gerbera is no longer experimental. *Vakblad voor de Bloemisterij*, **33**(15): 19.

Bik, R.A. and J.M.van Berg, 1984. Preliminary results of a gerbera trial. Peat appears to be more favourable than rockwool. *Vakblad voor de Bloemisterij*, **39** (II): 45.

Blomme, R. and P. Dambre, 1979. Cultural substract and fertilizer trials on gerberas. *Verbondsnieuws voor de Belgische Sierteelt*, **23**: 175-178.

Blomme, R. and P. Dambre, 1984. The culture of gerberas. Acidity and salt concentration as a function of the manual regime. *Verbondsnieuws voor de Belgische Sierteelt*, **28** (14): 684-687.

Bose, T.K. and B.K, Jana, 1978. Effect of different levels of nitrogen, phosphorus and potassium on growth and flowering of gerbera. *Indian J. Hort.*, **31**: 182-185.

Brickell, C. 1996. A-Z Encyclopedia of Garden Plants, pp. 467. Dorling Kindersley Limited, London.

Bruyan, J.W. de and A.N. Hulsama, 1972. Fluoride injury to cut gerberas. *Bedrijfsontwikkeling*, **3**(2): 209-211.

Buisman, J. 1985. Comparative production of gerbera at three plant spacings. *Vakblad voor de Bloemisterij*, **40**(3): 34-35.

Carpenter, W.J., E.R. Ostmark and J.A. Cornell, 1995. Temperature and seed moisture germination and storage of *Gerbera* seed. *HortSci.*, **30**: 98-101.

Charpentier, S., M. Laffaire, L.M. Riviere and E. Vidalie, 1986. Gerbera cultivation on rockwool. *Revue Horticole*, No. 271, pp. 47-54.

Choudhary, S., P. Mahanta, L. Paswan and M.C. Talukdar, 1998. Performance of some gerbera (*Gerbera jamesonii*) cultivars under the agro-climatic conditions of Jorhat, Assam. *Hort. J.*, 11(1): 109-114.

Chu, C.Y. and M.C. Huang, 1983. *In vitro* formation of *Gerbera hybrida* Hort. plantlets through excised scape culture. *J. Japanese Soc. hort. Sci.*, **52**(1): 40-50.

Das, P. and P.K. Singh Samanta, 1989. Gerbera. In: *Commercial Flowers* (eds Bose, T.K. and L.P. Yadav), pp. 601-622. Naya Prokash, Kolkata.

Deepak Kumar and Ramesh Kumar, 2000. Seasonal response on gerbera cultivars. *J. Ornam. Hort.* (N.S.), **3**(2): 103-106.

Dole, J.M. and H.F. Wilkins, 1999. *Floriculture Principles and Species*, pp. 356-361. Prentice Hall, New Jersey, USA.

Dufault, R.J., T.L. Phillips and J.W. Kelly, 1990. Nitrogen and potassium fertility and plant populations influence field production of gerbera. *HortSci.*, **25**: 1599-1602.

Erwin, J., R. Heins and W. Carlson, 1991. *Pot Gerbera Production*. Minnesota Flower Growers Association Bulletin, No. 40 (5), pp. 1-6.

Fischer, P. and N. Forchthommer, 1987. Copper fertilization of gerbera. *Deutscher Gartenbau*, **41**(II): 674-678.

Gutmane, L. and D. Grivina, 1975. The effect of different kinds and concentrations of fertilizers on the growth and development of gerbera and the nutrient content of the plants. *Izvestiya AN Lato SSR*, **II**: 22-26.

Hansen, H.B. 1999. A story of the cultivated *Gerbera*. *New Plantsman*, **6**(2): 85-95.

Hay, R. and K.A. Beckett, 1971. *Reader's Digest Encyclopaedia of Garden plants and Flowers*, pp. 301. The Reader's Digest Association, London.

Hedtrich, C.M. 1979. Production of shoots from leaves and propagation of *Gerbera jamesonii*. *Gartenbauwissenschaft*, **44** (1): 1-3.

Huang, J.M., Y.Y. Hi and M.H. Lin, 1987. The micropropagation of gerbera. *Acta Hort. Sinica*, **14**: 125-128.

Huang, H. and J. Harding, 1998. Quantitative analysis of correlations among flower traits in *Gerbera hybrida*, Compositae. III. Genetic variability and structure of principal component traits. *Theor. Applied Genet.*, **97**(1-2): 316-322.

Jacob, M. 1997. Beneficial organisms increase yield and resistance of gerbera. *TASPO Gartenbau Mag.*, **6**(12): 56-57.

Jawaharlal, M., K. Rajamani, K. Soorianathasundaram and G. Balakrishnamurthy, 1998. Evaluation of gerbera genotypes for certain floral characters, flower yield and vase life. *South Indian Hort.*, **46**(3-6): 291-293.

Kamel, H.A., A.L. Bishara, A.A. Ibrahim and M.A.K. Nada, 1975. Studies on the effects of N, P and K on the flower yield of *Gerbera jamesonii* L. var. Superba. *Agaric. Res. Rev.*, **53**(3): 97-103.

Kasten, P., K. Sommer and A. Papenhagen, 1990. Ammonium placement fertilizing of cut-flowers (German). *Deutscher Gartenbau*, **44**(26): 1710-1713.

Katinas, L. 1998. The Mexican *Chaptalia hintonii* is a *Gerbera* (Asteraceae, Mutisieae). *Novon*, **8**(4): 380-385.

Kaur, R., S. Chander and D. Sharma, 1999. Modified Murashige medium for micropropagation of gerbera. *Hort. J.*, **12**(1): 89-92.

Kacperska, I. 1985. The effect of quantity of NPK fertilizer doses on the yield of gerbera cv. Peter. *Prace Instytutu Sadownictwa I Kwiaciarstwa w Skierniewicach*, **B** (*Rosliny Ozdobne*), No. 10, pp. 105-114.

Koch, G.M. and E.J. Holcomb, 1982. The effect of control released fertilizer on growth and flowering of Aztec gerbera. *Bull. Pennsylvania Fl. Grs*, No. 340, pp. 4-5.

Kurzawiska, H. and E. Pacyna, 2000. Fungi isolated from substrates of gerbera plants and their effect on the growth of *Phytophthora cryptogea* and *Pythium ultimum*. *Phytopathologia Polonica*, **20**: 123-129.

Laliberte, S., L. Chretien and J. Vieth, 1985. *In vitro* plantlet production from young capitulum explants of *Gerbera jamesonii*. *HortSci.*, **20**(1): 137-139.

Lamanna, D., M. Castelnuovo and G. D'Angelo, 1991. Compost-based media as alternative to peat on ten pot ornamentals. *Acta Hort.*, No. 294, pp. 125-129.

Lê, C.L., C. Julmi and D. Thomas, 1999. *In vitro* regeneration and multiplication of *Gerbera jamesonii* Bolus Tschuy. *Revue Suisse de Viticulture, d'Arboriculture et d'Horticulture*, **31**(4): 207-211.

Lin, W.C. and C.J. French, 1985. Effect of supplementary lighting and soil warming on flowering of three *Gerbera* cultivars. *HortSci.*, **20**: 271-273.

Lisiecka, A. 1976. The effect of certain micro-elements on the quality of seedlings and flowering in gerbera. *Zesz. Probl. Postepow Nauk Rol.*, No. 179, pp. 323-326.

Lisiecka, A. 1991. Flowering of gerberas grown in an unheated plastic tunnel in relation to plant spacing. *Prace Z. Zakresu Nauk Rolniczych*, **71**: 59-64.

Lin, W.C. and C.J. French, 1985. Effect of supplementary lighting and soil warming on flowering of three *Gerbera* cultivars. *HortSci.*, **20**: 271-273.

Lupi, R., *E.G.* Accati and T. Schiva, 1977. Storage of cut gerbera flowers. Experimental results. *Annali dell Instituto Sperimentale per la Floricultur*, **8**(1): 1-13.

Macnish, A.J., R.T. Leonard and T.A. Nell, 2008. Treatment with chlorine dioxide extends the vase life of selected cut flowers. *Postharv. Biol. & Tech.*, **50**(2/3): 197-207.

Mahanta, P., S. Choudhury, L. Paswan and M.C. Talukdar, 1998a. Studies on variability and heritability of some quantitative characters in gerbera (*Gerbera jamesonii*). *South India Hort.*, **46**(1-2): 43-46.

Mahanta, P., S. Choudhury, L. Paswan, M.C. Talukdar and D. Sarma, 1998b. Correlation and path coefficient analysis in gerbera (*Gerbera jamesonii*). *Hort. J.*, **11**(2): 79-85.

Malavolta (Jr.), V.A., C.F. Robbs, O. Victor and J. Rodriguez Net, 1994. Bacterial blight of gerbera. *Biológico*, **56**(1-2): 29-30.

Manreet Sooch, Kushal Singh, Ramesh kumar and Premjit Singh, 2002. Effect of chemicals on vase life of gerbera. *Floriculture Research Trend in India* (eds Misra R.L. and Sanyat Misra), pp. 321-322, ISOH, New Delhi.

Mantrova, E.Z., T. Nikolaeva and V.V. Dvortsova, 1982. Effectiveness of supplying fertilizers to gerberas. *Byulleten' Glavnogo Botanicheskogo Sada*, No. 126, pp. 51-57.

Meera Manjusha, A.V. and V.S. Patil, 2002. Cobalt and postharvest quality of gerbera. *Floriculture Research Trend in India* (*Proc. Symp. on Indian Floriculture in the New Millennium*, held at Lal-Bagh, Bangalore, Feb. 25-27), pp. 319-320. ISOH, New Delhi.

Minuto, A., A. Pomè, M.L. Gullino and A. Garibaldi, 2000. Soil fumigaton for the control of *Phytophthora cryptogea* on gerbera. *Colture Protette*, **29**(2): 109-112.

Mitika Kalra, S.K. Sehrawat, Praveen Batra, Suresh Kumar, D.S. Dahiya and A.K. Gupta, 2008. Micropropagation studies in gerbera (*Gerbera jamesonii* Bolus). *Haryana J. hort. Sci.*, **37**(1-2): 78-79.

Moe, R., J.E. Erwin and W. Carlsson, 1996. Factors affecting *Gerbera jamesonii* early seedling branching and mortality. *HortTech.*, **6**: 59-61.

Mondal, T., Sanyat Misra, R.L. Misra and P. Naveen Kumar, 2007. Gerbera. AICRP on Floriculture Tech. Bull. No. 27. AICRP on Floriculture, ICAR, Division of Floriculture and Landscaping, I.A.R.I., New Delhi, 36 pp.

Mound, L.A. and S.H. Halsey, 1978. *Whitefly of the World*. British Museum of Natural History and John Wiley & Sons Publishers, Chichester, New York, 340 pp.

Mouninier, H. and M. Montarone, 1978. The nutrition of gerberas. *Pepinieristes Horticulteurs Maraichers*, **188**: 13-18.

Murashige, T., M. Serpa and J.B. Jones, 1974. Clonal multiplication of gerbera through tissue culture. *HortSci.*, **9**(3): 175-180.

Nowak, J. 1981. The effect of silver complex and sucrose on longevity of cut flower inflorescence stored for different period of time. *Rosliny Ozdobne*, **6**:83-88.

Nowak, J. and H. Plich, 1982. Ethylene synthesis during senescence of cut gerbera inflorescence as affected by different chemical pretreatments. *Proc. XXI intern. Hort. Congr.*, vol. II: 1736 (abstr.).

Nowak, J. and R.M. Rudnicki, 1990. *Postharvest Handling and Storage of Cut Flowers, Florist Greens, and Potted Plants.* Timber Press, Portland, Oregon.

Orlikowska, T., E. Nowak, A. Marasek and D. Kucharska, 1999. Effects of growth regulators and incubation period on *in vitro* regeneration of adventitious shoots from gerbera petioles. *Plant Cell Tissue Organ Culture*, **59**(2): 95-102.

Paradikovi, N. 1998. New possibilities for controlling the leaf miner, *Liriomyza trifolii* Burgess, on gerberas in glasshouses. *Poljoprivreda*, **42**: 87-90.

Parthasarathy, V.A. and V. Nagaraju, 1999. *In vitro* propagation in *Gerbera jamesonii* Bolus. *Indian J. Hort.*, **56**(1): 82-85.

Parthasarathy, V.A., V.I. Nagaraja and M. Mishra, 1997. Callus induction and subsequent plant regeneration from leaf explants of *Gerbera jamesonii. Folia Hort.*, **9**(2): 83-86.

Pasini, C., T. Berio, P. Curir and F. D'Aquila, 1996. Further characterization of *Rhizoctonia solani* isolated from carnation and other ornamental plants. *Informatore Fitopatologico*, **46**(6): 33-36.

Petru, E. and J. Matous, 1984. *In vitro* culture of gerbera (*G. jamesonii*) Bolus. *Zakradnictvi*, **II (4)**: 309-314.

Pierik, R.L.M., H.H.M. Steegmans and J.J. Marelis, 1973. Gerbera plantlets from *in vitro* cultivated capitulum. *Scientia Hort.*, **1**(1): 117-119.

Pierik, R.L.M., H.H.M. Steegmans, J.A.M. Verhaegh and A.N. Woeeters, 1982. Effect of cytokinin and cultivar on shoot formation on *Gerbera jamesonii in vitro. Netherlands J. agric. Sci.*, **30**(4): 341-346.

Pizzeti, I. and H. Cocker, 1975. *Flowers A Guide for Your Garden* (vol. I), pp. 524-526. Harry N. Abrams Inc., Publishers, New York.

Prabhat Kumar, 2002. Managing micronutrient deficiency in ornamental crops. *Indian Hort.*, **46**(4): 30-32.

Radice, S. and P.L. Macaroni, 1998. Clonación *in vitro* de diversos cultivares de *Gerbera jamesonii* a partir de capitulos florales. *Revista de la Facultad de Agronomía (La Planta)*, **103**(2): 111-118.

Rodriguez-Navarro, J.A., E. Zavaleta-Mejía, P. Sánchez-García and H. González Rosas, 2000. The effect of vermicompost on plant nutrition, yield and incidence of root and crown rot of gerbera (*Gerbera jamesonii* H. Bolus). *Fitopatologia*, **35**(1): 66-79.

Rogers, M.N. and B.O. Tjia, 1990. *Gerbera Production*, pp. 32-34. Timber Press, Portland, Oregon.

Roh, M.S.M. and R.H. Lawson, 1984. The graces of *Gerbera. Greenhouse Manager*, **3**(8): 79, 82, 86-88, 90, 92, 94-95, 98, 200.

Ruffoni, B. and F. Massaba, 1991. Tissue culture in *Gerbera jamesonii* hybrids. *Acta Hort.*, No. 289, pp. 147-148.

Sacalis, J.M. 1993. Gerbera. In: *Cut Flowers, Prolonging Freshness, Postproduction Care and Handling* (2nd ed.; ed. Seals, J.M.), pp. 61-63. Ball Publishing, Batavia, Illinois.

Salinger, J.P. 1985. Gerbera. In: *Commercial Flower Growing*, pp. 193-196. Butterworths Horticultural Books, Wellington, New Zealand.

Saniewska, A. and L.B. Orlikowski, 1994. Use of garlic homogenate in the context of some plant diseases. *Materialy sesji Instytutu Ochrony Rosliin*, **34**(1): 139-143.

Sheela, V.L. 2006. Gerbera. In: *Advances in Ornamental Horticulture*. Vol. 2. *Herbaceous Perennials and Shade Loving Foliage Plants* (ed. Bhattacharjee, S.K.), pp. 130-149. Pointer Publishers, Jaipur.

Shoemaker, C. and W.H. Carlson, 1985. Don't let iron chlorosis soil gerbera. *Greenhouse Grower*, **3**(12): 14-15.

Silas, R.A. and A.A. Ansari, 1999. *Gerbera maxima* (D. Don) Beauv. – a rare plant rediscovered after a century from Pauri Garhwal with notes on its uses. *Adv. Pl. Sci. Res. India*, **10**: 69-70.

Skalska, E. 1979. Effect of fertilization on young gerbera plants. *Sbornik UVTIZ, Zahradnictvi*, **5/6**(1/2):123-130.

Skalska, E. 1980. Gerbera nutrition according to season on the basis of substrate analysis. *Sbornik UVTIZ, Zahradnictvi*, 7(2): 145-150.

Skalska, E. 1985. Effect of nitrogen fertilization on gerbera yields. *Acta Prühonciana*, **50**: 43-54.

Soni, S.S., A.K. Godara, R.K. Goyal and S.K. Sehrawat, 2010. Leaf nutrient and chlorophyll content of gerbera varieties grown under greenhouse. *Haryana J.hort.Sci.*,**39**(3-4):308-309.

Soni, S.S., A.K. Godara, S.K. Sehrawat and R.K. Goyal, 2010. Effect of foliar application of micronutrient on nutrients and chlorophyll content of gerbera var. Winter Queen. *Haryana J. hort. Sci.*, **39**(1-2): 153-154.

Sonneveld, R. and W. Voogt, 1997. Effects of pH value and Mn application on yield and nutrient absorption with rock wool grown gerbera. *Proc. of the intern. Symp.on Growing Media and Pl. Nutr. in Horticulture* (ed. Roebee, R.V.). Freising, Germany; Sept. 2-7, 1996. *Acta Hort.*, No. 450, pp. 139-147.

Steen, J.A. Van der, 1975. Plant density for gerberas. *Vakblad voor de Bloemisterij*, **30**(22): 16-17.

Stolz, 1996. Investigations on the occurrence of myzid leaf miners and their natural enemies in greenhouse cut gerbera in Austria. *Bulletin* 01LB/SROP [*Proc. Meet. Integrated Control in Glasshouses* (ed. Lenteren, J.C. Von), held on May 20-25 in Vienna, Austria], No. 19(1), pp. 167-170.

Strojny, Z., M. Saniewski and E. Jesiotr, 1976. Nitrogen and magnesium nutrition of *Gerbera jamesonii* growing on composted pine bark. *Prace Instytutu Sadownictwa i Kwiaciarstwa w Skierniewicach*, **B. 2**: 87-95.

Suma, P. 1993. Effect of Time of Planting and Growth Regulators on Flowering and Vase Life of *Gerbera jamesonii* (M.Sc. Thesis). College of Horticulture (K.A.U.), Vellanikkara, Kerala.

Tjia, B., F.J. Marousky and R.H. Stamps, 1987. Responses of cut gerbera flowers to fluorinated water and a floral preservative. *HortSci.*, **22**:896-897.

Tjia, B. and M.N. Rogers, 1984. Growing *Gerbera*, learn to solve problems of bent neck, stem break. *Greenhouse Manager*, **2**(6): 67, 70, 73-75.

Tourjee, K.R., J. Harding and T.G. Byrne, 1994. Early development of *Gerbera* as a floriculture Crop. *HortTech.*, **4**: 34-40.

Tsujita, M.J. 1983. Greenhouse gerbera production. *Ontario Mininstry of Agricultural Fact Sheet* 83-006, pp. 1-4. Ontario, Canada.

Van Meeteren, U. 1978. Water relations and keeping quality of cut gerbera flowers. I. The cause of stem break. *Scientia Hort.*, **8**: 65-74.

Vidalie, H., M. Laffaira, L.M. Riviere and S. Charpentier, 1985. First results on the performance of gerbera cultivated on rockwool. *P.H.M. - Revue Horticole*, No. 262, pp. 13-18.

Voogt, S.J. 1976.Chlorosis of gerbera mainly caused by Mn deficiency. *Vakblad voor de Boemisterij*, **31**(36): 20-21.

Wernett, H.C., T.J. Sheehan, G.J. Wilfret, F.J. Marousky, P.M. Lyrine and D.A. Knauft, 1996. Postharvest longevity of cut flower *Gerbera*. I. Response to selection for vase life components. *J. Amer. Soc. hort. Sci.*, **121**: 216-221.

19

Gladiolus (Family: Iridaceae)

Y.C. Gupta, R.L. Misra, Nomita Laishram, Priyanka Sharma, Bhavya Bhargava and Sanyat Misra

[**Common names**: Autumn painted lady (*Gladiolus monticola*), Blue/Mauve Afrikaner (*G. carinatus*), Brown/Small brown Afrikaner (*G. maculatus*), Brownies (*G. mutabilis*), Caledon bluebell (*G. bullatus*), Christmas painted lady/Painted lady (*G. carneus*), Corn flag (*G. illyricus*), Gladiolus (pl. gladioli)/Sword lily (*Gladiolus* spp. and its varieties), Hermanus cliff gladiolus (*G. carmineus*), Large brown Afrikaner (*G. liliaceus*), Long-tubed painted lady (*G. angustus*), Marsh Afrikaner (*G. tristis*), New year lily/Waterfall gladiolus (*G. cardinalis, G. primulinus*, now known as *G. natalensis* ssp. *primulinus*), Painted lady (*G. carneus, G. debilis*), Pink bell (*G. ornatus*), Small brown Afrikaner (*G. hyalinus*), Riversdale bluebell (*G. rogersii*), Tulbagh bell (*G. inflatus*), Wit Afrikaner (*G. concolor*)]

Introduction and History

The name gladiolus was originally coined by Pliny the Elder (A.D. 23-79). The word gladiolus is derived from the Latin *gladius* meaning a sword, because of the sword-like shape of its foliage. As a genus it was first proposed by Tournefort and accepted in botanical literature by Linnaeus through his treatises 'Species Planterum' in 1753 and 'Genera *Plantarum*' in 1754. The gladioli are said to be cultivated since the days of ancient Greece. The genus *Gladiolus*, at present, comprises 255 recorded species (Misra, 1995; Goldblatt and Manning, 1998) with a wide distribution that includes Central Europe, the Mediterranean region, Central and South Africa, particularly majority of the species (more than 125) around the Cape of Good Hope as South Africa is the primary centre of its origin. Campbell and Bower (2003) mentions about one more species *Gladiolus scabridus* – a new species first recorded in 1978 from the mountains of northern KwaZulu-Natal and southern Swaziland, restricted to well-drained rocky habitat, and where it is endemic. This species has enormous commercial potential having large and bright pink flowers. So now number of recorded *Gladiolus* species are more than 260. The gladiolus was brought into cultivation first during the Greek period. Gladiolus has been documented in the literature since 1578, as evidenced from the record *Lyte's Nievve Herball*. According to the record it was first introduced into France and soon after spread to England, Germany, Holland and North America. In India, its cultivation dates back to 19th century as mentioned in the 'Firminger's Manual of Gardening in India' that Mr. Charles Gray of Conoor grew some gladioli from corms and seeds in his garden. The European Union gladiolus importing countries are Germany, Belgium, Luxembourg, Denmark, Spain, France, united Kingdom, the Netherlands, Italy and Portugal.

The development of gladiolus started only at the beginning of 18th century when William Herbert (1806) in England produced interspecific hybrids, but all proved to be sterile and were never commercialized. Later, in 1837, Hermann Josef Bedinghaus developed Gandavensis hybrids in Belgium which proved very helpful for the production of modern gladioli. In 1596, John Gerard, gardener to Lord Burleigh had two European species: *Gladiolus communis* L. from Mediterrenean region and *G. segetum* Ker from South Europe, as is evidenced by his catalogue of the plants he grew some 1,030 kinds. 'General Historie of Plantes' (1597), the famous Herbal by Gerard has mention of these two species. By 1862, John Parkinson introduced *G. byzantinus* Mill. into Europe from Mediterrenean region and published in his ' Paradisi in Sole Paradisus Terrestris'. *G. byzantinus* Mill. was bred in 1893 with *G. cardinalis* Curt. by Dammann & Co.,

Italy, resulting in *G. victorialis*, a slightly fragrant hybrid. No evidence is available for any species of gladiolus of South African origin being grown in Europe prior to 1700. *Gladiolus aethiopian* was first described by Cornutus from Cape region. In the middle of the18th century the South African species were introduced into Europe. *Gladiolus tristis* (fragrant) was introduced from Cape region in Chelsea Physic garden in 1745. *Gladiolus grandis* Thunb. was naturalized in Spain from South Africa. By 1816, French, Dutch and Belgium started raising hybrids, the first being *Gladiolus gandavensis* Van Houtte by crossing *Gladiolus cardinalis* and *Gladiolus psittacinus* in 1838 by M. Beddinghaus. Colvillei hybrid was produced in 1823 by Colville's Nursery, England by crossing *Gladiolus tristis* var. *'concolor'* with *Gladiolus cardinalis*. *Gladiolus primulinus* with its hooded characteristic of its central petals was collected from the bank of Zambesi River and has been extensively used in breeding programmes. Victor Lemoine of France used *Gladiolus primulinus* for producing Primulinus hybrids during 1906-1908. About 23 species of gladiolus seem to have contributed for the development of modern gladiolus (Misra and Kaicker, 1986), the latest one being *Gladiolus callianthus murielae* (formerly known as *Acidanthera bicolor murielae*). One related genus *Homoglossum watsonianum* has also been crossed with gladiolus in1935 by Collingwood Ingram but was never commercialized. In 1956, Mrs. John Wright in New Zealand, attempted the second bigeneric cross using cultivar 'Filigree' with *Acidenthera bicolor* and developed the famous fragrant cultivar 'Lucky Star'. Kazinori Suzuki *et al.* (2008) described that previously studied species and their floral scent types are *Gladiolus alatus, G. maculatus, G. recurvus* and *G. tristis* with linalool; *G. jonquilliodorus, G. orchidiflorus, G. patersoniae* and *G. scullyi* with geraniol or geraniol acetate, nerol and citronellol; *G. liliaceus* with eugenol; and *G. carinatus* & *G. virescens* with ß-ionone (Goldblatt *et al.*, 1998; Goldblatt and Manning, 2002). They had obtained a large qualitative variations in floral scents including benzenoids. Qualitatively and quantitatively, Suzuki *et al.* (2008) analyzed nine *Gladiolus* species (*G. gracilis, G. orchidiflorus, G. pritzelli, G. recurvus, G. trichonemifolius, G. tristis, G. uysiae, G. venustus* and *G. watermeyeri*) at Ibaraki Agrticultural Center, Japan for the floral scent these species emit. They collected the emitted scent compound from these species from 6 a.m. to 6 p.m. and from 6 p.m. to 6 a.m. for a 12 hour period when average temperature was 25 °C and relative humidity 25 per cent and analyzed using gas chromatography-mass spectrometry as per procedure of Oyama-Okubo *et al.* (2005) and detected 20 scent compounds. After analysis, they divided the species into four groups ~ 'linalool/benzenoid group', 'nerol' group, 'ionone' group and 'ocimene/caryophyllene' group and advocated *G. orchidiflorus, G. recurvus, G. tristis* and *G. watermeyeri* as potential genetic resources for fragrance.

Gladioli have almost all the colours except true green, blue and black though nearby colours are available. Gladioli are versatile as can be grown in the plains (absolutely tropical areas), in temperate areas during summer months, in the beds or borders, for mass effects and the smaller ones even in the pots and bowls. These are not suitable either for rock gardens or for naturalizing effects as the corms require to be lifted every year after the flowering.

Botany, Cytology and Genetics

Gladiolus is grown from corms and at the base of the corm, there is basal plate from which roots emerge. The corm is covered by 4-6 dry scales. The scale is attached to the corm at the node. Each node bears one or two quite visible buds which has the potential to grow into a shoot. The bud at the top is the largest and the first to sprout. Corm usually sprouts within a week after planting. The filiform roots start emerging from the base of the corm. When the plant reaches nearly 4-leaf stage, new contractile roots start emerging from the base of the growing stem where new corm starts developing. These new roots store water and orient the plants straight and deep into the soil. Leaves are ribbed, sword-shaped and clustered around the swollen stem base. In all, the plant produces 6-9 leaves. When the plant reaches 3-leaf stage, flower bud initiation starts. The process of flower differentiation is complete in 7-10 days. The inflorescence of gladiolus is a one-sided spike where widely funnel-shaped individual florets are directly attached to the central axis, and arise in an alternate fashion opposite one another, progressing from base to tip. Each flower bud is encased separately within its own spathe, which consists of two green bracts. Each bloom has a short curving tube and 6 petal-like lobes, the 3 upper ones usually larger, the 3 lower ones often blotched, and these are collectively known as perianth or segments. The new corm continues to develop and the old one below it gradually shrivels. At the base of the new corm arise the stolons, the tip of which swells and develop into the cormels.

Gladioli have basic chromosome number n=15. De Vilmorin and Simonet (1927) reported the chromosome number in *Gladiolus primulinus* var. 'La Meurthe' as n=15, and subsequently Ernst-Schwarzenbach (1931) reported *G. tristis* var. *concolor, G. cardinalis, G. colvillei* var. *roseus* and *G. cuspidatus* as 2n=30, *G. ramosus* with 2n=46 (3x+1), and *G. byzantinus, G. primulinus* vars. Souvenir, *G. gandavensis* varieties 'Pompée', 'Alexandre', 'Red Canna', *G. lemoinei* vars 'Catharina', 'Don Salluste', 'Mrs. Frank Pendleton', and *G. nanceianus* var. *desdémone* with 2n=60. Bamford (1935) worked out chromosome numbers of various species and varieties of gladiolus and reported that species in Europe and western Asia such as *G. byzantinus* and *G. atroviolaceous* had 2n=90, *G. anatolicus* and *G. segetum* had 2n=120, and *G. communis* 2n=138 (9n+3); species in Cape and tropical Africa such as *G. alatus, G. angustus, G. blandus, G. brevifolius, G. callistus, G. cardinalis, G.c. elegans, G. carmineus, G. colvillei alba, G.c. roseus, G.c. rubra, G. crassifolius, G. cuspidatus, G. debilis, G. gracilis, G. grandis, G. hirsutus, G. nanus* vars. 'Ackermanni', 'Blushing Bride', 'Peach Blossom', 'Robinhood', 'Roos van Dekama', 'Siren', 'Spitfire', *G. odoratus, G. oppositiflorus, G. pappei, G. permeabilis, G. recurvus, G. splendens, G. trichonemifolius, G. tristis, G.t. concolor, G. undulatus*, and *G. villosus*, all having 2n=30; *G. saundersii* with 2n=30, 45; *G. formosus, G. alatus* hybrid, *G. nanus* 'Liberty', 'Nymph', *G. orchidiflorus* & *G. tuberginii* vars. 'Charm' and

'Prunella', and a general var. 'Leeuwenhoek', all with 2n=45; *G. coccineus, G. nanus, G. nanus* 'Groenendaal', *G. primulinus, G. platyphyllus*, and certain other varieties such as 'Abbé Raucourt', 'Alda', 'Albatross', 'Alice Tiplady', 'Alsace Lorraine', 'Altar', 'Anthony Kunderd', 'Ave Maria', 'Baron Joseph Hulot', 'Blue Isle', 'Blue Triumphator', 'Break O'Day', 'Cardinal Prince', 'Catharina', 'Catherine Coleman', 'Cattleya Rose', 'Commodore', 'Contemplation', 'Desdémone', 'Dillenberg', 'Dr. F.E. Bennett', 'Encelade', 'Enchantress', 'Evelyn Kirtland', 'Francis King', 'Giant Nymph', 'Gloriana', 'Golden Dream', 'Golden Measure', 'Hyperion', 'Impressario', 'Indian Chief', 'Jane Addams', 'Joost v.d. Vondel', 'King of Oranges', 'L. von Beethoven', 'Los Angeles', 'Marshal Foch', 'Mary Jane', 'Meadow Lark', 'Minuet', 'Mr. Mark', 'Mr. W.H. Phipps', 'Mrs. F.C. Peters', 'Mrs. Frank Pendleton', 'Mrs. Leon Douglas', 'Mrs. P.W. Sisson', 'Mrs. Van Konynenburg', 'October', 'Pacha', 'P.C. Hooft', 'Pfitzer's Triumph', 'Picardy', 'Pierian', 'Porthos', 'Pride of Wanakah', 'Princeps', 'Prof. Donders', 'Prof. E.H. Wilson', 'Purple Glory', 'Rob Roy', 'Rose', 'Syncopation', 'Taurus' and 'Vermillion', all with 2n=60; *G. dracocephalus* and *G. quartinianus* with 2n=75; *G. papilio* and *G. psittacinus* with 2n=75, 90; and *G. dracocephalus* with 2n=90.

Shibata and Nozaka (1963) isolated seven anthocyanins, pelargonidin 3-*O*-rhamnosylglycoside, peonidin 3-*O*-rhamnosylglucoside-5-*O*-glucoside, pelargonidin 3-*O*-rhamnosylglucoside, peonidin 3-*O*-rhamnosylglucoside-5-*O*-glucoside, pelargonidin 3—*O*-diglucoside-5-*O*-glucoside, cyanidin 3,5-di-*O*-glucoside, delphinidin 3,5-di-*O*-glucoside, and delphinidin triglucoside from 10 gladiolus cultivars. Yatomi and Arisumi (1968) and Arisumi and Kobayashi (1971) reported 10 and 13 anthocyanins from 10 and 12 cultivars, respectively, and these were pelargonidin 3,5-di-*O*-glucoside, pelargonidin 3-*O*-rhamnosylglucoside, peonidin 3-*O*-rhamnosylglucoside, peonidin 3,5-di-*O*-glucoside, malvidin 3-*O*-rhamnosylglucoside, malvidin 3,5-di-*O*-glucoside, malvidin 3-*O*-rhamnosylglucoside-5-*O*-glucoside, petunidin 3,5-di-*O*-glucoside, petunidin 3-*O*-rhamnosylglucoside-5-*O*-glucoside, pelargonidin 3-*O*-glucoside, malvidin 3-*O*-glucoside, petunidin 3-*O*-glucoside, and petunidin 3-*O*-rhamnosylglucoside for the first time. Akavia *et al.* (1981) through HPLC, recorded 18 anthocyanins, including cyanidin 3-*O*-rhamnosylglucoside-5-*O*-glucoside, cyanidin 3-*O*-rhamnosylglucoside, delphinidin 3-*O*-rhamnosylglucoside, and delphinidin 3-*O*-rhamnosylglucoside from nine Israeli gladiolus cultivars. Takemura *et al.* (2005) from Japanese cv. 'Ariake' isolated malvidin 3,5-di-*O*-glucoside as the major anthocyanin together with two minor malvidin glycosides and three major flavonols, kaempeferol 3-*O*-sophoroside, kaempeferol 3-*O*-rutinoside, and quercetin 3-*O*-rutinoside. Takemura *et al.* (2008) through TLC, LC-MS and HPLC studied the anthocyanins of six *Gladiolus* cultivars, *viz.* 'Beijing', 'Ben Venuto', 'Fado', 'Juku-gaki', 'Marches', and 'Violetta' and isolated 11 anthocyanins, *e.g.* 3-*O*-rutinoside-5-*O*-glucosides of cyanidin, malvidin, pelargonidin and peonidin, and 3,5-di-*O*-glucosides of petunidin, malvidin, pelargonidin, cyanidin and peonidin, and pelargonidin 3-*O*-rutinoside and malvidin 3-*O*-glucoside. Of these the first four and pelargonidin 3-*O*-rutinoside have previously been characterized as 3-*O*-rhamnosylglucoside-5-*O*-glucoside and 3-*O*-rhamnosylglucoside though in this study they were, in fact, found to be 3-*O*-rutinoside-5-*O*-glucoside and 3-*O*-rutinoside, for the first time. After analyzing 84 cultivars through HPLC, as per anthocyanin pattern, it was divided into 18 groups, and from which it was inferred that major anthocyanins of purple flowers were malvidin glycosides together with petunidin 3,5-di-*O*-glucoside as a minor component but delphinidin glycosides were not detectable, red colouration of flowers was found due to pelargonidin glycosides, pink flowers due to various anthocyanins, pelargonidin, cyanidin, peonidin, petunidin and malvidin glycosides though were comparatively scarse compared with those of purple and red flowers, however, in many of the yellow and white cultivars no anthocyanins were detected but a few which had coloured spot or streak on perianth contained extremely small amount of anthocyanins (Takemura *et al.*, 2008).

Pragya *et al.* (2010a) while working with 54 gladiolus cultivars, employed nine morphological traits and 225 random amplified polymorphic DNA (RAPD) markers amplified with 25 arbitrary primers to discriminate between the cultivars and to evaluate the relatedness between them, and a total of 225 RAPD bands (250-3000 bp) were obtained out of which 21 (93.78 per cent) were polymorphic. The number of bands generated per primer ranged from 4 to 14 with an average of 9 bands. For morphological data, Euclidean distances were calculated and an unweighed pair-group method using an arithmetic average was used to construct a dendrogram. In addition, principal component analysis was carried out based on RAPD data. In both the cluster analyses, cultivar Pusa Lohit branched out from the dendrograms, confirming that it is quite different from all other genotypes. Furthermore, cultivar pairs 'Ripples' and 'Anglia', 'Friendship' and 'Yellow Jame', 'Creamy Green' and 'Morello', 'Pusa Suhagin' and 'Pusa Archana', 'The Barton' and 'Christian Jane', 'Plumtart' and 'Chantilar', and 'Peter Pears and 'Picardy' were very close in both the dendrograms indicating that they are quite similar to each other as also got support through PCA analysis though incongruence existed between these cluster analyses. Pragya Ranjan *et al.* (2010b) analyzed genetic relatedness in 54 gladiolus cultivars using amplified fragment length polymorphism (AFLP) through 9 primer sets out of 24 primer pairs initially tried by subjecting fluorescence-labeled amplification products to electrophoresis and then analyzed using an automated sequencer. They constructed a dendrogram by the unweighted pair group method using the arithmetic average (UPGMA). The number of AFLP fragments generated per primer set ranged from 10 to 151 with fragment sizes varying from 50 to 450 bp. A total of 660 fragments were detected, of which 658 (99.70 per cent) were polymorphic. All the primers except E-AGG/M—CTA displayed 100 per cent polymorphism. All cultivars were clearly differentiated by their AFLP profiles. The AFLP data were compared with previously obtained RAPD data and combined to generate a common dendrogram. The first cluster was dominated with indigenously bred cultivars while the second was dominated with exotic cultivars which shows that most of the exotic cultivars as well as indigenous cultivars are closely related with each other, however, two indigenous cultivars, *viz.* Pusa Suhagin and Pusa Archana share genetic similarity with exotic cultivars. Among all the cultivars studied, Pusa Gunjan was found as the most distinct genotype. The AFLP markers developed will help future *Gladiolus* cultivar identification, germplasm conservation and new cultivar development.

Neeraj *et al.* (2005) reported a high heritability (in broad sense) and high genetic advance (as a percentage of mean) for number and weight of cormels per plant and average weight of corms. High heritability with comparatively moderate genetic advance was obtained for number of corms per plant, length of spike, length of rachis and days to spike initiation. Heritability in these characters was attributed to additive gene effect. Ghimiray (2005) also reported that the phenotypic coefficients of variation were generally higher than the genotypic coefficients of variation for all the characters studied. Spike length, plant height and leaf length exhibited high heritability coupled with high genetic advance, whereas number of florets per spike showed low heritability but high genetic advance. This indicates the importance of additive gene effects in the control of these characters and their potential for direct selection. Non-additive gene effects were evident for the other characters, indicating the importance of heterosis breeding for the improvement of these characters. Kadam (2014) tried 26 gladiolus varieties and recorded earliest flowering in 'Hunting Song', maximum plant height and spike length in 'Amsterdam', maximum rachis length and number of florets per spike in 'Snow Princess', largest floret size in 'Snow Princess' and 'Priscilla', maximum number of corms in 'Yellow Stone', corm weight and corm size in 'Forta Rosa' though highest cormel weight in 'Argentina'. They recorded highest PCV and GCV values for cormel weight, *i.e.* 104.36 and 67.26, and number of cormels per plant, *i.e.* 99.82 and 71.58, respectively. The heritability was recorded maximum for days to first floret opening followed by rachis length and days to spike initiation. They recorded highest values for genetic advance for plant height and spike length though low genetic advance for days to sprouting followed by number of corms and leaves.

Classification, Species and Varieties

Bentham and Hooker (1883) placed the family taxonomically in the series Epigynae. Way back in 1888, Pax classified Iridaceae into Crocoideae, Iridoideae and Ixioideae, the latter one being further divided into Ixieae, Watsonieae and Gladioleae on the basis of the floral forms (zygomorphic or actinomorphic), styles (complete or divided into 2) and ovary type (mono- or bicarpellary) and the genus *Gladiolus* was placed under Gladioleae along with seven other genera. The genus *Gladiolus* further divides into four sections, *viz.* I: *Eugladiolus* (100 species of Tropical and South Africa, Europe and western Asia; II: *Habea* (12 species of Cape Peninsula of South Africa and Madagascar Island); III: *Schweiggeria* (only two species, *i.e. G. montana* and *G. arenarius*, both from Cape); and IV: *Homoglossum* (5-6 species of South Africa). In 1892, Baker also adopted the same system placing 109 species under *Eugladiolus*, 15 under *Habea*, 2 under *schweiggeria* and 6 under *Homoglossum*. Hutchinson in 1932 placed a separate order Iridales through his treatise *The Family of Flowering Plants II : Monocotyledons.* When this book was being revised, in 1959, the genus *Gladiolus* was placed in tribe 10-Gladioleae. However, Lewis *et al.* (1972) placed *Gladiolus* under four groups on the basis of leaf number and their emergence before or after the flowering, and these are being placed below.

I **Plurifoliati** containing 22 species [{red or scarlet coloured ones, *viz. G. cardinalis, G. cruentus, G. saundersii,. G. sempervirens*}; *G. opositiflorus* (white or pink); *G. elliottii* (white speckled purple); *G. sericeo-villosus* (pink and/or yellow arranged opposite); *G. varius* (pink); *G. ochroleucus* (cream to yellow); *G. ecklonii* (white, red and mauve); *G. buckerveldii* syn. *Antholyza buckerveldii, Petamenes buckerveldii* (yellow); *G. natalensis* (red, orange, yellow and/or greenish); *G. papilio* syn. *G. purpureoauratus* (yellow); *G. hollandii* (white speckled pink); *G. exiguous* (pink); *G. crassifolius* (white, pink and mauve); *G. densiflorus* (white spotted red); *G. invenustus* (white to mauve with blotches); *G. calcaratus* (white); *G. appendiculatus* (white marked mauve); and *G. pole-evansii* (pinkish-lilac, vegetative parts hairy)] all with 5-8 or more well developed leaves arranged distichously in the form of a fan.

II. **Paucifoliatii** containing 34 species [*G. stellatus* (fragrant whitish to pale-mauve); *G. gueinzii* (pink marked purple); *G. acuminatus* (fragrant, dull greenish-yellow); *G. lapeirousiodes* (white but lower lobes with 2 reddish blotches); *G. leptosiphon* (cream to cream-brown); *G. vigilans* (pale-pink); *G. carneus* (white, cream, pink or mauve); *G. macneillii* (pale-salmon-pink); *G. microcarpus* (whitish, pink or mauve); *G. buckerveldii* (also under Plurifoliati); *G. angustus* (white to cream-yellow); *G. floribundus* (fragrant white, cream, pink or mauve and flowers closing at night); *G. undulatus* (greenish to cream-white or pink); *G. caryophyllaceus* (strongly scented, pale to mauve); *G. lewisiae* (fragrant, creamy-white flushed pink); *G. involutus* (pale-pink or white); *G. scullyi* (fragrant, cream, yellow, lime-green, pink, mauve or maroon); *G. arcuatus* (fragrant mauve or purple); *G. salteri* (pink); *G. kamiesbergensis* (mauve); *G. permeabillis* (fragrant white to cream, pink, mauve or brown); *G. vernus* (pink or pale-magenta); *G. pretoriensis* (dull purple to pink); *G. marlothii* (fragrant mauve, lower lobes minutely speckled); *G. mostertiae* (pale-pink); *G. rufomarginatus* (white or cream-yellow densely flecked with red or purple); *G. orchidiflorus* (fragrant, grey-green, sheen purple); *G. watermeyeri* (fragrant, cream or greenish-cream, flushed dull purple); *G. virescens* (fragrant, greenish-yellow); *G. seresianus* (fragrant, dull or greenish-yellow); *G. uysiae* (fragrant, cream or greenish); *G. alatus* (fragrant, red, orange or salmon, lower lobes greenish); and *G. equitans* (fragrant, orange, red or vermilion, yellow-green patches on laterals and falls)] each with 2-5 well developed distichously or spirally arranged leaves.

III **Unifoliati** contains 29 species [*G. brevitubus* (pale-pink or red); *G. quadrangulus* (fragrant, white to pale-blue or pinkish-mauve); *G. citrinus* (bright yellow); *G. tenellus* (fragrant, yellow, cream or white, tinged purple); *G. oreocharis* (white to mauve); *G. inflatus* (pale, pink or mauve);*G. robertsoniae* (fragrant, mauve or white tinged pale-mauve); *G. cylindraceus* (pale- to creamy-pink, flushed salmon-pink); *G. debilis* (white to pale-pink, throat dotted red); *G. longicollis* (evening fragrant, white-cream or yellow); *G. tristis* (night fragrant, white to buff-cream or pale-yellow, tinged green); *G. liliaceus* (night-fragrant, yellowish to dull-yellow, flecked brown, pink, red or purple); *G. hyalinus* (mildly evening fragrant, yellow-brown, yellowish-cream or greenish-brown, spotted brown and purple); *G. recurus* (fragrant, pale-greenish-grey, greenish-mauve to pale-yellow or cream); *G. symonsii* (pink); *G. punctulatus* (pink to mauve); *G. carinatus* (fragrant, pale-greyish-blue to pale-pink); *G. viridiflorus* (fragrant, pale-green); *G. blommesteinii* (pale-pink or mauve); *G. ornatus* (pale to deep pink); *G. gracilis* (fragrant, pale-mauve, pink, mauve, white or yellow); *G. exilis exilis* (mildly fragrant, pale-blue, mauve or white, tinged pale-mauve); *G. mutabilis* (fragrant, grayish-mauve to cream, yellow or brown); *G. violaceo-lineatus* (mauve); *G. comptonii* (bright yellow with reddish streaks); *G. rogersii* (purple); *G. bullatus* (deep mauve); and *G. pritzelli* (fragrant, yellow, seldom tinged red or grey outside)] with one well developed spirally arranged basal leaf, following reduced, bract-like with short blades.

IV. **Exfoliati** contains 24 species [*G. stefaniae* (red); *G. stokoei* (scarlet); *G. nerineoides* (pale-salmon-pink to deep golden-red); *G. guthriei* (fragrant, deep pink or reddish); *G. carmineus* (deep rose-pink or carmine with white or cream median stripe); *G. monticola* (cream tinged pink); *G. maculatus* (fragrant, white, yellow or pink with brown, purple or red streaks and spots); *G. engysiphon* (strongly and sweetly fragrant, white with a crimson median line on falls and laterals); *G. bilineatus* (white or cream, shadowed light pink); *G. emiliae* (fragrant, yellow, speckled densely with red, brown or purple); *G. odoratus* (fragrant, dull or brownish-yellow, speckled and striped purple, red or maroon); *G. vaginatus* (fragrant, pale-mauve to grey-mauve, seldom white); *G. brevifolius* (fragrant, pale to pink, rarely white, mauve, brown or brownish-green); *G. pillansii* (fragrant, mauve or pink); *G. martleyi* (white suffused rose-pink); *G. subcaeruleus* (pinkish or mauve); *G. jonquilliodorus* (fragrant, cream or pale-yellow, flushed pink or mauve); *G. aurantiacus* (golden yellow or orange and yellow, streaked red); *G. brachyphyllus* (pink to mauve with red spots); *G. unguiculatus* (mauve with purplish blotches on lower lobes); *G. woodii* (dark brown); and *G. parvulus* (mauve, pink or white)], all having absent well developed leaf or leaves at flowering, the scape may bear 1-4 sheathing bracts with or without small free blades.

Goldblatt (1971) placed the genus *Gladiolus* under subtribe Gladiolineae of the tribe 3-Ixieae.

There had been 226 recorded *Gladiolus* species (Misra *et al.*, 2005) though now this number exceeds 260 (Takemura *et al.*, 2008). Goldblatt and Manning (1996) collected two new species from South Africa, *viz. Gladiolus povonia* (found in dolomite soil) and *G. serpenticola* (restricted to serpentine soil). Three additional species, *viz. G. miniatus* Eckl., *G. caeruleus* Baker and *G. variegatus* were recognized in western Cape Province, the first which was retained as subspecies under *G. floribundus* has now been reinstated at species rank, second species was previously treated as *G. gracilis* var. *latifolius* now holds species rank as *G. caeruleus*, and the third which was previously *G. debilis* var. *variegatus* is now a full rank species *G. variegatus*. A new species *Gladiolus somalensis* has also been identified in Sonang and Bazi regions of Somalia, an endemic species apart from four others (Goldblatt and Thulin, 1995). Some 34 South African species such as *G. acuminatus, G. alatus, G. arcuatus, G. brevifolius, G. brevitubus, G. carinatus, G. caryophyllaceus, G. ceresianus, G. engysiphon, G. equitans, G. exilis, G. floribundus, G. gracilis, G. guthriei, G. jonquilliodorus, G. lewisiae, G. liliaceus, G. longicollis, G. maculatus, G. marlothii, G. mutabilis, G. odoratus, G. orchidiflorus, G. permeabilis, G. pillansii, G. pritzelli* var. *pritzelli, G. recurvus, G. robertsoniae, G. tenellus, G. tristis, G. uysiae, G. vaginatus, G. viridiflorus* and *G. watermeyeri,* and one Ethiopian species *G. callianthus* are the fragrant species (Misra, 1995; Misra *et al.*, 2005). Misra (1995) mentions that out of a total number of 226 gladiolus species, Albania, Corsica, Cyprus, England, France, Ghana, Italy and Sardinia, Iraq, Israel, Ivory Coast, Labanon, Mali, Reunion Island, Sicily, Sierra Leone, Uganda, USSR and Yugoslavia, each contains one species of gladiolus. Countries like Botswana, Greece, Guinea, Mediterranean areas, Namibia, Portugal, Somali, Spain and Tropical Africa contain two species each. Cameroon, East Africa, Madagascar and Turkey contain three species each. Ethiopia, Iran, Kenya and Zambia contain four species each, Nigeria and Zimbabwe five species each, Mozambique 10 species, Zaire 11 species, Tanzania 13 species, Malawi 14 species and Angola contains 16 species though some 20 species exist of uncertain origin. It is only South Africa from where 114 species are recorded. Some of the most popular ones are being described here.

G. acuminatus. A species restricted to Cape coastal districts of Caledon and Bredasdorp where it flowers in August and September. Its corms are ovoid or subglobose with 0.8-1.5 cm diameter, stems are firm, slender or moderately slender, flexuose, simple to 1-2-branched, up to 60 cm long and basally ensheathed with a single hyaline obtuse or subacute sheath 2.5-7.0 cm long. Leaves 5-10, 2-5 basal short-sheathed, blades 1-2 mm wide and less in height than the stem. Spike secund, 2-10-flowered, bracts oblong to oblong-lanceolate and acute or acuminate and flowers tubular having erect tube with recurved dull greenish-yellow segments that are almost equal in size. Capsule is ellipsoid, 1 cm long, seeds small, dark brown with paler wings.

G. alatus. Widespread in the Cape winter-rainfall region from Namaqualand in the west to Cape Peninsula and along the SE coastal belt from Caledon to Bredasdorp where this flowers from August to October. This species grows up to 35 cm tall with slender, flexuous and simple to rarely branched stems where leaves are linear to lanceolate with prominent longitudinal ribs, and the spikes are 1-10-flowered. Its 6 cm wide scented flowers are bilabiate and small to medium. Upper segments reddish, salmon, orange or orange-red, lower tipped on lime-green or yellow background whereas laterals spathulate. In its native haunts it flowers from August to October.

G. angustus (*G. angustifolius* Salisb., *G. telifer* Stokes). From fairly moist areas of Cape Peninsula to Clanwilliam and Bredasdorp, this species bears flexuous, simple and slender stems up to 75 cm long with lax spike bearing 2-10 long-tubed funnel-shaped white, cream or dark creamy-yellow flowers, sometimes flushed pink. Tube is up to 7 cm long, and the segments obtuse or acute. Falls up to 2.5 cm long, 1 cm wide and with reddish-purple mark up to the middle. Upper laterals shorter than the hooded central, 2.5-3.5 cm long and 2 cm wide.

G. atroviolaceus (*G. aleppicus*). From sub-arid areas of Greece, Turkey and Iran, this species bears stiff, slender and up to 60 cm tall stems sheathed with linear leaves. Spike 5-12-flowered only at one side bearing dark to crimson-violet flowers, striped paler in the throat, Segments oblong-spathulate, central segment strongly curved, upper laterals about 2 cm long and little spreading, the falls are spathulate and about 3 cm long. It has been observed flowering there from April to July.

G. aurantiacus. From the moist hill areas of Natal, Transvaal and Swaziland, this species bears rarely branched stout stems about 70 cm long where 7-8 distichous basal leaves devlop only after flowering. Spike is 8-12-flowered, flowers tubular, golden-yellow or orange and yellow, up to 10 cm long, often streaked or marked red, perianth tube little curved and about 5 cm long, upper segment 4 cm long and 2 cm wide and little hoodeed, the lower small and elliptic, 2.0-2.8 cm long, 1.0-1.5 cm wide, and the laterals are either little shorter or longer than the falls. In its native haunts it flowers in October and November.

G. aureus (*Homoglossum aureum*). From Cape Peninsula, this species bears small globose corms with 15-30 cm long and hairy slender stems, and only two hairy leaves, upper entirely sheathed and the lower long-sheathed, narrowly linear and ribbed. Spikes 1-sided with bright yellow and funnel-shaped 1-2 flowers 2-3 cm long, and the segments oblong-acute, unequal and shorter than tube.

G. barnardii. From Zimbabwe (near Harare), this species is about 1 m tall with sturdy stems, having 5 distichous and up to 65 cm long basal leaves, and 6 cauline the upper one 11 cm long. Spike is 10-flowered and about 35 cm long, perianth tube curved, flowers pale orange-yellow with pinkish shade, segments gradually smaller from central upper to central falls where cental fall is twisted too. In Zimbabwe this flowers from May to July.

G. brevicaulis. From moist areas at an altitude of 1250 m of Angola and Zaire, this species bears globose corms of the diameter of 2.5 cm. Stems are terete, slender and up to 18 cm long with 1-2 reduced leaves. Leaves are linear, glabrous and 15-30 cm long. Spike is lax, up to 20 cm tall with many violet-purple flowers some 2.5 cm long but only at one side, perianth tube funnel-shaped and curved, and the segments are oblong-spathulate, the lower smaller.

G. brevitubus (*Tritonia ventricosa*). A native of mountain slopes of Caledon and Swellendam areas, this species bears some 50 cm long stem with 2-6-flowered short spikes, flowers almost regular, scented, small, pale to deep rose-pink, orange or brick-red, and rarely yellow in the centre. Perianth tube only 3 mm long, segments obtuse, outer a little larger and lower segments often with yellowish mark surrounded with red. It flowers there from September to December.

G. buckerveldii [*Antholyza buckerveldii* L., *Petamenes buckerveldii* (L. Bol.) N.E.Br.]. This species is found in Cape (Clanwilliam district) and close to the mountain streams up to an elevation of 1,000 m from Cedarberg (near Clanwilliam) mountains at Algeria where it flowers in January. Its corms are 1.5 cm in diameter, bearing up to 15 cm long 3 reddish prophylls, 5-6 sheathing leaves with up to 60 cm long and 1.5-3.0 cm wide, and the upper leaves are bract-like and about 12 cm long. Spike is about 60 cm long, comprising of closely set 30 greenish-yellow or creamy flowers 5.0-6.5 cm long with a 4 cm long floral tube, where falls have a small central purple spot. Segments are small and unequal, central lobe oblong to obovate while others ovate-acuminate and gradually smaller from upper to lower. Its capsules are 2.3 cm long and 0.9 cm wide with seeds broadly winged.

G. bullatus (*G. inflatus* sensu Klatt, *G. spathaceus* Pappe ex. Bak.). Its native haunts are rocky areas around Caledon and Bredasdorp districts of the Cape Province where it flowers from August to October. Its ovoid to subglobose corms are up to 2 cm in diameter. Its simple stems are slender up to 70 cm high, with two basal sheaths, inner one about 18 cm long. Its 3-4 leaves sheath the stem up to the rachis and then beyond short tail-like. Spike is secund and 1-2-flowered with curved, funnel –shaped and 1.0-1.3 cm long perianth tube. Flowers are inflated campanulate, some 5 cm long, deep blue to mauve-blue, and lower part of falls paler with deeper spots and a median stripe on lowest segment. Segments larger to smaller from upper to lower in the flower. Capsule is ellipsoid and 2 cm long, and the seeds are yellow-brown and small.

G. byzantinus. A native of central and eastern Mediterranean region (Spain and Italy), North Africa, Corsica and Malta, this species flowers from late spring to early summer. Leaves are 3-5 and form a fan at the base of the stem. Its spike is up to 75 cm tall with 2-3-sided and more than 15 florets up to 7.5 cm long with rich purplish-red but having a narrow red line along the centre of the lower tepal. In the florets, petals usually overlap each other. Also the 'Albus' is its white form though very rare.

G. callianthus (*Acidanthera bicolor, A. b. murielae, Gladiolus murielae, Ixia quartiniana, Sphaerospora gigantea*). A native of Ethiopia, Eritrea, Somali, Mozambique, Malawi and Tanzania up to 1,200-2,500 m elevations, is a half-hardy species with globose corms about 2.5 cm in diameter having reticulated tunic. Plants up to 60 cm tall with simple stem, leaves basal, glabrous and up to 30 cm long, two basal erect and linear while 1-2 semi-erect. Spikes simple, quite lax, 2-10-flowered, flowers 6-8 cm wide, long-tubed, creamy-white with dark reddish-brown dusting in the throat and fragrant. Perianth tube slender, up to 14 cm long, and segments oblong, acute and about 3 cm long.

G. cardinalis (*G. speciosus*; Waterfall gladiolus). A hardy species native to mountain areas of Paarl, Wellington, Worcester and along the stream banks in the SW Cape (South Africa) where it flowers in July, especially in the northern hemisphere. It is one of the major parents of the modern garden cultivars. Its stems are erect up to 1 m long and curved, leaves are 4-6, evergreen, sword-shaped and about 50 cm long, spikes 5-12-flowered, flowers scarlet or bright crimson and 7.5 cm in diameter, and the falls are with large white or cream marks surrounded with pinkish-mauve-red. It is one of a few gladioli which likes a little shade and regular moisture.

G. carinatus. It is native to SW Cape (South Africa) where it flowers from July to September. Its subglobose corms range from 1-3 cm in diameter. Stem is slender, more or less flexuose, 25-50 cm long with 2-3 basal sheaths. Leaves are usually three, the lowest with linear blade is as long as the stem and 1-8 mm wide grass-like but two upper ones are much shorter. Spike is fairly lax and secund, 1-9-flowered, flowers in main species fragrant, almost funnel-shaped, 4-6 cm long and pale-greyish-blue to deep blue or violet, mauve, pale-pink to reddish and pale-yellow to rich golden-brown with darker veining, lower tepals arching down and banded primrose-yellow.

G. carmineus. It has been recorded from the sandstone cliffs or low mountain slopes near the coast in the Caledon and Bredasdorp districts of southern Cape, flowering from February to April. Corm is subglobose with 2-3 cm diameter. Stem is simple, slender, usually flexuose, 20-35 cm high with 2-3 prophylls, basal leaf solitary the first year, two in the second year and sometimes three which appear after the flowering, and are glaucous, falcate, decumbent, up to 65 cm long, 0.6-1.7 cm wide and linear-lanceolate, however, cauline leaves are usually 3-6 and up to 13 cm long. Short spike is almost secund, 2-6-flowered, flowers deep rose-pink or carmine-red, funnel-shaped, up to 9 cm across, perianth tube may be little curved and up to 4.5 cm long, and the segments ovate, obovate or broadly elliptic. Capsule oblong-ellipsoid, 2 cm long, 1 cm wide and the seeds broadly winged.

G. carneus (*G. angustus* senus Thunb., *G. blandus* Ait, *G. trimaculatus* Lam., *G. ventricosus* Lam.). A native of western Cape Province (Cape Peninsula to Ceres and Riversdale up to 1,200 m elevation in moist and sandy areas), this tender species bears 4 leaves and produces 20-100 cm long spikes bearing 3-8 white, cream, pink or mauve campanulate flowers about 8 cm long with purple hues, and with or without markings on lower segments or in the throat. The segments are spreading, central slightly hooded and all with wavy margins.

G. caryophyllaceus (*G. ambiguous* et *G. mucronatus, G. rubromarginatus, G. similis, Antholyza caryophyllacea, Homoglossum caryophyllaceum, Watsonia amoena*). This was recorded from the SW Cape (South Africa) from sea level to about 1,000 m elevation where it flowers from August to October. The corms are subglobose or ovoid with 1.5-3.0 cm in diameter. Stems are stout, erect or flexuose, sometimes branched and 50-75 cm tall with two prophylls. Leaves usually 4-6, sword-ahaped and hairy, the lowest one about half the size of the stem with up to $^{1}/_{3}$rd ensheathing and 1-2 cm wide, second usually shorter and narrower and the upper 2-3 cauline leaves half or more sheathing with short erect and lanceolate blades, spirally twisted and usually with crenulate margins. Spike distichous or secund, 2-8-flowered, flowers have high carnation-scent, campanulate or broadly funnel-shaped, 5.0-7.5 cm across, with central petal slightly hooded, pale to deep pink or pinkish-mauve with a dark median line externally on the segments, and three lower ones variously pink- or red-spotted or streaked, and segments larger to smaller gradually from upper to lower.

G.* × *citrinus (*G. symmetranthus*). It is one of the earliest produced hybrids much used in the breeding programmes. A low growing hybrid up to 30 cm in height, flowering in August-September with pale-yellow flowers having maroon throat and long rounded segments.

G.* × *colvillei (*G. tristis* × *G. cardinalis*) and ***G.* × *nanus*** (*G. carneus* × *G. cardinalis*), both almost hardy, growing to 45 cm with about 25 cm long spikes, flowers loosely arranged, about 7.5 cm across, almost erect and scarlet with yellow throat, tepals pointed with lip yellow, and *G.* × *c.* 'The Bride' white. Some other promising hybrids and varieties are *G.c. albus* (white with a yellow stripe on the lower segments), *G.c. roseus* (soft pink), *G.n.* 'Amanda Mahy' (salmon-red, blotched violet), 'Atom' (red with creamy border), 'Comet' (bright red), 'Elvira' (pale-pink with yellow throat, blotched red), 'Fair Lady' (pink), 'Floriade' {salmon-pink}, 'Guernsey Glory' (deep pink, edges pale-purple-red, blotches creamy), 'Impressive' (orange-pink, blotched purple rose, edged brown), 'Nymph' (white, marked crimson), 'Prins Claus' (white, lower segments edged purple), 'Rose Marie' (bright pink, lower petals blotched scarlet), 'Sunmaid' (pink, interior rose with white centre), *etc.*

G. communis. A native of many parts of Mediterranean area, Spain, Italy, Greece, Corsica, Sicily, Sardinia, North Africa, *etc.*, this is a very perplexing species with so many synonyms. It bears slender and up to 1 m long stems, spikes sometimes branched and with up to 20 flowers, flowers pink, red or magenta-purple and segments almost equal in size. Subsp. *communis* from Spain to Greece bears rose-pink to purplish-red flowers where lower segments are blotched or streaked white or dark red. *G. communis* subsp. *byzantinus*, a native of throughout Mediterranean region is half-hardy and up to 75 cm tall, flowers rose-purple to magenta and 5 cm long, and tepals overlapping. *G. b. albus* is white.

G. crassifolius (*G. conrathii, G. dieterlenii, G. junodii, G. paludosus, G. rachidiflorus, G. tritoniaeformis*)**.** It is a native to eastern Cape throughout Natal, Lesotho and Orange Free State to the Transvaal and Krugersdorp in South Africa apart from Rhodesia and Mozambique, where it is found flowering in the autumn, *i.e.* February to April. Corm globose with 2-5 cm in diameter, cormlets sometimes sessile and this produces solitary up to 1 m high plants. Leaves are 5-11, long sheathing, strongly veined and fan-form, up to 70 cm long and 2 cm wide. Spikes just overtop the leaves, rarely with a short basal side spike, flowers up to 40, secund during anthesis, perianth narrowly campanulate, curved and 2.5-4.0 cm long in various shades of pink, red, orange, mauve, purple or white, two side-falls bearing a dark blotch below the apex, tube narrow, curved and 8-15 mm long, upper lobes little hooded and larger, capsules subglobose, up to 8 mm long and reddish-brown and seeds dark reddish-brown.

***G. cruentus*.** Endemic to Drakensberg summit around the Mont-aux-Sources and nearby peaks at an altitudes of 2,300-3,300 m, flowering in December-January. Corms are large and globose with simple, stout, erect or little curved stems 30-90 cm long, bearing 4-5 distichous leaves, which are 20-45 cm long, highly veined, glabrous or minutely pubescent and with ensiform or linear-ensiform 1.1-2.5 cm wide blades. Perianth tube of the flower 3.5-4.0 cm long and slightly curved, flowers funnel-shaped, scarlet but tube pale and the throat yellowish-white mottled with red, three lower perianth segments are smaller than upper.

***G. debilis*.** A native of Cape (South Africa), the corms of the species are ovoid and 1.0-2.5 cm in diameter, and flowering occurs from September to November. Simple stems are slender, flexuose and 30-65 cm long with short spikes that are 1-4-flowered with flexuose rachis. The leaves are usually three, stiff and 20-25 cm long. Small flowers are trumpet-shaped, white or pale-pink and patterned with deep red dots and blotches. Perianth tube is erect, very narrowly funnel-shaped or subcylindrical and 1-2 cm long, perianth segments ovate and subequal, three lower smaller than three upper, capsule is 1.7 cm long, oblong-ellipsoid and seeds ellipsoid and 8 mm long with a broad wing. *G.d.* var. *debilis* (*Geissorhiza albens* E. Mey) grows hardly more than 30 cm high, bearing three stiff leaves more than 20 cm long. Perianth tube is short, cylindrical and statight. Flowers are pale-pink or off-white and the lower petals are marked with rose stripes and blotches. *G.d. cochleatus* Sweet (*G. debilis* sensu Bak. pro parte, *G. lambda* Klatt, *Waitzia hastulifera* Herb.) bears white or pale-pink flowers with red throat and reddish markings on lower segments. *G.d.* var. *variegatus* Lewis produces only three leaves. Perianth tube is dark- or purple-red. Flowers are white or pale-pink patterned with dark red or purple spots and markings on lower petals which are almost equal and 3 cm long.

***G. densiflorus*.** This species is endemic to eastern Transvaal from around White River and the Pretoriuskop area in the Kruger National Park to Letaba at altitudes from 500-1,300 m, flowering from February to April. Corms are subglobose and about 3 cm in diameter, with sessile cormlets. Plants grow solitary or clump-forming, 1.3 m tall, secund, sometimes with a basal side spike and bearing 15-30 flowers per spike. Leaves are soft glaucous green, 9-12, broadly linear to ensiform, basal 6-8 forming a spreading fan, 25-50 cm long, 1.0-3.5 cm wide, 3-4 upper leaves are reduced, bract-like and ensheath the scape. Flowers are 3-4 cm long, obliquely funnel-shaped, whitish or grayish, spotted red and patterned with raised maroon, mauve or pink dots. Central segment the largest and somewhat hooded, capsule globose and 1 cm in diameter and seeds are oblong, winged and about 6 mm long.

G. dubius (probably synonymous with *G. italicus* Mill. or *G. communis* L.)**.** A native of CS Portugal and central Spain, flowering from April to May. It is possibly intermediate between subsp. *communis* and subsp. *byzantinus*. Its spike is about 35 cm long, bearing 5-8 pinkish- to purplish-red flowers.

***G. ecklonii*.** It is widespread in parts of eastern Cape Province, Natal, Transvaal, Orange Free State, Lesotho and eastern Botswana at varying altitudes to over 2,000 m, flowering from January to March. Corm is subglobose and 1.5-5.0 cm in diameter with numerous ovoid and sessile cormlets 2 mm or more long. Leaves are 6-10, variable in shape, the 4-6 ensiform-acuminate basal ones forming a spreading fan, lamina 16-24 cm long and up to 3 cm wide with prominent red or yellow margins, 1-2 upper cauline leaves are reduced, bract-like and glaucous green. Spike secund with 6-14 closely set funnel-shaped flowers up to 6 cm long. Perianth tube is curved and 3 cm long, segments oblong, central slightly hooded, laterals little larger and spreading and the falls smaller and curved. Capsule is oblong-ellipsoid, 1.5-3.0 cm long and seeds winged. *G.e.* subsp. *ecklonii* from Cape to eastern Transvaal bears ensiform and small leaves, 35-50 cm tall plants, and perianth speckled with red, maroon or pink. *G.e.* subsp. *vinoso-maculatus* is confined to central Transvaal and eastern Botswana. It bears linear and 70 cm long leaves, 1 m tall plants and perianth segments are plain or speckled with maroon. *G.e.* subsp. *rehmannii* from Botswana bears glaucous green, linear to narrow lanceolate, up to 70 cm long and 1.5 cm wide leaves. The perianth segments are plain, pale-mauve, greyish, pinkish or white with green or yellow blotch on the lower lateral lobes.

***G. elegans*.** It is found in Malawi at an altitude of 2,300 m, flowering in March and April. Plant is erect, terete, glabrous and about 30 cm tall, bearing erect, highly nerved, narrowly lanceolate, distichous and acuminate leaves. Spike is 6-flowered, flowers funnel-shaped and white, perianth tube curved, central petal oval and hooded, upper laterals oblong and narrower, lower laterals smaller and unguiculate and the lowest obtuse.

G. elliotii (*G. dehnianus* Merxm., *G. rigidifolius* Bak.)**.** It is found in parts of Transvaal, Natal, Orange Free State, Botswana and Zimbabwe at altitudes of 1,000-1,500 m, and flowers from December to late April. Corm is depresso-globose and about 2.5 cm in diameter, the plant is glabrous or somewhat hairy and up to 80 cm high, bearing usually 7 base-clasping, lanceolate-acute to linear-acuminate, in ascending rosette, up to 70 cm long and 1.5 cm wide leaves. Spike is distichous, as long as the leaves, rarely with a side spike, up to 25 closely set whitish-blue or very pale-mauve flowers which have median line of densely patterned dots of maroon, purple or pink while throat is yellow. Perianth tube is 2 cm long, segments 3.5 cm long, central hooded, and lower segments narrower and outward curving.

G. engysiphon. It is native to the foothills in the Riversdale area of South Africa, flowering in February and March. Corms are subglobose and 1.5 cm in diameter. Stem is slender, erect and up to 60 cm tall bearing 3-6 closely set fragrant and small white flowers with reddish median line on the segments except on the upper central. Floral tube is up to 5 cm long, segments are variable, central 1.6 cm long and 7 mm wide, laterals longer and wider while falls shorter and narrower.

G. floribundus. It is from the winter-rainfall area of the Cape, found up to 1,200 m elevation and flowers from September to November. The corm is ovoid and 1.5-4.5 cm in diameter, producing several cormlets 1-2 cm long on short to long stolons. Stem is sturdy, simple or branched and 20-50 cm long. Leaves are normally 5-7, the basal 2-4 are 12-40 cm long with short sheaths and sometimes mottled with red or purple, the blades are linear-lanceolate to ensiform, 0.4-1.7 cm wide, lower cauline-like basal and 1-2 upper much shorter. Spike secund or subdistichous, arching and 2-14-flowered. Funnel-shaped flowers are 4-10 cm long, occasionally scented, with curved to erect tube and variable lobes which close at night and the colour is cream to pink, salmon-pink, or mauve and variously flushed. Capsule is oblong-ellipsoid and 2.5-4.0 cm long and the seeds are broadly winged. Its five subspecies are recognized, *viz. floribundus, miniatus, milleri, fasciatus* and *rudis*.

G. × gandavensis (*G. psittacinus* × *G. cardinalis*). It is a late flowering hybrid bearing long flower spike with densely packed flowers of reddish-yellow or red having striped segments.

G. gracilis. Native to Cape Province (South Africa), it flowers from June to September. Corms are subglobose to ovoid and 1-2 cm in diameter. Stem is simple, slender, flexuose near the top and up to 75 cm high with spikes secund and 1-8-flowered. Leaves usually 4, somewhat cylindrical and rush-like, the lowest long-sheathing, linear and with short blade, and upper leaves much shorter with short blades. Spike lax, flowers funnel-shaped, 4-5 cm long, fragrant and pale-blue, mauve-blue or pale-pink with a band of cream or pale-yellow streaked with broken purple lines and dots on lower laterals. Sometimes streaking is with yellow and brown or white.

G. grandis (*G. liliaceus* Van Houtte, *G. versicolor* Andr.). A native to South Africa bearing 3-4 reddish-brown ruffle-edged flowers of carnation-like fragrance being emitted by the evening, it changes its reddish-brown colouration in the evening to pale-blue and just before morning these revert back, each changing taking about one hour and this activity continues till the flowers are fresh for up to 5-6 days, and for which there is no any scientific explanation as yet.

G. illyricus [*G. alpigenus* Koch, *G. byzantinus* Coss ex Ball., *G. communis* Cav., *G. imbricatus* Reut.ex Willk. & Lange, *G. glaucus* Heldr. Ex Halacsy, *G. germanicus* Jord., *G. reuteri* Boiss, *G. serotinus* Welw. Ex Boiss. & Reut., *G. sintenisii* Bak., *G. tenuiflorus* Koch (*G. italicus*), *G. vexillare* Martelli]. A native to Mediterranean regions and west Europe, Turkey, Lebanon, Israel and southern part of England where this flowers from April to June. This grows some 45 cm tall with slender stems bearing a few axillary branches and some 3-10 loosely-spaced, small, pink, purplish-pink, brilliant red, reddish-purple or purple flowers some 3-4 cm long with unequal segments. Central petal is oblong-elliptic and longer than others.

G. imbricatus. A native to many moist parts of East and Central.Europe (Greece, Romania, Bulgaria, Czechoslovakia, Yugoslavia, *etc.*), USSR, Turkey and Iran, is half-hardy and grows to 80 cm usually with three leaves. Spikes are dense and 4-12-flowered, flowers funnel-shaped and purple-red or violet-purple. Segments unequal, upper laterals 2 cm long and lanceolate, dorsal slightly longer and lower segments 3 cm long. Flowering occurs from late March to June.

G. inflatus. A native to Cape region of South Africa up to 3,000 m elevations, flowering from August to December. Its corms are ovoid or globose and 1-3 cm in diameter. Stem is simple, slender, erect or curved and up to 60 cm high, bearing 1-2 basal sheaths and 1-6-flowered closely set spikes. Leaves are 3, lowest sheathing below, as long as the spike or shorter, subterete and 0.5-1.0 mm in diameter, glabrous to hispid near the base, second leaf longer or shorter than lowest where sheath is longer but blade is shorter, and uppermost leaf is 1-10 cm long. Spikes 2-4, secund with closely set flowers. Flowers are narrowly funnel-shaped, 3.5-5.5 cm long, horizontal, pale to deep pink, pinkish-mauve to mauvish-blue, greyish or whitish, often with creamy-yellow markings, edged with reddish-purple on central fall-petal and variable markings on fall-laterals. Segments are unequal, central petal hooded, 2-3 cm long and about 2 cm wide, upper laterals up to 2.5 cm long and 1.8 cm wide and the lowest segments more or less equal and about 3 cm long. Its subspecies and botanical varieties are described here with. *G.i.* subsp. *inflatus* (*G. inflatus* Thunb., *G. hastatus* Thunb., *G. tristis* vars. Sensu Thunb. var. *hastatus, inflatus, violaceus, G. bolusii* Bak., *G. taubertianus* Schltr.) from Montague, Ceres, Worcester and Tulbagh (Cape, South Africa) produces short-tubed, curved, 2 cm long, pink flowers with somewhat closed segments. *G.i.* subsp. *inflatus* var. *intermedius* Lewis is from Montague and Laingsburg to Uniondale and Willowmore, flowering from September to December and produces greyish-mauve to blue flowers 2-4 cm long with short and little curved perianth tube. *G.i.* subsp. *inflatus* var. *louiseae* (L.Bol.) Oberm. (syn. *G. louiseae* L.Bol.) is recorded from Ceres, Laingsburg and Clanwilliam areas of Cape, flowering from September to November. Upper bract-like acute leaf which completely sheaths the stem is much smaller than other varieties and flowers are in varying colours.

G. italicus (*G. borneti* Ardoino, *G. byzantinus* Bieb., *G. caucasicus* Herb., *G. communis* L. (in part), *G. commutatus* Bouch., *G. dalmaticus* Tausch. Ex Reichb., *G. dubius* Guss, *G. guepini* Koch, *G. humilis* Stapf., *G. illyricus* Koch var. *Anatolicus* Boiss, *G. inarimensis* Guss, *G. infestus* Bianca, *G. italicus* Gaud., *G. leicanthus* Bouche, *G. ludovicae* Jan. ex Bertol, *G. mortonianus* Steud., *G. narbonensis* Bub., *G. pallidus* Bouche, *G. proteiflorus* Romano, *G. segetum* Ker, *G. segetalis* St. Lag., *G. spathaceus* Parl., *G. tenuiflorus* Koch). A native of the Mediterranean and southern Europe and northwards into west-central France, flowering from April to June. Its stems grow up to 80 cm tall, bearing 6-16-flowered spike with pale-pink to purplish-red flowers, each flower 4-5 cm long. Segments are unequal with upper central largest and broadest, the anthers are longer than filaments, and the seeds are without wings.

G. jonquilliodorus. A native of Cape Province of South Africa, this flowers during December to February. The corms are globose and 1.5-2.0 cm in diameter. Slender and up to 75 cm long stems are basally ensheathed with two prophylls 6.5-13.0 cm long. Basal terete leaves are generally two, 4-grooved, 60 cm long, 1.5-2.0 mm in diameter and appear after the flowering, though cauline leaves which are 2-3 are stem-sheathing, lowest 18-48 cm long and uppermost 2.5-8.0 cm long. Spike secund and closely set with 5-10 fragrant and cream or pale-yellow small flowers often flushed with pink to mauve. Perianth tube is curved, central petal is hooded, 2.0-2.5 cm long and 1.2-1.4 cm wide, laterals narrower and little shorter and the lowest ones are almost equal, 2 cm long, unguiculate and recurved.

G. liliaceus (*G. grandis* Thunb., *G. grandis* Thunb. var. *lucidus* Ingram, *G. tristis* L. var. *grandis* Thunb., *G. versicolor* Andr., *G. versicolor* var. *major* Ker, *Ranisia versicolor* Salisb.). It is widespread from Clanwilliam to Port Elizabeth in the Cape (South Africa) up to 1,800 m altitude, where this flowers from August to December. The corm is subglobose and 1.0-2.5 cm in diameter. Stem is simple, slender, nearly flexuose, 40-90 cm high with two prophylls at the base. Leaves 2-4, lowest mostly longer than the stem, subterete and sulcate or linear and up to 3 mm wide, second with longer sheath and shorter blade, and uppermost half-sheathing, 7-22 cm long with subulate or acicular blade. Spike is secund, laxly 1-5 night-fragrant flowers which are yellowish or dull pale-yellow, flecked densely with pale to dull brown or pinkish-brown, purplish-red or purplish streaks. Central and upper laterals sometimes have median line, the falls greenish-white and the throat sometimes reddish.

G. longicollis. This species is widespread in many eastern parts of South Africa, and there it flowers from October to December. Corms are globose to ovoid with sessile cormlets. It produces solitary plants growing from 40-100 cm high with two basal prophylls which sheath the scape. Leaves are reduced, scape-sheathing, lowest one the longest, linear to linear-subulate, up to 75 cm long, 2-5 mm wide, apex spinescent, 2 upper leaves reduced and stem sheathing with little of free blade, and the lower one often envelops the other leaves near to the top. Spike with flexuose rachis, 1-4-flowered, flowers evening-fragrant, white, cream or yellow, patterned with dark reddish-brown and purple spots and dark median lines, long-tubular (4-12 cm), little campanulate to trumpet-shaped, and sometimes the throat is greenish. Segments 2.5-4.0 cm long and have tapered tips, central petal slightly hooded and wider though others spreading.

G. maculatus. It is native to the Cape region of South Africa, found up to 1,000 m elevation and flowers during autumn (March to July). Corms are subglobose or ovoid and 1-2 cm in diameter. Stem is simple, slender, up to 80 cm high with 2-3 basal sheaths, two lower as small prophylls though upper ones up to or more than 15 cm long and turning into cauline-sheathing leaves, basal leaves absent or withering at flowering time, and cauline leaves are 3-5, ensheathing the stem up to one-third to three-fourth, blades linear or subulate, 2-10 cm long and 1-4 mm wide. Spike lax, secund, 1-4-flowered, flowers strongly sweet scented, 5-10 cm long, tubular-campanulate, dull to brownish-yellow, pink or white, with or without spots in the throat but are finely spotted and streaked with dark brown, purple or red. Perianth tube is curved, lower part slender but upper funnel-shaped or broadly cylindrical, lobes ovate to ovate-acuminate and shorter or equal to tube but with recurved tips, central largest and obovate though remaining five are subequal. Subspecies *maculatus* is similar to type species where perianth segments are streaked and spotted with dark brown, red or purple on a dull yellow colouration, tube 3.5 cm long, and lobes tapering to undulate with acuminate tips. Subspecies *meridionalis* perianth segments are pink, cream or yellow, throat spotted, tube 5 cm long and becoming broadly cylindrical above and the lobes are rounded and acute. Subspecies *eburneus* perianth segments are white or cream, without or with faint red-brown streaks in the throat, tube 4-5 cm long and narrowly cylindrical, and the lobes are ovate, acute or acuminate. Subspecies *hibernus* perianth-lobes are ovate-lanceolate *cum* attenuate, and the tips are undulate and as long as the lobes.

G. natalensis (Eckl.) Reinw. ex Hook. (*G. natalensis* Sweet, *G. adlami* Bak., *G. antholyzoides* Bak., *G. angolensis* Welw. ex Bak., *G. andogensis* Welw. ex Bak., *G. brachylymbus* Bak., *G. buettneri* Pax, *G. cooperi* Bak., *G. dracocephalus* Hook.f., *G. daleni* VanGeel., *G. fuscoviridis* Bak., *G. gallaensis* Vaup., *G. garuanus* Vaup., *G. katubensis* deWild., *G. kilimandscharicus* Pax, *G. leptophyllus* L. Bol., *G. megaliesmontanum* Bol.f., *G. nubicola* Bak., *G. nebulicola* Ingram, *G. occidentalis* A. Chev., *G. psittacinus* Hook., *G. psittacinus* var. *cooperi* Bak., *G. pageae* L., *G. platyphyllus* Bak., *G. primulinus* Bak., *G. quartinianus* A. Rich., *G. retrocurvus* Lewis, *G. saltetorum* Bak., *G. schlechteri* Bak., *G. schimperianus* Steud. ex Bak., *G. tayloriana* Rendle, *G. tysonii* Bak., *G. velutinus* Nob., *G. velutinus* deWild, *G. vogtsii* L. Bol., *G. welwitschii* Bak., *G. xanthus* Lewis, *Antholyza laxiflora* Bak., *Bobartia natalensis* Klatt, *Geissorhiza schlechteri* Bak., *Hesperantha schlechteri* (Bak.) R.C. Foster, *Homoglossum schlechteri* Bak., *Keitia natalensis* Regel, *Tritonia schlechteri* Bak., and *Watsonia natalensis* Eckl.). It is a more variable species from South Africa (Cape Province, Natal, Orange Free State, Transvaal), Swaziland, Lesotho, Mozambique, Botswana, Namibia, Zimbabwe, Ethiopia, Zaire, Nigeria, western Arabia, *etc.* This is a parent of many beautiful hybrids and varieties. Its corms are globose to depresso-globose, glabrous, 2-5 cm wide, cormels numerous and large formed on stolons or runners up to sometimes 30 cm long. Leaves 2-4 basal tubular prophylls, and 5-12 distichous, ascending, linear or ensiform, up to 60 cm long, 1-3 cm wide and arranged in fan-shape, stems robust, erect and up to 90 cm tall, spikes lax and about 30 cm long, sometimes with a side spike, having few to many up to 25 flowers with 3-6 remaining open at a time, flowers with curved tube, zygomorphic, secund, 10-12 cm long, colours blood-red to pale-yellow, upper segment large and hooded, lower smaller and deflexed, upper laterals about $^{1}/_{3}$rd the size of central, recurved, variously coloured, streaked and mottled with green, brown, orange, orange-yellow, yellow and red, mostly in combination, but lower lobes are generally lighter in the middle. Capsules cylindrical to terbinate, obtusely triangular, about 3 cm long and 1 cm broad, and golden-brown seeds are ovate to obovate or rounded, 1 cm in diameter and winged. ***G. n.* subsp. *primulinus*** is native to areas around Victoria Falls at Zambesi River in East Africa, growing some 90 cm in height and bears pale-yellow widely spaced flowers. ***G. n.* subsp. *psittacinus*** is native to southern Africa, particularly Rhodesia, grows to 90-120 cm, flower spike erect and stout with yellow flowers marked crimson and scarlet. ***G.n.*** cv. Hookeri (*G. psittacinus* var. Hookeri Stanford) has originated from the stoloniferous species from Mozambique or any other habitat and is very suitable for growing in warmer parts.

G. ochroleucus. This species is widespread in the eastern Cape, from Knysna to Pondoland and Lesotho from near the sea to the slopes of the Drakensberg at altitudes up to 2,000 m, flowering from February to April. Corms are globose and 1.5-4.0 cm in diameter often bearing sessile cormlets. Plants grow from 45 to 100 cm high. Leaves are 6-12, distichous, often evergreen, often free as do not form stem, usually ensiform, about 30 cm long, 1.5 cm wide and with 5-8 raised ribs. Spike up to 1 m tall, often with basal branching, bearing many distichous or secund flowers which vary in size and colour, funnel-shaped, and whitish, cream, yellowish, pink or mauve with throat and central veins usually deeper. Perianth tube curved, throat funnel-shaped with spreading lobes in obtriangular shape, lobes oblong to ovate, acute to acuminate, recurved and as long as the tube, central lobe widest and nearly hooded, upper laterals spreading, 2 lower laterals over the central longer and laxly recurved segment, capsule is oblong-ellipsoid and 1.2 cm long, and the circular 5 mm wide seeds have wide wings. Variety *ochroleucus* (*G. ochroleucus* Bak., *G. carneus* sensu Klatt, *G. kirkii* Bak., *G. masoniorum* C. H. Wright, *G. reductus* Bak., *G. stanfordiae* L. Bolus, *G. triangulus* Lewis, *G. zanguebaricus* Bak.) is native to the same regions as the type species bearing white, cream, yellowish or pink and about 5 cm long flowers. Variety *macowanii* (Bak.) Oberm. stat. nov. (*G. macowanii* Bak., *G. davisoniae* Bol.f., *G. massonii* Klatt ex Dur. & Schinz) resembles the flowers of *G. oppositiflorus* ssp. *salmoneus* but leaves are evergreen, glabrous, ensiform, short and acute, and being evergreen new leaves are regularly coming up with no dormancy so no reserve food, hence transplanting is difficult. Flowers up to 9 cm, larger than the type species, segments long and tapering and are pale to deep pink or mauve with darker throat. This is recorded from Somerset East to Barkly East on mountain slopes flowering from February to April.

G. odoratus. It is a species from western Cape Province at altitudes from 300-1,000 m, flowering from April to June. Corms are ovoid or subglobose and 2.5-4.0 cm in diameter. Stems are simple, slender, erect or subflexuose, 30-80 cm high and with two prophylls enclosing the stems at the base. By flowering the basal leaves are not present but after it is solitary, ensiform, up to 65 cm long, 1.2-2.0 cm wide, and spirally twisted with 2-3 basal sheaths as in case of stem, cauline leaves up to 4, lowest 13-35 cm long with 3/4th sheathing and the blades lanceolate, arching and glabrous, second with shorter and narrower blade while the uppermost sometimes sheathing to its half and sometimes not and 4.5-13.0 cm long. Spike secund, laxly to closely 3-13-flowered, flowers small, highly fragrant and yellowish ~ speckled and lined with purple, reddish-purple, red or maroon and rarely with yellowish margins on some segments. Perianth tube about 2.5 cm long and the segments almost equal to 3 cm long, central is little hooded and sometimes the upper laterals are slightly longer.

G. oppositiflorus. It is recorded from the eastern Cape, from near the East London to southern Natal and flowering in November and December along the coast but at higher altitudes usually during February to March. Corms are depresso-globose, and the plants up to 1.5 m tall often forming small clumps. Leaves usually 8, sheathing below, blades linear to linear-lanceolate, about 1 m long, 2-3 cm wide, clustered and forming a fan-shape, and covered with minute hairs which on the ribs face sideways. Spikes up to 1.5 m tall, rarely branched, rachis terete and glabrous, bearing 20-35, distichous or secund, imbricate and funnel-shaped night-closing flowers where some 15 remain open at a time. Perianth up to 10 cm long, white or in shades of pink, deep salmon-pink with dark median lines on segments and dark blotches in the throat, the lobes are spreading, ovate-lanceolate, acute to acuminate, mostly wavy and the central lobe slightly hooded. Its two subspecies are recognized. Subsp. *oppositiflorus* [*G. oppositiflorus* Herb., *G. mortonius* (Herb. ex Hook.) Bak., *G. mortanianus* Bak., *G. flabellifer* Tausch, *G. alicea* Hort.] is recorded mostly from near the coast in eastern Cape Province, bearing up to 30 flowers per spike in pink or white colour. Subsp. *salmoneus* (Bak.) Oberm. (*G. salmoneus* Bak., *G. blackwellii* L. Bol., *G. blandus* Ait. var. *erubescens* Bullock) is recorded from more inland areas up to 2,500 m elevation in eastern Cape Province and Natal. In this case, the plants are almost pubescent and there are only up to 20 salmon-pink flowers per spike, facing only to one side.

G. orchidiflorus (*G. alatus* sensu Jacq., *G. dregei* Klatt, *G. viperatus* Ker., *G. virescens* Thunb., *Hebea alata* Eckl., *Ophiolyza orchidiflora* Salisb., *O. viperata* Salisb., *Sisyrinchium viperatum* Pluk.). This species is frequently recorded from Cape and Orange Free State, flowering from August to October. Corms are often deep-seated, ovoid to globose, 1.0-2.5 cm in diameter and with sessile or stoloniferous cormlets. Leaves are short, erect to falcate and grass-like, distichous, clustered, linear-lanceolate to linear to filiform, up to 40 cm long, 5 mm wide and the midribs and veins are prominent. The plant at the base bears 1-2 pale, narrow, tubular, and short to long prophylls, and the spike is often branched, geniculate below, 5-15 secund to subsecund and evenly spaced fragrant flowers some 3-5 cm long which are bilabiate in various dull shades of green, grey or beige with a purplish sheen and purple median stripe, and the lower laterals are yellow to lime-green encircled with purple or with a purple blotch. Segments are clawed and unequal, central linear and arching with rounded apex, upper laterals acuminate, lower laterals spathulate, recurved and shorter than lowest lobe, and the lowest lobe is acuminate.

G. papilio (*G. brachyscyphus* Bak., *G. purpureo-auratus* Hook.f., *G. schlechteri* Bak., *G. spathulatus* Bak.,). A native of moist areas of South Africa at altitudes of 2,000 m, it flowers there from December to March. Corms are subglobose and 1-3 cm in diameter, bearing a few short stolons which terminate into cormels. It is hardy, stoloniferous, stout and slender up to 90 cm tall with 5-10-flowered spike. Flowers quite small (4-6 cm long), campanulate and yellow sometimes suffused mauve to purple. Upper segments broadly obovate and lower similar to upper but narrower and little spathulate.

G. quadrangulus (*G. biflorus* Klatt, *G. linearis* (L.f.) N.E. Br., *Geissorhiza quadrangula* (Delaroche) Ker, *Ixia linearis* L.f., *I. quadrangular* Delaroche). Endemic in the Cape Flats (South Africa) in the Isetes Vlei Nature Reserve, in the sandy ground around seasonal pools, flowering from August to October. Corms are ovoid or subglobose and with 1.0-1.5 cm diameter. Stem is quite slender, flexuose or erect, 15-30 cm high and with 2 ensheathing prophylls. Erect leaves are 3, lowest with long sheath, linear, as long as the spike and with 1-2 prominent veins, and 2 upper similar but much smaller. Spike lax with closely set 1-7 fragrant flowers which are actinomorphic with a short yellow tube, and the segments pale-blue, pale-lilac, pinkish-mauve or rarely white with darker or reddish-mauve spots basally. Perianth segments subequal, oblong, obtuse or subacute, 1.6-2.0 cm long and 0.7-1.0 cm wide. Capsule ellipsoid, pale-creamy-brown and 1.3-1.6 cm long, and yellowish-brown seeds broadly winged.

G.* × *ramosus 'Robinetta' (*G. cardinalis* × *G. oppositiflorus* seedlings). It was introduced in 1839 which is valued for its brilliant red flowers. Its leaves are sword-shaped and up to 60 cm long. Spike is stout, up to 1 m tall and bears many flowers during summer.

G. recurvus L. [*G. modestus* Ingram, *G. ramosus* Burm.f., *G. recurvus* Thunb. *Watsonia recurva* (L.) Pers.]. It is a native of SW Cape (South Africa) from Tulbagh and Ceres to Stellenbosch and Caledon, flowering from June to early October. This species is very closely allied to *G. maculatus* but differs in floral colouring and shapes, being narrower, more tapering, with recurved lobes and nondescript dull colouring so it could not become a horticultural favourite despite its delightful scent. The corms are ovoid to subglobose and 1.2-1.5 cm in diameter. Stem is simple, slender, flexuose, 30-50 cm high and with two ensheathing prophylls. Leaves are four, 2 lower up to or more than $^{2}/_{3}{}^{rd}$ sheathing with blades being 5-12 cm long and 1 mm in diameter though 2 upper similar but smaller and only ¼-$^{1}/_{2}$ sheathing. Spike is lax, secund, 1-4-flowered, flowers tubular to spreading lobes, tapering and recurved at the tips, pale-greyish-green or greyish-mauve to cream or pale-yellow, central petal 2.5-3.0 cm long, 1.3-1.6 cm wide and obovate with nearly concave base, and the upper laterals and 3 lower subequal, nearly oblong or obovate-oblong.

G. saundersii (*G. sandersii* Hort., *G. saundersonii* Hort., *G. cruentus* sensu Pole Evans, *G. spectabilis* Bak.). It is native of SW Cape Province and Lesotho, flowering from January to March. Corm is subglobose with 2.5-4.0 cm in diameter. Stem is simple, stout, straight, 40-90 cm high and with two prophylls. Leaves are 7-8, suberect, distichous, the 4-5 basal lanceolate or ensiform, 25-60 cm long, 0.6-2.6 cm wide and with several veins, upper cauline much shorter with short and acuminate blades. Spike is secund, laxly 3-12-flowered, flowers circular cup-formed, large, segments nearly deflexed, scarlet, red or orange-red, and lower segments at their below half are white or cream. Perianth tube is about 3 cm long. Central segment is slightly concave, upper laterals 3.5-7.0 cm long and 2.0-4.5 cm wide, and the segments in the falls are similar but shorter and narrower.

G. sempervirens (*G. splendens*). It is recorded from the Cape Province in the George, Knysna, Uniondale and Humansdorp districts at an altitudes ranging from 400 to 1,700 m, flowering from February to April. Corm is subglobose, small from 1.0 to 1.5 cm in diameter and nearly rhizomatic with several short slender rhizomes usually present. Stem is simple, slender, 40-50 cm long, ensheathed nearly up to the spike and with prophylls present only in the young shoots and not on the flowering stems. Leaves evergreen, 9-16, distichous, lanceolate or ensiform, basal leaf up to 30 cm long, 1-2 cm wide, the 2-3 outer shorter, narrower and almost brown at flowering time, lower cauline-like basal becoming shorter and 2 upper ones 11-15 cm long and lanceolate. Spike is secund, laxly 4-8-flowered, perianth tube curved, little twisted and 2.5-5.0 cm long, flowers up to 12 cm long, facing up, bright scarlet with white dagger-shaped marks in the lower half of the falls, segments acute and narrowed at the base, central segment not-hooded and obovate-elliptic, 3 upper 5.0-6.5 cm long, 2.5-3.3 cm wide, laterals ovate-elliptic and little wider than central, and the 3 lower elliptic, 5.0-5.5 cm long, connate for 3-4 mm and 1.8-2.5 cm wide.

G. sericeo-villosus. It is native to NE Cape Province to eastern parts of Orange Free State, Natal, Swaziland and Transvaal at altitudes of 100-2,100 m, flowering from December to June. Corms are globose with up to 4 cm diameter. Plants are 60 cm high with short spike or about as long as the leaves, closely set flowers up to 40, distichously and imbricately arranged, rarely with a basal side spike, and the rachis is flexuose and ridged. Leaves are 7 and arranged in an elongated fan, vaginate below, linear, up to 2 m long and 2 cm wide and with numerous raised yellow ribs. Flowers in the shade of cream, yellow, pink, lilac, lavender or maroon, sometimes lower segments are blotched with darker yellow or green shade. Perianth tube is curved and about 2 cm long, central petal hooded and lanceolate, and lower ones are smaller and narrower, and lowest one is curved downwards and longer than laterals. The segment size is usually 4.5 cm or up to 6 cm long. Its two forms are recognized. Forma *sericeo-villosus* (*G. sriceo-villosus* Hook.f., *C. sericeo-villosus* var. *ludwigii* Pappe ex Bak., *G. ludwigii* Pappe ex Bak., *G. ludwigii* Pappe ex Bak. var. *rubicundus* O. Kze., *Antholyza hirsuta* Klatt) is from NE Cape to Natal, the eastern Orange Free state, Swaziland and the southern Transvaal at altitudes from 500 to 1,500 m, flowering from November to May. It is a typical form where rachis, bracts and leaves are covered to a larger or lesser extent with minute hairs and bristles. Forma *calvatus* (Bak.) Oberm. comb. nov. (*G. ludwigii* Pappe ex Bak. var. *calvatus* Bak., *G. sericeo-villosus* Hook. f. var. *glabrescens* L. Bol.) is from Transvaal. It is without pubescence, mostly glabrous or with a few minute hairs and the bracteoles are usually tubular towards the base.

G. tristis. It is common in the winter rainfall region of the Cape (South Africa) from Clanwilliam to Humansdorp up to 1,800 m altitude, flowering from August to October and sometimes even in November to January. It is a tender and sweetly-fragrant species growing up to 45-150 cm (usually 50-70 cm) high, bearing subglobose corms 1-3 cm in diameter and numerous sessile cormlets with two ensheathing prophylls. Leaves are 2-4, the lowest almost as long as the stem with a short striate sheath and linear blade 1.5-5.0 mm wide spirally twisted towards the apex, second with longer sheath and shorter blade and the third much shorter, half or more sheathing. Spike is secund, lax and 1-20-flowered. Flowers are, erect, tubular-campanulate, up to 7 cm wide, highly night fragrant and colour white or cream to buff or sulphur-yellow, tinged green, and lower half of the lower segments with greenish-yellow zone. The flowers are finally flushed with, pink, dull red, brown or purplish-brown. Perianth tube is 4-6 cm long and curved, segments acute to acuminate and 2.0-3.3 cm long, central ovate to ovate-elliptic and 1.5-2.3 cm wide, other 5 subequal or sometimes lowest a little narrower than laterals. The capsule is spindle-shaped, up to 4 cm long, 1.2 cm wide and seeds 5 mm in diameter with a wide wing. *G.t.* var. *tristis* [*G. tristis* L., *G. tristis* var. Jacq. (1975), *Tilesia spiralis* Thunb. ex Steud., *Ranisia tristis* Salisb., *G. versicolor* var. *longifolius* Ker, *G. spiralis* Pers, *G. flavidus* Ingram, *G. fulvescens* Ingram] flowers from August to November with 1-8 flowers per spike, flowers being striped and spotted green or purplish-brown or mauve on the background colour of cream or yellow. *G.t.* var. *aestivalis* (Ingram) Lewis (*G. aestivalis* Ingram) flowers from December to January, growing up to 150 cm high, spikes with up to 20 cream, pale-yellow or deeper yellow flowers marked purple or brownish, the lower segments and throat greenish-yellow and the upper segments with reddish-brownish-purplish median line. *G.t.* var. *concolor* (Salisb.) Bak. [*G. concolor* Salisb., *G. tristis* Jacq. (1792), *G. tristis* var. Ker (1808), *Ranisia concolor* Salisb.] bears 1-8 sulphur-yellow or cream fragrant flowers, larger and compact than original species, often greenish in the throat overlaid with greenish median lines or purplish streaks; *G.* × 'Christabel' [*G. tristis concolor* × *G. virescens* (*G. bicolor*)] is similar to *G. tristis* with flowers which are of primrose-yellow colour and are slightly hooded where upper tepals are highly veined and shaded purple; *etc.*

G. undulatus (*G. affinis* Pers, *G. carneus* sensu Ker, *G. cuspidatus* Jacq., *G.c.* var. *longiflorus* DC, *ensiflorus* Bak. *breviflorus* Bak.). It is native to the moist conditions of the mountainous areas of Cape Peninsula to Clanwillam and Caledon. Flowering from October to January. This bears ovoid corm with up to 3 cm in diameter which produces numerous reddish-brown large cormlets. Stem is 30-95 cm tall with 2 prophylls. Leaves 5, variable in size and shape, short and ensiform to long and linear, sheathing only at the base, coriaceous, closely ribbed, 25-75 cm long and 5-20 mm wide. Spike is erect, rarely with a basal side spike, and distichous with laxly arranged 4-9 flowers. Flowers are long-tubed and up to 10 cm long with campanulate limb, the lobes are long-tapered, greenish-white, white, cream or pink with darker pink to red markings on lower lobes, lobes shorter than the tube, ovate to lanceolate with long tapering tips, central segment hooded, lower laterals recurved with faint markings, and midrib red-purple.

G. uysiae (*G. uysiae* L. Bolus, *G. ceresianus* sensu sealy). It is a rare species from the western Cape found in the Calvinia, Clanwilliam and Ceres districts, and in Transvaal, Mali, Nigeria, Ghana, Sierra Leone, *etc.*, flowering from August to October. Corm is depressed-globose and 1-2 cm in diameter. Stems are simple, slender, flexuose, up to 30 cm long with a single prophyll. Leaves are 3, lowest falcate and longer, 8.5-30 cm long, 0.4-1.0 cm wide, ¼ sheathing, second leaf similar but with longer sheath, shorter and erect, and the uppermost smaller with about half sheathing. Spike is subsecund, short, 1-6-flowered, flowers bilabiate, sweetly fragrant, central segment spathulate, cream or greenish with creamy margins, veined and suffused mauve and 3.6 cm long, upper laterals lobed, 1.6 cm long and 8 mm wide, all the lower ones are unguiculate, 8 mm long, 4 mm wide, the central of the falls is a little larger.

G. vaginatus. A species from Cape Province (South Africa), frowering from February to April. The corm is globose to ovoid and 1.5-3.0 cm in diameter. The stem is simple, erect, slender, 15-110 cm high, basally enclosed with 2-3 prophylls. Up to flowering, the basal leaves are absent but afterwards one, glabrous, up to 60 cm long and 0.5-1.5 mm in diameter appear, which at the base is enclosed in 1-2 obtuse sheaths 4-8 cm long, whereas cauline leaves are 2-3 where lowest one sheaths entirely the stem of its 3/4th region while remaining 1 or 2 are 2-6 cm long and acute or apiculate. Spike is secund, with 1-9 flowers where 2-3 open at a time, flowers are small, fragrant, pale-mauve to grey-mauve or grey-blue or rarely white, throat yellow-orange internally, and the lower lateral lobes with a pale-yellow median stripe encircled by purple streaks and dots or with 3 fine purple lines in the lower half. Capsule is rotund, and 1.0-1.3 cm long and the seeds are yellow-brown, 5-6 mm long and broadly winged. Subspecies *vaginatus* Bolus f. from Riversdale, west of Albertinia and Kirstenbosch in Cape bears 2-9 flowers per spike and the flowers are with very closely set lips. Subspecies *subtilis* oberm. bears 1-4-flowered spike having flowers with well apart lips which is recorded from Bredasdorp to Caledon of Cape.

G. virescens. It is widespread in the winter-rainfall region of Cape from Ceres and Worcester to Laingsburg, and along the coast from Caledon to Port Elizabeth, flowering from August to October. It has subglobose corms with 0.8-1.5 cm diameter. Stem is slender, sometimes branched, 10-30 cm long, flexuose with one prophyll up to 10 cm long. Leaves are 3 with glabrous to pubescent sheath, the blade of the lowest 20-60 cm long, 1.5-3.0 mm wide, linear to subterete or terete and sulcate, 2 upper leaves much shorter, 5.5-10.0, and 2.5-7.0 cm long with short sheaths and linear or subulate blades. Spike is secund, 2-7-flowered, flowers scented or scentless, bilabiate, yellowish, white, cream or greyish, veined in mauve, purple, pink or brownish, perianth tube about 1.3 cm long, segments unequal, central 3-4 cm long and 1.8 cm wide, upper laterals

2.7 cm long, 2.2 cm wide, spathulate, spreading with recurved tips, and lower segments up to 2.8 cm long, 1.5 cm wide and clawed and the lowest similar to upper laterals though smaller. Variety *virescens* (*G. luridus* Hornem, *G. bicolor* Eckl. ex Bak., *G. luteus* Klatt, *G. templemanii* Klatt, *Hebea bicolor* Eckl.) is from Caledon, Ceres, Worcester and Laingsburg areas with pubescent leaf sheaths, and yellow or greenish-yellow scented flowers. Variety *lepidus* Lewis (*G. pulchellus* Eckl. ex Klatt, *Hebea pulchella* Eckl.) is from Caledon to Port Elizabeth. Its fragrant flowers are smaller than the type species, and are grey, cream, pink, or brownish-pink. The upper segments and the tips of lower segments are veined mauve, maroon or purple, while the lower segments have central yellowish green blotch. Variety *roseo-venosus* Lewis from around Worcester bears flowers without scent in September. The flowers are 3-9 per spike in white or creamy-pink, veined deeper pink and leaf sheaths are glabrous.

G. viridiflorus. This species is from the hilly county, Namaqualand and Calvinia areas of South Africa, flowering from May to July. This is allied to *G. carinatus* but has larger corms and fragrant pale-green flowers, lower lateral lobes with an irregular yellow-green transverse band with purple outline near the middle, and the central segment of the falls with a spathulate mark of similar colouration. Corm is ovoid or subglobose with 2.0-2.5 cm diameter. Stem is simple, slender, highly flexuose, 18-50 cm long and with 2 prophylls at the base. Leaves are 3-4, two lower with half to three-quarter ensheathing, and the blades are glaucous green, soft, grass-like, linear-lanceolate, spirally twisted, 4-27 cm long, 2-3 mm wide with prominent midrib, uppermost cauline leaf is involuted at the base though not sheathing and 4-11 cm long. Spike is secund to subsecund with very flexuose rachis having 4-7 flowers. Perianth tube is curved and about 1.5 cm long. Segments are clawed, central one is hooded with recurved apex, up to 3 cm long and 1.6 cm wide, upper laterals smaller and spreading, lower laterals 2.7 cm long and 7 mm wide, while the lowest one is up to 1 cm wide.

G. watermeyeri. It is common around Nieuwoudtville in the western Cape, flowering from July end to September. The corms are globose or depressed-globose with 1-2 cm in diameter. Slender stem up to 40 cm long are stout, flexuose, rarely branched and with single prophyll. Leaves are 4, lowest erect to subfalcate, up to 44 cm long, 0.2-1.1 cm wide, 1/4th sheathing and the blades are linear and lanceolate or ensiform, 2nd and 3rd leaves smaller and erect while the uppermost is bract-like. Spike is secund, 2-6-flowered, flowers scented, bilabiate, upper segments cream to greenish-cream suffused pinkish-purple and the laterals are veined, lower segments are pale- to orange-green with creamy apex. Perianth tube is completely hidden in the bracts. Segments are unequal, central helmet-shaped, 3.5 cm long and 1.4 cm wide, upper laterals spreading, 2.5 cm long and 2.2 cm wide, lower laterals 2.8 cm long and 5 mm wide, and the lowest as long but 1.8 cm wide.

G. zambesiacus. This species is native to Malawi near Blantyre mountains east of Lake Nyasa. Its corm is small and globose, and the stem quite slender, 0-45 cm long with spike lax and only 8-10 cm long. Leaves are 3, glabrous, narrowly linear and up to 30 cm long. Flowers are pale-purple, 2.5 cm

Table 19.1: Size and Colour Classifications of Gladiolus as Designated by the N.A.G.C. (Wilfret, 1980)

Class[a]	*Designation*	*Floret Size (cm)*
100	Miniature	<6.4
200	Small or miniature	≥6.4 to <8.9
300	Decorative	≥8.9 to <11.4
400	Standard or large	≥1.4 to 14.0
500	Giant	>14.0

Colour[b]	*Pale*	*Light*	*Medium*	*Deep*	*Others*
White	00				
Green	02	04			
Yellow	10[c]	12	14	16	
Orange	20	22	24	26	
Salmon	30	32	34	36	
Pink	40	42	44	46	
Red	50	52	54	56	58 Black-red
Rose	60	62	64	66	68 Black-rose
Lavender	70	72	74	76	78 Purple
Violet	80	82	84	86	
Smokies	92	94	96		
Tan	90	98 Brown			

[a]The first digit indicates floret size in the five classes.

[b]The last two digits in the classification indicate floret colour and shade. An odd last digit indicates a conspicuous mark or blotch.

[c]Includes cream.

long, with 1.2 cm long curved perianth tube while segments are cuneate and 2.5 cm long.

Since gladioli hybridize easily, there are over 30,000 cultivars that have been bred. The **florist's classification** of varieties is furnished here with. There is no universal synchrony in its classification as its classification is based on the country's preferences.

American system is more scientific as they classify the varieties on 3-digit system which originated there during 1940's through North American Gladiolus Council as is given above in Table 19.1. First digit (from 1 to 5) represents floret size, *i.e.* from 100 (miniature, below 6.4 cm floral diameter) to 500 (giant, above 14.0 cm) *via* 200 (small, 6.4-8.9 cm), 300 (decorative, 8.9-11.4 cm) and 400 (large, 11.4-14.0 cm). Second digit (0 to 9) denotes basic hue or colour and since white does not fit in the series, hence it is placed first, 00 means pure white whereas 01 white with markings. Since there is no pure green and blues, hence no number is assigned to these, however, the depth of greenness is defined on the basic colour of 0, *i.e.* 02 (light green), 04 (medium green), etc. In the second digit, it is 1 for yellow, 2 for orange, 3 for salmon, 4 for pink, 5 for red, 6 for rose, 7 for lavender, 8 for violet (no blue) and 9 for tan, whereas third digit 0 to 9 depicts the depth of the colour, odd numbers standing for conspicuous markings though even numbers for self colouration. For example if it is '498 Brown Orchid' Baxter, 64, it means that the variety '400 Brown Orchid' was developed by Baxter in 1964 and is of large size (11.4-14.0 cm) with deep smoky (towards brown type) colouration, *i.e.* 098 where third digit 008 shows that the variety has inconspicuous markings.

Europeans use prefatory letters, P (Primulinus), B (Butterfly) or FU (Face-ups), and suffix letters such as E (early), M (mid-season), L (late) and V (very for extremes). New Zealanders and Aisans classify them on the basis of the size of their flowers, *i.e.* (i) Grandiflorus (large or exhibition type) which grows to a height of 90-150 cm with strong and erect stalks bearing 14-20 cm closely arranged triangular and symmetrical flowers that appear late in the season; (ii) Primulinus which are free-flowering and daintier than the grandiflorus, growing up to 100 cm height bearing 5-10 cm across soft-coloured florets, mostly with hooded central petals and blooming during the mid-season; (iii) Butterflys which grow up to 125 cm in height bearing medium-sized flowers spaced closely but marked with attractive blotches and throat markings; and (iv) Miniatures growing to a height of 75 to 90 cm, flowering early and bearing 5.0-7.5 cm across flowers similar in arrangement to the Primulinus types but frequently ruffled. **New Zealanders** also follow **Australian system** such as formals, informals, intermediates and miniatures.

Some of the commercially important cultivars belonging to different colour groups are listed below:

- ✰ **Pink**: Alfred Nobel, Amanda Mahy, America, Applause, Archana, Ben Venuto, Bo Peep, Charm Glow, Dancing Doll, Dawn Pink, Dr. Fleming, Dresden Doll, Enchantress, Evangeline, Flos Florium, Friendship, Frosty Pink, Howard, Jessica, Joyce, Kimberly, Legend, Maiden's Blush, Marjayne, Miss America, Miss Salem, Mme Butterfly, Musical, My Love, Palet, Peach Blossom, Pink Diamond, Pink Perfection, Pink Cheer, Pink Formal, Pink Prospector, Pink Triumph, Polynesia, Powder Puff, Priscilla, Rose Supreme, Rosy Frills, Rosy Maid, Spic and Span, Spring Song, Suchitra, Super Ruffles, Sweepstake, Sweet Debbie, Tea Party, Toulouse Lautrec, True Love, Uhu, Vicki Lin, Windsor, Wine and Roses, Zenith, etc.
- ✰ **Orange**: Acca Laurentia, Ackerman, Albert Schweitzer, Athalia, Autumn Glow, Autumn Gold, Bonaire, Cartage, Coral Seas, Fiesta, Foxfire, Graceful, Gypsy Dancer, Inca Chief, Kathleen Ferrier, Little Mo, Melodie, Orange Beauty, Orange Chiffon, Peter Pears, Picardy, President Kennedy, Red Star, Salmon Star, Setting Sun, Spitfire, Tangerine, Vivaldi, etc.
- ✰ **Red**: Advance, American Beauty, Apex, Applause, Aristocrat, Black Prince, Carthago, Decathalon, Delicious, Dixieland, Eclipse, Enticement, Eurovision, Fatima, Femina, Gerda, Happy End, Hawaii, Hunting Song, Jo Wagenaar, Life Flame, Little Slam, Mascagni, Music Man, Nautilus, New Europe, Oscar, Red Bantam, Red Beauty, Redeem, Richard Unwin, Rotterdam, Sans Souci, Sassy Willie, Scarlet Knight, Sequoia, Shirley Cole, Summer Fairy, The Barton, Uranus, Victor Borge, Victoria, etc.
- ✰ **Yellow**: Angelina, Anglia, Ares, Aurora, Brightsides, Chinese Lantern, Daily Sketch, Donald Duck, Fatima, Flowersong, Folksong, Forgotten Dreams, Golden Harvest, Golden Peach, Goldfield, Gold Standard, Green Bird, Green Willow, Green Woodpecker, Jacksonville Gold, Jester, Junior Prom, Lemon Ruffles, Limelight, Medusa, Mickey Mouse, Monanza, Morning Sun, Nova Lux, Nugget, Pactolus, Royal Gold, Souvenir, Sweet Fairy, True Yellow, Vinks Glory, Walt Disney, Yellow Stone, etc.
- ✰ **White**: Althena, Amsterdam, Bush Ballad, Classic, Cotton Blossom, Dream Girl, Eastern Star, Gambier Pearl, Ice Follies, Icicle, Ivory and Cream, Jennifer, Leif Ericsson, L'Innocence, Lipstick, Lunar Moth, Maria Goretti, Marjorie Ann, Maybride, Melissa, Mighty Mite, Moon Frost, Morning Bride, Pegasus, Prof. Goudriaan, Sancerre, Simplicity, Snowdrop, Snow Dust, Snow Princess, Super Star, The Bride, White Angel, White City, White Enchantress, White Friendship, White Goddess, White Oak, White Princess, White Prosperity, White Wonder, etc.
- ✰ **Lavender/Purple/Violet**: Abu Hassan, All Aglow, Angel Eyes, Anniversary, Ataturk, Bishop, Blue Bird, Blue Conqueror, Blue Isle, Blue Hawaii, Blue Mist, Blue Moon, Blue Ruffles, Blue Sky, Burgundy Beau, China Blue, Dawn Mist, Delphine, Elegance, Eternal City, Fidelio, Grock, Her Majesty, High Style, Keukenhof, Lavender Masterpiece, Linda Ruth, Mabel Violet, Mayur, Memorandum, Monte Negro, Pfitzer's Sensation, Plum Tart, Priscilla, Province, Purple Giant, Purple Moth, Pusa Sarang, Pusa Shringarika, Purple Giant, Pusa Urmi, Pusa Urmil, Royal Dame,

Rose Delight, Rose Spire, Royal Dutch, Royal Ruffles, Shalimar, Tropic Seas, Wind Song, etc.

- ☆ **Tan, smoky and brown:** Autumn Charm, Autumn Sensation, Aztec Chief, Blue Smoke, Brown Beauty, Chocolate Chip, Chocolate Dip, Cocolette, Cosmic, Hastings, Little Fawn, Little Tiger, Misty Eye, Misty Morn, Mystic Glow, Papoose, Piccolo, Table Talk, Tapestry, etc.

(i) Varieties Developed at IARI, New Delhi

Agnirekha, Anjali, Chandni, Mayur, Neelam, Pusa Archana, Pusa Bindiya, Pusa Chirag, Pusa Dhanvantari, Pusa Gulaal, Pusa Gunjan, Pusa Jyotsana, Pusa Kamini, Pusa Kiran, Pusa Lohit, Pusa Manmohak, Pusa Mohini, Pusa Rangmahal, Pusa Red Valentine, Pusa Sanjeevani, Pusa Sarang, Pusa Shabnam, Pusa Shagun, Pusa Shringarika, Pusa Shubham, Pusa Shweta, Pusa Suhagin, Pusa Sukanya, Pusa Sunayna, Pusa Swapnil, Pusa Swarnima, Pusa Urmi, Pusa Urmil, Pusa Urvashi, Pusa Vidushi, Suchitra, Suryakiran, Vandana.

(ii) Hybrids Developed at IIHR, Bangalore

Aarti, Apsara, Darshan, Kum Kum, Meera, Nazrana, Poonam, Sagar, Sapna, Shakti, Shobha (mutant), Sindur, etc.

(iii) Hybrids Developed at NBRI, Lucknow

Archna, Arun, Basant Bahar, Dhiraj, Gazel, Hans, Indrani, Jwala, Kalima, Kohra, Manhar, Manisha, Manmohan, Manohar, Mohini, Mridula, Mukta, Pitamber, Priyadarshini, Sada Bahar, Smita, Sanyukta and Triloki.

(iv) Hybrids Developed at HETC, Chaubattia

Chaubattia Ankur, Chaubattia Arunima, Chaubattia Shobhit, Chaubattia Tripti.

(v) Hybrids Developed at PAU, Ludhiana

Punjab Dawn, Punjab Morning, Shan-e-Punjab, etc.

(vi) Hybrids developed at GBPUAT, Pantnagar

Subhangini.

(vii) Hybrids Developed at MPKV, Pune

Shree Ganesh.

(viii) Hybrids Developed by Bajrang Bahadur Singh Bhadri in Himachal Pradesh

Bhadri (1963) developed some 150 varieties suitable for cultivation in Himachal Pradesh. Some of the important ones are Bhadri Blazing Star, Bhadri Blue Maiden, Bhadri Blue Marvel, Bhadri Bouquet, Bhadri Bright Red, Bhadri Carmine King, Bhadri Carnival, Bhadri Carpet, Bhadri Challenge, Bhadri Dazzler, Bhadri Delight, Bhadri Double Cream, Bhadri Early Peace, Bhadri Elite, Bhadri Favourite, Bhadri Fireball, Bhadri Fire King, Bhadri Flesh Pink, Bhadri Gaiety, Bhadri Giant Flowered, Bhadri Glitters, Bhadri Gold, Bhadri Indian Chief, Bhadri Jewel, Bhadri Jupiter, Bhadri Lemon Queen, Bhadri Lemon Supreme, Bhadri Lily Pink, Bhadri Little Yellow, Bhadri Lovliness, Bhadri Magenta, Bhadri Mauve Beauty, Bhadri May Blossom, Bhadri Nobility, Bhadri Orange Gleam, Bhadri Orange King, Bhadri Oriental Charm, Bhadri Peach Glow, Bhadri Pearl, Bhadri Pink Baby, Bhadri Pink Queen, Bhadri Pink Spice, Bhadri Primulinus Beauty, Bhadri Purple King, Bhadri Purple Sensation, Bhadri Queen of Pink, Bhadri Radiance, Bhadri Red, Bhadri Red Primulinus, Bhadri Red Wing, Bhadri Rose Glory, Bhadri Royalty, Bhadri Salmon Glory, Bhadri Salmon Glow, Bhadri's Baby Glow, Bhadri's Brownie, Bhadri's Liliput, Bhadri's Love Song, Bhadri's Red Prince, Bhadri Scarlet, Bhadri's Sungold, Bhadri Sunset, Bhadri Superb, Bhadri's Yellow Crown, Bhadri Symphony, Bhadri Tricolour, Bhadri White Exquisite, Bhadri Wonder, Bhadri Yellow Charming, etc.

From Netherlands, the large flowered gladiolus varieties being auctioned are 'Amsterdam', 'Angie', 'Applause', 'Atilla', 'Ben Travato', 'Ben Venuto', 'Bono's Memory', 'Campinas', 'Chapeau Bas', 'Chiquita', 'Cobra', 'Deciso', 'Early Yellow', 'Eurovision Elite', 'Eurovision', 'Fiancee', 'Fidelio', 'Friendship White', 'Friendship', 'Goldfield', 'Golden Game', 'Her Majesty', 'Hunting Song', 'Impale', 'Invitatie', 'Jacksonville Gold', 'Jessica', 'Joyeuse Entree', 'Leen Van der Mark', 'Madonna', 'Mascagni', 'May Bride', 'Memorial Day', 'My Love', 'Nicole', 'Nova Lux', 'Oscar', 'Peter

Dhanvantari

Gulaal

Jyotsna

Urmil

Urvashi

Phule Neelrekha

Pears', 'Picture', 'Princess Margaret', 'Priscilla', 'Rose Supreme', 'Sancerre', 'Saxony', 'Sittara', 'Spic & Span', 'Stardust', 'Superstar', 'Sweet Shadow', 'T 512', 'Teach In', 'Tenderesse', 'Trader Horn', 'Veerle', 'Victor Borge', 'Violetta', 'Waarland', 'White Prosperity', 'Windsong', 'Wine & Roses', etc.

Propagation

Gladiolus is propagated through corms, cormels or from corm division. It can also be propagated sexually by seed to evolve a new variety through hybridization. Gladiolus set seeds abundantly. Plants raised from seeds require 2-3 seasons to come into bloom under ordinary conditions. Six weeks after flowering, the capsules start turning brown and are ready for harvesting. The soil should be porous and rich in organic matter. After sowing, temperature should be maintained at 7 to 13°C. When the foliage starts withering, the cormels are lifted, cleaned and stored in a cool place which are planted in the following season. Some of these, especially the late type cormels may produce flowers in the same year while the others may take one or two seasons to come into flowering.

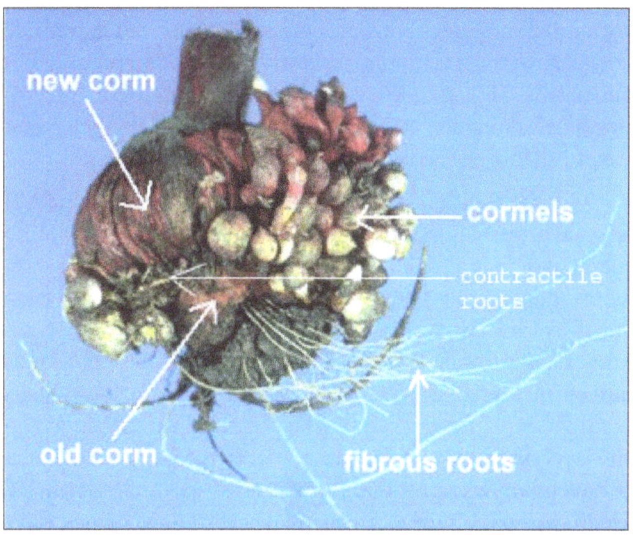

Corm and Cormlet Formation in Gladiolus.

Shredded contractile roots are attached below the new daughter corm. Remnant of old planted mother corm is attached at the basal-disc of the new corm. Many stolons coming out of the basal portion of the new corm, bear diminutive corms (cormlets = cormels) which have normally a very hard tunic though large daughter corm shredded ones.

The most commercial means for the propagation of gladiolus is by means of **corms** and **cormels**. Corm is a thickened underground perennating structure consisting of short vertical stem having many ring- like nodes which bear buds covered with dry scale leaves and a disc-like root zone at the base. Commercially corms are produced from **cormels** which grow in clusters on stolons between mother and daughter corms but are produced by the daughter corms. Yield and quality of commercial cut flower production in gladiolus is influenced by a number of factors. One such important factor is the size and weight of the corm to be planted. It is well known that bulbs below a certain size do not flower and if they do so, the quality remains poor. The critical size of the corm for flower production is 2.5 cm in diameter. According to the North American Gladiolus Council, corms and cormels can be classified into different grades depending upon their sizes (Wilfret, 1980).

Grades of Gladiolus Corms Developed by North American Gladiolus Council

Grades		*Diameter (cm)*
Large	Flowering stock	
	Jumbo	>5.1
	No.1	>3.8 to < 5.1
Medium		
	No. 2	>3.2 to <3.8
	No. 3	>2.5 to < 3.2
Small	Planting stock	
	No. 4	>1.9 to <2.5
	No. 5	>1.3 to <1.9
	No. 6	>1.0 to <1.3

The corms, on the basis of their spherical diameter, are classified into two categories, *viz.* the flowering stock (>2.5 to 5.0 cm or more in diameter) and planting stock (>1.0 to 2.5 cm in diameter). The flowering stock is used for production of cut spikes, whereas the planting stock is used for the production of flowering grade corms for the subsequent planting season. There exists a positive correlation between the weight of corm planted with plant growth and flower production. Larger the size better is the growth.

The rate of regeneration of corms may be increased by removing top eyes from the corms mechanically before sprouts get developed and by pinching the flowering spikes at the initial stage of development. However, response to this technique needs further study.

Singh (2000) compared the performance of nine grades of corms of gladiolus cv. Aarti and recorded that the largest grade-size (>6.0 to <6.5 cm diameter) gave better vegetative growth and superior quality flowers and higher weight and diameter of the daughter corms. The flowering percentage was reduced as corm size decreased. Singh (2001) conducted an experiment to determine the potential of lowest grade of planting material for flower production and multiplication in gladiolus cv. Pink Friendship. The corms of 3.2-3.8 cm diameter (average weight 14.7 g) was recorded with greatest plant height, highest number of leaves, spike and rachis length, floret size and also induced earlier flowering. Corm grade ranging from 1.9 cm to 3.8 cm diameter produced A grade flower spike. Propagation coefficient was reported to be highly increased with each reduction of corm grade size. Remotti *et al.* (2002) also reported that superior quality flowers were obtained when large corms were used as planting material. Galgate and Desai (2003) observed that large (>4.5 cm diameter) corms of gladiolus cv. Sancerre showed better vegetative characters, flower quality and corm production compared to medium (3.0 to 4.4 cm diameter) and small (< 3.0 cm diameter) corms. Sharma and Gupta (2003) also reported that plant height, number of leaves per plant, time taken for blooming, spike length and quality, number and size of corms and cormels produced were significantly improved by increasing the size of mother corms. However, different corm size treatment failed to exert any significant effect on size of florets and cormels.

In an experiment conducted on different cultivars of gladiolus, Dilta *et al.* (2004) observed that A grade corms (> 5.0 to 6.0 cm diameter) resulted in increased plant height as compared to B grade corms (> 2.5 to 3.5 cm diameter) and also took less days to flower and gave more number of florets per spike and larger floret size. Shalini *et al.* (2004) found that large sized corms (5.5 ± 0.5 cm) and wider spacing (20 x 30 cm) gave significantly superior results in respect of growth, flowering and corm production in gladiolus cv. Debonair.

Shinoda and Murata (2005) observed that planting of gladiolus corms weighing 0.4 to 1.4 g did not flower. However, 100 per cent flowering was observed in corms weighing 15.0 g and above. As the weight of the mother corm increased, the number of florets, scape length and number of daughter corms also increased significantly. Planting of large sized corms (5.1 to 6.0 cm diameter) in gladiolus cv. Pusa Jyotsna induced early flowering and recorded increased plant height, leaf breadth, length of spike and rachis, florets per spike and cormels per plant with corresponding increase in weight when compared to medium or small sized corms (Kumar and Yadav, 2006).

With the increase in popularity, the demand for planting material has also increased tremendously. However, some most preferred cultivars are shy-multipliers and produce less number of corms and cormels which limits the quantum of planting material. High percentage of storage rot is also one of the major problems which also limits the planting materials. In such a case, propagation may be done by **cutting** the corms into several pieces to maximize the number of planting materials. The segment of corm to be used as a propagule may have at least one visible eye and a portion of basal plate or root zone, but not necessarily. The corms are cut 7 to 10 days before planting. Small corms can be divided into 3 to 4 pieces while the large ones can be divided into 7 to 10 pieces (Gromov, 1972). Better results can be obtained when radial cuts are made. The corms are dehusked, cleaned and graded prior to cutting. After cutting, the propagules are sterilised by dipping in 0.2 per cent solution of Saaf (Carbendizim 12 per cent + Mancozeb 63 per cent WP) for 30 minutes and dried under shade for 6 days to suberize the wounds. Gromov (1972) reported that cutting of corms markedly increased the growth of the filial corms, the weight of the corms, and the number and weight of cormels in comparison with those from whole corms. MacKay *et al.* (1981) found increased flowering percentage by cutting the large corm of Jumbo, No.1 and No.2 grade-sizes into two pieces but reduced percentage when smaller corms (No.4 and No.5 grade) were divided. Inflorescence quality was also reduced by cutting the corms. The yield of new corms was, however, increased through division of large and medium sized corms. Lopez Oliveras *et al.* (1984) suggested that splitted and treated corms should be planted on well prepared warm and moist (not wet) soil media. Corm of cultivar Peter Pears when were divided into 4 or 8 sections and planted in 50 per cent peat and 50 per cent perlite substrate produced larger number of grade 1 corms (4.8 cm diameter) while soaking of corms for 24 h in 900 ppm GA_3 solution increased the cormel production.

Hatibarua (1989) reported that initial plant growth was delayed with increase in number of splits. Corms of all sizes produced flowering grade corms. Splitting of corms into two halves increased the number of corms over whole corms of all sizes. Substantial increase in number of cormels was also reported with the increase in number of corm splitting within the same size of corm in cv. Sylvia. Barman *et al.* (2006) conducted an experiment to study the effect of excised corm on corm and cormel production in cv. Jester in which they found a gradual decrease in spike length with increasing division of corms. Maximum plant height was recorded from whole corms which was at par with ½ excised corms. Largest diameter of corm was obtained from the whole corms, thereafter it decreased with subsequent division of corms. A similar pattern of variation was also recorded in case of number of florets per spike. Ramachandradu and Thangam (2006) cut Jumbo sized corms into three pieces and compared the performance with different grades of whole corms. They noticed that whole corms gave the maximum vegetative growth as compared to cut corms. However, cut corms of Jumbo grade (>5.1 cm) produced daughter corms with greater corm diameter and weight than whole corms of Jumbo and No.1 grade.

Clonal multiplication through conventional method of asexual propagation is not rapid to meet disease-free planting material demand. Callusing may be induced in inflorescence, stems, buds, perianth and anthers of gladiolus. Regeneration starts by formation of callus and root primordia on the basal end of a thin layer. This is followed by the formation of buds and cormels on distal end. Regeneration of the various organs from the explants was found to be polarized and depended on the level of growth substances added to the basal medium. It is now possible to produce foundation stock of virus indexed gladiolus cormels on a commercial scale. Such cormels have demonstrated the production of slightly earlier, uniform and high quality flowers. The protocol for ***in vitro*** production in gladiolus has been standardized at the PAU, Ludhiana. The buds on spike axes are ideal explants for propagation through tissue culture because they develop shoot buds directly without any intervening callus formation. On each spike there are 3 to 4 buds present in the axils of leaves, just below the florets. These buds can be excised at the spike emergence stage and used as explants. The explants should be thoroughly washed in running water and surface-sterilized with mercuric chloride (1 per cent) for two minutes. After surface-sterilization, the explant should be washed at least thrice with sterile water. The sterilized explants are cultured on the autoclaved MS medium containing sucrose (30g/l), myco-inositol (100mg/l), 6-benzyl amino purine (5 mg/l), and agar-agar (7g/l). Before autoclaving, pH of the culture medium should be adjusted to 5.7-5.9. The culture thus produced should be incubated at 20-25°C under 16 h light of 3000 lux intensity and 8 h dark cycles. The shoot bud cultures on the above mentioned medium containing BAP (5 mg/l) induce 4-6 shoot buds in four week's time. By repeated subculturing of bud clumps on to the same medium at monthly intervals, considerable number of shoot buds can be raised. The shoot buds thus formed are elongated on BAP-free MS medium. The bud clumps, each containing about 10 buds, should be cultured on this medium. These buds start elongating in 7-10 days and attain length of about 4-6 cm in 4 week's time. When *in vitro* raised shoots attain a length of 4-6 cm (after about 4 weeks), half strength MS medium containing elevated level of sucrose (6 per cent) and IBA (2 mg/l), should be poured in the culture vessels. IBA induces profuse root formation at the bases of the shoots, whereas high level of sucrose favours cormel formation. After 10-12 weeks of addition of half- strength MS medium containing IBA and sucrose, the shoots start drying and cormels turn brown, and the cormels which are mature by this time should be collected from the culture vessels. Agar-agar adhering to them should be removed. These cormels should be stored at low temperature (3-4°C) till planting in the next season. They develop into flowering-grade corm in 2-3 years. Sakkeer Hussain (1995) reported that best time for collection of corm axillary buds and cormel tips is from September to May in Kerala. Culture establishment of the corm axillary buds and cormel tips was found better in MS medium supplemented with BAP ranging from 1.0 to 4.0 mg l^{-1}, multiple axillary bud production was very high when MS medium was supplemented with BAP 1.0 mg l^{-1} and NAA 0.5 mg l^{-1} or BAP 2.0 mg l^{-1} and NAA 0.5 mg l^{-1}, and when NAA concentration was increased, callus proliferation initiated from the base of elongated shoots. These elongated shoots produced maximum number of roots in MS medium containing 1.0 mg l^{-1} IBA in darkness. Maximum plant survival was observed with 0.2 per cent Bavistin treated just after removing from culture vessels followed by treatment with 0.2 per cent mancozeb and norfloxacin at transplanting, and post-planting treatment with $^{1}/_{10}$th MS solution and drenching with triadimefon 20 mg l^{-1} at 3-day's interval inside improvised mist chamber. He recorded direct organogenesis from immature inflorescence segments in modified MS medium supplemented with 15 mg l^{-1} NAA and 3 mg l^{-1} BAP. Among the various explants tried for callus mediated organogenesis, callus index was found maximum (400) in immature inflorescence segments inoculated in MS medium supplemented with NAA 15 mg l^{-1} in 16 h photoperiod and the callus derived so differentiated into shoots in MS medium supplemented with 3 mg l^{-1} BAP. Koushik Dutta *et al.* (2010) cultured disinfected apical and axillary buds of gladiolus corms and apical buds of cormels on MS medium supplemented with BAP and kinetin and recorded 4 mg/l each of BAP and kinetin best for shoot regeneration in the apical buds, however, rooting was best at ½ MS medium supplemented with 4 mg/l NAA in cv. 'Fidelio'.

Cultural Practices

Gladiolus may be grown in wide range of **soils**. Sandy-loam soil is the best for its corm yield and quality cut flower production. The most suitable pH range is 5.6 to 6.5. Heavy soils cause poor drainage which results in poor stand, because of corm rotting. Availability of soil moisture affects the growth and development of plants which in turn affects flower, corm and cormel yields. Gladiolus growth and development are best at 60 to 90 per cent total soil moisture capacity. It has been reported that at low soil moisture content (30 per cent total soil moisture capacity) gladiolus plants contain most of the nutrients in corms and roots and at a high moisture content (90 per cent total soil moisture capacity) plants contain nutrients mainly in the leaves and inflorescences.

Gladiolus thrives best in temperate to sub-tropical climate. It is a summer crop in the temperate regions though a winter crop in the sub-tropical conditions. However, in tropical climates, it can be grown throughout the year. But, at all, there is no comparison of the quality of the spike as well as of the corms and cormels, their health and multiplication of the hill-grown crop because there the soil is normally acidic and sandy-loam and the summer temperature is normally 18-33 °C which is quite congenial for its growing. Since on the hills it is a summer-growing crop when the temperatures are not so drastic and only for a brief period it touches 33 °C when nights are still very pleasant so normally the corms are not attacked with *Fusarium oxysporum* f. sp. *gladioli*, the most serious disease of gladiolus in the plains. Likewise, during gladiolus growing season in temperate reagions the weather is never chilly so *Botrytis gladiolorum* infection, again a serious blight disease in the plains which burns whole of the crop during winter when weather is chilly, windy and humid, normally also does not occur. In India, the agro-climatic conditions differ from one part to another hence it can be grown throughout the year under suitable climate. **Solar radiation** is the most important factor for growth and development, flowering and yield of corms, particularly in winter flowering gladiolus. Thermal fluctuation and illumination are found to be necessary for induction of contractile roots. The light stimulus is perceived by the leaves and not the corms. Under low light intensity in winter, exposure of 1-4 °C increases the occurrence of flower blasting. The light requirement of gladiolus is usually too high for good growth of the plants under winter conditions, and the plants fail to flower because of flower-bud-blasting. Higher the greenhouse temperature the higher is the light requirement. On the basis of this principle, the earliest planting date at Bulb Reaserch Centre, Lisse (Netherlands) is 1 February at a greenhouse temperature of 12 °C, which gives flowering early in May. Nineteenth century miniature cultivars (*Gladiolus colvillei* and *G. nanus* hybrids) generally are not sensitive to photoperiod and mostly do not respond to supplemental lighting. Light affects even the dormancy of the gladiolus corms. The corms from mother plants grown under shade (30 or 80 per cent of natural light intensity) from 3-leaf stage sprout earlier than corms from plants grown under natural conditions. Better sprouting of corms is obtained under natural conditions and from plants whose leaves are given treatment with photosynthesis inhibitors DCMJ (diuron) or PAZ-428. Such corms also contain less quantity of growth inhibitors. Further, there is earlier sprouting and less inhibitors in corms obtained from plants whose leaf bases are wrapped with aluminium foil from 3-leaf stage. This chilling damage is observed at two stages, immediately after planting and at 7th leaf stage when spike emergence begins. **Temperature** plays most important role in dormancy. The dormancy in gladiolus corms can easily be broken with a 49-day treatment of corms at 5 °C temperature. Pre-planting of stored corms of early flowering gladiolus cultivars 'Lovely Melody' at 30 °C for 45 days accelerates the growth rate, resulting in anthesis 8 days earlier and flowering 11 days earlier than control stored under normal conditions. Storage at 5 °C for 30 days retards growth, and anthesis is delayed by 5 days but taller spikes with larger and more number of flowers are produced. Storing the gladiolus corms at 10 °C, followed by transferring them to 20 °C for last 4 weeks before planting, hastens emergence, advances flowering, and yields more corms. Temperature during planting affects sprouting behaviour. Planting cormels in January or later sprout at a wider range of temperatures (10-25 °C), and with April planting some cormels sprout at 30 °C. Bare cormels (removal of outer covering) sprout at wider temperature range than intact (without removal of outer covering) cormels. In spring planting, high percentage of sprouting has been recorded in gladiolus cormels which were taken from dry storage at room temperature after 17 months, but dormant cormels under moist conditions remained dormant for a year or even more. Atmospheric **humidity** also affects plant growth and development. High humidity affects flower quality adversely. Low humidity and low temperature and optimum soil moisture are important factors for production of quality flowers. Insufficient light leads to flower abortion. High temperature after planting affects the plant vigour negatively and favours *Fusarium* infections in contaminated soils.Low temperature affects flower development or causes frost injury (emerging spikes are particularly very sensitive).

The **land** must be brought to a fine tilth by ploughing and harrowing and the beds must be raised to a height of 20 cm from the ground level to allow proper drainage. Well decomposed FYM @ 5 kg/m^2 should be incorporated in the soil a week before planting. A fertilizer must be applied depending upon the soil fertility status. Half of the N and full dose of P and K should be applied as basal dressing. The remaining half of N should be applied as top dressing at 4-leaf stage. Malathion dust @ 0.05 per cent must be applied to the soil before planting to protect the young sprouts from garden cutworms and red ants. The corms are dehusked and cleaned prior to planting. Only the non-dormant corms should be **planted**. The emergence of root buds at the base of the corms shows that the corms are ready for planting. The corms should be suitably treated with fungicide before planting. When planting, the lower portion of the corm should be positioned so that bud at the top lies straight above. It is must to make sure that stem grows erect and does not show crooked growth. At planting time, the soil should contain sufficient moisture to facilitate uniform sprouting of corms. The corms are planted either on flat beds, on ridges or on raised beds. Flat planting is generally practiced in sandy soils, whereas in heavy soils planting should preferably be done on ridges or raised beds. This is because heavy soils are more prone to water-logging which creates anaerobic conditions leading to rotting of corms.

The **planting time** of corm varies with the geographical location. In North Indian plains, the best time for planting the corms is from September to November, optimum being October and for hilly areas from Febuary to April when danger of frost is over. Suneetha (1994) when planted three gladiolus cultivars at Vellayani conditions of Kerala from mid-August to mid-January, she found best performance in October planted ones, and though spike emergence in November planting was earlier but of inferior quality. Although gladiolus can be grown throughout the year in Bengaluru, June to November planting being most ideal. Staggered planting at an interval of 10-15 days is beneficial to get continuous flowering. The effects of planting date on the yield and quality of gladiolus (*Gladiolus grandiflorus*) corms and cormels were studied by Zubair *et al.* (2006). They observed that delay in planting date (1 November, 1 December and 1 January) reduced the number of daughter corms per mother corm (1.2, 1.2 and 1.0), number of cormels per mother corm (24.8, 20.9 and 19.5), and sizes of daughter corm (6.3, 5.6 and 5.1 cm) and cormels (1.5, 1.3 and 1.2 cm). Nijasure and Ranpise (2005) in Maharashtra evaluated the effect of planting dates (15 September, 1 October and 15 October) on growth, flowering and flower yield of gladiolus cv. American Beauty and reported that 15 October planting is superior with respect to growth, flowering time, flower quality, spike yield and the further vase life. September planting produced greater plant height and higher number of cormels per plant compared to October planting. Salvi *et al.* (2003) recommended Her Majesty for early planting. In Jammu region, Ravneet Kour (2009) planted the corms of gladiolus cv. White Prosperity on 15 September, 1 October and 15 October and found 15 October planting as the best one with respect to vegetative growth and flowering characters. Bagde *et al.* (2009) observed 13th October planting the best under Nagpur conditions for spike length, rachis length and florets per spike. They also observed that earlier planting dates produced maximum yield of corms and cormels while September to mid-October planting recorded the maximum weight of corms and cormels. Sheikh and Jhon (2005) reported that planting of corms on 31 March in Srinagar (J & K) resulted into maximum plant height (119.12 cm), spike length (98.06 cm), floret number (13.20) and floret diameter (9.77 cm) while minimum number of days for sprouting (10.88), days to slipping (74.27) and days to basal floret opening (86.15) were recorded for corms planted on 15 May. It was also observed that with successive delay in planting, the days taken to sprouting, slipping and basal floret opening declined. Late planting is subjected to relatively less time span available for accumulation and assimilation of photosynthates as a result of which growth and quality of flowers get deteriorated. **Planting density** varies with the type of soil, climatic conditions and the cultivar used. Very close planting should be avoided as it may adversely affect the plant vigour and spike quality due to increased competition for nutrients, water and light and also due to higher incidence of foliar diseases. At too high density, spike length will be reduced and flower quality impaired. Under poor light conditions, the density of corms should be reduced. Generally, 30x20 cm^2 (15 corms/m^2) is adopted. The effect of spacing (25x10, 25x20 or 25x30 cm) on the performance of gladiolus was studied in Sobour (Bihar). Widest spacing (25x30 cm) resulted in the greatest plant height, number of leaves per plant, number of spikes per plant, spike length, number of florets per spike, floret diameter, number of corms per plant and diameter of new corm, and the lowest number of days to first spike emergence and number of days to first floret opening (Anwar and Maurya, 2005). The corms (4.5 cm in diameter) of gladiolus cv. Red Beauty were planted at 15x30, 20x30, 25x30 and 30x30 cm across in Medziphema (Nagaland). Wider spacing (25x30 cm) significantly affected days to spike emergence, length and diameter of spike, number of florets per spike, and number of spikes per hectare (Singh and Bijimol, 2005). A spacing of 20x30 cm recorded significantly superior results in respect of growth, flowering and corm production in Maharashtra conditions (Pillewan *et al.*, 2004). The optimum spacing has been standardized as 25 cm x 30 cm for gladiolus cultivation in Bay Islands (Andamans) (Nair and Singh, 2004). In Meerut (U. P.), a spacing of 25x20 cm produced the tallest plants with the longest and widest leaves, earliest spike emergence, earlier opening of first flower, number of flowers per spike and size of spike (Singh and Singh, 2004). Sharma and Gupta (2003) studied the effects of spacing (10x40, 20x40, 30x40 and 40x40 cm) on growth and flowering of gladiolus at Haryana condition. They odserved that increasing spacing resulted in an increase in plant height, spike length, number of florets per spike and number of spikes per plant, number of corms per plant, corm weight and diameter and number and weight of cormels per plant. **Planting depth** varies with soil type and soil temperature, importantly with corm size and planting season. Medium corms are planted at 7-10 cm deep while large corms at 10-15 cm deep. A planting depth of 10 cm produced the highest plant height, number of leaves per plant, length of leaf and corm production in gladiolus cv. Candyman (Rana *et al.*, 2005).

Weeds compete with the crop for space, light, nutrients and water. Generally, **weeding** is carried out manually, *i.e.* hand weeding or through ploughing. Chemical weed control is essential for commercial cultivation and the selection of herbicide depends upon the nature of the weeds infesting the crop and the sensitivity of the crop towards the herbicide. Some pre-emergence herbicides reported for gladiolus are alachlor and metachlor (4.5 kg/ha), simazlne and atrazine (4 kg/ha) and oxyfluorfen (0.5 kg/ha). Stomp 30 EC @ 3 l/ha as pre-emergence is also very effective.

Mulching between rows is beneficial to conserve the moisture, to keep down the weeds which in turn will save the nutrients to help improving plant growth. Fresh manure, chopped straw, dried grass, clippings, saw dust, peat, husk, bark and strips of black polythene can effectively be used as mulching materials. Barman *et al.* (2005) reported that mulching with organic material further improves the chemical properties of soil. Organic dry weed is the best for growth and flowering of gladiolus cv. Jester. **Earthing up** (hilling up) is done to the plants up to the height of 10 cm when the plants have attained 20 cm height. This enables the plants to grow erect despite high winds and rains and reduces weed growth.

Especially large flowered varieties of gladioli grown outdoors are susceptible to lodging, hence need **staking**. Stems should be secured in position with bamboo or twig supports. *Salix* twigs make good supports if these are 75-100 cm in length. These are firmly pressed slightly away from the region of corm development at spike emergence stage and the spikes are gently tied. For large plantings, strong supports should be inserted at all the corners of the beds and then strings are stretched crosswise in 3-tier system so that plants remain in their position.

Feeding and Irrigation

Fertilizer requirement for rapidly growing plants vary with climatic conditions, irrigation methods, and soil types. In sandy soils, it is necessary to provide fertilizers frequently especially during rainy season. In some heavier loam soils, light or no fertilizer is required for flower production, as the large supply of inorganic and organic nutrients, present in large corms is sufficient.

At least, four applications of fertilizers are advisable:

1. Preplant incorporation
2. Side dressing at the 2nd to 3rd leaf stage
3. Side dressing at slipping stage when the inflorescence emerges from leaves
4. Side dressing about two weeks after flowering to develop new corms and cormels.

Chanda et al. (2000) in West Bengal conducted an experiment to study the influence of different levels of nitrogen, phosphorus and potassium on growth and flowering of gladiolus. Nitrogen promotes vegetative growth, due to the fact that N application increases more metabolic transport for growth and delays flowering. Out of phosphorus and potassium, potassium shows marked effect on flowering. Uday Sharma *et al.* (2008) studied the effect of 100 per cent recommended NPK dose, 100 per cent NPK + FYM, 100 per cent NPK + FYM + biofertilizers, 100 per cent NPK + vermicompost, 100 per cent NPK + vermicompost + biofertilizers, 75 per cent NPK + FYM + biofertilizers (*Azotobacter* + PSB), 75 per cent NPK + vermicompost + biofertilizers, and FYM + vermicompost + biofertilizers and obtained earlier sprouting of corms, maximum plant height, spike length and number of florets per spike, *vis-à-vis* corm and cormel production with the application of recommended doses of NPK along with vermicompost and biofertilizers. Baskaran *et al.* (2014) tried various bio-fertilizers (phosphorus solubilizing bacteria, *Azospirillum* and *Azotobacter*) and commercial formulations (Annapurna®, Prokissan®, Sumangala®, General liquid®, Chamak®, Twin®, Samaras®, Multinol® and Flower Booster®) and observed *Azotobacter* giving earlier corm sprouting with maximum rachis length, *Azospirillum* giving maximum number of corms per plant, phosphorus solubilizing bacteria giving maximum plant height and maximum diameter of florets, Annapurna inducing earlier flowering with maximum rachis length and maximum number of florets per spike, Sumangala recording increased spike length, and Flower Booster recording maximum weight of corms and cormels with maximum volume and diameter of corms and maximum propagation coefficient. *Bhalla et al.* (2006) studied the effect of biofertilizers and biostimulants on growth and flowering in gladiolus cv. 'Red Beauty'. They observed that common basal dose + 4 per cent Manchurian mushroom tea (Kombucha)

+ 6 per cent Panchgavya, common basal dose + 4 per cent Kombucha tea + 4 per cent Panchgavya, and common basal dose + 4 per cent Kombucha tea + 2 per cent Panchgavya to be the three best doses for growth and flowering parameters whereas common basal dose + 4 per cent Kombucha tea + 6 per cent Panchgavya, common basal dose + 6 per cent Panchgavya, and general medium + required fertilizer dose were found best for corm parameters.

In gladiolus, since both the spikes and corms & cormels are equally important, therefore, early and quick vegetative growth is necessary to obtain long and healthy spikes with more number of buds and to obtain more yield of corms in terms of size and number. Maximum spike length (89.7 cm) was observed in plants treated with 155gN+77.5gK in gladiolus cv. High Style (Butt, 2005). Application of NPK at 60:60:60 units significantly improved growth and quality of flowers and corm and cormel yields. Under West Bengal conditions, a fertilizer combination of N at 50 g/m^2, P at 10 g/m^2 and K at 20 g/m^2 produced highest spike weight, number of flowers per spike, flower diameter, number of open flowers at a time, size and weight of corms, and number of corms in *G. grandiflorus* cv. Tropic Sea (Kumar and Chattopadhyay, 2001). Hatibarua and Misra (1999) reported that among the different N sources, calcium ammonium nitrate (CAN) and ammonium sulfate performed better compared with urea. CAN at 30 g N/m^2 was best for increasing plant height, spike length and leaf number, while 40 g/m^2 was best for number of florets per spike. Coating corms either with rizobacterin (*Azotobacter chroococcum*; 1010 cfu/g) or cerealin (*Azospirillum brasilense*; 1010 cfu/g) increased the survival percentage. Rizobacterin was more effective in disease control at different irrigation periods than cerealin. Both the biofertilizers increased growth parameters, flower, corm and cormel yields with rizobacterin being nore superior to cerealin (Hilal *et al.*, 2001).

Sharma and Singh (2007) reported that application of $N_{40}P_{20}K_{20}$ g/m^2 significantly increased spike length, number of florets per spike, diameter of first floret, life of spike including durability of the whole spike under field conditions in gladiolus cv. White Friendship grown under mango orchard in Raipur. A gradual, steady and significant increase in corms produced per plant, corm size and weight including the number of cormels per plant were observed with increasing fertilizer levels, recording highest values at $N_{50}P_{25}K_{25}$ g/m^2. Srivastava and Govil (2007) are of the opinion that biofertilizers improve vegetative and floral characters of gladiolus. Vegetative growth was enhanced most effectively by Azotobacter treatment. However, for quality spike production, PSB was found more effective. It was found that treatment of corms with biofertilizers increased the total rhizospheric bacterial population. This indicates that the improvement in various characters of gladiolus is due to the activity of rhizospheric bacteria, which is enhanced by biofertilizer inoculation. According to Dongardive *et al.* (2007), application of 500:200:200 kg/ha NPK is optimum for growth and flowering of gladiolus cv. White Prosperity in Nagpur. Dongardive *et al.* (2009) under Nagpur conditions obtained maximum corm and cormels with treatment of 500:200:200 NPK kg/ha and in the treatment of vermicompost 8 t/ha+(Azotobacter 5 kg/ha)+PSB 5 kg/ha. Dalve *et al.* (2009) suggested use of biofertilizers (*Azotobacter* and *Azospirillum*) with reduced doses of nitrogen significantly influencing the growth, flowering and yield of gladiolus. N 75 per cent + P and K 100 per cent, *i.e.* 375:200:200 kg NPK ha^{-1} + *Azotobacter* + *Azospirillum* giving the best results, thus saving 25 per cent of nitrogenous fertilizer. Nitrogen at 30, 60 and 90 g/m^2, P_2O_5 at 15, 30 and 45 g/m^2 and K_2O at 10, 20 and 30 g/m^2 were when applied to gladiolus cv. 'Friendship' administering full dose of phosphorus and potassium and half dose of nitrogen along with 5 kg FYM/m^2 before planting of corms and remaining half dose of nitrogen at 4-leaf stage, Gupta *et al.* (2010) recorded maximum plant height at 60 g nitrogen though no obvious effects were observed with phosphorus and potassium applications. They recorded highest duration of flowering per spike under 20 g/m^2 K_2O, however, spike length, rachis length, number of florets per spike, diameter of florets diameter of corms and number of cormels were found significantly higher under 60 g N, 30 g P_2O_5 and 20 g/m^2 K_2O.

The soil should have sufficient moisture at the time of planting of corms and there after light irrigation is required just after sprouting. Later irrigation is done once in 4-6 days depending on soil and weather conditions. Irrigation should be withheld at least 4-6 weeks before lifting of corms. The stage of growth at which flower initiation takes place, *i.e.* 4th leaf stage, is more sensitive to water stress.

Control of Growth and Flowering

Control of flowering is only possible when corms of suitable size are grown under fovourable light conditions. For economic reasons, flower producers want to obtain advanced or retarded flowering. Use of different sizes of corms, temperature treatment to the stored corms, staggered plantings at 15-20 days intervals, planting of corms at varying depths (7.5, 10.0 and 15.0 cm), use of different cultivars and use of different chemicals and growth regulators can manipulate the growth and flowering in gladiolus. Corm and cormels of gladiolus undergo a period of **dormancy** which is more pronounced under warmer climates. Flowering corms for forcing are obtained by two temperature treatments. Freshly harvested corms and cormels undergo a period of dormancy which is due to the high content of growth inhibiting substances mainly the abscisic acid (apart from linolenic acid, linoleic, stearic and palmitic acids) and low content of growth promoting substances such as the ethylene in the tissue. Generally, the dormant corms require 3-4 months of cold storage at 3 to 7 °C (preferably at 4.5 °C) with 68-75 per cent RH in a well aerated room to break dormancy. A high temperature of 20-30 °C is then used to induce shoot and root development during storage and subsequent rapid growth of the plant. The length of this treatment varies from 4 to 6 weeks. Due to total duration of drying, cleaning and curing treatments and storage at low and then high temperatures, the corms cannot be planted earlier than 12 to 15 weeks. **Cormels** exhibit longer dormant period than the corms because they are enclosed in hard shell which is less permeable to water. Root bud swelling indicates that corms and cormels are ready for planting. Room should be opened for 1 hour once or twice a week to prevent building up of toxic gases generated from respiration which will prevent rotting of corms. At least, 1-3 weeks prior to planting, the corms are taken out from cold storage for acclimatization and to encourage root swelling. About a week prior to planting, the corms should be soaked in 0.2 per cent Captafol to kill the spores of the pathogens present on the surface. Cormels are also soaked in water and dried, then again this process is repeated twice so that their hard tunics crack and these sprout timely and uniformly. Corms and cormels after treatment will sprout within 3-4 weeks. An alternate method to break the dormancy is the exposure of corms and cormels to growth regulators and chemicals. Corms and cormels, following at least 1 week of cold storage are soaked in 1 litre container that contains 4 ml of 40 per cent ethylene chlorhydrin solution and are then held for 3-4 days at 23°C room temperature (Wilfret, 1980). Another method is to soak cormels in 3 per cent solution of ethylene chlorhydrin for 3-4 minutes and then to seal them in a tight glass container for 24 hours at 23°C. Growth of the plant is temperature dependent and through its manipulation **advanced flowering** may be obtained if light conditions are optimum. As long as elongating shoots are under the soil surface, soil temperature is critical. The lower threshold of activity is 7-9 °C. Once the inflorescence is above the soil level, air temperature is critical. Maximum light must be provided to the plants, *e.g.* using a low plant density, removal of axillary shoots to reduce mutual shading and the use of clear greenhouse. The utilization of large sized corms of low light requiring cultivars, *e.g.* Peter Pears and Friendship is recommended. During periods of very low light intensities, the temperature in the greenhouse must be lowered. Under limiting light conditions, flowering is promoted by supplementary lighting (4-5 hours lighting at 35 lux as day extension or continuous or cycle night break).

Through corm treatment, early sprouting but with prolonged blooming occurred under GA 50 ppm and early sprouting with maximum weight of corms under ethrel 200 ppm whereas over all performance including corm weight was better with 50 and 100 ppm GA and 100 ppm ethrel, ethrel 100 ppm giving even longer spikes though paclobutrazol resulted in significantly dwarfer plants (Vidya Gopinath, 1997). She recorded highest number and maximum weight of cormels as well as maximum number of florets opening at a time in 100 ppm GA treatment. Kirad *et al.* (2001) concluded from a field trial that GA_3 at 100 ppm induces earliest sprouting. The same concentration when applied as dipping + spraying resulted in highest leaf number and tallest plants in gladiolus cv. White Prosperity. Kumar *et al.* (2002) reported a significantly earlier sprouting, maximum size and weight of corms, best propagation coefficient with corm dipping in 400 ppm GA_3 for 24 hours prior to planting while days to 50 per cent heading, first floret showing colour, durability of the whole spike were found maximum by dipping + spraying of 400 ppm GA_3 at 40 + 65 days after planting. Karaguzel and Dosan (2000) stated that soaking the corms in 100 ppm GA_3 for 1-h 5 days before planting and application of 25 g/m² KNO_3 (5 times) as an additional fertilizer at weekly intervals after 3-4 leaf stage shortened the time from planting to harvest and increased flowering, spike length, florets per stalk and flower size. Leena Ravidas (1991) tried five gladiolus varieties, *viz.* 'Agnirekha', 'American Beauty', 'Friendship', 'Mansoer Red' and 'True Yellow' where she recorded least height with TIBA, CCC and salts of Ca but with increased spike diameter and extended vase life, reduced period in spike appearance with K_2SO_4 and GA_3 100 ppm, and increased spike length with K_2SO_4 0.5 per cent and $CaSO_4$ 0.5 per cent and GA_3 50 ppm. Datta *et al.* (2001) found GA_3 at 100 ppm giving maximum vegetative growth except for diameter of neck. It also induced earliest flowering, reduced periodicity of flowering, improved longevity of spike and flower, and increased length of spike, number of florets per spike and floret size in cv. American Beauty. Maurya and Nagda (2002) reported an increase in plant height, number of leaves per plant, spike length and number of florets per spike with foliar application of GA_3 at 100 ppm applied 45 days after planting in gladiolus cv. Friendship. Prasad *et al.* (2002) reported increased number of leaves under GA_3 at 500 ppm. GA_3 treatment induced earlier spike emergence irrespective of the cultivars used. GA_3 at 250 and 500 ppm has been suggested to be optimum treatments for producing maximum number of florets per spike and largest flower. Ram *et al.* (2002) found increased growth and development of corms and cormels in 100 ppm GA_3. Gaur *et al.* (2003) recorded 200 ppm GA_3 increasing the plant height and inducing earlier spikes with better spike length, floret size and floret longevity when compared to its 25 and 100 ppm treatments in gladiolus cv. Eurovision. Patil and Desai (2003) reported that gladiolus cv. Sancerre sprayed with 100 ppm GA_3 at 4 and 6 leaf stage produced maximum plant height, number and area of leaves, total dry matter content, more number of corms and cormels and heaviest corms with maximum diameter as compared with other plant growth regulators like NAA and ethrel. Sable and Havale (2003) found better vegetative growth with superior floral characters by preplanting soaking of gladiolus corms in 200 ppm GA_3. Shukla and Sagar (2003) in gladiolus cv. Snow Princess recorded improved plant height, spikes per plant, induced flowering with better flowering duration and number of florets per spike by spraying the plants with 200 ppm GA_3 at 30 days after planting. Sharma *et al.*(2004) obtained almost the same results when sprayed the plants of gladiolus cv. Friendship with 100 ppm GA_3 at 45 days after planting. Similarly, when Barman and Rajni (2004) dipped the corms for 24 h in GA_3, the results were almost the same as above but together they also recorded increased corm weight and cormel number per plant. In a field experiment conducted with gladiolus cv. White Prosperity, Ramachandrudu and Thangam (2007) observed that corms soaked in 150 ppm GA_3 for 24 hours prior to planting resulted in maximum plant height, advanced flowering and in prolonged flowering duration compared to the control. Kumar and Singh (2005) observed 17.96 days earlier sprouting in corms of gladiolus cv. Congo Song treated with 150 ppm GA_3. However, maximum weight per corm (52.88 g) was recorded in plants treated with 100 ppm GA_3 grown with sand application. Sharma *et al.* (2006) reported that GA_3 at 200 ppm given as corm soak treatment for 24 hours prior to planting gave earlier sprouting, maximum vegetative growth and superior floral characters. Number of corms and size of corms also increased significantly. GA_3 200 ppm corm dipping for 24 h induced earlier sprouting, 150 ppm dipping + spraying gave maximum plant height and leaf number, 200 ppm dipping + spraying caused earliest flowering but 150 ppm dipping + spraying brought about maximum duration of flowering, longest spike with more number of florets per spike and maximum spike weight, and maximum number of spikes per plant in gladiolus cv. 'Friendship' (Sharma *et al.*, 2008). However, 150 ppm soaking + spraying produced maximum number and weight of corms and cormels. In another experiment, Bhalla and Kumar (2008) reported that ethrel @ 1500 ppm was most effective in causing earlier sprouting, *vis-a-vis* earlier flowering.

Retarded flowering is obtained by delaying corm planting. The date of flowering is affected by planting dates, climatic conditions during growing, corm size and cultivars. For delayed flowering, dormancy is not a problem. However, after planting, these must be able to produce vigorous daughter plants. This can only be achieved if they are stored at a temperature that delays their physiological development. Corms used for delayed flowering must be put into cold storage (< 5°C) a few weeks longer, as per the requirement. They can be transferred to a mean temperature (15-20°C) one or two weeks before planting. Planting dates should be selected so that plants are not at the most sensitive stage (3 to 6 visible foliage leaf) when light conditions are favourable.

Postharvest Management

Gladiolus spikes can be **harvested** from 70 to 110 days after planting, depending upon the planted cultivar and season of planting. For export and distant market, harvesting is done when basal 1- 2 florets start showing colour. For local market, harvesting should be done when basal floret is half to fully opened. Spikes are cut with 4-5 leaves remaining on the stem intact as these are needed for development of the new corm and cormels. Harvesting is done in morning or evening hours. Cut spikes are immediately placed in water and kept indoors at 6 °C. Spikes are bundled in groups of 100 and sent to packing house for grading. They are transported in an upright position to prevent geotropic bending of spikes.

According to North American Gladiolus Council, USA, the **grades** of spikes are given below:

Grade	*Spike Length (cm)*	*Minimum Number of Spikes*
Fancy (Blue)	>107	16
Special (Red)	96-107	14
Standard (Green)	81 -96	12
Utility (Yellow)	<81	10

When graded, these are bundled in units of 10 and held together by rubber bands. They are then held upright in cold storage (4-6 °C) unitl packed. **Pulsing** treatments prior to storage of cut stem is generally given for 20-24 hours at moderate temperatures. Higher levels of sucrose are used along with various biocides. Pulsing of gladiolus for 8 h with 600 mg silver thiosulfate/litre improved the vase life (15 days) and floret opening (93 per cent) of gladiolus cv. Eighth Wonder. Further, holding of spikes in silver nitrate at 800 mg/litre delayed spike bending (7.20 days), improved floret opening (87 per cent) and vase life (13 days) (Barman *et al.*, 2004). Monika Singla *et al.* (2008) treated cut flowers of 10 gladiolus varieties, *viz.* 'Asia', 'Jessica', 'True Love', 'Amsterdam', 'Wine & Roses', 'White Friendship', 'Chapter', 'Happy End', 'Rose Supreme' and 'Dallas' with 4 per cent sucrose + 200 ppm 8-HQC and evaluated for their vase life. Cv. 'Chapter' recorded maximum per cent opening of florets, *vis-à-vis* vase life, followed by 'Wine & Roses' and 'Rose Supreme'.

Polyamines in vase solution significantly improved fresh weight, uptake of vase solution, flower opening and **vase life.** Spermine 100 ppm+4 per cent sucrose, spermine 500 ppm+4 per cent sucrose, putrescine 100 ppm+4 per cent sucrose, spermidine 100 ppm+4 per cent sucrose and spermidine 500 ppm+4 per cent sucrose significantly improved vase life over control and sucrose 4 per cent. Polyamines also delayed senescence and improved vase life of cut spikes by improving membrane stability (Dantuluri *et al.*, 2008). In vase solution treatment, combinations of GA_3 (50 mg/l) and BA (50 mg/l) with sucrose significantly increased the membrane stability index and enhanced the vase life as compared to sucrose alone or controls (Singh *et al.*, 2008). Suneetha (1994) reported sucrose 5 per cent + 8-HQ 300 or 600 ppm as best holding solution, followed by sucrose 5 per cent + silver nitrate (100 or 200 ppm). Holding solution containing silver nitrate (25 ppm) was reported to be the best for improving the quality and lengthening the life of cut gladiolus spikes. Aluminium sulfate (0.25 mM) was also good for improving the postharvest life of the cut spikes (Sharma and Devi, 2005).

The 10-spike bunches are **packed** in rectangular boxes of the size of 1.2 meter length, 60cm width and 30cm height. Fifteen to 24 bunches per box are wrapped in kraft paper or polyethylene for protection from sudden temperature fluctuation, bruising and moisture loss.These boxes are so designed that they fit well inside the refrigerated vans used for transporting the spikes. The size of boxes may differ in countries, depending upon the mode of transport and length of spikes. The packing boxes should be presentable and of good quality and strength to withstand the shocks during transit.

Sometimes it becomes necessary to store the spikes before transit, either because sufficient quantity of the produce is not available for the transportation or to regulate flower supply in the market to get a remunerative price. The best way is to hold the spikes under refrigerated conditions. Only the best quality spikes should be selected for storage. Boxes are then stored at 4 °C until transportation. CA storage with 1 per cent 0_2 and 5 per cent CO_2 if provided for 6-8 days, improves floral opening. Cut spikes can be stored for 7-10 days by keeping the spikes in bucket containing water or holding solution at 4-5 °C.

Lifting of Corms and Storage

Corms are **lifted** six weeks after flowering when 25 per cent cormels have become brown. One should not wait for lifting of the corms till yellowing of the leaves as it will become too late and certain pathogens will start entering the corms, *vis-à-vis* stolons will also rot so lifting of the cormels will become quite cumbersome. Lifting after 45 days of flowering ensures cent per cent cormel recovery. Lifting of corms is effected from September to November in the temperate regions while from March to May in subtropical regions of the country, depending upon the varieties planted, planting time and prevailing weather conditions. At the time of lifting the soil is loosened with forks or spades, the aerial stems are twisted off immediately while lifting and withered mother corms attached to the bottom of daughter corms are taken out with the help of thumbs and fingers. At the time of lifting the soil should be soft having sufficient moisture so that cormels remain intact with the daughter corms which can easily be collected, and the corms and cormels are properly cleaned of the adhering soil. While lifting, the corms should not be injured, however, injured and diseased ones should be discarded. The cleaned corms and cormels are shade-dried for 15-30 days at a well aerated place then should be dipped in 0.2 per cent solution of Captafol or Thiride for one hour and then again shade-dried for 15-60 days depending upon the convenience. More curing will ensure healthy crop in the following season. Spike removal and retaining of 4-5 leaves on the plant will ensure maximum propagation coefficient. Das (1998) and Ahmed and Siddique (2005) when removed the spike at slipping stage, and Chowdhury *et al.* (1999) when clipped four leaves, recorded highest number of corms and cormels with heavier weights.

Corms are **stored** at low temperatures (3 to 7 °C) with 68-75 per cent relative humidity.This slows down the physiological activities going on in the corms and synchronises the sprouting process so that these are timed for planting when proper environemental conditions appear. There should not be drastic fluctuations of temperature in the storage as too wide fluctuations (below 2 °C and above 10 °C) cause condensation of moisture which favours growth of micro-organisms. Temperature fluctuations may also lead to premature sprouting of corms. The air in storage rooms should also be circulated periodically to remove stagnant air which promotes infection by various pathogens. Many growers in our country store the corms in gunny bags. Insufficient movement of air in gunny bags may cause rotting of corms. The corms should preferably be stored in perforated plastic crates, wooden boxes, bamboo boxes or nylon mesh bags of convenient sizes which should be stacked in staged cool chambers. This will provide free air movement through the corms and keep them healthy. However, in the temperate regions where lifting is from September to November, and winter starts from November and continues till March, the corms are as such stored inside the open-window rooms at room temperature as the prevailing temperature during the period is quite congenial for their storage.

Insect-Pests, Diseases and Physiological Disorders

Gladiolus does not encounter any serious insect-pest. Only a few insects-pests attack this crop but none of them is so serious in India. Aphids (A*phis gossypi, Myzus persici*) are small, round-bodied insects which suck sap from tender plant parts. The infested plants become weak and may get deformed. The aphids excrete a sticky substance which attracts sooty mould and other insects such as ants. These are controlled by spraying the crop with Malathion 50 EC (0.15 per cent), Metasystox 25 EC (0.1 per cent) or Rogor 30 EC (0.1 per cent). Spraying of these insecticides may be repeated at a fortnightly interval, if required. **Thrips** (*Taeniothrips simplex)* suck the sap from individual cells, which in the process of drying become filled with air and reflect light, thus giving a glistering silvery or grey appearance. On flower petals, spots appear white in colour. These are controlled by weekly application of Rogor 30 EC (0.2- 0.3 per cent). **Borers** *(Helicoverpa armigera)* feed on leaves and unopened florets. They also bore into the seed capsules and damage the seeds. These may be controlled by spray applications of Thiodan 35 EC (0.5-0.8 per cent) or Ekalux (0.5- 0.8 per cent). Spraying should be done at first instar of the larvae if effective control of this pest has to be achieved. **Cut worms** (*Agrotis segetum*) feed on various parts of the plants during night and rest in the soil during the day. Methyl parathion 0.2 per cent spraying will control this pest. **Loopers** and **semiloopers** *(Pieris brassicae)* as well as **leaf caterpillar** *(Spodoptera litura,* Ranjith, 2013) feed primarily on leaves of gladiolus. Control measures are the same as for borers except that a lower dose of the insecticide (0.5 per cent) has to be employed. Nuvan (0.1 - 0.2 per cent) may also be used. **Seed corn maggots** (*Hylemya cilicura*) enter the capsules and feed on seeds, and on gladiolus shoots and leaves. It can be controlled through Thiodan 35 EC or Ekalux (0.5-0.8 per cent) spraying and also by dipping the corms in Thiram + insecticide solution 24 hours before planting. **Mealy bug** (*Ferrisia virgata,* Ranjith, 2013) is serious under dry conditions in the field as it feeds on the corms by sucking the sap. Drenching such plants with 0.3 per cent Metasystox will control this pest. **Mites** (*Tetranychus equatorius,* Ranjith, 2013; *Tetranychus urticae,* etc.) infest the leaves under warm and shady conditions and web the whole plant. **Lily bulb mite** (*Rhizoglyphus echinopus*) which is yellowish-white in colour with a pinkish tinge, feeds on the corms and roots that causes yellowing of leaves. Some effective miticide spraying will control this pest. **Slugs & snails** infest on gladiolus plants in warm, shady and humid situations, feed on the plants during nights and resting during day in the debris surrounding such plants and sometimes even on the underside of leaves. Its infestation batters the plants completely leaving only the veins. These are either light-trapped or controlled through poison baits (250 g wheat bran + 50 g jaggery + 6-8 g metaldehyde). Gladiolus roots are also damaged by root-galling **nematodes** (*Melodogyne incognita, Trichodorus* spp.), especially in well-drained sandy-loam soils. These are controlled through soil fumigation and HWT of dormant corms at 53 °C for 30 minutes.

There are many serious diseases which have been found infecting gladioli in India. **Fusarium wilt** (*Fusarium oxysporum* f. sp. *gladioli)* is a vascular disease which apart from corm rotting in the field as well as in storage, also causes dry rot or core rot of the corms and plant yellowing. This fungus is carried in corms and cormels in the storage which infects further planting, and also is soil-borne. The disease becomes very serious when temperature is high. The disease appears in the form of water-soaked lesions on the corms. Corms from infected plants may fail to sprout. Leaves of infected plants mostly become pale and sickle-shaped because of its infection in vascular system. The flower spikes remain stunted or get distorted. The disease develops rapidly in storage. The corms develop deep lesions on them and their surface dries and wrinkles in concentric zones. The rotted tissue becomes hard and mummified. Use of resistant varieties like Apricot Glow, Dhiraj, IARI Sel-1, Pusa Dhanvantari, Suchitra, Sylvia, White Friendship, White Prosperity, etc. is the best option. Use of healthy corms, clean cultivation, crop rotation, corm dipping in carbendazim (1g/l) or Captan (2g/l) for 15 - 20 minutes before planting is quite effective. Treatments of gladiolus corms with carbendazim + Thiram proved effective against *Fusarium* and *Botrytis* but Funaben 50 (benomyl) proved most effective (Niebisch and Kelling, 1986). Hot water treatment of corms at 38-40°C for about 30 minutes is also suggested. Fortnighly sprayings with Bavistin 0.1 per cent alternate with Captafol 0.2 per cent to the standing crop regularly from sprouting to spike cutting will keep this disease under control. However, corm dipping about a week before planting in Captafol 0.2 per cent for one hour and again after about 15 days after lifting of corms and cormels, together with regular spraying to the standing crop will keep this disease under complete check. Raj *et al.* (2005) reported that application of quintal, carbendazim or SAAF as corm dip followed by the drenching not only reduces the wilt incidence of gladiolus but also enhances growth of the crop. Mohamed and Gomaa (2000) suggested use of biological control agents (*Trichoderma harzianum* and *Bacillus subtilis*) as soil or corm-soaking treatments which were more effective for reducing this disease on gladiolus cv. Peter Pears as compared with Vitavax 75 WP and/or Rizolex-T50 treatment. Aqueous bulb extracts of *Allium sativum* and *Allium cepa* and the rhizome extract of *Zingiber officinale* showed better disease management potential than that of the recommended dose of carbendazim (Riaz *et al.*, 2010). Soil amendments with cabbage leaf residues followed by polyethylene mulching can effectively control the wilt pathogen. The combination of cabbage leaf residues with soil solarization caused 74.6 per cent reduction in mycelial growth of the pathogen compared to the unamended and unsolarized control. However, soil amendment of cabbage leaf residues in combination with soil solarization resulted in complete elimination of the pathogen at 5 cm soil depth (Raj and Upmanyu, 2005). Remotti *et al.* (1997) with gladiolus cv. 'Peter Pears' which is susceptible to *Fusarium oxysporum* f. sp. *gladioli* and whose infection produces one of the toxins, fusaric acid, when the cell-lines were challenged externally with this toxin *in vitro* up to 0.5 mM, they obtained two plants with altered DNA content which showed an increased tolerance to this toxin. Pathania and Misra (2003) multiplied gladiolus cvs 'Eurovision' and 'Wine & Roses' *in vitro* and treated the shoot clumps with ^{60}Co gamma rays and recorded that 50 Gy dose though decreased shoot survival but was sublethal, and those surviving ones were multiplied on GMM-7 (MS medium + 2 μM KIN) and GMM-8 (MS medium + 4 μM KIN) after being challenged with 20 per cent *Fusarium oxysporum* f. sp. *gladioli* culture filtrate and 1.0-1.5 mM fusaric acid *in vitro* which resulted into insensitive mutants at the end of third selection cycle. **Spongy** or core rot (*Botrytis gladiolorum*) develops when weather in cool, windy and humid and in its infection florets do not open properly. Symptoms on the leaves and stalk appear in the form of oval to circular pin-head type lesions with brown or grayish-brown centres. These may coalesce to form big blotches. On stems the lesions are somewhat elongated. The plants give burnt look if infection is severe. On flower petals the lesions develop as small, round, water- soaked areas. These may completely involve the flowers under congenial conditions and turn them into a drooping mass of tissue covered with grey mass of spores, which may get carried away by wind and initiate secondary infection. Good ventilation during storage and proper distance in planting minimizes the incidence of this disease. Removal of infected plants and corms, clean cultivation with less relative humidity in the field and dry conditions check the spread of this disease. Old flower spikes should, therefore be removed and destroyed to prevent further spread of disease. Spray with carbendazim, Maneb, benlate, or vinclozolin at 0.2 per cent weekly during inclement weathers will provide complete protection against this problem. **Dry or neck rot** (*Sclerotinia gladioli) causes* small, dark and superficial spots on stored corms which later turn brown from tips downwards. Corm treatment by dipping in cold water tor 24 hours and then for 30 minutes in hot water at 54.5°C will control this problem. Copper solution spraying twice or thrice at fortnight interval is also quite effective. **Curvularia blight** or leaf spots (*Curvularia trifolii f.*sp.*gladioli)* is also a soil-borne disease surviving in the soil for three years or more. The disease is very serious for the plants grown from cormels or seeds. This infects all the plant parts, viz. leaves, stalks, flowers, corms and roots. The disease becomes very serious at 18.3-32.2 °C temperature range and during humid weathers. First appearance is noticed in the form of leaf yellowing with browning of leaf tips, initially with circular to oval spots which afterwards may become irregular and extend lengthwise along the leaf blades. These pale-yellow spots turn tan or dark brown surrounded by a darker reddish-brown border, and then soon afterwards the central area is filled with black spores. Use of mancozeb and miltox (copper oxychloride + zineb) each at 0.2 per cent at fortnightly spraying will control this disease. **Alternaria leaf spots** *(Alternaria fasciculata, A. tenuis)* are noted in the form of dirty dark brown to black irregular spots in long patches, mostly on apical and marginal parts on the leaves. **Septoria leaf spot** *(Septoria gladioli)* causes circular, medium, brown to purple-brown spots with black spore bodies in the centre of older spots later in the season but only in the temperate areas. These spots coalesce and make shot-holes. Dithane 0.2 per cent spraying will control both these leaf spot diseases. **Storage rot** (*Penicillium gladioli, Rhizopus arrhizus) causes* black, brown, bluish or greenish new growths on corms, especially at the injured portion, during storage which under poor air circulation results in emission of foul smell. Corm dipping before planting and after lifting in Captafol 0.2 per cent, spraying the standing crop with Bavistin 0.1 per cent fortnightly alternate with Captafol 0.2 per cent, proper curing of the corms and cormels before storage, and proper aeration in storage room will check the storage rot completely. *Uromyces transversalis* **rust** on leaves, and *Urocystis gladioli* & *Tubercinia gladioli* **smuts** on leaves, stems and corms have although been recorded infecting gladioli but not in India yet. However, in their infection, the corm dipping for 30 minutes at 43.3 °C will kill the spores of these pathogens. Smut spores in the soil are killed by treating the soil with formalin.

Bacterial scab or neck rot (*Pseudomonas marginata*) and **bacterial blight and spots** (*Xanthomonas gummisudans*) which also causes neck rot, are serious in excessive humid weathers, especially in the crop grown from cormels and seeds in over-crowded plots. Bazzi *et al.* (1987) reported the bacterium *Pseudomonas gladioli* pv. *gladioli* infecting *Gladiolus* corm rot in Italy. Infection of *Pseudomonas* appears as red-brown specks and spots on leaves, becoming dark-brown later, and destroying parenchymatous tissues together and in severe attack leaves start drying from the tip, and if weather is quite humid the tissues at the base decay. On corm husks, the scab appears as elongated lesions with black rough margins, and on corms these lesions appear brown, circular and sunken with black raised margins, and this depression is then filled with a gummy exudation which glues husks to the corms and in dry weather these lesions become hard, brittle and shiny while causes soft rot during wet weather. *Erwinia carotovora* causes soft rot in gladioli, especially during a very cold and humid weathers. Dipping of corms in mercuric chloride before and after storage will control bacterial infections.

Many **viral** diseases occur in gladioli. CMV causes white discontinuous streaks on foliage and streak and colour break in flowers. It spreads through aphids so control of aphids through reflective surface (aluminium foil) will check spread of CMV also. BYMV is also spread through *Myzus persicae* aphid. TRSV is transmitted through mechanical process, certain aphids and grasshoppers by which leaves show brilliant chlorotic ring spots though sometimes symptomless. Tom R (tomato ring spot virus) is transmitted through *Xiphinema* nematodes and its infection causes plant and spike stunting. TRV is transmitted through sap and through *Trichodorus* and the symptoms appear as constriction with indentation on the margins of the leaves. Aster yellows is serious in India, especially in the temperate regions where gladioli are growing during summer season. Its infection causes precocious sprouting and grassy top, weak shoots, stunting, corkscrew malformation and spiral bending of spikes, straw-yellowing of younger leaves and spikes and virescent flowers, vascular tissues on the base of the corm showing discolouration and degeneration, and finally death from the tip down. *Microsteles sexnotatus* (six-spotted leaf hopper) transmits this disease. In China, Duan YongJia and Cai Hong (1998) observed brilliantly variegated gladiolus flowers infected with tulip breaking potyvirus. To get rid of viral diseases, gladioli are planted away from beans, cowpea, tobacco, tomato, cucurbits and the members of Compositae. Infected plants should be uprooted and burnt and the insects in the field should be thoroughly controlled.

The tips of gladiolus spikes show tendency to bent against gravity if placed horizontally for longer periods. This is primarily due to the lateral downward movement of auxin (IAA) and its accumulation on the lower portion of the spike. IAA causes asymmetrical elongation of cells in this region thereby causing the upward **bending of spike tips**. The bending of tip does not affect the vase life of the spike but reduces its market value considerably as the spikes do not fit well in the flower arrangements. It is suggested that the tips of the spikes should be clipped if it shows bending. To prevent bending of tips, the spikes should be held vertically in storage as well as during transportation. **Tip burning** (scorching) is caused by presence of fluorides as it is a fluoride pollution indicator (1 part/billion) in the atmosphere as well as in irrigation water. Fluoride accumulates at the leaf tips by which the drying or discolouration of the foliage is caused. Crop should be grown away from the polluted area, especially the industries to avoid fluoride toxicity. Use of irrigation water and phosphate fertilizers containing high fluoride, especially the rock phosphate, should be avoided. To overcome this problem, spraying with 5 per cent lime or magnesium sulphate, as well as Blitox 50 WP (0.3 per cent) once or twice will certainly decrease its injury. **Spike topple** and bud rot of spikes are caused due to low calcium content in stalk tissues when spikes are kept in vase solution. In this case $CaC0_3$ @ 0.2-0.3 per cent spraying is recommended. **Blindness** and **floral blasting** are due to unfavourable climatic conditions prevailing during growing period, especially the low light intensity.

References

Ahmed, M.J. and W. Siddique, 2005. Effect of leaf and spike clipping on production of corms and cormels of gladiolus. *Sarhad J. Agric.*, **21**(3): 335-338.

Anwar, S. and K.R. Maurya, 2005. Effect of spacing and corm size on growth, flowering and corm production in gladiolus. *Indian J.Hort.*, **62**(4): 419-421.

Akavia, N., D. Strack and A. Cohen, 1981. The coloration of Gladiolus. I. survey of anthocyanins in petals of *Gladiolus. Z. Naturforsch*, **36c**: 378-383.

Arisumi, K. and Y. Kobayashi, 1971. Studies on the flower colours in *Gladiolus* II. *Lab. Hort. Fac.Agric. Yamaguchi Univ.*, **22**: 157-170.

Bagde, M.S., V.J. Golliwar, B.M. Yadgirwar and M.N. Wankhede, 2009. Effect of planting dates on flower quality and yield parameters of gladiolus. *J. Soils and Crops*, **19**(2): 351-354.

Baker, J.G. 1892. *Handbook of Irideae*. London.

Bamford, R. 1935. The chromosome number in *Gladiolus. Br. Sci.*, **58**: 89-92.

Barman, D. and K. Rajni, 2004. Effect of chemicals on dormancy breaking, growth, flowering and multiplication in gladiolus. *J. Ornam. Hort.*, **79**(1): 38-44.

Barman, D, K. Rajni, Ram Pal and R.C. Upadhyaya, 2005. Effect of mulching on cut flower production and corm multiplication in gladiolus. *J.Ornam.Hort., 8(2):* 152-154.

Barman, D, K. Rajni, Ram Pal and R.C. Upadhyaya, 2006. Studies on corm and cormel production in gladiolus. *J.Ornam. Hort.*, **9**(2): 118-121.

Barman, D., K. Rajni and R.C. Upadhyaya, 2004. Postharvest life of gladiolus cut flower influenced by silver salts. *J. Hill Res.*, **17**(1): 6-10.

Baskaran, V., R.L. Misra, S.K. Singh and K. Abirami, 2014. Response of bio-fertilizers and commercial formulations on growth, yield and corm production of gladiolus. *Indian J. Hort.*, **71**(2): 237-241.

Bazzi, C., P. Minardi and U. Mazzucchi, 1987. Bacterial diseases of flower and ornamental plants in Italy (Italian). *Informatore Fitopatologico*, **37**(6): 15-24.

Bentham, G. and J.D. Hooker, 1883. *Genera Plantarum 3*, London.

Bhalla, R. and A. Kumar, 2008. Response of plant bio-regulators on dormancy breaking in gladiolus. *J. Ornam. Hort.*, **11**(1): 1-8.

Bhalla, R., P. Kanwar, S.R. Dhiman and R. Jain, 2006. Effect of biofertilizer and biostimulants on growth and flowering in glagiolus. *J. Ornam. Hort.*, **9**(4): 248-252.

Bhosale, A.M., G.J. Bankar, R.E. Kadam and S.P. Patinge, 2002. Postharvest life of gladiolus spikes influenced by sucrose and metal salts. *Ann. Pl. Physiol.*, **16**(1): 99-100.

Butt, S.J. 2005. Effect of N, P, K on some flower quality and corm yield characteristics of gladiolus. *J. Tekirdag agric. Fac.*, **2**(3): 212-214.

Campbell, T.B. and J.P. Bower, 2003. *Gladiolus scabridus* – the road to conservation and commercialization. In: *Elegant Science in Floriculture* (eds Blom, T. and R. Criley). *Proc. XXVI intern. hort. congr.*, held on August 11-17, 2002 in Toronto, Canada. *Acta Hort.*, No. 624, pp. 67-72.

Chanda, S., G. Barma and N. Roychowdhury, 2000. Influence of different levels of nitrogen, phosphorus and potassium on growth and flowering of gladiolus. *Hort. J.*, **13**(1): 64-65.

Chowdhury, S.A., R. Ara, F.N. Khan, M.S. Mollah and A.F.M.F. Rahman, 1999. Effect of leaf and spike clipping on corm and cormel production of gladiolus. *Bangladesh J. Scientific and Indust. Res.*, **34**(3/4): 428-431.

Dalve, P.D., S.V. Mane and R.R. Nimbalkar, 2009. Effect of biofertilizers on growth, flowering and yield of gladiolus. *Asian J. Hort.*, **4**(1): 227-229.

Dantuluri, V.S.R., R.L. Misra and V.P. Singh, 2008. Effect of polyamines on postharvest life of gladiolus spikes. *J. Ornam. Hort.*, **11**(1): 66-68.

Das, T.K. 1998. Effect of stage of spike removal and potassium application on corm and cormel production of some gladiolus varieties. *Indian J. Agron.*, **43**(4): 756-761.

Datta, R., J.P. Verma, G.K. Verma and D. Ram, 2001. Effect of plant growth regulators on vegetative growth of gladioli. *Ann. Agri. biol. Res.*, **69**(1): 81-84.

De Vilmorin, R. and M. Simonet, 1927. Nombre des chromosomes dans les genres *Lobelis, Linum* et chez quelques autres espéces végétales. *Compt. Rend. Soc. Biol.*, **96**: 166-168.

Dilta, B.S., S.D. Badiyala, Y.D. Sharma and V.K. Verma, 2004. Effect of corm size on performance of different gladiolus cultivars. *J. Ornam. Hort.*, **7**(2): 153-158.

Dongardive, S.B., V.J. Golliwar and S.A. Bhongle, 2007. Effect of organic manure and biofertilizers on growth and flowering in gladiolus cv. White Prosperity. *Plant Archives*, **7**(2): 657-658.

Dongardive, S.B., V.J. Golliwar and S.A. Bhongle, 2009. Influence of organic manure and biofertilizers on corm and cormel yield of gladiolus. *Ann.Pl. Physiol.*, **23**(1): 114-116.

Duan YongJia and Cai Hong, 1998. Studies on the variegated flowers of flowering plants (Chinese). *Acta Phytopath. Sinica*, **28**(1): 85-89.

Ernst-Schwarzenbach, M. 1931. Contribution a l'étude des chromosomes chez le genre *Gladiolus* L. *Ann. Sci. Nat. Bot.*, **13**(10): 345-351.

Galgate, V.S. and J.R. Desai, 2003. Effect of corm size on growth and flower production in gladiolus. *Symp. on Recent Advances in Indian Floriculture* (Abstracts), pp. 9-10. Symposium held at K.A.U., Vellanikkara, Thrissur (Kerala) in Nov. 12-14, 2003.

Gaur, G.S., T.C. Chaudhary and J.D. Trivedi, 2003. Effect of GA_3 and IAA on growth, flowering and corm production in gladiolus cv. Eurovision. *Farm Sci. J.*, **12**(1): 1-3.

Ghimiray, T.S. 2005. Studies on genetic variability in gladiolus. *J. Interacademicia*, **9**(3): 314-317.

Goldblatt, P. 1971. Cytological and morphological studies in the South African Iridaceae. *J. South African Bot.*, **37**: 317-460.

Goldblatt, P. and J.C. Manning, 2002. Evidence for moth and butterfly pollination in *Gladiolus* species (Iridaceae: Cocoideae). *Ann. Missouri bot. Gard.*, **89**: 110-124.

Goldblatt, P. and J.C. Manning, 1996. Two endemic species and taxonomic changes in gladiolus (Iridaceae) of southern Africa, and notes on Iridaceae restricted to unusual substrates. *Novon*, **6**: 172-180.

Goldblatt, P. and J.C. Manning, 1998. Gladiolus in Southern Africa. Fernwood Press, Vlaeberg, South Africa.

Goldblatt, P., J.C. Manning and P. Bernhardt, 1998. Adaptive radiation of bee-pollinated *Gladiolus* species (Iridaceae) in southern Africa. *Ann. Missouri bot. Gard.*, **85**: 492-517.

Goldblatt, P. and M. Thulin, 1995. *Gladiolus somalensis* (Iridaceae), a new species from northeastern Somalia. *Novon*, **5**: 325-328.

Gromov, A.N. 1972. *The World of the Gladiolus*, NAGC,USA, pp.98-102.

Gupta, R.B., J.R. Sharma and R.D. Panwar, 2010. Growth, flowering and corm production of gladiolus as affected by application of nitrogen, phosphorus and potassium. *Haryana J. hort. Sci.*, **39**(3-4): 286-287.

Hatibarua, P. 1989. Studies on corm production of gladiolus. M.Sc.(Agri.) Thesis, AAU, Jorhat.

Hatibarua, P. and R.L. Misra, 1999. Effect of nitrogen sources on vegetative and floral characters of gladiolus cv. Dhanvantari. *J. Ornam. Hort.* (New Series), **2**(2): 111-114.

Hilal, A.A., I.S. Elewa, S.E. Hassan and S.A.A. El-Malak, 2001. The effect of fertilization and irrigation on *Fusarium* disease development and yield components of gladiolus. *Egyptian J. Phytopath.*, **29**(2): 97-105.

Hutchinson, J. 1932. *The Families of Flowering Plants, II. Monocotyledons.* Oxford.

Kadam, G.B., Gunjeet Kumar, T.N. Saha, A.K. Tiwari and Ramesh Kumar, 2014. Varietal evaluation and genetic variability studies in gladiolus. *Indian J. Hort.*, **71**(3): 370-384.

Karaguzel, O. and J. Doran, 2000. Effect of GA_3 and KNO_3 on growth and flowering of gladiolus. *Ziraat Fakultesi Dergisi*, **13**(2): 123-132.

Kirad, K.S., R.N.S. Banafar, S. Barche, M. Billore and M. Dalal, 2001. Effect of growth regulators on gladiolus. *Ann. agric. Res.*, **22**(2): 278-281.

Koushik Dutta, Pushpa Kharb, G.S. Rana and S.K. Sehrawat, 2010. Studies on *in vitro* multiplication of gladiolus cultivars. *Haryana J. hort. Sci.*, **39**(1-2) 149-152.

Kumar, M. and T.K. Chattopadhyay, 2001. Effect of NPK on the yield and quality of gladiolus (*Gladiolus grandiflorus* L.) cv. Tropic Sea. *Environ. and Ecol.*, **19**(4): 868-871.

Kumar, R., R.K. Dubey and R.L. Misra, 2002. Effects of GA_3 on growth, flowering and corm production of gladiolus. In : *Floriculture Research Trend in India* (eds Misra, R.L. and Sanyat Misra), pp. 110-113. Indian Society of Ornamental Horticulture, IARI, New Delhi.

Kumar, R. and D.S. Yadav, 2006. Effect of different grades of mother corms and planting distances on growth, flowering and multiplication in gladiolus under Meghalaya conditions. *J. Ornam. Hort.*, **99**(1): 33-36.

Kumar, V. and R.P. Singh, 2005. Effect of soaking of mother corms with plant growth regulators on vegetative growth, flowering and corm production in gladiolus. *J. Ornam. Hort.*, **8**(4): 306-308.

Leena Ravidas, 1991. Effect of Growth Regulators and Nutrients on Spike Qualities of Gladiolus (M.Sc. Thesis). Kerala Agriculture University, Vellanikkara, Kerala.

Lewis, G.J., A.A. Obermeyer and T.T. Barnard, 1972. *Gladiolus – a Revision of the South African Species.* Purnell & Sons, Cape Province.

Lopez Oliveras, A.M., D. Lopez Perez and M. Pages Pallares, 1984. Studies on corm splitting of gladiolus. *Anales del Inst. Nacional de Investing Aciones Agrarias*, Agricola, **27**: 29-45.

MacKay, M.E., D.E. Byth and J.A. Tommerup, 1981. The effect of corm size and division of mother corm in gladioli. *Aust. J. exp. Agric. Anim. Husb.*, **21**: 343-348.

Maurya, R.P. and C.L. Nagda, 2002. Effect of growth substances on growth and flowering of gladiolus cv. Friendship. *Haryana J. hort. Sci.*, **31**(3-4): 203-204.

Misra, R.L. 1995. Genetic Resources of Gladiolus. In: *Advances in Horticulture*, vol. 12 –*Ornamental Plants* (eds Chadha, K.L. and S.K. Bhattacharjee), pp. 193-202. Malhotra Publishing House, New Delhi.

Misra, R.L., C.T. Sakkeer Hussain, N.S. Pathania, Sanyat Misra and Anjula Pandey, 2005. Ornamental Plants. In-*Plant Genetic Resources:Ornamental Crops* (eds Dhillon, B.S., R.K. Tyagi, S. Saxena and G.J. Randhawa), pp. 309-327. Narosa Publishing House Pvt. Ltd., New Delhi.

Misra, R.L. and U.S. Kaicker, 1986. Geohistorical development of gladiolus. *NAGC Bull.*, No. 165, pp. 43-45.

Mohamed, F.G. and A.O. Gomaa, 2000. Effect of some bio-agents and agricultural chemicals on *Fusarium* wilt incidence and growth characters of gladiolus plants. Ann. *Agric. Sci.*, Moshtohor, **38**(2): 883-906.

Monika Singla, A.K. Gupta, S.K. Sehrawat, Suresh Kumar and D.S. Dahiya, 2008. Studies on vase life of gladiolus cultivars. *Haryana J. hort. Sci.*, **37**(1-2): 99.

Nair, S.A. and D.R. Singh, 2004. Effect of varieties and spacing on growth and flowering of gladiolus in Andamans. *Indian J. Hort.*, **61**(3): 253-255.

Neeraj, H.P. Mishra and P.B. Jha, 2005. Variability study in gladiolus. *Haryana J. Hort. Sci.*, **34**(1/2): 70-72.

Niebisch, R.M. and K. Kelling, 1986. Results of chemical control of fungal diseases in ornamental plant production (German). *Gartenbau*, **33**(7): 215-218.

Nijasure, S.N. and S.A. Ranpise, 2005. Effect of date of planting on growth, flowering and spike yield of gladiolus cv. American Beauty under Konkan conditions of Maharashtra. *Haryana J. Hort. Sci.*, **34**(1/2): 73-74.

Oyama-Okubo, N., T. Ando, N. Watanabe, E. Marchesi, K. Uchida and M. Nakayama, 2005. Emission mechanism of floral scent in *Petunia axillaris. Biosci. Biotechnol. Biochem.*, **69**: 773-777.

Pathania, N.S. and R.L. Misra, 2003. *In vitro* mutagenesis studies in *Gladiolus* for induction of resistance to *Fusarium oxysporum* f. sp. *gladioli*. In: *Elegant Science in Floriculture* (eds Blom, T. and R. Criley). *Proc. XXVI intern. hort. congr.*, held on August 11-17, 2002 in Toronto, Canada. *Acta Hort.*, No. 624, pp. 487-494.

Patil, R. V. and J.R. Desai, 2003. Effect of plant growth regulators on gladiolus corms and cormel production. *Nat. symp. on Recent Adv. in Indian Floriculture* (Abstract), pp. 24. Symposium held at K.A.U., Vellanikkara, Thrissur (Kerala) on Nov. 12-14, 2003.

Pax, F. 1888. *Iridaceae* (Engler & Prantl.). *Die Naturliche Pflanzen-familien* 2.5.

Pillewen, S., S.K. Kohale, P.V. Belorkar and K. Madhuri, 2004. Effect of spacing and corm size on growth, flowering and corm production in gladiolus cv. Debonair. *J. Soil & Crops*, **14**(2): 394-396.

Pragya, K.V. Bhat, R.L. Misra and J.K. Ranjan, 2010a. Analysis of diversity and relationships among *Gladiolus* cultivars using morphological and RAPD markers. *Indian J. agric. Sci.*, **80**(9): 766-772.

Pragya Ranjan, K.V. Bhat, R.L. Misra, S.K. Singh and J.K. Ranjan, 2010b. Genetic relationships of gladiolus cultivars inferred from fluorescence based AFLP markers. *Scientia Hort.*, **123**: 562-567.

Prasad, A., R. Kumar, S. Arya and K. Saxena, 2002. Varietal response of gladioli corms to GA_3 dippings. *J. Ornam. Hort.* (New Series), **5**(1): 69-70.

Raj Harender, Anand Sachin and Upmanyu Sachin, 2005. Evaluation of fungicides for efficacy against *Fusarium* yellows in gladiolus. *J. Ornam. Hort.*, **8**(4): 320-321.

Raj, H. and S. Upmanyu, 2005. Integration of soil solarization with cruciferous leaf residues for the control of *Fusarium* wilt pathogen of gladiolus. *Integrated Plant Disease Management:Challenging Problems in Horticultural and Forest Pathology*, pp. 215-220, held on November 14 to 15, 2003 at Solan (H.P.).

Ram, R., D. Mukherjee and S. Maniya, 2002. Plant growth regulators affect the development of both corms and cormels in gladiolus. *Hort. Sci.*, **37**(2): 343-344.

Ramachandrudu, K. and M. Thangam, 2006. Effect of corm size on growth, flowering and corm production in gladiolus. *J. Ornam. Hort.*, **9**(4): 298-299.

Ramachandrudu, K. and M. Thangam, 2007. Effect of plant spacing on vegetative growth, flowering and corm production in gladiolus. *J. Ornam. Hort.*, **10**(1): 67-68.

Rana, P, J. Kumar and M. Kumar, 2005. Response of GA_3, plant spacing and planting depth on growth, flowering and corm production in gladiolus. *J.Ornam. Hort.*(New Series), **8**(1): 41-44.

Ranjith, A.M. 2013. Gladiolus. In: *Identification & Management of Horticultural Pests*, pp.139-140. New India Publishing Agency, Pitam Pura, New Delhi-110 088.

Ravneet Kour, 2009. Impact of date of planting on growth, flowering and spike yield of gladiolus cv. White Prosperity. *Intern. J. agric. Sci.*, **5**(2): 651-652.

Remoti, D., M. Deveehi and A. Schubert, 2002. Influence of the level of carbohydrate in gladiolus corms on the production of propagation material and flowering. *Italus Hortus*, **9**(4): 39-43.

Remotti, P.C., H.J. M. Löffler and L. van Vloten-Doting, 1997. Selection of cell-lines and regeneration of plants resistant to fusaric acid from *Gladiolus* × *grandiflorus* cv. 'Peter Pears'. *Euph.*, **96**: 237-245.

Riaz, T., S.N. Khan and Arshad Javaid, 2010. Management of corm-rot disease of gladiolus by plant extracts. *Natural Product Research*, **24**(12): 1131-1138.

Sable, A. S. and V.B. Havale, 2003. Effect of chemicals and growth regulators on sprouting, growth, flowering and corm production of gladiolus. *Nat. symp. on Recent Adv. in Indian Floriculture* (abstr.), pp. 30. Symposium held at K.A.U., Vellanikkara, Thrissur (Kerala) on Nov. 12-14, 2003.

Sakkeer Hussain, C.T. 1995. Response of Gladiolus to Rapid Cloning through *in vitro* Techniques (M.Sc. Thesis). Kerala Agricultural University, Vellanikkara, Kerala.

Salvi, J.S., S.R. Dalal, Anjali Mohariya and Renuka Deshpande, 2003. Effect of planting dates on growth, flowering and quality of gladiolus under shade-net. *Ann. Pl. Physiol.*, **17**(1): 88-91.

Shalini Pillewan, S.K. Kohale, P.V. Belorkar and Madhuri Kewte, 2004. Effect of spacing and corm size on growth, flowering and corm production in gladiolus cv. Debonair. *J. Soils and Crops*, **14**(2): 394-396.

Sharma, D.P., Y.K. Chattar and N. Gupta, 2006. Effect of gibberellic acid on growth, flowering and corm yield of three cultivars of gladiolus. *J. Ornam. Hort.*, **9**(2): 106-109.

Sharma, G. and J. Devi,. 2005. Effect of different holding solutions on postharvest quality of cut tuberose and gladiolus spikes. *Mysore J. agric. Sci.*, **39**(4): 447-451.

Sharma, G. and P. Singh, 2007. Response of N, P and K on vegetative growth, flowering and corm production in gladiolus under mango orchard. *J. Ornam. Hort.*, **10**(1): 52-54.

Sharma, J.R. and R.B. Gupta, 2003. Effect of corm size and spacing on growth, flowering and corm production in gladiolus. *J. Ornam. Hort.* (New Series), **6**(4): 352-356.

Sharma, J.R., R.B. Gupta and R.D. Panwar, 2004. Growth, flowering and corm production of gladiolus cv. Friendshiop as influenced by foliar application of nutrients and growth regulators. *J. Ornam. Hort.*, **7**(3-4):154-158.

Sharma, J.R., R.B. Gupta, R.D. Panwar and Surender Singh, 2008. Effect of gibberellic acid application on growth and flowering of gladiolus. *Haryana J. hort. Sci.*, **37**(1-2): 80-81.

Sheikh, M. Q. and A.Q. Jhon, 2005. Response of planting dates and genotypes on vegetative and floral characters in gladiolus. *J. Ornam. Hort.*, 8(3): 219-221.

Shibata, M. and M. Nozaka, 1963. Paper chromatography survey of anthocyanins in *Gladiolus* flowers I. *Bot. Mag. Tokyo*, **76**: 317-323.

Shinoda, K. and N. Murata, 2005. Effect of corm weight and low temperature treatment on the flowering of gladiolus. *Acta Hort.*, No. 673(vol. 2), pp. 495-499.

Shukla, P. K. and V. Sagar, 2003. Effect of GA_3 and succinic acid on growth and flowering in gladiolus cv. Snow Princess. *Nat. Symp. on Recent Adv. in Indian Floriculture* (Abstract), pp. 34. Symposium held at K.A.U., Vellanikkara, Thrissur (Kerala) on November 12-14, 2003.

Singh, A., J. Kumar and P. Kumar, 2008. Effects of plant growth regulators and sucrose on postharvest physiology, membrane stability and vase life of cut spikes of gladiolus. *Plant Gr. Regul.*, **55**(3): 221-229.

Singh, A.K. and G. Bijimol, 2005. Effect of spacing and nitrogen on gladiolus. *J. Ornam. Hort.* (New Series), **6**(1): 73-75.

Singh, A.K. and C. Singh, 2004. Effect of spacing and zinc on growth and flowering in gladiolus cv. Sylvia. *Progressive Hort.*, **36**(1): 94-98.

Singh, K.P. 2000. Growth, flowering and multiplication in gladiolus cv. Aarti as affected by grades of mother corm and cormel. *J. Applied Hort.*, **2**(2): 127-129.

Singh, K.P. 2001. Influence of corm and cormel grade-sizes on growth, flowering and multiplication of gladiolus. *Annals. Agri. Res.*, **22**(4): 523-527.

Srivastava, R. and M. Govil, 2007. Influence of biofertilizers on growth and flowering in gladiolus cv. American Beauty. *Acta Hort.*, No. 742, pp. 183-188.

Suneetha, S. 1994. Effects of Planting Dates and Floral Preservatives on Spike Qualities of Gladiolus (*Gladiolus grandiflorus*). M.Sc. Thesis. College of Agriculture (K.A.U.), Vellayani, Kerala.

Suzuki, K., N. Oyama-Okubo, M. Nakayama, Y. Takatsu and M. Kasumi, 2008. Floral scent of wild *Gladiolus* species and the selection of breeding material for this character. *Br. Sci.*, **58**: 89-92.

Takemura, T., Y. Takatsu, M. Kasumi, W. Marubashi and T. Iwashina, 2005. Flavonoids and their distribution patterns in the flowers of *Gladiolus* cultivars. *Acta Hort.*, No. 673, pp. 487-493.

Takemura, T., Y. Takatsu, M. Kasumi, W. Marubashi and T. Iwashina, 2008. Anthocyanins of *Gladiolus* cultivars and their contribution to flower colors. *J. Japan. Soc. hort. Sci.*, **77**(1): 80-87.

Uday Sharma, S.V.S. Chaudhary and Rajesh Thakur, 2008. Response of gladiolus to integrated plant nutrient management. *Haryana J. hort. Sci.*, **37**(3-4): 285-286.

Vidya Gopinath, 1997. Vegetative and Floral Characters of *Gladiolus* 'Friendship' as influenced by Corm Size and Growth Substances. M.Sc. Thesis. Kerala Agrticultural University, Vellanikkara, Kerala, Thrissur.

Wilfret, G. 1980. Gladiolus. In: *Introduction to Floriculture* (ed. Roy A. Larson), pp. 165-181. Academic Press, London.

Yatomi, T. and K. Arisumi, 1968. Studies on the flower colours in *Gladiolus* I.*Lab. Hort. Fac. Agric. Yamaguchi Univ.*, **19**: 1183-1196.

Zubair, M., F.K. Wazir, Gohar Ayub and Sohail Akhter, 2006. Planting dates affect production and quality of gladiolus corms and cormels. *Sarhad J. Agric.*, **22**(2): 249-258.

20

Gypsophila (Family: Caryophyllaceae)

Pragya Ranjan, J.K. Ranjan, Sanyat Misra and R.L. Misra

[**Common names**: Baby's breath/Chalk plant/Gypsum pink/Mist (*Gypsophila paniculata*), Soapwort]

Introduction and Origin

Gypsophila derives its name from the Greek *gypsos* meaning gypsum and *philos* meaning loving, owing to its habit of growing in gypsum rich soil. *Gypsophila* is actually a group of annual, biennial and evergreen perennial plants prized for its tiny soft blooms having star-like flowers with soothing effect. There may be some 126 species or more reported so far worldwide of which 55 are reported to be endemic to Turkey alone (Korkmaz and Özçelik Özçelik, 2011). The most commonly cultivated species are *G. paniculata* (perennial) and *G. elegans* (annual), however, other important species are *G. acutifolia, G. cerastioides, G. festucifolia, G. muralis, G. pacifica, G. perfoliana, G. repens, G. stevenii, G. tekyra,* etc. Recently, Hamzaoglu (2012) reported one more species from Turkey and that is *G. turcica*. The main centres of diversity is Turkey (here it is commonly known as Cöven, Cöven out, Helva kökü and Sabunotu), Caucasia, N. Iraq and N. Iran (Barkoudah, 1962). The main areas of production are Italy, Belgium, Germany and Holland. Gypsophilas are grown worldwide as a commercial filler in the cut flower industry. It is also adorned as potted plant, for floral arrangements and for dry flower arrangements. It is also used for mist-like effect in mixed borders, for bedding, for edging, in rockery plantings and in dry walls. The adolescent girls usually adorn their hairs with this flower during weddings in USA. The colour of the flowers ranges from white to carmine via pale-pink, pink, deep pink, rosy, red and violet with flower forms of single and double types. Its roots contain saponins of industrial interest and the detergent and expectorant are prepared from the saponins extracted from the roots of *G. paniculata* and *G. arrostii* (Hostettmann and Marston, 1998). *G. tekyra* produces saponins 4,920 kg/ha of fresh roots which is 39.3 per cent higher than *G. paniculata* (Todorov and Todorov, 1995; Srivastava and Kavita Kandpal, 2006).Turkish soaproot is widely obtained from six species of *Gypsophila, viz. G. arrostii* var. *nebulosa, G. bicolor, G. eriocalyx, G. graminifolia, G. perfoliata* var. *anatolica* and *G. venusta,* one species of *Ankyropetalum, i.e. A. gypsophiloides* and one species of *Saponaria, i.e. S. officinalis* (Özçelik and Muca, 2010). The rhizome liquid is mixed in the pudding (*haluwa*) for making it crispy, used in preparing ice creams, in the liquor and in the herby cheese (Korkamaz and Özçelik, 2011). Root extract is also used in gold-polishing, as cleaner and softener of delicate fabrics and as fire extinguisher (Kaminska *et al.,* 1996). *G. perfoliata* and *G. sphaerocephala* are boron hyperaccumulators (Korkmaz and Özçelik, 2011).

Botany

The plants are slender and lance-shaped with fleshy roots, herbaceous, about 1 m tall, stems highly branched or spreading type, foliage scant when in bloom, leaves opposite, entire, small, narrowly lanceolate and bluish-green in colour, annual species growing some 50 cm in height having delicate stems, border herbaceous perennials forming bushy dome-shape plants of the height up to 90 cm with slender flower stems, basal nodes congested but upper ones quite distant, flower stem is a sympodial dichasium and inflorescence a cymose multibranched panicle (panicled cyme) of 50-120 cm height bearing numerous small flowers being 3 per pedicel, and these flowers open in basipetal succession, flower small and 3-10 cm in diameter with 5 below united sepals, calyx naked at the base and 5-nerved, corolla with very small 5 clawed petals, stamens 10, styles 2 and pods 4-valved (Bailey, 1963). The basic chromosome number is n = 17 with most of the species being diploid (2n = 34), though *C. elegans* possesses 2n = 40 and *G. pacifica* is reported tetraploid (Pal, 1989).

Species and Varieties

The important cultivated species are *G. acutifolia*, a native of Caucasus, plant quite similar to *G. paniculata* though more greener, leaves narrower and 3-nerved, and the flowers white; *G. arietioides* (alpine species), a native of Iran, height 5 cm with a spread of 15-22 cm, leaves minute ovate, grey-green and forming a tight hard cushion, stemless, flowers white to pale-pink and 3 mm in diameter; *G. cerastioides* (alpine species), a native to Himalayas, tufted, basal leaves stalked, spathulate to oval, stems 10 cm tall and branched, flowers some 1 cm in diameter, white but veined red-purple; *G. dubia* (*G. repens, G. prostrata fratensis*, temperate to alpine species), a native to mountainous regions of CS Europe, stems arching to prostrate and forming low mounds, leaves usually glaucous and linear, 1-3 cm long, inflorescence corymbose and 20 cm tall, flowers 5-30 and some 1 cm in diameter, white to lilac or pale-purple; *G. elegans* (annual), a native of Iran, Turkey and Caucasus, growing up to 50 cm, stems highly branched, leaves oblong-spathulate to linear-lanceolate, flowers 1 cm in diameter, and flowers white, pink and carmine, *G. e.* 'Paris Market' bearing profusely though flower size smaller than the type and colour white, *G. e.* 'Covent Garden' (flowers large white), *G. e.* 'Grandiflora Alba' (large white), 'Rosea' (bright rose), and many other strains in white, pink and carmine colours; *G. muralis* (alpine annual most suitable for dry walls), a native of Europe, Asia Minor Caucasus and Siberia, a much branched little plant with wiry stems but spreading up to 40 cm, leaves opposite, minute, smooth, linear and bright green, and flowers numerous though quite small and of flesh colour; *G. oblanceolata*, an endangered and Turkish endemic species (Ekim *et al.*, 2000) where rhizomes are rich in saponins, having native haunts in hydromorphic soils (salty marshes) of Salt Lake (Konya, Aksaray and Nigde) and Sultansazliði in Central Anatolian part of Turkey (Hamzaoglu and Aksoy, 2009), and is valued in detergent, pharmaceutical and food industries,;*G. paniculata* (perennial), a native of C. Europe to C. Asia including Caucasus and Siberia, clump-forming, 90-120 cm tall, leaves lanceolate, grass-like and grey-green, flower diameter some 5 mm, panicles large and cloud-like and colour white to pink, *G. p.* 'Bristol Fairy' (double white), 'Compacta Plena' (double white), *G. p.* 'Flamingo' (semi-double pale-pink), 'Pink Star' (double pink), *G. p.* 'Rosy Veil' (syn. Rosenschleier, double rose-pink or paler than 'Pink Star', up to 40 cm tall); *G. perfoliata* (*G. scorzonerifolia*), a native to Mediterranean region, a tall perennial with thick, round and usually glabrous (sometimes hairy near the summit) stems, leaves perfoliate and 5-nerved, sepals a little shorter than the petals and flower colour purplish; *G. stevenii* (*G. glauca*), a native of Caucasus, plants perennial and glaucous-green with lesser height having smaller panicles but larger flowers than *G. paniculata*, leaves mostly radical, carinate and linear-lanceolate, and flowers white with shorter petals than the calyx; etc.

Other important varieties are 'Arbel' (double and insensitive to daylength), 'Blancavievens' (male sterile), 'Cassiopeia' (low light sensitivity), 'Dynamic Love' (low daylength sensitivity), 'Florello' (long daylength variety requiring 13-15 h light for flowering), 'Gilbosa' (semi-double, day neutral), 'Million Stars' (globally leading variety with small flowers most suitable for cutting and bedding, low sensitivity to daylength and with prolonged flush duration), 'Magic Golan' (semi-double large white flowers, day neutral, good vase life), 'My Pink' (flowers large, medium sensitivity to daylength), 'New Hope' (flowers large, medium sensitivity to daylength), 'Perfecta' (cloud-like double white large flowers), 'Romano 4' (double white), 'Snowball' (large double flowers with prolonged flush duration), 'Xlene' (flowers large and low sensitivity to daylength), 'Yukinko' (semi-double or double flowers and low sensitivity to daylength), etc.

Propagation

Annual species such as *G. elegans* and *G. muralis* are propagated only through seeds. Many perennial species are also raised through seeds but those producing double flowers are normally sterile. However, *G. paniculata* has sterility even in many of the single cultivars. Perennial species including the most commercial *G. paniculata*, are propagated through cuttings or through grafting onto the roots of *G. paniculata* raised from seeds, by cuttings or by means of root cuttings. **Seeds** can be sown during September-October either in the nursery beds or directly in the prepared field permanently to bloom there itself. Seeds which do not have dormancy are sown at the rate of 4-6 kg/ha in spring in temperate areas to bloom during summer or during mid-June to bloom from September to November. Seeds are covered barely 5-8 mm of fine sandy soil and then further covered with straw until germination. These tiny seeds germinate within 2-7 days with almost cent per cent germination. Planting in the prepared field is carried out one month after seed sowing at a distance of 25 × 25 cm in case of annuals while 45 × 60 cm in case of perennials. In case of **cuttings**, the defoliated cuttings taken should be 8-10 cm long from the basal vegetative shoot, treated with rooting hormone and then inserted in the porous rooting media in the mist chamber. These also become ready for planting in a month. When propagating through **root cuttings**, these should be treated with IBA at 6,000 mg/l (Sawwan, 1993). If the cuttings taken from the mid-portion of stem are stored for one month at 0 °C, there is about one fortnight advancement of flowering with greatest number of floral stem production (Diaz and Amezquita, 1990). *G. paniculata* and others not setting seeds, almost all the double forms and also those where seed setting is there but a particular colour is desired in bulk, these are grown through **tissue culture**. Use of thidiazuron induces shoot buds from stem explants in cv. 'Arbel' (Arhoni *et al.*, 1997). Zucker *et al.* (1997) got successs through leaf explants in the cultivars 'Arbel', 'Flamingo', 'Gilbosa', 'Golan', 'Perfecta' and 'Tavor'. Cv. 'Flamingo' does not grow freely on its roots, hence, it is grafted onto *G. paniculata* 'Bristol Fairy' rootstocks. Culturing of meristem tips (2 mm) of *Gypsophila paniculata* on half-strength MS medium supplemented with kinetin and IAA (0.5 mg/l) provides highest shoot length (Castro *et al.* (1996). This plant got international attention as now this is being flowered in winter by growing in the polyhouses and through tissue culturing of elite varieties.

Cultural Practices

Gypsophila can be grown from the mild **climate** to extreme cold climate if protection is provided, however, in the plains as well as in the temperate climatic conditions of India it is grown from September-Ocober (sowing/planting) to April (end of cropping) though on the hills

It can be grown as a perennial crop throughout the year but with winter protection. It can tolerate frost to some extent but persisting high temperatures are not congenial for its growing. Ideal growing temperature for gypsophila is from 16 to 30 °C, relative humidity 60-70 per cent and a light intensity of 5500-7000 lux. For a good growth and flowering, it requires sunny situation. Longer days reduce the number of nodes as Kusey *et al.* (1982) recorded total number of nodes from 12-23 under 14 h daylength while 12 h daylength formed 38 nodes, *vis-à-vis* high radiation of 80 klux at an ambient photoperiod of 13-14 h in May-June formed 18-20 nodes though 24 klux induced 30 nodes. At 38/20 °C (day/night) temperature at 12-14 h photoperiod it formed 18-20 nodes while 20/14 °C (day/night) temperature there were 28-30 nodes (Shillo, 1985). A well-drained sandy-loam **soil** with a pH range of 6.5 to 8.0 is most suitable for its growing, however, it requires to be fortified with lime. Repeated ploughings (at least thrice) is done at least to a depth of 30 cm to make the soil completely pulverized. At the time of ploughing, the soil should be incorporated with some 150 quintals of wellrotten farmyard or any other organic manure. Beds of convenient sizes are made when soil is thoroughly ploughed, irrigation channels are drawn and the beds are levelled properly. **Planting** is done at 25 × 25 cm in case of annuals but at 45 × 60 cm in case of perennials, followed by watering. However, a plant density of 4-5 for perennials whereas 10-15 plants for annual species will be optimum. Immediately after planting, a net to encourage tall straight stems should be held in the bed to a height of 20 cm. It may require pinching in the first year of cropping where central lateral is removed for better development of side shoots. Shoots may be **pinched** at 5th to 7th node and this stage is attained after five weeks of planting. Emerging **weeds** should immediately be weeded out so that these may not compete with the main crop. During summer season on *Gypsophila*, all the herbicides (Alachlor, Napropamide, Oxadiazon, Oxyfluorfen and chlorthal dimethyl) tested were found effective against >90 per cent of grasses and 67-100 per cent of broad-leaved weeds (Lamont, 1986). Pre-emergence application during summer of Alachlor (1.125 or 2.25 l), Oxadiazon (l, 2 or 4 kg), chlorthal dimethyl (7.5 or 15.0 kg), Oxyfluorfen (2 litres) and Napropamide (2.25 or 4.5 kg/ha) showed that only Oxadiazon and Oxyfluorfen provide 100 per cent control of all the weeds in *Gypsophila elegans* plantings and in direct-sown gypsophila crop, and in direct-sown crop Alachlor was found quite safe controlling 70 per cent of broad-leaved weeds and up to 90 per cent grasses during summer as well as in winter (Lamont and O'Connell, 1986). **Mulching** with organic matters or with black polythene films between the rows will also control the weeds.

For off season production and for **continuity of flowering** throughout, these require to be grown under temperature and light controlled polyhouses and by manipulating the planting time, *vis-à-vis* certain other cultural practices as suggested by Elgar (1998). Outdoor planting at 15-21 day's intervals from early September to late November of the cuttings planted in July on the benches and after rooting when kept in containers in the polyhouses will produce blooms from February to April. Seed-sown in October and planted in November outdoors will also flower from January to April, and this is the natural flowering period. Older or one-year old crowns left in the field, will start blooming as early as November, especially in the temperate regions, however, in the plains the crowns require polyhouse protection during summer and on the hills during December to February to facilitate crop growth and flowering from November to January-February. After flowering naturally in their natural flowering season, *i.e.* from November to February, these are headed back in early February and forced under protection and provided with the required temperature and light to flower from May to June or July. To get flowering from June to October, the crowns are lifted after removing the stems and leaves in January-February, crowns are dipped for a while in Captan (0.2 per cent) then placed in the polythene bags without sealing these, and held in the cool store at 2.5 °C for two months and then planted under protection but with proper lighting and temperature where these will flower from June to October. Also the cuttings taken from the vegetative stock plants during summer to early autumn, rooted and planted in the greenhouse in autumn, will provide the blooms throughout the winter months if proper lighting and temperature had been met with.

Seed yield per hectare is approximately 125 kg and the seeds have higher vigour and storage life at 5 °C.

Feeding and Watering

Gypsophilas are less demanding as liberal use of chemical **nutrients** may weaken the stems. If with the preparation of the soil, the organic manure has been incorporated in the soil at the rate of 150 quintals per hectare, there may not be any further requirement of nutrients provided soil is normal. Srivastava and Kavita Kandpal (2006) state that in *G. elegans alba* 'Covent Garden', it has been shown that nitrogen application alone provides far more better yield than those of P_2O_5 and K_2O as N alone at 40 kg/ha has been found most satisfactory. They further stated that for its soilless culture, 1:1 vermiculite:perlite mixture with a nutrient solution containing 17N, 4P, 8K and 4Mg (meq/l) along with the trace elements produced export quality flowers. One of the main external factors that influences anthocyanin accumulation is temperature, with low temperatures increasing and elevated temperatures decreasing pigment concentration but Nissim *et al.* (2007) showed that higher magnesium levels also increase anthocyanin accumulation in aster flowers grown at elevated temperatures, probably due to increased stability of the pigments. In pink-flowered *Gypsophila* 'Pinkolina', Mg treatment to whole plant, cut branches or detached flower buds caused significant increase in anthocyanin concentrations, with a stronger effect under elevated temperature regimes. Regarding **watering**, the pots or fields are watered immediately after planting. In case of seed-sown crop, the sowing should be done only when soil is sufficiently moist, however, in the nursery the beds are watered with fine rose can daily by the evening. In direct sowing, the watering is done only after germination of seeds. Watering is carried out almost every fourth day during summer and every 10 days during winter. Heavy soils require shallow watering though sandy-loam soil deeper. In the polyhouses, it would be better if it is fertigation. The gypsophilas are quite drought tolerant.

Growth and Development

Temperatures have greater role to play in gypsophila flowering. When grown under warm temperatures for months together, plants require below 11 °C treatment to bring the plants into flowering. Also, the night temperatures above 11 °C will induce flowering. Very low cold treatment (2-3 °C) is, at all, not required in gypsophila, albeit at forcing phase it is inversely related to flowering (Armitage and Garner, 1999), though this treatment for 6 weeks before and after shoot emergence brings down the floral malformation (12.5 per cent) and increases first branch number, number of nodes, cut flower yield to a tune of roughly $1^1/_2$ times and length of floral stalk (Cheong *et al.*, 1999). In heated greenhouses, GA_3 applications substitute the cool temperature requirement and BA the warm temperature requirement where some three crops per year are taken (Srivastava and Kavita Kandpal, 2006). For better flowering, *G. paniculata* 'Bristol Fairy' as is long day plant (critical being 8-10 h; Iwanami, 1987), requires above 11 °C night temperature so it is influenced by application of long day cycles (Hicklenton *et al.*, 1991). High temperature of 28 °C is normally not congenial for development of many of the cultivars, rapid growth occurs between 12 to 20 °C though there may be varietal differences (Hicklenton *et al.*, 1991) and the best growth and flowering occurs when the plants are exposed from 0 to 10 °C temperatures (Doi, 1992). An average of 10 °C minimum temperature causes the plants to flower in 120 days, though low temperature under supplementary night lighting retards flowering (Shillo and Halevy, 1982; Davies *et al.*, 1996). Short days and high night temperatures (18 or 24 °C) make the plants of *G. p.* 'Bristol Fairy' vegetative and rosetting though a temperature of 12 °C delays but induces cent per cent flowering with increased lateral branching and the flower yield if subjected at young vegetative phase, however, long days and high night temperatures enhance floral initiation, floral differentiation and bud development. Doi *et al.* (1984) for breaking the dormancy in the plants of cvs 'Bristol Fairy', 'Diamond' (white), 'Flamingo' and 'Red Sea' (both pink) and 'Perfecta', subjected them to natural low temperatures for 61 days during winter and then shifted them to polyhouse at temperatures ranging from 15 to 27 °C with an extended daylength of 16 h but results differed highly from variety to variety and even in different plants of the same variety. In 'Bristol Fairy', only a few plants flowered even after longest cold treatment, pink varieties followed by the white showing least requirement of the cold treatment, and 'Perfecta' did not flower at all. This shows that cold treatment does not have any significance for flowering in gypsophila. During summers, high temperatures inhibit or delay bolting though vegetative growth is promoted.

It is basically a long day plant requiring a **daylength** of 12-18 hours for flowering though there are genotypes which have very low sensitivity to daylength. Short days reduce internodal elongation in *G. repens* (Armitage and Garner, 1999). Photoperiod of 16 h to *G. paniculata* cvs 'Superwhite', 'Covent Garden Market' and 'Elegans Crimson' advanced their flowering by 35, 23 and 15 days, respectively, than those grown under natural photoperiods. *G. paniculata* is thermal- and photo-sensitive, and *G. elegans* is also a long day plant with a critical daylength requirement of 8-10 hours (Iwanami, 1987), however, long days with 16 h promoted flowering of cv. 'Bristol Fairy'. Karaguzel and Altan (1999) stated that flowering in plants of cv. 'Perfecta' grown under natural short days depends whether in the beginning these were subjected to natural long days. Floral induction becomes more effective when the plants have been subjected to 16 h photoperiod per day for 70 days after the plants were given vernalization for 45 days (Amezquita *et al.*, 1999). Doi *et al.* (1984) stated that in cv. 'Bristol Fairy', long photoperiod is effective only at higher temperatures of 27/22 or 22/17 °C (day/night), though at low temperature if night temperature is below 12 °C, the plants remain vegetative even in long days. If plants have been provided with less than 10 h photoperiod during growth period, these remain in rosette vegetative phase but at the end of the summer if supplemental lighting is provided the bolting in such plants is promoted, intermediate photoperiod of 12-14 h forces the plants to bolt with very short internodes but during longer photoperiods the degree of flowering is promoted. A continuous light regime promotes flowering. It becomes a critical factor during winter when days are quite short and then supplemental lighting is must. Shlomo (1984) obtained 100 per cent flowering under 80 or 24 klux though decreased radiation decreased the flowering percentage, however, in winter he observed 100 per cent flowrering at 58 klux while no flowering under 24 klux even under 24 h photoperiod. Shillo and Halevy (1982) reported that high light intensity during day time is important for producing high quality cut flowers with more florets, and where only a few blind shoots appear. Night break of 4 h even at the end of night with 60 to 100 lux has been found quite effective. Hicklenton (1986, 1987) reported that supplemental lighting (93 m mol $m^{-2}s^{-1}$) given from high pressure sodium lamps from 20 hours to 07 hours each day up to 42 or 63 days during September-February or January-June enhanced vegetative growth and flowering. Cumulative PAR up to about 745 mol m-2 provided up to 115-day period increased vegetative growth as well as flower number, more pronounced being when plants were subjected at early stage of development.

Growth regulators, especially GA (Shlomo *et al.*, 1985) and BA (Doi *et al.*, 1989) have been tried if these can alter flowering in gypsophila. Plants in short days (8 h) do not respond to GA_3 or GA_{4+7} (500-1000 ppm) at any stage even when treated frequently as it is not a substitute to long days, however, 500 ppm GA_3 treatment under marginal photoperiod of 12 h with high light intensities and high temperature in the tropics, the flowering was promoted. GA_3 at 125-500 ppm under long day and at low night temperatures (7-14 °C) had been found promoting flowering which otherwise prevents the realization of long days. In fact, GA_3 substitutes for the high night temperature required as a background for the photoperiodic induction but not for the long photoperiod itself. Application of BA at 300 ppm to the plants in the heated polyhouse under long day conditions promoted floral initiation and advanced the flowering by two weeks in the vegetative lines 'BF04', 'BF08' and 'BF13' but not in 'BF20'. BA substitutes the chilling requirement for floral induction. BA promoted flowering in the vegetative lines 'BF08' and 'BF13' in an unheated polyhouse under the natural daylength (Doi *et al.*, 1989). This makes possible the harvesting of gypsophila flowers from autumn to winter if BA treatment to the low chilling vegetative lines such as 'BF08' and 'BF13' are used (Doi *et al.*, 1989). Growth retardants such as CCC (4000 ppm) or B9 (5000 ppm) though suppress plant height, aggravating the situation under short days, but in long days promote flowering. Kaukovirta (1966) found marginal retarding growth with Phosphon-D, CCC and SADH. Bose (1965) obtained increased height but no effect on flowering with 100 ppm TIBA, however, MH at 1000 ppm suppressed plant height and flowering both. **CO_2** application just after planting in the polyhouse increased plant weight 2-3 times with more than double leaf area within a month of application, 50 per cent increase in branching and the shoots were also found elongated.

Postharvest

Gypsophila flowers are harvested when almost 50 per cent flowers have opened. The harvesting at the stage of 15-30 per cent floral opening, the inflorescence is treated with STS to protect them from external and internal ethylene, followed by pulsing in 7-10 per cent sugar and biocide till two-third of florets show opening. However, these can also be harvested at the stage when first buds are to open, or at once –over harvesting, and then treated with 2 per cent sugar + 200 ppm 8-HQC or 5-10 per cent sugar + 25 ppm $AgNO_3$ bud opening solution, latter being very effective for low temperature wet storage, however, dry storage should be avoided for gypsophila cut flowers. *Gypsophila paniculata* cut flowers were when held immediately after cutting in water or in various conditioning solutions before marketing, best results were obtained with Chrysal-avb which extended the vase life by 3.5 days, and the life was found further extended by 2.1 days when in the solution sugar was added at 50 g/l (Kalkman, 1983). These should be harvested before sunrise or even earlier when the weather is cool, sometimes even after sunset so that in the night these are treated, bundled and stored for sending to the market by next early morning. The stem of the cut flowers should be 25-60 cm as for Japan the total inflorescence length requirement is necessary (Salinger, 1985). After harvesting, the leaves from 10 cm from the lower end are removed and bunched in 25 stems per bunch. The quality of the flowers from late season harvest is reported to be better than the early season harvest, as late season harvest stores better and exhibits longer vase life (Hanchek and Cameron, 1995). While harvesting, the stems should be placed in water. These are now brought in the cold room, cut flowers are dressed as per the specified standard, pulsed with preservative solution containing silver thiosulphate at 4 mmoles per litre for 30 minutes, bunched and cold-stored for 1-2 days at 4 °C. These can be wet-stored up to two days at 4.4 °C. Salinger (1985) stated that gypsophila responds well to STS before or after bunching and can be cold-stored at 5 °C for up to three weeks in a vase solution. Since it is sensitive to ethylene, hence pretreatment with ethylene inhibitors and biocide will prolong its vase life. Normally, its cut flowers last for 5-7 days in vase. Certain holding solutions can double its life. Marousky (1978) recommended 400 mg DICA (dichloroisocyanurate) + 20 mg sucrose to per litre of water where its life is increased more than a week. Pulsing in STS, followed by holding solution containing 200 ppm Physan (a quaternary ammonium compound acting as bactericide) + 15 g/l sucrose (Newman *et al.*, 1998) may cause lasting of the flowers some two weeks. Miyamae *et al.* (2009) stated that treatment with sucrose (3 per cent) + antimicrobial compounds CMI/MI (an isothiazolinonic germicide) + aluminium sulphate promoted floret opening, increased the fresh weight regardless of harvest stage and extended the vase life. Benzalkonium chloride (510 ppm) when combined with GA_3 (300 ppm) at 21 °C shows synergistic effect on flower opening, and incorporation of 5 per cent sucrose in the solution causes 99 per cent flower opening and longest vase life (Son *et al.*,1995). At 23 °C and above temperature, its open flowers turn brown, wilt and inroll but pulsing with 0.2 mM STS + 0.26 mM 8-HQS, and continuous supply of 4 per cent sucrose to cut flowers prevents occurrence of floral browning. The effects of STS and 1-MCP were studied by Dole *et al.* (2005) who treated cut *Gypsophila* cv. 'Golan' stems and placed them either in deionized water (DI) and subjected to 1-MCP (740 nl l^{-1}) or ambient air for 4 h or DI + STS at 0.2 mM for 4 h, and then stems were removed, placed in polyethylene sleeves and stored either wet in DI water or dry in plastic-lined floral boxes at 5 °C in the dark for 4 days. After storage, bunches were placed in DI water under 12 h light (76 to 100 μmol $m^{-2}s^{-1}$) per day where it was found that STS increased the vase life though 1-MCP did not. Macnish *et al.* (2008) when added aqueous chlorine dioxide (ClO_2) at 2 or 10 μl l^{-1} in clean deionized water, the build up of bacteria in vase solution was prevented and the vase life was found extended in *Gypsophila paniculata* cvs 'Crystal' and 'Perfecta', however, the vase life in the cut flowers of cv. 'Perfecta' was found increased even when ClO_2 was added in water already contaminated with 1011 CFUL-1 aerobic bacterium. Gypsophila flowers can also be used as dry flowers, and its dry flowers can be stored for more than a year. Its flowers are transported through air in standard cardboard boxes which are securely insulated.

Insect-Pests and Diseases

Major **insect-pests** infesting this crop are leaf miners (*Liriomyza huidobrensis*). Torres *et al.* (1995) reported that this pest infests on stem biomass and the flowers. Two-spotted mites (*Tetranychus urticae*) also web and infest this plant damaging whole of the plant thus making the plant totally unsuitable for use. The nymphs and adults of flower thrips (*Frankliniella* sp.) suck the sap from flowers, leaves and the tender parts of the stalk. Lepidopterous larvae, especially beet armyworm (*Spodoptera* sp.), *Helicoverpa armigera* and certain others infest the whole plant. These all the pests can be controlled through spraying with 0.2 per cent Malathion combined with some miticide. Sanitation and timely removal of weeds will keep the attack to the minimum.

Major **pathogens** infecting this crop are *Agrobacterium gypsophilae, Pythium* sp. and *Rhizoctonia solani* which cause roots to rot. The visual symptoms are stunting, wilting, uneven plant growth, crown rot and finally the plant death. Crown gall caused by *Agrobacterium gypsophilae*, a major disease of *G. paniculata* whose attack is symptomised through soft nodular galls developing in the collar region of the grafted plant, therefore propagation of the plants bearing galls should be avoided, and plant and soil disinfection and streptomycin dip are the measures to control this problem. *Pythium* is soil-borne so proper field sanitation and drainage will keep this pathogen under check. Both these pathogens should be controlled by weekly spraying of the plants with 0.2 per cent carboxin + Captan. Application of *Trichoderma* isolates T13, T17 and T18 will also control these pathogens (Garces de Granada *et al.*, 1994). Damping off (*Pythium aphanidermatum*) of rooted cuttings is caused under high relative humidity and temperature. Prothiocarb 0.3 per cent weekly misting and soil treatment will kill this pathogen. Stem rot is caused by *Fusarium culmorum, F. solani, Penicillium, Mucor* and *Phytophthora nicotinae* var. *nicotinae*. *F. culmorum* occurs saprophytically but under optimum relative humidity and temperature conditions the shoots start rotting (Orlikowski *et al.*, 1991). Soil drenching with Fosetyl or sprayings with Fosetyl, mancozeb + metalaxyl, Oxadixyl + mancozeb or Ridomil is quite effective (Orlikowski and Wojdyla, 1991). Crown rot is caused by *Phytophthora cryptogea* where sudden wilting of plants occurs during heavy rains or due to excess moisture in the soil. Its control is similar to stem rot. Gypsophila blight is caused by *Botrytis cinerea* which can be controlled through Captan, ferbam or maneb. Powdery mildew is caused by *Oidium* sp. subtype *Erysiphe polygoni* which makes white mycelia colony on leaves and stems (Satou *et al.*, 1996). It normally occurs at bud emergence stage when temperature is around 20 °C which is optimum for its spread. Matsuura *et al.* (2000) recommended bitertanol, diflumetorim, fenarimol, kresoxim methyl and triflumizole to control powdery mildew effectively.

Witches broom is caused due to some mycoplasma where there is general chlorosis of the plants, their upright growth, *vis-à-vis* abnormal secondary shoot production, leaf reddening or bronzing, production of malformed flowers and finally dieing back (Kaminska *et al.*, 1996). Gera *et al.* (2006) recorded phytoplasma and/or spiroplasma disease on *Gypsophila* spp. in Israel. Such plants should immediately be lifted and destroyed to check further spread through the vectors. However, proper control of insect-pests will keep this problem to the minimum.

References

Arhoni, A., A. Zuker, Y. Rozen, H. Shetjman and A. Vainstein, 1997. An efficient method for adventitious shoot regeneration from stem-segment explants of gypsophila. *Pl. Cell Tiss.Cult.*, **49**(2): 101-106.

Amezquita, M.O.De, F. Cantor, J. Bernard, R. Cure and G. Fisher, 1999. Basic for growth stimulation in *Gypsophila paniculata,* including effects of photoperiod and vernalization. *Acta Hort.*, No. 482, pp. 33-37.

Armitage, A.M. and J.M. Garner, 1999. Photoperiod and cooling duration influence growth and flowering of six herbaceous perennials. *J. hort. Sci. & Biotech.*, **74**(2): 170-174.

Bailey, L.H. 1963. The Standard Cyclopedia of Horticulture. Macmillan & Co., New York.

Barkoudah, Y.I. 1962. A Revision of *Gypsophila, Bolanthus, Ankyropetalum* and *Phryna*. *Wendia*, **9**:1-203.

Bose, T.K. 1965. Effect of growth substances on growth and flowering of ornamentals and annuals.*Sci. & Cult.*, **31**: 34-36.

Castro, V.M., H.C. Dardel and R.G. Verdugo, 1996. *In vitro* propagation of *Gypsophila paniculata* L. *Agricultura Technia Santiago,* **56**(3): 224-228.

Cheong, D.C., J.M. Kim and H.B. Park, 1999. Effects of pruning, pinching and chilling treatment on flower malformation and rosette formation of *Gypsophila paniculata*. *J. Korean Soc. hort. Sci.*, **40**(4): 470-476.

Davies, I.J., P.R. Hicklenton and J.L. Catley, 1996. Vernalization and growth regulator effects on flowering of *Gypsophila paniculata* L. cvs Bristol Fairy and Bridal Veil. *J. Hort. Sci.*, **71**(1): 1-9.

Diaz Arteaga, A. and M. Orozco de Amezquita, 1990. Growth and development responses in *Gypsophila paniculata* in relation to cutting position on the mother plant and to low temperature treatments. *Agronomia Colombiana*, **7**(1-2): 54-60.

Doi, M. 1992. Studies on the chilling requirement of *Gypsophila paniculata* L. *Bull. Univ. Osaka Pref.* (Ser. B), *Agric. & Biol.*, **45**: 107-154.

Doi, M., Y. Takeda and T. Asahira, 1984. Differences in flowering response to low temperature among cultivars of *Gypsophila paniculata* L. and among vegetative lines of cv. Bristol Fairy. *Memoirs Coll. Agric., Kyoto Univ.*, **124**: 27-34.

Doi, M., Y. Takeda and T. Asahira, 1989. Effect of BA treatment on growth and flowering of some vegetative lines of *Gysophila paniculata* L. Bristol Fairy. *Japanese Soc. hort. Sci.*, **58**(1): 221-226.

Dole, J.M., W.C. Fonteno and S.L. Blankenship, 2005. Comparison of silver thiosulfate with 1- methylcyclopropene on 19 cut flower taxa. *Acta Hort.* (*Proceedings of the Fifth International Postharvest Symposium*, eds Mencarelli, F. and P. Tonutti), held at Verona, Italy, on 6-11 June, 2004; No. 682 (vol. 2), pp. 949-956.

Ekim, T., M. Koyunee, H. Vural, H. Duman, Z. Aytac and N. Adiguzel, 2000. The Red Book of Turkish Plants. Turkish Association for the Conservation of Nature. TTKD and Van Yuzuncu Yil Univ., Ankara, 245 p.

Elgar, J. 1998. HortFACT-Gypsophila for Cut Flowers – Propagation, Production and Harvesting. *HortRes.*, Mt. Albert, prepared for HortNET-November 1998.

Farr, D.F. and A.Y. Rossman, 2010. Fungal Databases. Systematic Mycology and Microbiology Laboratory. ARS, USDA. Retrieved from http://nt.ars-grin.gov/fungaldatabases/, October 27, 2010.

Garces de Granada, E., M. Orozco de Amezquita, P. Saray, L.M. Carvajal and C.I. Sarmiento, 1994. Management of *Pythium* sp. and *Rhizoctonia solani* Kühn in rooting banks of *Gypsophila paniculata* L. *Agronomia Colombiana*, **11**(2): 158-163.

Gera, A., L. Maslenin, A. Rosner, M. Zeidan and P.G. Weintraub, 2006. Phytoplasma diseases in ornamental crops in Israel. *Acta Hort.* (*Proceedings of the XIth Int. Symp. on Virus Dis. of Ornam. Plants*, held at Taichung, Taiwan, on March 9-14, 2004; ed. Chang, C.A.), No. 722, pp. 155-161.

Hamzaoglu, E. 2012. A new species of *Gypsophila* and a new name for *Silene* (Caryophyllaceae) from Turkey. *Turk. J. Bot.*, **36**: 135-139.

Hamzaoglu, E. and A. Aksoy, 2009. A phytosociological study on the halophytic communities of Sultansazlizi (inner Antolia-Turkey). *Ekoloji,* **18**(71): 1-14.

Hanchek, A.M. and A.C. Cameron, 1995. Harvest date and storage quality of herbaceous perennials. *HortSci.*, **330**(3): 573-576.

Hicklenton, P.R., L.J. Davies and S.M. Newman, 1991. Response of *Gypsophila paniculata* L. 'Bristol Fairy' and 'Bridal Veil' to different night temperatures and variable irradiance conditions. *New Zealand Controlled Environment Lab Report*, **454**: 215-218.

Hostettman, K. and A. Marston, 1998. Chemistry and Pharmacology of Natural Products : Saponins. Cambridge Univ. Press, pp. 326-327.

Iwanami, K. 1987. Studies on extending the cropping period of annual cut flowers. 1. Photoperiodic response of annual *Gypsophila*. *Scientific Reports Miyagi Agric. College,* **35**: 9-13.

Kalkman, E.C. 1983. Pre-treatment improves the quality of summer cut flowers (Dutch). *Vakblad voor de Bloemist.*, **38**(5): 26-29.

Kaminska, M., T. Malinowski, B. Komorowska and A. Langwals Rudzinska, 1996. 'Etiology of yellows and witches broom' symptoms in some ornamental plants. *Acta Hort.*, No. 432, pp. 96-106.

Karaguzel, O. 1996. Effect of GA_3 on cut flower yield and quality of *Gypsophila paniculata* L. 'Perfecta' under late winter and spring natural photoperiod. *Bahce,* **25**(1-2): 55-59.

Kaukovirta, E. 1966. Treaty group of plants with growth retardants. *Puutarha*, **69**: 16-18.

Korkmaz, M. and H. Özçelik, 2011. Economic importance of *Gypsophila* L., *Ankyropetalum* Fezl and *Saponaria* L. (Caryophyllaceae) taxa of Turkey. *African J. Biotach,* **10**(47): 9533-9541.

Kushy, W.E., T.C. Weiler, P.A. Hammer, B.K. Harbaugh and G.J. Wilfret, 1982. Seasonal and chemical influences on the flowering of *Gypsophila paniculata* 'Bristol Fairy' selections. *J. Am. Soc. hort. Sci.*, **106**: 84-88.

Lamont, G.P. 1986. Herbicide evaluation trials on cut flowers. *Australian Hort.*, **84**(7): 89-91.

Lamont, G.P. and M.A. O'Connell, 1986. An evaluation of pre-emergent herbicides in field-grown cut flowers. *Plant Prot. Quart.*, **1**(3): 95-100.

Macnish, A.J., R.T. Leonard and T.A. Nell, 2008. Treatment with chlorine dioxide extends the vase life of selected cut flowers. *Postharv. Biol. & Tech.*, **50**(2/3): 197-207.

Marousky, F.J. 1978. Control of bacteria in *Gypsophila* vase water. *Proc. Fla St. hort. Soc.*, **90**: 297-299.

Matsuura, S., S. Kajihara, T. Iwamoto, T. Kakou and T. Kohguchi, 2000. Environmental factors of disease incidence, host range and control of powdery mildew (*Oidium* sp.) of Baby's Breath (*Gypsophila paniculata* L.). *Bull. Hiroshima Perfectural Agric. Cent.*, **68**: 21-29.

Miyamae, H., H. Shinto and H. Kontani, 2009. Extension of vase life of cut gypsophila flowers by sucrose treatment during transport. *Hort. Res.* (Japan), **8**(4): 509-515.

Newman, J.P., L.L. Dodge and M.S. Reid, 1998. Evaluation of inhibitors for postharvest treatment of *Gypsophila paniculata* L. *HortTechn.*, **8**: 58-63.

Nissim Levi, A., R. Ovadia, I. Forer and M. Oren Shamir, 2007. Increased anthocyanin accumulation in ornamental plants due to magnesium treatment. *J. hort. Sci. & Biotech.*, **82**(3): 481-487.

Orlikowski, L.B. and A. Wojdyla, 1991. Diseases of *Gypsophila paniculata* in Poland and their control. II. Chemical control of stem and crown rot caused by *Phytophthora nicotinae* var. *nicotinae. Prace Instytutu Sadownictwa-i-Kwiaciarstwa-w-Skierniewicach*, Ser. B, *Rosliny Ozdobne*, **16**: 163-169.

Orlikowski, L.B., A. Wojdyla and C.Z. Skrzypezak, 1991. Diseases of *Gypsophila paniculata* in Poland and their control. I. Stem rot of *Gypsophila paniculata* incited by *Phytophthora nicotinae* var. *nicotinae* and *Fusarium culmorum*. *Prace Instytutu Sadownictwa-i-Kwiaciarstwa-w-Skierniewicach*, Ser. B, *Rosliny Ozdobne*, **16**: 157-162.

Özçelik, H. and B. Muca, 2010. Habitat characteristics and distribution in Turkey of species belonging to genus *Ankyropetalum* Fenzl (Caryophyllaceae). *J. Biol. Sci. Res.*, **3**(2): 47-56.

Pal, P. 1989. Gypsophila. In: Commercial Flowers. Naya Prokash, Kolkata, pp. 842-845.

Salinger, John P. 1985. Other Herbaceous Flowers – Gypsophila. In: Commercial Flower Growing. Butterworths of New Zealand (Ltd.), Wellington, New Zealand.

Satou, M., S. Matsuura, K. Tairako, F. Fukumoto and K. Abiko, 1996. Powdery mildew of Baby's Breath (*Gypsophila paniculata* L.). *Annals Phytopath. Soc. Japan*, **62**(5): 541-543.

Sawwan, J.S. 1993. Propagation of *Gypsophila paniculata* Bristol Fairy by stem-tip cuttings. *Adv. Hort. Sci.*, **7**(3): 103-104.

Shillo, R. and A.H. Halevy, 1982. Interaction of photoperiod and temperature in flowering control of *Gypsophila paniculata* L. *Sci. Hort.*, **16**(4): 385-393.

Shillo, R. 1985. *Gypsophila paniculata*. Handbook of Flowering, vol. III (ed. Halevy, A.H.), pp. 83-87. CRC Press, Boca Raton, Fla, USA.

Shlomo, E. 1984. Flowering in *Gypsophila paniculata*. M.Sc. Thesis. Hebrew University of Jerusalem, Rehobot, Israel.

Shlomo, E., R. Shillo and A.H. Halevy, 1985. Gibberellin substitution for high night temperature required for the long-day promotion of flowering in *Gypsophila paniculata* L. *Sci. Hort.*, **26**(1): 69-76.

Son, K.C., E.G. Gu and S.C. Chae, 1995. Effect of pulsing solutions on the flower opening and senescence of cut *Gypsophila paniculata* L. *J. Korean Soc. hort. Sci.*, **36**(2): 273-280.

Srivastava, R. and Kavita Kandpal, 2006. Gypsophila. In: Advances in Ornamental Horticulture (ed. Bhattacharjee, S.K.), pp. 305-322. Pointer Publishers, Jaipur.

Todorov, T. and M. Todorov, 1995. Investigation of the productivity and of the chemical composition of some species of the genus *Gypsophila* L. *Rasteniev dni Nauki*, **32**(5): 33-35.

Zucker, A., A. Ahroni, H. Shejtman and A. Vainstein, 1997. Adventitious shoot regeneration from leaf explants of *Gypsophila paniculata* L. *Pl. Cell Rep.*, **16**: 775-778.

21

Helianthus (Family: Asteraceae)

Ajit Kumar, Manish Kapoor, Meenu Kumari, Sanyat Misra, R.L. Misra, Sunder Pal, Mamta Bohra and Himanshu Trivedi

[**Common names**: Japanese sunflower/Miniature sunflower (*Helianthus cucumerifolius,* syn. *H. debilis*), Jerusalem artichoke/Sun choke/Girasole/Canada potato (*H. tuberosus*), Prairie sunflower (*H. maximillianii*), River sunflower (*H. decapetalus*), Silver leaf sunflower (*Helianathus argyrophyllus*), Sunflower (*H. annus*, syn. *H. lenticularis*), Swamp sunflower (*H. angustifolius*), Hindi – *surajmukhi*)]

Introduction and Origin

Helianthus derives from the Greek *helios* meaning 'sun' and *anthos* meaning 'flower', as it was thought that the flowers follow the course of the sun though it is incorrect. Its native haunts are North and South America, sunflower being from Mexico and Peru, from where it was introduced to Spain and grown in the Royal Gardens in Madrid in 1562, and by the end of the 16th century, it is thought to have been cultivated in Florence (Italy) where that time it was known by its Portuguese name meaning 'Indian Sun' or 'Sunflower'. From Spain, it was intoduced into England. Initially these were being valued for the ornamental qualities of their yellow flowerheads but now they are also considered as a major cooking oil yielding plant

since the start of its cultivation as oilseed crop in the 19th century in USSR, and so it was introduced into India in 1969 with many of the improved varieties having higher oil yield.

Sunflower (*Helianthus annus*) is an important summer season annual, which attracts attention in the garden due to its attractive flowers. It can also be used for its large, bright flowers in floral and wedding displays. Sunflower is also used for screening and as a background plant. Even before introduction of this crop in India from Russia for production of edible oil, the crop was being grown in British India but only for its aesthetic values as cut or loose flowers, but the specimens were only of primitive types and not the large-flowered improved ones which yield good cooking oil. Whether these are small flowered or large ones but in all the cases the flowers are highly ornamental. A bed or field full of sunflower when is in flowering with its majestic yellow and nodding flowers, it looks quite spectacular. *H. debilis* has showy flowers which attract butterflies and the plants make wonderful drought tolerant ground cover in landscaping. Most are tall enough for planting in clumps at the back of wide borders. In the language of flowers, sunflower is a symbol of adoration, pure and lofty thoughts and love and respect. It also serves as an ideal catch crop when land is left fallow.

The most common sunflower (*Helianthus annus*) produces good quality of seeds which after roasting and grinding serve as a beverage similar to coffee, seeds as such yield edible oil used for cooking, and they also serve as food for birds, especially parrots. The roasted seeds of sunflower are eaten as they contain many nutritional constituents such as vitamin D, B, niacin and protein. Vitamin A, D and certain valuable minerals such as iron and calcium are found in the sunflower oil which make it more effective for many other therapeutic uses. Seeds make the flour more nutritious when ground by mixing with wheat. A bitter liquor is prepared against fever in Europe by infusing its petals in alcohol. Its petals also yield an yellow dye. The roots of sunflower were used by the native American Indians to treat snake bites. Its oil extracted from flowerhead was used as a hair conditioner. Sunflower has been reported as an important resource of natural oil and lipid-rich nutrients for centuries. Its oil contains therapeutic properties as it contains 40-44 per cent high quality protein and used for curing skin diseases, haemorrhoids and leg ulcers. Bronchitis and rheumatism-like diseases were treated by flowerheads and leaves of sunflower by Russians. Certain primitive tribes eat raw tubers of *H. tuberosus* with pepper and salt or after boiling in water similar to Jerusalem artichoke or by roasting in mild fire. The flour prepared from the tubers of *H. tuberosus* is also used in the preparation of dietic and health foods. Some American Indian tribes relish the tubers of *H. maximilianii*, and use the seeds of *H. giganteus* to make a flour mixed with maize flour for making bread (Pizzetti and Cocker, 1968).

The worldwide cultivation of sunflower is more than 35 metric tonnes in the year 2010 (The Statistical Division of the Economic and Social Department, FAO Report, 2011). India is at 12th rank in sunflower production with the yield of 1.06 metric tonnes (FAO Report, 2010). In India, sunflower is cultivated in around 18 lakh hectares (10 per cent of the total global sunflower growing areas) and production is around 12.52 lakh tonnes (4 per cent of the total global production). The major sunflower producing states are Karnataka (54.86 per cent), Andhra Pradesh (20.83 per cent) and Maharashtra (14.58 per cent) (Anon., 2009). Since the people of this country are now fully aware about its potential for oil production, so commercial production of the crop for ornamental purpose is limited only for local use.

Botany and Breeding

These are hardy, erect, unbranced and fast-growing annuals and herbaceous perennials having height of 0.7-3.5 m. The perennial kinds of sunflower are tall and vigorous, fibrous-rooted, tuberous or rhizomatous, forming clumps or colonies of erect, thick and branched stems bearing alternate leaves (lower leaves opposite and cordate) which are rough on both the surfaces, ovate to lanceolate, irregularly toothed on the margins, and acute-acuminate with 3-veined, 10-30 cm long and 5-20 cm wide lamina. The flowering head is daisy-like on the terminal position on the main stem with 10-40 cm in diameter rotating to face the sun and something drooping in nature. This head consists of sterile outer ray florets and inner brown or black numerous disc florets spirally arranged. A definite pattern of florets in the head of a sunflower was given by H. Vogel in 1979 in polar coordinates. A single head has capability of producing 350 to 2,000 greyish-green seeds covered by an outer shell (pericarp) that appears black, white or striated. The length of the seed may be up to 1.5 cm long. Flowering in sunflower takes place in late summer and autumn bearing large, golden-yellow blooms. Ovary found in *Helianthus* is inferior with single basal ovule, achenes obovoid, compressed, slightly 4-angled, variable in size, usually from 1.0-1.5 cm long with fully coloured or striped. Sunflower is protandrous in nature, and lagging time between maturation of stamens and pistils is around 18-24 hours, favouring cross pollination.

The first step towards improvement in sunflower is through **introduction** of high-yielding cultivars worldwise. In India, sunflower was introduced in 1969 from USSR. Some varieties released through introduction were EC68414, EC68415 and EC69874. After its introduction in the country, it is being grown in Karnataka, Maharashtra, Tamil Nadu and Andhra Pradesh and is spreading fast in other states, especially Punjab and Haryana. Second step is genetic improvement by combining good characters of various parents through **hybridization** and this way the emphasis on developing the hybrids in India reached to sound footing after 1980, and now around 80 per cent of sunflower growing area in the country is under the hybrids. Sunflower is a highly cross-pollinated crop and is mainly pollinated through insects and to some extent by wind. Hand pollination increases seed setting up to 30 per cent if followed in morning hours. Firstly, flower head opens which is preceded by the centre of the head blooms. This whole procedure takes place in 5 to 10 days which depends upon size of flower and season. Occurrence of anthers usually takes place between 5 to 8 a.m. with viable pollen whose viability remains for 12 hours, and stigma remains receptive for 2 to 3 days. Emasculation in sunflower is done by two methods, *i.e.* by hand and by chemical means (application of GA_3 @ 100 ppm after bud initiation for atleast 3 days). The emasculated heads are bagged to ensure that no foreign pollen reaches over the emasculated heads afterwards, and when stigma becomes receptive hand pollination with the help of cotton or brush is effected and again bagged. For hybrid seed production, generally **cytoplasmic male sterile lines** (CMS) are used. Leclercq of France discovered the first usable source of cytoplasmic male sterility in a cross *H. petiolaris* Nutt. × *H. annus* L. and Kinman of the United States developed fertility restorer lines RHA 265 and RHA 266 which allowed practical development of sunflower hybrids. Likewise, BSH-1 (CMS 234 × RHA- 274) was the first hybrid of sunflower which was developed by private sector by using CMS line used in hybrid seed production and was released commercially in 1980. Since then the hybrid programme was boosted up and 29 productive hybrids were developed, 11 by private sector and 18 in the public sectors. Non-functional male gamete (pollen), *i.e.* male-sterility, but with functional female gamete in the flowers of a variety offers the best opportunity for crossing with desired males without going for emasculation. Tripathi and Singh (2013) studied the effect of various concentrations of benzotriazole ($C_6H_5N_3$), an hybridizing agent, on plant morphology, days taken for first flowering, pollen fertility and yield in *H. annus* var. MSFH-17 through male sterility, and recorded that 1-2 sprays at 1 per cent caused reduction in plant height, pollen fertility (almost complete pollen sterility) and yield components, *vis-à-vis* delayed flowering. Plants treated once and twice with 1.0 per cent benzotriazole showed 23.01 and 22.22 g total seed yield/plant, respectively, in comparison to 23.68 g total seed yield/plant in control. **Heterosis** may be defined as the superiority of an F_1 hybrid over both its parents in terms of yield or some other character. But in some cases, the hybrid may be inferior to the weaker parent. Ahmad *et al.* (2005) studied the effect of heterosis and inbreeding depression in sunflower (*H. annus*) with an objective of recording heterotic effect in F_1 hybrids, and inbreeding depression in F_2 population. The yield and leaf area showed highly significant heterosis in F_1 hybrids ranging from 102 to 309 per cent and 46.3 to 163.9 per cent, respectively, while inbreeding depression in the F_2 population ranged from 17-71 per cent and -9.7 to 4.3 per cent, respectively. However, leaves per plant showed low level of heterosis in F_1 hybrids (-0.9 to 39.7 per cent) whereas the effect of inbreeding depression in the F_2 population was comparatively high (1.1 to 22.2 per cent) for this character. The parent RHA-882 proved itself to be a good general combiner by making higher contribution towards heterosis both in F_1 hybrids and in F_2 populations. A breeding method based on repeated **backcrossing** of the F_1 and the subsequent generations to one of the parents is known as a method of further improvement. Sujata *et al.* (2008) studied a set of 250 distinct, stable and uniform backcross-derived inbred lines which were developed in sunflower through 5 interspecific cross combinations involving four wild diploid annual species (*H. argophyllus*, *H. petiolaris*, *H. annus* and *H. debilis*). The presence of wild-species genome in these inbred lines was confirmed through higher chromosome associations (tri- and quadrivalents) at diakinesis. Maximum structural rearrangements of chromosomes were observed in lines derived from *H. petiolaris*. There were 46 unique alleles and the number of unique alleles was highest in the lines derived from the cross involving *H. petiolaris*. The results indicate that the sunflower gene pool could benefit from introgression of novel alleles from the latent genetic diversity present in the wild species and particularly through exploitation of the diploid annual *H. petiolaris*.

Species and Varieties

Genus *Helianthus* comprises of some 80 species out of which 38 are perennial. The commonly cultivated species include *Helianthus angustifolius, H. annus, H. argophyllus, H. atrorubens, H. cucumerifolius, H. decapetalus, H. laetiflorus, H. maximillianii, H. mollis, H. salicifolius, H. tuberosus*. The two species, *Helianthus annus* and *H. tuberosus* are more commonly grown for their exquisite blooms and *H. annus* for oil content. All the annual species described below such as *H. annus, H. argophyllus* and *H. cucumerifolius* produce exquisite blooms, though remaining ones here are perennials producing beautiful flowers but are least cultivated.

Helianthus angustifolius: A rhizomatous herbaceous perennial from United States (New York to Florida and west to Texas) is always found growing in wet, boggy or swamp-like positions to a height of 2.1 m with flowering from August to October. This tends to spread by rhizomes to form clumps. Its all the parts are hairy, leaves are lanceolate and 15-18 cm long, flowers are solitary (rarely 3), terminal and 7.5 cm across, the petals (ligules) are yellow and the disc florets are purple.

Helianthus annus (syn. *H. lenticularis*): An annual from United States (Minnesota to Washington and California), growing from 1.8 to 3.0 m in height and flowering from July to September producing huge black-centred flowers. Apart from being ornamental, this is the major oil-seed crop with bright yellow giant flowers (up to 45 cm across flowerheads) grown in every part of India, especially in all the five states of South India with many of the improved varieties. Much-branched smaller flowered bushy types are in plenty in the country which are mostly used for aesthetic taste. The dwarf type much-branched forms are mostly used as 'cut-and-come-again'. A dwarf vigorous form but with short fibrous roots requires support to save the plants from falling off due to lodging and winds. The leaves are stalked, alternate, heart-shaped or elongated-oval, irregularly toothed, 35-40 cm long and are hairy on both the surfaces. The terminal and solitary flowerheads have numerous orange-yellow and oval-lanceolate ligules and radiate round the velvet-black central disc, however, later in the season, in the axils of the leaves, relatively insignificant smaller blooms are borne. The disc bears several small pale-yellow tubular florets and when young these dominate the central disc with their colour. The oil content in seed ranges from 22 to 36 per cent while the kernel consists of 45-55 per cent oil. The colour of oil is pale-yellow with mild and pleasant flavour. In 1910, Mrs. T.D.A. Cockerell from Colorado collected a form suffused with a beautiful chestnut-red tint instead of pure golden-yellow and named it var. *coronatus* which she crossed with *H. annus* and many other varieties, and obtained one with reddish petals tipped yellow which she named var. *bicolor*, one with a red circle which she named var. *zonatus*, and one more where all the petals were completely chestnut-red which she named var. *ruberrimus*, and when in 2014 she crossed var. *ruberrimus* with var. *primulinus* with primrose-yellow flowers, she obtained a wine-red form which she named var. *vinosus*. Apart from these she also developed var. *vinosissimus* (petals completely dark wine-red) and var. *niger* (black petals suffused red at the tips). Apart from the Cockerell varieties, there are numerous other cultivars such as 'Abendsonne' (head diameter 15 cm, petals purplish changing to chestnut-red or yellow and centre maroon), '*californicus*' (semi-double with bright yellow disc), '*chrysanthaeflorus*' (large double flowers with cut petals), '*citrinus*' (primrose-yellow), 'Gelber Knirps' (dwarf with only 1.35 m height, double flowered), 'Globe of Gold' (dwarf with up to 1.35 m height, double flowered), '*globosus fustulosus*' (enormous globose flowers), 'Goldener Neger' (much-branched, colour golden with dark centre), 'Indianerin' (much-branched, disc small, golden with dark centre), '*intermedius*' syn. *Helianthus hybridus* (as the original species but with smaller flowers), '*nanus flore pleno*' (height up to 1.25 m, double-flowered), '*purpureus*' (flowers suffused with pink to purple), 'Red' (red), 'Russian Giant' (very tall oil-seed crop), 'Sonnengold' syn. 'Sungold' (90 cm high, flowers quite double), '*variegatus*' (leaves variegated), *etc.* (Pizzetti and Cocker, 1968). Statewise recommended varieties for oilseed crop in India are 'Badshah', 'Divyamukhi', 'Haryana Surajmukhi-1' and 'Peredovik' for Punjab and Rajasthan; former three varieties and 'Jwalamukhi' for Haryana; 'APSH-11', 'Badshah', 'Divyamukhi', 'Laxmi', 'Mega-363', 'MSHF-8', 'MSHF-17', 'MSHF-30', 'Peredovik' and 'Sunrise Selection' for U.P.; 'Badshah', 'Divyamukhi', 'Peredovik' and 'Vinimik' for M.P. and Gujarat including 'MSHF-30' only for the latter state; 'APSH-11', 'CO-1', 'CO-2', 'Garden', 'MSHF-8', 'MSHF-17', 'MSHF-30', 'RSH-1', 'SS-56' and 'Surya' for Maharashtra; 'KBSH-1', 'MSHF-17', 'MSHF-30', 'Peredovik' and 'RSH-1' for Karnataka; and 'APSH-11', 'Armaverts', 'Armavirskij', 'Peredovik' and 'RSH-1' for Andhra Pradesh. ***H. annus*- Sunny Smile:** It is best known dwarf cultivar of *H. annus* which grows up to the height of 30 to 37.5 cm. It is perfect for growing inside the houses as no pollen is found. Growing this cultivar in the pot yields quality blooms. Some of the cultivars recently developed are 'Banilla Ice' (stems multi-branched, blooms nearly white and 12.5 cm across, suitable for cutting), 'Moonshadow' (height 1.2 m, male-sterile near white blooms with 10 cm across good for cutting and drying), 'Velvet Queen' (blooms orange and yellow but centre black, 20 cm across), 'Claret Hybrid' (1.5-1.8 m tall with side-branching, blooms male-sterile, deep red and 15 cm across), *etc.*

Helianathus argophyllus: An annual from North America growing up to 1.8 m and bears silky and soft hairy-clothed leaves giving silvery appearance. It generally produces 1.2-1.5 m long solitary stems which are usually soft grey and silky-hairy, especially the upper part, otherwise it is similar to *H. annus* with yellow flowers.

Helianthus atrorubens (syn. *H. sparsifolius*): A herbaceous perennial from United States (Virginia to Florida and westwards as far as southern Illinois, Missouri, Arkansas and Louisiana) found wild from light woodlands in dry soil, growing up to 2.1 m in height and flowering from August to October bearing large golden-yellow blooms. Its roots are wide-spreading, stems hairy and cylindrical, leaves mostly opposite, hairy on both the sides, oval-acuminate to lanceolate, dentate and prominently 3-veined, flowers several on naked stems with up to 15 cm long and 1.25 cm wide ligules of yellow colour, and the florets in the central disc purplish-brown. The species was one of the parents in evolving the most elegant and graceful autumn-flowering var. 'Monarch', a quite hardy bushy plant with height up to 2.0 m, flowerhead some 15 cm across with semidouble yellow flowers, which as cut flower lasts for a couple of weeks.

Helianthus cucumerifolius (syn. *H. debilis*): A much-branched annual from southern United States (Florida to Texas) is found growing up to 1.2 m with cucumber-like leaves, and free-blooming from July to October, bearing numerous small yellow, golden-yellow or yellow with dark eye flowers on long stalks. Its stems are generally reddish-blue spotted-yellow, leaves are cordate to triangular, dentate and some 10 cm long, disc florets purple-brown but other flowers yellow, solitary, terminal and axillary, 7.5 cm across, and the petals are narrow-lanceolate and spaced. They are excellent for cutting. Some of the most popular varieties are 'Dazzler' (orange, spotted-marron), 'Diadem' (large lemon-yellow but disc almost black), 'Herbstschönheit' (yellow-bronze, central disc ringed maroon), 'Orion' (golden-yellow with a little twisted petals), 'Perkeo' (free-flowering, blooms small and in various shades of yellow), '*plumosus*' (central disc changing from pale to intense yellow), '*purpureus*' (petals changing from pale-pink to intense purple), 'Rotstern' (flowers suffused with different shades of reddish-brown), 'Stella' (flowers intense golden-yellow with curled petals, disc dark violet-maroon), *etc.*

Helianthus decapetalus: A herbaceous perennial from North America (Maine and Quebec to Wisconsin, and as far as South Carolina and Missouri) is found in woodlands and in the vicinity of rivers. It is thought to have been introduced into cultivation in 1688. It grows up to 1.5 m with flowering from August to October. In mature plants the roots swell and become tuber-like. The stems are cylindrical and smooth but top slightly hairy, leaves thin and opposite, 10-20 cm long and 5.0-7.5 cm wide, oval to lanceolate, dentate, and little wrinkled above and smooth below, and the flowers are terminal, and yellow with some 2.5 cm long ligules. Some of the popular varieties are 'Flore Pleno' syn. 'Duplex'(fully double), 'Grandiflorus' (fully double), 'Major' (flowers large), 'Maximus' (flowers large and simple but ligules pointed), 'Simplex' (flowers like cactus dahlia), 'Soleil d'Or' (flowers as cactus dahlia), *etc.*

Helianthus laetiflorus (syn. *H. rigidus, H. scaberrimus*): A herbaceous perennial with spreading rhizomatous roots from North America (Minnesota, Indiana and Missouri, and westwards as far as Montana and New Mexico) has been found growing up to 2.4 m from August to September in dry prairieland. Stems downy-silky, leaves short-stalked, opposite towards base but alternate upward on the stem, oval-lanceolate, dentate and wrinkled, while flowers are short-stalked and several together, the ligules are yellow and 15-20 cm long, and the disc florets are purplish but sometimes yellow.

Helianthus maximillianii: A herbaceous perennial from North America growing up to 3 m in height with 7.5 cm diameter of flowerheads, and flowers during July to October. It has a tendency of spreading quickly. Leaves are trough-shaped, flower peduncles short and densely hairy, and the flowers are deep yellow with about 15-30 rays which are some 3.5 cm or more long.

Helianthus mollis: A frost-tender herbaceous perennial from United States (Florida to Midwest) which grows up to 3 m and flowers from September to October. It grows vigorously having been clothed throughout with soft white hairs. Leaves are 12.5 cm long, broadly lanceolate and dentate. Flowers some 7.5 cm across appear either solitary or in small groups in yellow colour but disc being darker. It is most suitable for planting at dry locations.

Helianthus salicifolius (syn. *H. orgyalis*): A herbaceous perennial from North America (Missouri and Kansas to Oklahoma, to as far east as Ohio), growing up to 3 m with flowering during September-October. It is found growing in dry prairielands in calcareous soils. Stems are reddish-brown, cylindrical, smooth and thickly covered with shortly-stalked, 15-40 cm long and some 8 mm wide, attractive, alternate, thin, lanceolate, willow-like and acuminate, drooping, slightly rough, leaves up to the crotch of the stem from where it has started producing pedunculate flowers though quite less on the portion of the raceme, *i.e.* stem bearing flowers. Flowers are numerous together on some 1.2 m long and branched spikes, 5 cm across, first greenish and then yellow, and the petals are lemon-yellow. This species does very well in poor, dry and sandy soils.

Helianthus tuberosus: A tuberous-rooted hardy herbaceous perennial from eastern United States extending to western rocky mountains, and is the state flower of Kansas. It was introduced into England possibly in 1616. It grows up to 3.6 m in height with 10-40 cm leaf length, and flowering during September-October. Apart from its ornamental value, its tubers are also edible having texture similar to potato and rich in insulin and carbohydrates, and can easily be digested by a diabetic patient. Mostly used for growing around the corners of a garden. The stems are erect, cylindrical, leafy and hairy. Leaves are stalked, opposite or alternate, 15-20 cm long and 5-10 cm wide, oval-acuminate, roughly wrinkled and coarsely dentate. Flowers are long-stalked, terminal, solitary, 6.0-7.5 cm across and yellow along with ligules. Its tubers are edible by human beings and are also used as fodder for cattle and pigs.

Some of the other promising cultivars are 'American Giant' (height up to 5 m, blooms golden-yellow and 25 cm across, suitable as background crop), 'Mammoth Russian' (very vigorous, orange-yellow), 'Prado Red' (height 1.2-1.8 m, floral colours various, vase life poor), 'Zohar' (improved version of 'Jerusalem Gold' (male-sterile, single-flowered, blooms 11.5 cm across, petals deep orange and centre dark brown, good for cutting), *etc.* All these are giant types which mature late with 1.8-4.2 m height bearing 30-50 cm across blooms. Semi-dwarf types are early maturing with 1.3-1.8 m plant height, flowerheads 17-23 cm across such as 'Jupiter' and 'Pole Star'. Dwarf types are also early maturing with 0.6-1.4 m height and 14-16 cm across flowerhead such as 'Advance' and 'Sunset'. Some other cultivars suitable for oil yielding as well as for flowers are 'PKVSH-27' developed by Krishi Vidyapeeth, Pune; and 'CO-2', 'CO-4' (TNAUSUF-7), and 'COSFV-5' by TNAU, Coimbatore. Some of the newly introduced F_1 hybrids suitable for home gardening are 'Full Sun' (large yellow head), 'Moonwalker' (attractively branched with pale heads), 'Sunbeam' (attractive golden-yellow with green eyes), and 'Valentine' (creamy-yellow with black disc). Krishi Vidyapeeth, Pune has developed three open pollinated varieties in sunflower namely 'SF9', 'Surya PKV', and 'TAS 82'. 'Sunny Smile' is excellent as pot plant.

Propagation

Propagation of sunflower is done through sexual (seeds) in case of all the annual types, as well as asexual (cuttings) through division or by soft cuttings in spring of annual as well as of perennial types. Usually seed propagation is preferred over cuttings as it is more convenient and the plants raised from seeds are tall, vigorous and heavy yielder. The **seeds** of *H. annus* and its various cultivars, *vis-à-vis* other species and their varieties, in most of the cases, are easily available and germinate quickly at temperature ranging from 15 to 25 °C, preferably at 16 °C. The seeds can be sown during spring, *i.e.* in mid-March in sub-tropical conditions, in April under sub-temperate and temperate conditions whereas in February under tropical conditions, at their permanent positions, setting 2-3 seeds at a place followed by thinning later on, in pots, boxes or on raised beds. Since the plants grow very fast so they do not relish seedling-transplanting as shock by these plants is unbearable. For winter season crop, sowing of seed is done during mid-September to mid-October. Similarly, for summer and rainy season crop, sowing of seed is done in January-February and mid-June, respectively. With utmost care and to take the early crop, the crop may be grown through seeds in the nursery beds, in polybags or in plugs and gently transplanted one month after sowing. However, seeds of perennial sunflowers should be sown in nursery beds, then pricking them off 15 cm apart and then these are transplanted to their permanent positions or in the beds one month after sowing. The seeds of hardy types are sown in September and transplanted in October though tender ones from February to April and transplanted from March to May.

Vegetative propagation is used to derive uniform plants for various vegetative and floral attributes of a particular variety. Tip-cuttings of perennials from shoots of about 5-10 cm length from apical portion of the purely vegetative phase are taken which give out roots adventitiously. Cuttings root best when planted in sand or vermiculite in the plugs under mist chamber or in a highly humid atmosphere. Root promoting hormones such as IBA and IAA can also be used. The rooted cuttings are successfully transplanted in beds or pots for flowering. Apart from this, perennial plants can also be multiplied by division of rootstock before or when have initiated sprouting from October to April and replanted direct in permanent positions.

Cultural Practices

Sunflower can be grown in a wide range of well-drained **soils** with a pH range of 6.0 to 7.5. Drought-prone soil can also be easily accepted by sunflower as it is a drought tolerant crop. Waterlogged and more acidic soils should always be avoided. However, there are certain species and varieties which can easily be grown in alkaline (calcareous) soils, and certain others even in the wet or boggy soil. In any case, it should be at an exposed site. Commercial production of sunflower is mainly seen in the warm temperate regions, but due to breeding approaches, its varieties are adapted to the wide range of environments. Sunflower is grown 2,500 m above sea level and maximum oil yield/ha is recorded below 1,500 m a.s.l. Usually, it is a **day-neutral** plant and its development is seen more rapid at 12 h daylight. Flowering is affected by **daylength** duration. High temperature drastically reduces its time of maturity. Sunflower holds great potential as it is short duration, photo-insensitive, widely adapted and drought-tolerant. Since it is a photo-insensitive crop, it can be grown in any season, *i.e. kharif*, *rabi* and spring-summer. All stages of growth including immature seed are likely to get damaged by frost. A frost-free period of 120 days is recommended for a commercial crop. The young plants of 4-6-leaf stage can withstand temperature of 5 to 6 °C for short period. Sunflower grows well within range of temperature of 20-25 °C. Temperature above 25 °C adversely affects the flowering, seed yield and the oil content though for ornamental purposes there is no problem of temperature as it grows luxuriantly even up to 40 °C temperature.

Sunflower crop raised by transplanting method gives higher yield and early maturity than that of seeded crop. Therefore, the crop should be raised by **transplanting the nursery**. About 100 m^2 of nursery is sufficient for transplanting one hectare area. Nursery for raising seedlings should preferably be made on raised beds due to more aeration and better drainage for healthy growth of seedlings for round the year cultivation of sunflower. Seed rate for raising nursery in one hectare area is 4.0 kg/ha. Nursery should be raised by sowing seed 30 days before transplanting in well prepared seed-bed after mixing 1.5 kg urea and 4.5 kg single superphosphate per 100 m^2 nursery bed area. Seeds should be spread thinly in the nursery beds and covered with fine soil/sand or FYM. Seeds should be sown in lines spaced at 6-8 cm and at a depth of 5-6 cm. Thick sowing of seed should be avoided as it results in week seedlings which take longer time to regain growth. After seed sowing, beds should be covered with mulch materials of grasses or dried leaves for better germination. The beds should be watered with a fine nozzle can and should be kept moist. When the seed starts germinating, mulch should be removed in the evening hours otherwise etiolation of seedlings will take place. To minimize mortality of plants in the nursery, the soil should be drenched with 0.2 per cent Captan. Land should be properly manured before transplanting of the seedlings. About one month old seedlings at 4-leaf stage should be transplanted at a distance of 30 cm between plant to plant and in rows 60 cm apart. Transplanting should be done during cooler hours of the day followed by irrigation. Uniformity in spacing ensures uniform growth and good yield. Ali *et al.* (2011) studied the performance of sunflower hybrids ('Hysun-38' and 'FH-331') at 17.5, 20.0, 22.5 and 25.0 cm spacings and recorded maximum achene's yield (1920 kg/ha) in spacing of 22.5 cm, whereas 17.5 cm plant spacing resulted in minimum yield.

For **direct sowing of the crop**, pre-treatment with Apron® XL and Maxim® 4FS fungicides available in Cruiser Maxx at the rate of 6 g per kg of seed offers protection against a broad-spectrum of fungal diseases such as 'damping off', 'seedling blight' and 'downy mildew' usually caused by fungi such as *Pythium*, *Rhizoctonia* and *Fusarium*. Seed treated with Cruiser Maxx demonstrates better stand and establishment, more vigour, faster early-season growth and higher yield potential. Pre-treatment of sunflower seeds with 10 mg IAA and 10 mg NAA/litre of water for 6 hours gives high yield also. Sowing period for *kharif*, *rabi* and *zaid* (spring season) crops is first fortnight of July, second fortnight of October and first fortnight of March, respectively. The seed should be sown at a depth of 4 cm in lines 60 cm apart. It takes about 10-12 days for germination. The crop should be thinned to maintain a plant to plant spacing of 20 cm. Normal recommended seed rate of sunflower is 8-10 kg per hectare. Seed rates of sunflower varieties are 7 kg/ha under rainfed conditions and 6 kg/ha under irrigated, whereas of the hybrids are 5 kg/ha under rainfed conditions and 4 kg/ha under irrigated conditions. **Earthing up** is an important operation in sunflower cultivation. It should be carried out when the crop attains a knee high stage to prevent the crop from lodging which generally occurs at heading stage in high speed windblown areas.

This crop responds well to the application of manures and fertilizers. Application of 25-30 tonnes per hectare of well rotten farmyard manure at the time of field preparation should be incorporated into the field. Generally, sunflower requires 75-80 kg N/ha under irrigated conditions and 50-60 kg N/ha for rainfed crop. Excessive application of nitrogen may reduce the oil content of sunflower and make it more prone towards diseases and lodging. Recommended rate of phosphorus and potash are 60-80 kg P_2O_5/ha and 50-150 kg K_2O/ha which should be provided only after soil analysis. Application of 50 per cent nitrogen along with full dose of phosphorus and potash should be given as basal dose. Remaining 50 per cent of the nitrogen should be given in two split doses, *i.e.* 25 per cent at buttoning and 25 per cent at flowering stage. Beneficial effects of sulphur application have been observed in sunflower. Recommended dose of 37.0 kg/ha is beneficial for qualitative and quantitative oil production in sunflower. As per recommendations of Indian Fertilizer Association in 1992, the crop producing 3.5 tonnes of seeds per hectare requires 131 kg N, 87 kg P_2O_5, 385 kg K_2O and 70 kg MgO per hectare and micronutrients like Fe 732 g, Zn 348 g, Mn 412 g, Cu 59 g and B 396 g/hectare. However, for ornamental crops the doses may be lowered to its half. Sunflower is a good water drinker though it varies from season to season. *Kharif* crop hardly needs any irrigation but if rainfall is uneven, atleast one irrigation per week should be given. For proper germination and good crop stand, pre-sowing irrigation is must for both *rabi* and *zaid* crops. *Rabi* season crop should be irrigated after around 40, 75, 110 days of sowing which will coincide with four to five leaf stage, flowering stage and grain filling stage, respectively. It should be kept in mind that sunflower is highly sensitive to water stress during flowering and grain filling stage, so one irrigation is necessary during this period. Abdou *et al.* (2011) studied the response of sunflower yield with irrigation schedule and found highest seed yield of 2,825 kg/ha by applying irrigation at 1.2 cumulative pan evaporation (C.P.E.), while the lowest seed yield of 1,490 kg/ha was obtained from 0.8 C.P.E.

Good **weed control** should be effected as sunflowers are poor competitors, especially in first four weeks of establishment. Application of Fluchloralin (Basalin) at 1.0-2.0 kg/ha should be applied as pre-plant incorporation (PPI) into soil or sprayed into soil before sowing of a crop followed by slight stirring of surface soil for incorporation into soil to reduce vaporization. Uses of Pendimethalin @ 1.5 kg a.i./ha 1-2 days after sowing (pre-emergence sprays) also show positive results. Weeds can also be effectively controlled by applying Oxadiazon @ 0.5 - 1.0 kg/ha or Alachlor @ 1.0-2.0 kg/ha as pre-emergence sprays. In case no pre-emergence weedicide is used, two hand weedings at monthly intervals should be practised to ensure weed-free conditions up to 60 days after sowing which results in higher yield. **Hoeing** should also be practised as it will not only kill the weeds but also aerate the soil.

Growth and Flowering

Plant growth regulators play an important role on growth and yield attributes of sunflower. Koutroubas *et al.* (2004) studied the effect of PGRs on plant height at various stages of growth. Chlormequat chloride at 3.8 kg/ha reduced plant height at all the three stages under observation. Conversely, mepiquat chloride at 0.025 kg/ha had a significant effect at both flowering (9.2 per cent reduction compared to untreated control) and maturity (11.7 per cent) stages. Split applications at 0.0375+0.0375 kg/ha and 0.05+0.05 kg/ha revealed the shortening effect at maturity (9.5 per cent) and 10 days after transplanting (6.5 per cent), respectively. Paclobutrazol at 0.0125 kg/ha had only a slight reduction effect at maturity (11.4 per cent), while no effect at the other stages was observed. Tahsin and Kolev (2006) investigated the response of sunflower hybrid 'Super Star' to various plant growth regulators, *viz.* H-40 (alfanaftilvinegar acid), M-2 (amid of dicarbonovy acids of benzolovid number), 31 (derivate of naphthalene with auxin efficacy), XP (flalaminova acid) and Agat 25 EK- 2.5 g/da application during pinhead and flowering stage. The results revealed that there was no significant difference between the seed yield (kg/da) and oil content (l/da) in pinhead period by the use of plant growth regulator 31 (1 cm^3/l water) and Agat 25 EK (2.5g/da). During the flowering period, the seed yield was increased by 15.3 per cent and oil content by 16.4 per cent with Agat 25 EK but it was decreased by 18.6 per cent by application of 31.

Postharvest of the Blooms

They are harvested early in the morning when outer florets in the heads have opened. If humidity on the plants is more at the time of cutting, better these should be harvested in the evening. Immediately after harvesting the cut-ends of the cut flowers should be put in distilled or mineral water. As a cut flower these last from one to two weeks depending upon the cultivars, growing conditions and prevailing weather conditions. Quality sunflowers are of uniform maturity and free from blemishes and any other defect. They should have straight stems and good quality foliage. Smaller flowered cultivars should be bunched in 10's or 12's but large flowered are packed individually. Pre-treatment of sunflower includes treating with detergent like Tween-20, Trilon X-100, dishwashing detergent in clean water for 15 to 30 minutes which avoids the premature wilting of sunflower in the vase. After that sunflower can be safely stored at 0-1° C and packed in standard horizontal flower boxes. *Helianthus* are also gravity-sensitive as horizontal packaging may permanently bend the flower heads upwards due to accumulation of auxins in the lower side of the stem which decreases the beauty of the flower stem. It is important to maintain cool temperature during transportation and storage of cut flowers. As the loose flowers also this is an important crop, and the ones only with smaller heads are used for the purpose. After harvesting these are filled in the baskets and taken to the market. Packing in gunny bags damages the flowers and these degenerate soon. If need be these may be cold-stored at 5-7 °C for 3-4 days. These as cut flowers are used for worshipping and for making garland.

Seed Harvesting, Curing and Value addition

When black flowerheads of sunflower become yellow, it is said to be physiologically mature. Maturation of crop takes about four months after sowing. The seeds are ready to **harvest** when colour of head turns brown on the back or when involucral bracts turn yellow and loosening of seeds takes place. High moisture content at time of harvesting reduces seed shattering and saves the plants from lodging. Moisture in the sunflower seeds should not exceed 20 per cent at the time of harvesting. All the heads are not ready for harvesting at a time, therefore, harvesting is carried out twice to thrice to avoid seed shattering. The harvested heads should be spread on the floor and **sun-dried**, followed by their threshing with a small stick. In commercial production, threshers can also be used after speed adjustment in the thresher. Cold air treatment can be used for reducing the moisture up to 15 per cent and can be said as the best method of drying. However, the moisture percentage of seeds should be brought down to 10 per cent before storage or oil extraction. If seed has to be used for long term storage then reduction of moisture to 9 per cent is necessary. On an average, one can get seed yield up to 20 q/ha by following improved package of practises. In case when the crop is harvested for fodder use or for making silage, it is done when the crop is in flowering.

Its flower petals are also used for **value addition**. Lutein extracted from the flower petals of sunflower is the best source of carotene for colouring food stuffs. Essential oil is also extracted from sunflower (*H. annus*) which is a potential source of protein in poultry diets. After extraction of oil from sunflower seeds, they are passed through a 1.4 mm screen which produces low-fibre sunflower residues and results in a feed that improves poultry performance (Mupeta *et al.*, 2001). North Dakota State University has done research on purple hulled sunflower which gave red food colorant extract which can be used for colouring the food stuff or as a dye. Since the petals of the sunflower are rich in coloured source compound, so an attempt has also been made to study its ability to quench reactive oxygen species. The petals from their flowers add bright colour and a spicy tang to tossed salads. The petals are chopped to make a tangy garnish for boiled eggs, steamed vegetables or fish dishes. Sunflowers show high tolerance to heavy metals, therefore, are used in phytoremidation studies (Pilon-Smits, 2005). It is also having the potential of removing heavy metals such as Zn or Cu as well as several radio nuclides from contaminated environment (Nehnovajova *et al.*, 2005).

Insect-Pests and Diseases

Sunflower crop is susceptible to various insect-pests and diseases which cause decrease in yield qualitatively and quantitatively. Various insect-pests that attack the crops are elaborated herewith. **Head borer** (*Helicoverpa armigera*) is a polyphagous insect and a severe pest of sunflower responsible for causing 20-25 per cent loss in yield under normal conditions. However, sometimes the damage is very severe and loss goes up to 40-70 per cent. Newly hatched and grown up larvae feed on leaves, buds and flowers for a short period of time and after making a hole in the disc may enter in it to feed the developing seed. The full grown larvae are greenish in colour and about 3.5 cm long. Deep ploughing of the field is helpful to kill the hibernating larvae. Spray of nuclear polyhedrosis virus (HaNPV) @ 350 LE (larval equivalent)/ha for the control of 1st and 2nd instar larvae and release of egg parasitoid, *Trichogramma* sp @ 50,000 adults/ha at weekly interval will keep the head borer at bay. **Hairy caterpillars** (*Spilosoma obliqua*) are polyphagous pests found throughout the year. Among various hairy caterpillars, Bihar hairy caterpillar is one of the most devastating causing severe damage to the sunflower crop. The female lays eggs in cluster on the lower surface of leaves. After hatching, the tiny larvae up to second instar feed gregariously on the chlorophyll content of the leaves. The infested leaves look like a dirty paper, which can be recognized from a distance. The eggs which are laid in cluster should be collected and destroyed manually and application of *Bacillus thuringiensis* (Bt) @ 1.0 kg/ha has been found effective in controlling hairy caterpillars. They can also be controlled by spraying of Quinalpos 25 EC @ 1.5ml/l or Chlorpyriphos 20 EC 1.0ml/l. The nymphs and adults of **jassids** (*Amrasca biguttula*) are important pests of sunflower as they suck cell sap. They are small, greenish-yellow and very active and can be seen in clusters on the lower surface of the leaves. The symptom is characterized by typical yellowish-white spots on the leaves, shrivelling of seeds and the drastic reduction in oil content. Balanced dose of fertilizers should be applied as excess nitrogen makes the plants more sappy and susceptible. Spraying of 5 per cent ***neem* seed kernel extract** protects the crop from insect damage and application of oxydemeton methyl 25 EC or Dimethoate 30 EC @ 1.0 ml/l water controls almost all the pests effectively. **Cut worm** (*Agrotis ipsilon*) is also a polyphagous pest causing serious problem with sunflower cultivation. Greasy brown coloured larvae feed on the young roots and basal portion of the plant below the ground and kill the plant by cutting at the base. Deep ploughing of the field should be done after harvesting of the crop in order to expose the pupal stage of pest so that these are destroyed being exposed from their safe places to open environments immaturely where these may be eaten by various insects and birds. Treating the seeds with Chlorpyriphos 20 EC @ 12 ml/kg seed will keep these worms away from being swallowed by them after sowing. Ants also feed on the seeds by taking these away to their burrows so the burrows should be dusted with pure and fresh turmeric powder or their accees to the field should be checked by drawing a line of powdered turmeric as a barrier around the field.

The **birds** like parrot, house sparrow and doves eat away seeds which are not fully covered at the time of sowing. They also cause damage to the crop at the time of seed formation. Seed treatment with Chlorpyriphos 20 EC @ 12 ml/kg seed protects the sown seeds from bird damage. Use of bird scarer is very important from the starting of seed development stage to harvest, *vis-à-vis* at the time of sowing.

Various diseases which infect sunflowers are elaborated here under. **Alternaria blight** (*Alternaria helianthi*) causes more than 80 per cent of yield losses under severe epiphytic conditions (Shankergoud *et al.*, 2006). Symptoms of alternaria blight first appear on lower leaves which later spread to middle and upper leaves. Appearance of dark brown irregular leaf spots with very dark border and grey centre are its major symptoms. This disease affects the quality of seeds by reducing the germination percentage and infection of the disease causes losses to an extent of 30 per cent. Till now, no blight-resistant variety is available. Therefore, it becomes inevitable to go for the use of fungicides for the management of this disease. Mesta *et al.* (2011) used Propiconazole and Hexaconazol @ 0.1 per cent sprayings fortnightly from seed germination stage and recorded maximum yield of 10.06 q/ha through Propiconazole, followed by 9.42 q/ha through Hexaconazol. The highest benefit cost ratio was obtained in Hexaconazol, though net returns were highest in Propiconazole.

Botrytis head rot or grey mould (*Botrytis cinerea*) is favoured by root-zone temperature of 15-25 °C and wet conditions. It usually develops on seedlings, stems, leaves and causes major destruction when it infects the head. Symptoms first appear on the back of the head as sunken brown spots. If infection is found in head in the late season, *i.e.* during harvesting period, maximum seed losses take place but both yield and quality get reduced during early infections. As there is no chemical options for controlling *Botrytis* for this crop, application of a desiccant in the crop should be applied when seed moisture reaches about 30 per cent. Zineb 0.2 per cent spraying when its first appearance is visible and then repeated 2-3 more times will control this problem. **Sclerotinia wilt**

and rot (*Sclerotinia sclerotiorum*) causes stem and head rot in sunflower. Stem rot is caused by the fungal threads known as mycelium. This leads to wilting and finally collapsing of the plant. An air-borne spore causes head rot which infects the plants during the growing season and finally results in total destruction of the head and plant. The attack of this disease is related to the number of crops grown in the rotation. If the interval between susceptible crops gets reduced, the risk of infection decreases. Till today there is no chemical control available in sunflower against this problem. However, proper sanitation and crop rotation are only the alternatives left. Among the **viral diseases** affecting the crop, the sunflower necrosis viral disease (SND) caused by 'tobacco streak virus' (TSV) belonging to the *Ilarvirus* class, and bud necrosis caused by 'tospovirus' (Lavanya *et al.,* 2005) are most important. Ravi *et al.* (2001) also reported that SND in India is caused by *Ilarvirus* related to TSV and in their study they observed that the symptoms of this disease comprised of chlorotic & necrotic ringspots, and leaf distortion. They also observed that in electron microscopy *Ilarvirus*-like particles were detected in crude sap of SND-affected sunflower inoculated with leaf extracts prepared from SND-affected sunflower plants. Back transmission to healthy sunflower seedlings with leaf extracts of systemically infected indicator plants resulted in identical symptoms of SND, hence confirming the *Ilar*-like virus as the causative agent of SND. In DAC-ELISA, SNV was detected up to 0.370μg/ml in purified virus preparation and up to 10-4 dilution in leaf sap extracted from virus infected sunflower leaves. In Dot-ELISA, SNV was detected up to 0.0370μg/ml in purified virus preparation and up to 10-5 dilution in leaf sap from infected sunflower leaves. Whereas in DAS-ELISA, the virus was detected up to 370μg/ml and 10-3 dilution in purified virus preparation and in leaf sap extracted from infected sunflower leaves, respectively. Hence, it is concluded that DAC-ELISA and Dot-ELISA were found to be the suitable tests for the detection of sunflower necrosis virus (Kotakadi *et al.*, 2012).

References

Abdou, S.M.M., K.E. Abad El-Latif, R.M.F. Farrag and K.M.R. Yousef, 2011. Response of sunflower yield and water relations to sowing dates and irrigation scheduling under middle Egypt conditions. *Advances in Applied Science Res.*, **2**(3):141-150.

Ahmad, S., M.S. Khan, M.S. Swati, G.S. Shah and I.H. Khalil, 2005. A study on heterosis and inbreeding depression in sunflower (*Helianthus annus* L.). *Songklanakarin. J. Sci. Technol.*, **27**(1): 1-8.

Ali, A., M. Afzal, I. Rasool, S. Hussain and M. Ahmad, 2011. Sunflower (*Helianthus annus* L.) hybrids performance at different plant spacing under agro ecological conditions of Sargodha Pakistan. *International Conference on Food Engineering and Biotechnology IPCBEE* vol. 9 (2011) @ (2011) IACSIT Press, Singapore.

Anonymous, 2009. Sunflower. Annual Report 2009-10, p. 225. All India Coordinated Research Project. Directorate of Oil Seed Research, Rajendranagar, Hyderabad.

Kotakadi, V.S., S.A. Gaddam and D.V.R.S. Gopal, 2012. Serological tests for detection of sunflower necrosis Tospo Virus causing necrosis disease of sunflower (*Helianthus annus* L.). *The Bioscan.*, **7**(3): 543-545.

Koutroubas, S.D., G. Vassiliou, S. Fotiadis and C. Alexoudis, 2004. Response of sunflower to plant growth regulators "New directions for a diverse plant", *Proceedings of the 4th International Crop Science Congress*, Brisbane, Australia, 26 September-1 October, 2004.

Lavanya, N., R.M. Ramiah, A. Sankaralingam, P. Renukadevi and R. Velazhahan, 2005. Identification of hosts for ilarvirus associated with sunflower necrosis disease. *Acta Phytopathol. Entomol. Hungarica*, **40**: 31-34.

Mesta, R.K., V.I. Benagi, S. Kulkarni and M.P. Basavarajappa, 2011. Management of alternaria blight of sunflower through fungicides. *Karnataka. J. Agric. Sci.*, **24**(2): 149-152.

Mupeta, B., J. Wood, F. Mandonga and J. Mhlanga, 2001. The performance of indigenous under improved feed management compared with the hybrid chickens in Zimbabwe. DFID project R7524. NR International, Aylesford, Kent, U.K.

Nehnovajova, E., R. Herziq, G. Fedcrer, K.H. Erismann and J.P. Schwitzguebel, 2005. Screening of sunflower cultivars for metal phyto extraction in a contaminated field prior to mutagenesis, *International Journal of Phytoremediation*, **7**: 337-349.

Pilon-Smith, E. 2005. Phytoremediation, *Ann. Rev. Plant. Biol.*, **56**, 15-39. Online at: arjournals. Annual reviews.org.

Pizzetti, I. and H. Cocker, 1968. *Helianthus*. In: *Flowers ~ A Guide for Your Garden*, pp. 557-565. Harry N. Abrams, Inc., Publishers, New York, USA.

Ravi, K.S., A. Buttgereitt, A.S. Kitkaru, S. Deshmukh, D.E. Lesemann and S. Winter, 2001. Sunflower necrosis disease from India is caused by an ilarvirus related to *tobacco streak virus. New Disease Reports*, **3**: 10.

Shankergoud, I., K.G. Parameshwarappa, H.T. Chandranath, P. Katti and R.K. Mesta, 2006. Sunflower and castor research in Karnataka. An overview, *Univ. Agric. Sci.*, Dharwad (India), p.1-41.

Sujatha, M., A.J. Prabakaran, S.L. Diwedi and S. Chandra, 2008. Cytomorphological and molecular diversity in backcross-derived inbred lines of sunflower (*Helianthus annus* L.). *Genome*, **51**: 282-293.

Tahsin, N. and T. Kolev, 2006. Investigation of the effect of some plant growth regulator on sunflower (*Helianthus annus*.L). *Journal of Tekirdag Agricultural Faculty*, **3**(2):229-232.

Tripathi, S. and K.P. Singh, 2013. Hybrid seed production in *H. annus* L. (doi: 10.4172/Scientificreports.645) through male sterility induced by Benzotriazole. **2**: 645.

22

Heliconia (Family: Heliconiaceae, formerly Musaceae)

T. Janakiram, Anuradha Sane and R.L. Misra

[**Common names:** Balisier/Firebird/Large lobster claw/Macaw flower/Wild plantain (*Heliconia bihai*), False plantain, Giant heliconia (*H. caribaea*), Hanging lobster claw (*H. rostrata*), Lobster claw (*H. bihai, H. stricta*), Parakeet flower/Parrot beak/Parrot's flower/Parrot's plantain (*H. psiattacorum*), Small lobster claw (*H. stricta*)]

Introduction and Origin

Heliconia is named so due to Mt. Helicon (in Greece), the seat of the mythological Muses. It may also have been named so by Linnaeus because of its close relationship with *Musa*. The genus is close to *Musa* in many aspects but differs chiefly in having a dry, mostly dehiscing 3-locular, capsular blue fruit containing only three seeds, lamina anatomy and as Musaceae blades have an irregular apex (Triplett and Kirchoff, 1991). It is also close to *Strelitzia* where perianth segments are fused and there are many ovules in each chamber while *Heliconia* perianth segments are only partly fused and each chamber contains a single ovule (Everard and Morley, 1970). *Heliconia* consists of over 250 species and some 350 varieties, majority of these native to Tropical America (Mexico to Central America and northern South America, the Caribbean Islands and some of the South Pacific islands), South East Asia and Polynesia (Everard and Morley, 1970). A nomenclatural study on heliconias based mainly on published articles and books published from 1900 to 2002, revealed 176 species of heliconias occurring in the Neotropical Region, and 6 species occurring in the Pacific Islands, giving a total of 182 species, out of which 94 are from Colombia, 60 from Ecuador, 56 from Panama, 47 from Costa Rica, 37 from Brazil, 32 from Peru, 26 from Venezuela, 22 from Nicaragua, 16 from Guatemala, 15 from Bolivia, 14 from Honduras and Mexico and 13 from Surinam, however, 94 of the 182 species are endemic to certain regions, therefore, it is necessary for botanists to draw out an international and/or national programme for the collection and conservation of its entire germplasm (Castro *et al.*, 2006). Most of the species are from moist or wet regions though some are also from seasonally dry areas. Although most heliconias are reported to thrive in the humid lowland tropics at elevations below 500 m, the greatest number of species is found in middle elevations rain- and cloud-forest habitats. Heliconias have gained importance because of being most attractive of all exotic tropical flowering plants with diverse flower forms and colours, *vis-à-vis* attractive foliage and has good potential as commercial cut flower. It is also used in creating boundaries and screens for privacy and for special mass effects in the gardens. Its planting also checks erosion. In Caribbean Islands and Mexico, it is used in thatching the roofs, leaves being used for wrapping the food, and roots and seeds of certain heliconias in Brazil being used in certain medicines. Sanabria *et al.* (1998) assayed the leaves, spathes, flowers, fruits, seeds, rhizomes and roots of *Heliconia bihai*, *H. latispatha* and *H. psittacorum* collected in Venezuela and recorded interspecific differences in the presence or absence of alkaloids, flavonoids, polyphenols, saponins, tannins, essential oils and anthraquinones. They detected saponins, flavonoids and polyphenols in all the species, tannins, essential oils and anthraquinones in none of the species, and alkaloids in all the organs of *H. bihai* except spathes, in leaves of *H. latispatha* and in leaves and seeds of *H. psittacorum*.

Its brilliant colours and long peduncles, *vis-a-vis* excellent postharvest life make it an outstanding flower for the florist trade. *H. psittacorum* and *H. latispatha* × *H. psittacorum* are excellent for floral arrangements. It is well adapted to all major agro-climatic zones of the country and there are species and cultivars suitable for cultivation up to an elevation of 3,000 to 4,000 metres. Criley (1989) mentions that in Hawaii, heliconia along with alpinia (red ginger) account for about 6 per cent of the floriculture industry, out of which heliconia comprises of about three times ($ 14,00,000 as compared to alpinia's only $485,000). Major heliconia producing nations are Barbados, U.S.A. (Hawaii), Brazil, Venezuela, the Netherlands and Germany. In India, it is West Godavari district of A.P. which produces half of the total heliconias in the country, followed by Karnataka and Kerala.

Botany

In general, there are three types of leaf arrangements in heliconias, *viz.* musoid (growing vertically with long petioled leaves), cannoid (angular growing with short to medium petioles) or zingiberoid (horizontal growing with short petioles) with banana-like foliage and have long-lasting (several days to several months) erect or pendulous terminal inflorescences (Anderson, 1985) composed of 3-30 boat-shaped and colourful bracts (mostly brilliantly coloured in hues of yellow, orange, pink and red or combination of these with variously splashed colours and green markings) arising from a central floral axis, bracts distichous or spiralling, usually large, carinate, waxy or leathery, each enclosing few to many flowers on short or long pedicels. In some species the bracts may be smooth though woody in certain others and arranged spirally on a central connecting stem, *i.e.* rachis. Each bract may hold several true flowers which while opening peep out gently or boldly. In appearance and relationship, its flowers are in between *Musa* and *Strelitzia*. Heliconias are normally large to giant evergreen and herbaceous perennials coming up from rhizomes, clump-forming and with often long stalked, paddle-shaped leaves, erect to pendulous inflorescences bearing colourful boat-shaped bracts mostly distichously arranged, and only the tips of the small and 6-tepalled clustered flowers protrude from the bracts in most species but not the all. Flowers opening after dawn, appearing in clusters normally below the leaves though sometimes in certain species from above the leaves, subtended by bracts, sepals showy, 3 ~ 2 fused to corolla except at apex and 1 adaxial (free), linear, perianth slender, cylindric, upcurved, petals 3, more or less connate, nectary bulbous at the base, stamens 5 (almost fertile) and staminodium 1 (modified into staminode), anthers surrounding the style and dehiscing longitudnally, stigmatic surface remaining receptive only for 4 to 5 hours after flower opening, ovary inferior and 3-locular, each locule with single ovule attached basally, and a dehiscing, 3-loculed and 3-seeded blue or violet drupe in American species (sometimes yellow, orange or red in South Pacific species). Depending on variety, the height varies from 60 cm to 6.0 metres, often with extensive rhizomatous growth. The morphological and anatomical features of the inflorescence bracts studied in *H. rivularis, H. spathocircinata* and *H. velloziana* (subgenus *Heliconia*) and in *H. angusta, H. hirsuta* and *H. subulata* ssp. *gracilis* (subgenus *Stenochlamys*) revealed that in the subgenus *Heliconia* the bracts are cymbiform and in *Stenochlamys* they are lanceolate-conduplicate, all the species having bracts with uniseriate epidermis possessing thin-walled cells, tetracytical stomata in *H. hirsuta* and *H. subulata* though polycytical in others, the vascular bundles and air canals being distributed in arc-shape and the fibre bundles situated close to the abaxial surface, the bract shape being associated with the distribution of the fibre bundles and the size of the air canals being different in both the studied subgenera, so on these features the identification key was prepared (Simao and Scatena, 2004).

The investigation on anther development, microsporogenesis and microgametogenesis in several species of *Heliconia* revealed that all the studied species present bithecate and tetrasporangiate anthers having fertile pollen grains but with only *H. rivularis* (a natural hybrid) presenting sterile pollen grains of variable size, the anther wall having an uniseriate epidermis and endothecium, the latter with helicoidal thickenings, although some cells of the middle layers also showed thickenings, the biseriate tapetum being of amoeboid non-syncytial type since the tapetum cells did not fuse together forming a true plasmodium, the microsporogenesis successively leading to isobilateral tetrads, the inaperturate pollen grains having a very reduced exine consisting of a thin, more or less continuous layer with small spines, the pollen grain shape being variable among the species, all of them presenting heteropolar pollen, except *H. angusta* with isopolar ones (Simao *et al.,* 2007). Simao *et al.* (2006) stated that *Heliconia* species have an ovule primordium with an outer integument of both dermal and subdermal origin, the archesporial cell is divided into a megasporocyte and a single parietal cell, which in turn is divided only anticlinally to form a single parietal layer, disintegrating later during gametogenesis, the embryo sac is fully developed prior to anthesis, in the developing seed the endosperm is nuclear with wall formation in the globular stage, a nucellar pad is formed during embryo development which later on becomes compressed, the ripe fruit contains seeds enveloped by a lignified endocarp that formed the pyrenes, with each pyrene having an operculum at the basal end, and the embryo was considered to be differentiated, and these all the characteristics are observed with other Zingiberales though the derivation of the operculum from the funicle and the formation of the main mechanical layer by the endocarp are unique to the Heliconiaceae. Out of the six cultivars of *H. psittacorum* selected for studies on pollen formation, pollination and their natural fruit-bearing ability under the tropical climatic conditions, 3 of them (Tay, Andromeda and Lady Di) were partially fertile with a very low rate of fruit set, ranging from 2.8 to 4.7 per cent though were diploid with 2n = 24 chromosomes, however, the process of pollen formation (microsporogenesis) and the pollen appearance were normal, *vis-a-vis* pollen grains in size were uniform, so the cause of poor seed set may have been due to poor pollen germination on the stigmas and not the self-incompatibility or poor pollination, however, other 3 cultivars (Petra, Sassy and Iris) were completely sterile as their pollen grains were of variable sizes and appeared to be abnormally fragmented and over 80 per cent of the pollen grains aborted 1-2 days before pollination because of these being triploid (2n = 3x =36) and so there had been irregular distribution of chromosomes during meiosis in microsporocytes (Lee *et al.,* 1994). Anthers of *H. aurantiaca, H. collinsiana* and *H. psittacorum* var. *subulata* were examined, following critical point drying in the SEM, and recorded calcium oxalate raphides, and long threads (as decay products of the walls separating the pollen chambers and the byproducts of stomium region while rupturing of the anther) within the extruded pollen mass (Rose and Barthlott, 1995). *Phaethornis idaliae* and *Glaucis hirsuta* hummingbirds visit heliconia flowers regularly as observed by Cruz *et al.* (2006) on the two species, *H. laneana* and *H. spathocircinata,* along with certain lepidopterous adults, bugs, bees and certain other insects, particularly *Trigona spinipes* bees, the latter *Heliconia* species inviting a large number of other visitors due to possessing quite large flowers. Some species are pollinated through bats even.

Breeding

While assessing genetic variability of 12 heliconia cultivars (Andromeda, Black Cherry, Deep Orange, Dirooji Red, Gayana, Golden Torch, Lady Di, Parakeet, Pascal, Petra Orange, Sassy and St. Vincent Red) for various characters, Sheela *et al.* (2005, 2006b, 2007) recorded phenotypic coefficient of variation (PCV) higher than genotypic coefficients of variation (GCV) for all the characters indicating role of the environment on expression of the genotype, however, higher PCV and GCV estimates were found for flower yield, bract size and leaf area, while high heritability with high genetic advance for flower yield, leaf area and bract size. Through analysis of variance, they observed significant differences among cultivars for all the studied characters which indicates that there is much scope for improvement in the crop through selection, and as high heritability coupled with high genetic advance was recorded for plant height and stalk length, number of flower shoots, days for flowering, flowering duration, bract size and vase life which indicate the preponderance of additive gene effects in the expression of these characters. The association between various characters and the direction and magnitude of different characters towards number of flowering shoots in 12 genotypes of *Heliconia* revealed that number of flowering shoots per year had a strong positive correlation with vase life, number of flowers per bract and size both at genotypic and phenotypic levels. In another study on genetic variability, heritability and correlations among floral stem characteristics in the cultivars and hybrids of *Heliconia psittacorum,* Costa *et al.* (2006) recorded highest heritability values for inflorescence emergence after shoot formation, stem length, inflorescence length, and stem floral mass while highest genetic coefficient values for inflorescence emergence, stem floral mass, and inflorescence length, and they further observed genetic correlation of inflorescence emergence after shoot formation with number of leaves at inflorescence emergence as 0.52 which may indicate the latter character as a marker for flowering. They observed stem length showing negative genetic correlation with inflorescence emergence after shoot formation and number of days before harvesting the inflorescence after its emergence (-0.72 and 0.81, respectively) which indicate that high stem length is observed in genotypes with less number of days from harvest. Eighteen heliconia cultivars were when studied for correlation studies, Pavan Kumar *et al.* (2011a) recorded higher genotypic correlations than phenotypic correlations with spike length having strong positively significant correlation with plant height, stalk length, leaf width, leaf length and flowers per bract both at genotypic and phenotypic levels. In a further study on genetic variability on 18 heliconia cultivars, Pavan Kumar *et al.* (2011b) recorded a wide range of genetic variability for several morphological and flowering characters in the genotypes. The PCV was recorded higher than GCV for all the characters but for spike length, number of bracts and leaf length, higher PCV and GCV estimates were found.

Kress (1983) in Costa Rica studied artificial hybridization among 14 species of neotropical heliconias where individuals in cultivation were used as parents in crosses primarily between species with pendent inflorescences that are normally distributed allopatrically or among the normally sympatric species with either pendent or erect inflorescences being crossed in their natural habitats, and the observations of pollen tube growth by means of fluorescence microscopy and seed set were used to determine the extent of crossability, and recorded that crossability barriers between the majority of species were strong and foreign pollen tubes were inhibited at the stigmatic surface within the stylar tissue or within the ovary, the site of inhibition being consistent for each pair of species and depended on the parentage and the direction of the cross, though additional isolating mechanisms such as pollinator specificity and phenological separation were present, pre-fertilization crossability barriers acted as the ultimate mechanism to prevent hybridization, and the type of barrier (stigmatic, stylar or ovarian) that existed between two species was not dependent upon the geographical distribution of the parental species or the specific types of pollinators that visit them but in some cases it might indicate taxonomic relationships. Melendez Ackerman *et al.* (2008) found *H. bihai* as self-compatible but various plants of the species showed variation as these were observed full to partially compatible, however, through artificial pollination more pollen tubes were observed growing inside the style.

In a mutational studies, Rodrigues *et al.* (2005) tried *in vitro* ^{60}Co gamma rays at 0, 10, 20, 30, 40 and 50 Gy to the shoot apices of *H. bihai* and they observed LD-50 between 30 and 40 Gy.

Species and Varieties

Heliconia is native of tropical America having more than 250 species mostly native to the New World tropics where they are found in open clearings. *Heliconia* subgenus *Heliconia* consists of four sections ~ section Episcopales (section nov.) with only one species, section Heliconia with 8 species, section Tenebria (section nov.) with 2 species, and section Tortex with 21 species. *Heliconia* subgenus *Taeniostrobus* has been redefined in 1990's with 3 species + 1 provisionally placed species (Sheela, 2006). Two new species, *Heliconia darienensis* and *H. rubigena* have been recorded further. Two new combinations are made, *viz. H. albicosta* and *H. undulata.* Kress (1991) also described new species ~ *H. fredberryana* (growing to 6-7 metres tall), *H. litana* (2 m) and *H. lutheri* (4 m) having colourful pendent inflorescences. A new subsepecies, *H. obscura* ssp. *dichroma* which grows some 4.0-4.5 m tall is also described. Two main types are found ~ erect heliconias with bracts standing straight pointing up and the pendent ones with hanging bracts pointing down. The abundance, population structure, and recruitment of a focal understorey herb (*Heliconia acuminata*) were compared in four secondary forest sites in Amazons (Brazil) which revealed that density of naturally emerging seedlings ranged from 0-6 per quadrat and there was a highly significant difference among the four stands, and thus Bruna and Ribeiro (2005) suggested that secondary forest type may be a poor predictor of its density or population structure, however, the significant differences were found between forest types in seed germination and seedling establishment.

Important species are *Heliconia acuminata*, a native of Brazil, Venezuela and Peru, habit musoid, height 0.5-3.0 m, leaves 15-70 cm, inflorescence 25 cm, erect, red, orange or yellow, bracts distichous, usually alternate on flexuous rachis, 6-25 cm, 4-6, red, orange or yellow, usually with dark green spots towards apex or olive-green with white or pale-orange apices, keeled, triangular in section, fruit 7-9 mm, and its varieties are 'Cheri R', 'Ruby', 'Taruma', 'Yellow Waltz', etc. *H. angusta*, a native of South America (SE Brazil), habit musoid, growing from 1-3 m, leaves 0.25-1.00 m, inflorescence 35 cm and erect, boat-shaped bracts 6-22 cm, 3-10, distant distichous, alternate, lanceolate, yellow or orange to vermilion or scarlet, flowers creamy-white, and its varieties are 'Holiday', 'Orange Christmas', 'Yellow Christmas', etc. *H. aurantiaca*, a native to Mexico to Costa Rica, habit zingiberoid, stems 2 m, inflorescence erect, distichous to loosely spiralled, peduncle 2-12 cm with 1-2 leafy bracts, bracts 4-9 cm, 2-5, broadly carinate, red to orange or yellow, seldom tipped or wholly green, flowers 5-8 per bract, pale-yellow or cream to orange, mostly dark tipped, pedicels red, and fruit roughly 7 mm. *H. bihai* (*H. distans*) having tight and compact growth habit with highly variable bract colour is mostly confused with *H. caribbaea, H. humilis* and *H. wagneriana*, a native of Northern South America and Lesser Antilles, habit musoid, grows some 3-5 m in height, leaf blades 90 cm or more, smooth-textured, oblong-oval and pointed, enormous inflorescence erect and some 60 cm long and 30 cm wide bearing a series of scarlet and yellow spathes (bracts) arranged in two ranks but the flowers are inconspicuous and in whitish shade, flowers greenish and about six together, its forms are *H. b. aureo-striata, H. b. illustris, H. b. illustris* var. *rubricaulis*, etc. and the varieties are 'Arawak', 'Aurea', 'Balisier', 'Banana Split', 'Chocolate Dancer', 'Emerald Forest', 'Five A.M.', 'Giant Lobster Claw', 'Hatchet', 'Jade Forest', 'Kamahamcha', 'Kuma Negro', 'Lobster Claw One', 'Lobster Claw Two', 'Nappi', 'Purple Throat', 'Schaefer', 'Swish', 'Yellow Dancer', 'Cinnamon Twist', etc. *H. caribaea*, a West Indian species (Jamaica and Cuba to St. Vincent), habit musoid, is distinguished by its red spathes which contrasts sharply with its whitish flowers and blue fruits, stems 2.0-5.5 m, leaves 60-130 × 40 cm, inflorescence 20-40 cm, sessile, erect, rachis straight, bracts up to 25 cm, 6-15, distichous in 2 overlapping rows, broadly triangular, red or yellow, keels and apias sometimes green or yellow, flowers 9-12 per bract, sepals straight or slightly curved, base white, apex green and size 4.8-6.0 cm, fruit 1.0-1.4 cm and seeds 8-13 mm in size, and its varieties are 'Barbados Flat', 'Black magic', 'Chartreuse', 'Cream', 'Flash', 'Gold', 'Jacquinii', 'Purpurea', etc. and *H. caribaea* × *H. bihai* varieties are 'Carib Flame', 'Criswick','Grand Etang', 'Green Thumb', 'Jacquinii', 'Kawandri', 'Vermilion Lake', 'Yellow Dolly', etc. *H. collinsiana* (syn. *H. pendula*), a native of Mexico to Nicaragua, habit musoid, bears 72 cm long pendent inflorescence below the leaves, 7-18 spirally arranged yellow-pink to red inside and red to orange-red outside spreading bracts covered with waxy powder which is yellowish towards the tip and 4.6-6.0 cm cream flowers that are 10-20 per bract, and sepals orange to yellow edged red. *H. hirsuta*, a native of Antilles and Belize to Bolivia, N. Paraguay and E. Brazil, zingiberoid, grows from 0.5-5.0 m tall, leaves 15-50 cm, ovate to oblong, petioles to 1.8 cm or absent, inflorescence 5-20 cm, erect and compressed, peduncle 5-25 cm, bracts 3-9 per inflorescence, upcurved, 2-ranked and orange to red, flowers 2-5-5.2 cm on 2-cm pedicel, yellow to orange or red, free sepals and petals have dark spots on apices and fruit 7-10 mm, its varieties are 'Alicia' (bracts orange, 3-6, sepals orange with green banding at apex), 'Costa Flores' (bracts 6-8, red, like lobster's claw, sepals yellow banded black, tip white), 'Halloween' (bracts 6-7, pale-orange to deep peach, sepals orange banded black and tipped white), 'Pancoastal' (red bracts are 6-8, sepals yellow banded black), 'Roberto Burle Mars' (red to tangerine bracts 4-6, sepals pale-yellow banded black but tipped white), 'Trinidad Red' (tangerine to fuschia-pink bracts are 8-10, flowers orange to scarlet, sepals orange darker upward), 'Yellow Panama' (yellow-green bracts 5-10, sepals sulphur-yellow tipped dark blue-green), etc. *H. humilis*, a native of Brazil and Trinidad, leaves shiny green, inflorescence erect with salmon-red boat-shaped bracts, changing to green towards the tip and the flowers greenish-yellow. *H. indica*, a native of Papua New Guinea, musoid, stems up to 5 m, leaves up to 2.5 m, oblong-elliptic, inflorescence erect and 1 m, bracts up to 35 cm, green, distichous, carinate and usually with leafy structures, sepals green and its varieties are 'Eswardus Rex', 'Illustris', 'Rubra', 'Rubricaulis', 'Aureo-Striata', etc. *H. latispatha*, a native of South Mexico to Colombia and Venezuela, musoid, grows up to 3 m in height, leaf blades some 90 cm long and broadly oblong-oval, and sometimes edged red, inflorescence 30-50 cm, erect, rising above the leaves, well separated floral bracts up to 50 cm, 10-20, carinate, orange or sometimes yellow or red at the base, red towards the tip and some 15 cm long which are arranged spirally (but not overlapping) making the inflorescence more attractive with greenish flowers, sepals yellow edged and tipped green, and the var. 'Orange Gyro' is also orange in colour. *H. metallica*, a native to Honduras to NW Colombia and Bolivia, cannoid in habit, is a tufted perennial growing 0.5-4.0 m, leaves to 0.25-1.4 m, oblong-obovate, base cuneate, long stalked, velvety-green above with silver midribs and veins, and metallic-purple beneath, petioles to 28 cm, inflorescence 7-25 cm (peduncle 20-62 cm), erect with tubular, glossy, greenish-white tipped, red flowers subtended by narrow, boat-shaped and 5-14 cm green to red bracts that are 3-8 in an inflorescence, bracts distichous, linear-lanceolate, sepals 3-5 cm, rose-pink to purple tipped white, pedicels to 0.5 cm, white and fruit 8-9 mm. *H.* × *nickeriensis* (*H. psittacorum* × *H.marginata*), found in Guyana and Surinam, musoid, stems 1-4 m, leaves 25-84 cm, petioles 8-60 cm, inflorerscence 8-20 cm, peduncle 25-40 cm, orange, bracts 5-14 cm, 3-7, red with a broad yellow to orange scarious margin, sepals orange to yellow-orange. *H. psittacorum*, a native of Caribbean Islands and the Coast of Guyana, musoid and tends to be a runner, is small (90 cm-2.0 m), floriferous throughout the year under tropical humid climates though it can also be grown in cooler climates, daintier than others, leaves 10-50 cm, long-stalked and lance-shaped, and when in bloom appears as *Strelizia*, petioles 0.2-24 cm, often tinted red, flower heads erect, 7-18 cm (peduncle 6-58 cm) and brilliantly coloured green-tipped orange, red or yellow-green flowers having black spots at the tips, the flowers 3.2-5.4 cm the bracts 4-16 cm, 2-7, pink, orange or orange-red, narrow and glossy, fruit 7.5-10 mm, ovoid, deep metallic-indigo, under this species some more than 1000 distinct variants (ecotypes) have been recorded with prominenat varieties being 'Alan Carter', 'Andromeda' (bracts reddish-orange), 'Blush' (bracts distally light pink and proximally pale-cream-green), 'Choconiana' (orange), 'Fuschia', 'Golden Torch' (probably *H. latispatha* × *H. psittacorum*, orange-yellow), 'Golden Torch Adrian', 'Lady Di' (dark rose-red bracts and cream-yellow sepals, probably most beautiful of other psittacorum cultivars), 'Lilian', 'Nickeriensis' (pink-red and golden-yellow), 'Rosi' (bracts distally rosy-pink and proximally light pink), 'Parakeet', 'Peter Bacon', 'Sassy' (dwarf, bracts pale-green or cream at base and reddish-pink above), 'St. Vincent Red' (deep red bracts), 'Shamrock', 'Strawberries and Cream', 'Tay' (orange-red), 'Tortuga' (fiery orange-red at base and with yellow-gold ends splashed red), etc. and for this species, Geertsen (1988) when studied various species in the greenhouse, he found this species of immense value for cut flower use due to its erect stems, vigorous growth, prolonged flowering season and long lasting inflorescences. *H. rostrata*, a native of Peru to Argentina, musoid in habit, grows to 90 cm to 5.4 m tall, quite hardy, flowers appearing on pseudostems developed the previous year, and blooming throughout the year under full sun to semi-shade conditions, leaves 60-120 × 15-30 cm, bears magnificent pendent inflorescences of 30 to 60 cm having alternating scarlet-red bracts tipped yellowish, and each bract 6-10 cm long, distichous, carinate, base red, apex yellow, margins green and flowers yellow-green. *H. schiedeana*, a native of S. Mexico, musoid, is though quite similar to *H. latispatha* but the inflorescence appears below the leaf blades, stem length 1.5-3.0 m, leaves to 1.5 m, oblong, inflorescence 15-35 cm, erect and red but twisting at maturity, peduncle 5-15 cm, bracts narrower, 6-21 cm, 5-14, reflexed, spiralling red or red-orange with small yellow flowers, sepals 3.0-3.8 cm, yellow-green at apex, pedicels 1-2 cm and fruits 1.0-1.2 cm. *H. spathocircinata*, a native to E. Venezuela, musoid, stems up to 2.0 m, leaves up to 80 cm and oblong, inflorescence red and erect, peduncle stout, bracts 7-15 cm, 8-10, spiralling, red with yellow margins, flowers to 5 cm, 14 per bract, yellow and pedicels 8 cm long. *H. stricta*, a native of S. Venezuela and Surinam to Bolivia and Ecuador, musoid, appearing as lobster claws of various kinds put together and their colour remains as such for weeks together, most suitable for floral arrangements due to its characteristic flat bracts, stems 0.5-4.0 m, leaves 40-150 cm, oblong, petioles 20-75 cm, many variants with various plant sizes having erect inflorescence, size ranging from 12 to 30 cm, *vis-a-vis* colours of green, gold, orange, red, maroon, singly or in combination, bracts 7.7-11.2 cm, 3-10 per inflorescence, distichous, overlapping, carinate, red or orangish, keel and margins yellow tipped green, sepals 4.7 cm, 6-25 per bract, bract mostly flat but in a few even rounded, white or pale-yellow at base and apex ornamented with a green banding, pedicels 0.6-1.2 cm, fruit 11-12 mm, and its most prominent varieties are 'Ali', 'Bucky', 'Carl's Sharonii', 'Catanza', 'Dimples', 'Dorado Gold', 'Dwarf Jamaican' (good as a ground cover, foliage attractive, most suitable for pot culture with 45-90 cm tall plant and 12 cm long rose-coloured inflorescence with pale to deep hues), 'Dwarf Wag', 'Fire Bird', 'Lee Moore', 'Petite', 'Sharonii' (height up to 1 m, stalks red and foliage red-wine, erect inflorescence with red and yellow colouration, can grow up to 80 per cent shade), 'Tagami', etc. *H. wagneriana* (sometimes wrongly called as *H. humilis*), a native of Brazil and Trinidad, musoid, grows some 1.2-4.0 m in height, leaves 1.2-2.0 m, margins undulated, petioles up to 45 cm, inflorescence 20-45 cm, erect and sessile, bracts red and green, 13-14 cm, 6-13, per inflorescence, distichous, overlapping, pink to crimson or orange, keel cream or yellow and edged green, flowers 5.3-7.4 cm, 10-31 per bract, base white, apex dark green, pedicel 9-15 mm and fruit 1.0-1.3 cm, etc. Yamakawa *et al.* (2009) reported three new heliconia cultivars which are 'Kauai Christmas' having musoid habit, leaves glabrous but petioles slightly pubescent, rachis long, erect and white, bracts green at upper margins and the tips with a narrow black border while flowers bright red; 'Kauai Morning Sun' with musoid habit, rachis medium, erect and pale-yellow, bracts compact, relatively non-pubescent, light rosy-red, upper margins and tips green but with a black narrow strip; and 'Kauai Sunset' also has musoid habit bearing compact, erect and light yellow rachis with deep red appearing as deepening sunset and pubescent bracts where upper margins are green but tips below have a flat-black border.

Thirty *Heliconia* spp. were evaluated through the point-scoring system for the purpose of diversifying the number of species available to the local cut flower market on the basis of their plant stand, stem rigidity, flowering period, postharvest durability, handling, packaging and transport, and market value and the most suitable ones scoring more than 35 points in order of merit were *H. orthotricha, H. bihai, H. stricta, H. psittacorum, H. aurea* and *H. wagneriana* while others scoring 25 to 35 points were *H. hirsuta, H. bourgaeana, H. foreroi, H. rauliniana, H. angusta* var. *flava, H. caribaea, H. chartacea, H. latispatha, H. angusta* cv. Holiday, *H. episcopalis, H. rostrata, H. angusta* var. *aurorea, H. sampaloana, H. librata* and *H. Velloziana* (Castro *et al.*, 2006). They further reported that short heliconia types, *viz. H. psittacorum* hybrids and cultivars produce largest number of shoots compared to large genotypes. They had compared *H. psittacorum* cv. 'Red Opal', *H.* × *nickeriensis* and *H. orthotricha* under sun and *H. bihai* cv. 'Nappi Yellow' and *H. stricta* 'Fire Bird'. Criley *et al.* (2003) stated that medium flower size, bold colouration, good keeping quality and the long blooming season characterize the selections of *H. orthotricha*, a species relatively new to the palette of bold tropical cut flowers and when they evaluated four heliconia cultivars continuously for two years on the basis of their growth habits from shoot emergence to flowering and the season of flowering, they observed that time required for flower harvesting from the date of planting was highly variable as 'Garden of Eden' took 125-254 days, 'Candy Cane' 131-248 days, 'Macas Pink' 156-291 days and 'Eden Pink' 170-283 days, and after 248 days of planting, *i.e.* during summer months some 50 per cent of the inflorescences were harvested though each variety continued producing some flowers throughout the year, the number of flowers being produced during the first 12 months was 57 in the foremost variety, 54 in the second and third varieties and 40 in the last variety though, in all, these had produced more than 200 flowers in a 2-year period, however, the flowering took maximum days when shoot emergence happened to be during autumn than those sprouting during spring, and so they recommended these four varieties as the potential export cultivars. Cvs 'Andromeda' and 'Golden Torch' which are commercially propagated through tissue culture or their rhizomes, in outdoor evaluation in Florida, Broschat *et al.* (1984a) stated that both the cultivars generally start flowering in April or May, the peak being between July and September but drastic decline in October and November when temperatures drop below 10 °C, and at 23 °C the vase life they observed in deionized water was 14-17 days. Ramachandrudu and Thangam (2012) evaluated three heliconia varieties, *viz.* 'Choconiana', 'Golden Torch' and 'Lady Di' in Goa under partial shade of coconut garden (45 per cent shade), *vis-a-vis* open conditions and recorded better performance for most of the characters such as plant height, shoots per plant, spikes per clump, stalk girth, earlier blooming, vase life and rhizome yield in open field conditions, and though the varieties 'Golden Torch' and 'Lady Di' were found statistically almost at par but over all performance was recorded better in 'Golden Torch' while quite poor in 'Choconiana'.

Many *Heliconia* spp. are polymorphic with a large number of cultivars. Goh *et al.* (1995a) used RAPD analysis and recorded *H. rostrata* having 3 prominent bands that were absent in all the 19 *H. psittacorum* cultivars using 3 different primers, three cultivars of *H. psittacorum* × *H. spathocircinata* (Golden Torch, Red Torch and Alan Carle) had slightly different RAPD profiles which were distinct from that of *H. psittacorum* × *H. marginata,* and of the 3 triploid cultivars of *H. psittacorum* tested, 'Iris' and 'Petra' had identical profiles with all the primers though a prominent band identified in these two cultivars was absent in 'Sassy', so conclusively it can be said that 'Iris' and 'Petra' are the same genotype. Data from 11 primers suggested that the RAPD technique can be used to distinguish species and cultivars of *Heliconia* through DNA extracted from the leaves, and using a single 10-mer primer (OPA18), distinct RAPD profiles were determined for 16 cultivars of *H. psittacorum* grown at Heliconia Society International Depository, Jurong Bird Park, Singapore (Kumar *et al.,* 1998). The phylogenetic tree derived from the RAPD data showed that all of the 16 cultivars examined are closely related to each other, providing the first genetic evidence that this large group of cultivars has a common genetic background. Moreover, 2 triploid cultivars of H. *psittacorum* (Iris and Petra) showed identical RAPD profiles with 10 different primers in agarose and polyacrylamide gels suggesting that they are of the same genotype. Sheela *et al.* (2006a) analysed 17 *Heliconia* species and varieties through RAPD markers using 8 primers which produced the highest number of bands for DNA amplification. The genetic similarity matrix constructed with Jaccard's coefficient using RAPD marker scores showed that the highest value was between 'Petra Orange' and 'Parakeet' while the lowest was between 'Golden Torch' and *H. humilis.* Seventeen species and varieties of *Heliconia* formed nine distinct clusters at similarity coefficient value of 0.42, implying a strong parallelism between genetic and morphologic/ taxonomic variability. 'Petra Orange', 'Deep Orange', 'Parakeet', 'Pascal', and 'Alan Carle' formed a big cluster within which 'Petra Orange' and 'Parakeet' formed a more cohesive entity.

Propagation

Increased interest in tropical cut flower export in developing nations has increased the demand for clean planting stock. **Seed**-grown plants may not flower true to the type. Seed setting may be poor due to specific character of the species or varieties or due to lack of pollinators (Montgomery, 1986). Study of the morphology of seed and vegetative propagation of *Heliconia velloziana* with a large aerial part revealed that **seeds** take four months to germinate and sometimes even three years and that too with unpredictable germination rates (Criley, 1988; Carle, 1990) and in this case even acid scarification is not effective (Kress and Roesel, 1987). Delay in germination is probably due to the undifferentiated embryo and hard endocarp. Before sowing the seeds, these may be rubbed with sand paper to expose the endosperms so that these may absorb the water for quick germination. While germinating, the seedlings have two first foliar structures as scale leaves and development of the primary root, but adventitious roots being more conspicuous, though vegetative propagation by **rhizomes** showing the faster rate than sexual propagation where roots develop after one month and the young shoots after 4-6 weeks, the young aerial shoots presenting a variable number of cataphylls and two protophylls (Simao and Scatena, 2003) and the plants also appear true to the type. Size and weight of *Heliconia* rhizomes differ from species to species and varieties to varieties. *H. psittacorum* rhizomes weigh 50-70 g while *H. caribaea* and its varieties from 300-700 g so while propagating the size influences the vigour of the plants. However, *Heliconia stricta* cv. 'Dwarf Jamaican' records most irregular pattern of germination (Maciel and Mogollon, 1997). They are propagated through rhizomes, side shoots and the suckers being produced by the mother plant. Both terminal and axillary buds are borne by the rhizomes so while propagating through rhizomes, care should be taken that the bits should contain at least either type of one bud and the division should be done only in early spring. Side shoots and suckers should also be taken out in spring and planted. Collected seeds should, however, be started in late winter to early spring.

Dias and Rodrigues (2001) described about the endophytic contamination through *in vitro* propagation whereas Sato *et al.* (1999) stated the system being more popular due to pathogen-free rapid multiplication of the elite germplasm. Callus initiation in heliconia was carried out using different plant tissues, *viz.* shoot apices, primordial leaves and rhizome sections as explants, *vis-à-vis* zygotic embryos as these are derived from a sterile environment within the ovary walls. Both, shoot apices and primordial leaves produced good callus on MS medium with 2.4-D (0.15 mg/l), NAA (1.96 mg/l) and kinetin (4.3 mg/l) whereas rhizome sections failed to induce callus. However, the calli thus derived remained non-organogenic on the regeneration medium (Siddique and Paswan, 1997). Shoot explants from *Heliconia psittacorum* rhizomes were surface-sterilized and cultured on modified MS medium comprising 0.25 mg IAA/litre and benzyladenine (BAP) at 4, 6, 8 or 10 mg/l and the creamy-white buds turned green within 10-14 days of culturing and subsequently increased in size, best being with 6 mg BAP/l recording 59.0 per cent regeneration though up to 2-3 months shoots did not proliferate, however, in other media the buds remained green, retained their size and turned brown thereafter and died. Treatment with 0.25 mg IAA/l + 2.0 mg BAP/l recorded highest shoot proliferation within 4-5 weeks of culture with a mean shoot length of 5.4 cm having mean number of leaves as 4.2 per culture. Higher rooting with faster root growth was initiated when in the medium, 1.5 mg IAA/l was added (Talukdar *et al.*, 2002). Through terminal shoot tip explants in *H. psittacorum* cv. 'Rhizomatosa', Shiau YihJuh *et al.* (1999) obtained lower rates of contamination after pretreatment of explants in liquid MS medium supplemented with 500 mg/l cefotaxime for 3 days and the well developed shoots were successfully obtained after 45 days of culture when MS medium was supplemented with 4-8 mg/l BA and 0.5 mg/l NAA. When they added 0.01-0.5 mg/l thidiazuron the shoot tip explant growth was found rapid. Bora and Paswan (2003) while culturing *H. psittacorum* axillary buds on MS medium obtained increase of more than three-fold length and double the breadth when medium was supplemented with 2.5 mg BA, 0.2 mg IAA, and 2.5 mg GA_3/l where they also obtained ideal initiation of multiple shoots and their multiplication rates, and the medium supplemented with 0.75 mg IAA and 0.05 mg NAA/l gave maximum rooting response (85 per cent), earliest root initiation (within 21 days), and maximum number of primary (3.70) and secondary (3.90) roots. Goh *et al.* (1995b) with *H. psittacorum* developed a further rapid propagation system using 1 mm thin sections of the base of *in vitro* plantlets on MS medium where they found rooted plantlets in presence of 10 micro M 2,4-D. 2,4-D at 40-80 micro M induced the formation of highly morphogenic callus. When subcultured on basal medium, this callus produced protocorm-like bodies that developed into plantlets. Nathan *et al.* (1992) developed a protocol for *in vitro* propagation of *H. psittacorum* by culture of terminal and axillary buds of rhizomes on modified MS medium containing 40 micro M BA, 150 ml coconut water, 30 g sucrose and 2 g Gelrite/l. Shoot multiplication was achieved in this medium without coconut water but when supplemented with 10 micro M BA. Rooted shoots on MS basal medium were successfully acclimatized to greenhouse conditions. New buds obtained by dark culture from the rhizomes of 10 species, hybrids and cultivars of *Heliconia* were used as explants on MS medium supplemented with 2-10 mg BA l^{-1} and 0.2-1.0 mg NAA l^{-1} where the explants could induce shoots and proliferate, and the best results for rooting was observed on MS medium supplemented with 0.5 mg NAA l^{-1} + 0.5 per cent active carbon (Zeng SongJun *et al.*, 2004). In *Heliconia rostrata*, Torres *et al.* (2005) succeeded in micropropagation through embryo excision from mature fruits and culturing in basal medium with MS salts, vitamins and sucrose (1, 2 and 3 per cent w/v sucrose essential for embryo development) as without sucrose the embryos die. In embryogenic callus of *Heliconia chartacea* cvs 'Sexy Pink' and 'Sexy Scarlet' from ovary sections, Ulisses *et al.* (2007) developed a protocol for *in vitro* culturing in MS medium supplemented with IAA 1 mg/l + 2,4-D 10 mg/l and IAA 2 mg/l + 2, 4-D 5 mg/l, getting 100 per cent callus formation though callus formation in former cultivar was of higher frequency. Ulisses *et al.* (2010) cultured embryos derived from mature fruits of *H. bihai* after inoculation in ½ MS medium + carbohydrates 30 g/l and obtained differentiation 15 days after inoculation and these were acclimatized in a washed sand substrate under 80 per cent shadenet.

Zeng SongJun *et al.* (2004) transferred the tissue-cultured heliconia plantlets to pots containing sand/turf mixture (1:1), sand/perlite (1:1) or only sand with a survival rate of more than 90 per cent, and out of three organic substrates [carbonized rice hull vs. dry or green coir dust, green vs. dry coir dust, and two fertilizers, *viz.* Vitasolo vs. earthworm refuse at 3:1 (v/v)] on the acclimatization of plantlets of *Heliconia bihai* obtained by *in vitro* culture were studied, and when after 75 days, plant height, pseudostem diameter, leaf number, and third leaf area were evaluated, it was found that carbonized rice hull had a more beneficial effect than coir dust (green or dry), however, the earthworm refuse was more efficient in the enhancement of plant growth than Vitasolo.

Of *Heliconia bihai* var. 'Lobster Claw I', 200 plants obtained from rhizomes, and from *in vitro* multiplication were planted in alternate rows (20 plants per row) in the field. Evaluation was carried out by the number of suckers, time for flowering initiation and cut flower productivity. Flowering was initially observed after seven months in the field in rhizome-derived plants but the number of suckers, however, was higher in micropropagated plants resulting in higher productivity with better quality of flowers (Rodrigues *et al.*, 2006).

The presence of endophytic microorganisms such as *Pseudomonas* sp. and *Klebisiella* sp. are stumbling block for *in vitro* propagation of *Heliconia* species, hybrids and cultivars and when stem apices of *H. rauliniana* were *in vitro* cultured the presence of their infection was confirmed, so for their control antibiotics such as chloramphenicol, cefotaxime and the association of both were tested by applying in MS medium supplemented with 3.5 mg/l 6-BA where after a 50 day-period 500 mg/l cefotaxime was found most effictive against the infection though chloramphenicol despite being effective proved, in fact, cytotoxic to the explants (Rodrigues, 2005). Micro-organism contamination was obscured in shoot tip explants when cultured in liquid MS medium supplemented with 4-8 mg/l BA, 0.5 mg/l NAA and 500 mg/l cefotaxime, and after 45 days of culturing, well developed shoots were successfully obtained, however, addition of 0.01-0.5 mg/l thidiazuron accelerated the growth process (Shiau YihJuh *et al.*, 1999). Heliconia plantlets obtained by *in vitro* culturing on MS medium were inoculated with three VAM (*Glomus clarum, G. etunicatum* and *Gigaspora margarita*) species, alone and in combination, the plantlets were acclimatized in a mixture of soil:sand:'torta de filtro' in the ratio of 3-2-5, and at 90 days of inoculation the fresh and dry weights of the aerial and root parts were recorded and percentage of mycorrhizal colonization of roots was determined but no effect on heliconia was observed despite high root colonization (Sato *et al.*, 1999).

Cultural Practices

It can be grown successfully from full sun to 80 per cent shade (depending upon the species, hybrids and cultivars) under humid climates where **temperature** does not persist for longer below 18 °C, however, during summer the temperature may rise up to 35 °C and if the atmosphere remains humid there is no danger of desiccation even at higher temperatures. Geertsen (1989) found increase in number of flowers from 25 to 60 when temperature was increased from 15 to 21 °C in *H. psittacorum* 'Tay' and the increase in growing temperatures increased overall growth rate in *H. aurantiaca, H. psittacorum* and its var. 'Golden Torch' (Criley, 1989; Geertsen, 1989) though decrease in minimum temperature from 15 °C to 10 °C, growth and flower production had been poor *vis-à-vis* appearance of the black spots on rachis at bract emergence point occurred due to cold injury. However, only a very few species survive 10 °C and below temperatures. There is report that most heliconias produce more flowers under full **sun** than under partial shade as these have been found growing naturally in bright clearings, except *H. pendula* and *H. stricta* var. 'Carli's Sharonii' which are shade loving; *H. aurantiaca, H. stricta* 'Dwarf Jamaican' and *H. wageneriana* and its var. 'Turbo' initiate flowering under short days (Criley and Kawabata, 1986; Criley and Lekawatana, 1990); though species such as *H. bihai, H. chartacea, H. episcopalism, H. hirsuta, H. nickeriensis, H. psittacorum* and its var. 'Golden Torch', and certain cultivars of *H. stricta* are considered day-neutral or non-photoperiodic as these flower all the year round under suitable light intensisties (Broschat *et al.*, 1984c). However, a 6 hours of morning sunlight should certainly be provided for better results. Three parts compost, two parts fibrous loam, two parts soil and one part each of leafmould, peat and sand is the potting medium or **substrate**. For field planting the soil rich in organic matter is suitable for its cultivation. Four kilogrammes of well-rotten farmyard manure per square metre at the time of soil preparation and 25 g each of N, P_2O_5 and K_2O at the time of final bed preparation should be incorporated in the soil, and when shoots emerge the crop is top-dressed with 25 g nitrogen per square metre. Since it flowers throughout the year, especially from April to December in previously planted crop so it requires regular watering during active growth and flowering period, however, in the first year of its planting it may flower from September to December only when planted in January. The study on the effects of culture medium (field soil, yellow soil + peat + mushroom compost, yellow soil + mushroom compost + sand, and yellow soil + peat) on growth, cut flower yield, leaf structure and ultramicroscopic structure of *H. psittacorum* revealed that it produced green leaves with normally developed cells and high cut flower yield when grown in yellow soil + peat. The mixture of yellow soil + peat + mushroom compost or yellow soil + mushroom compost + sand inhibited the growth of the plants and gave plants with leaves exhibiting chlorosis, densely arranged cells and damaged chloroplasts (Chen GuoJu *et al.*, 2005).

When *H. stricta* and *H. rauliniana* were tried outdoors at **spacings** of 0.5 × 1.0 m, 1.0 × 1.0 m, 1.5 × 1.0 m or 2.0 × 1.0 m, Ibiapaba *et al.* (1997) recorded first inflorescence emergence at 180 days under each spacing in both the varieties but the number of inflorescences per plant was highest at 0.5 × 1.0 m (8.6 in *H. stricta* and 1.5 in *H. rauliniana*), stem lengths being commercially acceptable in *H. stricta* at 0.5 x 1.0 m and 1.0 x 1.0 m spacings and inflorescence size at all spacings while stem lengths and inflorescence size both were commercially acceptable at all spacings in the second variety but plants at 1.5 × 1.0 m and 2.0 × 1.0 m had high number of discoloured inflorescences. Lalrinawani and Talukdar (1999) when studied the rhizome size (10, 20, 30 or 40 g) of *H. psittacorum* with 20 × 20 and 40 × 40 cm spacings (traditionally being planted at 40 × 40 cm), they obtained highest values for rhizome number and yield (weight) per clump, girth and spread though per square metre the number and yield of rhizomes were found much better in closer spacings. The study on the effect of two spacings (20 and 40 cm) and four rhizome sizes (10, 20, 30 and 40 g per bit) on growth, number of inflorescences/m^2, and other flower characters of *H. psittacorum* revealed that wider spacing, and also the increasing rhizome sizes yielded taller plants with larger inflorescence, maximum (116.80 cm) being under 40 g size compared to 91.77 cm under 10 g, and so had been the case even with the number of shoots and inflorescences per clump though with reduced inflorescence size, however, smallest rhizome produced the largest inflorescence of 18.56 cm size (Lalrinawani and Talukdar, 2000). *H. psittacorum* cv. 'Sassy' and 'Andromeda' when grown at spacings of 0.50 × 0.25, 0.50 × 0.50, 0.50 × 0.75 or 0.50 × 1.00 m in Brazil, Ibiapaba *et al.* (2000) recorded first shoot emergence under all the spacings 20-30 days after transplanting and the first inflorescences after about 120 days with lasting ability of about 14 days though both the cultivars produced largest number of buds and inflorescences at a spacing of 0.50 × 0.25 m, more being in 'Sassy', however, 'Andromeda' even after showing best floral stem length under 0.50 × 0.50 m spacing, no treatment affected inflorescence length or width in any studied variety.

Feeding, Watering and Humidity

Though water-logged conditions are at all not suitable for heliconias, but they perform well under conditions where atmosphere remains humid throughout. During active growth and flowering it should be watered sufficiently but during off season though soil or medium should be moist but watering is not required. The study of *H. psittacorum* × *H. spathocircinata* cv. 'Alan Carle' in a greenhouse to estimate the evapotranspiration through use of soil water balance method, *vis-à-vis* class A pan evaporation to estimate the reference evapotranspiration and based on these data determination of the crop coefficient for each crop development stage revealed that crop evapotranspiration average was 2.2 mm/day, the crop coefficients from evapotranspiration in greenhouse were higher than those from the estimated agrometeorological station situated nearby (Gomes *et al.*, 2006).

If at the time of soil preparation for planting, the soil has been incorporated with FYM and NPK fertilizers as described under **cultural practices**, there is no further need for any other fertilizer afterwards for whole of the year but the same quantity of FYM and fertilizers are to be applied year after year at the start of the crop. The works carried out on the crop on its nutritional requirement is therefore being elaborated herein. Finger and Barbosa (2006) studied morphological and physiological characteristics as well as postharvest longevity of the first floral stems under macronutrient omissions in *Heliconia psittacorum* × *H. spathocircinata* cv. 'Golden Torch' under greenhouse conditions and recorded that inflorescences under N omission showed a pale orange colour and floral stem deformation; N, P or K omission reduced stem length and diameter as well as inflorescence length; N or K omission reduced floral stem dry matter and postharvest longevity. In an outdoor trial at Florida with *H. latispatha* × *H. psittacorum* cv. 'Golden Torch', Broschat *et al.* (1984b,c) recorded flower production peaking from July to September under full sun (at a temperature range of 21-35 °C) with an average yield of 84 flowers per square metre per year with application of 3.6 kg 18-6-12 NPK/m^2, however, the flower production was found reduced to its half when the crop was grown under 63 per cent shade. In *H. psittacorum* cv. 'Andromeda', the flower production peaked from July to September with 130 flowers/m^2 in the first year and 160 the second year under full sun and at the same temperature range while under 63 per cent shade there was reduction in flower yield to about 35/m^2 per year in the first year and to 65 to 75/m^2 per year in the second year. *Heliconia psittacorum* cv. 'St. Vincent Red' and *H. psittacorum* × *H. spathocircinata* cv. 'Golden Torch' in an outdoor study were supplied with N (200, 400 and 600 kg/ha), P (80, 169 and 240 kg/ha) and K (80, 160 and 240 kg/ha) fertilizers, where Ferreira and Oliveira, (2003) recorded increased pseudostem length, floral stem length and diameter, and flower production in 'Golden Torch', and flower production in 'St. Vincent Red' with increasing rates of N, but increasing rates of P and K resulted in decreasing values of the parameters measured, optimum rates for flower production being 600:160:160 and 400:240:160 kg/ha NPK, respectively, for both the cultivars. Containerized plants of *H. psittacorum x H. spathocircinata* 'Golden Torch' were grown in the greenhouse for eight months from early summer to winter under selected combinations of N, P and K ranging from 0 to 1.5-2.0 times of the optimum doses as reported by Clemens and Morton (1999). High N, NP and NK ratios maximized biomass variables (vegetative and inflorescence dry weight, and leaf area), though the number of shoots and flowers produced per rhizome was maximum at lower NK ratios, however, heavier the rhizome planted the shorter the time for shoot emergence and flowering occurred, *vis-à-vis* greater the number of flowers harvested. However, rhizome weight had no effect on number of shoots to emerge (Clemens and Morton, 1999).

Control of Growth and Flowering

The natural flowering season for *Heliconia* species in their natural habitats is sometimes indicated in the taxonomic literature, but it may be influenced locally by rainfall and drought periods as well as by photoperiod and may not be reliable in indicating production periods elsewhere. Wild-collected plants are often introduced with little or no information on their season of flowering. With more than three dozen species of *Heliconia* grown and exported in international trade, the seasonality of flowering is important to the supply and marketing of this bold tropical flower. In Hawaii, sales records from several commercial growers were collected over several years to identify the peak seasons of flowering as well as the duration of the flowering seasons. Such information is helpful in coordinating with the flower markets. *Heliconia* species of commercial interest with strong seasonal flowering periods are *H. angusta*, *H. bihai*, *H. caribaea*, *H. caribaea* × *H. bihai*, *H. collinsiana*, *H. lingulata*, *H. rostrata*, *H. stricta* and *H. wagneriana* as well as species with longer flowering periods such as *Psittacorum* cultivars and hybrids (Criley, 2000).

Catley and Brooking (1996) studied the flowering responses of *H. psittacorum* × *H. spathocircinata* cv. 'Golden Torch' at temperatures of 32 °C day/20 °C night, and 24 °C day/20 °C night and photosynthetic photon flux (PPF) 475 and 710 micro mol m^{-2} s^{-1} under controlled environmental conditions and recorded that temperature did not have any significant effect on new shoot production (with an average of 9.3 shoots per plant being produced over the period of 248 days of treatment) though more shoots were produced at the higher PPF level (10.1 shoots, compared with 8.3 shoots at the lower light level), and the proportion of shoots that initiated flowers (85 per cent) was almost the same in all the treatments. The time from shoot emergence to inflorescence emergence was significantly shorter at 32/20 °C than at 24/20 °C (140 and 146 days, respectively) though was unaffected by PPF, however, second shoots to emerge had the shortest duration from shoot emergence to inflorescence emergence. Higher temperature caused increase in the number of leaves subtending the inflorescence and decreased as shoot order increased but was unaffected by PPF though levels of temperature and PPF influenced total leaf area at flowering, with highest areas being achieved in the high temperature and low PPF combination. Acceptable flower quality with at least two well-formed and properly coloured bracts opened was obtained in all the treatments though with taller and thicker flower stems at 32/20 °C and these dimensions increased further with increasing order of shoot appearance, however, stems were thinner at the lower PPF level. Though temperature, *vis-à-vis* developmental factors associated with order of shoot appearance played a more dominant role than light in influencing production and quality of flowers, and the cut flower production may be possible in cv. 'Golden Torch' in temperature controlled greenhouses in temperate regions with mean air temperature of 20 °C. The leaves of *Heliconia* cv. 'Golden Torch' grown in full sunlight or full sunlight + low levels of fertilizers than being grown under shade caused reduction in photosynthesis and chlorophyll content while plants grown at high nutrient levels showed higher values for all these parameters and when plants were grown under intermediate or deep shade, there was no significant difference in any of the parameters irrespective of nutrient supply, however, in the recovery experiments, when the plants without fertilizer were refertilized weekly, photosynthetic rates and chlorophyll content were found increased maximally but gradually after growing them in full sunlight though no significant changes in these parameters were observed when plants were grown under intermediate or deep shade over the same period and photosynthetic rates and chlorophyll content showed a clear linear correlation with total leaf N in plants grown in full sunlight while there was no clarity under intermediate or deep shade suggesting that acclimatization of *Heliconia* under full sunlight could be achieved by high nutrient levels (Jie He *et al.*, 2000).

Although *Heliconia* is a tropical genus, many species exhibiting highly seasonal patterns of flowering, which was being thought due to seasonal rainfall but studies reveal it otherwise that *H. wagneriana* and *H. stricta* cv. 'Dwarf Jamaican' are short day species, the latter initiating its flowers after its pseudostem attains three unfurled leaf blades and it reaches the anthesis in 15-19 weeks after start of SD, while *H. angusta* the long day species requiring above 13.3 h of daylengths and then 15-17 weeks such daylengths for anthesis, and such studies may be used to manipulate the time of flowering than the usual flowering time (Criley *et al.*, 1999). *H. wagneriana* being a SD species with natural floral initiation in Hawaii usually being in October but through artificial short (8-9 h) and long days treatments looking into the market demand the flowering can be manipulated as 120-150 days are required from the onset of inductive SD to inflorescence emergence. Grower shipping records have been used to identify other species with marked seasonality. Influence of habitat on physiological and structural characteristics was investigated in *H. imbricata*, *H. latispatha*, *H. pogonantha* and *H. wagneriana* occurring in open sites receiving full sunlight, *H. mathiasiae* found at the forest edge under partial sunlight, and *H. irrasa* ssp. *undulata* and *H. umbrophila* occurring in the forest understorey under deep shade where light response curves showed a clear gradient with respect to light saturation and rates of maximum net assimilation (Amax) as *H. latispatha* showed saturation at higher photon flux densities (PFD, 1400 micro mol m^{-2} s^{-1}) and higher Amax (14-16 micro mol m^{-2} s^{-1}) than *H. mathiasiae* (PFD 1000 micro mol m^{-2} s^{-1}; Amax 7.5-8.5 micro mol m^{-2} s^{-1}) and *H. irrasa* (PFD 250 micro mol m^{-2} s^{-1}; Amax 3.5 mol m^{-2} s^{-1}), however, leaf blade areas were greatest in open sites coupled with more leaf thickening due to more chlorenchyma, *vis-à-vis* significantly higher leaf specific mass, but leaf support efficiency was highest in understorey species whereas deep-shade species had very thin leaves and low stomatal densities (Rundel *et al.*, 1998). Year-round flowering of *H. chartacea* is potentially possible in Hawaii as new shoots develop regularly though there is poor flowering from late-March to early-June because the shoots which should flower at this time are produced 40-48 weeks earlier in June-August and for floral initiation there should be at least four unfurled leaves (2-3 still within the enclosing pseudostem) and so the flower initiation is delayed to October-December time-frame for these shoots. In contrast, autumn-winter shoots which bear normally one leaf less than summer shoots, initiate flowers during January-March which suggests that greater leaf count and low flowering percentage of spring shoots are not favourable for floral initiation but autumn shoots with less leaves initiate flowering, though the attempt to promote flower initiation during autumn with light-break lighting is also thwarted by *Phytophthora* disease. Growth and development of *H. bihai* cv. 'Lobster Claw I' and *H. latispatha* were studied under three shade levels (0, 40 or 60 per cent) for 20 months by Maciel *et al.* (1994) during which five generations of shoots developed where plants without shade produced maximum pseudostems though number per clump being greater in *H. bihai* than *H. latispatha* and in the first four generations of all the plants the abortive and generative shoots developed though *H. latispatha* showing a higher percentage of aborted shoots than the other species, however, in successive generations the increase in the shade levels increased the height of generative shoots but with decreased number of leaves. In *H. bihai* the first three generations flowered simultaneously after the clumps attained the age of 12 months and continued the flowering for six months with a peak during March-June when 95 per cent of the flowers developed, however, in *H. latispatha* the flowering began after the clumps were 10 months old and showed an irregular pattern during the cycle with the peak (82 per cent) occurring during July-August. The generative stage could begin at the apex when the number of unfurled leaves on the pseudostem was 4-5 in *H. bihai* and 5 in *H. latispatha* (Maciel and Rojas, 1994).

When plants of *H. stricta* cv. 'Dwarf Jamaican' were grown for one year in 10-litre containers under full sun produced more inflorescences than those grown under 50 per cent shade though these shaded plants treated with DCPTA (3,4-dichlorophenoxy triethylamine) produced more and taller pseudostem, and the plants of *H. caribaea* cv. 'Purpurea' when grown in the open field for 2.5 years after soaking the rhizomes for 1 h before planting or spraying the plants at 2-leaf stage with 30 micro M DCPTA, the plants produced more pseudostems during first year, but differences in the number of pseudostems and inflorescences during subsequent years were not observed significant (Broschat and Svenson, 1994).

Postharvest

The flowers are harvested early in the morning, *i.e.* before 8 A.M., near the ground with peduncles of 70-100 cm length, at desired stage of opening, *i.e.* semi-mature stage since further opening is little. The yield of the stems is 60 to 120 stems per year per square metre of area. At Gujarat conditions when the spikes of the heliconia var. 'Golden Torch' were harvested at unopened stage, 1 bract open stage, 2 bracts open stage, 3 bracts open stage and 4 bracts open stage, Mangave *et al.* (2011) recorded best postharvest life under 3 bract open stage. Inflorescences last longer if harvested from well irrigated plants than from plants under stress. Harvesting at mid-day hampers the cut flower life by about one week because after harvesting these do not take water well and desiccation is fast. Immediately after cutting the cut ends are placed in water and after every 2-3 days these cut ends are recut for proper absorption of water or solution. Finger and Barbosa (2006) stated that for most flowers the xylem blockage of a cut flower is reduced by recutting the stem, reducing the pH of the vase water to 4 and by applying antibacterial substances to the water. The use of sucrose in pulsing solution or as constituent of vase solution may extend the vase life of the flowers by improving the water balance, stimulating flower opening or by delaying the senescence due to lower synthesis of ethylene, as observed in carnation. Flowers can be classified as sensitive, low-sensitive or ethylene-insensitive based on the responses to the growth regulator. Cytokinins and gibberellic acid may have positive effect on vase life of some flowers, specially in those with low or ethylene-insensitive ones like *Heliconia* sp. The reduction of temperature drastically reduces CO_2 and ethylene production as well as its action. However, chilling-sensitive flowers such as bird-of-paradise and *Heliconia* cannot be stored below 10-13 °C due to intense tissue discolouration.

After trimming, the whole inflorescence is given an insecticide dip (5 minutes) and subsequently rinsed. Manual cleaning is necessary on some inflorescence to dislodge insect and remove dead flowers. The stalks are held in water to prevent drying of inflorescence before packing. Heliconias (cvs Andromeda, Golden Torch, St. Vincent Red, etc.) have a postharvest life of 14 to 15 days in deionized water at 23 °C and flowers are damaged when stored at temperatures below 10 °C. However, Zimmer and Carow (1977) reported that heliconia inflorescences when kept at temperatures at 0-12 °C under reduced pressure, all the species stored well at least up to four weeks. Bredmose (1987) reported that without additives, heliconia inflorescences lasted for 2-4 weeks, however, cold temperature hampered the life. BA delays bract darkening and abscission and increases the vase life from 1.2 to 1.7 times (Whittaker, 1993).When BA at 100 mg/l was applied as spray or dip, the cut flower life in case of *H. chartacea* cvs 'Sexy Pink' and 'Sexy Red' as well as in *H. psittacorum* cv. 'Andromeda' was found increased (Paull and Chantrachit, 2001). BA treatment gives 2.4 times increase in vase life of *Heliconia* 'Andromeda' leaves and flowers. Mangave *et al.* (2011) when sprayed the cut inflorescences of heliconia 'Golden Torch' immediately after harvesting in the laboratory and again after two days with sucrose 3 per cent, sucrose 3 per cent + 8-HQC at 250 mg l^{-1}, GA_3 at 100 and 50 mg l^{-1}, BA at 100 and 50 mg l^{-1}, BSA at 50 mg l^{-1}, alar at 50 mg l^{-1} and $KMnO_4$ at 50 mg l^{-1}, they recorded GA_3 at 50 mg l^{-1} as the most effective chemical for maintaining TSS, higher carotene pigment (carotene) retention, lower lipid peroxidation, low electrolytic leakage and highest vase life. Mangave *et al.* (2014) tried 8-HQC, sucrose, calcium chloride, α-lipoic acid, sodium benzoate, spermine, citric acid and commercially available surfactant and recorded that α-lipoic acid 100 mg/l + 8-HQC 250 mg/l + sucrose 3 per cent, and spermine 100 mg/l + 8-HQC 250 mg/l + sucrose 3 per cent were effective in retaining fresh weight and maintaining carotene pigmentation in the bracts, and in the former combination gave maximum vase life in heliconia flowers, followed by spermine 100 mg/l + $Cacl_2$ 250 mg/l along with 8-HQC 250 mg/l + sucrose 3 per cent.

The large heliconias are usually cut to fit a 150 cm long box while the *H. psittacorum* types are trimmed to lengths of 60 to 90 cm. They are often packed in a bunch of 10, and sleeved in plastic film or open weave netting. Up to 25 such bunches may be packed in a 150 × 50 × 25 cm box. Medium sized heliconias such as *H. bihai* and *H. stricta* may be packed 20 to 50 per box while the large *H. caribaea* inflorescences are packed as 10 to 5 pieces. While transporting the cut heliconias, the packing material should not be too moist and storage temperature not below 10 °C (Jeroenkit and Paul, 2003), and the spikes stored at 12 °C and 19 °C there was no change in bract colour.

Tropical flowers such as *Heliconia psittacorum* have a vase life that is limited by inhibition of water uptake due to exudation of mucilage at the cut surface. Within an hour of *H. latispatha* inflorescence cutting, spraying with benzyladenine at 'run off' stage at 100, 200 or 300 mg/l to the inflorescences harvested at stage I having 1-2 bracts fully open or at stage II with 3-4 bracts fully open' though did not affect the average number of open bracts throughout the vase life (2.5 for stage 1 and 4.2 for stage 2) but floral longevity was found prolonged at both the stages, higher the concentration more was the vase life and 300 mg/l increased the floral longevity by at least 1.85 times compared to control (Moraes *et al.*, 2005).

The sensitivity of cut flowers and foliage to irradiation used for insect disinfestation varies from species to species and to a lesser extent variety to variety, highly sensitive being *Heliconia* 'Keanae' to 250 Gy though a hot water treatment of 45.0-47.5 °C for 40 minutes effectively reduces the irradiation injury and medium sensitive being *H. stricta* cv. 'Red Stricta' (Sangwanangkul *et al.*, 2008).

To get rid of spathe browning in *H. rostrata*, the optimum harvest time was found at middle flowering bud **stage** and subsequent whole inflorescence immersion for 1 h in a solution containing 500 mg H_3BO_4 or 500 mg/l $MgSO_4$ then placing in a low pH vase solution containing 2 mg BA + 200 mg aluminium sulphate, and then spraying with 0.1 per cent polyvinyl alcohol which controlled the browning and where vase life was recorded 11.5-12.0 days (Sheng-Aiwu and Liu-Nian, 2006).

Insect-Pests and Diseases

Snails (*Achatina fullica*) sometimes feed on the plants during night hours by making irregular holes on the leaves and batter them. Handpicking and killing by dropping snails in 5 per cent salt solution, sprinkling of salt crystals in the paths or around pots, spreading 'Snail Kill' (3 per cent metaldehyde pellets) in the field or neem oil 10 ml/l and soap nut extract 60 g/l spraying will save the plants from their infestation and reduce the population. Some **chewing insects** while robbing the nector in *H. wagneriana* damage the bracts for which Wootton and Sun (1990) suggested 'bract liquid' as a safequard. Root-knot **nematode** (*Meloidogyne* sp.) have been found infesting heliconia rhizomes which can be controlled by dipping the rhizomes in hot water for one hour at 48 °C or for 30 minutes at 48 °C (Criley, 1988). Apart from root-knot, there are burrowing, spiral, lesion and reniform nematodes that infect heliconias, burrowing nematodes being most common in Hawaii (Holzmann and Wong, 1986) while spiral and lesion nematodes in California. Chemical fumigation of soil before planting and planting only nematode-free or HWT treated rhizomes will manage this problem (Holzmann and Wong, 1986).

Chowdhry *et al.* (1983) recorded *Cercospora heliconiae* from diseased plants of *H. caribaea*. During 1990's in Venezuela (Carabobo state), a *Myrothecium* sp. on *H. psittacorum* × *H. spathocircinata* cvs 'Golden Torch' and 'Tropic' with dark brown lesions on inflorescence spathes and causing about 30 per cent loss was recorded. In Venezuela, Madriz *et al.* (1991) while surveying the heliconia growing areas, isolated *Alternaria alternata, Curvularia* sp., *Drechslera musae-sapientum, Fusarium oxysporum, Gloeosporium musarum* (*Colletotrichum musae*), *Glomerella cingulata, Guignardia musae, Mycosphaerella musicola, Pestalotiopsis* sp. and *Phyllosticta musarum* (*P. musae*) from the lesions found on leaves and inflorescences, as pathogenic especially on *H. caribaea* cv. 'Riqui-riqui', *H. latispatha* cv. 'Platanillo' and *H. psittacorum* cv. 'Avecilia'. Herrera Isla (1994) in Cuba recorded a new species of pathogen on heliconias as *Glomerularia heliconii*. *Phytophthora* has been recorded causing heliconia root rot and *Pythium* stem rot. These all the diseases can be controlled through use of Bavistin 0.1 per cent alternate with Captan 0.2 per cent sprayings to the plants fortnightly. However, leaf spot diseases can be controlled through spraying with Dithane M-45 or Dithane Z-78. *Ralstonia solanacearum* also infects heliconias. In excessive humidity and chilly weathers the crop is also infected with *Erwinia* bacterium. Shiau YihJuh *et al.* (1998) took explants from different parts of *H. psittacorum* to study the source and control of contaminants (bacteria such as *Agrtobacterium radiobacter* syn. *A. tunefaciens, Pseudomonas diminuta, P. putida, Serratia marcescens, Sphingomonas pancimobilis* and *Xanthomonas campestris*) in tissue cultured plants, and found least (57.7 per cent) contamination after 21 days of culture in the explants taken from the leaf tips though it ranged from 80-100 per cent in the explants taken from petioles, leaf bases, ovaries, peduncles and terminal buds, however, the contamination was quite low when explants were grown in soilless media in the greenhouse. They observed 40-46 per cent contamination rates from ovary explants cultured in a liquid MS medium containing 100 mg/l streptomycin sulphate and 50 mg/l gentamycin sulphate for 24 or 48 hours. Sometimes the crop is also found infected with cucumber mosaic virus. The virus-infected plants should be uprooted and burnt immediately.

References

Anderson, L. 1985. Musaceae. In: *Flora of Ecuador*, No. 22 (eds Harling,G. and B. Sparre), pp. 1-87. Swedish Research Council Publishing House, Stockholm, Sweden.

Bredmose, N. 1987. Keeping quality of some new flowers for cutting. *Gartner Tidende*, **103**(6): 146-147.

Bora, S. and L. Paswan, 2003. *In vitro* propagation of *Heliconia psittacorum* through axillary bud. *J. Ornam. Hort.* (N.S.), **6**(1): 11-15.

Broschat, T.K., H.M. Donselman and A.A. Will, 1984a. 'Andromeda' and 'Golden Torch' heliconias. *HortSci.*, **19**(5): 736-737.

Broschat, T.K., H.M. Donselman and A.A. Will, 1984b. Golden Torch, an orange heliconia for cut-flower use. *Circ. Agricultural Experiment Stations*, University of Florida, No. S-308, p. 4.

Broschat,T.K., H.M. Donselman http://129.200.9.213:8595/webspirs/doLS.ws?ss=Donselman-H-M+in+AU and A.A. Will, 1984c. Andromeda, a red and orange heliconia for cut-flower use. *Circ. Agric.Experiment Station*, University of Florida, No. S-309, p. 5.

Broschat,T.K. and S. Svenson, 1994. EDCPTA enhances growth and flowering of heliconias. *HortSci.*, **29**(8): 891-892.

Bruna, E.M. and M.B.N. Ribeiro, 2005. Regeneration and population structure of *Heliconia acuminata* in Amazonian secondary forests with contrasting land-use histories. *J. Trop. Ecol*, **21**(1): 127-131.

Carle, A.W. 1990. Heliconias by seed. *Bull. Heliconia Soc. int.*, **4**: 6.

Castro, C.E.F. de, A. May and C. Goncalves, 2006. Heliconia's species for cut flowers. *Revista Brasilieira de Horticultura Ornamentale*, **12**(2): 87-96.

Catley, J.L.and I.R. Brooking, 1996. Temperature and light influence growth and flower production in *Heliconia* 'Golden Torch'. *HortSci.*, **31**(2): 213-217.

Chen GuoJu, Chen RiYuan, Wu XiaoYing, Huang DingHua, and He ShuiHua, 2005. The effect of different culture mediums on growth and cell structure of *Heliconia psittacorum*. *J. South China agric. Univ.*, **26**(1): 28-30.

Chowdhry, P.N., D. Gupta and B. Padhi, 1983. Some new species of *Cercospora* from ornamental plants of India. *Indian Phytopath.*, **36**(4): 624-625.

Clemens, J. and R.H. Morton, 1999. Optimizing mineral nutrition for flower production in *Heliconia* 'Golden Torch' using response surface methodology. *J. Amer. Soc. hort. Sci.*, **124**(6): 713-718.

Costa, A.S. da, V. Loges, A.C.R. de Castro, A.L. Verona, C. de O. Pessoa and V.F. dos Santos, 2006. Number of shoots and area per clump of heliconia. *Horticultura Brasileira*, **24**(4):460-463.

Criley, R.A. 1988. Propagation of tropical cut flowers: *Strelitzia, Alpinia* and *Heliconia*. *Acta Hort.*, No. 226 (vol. II), pp. 509-517.

Criley, R.A. 1989. Development of *Heliconia* and *Alpinia* in Hawaii ~ cultivar selection and culture. *Acta Hort.*, No. 246, pp. 247-258.

Criley, R.A. 2000. Seasonal flowering patterns for *Heliconia* shown by grower records. *Acta Hort.*, No. 541, pp.159-165.

Criley, R.A. and O. Kawabata, 1986. Evidence for a short-day flowering response in *Heliconia stricta* 'Dwarf Jamaican'. *HortSci.*, **21**(3): 506-507.

Criley, R.A. and R. Lekawatana, 1990. Environment influences seasonality in flowering of *Heliconia* (abstr.). *XXIIIrd int. hort. Congr.*, **1**: 376.

Criley, R.A., W.S. Sakai, S. http://129.200.9.213:8595/webspirs/doLS.ws?ss=Lekawatana-S+in+AU and E. Kwon, 1999. Photoperiodism in the genus *Heliconia* and its effect upon seasonal flowering. *Acta Hort.*, No. 486, pp. 323-327.

Criley, R.A., J.Y. Uchida and Z.Q. Fu, 2003. Productivity and periodicity of flowering in *Heliconia orthotricha* cultivars. *Acta Hort.*, No. 624, pp. 207-212.

Cruz, D.D. da, M.A.R. Mello and M. van Sluys, 2006. Phenology and floral visitors of two sympatric *Heliconia* species in the Brazilian Atlantic forest. *Flora Jena*, **201**(7): 519-527.

Dias, M.A. and P.H.V. Rodrigues, 2001. Fontes de explantes e contaminates isolados em cultivo *in vitro* de heliconia. *Revista Brasileira de Horticultura Ornamentale*, **7**: 165-168.

Everard, B. and Brian D. Morley, 1970. *Wild Flowers of the World*, pp. 189. Peerage Books, London.

Ferreira, L.B. snd S.A.Oliveira, 2003. Study of NPK fertilization in the growth variable and inflorescences productivity of *Heliconia* sp. *Revista Brasileira de Horticultura Ornamentale*, **9**(2): 121-127.

Finger, F.L. and J.G. Barbosa, 2006. Postharvest physiology of cut flowers. *Advances in Postharvest Technologie for hort. Crops*, pp. 373-393 (*http:*//129.200.9.213:8595/webspirs/doLS.ws?ss).

Geertsen, V. 1988. Development of new cut flower crops. *Gron Viden. Havebrug*, No. 123, p. 4.

Geertsen, V. 1989. Effect of photoperiod and temperature on the growth and flower production of *Heliconia psittacorum* 'Tay'. *Acta Hort.*, No. 252, pp. 117-122.

Goh, C.J., P.P. Kumar and J.C.K. Yau, 1995. Genetic variations detected with RAPD markers in *Heliconia*. *Acta Hort.*, No. 420, pp. 72-74.

Goh, C.J., M.J. Nathan, P.P. Kumar, R.A. Criley and S. Lekawatana, 1995b. Year around production with high yields may be a possibility for *Heliconia chartacea*. *Acta Hort.*, No. 397, pp. 95-101.

Gomes, A.R.M., J.H.T. D'Avila, R.S. Gondim, F.C. Bezerra and F.M. Bezerra, 2006. Evapotranspiration and crop coefficient of *Heliconia psittacorum* L x H. spathocircinada (Arist) under greenhouse conditions. *Revista Ciencia Agronomica*, **37**(1): 13-18.

Herrera Isla, L. 1994. A new disease of *Heliconia* sp. caused by *Glomerularia heliconi* sp. nov. *Centro Agricola*, **21**(1): 95-96.

Holtzmann, O.V. and Y.M. Wong, 1986. Nematodes in tropical cut flower and their control. *Hort. Digest* (Hawaii), **80**: 5-6.

Ibiapaba, M.V.B. da, J.M.Q. Luz and R. Innecco, 1997. Performance of two *Heliconia* species at different planting spacings in Fortaleza (CE). *Revista Brasileira de Horticultura Ornamentale*, **3**(2): 74-79.

Ibiapaba, M.V.B. da and J.M.Q. Luz, 2000. Evaluation of spacing in *Heliconia psittacorum* L. cv. Sassy and Andromeda. *Ciência e Agrotecnologia*, **24**(1): 181-186.

Jeroenkit, T. and R.E. Paul, 2003. Reviews on postharvest handling of heliconia, red ginger and bird of paradise. *Hort. Techn.*, **13**: 259-266.

Jie He, Lay PhengTan and ChongJin Goh, 2000. Alleviation of photoinhibition in *Heliconia* grown under tropical natural conditions after release from nutrient stress. *J. Pl. Nutr.*, **23**(2): 181-196.

Kress, W.J. 1983. Crossability barriers in neotropical *Heliconia*. *Ann. Bot.*, **52**(2): 131-147.

Kress, W.J. and C.S. Roesel, 1987. Seed germination trials in *Heliconia stricta cv.* Jamaica. *Bull. Heliconia Soc. int.*, **2**(2): 6-7.

Kumar, P.P., J.C.K. Yau and Goh ChongJin, 1998. Genetic analyses of *Heliconia* species and cultivars with randomly amplified polymorphic DNA (RAPD) markers. *J. Amer. Soc. hort. Sci.*, **123**(1): 91-97.

Lalrinawmi and M.C. Talukdar, 1999. Effect of spacing and size of rhizome on the rhizome production of heliconia (*Heliconia psittacorum* L.). *Hort. J.*, **12**(1): 45-49.

Lalrinawani and M.C. Talukdar, 2000. Effect of spacing and size of rhizome on the flower production of heliconia (*Heliconia psittacorum* L.). *J. agric. Sci. Soc. NE India*, **13**(1): 48.

Lee, Y.H., N.Y. Ng and C.J. Goh, 1994. Pollen formation and fruit set in some cultivars of *Heliconia psittacorum*. *Sci. Hort.*, **60**(1/2): 167-172.

Maciel, N and N. Mogollon, 1997. Germination in six ornamental Zingiberales. *Proc. interamerican Soc. Trop. Hort.*, **41**: 55-61.

Maciel, N., E. Rojas and R.J. Campbell, 1994. Growth and development of *Heliconia bihai* and *H. latispatha* under different levels of shade. *Proc. interamerican Soc. Trop. Hort.*, **38**: 257-263.

Madriz, R., B.G. Smits and R. Noguera,.1991. Main pathogenic fungi affecting some ornamental species of the genus *Heliconia*. *Agronomia Tropical* (*Maracay*), **41**(5-6): 265-274.

Mangave, B.D., Alka Singh, S.L. Chawla and B.K. Dhaduk, 2011. Effect of harvesting stage and postharvest handling on quality and vase life of heliconia inflorescence. *J. Ornam.Hort.*, **14**(1-2): 48-52.

Mangave, B.D., Alka Singh, Sanjay Jha and S.L. Chawla, 2014. Postharvest physiology and quality of heliconia inflorescence cv. Golden Torch as influenced by antioxidants. *Indian J. Hort.*, **71**(2): 232-236.

Melendez-Ackerman, E, J. Rojas-Sandoval and S. Planas, 2008. Self-compatibility of microgametophytes in *Heliconia bihai* (Heliconiaceae) from St. Lucia. *Caribbean J. Sci.*, **44**(2): 145-149.

Montgomery, R. 1986. Propagation of *Heliconia* from seeds. *Bull. Heliconia Soc. int.*, **1**: 6-7.

Moraes, P.J. de, F.L. Finger, J.G. Barbosa, P.R. Cecon and L.P. Cesa, 2005. Influence of benzyl adenine on longevity of *Heliconia latispatha* Benth. *Acta Hort.*, No. 683, pp. 369-373.

Nathan, M.J., C.J. Goh and P.P. Kumar, 1992. *In vitro* propagation of *Heliconia psittacorum* by bud culture. *HortSci.*, **27**(5): 450-452.

Paull, R.E. and T. Chantrachit, 2001. Benzyladenine and the vase life of tropical ornamentals. *Postharvest Biol. & Tech.*, **21**(3): 303-310.

Pavan Kumar, P., T. Janakiram, D.P. Kumar and R. Venugopalan, 2011a. Correlation studies in *Heliconia* genotypes. *J. Ornam. Hort.*, **14**(1-2): 21-23.

Pavan Kumar, P., T. Janakiram, D.P. Kumar and R. Venugopalan, 2011b. Genetic variability studies in *Heliconia* cultivars. *J. Ornam. Hort.*, **14**(3-4): 10-13.

Ramachandrudu, K. and M. Thangam, 2012. Performance of heliconia under coconut garden and open field conditions. *Indian J. Hort.*, **69**(3): 450-453.

Rodriguez, P.H.V. 2005. *In vitro* establishment of *Heliconia rauliniana* (Heliconiaceae). *Scientia Agricola*, **62**(1): 69-71.

Rodriguez, P.H.V., A. Tulmann Neto and M. de F.B. Dutra, 2005. Gamma radiation LD^{5o} determination in *Heliconia bihai* (Heliconiaceae) explants. *Revista Brasileira de Horticultura Ornamentale*, **11**(1): 79-81.

Rodrigues, P.H.V., G.M.B. Ambrosano, A.M.L.P. Lima, B.M.J. Mendes and A.P.M. Rodriguez, 2006. *Heliconia bihai* var. Lobster Claw. I. Cut flower field production from micropropagated versus rhizome-derived plants. *Floriculture, Ornamental and Pl. Biotech.*, pp. 558-560 (*http*:1239.200.9.213:8595/webspirs/doLS.ws?ss).

Rose, M.J. and W. Barthlott, 1995. Pollen connecting threads in *Heliconia* (Heliconiaceae). *Plant Syst. Evol.*, **195**(1/2): 61-65.

Rundel, P.W., M.R. Sharifi, A.C. Gibson and K.J. Esler, 1998. Structural and physiological adaptation to light environments in neotropical *Heliconia* (Heliconiaceae). *J. Trop. Ecol.*, **14**(6): 789-801.

Sanabria, M.E., O. Crescente and A.E. de Caserta, 1998. Secondary metabolites as a chemotaxonomic character for three species of the genus *Heliconia* L. (Heliconiaceae) from the State of Sucre, Venezuela. *Ernstia*, **8**(1): 27-39.

Sangwanangkul, P., P. Saradhuldhat and R.E. Paull, 2008. Survey of tropical cut flower and foliage responses to irradiation. *Postharvest Biol. Techn.*, **48**(2): 264-271.

Sato, A.Y., D. de carvalho Nannetti, J.E.B.P. Pinto, J.O. Siqueira and M. de F.A. Blank, 1999. Application of arbuscular mycorrhiza to micropropagated heliconia and gerbera plants during acclimatization period. *Horticultura Brasileira*, **17**(1): 25-28.

Sheela, V.L. 2006. Heliconia. In: *Advances in Ornamental Horticulture* (vol. 2) : *Herbaceous Perennials and Shade Loving Foliage Plants*, pp. 150-165. Pointer Publishers, Jaipur (India).

Sheela, V.L., T.S. George, R. Rakhi and P.R.G. Lekshmi, 2007. Variability studies in cut flower varieties of heliconias. *Indian J. Hort.*, **64**(1): 109-111.

Sheela, V.L., P.R.G. Lekshmi, C.S.J. Nair and K. Rajmohan, 2006a. Molecular characterization of *Heliconia* by RAPD assay. *J. Trop. Agric.*, **44**(1/2): 37-41.

Sheela, V.L., C.S.J. Nair, R. Rakhi and P.R.G. Lekshmi, 2006b. Correlation studies in heliconia. *J. Ornam. Hort.* (New Series), **9**(1): 29-32.

Sheela, V.L., R. Rakhi, C.S.J. Nair and T.S. George, 2005. Genetic variability in *Heliconia*. *J. Ornam. Hort.* (New Series), **8**(4): 284-286.

Sheng Aiwu and Liu Nian, 2006. Control of browning of spathe tissues of cut *heliconia rostrata*. *Acta Hort. Sinica*, **33**(4): 898-900.

Siddique, A.B. and L. Paswan, 1997. Callus initiation in heliconia (*Heliconia psittacorum*). *J. agric. Sci. Soc. NE India*, **10**(1): 150-152.

Shiau YihJuh, Hseu ShiowHuey, Wang TsaiYih and Tsay HsinSheng, 1998. Studies on tissue culture of *Heliconia psittacorum* 'Rhizomatosa'. I. Identification and control of bacterial contamination of explants cultured *in vitro*. *J. agric. Res. China*, **47**(4): 346-363.

Shiau Yihjuh, Wang TsaiYih and Tsay HsinSheng, 1999. Studies on tissue culture of *Heliconia psittacorum* cv. Rhizomatosa. III. Effect of cefotaxime and plant growth regulators on growth of shoot tip explants *in vitro*. *J. agric. Res. China*, **48**(4): 42-51

Simao,D.G and V.L. Scatena, 2004. Morphology and anatomy of the bracts in *Heliconia* (Heliconiaceae) occurring in the Sao Paulo State, Brazil. *Acta Botanica Brasilica*, **18**(2): 261-270.

Simao, D.G., B.L. Scatena and F. Bouman, 2006. Developmental anatomy and morphology of the ovule and seed of *Heliconia* (Heliconiaceae, Zingiberales). *Pl. Biol.*, **8**(1): 143-154.

Simao, D.G., V.L. Scatena and F. Bouman, 2007. Anther development, microsporogenesis and microgametogenesis in *Heliconia* (Heliconiaceae, Zingiberales). *Flora Jena*, **202**(2): 148-160.

Talukdar, M.C., Lalrinawmi and S. Singh, 2002. *In vitro* propagation of heliconia. *J. Ornam. Hort.* (New Series), **5**(1): 45-46.

Torres, A.C., F.D. Duval, D.G. Ribeiro, A.F.F. Barros and F.A.D. Aragao, 2005. Effect of sucrose, kinetin, isopentenyl adenine and zeatin on the development of embryos of *Heliconia rostrata in vitro*. *Horticultura Brasileira*, **23**(3): 789-792.

Triplett, J.K. and B.K. Kirchoff, 1991. Lamina architecture and anatomy in the Heliconiaceae and Musaceae (Zingiberales). *Canad. J. Bot.*, **69**(4): 887-900.

Ulisses, C., T. Camara, L. Willadino and C. Albuquerque, 2007. Induction of somatic embryogenesis in *Heliconia chartacea* Lane ex Barreiros cv. Sexy Pink and cv. Sexy Scarlet from sections of ovaries. *Revista Brasileira de Ciencias Agrarias*, **2**(4): 282-284.

Ulisses, C., T. R. Camara, L. Willadino, C. C. de Albuquerque and J. Z. de Brito, 2010. Early somatic embryogenesis in *Heliconia chartacea* Lane ex Barreiros cv. Sexy Pink ovary section explants. *Brazilian Archives of Biol. & Techn.*, **53**(1): 11-18.

Whittaker, J.M. 1993. *Postharvest Habndling Procedures for Jamaican Grown Cut Flowers*. M.Sc. Thesis, Univ. of Florida, Gainesville, USA.

Wootton, J.T. and I.F. Sun, 1990. Bract liquid as a hervivore defence mechanism for *Heliconia wagneriana* inflorescences. *Biotropica*, **22**(2): 155-159.

Zeng SongJun, Guo ShaoCong, Wu KunLin, Zhang YiQi, Chen GuoHua and Duan Jun, 2004. *In vitro* propagation of *Heliconia* plants. *J. Trop. & Subtrop. Bot.*, **12**(2): 153-158.

Zimmer, K. and B. Carow 1977. Possibilities of partial vacuum storage. *Deutscher Gartenbau*. **31**(3): 87-89.

23

Iberis (Family: Brassicaceae)

Ajit Kumar, Kiran Kumari, Manish Kapoor, Ranjan Srivastava, Sanyat Misra and R.L. Misra

[**Common names**: Bitter candytuft/Candytuft/Down's mustard/Hyacinth candytuft/Rocket candytuft (*Iberis amara*, syn. *I. coronaria*), Common annual candytuft (*I. amara, I. umbellata*), Candia mustard/Common purple candytuft/Globe candytuft (*I. umbellata*), Dune candytuft (*I. procumbens*), Fragrant/Sweet-scented candytuft (*I. odorata*), Gibralter candytuft (*I. gibraltarica*), Perennial candytuft (*I. sempervirens*), Pinnate-leaved candytuft/Spanish tuft (*I. pinnata*), Winter candytuft (*I. semperflorens*), *etc.*]

Introduction and Origin

The word candytuft comprises of two words, *i.e.* candy (derives from Candia the former name of Iraklion on the Island of Crete) and tuft (originating from the clustered flowers or tufted plant growth). *Iberis amara* was thought to be a species of *Thalaspi* or wild mustard. The seeds of candytufts are very pungent and were used as a substitute for mustard. The generic name *Iberis* derives from Iberia which is ancient name of Spain as most of the species of candytuft were discovered in Spain. Specific name *amara* comes from the Latin *amarus* which refers to its bitter flavour. *Iberis semperflorens* is a Latin word where *semper* means always and *virens* means alive, denoting

to its evergreen habit as it is perennial. *Iberis* is indigenous to the Mediterranean region, but with a range also extending northwards to Central Europe and eastwards to Asia Minor, at an altitude ranging from 180 to 2,100 metres above mean sea level in the arid, stony and calcareous soil. Several species of the genus can be found growing wild in Spain. However, the perennial species are native to temperate climates. Among the various species of the genus, *I. amara* and *I. umbellata* are commonly cultivated as annual bedding plants whereas *I. sempervirens* and *I. gibraltarica* are grown as garden perennials (Armitage, 1989; Dole and Wilkins, 2005). *Iberis sempervirens* is native to Southern Europe and Western Asia. Most of the species of the genus *Iberis* are native to Iberian Peninsula.

Candytuft is an important winter season annual with versatile uses. Very free flowering, the sweet scented blooms are produced in abundance throughout the summer in the temperate regions though only during winter-spring in subtropical and tropical climates. Extremely easy to grow, and probably because of this, it has been somewhat neglected of late but is now regaining much of its earlier popularity again. It is useful as cut flower in bouquets and flower arrangements, bedding purpose, border and edging along paths. Candytufts are suitable for growing in rock garden as the plants do well in gravelly soil. Because of variable height of plants and floriferous nature, they are ideal choice for growing in pots also. *Iberis* is tolerant to drought and often used as non-invasive ground cover. They are tolerant of smoke and grime. The white candytuft flowers are compatible with intermediate annuals and perennials and can be used to cover tall bloomless stems of flowers. Various parts of the plants such as leaves, seeds, roots and stems possess medicinal properties, seeds being more potent, and these can be used in the treatment of digestive problems as well as in the treatment of rheumatism, gout, asthma, bronchitis and cardiac hypertrophy in doses of from 1-3 grains (0.065-0.2 g) of the seed. In curing dyspepsia, candytuft is one of the primary ingredients in herbal-combination therapy.

Botany, Cytology and Breeding

Candytufts are hardy and half-hardy annuals, evergreen perennial herbs and sub-shrubs growing up to 15-45 cm. The plants are more or less woody. Most of the species have bristly stem, short-haired and have branching that arise from crown. Bristles that are present on stem are descending oblique. Leaves are alternate, middle and upper ones stalkless, linear to narrowly spathulate or obovate, sparse-toothed at the tips (sometimes with short margins), margins hairy and 3-veined. Flowers are roughly cross-shaped, white, pink, carmine-rose, purple and lilac, having inflorescence initially dense then broadly corymbose or lengthening raceme and becoming sparser in fruiting stage. Corolla is zygomorphic (irregular), with 4 petals out of which 2 outermost are longer (6 mm) while 2 innermost shorter (3 mm). Total sepals present are four and stamens are six. Gynoecium is fused consisting of a single carpel. Fruits are silicula with long wings, wings widest at tip and notched with approximately 1 mm bristle. Seeds are round and flat narrowly winged and of the size from 4-6 mm.

A little **cytological** work is available on this plant but what have been compiled about its chromosome number by Manton (1932), and Darlington and Wylie (1955) are furnished here with in Table 23.1.

Table 23.1: Chromosome Numbers of *Iberis* Species Based on the Tables by Manton (1932), and Darlington and Wylie (1955) (X = 7, 8 & 11)

Sl.No.	Species	Chromosome Number
1.	*Iberis pinnata*	16
2.	*I. amara* L.	16
3.	*I. amara* L.	14
4.	*I. affinis* Jord.	14
5.	*I. gibraltarica*	14
6.	*I. lagascana*	14
7.	*I. odorata*	14
8.	*I. pectinata*	14
9.	*I. taurica*	14
10.	*I. umbellata*	14
11.	*I. umbellata*	16
12.	*I. jordani*	22
13.	*I. nana*	22
14.	*I. pruitii*	22
15.	*I. semperflorens*	22, 44
16.	-do- var. *garrexiana*	22+1 fragment
17.	*I. correaefolia*	44
18.	*I. garrexiana*	44
19.	*I. saxatilis* L.	22+2 fragments
20.	*I. corifolia*	44 (fragmented)
21.	*I. ciliata*	—
22.	*I. bernadiana*	—
23.	*I. welwitschii*	—

Ene (1973).

Bateman (1954, 1955) stated that two members of the Cruciferae, *viz. Iberis amara* and *Brassica campestris*, have a sporophytic incompatibility system where pollen behaviour is determined by the genotype of the pollen parent.

Occurrence of **polyploidy**, *viz.* trisomics, tetrasomics and occasionally double trisomics, has been found in some species of the genus *Iberis* (Ene, 1968). However, seeds are not produced in triploid, tetraploid or in plants having higher ploidy levels as in *I. amara* and *I. pectinata*, which indicates the presence of some intrinsic factors which keep these species in diploid state in nature. In *I. umbellata*, the diploid seeds obtained from 2n female crossed with 4n males must have arisen from haploid ovum fertilized by haploid pollen which was produced by a spontaneous break-up of diploid genome to haploid ones during gametogenesis in tetraploid plants (Ene, 1973).

The main objective of **hybridization** is to create genetic variation, therefore, various species or varieties have been crossed for production of F_1 hybrids as they are uniform in flowering with large blooms and high in yield. In the genus *Iberis*, interspecific hybridization was successful between *I. pectinata* as female parent and *I. amara* as male parent (Ene, 1968). Interspecific and intraspecific crosses were made between four varieties of *I. amara, viz.* 'Coronaria', 'Grandiflorum', 'Hyacinthiflora' & 'Iceberg', and the species *I. pectinata* & *I. amara*. The F_1 hybrids showed hybrid vigour, bushy plant growth, large flowers and more resistance than either parent to grey mould disease caused by *Botrytis cinerea*. However, F_1 hybrids were highly sterile and were morphologically intermediate between the two parents except in siliqua shape where these resembled the maternal parent more closely. The parent *I. pectinata* and *I. amara* had 96 per cent and 98 per cent pollen fertility, respectively, compared to 0 per cent in the F_1 hybrids. Reciprocal crosses made to the original male parent *I. amara* aggregate failed to set any seed, but the reciprocal back crosses to the female parent *I. pectinata* set viable seeds (Table 23.2).

Species, Cultivars and Classification

The genus *Iberis* comprises some 50 species of which the commonly cultivated ones are *Iberis amara, I. garrexiana, I. gibraltarica, I. odorata, I. pectinata, I. pinnata, I. pruitii, I. saxatilis, I. semperflorens, I. sempervirens, I. tenoreana* and *I. umbellata*.

Iberis amara (syn. *I. coronaria*). This is an annual species native to Central and Southern Europe, and in the garden it is being grown since the 16th century. The species grows up to 45 cm tall though its varieties from 10 to 75 cm. Its whole plant tastes bitter. The stem is erect, branched, and bristly with short hairs, forked at the summit and forms a large corymb of vegetation. Leaves are alternate, slightly dentate and hairy, 5-10 cm long and wider, oblanceolate to spathulate, widely pinnatifid or toothed and obtuse. Its 4-petalled (2 outermost larger some 6 mm long and 2 innermost smaller some 3 mm long) flowers resemble the hyacinth, are scented white (occasionally light purple) and are borne as umbel on 15 cm long spikes. The plants have medicinal uses in cardiac diseases as they contain amines, cucurbitacins, flavonoglycosides and mustard oil glycosides. There are several popular varieties with *Iberis umbellata* such as 'Hyazinthenblutige Riesen', 'Imperialis' both being hyacinth-flowered, 'Empress', 'Giant Snowflake', 'Iceberg', 'Spiral White', *etc.* all tall types, and 'Little Prince', 'Tom Thumb', *etc.* the dwarf types.

Iberis garrexiana. It is native to Piedmont and Pyrenees, and is intermediate between *I. saxatilis* and *I. sempervirens*, having the habit of the former. Sometimes it is described as *I. sempervirens* var. 'Garrexiana'. Its leaves are narrow at the base, oblong and glabrous. Flowers are white, small and the racemes much elongating.

Iberis gibraltarica. This moderately hardy evergreen shrub from Iberian Peninsula (Gibralter and South Spain) and Morocco is woody-based forming hummocks and one of the most attractive of the genus, growing some 30 cm high with a spread of 30-45 cm, though often capricious. Though it is perennial but when grown as an annual it flowers the same year if the seeds are sown at the site of flowering in March-April in the temperate region. As a biennial, it should be sown in summer, the potting of young seedlings in 5-7.5-cm pots and then keeping them in cold frames until flowering time in the following spring. Adult plants are wide-spreading and a little unruly. Stems are red and little pubescent, leaves are dark green, thick-textured, 2.5-5.0 cm long, narrowly oblong or broadly spathulate, cuneate and irregularly dentate at the tips. Light pinkish-violet flowers appear in flattish corymbs where outer individual flowers are 3-4 times larger than the inner ones.

Iberis odorata. A dwarf annual species growing from 10 to 30 cm tall is native to Crete (Greek island) and Syria. Stem is erect, toothed, bluntly angular, white-hairy and little branched. The leaves are alternate, succulent, slender but wider towards the top, fringed at the base and toothed. Sweet-scented flowers are white, corymbose, small and loose though pods are acutely lobed.

Iberis pectinata (syn. *I. affinis* Hort.). A native of Spain, it is confused with *I. odorata* but the petals are four times as long as the calyx and the pods have short hairs though in *I. odorata* the petals are $1^1/_2$ times as long as the calyx and the pods are glabrous. Leaves are pectinate and the flowers are white.

Table 23.2: Seed Set in the Reciprocal Crosses between *I. pectinata* and Varieties of *I. amara*

Sl.No.	Crosses Female × Male	No. of Crosses	Per cent Success	No. of 20 Seed Samples	Av. wt. in mg of 20 Seed Samples	SE
1.	*I. pectinata* × *I. amara*					
	a. –do– × var. 'Iceberg'	300	16	4	16.7	0.74
	b. –do– × var. 'Hyacinthiflora'	25	10	1	18.2**	–
	c. –do– × var. 'Coronaria'	25	8	1	17.6**	–
	d. –do– × var. 'Grandiflorum'	25	8	1	17.1**	–
2.	*I. amara* × *I. pectinata* 4 Varieties	300	0	–	–	–
3.	*I. amara* × *I. amara*					
	var. 'Iceberg' x var. 'Iceberg'	300	72	5	61.5	1.05
	var. 'Hycinthiflora' × 'Iceberg'	100	66	5	62.3	1.03
	var. 'Iceberg' × var. 'Hyacinthiflora'	100	64	5	60.2	1.52
4.	*I. pectinata* × *I. pectinata*	100	50	5	38.4	1.35

* Ene (1971); ** Calculated from less than 20 seed samples.

Iberis pinnata. An annual or biennial fragrant species, native to Spain, South France and Italy, grows 10-20 cm high bearing pinnate leaves. It was being grown in the gardens of Southern Europe since 1700. Stems are either pubescent or smooth and branched at the summit. Pinnae (segments) of the leaves have feathered-arrangement and are little velvety. White flowers appear in flat and compact corymbs, the calyx slightly tinged purple, petals small and irregular, two basal petals on the outer flowers larger and tongue-shaped, the two inward facing petals are small, pure white, round and erect. It requires careless culture for keeping beauty of the flowers intact as by providing good cultural practices flowers elongate and mar the beauty. Its two beautiful varieties 'Affinis' and 'Pandurata' with fragrant white flowers grow only 10-15 cm high and form compact cushions up to 25 cm across. The flowers in the bud are lilac-coloured.

Iberis pruitii. A 15 cm tall growing perennial with woody base is native to Sicily. Leaves are glabrous, entire or subdentate, and obovate-spathulate. White flowers are borne in compact clusters and the pods are merely notched at the apex.

Iberis saxatilis. A hardy evergreen perennial semi-procumbent sub-shrub which does not grow more than 15 cm in height with a spread of about 30 cm is native to the Southern Europe, found frequently in the mountainous areas. It is a neat alpine bush. Leaves are dark green, fleshy, linear, semi-cylindrical, 18-20 mm long and occasionally dentate. Flowers appear on flat terminal heads some 5.0 cm across in compact umbels on some 7-15 cm tall stems and are white, sometimes tinged-purple.

Iberis semperflorens. A native of Southern Europe, most common being in Italy (Island of Sicily), it is a rare evergreen and ever-flowering sub-shrub with 30-70 cm plant height, which is winter-flowering. Iapichino and Bertolino (2009a) described it endemic to southern Italy and Sicily. Though it is hardy but not tolerant of severe frost. The leaves are entire, smooth, thickish and cuneate to spathulate. Its flowers are large, attractive white and fragrant which appear in corymbs from autumn to spring, *i.e.* October to April (Iapichino and Bertolino, 2009a). It is most suitable in pots for indoor use. Its var. *plena* is double-flowered.

Iberis sempervirens. A spreading and much-branched but very compact evergreen perennial and semi-procumbent sub-shrub which is herbaceous with woody base, grows some 30 cm in height and 40 cm in spread, and is native to the mountainous regions of Southern Europe, especially Mediterranean region. This species is drought tolerant. On touching the ground, the stems root and create new plants which can be allowed to flower there or transplanted elsewhere. Leaves are numerous, dark green with bluish cast, leathery, some 2-5 cm long and 3 mm wide, smooth, oblong-obtuse and narrowing at the base. Small flowers are bright white (turning pinkish when nights are cold) and cloud-like, appear in flattish terminal raceme up to 4.0 cm across on a 10 cm long stem, and on maturity these flowers become elongated. The blooms appear early in the spring and continue blooming for up to 10 weeks, *i.e.* up to mid-spring. After the flowering is over, the plants are cut back leaving only 5-8 cm from the ground so that these remain full and lush. However, in case seeds are to be harvested, these should not be pruned. This small evergreen perennial is perfect for edging or cascading over rocks and walls during spring season. Var. 'Superba' with very large blooms is the best among all. Vars. 'Little Gem', 'Nana' and 'Pygmaea' are very attractive dwarf forms. Var. 'Foliis Variegatis' has variegated foliage. Var. 'Rosea' bears pink flowers and var. 'Plena', the double form. 'Autumn Beauty' blooms in the spring with a repeat during autumn. Var. 'Whiteout' from Kleft Seeds (USA) bears bright white flowers appearing during early spring, which has superseded all other varieties under this species in its consistency of uniform flowering.

Iberis tenoreana. This hairy perennial from Naples is little shrubby at base, grows up to 15 cm high, lower leaves obovate and narrowed at base, upper ones oblong-linear, flowers are whitish or purplish and are borne in flattish clusters, and the pods are bifid. Var. *petrea* bears white flowers tinged red in centre and is most suitable for rock gardens.

Iberis umbellata. A native of Candia (Southern Europe, Mediterranean) is a robust and hardy annual species with erect branching habit, lower branches being larger than those at the summit. It was brought to England by Lord Edward Zouche around 1590. The plant is 15-40 cm tall but its varieties grow from 20 to 75 cm high. Leaves are mid-green, 1.5-7.0 cm long, alternate, and pointed, the basal ones lanceolate to linear-lanceolate though others only linear. Broad umbellate (some 40 cm across) and compact clusters of small, violet-lilac suffused pink or ice-white flowers in the shape of umbrella appear in large heads. Also there are forms with yellow flowers. Its varieties have a wide colour range of white, pink, carmine-rose, lilac and purple where petal apices are rounded. Fruits are 7-10 mm long siliques. Being a cool season annual it does not like hot weather so flowering period occurs from February to April whether it is sub-tropical area or temperate one, and therefore, its seeds will have to be sown during autumn. Its varieties are variously coloured and very popular. 'Alba' (white), 'Carminea' (carmine), 'Formosa Purpurea' (purple-red or violet-wine), 'Liliacina' (purple), 'Königin von Italien' (lilac), 'Purpurea' (dark purple), 'Red Cardinal' (red), 'Red Flash' (vivid carmine), 'Vulcan' (tetraploid with large carmine flowers) are some of the popular varieties. Its dwarf type varieties are 'Dwarf Hybrid', 'Dwarf Fairy', 'Dwarf Fairy Mixed' (white, lavender, rose-pink and red), 'Rose Cardinal' (rose-scarlet), etc. which grow mound-like completely covered with a multitude of attractively coloured flowers.

Various other popular cultivars of candytufts are 'Autumn Beauty', 'Autumn Show', 'Little Gem', 'Mauve Ice', 'Nana' and 'Pygmaea', 'Snowflake', 'Snow White', 'Whiteout', *etc.*

Two **classes** are defined, (i) **Hyacinth-flowered Series** which are fully hardy, fast-growing, upright and bushy annuals as varieties of *Iberis amara*, growing up to 30 cm high with a spread of 15 cm, leaves are mid-green and lanceolate, and the large flowers which appear on flattish heads during summer are of various colours and scented; and (ii) **Fairy Series** are, in fact, 20 cm high varieties of *Iberis umbellata*, and these are upright and fast-growing bushy annuals with mid-green lanceolate leaves, and the flowers which appear in shades of white, pink, red and purple appear in summer and early autumn on small heads.

Propagation

Usually annual candytufts are propagated mainly through seeds, whereas perennials are propagated through seeds as well as cuttings. **Seed propagation** is the easiest method of propagation. The seeds germinate at 16-20 °C temperatures. Seeds are sown either directly into the field or the seedlings can be raised in pots, boxes or in nursery beds depending upon the requirement of the seedlings. For winter season crop as in case of tropical and sub-tropical regions of the country, sowing of seeds is done in mid-September and seedlings are transplanted after about a month. In temperate areas the seeds can be sown in the flowering site from September-October, March, April and May in order to extend the flowering season. In pots or boxes the seeds should be sown in seed compost under polyhouse in February and then pricked out in potting compost and hardened off in a frame and in April these may be planted out at their flowering site. In case of **vegetative propagation**, soft-wood terminal cuttings of perennial species measuring some 14-15 cm, having 8-15 nodes (Iapichino and Bertolino, 2009a) are taken from non-flowering shoots from June to August in the temperate regions and from October to February in the sub-tropical and tropical regions. This method is used to derive uniform plants for various vegetative and floral attributes of a particular variety. However, the size of cuttings may be even from 5-9 cm with 3-5 nodes. Cuttings should be dipped in IBA solution @ 500 ppm for 1 minute (Iapichino and Bertolino, 2009a) and planted in a mixture of peat (3 parts) and perlite (1 part) or peat and sand in equal parts (v/v) at 22-±2 °C. Iapichino and Bertolino (2009a) suggested planting of *Iberis semperflorens* cuttings in 16-cm pots filled with peat-perlite medium and 3 kg/m^3 Osmocote containing 16N-8P-10K for better results. The rooted cuttings should be transplanted in beds or pots for flowering, and if this is not the proper season of planting the rooted cuttings may be over-wintered in a cold frame and in the following March or April the plants may be set out in their flowering beds.

Taking MS medium for **micropropagation** of *Iberis semperflorens*, Iapichino and Bertolino (2009a) when tried nodal explants, they stated that axillary shoots were visible within two weeks in medium containing 0.88 μM BA and were about 1.5 cm tall after five weeks. On excision and subculturing of these shoots onto the same medium, it continued to develop new axillary shoots. They obtained 50 per cent rooting when the microcuttings were placed on MS medium augmented with 2.85 μM (0.5 mg/l IAA). They further stated that 80 per cent of the acclimatized plants survived in the soil.

Cultural Practices

Candytufts require light, well drained garden **soil** containing lime. The soils having pH range of 5.5 to 6.2, *i.e.* slightly acidic are ideal for seed germination, though for successful growing of plants, the soil should be slightly alkaline, *i.e.* pH 7.2 to 8.5, therefore, acidic soils need liming. pH above 7.2 keeps away the clubroot problem of the plants. Acidic soils need liming. If the soil is heavy clay or poorly drained, planting should be done in raised beds as wet and poorly drained soil may lead to crown rot and clubroot disease. Candytufts require mild climate and **sunny situations** for luxuriant growth and flowering. *Iberis* is a drought tolerant and requires 8-10 weeks of vernalization for flowering. Flowering behaviour in *I. sempervirens*, a spring flowering species is modulated by vernalization and by attaining a certain plant maturity (Hamrick, 2003). Shady garden areas are at all not suitable for its cultivation. There are two methods of its seed sowing, (i) *in situ* sowing directly on the raised flowering beds where these can be thinned out when attain some 7-8 cm height, and (ii) sowing in the raised nursery beds and then transplanting in its permanent position on the raised beds. As in India, it is grown as winter-season ornamental so the seeds can be sown in October-November. In heavy clayey soils, these should preferably be sown or transplanted on raised beds due to more aeration and for better drainage for healthy growth of seedlings or plants. Seeds should be spread thinly in the nursery beds and covered with fine soil, sand or FYM. Seeds can germinate in light as well as dark conditions with soil temperatures of 16-21 °C. At 20- 21 °C, the seeds germinate within two weeks (Tomar, 2001). However, seed germination in *I. pectinata* is quite low in darkness than at an illuminated place. Seeds should be sown in lines spaced at 5-6 cm apart and at a depth of 2 cm. Thick sowing of seeds should be avoided as it results in week seedlings which then take longer time to regain proper growth. After seed sowing, beds should be covered with dried leaves or grasses to conserve moisture for better germination. The beds should be watered lightly daily with a fine nozzle can and should be kept moist regularly. When the seeds start bulging of the upper crust of the soil in the process of germination, mulch should be removed in the evening otherwise seedlings will etiolate. To minimize mortality of plants in the nursery, the soil should be drenched with 0.2 per cent Captan or Thiram. *Iberis* seedlings when attain 5-8 cm of height, *i.e.* roughly four weeks of sowing or when there are 3-4 true leaves, these can be **transplanted** in pots or in the well prepared field during cool hours of the day, preferably just before the evening or on cloudy days for maximum survival, followed by light irrigation to settle the roots. Proper spacing between plants is required for better development of plants. For commercial cultivation, *Iberis* should be transplanted at a spacing of 15 cm × 20 cm, 20 × 20 cm, 20 × 25 cm or 25 × 25 cm looking into the spread of the varieties or species. In poorly drained and heavy clay soils, planting should be done always on raised beds.

For producing good vegetative growth, *vis-a-vis* quality flower production, well rotten **farmyard manure** should be incorporated @ 20-30 tonnes per hectare at least one month before direct sowing of seeds, transplanting of seedlings or rooted cuttings. Kaur and Kumar (2001) reported that nitrogen application @ 30 g/m^2 resulted in tallest plants (38.89 cm), widest spread (48.27 cm), more number of branches (51.41/plant), longest flowering duration (58.68 days) and maximum seed yield (39.75 g/m^2). However, phosphorus application up to 10 g/m^2 resulted in significantly better growth and seed yield. In vegetative propagation when true leaves have appeared, fertilizers @ 75-100 ppm (N from a balanced nitrate form) should be applied, and when 2-3 sets of true leaves have developed, the level of fertilizers should be increased to 100-175 ppm but the fertilizer should have low phosphorus levels. Immediately after transplanting, light **irrigation** should be followed for proper establishment of seedlings. Thereafter, irrigation should be carried out at 10-12 day's intervals during winter season though during hot weather and in summer, it should be effected at weekly intervals. Since candytuft plants can tolerate long periods of drought so any delay in irrigation for 1-3 days is not going to harm the plants significantly though persistent drought is harmful, however, excessive moisture or shaded place is not congenial for its growing. Frequency of watering depends upon the soil type, season of growing, prevailing weather conditions, *etc.* Inadequate moisture level in the soil during plant growth and flowering adversely affects the flower yield. For pot plants, care should be taken not to drench the tender seedlings as drenching may cause damping off. Once the topsoil appears dry, plants should be watered thoroughly. Underwatering is preferred over overwatering in candytuft.

Weeding is an important operation for quality flower production in candytuft as weeds not only compete with the main crops for space, nutrients, light and water but also harbour various pests and diseases which affect the plant growth and flower production significantly. Moreover, weeds also become problematic during harvesting of flowers as well as seeds. Removal of weeds manually, as involves hoeing of field so soil is facilitated for aeration. In general, to keep the field weed-free, 4-6 manual weedings are required during the entire cropping period.

Pot Plant Production

Iberis semperflorens has good potential for pot cultivation and outdoor decoration due to its abundance of flowers and long-flowering period during winter months. Plants for pot plant production can be raised either through seeds or through softwood terminal cuttings (Iapichino and Bertolino, 2009b). For seed germination, seeds are sown in September when the soil temperature is around 20 ± 2°C. Seeds start germinating within two weeks, and then after proper cultural methods should be adopted for seedling growing. After one month of sowing or when seedlings have attained 3-4 true leaves, these are transplanted into polystyrene plug trays filled with 3 peat: 1 perlite (v/v) growing medium or in small polybags filled with soil, sand and FYM (1:1:1) and maintained in the same growing environment. Two parts ordinary garden soil, 2 parts fibrous loam, 1 part fresh and coarse sand and 1 part well-rotten farmyard manure or compost, all by volume, when used as planting mixture, the plants feel quite comfortable. In soft wood terminal cutting method, cuttings from vegetative shoot of almost 5-9 cm length having 3-5 nodes should be taken for rooting with upper leaves intact but lower removed up to 1-2 nodal lengths, when the plant is not in flowering. Cuttings should be treated with IBA solution (500 ppm) as a quick dip method. The treated cuttings should be inserted into plug trays filled with potting mixture as elaborated above. The plants thus raised from seeds (10-12 leaves) as well as cuttings (15-20 leaves) are transplanted into plastic pots (14-16 cm across).

Seed Production

For seed production in candytuft, healthy seedlings are transplanted on slightly raised beds at 40 cm × 30 cm distance. The best time for transplanting seedlings for seed production under North Indian condition is early November. Being a cross-pollinated crop, candytuft requires an isolation distance of 400-1,000 m for maintaining purity. Seeds become ready for harvesting in May when seed colour turns brown. Whole plants are removed manually and spread over tarpaulin in a well-ventilated room for a week and reshuffled daily for proper drying. The seeds are threshed manually and sieved with various types of mesh, and then cleaned finally by hand-winnowing or using a table fan which will separate out light and unviable seeds, *vis-a-vis* dust particles. Seeds are dried in shade under well-ventilated place to a safe moisture level. Seed yield depends mainly on species or varieties and management practices. One gramme of *Iberis umbellata* seed contains 350-400 seeds and seeds remain viable up to five years under refrigerated conditions (Rao, 1991).

Growth and Development

In *Iberis* species, there is limited application of plant growth regulators in crop production. In perennial species of the genus, rooting can be enhanced through application of IBA @500 ppm. In *I. sempervirens*, height management can be done through application of paclobutrazol @ 15 ppm or uniconazole @ 2.5 ppm. In *I. odorata*, GA_3 @ 50-100 ppm promotes plant growth and stem elongation when applied as foliar spray on plants with 10-20 cm height. Seed treatment with lower doses of gibberellic acid improves germination percentage, seedling vigour, final population stand, *etc.* Attoa *et al.* (2002) reported increased growth and photosynthetic pigments in *I. amara* by spraying the plants with tryptophan.

Postharvest

Candytuft flowers should be harvested at 50 per cent open stage in a tuft during cool hours of the day and then the stems should immediately be placed in a bucket containing neutral water for maximum vase life of the flowers. Good grading and bunching can enhance the quality of flowers and fetch better price. In cold storage, candytuft flowers should not be stored at temperature lower than 4.4°C. Humidity is an important factor affecting the life of cut flowers. A relative humidity of approximately 80 per cent is satisfactory. According to Patil and Dhaduk (2007), flowers harvested at 50 per cent open flower stage exhibit maximum vase life (10.42 days) and good quality of flowers for longer time. Among the various preservatives tried, $AgNO_3$ @ 25 mg/l + sucrose 2 per cent was found best with maximum vase life (13.67 days) along with excellent flower quality, high turgidity and freshness.

Patil and Patil (2008) tested various edible dyes (yellow, orange-red, phalsa-blue, apple-green, pink-rose, tomato-red and kalakhatta) for tinting *Iberis umbellata* cut flowers by dipping the cut ends in their 0.5 per cent, 1 per cent and 1.5 per cent dilutions for the duration of 30 to 180 minutes, with an increment of 30 minutes each, and recorded that all the dyes tinted the white flowers in the colours of the dyes used, the concentration and durations increasing the shades but with no any adverse effect on vase life though the shades persisted throughout the vase life.

Insect-Pests and Diseases

Insect-pest problem is little in candytuft due to presence of primary and secondary plant metabolites such as cucurbitacins and glucosinolate compounds (Kamel and El-Gengaihi Souad, 2008) but a few pests attacking candytuft are furnished here under. **Oystershell scale** (*Lepidosaphes ulmi*) sporadically infests the hardy candytuft and can be controlled by spraying the plants with Malathion or Sevin. **Diamond back moth** (*Plutella xylostella*) larvae feed on plants resulting in complete removal of foliar tissues except the leaf veins. The pest can be effectively controlled through the application of Coragen 20 SC (chlorantraniliprole) @ 150 ml/ha. **Root knot nematodes** (*Meloidogyne incognita* and *M. arenaria*) damage the plant root system and reduce its ability to absorb nutrients and water from soil. In severe case, aboveground symptoms include yellowing, wilting, stunting and ultimate death of plants. Various cultural methods such as soil solarization, organic amendments such as manure, green manure and compost are recommended to discourage growth of nematodes in the soil. Soils infested with nematodes can also be corrected by growing marigold and ploughing it in the field at flowering stage. Other control measures include application of neem cake @ 1 tonne per hectare and Furadan 3 per cent (Carbofuron) in the nematode infested areas.

Club root is a soil borne disease caused by *Plasmodiophora brassicae*. The fungus attacks roots and such roots are deformed and stunted. The disease can be controlled by growing the crop in neutral to alkaline soil as acidic soils favour its growth. Soil solarization as well as application of PCNB (Terraclor) will control this pathogen. **Damping off** is a serious disease of seedlings in the nursery and is caused by various pathogens such as *Rhizoctonia solani*, *Pythium ultimum* and *Pellicularia filamentosa*. The pre-emergence symptoms appear as brown necrotic spots, girdling the radicle though later on extend to plumule. The infection causes pre-emergence mortality of seedlings. Under post-emergence symptoms, lower part of hypocotyl has water-soaked brown necrotic ring leading to death of seedlings. When such seedlings are pulled, the root system appears partially to fully decayed. The fungi cause stems to curl and plants to collapse. Humid weather conditions favour the occurrence and spread of this disease. To control this disease, proper drainage in nursery beds should be ensured and drenching of nursery beds with Brassicol @ 0.3 per cent is quite effective. **Leaf spot** is caused by *Alternaria brassicaecola*. The symptoms appear as dark blackish, circular, elongated or irregular spots which appear on leaves, stems and inflorescences and enlarge at later stage of infection. To control this disease, spray the plants with Dithane M-45@ 0.2 per cent or Difolatan @ 0.3 per cent. **Powdery mildew** is another serious disease caused by *Erysiphe polygoni*. The symptoms appear as whitish, tiny and superficial floury patches on the leaves which later spread to whole aerial part of the plant. It is effectively controlled by spraying the plants with Karathane (0.025 per cent) or dusting the plants with wettable sulphur (0.2 per cent) at fortnightly interval. **Botrytis blight** is caused by *Botrytis cinerea*, a fungus that colonizes dead, dying and wounded plant parts from which infection spreads to healthy tissues also. Flower petals are most susceptible to its infection. Grey brown spore masses may develop on the dead tissues during moist conditions. To control this disease, various cultural methods should be followed such as avoiding overhead irrigation, removing all dead and dying plant parts, spacing plants for good air circulation and planting in neutral to alkaline soil. Other diseases which are not so seriously found in this crop are downy mildew (*Peronospora parasitica*) and white rust (*Albugo candida*). *Iberis sempervirens* cvs 'Tahoe' and 'Fish Back' plants were found to be infected with **phytoplasma** disease related to 16SrX-A group, exhibiting stunting, yellowing and witches broom symptoms. To control this disease, infected plants are uprooted and destroyed.

References

Armitage, A.M. 1989. Herbaceous Perennial Plants. *Varsity Press, Inc.*, Athens.

Attoa, G.E., H.E. Wahba and A.A. Frahat, 2001. Effect of some amino acids and sulphur fertilizers on growth and chemical composition of *Iberis amara* L. plant. *Egypt. J. Hort.*, **29**(1): 17-38.

Bateman, A.J. 1954. Self-incompatibility systems in Angiosperms. II. *Iberis amara*. *Heredity*, **8**: 305-332.

Bateman, A.J. 1955. Self-incompatibility systems in Angiosperms. III. *Cruciferae*. *Heredity*, **9**: 53-68.

Darlington, C.D. and A.P. Wylie, 1955. Chromosome Atlas of Flowering Plants. *Allen and Unwin*, London.

Dole, J.M. and H.F. Wilkins, 2005. Floriculture Principles and Species. *Prentice Hall International, Inc.*, London.

Ene, L.S.O. 1968. Cytogenetics of trisomics and tetrasomics in some species of *Iberis* L. (Cruciferae). *Cytologia*, **33**: 82-93.

Ene, L.S.O. 1971. A cytogenetic study of interspecific and allopolyploid hybrids of *Iberis amara* L. and *I. pectinata* Boiss (Cruciferae). *Cytologia*, **36**: 93-103.

Ene, L.S.O. 1973. Polyploids in the genus *Iberis*. *Cytologia*, **38**: 699-706.

Hamrick, D. 2003. Ball Red Book, Crop Production, Vol.2. *Ball Publishing*, Batavia, Illinois.

Iapichino, G. and M. Bertolino, 2009a. *Iberis semperflorens* L., an attractive Italian endemic shrub with high potential as flowering potted plants. *Acta Hort.*, No. 813, pp. 329-334.

Iapichino, G. and M. Bertolino, 2009b. Propagation techniques for *Iberis semperflorens* L. *Acta Hort.*, No. 813, pp. 427-434.

Kamel, A.M. and E. El-Gengaihi Souad, 2008. Secondary and primary plant metabolites as chemical markers for resistance of bitter candytuft (*Iberis amara*) plant against insect attack. *Not. Bot. Hort. Agrobot. Cluj*, **36**(2): 80-87.

Kaur, K. and R. Kumar, 2001. Effect of nitrogen and phosphate on seed yield and its parameters in candytuft (*Iberis amara*). *Proc. 4th Punjab Science Congress*, February 9-20, held at PAU, Ludhiana (abstr. AAMF 45).

Manton, I. 1932. Introduction to the general cytology of the *Cruciferae*. *Ann. Bot.*, **46**: 509-556.

Patil, S.D. and B.K. Dhaduk, 2007. Postharvest behaviour of candytuft cut flowers (*Iberis umbellata* L.) as affected by harvesting stages, chemical preservatives and colouring solutions. *Asian J. Hort.*, **2**(2): 256-265.

Patil, S.D. and H.E. Patil, 2008. Value addition of candytuft (*Iberis umbellata* L.) cut flowers coloured with edible dyes. *Asian J. BioSci.*, **3**(1): 163-167.

Rao, K.M. 1991. Textbook of Horticulture, pp. 59. Macmillan India Limited, New Delhi.

Tomar, B.S. 2001. *Seed Production of Annual Flowers*. ***In***: Commercial Flower Production (eds Sindhu, S.S. and M.L. Choudhary), pp. 61-67. Division of Floriculture and Landscaping, IARI, New Delhi.

24

Lilium (Family: Liliaceae)

Sanyat Misra, Ranjan Srivastava, Pragya Ranjan, R.L. Misra, Narendra Bhandari, Ajit Kumar and Yalek Messar

[**Common names**: Alpine lily/Sierra lily/Small tiger lily (*Lilium parvum*), American lily/Flame lily/Flame lips lily/Huckleberry lily/Orange cut lily/Red lily/Wild orange-red lily/Wood lily (*L. philadelphicum*), American Turk's cap lily/Lily Royal/Southern swamp lily/Wild tiger lily/White lily of the east (*L. superbum*), Asiatic hybrid lilies/Annunciation lily/Bamboo lily/Japanese lily (*L. japonicum*), Bourbon lily (*L. szovitsianum*), Canada lily/Meadow lily/Wild yellow lily/Yellow belled lily/Yellow lily (*L. canadense*), Candlestick lily (*L.* × *hollandicum, L. pennsylvanicum*), Carolina lily/Turk's cap lily (*L. michauxii*), Caucasian lily (*L. monadelphum, L. szovitsianum*), Chamise lily/Chaparral lily/Redwood lily (*L. rubescens*), Chinese white lily (*L. leucanthum*), Coast lily (*L. maritimum*), Cold-banded lily/Golden-rayed lily/Mountain lily/Yama-Yuri (*L. auratum*), Columbia lily/Oregon lily (*L. columbianum*), Coral lily (*L. pumilum*), Easter lily/Church lily/Trumpet lily/White lily/White trumpet lily (*L. longiflorum*), Eureka lily/Western lily (*L. occidentale*), Gray's lily/Orange-bell lily/Roan lily (*L. grayi*), Humboldt lily (*L. humboldtii*), Japanese lily/Rubrum lily/Showy Japanese lily/Showy lily (*L. speciosum*), Japanese Turk's cap lily (*L. hansonii*), Korean lily (*L. amabile*), Lemon lily of California (*L. parryi*), Leopard lily/Pine lily/Southern red lily (*L. catesbaei*), Lesser Turk's-cap lily/Little Turk's cap lily/Minor Turk's cap lily/Turban lily (*L. pomponium*), Lily, Madonna lily/St. Joseph's lily/White lily of the east (*L. candidum*), Manipur Nomocharis (*L. mackliniae*), Martagon lily/Turban lily/Turk's cap/Turk's-cap lily (*L. martagon*), Michigan lily (*L. michiganense*), Nankeen/Nankin lily (*L.* × *testaceum*), Nodding lily (*L. cernuum*), Orange lily (*L. bulbiferum, L. croceum*), Orange speciosum (*L. henryii*), Oriental hybrid lilies/Leopard lily/Panther lily (*L. pardalinum*), Port-of-gold lily (*L. iridollae*), Red Martagon/Scarlet Turk's cap (*L. chalcedonicum*), Regal lily/Royal lily (*L. regale*), Star lily (*L. concolor*), Devil lily/Kentan/Tiger lily (*L. lancifolium* syn. *L. tigrinum*), Turk's cap lily (*Lilium* species having much recurved flowers), Washington lily (*L. washingtonianum*), Wheel lily (*L. medeoloides*), Yellow Turk's cap lily (*L. pyrenaicum*), *etc.*]

Introduction and Origin

The flower, which is most beautiful part of the plant, is inseparable from social fabric of human life. *Lilium* is one of the most important cut flowers and pot plants in the global market. *Lilium* is a Latin name but is derived from the Greek word *leirion* used by Theophrasthus for the Madonna lily (*Lilium candidum*). It ranks 5th after rose, tulip, spray chrysanthemum and gerbera for total sales at the flower market of Aalsmeer (Ballarin *et al.,* 2009). The genus *Lilium is* herbaceous flowering plant normally growing from bulbs, comprising of about 130 species in the family Liliaceae, all originating from the Northern Hemisphere (Beattie and White, 1993). It is native to the northern temperate Hemisphere though a few species require greenhouse protection, however, the distribution is worldwide. *Lilium* is indigenous to a vast areas extending throughout north temperate zones of both hemispheres, and the area of distribution stretches through Europe and Asia across the Bering Strait, and from Vancouver eastwards across North America (Sanyat Misra and Misra, 2013). Many species are native to North America (some 24), Asia (50-60) and Europe (some 12 species), while it is only *Lilium martagon* which is distributed across Asia and Europe (Sanyat Misra and Misra, 2013). Fourteen species are native to Japan though six of them, *viz. Lilium alexandrae, L. auratum, L. japonicum, L. nobilissimum, L. pratyphyllum* and *L. rubellum*, are endemic. The primary centre of origin of the most important horticultural species *Lilium longiflorum* and *L. speciosum* is Japan but Taiwan and China are their secondary centre of origin. The remaining six *Lilium* species native to Japan are *L. callosum*, *L. concolor*, *L. dauricum*, *L. maculatum*, *L. medeoloides* and *L. leichtlinii* var. *maximowiczii*. From among the native Japanese lilies as is mentioned in one of Kaempfer publications in 1712 that he introduced *L. japonicum*, *L. maculatum* and *L. speciosum* into Europe. Subsequently, Thunberg introduced *L. longiflorum* and *L. speciosum* in 1794. *L. japonicum* and *L. maculatum* were introduced in 1804 and of *L. longiflorum* in 1819. Jeong *et al.* (1989) studied Korean native lilies, *viz. Lilium tigrinum* (*L. lancifolium*), *L. leichtlinii* var. *tigrinum*, *L. amabile*, *L. concolor* var. *parthenion*, *L. callosum*, *L. cernum* (*L. cernuum*), *L. distichum*, *L. tsingtauense* and *L. hansonii* for their shade conditions, altitude, the sites, soil characteristics and soil moisture content of the locations where these were recorded in the wild.

Beattie and White (1993) mention that **North America** is native for *L. bolanderi*, *L. canadense*, *L. carolinianum*, *L. catesbaei*, *L. columbianum*, *L. fairchildii*, *L. fortunofulgidum*, *L. gazarabrum*, *L. grayi*, *L. humboltii*, *L. iridollae*, *L. kelleyanum*, *L. kelloggii*, *L. ledebouri*, *L. maritimum*, *L. mary-henryae*, *L. michauxii*, *L. michiganense*, *L. nevadense*, *L. occidentale*, *L. oscellatum*, *L. pardalinum*, *L. parryi*, *L. parviflorum*, *L. parvum*, *L. philadelphicum*, *L. pikinense*, *L. rubescens*, *L. superbum*, *L. vollmeri*, *L. washingtonianum*, and *L. wigginsii*; **Europe and Asia Minor** are native for *L. albanicum*, and *L. armenum*; **Western Asia and Caucasus** are native for *L. aurantiacum*, *L. bosniacum*, *L. bulbiferum*, *L. candidum*, *L. carniolicum*, *L. caucasium*, *L. chalcedonicum*, *L. georgicum*, *L. heldreichii*, *L. janke*, *L. kesselringianum*, *L. ledebourii*, *L. martagon*, *L. monadelphum*, *L. polyphyllum*, *L. pomponium*, *L. ponticum*, *L. pyrenaicum*, *L. rhodopaeum*, and *L. szovitsianum*; and **Asia (China, Japan, Korea, Pacific Islands, India, Tibet, Myanmar and Himalayas)** is native for *L. alexandrae*, *L. amabile*, *L. amoenum*, *L. arboricola*, *L. auratum*, *L. bakerianum*, *L. brownie*, *L. callosum*, *L. cernuum*, *L. concolor*, *L. cordatum*, *L. cordifolium*, *L. dauricum*, *L. davidii*, *L. distichum*, *L.duchartrei*, *L. euxanthum*, *L. fargesii*, *L. formosanum*, *L. hansonii*, *L. henryi*, *L. japonicum*, *L. lancifolium*, *L. lankongense*, *L. leichtlinii*, *L. leucanthum*, *L. longiflorum*, *L. mackliniae*, *L. maculatum*, *L. medeoloides*, *L. miquelianum*, *L. nanum*, *L. neilgherrense*, *L. nepalense*, *L. nobilissimum*, *L. oxypetalum*, *L. papilliferum*, *L. paradoxum*, *L. pennsylvanicum*, *L. philippinense*, *L. poilanei*, *L. polyphyllum*, *L. primulinum* var. *ochraceum*, *L. pumilum*, *L. regale*, *L. rubellum*, *L. sargentiae*, *L. sempervivoideum*, *L. sherriffiae*, *L. souileie*, *L. speciosum*, *L. stewatianum*, *L. sulphureum*, *L. taliense*, *L. tenuifolium*, *L. tsingtauense*, *L. wallichianum*, *L. wardii*, and *L. wilsonii*. The species of **unknown origin** are given as *L. buschianum*, *L. croceum*, *L. maximowiczii*, *L. pavium*, *L. sachlinense*, *L.* spp. with much recurved flowers (Turk's cap type), *L. tabasco*, *L.* × *hollandicum* and *L.* × *testaceum*. In all, these are 121 known species and one unknown one, but certain species mentioned here are, in fact, botanical forms/varieties or natural hybrids of other species.

The species in this genus are the true lilies while the other plants with lily as the common names are related to the other groups of plants. Certain lilies are also edible such as tiger lily (*Lilium tigrinum*) in particular is consumed in China. A flavanoid alkaloid called lilaline is also isolated from the flowers of *Lilium candidum* (Jana and Roychoudhury, 1989). Most of the *Lilium* species and their varieties are fragrant though a few having quite unpleasant smelling. They may be of various colours and hues such as white, yellow, orange, pink, red to flaming red, maroon, and deep purple and mixed ones but not the exact blue. The floral diameter may be from 2.5 cm to 30 cm and in the shape of long trumpets to bowl-shaped, arranged on the erect stems up to the top. They are suitable for massing in mixed borders, in the shrubberies for enchantment when flowering in shrubs is over, in decks and patios as pot plants, in large rock gardens, in the wild or woodland gardens through selection of right species and varieties, for forcing in pots and for flower harvesting throughout the year especially in the temperate regions, and for excellent cut flowers

Botany, Genetics and Breeding

Lilies are usually erect and leafy-stemmed perennial herbs. These have non-tunicated fleshy scales underground, conforming to true bulbs. Fleshy scales overlap one another loosely or tightly. The bulbs may be of five types, *e.g.* (i) **concentric** where scales overlap each other and cluster firmly around a central growing point such as *L. auricum*, *L. medeloides* and other European & Asiatic lilies, (ii) **sub-rhizomatous** with declining basal plate and where these scaleless rhizomes elongate horizontally only to one direction, setting the new growing point away from the planted (old) ones such as *L. canadense*, *L. michiganense*, *L. superbum* and *L. washingtonianum*, (iii) **branched rhizomes** covered with pointed scales forming extensive mats such as *L. pardalinum*, (iv) **stoloniferous** (wandering stems) developing new concentric bulbs at apices of one or more horizontal stolons during the growing season such as *L. canadense*, *L. lankongens*, *L. leichtlinii* and *L. wilsonii*, and (v) **stoloniform** where the stem wanders some distance underground before emergence of shoots from the ground and the new bulbs often arise at intervals on the sections below ground, the example being *L. nepalense* and *L. wardii*. Most of the hybrid lilies produce concentric bulbs though ones developed from West American species which are generally rhizomatous, and some Asiatic hybrids have stoloniferous stems. Though bulb formation, by and large, is dependent upon place of origin, these form bulblets underground and a few species form bulbils in the aerial stems either in the axils of leaves or at lower leafless stem portion. Contractile and feeder roots are produced by the bulbs. Floral stalks are usually unbranched, erect and green but sometimes tinged purple or brown and always clothed with linear or lanceolate and sessile leaves (short pedicels in two species, *L. cordifolium* and *L. giganteum*) either scattered to their whole length such as majority of Asiatic lilies which are most suitable for growing in the modern greenhouses or in whorls such as *L. canadense*, *L. hansonii*, *L. martagon*, *etc.* In a few species the stem bases where these are attached with bulbs, form roots underground. Greller (1969) studied the patterns of spirality in developing and mature plant parts to relate with the evolutionary development of inflorescence in *Lilium* and other related taxa, and stated that mature inflorescences of *L. tigrinum* fitted the definition of a raceme but the ancestors of this species probably had well developed cymose branching patterns in the inflorescence. The large flowers have three petals along with three petal-like sepals. Flowers are borne in the form of an umbel or a terminal raceme, and are erect (cup-shaped) when solitary but when it is more than one, these are pendulous (bell-shaped), inclined or horizontal (funnel- or bowl-shaped) with six separate segments, which are scarcely differentiated as between sepal-like and petal-like organs, each bearing a nectar groove or furrow at the base. Stamens are six in number, hypogynous or slightly adherent to perianth mostly shorter than the segments, anthers are versatile, filaments slender, pistil one with one long style and three-pronged stigma, fruit dry and a 3-chambered capsule, and seeds many in two rows in every chamber.

For more than 100 years in Japan, bulbs of *L. longiflorum* are being grown and in the recent past they even evolved certain *Longiflorum* hybrids though as such *L. longiflorum* is self-incompatible, however, this barrier is broken by giving 45-49 °C heat treatment in distilled water to the style for 5-10 minutes (Sanyat Misra and Misra, 2013). A comparison between the bulb production in the Netherlands and Japan was made by Matsuo and van Tuyl (1986). Despite a long history of cultivation, a large number of *Lilium* cultivars are diploid (2n=2x=24) though *Lilium tigrinum* (*L. lancifolium*) is normally a triploid (2n=36) apart from having even diploid forms, and now the polyploid cultivars are also increasing (Schmitzer, 1991) though in most cases tetraploid lilies are also sterile. When diploid and tetraploid forms of *L. longiflorum* are crossed either way, the resultant hybrids produced are triploids with 2n = 36, however, diploid *L. formosanum* through colchicine treatment at stem apex resulted into tetraploids (Sanyat Misra and Misra, 2013). *L. longiflorum* × *L.* 'Utagoe' and *L. longiflorum* × *L.* 'Karakoma' hybrids are allotriploids, *i.e.* 2n = 36, 37; *L. longiflorum* × *L. pumilum* hybrids are also allodiploids with 2n = 24, 25; though *L. longiflorum* × *L. amabile* or *L. longiflorum* × *L. candidum* hybrids are diploid with 2n = 24. The *Lilium* varieties 'Sasatame' with *L. henryi*, 'Shikayama' with *L. nobilissimum* and '606', '6542' with *L. candidum*, 'Royal Gold' with *L. speciosum*, *L. auratum platyphyllum* with *L. henryi* and *L. regale* with *L. leichtlinii maximowiczii* also produced hybrids with 2n = 24. Bennici (1979) reported the mixoploid condition (chromosome mosaicism) when he regenerated *L. longiflorum* diploid form (2n = 24) tissues *in vitro*. In addition to having robust stems, large flowers, thicker and larger leaves, polyploids can also serve to combine desirable characters from species of different taxonomic sections; not only from the cultivated groups, *viz.*, Sinomartagon, Archelirion and Leucolirion, but also from sections that include non-cultivated species. Since the F_1 hybrids between the species of different taxonomic sections are highly sterile, it is imperative that breeding has to be carried out at the polyploidy level. This is based on the results on using intersectional interspecific hybrids of *Longiflorum* × Asiatic (LA) and Oriental × Asiatic (OA) groups of lilies and their polyploid backcross progenies (BC_1, BC_2 and BC_3) (Lim *et al.*, 2003). In order to cross genotypes of different ploidy levels, *i.e.* interploidy crosses, knowledge of embryo and endosperm ploidy levels is valuable because, unlike in most other plant species, lily has very large chromosomes and the embryo sac formation is of tetrasporic 8-nucleate type. In order to highlight the differences, a comparison is made with the most commonly occurring, monosporic 8-nucleate type of diploid potato, with 24 chromosomes (same as lily). The species used in interspecific hybridization programme (Van Tuyl *et al.*, 1988, 1996) were chosen on the basis of their respective characters such as *L. candidum* (section Lilium) for its whiteness, fragrance and low light and temperature tolerance; *L. longiflorum* (section Leucolirion) for its forcing ability and growth vigour; *L. henryi* (section Archelirion) with resistance to virus diseases, and bulb rot caused by *Fusarium oxysporum*, *Penicillium*, *Pythium* and *Rhizoctonia*, and *L. pumilum* for its earliness and bright red colourattion (Löffler *et al.*, 1996). Apart from the above, *L. rubellum* contributes earliness and therefore is being used to produce early Orientals and Asiatic hybrids (Sanyat Misra and Misra, 2013). Breeding is for major issues, *viz.* long vase life, fragrance, disease resistance, and development of male sterile lines especially in Orientals as these produce abundance of pollen which stain the clothes permanently. The reasons for using polyploidy in lily breeding are the larger flowers and the stronger stems of polyploid plants (especially important for forcing under low light conditions during the winter period; Van Tuyl *et al.*, 1985). In interspecific hybridization, the F_1-sterility at the diploid level is restored through tetraploidy though mostly the tetraploid lilies turn out sterile. Tetraploids can be obtained by colchicine treatment of mitotic cells (mitotic polyploidization). Restoration of F_1-sterility by doubling the number of chromosomes was performed successfully in *L. henryi* × *L. candidum*, *L. longiflorum* × Asiatic hybrids, *L. longiflorum* × *L. candidum*, *L. longiflorum* × *L. concolor*, *L. longiflorum* × *L. henryi* and *L. longiflorum* × *L. dauricum* using oryzalin. Using these tetraploids, backcrossing were performed on Asiatic, Oriental and Longiflorum hybrids (Zhou *et al.*, 2008). In the near future, the tetraploid *Lilium*-hybrids will originate from a range of different genotypes such as *L. longiflorum*, *L. henryi*, Asiatic hybrids, Oriental hybrids and so on, which could not be combined up till now, and this will open a complete new and promising possibility for innovating the lily assortment (Lim *et al.*, 2003).

Classification, Species and Varieties

The introduction of numerous beautiful and robust hybrids, notably from the Oregon Bulb Farm, has led to a classification of lilies into **IX Divisions**, and several sub-divisions. This classification has been adopted by the Royal Horticultural Society and North American Lily Society. This classification is particularly used by specialist nurserymen and for exhibition purposes. **Division I** (**Asiatic hybrids**) having three sub-divisions, **Division VI** (**Trumpet and Aurelian hybrids**) & **VII** (**Oriental hybrids**) with four sub-divisions each, and **Division IX** (**all true species and their botanical forms and varieties**) with eight sub-divisions whereas **Division II** (**Martagon hybrids**), **III** (**Candidum hybrids**), **IV** (**American hybrids**), **V** (**Longiflorum hybrids**), and **VIII** (**all hybrids not belonging to any other division**) have no further divisions. From academic point of view, a **Division X**

(those species belonging to other genera and not the *Lilium* though in the long past were considered as *Lilium*) is also sometimes found in certain literature sources though now it is obsolete. The Asiatic (Division I), Oriental (Division VII) and *Longiflorum* hybrids (Division V) are very popular in the trade and each of these with their specific positive and negative characteristics are furnished below along with other divisions systematically, following RHS and NALS specifications.

Division I (Asiatic Hybrids)

The commonly grown hybrid lilies forced in greenhouses are **Asiatic lilies** and their garden forms developed through 12 Asiatic species (China and Japan) and their hybrid groups [*L. amabile, L. bulbiferum, L. cernuum, L. concolor, L. dauricum, L. davidii, L.* × *hollandicum, L. leichtlinii, L.* × *maculatum, L. maximowiczii, L. pumilum* (syn. *L. tenuifolium*) and *L. tigrinum* (syn. *L. lancifolium*)]. Only because of this reason these are called as Asiatic hybrids. Mainly it is only the parents which differentiate Asiatic hybrids from those of Oriental hybrids. The Asiatic hybrids are known for their wide array of colours, generally profuse flowering and their smaller bulb size in comparison to the Oriental hybrids (Roberts *et al.,* 1985). On the other hand, the flowers of Asiatic hybrids appear smaller and less exotic when compared to the other two groups and some of its cultivars are susceptible to leaf scorch. The species used in the evolution of Asiatic hybrids have a diversity of shapes, *i.e.* dainty upright flowers of *L. concolor* and the nodding tiny Turk's caps of *L. pumilum*, to the large Turk's cap reflexed flowers of *L. tigrinum*, resulting into prolonged flowering period with various colours and shapes but these hybrids are odourless; and floral colouration, *i.e.* mostly orange and red and sometimes yellow. These have a wide range of flowering periods. Further these have three **sub-divisions**, *viz.* **IA** where the hybrids or varieties are early flowering, upright, flowers are either solitary or in umbels, and representative varieties are 'Alpenglow', 'Bravo', 'Butternut', 'Charisma', 'Cherub', 'Chinook', 'Cinnahar', 'Connecticut King', 'Croesus', 'Destiny', 'Edith', 'Enchantment', 'Endeavour', 'Fireband', 'Firecracker', 'Golden Pixie', 'Halloween', 'Heritage', '×*hollandicum*', '×*maculatum*', 'Mid-Century Hybrids', 'Pirate', 'Rosefire', *etc.*; **IB** where flowers face outwards such as 'Connecticut', 'Fighter Group', 'Lemonglow', 'Ming Yellow', 'Orange Glow', 'Preston Hybrids', 'Prosperity', 'Stenographer Group', *etc.*; and **IC** with pendant flowers such as 'Burgundy', 'Citronella', 'Connecticut Yankee', 'Debutante', 'Fiesta Hybrids', 'Halequin', *etc.*

Division II (Martagon Hybrids)

This includes hybrids of *L. martagon* and *L. hansonii*, and the varieties/hybrids are 'Achievement', 'Backhouse hybrids', '×*dalhansonii*', 'Gaylights', 'Marhan', 'Paisley hybrids', 'Sonata', *etc.*

Division III (Candidum Hybrids)

This is derived from *L. candidum, L. chalcedonicum* and other related European species such as *L.* × *testaceum* with exception of *L. martagon*. The varieties are 'Aries', 'Apollo', 'Artemes', 'Prelude', '×*testaceum*', *etc.*

Division IV (American Hybrids)

It is derived from American species. The varieties/hybrids include 'Bellingham Hybrids', 'Bellmaid Hybrids', 'Buttercup', 'Shuksan', *etc.*

Division V (Longiflorum Hybrids)

It is derived from *L. longiflorum* and *L. formosanum* such as *L.* × *flormolongi*, but excluding forms and polyploids of either species. The *Longiflorum* hybrids can be distinguished by their large, calyx-shaped flowers, their need for a shorter cold period and their good forcing characteristics. Their unpleasant characteristics are their very limited range of colours and being susceptible to viral diseases. The varieties include 'Formobel', 'Formolongi', *etc.* **LA hybrids** were derived in 1970 by crossing *Longiflorum* with Asiatic hybrids.

Division VI (Trumpet and Aurelian Hybrids)

These hybrids are derived from *L. henryi, L. sergentiae* and other Asiatic species but not *L. auratum, L. japonicum, L. rubellum* or *L. speciosum*. Further these have four **sub-divisions**, *viz.* **VIA** (**Trumpet type**): These include 'Aurelian group' derived from *L. sergentiae* × *L. henryi* × *L. leucanthum*, ×*aurelianense*, 'Black Dragon', 'Black Magic Strain', 'Golden Clarion group', 'Black Dragon', 'Damson', '×Golden Clarion', 'Golden Splendor', 'Green Dragon', 'Green Emerald', 'Green Magic', 'Honeydew', '×*imperiale*', 'Limelight', 'Olympic Hybrids', 'Pink Perfection', *etc.*; **VIB** where flowers face outward and are bowl-shaped such as 'Heart's Desire', '×First Love', 'Good Hope', 'New Year', 'Stardust', *etc.*; **VIC** with pendant flowers such as 'Christmas Day', 'Golden Showers', 'Summer Song', *etc.*; and **VID** (Sunburst type) where flowers open flat and are star-shaped such as 'Bright Star', '×Golden Sunburst', 'Thunderbolt', *etc.* **OT hybrids** were derived in 1980 by crossing Oriental hybrids with Trumpet lilies.

Division VII (Oriental Hybrids)

After Asiatic hybrids, this is the other group which is grown in greenhouses for forcing, and is derived using other four species excluded from Division VI. These are developed using four species, *viz. L. auratum, L. japonicum, L. rubellum* and *L. speciosum* (Sanyat Misra and Misra, 2013), native to China and Japan, *vis-à-vis* their crosses with *L. henryi* native of China, and these are late in flowering. Corr and Widmer (1984) also stated that Oriental hybrids were developed using *L. auratum, L. rubellum* and *L. speciosum*. The Oriental hybrids produce large flowers with a beautiful shape, require less light but take longer time to produce flowers, have strong fragrance, offer less colour variation (mostly white and pink), and are susceptible to various diseases. Further this has four **sub-divisions**, *viz.* **VIIA** with trumpet-shaped blooms, but now not in cultivation, however, this is in the process of giving some varieties; **VIIB** has rooting stems, produces bowl-shaped flowers and the varieties are 'Crimson Beauty', 'Empress of China', 'Empress of India', 'Empress of Japan', '×Parkmannii', 'Sunday Best', *etc.*; **VIIC** has rooting stems, bears flat and star-shaped flowers with recurved petal-tips, and the varieties are 'Allegra', 'Jillian Wallace', '×Imperial Crimson', '×Imperial Gold', *etc.*; and **VIID** has stem-rooting and bears recurved blooms such as '×Jamboree', 'Potomac hybrids', *etc.* **LO hybrids** were derived in 1970 by crossing *Longiflorum* with Oriental hybrids, **OA hybrids** were derived in 1995 by crossing Oriental with Asiatic hybrids, and **OT hybrids** in 1980 by crossing Oriental hybrids with Trumpet lilies.

Division VIII

Contains hybrids and forms not listed above or below in any of the divisions.

Division IX (All true species and their botanical forms and varieties)

It is further divided into eight **sub-divisions**, *viz.* **IXA** which contains the forms similar to *L. martagon* with reflexed flowers such as *L. cernuum, L. chalcedonicum, L. davidii, L. duchartrei, L. hansonii, L. martagon, L. monadelphum* and *L. waardii* but excluding those of American origin; **IXB** with upright flowers such as *L. bulbiferum, L. dauricum* and *L. tsingtauense*; **IXC** of American origin; **IXD** includes forms and polyploids of *L. formosanum, L. longiflorum* and *L. philippinense*; **IXE** bears bowl-shaped or trumpet flowers such as *L. regale,L. sargentiae* and *L. sulphureum* but excluding those mentioned under IXD; **IXF** includes those Asiatic lilies that have short trumpet or slightly recurved pendulous flowers such as *L. bakerianum, L. nepalense, L. primulinum, etc.*; **IXG** includes forms and varieties of *L. auratum* and *L. speciosum*; and **IXH** which includes dwarf ones of Asiatic lilies having affinity to *Nomocharis* such as *L. henrici, L. mackliniae, L. nanum, L. sheriffiae, etc.* Most important ones, out of a total of 80-100 *Lilium* **species**, are being described here below.

Lilium amabile (syn. *L. fauriei*), a native of Korea is one of the earliest lilies to cultivate. It comes up very well even with seeds, grows up to 1.2 m high and flowers even from small-sized bulbs, bearing unpleasantly fragrant, 6-8, racemose, pendant and deep red flowers but spotted black on the recurved petals though its form *luteum* bears yellow flowers. It produces numerous bulbils. Its leaves are narrow, lanceolate and irregularly distributed along the stem. It is tolerant to calcareous (alkaline) soil, prefers semi-shade and is most suitable for naturalizing in the woodlands.

L. auratum (syn. *L. dexterii*), a native of Japan is probably the most charming one in the world of lilies; therefore, it is a parent for many of the present day varieties. It requires planting distance of 30 × 30 cm, is stem-rooting, grows up to 1.5-2.4 m high where stems are purple-green and flowers from August to September bearing bowl- to funnel-shaped 10-30 highly fragrant white flowers of the diameter of 20-30 cm. The flowers are white with golden-yellow ray or band, and deep purple-wine petal-interior. Leaves are narrow, lanceolate and irregularly distributed on the stem. It is an excellent lily for pot cultivation. It does not like alkaline (calcareous) soil. Its var. 'Apollo' (*rubro-vittatum*) is red; var. *pictum* bears more intense markings of crimson; var. *platyphyllum* grows up to 3.0 m high bearing outward and downward facing wax-white flowers marked crimson; var. *rubrum* with red flowers; var. *virginale* bears white- and pale-yellow flowers. Apart from these there are many other important varieties under this species. Its var. 'Tom Thunb' is also very charming.

L. bolanderi is native to California, growing up to 90 cm tall, leaves are basal, whorled and greenish-blue. The floral stalks bear 1-4 small and reddish-purple flowers whose interior is spotted darker red. It flowers during June-July.

L. brownie is native to China which is the first Chinese lily introduced into Europe around 1835. It is most spectacular species bearing large trumpet-shaped slightly fragrant flowers generally in clusters of four. It does not like alkaline soil, is half-hardy, grows up to 90 cm in height and flowers in July but there is no seed-set. Leaves are dark, glossy green and lanceolate which are arranged irregularly on the stems. Petal texture is waxy, interior white though purple-mahogany exterior

L. bulbiferum (syn. *L. aurantiacum, L. buchenavii, L. croceum, L. humile, L. latifolium*), a native of Western and Central Europe (South Germany, Italy, Pyrenees, Switzerland and Yugoslavia) is stem-rooting, hardy, requiring 15-20 cm apart planting distance and 10-15 cm depth of planting, grows 0.9 m high and bears minute bulbils in its leaf axils. Its 6-7 cm long trumpet-shaped upright flowers are orange-red, fire-red or tangerine-orange spotted purple and has many forms such as *L. b. croceum* (syn. *L. aurantiacum, L. b. aurantiacum, L. croceum, L. humile*) which is native to South Germany, north-central Italy, South France and Corsica, grows up to 1.35 m high and bears some 11 cm diameter of light orange clustered flowers; *L. b. chaixii* is native to Maritime Alps, shorter and fewer orange-yellow flowers than former, tipped orange-red and spotted reddish-purple, *etc.* Certain of its forms bear bulbils.

L. canadense (syn. *L. canadense* var. *canadense, L.c.* ssp. *typicum, L. penduliflorum, L. pendulum, L. pulchrum*), is a native of eastern coast of North America from Quebec, New Brunswick and Nova Scotia, and in the south of West Virginia, Pennsylvania and Alabama. Its bulbs form 2-5 cm long stolons where at the end the bulbs are formed making a clump. It grows up to 1.5 m tall with whorled leaves and 7-10 cm diameter some 20 bell-shaped drooping lemon-yellow flowers, which are speckled with crimson-black spots internally. Its other forms such as *L. c. editorum, L. c. coccineum, L. c. flavum, etc.* are quite popular.

L. candidum (syn. *L. album, L. candidum* ssp. *candidum*) is though found in various countries around the Mediterranean such as Lebanon, Macedonia, Turkey and the eastern part of Israel, the Balkans, W. Greece and the Middle East but its actual habitat is not established. It grows 1.2-1.8 m high having stout stem with lanceolate and irregularly arranged whorled-like leaves and about 20 trumpet-shaped white, deliciously fragrant and 7-10 cm diameter of flowers where anthers are golden. It is a best known and most liked of all lilies in the old-fashioned ones. Once after senescence of stems again new rosette of leaves appear at the onset of winter. For planting it requires barely covered bulbs some 20-25 cm apart, is tolerant of alkaline soil and these rarely produce seeds. It is highly susceptible to *Botrytis* when there is poor aeration. It has various forms such as *L. cernuum* (syn. *L. cernuum* var. *atropurpureum, L. graminifolium, L. palibinianum*) native to Korea and Manchuria, requires planting distance of 15-20 cm and depth of 8-10 cm though its bulbs are quite small, is stem-rooting, comes up very well through seeds where these flower 2nd or 3rd year, grows up to 45-90 cm high bearing up to 12 nodding Turk's cap-shaped, pale-purplish-pink and fragrant flowers heavily marked with carmine; *L. c. salonikae; etc.*

L. cernuum, a native to Korea and Manchuria, it is stem-rooting lily growing up to 0.5 m in height, requiring planting distance of 18-25 cm and depth of 10 cm, and bearing mid-green linear to narrowly lanceolate leaves. The fragrant pale-pink flowers, appearing 5 cm across, having red-purple spots, appear in the shape of Turk's cap in June. It can be planted at sunny site or in partial shade.

L. chalcedonicum (syn. *L. pomponium*) is native to Greece. It is similar to Martagon lilies and is most intensely coloured. Bulbs are quite large (some 7.5 cm in diameter), ovoid and yellowish. It grows up to 1.2 m high, stalks clothed with leaves and bears some 10 brilliant red pendulous flowers covered with a lacquer-like sheen, may or may not be spotted, the petals are thick, recurved and waxy,and it is a seed parent which when was crossed with *L. candidum*, produced renowned cross *L.* × *testaceum*. The leavers are green, edged with silver, linear and are produced on the lower part of the stalk. It prefers alkaline (calcareous) soil.

L. concolor (syn. *L. concolor* var. *sinicum, L. c.* var. *typicum, L. mairei, L. sinicum*) is native to North-east Asia, mainly Central China. It grows up to 90 cm in height and bears up to 10 upright and slightly fragrant, scarlet and spot-free bowl-shaped flowers. Its forms *L. c.* var. *coridion* (syn. *L. c.* var. *luteum, L. c.* var. *partheneion* subvar. *Coridion. L. c.* var. *pulchellum* f. *coridion*) bears bright yellow spotted carmine flowers, and var. *pulchellum* (syn. *L. buschianum, L. c.* var. *buschianum, L. pulchellum*) bears vermillion to apricot and orange-red flowers on stout stems.

L. dauricum is native to Asia and Siberia, and is distributed from Mongolia to Korea. It is one of the first Asiatic lilies to be grown in Europe. It is very hardy and sun-loving species but is susceptible to viral diseases. It is planted some 10 cm deep and requires planting diatance of 20 cm apart. This species is hardy, stem-rooting and grows up to 90 cm tall, bearing upright, cup-shaped and bright red, spotted brown flowers some 12.5 cm in diameter. Its var. *luteum* bears deep yellow flowers.

L. davidii is native to China and is one of the most beautiful Chinese lilies which require 10-15 cm deep planting and 20-25 cm apart planting distance. It is hardy stem-rooting species which grows up to 1.8 m high, bearing up to 20 Turk's cap-shaped semi-pendulous flowers which are orange-red, spotted darker. Its leaves are prolific, narrow and linear with small tuft of white hairs at the base. Its particularly two good varieties are 'Maxwell' (*L. davidii* × *L. davidii* var. *willmottiae*) which bears orange-red pendulous flowers, and var. *willmottiae* which bears up to 40 dark orange-red spotted flowers.

L. formosanum is wrongly referred to as *L. longiflorum* var. *formosum* or *L. philippinense* var. *formosanum*. It is native to Taiwan, is half-hardy and stem-rooting. It requires 12-15 cm deep planting with 30 cm across planting distance. It easily comes up through seeds and flowers the same year after nine months hence can be considered as an annual lily. Leaves are dark green, narrow and linear and appear on the stem. It grows 1.2-1.8 m tall and bears 4 or more, 15 cm long, fragrant and white flowers which are suffused crimson-purple. The flowers are wide-mouthed and horizontally trumpet-shaped. It is widely grown as commercial cut flower for forcing.

L. hansonii (syn. *L. avenaceum, L. maculatum*) is native to the island of Pagelet off the coast of Korea. It is hardy, stem-rooting and requires 15-20 cm deep planting at a distance of 25 cm across. It grows 1.2 m in height with whorled leaves, bearing up to 12 pendulous orange flowers splashed yellow and spotted brown. It crosses well with *L. martagon* and has produced many notable hybrids such as 'Backhouse Hybrids'.

L. henryi is native of China and bears giant bulbs (diameter 20 cm or even more and polar diameter some 15 cm). It is hardy, stem-rooting, suitable for planting in alkaline or acidic soils and after it is once established, it naturalizes through rapid multiplication of its bulbs. The bulbs produce numerous bulblets, and the purple-brown stems which grow up to 2.4 m high are arching hence require staking and bear bulbils. Leaves are many, lower ones lanceolate while upper ones ovate. Its planting distance is 30 cm apart and 15-20 cm deep. Pale apricot-yellow flowers are spotted red, pendulous, recurved, solitary or 2-3 together while whole stem may bear up to 20 flowers, 7-9 cm long and 8-9 cm in diameter, Martagon-type and makes excellent cut flower. Its var. *citrinus* bears lemon-yellow flowers.

L.* × *hollandicum is, in fact, a group of lilies varying in colour from yellow to orange-red and is perhaps a cross between *L. bulbiferum* and *L. elegans* or *L.* × *maculatum*, and is also commercially referred to as *L. umbellatum*. Its stout stems grow some 75 cm tall bearing several cup-shaped and upright flowers. The flowers are in varying colours such as yellow to orange-red. The notable varieties under this complex are 'Erectum' (orange-red), 'Golden Fleece' (golden-yellow), 'Moonlight' (pale-yellow) and 'Vermillion Brilliant' (brillianat vermillion-scarlet), *etc.*

L. humboldtii is native to Sierra Nevada Mountains in California. Leaves appear on stems in whorls. It is suitable for alkaline or acidic soil and is semi-shade loving. It grows up to 1.8 m and bears 15 or more pendulous flowers which are 8-9 cm long, orange-red and spotted with maroon or dark purple.

L. japonicum (syn. *L. belladonna, L. japonicum* f. *roseum, L. krameri, L. kramerianum, L. makinoi*) is native to Japan where it is found growing among the bamboos and is one of the aristocrats of the genus. It is stem-rooting, grows up to 90 cm tall and requires planting distance of 20-25 cm across and 15 cm deep and is suitable for pot cultivation in the semi-shaded condition. It bears a few dark green, narrow, lanceolate and 15 cm long leaves. Its stems bear 5 trumpet-shaped, fragrant, some 10 cm in diameter and white shaded pink or shell-pink flowers with slightly recurved petals.

L. kelloggii is native to Pacific Coast of the United States and bears 0.9 to 1.2 m high brownish stems. The stems are clothed with lanceolate leaves. Flowers are Martagon-type, 5 cm long, pinkish, spotted red and the petals recurved.

L. leichtlinii is native to Japan. The bulbs are stoloniferous, floral stalk some 1.2 m tall bearing 1-6 pendulous and Turk's cap-shaped lemon-yellow flowers which are spotted maroon. Its notable variety is *maximowiczii* (syn. *L. maximowiczii, L. pseudotigrinum, L. leichtlinii* var. *tigrinum*) with 1-2 bright orange-red spotted purple-brown flowers.

L. leucanthum (syn. *L. leucanthum* var. *primarium, L. formosum*) is native to Western China (Hupeh Province). It is hardy, tolerant to alkaline soil, and grows 1.2 m tall, bearing numerous linear-lanceolate leaves, four or more funnel-shaped (trumpet-type), 12.5 cm long, fragrant, cream-white flowers suffused pale-yellow at base and each petal is striped greenish. Its more popular variety is *L. l.* var. *centifolium* (better known as *L. centifolium*) which grows up to 2.7 m tall with stout stems which bears up to 18 trumpet-shaped flushed-green and pinkish-purple flowers.

L. longiflorum is native to Japan which is semi-hardy and stem-rooting. Its planting distance is 20-25 cm across and depth of planting 15 cm. Floral stalk length is up to 90 cm with 4-5 trumpet-shaped slightly fragrant white flowers some 15 cm long. Numerous lanceolate leaves are irregularly placed on the stem. It is easily grown through seeds and flowers within a year, and as it is highly susceptible to viral infection so each year it can be grown through seeds to get virus-free material. Its varieties are numerous such as 'Croft', 'Estate', 'Eximium' (a well known Easter lily), 'Holland's Glory', 'Slocum's Ace', 'White Queen', *etc.*

L. mackliniae is native to Manipur (India) and is hardy and stem-rooting. It grows from 25 to 90 cm tall and requires planting depth of 15 cm and planting distance of 20 cm apart. The 2-3 or up to 8 bell-shaped and nodding flowers are borne on each stem, which are 5 cm across, rose-purple outside and white within. This species is easily raised from seeds, and the plants do well under semi-shade in moist but well-drained soil.

L. × maculatum (syn. *L. atrosanguineum, L. batemanniae, L. elegans, L. fortunei, L. thunbergianum*) is native of Japan and is considered a cross between *L. dauricum* and *L. concolor*. Its bulbs are concentric and are stem-rooting. It grows to 60 cm and bears cup-shaped and erect yellow, orange or red flowers which are variably spotted. Its varieties are 'Alutaceum' (deep apricot, spotted purple-black), 'Aureum' (orange-yellow spotted black), 'Bicolor' (orange with few spots and margins red), 'Biligulatum' (chestnut-red), 'Blush' (pink), 'Sanguineum' (stems up to 40 cm, flower orange-red and solitary), 'Wallacci' (apricot with maroon spots), *etc.*

L. martagon is native to Albania, most of Europe, especially the eastern part, Northern Asia Minor, Siberia, N. Turkestan and Mongolia. Kim and Lee (1990) made SEM studies of the epidermis of leaves and perianths in seven *Lilium* species and two botanical varieties and found that Korean representatives of genus *Lilium* belong to two sections, *Martagon* and the more advanced *Sinomartagon*. It is basal-rooting and hardy species, requiring a planting distance of 25 cm across and 10 cm deep, and grows up to 1.5 m tall with quite stout and greenish-purple stems, bearing oval-lanceolate leaves in whorls of six at intervals and numerous waxy-textured pendulous (Turk's cap-shaped) blooms. It is tolerant to alkaline soil and is mostly found growing in light woodlands. This species represents an entire class of its own. Unpleasantly fragrant flowers are 3-4 cm long, recurved and rose-purple with darker spots. It is easily raised from seeds. Its certain notable varieties are 'Album' (white), 'Backhouse Hybrids' (yellow, orange, and orange-pink flowers), 'Blush' (ruby red), *L. m. cattaniae* (dark purple), 'Dalhansonii' (dark reddish-maroon and spotted), 'Gleam' (black, spotted red), 'Marhan' (orange, spotted reddish-brown), *etc.*

L. michauxii is native to south-eastern United States which grows up to 75 cm in length; bearing whorls of seven or so, blue-green and lanceolate leaves; and Martagon-type (Turk's cap form) pendulous flowers some 75-12.5 cm long. Flower colour is orange-red, spotted purple.

L. monadelphum, a native of Caucasus is stem-rooting and hardy. It grows up to 1.3 m high and requires planting distance of 25 cm across and depth of 10-12 cm. This species tolerates alkaline-calcareous soil and prefers partially shaded position. The leaves are scattered along the stem, are up to 13 cm long and lanceolate to oblanceolate. The pendant flowers are Turk's cap-shaped (bell-shaped) with recurved petals, fragrant, 6.2-9.0 cm long, pale-canary or deep yellow dotted lightly with purple and petal-bases tinged red. *L. m. szovitsianum* (syn. *L. szovitsianum*) is quite similar to the type species.

L. nepalense grows some 1.5 m tall with dark purple stalk, bearing only a few scattered leaves. It is quite hardy and is found growing naturally in Nepal at an altitude of 2,400 metres. The flowers are large, beautiful, pendulous, fragrant and greenish-yellow but suffused purple in the throat.

L. ochraceum (syn. *L. primulinum*) is native to western China and northern Myanmar and is quite hardy. It is a very beautiful species but with endangered habitat. It grows up to 2.7 m high with purple-green stout stems, bearing seven or more large, pendulous and fragrant blooms which start appaearing late in the season, *i.e.* from September to November in its natural habitat. Its irregularly placed leaves are dark green and broadly lanceolate. Flower colour of half of the basal segment inside is bright reddish-purple and upper half inside is pale citron-yellow while exterior is greenish-yellow with purple-red spots. It is closely allied to *L. nepalense*. This is a variable species with at least three different forms.

L. pardalinum (syn. *L.p.* var. *angustifolium, L. p.* var. *pallidifolium, L. californicum, L. harrisianum, L. roezlii, L. superbum* var. *pardalinum*) is native to coastal belt of California (USA), a very easily cultivated species, whether it be sun or shade and is particularly suited to damp sites which is not waterlogged. Its bulb is quite irregular in shape and is a more branching scaly rhizome. The leaves are linear or lanceolate and whorled on the stems. It is planted at 30 cm apart, 10-15 cm deep and grows 1.5-2.4 m tall with Turk's cap-shaped 1-30 pendulous and fragrant blooms per stalk. The floral diameter is about 10 cm, the colour leopard-spotted dark crimson-brown and the petal-tips are crimson. In form and colour, it is a very variable species. Out of many varieties it has produced, the varieties 'Red Giant' produces quite large flowers and 'Sunset' are quite beautiful.

L. parvum is native to United States from Oregon to California at the Pacific Coast. It grows to a height of 1.5 m bearing orange or yellow and almost erect flowers where base is spotted dark purple. The flowers are small, some 3.8 cm long and trumpet-shaped.

L. pumilum (syn. *L. linifolium, L. sinensium, L. tenuifolium*) is native to China, Mongolia, Manchuria, North Korea and Siberia. It is a dwarf type, stem-rooting, easily raised through seeds, bulb planting depth 10 cm and distance 15 cm across, and grows 60-75 cm tall bearing about 20 Martagon-type (Turk's cap) nodding flowers about 3-5 cm in diameter with recurved petals. Flowers are brilliant red with no or a few black spots at the segment tips. Its few of the many varieties are 'Golden Gleam' (golden-yellow), 'Red Star' (red), 'Yellow Bunting' (yellow), *etc.*

L. pyrenaicum is native to Pyrenees and northern Spain, easy in cultivation and most suitable for naturalizing. It grows up to 90 cm high bearing prolific and narrowly lanceolate leaves, and up to 12 fox-like unpleasantly fragrant pendulous flowers which are Turk's cap-shaped with recurved petals, and yellow spotted dark greenish-yellow. Its var. *rubrum* is very attractive and bears 5 cm long reddish-orange spotted maroon flowers.

L. regale (syn. *L. myriophyllum*) is native to central-west China (Szechwan Province) and is finest of the trumpet-shaped lilies, easy in cultivation, tolerates even alkaline-calcareous soils, easily raised from seeds which flower the second year, and is widely grown commercially for cutting. It is robust-growing, hardy, stem-rooting and bears numerous scattered and linear leaves. It requires planting distance of 30 cm apart and 15-20 cm deep. It grows up to 1.8 m high bearing up to 30 large funnel-shaped, 12.5 cm long, fragrant and white flowers suffused maroon, dark pink at the base, and throat yellow.

L. rubellum (syn. *L. japonicum* var. *rubellum, L. makinoi* var. *rubellum*) is native to Japan and is stem-rooting. It requires a planting distance of 20-25 cm apart and 15 cm deep, and grows 30-75 cm tall, bearing 1-9 sweetly-scented, up to 7.5 cm wide, bell-shaped and rose-pink flowers with golden-yellow anthers.

L. rubescens is a native to United States from Oregon to California along the Pacific Coast and grows up to 1.8 m high. It prefers partial shade though not easy to cultivate. It is most beautiful species and presents three colour changes on the same stem: just emerged white ones which are waxy textured, pinkish ones when fully expanded and almost purple when old. Flowers are fragrant, 5 cm long, tubular and upright with wide flat star-like mouth, and have similarity to *L. washingtonianum*.

L. sargentiae (syn. *L. brownie* var. *leucanthum, L. leucanthum, L. myriophyllum*) is a late-flowering *Lilium* native to China. It is stem-rooting and bears many bulbils on the stem in the axils of the foliage. The leaves are dark green, broad and linear-oblong. It requires a planting distance of 30 cm and 15-20 cm deep planting. Floral stalk grows 1.2-1.8 m tall, bearing trumpet-shaped up to ten some 15 cm long and 10-12 cm wide and fragrant white flowers shaded purple-brown. The mouth of flower is quite wide. It is semi-hardy and difficult to cultivate.

L. speciosum (syn. *L. lancifolium, L. superbum*) is native to Japan. It is a half-hardy stem-rooting lily with stout stems with dark green and linear leaves, requiring planting distance of 30 cm and depth of 15 cm, growing 1.2-2.1 m tall, bearing up to 12 or even more bowl-shaped, 7.5-12.5 cm long, fragrant and white nodding flowers heavily shaded with crimson. Though prefers sunny situation but does not like alkaline-calcareous soil. It continues flowering for nearly three months continuously. Petals are recurved, thick and waxy and the edges are undulating. It has produced numerous notable varieties, a few being *album* (pure white), 'Gilrey', *magnificum* (more intensive and vivid colour markings), *rubrum* (red and crimson-scarlet), *etc.*

L. sulphureum (*L. myriophyllum*) is native to northern and southwestern China and north Myanmar, and is a very beautiful species. It grows up to 2.5 m tall preferring sunny places and bears numerous linear-lanceolate and up to 20 cm long leaves with bulbils being produced in their axils. Floral stalk bears up to 15 funnel-shaped highly fragrant blooms measuring some 25 cm in length and 20 cm wide (diameter), the largest in the genus and the colour of the bloom is sulphur-yellow, suffused pink, chocolate-brown and green exterior while paler-yellow interior, and the throat is suffused with dark yellow.

L. superbum (syn. *L. superbum* var. *superbum*) is native to eastern North America (eastern United States, *i.e.* from Massachusetts to Florida and from northern Alabama to southern Indiana). It is an excellent hardy lily with stem-rooting and large bulbs, likes partial shade as well as damp positions but not the calcareous soils, easily raised through seeds and requires a planting distance of 20-30 cm apart and depth 15-20 cm. Stems have whorls of lanceolate leaves. It grows 1.2-2.4 m tall and on pyramid-shaped raceme it bears up to 40 pendulous, 7.5-10.0 cm long and 10.0 cm wide Martagon-type (Turk's cap) recurved flowers of orange-red colouration where petals are tipped red, and the centres are red-spotted.

L. taliense is native to Tali Mountain Range in Yunnan (China) hence its specific name *taliense*. Its seeds were collected in the mid-1930 by George Forrest and taken to Europe where it flowered afterwards. Its stem is basal rooting. It grows from 0.9 to 3.0 m in height and bears numerous irregularly spaced narrow leaves on the stem, each with three faint veins. The inflorescence bears 1 to 10 nodding, fragrant and purple-spotted white flowers.

L. × testaceum (a hybrid between *L. candidum × L. chalcedonicum*, syn. *L. isabellinum, L. excelsum*) is an oldest esteemed lily-hybrid producing no seeds, and which grows 1.8 to 3.0 m high with stout and purple-green stems where leaves are small, prolific, thick in texture, lanceolate and silvery-margined. It likes sunny situation and is tolerant to alkaline-calcareous soil. The stems bear up to 12 large fragrant, pendulous, apricot-yellow, *i.e.* orange reddish-yellow (almost unspotted) flowers with recurved petal tips.

L. tigrinum (syn. *L. t. sinense, L. lancifolium, L. speciosum*) is native to East China, Japan and Korea. The stems are robust and clothed with dense woolly hairs. A stem-rooting species though produces no seeds but numerous black bulbils are formed mostly in threes in the leaf axils. Leaves are numerous, green and linear-lanceolate. It is one of the best known garden lilies though does not prefer alkaline-calcareous soils. It grows 0.6 to 1.5 m tall bearing Turk's cap-shaped, 12.5 cm wide, 25-40 pendulous and recurved flowers when opening but spreading afterwards. Flowers are bright orange-red and freely black-spotted. The fleshy bulb scales in China and Japan are peeled and cooked with sugar water and eaten as this is thought to act as a tonic and purifier. It has several excellent varieties such as *flaviflorum* (lemon-yellow, spotted purple), *flore-pleno* (syn. *plenescens*, double), *folis variegalis* (variegated foliage), *fortunei* (stems clothed with dense white wool), *splendens* (vigorous, rich fire-red, late-flowering), *etc.*

L. tsingtauense from China is hardy and stem-rooting, grows up to 90 cm in height, requires planting distance of 25 cm across and depth 10 cm, and grows well in lime-free soil. The stem bears a few-flowered widely funnel-shaped cluster with bright orange flowers some 5 cm long which appear in July.

L. wardii from Tibet is a small-bulbed stem-rooting hardy species which requires lime-free soil and partial shade with a planting distance of 30 cm across and 10-12 cm deep. Its Turk's cap-shaped fragrant flowers some 6.2 cm wide appear in pink colour, spotted purple in July and August.

L. washingtonianum is native to United States from Oregon to California in the Pacific Coast, found along with *L. kelloggii, L. parvum* and *L. rubescens.* It is a quite prominent species in its native habitat, bears stout stems with horizontally nodding and tubular-shaped fragrant flowers, growing up to 1.8 m high but is difficult to cultivate. The flowers are some 10 cm long with a wide flat mouth and with colouration of white, spotted purple.

Division X

Though this division does not exist as per RHS or NALS classification system but formerly was in existence when certain species, varieties and forms which now rank as *Cardiocrinum*, were then a part of *Lilium*. Here this division has been included from academic point of view. Woodcock and Steam (1950) suggested certain new groups as standards to be included within the lily assortment.

Commercial **varieties** should be selected on the basis of certain characters to ensure good domestic as well as international marketing. The characters should be internationally accepted for fetching a good market price. Market preference is for Asiatic varieties as these have wide colour range than other groups but are not fragrant. Stout and longer stalks and that too with 3-5 buds (3-4 when flowers are quite large otherwise 5) in Asiatic and Orientals are more preferable but then these require longer growing period and such spikes are more susceptible to light-deficient conditions so are not suitable for growing in winter, however, dwarf types are suitable for pot-growing only. The growth of lilies is slow during autumn and winter-growing though better when grown during summer. Growing of the varieties susceptible to leaf scorch and iron deficiency such as 'Star Gazer', 'Sterling Star', 'Connecticut King' and '*Longiflorum* group' should be avoided. However, those having larger bulbs are also susceptible to leaf scorch though larger bulbs are directly correlated to more height and larger flowers. Those susceptible to bud drop should not be grown during winter when there is poor lighting, however, these should get full light conditions to minimize this problem. Only those varieties should be selected which have longer vase life. Varieties having upward facing flowers have more acceptability than outward and nodding types. The popular lily varieties of the **Asiatic lilies** are 'America', 'Black Out', 'Brunello', 'Cinnabar', 'Compass', 'Corsage', 'Discovery', 'Elite', 'Enchantment', 'Fireflame', 'Grand Cru', 'Iberflora', 'Menton', 'Minstreel', 'Mona', 'Navona', 'Nello', 'Nerone', 'Nove Cento', 'Pollyanna', 'Prato', 'Prince Charming', 'Pronto', 'Prosperity', 'Romano', 'Sancerre', 'Shiraz', 'Tresor', 'Val di Sole', 'Vermeer', 'Vivaldi', 'Yellow Baby', 'White Pixels', *etc.*; of the **Oriental hybrids** are 'Allegra', 'Acapulco', 'Baccardi', 'Bergamo', 'Berlin', 'Bernini', 'Casa Blanca', 'Con Amore', 'Crimson Beauty', 'Crystal Blanca', 'Empress of India', 'Le Reve', 'Louvre', 'Marco Polo', 'Mero Star', 'Miditerannee', 'Merostar', 'Mother's Choice', 'Noblesse', 'Pompei', 'Rialto', 'Siberia', 'Snow Queen', 'Star Gazer', 'Sunday Best', 'Tiber', *etc.*; of the **LA hybrids** are 'Aatistero', 'Adoration', 'Arcachon', 'Bach', 'Bright Diamond', 'Brindisi', 'Busseto', 'Ceb Dazzle', 'Couplet', 'Courier', 'Indian Diamond', 'Indian Summerset', 'Litouwen', 'Menorca', 'Pavia', 'Rodeo', 'Royal Fantasy', 'Sulpice', *etc.*; and of the **OT hybrids** are 'Conco d'Or', 'Holland Beauty', 'Nymph', 'Robina', 'Shocking', 'Yelloween', *etc.*

Propagation

Lilies are commonly propagated only vegetatively through bulb splitting and simple division. In India, lily is a subject of temperate regions where these flower and form **bulbs** naturally, and in case when its chilling requirement is complete, it can also be flowered in the sub-tropical and tropical conditions only during winter season but there these may not form bulbs. In temperate regions, it is planted from September to March and the flowering is normally during summer and in some cases continuing till autumn. In the plains the lilies being winter crop where Easter lilies may form bulbs, to some extent, but others not, and for bulbing these require temperate conditions. However, in the process of growth and development in temperate and sub-temperate regions there is bulb splitting along with formation of one or more **bulblets** underground, a few species also form **bulbils** (aerial bulbs that are black-purple or green) in the axils of leaves which maintain the varietal character and can save one complete year for flowering as compared to seed-grown ones, and such lilies are *L. bulbiferum, L. sargentiae* and *L. tigrinum*. Other lilies may also be stimulated to form bulbils through foliar use of BA. Iizuka *et al.* (1978) applied 6-benzyladenosine, 6-benzylamino-9-2-methylpurine or 6-benzylamino-9-2-ethylpurine on stems of no-bulbil-forming species but bulbils were formed in 21 days with *L. longiflorum*, and in 30-35 days with *L. speciosum, L. platyphyllum* and *L. hansonii*. KT-30 (kinetin) and BA at 4, 20 and 100 ppm accelerated bulbil formation on stems of *L.* × *formolongi* and *L.* × *longiflorum*, maximum being 3 bulbils per leaf though BA was less effective (Watanabe, 1989), however, KT-30 treatment increased bulblet formation 6-30 times in *L.* × *elegans* (*L. maculatum*) though with BA it was only 1.3 to 1.6 times. Once, twice or three times sprayings at 2-week interval with PBA at 0, 50, 100, 500 or 1,000 ppm to *L. longiflorum* plants when they were 5 and 15 cm high, Nightingale (1979) obtained greatest number of bulbils when 15-cm plants were sprayed thrice with 500 ppm. Bulbils at lifting should be removed and kept at 4.5 °C for three months for physiological maturity. For current season flowering, the bulbs and for further growth and next or next to next season flowering the bulblets and bulbils are planted in well prepared field. Some stem-rooting lilies also form small bulbs on the stem just below the ground levels so when the daughter bulbs are mature for lifting, the soil near the stem is gently and carefully removed to recover these **offset-bulblets** for planting in the prepared field. Okada (1951) found significant negative correlation between the weight of mother bulbs and the number of bulblets produced. Bulblets, offset-bulblets and bulbils may attain the flowering size in one to three seasons. **Broken-up stems** before flowering should not be discarded but buried horizontally at 16 °C, and these stems may form bulbils at the base of the stems or in the leaf axils. In *L. longiflorum*, the **leaves** can be detached from the upper parts of the stem (leaves from lower part are less regenerative) and placed 1.5 cm deep in coarse vermiculite at 21 °C and 17 h photoperiod with misting, these develop adventitious bulblets in 4-6 weeks (Roh, 1982). Suzuki (1989) studied leaf cuttings of the cvs 'Hinomoto' and 'Georgia' by taking these as (i) 5 mm above the leaf base, (ii) whole leaf by hand removal, (iii) cutting at the base of the leaf leaving a small portion of the stem epidermis attached, and (iv) as the (iii) but leaving additionally a portion of stem parenchyma attached and then these were planted in soil where after 45 days the respective treatments produced bulblets at the leaf base at 0, 74, 69 and 97 per cent with 0, 3.1 (total weight being 0.35 g), 3.2 (0.15 g) and 2.9 (0.61 g) bulblets per leaf, respectively. When the cuttings were taken one week before flowering, produced more bulblets compared to those taken 3.5 weeks after flowering, and there was greater bulblet yield when cuttings from upper part of the stems were taken. Histological examination revealed that bulblet primordia were produced in the tissues between leaf blade and stem. 'Georgia' detached leaves treated with a range of concentrations of IAA, IAA-amino acid conjugates or BA, both alone and in combination, revealed that 25 mg IAA and 25 or 50 mg BA/l formed highest number of bulblets (Roh, 1990a). Oglevee *et al.* (1986) advocated programming of flowering in *L. longiflorum* through **leaf node cuttings** for producing bulblets, and for flowering without lifting of bulbs or applying cold treatment. This way the plants grow continuously. Flokstra (1969) when tried 3 cm long 10-15 leaf-bud cuttings with one leaf attached from each 3-month old plant of *L. longiflorum* selection 'Arai No. 5' derived from cv. 'Georgia' when flowering was over, their bases were dipped in rooting powder before insertion in a 1:1 peat and sand medium, and the propagating case was held at 25-30 °C with 16 h of lighting daily, he found strong callusing with some root formation after three weeks. When now the cuttings were moved into pots containing a mixture of leaf mould, clay, peat and sand and temperature was lowered by 5 °C, after further 3-weeks 2-3 bulbils were visible at the base of each cutting and not at the leaf axil. After these produced 2-3 roots and 3-4 cm long leaves, they were removed and planted out in boxes and then after some growth in the individual pots where these reached full size after 8-months from propagation.

During September-October or March-April, healthy dormant **scales** from the bulbs still in the field may be gently pulled off from the ground (bulbs of hybrid lilies underground are non-tunicated composed of fleshy, scale-like leaves formed in continuous concentric layers closely around each other), treated with some effective fungicide and then their half of the length is inserted in prepared beds obliquely in peat and sand mixture and provided with 10-13 °C soil temperature for establishment and bulbing at the scale-bases within 4-8 weeks. During this period, the temperature in no case should rise above 23 °C. If the scales not touching each other are scattered uniformly across the surface of the medium in boxes between alternate layers of moist peat and vermiculite, and the crates are stacked in a dark room having proper aeration, high humidity and optimum temperatures, the scales will come up very well, and the performance of scales taken in spring proves better than those taken in autumn and winter. 'Hinomoto' lily scales when were held at 25°, 15° or 10 °C in darkness or with upper part exposed to continuous red, orange or blue lights from November to April in Japan, no difference was recorded in bulblet initiation, growth or death of the mother scales at 25° or 15° irrespective of the colour of the light, but at 10 °C the bulblet initiation and rooting were more quickly in coloured light than in darkness and most rapidly in red light (Matsuo, 1974). Lin and Roberts (1970) stated that scales can perform inhibitory and promotive roles at various times as in case of cold-treated Easter lily bulbs where scales were removed up to 100 per cent. They stated that scales were not necessary for flower induction though number of leaves and flowers initiated were proportional to the number of scales retained, and the daughter scale removal accelerated daughter bulb sprouting by increasing internode elongation, reduced the number of leaves and flowers initiated, *vis-à-vis* delayed anthesis due to slower expansion of these organs, however, there was subsequent reduction in the rate of organ formation and expansion. Matsuo and Arisumi (1978a) when planted pre-propagated small scale pieces of parent bulbs of lily cv. 'Hinomoto', they obtained increased number of hypogeous-type plants (HTPs, where first leaves are foliage leaves) and hypo-epigeous-type plants (HETPs, where first foliage leaves develop after spring when bolting occurs) though decreased number of epigeous-type plants (ETPs, which produce a stem first) compared with undivided parent scales. However, scales from 2-year old bulb produced more ETPs and fewer HTPs than among those derived from 1-year-old bulbs. One-year old bulbs of less than 10 g and the parent scales producing less than 0.4 g did not produce ETPs, though smaller the parent bulb the greater was the percentage of HTPs. In another studies, Matsuo and Arisumi (1978b) recorded increased proportion of scales producing two or more bulblets following HWT for 30 or 60 minutes to parent bulbs prior to scale propagation. The results were highly pronounced on scales taken from inner or middle parts of the bulbs compared to outer parts. The small and larger bulbs of lily cv. 'Hinomoto' were when subjected to HWT at 45 °C for 30 minutes before scaling and the scales planted on August 7, 1975 or July 27, 1976, after 4½ months the leaves were found emerging more rapidly on scales of small parent bulbs compared to large ones though rate of leaf emergence from scale bulblets of larger bulbs was found increased (Matsuo *et al.,* 1978). The bulblets from inner scales when scaled in light was earliest though from outer scales it was latest from bulblets, and likewise in dark scaling the middle scale bulblets produced earliest leaves and inner scales the latest. Scaling in light through deeper planting of Easter lily scales (Matsuo *et al.,* 1982) found more epigeous-type plants and fewer hypo-epigeous and hypogeous types with more frequent occurrence of premature leaf and/or shoot development on the daughter axis of scale-propagated plantlets on epigeous type plants than hypo-epigeous types. Steffen and Bünger (1955) planted two layers of bulb scales in first week of November in boxes with moderately moist peat and other media at 18-20 °C. After 70 days when 80 per cent of scales formed bulblets the boxes were shifted to a cool place up to March and then the rooted scales were planted out in beds, however, they suggested that in case of *L. candidum* the scales should be planted in boxes earlier, preferably in July or August. Baardse (1977) recommended one volume of scales to 2.5 volumes of wet vermiculite (per 10 litre of vermiculite is wet with 2 litres of water). Matsuo *et al.* (1982) obtained three types of premature leaf and/or shoot development (PLS) from daughter axes of scale propagated Easter lily cv. 'Hinomoto' six months after scale planting in late July in boxes outdoors. They recorded enhanced PLS when applied slow-release NPK fertilizer and planted the scales with upper third above the soil as compared to planting 3 cm below the surface. Matsuo *et al.* (1983) removed middle scales from the basal plate of Easter lily bulbs of cv. 'Hinomoto', soaked in 0.2 per cent benomyl for 30 minutes and planted in boxes filled with sand 2 parts and loam 1 part (v/v) and then the boxes were covered and planted outdoors. With the application of slow-release fertilizer (13N-3P-11K) at 20 g per box of 40 scales + liquid fertilization with a 0.1 per cent solution of Hyponex (7N-6P-19K) every two weeks, they recorded earlier bolting, great number of stem roots and earlier root development though low winter temperatures retarded stem root development. At bulblet-forming stage for 8-12 weeks the temperatrure should be maintained at 23 °C, and then gradually brought to 17 °C for 8-12 weeks when these form stems (Beattie and White, 1993). These bulblets continue developing through whole winter by utilizing the starch accumulated at scale apex (Miller, 1990). Soluble sugars (glucose, fructose, and later sucrose and mannose) accumulate in scale regions due to active starch disintegration (Miller, 1993). Production of bulblets is affected by the genetic constitution of the species or varieties, the temperature during the scaling process, scale position on the mother bulb and bulb storage temperature before scaling (Beattie and White, 1993). Van Tuyl (1984) in *L. longiflorum* 'White Europe' found increased bulblet production with initial increase in temperature for holding the scales though it was found decreased in case of Asiatic hybrid 'Enchantment'. In most of the cases of Asiatic lilies for scale-bulb production, the initial temperature is maintained at 23 °C to encourage bulblet initiation, followed by gradually lowering to 17 °C for stem initiation, though McRae (1986) reported that some scales initiate bulblet formation within 15 days at 35 °C temperature, white-coloured bulb-scale types as in 'Enchantment' form bulbs in 4-6 weeks, *L. regale* 'Pink Perfection' with purple scale types in 6-10 weeks and other types such as *L. auratum* or the Oriental lilies may require 12-14 weeks. When *L. longiflorum* 'White American' bulbs were stored at 30 °C, their scales produced more scale bulblets than those stored at 0 or 10 °C, however, the scale bulblets from dormant mother bulbs were found sprouting more rapidly, and outer scales were observed producing more bulblets than inner ones (Matsuo and Van Tuyl, 1984, 1986). In commercial practice the scales are not treated with growth regulators for bulblet initiation though IAA, IBA and NAA may be tried which will certainly initiate good rooting as well as bulbing as has been recorded by Emsweller (1986) with NAA. Ivanova (1983) who tried several lily cultivars and species including *Lilium regale* by soaking for 6 h in several growth regulators and recorded greatest number and size of bulbils with 100 mg/l succinic acid, followed by 25 mg/l of NAA where though bulbil numbers were almost the same but sizes were smaller. However, CCC at 100 mg/l inhibited bulbil formation. On the scales two or more bulblets are formed on the inner surface in the lower part where these are attached with stem axis, though these are found formed along the scale-edges in *L. tigrinum* and its hybrids. In fact, every part of *Lilium* is regenerative, and scale bases are more regenerative with meristematic tissues in these areas so these easily differentiate into bulb tissues. The scale-bulblets after attaining satisfactory size should be shifted to 4-5 °C until April and then planted out in the prepared beds outside where these may take two years to attain proper size for flowering. Kachroo (1949) while studying vegetative propagation of *Lilium longiflorum* reported that small vegetative buds were formed on the inner and outer surfaces and edges of scale leaves, as well as in the normal axillary position, and a single scale leaf produced as many as three such buds which made their appearance as small papillae consisting of two overlapping opposite scale leaves, which later assumed the form of normal bulbs. Tamura (1949) found average moisture of the medium and 21 °C temperature as the best for number and development of bulblets, *vis-a-vis* root growth and though there was no difference of growing between light and darkness with regard to number of bulblet formation but those grown in darkness produced larger bulblets. Hosaka and Yokoi (1959) stated that early scaling from inner part and outer parts of 10-20 days early lifted bulbs resulted into rapid and much larger bulblets when grown at 15-25 °C with 30-60 per cent soil moisture, and the bulblet and root formation were found quite rapid in soil or river sand than in vermiculite or sawdust. *Lilium longiflorum* 'Chotaro', a variegated-leaved variety is most suitable for pot-planting and when this is conventionally scale-propagated, a sizeable number of plants appear with entirely green foliage, *viz.* 7 per cent from middle, inner or innermost scales but being significantly higher in outer scales though when basal plate of the parent bulb is nail-wounded, the number of daughter bulb production is almost 3 times more with all the plants coming up variegated (Matsuo *et al.*, 1989).

However, for getting virus-free planting material and for quick multiplication of elite germplasm, the lilies are propagated through **tissue culture method** by taking explants from all the plant parts as every lily plant part is regenerative. *L. longiflorum* has been successfully micro-propagated by taking explants from shoot apices (Sheridan, 1968), pedicels (Liu and Burger, 1986), leaves (Stenberg *et al.*, 1977) and bulb scale sections (Gupta *et al.*, 1978; Stimart and Ascher, 1978; Takayama *et al.*, 1982). Coquen and Astié (1977) used sepals, petals, stamens and carpels cut from the floral buds above their point of insertion on the receptacle for culturing on nutrient agar + NAA, their basal end penetrating 5-10 mm deep into the medium. After some days callus tissue began to appear near the surface of the agar on the adaxial face of the flower part and under the epidermis. Meristems arising in this callus developed into bulbils on the outer surface. They found bulbils appearing faster from sepals and petals but most prolifically from carpels. Takayama and Misawa (1979) while working with *L. aurantium* and *L. speciosum* stated that best material for bulblet formation is scale from intact or *in vitro* produced bulblets, 20 °C temperature and media-pH 6 where leaf emergence could be initiated by irradiance. They found high kinetin concentration stimulating the formation of many bulb scales, high sucrose stimulating the organ formation, increase in the strength of the MS medium stimulating bulblet formation and growth though suppressing root elongation, and high NAA inducing root formation but kinetin reversing this effect. Aguettaz *et al.* (1990) harvested scale-explant *in vitro* generated bulblets of *L. speciosum* cv. Rubrum No. 10 after 11 weeks of culture, and without any cold-treatment planted in soil where they measured dormancy of the bulblets through their sprouting in the soil. Bulblets obtained by *in vitro* culturing at 15 °C showed no dormancy though those cultured at 25 °C induced a high level of dormancy; however, growing cultures in short or long days, light or darkness, up to 1 mg ABA/l or 0.5 mg BA/l or 5 mg NAA/l or 10 mg IAA/l or 1 mg TIBA/l or a MS macro- and microelement mixture did not affect the dormancy status; but addition of 1 g/l sucrose or 1 mg/l GA reduced the level of dormancy. MS medium-cultured bulblets did not produce any leaf when planted, those of such bulblets chilled at 4 °C for 12 weeks and then planted led to emergence of leaves 20 days later, but those chilled for <10 weeks depended on growth temperature after transplanting, however, immersion of bulblets in 250 or 500 mg GA_3/l substituted the chilling requirement but only in partly- and fully-chilled bulblets (Niimi *et al.*, 1988). Novak and Petru (1981) with 'Crimson Beauty' of Oriental hybrid regenerated 25 bulblets per scale explant after 60 days in Linsmaier & Skoog medium with 5 µM BA+1µM NAA. They induced bud differentiation on the same medium on ovary- and leaf-derived explants and in callus culture. When bulb scale segments from freshly harvested bulbs of *L. martagon* were cultured on MS medium with 0.01 or 0.1 mg NAA and/or 0.01 or 0.1 mg BA/l, bulblets started forming after 3rd week and developed fully by the next 5 weeks and after 15 weeks of culture bulblet production per explants was found highest in media with 0.1 mg/l NAA, followed by 0.01 mg BA/l (Rybczynska and Gomolinska, 1989). Takayama and Misawa (1983) estimated that from a medium size bulb in one year the *L. speciosum* to a tune of 1.2×10^{10} and *L. auratum* to a tune of 3.2×10^{12} can be multiplied. The success was achieved by Niimi and Onozawa (1979) from leaves of *L. rubellum*; Maesato *et al.* (1991) from shoot apices of *L. japonicum*; and Shoyama *et al.* (1988) from bulblets of *L. japonicum*. Emsweller (1963) for the first time applied embryo culture technique in Oriental lilies. Van Tuyl *et al.* (1990) when studied influence of sucrose, NAA, pH and the presence of anthers on the media where ovules were cultured, stated to have recorded beneficial effects by the presence of anthers, high sugar concentrations, NAA and high pH and obtained a much higher percentage of interspecific hybrids compared to usual rescue procedures. *Lilium* species have also been micro-regenerated through bulb scale sections (Robb, 1957; Van Aartrijk and Blom-Barnhoorn, 1980), anther culture (Sharp *et al.*, 1971), filamentous tissues (Montezuma-de-Carvalho and Gulmaraes, 1974) and callus (Simmonds and Cumming, 1976; Kato and Yasutake, 1977).

A few of the species which breed true are raised through **seeds** if gestation period from seed-sowing to flowering is not considerably long. However, for raising new varieties and hybrids, these are grown only through seeds. This method is cheap and easy and is used generally by breeders for evolution of new varieties. Seeds are sown during winter in pots and are planted outdoors in the spring. Ideal soil mixture should have loam, peat and sand in the ratio of 7:3:2. Nagamura (1983) in Japan when had sown *L. concolor* seeds on various dates for bulb production, recorded suppression of germination at 10 and 35 °C and optimum germination at 20-25 °C temperatures, maximum bulb yield from April and May sowings but largest bulbs from June-sowing and found hotter months of July and August most unsuitable for seed sowing. Together he recommended planting of small bulbs 10 cm deep at 5 × 5 cm distance in a sawdust bed, administered with Nutricote at 450 or 600 g/m^2. Moisture level should be maintained in the soil so that seeds do not suffer a dry spell. The only demerit in this method is that it involves a considerable time lapse. Seeds with delayed hypogeal germination normally do not produce top growth in the first season as in case of *L. auratum, L. canadense, L. martagon, L. monadelphum, L. speciosum* and *L. szovitsianum*. Therefore for best results these should be sown in June-July in the boxes filled with sterilized mixture of 7 parts loam:3 parts peat:2 parts sand and throughout the growing period the mixture should remain moist. Payne (1960) has advocated vernalization of slow-germinating types of lily seeds. He stated that either seeds should be dusted with a fungicide along with damp sand and sphagnum moss or vermiculite in glass jars or polythene bags which after sealing should be kept at a warm place until appearance of bulblets and then these are either planted in pots and are left out during the cold weather or kept in the jars in a refrigerator for a month before planting out in the open, or the seeds should be sown in a damp mixture of leaf mould and loam in containers and then kept in a warm and dark place for six weeks and then transferred to a refrigerator or cold place for one month, and then the pots are placed in a cold frame where germination takes place within a few weeks. In **hypogeal germination**, there is no cotyledon but true seed leaves emerge above the ground but after transference of the food from the seed to a point which is underground in between the root tip and true seed leaves so the point swells as if a miniature bulb. If the growth cycle is continuous it is called **hypogeal immediate** but if incubation period is necessary after the initial germination, it is called **hypogeal delayed**. In case of **epigeal germination** a cotyledon emerges above the ground where developing seedling is nourished by the endosperm and just above the root tip apparently a node is formed which afterwards becomes a bulb. True leaves are produced only when this node has sufficient strength. **Epigeal immediate** types are immediately sown after harvesting though is not always practical, however, **epigeal delayed** requires a definite cold period before germination. Asiatic and trumpet lilies have epigeal germination. In March the seed-sown boxes are placed in cold frame where leaf growth should appear. Jackson (1958) stated that sowing of seeds of *L. formosanum* var. 'Pricei' in heat under glass in January and moving the seedlings to flowering house in March caused flowering in August and September with capsule ripening in October which fetched a quite handsome price in the market. This way, certain lilies can also be grown as an annual crop. Rohde (1970) studied six *Lilium* species for their germinability over a wide range of temperature range (between 1 and 25-30 °C) and he found optimum as 15-20 °C though in case of *L. pumilum* the optimum was found 2-4 °C higher. Germination was found reduced or delayed by a long period of temperature deviation from the optimum. Soaking of *Lilium* seeds for 6-16 h in a 0.005 per cent sodium humate solution with and without addition of minor nutrients increased germination between 20 and 112 per cent with varying treatment durations (Gindina *et al.*, 1974). *L. speciosum* × *L. auratum* seeds when were washed for 10-14 h in running water to remove water-soluble acids including ferulic (a growth inhibitor), *p*-coumaric and sinapic acids (Emsweller *et al.*, 1963).

Cultural Practices

The production of lilies for use as cut flowers usually takes place in the border soil or in containers under glass, plastic or shaded greenhouses. In this way, the growers will have fewer problems due to unfavourable weather conditions, will be able to control the climate for lily cultivation, and will have a way of producing lilies year round. The planting of lilies outdoors

***Lilium*: Shoots and Plants in Beds, Harvesting and Preparation for Packing.**

can be done only in regions that provide a favourable climate throughout the cultivation period. Before starting outdoor cultivation, one should consider the possible problems including *Botrytis* that can occur as a result of heavy or prolonged rainfall, hailstorms, heavy wind and occurrence of frost. Intense sunlight will produce lilies with short stems. For outdoor cultivation, it is particularly important to have rich, moisture-retentive, well-drained soil; an effective irrigation system; and screening to protect the plants from excessive wind and sunlight. The last factor is important for producing long stems possessing more buds during summer months.

Lilies can be flowered in almost any type of **soil** which has good structure to permit aeration, retains moisture throughout the growing season and is not water-logged. Heavy loam and clay soils are less suitable for cultivation of Oriental hybrids. For producing other groups of lilies, these soils can be improved by working in substrates containing humus and coarse sand to a depth of 30-40 cm. Maintaining proper pH in the growing layer is essential for root and bulb development of lily plants and for proper absorption of nutrients. Soil with an excessively low **pH** can result in excessive absorption of such elements as manganese, aluminium and iron; an excessively high pH leads to an insufficient absorption of such elements as phosphorus, manganese and iron. For growing Asiatic, LA and *Longiflorum* hybrids, maintaining a pH of 6.0 to 7.0 is recommended; for the Oriental, OA, LO and OT hybrids, a pH of 5.0 to 6.5 should be maintained. In general, the pH of the soil should be 6.5-7.5. *Lilium philadelphum* prefers acidic soil, *L. candidum* alkaline while Easter lilies non-acidic soils with high Ca content, *i.e.* pH 6.5-7.0. However, the pH in soilless media should be maintained at 6.1-6.5 so that fluoride leaf scorch is avoided to the maximum. To reduce the pH, pH-reducing materials such as peat products should be worked into the upper layer of the soil. When using artificial fertilizers, pH-reducing fertilizers such as the ones containing ammonium and urea are preferable. To increase the pH, liming materials with or without Mg can be incorporated in the soil.

It is essential to provide an initial **temperature** of 12 to13 °C to all lily groups until stem roots have developed. Always this level of temperature may not be possible but somehow the soil temperature will have to be adjusted somewhere below 20-25 °C (Erwin and Heins, 1990). To some extent requirement of cold temperature can be achieved through shading for long after emergence, proper ventilation, application of cold ground-water and mulching with reflective material. During the cultivation stage, the optimum daily temperature requirement is between 15 to 22 °C (Roh and Wilkins, 1973). Temperature below 15°C can result in bud drop and yellowing of the foliage in Oriental hybrids. Choi (1983) reported that low temperature treatment for 40 days at 0.5 °C inhibited leaf emergence in bulblet during propagation of Easter lily cultivars 'Georgia' and 'Hinomoto'. Van Tuyl (1984) advocated that higher storage temperature promotes leaf emergence from scale bulblets of *L. longiflorum* cv. 'White American'. Higher storage temperature produces more epigenous type plants, especially from outer and middle scales. The variation in temperatures also affects the bulb formation. The wider temperature variation gives better results than constant high temperature which fails to induce bulb formation (Gries, 1971). Improper **ventilation** in polyhouses becomes suicidal during summer as temperature inside shoots up soon, hence the polyhouses should have top and side vents for controlling the temperature.It is advisable to build a structure with a top ventilation gap of minimum 3 feet to have proper ventilation inside the greenhouse. Depending on design and size of the greenhouse, a provision for side ventilation as well can be made. If the distance from side to centre of the greenhouse is less than 9.14 m, side ventilation in combination with top ventilation can be recommended. The sides should not be closed completely to allow free exchange of fresh air.

Greenhouse during growing of lilies should have 80-85 per cent **relative humidity**, and wide fluctuations will cause stress and leaf scorch in susceptible varieties. Optimum **CO_2** concentration for *Longiflorum* hybrids is ±2,000 ppm though for other lilies it is 800-1,000 ppm. Raising of the CO_2 concentration to 0.1 per cent in winter before flowering at the time when continuous supplementary lighting was being provided, lessened the bud drop in glasshouse-grown 'Enchantment' lilies and produced though a little delayed but better flower quality than illumination alone did (Durieux, 1978a). For the best results, it is suggested that light intensity should not be less than 7,500 mW/m^2 and due to reduction in bud drop the temperature during the illuminated period could be raised from 18 to 20 °C to shorten the forcing **period.** Kohl and Nelson (1963) reported that low **light intensity** increases plant height but reduces percentage of dry weight in *L. longiflorum*. Schenk (1975) found that bulblet formation was greater on basal tissues than on distal tissues in light intensity of 1,000 lux, whereas, those formed in dark conditions were larger. Depending on the season and time of the year, location of growing, the genotype and greenhouse structuring, the requirement of light may differ. Insufficient minimum light intensity in the greenhouse for Asiatic hybrids is 190 Joules/cm^2 and such light conditions result in inadequate growth and bud drop, more susceptible being Asiatic hybrids as compared to Oriental and *Longiflorum* types. Heins *et al.* (1982a) found reduced plant height in forced lilies under 8 h photoperiod than those forced at natural day light. Plant heights of cvs 'Ace' and 'Nellie White' were reduced by 29 and 45 per cent, respectively when forced under shortdays from emergence to flowering. Receptive light/dark cycles of 4, 6, or 12 h had no effect on flower bud development rate from the time buds were 6-12 cm long to anthesis. Kawabata *et al.* (2002) reported the cause of poor floral pigmentation in lily cv. 'Acapulco' due to low light intensity because of reaction mediated by photoreceptors located in the petal and the reaction mediated by sugar supply from leaves or stems, but to find out more effective one out of the two by shading flowers with cheesecloth (PPFD<17 µmol m^{-2}sec^{-1}) or with aluminium foil (0 µmol m^{-2}sec^{-1}), the anthycyanin concentration was found reduced which suggests that anthocyanin production is a photoreceptor-mediated reaction, however, shading the whole plant with cheesecloth lowered anthocyanin concentration compared to flower-shaded ones due to reduced total sugar concentration from 3.0 to 1.6 per cent. **Shading nets** not only reduce temperatures but also avoid leaf scorching. The percentage of shading depends on the light conditions at the site and time of the year. The movable shading net is ideal, one that opens or closes depending on the light conditions. A cheaper option is to go for a 50 per cent agro-white shadenet which should be fixed on top of the polyfilm. During monsoon, the same can be removed depending on the light conditions at that time.

Soil sterilization by thorough drenching some 10-15 cm top layer of soil with chloropyriphos (0.2 per cent) and Thiride (0.2 per cent) solution should be carried out after incorporating farmyard or any other organic manure in the bed at 3 kg/m^2 (300 quintals/ha), however, at planting again the fungicides may be used. For **field preparation**, the soil should have been dug out thrice one by one at a depth of 25-30 cm followed by planking every time, all the foreign material such as woods, alkathene pieces, glass, stone, pebbles or any other material which may hinder plant growth and perennial rootstocks of grasses should be collected either manually or through forks and then soil should be made fully pulverized to work with. Before final or third digging or ploughing, it would be better to flood the field so that at planting the field should have sufficient moisture in the soil which may support sprouting of the bulbs properly and before sprouting the crop may not require watering. After third ploughing, the field is levelled, and final lay out is made by demarcating bunds and irrigation channels and their tributaries. Now beds are prepared to any width, preferably 2 m and to any length but the topography should be even and properly levelled so that in case of flood irrigation the water spread is even. Two metres wide beds will facilitate cultural operations easier as by sitting to both sides of the channels one can operate earthing, weeding, hoeing and other practices. Flowering size bulbs are taken out from cold storage at least one week before planting. The imported frozen bulbs before planting are stored at 0 to 2 °C with a relative humidity of 95-98 per cent to the maximum of two weeks in well aerated store rooms though unfrozen ones stacked only at aerated room temperature of 10-15 °C in a single layer. Once the bulbs are unfrozen, these should not be stored at 2 °C as these may be damaged by frost. In unfrozen bulbs where shoots have started emerging, either these may be rooted in the crates at a temperature of 10-12 °C or planted directly in the prepared field immediately without damaging the emerging shoots. It is best to use smaller bulb sizes from various lily groups as large bulb sizes in the Asiatic, LA and Oriental hybrids result in higher risk of leaf scorch. The bulb size chosen also depends on the desired number of buds per stem. However, smaller bulbs produce smaller stems with only a few buds per stem so that stems do not lodge. Following list indicates the bulb sizes for each lily group that can be used.

Group	Bulb Size
Asiatic & LA hybrids	10/12 cm, 12/14 cm, 14/16 cm, 16/18 cm and 18 cm+
Oriental, OT, LO & OA hybrids	12/14 cm, 14/16 cm, 16/18 cm, 18/20 cm, 20/22 cm & 22 cm+
Longiflorum hybrids	10/12 cm, 12/14 cm, 14/16 cm, 16/18 cm & 18+

Planting is usually carried out during October- November in the plains (tropical and sub-tropical areas) and in February to April in the hills (temperate and sub-temperate areas). Once a lily bulb is planted, for the first three weeks for its intake of water, oxygen and nutrients, it depends only on the bulb roots it has already developed before planting though during this period the bulbs further develop stem-roots underground at the stem bases just above the bulbs. These stem-roots soon take over some 90 per cent function of the bulb roots for transport of water and nutrients. Therefore, for production of quality lilies, proper development of stem-roots should be ensured by deep planting (some 10 cm or even more) only healthy bulbs in the healthy (disease-free), moist and cool soil on cold days or only in the morning, and in the process of planting care should be taken to prevent drying out of the bulbs or its roots, and after planting the beds should be **mulched** with rice hulls (20-30 kg/100 m^2), sawdust (both softwood and hardwood whether fresh or rotted), straw, chaffed bagasse pieces, cocopeat, dried leaves, leaf-mould, compost, vermicompost, *etc.* with some 2.5 cm thick layer for moisture conservation and to prevent soil desiccation, but 5.0 and preferably 7.5 or 10.0 cm for suppressing weeds of any type, to maintain soil structure, and to increase soil temperature especially during winter in autumn-planted crop. Box (1963) with a vigorous strain of 'Georgia' lily in pots with an usual soil pH of 6.5-7.0, when mulched up to the rim of the pots with sawdust, and left in the open from the time of emergence in October except night-covering of the plastic over plants during winter to save them from dropping temperatures, and then keeping as such up to the forcing. He suggested that forcing should begin 90, 85 or 80 days before Easter when temperatures of 15.6 °, 18.3 ° or 21 °C, respectively, are maintained. He for all the three treatments advocated a transition temperature of 10 °C for two weeks, and for forcing at 18.3 ° or 21 °C an additional conditioning period of 1-2 weeks at 15. 6 °C. However, if pendimethalin (Stomp) is to be used as pre-emergence weedicide, some 2 litres of stomp per hectare of land should be dissolved in tanker-water to wet upper crust of the soil by spraying so that whole field is wet. Spraying of the Stomp and spreading of mulches should go on one by one and bit by bit so that upper layer of soil is not disturbed after Stomp application. It would be better to plant the bulbs in double-rows spaced 30 cm and from 10 to 25 cm deep, deeper for stem-rooting species and varieties though *L. candidum* requires shallow planting. Safe enough is planting the bulb 2½-3 times the height of the bulb, however, in case of *L. candidum* the nose of the bulb should be just below the surface of the soil. Roh (1979) has recommended planting of *L. lancifolium* bulbs with circumference of 15 cm, weighing 13-14 g, and at 10 × 10, 15 × 15 and 20 × 20 cm spacings. Commercially, larger bulbs are given more spacing, *i.e.* 16 × 18 cm accommodating some 36 plants/m^2 while smaller flowering- bulbs 12 × 12 cm, accommodating 64 plants/m^2. In a survey with Dutch nurseries, Rotten (1978) found that the planting density for 'Enchantment' lily ranging from 46 bulbs (size-grade 14-16) to 87 bulbs (size-grade 8-10) per net square metre and peak time for flower supply being April and May. **Planting distance** and **depth** depend on the bulb size, plant spread, soil type and availability of sun. The varieties and species having limited growth can be given 100 cm^2 spaces though others up to 400 cm^2. Young lily bulbs have contractile roots which pull them down to their proper depth. Following table indicates the maximum and minimum planting densities per m^2 area.

Bulb Size	*Bulbs/m^2*	*Planting Distance*
8-10 cm	49	15 × 15 cm
10-12 cm	42	16 × 15 cm
12-16 cm	36	16 × 18 cm

Tropical and sub-tropical regions encounter winter **weeds** though temperate regions summer and rainy season weeds. Round Up (Glyphosate) can be used when leaves in lily crop have senesced. Diuron at 1.1 kg/ha or Neburon at 1.6 kg/ha will control all the pre- or post-emergence weeds in lily field. Gentner *et al.* (1958) when used with Easter lily planting the pre-emergence weedicides dalapon (sodium) at 6.7-18.0 kg/ha, CIPC at 6.7, 9.0 and 18.0 kg/ha, monuron at 2.25 kg/ha, diuron 1.12 and 2.25 kg/ha, 2,3,6-TBA (sodium) 1.12 and 2.25 kg/ha or dinoseb (alkanolamine) at 9.0 and 13.5 kg/ha, they obtained satisfactory weed control for more than two months without any ill-effect. However, when they applied post-emergence weedicide such as 2,4-DES 4.5 kg/ha, neburon 4.5 kg/ha, CIPC at 13.5 kg/ha and 2,4-D at 2.8 kg/ha, though there was no injury with 2,4-DES but others caused slight injury when used as directed sprays or as broadcast-spray. Chlorozuron (Tenoron) at 5 g/l of water for 400 m^2 area of lily planting is quite effective when applied first after six weeks after planting with a repeat two weeks before harvest. Boer (1972) applied Tenoran (50 per cent chloroxuron) at 0, 125 g, 250 g or 375 g/ha to glasshouse planted Easter lily 'White Lady', *L.* Fiësta hybrid 'Citronella', *L. hollandicum* hybrid 'Fire King', *L.* Mid-Century hybrid, *L. speciosum rubrum* and *L. tigrinum* and found satisfactory weed control with no plant injury, but suggested for deep planting of bulbs, weedicide treatments from 15 days to three months of planting and sprinkler irrigation following weedicide sprayings. Chlorpropham (chlor-IPC) 40 ml/100 m^2 along with Paraquat (30-40 ml/100 m^2) can be used when field is full with weeds but lily crop is yet to emerge, Lenacil at emergence and Simazine 3-4 weeks after emergence. Adamson (1959) suggested use of monuron at 2.8-3.4 kg/ha + chlorpropham at 4.5 kg/ha during autumn as pre-emergence for controlling broad-leaved weeds, grasses and volunteer cereals in Easter lilies but where only grasses and cereals are a problem he suggested pre-emergence autumn spraying with dinoseb (amine) at 5.0 kg/ha along with chlorpropham at 4.5 kg/ha. As per Dutch recommendations, the chlorpropham together with paraquat should be sprayed pre-emergence when weeds are actually present but lily shoots are at least 1 cm below the surface, however, lenacil should be applied at emergence and simazine 3-4 weeks after emergence (Anon., 1976). Maximova (1970) recommended 2 kg a.i./ha atrazine application between the rows in autumn which controls annual weeds for two years along with a number of perennial weeds also to some extent, though monuron at 2 kg/ha or propazine at 4 kg a.i./ha when applied in the spring gave excellent control of perennial weeds during the second year of treatment, without any adverse effect on main crop in any form. The weeds in small holdings should be managed manually.

A plant support system may be necessary depending on the cultivation period and cultivar. Since winter crop is vigorous so this may, in most cases, require **staking**. Plants attaining the height of 80 cm and above or those producing more number of larger flowers should certainly be staked otherwise due to lodging even slight wind or rain may break such plants. In greenhouses, 3-tier strings may be used at every 25-30 cm height. Harvesting of spikes will involve pulling of the stems rather than cutting.

Feeding and Watering

Fertilizers such as calcium ammonium nitrate (50 g), diammonium phosphate (25 g) or single superphosphate (20 g) and muriate of potash (50 g) each per m^2 should be applied and mixed thoroughly in the upper layer of the beds before planting. During first three weeks of planting, good root development should occur, so at this time any salt damage should be prevented. For this reason, the weekly alternating applications of the quantities of calcium nitrate and potassium nitrate will have to wait until three weeks following planting. Magnesium sulphate (0.15 to 0.20 kg/100 m^2) can be applied if any yellowish discolouration occurs in the lower foliage. Sanyat Misra and Misra (2013) mention that most of the lilies are comfortable at 140 kg N, 280 kg P_2O_5 and 200 kg K_2O per hectare. Roughly one-third of N and whole of P and K should be used as basal dose while remaining two-third of N in further two doses, once when the plants are 20-30 cm tall and the second when spike is going to emerge. Gindina (1976) in a fertilizer trial with *L. davidii* stated that when he supplied 45 kg N + 60 kg P_2O_5/ha at the start of the growth, 45 kg N + 40 kg K_2O during the peak of the growth, and 45 kg N + 60 kg P_2O_5 + 40 kg K_2O at bud development, the ornamental value of the crop was found improved only in the third year of application. Giustiniani *et al.* (1988) through a study in the autumn in Italy on 'Mid-Century' lilies stated that 16 and 42 g each of N and K_2O/m^2 brings about quality blooms. It would always be better to use the fertilizer only after soil analysis. Triple superphosphate in glasshouse Easter lily crop should be avoided as it contains fluoride, however, in its damage Ca application will reduce it (Nederpel, 1979). The requirement of calcium is more by *L. longiflorum*, so lime as pre-plant and calcium nitrate during active growth should be applied. Absorption of nitrogen increases during inflorescence development so at this time nitrate nitrogen to *L. longiflorum* should be applied. Eastwood (1952) stated that low concentrations of soil nitrate ranging from 2 to 10 ppm produce better 'Creole' lily with a greater yield of better quality flowers than do high level of nitrate in the range of 50 to 80 ppm as quality appears to deteriorate at 15-20 ppm. In Easter lily plantings at five sites along the coasts of South Oregon and North California, monthly leaf analysis showed that leaf N, P, K and Zn fell while Ca, Mg and Cu rose from March to October; Mg and B were found stable till June-July and then rose till October; and Fe fell between March and April and then remained constant (Chaplin and Roberts, 1981) and then they suggested for early leaf analysis to adopt proper corrective fertilizing. Prince and Cunningham (1989) when terminated liquid fertilization to Easter lilies 'Nellie White' plants at the visible bud stage or 2 weeks later during potted lilies when stored for 3 weeks at 2 °C in the dark, increased foliar chlorosis than those which were fertilized up to stem harvest.

The bulbs should not be planted in soil that is too dry; however, planting should be done only when soil is sufficiently moist so that crop sustains up to sprouting without watering. After sprouting, first **irrigation** sufficient to wet the root zone is carried out. The EC of irrigation water should be 0.5 mS/cm or even lower. The maximum acceptable chlorine level of irrigation water used for greenhouse irrigation is 200 ppm. If enough water is not provided, the result will be slow growth, uneven development, shorter stems and early flower bud desiccation. Too much water should be avoided as well since this will reduce the presence of oxygen around the roots and limit their development, *vis-à-vis* weakens the roots which will become susceptible to *Pythium* and *Phytophthora*. An excessively wet soil during the period when the stems are growing rapidly will also result in limping of plants due to excessive cell enlargement. During dry periods, water consumption can reach 8 to 9 litres/m^2/day. The following tests have been recommended to check the correct quantity of water: squeeze some soil tightly in hand, if the moisture in the soil almost but not quite drips from hand, this indicates the proper degree of moistness. Also, the water distribution of irrigation system should be checked on a regular basis.

Pot Culture

Lilium can be grown successfully in the pots even for forcing. Goldsberry (1966) reported 'Cinnabar', 'Enchantment', 'Fireflame', 'Formosanum Pricei', 'Harmony', 'Joan Evans', 'Sunspot', and 'Tobasco', whereas Seeley (1982) reported 'Connecticut King', 'Connecticut Lemon Glow', 'Enchantment', 'Sunkissed' and 'Sunray' as pot plant lilies. During autumn the bulbs may be set in 15-20 cm pots singly or three bulbs to a 25 cm pot containing 1 part loam, 1 part well-decomposed organic manure and ½ part coarse sand, all by volume. Boodley and Sheldrake (1963) with forcing of 'Ace' lilies recorded 2:1:1 (v/v) combination of peat moss, vermiculite and perlite as the best potting mixture compared with peat moss + vermiculite and peat moss + perlite, and all the three superior over soil mixture. Stem-rooting species and varieties are planted deeper in the pots. Potted ones may be stacked at dark and aerated place at a temperature of 12-13 °C for proper root initiation and development. After appearing of the sprouts these are brought to full sunlight but at a cooler place for further growth and flowering. At no time the potting medium should become dry otherwise results will be disastrous. Such pots after flowering may be shifted outside at a shaded spot for senescence of aerial growth on its natural way though at this time also the pots should be regularly watered to prevent the bulbs from drying. These are again repotted during autumn anywhere or during March-April in temperate areas. Oriental lily 'Sans Souci' is usually grown as cut flower in America but if its height is effectively controlled it may have an excellent market there as a new potted plant. In this direction, Holcomb *et al.* (1989) studied ancymidol and XE-1019 (uniconazole) as bulb dipping, foliar spray, soil drenching, chemigation and their combinations. XE-1019 sprays at 1 to 25 ppm and drench at 0.1 to 0.5 mg a.i. per plant proved effective in reducing stem length. Ancymidol applied at 66 and 132 ppm as foliar spray and 0.25 and 0.5 mg per plant as drench also proved quite effective. Spraying or drenching of ancymidol at 0.25 g per plant reduced the height by 6 per cent and 9 per cent, respectively, whereas 0.5 mg per plant drenching reduced the height by 12 per cent. Bulb dipping or chemigation with 100 ppm ancymidol brought about 21 per cent and 22 per cent, respectively, plant height reduction. They recorded further reduction when bulb dipping was combined either with spray or drench. In another experiment with the same variety, White *et al.* (1989) cooled the bulbs for 42 days starting from October 28 at 4 °C, followed by 28 days at 2 °C, potted in a peat-perlite medium and grown in a greenhouse at 17 °(night)/22 °C (day) temperatures. They studied ancymidol and uniconazole as bulb dipping, media drenching at shoot emergence when shoot height was 7-8 cm, as foliar spraying at 7-8 cm or 14 cm plant height or daily application through irrigation water, and recorded reduced height with all the treatments, particularly those treated with uniconazole and that too when applied as drench, though those given combined treatments of bulb dip, drench as well as spray were very effective in dwarfing the plants.

Out of all the dwarfing chemicals, α-cyclopropyl-α(methoxyphenyl)-5-pyrimidine methanol (Ancymidol, A-Rest, EL-531, Quel, Reducynol) has extensively been used for reducing the height of stems in *Lilium* (White, 1971a, b; 1972a, b; Johnson, 1973; Dicks *et al.*, 1974; Hammer, 1974; Marshall *et al.*, 1974; Besemer, 1975; Kiplinger *et al.*, 1975a, b, c; White, 1976; Tjia *et al.*, 1976; Roh and Wilkins, 1976, 1977; Lewnes *et al.*, 1977; Weiler, 1977, 1978; Staden and Maas, 1978; Tsujita *et al.*, 1978, 1979, 1980a, b; Staden, 1979; Lewis and Lewis, 1980, 1981, 1982; Parivar *et al.*, 1985a, b; Adzima, 1986; Wilkins *et al.*, 1986; Grueber *et al.*, 1986; Lewis and Gilbertz, 1987; Larson *et al.*, 1987; Gilbertz and Lewis, 1990; etc.). This has been used as bulb dipping in 28 to 264 mg/l from 5 seconds to 60 seconds, or up to 12 h; as vacuum bulb soak; as soil drench at 0.25-1.0 mg and 0.313 to 5.0 ppm (or with repeated use) per 12.5 cm pot when the shoots were 7.5, 15.0 and 22.5 cm high where 7.5 cm high shoots gave good response; as foliar spray with 50-100 ppm when plants were 7.5 to 30 cm tall; and with irrigation water at 40 to 100 ppm. Kiplinger *et al.* (1971, 1972a, b) applied EL-531 (0.025 per cent a.i.) at 1 mg in 6 fluid ounces per 15-cm pot of 'Ace' and 'No. 44' lilies, when shoots had just appeared or were 7.5 cm long, they recorded significant reduction in stem length. However, when Phosfon drenches at 1:96 solution at 6 fluid ounces per pot, EL-531 drenches and sprays at 2.2 ml/l and CHE-8913 foliar sprays at 500 ppm were applied at three stages, *i.e.* pot-planting, 15 cm tall and 30 cm tall, they obtained reduced plant height with Phosfon drench when shoots were 15 cm tall. Dicks *et al.* (1974) applied a 15-ppm soil drench of ancymidol at 750 μg per plant on three different dates, *viz.* at inflorescence initiation, and then 10 and 20 days later, where they recorded shortest plants when the plants were sprayed on the second date because of the good absorption of the chemical by quite developed root system. However, when they used aqueous solutions of GA_3 to the shoot tip with 20, 200 and 2,000 µg per plant to the already ancymidol-treated plants (500 µg/plant as a 10 ppm soil drench) and in the absence of ancymidol, 2,000 µg GA_3 per plant, it significantly promoted stem extension.

Apart from ancymidol, various other chemicals and/or their formulations were also tried by various workers. Struckmeyer and Beck (1952) used diethanolamine salt of maleic hydrazide (1, 2-dihydrophyidazine-3-6 dione) in the concentrations of 0.025, 0.05, 0.1, 0.2, 1 per cent and higher, sprayed once or twice at four stages of growth and recorded decreased height with increased doses along with abnormal flowers. Struckmeyer (1953) with 0.2 per cent MH recorded only deleterious effects on 'Croft' Easter lily. Stuart *et al.* (1961) used Phosfon (as drench at 2.25 g/pot or as a soil amendment) and recorded about 50 per cent reduction in forced 'Georgia' Easter lily with soil amendment. Wilkins *et al.* (1963) recorded Phosfon L (a liquid formulation) at 4 and 8 per cent more effective compared to Phosfon D (a dry material) or CCC (Chlormequat, Cycocel) in case of 'Croft' lilies. Pope and Baum (1965) advocated Phosfon treatments as pot soak at low concentrations because it results in more blasted buds and leaf damage in 'Georgia'. Easter lilies, *Lilium longiflorum* 'Georgia', 'Enchantment' and *L. speciosum* 'Rubrum' plants when a few cm tall in 12-cm pots were sprayed with 50 cc of 5 per cent CCC per pot, which reduced stem length some 15 cm in 'Enchantment' but not in others (Systema, 1968). Adelman *et al.* (1970) recommended 1:96 Phosfon and water application when 'Ace' and 'Nellie White' lilies were 30 cm in height though still there were some plant damages. Kneissl (1971) recommended thrice 5 per cent CCC watering to the cvs 'Arai No. 5', 'Destiny', 'Georgia', 'Hinomoto' and *L. speciosum* when shoots were 1-3 cm, 8-12 cm and 15-25 cm long, however, spraying twice either with CCC or B9 was not found effective. When pot-grown Mid-Century hybrids 'Enchantment' and 'Joan Evans' were administered single soil drenches with CCC at 25,000 ppm, Phosfon and ethrel at 1,000 ppm, and EL-531 at 5-15 ppm, Dicks and Rees (1972) recorded height reduction by 21-37 per cent in CCC and ethrel and 40-57 per cent under EL-531 at 10 and 15 ppm, though ethrel reduced flower number in 'Enchantment'. CCC as drench was when applied at 1, 2, 3 or 4 per cent to the 4 cm tall shoots in 12.5-cm pots, out of six cultivars three of them made an attractive and compact pot plants (Pearse, 1972).

Hanks and Menhenett (1980) compared some new growth retardants such as piproctanyl bromide, dikegulac-sodium, PP 528 (chlorophenyl tetrazole acetate) and 2, 3-dihydro-5, 6-diphenyl-1,4-oxathiin (UBI-P293) with already being used ones such as chlormequat chloride (CCC), ancymidol and chlorphonium chloride (Phosfon) taking Mid-Century Hybrid lily cvs 'Enchantment' and 'Joan Evans'. They provided single compost drenches 2-3 weeks after plants were transferred to the glasshouse and obtained piproctanyl bromide at 2,000 mg/plant, chlorphonium chloride 250 mg/plant and CCC at 1,250 mg/plant at par, giving some dwarfing, though ancymidol at 0.25-0.50 mg/plant was found quite effective. Remaining chemicals though were also effective in reducing stem length but these brought about poor quality blooms. When a new retardant PP 333 was compared with ancymidol as compost drench, its 5 mg/15-cm diameter pot made substantial reduction of height as to that of 0.5 mg ancymidol, though some zones in the lower leaves were dead (Menhenett and Hanks, 1982). Wulster *et al.* (1987) tried ancymidol, propiconazole (Banner), triadimefon (Bayleton) or Mobay RSW0411 [ß- (cyclohexylmethylene) – α – (1, 1-dimethylethyl) – 1*H*-1, 2, 4-triazole–1–ethanol] as pre-plant bulb-soak or soil drench, and recorded that the chemicals Banner and Bayleton reduced some plant height but Mobay RSW0411 and ancymidol were found quite effective in reducing the stem length in *L. longiflorum*. Hendriks and Kok (1989) dwarfed the plants of 'Aristo' and 'Sunray' by applying 0.25 ml in 100 ml of water/12-cm pot Bonzi (daminozide, paclobutrazol) and for cvs 'Connecticut King' and 'Star Gazer' 0.75 ml/pot, while its double concentrations brought about the same dwarfing when Reducymol (ancymidol) was used. Jiao *et al.* (1991) with uniconazole soil drench (0.1-0.3 mg/plant) to *L. longiflorum* recorded 20-50 per cent reduced plant height, with foliar spray treatments (single or double applications) at 0.1, 0.15 or 0.2 mg/plant recorded 49-70 per cent height reduction, and with foliar application of GA_{4+7} at 200 mg/l recorded counteracting the inhibitory effect of uniconazole on shoot growth and the rate of flower bud development.

Growth and Development

For forcing of Mid-Century Hybrids, *L. auratum, L. longiflorum* and *L. speciosum*, these are planted in the pots and when sprouts start appearing, the pots are taken to cellars and kept at a temperature of 16 °C but when buds start forming the temperature is raised to 18-21 °C and the plants are fed fortnightly with liquid manure. When buds start showing colour the liquid feeding is stopped and temperature is again brought down to 16 °C. Such plants flower six weeks after appearance of the buds. Flower opening at higher temperature is earlier as compared to lower temperatures. Miller and Langhans (1989) studied the changes in the carbohydrate types and quantities in Easter lily 'Nelliw White' bulbs, stems, leaves and buds forced under ambient or reduced irradiance conditions. It was revealed that sucrose is the dominant soluble carbohydrate in bulb tissues, with glucose, fructose and mannose being in quite lower concentrations. In the process of developing inflorescence, mother bulb reserves are preferentially utilized irrespective of greenhouse irradiance. From the 40th day of planting, there was a steady decrease in mother bulb starch concentration until anthesis, *i.e.* up to 70 days later. On the basis of the flower bud differentiation, Ohkawa (1989) in Japan studied 18 native species and had put them into four groups. Group I consists of those where buds are initiated in early autumn and complete their development in early winter before sprouting of the bulbs such as *L. rubellum* and *L. dauricum* (*L. pennsylvanicum*), group II where flower buds are initiated in late autumn and complete their development in the spring when the bulbs sprout, group III where floral buds are initiated just after sprouting, while in group IV the floral buds initiate one month after sprouting such as *L. lancifolium* and *L. henryi*. Niimi and Oda (1989) through their studies with field-grown *L. rubellum* observed initiation and development of a new bulb at the base of the mother axis for the first time in May, and both scales and leaves were being formed continuously through late August on the slightly convex-shaped apices which consisted of several tunica layers. By early September the external surface of the apices had become more bluntly convex and broadened, and the tunica layers were decreased in number which was indicative for the start of the early reproductive stage and then after the floral primordia were observed rapidly developing into bracts, perianths, stamens and carpels in November and December but pollen grains in the androecia and ovules in the gynoecia could be found before sprouting of the floral buds in the spring. Mid-Century Hybrids and *L. longiflorum* are more amenable to year round flower production in temperate areas through proper regulation of heat and light. In these cases the bulbs are stored at 4-7 °C until required for forcing. 'Enchantment' plants were given assimilation lighting with high pressure sodium discharge (HID) lamps for 16 h a day at 2500 lx or for 24 h a day at 2000 lx up to picking, *i.e.* for 7 weeks from February 4, which resulted with heavier and longer stems, and under 24 h a day continuous lighting the plants flowered a few days earlier (De Groote, 1981). When 'Nellie White' plants were exposed to natural daylight (ND), 50 per cent ND, ND + 16 h incandescent (Inc) light or ND + 16 h HID lamp light, Heins *et al.* (1982b) reported that all lights increased plant height over the ND plants, but only incandescent or HID light, when given from emergence to flowering, hastened flowering by 5-8 days compared to ND plants though incandescent lighting increased flower bud abortion. At Lisse, Boontjes (1980a, b) advocated retarding of lily bulbs for up to 13 months, *i.e.* up to mid-January for late planting under glass, by storing them at -2 °C. For 'Enchantment' he recommended -2 °C storage from January to November, packed in boxes filled with up to 24 litres of moist black peat and the box enclosed in a perforated polyethylene bag and then forcing for cut flowers for getting increased stem length and weight, and number of flowers per stem. Hong *et al.* (1978) recorded broken bulblet dormancy of *L. lancifolium* after subjecting them to 10 °C temperature for 41 days, and these bulblets flowered two years after planting, and when planted in August than usual September these produced slightly larger bulbs. The 'Ace' bulbs potted in December and (i) cooled for 6 weeks at 4.4 °C and forced at 15.6 °C, (ii) cooled for 6 weeks at 4.4 °C and forced at 12.8 °C, (iii) cooled for 7 weeks at 4.4 °C and forced at 15.5 °C, (iv) cooled for 7 weeks at 4.4 °C and forced at 12.8 °C, or (v) potted in October and kept for 3 weeks at 15.5-18.3 °C, cooled for 6 weeks at 4.4 °C and forced at 15.5 °C, and all these flowered after 118, 116, 111, 115 and 158 days, producing 9.1, 8.8, 8.3, 7.0 and 9.8 flowers, respectively (Hammer and Kirk, 1981). Van Nes (1981) stated storing the bulbs of *L. longiflorum* cv. 'Arai No. 5' and 'White Europe' at -2 °C either from mid-December or from 1 April (the latter after storage at 0.5°), and planting them from July to October for forcing as cut flower with an advise to keep the bulbs first in woodwool packing at 0.5° before transferring them to potting soil at -2 °C so that no freezing damage occurs. Pergola (1986a) when raised the bulbs in Pescia (Italy) and lifted on October 10, or imported from the Netherlands, stored them for 12-24 weeks at 2°, -2°, 18-20 °C (ambient) or at combinations of these before planting in a moderately heated (6°) greenhouse, noted that storing the bulbs at 2 °C for 8 weeks followed by -2 °C for remaining time provided best results with short growing period and most flowers per stem though those stored at 2° took less time for growth but the flower numbers kept on decreasing with increase of the storage period and those stored at ambient temperature took longest growing time with poorest inflorescence quality. In another study, Pergola (1986b) lifted the bulbs in July when soil temperatures were above 22 °C, stored at 4° for 0-11 weeks and then planted in growth chambers at 20/23 °C night/day temperatures, only vernalized bulbs sprouted and the time taken to produce visible buds was found inversely proportional to the length of cold storage period but non-vernalized bulbs did not sprout. Bulbs once forced should not be forced again, and spent pots are then taken out and the bulbs planted outdoors (Hay and Beckett, 1971).

Oelker (1958) tried storage of *L. auratum* seeds at 20 °C which did not alter germination of seeds even well after 6 months though germination power was highest immediately after capsule ripening, and sowing even when temperature is -16 °C does not have any adverse effect provided there was no any previous warm spell to start the germination. Most species require below 20 °C temperature and even a few even below 15 °C for seed germination. He suggested sowing of seeds of those species which form bulbs before producing aerial growth, *e.g. L. martagon, L. canadense, L. auratum, L. superbum* and *L. japonicum* in late February or early March for emergence the following spring, though late-sowing delays the emergence until second spring after sowing. Likewise, Matsuo and Arisumi (1979) stored the mother bulbs of cv. 'Hinomoto' for 5 weeks either at 25 or 10 °C throughout or at 10 °C for 1-4 weeks followed by 25 °C for the rest of the time and then the middle scales and the inner core of each bulb was planted in the open. They observed increased scale bulblet number with an increase in 10 °C treatment duration. Chilling increased the production of epigeous-type plants and reduced that of hypogeous-type plants. Leaf emergence from the middle scale bulblets was delayed as chilling duration for the mother bulbs increased. With 4-5 week's chilling most of the bulbs derived from inner core material rapidly produced tall plants that initiated flower buds whereas with 0-2 week's chilling bulbs were slow and produced short and non-flowering plants.

In summer-forcing when night greenhouse temperatures for Asiatic lilies are not maintained below 18 °C, bud loss during early shoot development occurs (Beattie *et al.*, 1985) as has also been reported by Boontjes (1982) with 'Connecticut King', and Roh (1990b) with 'Cherub', 'Connecticut Lemonglow', 'Red Carpet' and 'Sunray' though he also reported that 26/24 °C accelerated the flowering compared to 16/13 °C (day/night) temperatures but number of aborted flowers was found higher at higher temperatures. Ethylene is reported to cause flower bud abscission in *Lilium* 'Enchantment' (Durieux *et al.*, 1982/1983) and other lilies more effectively, which occurs only during a critical stage of floral bud development, though bud abortion occurs throughout inflorescence development.

Abscission coincides with the peak in the endogenous of ethylene by the buds and with the end of the meiotic phase of the stamens. 'Nellie White' and 'Georgia' Easter lilies were when treated with ethylene and ethephon at various developmental stages, Lee and Roh (1985) reported that ethylene hastened flower bud opening, *vis-à-vis* flower senescence though both the chemicals decreased petal length and increased floral deformity. Van Meeteren and Slootweg (1986a) stated that the capacity of flower buds to convert ACC (ethylene precursor) into ethylene was independent of the length of buds if they were in between 1-5 cm long. In a greenhouse experiment of *Lilium* × 'Enchantment', Van Meeteren and Slootweg (1986b) recorded reduced floral bud abscission to about 20 per cent by foliar spraying of STS.

Growth regulators have influence on almost all the stages of growth and development of lilies but the effect varies with the type of growth regulator used. NAA and IBA at 300 and 100 ppm, respectively, promote bulblet formation. α-NAA at 5.4 μM results in bulblet formation while 2,4-D at 2.2 μM results in root formation. In *Lilium longiflorum,* increased number of flowers is observed with CCC at 1,000 ppm and 2,500 ppm; accelerated bulb formation with IAA at 100 ppm and TIBA at 500 ppm; vegetative growth with GA_3 and NAA at 100 ppm, and beneficial response on flowering with GA_3 at 10 ppm. Potted 'Ace' lily bulbs were soil-drenched with 1,000 ppm either with gibberellins A_3 or gibberellins A_{4+7} using 200 ml/pot in January and De Hertogh and Blakely (1972) recorded induced flowering as compared to GA_{4+7} though both the GAs increased the stem length and reduced the number of primary flowers initiated and completely inhibited the initiation of secondary flowers, and GA_{4+7} increased the number of primary flowers which aborted shortly after initiation but not in GA_3. Sajid *et al.* (2009) reported that mean shoot length of the plants sprayed with both the hormones and nutrients increased by 33 per cent, nodes per stem 71 per cent and bud number 92 per cent whereas those with only nutrients the shoot was found increased by 6 per cent, nodes per stem 42 per cent, number of buds 28 per cent and mean number of flowers 14 per cent more. Foliar application of GA_3 and nutrient application was found to have improved the productivity and quality of lily cut flowers.

Postharvest

It is very important to harvest the inflorescences early in the morning by pulling them rather cutting when the buds are properly coloured but yet not open in case of Asiatic, Oriental and Easter lilies. In case where bulb harvesting for next planting is also desired, it would be better to cut the stems 15-20 cm above the ground. However, for local markets these should be harvested when one floret is opening or opened. The earliest harvest stage for stems with five to ten buds is when 2-3 buds display colour. Harvesting of less mature flowers will cause unopening of a few buds and the opened flowers will also be small and pale (Wilkins and Dole, 1997). To avoid desiccation, the spikes should not be stored dry in the storage for more than 30 minutes. Stems are graded as per stem length & firmness and number of flower buds per stem. During bunching, foliage up to 10 cm from the stem end should be removed, flowers sleeved and after bunching, the flowers should be placed in cold water in cold storage at 2-3 °C with 2 per cent sucrose and 100 ppm GA_3 as preservative agents. The stems are packed in perforated cardboard boxes and carried in refrigerated vans. Average yield of lily is one stem per plant. Vase life in *Lilium* is 7-14 days in Asiatic, 8-10 days in Easter and 10-15 days for Orientals from cutting stage. Orlov (1957) stated that the buds of *L. regale* take 50-70 hours to open and the inner petals some 2 hours, flowers taking 5 days to remain fresh though wilting takes 2 days. Swart (1982) when immersed the lily cv. 'Enchantment' bulbs in a silver thiosulphate solution before planting, recorded improved quality of the harvested flowers. Foliar STS-sprayed Easter lily 'Enchantment' flowers were when cut and placed in water the floral bud opening was greatly improved similar to that of STS-pulsing, and even where no supplementary illumination was done or even after a dry storage in the air with 0.5 ppm ethylene at 20 °C for 24 hours (Van Meeteren and Slootweg, 1986b).

Pulsing consists of placing the cut ends of flowers in a concentrated solution of sucrose and germicides. The vase life of freshly harvested lilies can be extended for up to two weeks if they are pulsed and then placed in a solution containing floral preservative. Floral preservatives should be used for fresh arrangements in order to promote bud opening and to prolong the vase life, storage in water prior to long distance transportation or upon receipt and pulsing at retail level. Floral preservatives that may be used include Chrysal Select, Chrysal Universal, Crystal Clear, Floralife, Floralife Special Blend Pure (for high quality water), Floralife Special Blend Hard (for hard water), Oasis Floral Preservative, Prolong and Rogard. Ethylene inhibitors, which may be used for Asiatic lilies, include Silflor and Silgard RS. Silgard is designed specifically for mixing with Rogard. Chrysal LVB is a STS product designed for lilies. Silver thiosulphate (STS) inhibits the activity of ethylene in flowers and retards the rate of senescence. Asiatic and LA hybrids respond well to STS though Oriental hybrids not. Varieties 'Annabella', 'Cannes', 'Chianti', 'Concorde', 'Connecticut King', 'Cote d'Azur', 'Eurovision', 'Festival', 'Florence', 'Gran Sasso', 'Hilde', 'Jolanda', 'Madras', 'Mont Blanc', 'Monte Rose', 'Navona', 'Prominence', 'Pronto', 'Rodeo', 'Sterling Star', 'Toscane', 'Vivaldi', and 'White Cloud' need not be protected with STS as per the specifications of Dutch Auction Centre. Freshly harvested lilies can be pulsed in a solution containing sucrose and 4mM STS for 20 minutes at 20 °C or 2 hours in 1mM STS at room temperature. Flowers may also be pulsed overnight for 18 hours at 2-5 °C in 1 mM STS. The vase life of 'Enchantment' lilies improves when pulsed for 24 hours in a solution of 1.6 mM STS and 10 per cent sucrose followed by transfer to water. Other fresh lilies should also be pulsed with 1.6 mM STS and 10 per cent sucrose solution for 24 hours before wrapping them in tightly sealed moisture –retentive polyethylene film to reduce water loss. 'Enchantment' Easter lily flowers harvested before correct stage last better if are stood for 24-48 hours in a solution containing AAdural AK at 30 g/l at a temperature above 10 °C than being kept in water alone before marketing (Barendse *et al.*, 1978). 'Enchantment' cut flowers pre-treated by standing in a solution of a compound containing silver nitrate and STS for 24 h at 2 °C improved the quality of the flowers and vase life even when they were subsequently subjected to 72 h of dry storage or exposed to ethylene though some flowers were misshapen (Swart, 1979). Swart (1981) treated standing cut flower of hybrid lily 'Enchantment' in water containing carnation flower preservative Bendien Anjer VB for 4-36 hours immediately after cutting and before dry storage (prior to transport) and recorded increased percentage of normal flower opening, prolonged vase life and prevented damage by ethylene being emitted by smaller buds (a specific problem with this cultivar). Lilies may be stored up to 4 weeks in this manner. This will improve the keeping quality of the lilies during distribution stage, as they will be less susceptible to ethylene. Nowak and Mynett (1985) when continuously used 30g/l sucrose for inflorescences of cv. 'Prima' cut at bud stage, recorded satisfactory bud opening and increased longevity. They were able to store the flowers for 4 weeks at 1 °C without any potential vase life loss provided the cut inflorescences were pre-treated with STS+100 g/l sucrose for 24 hours and kept in a cold room in a solution containing 50 mg/l silver nitrate, and then held in a solution with 30 g/l sucrose and 200 mg/l 8-HQC. The effects of STS and 1-MCP were studied by Dole *et al.* (2005) on cut *Lilium* cv. 'Polyana' by treating the unpacked cut stems and placing them either in deionized water (DI) and subjected to 1-MCP (740 nl l^{-1}) or ambient air for 4 h or DI + STS at 0.2 mM for 4 h. After treatment, stems were removed, placed in polyethylene sleeves and stored either wet in DI water or dry in plastic-lined floral boxes at 5 °C in the dark for 4 days. After storage, bunches were placed in DI water under 12 h light (76 to 100 µmol $m^{-2}s^{-1}$) per day where it was found that both STS and 1-MCP increased the vase life, more being under STS.

After pulsing, lilies can be transferred to a **holding solution** containing a floral preservative or 200 ppm of hydroxyquinoline citrate and 3 per cent sucrose solution. This may be followed by pulsing with STS and sucrose with 200 ppm of GA_3 for 24 hours or adding 50 ppm GA_3 in the vase water. GA_3 does not profoundly affect the cut flower life of lilies but it does not inhibit the loss of chlorophyll from lily leaves and may be added at the discretion of the grower. Nowak and Mynett (1985b) when used 20-2000 mg/l of GA, recorded increased longevity and floret size and retarded foliage discolouration, even in those cases where inflorescences were stored at 1 °C for 3 weeks, though BA, kinetin, CCC or Alar were not found effective. Macnish *et al.* (2008) when added aqueous chlorine dioxide (ClO_2) at 2 or 10 µl l^{-1} in clean deionized water, the build up of bacteria in vase solution was prevented and the vase life was found extended in *Lilium asiaticum* cv. 'Vermeer'.

Freshly cut lilies which are pre-treated with pulsing solution may be **stored** dry in tightly sealed moisture-retentive polythene foil at 1-5 °C for 4-6 weeks. Before wrapping the lilies in foil bags for dry storage, the bottom 1/3rd of the foliage should be cut; stems should be graded according to their size and bunched into desired quantities. Pulsed lilies can be stored loose for a period of 4 weeks at 0-2 °C in deionized water having a pH of 3.5. *Botrytis* infection may be a cause for concern during wet or dry storage.

Cold storage of cut flowers is necessary to limit desiccation and to enhance bud development, therefore, to remove field heat these are taken to the cold storage as soon as possible, following harvest, and retained there in containers filled with clean water for at least 3-4 hours (no longer than 48 hours) at 1-2 °C when other activities may be performed such as dressing, grading, bunching and so on. Pre-treatment agents such as silver thiosulphate + GA (*e.g.* 6 ml of Chrysal AVB + 1 tablet of SVB for each 3 litres of water) should be added to the clean water filled in clean containers for Asiatic and LA hybrids. This will nullify adverse effect of ethylene gas during the distribution process.

Following product-cooling, the stems are **graded** as per health, stem length, stem sturdiness, number of flower buds or any other damage or disorder affecting leaves and buds. The lily stems are then bunched, part of this process including the trimming off of the lower 10-20 cm of stem. This can be done by hand or by using a special leaf-stripping shears. Stripping the leaves improves product presentation and, because it reduces the build-up of bacteria in the water, it improves the keeping quality of the lily stems. After bunching, the stems are cut to equal lengths and then wrapped in sleeves that protect both the flower buds and the leaves. To substantially reduce processing time, grading and bunching can also be done mechanically on a flower processing line, and keeping processing time to within one hour will prevent desiccation of stems. Another point to remember is that people operating a flower processing line must be able to do so ergonomically.

After grading and bunching, the lilies can be **stored** in the cold storage room without placing them in water. The best storage temperature for cut lily stems (except for some such as 'Star Gazer'), when harvesting takes place under warm weather conditions, is 1-2 °C. In addition, storage should always be kept as short as possible since the best storage period is always the shortest. Lilies should be **transported** in perforated boxes. The perforations are necessary to prevent the accumulation of excessive concentrations of ethylene as it causes acceleration in maturation process that leads to bud drop and reduced keeping quality. To prevent the premature forcing of flowers and the development of fungi, make sure when packaging that the product goes into the box dry. Low transport temperatures (preferably cooled to between 1 and 2 °C) are necessary for lilies to prevent flower bud development as well as the adverse effects of ethylene. For a lengthy period of transportation, the boxes should be pre-cooled before dispatch. Upon arrival at the wholesaler and/or retailer, the lilies should again be trimmed, placed in clean water, and stored at 1 to 5 °C.

Prince *et al.* (1987) studied post-storage and post-shipping quality of potted Easter lily cv. 'Nellie White' under simulated interior environment after whole plant spraying with STS complex (0.5 to 2.0 mM Ag) or phenidone, and recorded that STS reduced pre-marketing storage-induced bud abortion and increased floral longevity but did not reduce foliar chlorosis, while bud abortion and foliar chlorosis increased and floral longevity decreased with increasing storage period from 0 to 4 weeks in dark at 2 °C.

Bulb Harvesting and Storage

While lifting the bulbs from the field, neither the soil should be dry enough to damage outer scales or to hinder lifting, nor so wet as to invite bulb rotting pathogens, so the soil should have proper moisture at the time of lifting. This will ensure proper instant cleaning of the bulbs and bulblets as adhering soil will be easily removed through shaking and the stems twisted off. The lifted bulbs are cured through aerated shade drying for 4 to 5 weeks. The bulbs may now be treated for one hour in a 0.2 per cent Captan solution and then again air-dried in shade. Too much drying leads to loosening and wrinkling of scales which causes poor performance of the ensuing crop. Immediately after air drying, the bulbs are packed in plastic crates with sterilized moist cocopeat or sawdust wrapped with perforated plastic sleeves. The crates are stored at 2 °C for 2 weeks and then at -1°C for 6 weeks. Easter lily bulbs packed either in damp peat moss (by weight 68 per cent moisture) or in dry peat moss (8 per cent moisture) were pre-cooled at 7.2 to 10 °C for 0, 2, 4 or 6 weeks, then soaked in 900 ppm GA for two hours, and then planted in a glasshouse at 2-week intervals (Laiche and Box, 1970). Those pre-cooled for 4 or 6 weeks flowered much earlier but with reduced stem length and number of flowers where GA did not prove effective, and those pre-cooled at 0 or 2 weeks, flowering was delayed but GA treatment hastened flowering by about 4 weeks with little reduced stem length and usual number of flowers, and storage in damp peat caused earlier flowering than in dry peat. Prince and Cunningham (1990) advocated lining of shipping cases with low-density polyethylene which reduces moisture loss greatly from packing media and the bulbs of *L. longiflorum* cv. 'Nellie White' during shipping, handling and case-vernalization. Use of polyethylene liners accelerate floral sprout emergence above the growing medium, floral bud initiation and flowering. When *L. speciosum rubrum* bulbs were stored for 60 days at 5 °C as the common practice for breaking dormancy, hastening shoot emergence, further elongation and flowering, Ohkawa (1976) recorded dramatic decrease of endogenous GA-like substances, while contrary to this when the bulbs were stored for 60 days at 21 °C, the bulbs were observed with increased GA-like substances though temperature at 21 °C was in no case effective in breaking dormancy, hastening shoot emergence and flowering, and in these bulbs neither GA_3 nor GA_{4+7} could induce flower differentiation, though flower inhibition caused by applying ancymidol was found reversed.

L. lancifolium bulbils harvested at two stages of maturity and after cold treatment of 6 °C were planted but when harvested in 1979 these were further given cold-treatment and then forced, to find out their flowering responses in 1979 and 1980 to test their harvest-maturity of 1978 (Roh, 1981). Immature bulbils though delayed emergence of scale leaves but once developed into bulb of sufficient size, the effect of bulbil maturity of 1978 was not expressed.

Insect-Pests, Diseases and Physiological Disorders

Insect-pests that attack lilies are furnished herewith. **Aphids** (*Acyrthosiphon solani, Aphis gossypii, A. solani, Aulacortum circumflexum, Macrosiphum euphorbiae, Myzus persicae*) feed normally clustering on young leaves, more severely on the lower side, on buds and floral parts by sucking their sap and leaving honey dew which attracts other insects and sooty mould. Its infestation causes curling and malformation of leaves and floral parts by which growth of the plant is arrested, and buds either do not appear, or are malformed. These also transmit viruses from one plant to the other. Proper sanitation of the field will minimize the aphid incidence. Giving jet-spray with simple water will also kill them. Spraying with nicotine sulphate or any weak insecticide at 0.2 per cent will control this pest. A good aphid control with healthy crop growth was obtained through granular Di-Syston application on lily bulbs over furrows at 2.25 and 4.50 kg/ha and phorate at 9.0 and 13.5 kg/ha against *Macrosiphum* and *Acyrthosiphon* aphids but with reduced plant height and bulb weight (Swenson, 1963); demeton at 18.0 kg a.i./ha soil application against *Myzus solani* (Streu, 1964); soil treatment either at planting time, at shoot emergence or when foliage initiated with ethylthiometon granules at 0.75-1.50 kg/ha (Eguchi and Matsui, 1968) or foliar sprays with Pirimor (pirimicarb) 0.15 per cent (Appleby and Carbonneau, 1976) against green peach aphid (*Myzus persicae*); and soil application in granular form or spraying the plant crowns with AC 47031, Temik and Furadan against *Aphis gossypii* though AC 47031 causing extreme leaf scorch and stunting (Morishitsa and Jefferson, 1969). **Thrips** (*Liothrips vaneecki, Taeniothrips simplex, Xylaplothrips subterraneus*) suck scale's sap causing rusted sunken areas on scale bases and on planting such plants become stunted. HWT at 45 °C and bulb storage at 10 °C will control damage by these pests. Bulb dipping for 15 minutes in a solution of 100 ml 25 per cent parathion in 100 litres of water and then storing the bulb for 1-2 days in a heated room has been found to control *Liothrips vaneecki* and *Xylaplothrips subterraneus* (Franssen and Mantel, 1962). **Beetles** (*Lilioceris lilii* and *L. merdigera*) feed on stem below the ground whose infestation can be brought to a minimum through field sanitation and by use of 0.2 per cent methyl parathion. Rode (1966) obtained full control of both the beetles through 1-2 sprayings of 0.2 per cent Bi 58 (a dimethoate preparation) starting in early June. Heladze (1964) had a good control of beetles by dusting the plants in early May with BHC or DDT and spraying with 0.3 per cent anabasine sulphate + o.4 per cent soap or 0.1 per cent parathion in late May, and at 10 day intervals application of DDT throughout the June. **Iris borer** (*Macronoctua onusta*) was fully controlled by four applications at 6-day intervals of 5 per cent DDT dust starting from April 28, or 3 per cent endosulfan dust or 4 per cent malathion dust applied 1-4 times from April 19 (Schread, 1966). **Lily weevil** (*Agasphaerops nigra*) larvae feed on bulbs and underground portion of stem and is a serious pest of lily in certain countries which has successfully been controlled in its adult stage by five sprayings at 10-day intervals, starting from March end, at 'run off' stage with a stomach poison lead arsenate at 0.4 per cent along with a spreader (Doucette and Latta, 1946). Rode (1966) when applied 1-2 sprays of 0.2 per cent Bi 58 from May end to June beginning, recorded control of *Contarinia* larvae **lily fly** (*Liriomyza urophorina*). Roh and Vorsatz (1966) recommended DDT-lindane or through a systemic insecticide based on demeton-methyl and dimethoate, spraying of DDT-lindane first before appearance of the pest and the second at 5- to 7-day intervals though of the systemic demeton-methyl or dimethoate should be applied one week after the beginning of swarming at 'run off' stage.

Mites (*Rhizoglyphus echinopus, R. bovini, R. robini*) infest on bulbs causing basal cracking, and on stems, by sucking their sap and cause yellowing of the leaves. These can be controlled through HWT at 45 °C from 15-60 minutes and storing the bulbs at 1.6 °C, methyl demeton at 0.1 per cent, and Tedion at 0.2 per cent, and the application of methyl demeton and trimethyl tauryl ammonium 2, 4, 5-trichlorophenoxide to the bulbs were found effective in controlling even rats apart from mites though there was increase of anthracnose problem (Ueno and Namikawa, 1963). Oakawa and Sugimoto (1966) obtained good control of *Rhizoglyphus echinopus* with application of 1 g granules of Di-Syston per bulb at planting whereas Ascerno *et al.* (1983) found significant reduction in the population of *Rhizoglyphus robini* by using aldicarb and dicofol.

Slugs (*Arion fasciatus*; Eaton and Tompsett, 1976) and **snails** infest the crop including the bulbs when there is excessive humid condition. These attack during night though by morning they hide themselves in the plant debris around such plants or sometimes can be seen on the lower side of the leaves. Their infestation batters the aerial plant completely and their movement leaves silvery trails. These are controlled through metaldehyde baits. Collection and burning of debris around such plants will kill them completely.

Plants produced from **leaf nematode** (*Aphelenchoides fragariae*, strawberry leaf nematode; *A. ritzemabosi*, chrysanthemum leaf nematode) infested bulbs are slow to develop; leaves become malformed, round and cupped (especially among Oriental hybrids), irregularly and densely arranged and thickened, especially towards the top and their severe infestation does not allow the plants producing flowers. Moist conditions favour their spread. Symptoms often develop in the middle of the stem, at first in leaf axils, or on the leaf tips and/or surfaces of downward hanging leaves. Uniform bronzy-green to brown discolourations develop on finely-veined leaves which causes premature withering and falling off, though thickly-veined leaves first show yellowing which later on turn into brownish patches and then these become discoloured on one side first and then on both the sides. Another symptom that sometimes occurs is the appearance of white speckles on curled leaves. Under moist conditions, leaf nematodes emerge through the stomata from plants grown from infested bulbs and can then easily be spread by splashing water or wind. If a crop remains wet for a prolonged period, the infestation can spread at an explosive rate. Practically no spreading takes place in dry conditions, either in the greenhouse or outside. Slootweg (1960) obtained marked reduction in the population of *A. fragariae* by 6-8 sprayings with 0.5 per cent solution of 25 per cent parathion emulsion, however, HWT for 1 h at 43½ °C gave full control but caused injury to the bulbs. HWT of *Aphelenchoides*-infested lily bulbs for an hour at 42 °C gave full control of nematodes, however, Muller (1966) recorded 36 °C for 6 h, 37 °C for 5 h, 38 °C for 3 h, 39 & 40 °C for 2 h or 41 °C for 1 hour giving full control of leaf-nematodes without any injury to the bulbs. **Root-lesion nematodes** (*Pratylenchus penetrans, P. pratensis, P. vulnus*) are the serious pests of lilies. To keep this pest at bay, field sanitation is a must as these nematodes have more than 600 host plants. Immediate removal and burning of those plants should be done which have started showing its symptoms. No excess humidity in the crop should be maintained at any time. Steam-sterilization or solarization of field will kill the population of these nematodes. Removal of root remnants of the bulbs after cold storage reduces the number of infested bulbs up to 84 per cent without affecting forcing quality and is quite less injurious than chemical soaks, vacuum fumigation or hot- water treatment (Caveness and Jensen, 1955), however, each year soil treatment with chloropicrin, DD or steam is the best way in managing these pests (Ohkawa and Saigusa, 1972). D'Herde *et al.* (1959) recommended HWT at 46 °C for 20 minutes as an effective control. Overman (1961) recommended pre-storage dip-treatment of lily bulbs for 15 minutes in parathion solution or through in-furrow granule application at planting time which hastened bulb emergence, and increased the number of open buds per stem, *vis-à-vis* bulb weight. Out of phorate granules, dusts and dips, 10 per cent phorate granule application at 13.5 kg a.i./ha at planting time was recorded as the best treatment for controlling *Pratylenchus penetrans* and *P. pratensis* both (Jensen, 1966; Jensen and Doucette, 1968), though Maggenti *et al.* (1967) recommended 1, 3-D (1, 3-dichloropropene) as split application at 100 gal/ha quite effective and granular phorate least effective. Boontjes *et al.* (1978) when applied aldicarb 30 kg/ha in the furrows and then planted lily bulbs in previously fumigated soil with their roots infested with *P. penetrans*, they obtained a very satisfactory result in a 2-year lily crop. Out of fungicidal dips, HWT, root washing and pruning, 0.5 per cent a.i. phorate pre-planting dip, granular phorate at 14.5 kg a.i./ha applicatrion at planting in nonfumigated, or in soil fumigated with 1, 3-dichloropropene, Hart *et al.* (1967) found only phorate dip quite effective in minimizing the nematode population.

Various **diseases** which infect lily crops are described here with. **Bulb and scale rot** disease (*Fusarium oxysporum* f. sp. *lilii, Cylindrocarpon destructans, Nectria radicicola*) infects the plants from the wounds underground. This causes root and scale rot, weaker growth of plants and their stunting, leaves turning pale-green, bud desiccation, and failure of shoots to emerge. Brown spots may be noticed on the tips and sides of the scales as well as at the bases where these are attached with the bulbs. Finally this causes rot of the basal plate and scale bases. Smith and Maginnes (1969) identified one quite *Fusarium*-resistant cv. 'Rose Cup'. **Stem spot disease** (*Cylindrocarpon destructans*) is symptomized by premature yellowing of lower leaves, later on turning brown and falling off. On the stems underground, orange to deep brown spots appear, which coalesce later on and enter the stems inside which causes stem rotting. Its light brown, dry spots and patches on stems and bulbs are obviously seen. **Glasshouse edge disease** (*Nectria radicicola, Fusarium oxysporum*) can be controlled by disinfecting the bulbs in 1.5 per cent Difolatan + 0.3 per cent Benlate when preceded with HWT. Meij *et al.*

(1978) obtained best results with HWT followed by soaking the bulbs in 0.4 per cent benomyl + 0.75 per cent trichlophenol against *Fusarium oxysporum* basal rot. Sometimes such symptoms also appear due to infection of *Fusarium*. Conditions that encourage infection are high soil temperatures, excessively wet soil and excessive fertilizing. Soils infected with pathogens should be solarized, sterilized through formalin or should be drenched with a contact fungicide. All the infected bulbs should be discarded. Growing of lilies at higher temperatures should be avoided as these favour infection of these diseases. Moreover, excessive wetting of soil and fertilizers should also be avoided. ***Botrytis* blight** (*Botrytis cinerea*) is probably the most serious disease of field grown lilies and occurs at 10-18 °C temperatures. Initially reddish-brown spots (red streaks in case of Asiatic lilies) on leaves appear but in Easter lily the streaks are yellow. Its infection causes grey mould on the flowers. Symptoms of *Botrytis elliptica* appear on the leaves as grey-brown to dark brown speckles, sometimes with a dark green edge, that measure 1 to 2 mm across. Under moist conditions, the speckles can quickly expand and coalesce to become larger round or oval. The infection begins in the middle of the leaf surface or at the edge where it is crescent-shaped which results in stunted and malformed leaves and finally engulfing the whole plant including the buds and flowers and shows a burnt appearance. On necrotic tissues, the fungus produces large quantities of light brown to grey-brown spores that easily disperse upon slight contact or falling water droplets. On a dry crop, however, the spores cannot germinate, so the lack of water will keep an infection at bay. At the end of the season, the fungus on infected and necrotic tissue will form round black sclerotia 2-3 mm in diameter that can survive in the soil for many years. The Asiatic and *Longiflorum* hybrids are much more susceptible than Oriental hybrids. Within the Asiatic hybrids, the white and pink cultivars are especially susceptible. *Botrytis* blight and fire can be managed by keeping the crop and field dry, by reducing planting density for proper aeration, watering the soil in the morning and providing proper ventilation, use of fans for extra air circulation, avoiding sprinkler irrigation, preventing condensation in the morning by increasing the temperature about one hour before sunrise and through proper sanitation of the field. Maneb and zineb fungicides at 0.2 per cent should be sprayed weekly when weather is windy and cloudy. Benlate is also quite effective in checking this disease. Anahosur *et al.* (1971) also reported a serious **leaf blight** on lily caused by *Botryodiplodia theobromae*. ***Phytophthora* foot rot** or **stem rot** (*Phytophthora parasitica, P. nicotianae, P. cryptogea) causes* plants not developing properly and yellowing of the plants starts from below and suddenly the plants wilt. This fungus thrives in moist conditions. Stem bases show soft rot and become deep green to dark brown (sometimes purplish-brown with streaking upward to the aerial plant parts) and then bending to falling over. If the plants are infested late in the season, the plants will not fall but the diseased stem tissue will become desiccated. This results in a hollow interior within which fungal wefts may develop. Finding this kind of soft rot in the aerial part of the stem is not unusual either; here it is found just beneath the top of the plants that are not completely developed. In this case, the top of the plant turns black which leads to localized leaf yellowing and/or a bent stem. The pathogens are found in soils previously cultivated with tomatoes and gerbera and can survive several years in moist soil. Excessively wet soil or a combination of a wet crop and high temperatures (20 °C or more) encourage the development of the disease. The fungus spreads by means of zoospores that are distributed by means of soil particles and splashing water. It can be managed by avoiding irrigation water standing for long in the field and by keeping the soil well-drained. Drenching the affected plants with Thiride (Thiram) 0.2 per cent will control this disease. ***Pythium* root rot** (*Pythium ultimum, P. splendens* and certain other species) is observed in the field here and there in patches. In its attack the roots develop poorly, plants are stunted, their bottom leaves turn yellow though upper ones are narrower, drabber & somewhat drooping, particularly during periods of heavy transpiration, more bud desiccation, during winter more bud drop, *vis-à-vis* bulb rotting. The flowers often remain smaller, frequently fail to open entirely and if open they fail to colour properly. When the plants are removed from the soil, the bulb and stem roots display glassy, light browning and soft-rotting, leaving only an empty membrane-like shell that can easily be detached from the core. In general, these fungi thrive under moist conditions and grow best at 20-30 °C. Development of fungi is promoted by too wet soil, poor soil structure, or very high EC. Therefore it is necessary to have good soil with proper drainage facilities so that these pathogens may not get proper environments for development. However, as a precaution the soil should be disinfected. A low soil temperature at the start of cultivation need to be maintained and good cultural practices should be followed. As and when its infection is observed, the greenhouse climate should be made as cool as possible to limit transpiration in the lily plants. Slight infection of ***Rhizoctonia solani*** will be limited to leaves and leaf scales, more pronounced being in the lowest leaves of the young shoot, and its infection will display light brown spots in the form as if eaten by a pest. Often hanging from the affected leaves are hyphae with soil particles clinging to them. Though plants continue growing in its infection but growth is slower. In severe infection, the plant emergence is delayed, lowest leaves wilt and fall off leaving a brown scar on the stem, growth points and younger leaves look sickly, underground stem parts display brown stripes and spots that are usually elongated, emergence of stem roots is suppressed and delayed, stems become susceptible to breakage, and the floral buds either dry out at an early stage or flowering is very poor. It is a soil-borne fungus which is favoured by moist conditions and when temperature is above 15 °C. After sprout emergence, generally its further infection is stopped leaving with its after-effects only, though such plants may recover in the long run. Pasini *et al.* (1996) in Italy isolated *Rhizoctonia solani* from *Lilium* causing collar or foot rot.To manage this problem, it would be better to pre-root the bulbs before planting. Such soils should be disinfected with formalin before planting and after planting the field should be kept clean. In absence of formalin treatment, the soil should be tilled up to a depth of 10 cm with 5-10 g/m^2 Rizolex (50 per cent tolclophosmethyl) when soil temperature during growing season is expected to go beyond 16 °C. ***Sclerotium* crown rot** (*Scleritium rolfsii* var. *delphinii*) and **bulb and basal bulb rot** (black leg disease) (*Sclerotium wakkeri*) are also soil-borne fungi whose spread is favoured when soil temperature is higher than 18 °C, and in heavily infected soil, the shoots in some areas of the greenhouse will either grow quite slow or will barely emerge. Those leaves coming into contact with the soil will wilt and start to rot and brown spots will appear at the stem bases which will also ultimately rot and collapse. Very characteristic of this fungus is the presence of white strands of hyphae, and later the formation of white and round sclerotia on the diseased tissues and surrounding soil, which later on turns bright brown to golden-brown, sometimes forming together a crust. In the process the bulbs are also infected and rot. In slightly infected soil, sprouts usually appear normal but afterwards their further growth becomes poor when stem bases are infected, leaves yellow and the plants topple over. ***Penicillium* brown rot** (*Penicillium hirsutum)* develops in storage through wounds inflicted on the bulb scales. Its infection shows brown rotten spots covered with a white fungal weft that later turns blue-green and is accompanied by masses of spores. Once established, the rot slowly spreads throughout the storage period, when even low temperature of -2 °C cannot stop its spread. After a fairly lengthy period of time, the fungus can penetrate the basal plate and then afterwards to other scales. Such scales detach from the basal plate and then cannot participate in plant growth so plant growth becomes poor. The disease is not transferred from soil to the stem. An excessively high temperature coupled with excessively low RH during storage promotes the problem. Bulb dipping in 0.2 per cent Captan after curing and before planting, storing only disease-free and uninjured bulbs, by keeping storage humidity proper, by storing the bulbs at the lowest possible temperature, and proper sanitation and fumigation of the store will keep this disease under control. The substrate mixed with calcium hypochlorite powder (20-27 per cent available chlorine) @ 7 g/kg substrate gives excellent control of *Pecicillium* rot along with *Rhizoglyphus echinopus* mite (O'Leary and Guterman, 1937). Hendriks and Koster (1988) advocated bulb dipping from 15 seconds to 15 minutes in 1 per cent Captan + 0.2 per cent benomyl (50 per cent) or 0.2 per cent carbendazim (50 per cent) or 0.3 per cent Topsin M (Thiophanate methyl) as preventive regime against *Penicillium* infection. **Anthracnose scale rots** (*Colletotrichum lilii, C. dematicum*) appear as black scales on the bulbs, starting from scale tips, and on flowers at 22-26 °C or above soil temperatures, and this can be controlled successfully through postharvest bulb dusting or as a pre-planting soak for 30 minutes with Daconil 2787 (Magie, 1966).

Though **bacterial diseases** are not a problem but sometimes *Pectobacterium carotovorum* causes bulb decay along with *Bacillus lilii. Pseudomonas gladioli* pv. *gladioli* sometimes causes definite but restricted stem lesions and scale tip rot which can be controlled through Agrimycin 100 (streptomycin + oxytetracycline) as pre-plant dips (Bald *et al.,* 1979). *Corynebacterium fascians* causes leaf galls mainly in the bulblets (not the bulbils) and the scales get deformed, sometimes pointed or rounded, and beneath these clusters a thickened ridge of yellowish gall-like tissue is often found but only on the underground parts including the whole length of underground stem (Miller *et al.,* 1980). Also Weststeijn (1978) recorded a bacterial disease in Holland where numerous malformed bulbils are formed on the stems which consequently causes poor development of the main bulb, and he named this malady a 'proliferation disease'. Proper field sanitation should be adopted and injured bulbs should never be stored or planted. Boontjes (1973) and Kruyer & Boontjes (1982) recommended bulb disinfection first with HWT + 0.5 per cent commercial formalin for two hours at 39 °C (beyond 39 °C, it is quite harmful to the bulbs) and then fungicidal dips in a 2 per cent Zn-Mn carbamate or 2 per cent captafol in the first and third years, and a 0.75-1 per cent phenol derivative dip in the second year, but for late treatment the HWT should be omitted and all the three years the bulbs should be dipped in 2 per cent captafol to get rid of pathogens such as *Fusarium oxysporum, Nectria radicicola, Corynebacterium fascians* and *Pythium* spp., and *Pratylenchus penetrans* nematode, *etc.* PCNB (quintozene) + ferbam is quite effective for controlling most of the pathogens including *Fusarium* and *Pseudomonas*, and its effectivity further increases if thymol is also added to the dip. Benomyl at 0.2-0.4 per cent applied as a 30-minute soak at 10-20 °C is quite satisfactory in case of fungal diseases, particularly *F. oxysporum* and *Nectria radicicola* (Boontjes, 1970).

Lily **viruses** are transmitted largely by aphids. Virus infected plants show irregular mottling and flecking on the leaves, reduction in plant height, distorted growth, colour-breaking on the flowers and leaves, and brown ring patterning on bulb scales. '**Lily symptomless carlavirus**' or 'lily latent virus' [transmitted mechanically and through aphids (*Aphis gossypii, Macrosiphum euphorbiae, Myzus persicae*)] is transmitted to tulips and *vice versa*, and '**tulip breaking potyvirus**' causes deformities in flowers, showing virescens and alterations of the reproductive structures and leaf pigment anomalies (Bertaccini and Marani, 1982); '**lily X potexvirus**' (Asjes, 1991); '**little mottle virus**' produces yellow mottling, mosaic and vein-clearing on the leaves and causes plant stunting; '**cucumber mosaic virus**' or 'tulip mosaic virus' for which vector is *Myzus persicae*, exhibits mosaic patterns, necrosis, deformity and stunting on the plants, *vis-a-vis* 'brown ring formation' and leaf drop where bulbs exhibit brown rings on the outer bulb scales (Bakker *et al.,* 1976) and it is transmitted by aphids; '**lily ringspot virus**' which shows a faint mottling, stunting and deformities, dead area on the leaves, and finally killing the growing points; '**lily rosette virus**' which causes necrosis of vascular tissues usually in the phloem elements and sheath cells; '**necrotic fleck disease**' which produces elongated virus-like particles on leaves; '**tobacco rattle virus**'; *etc.* Infected lily plants should be uprooted and destroyed along with bulbs and their scales. Planting of lilies should be avoided along with *Tulipa grinum* and aphids and other visiting pests should be controlled with regular sprayings of insecticides and destroying the infected plants. Brierley (1962) recommended propagation of lilies from mature outer scales which eliminates CMV in some plants raised from recently infected stocks though does not eliminate lily mottle virus at the same time. Spraying with 1 per cent mineral oil emulsions once or twice per week for three months from sprouting reduces the virus incidence to the minimum as the oil makes feeding of the aphids on the plants difficult hence the spread of the viruses are restricted. Asjes (1978) recommended 5 per cent Albolineum mineral oil or 3 per cent Luxan oil H or 7.5 per cent Schering 11E @ 400 litres solution of the former two and 200 litres the latter one per hectare spray. Allen (1974) obtained virus-free lilies by culturing 5-mm^2 pieces on inner bulb scales on Sheridan's medium, and the plantlets were planted into a pasteurized soil-peat mixture, and so from a 3-cm diameter bulb, he obtained 787 virus-free plants within a year.

Physiological disorders include leaf scorch (leaf tip burn); bud blast (bud abortion); leaf yellowing *cum* shedding; paling of entire leaf (nitrogen deficiency); dwarfing of plants with poor root system, leaves becoming light to very pale-green (sometimes with white spots) and downward bending of leaf tips which sometimes turn brownish (calcium deficiency); and interveinal leaf yellowing (iron deficiency) are mainly physiological disorders. **Leaf scorch** occurs just before the visibility of flower buds, usually on the upper leaves and when buds open the petals also show such symptoms. The young leaves first curl inward slightly and a few days later display yellow-green to whitish spots. On Oriental hybrids, leaf scorch occurs primarily on the edges of the leaves in the form of brown spots. In severe cases, the white spots turn brown in places, the leaves curl at the damaged portions and sometimes these are lost and young flower buds are destroyed so the plants stop developing, grow crookedly and may turn blackish-brown. Leaf scorch is caused by a disruption of the balance between the quantity of water being absorbed by the roots and the amount of water leaving it through the aerial parts of the plant. This occurs when the plant lacks ways to absorb and emit water by means of transpiration. The result is a deficiency of calcium in the cells of the leaves. These cells then collapse and die. Leaf scorch involves many factors that impact the growth rate, water intake and transpiration of plants. Yet one factor that has the greatest impact on this process is an abrupt change in the RH in the greenhouse. In such cases, mostly EC of the soil is excessively high which causes poor root system. Newly harvested bulbs if planted early in the season as compared to summer planting are less susceptible to leaf scorch. Plants grown outside suffer less than those grown in an average greenhouse climate. 'Croft' and 'Giganteum' Easter lily plants when grown in soil and sand cultures, high Ca, high N, low P and low K concentrations reduced leaf scorch or tip-burn injury probably due to absorption of Ca and its translocation in the plants (Dunham and Crossan, 1959), and in soil cultures the application of lime reduced tip-burn, and the regular fertilization of plants where 75 per cent N is derived from calcium nitrate had very little tip-burn as role of Ca is related to internal moisture distribution of plants. Mastalerz (1962) stated that the intensity of leaf scorch is directly related to the quantity of P, and inversely related to the amount of Ca, incorporated into the soil before planting. High level of superphosphate, i.e 2.5 kg/m^2 could be offset by using same quantity of ground limestone, however, he observed severe leaf scorch in a soil : peat : perlite mixture (1:1:1) than those of 3:1:1 or 8:1:1 mixtures. Furuta *et al.* (1963) described lithium toxicity as the cause of leaf scorch which could be prevented due to presence of Ca in the plants and continuous foliar misting. Marousky and Woltz (1977) when studied 'Croft', 'Ace' and 'Nellie White' lilies in fluoride-amended or lithium-amended sand cultures which had injured leaves, they recorded 'Croft' developing as a semi-circular necrotic areas at leaf margins near its apex on entire leaf mass due to fluoride injury which later on covered entire leaf tips and margins though in case of 'Ace' and 'Nellie White' there was semi-circular necrosis pattern to chlorotic or necrotic leaf margins only on the lower leaves, whereas lithium though injured all the three cultivars but the injury was comparatively quite less than fluoride. Lithium injury developed as chlorotic and necrotic leaf margins only on lower leaves. When they grew lilies in soil amended with dicalcium phosphate or superphosphate at both low and high lime rates and fertilized with NH_4-N or NO_3-N nitrogen, dicalcium phosphate with NH_4-N or at any lime rate there was no leaf injury though those grown with superphosphate had injured leaves, especially at low lime rates with NH_4-N. Susceptibility to leaf scorch varies widely depending on the cultivar and bulb size. Larger bulbs are more susceptible than smaller ones. Susceptible cultivars in Asiatic hybrids are 'Brunello', 'Dreamland', 'Navona' and 'Umbria' whereas in Oriental hybrids are 'Acapulco', 'Hit Parade', 'Kyoto' and 'Star Gazer' (De Hertogh and Le Nard, 1993). To get rid of this problem, only resistant cultivars or the smaller ones with healthy root system should be planted deeper in the soil, bulb-top being covered at least 8 cm in moist soil. RH of the greenhouse should be maintained around 75 per cent and the environment should be kept cooler, *i.e.* 10-12 °C for the first four weeks in case of Asiatic and LA hybrids, and 15 °C for the first six weeks in case of Oriental hybrids. Plants should transpire usually but in case of excessive transpiration the plants require to be shaded or in high sunny days water should be sprinkled lightly but frequently. Boron-injury is shown as **chlorotic and necrotic leaf tips** without affecting the basal leaves. **Flower bud drop** (flower bud abscission or desiccation) can occur from the time when buds attain 1 cm length and when crop receives insufficient light or/and when the buds are exposed to high concentration of ethylene. Under low light conditions the stamens in the buds start producing ethylene which causes buds to abscise. High temperatures also encourage bud drop. The buds going to drop first become light green and at the same time the stems become constricted at base of the bud, and then buds fall down. During spring, the lowest buds drop first while in autumn the upper buds drop first (Mason and Miller, 1991). Flower bud desiccation can occur during any stage of bud development. If it occurs early, the plants will remain short and the leaves will be dull green, short and narrow and arranged closely next to the stem but will display no symptoms of leaf scorch. Early bud drop is due to poor soil structure, poor or damaged root system, shallow planting, excessively high soil temperature, insufficient moisture level in the soil, and high level of salts present in the soil though late bud drop is associated with insufficiency of available nutrients due to poor lighting, and planting of large bulbs. Some or all the flower buds will desiccate during an early stage of growth and will later appear in the axils of the top leaves as small white specks. If flower bud drop occurs later in the plant's development, the plants usually develop with a normal root system and with flower buds that are already clearly visible. Later, however, the buds will turn light green and shrivel. Flower buds already in process of displaying colour will turn paler and dry up completely but will usually not fall off. The upper buds in the inflorescence will be the first to desiccate. Meeteren and Proft (1982) with lily 'Enchantment' stated that when daylength was shortened from 12 to 4 h the ethylene evolution from the 3 cm long buds increased about 4-fold, *vis-à-vis* when plants were kept in the dark up to 5 days and the bud length had been 2.0-3.5 cm. They advocated that flower bud abscission induced by darkness could be prevented by injecting flower buds either with 0.2 mM STS at the beginning of the dark period or when the bulbs were grown pre-treated with STS. They opined that light-controlled flower bud abscission is mediated by ethylene, and that an increase in ethylene evolution in darkness is a consequence of the influence of light on ethylene action rather than a direct influence on ethylene biosynthesis. Providing the supplementary lighting for one month or so before flowering, *vis-à-vis* supplementary CO_2 to bring glasshouse concentration up to 0.1 per cent will minimize bud drop incidence (Durieux, 1978b). **Bud blasting** (withered flowers) is an entirely different phenomenon from bud abscission which is also evoked by darkness or ethylene or high temperatures which causes rapid depletion of carbohydrates, and occurs throughout inflorescence development though bud abscission occurs only at a critical stage of bud development. **Blindness** (non-flowering plants) also mostly occur in the planting due to inappropriate forcing temperature. **Leaf yellowing and shedding disorders** are encountered with Easter lilies and some of the Oriental hybrids due to over-watering which results into root-rotting, moisture stress, high soluble salt accumulation, ethylene evolution, excessive or inadequate fertilizer use and insufficient light reaching to the plant bases due to more plant density. Spray of 50 ppm BA+GA (10:1, commercially known as Accel) only to lower leaves, first before 10-14 days of bud visibility and second (last) 10-14 days later, or spraying 100 ppm of BA+GA (1:1, commercially known as Promalin) to the lower foliage after two months of shoot emergence will control this problem. Stumm-Tegethoff (1986) when treated plants of *L. henryi* with *Ditylenchus dipsaci* extract, all these plants and also those receiving IAA at 10^{-4} and 10^{-5} M or kinetin at 10^{-4} along with extract were fasciated, and gibberellins at 10^{-4} M also produced fasciated stem.

Plants become weak and entire leaf turns pale in case of **nitrogen deficiency**, which becomes more conspicuous when the plants are in flowering stage. The situation is aggravated in summer cultivation of crops. Flower stems produced from nitrogen-deficient soil are lighter in weight with smaller leaves and produce fewer flower buds. When put in vase, the leaves turn yellow faster than normal ones. Only after getting analysed the soil, the fertilizers should be applied as required. If nitrogen deficiency is observed during cultivation, application of calcium nitrate, urea or potassium nitrate is immediately recommended, preferably with irrigation water or application followed with watering. Dwarfing of plants with poor root system, leaves becoming light to very pale-green (sometimes with white spots) and downward bending of leaf tips which sometimes turn brownish are due to **calcium deficiency**. Soil liming with calcium carbonate before planting, and to some extent even through application of magnesium carbonate, magnesium oxide and magnesium hydroxide the abnormalities may be checked from occurring. In **iron deficient** lilies, the mesophyll between the veins of young leaves turns yellow-green among rapidly growing plants, and intensity of yellowing increases with high iron deficiency in the soil. The disorder occurs more often in the soils having high pH and where soils are light, calcareous and susceptible to slaking, more pronounced when soil temperature is too low. Lack of iron in plant available form is major cause of its deficiency. A slight yellowing during cultivation, however, usually disappears as the harvest period approaches. Lily groups and cultivars susceptible to iron deficiency can be found among the Oriental and *Longiflorum* hybrids. To have control over iron deficiency the soil should have low pH and in case when this increases above 6.5, chelated iron should be applied, should have proper drainage facility, and the plants should have well developed root system. Chelated Fe-EDDHA 6 per cent can be used in soils with a pH as high as 12 and can be applied up to a few weeks previous to flowering. Fe-DTPA can be used only on soils having a pH of around 7 or less and can be applied only until the flower buds are visible. The application of too much Fe-DTPA can result in black spots appearing on the leaves. Applying chelated iron too late in the cultivation process can leave reddish-brown spots on the flowers. The addition of a wetting agent will help to prevent this. This can be mixed in the soil at 2 to 3 g/m^2 before planting but after planting the application should not exceed 2 grammes, followed by a second post-planting application not exceeding 1 to 1.5 g/m^2. In case no application is made before planting but a slight yellowing is noticed, this may be applied 2-3 g/m^2 through sprinkler circuit or mixed with dry sand and scattered beneath the leaves, however, in case of severe yellowing, 5 g/m^2 can be applied in a single application, followed by sprinkler irrigation.

References

Adanson, R.M. 1959. Weed control practices in bulbs and ornamentals. *Res. Rep. Joint Mtg – 16th Cent.* and *10th West Canadian Weed Control Conf.*, pp. 14-15.

Adelman, H., H.K. Tayama and D.C. Kiplinger, 1970. An explanatory study of the effects of Phosfon on 'Ace' and 'Nellie White' Easter lilies. *Ohio Flor. Ass. Bull.*, No. 484, pp. 8-9.

Adzima, R. 1986. The influence of bulb-dips on the height of hybrid lilies grown in warm temperatures. *Connecticut Greenhouse N.L.*, No. 135, pp. 15-18.

Allen, T.C. 1974. Production of virus-free lilies. *Acta Hort.*, No. 36, pp. 235-239.

Anahosur, K.H., K. Fazalnoor and B.C. Narayanaswamy, 1971. A new *Botryodiplodia* leaf blight of lily. *Sci. Cult.*, **37**(10): 480-481.

Anonymous, 1976. Nieuwe mogelijkheden van onkruidbestrijding in lelies (Dutch). *Bloembollencultuur*, **86**(38): 775.

Appleby, J.E. and M.C. Carbnneau, 1976. Green peach aphid control on Easter lily and hydrangea. *Ill. St. Florist's Assoc. Bull.*, No. 363, p. 5.

Aquettaz, P., A. Paffen, I. Delvallée, P. van Der Linde and G.T. De Klerk, 1990. The development of dormancy in bulblets of *Lilium speciosum* generated *in vitro*. I. The effects of culture conditions. *Plant Cell, Tissue Org. Cult.*, **22**(3): 167-172.

Ascerno, M.E., F.L. Pfleger, F. Morgan and H.F. Wilkins, 1983. Relationship of *Rhizoglyphus robini* (Acari: Acaridae) to root rot control in greenhouse- forced Easter lilies. *Environmental Ent.*, **12**(2): 422-425.

Asjes, C.J. 1978. Mineral oil on lilies to prevent virus spread (Dutch). *Bloembollencultuur*, **88**(42): 1046-1047.

Asjes, C.J. 1991. Control of air-borne field spread of tulip breaking virus, lily symptomless virus and lily virus X in lilies by mineral oils, synthetic pyrethroids, and a nematicide in the Netherlands. *J. Pl. Path.*, **97**(3): 129-138.

Baardse, A.A. 1977. Groot Lelie Boek. Drukkerij, 152 pp. West-Frieseland, Hoorn, the Netherlands.

Bakker, J., L. Barendse and P.C. Schenk, 1976. The effect of brown ring disease on the flower crop of *Lilium* cv. Enchantment (Dutch). *Vakblad Bloemist.*, **31**(27): 17.

Bald, J., J. Lenz and A.O. Paulus, 1979. Stem lesion of Easter lilies – a complex disease. *California Agric.*, **33**(3): 12-13.

Ballarin, A., D. Prisa, A. Grassotti, F. Pierandrei and G. Burchi, 2009. Leaf treatments to improve cut flower quality in Easter lily and in Asiatic and Oriental hybrid lily cultivars. *Acta Hort.,* No. 847, pp.369-376.

Barendse, L., C. Slootweg and W. van Den Broek, 1978. Improving the keeping quality of Enchantment lily that has been picked too soon (Dutch). *Vakblad Bloemist.*, **33**(27): 21.

Beattie, D., E.J. Holcomb and J. Iles, 1985. Steps to successful hybrid lily production. *Greenhouse Grower*, **4**: 16.

Beattie, D.J.and J.W. White, 1993. Lilium-Hybrids and Species. In: *The Physiology of Flower Bulbs* (eds De Hertogh, A.A. and M. Le Nard), pp. 423-454. Elsevier Science Publishers B.V., Amsterdam, The Netherlands.

Bennici, R. 1979. Cytological chimeras in plants regenerated from *Lilium longiflorum* tissues grown *in vitro*. *Zeitschrift für Pflanzenzüchtung*, **82**(4): 349-353.

Bertaccini, A. and F. Murani, 1982. Electron microscopy of two viruses and mycoplasma-like organisms in lilies with deformed flowers. *Phytopathologia Mediterranea*, **21**(1): 8-14.

Besemer, S.T. 1975. Japanese Georgia lily response to A-Rest. Progress Report. *Fl. Nursery Rept*, pp. 2-3.

Boer, W. Den, 1972. Chemical weed control in lily bulbs under glass. *Vakblad Bloemist.*, **27**(35): 15.

Boodley, J.W. and R. Sheldrake, 1963. Forcing Ace lilies in light weight media. *Bull. N.Y. St. Fl. Grs*, No. 216, pp. 1, 4.

Boontjes, J. 1970. Control of scale rot and bulb rot in lilies. *Praktijkmeded. Lab. BloembollOnderz., Lisse*, **32**: 8.

Boontjes, J. 1973. Disinfection of planting stock of *Lilium speciosum*. *Bloembollencultuur*, **83**(26): 662, 664.

Boontjes, J. 1980a. The condition of the lily bulb determines the risk of frost damage (Dutch). *Vakblad Boloemist.*, **35**(49): 33.

Boontjes, J. 1980b. Should peat be used in the storage of lilies in ice? (Dutch). *Bloembollencultuur*, **91**(15): 375.

Boontjes, J. 1982. 'Blinden' bij 'Connecticut King', nog steeds een problem. *Weekblad voor Bloembollencultuur*, **92**:1230-1231.

Boontjes, J., P. Mantel and H. Brinkman, 1978. The effect of aldicarb (Temik 10G) against root- lesion nematodes in lily planting stock (Dutch). *Bloembollencultuur*, **88**(41): 994-995.

Box, C.O. 1963. Natural cooling of Georgia lilies. *Bull. Miss. Agric. Exp. Stat.*, No. 675, p. 15.

Brierley, P. 1962. Easter lilies freed from cucumber mosaic virus by scale propagation. *Plant Dis. Reptr*, **46**: 627.

Caveness, F.E. and H.J. Jensen, 1955. Investigations of various therapeutic measures to eliminate root-lesion nematodes from Easter lilies. *Plant Dis. Reptr*, **39**: 710-715.

Chaplin, M.H. and A.N. Roberts, 1981. Seasonal nutrient element distribution in leaves of 'Ace' and 'Nellie White' cultivars of the Easter lily, *Lilium longiflorum* L. *Comm. in Soil Sci. and Pl. Analysis.*, **12**(3): 227-237.

Choi, S.T. 1983. Effects of low temperature and hot water treatment of dormant bulbs on leaf emergence in bulblets during propagation of Easter lily (*Lilium longiflorum* Thumb.) from scales. *J. Korean Soc. hort. Sci.*, 24:42-48.

Coquen, C. and M. Astié, 1977. Ontogenèse de bulbilles végétatives obtenues par culture de pieces florales isolées chez le *lilium candidum* L. *Bull. De la Sociéte Botanique de France*, **124**(1/2): 51-59.

Corr, B.E. and H.F. Wilkins, 1984. Hybrid lily forcing. *Minnesota State Florist's Bull.*, No. 33, pp. 1-4.

De Groote, A. 1981. The use of artificial light for the culture of lilies (Dutch). *Verbondsnieuws voor de Belgische Sierteelt*, **25**(16): 712.

De Hertogh, A.A. and N. Blakely, 1972. Influence of gibberellins A_3 and A_{4+7} on development of forced *Lilium longiflorum* Thunb. cv. Ace. *J. Amer. Soc. hort. Sci.*, **97**(3): 320-323.

De Hertogh, A.A. and M. Le Nard, 1993. Physiological Disorders. In: *The physiology of Flower Bulbs* (eds De Hertogh, A.A. and M. Le Nard), pp. 155-160. Elsevier Science Publishers B.V. Amsterdam, The Netherlands.

D'Herde, J., J. van Den Brande and A. Gillard, 1959. Control of *Pratylenchus penetrans* (Cobb), causal agent of root rot in lilies. *Nematologia, Suppl. II, Rep. 5th int. Symp. Plant Nemat.*, Uppsala (Belgium), pp. 64-67.

Dicks, J.W., J. Mc D. Gilford and A.R. Rees, 1974. The influence of timing of application and gibberellic acid on the effects of ancymidol on growth and flowering of Mid-Century Hybrid lily cv. Enchantment. *Scientia Hort.*, **2**(2): 153-163.

Dicks, J.W. and A.R. Rees, 1972. US growth retardant eases lily pot plant production. *Grower*, **78**(2): 81.

Dole, J.M., W.C. Fonteno and S.L. Blankenship, 2005. Comparison of silver thiosulfate with 1- methylcyclopropene on 19 cut flower taxa. *Acta Hort.* (*Proceedings of the Fifth International Postharvest Symposium*, eds Mencarelli, F. and P. Tonutti), held at Verona, Italy, on 6-11 June, 2004; No. 682 (vol. 2), pp. 949-956.

Doucette, C.F. and R. Latta, 1946. The lily weevil, a potentially serious pest in the Pacific Northeast. *Circ. U.S. Dept. agric.*, No. 746, p. 24.

Dunham, C.W. and D.F. Crossan, 1959. Factors in tip-burn injury on leaves of Easter lilies. *Proc. Amer. Soc. hort. Sci.*, **74**: 704-710.

Durieux, A.J.B. 1978a. CO_2 to prevent bud drop in the culture of Enchantment with supplementary lighting (Dutch). *Vakblad Bloemist.*, **33**(17): 24-25.

Durieux, A.J.B. 1978b. Carbon dioxide to prevent bud drop when growing Enchantment with supplementary lighting (Dutch). *Boembollencultuur*, **88**(43): 1048-1049.

Durieux, A.J.B., G.A. Kamerbeek and U. Van Meeteren, 1982/83. The existence of a critical period for the abscission and a non-critical period for blasting of flower-buds of *Lilium* 'Enchantment'; influence of light and ethylene. *Scientia Hort.*, **18**: 287-297.

Eastwood, T. 1952. Forcing Creole lilies at different levels of soil nitrate. *Proc. Amer. Soc. hort. Sci.*, **59**: 531-541.

Eaton, H.J. and A.A. Tompsett, 1'976. Avoiding slug damage to lily bulbs at Rosewarne Experimental Horticulture Station, Camborne, Cornwall. In: *Lilies and Other Liliaceae*, pp. 63-65. RHS, London, U.K.

Eguchi, T. and Y. Matsui, 1968. The effect of soil treatment with ethylthiometon granules on the control of aphids infesting white trumpet lilies. *Res. Bull. Plant Prot. Serv. Japan*, No. 5, pp. 66-70.

Emsweller, S.L. 1963. Propagation of lilies. *The Lily Yrbk of the North Amer. Lily Soc.*, **16**:143-155.

Emsweller, S.L. 1986. Propagation of lilies. *The Lily Yrbk of the North Amer. Lily Soc.*, **39**: 35-40.

Emsweller, S.L., S. Asen and J. Uhring, 1963. Breeders repeat age-old lily cross. *Agric. Res., Wash.*, **11**(9): 15-16.

Erwin, J.E. and R.D. Heins, 1990. Temperature effects on lily development rate and morphology from the visible bud stage until anthesis. *Amer. Soc. hort. Sci.*, **115**: 644-646.

Flokstra, W. 1969. Leaf cuttings of *Lilium longiflorum* (Dutch). *Vakblad Bloemist.*, **24**: 1744.

Franssen, C.J.H. and W.P. Mantel, 1962. *Liothrips vaneecki* and *Xylaplothrips subtennarneus*, two thrips harmful to lilies (Dutch). *Tijdschr. PlZiekt.*, **68**: 285-288.

Furuta, T., W.C. Martin, Jr. and F. Perry, 1963. Lithium toxicity as a cause of leaf scorch on Easter lily. *Proc. Amer. Soc. hort. Sci.*, **83**: 803-807.

Gentner, W.A., W.C. Shaw and F.F. Smith, 1958. An evaluation of several chemicals for weed control in Easter lilies. *Weeds, N. York*, **6**: 203-210.

Gilberte, D.A. and A.J. Lewis, 1990. Effect of planting-date and application method on ancymidol response of hybrid lilies. *Scientia Hort.*, **45**(1-2): 159-165.

Gindina, S.R. 1976. The effect of fertilizers on the growth and development of *Lilium davidii* (Russian). *Okhrana Sredy I Ratsional'n. Ispol'z. Rastitel'n. Resursov. Moscow, USSR; Nauka*, pp. 232-233.

Gindina, S.R., O.F. Mikhailov and Yu. I. Kara, 1974. The effect of water-soluble humates on seed germination and vegetative propagation of lilies by scales (Russian). *Adaptsiya Rastenii v Usloviyakh Sredy. Dnepropetrovsk, Ukrainian SSR*, pp. 52-57.

Giustiniani, L., E. Moschini and A. Graifenberg, 1988. Nutrient requirements of *Lilium* cultivar Mid- Century (Italian). *Colture Protette*, **17**(11): 81-87.

Goldsberry, K.L. 1966. Forcing lilies as pot plants. *Colorado Fl. Grs' Ass. Bull.*, No. 190, pp. 3-5.

Greller, A.M. 1969. Spiral development patterns in the stem and inflorescence of *Lilium tigrinum. Amer. J. Bot.*, **56**: 575-583.

Gries, H. 1971. Observations on lily propagation from bulb scales (Dutch). *Iris and Lilien*, Bull. No. 4, pp. 193-195. Published from International Flower Bulb Center, Hillegom, Holland.

Grueber, K., W. Healy, H.B. Pemberton and H.F. Wilkins, 1986. Minimum fluorescent light requirements and ancymidol interactions on the growth of *Lilium longiflorum* Thunb. 'Nellie White'. *Acta Hort.*, No. 177, vol. I, pp. 241-248.

Gupta, P.P., A.K. Sharma and H.C. Chaturvedi, 1978. Multiplication of *Lilium longiflorum* Thunb. by aseptic culture of bulb scales and their segments. *Indian J. exp. Biol.*, **16**(8): 940-942.

Hammer, A. 1974. Lily height control. *Focus Floric.*, **2**(4): 11-16.

Hammer, P.A. and T. Kirk, 1981. Easter lily results 1981. *Focus on Floriculture*, **9**(4): 4-6.

Hanks, G.R. and R. Menhenett, 1980. An evaluation of new growth retardants on Mid-Century Hybrid lilies. *Scientia Hort.*, **13**(4): 349-359.

Hart, W.H., A.R. Maggenti and J.V. Lenz, 1967. Bulb treatments for the control of root-lesion nematode, *Pratylenchus penetrans*, in Easter lily. *Plant Dis. Reptr*, **51**: 978-980.

Hay, R. and K.A. Beckett, 1971. *Lilium*. In: *Reader's Digest Encyclopaedia of Garden Plants and Flowers*, pp. 399-409. The Reader's Digest Association Limited, London. Great Britain.

Heins, R.D., H.F. Wilkins and W.E. Healy, 1982a. Influence of photoperiod and light stress on flower number, height and growth rate. *Amer. Soc. hort. Sci.*, **107**: 335-338.

Heins, R.D., H.B. Pemberton and H.F. Wilkins, 1982b. Influence of light intensity on plant development. *J. Amer. Soc. hort. Sci.*, **107**(2): 330-335.

Heladze, V.S. 1964. Leaf beetles on lilies (Russian). *Zašc. Rast. Vred. Bolez.*, **9**(12): 34.

Hendriks, C.H.M. and B.J. Kok, 1989. Growth regulators dwarf lilies grown in pots (Dutch). *Vakblad Bloemist.*, **44**(7): 48-51.

Hendriks, C.H.M. and A.T.J. Koster, 1988. New advice on disinfection gives reasonable control of *Penicillium* (Dutch). *Bloembollencultuur*, **99**(21): 31.

Holcomb, E.J., J.W. White and D.J. Beattie, 1989. Adaptation of 'Sans Souci' lilies to potted plant culture. *Acta Hort.*, No. 252, pp. 159-171.

Hong, Y.P., C.S. Lee and J.K. Choi, 1978. Studies on the morphological characteristics, bulblet dormancy and growth retardant responses of *Lilium lancifolium* Thunb. *J. Korean Soc. Hort. Sci.*, **19**(2): 147-153.

Hosaka, H. and M. Yokoi, 1959. Studies on the scale propagation of lilies (Japanese). *Tech. Bull. Fac. Hort. Chiba*, No. 7, pp. 45-55.

Iizuka, M., K. Takeuchi, J. Watanabe and S.H. Bo, 1978. Artificial induction of bulbils in *Lilium* species. *HortSci.*, **13**(6, sect. 1): 666-667.

Ivanova, N.V. 1983. Effect of growth regulators on the propagation of lilies (Russian). *Byulleten' Glavnogo Botanicheskogo Sada*, No. 127, pp. 62-64.

Jackson, A.A. 1958. The Formosa lily as a commercial crop. *Agriculture, London*, **64**: 554-555.

Jana, B.K. and N. Roychoudhury, 1989. *Lilium*. In: *Commercial Flowers* (eds Bose, T.K. and L.P. Yadav), pp. 789-825. Naya Prokash, Calcutta, India.

Jensen, H.J. 1966. Phosphate pesticides control root-lesion nematodes in Oregon Easter lily plantings. *Plant Dis. Reptr*, **50**: 923-927.

Jensen, H.J. and C.F. Doucette, 1968. Root-lesion nematodes can be controlled in Easter lily planting. *Lily Yrbk N.Amer. Lily Soc. Inc.*, No. 21, pp. 53-56.

Jeong, J.H., G.S. Kim, D.Y. Yeom and Y.P. Hong, 1989. Studies on the distribution and eco-morphology of Korean native lilies. Paper presented at the 27th meeting of the Korean Society of Horticultural Science, held at Kyung Hee University on February 1, 1989. *Korean Soc. hort. Sci.*, **7**(1): 180-181(abstr.).

Jiao, J., X. Wang and M.J. Tsujita, 1991. Antagonistic effects of uniconazole and GA_{4+7} on shoot elongation and flower development in 'Nellie White' Easter lily. *Scientia Hort.*, **46**(3-4) 323-331.

Johnson, C.R. 1973. Effectiveness of ancymidol on reducing height of Easter lilies grown under different environments. *Proc. Fla St. hort. Soc.*, **86**: 380-382.

Kachroo, P. 1949. A note on vegetative propagation of *Lilium longiflorum* Wall. *Curr. Sci.*, **18**: 299.

Kato, K. and Y. Yasutake, 1977. Plantlet formation and differentiation of epidermal tissues in green callus cultures from excised leaves of *Lilium*. *Phytomorph.*, **27**: 390-396.

Kawabata, S., Li YuHua, M. Adachi, H. Maruyama, E. Ozawa and R. Sakiyama, 2002. Role of photoreceptor-and sugar-mediated reactions in light dependent anthocyanin production in lily and stock flowers. *J. Japanese Soc. hort. Sci.*, **71**(2): 220-225.

Kim, Y.S. and W.B. Lee, 1990. A study of epidermal characters of the genus *Lilium* L. in Korea by SEM. *Korean J. Pl. Taxon.*, **20**(4): 195-207.

Kiplinger, D.C., H.K. Tayama and G. Staby, 1971. EL-531 on height of Ace and No. 44 lilies. *Ohio Florists' Assocn Bull.*, No. 505, p. 2.

Kiplinger, D.C., H.K. Tayama and G. Staby, 1972a. Arai and Georgia lilies as affected by growth retardants and short days. *Ohio Florists' Assocn Bull.*, No. 509, p. 6.

Kiplinger, D.C., H.K. Tayama and G. Staby, 1972b. Garden lilies as affected by growth retardants. *Ohio Florists' Assocn Bull.*, No. 509, pp. 5-6.

Kiplinger, D.C., H.K. Tayama and G. Staby, 1975a. A-Rest on Mid-Century hybrid lilies. *Ohio Florists' Assoc. Bull.*, No. 543, p. 5.

Kiplinger, D.C., H.K. Tayama and G. Staby, 1975b. A-Rest on a *rubrum*-type garden lily. *Ohio Florists' Assoc. Bull.*, No. 543, p. 7.

Kiplinger, D.C., H.K. Tayama and G. Staby, 1975c. A-Rest on Ace and Arai lilies. *Ohio Florists' Assoc. Bull.*, No. 543, p. 9.

Kneissl, P. 1971. Little-grown pot lilies. *Deutscher Gärtnerbörse*, **26**: 553-554.

Kohl, H.C., Jr. and R. L. Nelson, 1963. Day length and light intensity as independent factors in determining height of Easter lily. *Proc. American Soc. hort. Sci.*, **83**: 808-810.

Kruyer, C.J. and J. Boontjes, 1982. Hot water treatment of *Lilium longiflorum* bulbs. *Bloembollencultuur*, **93**(25): 622-623.

Laiche, A.J. Jr. and C.O. Box, 1970. Response of Easter lily to bulb treatments of precooling, packing, media, moisture, and gibberellins. *HortSci.*, **5**: 396-397.

Larson, R.A., C.B. Thorne, R.R. Milks, Y.M. Isenberg and L.D. Brisson, 1987. Use of ancymidol bulb dips to control stem elongation of Easter lilies grown in a pine bark medium. *J. Amer. Soc. hort. Sci.*, **112**(5): 773-777.

Lee, J.S. and S.M. Roh, 1985. Effects of ethylence on flowering and senescence of Easter lilies. *J. Korean Soc. hort. Sci.*, **26**(2): 145-149.

Lewis, A.J. and J.S. Lewis, 1980. Response of *Lilium longiflorum* to ancymidol bulb-dips. *Scientia Hort.*, **13**(1): 93-97.

Lewis, A.J. and J.S. Lewis, 1981. Improving ancymidol efficiency for height control of Easter lily. *HortSci.*, **16**(1): 89-90.

Lewis, A.J. and J.S. Lewis, 1982. Height control of *Lilium longiflorum* Thunb. 'Ace' ancymidol bulb-dips. *HortSci.*, **17**(3): 336-337.

Lewis, A.J. and D.A. Gilbertz, 1987. Hybrid lily response to various methods of ancymidol application. *Acta Hort.*, No. 205, pp. 237-240.

Lewnes, M.A., H.K. Tayama and D.C. Kiplinger, 1977. Effective height control of 'Georgia' Easter lilies with A-Rest. *Ohio Florists' Assocn Bull.*, No. 576, p. 2.

Lim, K.B., M.S. Ramanna, J.H. De Jong, E. Jacobsen and J.M. van Tuyl, 2003. Evaluation of BC2 progenies derived from 3x-2x and 3x-4x crosses of *Lilium* hybrids: a GISH analysis. *Theor. Appl. Genet.*, **106**: 568-574.

Lin, P.C. and A.N. Roberts, 1970. Scale function in growth and flowering of *Lilium longiflorum* Thunb. 'Nellie White'. *J. Amer. Soc. hort.Sci.*, **95**: 559-561.

Liu, L. and D.W. Burger, 1986. *In vitro* propagation of Easter lily from pedicels. *HortSci.*, **21**: 1437-1438.

Löffler, H.J.M., H. Meijer, Th. P. Straathof and J.M. van Tuyl, 1996. Segregation of *Fusarium* resistance in an interspecific cross between *Lilium longiflorum* and *Lilium dauricum*. *Acta Hort.*, **414**: 203-208.

Macnish, A.J., R.T. Leonard and T.A. Nell, 2008. Treatment with chlorine dioxide extends the vase life of selected cut flowers. *Postharv. Biol. & Tech.*, **50**(2/3): 197-207.

Maesato, K., K.S. Sarma, H. Fukui and T. Hara, 1991. *In vitro* bulblet induction from shoot apices of *Lilium japonicum*. *HortSci.*, **26**: 211.

Maggenti, A.R., W.H. Hart and J.V. Lenz, 1967. Soil treatments for the control of root-lesion nematode (*Pratylenchus penetrans*) on Easter lily. *Plant Dis. Reptr*, **51**: 549-552.

Magie, R.O. 1966. Control of anthracnose scale rot on *Lilium longiflorum* Thunb. ('Georgia'). *Proc. Fla St. hort. Soc.*, **79**: 471-477.

Maksimova, E.V. 1970. Herebicides in lilies. *Zašc. Rast.*, **15**(2): 26.

Marshall, P.A., J.N. Joiner and W.T. Witte, 1974. Effect of media, fertilizer rate and ancymidol on Mid-Century hybrid lily 'Enchantment'. *Proc. Fla St. hort. Soc.*, **87**: 493-495.

Marousky, F.J. and S.S. Woltz, 1977. Influence of lime, nitrogen, and phosphorus sources on the availability and relationship of soil fluoride to leaf scorch in *Lilium longiflorum* Thunb. *J. Amer. Soc. hort. Sci.*, **102**(6): 799-804.

Mason, M.R. and W.B. Miller, 1991. Flower bud blast in Easter lily is induced by ethephon and inhibited by silver thiosulfate. *HortSci.*, **26**: 1165-1167.

Mastalerz, J.W. 1962. Leaf scorch of Croft lilies as influenced by calcium-phosphorus levels and soil mixtures (abstr.). *Mic. Flor.*, No. 371, p. 24.

Matsuo, E. 1974. Studies on growth and development of Easter lily (*Lilium longiflorum*) bulbs. II. Effect of red, orange or blue fluorescent light or darkness during scaling on the growth of bulblets. *Sci. Bull. Fac. Agric., Kyushu Univ.*, **28**(4): 197-201.

Matsuo, E. and K. Arisumi, 1979. Studies on the growth and development of bulbs in the Easter lily, *Lilium longiflorum*. VIII. Effect of bulb cold treatment on leaf development in the bulb and the scale bulblet. *J. Japanese Soc. hort. Sci.*, **48**(2): 224-230.

Matsuo, E. and K. Arisumi, 1978a. Studies on the leaf development of the scale bulblet in the Easter lily (*Lilium longiflorum* Thunb.). II. Relationship between the size of the parent bulb or the parent scale and the type of leaf development (plant type)(Japanese). *Bull. Fac. Agric., Kagoshima Univ.*, No. 28, pp. 1-8.

Matsuo, E. and K. Arisumi, 1978b. Studies on growth and development of bulbs in the Easter lily (*Lilium longiflorum* Thunb.). VII. Scale position-dependence of the effect of the hot water treatments on the growth behavior of the scale bulblets. *Memoirs Fac. Agric., Kagoshima Univ.*, **14**(23): 69-76.

Matsuo, E., K. Arisumi and H. Kawashima, 1982. Cultural practices influencing premature daughter leaf and/or shoot emergence in scale-propagated Easter lily. *HortSci.*, **17**(2): 196-198.

Matsuo, E., K. Arisumi, K. Obmachi and Y. Sakata, 1982. Effect of scale planting depth on leaf and shoot emergence in scale propagated Eater lily. *Hort.Sci.*, **17**(5): 806-807.

Matsuo, E., K. Arisumi, K. Satoh and Y. Sakata, 1983. Factors influencing stem root emergence during scale propagation in the Easter lily. *HortSci.*, **18**(1): 78-79.

Mastuo, E., M. Matsuzawa, Y. Sakata and K. Arisumi, 1989. Asexual propagation of variegated *Lilium longiflorum* 'Chotaro'. *Scientia Hort.*, **39**(4): 349-354.

Matsuo, E., A. Nonaka and K. Arisumi, 1978. Studies on the leaf development of the scale bulblet in the Easter lily (*Lilium longiflorum* Thunb.). I. Some factors influencing the time of leaf emergence and the survival of the parent scale (Japanese). *J. Jap. Soc. hort. Sci.*, **46**(4): 515-520.

Matsuo E.and J.M. Van Tuyl, 1984. Effect of bulb storage temperature on leaf emergence and plant development during scale propagation of *Lilium longiflorum* 'White American'. *Sci. Hort.*, **24**: 59-66.,

Matsuo, E. and J.M. Van Tuyl, 1986. Early scale propagation results in forcible bulbs of Easter lily. *HortSci.*, **21**(4):1006-1007.

McRae, E. 1986. Scale propagation of lilies. *The Lily Yrbk of the North Amer. Lily Soc.*, **39**: 41-43.

Meeteren, U. van and M. De Proft, 1982. Inhibition of flower bud abscission and ethylene evolution by light and silver thiosulphate in *Lilium. Physiol. Plant.*, **56**(3): 236-240.

Meij, H. van Der, J. Boontjes and C.J. Kruyer, 1978. Hot-water treatment and disinfection of lily planting stock (Dutch). *Bloembollencultuur*, **89**(24): 580, 582.

Menhenett, R. and G.R. Hanks, 1982. New retardant shows promise for pot-grown lilies and tulips. *Grower*, **97**(20): 17, 19, 20.

Miller, H.J., J.D. Janse, W. Kamerman and P.J. Muller, 1980. Recent observations on leafy gall in Liliaceae and some other families. *Netherlands J. Pl. Path.*, **86**(2): 55-68.

Miller, W.B. 1990. Localization of reserve mobilization during scalet formation on Easter lily scales. *Acta Hort.*, No. 266, pp. 95-100. *Fifth intern. Symp. on flower bulbs*, held at Seattle, Washington in July 10-14, 1989.

Miller, W.B. 1993. *Lilium longiflorum*. In: *The Physiology of Flower Bulbs* (eds De Hertogh, A.A. and M. Le Nard), pp. 391-422. Elsevier Science Publishers B.V., Amsterdam, The Netherlands.

Miller, W.B. and R.W Langhans, 1989. Carbohydrate changes of Easter lilies during growth in normal and reduced irradiance environments. *J. Amer. Soc. hort. Sci.*, **114**(2): 310-315.

Montezuma-de-Carvalho, J. and M.L.L. Gulmaraes, 1974. Production of buds and plantlets from the stamen's filament of *Lilium regale* cultivated *in vitro. Biol. Plant.*, **16**: 472-473.

Morishita, F.S. and R.N. Jefferson, 1969. Granular formulations of systemic insecticides for control of aphids on Easter lilies. *Calif. Agric.*, **23**: 16-17.

Muller, P.J. 1966. The incidence and control of leaf eelworms in lilies (Dutch). *Mededs. Rijksfacult. Landb-Wetensch. Gent*, **31**: 666-671.

Nagamura, S. 1983. Bulb propagation of *Lilium concolor* Salisbury in a sawdust bed separated from the ground. *Bull. Nara agric. Exp. St.*, No. 14, pp. 56-65.

Nederpel, W.A.C. 1979. Be careful with fluorine for lilies (Dutch). *Vakblad Bloemist.*, **34**(37): 44-45.

Nightingale, A.E. 1979. Bulbil formation on *Lilium longiflorum* Thunb. cv. Nellie White by foliar applications of BPA. *Hort.Sci.*, 14(1): 67-68.

Niimi, Y., Y. Endo and E. Arisaka, 1988. Effects of chilling- and GA_3 treatments on breaking dormancy in *Lilium rubellum* Baker bulblets cultured *in vitro. J. Japan Soc. hort. Sci.*, **57**(2): 250-257.

Niimi, Y. and T. Onozawa, 1979. *In vitro* bulblet formation from leaf segments of lilies, especially *Lilium rubellum* Baker. *Sci. Hort.*, **11**: 379-389.

Niimi, Y. and M. Oda, 1989. Time of initiation and development of flowerbuds in *Lilium rubellum* Baker. *Scientia Hort.*, **39**(4): 341-348.

Novak, F.J. and E. Petru, 1981. Tissue culture propagation of *Lilium* hybrids. *Scientia Hort.*, **14**(2): 191-199.

Nowak, J. and K. Mynett, 1985a. The effect of sucrose, silver thiosulphate and 8- hydroxyquinoline citrate on the quality of *Lilium* inflorescences cut at the bud stage and stored at low temperature. *Scientia Hort.*, **25**(3): 299-302.

Nowak, J. and K. Mynett, 1985b. The effect of growth regulators on postharvest characteristics of cut *Lilium* 'Prima' inflorescences. *Acta Hort.*, No. 167, pp. 109-116.

Oelker, G. 1958. Moltiplicazione per seme di alcune specie di *Lilium. Sementi Elette*, **4**: 44-48.

Oglevee, J.R., J.F. Tammen and W. O'Donnovan, 1986. Lily process. United States Patent No. 4,570,379.

Ohkawa, K. 1989. Time of flower bud differentiation in lilies native to Japan. *J. Jap. Soc. hort. Sci.*, **57**(4): 655-661.

Ohkawa, K. and T. Saigusa, 1972. Studies on the nematode damage caused by continuous cropping of *Lilium speciosum rubrum* (Japanese). *Bull. Kanagawa Hort. Exp. St.*, No. 20, pp. 103-107.

Okada, M. 1951. Propagation studies on Easter lily bulbs. I. Correlation between growth and bulblet production (Japanese). *J. hort. Ass. Apan*, **20**: 125-128.

O'Leary, K. and C.E.F. Guterman, 1937. *Penicillium* rot of lily bulbs and its control by calcium hypochlorite. *Contr. Boyce Thompson Inst.*, **8**: 361-374.

Ookawa, K. and M. Sugimoto, 1966. Studies on methods of controlling the bulb mite *Rhizoglyphus echinopus* on *Lilium speciosum* var. *speciosum* (Japanese). *Bull. Kanagawa hort. Exp. Stat.*, No. 14, pp. 81-88.

Orlov, M.I. 1957. The life and unfolding time of lily flowers. *Bjull. Glav. Bot. Sada*, No. 28, pp. 121-122.

Overman, A.J. 1961. Pre-storage treatment of lily bulbs with nematicides. *Proc. St. hort. Soc.*, **74**: 386-388.

Parivar, F., J.E. Preece and G.D. Coorts, 1985a. Height control of Mid-Century hybrid lilies. *Illinois St. Florists' Assoc. Bull.*, No. 422, pp. 26-28.

Parivar, F., J.E. Preece and G.D. Coorts, 1985b. The effects of ancymidol concentrations and application methods on cultivars of Mid-Century Hybrid lily. *J. hort. Sci.*, **60**(2): 263-268.

Pasini, C., T. Berio, P. Curir and F. D'Aquila, 1996. Further characterization of *Rhizoctonia solani* isolated from carnation and other ornamental plants. *Informatore Fitopatologico*, **46**(6): 33-36.

Payne, G.O. 1960. Vernalization of lily seeds. *N.Z. Plants Gdns*, **3**: 320-322.

Pearse, H.L. 1972. Beautiful as pot plants. Dwarfed lilies. *Fmg South Africa*, **48**(6): 75-76.

Pergola, G. 1986a. The preparation of *Lilium* bulbs for the programmed production of cut flowers. Results of some preliminary trials. I. The effect of cold storage at 2 °C and -2 °C compared with storage at ambient temperature on bulbs of the cultivar Enchantment (Italian). *Annali dell'Istituto Sperimentale per la Floricoltura,* **17**(1): 127-139.

Pergola, G. 1986b. The preparation of *Lilium* bulbs for the programmed production of cut flowers. Results of some preliminary trials. Part II. Vernalization requirements to break dormancy and the assessment of bulb maturity. *Annali dell'Istituto Sperimentale per la Floricoltura,* **17**(1): 141-146.

Pope, T.E. and B.L. Baum, 1965. Effect of cooling method and Phosfon on *Lilium longiflorum* "Georgia". *Proc. Amer. Soc. hort. Sci.*, **87**: 510-514.

Prince, T.A. and M.S. Cunningham, 1989. Production and storage factors influencing quality of potted Easter lilies. *Hort. Sci.*, **24**(6): 992-994.

Prince, T.A. and M.S. Cunningham, 1990. Response of Easter lily bulbs to peat moisture content and the use of peat or of polyethylene-lined cases during handling and vernalization. *J. Amer. Soc. hort. Sci.*, **115**(1): 68-72.

Prince, T.A., M.S. Cunningham and J.S. Peary, 1987. Floral and foliar quality of potted Easter lilies after STS or phenidone application, refrigerated storage, and stimulated shipment. *J. Amer. Soc. hort. Sci.*, **112**(3): 469-473.

Robb, S.M. 1957. The culture of excised tissue from bulb scales of *Lilium speciosum* Thunb. *J. exp. Bot.*, **8**: 348-352.

Roberts, A.N., J.R. Stang, Y.T. Wang, W.R. McCorkle, L.R. Riddle and F.W. Moeller, 1985. *Easter Lily Growth and Development. Oregon St. Univ. Agr. Exp. Stn. Tech. Bull.* No. 148, 74 pp.

Rode, H. 1966. Several dipterous insects as pests on *Lilium martagon* (German). *Rev. Roumaine Bio. Ser. Bot.*, **11**(1/3): 185-190.

Rode, H. and E. Vorsatz, 1966. The biology and control of the lily fly, *Liriomyza urophorina* Mik. (German). *NachrBl. Dtsch. PflSchDienst, Berlin*, **20**: 107-112.

Roh, S.M. 1979. Various factors influencing the growth and flowering of *Lilium lancifolium* Thunb. *J. Korean Soc. hort. Sci.*, **20**(1): 72-83.

Roh, S.M. 1981. Dormancy and maturity in the bulbil of *Lilium lancifolium*. II. The influence of temperature treatment on growth and flowering response. *J. Korean Soc. hort. Sci.*, **22**(3): 199-208.

Roh, S.M 1982. Propagation of *Lilium longiflorum* Thunb. by leaf cuttings. *HortSci.*, **17**: 607-609.

Roh, S.M. 1990a. The effects of growth regulators on bulblet formation from Easter lily leaves. *Plant Gr. Regulator Soc. of America Quart.*, **18**(3): 140-146.

Roh, S.M. 1990b. Effect of high temperature on bud blast in Asiatic hybrid lilies. *Acta Hort,* No. 266, pp. 142-146.

Roh, S.M. and H.F. Wilkins, 1973. Influence of temperature on the development of flower buds from the visible stage to anthesis of *Lilium longiflorum* Thunb. cv. Ace. *HortSci*, **8**: 129-130.

Roh, S.M. and H.F. Wilkins, 1976. The study on the timing of ancymidol spoil drenches and extraction of ancymidol-like substances in potted *Lilium brownii* lily. *J. Korean Soc. hort. Sci.*, **17**(1): 93-99.

Roh, S.M. and H.F. Wilkins, 1977. The influence and interaction of ancymidol and photoperiod on growth of *Lilium longiflorum* Thunb. *J. Amer. Soc. hort. Sci.*, **102**(3): 255-257.

Rohde, J. 1970. The relationship between temperature and germination of *Lilium* spp. (German). *Gartenbwuwissenschaft*, **35**: 347-352.

Rotten, L.A.J.M. van De, 1978. Yields and returns for flower production with the lily cv. Enchantment in 1975 and 1976 (2) (Dutch). *Boembollencultuur*, **89**(2): 30-31.

Rybczynska, J.J. and H. Gomolinska, 1989. 6-Benzyladenine control of the initial bulblets formation of wild lily *Lilium martagon* L. *Acta Hort.*, No. 251, pp. 183-189 (Proceedings of a Symp. on Growth Regulators in Ornamental Horticulture, held at Skierniewice, Poland, on 5-10 Sept., 1988).

Sajid,G.M., Mahmoona Kaukab and Zahoor Ahmad, 2009. Foliar application of plant growth regulators (PGRs) and nutrients for improvement of lily flowers. *Pakistan J Bot.*, **41**(1): 233- 237.

Sanyat Misra and R.L. Misra, 2013. *Lilium*. In: *Commercial Ornamental Bulb Science*, pp. 84-99. Westville Publishing House, 47, B-5, Paschim Vihar, New Delhi-110 063.

Schenk, P.C. 1975. Periodicity in lilies. *Vakblad voor de Bloemisterij*, **29** (14):10-11, 13.

Schread, J.C. 1966. Iris borer and its control. *Circ. Conn. Agric. Exp. Stat.*, No. 202, p. 5.

Schmitzer, E. 1991. A survey of named polyploid lilies of the Asiatic section. *Quarterly Bulletin of the North American Lily Society*, **45**: 6-12.

Seeley, J.G. 1982. Colorful garden lilies for potted plants. *N. Y. St. Fl. Industries Bull.*, No. 137, pp. 1-5.

Sharp, W.R., R.S. Raskin and H.E. Sommer, 1971. Haploidy in *Lilium*. *Phytomorph.*, **21**: 334-347.

Sheridan, W.W. 1968. Tissue culture of the monocot *Lilium*. *Planta*, **82**: 180-192.

Shoyama, Y., N. Hasegawa and I. Nishioka, 1988.*In vitro* propagation of *Lilium japonicum* by culture of bulblets. *Shoyakugaku, Zasshi*, **41**: 352-355.

Simmonds, J.A. and B.G. Cumming, 1976. Propagation of *Lilium* hybrids. II. Production of plantlets from bulb-scale callus cultures for increased propagation rates. *Sci. Hort.*, **5**: 161-170.

Slootweg, A.F.G. 1960. Bulb diseases caused by *Aphelenchoides* species (Dutch, abstr.). *Tijdschr. PlZiekt.*, **66**: 126.

Smith, J.D. and E.A. Miginnes, 1969. Scale rot tests of hardy hybrid lilies. *Canad. Plant Dis. Surv.*, **49**: 43-45.

Staden, O.L. 1979. Pot plants from treated bulbs of lily cultivars (Dutch). *Vakblad Bloemist.*, **34**(39): 46-47.

Staden, O.L. and W. Maas, 1978. The lily cv. Enchantment, from bulb to pot plant (Dutch). *Bloembollencultuur*, **89**(13): 310-311.

Staden, O.L. and W. Maas, 1979. Growing the lily cv. Enchantment as a pot plant by applying growth retardant (Dutch). *Vakblad Bloemist.*, **34**(9): 32-33.

Staden, O.L. and W. Maas, 1980a. Retarding lily bulbs results in good pot plants (Dutch). *Vakblad Bloemist.*, **35**(48): 40-41.

Staden, O.L. and W. Maas, 1980b. Pot plants from bulbs of various lily cultivars (Dutch). *Vakblad Bloemist.*, **35**(48): 43.

Steffen, L. and C.H. Bünger, 1955. Propagation of lilies from bulb scales (German). *Gartenwelt*, **55**: 186-187.

Stenberg, N.E., C.H. Chen and J.G. Ross, 1977. Regeneration of plantlets from leaf culture of *Lilium longiflorum* Thunb. *Proc. South Dakota Acad. Sci.*, **56**: 152-158.

Stimart, D.P. and P.D. Ascher, 1978. Tissue culture of bulb scale sections for asexual propagation of *Lilium longiflorum* Thunb. *J. Amer. Soc. hort. Sci.*, **103**: 182-184.

Streu, H.T. 1964. Find systemic control of aphids on Easter lilies. *N. J. Agric.*, **46**(2): 5-8.

Struckmeyer, B.E. 1953. The effect of maleic hydrazide on the anatomical structure of Croft Easter lilies. *Amer. J. Bot.*, **40**: 25-29.

Struckmeyer, B.E. and G.E. Beck, 1952. The effect of maleic hydrazide on the growth and flowering of Croft Easter lilies. *Proc. Amer. Soc. hort. Sci.*, **59**: 542-548.

Stuart, N.W., D.L. Gill and M.G. Hickma, 1961. Height control of forced Georgia Easter lilies. *The Exch.*, **136**(52): 22-23, 42.

Stumm-Tedgethoff, B.F.A. 1986. Plant growth hormones as intermediates in stem fasciation caused by nematodes in *Lilium henryi. Nematologica.*, **32**(2): 234-235.

Suzuki, S. 1989. Bulblet formation from leaf cuttings of *Lilium longiflorum* Thunb. (Japanese). *J. agric. Sci., Tokyo Nogyo Daigaku*, **33**(3): 232-238.

Swanson, K.G. 1963. Systemic insecticides for aphid control in garden lily production. *J. econ. Ent.*, **56**: 680-681.

Swart, A. 1979. The quality of the lily cv. Enchantment as a cut flower (Dutch). *Boloembollencultuur*, **89**(34): 902-903.

Swart, A. 1981. Bij lelie Enchantment is voorbehandeling noodzakelijk. *Boembollencultuur*, **92**(16): 398-399.

Swart, A. 1982. Improvement of the bulb flower quality as a consequence of pretreatment by chemical (abstr. No. 1764). *Abstracts, XXIst intern. Hort. Congr. Hamburg, FRG; intern. Soc. hort. Sci.*, Vol. II.

Systema, W. 1968. Lilies as pot plants. *Vakblad Bloemist.*, **23**: 1703.

Takayama, S. and M. Misawa, 1979. Differentiation in *Lilium* bulb scales grown *in vitro*, effect of various cultural conditions. *Physiol. Plant.*, **46**(2): 184-190.

Takayama, S., M. Misawa, Y. Takashige, H. Tsumori and K. Ohkawa, 1982. Cultivation of *in vitro*-propagated *Lilium* bulbs in soil. *J. Amer. Soc. Hort. Sci.*, **107**: 830-834.

Takayama,S. and M. Misawa, 1983. A scheme for mass propagation of *Lilium in vitro. Sci. Hort.*, **18**: 353-362.

Tamura, T. 1949. Lily propagation by scales (Japanese). *J. hort. Ass. Japan*, **18**: 233-236.

Tjia, B., L. Stoltz, M.S. Sandhu and J. Buxton, 1976. Surface active agent to increase effectiveness of surface penetration of ancymidol on hydrangea and Easter lily. *HortSci.*, **11**(4): 371-372.

Tsujita, M.J., D.P. Murr and A.G. Johnson, 1978. Influence of phosphorus nutrition and ancymidol on leaf senescence and growth of Easter lily. *Canadian J. Pl. Sci.*, **58**(1): 287-290.

Ueno, K. and O. Namikawa, 1963. Bulb disease of *Lilium speciosum* and its control (Japanese). *Bull. Kanagawa hort. Exp. Stat.*, No. 11, pp. 67-72.

Van Aartrijk, J. and G.J. Blom-Barnhoorn, 1980. Effect of sucrose, mineral salts, and some organic substances on the adventitious regeneration *in vitro* of plantlets from bulb-scale tissue of *Lilium speciosum* 'Rubrum'. *Acta Hort.*, No. 109, pp. 297-302.

Van Meeteren, U. and G. Slootweg, 1986a. On the role of ethylene biosynthesis in flower-bud abscission of *Lilium* × 'Enchantment'. *Acta Hort.*, No. 177, pp. 641-644.

Van Meeteren, U. and G. Slootweg, 1986b. The effects of foliar sprays with STS during forcing of *Lilium* × 'Enchantment' on flower-bud abscission and opening. *Acta Hort.*, No. 181, pp. 473-476.

Van Nes, C.R. 1981. Possibilities with freezing *Lilium longiflorum* bulbs (Dutch). *Bloembollencultuur*, **92**(6): 142.

Van Tuyl, J.M. 1984. Effect of temperature treatments on the scale propagation of *Lilium longiflorum* 'White Europe' and *Lilium* × 'Enchantment'. *HortSci.*, **18**(5): 754-756.

Van Tuyl, J.M. 1988. Dutch-grown *Lilium longiflorum* a reality. *The Lily Yearbook of the North American Lily Society*, **41**: 33-37.

Van Tuyl, J.M. 1996. Interspecific hybridization of flower bulbs : A review. *Seventh International Symposium on Flower Bulbs, March 10-16, Herzliya, Israel. Acta Hort*, No. 332, pp. 85-89.

Van Tuyl, J.M., C.J. Keijzer, H.J. Wilms and A.A.M. Kwakkenbos, 1988. Interspecific hybridization between *Lilium longiflorum* and the white Asiatic hybrid 'Mont Blanc'. *The Lily Yearbook of the North American Lily Society*, **41**: 103-111.

Van Tuyl, J.M., J.E. van Groenestijn and S.J. Toxopeus, 1985. Low light intensity and flower bud abortion in Asiatic hybrid lilies. I. Genetic variation among cultivars and progenies of a diallel cross. *Euphytica*, **34**: 83-92.

Van Tuyl, J.M., K. van de Sande, R.van Dien, D. Straathof and H.M.C.van Holtejn, 1990. Overcoming interspecific crossing barriers in *Lilium* by ovary and embryo culture. *Acta Hort*., No. 266, pp. 317-322. Proc. of the Fifth intern. Symp. on Flower Bulbs, held in Seattle, Washington, USA on July 10-14, 1989.

Watanabe, H. 1989. The use of growth regulators applied to stem cuttings of lilies. *Bull. Nara agric. Exp. St*., No. 20, pp. 67-71.

Weiler, T. 1977. Light and A-Rest influence Easter lily height. *Focus on Floric*., **5**(4): 9-12.

Weiler, T.C. 1978. Shade- and ancymidol-altered shape of potted *Lilium longiflorum* 'Ace'. *HortSci*., **13**(4): 462-463.

Weststeijn, G. 1978. Proliferation disease in lilies (Dutch). *Bloembollencultuur*, **89**(21): 515.

White, J.W. 1971a. The response of Mid-Century hybrid lilies to Quel, a new growth regulating chemical. *Bull. Pennsylvania Fl. Grs*, No. 242, pp. 3-5, 13.

White, J.W. 1971b. Progress Report II. Easter lily height control. *Bull. Pennsylvania Fl. Grs*, No. 246, pp. 3-5.

White, J.W. 1972a. Progress Report III. Easter lily height control. *Bull. Pennsylvania Fl. Grs*, No. 247, pp. 1-3, 12.

White, J.W. 1972b. Easter lily height control. *Bull. Pennsylvania Fl. Grs*, No. 249, pp. 6-7.

White, J.W. 1976. *Lilium* sp. 'Mid-Century Hybrids' adapted to pot use with ancymidol. *J. Amer. Soc. hort. Sci*., **101**(2): 126-129.

White, J.W., E.J. Holcomb, K. Shumac, M. Rose and T. Morgart, 1989. Oriental hybrid lilies as pot plants. *Bull. Pennsylvania Fl. Grs*, No. 395, pp. 1-5.

Wilkins, H.F., G.D. Coorts and J.B. Gartner, 1963. The effects of several growth retardants on Croft lilies. *Ill. St. Flor. Ass. Bull*., No. 234, pp. 1, 3-4.

Wilkins, H.F. and J.M. Dole, 1997. The physiology of flowering in *Lilium*. *Acta Hort*., No. 430, pp. 183-188.

Wilkins, H.F., K. Grueber, W. Healy and H.B. Pemberton, 1986. Minimum fluorescent light requirements and ancymidol interactions on the growth of Easter lily. *J. Amer. Soc. hort. Sci*., **111**(3): 384-387.

Woodcock, H.B.D. and W.T. Stearn, 1950. *Lilies of the World: Their Cultivation and Classification*, pp. 15-22. Country Life Ltd., London.

Wulster, G.J., T.J. Gianfagna and B.B. Clarke, 1987. Comparative effects of ancymidol, propiconazol, triadimefon, and Mobay RSW0411 on lily height. *HortSci*., **22**(4): 601-602.

Zhou, S., M.S. Ramanna, R.G.F. Visser and J.M. van Tuyl, 2008. Genome composition of triploid lily cultivars derived from sexual polyploidization of *Longiflorum* × Asiatic hybrids (*Lilium*). *Euphytica*, **160**: 207-215.

25

Limonium (Family: Plumbaginaceae)

M. K. Singh, Sanyat Misra, R.L. Misra and Sanjay Kumar

[**Common names**: German statice/Notch-leaf/Sea lavender/Winged statice (*Limonium sinuatum*), Russian/Rat's-tail statice (*L. suworowii*), Statice (*Statice*)]

Introduction

Allied to *Acantholimon*, the genus *Limonium*, formerly known as *Statice* (includes the plants of the genera now known as *Armeria, Limonium* and *Statice*), of some 150-300 species (depending on classifier) of herbaceous annuals, perennials, and deciduous shrubs and sub-shrubs of cosmopolitan distribution, *i.e.* with some representation in most continents but especially found in eastern Mediterranean region and central Asia, especially in salt marshes and steppes in the saline regions both coastal and semi-desert. The perennial species are not all hardy and some are treated as annuals. It takes its name from the Greek *leimon* 'a meadow', an allusion to its salt-marsh habitat. For many years these have been grown to provide dried flowers but only recently the use as a fresh flower has been realized. Florists prefer yellow and clear blue as fresh flowers. They are suitable for beds and borders, rockeries, pot culture, for cutting and for drying. These are probably the most popular of the entire everlasting annual flowers. The texture of the flower is rather crispy tissue paper like. They grow on flat looking dark green stems, the main one is stiff and branches near the top with flowers appearing on the top of each stem. The flower sprays are white, lemon-yellow, golden, pale-orange, rosy-pink, blue, deep purple and in various combinations, and are borne in comb-like clusters. Individual flowers are tubular and tiny. They retain their attractive petal shades for a long time when dried and sold as everlasting flowers for flower arrangement. They can tolerate a great diversity of climate and soil such as saline soil as well as humid coastal salt marshes, sea cliffs and semi-desert, and desert climates and because of this reason, *vis-à-vis* its high demand as cut or dry flowers, it is grown all over the world, especially in the Netherland, Ecuador, Korea, Kenya, United Kingdom, USA, Colombia, Canary Islands, China, Canada, India, Israel, Italy, Japan, Spain, Algeria, Germany and Mediterranean region. Kenya has become a successful cut flower exporter of *Limonium*.

Limonium has turned into a multi-purpose plant. Besides its importance as a cut flower crop, it is also ideal for making dry-flower products. It can be painted, coloured and dyed. Various floral products such as cards, pictures, wall hangings, arrangements and pot-pourris can be prepared out of them. They need little care and are everlasting. Its products are sold at very high prices so there is a great potential to start a cottage industry based on *Limonium* products. By growing this plant, the economic condition of small flower growers and home makers will improve considerably by selling it in domestic as well as export market.

Antioxidant metabolites from roots of *Limonium brasiliense* have been reported for the first time.These active compounds are myricetin 3-*O*-alpha-rhamnopyranoside, (-)-epigallocatechin 3-*O*-gallate, (-)-epigallocatechin, (+)-gallocatechin and gallic acid (Murray *et al.*, 2004). *L. sinense* possesses a hepatoprotective effects against carbon tetrachloride and beta-D-galactosamine intoxication in rats (Chaung *et al.*, 2003). Free radical scavenging action of the medicinal herb *Limonium wrightii* has been found and gallic acid was identified as the active component to these actions (Aniya *et al.*, 2002). Extract from the roots of *Limonium sinense* is used to treat fever, hemorrhage, hepatitis and other disorders in a Chinese folk medicine (Tang *et al.*, 2008). Palmitic acid, a predominant volatile constituent from the flowering parts of *Limonium echioides* was found to inhibit the visible growth of *Escherichia coli, Micrococcus luteus, Salmonella typhimurium,*

Staphylococcus aureus and *S. epidermidis* (Saidana *et al.*, 2008). Ethanolic extracts of *Limonium vulgare* showed significant cytotoxic activity against *Artemia salina* and *Daphnia magna*. The extracts of this plant also showed antineoplastic activities in the potato disc assay (Lellau and Liebezeit, 2003). Butanol crude extracts of the aerial parts of *Limonium axillare* exhibited a superior antifungal activity when compared with standard chloramphenicol, tetracycline and nalidixic acid (Mahasneh, 2002).The methanol extract of *Limonium stocksii* roots was found effective against gram-positive bacteria (*Staphylococcus aureus* and *Bacillus subtilis*), gram-negative bacteria (*Escherichia coli* and *Pseudomonas aeruginosa*), and one of 3 pathogenic fungi (*Aspergillus terreus, A. flavus* and *Candida albicans*) (Bashir *et al.*, 1992). Fried powdered leaves of *Limonium macrorhabdos* is used in making traditional Ladakhi preparations like staspakchek (Navchoo and Buth, 1990). *Limonium sinuatum* pollen sometimes causes allergic symptoms (Vidal *et al.*, 2007).

Botany, Cytogenetics and Breeding

The most frequently grown one is the perennial species *Limonium sinuatum* which is treated as an annual and is available in all the colours, *viz.* white, yellow, pink, lavender, mauve and blue while majority of other species are basically of only one colour. Other species either may be annual, biennial or perennial, herbaceous or sub-shrubs, sometimes with woody rootstocks. They are tufted (mostly in case of herbaceous ones) or clump-forming with often basal rosette mostly radical, obovate to oblanceolate (in shrubby species mostly linear, spatulate-oblong or obovate), and mostly pinnately lobed but sometimes dissected leaves, from the centre of which normally erect, thin and little- or much-branched (sometimes from the base itself) floral stems (scapes) appear and terminate in the heads of small scale-like flower clusters which are somewhat clasping and where bracts subtend the flowers though the plumes of small flowers appear in case of *Limonium caspium*. Floral stems of good varieties are almost wingless (leafless) though in little improved varieties there are normally leaf-like growths along the stem length. Stems are branched into stiff panicles or corymbs comprising of small spikelets mostly packed densely with flowers, each flower with a persisting tubular and mostly coloured calyx and 5-lobed corolla which may also have various colours. The attractive long-lasting part of the flower is the calyx which can be coloured white, yellow, orange, pink, mauve, blue or purple. Its seeds are whole fruit which on sowing may produce 1 to 7 seedlings. Therefore, when raising from seeds, these require to be decorticated so that each seed is separated and sown individually for good results. A new tetraploid agamic species, *Limonium vigoi*, is described from coastal populations of the northeast of Spain (Ebro delta). The new species is related, on morphological grounds, to *L. girardianum* and *L. grosii*, from which it could be easily discriminated by its retuse leaves, the basal ones usually withered at anthesis, the very short (or even absent) leaf apiculum, the denser and longer (up to 0.7 mm) hairs of the calyx tube and the deeper colour of the corolla. In addition, *L. girardianum* is triploid, whereas *L. vigoi* and *L. grosii* are tetraploid. *Limonium girardianum* and *L. grosii* show the same pollen/stigma combination (cob-type) which differs from that exhibited by *L. vigoi* (papillate-type). The types of *Limonium girardianum* f. *retusum* and *L. glaucophyllum*, two taxa described from the Ebro delta, could not be distinguished from *L. girardianum*, but they clearly differed from *L. vigoi* (Saez *et al.*, 1998).

Asen *et al.* (1973) isolated anthocyanin delphinidin 3,5-diglucoside from flowers of *Limonium* cvs 'Twilight Lavender' and 'Midnight Blue', and delphinidin 3-glucoside from 'Blue Bonnet' and 'American Beauty'. The major flavonoid co-pigments in all the four cultivars were luteolin and its 6-C-glucoside (iso-orientin). These co-pigments were also present in white 'Iceberg' and yellow 'Gold Coast'. The range of colours from reddish-purple to blue for various cultivars was directly related to the pH of the tissues. Seeds of some species of genus *Limonium* exhibited steroid antagonist activity in the *Drosophila melanogaster* BII cell line bioassay (Whiting *et al.*, 1998). Some volatile compounds were also found in *Limonium* × *altaica, L. latifolia* and *L. perezii*. The essential oils have also been hydro-distilled from flowers of *L.* × *altaica* cv. 'Emille', *L. latifolia*, cvs 'Betlaard' & 'St. Pierre' and *L. perezii* cv. 'Violet'. 'Violet' contained a high content of carbonyl compounds including nonanal (9.8 per cent), phenylacetaldehyde (2.7 per cent) and decanal (1.5 per cent) (Soriano *et al.*, 1998). *L. axillare* is used to treat wounds and inflammation in the UAE. Using chromatographic techniques, sitosterol, stigmasterol, sitosterol glucoside, and 6 flavonoids (kaempferol, luteolin, myricetin, apigenin, luteolin-7-*O*-glucoside and apiin) were isolated from a methanol extract of aerial parts of *L. axillare*. This is the first report of apiin in the Plumbaginaceae. The antibacterial and antifungal activities of methanol and chloroform plant extracts, and the isolated flavonoids were investigated against *Bacillus subtilis, Escherichia coli, Pseudomonas aeruginosa, Staphylococcus aureus, Aspergillus terreus, A. flavus* and *Candida albicans*. The methanol extract was more active than chloroform extract, and all the six flavonoids inhibited the growth of the test organisms; apigenin and apiin exhibiting the greatest activity (Bashir *et al.*, 1994).

Plavcova (1976) in *L. sinuatum* cvs 'Vencovka', 'Bezovka' and 'Safir' and in *L. bonduellii* cv. 'Yellow' recorded the chromosome number of 2n=16. Dawson (1990) when studying cytologically the root tip mitoses of the two species, *L. humile* and *L. vulgare*, recorded former being a tetraploid with 2n = 4x = 36 and the latter a hexaploid with 2n = 6x = 54 though with a few variations at the sites of growing of these two species closely having chromsosome counts of 2n = 36 to 54. Artelari and Georgiou (2002) reported that the Kithira, Antikithira and the surrounding islets of Greece are represented by nine species, *viz. L. aphroditae* (2n=3x=27) and *L. cythereum* (2n=6x=52) from Kithira; *L. runemarkii* (2n=5x=43) and *L. ocymifolium* (2n=5x=43) endemic to Greece, and here they proposed *L. ocymifolium* a synonym to *L. pigadiense* due to their studied characteristics; *L. graecum* (2n=5x=43) and *L. sieberi* (2n=6x=61) distributed in East Mediterranean; and *L. virgatum* (2n=3x=27), *L. sinuatum* (2n=2x=16) & *L. echioides* (2n=2x=18) having a wider Mediterranean distribution. *L. runemarkii* which was considered so far as endemic to Southeast Evvia, a new extended distribution range is reported from Evvia to Northwest Kriti, through Peloponnisos and Kithira, showing the phytogeographic relationship of these areas. Concerning pollen-stigma combination, they reported that *L. aphroditae, L. ocymifolium* and *L. virgatum* are monomorphic with the self-incompatible combination B, while *L. cythereum* and *L. runemarkii* are monomorphic with the self-incompatible combination A which shows that these species may be apomictic. The facultative apomictic *L. graecum* was found monomorphic with the combination A, while *L. sieberi* had combination A or B within different populations. The autogamous species *L. echioides* had the self-compatible combination C, while the allogamous *L. sinuatum* was dimorphic having both combinations A and B in each population. Febles *et al.* (2004) studied the karyotypes of nine species and one subspecies of the genus *Limonium*, section Pteroclados, subsection Nobiles which is endemic to the Canary Islands, and which was found to be diploid with 2n=14 (basic number being 7 which is exclusive to this sub-section). The karyotype is highly symmetrical, made up of 7 pairs of metacentric (7 m) chromosomes in all the taxa except for the endemics of the Eastern Islands *L. bourgeaui* and *L. puberulum* with 6 m and 1 sm (pair 7). Georgakopoulou *et al.* (2006) recorded *L. palmare* and *L. roridum* as facultatively apomictic with 2n=4x=34 and 2n=5x=43, respectively; *L. graecum* as apomictic with 2n=6x=52; *L. virgatum* as also apomictic with 2n=3x=27; and *L. narbonense* being sexual with 2n=6x=54 and 2n=8x=72, the latter number being a new record and the highest number reported so far in the genus. Castro *et al.* (2007) reported somatic chromosome numbers, conventional karyotype features and idiograms for 27 *Limonium* species (*L. marisolii* with 2n = 54 as a new cytotype; *L. migjomense* with 2n = 50; *L. barceloi* and *L. geronense* with 2n = 36; *L. alcudianum, L. bonafei, L. camposanum, L. companyonis, L. dufourii* and *L. inexpoectans* with 2n = 26; *L. scopulorum* with 2n = 25; *L. ejulabilis, L. inexpectans,* and *L. interjectum* with 2n = 24; and *L. pseudodictyocladon* with 2n = 16 as a new cytotype) inhabiting the Western Mediterranean basin (Iberian Peninsula and the Balearic Islands). A group of polyploid species showed karyotypes comprising homologous chromosomes in groups of three (*L. antonii-llorensii, L. ejulabilis, L. interjectum, L. virgatum,* and *L. wiedmanii*), four (*L. geronense*) or six (*L. marisolii*), which suggests an autopolyploid origin. Other polyploid species were characterized by the presence of two different chromosome sets (x=8 and x=9) in the genome. The species *L. alcudianum, L. bonafei, L. camposanum, L. companyonis, L. dufourii, L. gibertii, L. girardianum, L. inexpectans, L. leonardi-llorensii, L. magallufianum, L. migjornense, L. minoricense,* and *L. scopulorum* showed various combinations of paired and unpaired x=8 and x=9 chromosome sets, suggesting that they are allopolyploids. Evliyaoglu *et al.* (2008) made a karyological analysis on three naturally growing *Limonium* species, *viz. L. iconicum* with 2n = 34, *L. lilacinum* with 2n = 36 and *L. globuliferum* with 2n = 18 chromosome numbers.

Agamospermous species account for a large proportion of the species of *Limonium*. Agamospermy is indicated by uneven polyploidy or aneuploidy, low pollen stainability and by the presence of monomorphic self-incompatible populations. The taxonomic treatment of agamospermous taxa varies from recognition of all of them at the same specific rank or by utilising a range of ranks in the taxonomic hierarchy. The influence of evolutionary hypotheses on taxonomic systems is considered. Molecular data provide a means of measuring the genetical relationships of taxa and establishing groups in a taxonomic hierarchy (Cowan *et al.*, 1998). Morphological characteristics of 56 protoplast-derived plants (protoclones) of *L. perezii* were compared with meristem-derived plants (controls). Some protoclones had smaller leaf shape index (length/width), and shorter flower stalks and calyces than the controls. However, no distinct somaclonal variations were observed in other characteristics such as colour of calyx, and number of petals and stamens, except in protoclone 1-7 which had flowers with abnormal petals and low pollen fertility (25 per cent). Microscopic observation revealed that all of the 17 protoclones examined, including 1-7, had the normal diploid chromosome number of 14 (Kunitake *et al.*, 1995).

Hamilton and Rand (1996) identified hyper-variable marker based on oligo (GATA)4-hybridization of DpnII-cut genomic DNA from *Limonium carolinianum*, and recorded somatically stable banding patterns though highly variable among unrelated individuals. They estimated band molecular weight sizing errors (as a percentage of band molecular weight) at 0.44 per cent +or-0.003 within gels and 0.76 per cent +or-0.964 between gels, and these sizing errors defined a 99 per cent confidence bin of +or-0.95 per cent (1.90 per cent total) of molecular weight. Band-sharing among nine unrelated individuals (theta) was 0.198+or-0.011. Experimental pollinations designed to produce selfed, full- and half-sib progeny groups led to five selfed progeny groups and no outcrossed progeny (mean band-sharing, S = 0.468+or-0.074). A linear regression between band-sharing (S) and relatedness (r) assuming 17 per cent inbreeding was r=0.006+0.914*S (R superscript 2=0.973) and established the maximum amount of inbreeding. S (0.392+or-0.022) estimated from wild pollinated seeds from four maternal families was intermediate to unrelated individuals and experimental selfed progeny, giving evidence for mixed mating in wild plants. A highly endemic triploid species, *L. dufourii* from the coasts of eastern Spain, which is a highly endangered species, was investigated by Palop Esteban *et al.* (1997). In a sample of 122 individuals collected from six extant populations, 65 alleles from 13 microsatellite regions were amplified which showed consistent microsatellite patterns with its triploid nature. The alleles were unambiguously assigned to two different parental subgenomes in this hybrid species and the greater contribution of the diploid parental subgenome was confirmed.

L cavanillesii, an extremely endangered species endemic to the east Mediterranean region of Spain, was regarded as extinct for several years but Placios and Gonzalez-Candelas (1997) when discovered it again with some 29 individuals only, they analysed its genetic variation among entire 29 plants through RAPD with 11 different primers which produced 131 monomorphic bands, apparently the lowest level of genetic variation. Palacios *et al.* (1999) studied the genetic variation and population structure through AFLP markers in *Limonium dufourii*, a triploid species endemic to the East Mediterranean coast of Spain, with obligate apomictic reproduction. Three different primers provided 252 bands; of these, 51 were polymorphic among the 152 individuals analysed. These polymorphic bands were able to define 65 different phenotypes, of which all but two were present in only one population. The comparative analyses of data from AFLPs with those from RAPDs showed a high degree of concordance. Relationships among different AFLP patterns and the estimates of population genetic parameters obtained with this evolutionary distance are in good agreement with previous results. These analyses showed that substantial genetic variability and differentiation exist within and among populations of *L. dufourii*. Their higher reproducibility and the possibility of obtaining estimates of nucleotide divergence make AFLPs a much better DNA fingerprinting technique. Using DNA fingerprint markers within species and populations of wild plants requires information on the relationship between fingerprint similarity and relatedness. Burchi (2002) used RAPD markers for genetic fingerprinting of *Limonium* spp., *Alstroemeria* spp., *Dianthus caryophyyllus, Gerbera* spp., *Rosa* spp., and *Prunus* spp. In *Limonium*, RAPD analysis was used for verification of hybridity in progenies of interspecific crosses and for the study of taxonomic relationships of seven putative interspecific hybrids and their parents where putative parentage in two of the hybrids was confirmed though five of the genotypes were found derived from self-pollinations. Bruna *et al.* (2004) when tested 13 wild species with 10 primers, they scored a total of 244 bands and the dendrogram drawn from cluster analysis showed high similarity among three species, *viz. L. bellidifolium, L. caspium* (*L. caspia*) and *L. otolepis* which some authors report as synonymous, therefore, several other genotypes belonging to these species were further analysed where new dendograms showd a score of 151 RAPD bands which shows that the genotypes did not group in clear clusters, as also confirmed by analysis of molecular variance (AMOVA), genotypes showing highest genetic variation though only 6.58 per cent variation resulted among the species. These results suggest that the species can be considered synonymous. Bruna *et al.* (2005) used 10 selected RAPD primers to analyse the genetic similarity in 11 wild species and 10 commercially cultivated varieties of *Limonium* spp. and for verification of hybridity in progenies of interspecific crosses that scored a total of 254 bands which were used for calculating genetic distances among entries. The largest cluster grouped five varieties and *L. latifolium*; the species *L. bellidifolium, L. otolepis* and *L. caspia* were clustered together; two cultivars were grouped together with *L. sinensis*; though the wild species *L. tataricum, L. perezii, L. sinuatum* and *L. longifolium* were clearly separated from all of the other genotypes. Seven putative interspecific hybrids and their parents were characterised by means of nine selected RAPD primers. The putative parentage was confirmed for only two hybrids; the remaining five genotypes, on the basis of RAPD patterns, were identified as apomixis and/or self-pollinations.

In *Limonium* the flowers are generally very small, thus (i) complete flower emasculation is difficult; (ii) anther dehiscence starts immediately after or also before flower opening; and (iii) anthesis of flowers on the same inflorescence branch does not occur at the same time. The studies of Zhu-Jin Wen *et al.* (1997) revealed that pollen grains and stigmas of *Limonium* spp. can differ in appearance between species and individuals of the same species, and can be classified into 2 types (A and B), and stigmas into 3 types (head-like, cob-like and nipple-like). Head-like stigmas remain fertile after pollination with both pollen grain types. Cob-like stigmas remain fertile after pollination with A type pollen, but not with type B, whilst nipple-like stigmas are successfully fertilized with type B, but not with type A pollen. These results were utilized in the crossing of *L. latifolium* and *L. caspium*. Hybrids with superior characters from both parental species were obtained and propagated *in vitro*. Dulberger (1975) stated that in dimorphic members of the Plumbaginaceae the floral morphs differ in the wall structure at the apical region of the stigmatic papilla. He examined many genera and species and recorded that intramorph pollinations are incompatible and inhibition of the pollen occurs at the stigma surface. Structural stigma dimorphism is probably involved in the incompatibility mechanism in Staticeae.

Exclusive *Limonium* breeding programme is going on in Sanremo and Pescia (Italy) for stress tolerance, free-flowering habit requiring little energy consumption, low input rerquirement, good stem architecture with precoecious branching, a wide colour combination of calyx and corolla and high yield with strong stems taking many species including *L. bellidifolium, L. bonduelli, L. caspia, L. gmelinii, L. latifolium, L. otolepis, L. serotinum, L. sinuatum* and *L. tataricum* (Burchi *et al.*, 2006). They studied the breeding capability among some species through diallele crosses, some through embryo rescue technique where the crosses were incompatible, some for very specific traits and some others through open pollinations, and obtained encouraging results. 'Chorus Magenta' with stem in excess of 60 cm was selected from *L. perigrinum* × *L. purpuratum*, but subsequent breeding gave four more selected hybrids suitable for pot growing in New Zealand and Australia (Morgan *et al.*, 2001). Through embryo rescue technique, *L. perezii* × *L. sinuatum* produced hybrids but they were sterile. The use of spindle toxins resulted in a doubling of the nuclear DNA contents of the hybrids and a large proportion of the pollen staining with Alexander's stain. Reciprocal crosses between the DNA tetraploids and their diploid parent species have produced embryos. The individual seedling plants which responded to one week exposure to 11 °C by early flowering were selected and allowed to cross-pollinate with the entire population of selected individuals, and their seeds collected and planted as half-sib families and recurrent selection was applied within and between the families for five generations. The response of the selected population to vernalization after five years of selection was compared to that of the original population, and then it was realized that non-vernalized plants of the selected strain flowered three months before non-vernalized plants of the original population. Yield curves of the non-vernalized selected plants were similar to those of the non-selected population that had been exposed to six weeks of vernalization (Cohen *et al.*, 1995).

The morphogenetic potential of leaf explants of 12 *Limonium* genotypes was when examined *in vitro* (the full strength MS salts and vitamins, 30 g sucrose and 7.5 g agar/l together with 0.6 mg BA or 2 mg IAA + 1 mg BA/l), 11 genotypes produced many shoots after 30-60 days of culture though only *L. perezii* failed to regenerate. *L. serotinum*, and *Limonium* hybrids '3060' and '3070' were transformed with *Agrobacterium tumefaciens*. Several plantlets showing kanamycin resistance and expressing the GUS marker gene were obtained (Mercuri *et al.*, 1999). In case of *L. sinuatum* using six cultivars (PF142, PF157, PF167, PF1108, PF1124 and PF1127), Kimizu *et al.* (2001) developed an *Agrobacterium*-mediated transformation system and cultured the shoots on TDZ-containing medium on which the cultivar PF142 explants regenerated to a tune of 40 per cent when TDZ was added 0.1 mM in the medium. Luciferase (LUC) and beta-glucuronidase (GUS) genes were introduced and transient expression of these genes were confirmed, along with stable expression of the GUS gene, but shoot formation was strongly inhibited by hygromycin B, of which resistant gene was used as a selection marker. Mercuri (2002) and Mercuri *et al.* (2003) got transformation of genotype 'L116' (*Limonium otolepis* × *L. latifolium*, sterile hybrid) and *L. gmelinii* species using the disarmed strain LBA4404 of *Agrobacterium tumefaciens* containing the binary plasmid pBIN19 harbouring rolA, rolB and rolC genes (EcoRI 15 fragment) of *Agrobacterium* rhizogenes. The genetically modified genotypes showed different morphological traits with drastic plant size reduction and earlier flowering, *vis-à-vis* three types of compactness (super-compact for ready-to-get bouquet and pot-growing, compact and semi-compact) with increased postharvest life. Smith *et al.* (2006) has stated that the rolC gene is found on the RI plasmid of *Agrobacterium* rhizogenes. Plants expressing this gene have significant changes in phenotype, including reductions in height, apical dominance, internode length, and leaf size. These changes can be beneficial to plant breeders attempting to create novel, smaller versions of existing cultivars. The rolC gene has been introduced into several ornamental species, including *Antirrhinum majus, Begonia tuberhybrida, Betula pendula, Chrysanthemum morifolium, Datura arborea, Datura sanguinea, Dianthus caryophyllus, Eustoma grandiflorum, Lilium longiflorum, Limonium otolepis* × *Limonium latifolium, Malus prunifolia, Osteopermum ecklonis, Pelargonium* × *domesticum, Petunia axillaris* × (*P. axillaris* × *P. hybrida*) and *Populus tremula* × *P. tremuloides*. These species exhibited changes in morphological traits such as dwarfness, increased branching, increased flower number, and unique leaf morphology. These traits are heritable from one generation to another, and the degree of phenotypic change is affected by the amount of rolC produced under a specific promoter. The introduction of rolC into plant species is an effective way to produce dwarf plants and has potential for breeding unique ornamental cultivars. Li-HongYan *et al.* (2007) cloned two metallothionein (MT) genes from *L. bicolor*, named LbMT1 (length 569 bp including 23 bp of 5' untranslated region and 300 bp of 3' untranslated region) having an open reading frame (ORF) of 246 bp, encoding a protein of 81 amino acid residues, with protein molecular weight of 8070 and isoelectric point of 4.7; and LbMT2 gene (length 523 bp, including 61 bp of 5' untranslated region and 216 bp of 3' untranslated region) having an ORF of 246 bp, encoding 81 amino acid residues with protein molecular weight of encoding protein as 8000 with the isoelectric point of 5.0. Both LbMT1 and LbMT2 have 14 Cysresidues (Cys) that have conserved sequence position at N terminal and C terminal of protein, ranking in the forms of CXC and CXXC. Hydrophobicity analysis showed that there was a high hydrophobic region between 35 and 48 residue positions. Multiple sequence alignment of MT proteins from some plants revealed that the MT gene family shared low identities in sequence, but were conserved in Cys residue positions and the number of Cys residues, implying that Cys residues may play very important roles in the function of MT proteins. These two genes were accepted by Genebank, the accession number of LbMT1 is EF103574 and of LbMT2 is EF103575.

Ban-QiaoYing *et al.* (2008) cloned, from a cDNA library of *L. bicolor*, the full length cDNA of a novel metallothionein (LbMT2) gene with an ORF of 249 bp in length, encoding a protein of 82 amino acid residues with the molecular mass of 8.1 kDa and theoretical pl of 4.71. The expression of LbMT2 gene was induced in *L. bicolor* by CuSO4, CdCl2, NaCl and cold though was inhibited by PEG stress. LbMT2 gene was inserted into a prokaryotic expression vector (pGEX-4T-2) to produce the recombinant expression vector pGEX-LbMT2. The expression of LbMT2 in *E. coli* BL21 was induced with IPTG, which produced a protein band with expected size of 35 kDa on SDS-PAGE.

Shoots of *L. sinuatum* were exposed to different doses of ^{60}Co gamma rays (10, 20, 30, 40 and 60 Gy) and the treated shoots were raised *in vitro*, and thereafter acclimatization to *in vivo* conditions they were transplanted in pot filled with peat-based medium and transferred into a greenhouse where after 75 per cent flowering of the plants, it was observed that in 10 Gy treatment the stem height and leaf number were affected so it was stipulated that concerted efforts will certainly bring out desirable genetic changes (Cardarelli *et al.*, 2002). The degraded alginate product with a weight-average molecular mass (Mw) of ~ 9.04x105 Da as growth promoter was irradiated at 10-200 kGy in 4 per cent (w/v) aqueous solution and used in tissue culturing of *L. latifolium*, where alginate irradiated at 75 kGy with an Mw of ~ 1.43x104 Da had the highest positive effect on the growth, though irradiated alginate at 30-200 mg/l increased the shoot multiplication rate from 17.5 to 40.5 per cent compared to control (Luan *et al.*, 2003). In plantlet culture, supplementation with 100 mg/l irradiated alginate enhanced shoot height (9.7-23.2 per cent), root length (9.7-39.4 per cent) and fresh biomass (8.1-19.4 per cent).

Morgan *et al.* (1995, 1998) succeeded in rescuing 12 to 16 day old embryos through ovule culture in interspecific crosses, *Limonium peregrinum* × *L. purpuratum* and *L. perezii* × *L. sinuatum*. Fifteen *L. perigrinum* × *L. purpuratum* embryos were successfully rescued when *L. perigrinum* was the female parent though none in the reverse cross which all produced intermediate floral colours with stems of 25-60 cm. Early confirmation of *L. perezii* × *L. sinuatum* hybrids was provided by flow cytometric analysis (Burge *et al.*, 1995).

Species and Varieties

Some 300 species of *Limonium* exist but only a few are ornamentals as furnished here with.

Limonium bellidifolium (perennial). A native of Europe to E. Asia, it is a woody species with much branched rootstock. The leaves are 1.5-4.0 cm long, obovate to spathulate and normally senescing before flowering ceases. Flowers are pale-lilac, 5 mm long, paniculate and the panicles are formed on decumbent stems. Panicles some 7.5 cm long and radiating outwards.

L. bonduellii (biennial to perennial). A native of Algeria, and is treated as hardy annual. Its leaves are pale-green, ovate, deeply lobed and are borne in a basal rosette. It is similar to *P. sinuatum* in growth and flowering. It grows to 30 cm in height. The stems are mostly winged though good varieties bear wingless stems. Its stems are erect and carry loose clusters some 7.5 cm across with yellow flowers. Its inflorescences are excellent for cutting as well as for drying.

L. brassicaefolium (perennial). A native to Canary Islands is a sub-shrubby species growing up to 45 cm high with broadly 2-winged stems which run out into large auricles below the forks (crotch), the branchlets 3-winged and the wings are dilated from the base upward and running out into short and falcate auricles. Leaves rosetted at base, ovate-rotund, irregularly lobed, petioled with small leafy lobes at intervals, shorter than plants and velvety with somewhat ciliate margins having largest terminal lobe. The scape is angled, paniculate-corymbose above, spikelets 2-flowered with 2-3 fascicles at the branch ends, calyx purple and the corolla yellowish-white. It flowers from late summer to early autumn.

L. caesium (perennial). It is native of Spain and grows 30-90 cm in height with white powdery coating throughout the plant having subtuberculate calcareous dots. Leaves are glaucous grey-green, radical, obovate or retuse and small. It bears numerous stout scapes having precocious branching at the base with sterile lower branches while the upper ones with profuse branching and paniculate. Sprays are composed of grayish-white calyx and rosy corolla, spikelets erect, quite slender and 1-flowered and are arranged in a few-flowered spike.

L. incanum (syn. *Goniolimon callicomum*; perennial). A woody-based hardy perennial native of Siberia, growing 30-60 cm tall. Leaves are mid-green, narrowly lanceolate to oblong-elliptic, about 5.0 cm long and mostly close to the ground. Spikes are erect, branched, some 20-25 cm long and bear small flowers with a white calyx and violet-rose to red corolla. Its var. *dumosum* is a good pink.

L. latifolium (perennial). A hardy perennial with a woody rootstock is native from SE Europe to USSR, *i.e.* southern Russia and Bulgaria. This species is most popular for general garden planting and most suitable for cutting and drying. The mid-green basal evergreen rosetted leaves are downy, up to 25 cm long and oblong-elliptic which appear in a large rosette having rounded-triangular shape. The bright cobwebby masses of pale- to lavender-blue flowers some 6 mm long appear in diffuse panicles up to 23 cm long in the form of large branching inflorescence on 30-60 cm high stems. The flowerheads initially appear in a flat fan-shape. Its var. *album* is white, 'Blue Cloud' is light lavender-blue and *violetta* is dark blue.

L. perezii (perennial). It is a native of Canary Islands and is more frost-tender so requires frost protection in cold areas. It is a low shrub growing 60-90 cm high, bearing long petioled, rhomboid-ovate or broadly triangular leaves which are truncate at the base. Calyces are rich purplish-blue and corolla pale-yellow. Flowering terminates by autumn to early winter.

L. sinuatum (biennial to perennial). It is a half-hardy circum-Mediterranean species flowering in the wild at its native haunts between March and July amongst rocks or in the waste places near the coast. It is a most popular perennial species cultivated as an annual for summer blooming in beds and borders and for drying for indoor enjoyment during winter. It grows up to 45 cm height with winged stems. Its leaves are mid to dark green, harp-shaped or oval (obovate-lanceolate), deeply pinnately lobed, waved on the broadly winged stalk and 10-20 cm long. It is frequently grown for ornamental purposes due to more durable flowers in 7.5-10.0 cm long clusters of blue and cream flowers some 10-12 mm long, appearing from funnel-shaped calyces, flowers surrounded by green bracts. Its inflorescences are in the form of much-branched dense panicles. It is often grown for its conspicuous bluish-mauve calyces which retain their colour many months after drying. It has many forms and varieties in shades of white, yellow, pink, red, violet, blue and others in combination. Flowers appear from a 1.0-1.2 cm long & purple, and funnel-shaped calyx. It has numerous garden cultivars including 'New Art Shades' (a blend of colours with yellow, orange, salmon, rose-pink, red, carmine, blue and lavender).

L. suworowii (syn. *Psylliostachys suworowii*; annual). It is half-hardy annual native from Iran to Central Asia and Caucasus, growing up to 45 cm tall. It was first discovered by Albert von Regel near Leningrad in NW Tadzhikistan (part of Turkestan) and was named so to commemorate Ivan Petrowitsch Suworow, the Inspector of Military Hospitals in that area at that time. It is the most common and usual garden species. In the wild it has been recorded growing up to 1.5 m, and is quite effective and striking species. Its leaves are mid-green, oblanceolate, lobed, waved, 15-25 cm long and forms a basal rosette. It produces narrow, candle-like upright rose-pink (purple-lilac) flowers some 4 mm long, densely packed in narrow plume-shaped (fingered) panicles where tips bend and twist little, so are most suitable for flower arrangements, especially informal line arrangements. Individual flowers are some 4 mm long and 2-3 are grouped together in a cup of chaffy bracts, each cluster (group) placed close together to form a continuous cylinder of flowers, forming narrow finger-like paniculate spikes, and the inflorescence remains in flowering for about eight weeks.

L. tartaricum (syn. *Goniolimon tartaricum*; perennial). It is a sub-shrub native to salt marshes and steppes of Caucasus Mountains, *i.e.* Dalmatia and Hungary, eastward to the Crimea and north Caucasus and was introduced into Britain about 1731 by Philip Miller. Its leaves are oval, widest at the ends and up to 15 cm long. It is a dwarf species growing up to 30 cm tall though in the wild has been found growing up to 91 cm across with ruby coloured flowers surrounded by chaffy sepals. In cultivation it is quite dwarf, so is more effective for planting in small beds, narrow borders and in the rock garden. Its stems are quite stiff and bear masses of minute brilliant pink and white flowers in the form of corymb. The calyx of the flowers are white though corolla in various other colours.

L. vulgare (perennial). A native of Europe, North Africa, Asia Minor and Syria usually along the seaside and grows some 25-40 cm tall. It prefers slightly moist soil for its growing at a sunny position. Its foliage is 10-15 cm long with long petiole, oblong-lanceolate, 1-nerved and obtuse. Scape is long and terete though paniculate-corymbose above with 1- to 3-flowered spikelets which are densely congested on the spike. Calyx in masses of minute white to lilac-purple (bluish), and corolla bluish-lilac, which continue appearing throughout the full summer in the temperate areas and from January end to March-April in the sub-tropical areas of India. Its flowers are most suitable for cutting and for dry arrangements. Its var. *alba* grows up to 30 cm tall and bears white calyx while var. *macrocladum* possesses much-branched and spreading panicles and is found in Mediterranean region and Syria.

Guda *et al.* (1998) described *L. otolepis* as a new *Limonium* species from Italy. Rizzotto (1999) stated that geographical isolation obviously prevents gene exchange between populations which depicts a case of slow divergent speciation of karyologically rather homogeneous populations where heterostyly plays a significant role. He further mentions that each island hosts a different, though little differentiated species such as *L. dianium, L. doriae, L. gorgonae, L. ilvae pignatti* and *L. planesiae*, whereas *L. sommierianum* is typified and its presence has been found limited to the Giglio Island. He also described two new species, *L. montis-christi* and *L. caprariae*. Saez and Rossello (1999) described a new triploid apomictic species (*L. perplexum*) from a single coastal locality in eastern Spain. The new species has been previously confused with *L. cavanillesii*, which is only known from a few herbarium specimens and it is believed to be extinct. Overall morphology, chromosome number and the pollen/stigma combination suggest that *L. perplexum* belongs to the *L. duriusculum* complex. Based on morphological observations it is suggested that *L. cavanillesii* could have developed from a cross between *L. perplexum* and *L. dufourii*. Erben (2001) described eight new species such as *Limonium portovecchiense* from Corsica; *L. caralitanum, L. capitis-eliae, L. carisae, L. malfatanicum* and *L. multifurcatum* from Sardinia; and *L. comosum* and *L. fesianum* which are locally distributed within Morocco and Tunisia; *L. coralliforme* as not differing from *L. ursanum* so being reduced to a synonym of the latter; *Statice bollei* is being recombined as *Limonium bollei*; and the typification of *L. dubium* and *L. tunetanum* is completed by the selection of lectotypes. In a survey of *Limonium* Mill. sect. Pteroclados (Boiss.) Bokhari (Plumbaginaceae Juss.), Karis (2004) found two previously recognized subsections Nobiles and Odontolepidae constituting two sister groups, *Limonium mouretii* (Pit.) Maire sharing synapomorphies with both subsections and was positioned as sister group of *Limonium sinuatum* (L.) Mill. s.l. plus *L. lobatum* (L. f.) Kuntze but with weak support, therefore, based on the results of the cladistic analyses, a single introduction of a common ancestor to *Limonium* sect. Pteroclados subsect. Nobiles into the Canary Islands is postulated. He found minor morphological differences between *L. sinuatum, L. bonduellei* (T. Lestib.) Kuntze and *L. beaumierianum* (Maire) Maire, hence he suggested these to be recognized as subspecies under *L. sinuatum*. Thulin (2004) stated that *L. maurocordatae* (*Statice maurocordatae*), originally described from eastern Ethiopia, is conspecific with *L. distichum*, a species distributed in eastern Ethiopia, Somalia and north-eastern Kenya, therefore, he proposed it a neotype *L. maurocordatae* though reduced *L. distichum* to its synonymy. Aparicio (2005) recorded *L. silvestrei* from southern Spain which he stated that based on the single A pollen/cob stigma combination, male sterility, jumbled 3x karyotype and high seed set and germination, this should be regarded as a new agamospecies in *Limonium*. The number of long metacentric chromosomes observed in the karyotype of this species is in conflict with Erben's theory about the evolution in the genus. Based on the morphology and chromosome number (2n=18), Erben and Aran (2005) described that *L. mateoi* sp. nov., an endemic species from the gypsaceous soils of the Huete-Buendia area in Cuenca (Central Spain), seems to be related to *L. dichotomum* and *L. erectum* and also the distribution areas of these three species being quite close though they do not overlap. Matsumura *et al.* (2006) recorded six morphs of interspecific colour variations (pink, orange, ivory, yellow, white, and cream) in *L. wrightii* on 36 islands in the NW Pacific Islands, two colours, *viz.* pink and yellow being most common. Orange would have resulted by natural crossing of pink with yellow. *Limonium failachicum* (2n=18), a new taxon so far the only known endemic species from Arabia (Kuwait) is reported by Erben and Mucina (2006) which is taxonomically related to *L. iranicum* and *L. carnosum*. Wu-YuHu and Yang-YongChang (2006) described *Limonium aureum* var. *maduoensis* var. nov. from Qinghai province of China, along the plateau area with an altitude of more than 4,000 m and grows in the *Stipa purpurea* alpine steppe on saline soil. While preparing a new account of *Limonium* as an initial part of a taxonomic revision of the family Plumbaginaceae in Turkey, Akaydin (2007) describes a new species *L. smithii* sp. nov. from salt steppe of Central Anatolia. *Limonium gueneri* Dogan, Duman & Akaydm sp. nova is described from Patara (Antalya, Turkey), where it grows on calcareous slopes on the coast, which is probably closely related to *L. ocymifolium* (Poir.) Kuntze, an East Mediterranean species (Dogan *et al.*, 2008).

Certain commercial *Limonium* cultivars are:

'Ballerina Rose'. It is regarded as the first commercially useful *Limonium peregrinum* variety for cut flower production. It was derived by selection from a range of plants collected from gardens in New Zealand. It is a semi-erect perennial shrub (60 cm high) of sympodial growth habit. It produces terminal inflorescences up to 45 cm long which bear 200-400 individual flowers, some 16.9 mm in diameter in spikelets of 10-75 flowers (Anon., 1994).

'Beltlaard'. Developed in Japan by crossing *Limonium caspia* with *L. latifolium*. More or less this is perpetual and resistant to heat and frost. Its inflorescences are more than 80 cm tall having spread of about 40 cm with open and spreading branching. Its petal colour is 87A/86C.

'Chorus Magenta'. 'Chorus Magenta' is derived by crossing *Limonium perigrinum* with *L. purpuratum* and then back-crossing with *L. purpuratum* as the pollen parent. This variety has acropetal opening of the florets and a panicled inflorescence bears some 400-700 flowers. Flowers are pink (RHS 72C) and some 8 mm long. It is relatively resistant to many pests and diseases. Cut inflorescences can be used both fresh and for drying (Seelye *et al.*, 2000).

'Cosita'. A variety of *Limonium perezii* derived from LIM94-2 × LIM94-15 which is vigorous, compact, and free-flowering, producing clusters of 2-4 flowers on branching spikes, with an average of 16 spikes per plant. Sepal colour at flowering is light purple (RHS 87C) and petal colour cream (RHS 1D) (Cherry, 1999).

'Daicean'. Developed by crossing *Limonium latifolium* with *L. bellidifolium*, and bears rigid stems with year round flowering (Anon., 1993a).

'Emille'. It is an open-pollinated Japanese variety of *Limonium altaica* with darker blue flowers and higher yields for autumn flowering. Its open and spreading inflorescences are 50-60 cm long with some 45 cm spread (Anon., 1993b).

'Oceanic Blue'. A variety developed by crossing *Limonium latifolium* with *L. dumosa* whose inflorescence grows with spreading branching, inflorescence attaining the height of 92.0 cm and spread of 52 cm, and the flowers are 5 mm long with blue (RHS 85A) petals (Anon., 1993c).

'Oceanic White'. A Japanese mutant from cv. 'Oceanic Blue' evolved through X-ray irradiation, which produces upright, open and spreading inflorescence with the length and spread of 95 and 55 cm, respectively. Its calyx is white and corolla RHS 155A (Anon., 1997).

'Pink Emille'. A natural mutant of Japanese var. 'Emille', bearing inflorescence some 50 cm long 40 cm in spread with open and spreading branching, having 6 mm long flowers with petal and calyx colour of pale pink/white (RHS 76D) with a pale red stripe (Anon., 1993).

'SOI 50'. A variety released at 50th anniversary of the Italian Society of Horticulture, was developed through selection in *Limonium tataricum* seedlings. It is characterized by a good production of flowered stems with quite attractive and high number of flowers per stem. Calyx and corolla both are white, unlike other cultivars of *L. tataricum* whose corolla is commonly pink or red-violet (Burchi *et al.*, 2003).

'Tall Emille' (LC.00281I). A spontaneous mutant of 'Emille' with open and spreading inflorescence which grows to a length of 60-65 cm, bearing 5 mm long flowers with petal colour of RHS 85B (Anon., 1996).

Other commercial colourwise *Limonium* cultivars are

White. 'Forever White', 'Fortress White', 'Iceberg', 'Qis White', 'Snow Queen', 'Soiree White', 'Turbo White', 'White Dream', *etc.*

Yellow. 'Bonduellii', 'Excellent Yellow', 'Forever Yellow', 'Fortress Yellow', 'Gold Coast', 'Sunset Yellow', 'Turbo Yellow', *etc.*

Apricot. 'Apricot Beauty', 'Excellent Salmon', 'Fortress', 'Sunset Apricot', 'Sunset Salmon', *etc.*

Pink. 'Excellent Pink', 'Forever Pink', *etc.*

Rose. 'Chamois Rose', 'Excellent Rose', 'Fortress Rose', 'QIS Rose', 'Rosea Superba', 'Rose Strike', 'Soiree Rose', 'Sunset Rose', *etc.*

Violet/lavender. 'Blauschleier', 'Blauwolke', 'Lavandin', 'Lavender Emille', 'Lavender Queen', 'Twilight', 'Violetta', *etc.*

Purple. 'Excellent Purple', 'Forever Purple', 'Fortress Purple', 'Soiree Purple', 'Turbo Purple', *etc.*

Blue. 'Blue Bonnet', 'Blue Cloud', 'Blue Perfection', 'Blue Symphonet', 'Charm Blue', 'Early Blue', 'Forever Blue', 'Marin Blue', 'Market Grower Blue', 'Midnight Blue', 'MistyBlue', 'Oriental Blue', 'Super Blue', 'True Blue Extra Select', 'Turbo Blue', *etc.*

'Art Shades', 'Pastel Shades', 'Avignon', 'Saint Pierre', *etc.* are also quite beautiful varieties.

Propagation

Usually the plants are raised through **seeds** whether these are annuals or perennials. The commercial seed is, in fact, a whole fruit producing 1-7 seedlings, therefore, only scarified, decorticated and clean seed should be purchased as it germinates quickly and produces individual plants. One gramme of decorticated seeds contains some 450 seeds. Hardy perennial species are sown in seed compost during March-April while half-hardy annuals in May in temperate areas of the country whereas in August-September (late summer) in the sub-tropical and trtopical areas of India to flower during spring through overwintering at 8 °C or sown in trays at 20-22 °C temperatures in the open or greenhouses. In polyhouses, these may be sown in February-March at a temperature of 13-16 °C. After appearance of the true leaves these should be pricked out and when a rosette of three leaves has appeared, these are placed for necessary chilling at 8-13 °C for at least three weeks though optimum is six weeks (Shillo, 1976). It would be better if during growth and development these are provided with 22-27 °C day and 12-16 °C night temperatures (Salinger, 1985). Perennial statice plants are cut down in October to November on the hills to encourage new growths. Vitanova and Kaninski (1998) studied flowering and seed production in *L. sinuatum* and *L. bonduellii* by sowing the seeds in Bulgaria on 15 February, 30 March or 15 September where floral stalks were formed earlier in both the species from sowing on 15 February, however, *L. bonduellii* flowered even earlier but with lower seed yield, though in *L. sinuatum* it was recorded highest (14.5 g/plant) from sowing on 15 February and in *L. bonduellii* (8.2 g/plant) from sowing on 15 September. The laboratory germination of the seeds was high, exceeding that of first class seeds according to the Bulgarian National Standard in both the species.

NaCl and iso-osmotic PEG had inhibitory effect on *Limonium bicolor* seeds, the result being more pronounced under PEG, though the inhibitory effect of NaCl decreased with delay of treatment time (Zhang-WanJun *et al.*, 2001). The drying of *L. aureum* seeds in a desiccating container with silica gel from 8.92 to 2.88 per cent moisture content (ultra drying) and keeping them for one month at 50 °C tested higher for their dehydrogenase, peroxidase, superoxide dismutase, glutathione reductase, ascorbate peroxidase and catalase activities, while volatile aldehydes and malondialdehyde were found lower than control (Li-Yi *et al.*, 2007). *L. stocksii* seeds were either pre-treated with sodium hypochlorite (NaOCl) or it was added to a medium with and without salinity (0-400 mM NaCl) at various temperature regimes (10-20, 15-25, 20-30 and 25-35 °C). Pretreatment for one minute in a 10 per cent sodium hypochlorite solution appeared to be the most effective treatment in alleviating salinity-induced dormancy, and at lower temperatures, it was beneficial for seed germination while at higher temperatures its effect was found reduced (Khan and Sabahat Zia, 2007). Redondo Gomez *et al.* (2008) in a laboratory experiment found delayed and poor germination when seeds of *E. memarginatum* (an endangered and endemic halophyte of the Strait of Gibraltar) were treated with 2, 4 or 6 per cent NaCl, effect being more pronounced at increased salinity levels because above 2 per cent NaCl completely inhibited the germination though in fresh water the germination was 60-70 per cent, however, salinity pretreatments had a stimulatory effect since germination speed was higher for the recovery experiment than for the seed germination experiment, and it was found that transition between germination and seedling establishment was a critical phase, given that less than 50 per cent of seedlings survived in distilled water and only 5 per cent at 2 per cent salinity. Seedlings of *L. aureum* and *L. gmelinii* were when treated with NaCl at 150 mmol/l, Zhang-ChaoQiang *et al.* (2007) recorded significant decrease in chlorophyll and increased amount ot malondialdehyde contents though in the leaves of *L. aureum* the soluble protein content was found increased while in *L. gmelinii* the same was decreased. When the seeds of two halophytes, *L. aureum* (Linn.) Hill and *L. gmelinii* (Willd.) Kuntze were treated with different NaCl concentrations, Yang-YingLi *et al.* (2008) recorded inhibition of germination and decrease in chlorophyll contents under 50, 150 and 300 mmol/l NaCl treatment though at lower level the soluble protein content in leaves was found enhanced but at higher levels there was inhibition. Li-Yan (2008) with *L. sinense* recorded seed germination even at higher level of salt concentration (400 mmol/l) with increase in the radical hypocotyls which shows that the crop has high tolerance in shoot growth under NaCl stress which may be reason for its wide utilization for saline soil rehabilitation. Three germination (16/13, 21/18 and 27/24 °C; day/night) and two growing temperatures (24/16 and 26/24 °C; day/night) on the flowering of long day grown plants when were studied for a 24-week period, Semeniuk and Krizek (1973) recorded 24/16 °C producing earliest and greatest number of flower stalks with good quality in nearly all the plants (exception 'Gold Coast') irrespective of the germination temperature though at 26/24 °C temperature the flowering percentages varied greatly depending upon the cultivars and the germination temperature, however, in the cv. 'Blue Bonnet' flowering percentage was lower at 26/24 °C irrespective of the germination temperature while in the cvs 'Iceberg', 'Midnight Blue' and 'Twilight Lavender' there was higher percentage of flowering whether it was 16/13 or 21/18 °C growing temperature, and 'Gold Coast' flowered irrespective of temperature treatment. The seeds of three statice cultivars, *viz.* 'Early Blue, (early flowering), 'Midnight Blue' (midseason) and 'Super Blue' (late flowering) when were sown in autumn started flowering in the following spring only after experiencing some chilling in the winter, and where plants from seeds were vernalized at 2-3 °C for 30 days flowered earlier with more cut flowers but with less number of foliage leaves than those non-chilled, best temperature being 2 °C (Azuma *et al.*, 1983). Early and midseason cultivars responded well when vernalized at 2 °C for 20 days and the late ones for 40 days but transfer of vernalized seedlings directly to high temperature (>25 °C) for five days the effect of vernalization was found nullified. The efficacy of kinetin (0.05 mM), ethephon (5 micro M), GA_3 (0.3 mM), proline (0.1 mM), betaine (0.1 mM), nitrate (10 mM) and thiourea (5 mM) in the alleviation of salinity-induced (0, 100, 200, 300, 400 and 500 mM NaCl) seed dormancy in *L. stocksii* under a 12-h photoperiod and the complete darkness was evaluated, where Sabahat Zia and Khan (2003) observed that only kinetin and ethephon alleviated salinity-induced seed dormancy, kinetin being more effective than ethephon, whereas under complete darkness the seed germination was substantially inhibited by salinity but here too kinetin and ethephon improved seed germination. At low salinity, there was no effect of other tried chemicals but at salinity these inhibited germination. *Limonium* seeds of different species and populations in Sicily (Italy) were sterilized following micropropagation protocols and placed in petri dishes with MS media without growth regulators to compare *in vitro* germination percentage and mean germination time (MGT), where Airo *et al.* (2004) observed MGT varying from 3.4 to 19.5 days and the germination percentage among the species from 28.1 and 97.5.

L. brasiliense, a native to Brazil shows two reproductive systems; sexual (seed) and asexual (**rhizomes**), so Lopes *et al.* (2003) tried 5-8 cm long rhizome fragment from apical (young portion) and non-apical (middle region) rhizomes under different substrates (marisma beach sand, original sand substrate and commercial substrate) where they recorded apical portion responding better to non-apical though no difference was observed in different substrates used.

Subsequent selection and propagation of good clones can be effected through **root cuttings**, crown divisions and tissue culture. Division of root, though a tedious job, is carried out in February-March. Cuttings are inserted in sandy soil in a cold frame to root. When these have attained 3-4 leaf sizes, these are set out in the nursery beds, and then these are grown and planted out in the autumn in case of temperate areas though in the sub-tropical and tropical areas these can be raised in September-October at 13-16 °C, and after appearance of leaves these are overwintered at 8-13 °C in greenhouses, and finally planted in March for harvesting of the flowers.

Multiplication even of the perennial type is not a problem as these come up easily through seeds but in order to overcome pollination and post-pollination barriers. Different *in vitro* culture techniques have been used for rapid plant propagation. *Limonium* hybrids are produced using ovule-embryo rescue techniques through **micropropagation** to overcome post-pollination barriers. Morgan *et al.* (1995) rescued the 12 to 15 days old embryos obtained from *Limonium peregrinum* × *L. purpuratum* and cultured within their embryo sacs on modified B5 or KM medium. After 2-3 days the embryos were excised from their embryo sacs and replated on fresh medium. When the embryo-derived plantlets had attained a length of 1 cm they were transferred to a modified MS medium containing BA and NAA for shoot proliferation. Plantlets were transferred to modified MS medium supplemented with IBA for 24 h for root initiation then to a modified growth-regulator-free MS medium for root growth. After a further 28 days the plantlets were transferred to soil-less medium for acclimatization. The hybrid characteristics of one of the 15 embryo-derived plants were determined by flow cytometry and by examination of morphological features. The mean DNA contents of 2C nuclei from *L. peregrinum*, the hybrid and *L. purpuratum* were 13.98 pg, 16.81 pg and 19.37 pg, respectively. Mitotic and meiotic chromosome counts from *L. peregrinum* and *L. purpuratum* showed that both parents and their hybrids had identical chromosome numbers (2n = 24), and that the species were closely related. Morgan *et al.* (1998) cultured 12-16-day-old embryos within their ovules of *L. perezii* × *L. sinuatum* on modified B5 medium and later excised and re-plated onto fresh medium. Embryos grew for varying lengths of time eventually reaching lengths of up to 10 mm, before ceasing growth and dying off. Transfer to modified MS medium containing 3 mg TDZ/l for 24 hours induced callus growth in the embryo-derived tissues from which shoots developed. Shoots were proliferated on modified MS medium containing IBA, BA and GA_3. Shoots transferred to modified MS medium with added IBA for 6 days initiated roots which continued development after transfer of the explants to modified growth-regulator-free MS medium. Further, after 28 days, the plantlets were transferred to soil. The hybrid nature of one of the embryo-derived plants was determined by flow cytometry and by examination of morphological features. The mean DNA contents of 2C nuclei from *L. perezii*, the hybrid and *L. sinuatum* were 8.69 pg, 7.59 pg and 6.42 pg, respectively. Mitotic and meiotic chromosome counts from *L. perezii* and *L. sinuatum* and their hybrid revealed that they had chromosome numbers of 14, 16 and 15, respectively. In order to overcome pollination and post-pollination barriers, *Limonium* hybrids were produced using ovule-embryo rescue techniques. Fifteen *L. perigrinum* × *L. purpuratum* embryos were successfully rescued when *L. perigrinum* was the female parent though none in the reverse cross which all produced intermediate floral colours with stems of 25-60 cm. Early confirmation of *L. perezii* × *L. sinuatum* hybrids was provided by flow cytometric analysis (Burge *et al.*, 1995).

Pathogen-free plants have been produced using techniques such as seed culture, meristem culture, micropropagation using axillary or adventitious shoot buds, and somatic embryogenesis.

Bach and Pawlowska (1992) used apical buds, leaf explants and axillary shoots for *in vitro* culture of *L. latifolium* and *L. tartaricum* on MS basal medium supplemented with 1 mg NAA/l and 3 per cent sucrose, and after eight weeks the explants were transferred to media containing NAA, IAA, BA or kinetin at 0.1 to 2.0 mg/l where both the species propagated well from each of the three explant materials, shoot multiplication being highest in *L. tartaricum* in kinetin at 0.5 mg/l and NAA at 0.1 mg/l, and in *L. latifolium* on kinetin at 0.5 mg/l and NAA at 1.0 mg/l. When shoots were obtained these were placed on MS rooting media without growth regulators or with IAA, NAA or IBA, each at 0.05-0.8 mg/l, and given 16 h light/day at 23 ± 2 °C, and then rooted explants were planted in boxes in a 3:1:1 peat:perlite:sand mixture and kept in a greenhouse. *L. latifolium* rooted well without growth regulators or in the presence of IAA, IBA or NAA with 98 per cent survival, whereas *L. tataricum* rooted only in the presence of NAA with 75 per cent survival. Lledo *et al.* (1993) reported 25 threatened *Limonium* species in the Valencian area of eastern Spain, out of which four species such as *L. dufourii, L. parvebracteatum, L. riguallii* and *L. santapalense* all having ornamental importance have been established through *in vitro* cultivation (axillary buds from leaf rosettes, inflorescence nodal segments and seeds). *In vitro* growth on ½ MS plug liquid medium (as compared to full MS plug liquid medium), followed by *ex vitro* acclimatization under 25 per cent shading in the glasshouse, of the plantlets cultured autotrophically of *Limonium* 'Misty Blue', was found superior to that cultured mixotrophically on MS agar (Lee EunJu and Jeong-ByoungRyong, 1999b). Huang *et al.* (2000) cultured *in vitro* the primary and lateral shoot tips, leaf and inflorescence node explants of an endangered medicinal plant, *L. wrightii* on MS medium supplemented with 8.87 mM N6-BA and 1.07 mM NAA where adventitious shoots were formed after 60 days and then these were sub-cultured and rooted on basal MS medium with 4.92 mM IBA, and then after acclimatizastion for one month these were transferred to the field where these grew normally. In *L. cordatum*, a Ligurian endemism of the northern Mediterranean area, Casazza *et al.* (2002) suggested low BA concentration to increase production rate of new shoots though it inhibits rooting, however, auxin encourages rooting but inhibits shoot development so simple MS medium without use of growth regulators should be adopted in its *in situ* growth and once these are established there won't be any problem in transplanting these *in vivo*. Ruffoni *et al.* (2002) took fragments from 24-hour old immature inflorescences of *L. perezii*, rinsed and sterilized with ethanol (70 per cent) and sodium hypochlorite (2.3 per cent) and cultured in presence of zeatin, isopentenyladenine or benzyladenine where shoots developed after 40 days with differing behaviours related to the different cytokinins used, and all such shoots proliferated in normal medium with high multiplication rates, and when 0.3 mg IAA/l was added in the medium there had been high percentage of rooting. On planting in the cold greenhouse these flowered as usual. In the *in vitro* multiplication of *Limonium* 'Misty Blue', out of various concentrations of cytokinins (KIN, BAP and TDZ) and auxins (IAA and IBA) used, the best results were obtained using MS medium supplemented with 0.5 mg KIN or BAP for a period of four weeks, whereas high concentrations of cytokinins induced cellular dedifferentiation and reduced multiplication rate, especially at 0.1 mg TDZ/l, though *in vitro* rooting was found induced when MS medium was supplemented with IBA at 0.5 and 1.0 mg/l for three weeks, however, IAA at any dilution did not initiate rooting (Chamorro *et al.*, 2007). Rathinasabapathi and Aly (2000) for the first time developed a protocol to induce somatic embryogenesis from 5-day old cotyledon and hypocotyl explants of *L. latifolium* cultured on agar-solidified MS medium supplemented with 3 per cent (w/v) sucrose, 1 mg/l 2,4-D and 0.1 mg/l kinetin where they obtained globular embryogenic callii in two weeks which differentiated into cotyledonary stage embryos. Aly *et al.* (2002a, b) succeeded in inducing somatic embryogenesis from cotyledon explants in *L. bellidifolium* (*Statice caspia*), *L. aureum, L. latifolium* and *L. sinuatum* on modified MS medium supplemented with 4.5 mM 2, 4-D and 88 or 118 mM sucrose which developed full shoots afterwards. By taking leaf segment explants of *in vitro*-cultured seedlings of *L. sinuatum* cv. 'Early Rose', when Igawa *et al.* (2002) cultured these on 0.25 per cent gellan gum-solidified half-strength MS medium containing 1.0 mg l^{-1} (4.14 mM) picloram, initially friable callii were induced which were maintained as cell suspension cultures where these showed high proliferation ability with about 80 per cent fresh weight increase in 2-week interval of sub-culturing, then after addition of zeatin shoot regeneration was achieved whereas after detachment from the callus these shoots rooted readily after one month of transfer onto 0.5 per cent gellan gum-solidified half-strength MS medium lacking plant growth regulators. These plantlets were successfully transferred to the greenhouse after acclimatization where these grew normally. Liu-TsuHwie *et al.* (2004) with *L. sinuatum* tissue culture recorded greater shoot regeneration and lower vitrification in the agar solidified medium compared to gelrite medium. Likewise, the proliferation rate was higher in C8 medium (1.1 mg BA + 0.5 mg NAA/l) than in the C1 medium (4.0 mg IAA + 2.5 mg kinetin/l) though in C1 medium the vitrification was lower, however, in the GA-7 vessels with parafilm and L-type vessels there was highest vitrification and lowest proliferation, may be due to poor ventilation of the culture containers.

Optimum proliferation in statice with axillary buds on Linsmaier & Skoog (LS) medium containing 0.6 mg/l BA, growth and rooting on half LS medium containing 0.5 mg/l NAA were obtained (Harazy *et al.*, 1985; Lesham and Cohen, 1983). *L. sinuatum* 'Midnight Blue' buds were cultured on basal MS medium supplemented with 36 combinations of BA (0.5 or 1.0 mg/l), calcium pantothenate (1.0 or 2.0 mg/l) and sucrose (3, 4 or 5 per cent), with 0.6, 0.7 or 0.8 per cent agar where vitrification was controlled in media supplemented with 2.0 mg calcium pantothenate/l and 4 per cent sucrose. The highest number of non-vitrified shoots was observed in the medium supplemented with 0.5 mg BA/l, 2.0 mg calcium pantothenate/l and 4 per cent sucrose, with 0.7 per cent agar (Pedroza *et al.*, 1997). Miyama *et al.* (1998) cultured lateral floral stalk buds of *L. sinuatum* cv. 'Early Blue' *in vitro* at 20 or 27 °C on a ½ strength MS medium containing 0.01 mg NAA/l, 1.0 mg BA/l, 30 g sucrose/l and 8 g agar/l, the emerging shoots were excised from clumps of multiple shoots and subcultured for 4 weeks at 20 °C on a hormone-free rooting medium and the rooted plantlets were transplanted into pots and grown at 20 °C. High temperature and repeated sub-culturing during *in vitro* multiplication caused a reversion of flowering, and that low temperature during sub-culturing restored it. The vegetative and reproductive buds of a deep blue selected clone of *L. sinuatum* 'Midnight Blue', having good demand by the growers, were used for *in vitro* culture where vitrification was solved by increasing the concentration of sucrose to 4 per cent, agar to 0.7 per cent and calcium pantothenate to 2 mg/l. Such plants showed complete colour homogeneity (Pedroza *et al.*, 1999). Martin and Perez (1995) successfully regenerated *L. dufourei, L. calaminare, L. gibertii, L. dichotomum* and *L. catalaunicum* from nodal segments excised from young seedlings (raised from seeds) on MS medium supplemented with IBA (1 mg/l) and BAP (0.1 mg/l), and rooting was obtained by transferring the isolated shoots on fresh MS medium devoid of plant growth regulators.

Leaf explants (5×5 mm sections) from mature plants of *L. altaica* cv. 'Emille' were *in vitro* cultured on MS medium supplemented with N6-BA and TDZ individually and in combination with IAA and NAA where most of media showed direct prolific adventitious shoot regeneration, best (99.5 per cent frequency of shoot regeneration and 112 shoots per explant) being on medium supplemented with 2.85 mM IAA and 1.14 mM TDZ, rooting in ½ strength MS medium and MS medium with IAA, and such plants survived more than 95 per cent when transferred to the soil (Jeong *et al.*, 2001). Hsu-NaiWen *et al.* (2000) used *in vitro* culturing of *L. wrightii* leaf explants on solid medium containing MS basal salts, vitamins, 2 mg/l BA and 0.5 mg/l NAA where callus was induced after two weeks, adventitious shoots were developed from callus within six weeks, and then the explants were sub-cultured on Lim1 (2 mg/l BA and 0.2 mg/l NAA), JP2 (0.5 mg/l BA and 0.1 mg/l IAA) and JP3 (0.5 mg/l BA and 0.2 mg/l NAA) media where multiplication rate of proliferated shoots at regular sub-culture intervals was 10-, 6.5- and 5.4-fold, respectively which rooted well after transferring them to MS medium with 0.2-0.8 mg/l IBA and then these were transplanted in the greenhouse where these grew well. Amo Marco and Ibanez (1998) sterilized 20 mm long basal immature inflorescence segments of *L. carvanillesii* (a threatened and endemic species of eastern Spain) and established *in vitro* on MS medium containing 2-5 mg/l kinetin, 5 mg 6-gamma isopentenyladenine or 0.1 mg/l 6-BA where they obtained best shoot formation, *vis-a-vis* rooting in four weeks in 80-85 per cent shoots without or with IBA or IAA at 0.1 or 0.5 mg/l and when such plantlets were planted under greenhouse conditions there was 90 per cent survival in four weeks. Hosni *et al.* (2000) took 2-3 mm segments of the imported *L. sinuatum* 'Citron Mountain' inflorescence, sterilized and cultured on MS basal medium supplemented with 0.0, 0.5, 1.0 or 1.5 mg/l NAA, where they recorded best bud break and shoot development with 1.0 mg/l NAA, best shoot multiplication at 1.5 mg/l of BA, followed by kinetin, and with addition of activated charcoal in the medium there was poor tendency of callii formation, and improvement in the length and colour of shoot, and high rooting (25-38) per shoot within four weeks with addition of 1.5 mg/l IBA in the medium augmented with or without activated charcoal. Fior *et al.* (2000) successfully reported *in vitro* culture of immature inflorescence of *L. latifolium* on MS medium for 35 days added with 0.7 mg BA/l, rooting in 30 days when 1 mg IBA/l was added to MS medium, and optimum survival after *ex vitro* transfer following acclimatization, and this way they could produce 15-30 plantlets in 100-120 days from one explant. Of the three vegetative and one floral stage of development in *L. sinuatum* 'Midnight Blue' cultured *in vitro* on LS medium supplemented with 0.0, 0.6 or 1.2 mg BA/l, Oviedo and Guevara (1988) observed satisfactory establishment only under 0.6 mg BA/l and that too with the axillary buds taken at 4th stage where BA induced a reversion from floral to vegetative growth, and they observed some 43 per cent root development when 0.5 mg/l NAA was added to the medium. Proliferation of the cultures was initially accelerated by 1.2 mg/l BA but the treatment induced growth malformation with repeated subcultures. Fang-XiaoFeng *et al.* (2007) used sterile leaves, petioles and young roots from the sterile seedlings of *L. aureum* and recorded higher induction rates in petioles and leaves than those of roots in MS media supplemented with NAA + 0.5 mg/l, 2,4-D at 1.0 mg/l or 6-BA at 0.3-0.5 mg/l + 5-2.0 mg/l NAA. During 14-28 days of culture, the explants produced milky-white, greenish-grained or compact calluses, with induction rates of 70 per cent. After transferring 1-2 times, red- and greenish-grained calluses were induced to produce clusters of adventitious buds on MS medium with 6-BA at 0.3 mg/l + NAA at 1.0 mg/l. The adventitious roots were produced when buds were cut from 3 cm height cluster buds, and cultured in ½ MS medium with IAA at 0.5 mg/l or IBA at 0.5 mg/l, and finally intact plants were obtained.

Leaf protoplasts cultured in Gellan gum-solidified 50 per cent MS medium containing 1 mg 2,4-D, 1 mg NAA, 1 mg BA and 250 mg casein hydrolysate/l, 3 per cent sucrose and 0.5 M mannitol, started to divide within 2-3 days of culture and formed colonies (0.5-1 mm) after 2 months of culture, and on transfer of these colonies to MS medium containing 1 mg 2,4-D, 3 per cent sucrose and 0.2 per cent Gellan gum, callus proliferation occurred, and these were then transferred for shoot regeneration to MS medium with or without 2 mg zeatin/l. Shoots were rooted on MS medium containing 0.5 mg IBA/l, 3 per cent sucrose and 0.2 per cent Gellan gum and transferred to pots where these grew normally (Kunitake and Mii, 1990). No abnormalities in leaf shape and growth habit were observed.

Cultural Practices

Limonium species show excellent potential as drought-resistant ornamentals. Since statice originates in dry and infertile locations, they grow best on well-drained soils, preferably sandy-loam with only a moderate level of fertility, at sunny location. The ideal soil pH is 6.5-8.5 with EC being 2.5. They are best grown on raised beds or ridges in the open though quality is certainly better when grown under cover, *i.e.* unheated greenhouses. These require wind protection, and simple stretching of wire along the length and breadth of the beds would be sufficient to support the stems. Netting is not preferred as flower heads tangle with the nets. They flower from seed within first year, earliest varieties taking 120-150 days from the date of sowing such as 'Gold Coast' and 'Roselight' though the varieties such as 'Heavenly Blue' and 'Iceberg' are high-yieldding ones. They are sown in late summer outdoors to overwinter and flower in late spring or sown in trays in a greenhouse at 20-22 °C temperature, where pricking out is carried out when true leaves have appeared, and when have attained the rosette of three or more leaves they should be shifted to a cool location at 11-13 °C for necessary chilling from 3-6 weeks and then these are planted outdoors at 30 × 30 cm spacing where these grow and flower usually if growing temperatures have been maintained 22-27 °C (day) and 12-16 °C (night). Field studies were conducted on a sandy loam soil to determine the potential of annual statice as an outdoor cut flower crop. Data were collected on seven cultivars in first year and 42 cultivars in the second year. In first year, total stem fresh weight, stem count and average stem weight differed significantly among cultivars. Yellow cultivars had more stems harvested than rose, apricot and blue cultivars, but individual stems of the yellow cultivars weighed less. The number of stems harvested over time tended to be concentrated in the first eight weeks after flowering began. In second year, the average stem fresh weight varied significantly between the apricot, blue and rose cultivars, but the number of stems harvested was significantly different only between the blue and rose cultivars (Whipker and Hammer, 1994). *L. latifolium* × *L. bellidifolium* cvs 'Misty Blue', 'Misty Pink' and 'Misty White' were grown in pumice (soilless) in a cold greenhouse having polythene covering where the plants started flowering three months after planting with better quality stems than those grown in normal soil, 'Misty White' producing 12 in pumice and 11.5 stems per plant in soil, while 'Misty Blue' 106.6 cm long stems with 73.5 g weight in pumice and 90 cm with 46.4 g in traditional soil (Zizzo *et al.*, 2003). Seeds of *L. virgatum* from calcareous, tufa, silt and marl soil types were germinated on beds of peat and sand and grown on in pots of the same mixture and when the seedlings had attained 8-12 leaves, they were planted in an open field at Palermo (Italy) with drip irrigation, where Zizzo *et al.* (2004) recorded 33 to 160 stems per plant with stem height of 53 to 86 cm where inflorescence initiated from mid-June to early August, may be due to their differing original locations. Since flower stems tended to lodge and the leaves to dry and lose greenness, none of the ecotypes were suitable for outdoor decoration but they have potential for the cut flower trade.

L. sinuatum seeds (cvs 'Bonduely Yellow' and 'Sophia') were sown at low altitude on February 20, and March 10 & 30 prior to summer season production in alpine areas (800 m a.s.l.) and these both the cultivars started flowering by early July when sown on February 20, the middle of June when sown on March 10 and late July for the last planting date with cut flower production being much higher for the first two earliest plantings though quality inflorescence production peaked from mid-July to ealy-August, however, quality spike production peaked in late August when sown on March 30 though with poor yield (Yoo *et al.*, 1994). One-month-old seedlings of six *L. sinuatum* cultivars were transplanted on 15 November or 15 December to a sandy loam soil at Ludhiana (India) at spacings of 60 × 30, 60 × 45 or 60 × 60 cm where Ramesh Kumar and Kiranjeet Kaur (1997) recorded greatest plant height (55.33 cm), spread (41.16 cm) and fresh weight in cv. 'Forever Purple' at 60 × 30 cm spacing when planted in November, likewise, November-planted 'Forever White' also produced highest fresh weight of flowers (1.072 kg). In a subsequent trial, Ramesh Kumar *et al.* (1998) studied five *L. sinuatum* cultivars such as 'Forever Blue', 'Forever Pink', 'Forever Purple', 'Forever White' and 'Forever Yellow' for planting in November and December at spacings of 60 × 30 cm, 60 × 45 cm and 60 × 60 cm and recorded intermediate spacing the best where stem heights were 30.88, 23.11 and 18.44 cm, respectively in 'Forever White', 'Forever Pink' and 'Forever Purple' in November-planted crop though the same spacing produced highest number of stems (23.05) with December planting, however, 'Forever Purple' bolted in 87.44 days (the earliest) while 'Forever White' in 104.66 days (the latest). The effects of spacing between plants (15, 30, and 45 cm) and rows (30, 45, and 60 cm) on 30-day-old *L. sinuatum* seedlings under Indian conditions of Pune revealed that there was an increase in height after 60 and 90 days after sowing but decrease in plant spread with decrease in the spacing between plants and rows, and a spacing of 60 × 45 resulted in the earliest flowering (47.43 days) and highest yield in terms of the weight of flower stalks per plant (871.36 g), flower stalk length (82.55) and flower weight per hectare (47.27 t/ha), however, in general, a plant spacing of 60 × 45 cm resulted in superior growth and flower quality but poor yield though yield was highest under 45 × 15 cm with poor quality though 45 × 30 cm spacing gave satisfactory flower yield and quality (Deshpande *et al.*, 2001).

Two clones of *L. perezii* tried in San Remo (Bari University, Italy) by Talia *et al.* (2004), resulted into strong and free-growing plants when grown in the open field than in the cold greenhouse.

In a 3-year trial near Sanremo (Italy), Farina *et al.* (1995) grew *L. perezii* from seed sown in July 1991 or February 1992 in greenhouses kept at a minimum temperature of 14 °C during germination and seedling growth, and these were transplanted into greenhouse beds in February or August where these started flowering after four months in both cases and continued without interruption until the end of the third year, though with highest yield in the second year, peak being in April and August though with lowest yield in December and January. In Sicily (Italy) with different roof opening systems (gull-wing vents equipped with 2-head ventilators, and with total roof opening by complete removal of the plastic film cover), Fascella and Zizzo (2004) evaluated two *Limonium* hybrids, 'Misty Blue' and 'Misty White' where they recorded no change in yield in both the varieties though there was quality improvement, better being in 'Misty White' and the yields were 16.7 and 26.4 stems per plant, respectively. Of the three species, *L. bellidifolium*, *L. perezii* and *L. sinense* (*L. tetragonum*) grown in cold greenhouse, *L. perezii* showed the highest yield but with poor quality, followed by *L. bellidifolium* with a lower yield but with good quality, and *L. sinense* that was the least productive but the best in quality (Talia and Pacucci, 2005).

Weed Control

Gilreath (1985) through pre- and post-transplanting applications of 4.48 kg oxadiazon, 3.36 kg EPTC and 8.97 kg/ha chlorthal-dimethyl on *Limonium sinuatum* field obtained marketable panicles comparable to the controls. Alachlor, napropamide and oryzalin caused stunting and reduced flower yields. The spring-planted *Limonium sinuatum* cv. 'Midnight Blue' crops in the soils treated four times the previous autumn with 9.0 kg DCPA (chlorthal-dimethyl), 560 g oxyfluorfen, 2.25 kg pronamide (propyzamide), 2.2 kg alachlor, 4.5 kg oxadiazon or 2.25 kg/ha oryzalin though certain chemicals reduced early season plant vigour but no herbicide residue was detrimental to flower production (Gilreath, 1986). Use of alachlor, chlorthal-dimethyl, napropamide, oryzalin and chloroxuron in winter season *Limonium* crop, though all were found effective in controlling >80 per cent of weeds except chloroxuron, however, the best and safest treatment had been alachlor for controlling 80 per cent broad-leaved weeds, though it was not suitable for some species when they were sown directly (Lamont, 1986). A greenhouse evaluation with annual statice (*L. sinuatum*) using chlorpropham, napropamide, EPTC, oxadiazon and trifluralin indicated that pre-emergence applications of three latter herbicides result into most effective weed control and least statice injury, oxadiazon being the most effective for broadleaved weeds and in increasing the greatest quality yield even better than hand-weeded control, however, the annual statice was found most sensitive to trifluralin (Hatterman *et al.*, 1987). Lamont and O'Connell (1986) found Chloroxuron (2.25 or 4.5 kg/ha) and Oryzalin (2.25 or 4.5 kg/ha) quite safe only on transplants of *Limonium sinuatum*, the former controlling about 50 per cent of weeds though latter 70 per cent of grasses and 60 per cent of broad-leaved weeds, however, when Oryzalin was tried on direct-sown crop, it was of no value. Sethoxydim, BAS 517-02-H (cycloxydim), fluazifop-butyl and AXF 1294 provided total control of field grasses without any injury to statice crop, *vis-a-vis* prodiamine and napropamide at 6.75 kg/ha gave excellent broad-spectum weed control (Kühns *et al.*, 1986). The efficiency and selectivity of 6 herbicide systems, *viz.* (i) pendimethalin (1320 g/ha) + chloroxuron (4500 g/ha)/oxadiazon (1000 g/ha); (ii) metolachlor (2000 g/ha) + oxadiazon (1000 g/ha)/fluazifop-butyl (375 g/ha); (iii) metolachlor (2000 g/ha) + chloroxuron (4500 g/ha)/oxadiazon (1000 g/ha); (iv) pendimethalin (1320 g/ha) + oxadiazon (1000 g/ha)/fluazifop-butyl (375 g/ha); (v) trifluralin (960 g/ha) + oxadiazon (1000 g/ha)/fluazifop-butyl (375 g/ha); and (vi) trifluralin (960 g/ha) + chloroxuron (4500 g/ha)/fluazifop-butyl (375 g/ha) + oxadiazon (1000 g/ha) applied to *Limonium sinuatum* were when tested, these were found very effective even under very high weed density, including annual grasses and dicotyledonous weeds, and the cut flower yield was also higher compared to control, except in the treatment trifluralin (960 g/ha) + oxadiazon (1000 g/ha)/fluazifop-butyl (375 g/ha) (Ivanova, 1999a). Out of the six herbicides (pendimethalin was applied at 1,320 g/ha, metolachlor at 2,000 g/ha, trifluralin at 960 g/ha, oxadiazon at 1,000 g/ha, metobromuron at 750 g/ha and chloroxuron at 4,500 g/ha) applied in *L. sinuatum* crop, first three proved very effective against the annual weeds *Echinochloa crus-galli*, *Digitaria sanguinalis* and *Setaria viridis* though latter three were effective against the annual weeds *Amaranthus retroflexus*, *Chenopodium album*, *Galinsoga parviflora*, *Stellaria media*. The soil-applied herbicides (except for metobromuron) also had a positive effect on the quality and quantity of *L. sinuatum* cut flowers (Ivanova, 1999b).

Feeding and Irrigation

Though most *Limonium* species are drought tolerant to a great extent but after all these require **water** for drawing nutrients from the soil. Immediately after planting these are irrigated for plant establishement. Watering also depends upon season, soil types and weather condition. Over-head watering after rosette formation is dangerous to these plants as flowers and leaves are affected with various diseases, especially *Botrytis cinerea*, therefore, only drip system of irrigation is best. Watering should be done only when plants express such signs. Excess of watering may cause rotting of rosettes. Alarcon *et al.* (1999) studied the effect of NaCl (140 mM for 20 days) with respect to growth, water relations and feeding in *L. latifolium* 'Avignon' and *L. caspia* × *L. latifolium* 'Beltlaard' in hydroponics where 'Beltlaard' showed reduced shoot dry weight and leaf area due to salinity as there had been higher concentrations of Na^+ and Cl^-, and changes in the contribution of K^+ and NO_3- to leaf osmostic potential, though 'Avignon' was not affected, however, leaf water relations under saline stress were not different between both the genotypes, indicating similar osmotic effect in both the cultivars as in the end of the experiment both the cultivars were able to achieve the same degree of osmotic adjustment, though in 'Beltlaard' the maintenance of turgor did not permit the maintenance of growth rate. The rates of NaCl absorption by roots and the net transfer from roots to shoots were higher in 'Beltlaard' than in 'Avignon', suggesting that radical exclusion mechanisms are more operative in 'Avignon'. Emily cultivars of *Limonium* showed high resistance to salinity and the **irrigation** water with an EC of up to 11.5 dSm^{-1} (Shillo *et al.*, 2002). *L. cossonianum*, a Mediterranean species was pot-grown from seeds for 57 days in a greenhouse in Spain and was drip-irrigated six days a week (T_1; 5.1 litres/plant during whole nursery period), three days a week (T_2; 2.6 litres) and twice a week (T_3; 1.8 litres) to reach the water-holding capacity and then T_1 and T_3 plants were transplanted to the field to study for 435 days the development of the aerial part and the root system dynamics, the latter using minirhizotrons (Franco *et al.*, 2002). The regime involving the least amount of water in the nursery period produced the plants best adapted to stress after transplantation, as manifested in their greater root length to leaf area ratio and a higher percentage of brown roots. After transplanting to field, T_3 and T_1 plants showed similar root length density (RLD) in the upper (0-46 cm) and in the deepest (115-160 cm) layers of soil, but RLD was greater for T_3 plants in the intermediate levels of the soil profile (46-115 cm deep), where the moisture content was low but sufficient for root development. The aerial part of T_3 plants developed more and had a more compact appearance; they also had longer flowering stems than T_1 plants. Root development was substantially lower when there was no establishment irrigation than when one establishment irrigation of 50 litres/m^2 was applied. In a recovery experiment 415 days post-transplanting and after 76 days without rain, root growth was reactivated by re-irrigation more rapidly in T_3 than in T_1 plants, confirming the positive effect of hardening in the nursery. Plants of *L. pectinatum* and *L. latifolium* × *L. caspium* cv. 'Beltlaard' grown under greenhouse conditions were irrigated with 200 mM NaCl solution for four months (saline treatment) which reduced plant dry weight, *vis-à-vis* inflorescence height, number and dry weight but not in *L. pectinatum* though leaf water relations under saline conditions showed a similar behaviour in both the genotypes, achieving the same degree of leaf osmotic adjustment through the accumulation of Na+ and Cl- (Morales *et al.*, 2001). They recorded though highest levels of these ions in treated *L. pectinatum* plants, but even the lower accumulation of salts in the tissues of 'Beltlaard' plants generated major toxicity and nutritional disturbances, which negatively affected the growth. The effect of sea water and NaCl at 0, 10, 20, 30, 40 and 50 dS/m on *L. stocksii* germination under a 12-h photoperiod and in complete darkness was investigated by Sabahat Zia and Khan (2002) where they found decreased seed germination with increasing salinity, and only a few seeds germinating above 30 dS/m in both sea water and NaCl, inhibition being more in sea water and synergistically under darker conditions, and when all these ungerminated seeds were kept for 20 days under saline conditions they readily germinated on transfer to distilled water. Carter and Grieve (2006) investigated into the salinity tolerance of cut flowers of *Limonium perezii* 'Blue Seas' and *L. sinuatum* 'American Beauty' at the U. S. Salinity Laboratory to determine marketability based on stem length, where plants were exposed to differing water ionic compositions and salinity levels which though caused reduction in plant height with increase in the levels but the flowers were of marketable quality at moderate salinity levels. Ding Feng and WangBaoShan (2006) recorded significantly increased dry and fresh weights of *L. sinense* plants following treatment with 100 mmol NaCl/l but these gradually declined; the secretory cells and the average diameter of salt glands did not significantly vary between the control and 300 mmol NaCl/l but significantly varied between the control and NaCl at 100 and 200 mmol/l and these concentrations as well as 300 mmol significantly enhanced the development of salt glands and the salt-secretng capacity, thereby increasing significantly the salt secretion rate in salt glands and leaves with increase in the NaCl concentration. Li (2008) found the seedlings of *L. bicolor* growing best under 100 mmol/l NaCl salt concentration, fresh weight and dry weight reaching the maximum, and parameters of oxidative stress of shoots-malondialdehyde (MDA) concentration slightly decreasing under 100 mmol/l salt stress, then increasing with increased NaCl concentration. The activities of enzymes (superoxide dismutase, peroxidase and catalase) increased with increase of the concentration of NaCl in the shoots.

Limonium remains comfortable in soils having medium nutrients as these are not heavy feeder, and feeding also depends largely on soil types and climatic conditions. If at the time of soil preparation the soil has been incorporated with up to 50 quintals per hectare of farmyard manure or compost, there is no further requirement of any feeding. During the period of growth if plant stand is poor, the crop may be fertilized only after soil analysis. Raulston and Geraldson (1971) had grown statice successfully under Saran cloth-shaded structures in ground beds receiving a single pre-plant application of 6-8-8 **fertilizer** or Osmocote (14-14-14) at 335-1,000 kg N/ha though mulching with polythene or paper made little difference to crop quality. Arisco and Koths (1984) planted *Limonium* seedlings in three media (Pro-Mix BX containing vermiculite, perlite and limestone; UConn Mix containing compost, peat and sand 3:2:1; and Arisco Mix containing loam, sphagnum peat, sand and perlite 3:2:1:1), each medium containing nutrients either 15-0-18 fertilizers (including 3:2, Ca and K nitrates) or 20-20-20, and under continual fertilization plants were irrigated daily at 150 ppm N from two weeks after planting; with pulse fertilization the plants receiving a weekly application at 8,000 ppm N, and where plants received 15-0-18 fertilizer the plants were darker green in the Arisco medium than in other two media though no significant differences in plant size were observed. A salt-tolerant *L. tartaricum* plants were when fed with supplementary K top-dressing as potash magnesia (40 g K/m^2) in early April increased the metabolic activity of enzymes involved in cellulose formation resulting in increased fresh and dry weights of inflorescences (Harm and Maync, 1985). When *L. sinuatum* cvs 'Iceberg' and 'Kampf's Blue' and *L. bonduellei* cv. 'Gold Coast' were grown in a fertile silty clay-loam field and fertilized with varying amounts of granular 12N:5.3P:10K, Paparozzi and Hatterman (1988) recorded significant increase in stem weight, stem number, and total fresh weight though there were varietal differences, 'Kampf's Blue' showing the highest and 'Gold Coast' the lowest values, 45.4 to 68.1 kg N/ha increasing the total fresh weight in 'Iceberg' and 'Kampf's Blue' but not in 'Gold Coast', and the number of subsequent fertilizer applications beyond the initial application was not found as critical as the total amount of fertilizer applied. *L. perigrinum* (*L. peregrinum*) cv. 'Ballerina Rose' plants were when grown in the greenhouse in 1.8-litre plastic planter bags with N (250, 500 or 750 mg/l), P (60 or 120 mg/l) and K (150 or 300 mg/l) fertilizers incorporated into the potting mix, Lewis *et al.* (1994) recorded better shoot dry weight, stem elongation, and inflorescence number & size with the potting mix containing 750 mg N, 60 mg P and 150 mg K/litre, the greatest response being with nitrogen than phosphorus and potassium and the leaf nutrient concentrations associated with the greatest growth found were 2-3 per cent N, 0.2-0.3 per cent P and 0.5-1.5 per cent K. Lee-EunJu and Jeong-ByoungħRyong (1999a) studied effect of P on the plantlet growth of *Limonium* cv. 'Misty Blue' under *in vitro* autotrophic and mixotrophic culture conditions in the medium having 55.7-5.85 pH with no organic materials, which most effectively promoted the growth under autotrophic conditions, however, in mixotrophic culture the pH was 5.29 and in the end the EC was 1.60 mS/cm, higher than that in autotrophic culture, suggesting fewer uptakes of nutrients by plantlets. Further addition of P in the MS basal medium enhanced further plantlet growth under autotrophic culture conditions. Optimum P levels were found to be 2.665 meq/l for greatest number of leaves, leaf area, leaf chlorophyll concentration, fresh weight, and length of the longest root; and 4.080 meq/l as initial concentration for plantlets with the greatest height, dry weight and percentage dry matter only under autotropic conditions. One of the main external factors that influences anthocyanin accumulation is temperature, with low temperatures increasing and elevated temperatures decreasing pigment concentration but Nissim *et al.* (2007) showed that higher magnesium levels also increase anthocyanin accumulation in aster flowers grown at elevated temperatures, probably due to increased stability of the pigments. In blue-bracted *Limonium* 'Blue Night', Mg treatment to whole plant, cut branches or detached flower buds caused significant increase in anthocyanin concentrations, with a stronger effect under elevated temperature regimes.

Growth and Development

Statice cvs 'Iceberg' and 'Midnight Blue' were sprayed with **GA** at 0-2,000 ppm 110 days after sowing where GA at 200 or 300 ppm in 'Midnight Blue' accelerated flowering by 30 days and at 400 ppm in 'Iceberg' by about 55 days, and with better flower yield in GA 200 ppm and above, optimum being 400-500 ppm (Wilfret and Green, 1975). Single applications of GA_3 at 500 ppm made at 87, 101 and 115 days after seed sowing on statice cvs 'Iceberg' and 'Midnight Blue' reduced the time taken for flowering by 90, 72 and 45 days, respectively though repeated applications did not have any such effect and rather delayed the flowering, and that single spray made at 129 days of sowing did not have any effect (Wilfret and Raulston, 1975). *L. sinuatum* cvs 'Early Blue' and 'Sophia' flowered earlier and produced increased number of flower stems per plant in October and November but without any change in the flower quality when GA_3 (500 ppm) was applied three weeks after planting in the cropping season of June to March (Huh *et al.*, 1995). Its 400 ppm single direct spray at various times between 9-52 days after planting to the plants of cv. 'Misty Blue' brought about accelerated flowering in all the treatments, more prominent being when treated at four weeks of planting, though increased cut flower yield was recorded when applied nine days after planting (Garner and Armitage, 1996). Micropropagated plants of *L. gmelinii* clones transplanted in unheated greenhouse in summer and which after summer flowering were subjected to GA_3 applications and/or 16-h photoperiod in next year September at an interval of 20 days, then maintained under natural photoperiodic condition (ND) or 16-h photoperiod (LD) (Farina *et al.*, 2003); and in the subsequent experiment the plants obtained in September were acclimatized in spring and planted in a greenhouse in

May and then were subjected to no GA_3application followed by ND condition, two-500 mg/l GA_3 applications in September followed by ND condition, and two-500 mg/l GA_3 applications in September followed by LD condition. The response of the two-GA_3 or of LD treatments was clone-specific though LD condition applied to the GA_3-treated plants decreased the promotive effect of the growth regulator on autumn flowering of clones. A clone failed to flower in autumn even after GA_3 and/or 16-h photoperiod treatment while a clone gave a certain flower yield in autumn even without growth regulator application and/or photoperiodic treatment. Flower quality was sufficient and in general poorly affected by GA_3 applications or by photoperiodic treatment though yield was recorded more with higher stem height in summer flush.

Statice plants grown under long days (16-h) started to flower about 22 days earlier with 98 per cent flowering than those grown under short (8-h) days where flowering could be achieved only 52 per cent and where cv. 'Blue Bonnet' did not flower at all, however, the falling of night temperature from 21 to 13 °C reduced the time to flowering by about one week while flowering percentage increased by about 34 days (Semeniuk and Krizek, 1972). In Barcelona (Spain), when Lopez *et al.* (1996) had sown the seeds of *L. sinense* (*L. tetragonum*) on 10 January and grown under natural **photoperiod** without heat or in a greenhouse heated to ±13 °C with a 16-h photoperiod and continuous fertigation, heating the greenhouse advanced the onset of flowering by six weeks with resulting increase in inflorescence width but with no change in quality.

Zimmer (1981) stated that in Germany *L. sinuatum* plants are usually grown under plastic houses, the seeds are germinated at 16/13 °C (day/night) and so such seedlings produce only fewer leaves as compared to the excessively leafy growth when seeds are raised at 27/24 °C, the long days (16 h) advance flowering with exception of 'Gold Coast' (yellow) and 'Iceberg' (white) that flower much earlier and most abundantly at 26/24 °C when compared to blue and pink cultivars, though 'Blue Bonnet' does not flower at all under such conditions as it requires low temperature treatment. The seedlings at 2, 4 or 8 true leaves of *L. gmelinii*, a perennial species flowering in spring-summer, were subjected to a fixed photoperiod of 12-h and PPFD of 50 mM m^{-2} and sec^{-1} by HPS lamps and low temperatures of 6, 11, 16 and 21 °C for 5 weeks and then grown in cold greenhouse with 16-h photoperiod and a minimum temperature of 16 °C but in no case the plants flowered within 87 days (5 August-31 October); in the second experiment, the seedlings with 16-20 true leaves or the adult plants were subjected to 6 or 24 °C for 5 weeks and then these were transferred to a greenhouse and grown under ND conditions (23 August-30 November), none of the plants subjected to 24 °C for 5 weeks flowered independently of growth stage but on the contrary both the adult plants and the plants with more than 16 leaves coming from the low temperature treatment reached flowering; while in the third experiment, LD conditions (16-h) enhanced flowering rate of vernalized adult plants (Enrico *et al.,* 2000). In a further experiment on adult plants a succession of four weeks at 6 °C or at 32 °C, and two weeks at 6 °C was compared with that of six weeks at 6 °C, the higher temperatures increased the flowering rate thus it was concluded that a low temperature exposure of the plants in a suitable growth stage is necessary for flowering, LD conditions enhance flowering rate after fulfilling the chilling requirement, five weeks at 6 °C applied to plants with more than 16 leaves are sufficient for flowering and the devernalization was not observed after exposure of the vernalized adult plants to constant 32 °C for 4 weeks (Enrico *et al.,* 2000).

Krizek and Semeniuk (1972) when exposed 'Midnight Blue' statice to 16/13 °C **temperature** this severely limited seedling growth as measured by leaf number and fresh and dry weight of tops; vegetative growth at 21/18 °C was markedly greater than at 16/13 °C but when temperature was increased to 27/24 °C, it had no appreciable effect on leaf number, but helped producing further increases in fresh and dry weights of plant tops. Despite the minimum amount of vegetative growth produced at 16/13 °C after 42 days, this treatment resulted in the greatest number of plants flowering in the shortest time which indicates that the cultivar has a low-temperature requirement for flower initiation. The stimulatory effects of a cool temperature or of a high temperature on flower initiation persisted after the seedlings were removed from the growth chambers and placed in the greenhouse under natural long photoperiods. It would thus seem possible to use cool-temperature treatment at the seedling stage to promote flowering in this cultivar. The plants of *Limonium sinuatum* require a period of low temperatures to induce flowering (Shillo, 1976) though other perennial species start flowering in autumn and once the flowering is induced, the long days enhance this. Zimmer (1982) studied chilling requirements of 27 *L. sinuatum* cultivars by growing pricked-out seedlings at 20 °C (warm) or 15/12 °C (cold) where 24 coloured cultivars were classified into four groups according to their chilling requirements. Chilled plants always produced more and earlier inflorescences than warm-grown plants, blue cultivars produced significantly fewer flowers per plant than pink ones while the three yellow cultivars ('Bonduellei', 'Gold Coast' and 'Marktgartner Gelb') showed little varietal differences but their chilled plants developed inflorescences quicker with more flower stems. With a blue and a white cultivar of *Limonium sinuatum*, Farina *et al.* (1993) when vernalized seedlings at 6 °C for 5 weeks and planted on November 22, recorded quite advanced flowering without affecting flower quality or yield, though in a second trial only the white-flowered one planted on August 5 produced 7-8 flower stems per plant from mid-October to end of November. To assess possibility for earlier flowering than in April-May in Sicily (Italy) in *L. sinuatum* blue and white cultivars, Farina *et al.* (1993) produced flowers of usual quality and yield by February end by vernalizing young plants at 6 °C for five weeks before planting on November 22, while in the 2nd trial planted on August 5, only the white-flowered cultivar gave a reasonable yield (7-8 flowering stems per plant) from 15 October to 30 November where yield was not affected when natural light intensity of 30 per cent was reduced between August 5 and September 25. When 15th July-propagated plants were raised in open field, in a filtered light condition or in a cool and controlled temperature room, they produced more cut flowers of higher quality than plants raised in a plastic greenhouse (Goto *et al.*, 1996). *L. sinuatum* cvs 'La Mer' and 'Sunday Pink' propagated from cuttings on 16 May, 15 June or 15 July (after which all flowering stalks were removed for the first 2 months) started producing cut flowers in October, best quality being under 15 July propagated ones, while those propagated on 14 August did not flower until December. Low temperature seed treatment of *L. sinuatum* cv. 'Sophia' reduces the period from sowing to flowering and increases cut flower yield, particularly between November and February and that too under highland cultivation than low land (Yoo *et al.*, 1998). Chilling of *L. sinuatum* seedlings by storing them at 2-3 °C for 50 days, combined with illumination, produces a greater number of cut flowers, and chilling at 15-expanded-leaf stage provides highest cut flower yield (Katsutani *et al.*, 1998).

Postharvest Management

The inflorescence of statice is comprised of many individual flowers, whose sepals remain open and have decorative value long after the petals have deteriorated. Inflorescence becomes ready for harvesting generally after 3-4months of planting, depending upon the varieties and climatic conditions. This results in a long natural vase life of 30-40 days immediately after the inflorescence has been harvested. They are harvested either in the morning when stems are fully turgid or by the evening when the inflorescences have accumulated maximum of food material in them, and at the stage when calyces of individual flowers are mostly open and showing colour, i.e when roughly $^{1}/_{5}{}^{th}$ of the florets have opened and then these are sold as fillers similar to gypsophila. Amos (1980) mentions that before harvesting all the flowers must have opened as the tips wilt if cut when immature, though now it is not necessary as cutting when $^{1}/_{5}{}^{th}$ florets are open and putting them in floral preservatives all the unopened buds are encouraged to open. After cutting, the inflorescences can be stored for 2-3 weeks at 2 °C. The leaves and leafy growths at the base are cleanly removed. The cut flowers usually last for more than a fortnight and dry flowers for more than a year. For dry flower use, usually the flowers of statice are air-dried by hanging downwards in a dry, airy and dust-proof shed.

When assessing the effects of physiological maturity (3-, 6-, 9-, or 12-leaf axils) and type of cutting (whole cutting, top-half, or bottom-half) on the yield, time to flower, and flower and stem characteristics of *Limonium* 'Chorus Magenta', it was recorded that cuttings struck at an increased maturity (9- and 12-leaf axils) with apical meristems retained (whole and top-half types) had the potential to produce at least a 20 per cent higher flower-stem yield in the first 12 months of production, flowered earliest (142 days), and had reduced standard deviations in flowering time (less than 20 days compared with up to 106 days). However, these cuttings had 19 cm shorter stems than bottom-half cuttings. Panicle diameter increased up to 5 cm with increased maturity for both whole and bottom-half cuttings. However, as the number of new leaves on bottom-half cuttings increased, the time to flowering increased exponentially. Bottom-half cuttings typically took longer to flower (up to 214 days) and produced higher number of new leaves (up to 18) than the other cutting types (Funnell *et al.*, 2003). *Limonium* cv. 'Misty Blue' in bouquets wrapped in plastic films that differ in their absorption of ethylene were kept at an average temperature of 23 °C, light of 3 micro m m-2 s-1, and RH of 42.8 per cent, where it had lost its shape, colour and fresh weight more rapidly though its aesthetic characteristics remained intact. It was found that the use of plastic film helps to conserve the fresh weight and characteristics. Plastic wrap PEAKfresh best conserved the fresh weight of the bouquets and reduced the level of ethylene present in their microenvironments only during the first few days (Colinas *et al.*, 2002). There was a marked decrease of stem fresh weight in *Limonium sinuatum* and *L. gmelinii* flowers, starting after 2 days of storage. The dry transportation of these flowers causes a marked increase in transpiration, which cannot be compensated for by immersion in water, and another reason could be poor capacity for water uptake (Vigna *et al.*, 1999). Chlorophyll loss, a marker process of senescence, was accelerated when flower stalks of *L. sinuatum* were kept in the dark though exposure to continuous white light or to brief red irradiations inhibit chlorophyll degradation, both in similar manner, but it could be reversed if each red irradiation is followed by a brief far-red irradiation. Infact the phytochrome is present in the flower stalks and mediates the retardation by light of flower stem senescence. Treatment of excised flower stalks with BA or GA_3 significantly retards the destruction of chlorophyll in stems kept in the dark (Steinitz *et al.*, 1980).

Flowering stems, each with 5 subspikes, of the hybrid *Limonium* cv. 'Blue Fantasia 100' were cut from the inflorescence and placed in solutions of sucrose (0, 10 or 20 g/l), alpha-aminoisobutyric acid (AIB, 0, 5 or 10 mM) and/or aminooxyacetic acid (AOA, 0, 1 or 2 mM) for 24 hours. Thereafter the stems were held in distilled water under a 12-h day/night photoperiod at 20 °C. Sucrose more or less than 10 g/l promoted bud opening. AIB at more or less than 5 mM extended floret longevity. AOA had little effect on vase life. Treatment with 10 mM AIB in combination with 20 g/l sucrose promoted floret opening and inhibited floret senescence. Concentrations of glucose, fructose and sucrose in florets of stems treated with sucrose alone (20 g/l) or sucrose (20 g/l) + AIB (10 mM) were higher than those in the control florets. These florets produced less ethylene than control florets. The results suggest that promotion of floret opening and extension of longevity by sucrose and AIB are due to increase in the concentration of sugars in tissue and inhibition of ethylene production. In trials cut with whole inflorescences, a pulse treatment with sucrose + AIB improved the vase life by 5 days (Shimamura *et al.*, 1997). Sugars play important roles in the keeping quality of cut flowers because the amount of sugar contained in cut flowers is limited. Although pulsed treatment with sucrose (20 g/l) was fairly effective at promoting floret opening in cut hybrid *Limonium* cv. 'Blue Fantasia 100', it hardly improved the vase life. Pulsed treatment with sucrose in combination with 10 mM alpha-aminoisobutyric acid (AIB), an ethylene biosynthesis inhibitor, markedly extended the vase life (Ichimura, 1998). Water uptake and stem weight of harvested inflorescences of *Limonium* cv. 'Chorus Magenta' declined rapidly when the stems were placed in water or a biocide solution. Florets remained open for only five days after harvest and there were no opening of the new florets. The decline in water uptake and stem weight was greatly reduced by a 24-hour wetting-agent pulse treatment with Agral LN or Tween 20 at 0.1 or 1 ml/l. These treatments extended the period over which flower buds continued to open by 5 days. The addition of 50 or 100 g/l sucrose to the 24-hour wetting-agent pulse treatment extended flower bud opening to 16 days. The number of open flowers was greater when the wetting-agent pulse treatment was followed by a 20 g/l sucrose vase-solution treatment (Burge *et al.*, 1998).

Flower bud opening in *L. sinuatum* was promoted when GA_3 (10^{-6}M or 10^{-4}M) in holding solution was taken up through the cut stem but it was not promoted when GA_3 was applied directly to closed buds. Addition of 4 per cent sucrose to the solution neither increased bud opening nor altered the effectiveness of GA_3 (Steinitz and Cohen, 1982). Regardless of their maturity at harvest, the vase life of cut inflorescences of the hybrid *Limonium* cv. 'Fantasia' placed in deionized water was 4-5 days. A vase solution containing Physan (a quaternary ammonium disinfectant solution) at 200 micro l/l and 20 g/l sucrose not only prolonged the longevity of individual florets but also promoted bud opening so that the vase life of cut inflorescences was extended to 17 days. Pulse treatment with 100 g/l sucrose in combination with Physan at 200 micro l/l for 12 hours partially substituted for a continuous supply of sucrose. Inclusion of 30 mg GA/l in the vase solution had no benefit (Doi and Reid, 1995). The postharvest response of *L. peregrinum* cv. 'Ballerina Rose' inflorescences to cool storage and preservative solutions was examined as part of the development of a new cut flower crop. Chrysal or sugar preservative solutions increased the number of flower buds that opened and the length of time that flowers with open petals were present on an inflorescence after harvest but reduced vase life to 10-20 days. If inflorescences were stored for 72 hours after harvest, display life was reduced to 20-25 days. Cool storage (5 °C) and rehydration of inflorescences after storage in a sugar solution increased the time that flowers with open petals were present but did not increase overall vase life. It is concluded that the use of preservative solutions may not be appropriate given the length of the normal display life (Lewis and Borst, 1993). In an experiment to study the application of 1-methylcyclopropene (1-MCP) and to find the optimum application temperature, rate and duration alongwith its competition to ethylene and its synergistic effect with STS on *Limonium* cv. 'Beltlaard', the combined use of 1-MCP and STS resulted in the highest quality. This combined treatment also enabled storage under sea transport conditions (8 days at 2 °C). The results suggest that 1-MCP may be very useful in preserving quality of various ornamentals and that it protects against ethylene effects (Philosoph *et al.*, 2005). Five inflorescences are bunched in one bunch and such five bunches in one bundle and then these are pre-cooled and maintained in the cool chain at/or below 5 °C.

For **drying**, statice floral stems are cut just before the flowers are fully opened. These stems are tied in the bundles and hanged upside-down in shade at an airy shed protected by dusts. *L. sinuatum* harvested cultivars in the first year were soaked in varying concentrations of glycerol:water solutions for 24 and 48 hours and then subjected to microwave treatment for 0, 1, 3 or 5 minutes. Half of the stems were measured for flexibility upto 3 weeks, and the remainder was assessed a year later. Stems harvested in second year were wet- and dry-stored at 3 °C for varying lengths of time and then preserved. Preservation was best when it was preserved immediately. Cold storage decreased preserved statice flexibility, but was better than air-drying. It is suggested that fresh cut stems, up to 34 cm long, should be preserved by soaking in a 1:2 or 1:3 glycerol:water solution for 48 hours followed by microwave treatment for 1 minute at 34 °C (Paparozzi and McCallister, 1988). *Limonium bonduellii, L. otolepis, L. perezii, L. sinuatum* and *L. suworowii* when grown in the open or in the unheated greenhouse, all the species produced flowers with good stem length and fresh weight and were sold out as fresh cut flowers but the left out ones were dried in a dark ventilated room at 30 °C in the first week and at 20 °C for next 2 to 3 weeks to maintain their colour and quality characteristics, and apart from these the *L. otolepis* was also found to be used as cut green in flower arrangements (Chimonidou, 2001). A study of *Limonium sinuatum* cv. 'White Flowers' for flower drying to obtain high quality dried flowers, air drying, glycol preservation and desiccants were used to assess petal colours, dryness of petals and overall appearance of the flower, where it was found that air drying is the best method for highest scoring which was achieved in the 4th week (Kumari and Peiris, 2000).

Insect-Pests and Diseases

The **aphids** are usually found on young leaves, tender shoots and flower buds. It sucks the sap of foliage and bud and causes retarded growth of the plant with poor quality flowers. It also acts as a vector for viral diseases. Two spray of Malathion or Metasystox at 1.0-1.5 ml/l water at 15-20 days interval or spraying with Dicofol 1.5 ml or Pirimor 50 g/100 litres of water will control this pest. The **thrips** are usually found on young leaves, tender shoots, floral buds and petals. While feeding on flowers, it causes streaks and makes the flowers unsuitable for marketing. The affected plant parts turn yellow and patchy with black dots and slight crinkling. While controlling the sphids, it will control this pest also. **Cutworms** are very serious if these attack the plants at seedling stage in the beds. **Armyworms** feed on the foliage. Cutworms feed on the plants at any stage and mostly these are highly devastating if not controlled in time. Spraying with Rogor at 0.03 per cent will control these pests. Larvae of *Phyllophaga ilhuicaminai* beetles which are nocturnal in nature, have been found causing damage to the root system of *Limonium sinuatum* plants (Aragon and Moron, 2000). These beetles are blackish or reddish-brown without any prominent marking, and their size ranges from 1.2 to 3.5 cm. These are attracted to light in great numbers. Adult chafers eat leaves and flowers. However, white grubs live in the soil and feed on plant roots. Infested plants may show stunted growth, premature shedding of leaves or sudden wilting. Wasps from certain families (Pelecinidae, Scoliidae and Tiphiidae) parasitoid the *Phyllophaga* grubs. The female flies from Pyrgotidae family are endoparasitoids of these beetles.

High CO_2 atmospheres combined with high temperature was effective for controlling **Indian meal moth** (*Plodia interpunctella*) pupae on *L. sinuatum* dry flowers or preserved floral products. Pupae were exposed to atmospheres of 60, 80, or 98 per cent CO_2 in N_2, or 60 or 80 per cent CO_2 in air at temperatures of 26.7 or 32.2 °C and 60 per cent RH. Controlled atmosphere treatments at 32.2 °C controlled pupae faster than the same treatments at the lower temperature. At both temperatures, high CO_2 concentration treatments combined with nitrogen killed pupae faster than high CO_2 concentration treatments combined with air. Exposure to 80 per cent CO_2 mixed with nitrogen was the most effective treatment, causing 100 per cent mortality in 12 h at 32.2 °C and 93.3 per cent mortality in 18 h at 26.6 °C, with no adverse effects on quality of preserved floral products (Sauer and Shelton, 2002).

Spider mites are scarlet-red tiny pests and have two dark spots on their back. They feed on the under surface of the leaves, suck the sap and cause curling and leaf rolling. In severe attack, plants become stunted, and infested flowers become unsuitable for marketing. Regular spray of Propargite at 0.5ml or Dicofol at 1.0 ml/l water will control these.

Nematodes reported on *Limonium* are *Meloidogyne incognita* and *Heterodera marioni*. These live and feed in the tissues of the root, producing galls varying in size. Often these galls coalesce to form large deformed swellings. These galls or cysts containing 300 to 600 eggs, after hatching they escape into the surrounding soil and infest new hosts. In severe attack the plants wilt, turn yellow and eventually die. Control measures include steam or formaldehyde (2 per cent) soil sterilization before planting and incorporation of Furadan in the soil. Infected plants should be uprooted gently along with the soil and burnt. Infested soils should be restricted at their own places and treated to check further spread, *vis-à-vis* tools should also be sterilized. Nematodes can also be controlled through biocontrol agent *Paecilomyces lilacinus*. Application of neem-based product and crop rotation with marigold can also minimize the nematode infestations.

Damping off (*Botrytis cinerea*) of *Limonium sinuatum* has been observed in seed-beds sown with non-decorticated seeds. The fungus persists in the old flower parts surrounding the seed and resumes the growth with the addition of moisture at the time of seeding. Soaking the seeds in hot water at 52 °C for 30 minutes, or in 0.52 per cent sodium hypochlorite for 2 minutes, or in a 0.1 per cent benomyl slolution for 1 h, or a soil drench with 0.2 per cent benomyl at pre-planting eliminates the subsequent damping-off without reducing germination. Non-decorticated seeds of five statice cultivars withstood HWT at 58 °C for 30 minutes with little or no loss in germination (Strider, 1973). To control *B. cinerea* on *L. sinuatum* cv. 'Midnight Blue', Diaz *et al.* (1999) harvested highest number of exportable bunches under the integrated method of control, *i.e.* Iprodine or Captan + *Trichoderma hamatum* at 6.5×106 spores/ml + cultural management instead of the individual treatments.

Leaf spot (*Cercospora insulana*) is a very serious disease of annual statice (*L. sinuatum* and *L. bonduelli*) at temperatures of 16-24 °C (Engelhard, 1975). In India *C. insulana* producing spots on leaf shoot was recorded long back (Ramakrishnan *et al.*, 1955). Braun *et al.* (2003) recorded two species of *Cercospora apii* (*C. statices* Pesante) and *C. insulana* (*C. statices* Lobik), and Nicoletti *et al.* (2003) recorded *C. insulana* causing circular brown spots on leaves and on the wings of scapes (not the scape proper) with darker edges surrounded by orange to reddish halo on *Limonium* spp. *C. insulana* is comparatively more menacing to plants bearing yellow floral colours. Such leaves yellow and senesce prematurely, growth of the plant is retarded and a part of panicle may also die in its severe attack. Maximum disease levels were recorded at 16 ° C. In a field trial to control the disease, Clorothalonil (Daconil W.P.) at 75 per cent gave the most effective control, followed by Propiconazole (Tilt E.C.) (Liu, 1991). From Punjab (India), Saini *et al.* (1989) recorded *Alternaria tenuis* (*A. alternata*) on *Limonium sinuatum* leaves. A new **crown rot, leaf and scape spot disease** (*Colletotrichum* sp.) on *Limonium sinuatum* has been found problematic and by which in the open or in the greenhouse there had been cent per cent crop loss of var. 'Gold Coast', a yellow flowering statice. Though it is highly menacing in the yellow type of statice, but also causes serious losses in white, pink and blue flowering varieties (Engelhard *et al.*, 1972). Spray of copper oxychloride (0.3 per cent) and maintaining proper aeration between the plants could minimize the disease incidence. ***Sclerotium rolfsii*** (*Corticium rolfsii*) also infects *Limonium*, and in its infection the plants either gradually show fire blight or may suddenly wilt and collapse even within 24 hours of detection. In its infection, there is complete leaf necrosis, casualty being up to 90 per cent but if crown and roots are not affected the plants may recover after treatment (Cox, 1972). Dithane M-45 or Z-78 spray at 0.2 per cent at its first appearance and then next at 5-7 days will control this disease. **Powdery mildew** (*Oidium* sp.) on perennial statice appears as numerous white, powdery mycelial colonies on the leaves, stems and branches of the plants (Hagiwara *et al.*, 1998). Karathane spraying at 0.05 per cent may control this problem. First occurrence of **downy mildew** (*Peronospora statices*) of statice (*Limonium bellidifolium* × *L. latifolium*) was reported by Koike *et al.* (1998). Its spread is favoured by copious rains in spring end. Species and cultivars of *Limonium* may differ in their susceptibility to the pathogen (Rapetti and Garibaldi, 1995). Symptoms observed in *L. sinuatum* are yellow chlorotic spots on the upper side of the lower leaves, followed by a mould on the lower side which is initially grey but grows darker. Infected leaves become necrotic and desiccated but remain attached to the stem. The compact cluster of leaves at the base of the stem impedes the penetration of pesticides whereas temperature range in the tunnels favours fungal development. It is advised to remove a few such basal leaves to reduce the humidity in the tunnel, followed by application of Dimethomorph alternately with copper solutions and dithiocarbamates and Cymoxanil (Polizzi, 2002). In a fungicide trial on *Limonium tataricum* to control the *Peronospora* disease, Ridomil Plus 50 WP (copper oxychloride + metalaxyl), Mikal C 64 WP (fosetyl) and Sandofan C (copper oxychloride + oxadixyl) gave moderate disease control (Szabo and Viranyi, 1990). In a study to explore the possibilities of downy mildew control on field grown German statice (*Limonium sinuatum*), only mixture of aluminium fosetyl (fosetyl)+fenamidone used at weekly intervals significantly inhibited the disease development (25 per cent of the plants had no disease symptoms). Mixture of Cymoxanil and Famoxate (famoxadone) also gave satisfactory results (Skrzypczak, 2006). Fungicide Furalaxyl has also been found to control an oospore-forming strain of this potentially highly destructive fungus (Hall *et al.,* 1997). A **stem canker** (*Phomopsis limonii* sp. nov.; Harvey *et al.,* 2000) of *L. perigrinum* × *L. purpuratum* cv. 'Chorus Magenta' was identified where initially the leaves turn yellow often only to one side of leaf blades. Though appearance of the disease may be noticed even at 10 °C, but higher temperatures (20-30 °C) are highly favourable for its rapid spread, and wound in the plants also helps entering this pathogen in the tissues. The disease develops blue-black colouration with dark fruiting bodies located below the epidermis of cankers on dying plants. In controlling this disease, no difference was found in the efficacy of Prochloraz and Tebuconazole. Overall, control of the disease was poor in all treatments although symptom development was delayed most when the fungicides were applied 1-2 h after inoculation (Long *et al.,* 2001). New occurrence of *Pestalotiopsis gracile* on statice (*Limonium* spp.) was first reported by Sato and Sawamoto (1997). **Rust** (*Uromyces limonii*) is serious on many of the statice species in greenhouses. *U. savulescui* has also been reported infecting *L. sinuatum* (Motokura *et al.,* 2001). Rust infections on leaves and stems usually appear as orange-brown pustules with swellings and deformations. Different species and cultivars of *Limonium* were tested for their reaction to the *U. limonii* isolate and all were found to be susceptible though no symptom appeared on *L. latifolia*, *L. perezii*, *L. otolepis* and *L. tataricum* (*Goniolimon tataricum*) (Longo *et al.,* 2003). To control the rust, preparations of dithiocarbamates, dodine, Bayleton 100 EC (triadimefon), Plantvax (oxycarboxin) and Funginex (triforine) are recommended (Bedlan, 1980). Spraying acibenzolar-S-methyl, azoxystrobin, bitertanol, kresoxim-methyl, myclobutanil, penconazole, trifloxystrobin, triforine and an isolate of the antagonistic fungus *Aphanocladium album* at weekly intervals showed that the new strobilurin fungicides gave good results, ergosterol biosynthesis inhibitors fungicides showed similar effect in reducing damage, and the new product acibenzolar-S-methyl & the isolate of *A. album* only partially controlled the disease with no phytotoxicity symptoms whatsoever (Pasini *et al.,* 2003).

Anthracnose (*Colletotrichum dematicum,* Hong *et al.,* 2006; *C. gloeosporioides*, Kagiwata, 1986) is the limiting factor in the commercial production of annual statice (*L. sinuatum* and *L. bonduelli*) (Engelhard, 1975), though *L. sinuatum* cvs 'Iceberg', 'Midnight Blue' and 'Roselight' are only slightly susceptible. It causes circular to irregularly-shaped reddish-brown lesions on leaves, stems and flowers, as well as crown rot in *Limonium bonduellii* cv. 'Gold Coast' (Sobers and Cox, 1973), *L. bellidifolium, L. sinuatum* and others. The disease was effectively controlled by spraying with benomyl 50 per cent W.P. at 0.25 per cent applied weekly (Sobers and Cox, 1973). In an experiment to study the benomyl resistant and sensitive populations of *Colletotrichum gloeosporioides* from statice, 64 isolates were isolated from 12 different locations where it was found that all the isolates belong to *C. gloeosporioides* complex. Of these isolates, 46 were resistant to benomyl at 10 micro g/ml and 18 were sensitive to this concentration of fungicide. Based on arbitrarily primed polymerase chain reaction of all isolates and internal transcribed spacer-1 sequence analyses of 12 selected isolates, the benomyl resistant and sensitive populations belonged to two distinct genotypes. Sequence analysis of the beta-tubulin genes, TUB1 and TUB2, of five sensitive and five resistant representative isolates of *C. gloeosporioides* from *Limonium* spp. revealed that the benomyl-resistant isolates had an alanine substitute instead of a glutamic acid at position 198 in TUB2. All data suggest that the resistant and sensitive genotypes are two independent and separate populations (Maymon *et al.,* 2006). **Root and basal rot** (*Phytophthora nicotianae, Sclerotinia sclerotiorum*) has been reported to infect *L. sinensis* (*L. tetragonum*) cv. 'Diamond' and *L. sinuatum* plants from the soil, causing leaf wilting and then collapse of the plant. Leaf bases initially show progressive necrotic areas, later turning dark brown to black, stem and root cortex become water-soaked and dark brown to black, rotted root tissues become brown with a characteristic black margin, and pith parenchyma turns grey-brown having a firm and wet rot. In plants with advanced disease symptoms, a cavity in the stem parenchyma is observed. On the basis of morphological features and isoenzyme genotyping, the causal organism was identified as *Phytophthora nicotinae* (Ilieva *et al.,* 2001). Basal rot of *Limonium sinensis* is also caused by *Sclerotinia sclerotiorum* (Abagnale and Nanni, 2007). Root and basal rot disease could be minimized by the spray of carbendazim or Ridomil at 0.05 per cent.

Bacterial wilt (*Pseudomonas caryophylli*) of *Limonium sinuatum* is symptomized as crown and leaf rot during hot season. In its infection, the affected leaves show chlorosis or wilting of entire leaf. The mid- and lateral veins of affected parts often turn red. Diseased plants yield isolates of a non-fluorescent, aerobic, gram negative bacterium with characteristics typical of *P. caryophylli* which affects carnation (Nishiyama *et al.,* 1988). **Bacterial leafspot** (*Pseudomonas andropogonis*) was for the first time recorded as the cause of lesions on *Limonium sinuatum* (Moffett *et al.,* 1986). Leaf and flower scape infected with bacterial leaf spot express water-soaked lesions which become reddish-brown, often coalesce and girdle the scape, resulting in death of the inflorescence (Anderson and Tisserat, 1994). **Leaf-tip necrosis** of micropropagated statice plantlets is a serious problem in commercial laboratories. Endophytic bacteria were detected in flower stalks collected from four different statice farms at frequencies ranging from 61 to 100 per cent. All plantlets regenerated from flower-stalk explants that tested free of endophytic bacteria as did not develop leaf-tip necrosis. The most frequently detected endophytic bacteria were *Alcaligenes* sp., *Pasteurella multocida* and *Stenotrophomonas maltophilia*, and most of such endophytic bacteria in statice plantlets were eliminated by the subculturing of plantlets on medium with augmentin, cefotaxime, or augmentin + cefotaxime. Those plantlets freed from endophytic bacteria by subculture on antibiotic-amended medium did not develop leaf-tip necrosis (Liu *et al*, 2005).

Mycoplasma-like organisms (MLO) were found in the phloem cells of ultra-thin sections of stems, leaves and pedicels of annual statice (*Limonium sinuatum*). The symptoms included yellowing and malformation of young leaves, leaf-reddening in older rosettes, bunching of floral stalks, phyllody and other floral abnormalities. Fourth- and 5th-instar nymphs of *Macrosteles fascifrons* were able to transmit the MLOs from diseased statice plants to healthy plants (Baker *et al.*, 1983). The other known phytoplasma vectors present in *Limonium* hybrid crops were *Circulifer haematoceps* (*Neoaliturus haematoceps*) complex, *Circulifer tenellus* complex, *Exitianus capicola* and *Orosius orientalis* (Weintraub *et al.*, 2004). In Israel, the phytoplasma and/or spiroplasma disease was identified in *Limonium* spp. causing leaf yellowing *cum* excessive branching coupled with precocious formation of long and narrow leaves, occasional production of leaf-like structures in place of flowers and production of small and/or white flowers, wherein six leafhoppers [*Austroagallia sinuata, Circulifer haematoceps* (*Neoliturus haemotoceps*), *C. tenellus, Exitianus capicola, Orosius orientalis* and *Psammotettix* spp.] suspected to be vector were trapped, out of which *C. haematoceps, C. tenellus* and *O. haematoceps* though tested positive for phytoplasma but as vector *O. orientalis* was confirmed transmitting mycoplama in *Limonium* (Gera *et al.*, 2006). A survey of *L. sinuatum* cultivars planted around for tolerance to MLOs revealed that 'Golden Coast' is most tolerant (Chang *et al.*, 1994). Changes in anatomy and cytology of conducting tissues of *Limonium sinuatum* plants affected by aster yellows phytoplasma were investigated. In the phloem tissues of affected plants, stem necrosis takes place. In necrotic regions, no sieve tubes are observed. The sieve tubes present on the border of necrosis show collapsed walls. Phytoplasma cells were observed in sieve tubes present in non- necrotic regions of the phloem. Various structural changes in sieve elements were investigated. The endoplasmic reticulum cisternae were often localized in the lumen of the sieve element without contact with the walls. Such localization of endoplasmic reticulum was never observed in healthy plants. Vesicles of different size, fuzzy material and clumping of p-proteins were characteristic for sieve elements from non-necrotic part of phloem. In infected plants there was a remarkable increase in phloem parenchyma cells with spiny vesicles (SV). Also the SV itself had not only a vesicular but also a tubular or extended cistern shape (Rudzinska and Kaminska, 2001a). In *L. sinuatum* plants with severe symptoms of aster yellows infection, phytoplasmas were present not only in the phloem but also in some cortex parenchyma cells. These parenchyma cells were situated at some distance from the conducting bundles. The phytoplasmas were observed directly in parenchyma cell cytoplasm. The cells with a small number of phytoplasmas show little pathological changes compared with the unaffected cells of the same zone of the stem as well with the cells of healthy plants. The cells filled with a number of phytoplasmas had their protoplast very much changed. The vacuole was reduced and in the cytoplasm a reduction of the number of ribosomes was noted and regions of homogenous structure appeared. Mitochondria were moved in the direction of the tonoplast and plasma membrane. Compared to the cells unaffected by phytoplasma, the mitochondria were smaller and had an enlarged cristae internal space. The chloroplasts from affected cells had a very significant reduction in size and the thylakoids system had disappeared (Rudzinska and Kaminska, 2001b). MLOs could be controlled by clean cultivation, removal of infected plants and regular spray of insecticides to check the vector population.

There are number of **viruses** infecting *Limonium* species. Two strains of a potyvirus from statice were identified as 'clover yellow vein virus' (**CYVV**). The host ranges of both isolates were more similar to the white clover isolate of CYVV from Pratt (CYVV-P) than to bean yellow mosaic virus from gladiolus (BYMV-G, BYMV-G82-25) and red clover (BYMV-204-1). CYVV-St was partially purified from *Chenopodium quinoa* with chloroform, carbon tetrachloride extraction, and differential centrifugation (Lawson *et al.*, 1985). Characterization and identification of 'cucumber mosaic cucumovirus' (**CMV**) on *L. sinuatum* have been reported by Matsumoto *et al.* (1997) but the transmission of mosaic is not via seeds (Vovlas and Cillo, 1996). Chlorotic and white necrotic spots on leaves appear together, showing mosaic pattern and such spots and pattern multiply naturally on the same plant (Inouye *et al.*, 1991) and due to infection such plants show leaf mosaic, plant stunting and absence of flowers. In its infection plants become stunted, flowers are either absent or a few and freakish and leaves show mosaic symptoms. '**Statice virus Y**' infects *L. sinuatum*. This weakly immunogenic virus was serologically related to 'bean yellow mosaic' (BYMV), 'clover yellow vein (CYVV), 'bean common mosaic', '*Gloriosa stripe mosaic virus*' and to 'potato virus Y'. Properties of the virus are intermediate to BYMV and CYVV (Lesemann *et al.*, 1979).'Tobacco rattle virus' (**TRV**) also infects statice. This virus is isolated from plants with bright yellow or red line patterns and ringspots on the leaves. Its serological relationship was found with TRV isolates from Europe and Brazil (Dijkstra and Van Dijke, 1981). Incidence of 'tomato ringspot nepovirus' (**ToRSV**) on *Limonium* spp. is very high. It was isolated and identified by the methods of test-plants, electron microscopy and ELISA indirect double sandwich technique (Navalinskiene and Samuitiene, 2000). 'Tomato ringspot virus' (**TomRV**) was isolated and identified from *Limonium* sp. (Navalinskiene and Samuitiene, 1999). 'Tomato spotted wilt tospovirus' (**TSWV**) was recorded from *L. tartaricum* (Ramasso *et al.*, 1994; Samuitiene *et al.*, 2003) and *L. sinuatum* (Jorda *et al.*, 1995). The incidence is commonly associated with 'Western flower thrips' which can be kept at bay by removing or avoiding the TSWV sources. '***Tombusvirus***' is a new soil-borne viral disease reported by Krczal *et al.* (1992). It was initially isolated from *Limonium sinuatum*. Infected plants show poor growth, distortion, mosaic and necrosis of leaves and poor flowering. The immuno-electron microscopy using antisera to 14 different *tombusviruses* or tombuslike-viruses, four groups could be distinguished. Subsequent sequence analysis of RT-PCR products covering the coat protein gene provided additional data which indicated that six of the isolates belong to one of the following *tombusvirus* species: Petunia asteroid mosaic virus, Grapevine Algerian latent virus or *Cymbidium* ringspot virus. The seventh isolate represents a new virus species, for which the name *Limonium* flower distortion virus is proposed (Verhoeven *et al.,* 2006). The two viruses, '**turnip mosaic virus**' and '**cucumber mosaic virus**' identified by host plant reaction, serology and electron microscopy, were among at least four isolated from *Limonium sinuatum* with pronounced symptoms (Hein *et al.,* 1976). In a 2-year field studies, turnip mosaic virus was transmitted from statice to statice by four aphid species. Disease incidence was high (67 per cent) in the first year and low (6 per cent) in the second, the differences being due to differences in aphid vector populations, *vis-à-vis* the proximity to severely diseased fields (Laird and Dickson, 1972). 'Bean yellow mosaic virus' (**BYMV**) has also been recorded infecting statice. Proper field sanitation by weed removal, proper control of aphids, thrips and other insect-pests, immediate rogueing of virus-infected plants from the field and their burning, and use of planting material from certified virus-free mother stocks are some of the measures which may keep the viruses at bay.

References

Abagnale, A. and B. Nanni, 2007. *Limonium sinensis,* new host of *Sclerotinia sclerotiorum. Colture Protette,* **36**(6): 91-92.

Airo, M., S. Aprile, G. Zizzo and A. Castelli, 2004. Seed germination of different *Limonium* species of the Sicilian flora (Italian). *Italus Hortus,* **11**(4): 163-165.

Akaydin, G. 2007. A new species of *Limonium* Mill. (Plumbaginaceae) from the Central Anatolian salt steppe, Turkey. *World App. Sci. J.,* **2**(4): 406-411.

Alarcon, J.J., M.A. Morales, A. Torrecillas and M.J. Sanchez-Blanco, 1999. Growth, water relations and accumulation of organic and inorganic solutes in the halophyte *Limonium latifolium* cv. Avignon and its interspecific hybrid *Limonium caspia* x *Limonium latifolium* cv. Beltlaard during salt stress. *J. Pl. Physiol.,* **154**(5/6): 795-801.

Aly, M.A.M., B. Rathinasabapathi and S. Bhalsod, 2002a. Somatic embryogenesis in members of the Plumbaginaceae ornamental statice *Limonium* and sea thrift *Armeria maritima. HortSci.,* **37**(7): 1122-1123.

Aly, M.A.M., B. Rathinasabapathi and K. Kelley, 2002b. Somatic embryogenesis in perennial statice *Limonium bellidifolium,* Plumbaginaceae. *Plant Cell Tissue Org. Cult.,* **68**(2): 127-135.

Amo Marco, J.B. and M.R. Ibanez, 1998. Micropropagation of *Limonium cavanillesii* Erben, a threatened statice, from inflorescence stems. *Pl. Gr. Reg.,* **24**(1): 49-54.

Amos, J. 1980. Annual cut flowers, commercial production. Ministry of Agriculture and Fisheries. Media Services. *Aglink* HPP 196.

Anderson, D. and N. Tisserat, 1994. Bacterial leaf spot of statice caused by *Pseudomonas andropogonis. Plant Dis.,* **78**(12): 1218.

Aniya, Y., C. Miyagi, A. Nakandakari, S. Kamiya, N. Imaizumi and T. Ichiba, 2002. Free radical scavenging action of the medicinal herb *Limonium wrightii* from the Okinawa islands. *Phytomedicine,* **9**(3): 239-244.

Anonymous, 1993. Variety: 'Pink Emille'. Application no. 92/128 (Burbank Biotechnology Pty Ltd.). *Pl. Var. J.,* **6**(4): 23-24, 37.

Anonymous, 1993a. Variety: 'Daicean' synonym: 'Ocean Blue'. Application no. 92/057 (Daiichi Seed Co. Ltd.). *Pl. Var. J.,* **6**(4): 20, 36.

Anonymous, 1993b. Variety: 'Emille'. Application no. 91/028 (Miyoshi & Co. Ltd.). *Plant Varieties Journal,* **6(4)**: 10-12, 33.

Anonymous, 1993c. Variety: 'Oceanic Blue' filed in the Netherlands as 'Misty Blue'. Application no. 92/058 (Daiichi Seed Co. Ltd.). *Pl. Var. J.,* **6**(4): 20-21, 37.

Anonymous, 1994. Variety: 'Ballerina Rose'. Application no. 90/056 (New Zealand Ministry of Agriculture). *Pl. Var. J.,* **7**(3): 9-10, 25.

Anonymous, 1996. Variety: 'Tall Emille' syn LC.00281I. Application (Miyoshi & Co.Ltd.) no: 94/154. *Pl. Var. J.,* **9**(3): 34.

Anonymous, 1997. Variety: 'Oceanic White'. Application (Daiichi Seed Co. Ltd.) no: 92/059. *Pl. Var. J.,* **10**(4): 31.

Aparicio, A. 2005. *Limonium silvestrei* (Plumbaginaceae), a new agamospecies from southern Spain. *Annales Botanici Fennici,* **42**(5): 371-377.

Aragon, A.and M.A.Moron, 2000. Description of third-instar larva of two species of *Phyllophaga* (Coleoptera: Melolonthidae).*Canadian Entomologist,* **132**(3): 323-332.

Arisco, M. and J.S. Koths, 1984. Pulse and continual fertilization of bedding plants. *Connecticut Greenhouse N.L.,* No. 121, pp. 12-15.

Artelari, R. and O. Georgiou, 2002. Biosystemic study of the genus *Limonium* (Plumbaginaceae) in the Aegean area, Greece. III. *Limonium* on the islands Kithira and Antikithira and the surrounding islets. *Nordic J. Bot.,* **22**(4): 483-501.

Asen, S., K.H. Norris, R.N. Stewart and P. Semeniuk, 1973. Effect of pH, anthocyanin, and flavonoid co-pigments on the color of statice flowers. *J. Amer. Soc. hort. Sci.,* **98**(2): 174-176.

Azuma, A, J. Shimasaki and S. Inubushi, 1983. Acceleration of flowering of statice (*Limonium sinuatum* Mill.) by seed vernalization (Japanese). *J. Japanese Soc hort.Sci.,* **51**(4): 466-474.

Bach, A and B. Pawlowska, 1992. Micropropagation of perennial species of limonium (*Limonium tataricum* L., *L. latifolium* (Sm.) O. Ktze) (Polish). *Zeszyty Naukowe Akademii Rolniczej im Hugona Kollataja w Krakowie, Ogrodnictwo,* Bull. No. 20, pp. 3-13.

Baker, K.K., S.K. Perry, T.M. Mowry and J.F. Russo, 1983. Association of a mycoplasma-like organism with a disease of annual statice in Michigan. *Plant Dis.*, **67**(6): 699-701.

Ban-QiaoYing, Liu-GuiFeng, Wang-YuCheng, Zhang-DaWei, and Jiang-LiLi, 2008. Cloning and expression of a novel metallothionein gene LbMT2 from *Limonium bicolor* (Chinese). *Hereditas Beijing*, **30**(8): 1075-1082.

Bashir, A.K., E.S. Hassan, A.A. Abdalla and I.A. Wasfi, 1992. Antimicrobial activity of certain plants used in the folk-medicine of United Arab Emirates. *Fitoterapia*, **63**(4): 371-375.

Bashir, A.K., A.A. Abdalla, I.A. Wasfi, E.S. Hassan, M.H. Amiri and T.A. Crabb, 1994. Flavonoids of *Limonium axillare*. *Intern. J. Pharmac.*, **32**(4): 366-372.

Bedlan, G., 1980. Rust fungi of some ornamental plants. *Pflanzenarzt*, **33**(12): 118-119.

Braun, U., C.F. Hill and M. Dick, 2003. New cercosporoid leaf spot diseases from New Zealand. *Australasian Pl. Path.*, **32**(1): 87-97.

Bruna, S., L. de Benedetti, A. Mercuri, T. Schiva, G. Burchi, N. Pecchioni and C. Agrimonti, 2004. Use of RAPD markers for the genetic characterization of *Limonium* species. *Acta Hort.*, No. 651(2), pp. 155-160.

Bruna, S., G. Burchi, L. de Benedetti, A. Mercuri, N. Pecchioni, C. Bianchini and T. Schiva, 2005. Molecular analysis of *Limonium* genus through RAPD markers. *Agricoltura Mediterrfanea*, **135**(1): 52-58

Burchi, G. 2002. Genetic characterization of ornamental germplasm through biotechnological approaches (Italian). *Italus Hortus*, **9**(5): 28-33.

Burchi, G., A. Mercuri, S. Bruna, L. de Benedetti, C. Bianchini, R. Bregliano, G. Foglia and T. Schiva, 2003. 'SOI 50', a new variety of *L. tataricum* obtained by the Experimental Institute for Floriculture, to celebrate the 50th Anniversary of the Italian Society of Horticulture (S.O.I.). *Italus Hortus*, **10**(4): 56-59.

Burchi, G., E. Mercatelli, M. Maletta, A. Mercuri, C. Bianchini and T. Schiva, 2006. Results of a breeding activity on *Limonium* spp. *Acta Hort.*, No. 714, pp. 43-49.

Burge, G.K., E.R. Morgan, J.F. Seelye, J.E. Grant, C. Zhang and M.E. Hopping, 1995. Generation of novel forms of *Limonium*. *Acta Hort.*, No. 420, pp. 78-80

Burge, G.K., E.R. Morgan, I. Konczak and J.F. Seelye, 1998. Postharvest characteristics of *Limonium* 'Chorus Magenta' inflorescences. *N.Z. J. Crop & hort. Sci.*, **26**(2): 135-142.

Cardarelli, M., M. Temperini, D. Vitti and F. Saccardo, 2002. Gamma rays induced mutagenesis in *Limonium sinuatum* (Italian). *Italus Hortus*, **9**(5): 85-87.

Carter, C.T. and C.M. Grieve, 2006. Salt tolerance of floriculture crops. In: *Ecophysiology of High Salinity Tolerant Plants*, pp. 279-287(eds Ajmal Khan, M. and D.J. Weber). Springer Science +Business Media, Dordrecht, Netherlands.

Casazza, G., M. Savona, S. Carli, L. Minuto and P. Profumo, 2002. Micropropagation of *Limonium cordatum* (L.) Mill. for conservation purposes. *J. hort. Sci. Biotech.*, **77**(5): 541-545.

Castro, M. and J.A. Rossello, 2007. Karyology of *Limonium* (Plumbaginaceae) species from the Balearic Islands and the westertn Iberian Peninsula. *Bot. J. Linnean Soc.*, **155**(2): 257-272.

Chamorro, A.H., S.L. Martinez, J.C. Fernandez and T. Mosquera, 2007. Evaluation of different concentrations of some plant growth regulators on *in vitro* multiplication and rooting of *Limonium* var. Misty Blue (Spanish). *Agronomia Colombiana*, **25**(1): 47-53.

Chang, K.F., S.F. Hwang and M. Mirza, 1994. Symptomatology and the occurrence of mycoplasma- like organisms disease of statice (*Limonium sinuatum*) in Alberta, Canada. *Zeitschrift für Pflanzenkrankheiten und Pflanzenschutz*, **101**(2): 113-123.

Chaung,S.S., C.C. Lin, J.L. Lin, K.H. Yu, Y.F. Hsu and M.H. Yen, 2003. The hepatoprotective effects of *Limonium sinense* against carbon tetrachloride and beta -*D*-galactosamine intoxication in rats. *Phytotherapy Res.*, **17**(7): 784-791.

Cherry, R.J., 1999. Variety: 'Cosita'. Application no. 97/233. *Pl. Var. J.*, **12**(1): 34-35.

Chimonidou, P.D. 2001. Cultivation of four *Limonium* species for fresh and dry flower production. *Technical Bulletin Cyprus Agricultural Research Institute*, No. 211, p. 8.

Cohen, A, A., H.D. Harazy, Rabinowitch and R. Stav, 1995. Selection for early flowering in blue statice (*Limonium sinuatum* Mill.). *Acta Hort.*, No. 420, pp. 118-124.

Colinas,L. M.T., R.R. Garnica and R.A. Curiel, 2002. Plastic films in six ornamental species in bouquet:changes in appearance, fresh weight and ethylene production. *Revista Chapingo Serie Horticultura*, **8**(2): 211-221.

Cowan, R., M.J. Ingrouille and M.D. Lledo, 1998. The taxonomic treatment of agamosperms in the genus *Limonium* Mill. (Plumbaginaceae). *Folia Geobotanica*, **33**(3): 353-366.

Cox, R.S., 1972. *Sclerotium rolfsii* on statice. *Plant Dis. Reoprter*, **56**(8): 656-657.

Dawson, H.J. 1990. Chromosome numbers in two *Limonium* species. *Watsonia*, **18**(1): 82-83.

Deshpande, G.M., P.C. Sonawane and Manjul Dutt, 2001. Effect of plant density on growth, flowering and yield of statice (*Limonium sinuatum*). *J. Appl. Hort., Lucknow*, **3**(2): 98-99.

Diaz, N.C., M.J. Barrera.and *E.G.*de Granada, 1999. Controlling *Botrytis cinerea* Pers. in statice (*Limonium sinuatum* Mill.) 'Midnight Blue' cultivar. *Acta Hort.*, No. 482, pp. 235-238.

Dijkstra, J. and H.D.Van Dijke, 1981. Line pattern in *Limonium latifolium* caused by *tobacco rattle virus. Netherlands J. Pl. Path.*, **87**(2): 35-44.

Ding Feng and Wang BaoShan, 2006. Effect of NaCl on salt gland development and salt-secretion rate in the leaves of *Limonium sinense* (Chinese). *Acta Botanica Boreali Occidentalia Sinica*, **26**(8): 1593-1599.

Dogan, M., H. Duman and G. Akaydn, 2008. *Limonium gueneri* (Plumbaginaceae), a new species from Turkey. *Annales Botanici Fennici*, **45**(5): 389-393.

Doi, M.and M.S. Reid, 1995. Sucrose improves the postharvest life of cut flowers of a hybrid *Limonium. HortScience*, **30**(5): 1058-1060.

Dulberger, R. 1975. Intermorph structural differences between stigmatic papillae and pollen grains in relation to incompatibility in Plumbaginaceae. *Proc. roy. Soc. London-B*, **188**(1092): 257-274.

Engelhard, A.W., 1975. Etiology, symptomatology, and economic importance of the diseases of annual statice (*Limonium* sp.). *Pl. Dis. Reptr*, **59**(7): 551-555.

Engelhard, A.W., C.M. Howard and G.J. Wilfret, 1972.A new crown rot, leaf and scape spot disease of statice (*Limonium sinuatum*) incited by *Colletotrichum* spp. *Pl. Dis. Reptr*, **6**(10): 894-895.

Enrico, F., D.G. Carla and S. Eleonora, 2000. Effects of low temperatures and photoperiod on flowering of *Limonium gmelinii. Acta Hort.*, No.541, pp. 193-199. Paper presented in the IVth intern. Symp. on New Floricultural Crops, held on May 22-27, 1999 at Chnia, Crete (Greece).

Erben, M. 2001. Comments on the taxonomy of the genus *Limonium*. VII. (German). *Sendtnera,* Mitteilungen der Botanischen Staatssammlung und des Instituts für Systematische Botanik der *Universitat Munchen*, **7**: 53-84, held at Munchen University, Germany.

Erben, M. and V.J. Aran, 2005. *Limonium mateoi* (Plumbaginaceae), a new species from Central Spain. *Annales del Jardin Botanico de Madrid*, **62**(1): 3-7.

Erben, M. and L. Mucina, 2006. *Limonium failachicum* (Plumbaginaceae) - new and so far the only endemic plant from Kuwait. *Folia Geobotanica*, **41**(2): 229-235.

Evliyaoglu, N., M. Kargoglu, E. Martin, M. Temel and O. Cetin, 2008. The karyotype of three *Limonium* Miller species in the family Plumbaginaceae conducted using Image Analysis System. *Int. J. Bot.*, **4**(2): 213-218.

Fan-XiaoFeng, Yang-YingLi, Liu-JunMei, Wang-Lai and Li-KeWen, 2007. Callus induction and plant regeneration of *Limonium aureum* (Chinese). *Acta Botanica Boreali Occidentalia Sinica*, **27**(2): 257-261.

Farina, E., M. Assenza and G. Donzella, 1993.Extending the production period for statice. *Colture Protette*, **22**: 4, 69-72.

Farina, E., C. dalla Guda, E. Scordo, S. Castello, T. Paterniani and M. Palagi, 2003. Advanced production techniques for *Limonium gmelinii. Acta Hort.*, No. 614(vol. 1), pp. 109-113.

Farina, E., T. Paterniani and M. Palagi, 1995. New ornamental crops of *Limonium*: I. Bioagronomic evaluation of *Limonium perezii* (Italian). *Italus Hortus*, **2**(5): 58-62.

Fascella, G and G.V. Zizzo, 2004. Influence of greenhouse roof opening system on internal climate and limonium yield control. *Acta Hort.*, No. 659, vol. 1, pp. 171-176.

Febles, R. and E. Perez-Rodriguez, 2004. Karyotypic analysis of *Limonium* Mill. Section Pteroclados Boissd. Subsection Nobiles Boiss. (Plumbaginaceae) (Spanish). *Botanica Macaronesica*, **25**: 79-94.

Fior, C.S., L.R. Rodrigues and A.N. Kampf, 2000. *In vitro* propagation of *Limonium latifolium* Kuntze (Plumbaginaceae) (Portuguese). *Ciencia Rural*, **30**(4): 575-580.

Franco, J.A., V. Cros, S. Banon and J.J. Martinez Sanchez, 2002. Nursery irrigation regimes and establishment irrigation affect the postplanting growth of *Limonium cossonianum* in semiarid conditions. *Israel J. Pl. Sci.*, **50**(1): 25-32.

Funnell, K.A., M. Bendall, W.F. Fountain and E.R. Morgan, 2003. Maturity and type of cutting influences flower yield, flowering time, and quality in *Limonium* 'Chorus Magenta'. *New Zealand J. Crop & hort. Sci.*, **31**(2): 139-146.

Garner, J.M. and A.M. Armitage, 1996. Gibberellin applications influence the scheduling and flowering of *Limonium* x 'Misty Blue'. *HortSci.*, **31**(2): 247-248.

Georgakopoulou, A., S. Manousou, R. Artelari and O. Georgiou, 2006. Breeding systems and cytology in Greek populations of five *Limonium* species (Plumbaginaceae). *Willdenowia*, **36**(2): 741-750.

Gera, A., L. Maslenin, A. Rosner, M. Zeidan and P.G. Weintraub, 2006. Phytoplasma diseases in ornamental crops in Israel. *Acta Hort.* (*Proceedings of the XIth Int. Symp. on Virus Dis. of Ornam. Plants*, held at Taichung, Taiwan, on March 9-14, 2004; ed. Chang, C.A.), No. 722, pp. 155-161.

Gilreath, J.P. 1985. Response of statice to selected herbicides. *Proc. Southern Weed Science Soc.*, 38th annual meeting, p. 135.

Gilreath, J.P., 1986. Response of gladiolus, statice and gypsophila to residues of preemergence herbicides. *Proc. Fl. St. hort. Soc.*, **99**: 275-278.

Goto, T., C. Yamato, Y. Kageyama and K. Konishi, 1996. Forcing culture of *Limonium sinuatum* Mill. using cutting plants (cuttings). *Scientific Reports of the Faculty of Agriculture*,Okayama University,No. 85, pp. 31-37.

Guda, C. dalla, B. Ruffoni, C. Cervelli and E. Farina, 1998. *Limonium otolepis* as a new ornamental crop. *Acta Hort.*, No. 454, pp. 289-296.

Hagiwara, H., K. Abiko, S. Izutsu, A. Tomikawa, K. Kuroda and J. Okamoto, 1998. Powdery mildew of perennial statice (*Limonium* spp.) in Japan. *Ann. phytopath. Soc. Japan*, **64**(5): 506-509.

Hall, G.S., C.R. Lane and J.R. Mellor, 1997. An oospore-forming strain of *Peronospora statices* on cultivated *Limonium* in the UK, the Netherlands and Italy. *European J. Pl. Path.*, **103**(5): 471-475.

Hamilton, M.B. and D.M. Rand, 1996. Relatedness measured by oligonucleotide probe DNA fingerprints and an estimate of the mating system of sea lavender (*Limonium carolinianum*). *Theor. Appl. Genet.*, **93**(1/2): 249-256.

Harazy, A., B. Leshem, A. Cohen and H.D. Rabinowitch, 1985. *In vitro* propagation of statice as an aid to breeding. *HortSci.*, **20**(3): 361-362.

Harm, U. and A. Maync, 1985. A study on increased potassium for sea lavender (German). *Deutscher Gartenbau*, **39**(24): 1174-1175.

Harvey, I.C., E.R. Morgan and G.K. Burge, 2000. A canker of *Limonium* sp. caused by *Phomopsis limonii* sp. nov. *New Zealand J. Crop & hort. Sci.*, **28**(1): 73-77.

Hatterman, H.M., P.J. Shea and E.T. Paparozzi, 1987. Weed control in statice. *Weed Sci.*, **35**(3): 373-376.

Hein, A., R. Koenig, D.E. Lesemann and G. Querfurth, 1976. Turnip mosaic virus and cucumber mosaic virus in cultivated statice in south Germany. *Zeitschrift für Pflanzenkrankheiten und Pflanzenschutz*, **83**(4): 229-233.

Hong, C.F., P.F.L. Chang, J.Y. Chang and J.W. Huang, 2006. Identification for the causal agent of caspia anthracnose and its pathogenicity tests. *Plant Path. Bull.*, No. 15(4), pp. 241-249.

Hosni, A.M., Y.A. Hosni and M.A. Ebrahim, 2000. *In vitro* micropropagation of *Limonium sinuatum* 'Citrom Mountain', a hybrid statice newly introduced in Egypt. *Ann. agric. Sci. Cairo*, **45**(1): 327-339.

Hsu-NaiWen, Liu-TsuHwie, Huang-ChaoLang and Wu-ReyYuh, 2000. The micropropagation of *Limonium wrightii* (Hance) O. Kuntze (Chinese). *J. Chinese Soc. hort. Sci.*, **46**(3): 277-286.

Huang, C.L., M.T. Hsieh, W.C. Hsieh, A.P. Sagare and H.S. Tsay, 2000. *In vitro* propagation of *Limonium wrightii* (Hance) Ktze. (Plumbaginaceae), an ethnomedicinal plant, from shoot-tip, leaf- and inflorescence-node explants. *In Vitro Cell. Dev. Biol. Plant*, **36**(3): 220-224.

Huh, K.Y., W.S. Kim, S.J. Jang and H.K. Lee, 1995. Flowering acceleration of chilled statice (*Limonium sinuatum*) by gibberellic acid. *RDA J. agr. Sci., Hort.*, **37**(2): 427-431.

Ichimura, K. 1998. Improvement of postharvest life in several cut flowers by the addition of sucrose. *Japan agr. Res. Quart.*, **32**(4): 275-280.

Igawa, T., Y. Hoshino and M. Mii, 2002.Efficient plant regeneration from cell cultures of ornamental statice, *Limonium sinuatum* Mill. *In Vitro Cell. dev. Biol. Plant*, **38**(2): 157-162.

Ilieva, E., W.A.M. Veld, B.B.F. Wessels and R.P. Baayen, 2001. First report of *Phytophthora nicotianae* on *Limonium* in Europe. *Plant Dis.*, **85**(4): 445.

Inouye, N. T. Maeda, H. Huttinga *and K. Mitsuhata*, 1991. Comparison of two strains of cucumber mosaic virus isolated from a single statice plant. *Nogaku Kenkyu*, **62**(3): 209-223.

Ivanova, I., 1999a. Effect of some herbicide systems on flower production of statice (*Limonium sinuatum*). *Rasteniev"dni-Nauki*, **36**(4): 212-215.

Ivanova, I., 1999b. Weed control in *Limonium sinuatum*. *Rasteniev"dni-Nauki*, **36**(3): 182-185.

Jeong, J.H., H.N. Murthy and K.Y. Paek, 2001. High frequency adventitious shoot induction and plant regeneration from leaves of statice. *Plant Cell Tissue Org Cult.*, **65**(2): 123-128.

Jorda, C., A. Ortega and M.Juarez, 1995. New hosts of tomato spotted wilt virus. *Plant Dis.*, **79**(5): 538.

Kagiwata, T. 1986. An anthracnose of statice (*Limonium sinuatum*) caused by *Colletotrichum gloeosporioides* Penzig. *J. agr. Sci.*, **31**(2): 101-110.

Karis, P.O. 2004. Taxonomy, phylogeny and biogeography of *Limonium* sect. Pteroclados (Plumbaginaceae), based on morphological data. *Bot. J. Linnean Soc.*, **144**(4): 461-482.

Katsutani, N., S. Kajihara, Y. Kawamoto and H. Hara, 1998. Forcing of statice (*Limonium sinuatum* Mill.) by low temperature treatment of seedlings with illumination to prevent their rot. *Bulletin of the Hiroshima Prefectural Agriculture Research Center*, No. 66, pp. 53-59.

Khan, M.A. and Zia Sabahat, 2007. Alleviation of salinity by sodium hypochlorite on seed germination of *Limonium* seeds. *Pakistan J. Bot.*, **39**(2): 503-511.

Kimizu, M., Y. Yamamoto, T. Iwasaki and S. Ohki, 2001. *Agrobacterium* mediated transient expression of beta -glucuronidase and luciferase genes in *Limonium sinuatum*. *Acta Hort.*, No. 560, pp. 181-184.

Koike, S.T., P.A. Nolan, S.A. Tjosvold and K.L.Robb, 1998. First occurrence of downy mildew of statice, caused by *Peronospora statices*, in California and the rest of the United States. *Plant Dis.*, **82**(5): 591.

Krczal, G., M. Beutel, V. Mokra (ed.), A. Brunt (ed.), T. Derks (ed.) and A. van Zaayen, 1994. A new soil-borne virus disease in statice (*Gonolium tartaricum*). *Eighth int. Symp. on Virus Diseases of Ornam. Plants*, held in Prague (Czech Republic) on Aug. 24-28, 1992. *Acta Hort.*, No. 377, pp. 115-122.

Krizek, D.T. and P. Semeniuk, 1972. Influence of day/night temperature under controlled environments on the growth and flowering of *Limonium* 'Midnight Blue'. *J. Amer. Soc. hort. Sci.*, **97**(5): 597-599.

Kühns, L.J., C. Haramaki, J. Becker and G. Lyman, 1986. Preemergence and postemergence grass herbicides on annual flowers. *Proc. 40th ann. Meet. NE Weed Sci. Soc.*, pp. 260-262.

Kumari, D.L.C. and S.E. Peiris, 2000. Preliminary investigation of preservation methods to produce dried flowers of rose and statice. *Tropical agr. Res.*, **12**: 416-422.

Kunitake, H. and M. Mii, 1990. Plant regeneration from cell culture-derived protoplasts of statice (*Limonium perezii* Hubbard). *Plant Sci. Limerick*, **70**(1): 115-119.

Kunitake, H., K. Koreeda and M. Mii, 1995. Morphological and cytological characteristics of protoplast-derived plants of statice (*Limonium perezii* Hubbard). *Scientia Hort.*, **60**(3/4): 305-312.

Laird, E.F., Jr. and R.C. Dickson, 1972. Turnip mosaic virus vector relationships in field grown statice, *Limonium perezii*. *Plant Dis. Reptr*, **56**(8): 722-725.

Lamont, G.P. 1986. Herbicide evaluation trials on cut flowers. *Australian Horticulture*, **84**(7): 89-91.

Lamont, G.P. and M.A. O'Connell, 1986. An evaluation of pre-emergent herbicides in field-grown cut flowers. *Plant Prot. Quart.*, **1**(3): 95-100.

Lawson, R.H., M.D. Brannigan and J. Foster, 1985. Clover yellow vein virus in *Limonium sinuatum*. *Phytopath.*, **75**(8): 899-906.

Lee-EunJu and Jeong-ByoungRyong, 1999a. Effect of initial phosphorus level in the medium on the in vitro plantlet growth of *Limonium* spp. 'Misty Blue' (Korean). *J. Korean Soc. hort. Sci.*, **40**(2): 253-256.

Lee EunJu and Jeong-ByoungRyong, 1999b. Growth of *Limonium* 'Misty Blue' as affected by culture environment *in vitro* and level of shading during *ex vitro* acclimatization (Korean). *J. Korean Soc. hort Sci.*, **40**(5): 623-626.

Lellau, T.F. and G. Liebezeit, 2003. Cytotoxic and antitumor activities of ethanolic extracts of salt marsh plants from the lower Saxonian Wadden Sea, Southern North Sea. *Pharmaceut. Biol.*, **41**(4): 293-300.

Lesemann, D.E., R. Koenig and A. Hein, 1979. Statice virus Y - a virus related to bean yellow mosaic and clover yellow vein viruses. *Phytopathologische Zeitschrift*, **95**(2): 128-139.

Leshem, B. and A. Cohen, 1983. *In vitro* propagation of statice as an aid to breeding. *Hassadeh*, **64**(2): 318-319.

Lewis, D.H. and N.K. Borst, 1993. Evaluation of cool storage and preservative solution treatments on the display life of *Limonium perigrinum* inflorescences. *New Zealand J. Crop hort. Sci.*, **21**(4): 359-365.

Lewis, D., M. Prasad, N. Borst and M. Spiers, 1994. Effect of N-P-K fertiliser on the growth of *Limonium perigrinum* 'Ballerina Rose'. *N.Z. J. Crop & hort. Sci.*, **22**(2): 217-220.

Li-HongYan, Yang-ChuanPing, Wang-YuCheng and Wang-BingFeng, 2007. Cloning and sequence analysis of two metallothionein genes from *Limonium bicolor* (Chinese). *J. NE Forestry Univ.*, **35**(8): 1-5.

Liu, H.L. 1991. Occurrence of *Cercospora* leafspot of statice and its chemical control. *Bull. Taichung District agric Impr. St.*, No. 33, pp. 25-35.

Liu, T.H.A., N.W. Hsu and R.Y. Wu, 2005. Control of leaf-tip necrosis of micropropagated ornamental statice by elimination of endophytic bacteria. *In Vitro Cell. dev. Biol., Plant*, **41**(4): 546-549.

Liu-TsuHwie, Hsu-NaiWen and Wu-ReyYuh, 2004. Factors effecting vitrification of tissue-cultured statice (Chinese). *J. Chinese Soc. hort. Sci.*, **50**(2): 167-173.

Li-Yan, 2008. Effect of salt stress on seed germination and seedling growth of three salinity plants. *Pakistan J. biol. Sci.*, **11**(9): 1268-1272.

Li, Y. 2008. Kinetics of the antioxidant response to salinity in the halophyte *Limonium bicolor*. *Plant Soil Environ.*, **54**(11): 493-497.

Li-Yi, Feng-HuYuan, Chen-Tuo, Yang-XiaoMing and An-LiZhe, 2007. Physiological responses of *Limonium aureum* seeds to ultra-drying. *J. Integrative Pl. Biol.*, **49**(5): 569-575.

Lledo, M.A., M.B. Crespo and J.B. del Amo-Marco, 1993. Preliminary remarks on micropropagation of threatened *Limonium* species (Plumbaginaceae). *Bot. Gdns Microprop. News*, **1**(6): 72-74.

Long, P.G., K.A. Funnell, W.F. Fountain, M. Bendall and E.R. Morgan, 2001. Control of a stem canker caused by *Phomopsis limonii* on *Limonium* 'Chorus Magenta'. *New Zealand J. Crop & hort. Sci.*, **29**(4): 247-253.

Longo, O., A. Ambrico, A. Siniscalco and F. Ciccarese, 2003. Reaction of *Limonium* spp. towards rust caused by *Uromyces limonii*. *Italus Hortus*, **10**(6): 49-51.

Lopes, M.S., E.R.T. Stumpf and F.I.F. de Carvalho, 2003. Substrate effect on the asexual reproduction of *Limonium brasiliense* (Boiss.) O. Kuntze (Portuguese). *Revista Brasileira de Agrociencia*, **9**(4): 421-424.

Lopez, D., P. Cabot and R. Molina, 1996. *Trachelium*, *Limonium* and *Lisianthus*: study on the advancement of flowering (Spanish). *Horticultura Revista de Hortalizas Flores y Plantas Ornamentales*, No. 116, pp. 118-120.

Luan, L.Q., N.Q. Hien, N. Nagasawa, T. Kume, F. Yoshii and T.M. Nakanishi, 2003. Biological effect of radiation-degraded alginate on flower plants in tissue culture. *Biotech. App. Biochem.*, **38**(3): 283-288.

Mahasneh, A.M. 2002. Screening of some indigenous Qatari medicinal plants for antimicrobial activity. *Phytotherapy Res.*, **16**(8): 751-753.

Martin, C. and C. Perez, 1995. Micropropagation of five endemic species of *Limonium* from the Iberian peninsula. *J. hort. Sci.*, **70**(1): 97-103.

Matsumoto, J.I., N. Okamura and S.T. Ohki, 1997. Cucumber mosaic and broad bean wilt viruses isolated from

Limonium sinuatum and *Rudbeckia × hybrida*. *Ann. phytopath. Soc. Japan*, **63**(1): 13-15.

Matsumura, S., J. Yokoyama, Y. Tateishi and M. Maki, 2006. Intraspecific variation of flower colour and its distribution within a sea lavender, *Limonium wrightii* (Plumbaginaceae), in the northwestern Pacific Islands. *J. Pl. Res.*, **119**(6): 625-632.

Maymon, M., A. Zveibil, S. Pivonia, D. Minz and S. Freeman, 2006. Identification and characterization of benomyl-resistant and-sensitive populations of *Colletotrichum gloeosporioides* from statice (*Limonium* spp.). *Phytopath.*, **96**(5): 542-548.

Mercuri, A. 2002. Genetic transformation of ornamental germplasm: results obtained with the biotechnological approach (Italian). *Italus Hortus*, **9**(5): 16-23.

Mercuri, A., L. Anfosso, G. Burchi, S. Bruna, L. de Benedetti and T. Schiva, 2003. Rol genes and new genotypes of *Limonium gmelinii* through *Agrobacterium*-mediated transformation. *Acta Horticulturae.*, No. 624, pp. 455-462.

Mercuri, A., M. Antonetti, G. Burchi, C. Bianchini, P.L. Pasqualetto and T. Schiva, 1999. *In vitro* manipulation of *Limonium* (Italian). *Colture Protette*, **28**(1): 89-93.

Miyama, T., K. Inamoto, M. Doi and H. Imanishi, 1998. Influences of temperature and subculturing *in vitro* on subsequent flowering of *Limonium sinuatum* Mill. (Japanese). *J. Japanese Soc. hort. Sci.*, **67**(4): 632-634.

Moffett, M.L., A.C. Hayward and P.C. Fahy, 1986. Five new hosts of *Pseudomonas andropogonis* occurring in eastern Australia: host range and characterization of isolates. *Plant Path.*, **35**(1): 34-43.

Morales, M.A., E. Olmos, A. Torrecillas, M.J. Sanchez-Blanco and J.J. Alarcon, 2001. Differences in water relations, leaf ion accumulation and excretion rates between cultivated and wild species of *Limonium* sp. grown in conditions of saline stress. *Flora Jena*, **196**(5): 345-352.

Morgan, E.R., G.K. Burge and J.F. Seelye, 2001. *Limonium* breeding: new options for a well known genus. *Acta Hort.*, No. 552, pp. 39-42.

Morgan, E.R., G.K. Burge, J.F. Seelye, J.E. Grant and M.E. Hopping, 1995. Interspecific hybridization between *Limonium peregrinum* Bergius and *Limonium purpuratum* L. *Euphytica*, **83**(3): 215-224.

Morgan, E.R., G.K. Burge, J.F. Seelye, M.E. Hopping and J.E. Grant, 1998. Production of inter- specific hybrids between *Limonium perezii* (Stapf) Hubb. and *Limonium sinuatum* (L.) Mill. *Euphytica*, **102**(1): 109-115.

Motokura, Y., H. Kon and Y. Kobayashi, 2001. Rust disease of statice caused by *Uromyces savulescui* intercepted in plant quarantine. *Res. Bull. Pl. Prot. Serv., Japan*, No.37, pp. 87-91.

Murray, A.P., S. Rodriguez, M.A. Frontera, M.A. Tomas and M.C. Mulet, 2004. Antioxidant metabolites from *Limonium brasiliense* (Boiss.) Kuntze. *Biosci.*, **59**(7/8): 477-480.

Navalinskiene, M. and M. Samuitiene, 1999. Virological state of field floricultural plants in Lithuania. *Sodininkyste ir Darzininkyste*, **18**(1): 17-27.

Navalinskiene, M. and M. Samuitiene, 2000. Natural occurrence of tomato ringspot nepovirus in ornamental plants in Lithuania.*Trans. Estonian agric. Univ., Agron.*, No. 209, pp. 140-143.

Navchoo, I.A. and G.M. Buth, 1990. Ethnobotany of Ladakh, India: beverages, narcotics, foods. *Econ. Bot.*, **44**(3): 318-321.

Nicoletti, R., F. Raimo, C. Pasini and F. D'Aquila, 2003. Occurrence of *Cercospora insulana* on statice (*Limonium sinuatum*) in Italy. *Plant Path.*, **52**(3): 418.

Nishiyama, J., T. Kobayashi and K. Azegami, 1988. Bacterial wilt of statice caused by *Pseudomonas caryophylli*. *Ann. Phytopath. Soc. Japan*, **54**(4): 444-452.

Nissim Levi, A., R. Ovadia, I. Forer and M. Oren Shamir, 2007. Increased anthocyanin accumulation in ornamental plants due to magnesium treatment. *J. hort. Sci. & Biotech.*, **82**(3): 481-487.

Oviedo, Y. del C. and E. Guevara, 1988. *In vitro* propagation of statice (*Limonium sinuatum*) cv. Midnight Blue (Spanish). *Agronomia Costarricense*, **12**(1): 113-122.

Palacios, C. and F. Gonzalez-Candelas, 1997. Lack of genetic variability in the rare and endangered *Limonium cavanillesii* (Plumbaginaceae) using RAPD markers. *Mol. Eco.*, **6**(7): 671-675.

Palacios, C., S. Kresovich and F. Gonzalez-Candelas, 1999. A population genetic study of the endangered plant species *Limonium dufourii* (Plumbaginaceae) base don amplified fragment length polymorphism (AFLP). *Mol. Ecol.*, **8**(4): 645-657.

Palop Esteban, M., J.G. Segarra Moragues and F. Gonzalez-Candelas, 2007. Historical and biological determinants of genetic diversity in the highly endemic triploid sea lavender *Limonium dufourii* (Plumbaginaceae). *Mol. Ecol.*, **16**(18): 3814-3827.

Paparozzi, E.T. and H.M. Hatterman, 1988. Fertilizer applications on field-grown statice. *HortSci.*, **23**(1): 157-160.

Paparozzi, E.T. and D.E. McCallister, 1988. Glycerol and microwave preservation of annual statice (*Limonium sinuatum* Mill.).*Scientia Hort.*, **34**(3/4): 293-299.

Pasini, C., P. Curir, F.D'Aquila and G. Brofiga, 2003. Evaluation of some products for control of *Limonium* rust. *Informatore Fitopatologico*. **53**(11): 32-34.

Pedroza, J., A. Angarita and G. Corchuelo, 1997. Control of vitrification in micropropagation of statice (*Limonium sinuatum* Mill.) cv. Midnight Blue (Spanish). *Agronomia Colombiana*, **14**(1): 22-27.

Pedroza, J., G. Corchuelo and A. Angarita, 1999. Quantitative and qualitative analysis of the phenotypic homogeneity of micropropagated and seed derived statice (*Limonium sinuatum* Mill) cv. 'Midnight Blue' plants. *Acta Hort.*, No. 482, pp. 321-327.

Philosoph, H.S., O. Golan, I. Rosenberger, S. Salim, B. Kochanek and S. Meir, 2005. Efficiency of 1-MCP in neutralizing ethylene effects in cut flowers and potted plants following simultaneous or sequential application. *Acta Hort.*, No. 669, pp. 321-328.

Plavcova, O. 1976. Chromosome numbers in *Limonium sinuatum* and *L. bonduellii* (Czech). *Sbornik-UVTIZ, Zahradnictvi*, **3**(6)(3/4): 263-265.

Polizzi, G., 2002.Serious attacks by *Peronospora statices* in crops of *Limonium* spp. *Informatore Agrario*, **58**(46): 63-64.

Ramakrishnan, T.S. and U.V. Sundaram, 1955. Additions to fungi of Madras-XVIII. *Proc. Indian Acad. Sci.*, **42B**:58-64.

Ramasso, E., G. Dellavalle, P. Roggero and V. Lisa, 1994. New hosts of tospoviruses in ornamental plants and vegetables in Liguria. *Informatore Fitopat.*, **44**(1): 44-48.

Ramesh Kumar and Kiranjeet Kaur, 1997. Effect of planting date and spacing on growth parameters of different cultivars of statice. *J. Ornam. Hort.*, **5**(1/2): 20-25.

Ramesh Kumar, Kiranjeet Kaur and Mandhir Singh, 1998. Effect of time of planting and spacing on flower production behaviour of statice (*Limonium sinuatum*) cultivars. *Indian J. Hort.*, **55**(2): 172-176.

Rapetti, S. and A.Garibaldi, 1995. Damage by *Peronospora* on *Limonium* in Tuscany. *Colture Protette*, **24**(12): 53-54.

Rathinasabapathi, B and M.A.M. Aly, 2000. Somatic embryogenesis in *Limonium latifolium* (SM.) O. Kuntze, Plumbaginaceae. *Proc. Fla St. hort. Soc.*, **113**: 170-172.

Raulston, J.C. and C.M. Geraldson, 1971. Production of cut-flowers on subirrigated mulched beds with a single pre-plant fertilizer application. *Proc. Fla St. hort. Soc.*, **84**: 403-407.

Redondo Gomez, S., E.M. Naranjo, O. Garzon, J.M. Castillo, T. Luque and M.E. Figueroa, 2008. Effects of salinity on germination and seedling establishment of endangered *Limonium emarginatum* (Willd.) O. Kuntze. *J. Coastal Res.*, **24**(1A, Supplement): 201-205.

Rizzotto, M. 1999. Research on the genus *Limonium* (Plumbaginaceae) in the Tuscan Archipelago (Italy). *Webbia*, **53**(2): 241-282

Rudzinska L.A. and M.Kaminska, 2001a.Ultrastructural changes in aster yellows phytoplasma affected *Limonium sinuatum* Mill. plants. I. Pathology of conducting tissues. *Acta Societatis Botanicorum Poloniae*, **70**(3): 173-180.

Rudzinska L.A. and M.Kaminska, 2001b.Ultrastructural changes in aster yellows phytoplasma affected *Limonium sinuatum* Mill. plants. II. Pathology of cortex parenchyma cells.*Acta Societatis Botanicorum Poloniae*, **70**(4): 273-279.

Ruffoni, B., M. Savona, C. Mascarello, M. Pamato, M.C. Talia, C. Pacucci and E. Farina, 2002. *In vitro* culture and neomorphogenesis from immature floral scapes of *Limonium perezii* (Italian). *Italus Hortus*, **9**(5): 57-61.

Sabahat Zia and M.A. Khan, 2002. Comparative effect of NaCl and seawater on seed germination of *Limonium stocksii*. *Pakistan J. Bot.*, **34**(4): 345-350.

Sabahat Zia and M.A. Khan, 2003. Effect of germination-regulating chemicals in alleviating salinity-induced germination inhibition of *Limonium stocksii* seeds. *Pakistan J. Bot.*, **35**(5): 939-948.

Saez, L., A. Curco and J.A. Rossello, 1998. *Limonium vigoi* (Plumbaginaceae), a new tetraploid species from the Northeast of the Iberian Peninsula. *Anales del Jardin Botanico de Madrid*, **56**(2): 269-278.

Saez, L. and J.A. Rossello, 1999. Is *Limonium cavanillesii* Erben (Plumbaginaceae) really an extant species? *Anales del Jardin Botanico de Madrid*, **57**(1): 47-55.

Saidana, D., S. Mahjoub, O. Boussaada, J. Chriaa, M.A. Mahjoub, I. Cheraif, M. Daami, Z. Mighri, and A.N. Helal, 2008. Antibacterial and antifungal activities of the essential oils of two saltcedar species from Tunisia. *J. Amer. Oil Chemists' Soc.*, **85**(9): 817-826.

Saini, S.S., S. Kumari and J. Kaur, 1989. Fungi of Punjab - III - new host records of *Alternaria*. *Indian Phytopath.*, **42**(4): 599-600.

Salinger, J.P. 1985. Commercial Flower Growing, pp. 200-201. Butterworths of New Zealand (Ltd), Wellington, New Zealand.

Samuitiene, M., M. Navalinskiene and E. Jackeviciene, 2003. Detection of tospovirus infection in ornamental plants by DAS-ELISA. *Vagos*, **57**: 38-42.

Sato, S. and T. Sawamoto, 1997. New occurrence of Pestalotia disease of statice (*Limonium* spp.) by *Pestalotiopsis gracile*. *Proc. Assoc. Pl. Prot., Kyushu*, **43**: 57-59.

Sauer, J.A. and M.D. Shelton, 2002. High-temperature controlled atmosphere for postharvest control of Indian meal moth (Lepidoptera: Pyralidae) on preserved flowers. *Journal Econ. Entom.*, **95**(5): 1074-1078.

Seelye, J.F., G.K. Burge and E.R.Morgan, 2000. 'Chorus Magenta' *Limonium*. *HortSci.*, **35**(6): 1179.

Semeniuk, P. and D.T. Krizek, 1972. Long days and cool night temperature increase flowering of greenhouse grown *Limonium* cultivars. *HortSci.*, **7**(3): 293.

Semeniuk, P, and D.T. Krizek, 1973. Influence of germination and growing temperature on flowering of six cultivars of annual statice (*Limonium* cv.). *J. Amer. Soc. hort. Sci.*, **98**(2): 140-142.

Shillo, R. 1976. Control of flower initiation and development of statice (*Limonium sinuatum*) by temperature and daylength. *Acta Hort.*, **64**: 197-203.

Shillo, R., Mu Ding, D. Pasternak and M. Zaccai, 2002. Cultivation of cut flower and bulb species with saline water. *Scientia Hort.*, **92**(1): 41-54.

Shimamura, M., A. Ito, K. Suto, H. Okabayashi and K. Ichimura, 1997. Effects of alpha-aminoisobutyric acid and sucrose on the vase life of hybrid *Limonium*. *Postharvest Biol. & Techn.*, **12**(3): 247-253.

Skrzypczak, C., 2006.Harmfulness and possibilities of downy mildew control on some ornamental plants in Poland. *Prog. in Pl. Prot.*, **46**(1): 121-126.

Smith, A.G., K.E. John and N. Gardner, 2006. Dwarfing ornamental crops with the rolC gene. *Floriculture, Ornamental and Plant Biotechnology*, pp. 54-59. Global Science Books, Ltd., Middlesex, U.K.

Sobers, E.K. and R.S. Cox, 1973.Anthracnose of statice in southern Florida. *Phytopath.*, **63**(1): 193-194.

Soriano Cano, M.C., E. Go Plaza, R. Gil Munoz and A Gonzalez Benavente, 1998. Volatile compounds of three *Limonium* species: *L. latifolia*, *L.* × *altaica* and *L. perezii*. *J. Essential Oil Res.*, **10**(1): 67-69.

Steinitz, B., A. Cohen and B. Leshem, 1980. Factors controlling the retardation of chlorophyll degradation during senescence of detached statice (*Limonium sinuatum*) flower stalks. *Zeitschrift für Pflanzenphysiologie*, **100**(4): 343-349.

Steinitz, B. and A. Cohen, 1982. Gibberellic acid promotes flower bud opening on detached flower stalks of statice (*Limonium sinuatum* L.). *HortSci.*, **17**(6): 903-904.

Strider, D.L., 1973. Damping-off of statice caused by *Botrytis cinerea* and its control. *Plant Dis. Reptr*, **57**(11): 969-971.

Szabo, T. and F. Viranyi, 1990. *Peronospora* disease of *Limonium tataricum*. *Novenyvedelem*, **26**(11): 508-511.

Talia, M.A.C., C. Pacucci and B. Pace, 2004. *Limonium perezii*, influence of the growing environment on quantity and quality (Italian). *Colture Protette*, **33**(9): 129-133.

Talia, M.A.C. and C. Pacucci, 2005. Bioagronomic assessment of three *Limonium* species (*L. bellidifolium* Dumort, *L. perezii* Hubb., *L. sinensis* Kuntze) (Italian). *Colture Protette*, **34**(9): 127-130.

Tang, X.H., J. Gao, J. Chen, L.Z. Xu, Y.H. Tang, X.N. Zhao and L. Michael, 2008. Mitochondrial modulation is involved in the hepatoprotection of *Limonium sinense* extract against liver damage in mice. *Journal of Ethnopharmacology*, **120**(3): 427-431.

Thulin, M. 2004. The identity of *Statice maurocordatae* (Plumbaginaceae). *Kew Bull.*, **59**(1): 163-164.

Verhoeven, J.T.J., J.W. Roenhorst, D.E. Lesemann and R. Koenig, 2006. Identification of tombusviruses isolated from *Limonium sinuatum* on the basis of host range, serological and molecular studies. *Acta Hort.*, No. 722, pp. 201-207.

Vidal, C., B. Bartolome.and A.Gonzalez, 2007. Occupational asthma to fresh sea lavender and cross- reactivity to sweet vernal grass pollen. *Ann. Allergy, Asthma & Immunology*, **99**(6): 576-577.

Vigna, R., M. Devecchi and E. Accati, 1999. Keeping cut flowers of *Delphinium* and *Limonium*. *Colture Protette*, **28**(10): 71-78.

Vitanova, G. and A. Kaninski, 1998. Flowering and seed production of two *Limonium* species. *Rasteniev"dni-Nauki*, **35**(6): 453-457.

Vovlas, C. and F. Cillo, 1996. Infections of cucumber mosaic cucumovirus (CMV) on statice in Salento. *Informatore* Fitopatologico, **46** (12): 7-8.

Weintraub, P.G., S. Pivonia, A. Rosner and A. Gera, 2004. A new disease in *Limonium latifolium* hybrids. II. Investigating insect vectors. *HortSci.*, **39**(5): 1060-1061.

Whipker, B.E. and P.A. Hammer, 1994. Growth and yield characteristics of field-grown *Limonium sinuatum* (L.). *HortSci.*, **29**(6): 638-640.

Whiting, P., T. Savchenko, S.D. Sarker, H.H. Rees and L.Dinan, 1998. Phytoecdysteroids in the genus *Limonium* (Plumbaginaceae). *Biochem Systemat. Ecol.*, **26** (6): 695-698.

Wilfret, G.J. and J.L. Green, 1975. Optimum gibberellic acid (GA_3) concentration to accelerate flowering and increase yield of statice. *Proc. Fla St. hort. Soc.*, **88**: 527-530.

Wilfret, G.J. and J.C. Raulston, 1975. Acceleration of flowering of statice (*Limonium sinuatum* Mill) by gibberellic acid (GA_3). *HortSci.*, **10**(1): 37-38.

Wu-YuHu and Yang-YongChang, 2006. A new variety of the genus *Limonium* from Qinghai Province, China (Chinese). *J. Wuhan bot. Res.*, **24**(4): 323.

Yoo, D.L., W.B. Kim, C.W. Nam, B.H. Hahn, B.H. Kim, J.K. Kim and K.S. Choi, 1994. Flowering and yield of statice (*Limonium sinuatum* Mill.) by sowing date for summer season production in alpine area(Korean). *RDA J. agric. Sci., Hort.*, **36**(1): 435-443.

Zhang-ChaoQiang, Yang-YingLi, Wang-Lai, Zhang-SaiNa, Ma-XuJun and Sun-Kun, 2007. Physiological responses to NaCl treatment in seedlings of two *Limonium* species (Chinese). *Acta Botanica Boreali Occidentalia Sinica*, **27**(11): 2245-2250.

Zhang-WanJun, Wang-DouTian, Fan-Hai, Feng-LiTian and Zhao-KeFu, 2001. Characters and physiological basis of halophytic seeds germination (Chinese). *Chinese J. Appl. environ. Biol.*, **7**(2): 117-121.

Zhu-JinWen, Ni-YueYuan and Zhao-XiaoYi, 1997. A study on interspecies hybridization of *Limonium* sp. (Chinese). *Acta Hort. Sinica*, **24**(2): 175-178.

Zizzo, G., S. Aprile, S. Agnello, A. la Mantia and S. Sortino, 2004. *Ex situ* cultivation of *Limonium virgatum* Willd. from different Sicilian localities (Italian). *Italus Hortus*, **11**(4): 160-162.

Zizzo, G.V., G. Fascella, G. Costantino and S. Agnello, 2003. First evaluation of *Limonium* suitability for soilless cultivation. *Acta Hort.*, No. 614(vol. 1), pp. 235-239.

Yang-YingLi, Zhang-ChaoQiang, Li-KeWen, Wang-Lai, Fan-XiaoFeng and Sun-Kun, 2008. Comparison of seed germination and resistance in two *Limonium* seedlings under NaCl treatment (Chinese). *Bull. Bot. Res.*, No. 28(1), pp. 73-78.

Yoo, D.L., W.B. Kim, C.W. Nam, B.H. Hahn, B.H. Kim, J.K. Kim and K.S. Choi, 1994.Flowering and yield of statice (*Limonium sinuatum* Mill.) by sowing date for summer season production in alpine area (Korean). *RDA J. agric. Sci., Hort.*, **36**(1): 435-443.

Yoo, D.L., S.Y. Ryu, W.B. Kim, C.W. Nam, Y.E. Choi, B.H. Kim and J.K. Kim, 1998. Effects of highland raising and seed low temperature treatment on the yield of statice (*Limonium sinuatum*) cut flowers under structures during winter season in lowland. *RDA J. hort. Sci.*, **40**(1): 150-154.

Zimmer, K. 1981. Growing *Limonium* (statice) (German). *Deutscher Gartenbau*, **36**(9): 318-320.

Zimmer, K. 1982. Yield investigations of *Limonium* (statice) (German). *Deutscher Gartenbau*, **36**(45): 1882, 1884.

26

Matthiola (Family: Cruciferae)

M. K. Singh, Sanyat Misra, R.L. Misra and Sanjay Kumar

[**Common names**: Annual stock/Biennial stock/Evening stock/Night-scented stock (*Matthiola bicornis*), Autumn-blooming stock/Brompton stock/Column stock/Common stock or stock/Gillyflower/Intermediate stock/Perennial stock/Queen stock/Summer-blooming stock/Ten-week stock (*M. incana, M. incana* var. *annua* or *graeca*), Night-scented stock (*M. bicornis, M. longipetala* ssp. *bicornis*), Perennial stock (*M. coronopifolia*), Sea stock (*M. sinuata*), Three-horned stock (*M. tricuspidata*), etc.]

Introduction

Matthiola was named for Pierandrea (Pietro Andrea) Mattioli (1500-1577), an Italian doctor and botanist. It is native to Azores and Canary Islands westwards through the Mediterranean region to central Asia. It is a genus of 55 species of annuals, biennials, perennials and sub-shrubs. They are excellent subjects for cut flowers, for beddings and borders, for sweet fragrance, and for growing in the pots (non-branching types). Single-flowered stocks are ideal for a bank, the edge of a wood or orchard or to any semi-wild position, at the edge of the vegetable garden near the house and near the windows for fragrant breeze with lily-carnation scent which *M. bicornis* or *M. incana* emit, more by evening than the daytime. The Brompton stocks and the ten-week stocks have their origin in *M. incana* which occurs from the northern Mediterranean to Egypt and Arabia and because of their tendancy to produce double flowers they were selected. The doubling of the flowers is due to repeated production of petals and sepals at the expense of stamens and carpels that is why most of the double forms do not yield seeds. However, for double types the selection is made at the seedling stage, and in the good strains some 70 per cent of selected seedlings may produce double flowers.

Two new kaempferol glycosides, kaempferol 3-0-(2”-α-l-rhamnopyranosyl)-ß-l-arabinopyranoside-7-0-α-l-rhamnopyranoside and kaempferol 3-0-(2”-α-l-rhamnopyranosyl)-ß-l-arabinopyranoside-7-0-α-l-rhamnopyranoside-4’-0-ß-d- glucopyranoside were isolated along with four known flavonols from *Matthiola longipetala* (ssp. *livida*) growing in Egypt and their cytotoxic activity of the alcoholic extract against cervix (HELA) and colon (HCT116) human cell lines showed IC50 values of 22.1 and 53.7 µg/ml, respectively though no activity was found against liver (HEPG2) carcinoma cells (Maezouk *et al.*, 2008). The chemical composition of the volatile parts of *Matthiola longipetala* flowers revealed some 46 constituents in the essential oil, most representative compounds being eugenol (19.93 per cent), bicyclogermacrene (13.60 per cent), heptacosane (7.85 per cent) and tetradecanoic acid (5.57 per cent) (Hammami *et al.*, 2006). Insecticide activity of the essential oil against *Tribolium confusum* larvae was evaluated using direct contact application method which showed a significant inhibitory effect of the test material with adult mortality of 35 per cent at 8 days after treatment. *Matthiola incana* extract yielded an herbicidal compound identified as sulforaphene (4-methylsulfinyl-3-butenyl isothiocyanate) (Brinker and Spencer, 1993). *Matthiola incana* produces seeds some 5-6 quintals per hectare with 20-25 per cent seed oil and this oil yields some 50 per cent of edible pure α-linolenic acid which can be used as a dietary supplement (Yaniv *et al*, 1997).

Botany, Breeding and Biotechnology

The species may be annuals, biennials, perennial herbaceous plants and evergreen sub-shrubs bearing alternate or rosetted, oblong to linear leaves, usually undivided though sometimes pinnatifid, and covered with starry or branched hairs, and the flowers are 4-petalled, borne on the erect racemes. They are fully hardy to frost-tender. The most of the annuals and biennials have highly fragrant flowers, most suitable for cutting. These prefer sun or semi-shade for growing.

Cytological studies carried out by Khosravi and Maassoumi (1998) revealed *Matthiola afghanica* and *M. alyssifolia* containing 2n = 12 and 2n = 14 chromosome counts. When Lan ZeQu *et al.* (1999) studied the root tips of *Matthiola incana* cytologically at mitosis, they recorded a karyotype of 2n = 14 = 6m + 8sm with no satellite, while at meiosis, a configuration of 7 bivalents were distinctly observed at diakinesis with a rod, ring or cross shape, rod-shaped bivalents increased and ring-shaped bivalents decreased in number in metaphase I, and chromosome bridges appeared at anaphase I and anaphase II.

A cDNA clone (pcM12) of the chalcone [naringenin-chalcone] synthase (CHS) of *Matthiola incana* was isolated from a cDNA library, sequenced and analysed, and this comprised the whole coding sequence for the CHS and 5' and 3' untranslated regions (Epping *et al.*, 1990). The deduced amino acid sequence showed that *Matthiola* CHS consisted of 394 amino acid residues, and comparison with CHS amino acid sequences of other plants indicated >82 per cent homology. Protoplast isolated from hypocotyls of *Brassica napus* cv. 'Xiangyou 15' and *Matthiola incana* were fused by the PEG-high Ca_2+-high pH method and the highest fusion frequency (17 per cent) was obtained with a fusion solution containing 30 per cent PEG and fusion was conducted at a density of 5.0 × 105 protoplasts/ml for 25 minutes, then the fusion-protoplasts were cultured by the 'thin liquid layer' method in improved KM8p medium supplemented with 1.5 mg 2,4-D, 0.5 mg NAA and 0.5 ml benzyladenine, 0.3 mol sucrose and 0.1 mol glucose/l used as osmotic regulator where fusion-protoplasts started to divide after 3-6 days and the division frequency was 13.8 per cent after being cultured for 7 days, and the cell-masses and small calluses could be obtained after 35 days, and the plating efficiency was 6.5 per cent, then transferred to the proliferation MS medium supplemented with 0.1 mg 2, 4-D and 0.2 mg benzyladenine/l and the calluses of 5 mm in diameter were transferred on MS differentiation medium supplemented with 0.5 mg BA, 0.1 mg NAA and 0.2 mg GA <sub>>3< sub>/l, and whole plants were obtained after transferring 1-2 cm shoots to 1/2 MS medium supplemented with 0.5 mg IBA/l for 2 weeks, which through cytological studies confirmed this as a true hybrid (Lin LiangBin, 2008).

Sheng XiaoGuang *et al.* (2008) through intergeneric somatic hybridization obtained hybrids between *Brassica oleracea* (2n=18) and *Matthiola incana* (2n=14). Random amplified polymorphic DNA (RAPD), sequence-related amplified polymorphism (SRAP), and cytological analysis confirmed somatic hybrids in the progenies, and on cytological studies of the root tips it revealed that hybrid cells contained 28 chromosomes. A cell-free extract of stock flowers catalysed a NADPH-dependent stereospecific reduction of (+)-dihydrokaempferol to 3,4-cis-leucopelargonidin (5,7,4'-trihydroxyflavan-3,4-cis-diol) with an optimum pH reaction of 6.0, the rate of reaction with NADH was about 50 per cent of that found with NADPH, and (+)-dihydroquercetin and (+)-dihydromyricetin were also reduced by the enzyme preparation to the corresponding flavan-3,4-cis-diols, which suggests that the correlation between the genotype and the presence of dihydroflavonol 4-reductase is strong evidence that this enzyme is involved in anthocyanin biosynthesis (Heller *et ai.*, 1985). During flower development of a line (line 07) with coloured petals, a clear pattern of chalcone synthase (CHS) enzyme production (the key enzyme of flavonoid biosynthesis) could be observed (Rall and Hemleben, 1984), with a large increase of the amount of CHS enzyme protein in the petals occurring at the stage of bud opening, differing from the results obtained with petals of the white-flowering mutant line 18 bearing a genetic defect in the gene f coding for CHS, *vis-à-vis* a reduced and nearly constant level of CHS enzyme protein occurring during flower development, though other white-flowering mutant lines [line 17, genotype ee; and line 19 (gg)] with a late block in the flavonoid biosynthesis pathway, exhibited a concomitant reduction of CHS enzyme activity and protein content compared with anthocyanin-producing lines with the f+f+e+e+g+g+-genotype. Line 18 was most convenient for studying the regulation of enzyme synthesis because under stress conditions a slight colouring of the flower petals occurred and was uniformly distributed and line-specific which suggests that the mutant produced a rather inactive enzyme protein at a low level. This protein may, however, regain enzyme activity under certain environmental conditions. Preliminary investigations suggest a high level of CHS mRNA transcription at the bud opening stage. All the *Matthiola incana* genotypes with different flower colours were found to contain a xylosyltransferase which catalyzed the transfer of the xylosyl moiety of uridine 5'-diphosphate-xylose to the glucose of cyanidin 3-glucoside (Teusch, 1986), though no significant differences in enzyme activity were detected amongst lines that were either dominant or recessive for gene u, indicating that this gene is not responsible for xylotransferase activity. The 3-glucosides of pelargonidin and delphinidin, cyanidin 3-(p-coumaroyl)-glucoside, and 3-(caffeoyl)-glucoside were also substrates for this enzyme, enzyme having optimum pH of 6.5 and its activity being depended on the stage of bud and flower development, and accumulation of cyanidin 3-glucoside during flower development was correlated with xylosyltransferase activity. Saito *et al.* (1995) isolated four triacylated cyanidin 3-sambubioside-5-glucosides (anthocyanins in variable amount as their main pigment) from 15 purple-violet cultivars of *Matthiola incana*. The presence of hydroxycinnamic (coumaric) acids in the acylated anthocyanins, in particular the presence of sinapic and ferulic acids, and sinapic and p-coumaric acids, are thought to produce the bluish colours in these flowers.

Basic helix-loop-helix (bHLH) proteins as in mammalian MYC transcription factors, regulate the anthocyanin biosynthetic pathway in both monocots and dicots, the two *Arabidopsis* bHLH genes, GLABRA3 (GL3) and MYC-146, encode proteins that are similar throughout the predicted amino acid sequence to R and DELILA, which regulate anthocyanin production in maize and snapdragon, respectively, and the northern blot analysis indicated that

MYC-146 is most highly expressed in flower buds and flowers (Ramsay, *et al.,* 2003). They found that expression of a MYC-146 cDNA from the CaMV 35S promoter was unable to complement the anthocyanin deficiency in a ttg1 mutant of *Arabidopsis* and resulted in no obvious phenotypic change in Columbia plants, however, transient expression of GL3 and MYC-146 upon microprojectile bombardment of petals of a white-flowered mutant of *Matthiola incana* was able to complement anthocyanin deficiency. The lack of anthocyanin-deficient *Arabidopsis* mutants mapping to the locations of GL3 and MYC-146 suggested that the two bHLH proteins may be partially redundant and overlap in function. For *Matthiola incana*, used as a model system to study biochemical and genetical aspects of anthocyanin biosynthesis, several nearly isogenic coloured wild type lines and white-flowering mutant lines are available, each with a specific defect in the genes responsible for anthocyanin production (genes e, f, and g). For gene f supposed to code for chalcone synthase (CHS; EC 2.3.1.74), the key enzyme of the flavonoid/anthocyanin biosynthesis pathway belonging to the group of type III polyketide synthases (PKS), the wild type genomic sequence of *Matthiola incana* line 04 was determined in comparison to the white-flowering CHS mutant line 18. The type of mutation in the chs gene was characterized as a single nucleotide substitution in a triplet AGG coding for an evolutionary conserved arginine into AGT coding for serine (R72S). Northern blots and RT-PCR demonstrated that the mutated gene is expressed in flower petals. Heterologous expression of the wild type and mutated CHS cDNA in *Escherichia coli*, verified by western blotting and enzyme assays with various starter molecules, revealed that the mutant protein had no detectable activity, indicating that the strictly conserved arginine residue is essential for the enzymatic reaction.

Crossing-over between a very small fragment and a whole chromosome in a stock plant that had 2 genes for doubleness (S) and an unstable fragment (Su) derived from breakage of a ring produced one single-flowered diploid plant in progeny consisting of 41 double-flowered to 10 variegated plants (Lesley, 1973). The selfed progeny of the single-flowered diploid plant reversed this, giving 33 diploid single-flowered to 10 double-flowered plants, and these double-flowered were always sterile, though single-flowered plants were of two types; 23 normal except for a variable number of bent or curled capsules, and 10 larger dark-green basal leaves and a central growing tip wholly or partly sterile, therefore, he opined that the normal looking plants are Sus and the more abnormal ones SuSu. Crossing of line A with a dominant gene for single flowers and B with paired recessive genes for double flowers, where marker genes influencing plant pigmentation are linked to these genes, the progeny from the cross included yellowish seedlings & green seedlings where 93 per cent of the latter gave double flowers though in the original method the double flowers were produced on somewhat inferior yellowish plants (Fehleisen and Schnack, 1972). Skiebe (1985) when studying the inheritance of doubleness in stocks, reported that the alleles of 3 gene sites exhibited to be linked but this link was not very close and was likely to be broken, and the linkage configuration could also be modified by mutation. Anthocyanidin 3-glucosides and 3-sambubiosides acylated with 4-coumarate or caffeate were identified in flower extracts of the lines of *Matthiola incana* with wild-type alleles of gene 'u' and an enzyme activity was demonstrated catalysing the acylation of 3-glucosides and 3-sambubiosides with 4-coumarate or caffeate using the respective CoA esters as acyl donors (Teusch *et al.,* 1987). Jana and Seyffert (1972) used six groups of stock breeding material, each with two loci both having two alleles for anthocyanins, to set up six ideal 4 × 4 diallel crosses in which all conditions except possibly that requiring the independent action of nonallelic genes were satisfied, and through diallel analysis of anthocyanin content in the flower tissues indicated that in the presence of epistasis, the dominance ratio (H1/D) ceases to be a reliable measure of the average degree of dominance, so in such situations additional genetic information obtained from the diallel analysis is not in agreement with the expectations on the basis of the already available genetic information on the materials. The estimator H2/4 H1, a measure of average value of the product of alleles with positive and negative effects seemed to remain unaffected by epistasis, and the Wr/Vr regression analysis does not always permit the detection of nonallelic gene interactions, therefore, it is inferred that duplicate interaction may escape detection by the regression analysis. Ecker *et al.* (1991) when analysed the fatty acid composition of the seed-oil of breeding lines and F_1 hybrids of stock, it showed that breeding lines differed significantly with respect to the levels of palmitic, oleic, linoleic and linolenic acids, and linolenic acid content was negatively correlated with both oleic acid content (r = - 0.85) and linoleic acid content (r = - 0.66). Ecker *et al.* (1994a) when studied inheritance of flowering time (FT) in a cross between 70 early-flowering (P1) and 120 late-flowering (P2) genotypes, they recorded distribution of FT in F_1 and BC_1 generations having additive genetic control with earliness alleles partially dominant, particularly in double-flowered plants. They found single-flowered plants flowering earlier than double-flowered ones, however, the mean difference between singles and doubles was 16 days for late-flowering (P2). Through F_3 family means analysis, this flower-doubleness related delay in FT was found to be heritable. In a study on the inheritance and linkage relationships of a leaf morphology gene of *Matthiola incana*, Ecker *et al.,* 1994) recorded that the allele for sinuate leaf shape 'c' was recessive to the allele for normal entire leaf 'C' and the 'c' allele was tightly linked to the recessive allele for double flowering 's'. The recombination frequency between the two loci was close to zero. The mode of inheritance of the C gene was in accordance with the hypothesis that a pollen lethal gene is responsible for the constant 1:1 segregation ratio of double-flowered (= male sterile) to single-flowered (= fertile) plants in most breeding lines. The sinuate leaf allele seemed to reduce the frequency and delayed the flowering of double-flowered plants. Ardelean *et al.* (2008) used various *Matthiola incana* cultivars (homogeneous in their characters) for incomplete diallel crosses, and assessed the F_1s for six quantitative characters of high ornamental value, *viz.* plant height, number of inflorescences per plant, number of flowers per inflorescence, diameter of individual flower, start of flowering and persistence of flowering. Additive effects of polygenes, emphasized by high GCA values in several characters, suggested that for most of these characters there were groups of polygenes with positive effects and other groups with negative ones. For SCA, significant values were found for most of the characters, which could be considered as a preliminary proof that there might be fair chances of developing commercial hybrids with an obvious phenotypic expression of the desired heterosis (negative or positive). Six parental patterns (P1-P6) of *Matthiola incana* cultivars and 10 direct and mutual hybrid combinations (obtained through cyclic crossbreedings using P1 as a tester) were when analysed, Pui and Ardelean (2007) recorded that in simple flowering F_1s, the majority or all of the progenies showed positive heterosis only in case of the flower diameter, siliqua length and number of inflorescences per plant when heterosis was calculated based on the parental mean. When heterosis was calculated in relation to Pmax, the other combinations had both positive heterosis and negative heterosis regardless of the way of its calculation. In descendants with plants presenting composite flowers, only the height of plants and the flower diameter had predominantly positive heterosis for both ways of calculation, the other characters having negative values. We assume that negative heterosis in this case, could be rather designated as the result of genetic transgressions. Six *Matthiola incana* cultivars were when tested through direct and indirect diallel crosses, Pui *et al.* (2007) in 10 of the hybrid combinations recorded, high values in wide sense (H) were accompanied by medium values of narrow sense (h2). It is quite reasonable to expect a high efficiency of phenotypic selection for such traits since additive effects seem to play a major role in their inheritance.

Classification, Species and Varieties

There are various series in stocks derived only from *Matthiola incana* as detailed here under:

There are 55 species under the genus *Matthiola* and the most important ones of ornamental value are described below:

M. bicornis (*M. longipetala bicornia*). It is a hardy annual native of central and south Greece and Aegean, and grows up to 40 cm tall and expansive, leaves are short-stalked, linear or lanceolate, grey-green and some 7.5 cm long, lower leaves are narrowly lanceolate, sinuate-toothed to pinnatifid and the upper ones are shorter, narrower and few-toothed, flowers are single, deliciously night fragrant, 2 cm across, widely spaced in the raceme, lilac-pink to purple and petals dull which droop during daytime and the raceme is up to 25 cm long. The seed-pods bear twin horns hence the specific name.

M. incana. It is a woody-based hardy perennial of bushy habit native to coastal areas of SW Europe, Asia Minor, North Africa, and the Canary Islands, though grown as biennial or annual. It grows 30-90 cm tall bearing grey-green, 5-10 cm long, linear to oblong-lanceolate leaves that have grey-white felt, and the fragrant flowers are 2-3 cm wide and white, pink or purple though in cultivated types these may also have purple-blue, red or yellow, which are borne on 15-25 cm tall inflorescences. The original species was in cultivation early in the 16th century though double-flowered one was introduced into Europs in 1570. For the sake of convenience they are broadly divided into **Brompton Stock** grown generally as biennials, **Intermediate Stocks** which contains the best one as East Lothian strain is treated as an annual, and **Ten-Week Stocks** which is grown as an annual, however, there are many groups and strains as described below.

The species *M. incana* is the parent of most of the garden hybrids and varieties including various series such as **Ten-Week** stocks that are fully hardy, fast-growing, branched and upright bushy biennials and short-lived perennials though grown as annuals with height and spread of some 30 cm, the leaves are greyish-green and lance-shaped, the flowers are single (only a few dwarf ones being doubles), fragrant and in wide range of colours which are borne on a spike more than 15 cm tall [Dwarf Large Flowering in shades of white, yellow, pink, crimson and lavender which grows up to 30 cm tall; Excelsior or Column in all the shades as above except yellow and grows up to 75 cm tall with densely flowered inflorescence; Giant Imperial grows up to 60 cm in height bearing flowers in shades of yellow, gold and copper; the Giant Perfection strain grows up to 75 cm tall and its popular varieties are 'Princess Alice' (white) and 'Queen of the Belgians' (silver-lilac); The Mammoth or Beauty strain grows up to 45 cm tall and its popular varieties are 'Beauty of Nice' (pink), 'Crimson King' (deep crimson), 'Heatham Beauty' (rose-mauve), 'Mont Blanc' (white), 'Parma Violet' (purple), 'Yellow of Nice' (pale-yellow), *etc.*]; second the **Perpetual Fowering or All the Year Round** stocks are a group of short-lived perennials or biennials but grown as annuals that are vigorous dwarf plants growing up to 40 cm in height bearing -*Cheiranthus*-like green foliage and dense spikes with double, large and single to white flowers; the third one is **Brompton** stocks which grow up to 45 cm tall with about 30 cm spread, well-branched bearing single to double highly fragrant flowers in shades of white, yellow, pink, red and purple, are fast-growing, erect and bushy biennials though grown as annuals, the leaves are lance-shaped and greyish-green, and spikes quite tall; fourth the **East Lothian or Intermediate series**, a group of fully hardy, fast-growing, bushy and upright biennials and short-lived perennials though grown as annuals, growing up to 40 cm in height and spread, leaves greyish-green and lance-shaped, and the spikes are more than 15 cm tall bearing fragrant and single or double flowers in shades of white, yellow, pink, red or purple; rhe **Park Series** that are fully hardy fast-growing bushy biennials and short-lived perennials though treated as annuals, the leaves are greyish-green and lance-shaped, and grow some 30 cm in height and spread, are fast-growing, upright with about 15 cm tall spikes bearing 4-petalled single and fragrant flowers having wide range of colours; the sixth one is **Trysomic** which is fully hardy, fast-growing, upright and bushy biennial or short-lived perernnial though grown as annuals, growing some 30 cm tall in height and spread, the leaves are greyish-green, lance-shaped, spikes more than 15 cm long and the flowers are mostly double (85 per cent) in various colours; and the last one is **All Double** which are fully hardy, 100 per cent fully double, fast-growing, upright and bushy biennial or shorter-lived perennials though grown as annuals, and grow some 30 cm tall and wide, leaves are greyish-green, lance-shaped and the flowers in various colours on more than 15-cm tall racemes.

M. sinuata. It is a perennial or biennial species with much-branched stems, growing up to 50 cm, and is native to the shores of northern Mediterranean, Asia Minor (especially Egypt) and Algeria (Arabia). Whole plant is covered with silky hairs, the leaves are thick-textured, margins grey-edged and dentate while those at the base are undulated, racemes are erect, bearing suffused blue or lilac and fragrant flowers some 2.5 cm across.

There are many other species of ornamental values, and in India Chandrasekar *et al.* (2004) recorded *Matthiola flavida* in Pin Valley National Park, Lahaul-Spiti (H.P.). Dirmenci *et al.* (2006) recorded *M. trojana*, a new species from Turkey.

Loeser (1986) tabulated characteristics of 64 cultivars from 8 races of *Matthiola* in a polyhouse and stated that dark blue form of 'Buchholzer Riesen Treib' had the shortest panicles (13 cm) having 67 per cent of the panicles with single flowers and 33 per cent with double flowers, the pink form of 'Super Giant' with longest panicles (26 cm), cvs 'Royal Treib' and 'Treibwunder' only with single flowers though earliest in flowering (20 May), 'Buchholzer Riesen Treib' the latest (12 June), while 'Exelsior Mammut' was the tallest (81 cm) and 'Royal Treib' the shortest (46 cm).

Promising globally important varieties of stocks in various colours (white, yellow, orange, light to dark pink, rose, red, lavender, lavender-blue, dark blue, carmine and maroon; single and double; and branching and non-branching types) are 'Akinobeni', 'Akinomurasaki', 'Asanami', 'Banrei', 'Beauty of Nice', 'Benihime', 'Brompton Stocks' (various colours), 'Buchholzer Riesen Treib' (various colours), 'Centrum Hybrids' (various colours including white, dark pink, red, lavender blue, dark blue, *etc.*), 'Cheerful White', 'Chohong', 'Christmas Blue', 'Christmas Rose', 'Christmas White', 'Cinderella Lavender', 'Cinderella Red', Cinderella Yellow', 'Crimson King', 'Dwarf Ten-week', 'Excelsior Mammut', 'Fressia', 'Frolic Carmine', 'Fujimusume', 'Giant Excelsior Early Original Hansen Selections' (yellow, pink, dark red, dark blue, *etc.*), 'Giant Excelsior Original Hansen Selections' (various colours including white, pink, red, pale to dark blue, *etc.*), 'Ginsho', 'Heatham Beauty', 'Iwaiaka No.1', 'Iwaiaka No.2', 'Julliette', 'Kanchidori', 'Kanshio', 'Lavender', 'Littlejem Yellow', 'Mammoth Column', 'Maria', 'Matsudoaka', 'Midget Red', 'Mont Blanc', 'Parma Violet', 'Pearl Pink', 'Princess Alice', 'Queen of the Belgians', 'Royal Treib', 'Rubin', 'Ruby Red', 'Senshonoyuki', 'Sentinel White', 'Seven-Week Trysomic Dwarf Double', 'Snow Wonder', 'Sourei' 'Super Giant', 'Trysomic Hi-Double Lilac-Lavender', 'Tsukinoyosooi', 'Wakazakura', 'White Beach', 'White Dorse', 'Yellow of Nice', 'Yukimatsuri', 'Zuisei', *etc.*

Propagation

For winter flowering, the seeds should be sown outside in July in the temperate regions of the country and in August-September in the plains of India, and when seedlings are some 4-week old these are transplanted in 15 cm pots individually and transferred to glasshouse or in flat beds inside the glasshouse at 7 °C, and for succession of blooming a lot can be maintained in the cold frame to be shifted in the glasshouse subsequently. For summer flowering, seeds are sown during October-November at a temperature of 13-15 °C, should be taken out in the boxes and hardened off in a cold frame before planting out inside the glasshouse or in the open. April-May planting is recommended only in the temperate regions. Trysomic strains and other double-flowered ones should be grown at 13-15 °C and then before pricking out the temperature for a week should be lowered to 10 °C to differentiate the colours of the seedlings, the yellow double plants with those of green singles so latter may be discarded. An automatic method for recognizing and evaluating double flowered stock seedlings was investigated to develop an automatic transplanting system, where three basic characteristics (area, degree of complexity and saturation of the cotyledon colour) of the seedling image were measured by a colour image processing system to fix the membership function of fuzzy sets, and so an algorithm based on the fuzzy evaluation theory was developed to discriminate between double and single flowering stock in the seedling stage (Koku *et al.*, 1999; Dohi and Fujiura, 1999) in the cvs 'Asanami' and 'Sourei' where accuracy of discrimination was recorded 85-90.8 per cent (much higher than that achieved manually) and the availability of the seedling was 39.4 per cent. Though morphological characterization of doubleness was not so promising in stock, which could be determined in the seeds by using the radiospectrometer RE 1301 by placing the seeds in a thin walled molybdenum tube to register the electronic paramagnetic resonance, and doubleness should be determined by the strength of the signal (Gladunov *et al.*, 1974). The soil for seed sowing should be a good garden sandy-loam soil fortified with wellrotten farmyard manure or compost and coarse sand all the three in equal quantity (w/w).

Dressing seeds with acrylic gels increased stock seed germination, results being best with a wet seed dressing of Akrygel KM and 10 or 25 per cent graphite and sowing into growing medium treated with Akrygel RP (Hetman *et al.*, 1996). Seeds of *Matthiola bicornis* (*M. longipetala* var. *bicornis*) were when subjected to accelerated aging at 42 °C and 100 per cent RH and then subjected to laboratory testing either using the tetrazolium test or Germ's method, and in a greenhouse by monitoring seedling emergence at 20 °/15 °C, and so Houbowicz *et al.* (1997) found aging reducing the seed viability by 19-42 per cent, the tetrazolium test giving good results, Germ's method being least useful but the seedling emergence in the greenhouse remains best. The effects of abscisic acid (ABA) priming singly and in combination with GA_3, kinetin (KT) and NAA, on the germination of *Matthiola bicornis* seeds, Zhang ZhiSheng and Doma (2005) recorded optimum priming method as seed treatment in a 2.5 × 10-5 mol ABA/l solution at 20 °C for 7 days which reduced the mean germination time and increased the germination index, though ABA + NAA treatment gave no further effect on the germination of seeds, but other treatment combinations significantly increased the percentage, however, KT + NAA increased the germination index compared with ABA or PEG priming. *Matthiola livida* and *M. longipetala* seeds were when germinated at 10, 15, 20, 25 or 30 °C, Ishak (1989) in Egypt recorded the lowest germination rates and percentages at 10 °C and highest at 20-25 °C (*i.e.*optimum sowing time being in spring and late autumn) though at 30 °C it was only 64-74 per cent. Three-horned annual stock (*M. tricuspidata*) of the Mediterranean sandy shores whose seeds are dark-germinating and negatively photosensitive, showed full germination at a wider range of temperatures (5-25 °C) in the dark, the inhibition of germination by white (fluorescent), blue and far-red light, applied either continuously or intermittently, consistently showed a linear dependence upon the logarithm of the flux density of the irradiation, and continuous blue or far-red irradiations, both establishing a similar phi-value (0.26), resulted in similar regression curves, thus favouring the hypothesis that phytochrome is the single photoreceptor in the photoinhibition of seed germination (Thanos *et al.*, 1994).

Multiple shoot buds differentiated from cotyledonary explants of *Matthiola incana* cultured on MS basal medium supplemented with BA (0.1-0.8 ppm) where only a few shoots developed on explants reared on MS + kinetin (0.1 p.p.m.) or MS + adenine (1 ppm), though roots initiated when subculturing was done on MS + NAA (1 ppm) medium, however, rooting was profuse on NAA (1 and 4 ppm) irrespective of explants type (cotyledonary or hypocotyls), whereas 2,4-D (2 ppm) mainly induced callusing with sporadic rooting in cotyledonary explants (Gautam *et al.*, 1983). Mizozoe *et al.* (1992) through 75 per cent shading and half strength Ootsuka liquid fertilizer application to stock mother plants found enhanced shoot formation in leaf explants of cv. 'Ginsho' but not in cv. 'Yukimatsuri'. Reduction in the NH_4^+:NO_3^- ratio in the culture medium increased shoot formation in both these cultivars, but with explants in cvs 'Littlejem Yellow' and 'Yukimatsuri', shading of mother plants for two weeks after sowing resulted in fewer shoots and the optimum NH_4^+:NO_3^- ratio in terms of shoot formation (85 per cent) and shoots per explant was found 1:10 and 1:2 for both the varieties, respectively. *In vitro* callus growth of *Matthiola incana* was promoted by NAA and 2, 4-D, best being with 10 μM NAA and 1 μM 2,4-D, slightly more by auxins along with cytokinins than in the auxins alone, and initially the adventitious shoot formation was poor from single stocks than the double ones but after 10 weeks no difference was recorded (Okuse *et al.*, 1996), though root formation occurred without addition of any of the growth regulators, and instead, addition of cytokinins (kinetin, BA, zeatin and 2iP) inhibited it. Of all the cytokinins tested, adventitious shoots formed on medium supplemented with 3-30 μM cytokinin, optimum being 10 μM. Mesuali Sodi *et al.* (1994) from young (apical microcuttings and leaves of *in vitro*-grown seedlings) and older explants (leaves of greenhouse-grown plants), recorded only adventitious shoot formation either indirectly from compact nodular callus at the basal end of the microcuttings or directly from nodular structures on leaf explants of cv. 'White King', though presence of BA for shoot regeneration in the medium was critical, being 1 mg/l for young explants and 10 mg/l for old ones. Siemens and Sacristan (1995) isolated the protoplasts from leaves of 21-28 days old plants of wild-type line MSC and mutant lines MChl (chlorophyll deficient) and MutF (petioleless leaves and stalkless flowers), with yields of 1.4 × 106 from MSC and MChl, and 5 × 104/g fresh weight from MutF, with high division frequencies and plating efficiencies (0.9-2.2 per cent) being obtained from protoplasts of MSC and MChl embedded in sodium alginate and cultured in MI medium at plating densities of 2-3 × 105/ml medium. Colonies with >1.5 mm diameter were transferred to 4 shoot regeneration media, obtaining highest shoot regeneration rates of 41, 37 and 27 per cent for MChl, MSC and MutF, respectively, on AIIR medium, which on transfer to G1 medium, roots developed in 15, 13 and 11 per cent shoots of MutF, MSC and MChl, respectively, and then these were transferred to soil in the greenhouse. When Xu HuaiLiang *et al.* (2002) studied the explants of cotyledons, hypocotyls and cotyledon petioles of common stocks as a new source of omega-3-linolenic acid, by inoculating onto MS medium supplemented with different combinations of plant growth regulators, they obtained induction of calli, *vis-a-vis* plantlet regeneration via organogenesis, best being through explants taken from cotyledons and hypocotyls for callus induction on MS+0.1 mg/litre NAA, with a callus formation frequency of 100 per cent, and on MS+0.1 mg/l 6-BA shoots were readily differentiated from the calli induced from hypocotyls with a frequency of 67 per cent, though rapid root induction was obtained under shoots on 1/2MS+0.2 mg/l IBA with a frequency of 86 per cent.

Cultural Practices

Stocks do well in any good garden **soil**, *i.e.* the fertile medium loam at fully exposed site. For its cultivation the pH of soil should be from 7.0 to 8.5. In poor soils, incorporation of well rotten farmyard manure, compost or any organic matter is recommended so that soil becomes fertile and porous. From planting to flowering it takes about 80-110 days in autumn-planted winter-flowering stocks while 50-60 days in spring to early summer-planted ones and continue flowering for up to three weeks. Winter-flowering stocks require some six months from sowing to seed maturity.

Land preparation should start a month before the planting out and three ploughings should be done, After each ploughing, all the perennial rootstocks of the weeds and any other foreign material such as brick pieces, stones, polyethylene pieces, wooden pieces, *etc.* should be taken out with the fork every time, and then after third ploughing the lay out should be done properly by earmarking the irrigation channels, the paths and the beds, and then the beds are prepared and levelled properly. At the time of first ploughing the soil should be incorporated with farmyard manure at the rate of 3kg/m^{2}and thoroughly mixed. If the beds are not more than 1.6 m in width that is most convenient for tillage operations as by sitting to both sides of the beds all the manual operations can be carried out. The planting is carried out in October-November in the open at 30 × 30 cm distance, followed by irrigation so that plants do not desiccate. The proprietary multi-unit container Vacapot 15 produces the most acceptable plants, and direct spaced seeding in boxes has been found as satisfactory as pricking out (Powell, 1974). It would be better if planting is either carried out in cloudy days or in the afternoon. After planting all the cares are to be taken that when required the beds should be watered, weeds if any should be taken out. For first two months of planting, the beds are required to be cleaned of weeds at every three weeks. If required, the plants should also be supported with stakes.

Bondt (1978) used incadescent lamps, mercury vapour lamps, iodine lamps and sodium lamps to extend the 8-h **daylength** by a further 8-h from 22 March to 13 April for winter-flowering stocks planted out under glass on 13 January, he recorded taller plants under incandescent and sodium lamp treatments, and 8 days later flowering in case of non-illumination. Groote (1980) when gave supplementary **lighting** under glass for 8 or 16 h/day to improve photosynthesis to *Matthiola annua* (syn. *M. incana*) which is a long-day plant, from 20 February to 11 April through HPS and incandescent lamps for both the periods, as well as HPS lamps for only the 8-h treatment, where sodium lamps gave best results with respect to stem length, length of flower truss and earliness and resulted into 7-14 day advancement in flowering, however, the sodium lamps at 8-h treatment gave highest quality blooms than a 16-h or no illumination though in 16-h lighting the crop was earliest. Effects of single-layer glass and double-layer antifog polyethylene films on growth and flowering of stock in Canada revealed that stock produces more buds per spike with shorter but thicker stems under single-layer glass [with higher photosynthetic capacity (PC)] and under antifog 3-year polyethylene (Densereau *et al.*, 1998) though latter treatment was energy-efficient with good plant quality, PC being lower under all polyethylene covers than under single-layer glass irrespective of light regimes, though there were no differences among covering treatments in the net photosynthetic rate (PN) at low light level (canopy level). Stock which is a long-day plant, when was exposed to high red/far red photon flux ratios during the day and irradiated with red light at night, flowered late and developed short internodes, while the opposite occurred in plants grown under a low red/far red photon flux ratio during the day and irradiated with far red light at night, and the contrast between the two lighting conditions was especially apparent between autumn-sown (winter-cultivated plants) and summer-sown (autumn-cultivated plants) because temperatures were lower during the winter flowering period, though spring-sown and summer-grown plants had shorter internodes and more leaves with advanced anthesis compared with plants sown at other times of the year (Yoshimura *at al.*, 2002). Hisamatsu *et al.* (2002) studied the effects of red (R) & far red (FR) light, and day/night temperatures (20/15 and 25/20 °C) on stem extension and flowering of stock cultivars ('Sousei', 'Christmas Apricot No. 2' and 'Banrei') and recorded marked suppression of stem extension (suppression of GA_4 response) by R-rich spectrum treatment until the flower buds became visible, whereas it was slightly promoted or was not affected by the FR-rich spectrum treatment. The FR-rich spectrum treatment accelerated flowering while the R-rich spectrum delayed it under weakly flower-inducing temperatures but these phenomena were eliminated under strong flower-inducing temperatures, however, the dark-treated plants did not initiate flower buds in spite of cold treatment, which indicate that light quality is involved in regulation of stem extension and flowering but temperature is rather more responsible than light quality in stock. Kawabata *et al.* (2002) reported the cause of poor floral pigmentation in stock cv. 'Pigmy Rose' due to low light intensity because of reaction mediated by photoreceptors located in the petal and the reaction mediated by sugar supply from leaves or stems, but to find out more effective one out of the two by shading flowers with cheesecloth (PPFD<17 μmol $m^{-2}sec^{-1}$) or with aluminium foil (0 μmol $m^{-2}sec^{-1}$), the anthycyanin concentration was found reduced which suggests that anthocyanin production is a photoreceptor-mediated reaction, however, shading the whole plant did not affect pigmentation compared to flower-shaded ones which suggests that the amount of available sugar is not a limiting factor for anthocyanin production, and the placing of detached flowers on sucrose solutions reduced anthocyanin concentration due to decrease in the hexose concentration in the petal, especially under 2 per cent, which indicates that though soluble sugars in the petal affected anthocyanin production, their high concentration prevented fading of flower colour even under low light conditions. The flowering of summer-sown (autumn-cultivated) stock when at flowering the daylength in Japan is short and temperature is cool, was promoted and by increasing the irradiance further accelerated by irradiation of compact self-ballasted fluorescent lamps of far red light, followed by incandescent lamps (Yoshimura *et al.*, 2006). In the process of light filtering by leaves, the leaves allow more far red light (720-740 nm) to pass through than red light (660-680 nm), thus alter the red:far red ratio by altering light quality, so the plants perceive canopy shading *via* alterations in phytochrome photoequilibria that can result in increased stem elongation and reduced branching among other responses (Erwin *et al.*, 2006), and in this process, in contrast to red:far red light filtering, leaves absorb much of the photosynthetically active radiation (PAR) incident on the leaf surface, but allow some to pass through. They observed highest red:far red light ratio below a leaf in stock.

Characteristics of (a) ethylene venyl acetate (EVA) 'foamed' film (210 μm) prepared by Polimeri Europa (Italy), containing scattered microscopic air bubbles, and (b) standard film (200 μm) were compared on tunnels with the aim of obtaining a thermic covering material able to limit excessive increase of the daytime **temperature** in greenhouses, where Magnani *et al.* (1997) recorded total transmittance of shortwave IR being slightly lower and direct transmittance in the bands 200-1100 nm almost 15 per cent lower for (a) *vis-a-vis* transmittance of longwave IR radiation being also lower and with some 10 days earlier stock flower harvesting under 'foamed' film with higher stem length, fresh and dry weights, and floral length. As the angle of incidence of the luminous radiation increased above 45 °, there was a greater decrease in transmittance for (a) so that when the angle of incidence was 75 ° inclination of the source of radiation transmittance was 77 per cent for (a) and 88 per cent for (b), and air temperature in small tunnels (1.5 m wide, 1.5 m high and 10 m long) in the first 3 weeks of May increased less with (a) than (b) under clear skies, but there was no difference between the films in temperature performance under cloudy skies, however, night temperatures in the tunnels were 2 °C higher under (b).

Night **temperatures** higher than those normally used in commercial production systems promoted vegetative growth in *Matthiola incana* seedlings; in the short (8 h) daylength regimes and with a day temperature of 16 °C, cool nights (8

°C) had a detrimental effect, especially when imposed for 6 weeks from pricking-out rather than later; and the night temperatures between 8 ° and 16 °C did not substitute for the long-day requirement for flowering in *Matthiola* after 11 weeks of growth, and there was no advantage with respect to vegetative growth in maintaining night temperatures above 12 °C (Summerfield *et al.*, 1977). Studies on the effects of differences (DIF) between day and night temperatures, and constant temperature regimes (15, 20 or 25 °C) on the elongation and development of *Matthiola incana* seedlings revealed that 25/15 °C (day/night, *i.e.* positive DIF) compared with constant 20 °C (zero DIF) regime treatment promoted plant height, internode, leaf elongation, leaf unfolding rate and petiole lengths during rapid extension phase (13 days after treatment) though final values (10 days later) were found non-significant, whereas 15/25 °C (negative DIF) suppressed growth compared with zero DIF, however, leaf length and leaf unfolding rate responded significantly different to different constant temperatures, while not being significantly affected by DIF treatments, though switching from negative DIF regime to positive regime or *vice versa*, on day 13, final differences in final plant height, 1st internode length and 1st leaf petiole length between the corresponding treatments were observed, therefore, it is inferred that DIF effects may be advantageous to facilitate production of compact seedlings but that short-term control of elongation may be more easily achieved by altering average temperature than by changing DIF, however, the DIF during early seedling stage had a small effect on subsequent growth (Ito *et al.*, 1997a, b). *Matthiola incana* plants were grown in raised sandy loam beds in double-layer polyethylene-covered greenhouses which in year 1 were either unheated or had a minimum night temperature of 13 °C and in year 2 had a minimum night temperature of 2 or 10 °C, where warmer greenhouses each year reduced the production time though colder greenhouses increased stem lengths (Cavins *et al.*, 2000).

In an ordinary polyhouse when Loeser (1991) grew stock in two batches, the flowers produced were of good quality in a cellular potting system with 60 plants/m^2 and 34 litres of substrate/m^2 or with 96 plants/m^2 with 34 litres substrate/m^2. Equal parts by volume in the **potting media** of peat + sand or peat + sand + sawdust with either Osmocote (14:6:12), Floranid (20:2:8) or Nitrolime (26:0:0 + CaCO3) at 300 or 420 g N/m^3 were evaluated for growing *Matthiola incana* where foliage growth and dry weight were best in the peat + sand medium + Osmocote, followed by the peat + sand + sawdust medium + Osmocote (Thomas *et al.*, 1980). Potting mixes in greenhouse for stock composing of ground pine bark or composted hammer milled softwood and sand which were when **fertilized** *via* liquid feeds containing similar concentrations of N, but with S (as sulphate) in concentrations ranging from 0.01 to 70 mg l^{-1}, Hendreck (1986) reported that stock needed 25-27 mg l^{-1} S in the liquid feed.

Feeding and Irrigation

A process has been developed which converts composts into highly effective growing media by addition of phosphoric acid (to add P and reduce pH) and ammonium nitrate (to add N), followed by dilution using a low-nutrient substrate such as coir depending upon the chemical analysis and the intended end use, and so samples taken from some waste composting operations were transformed into operational growing media by applying 1.7-3.0 g ammonium nitrate and 50 ml phosphoric acid per litre of compost and then dilution with coir in ratio of 1 to 5 parts by volume, where when stock was sown directly, Rainbow and Wilson (1998) recorded that in some cases untreated compost killed seedlings though in many cases treated composts performed well. Slow-release compound Enmay is quite effective in stocks, ammonium sulphate is not suitable for use in composts and ammonium nitrate should not be applied at rates above 1.5 g/l whereas liquid feeding with 200 ppm N at each watering is preferable to a single base application of 2.25 g/l of Osmocot 14:14:14 (Powell, 1974). Holmes *et al.* (1981) applied four levels of K at weekly intervals to two plantings of stock plants grown to flowering in three media, *viz.* bark, peat-lite and soil where at all the levels of K, the plants grown in bark medium had greatest height, spike length, dry weight and shortest time to flowering, whereas in soil or peat-lite medium the optimum level of K was found as 100 ppm, however, the differences in plant performance were found associated more with media than with K levels. Stocks were grown in 1:1 peat:sand with 5 levels of N, P, K where there was little demand for P and K deficiency symptoms occurring at low K application though N strongly promoted plant growth and was further enhanced by high K application (Thomas and Leong, 1980). Through the use of ammonium fertilizer in greenhouse stock, Kasten *et al.* (1990) recorded 60 per cent reduction in nitrate content of the soil while placement of straw fertilizer together resulted into 80 per cent reduction, and there was less fertilizer leaching in the groundwater. In stock cv. 'Asanami' when grown in sand cultures supplied with 130-200 ml of nutrient solution daily, Yamanouchi and Yang XiuZhen (2000) in experiment I examined the effects of nitrogen treatments (10, 20, 40, 60, 100 and 180 g N kg^{-1}) on the concentrations of leaf and stem P, K, Ca and Mg; and in experiment II the influence of N supply period (solution without N on 17, 25, 32, 39, 45, 51 and 60 days after transplanting) in contrast to a normal solution which contained 100 g N kg^{-1}, on growth and cut flower quality. The minimum level of nitrogen concentration that resulted in good cut flower quality was ~30-32 g kg^{-1} in leaves at the upper 11th position, which was sufficient for nitrogen absorption to continue until visibility of flower bud. They recorded a high positive correlation between N concentration and K (in stem only), Ca (except 180 N), P and Mg concentrations in the leaves and stem, respectively, but K concentration in the leaves decreased when N increased above 40 g kg^{-1}.

If tolerant to salinity, the oilseed plants such as stocks, rich in essential polyunsaturated fatty acids (linoleic and α-linolenic acid), may be considered highly economic substitute for common field crops **irrigated** with fresh water, and the stocks which are highly tolerant to salinity may be irrigated with saline water though to some extent this affects oil content and composition (Heuer *et al.*, 2002). Heuer *et al.* (2005) tried five salinity levels (water EC 1.0 dS/m as control, and EC with 2, 4, 6 & 8 dS/m as saline water treatments) starting from 22 days after transplanting and up to four months, and then when after two months the plants were harvested there was decrease in yield with every increase in the salinity level, negatively significant being at EC 8 dS/m for total yield and seed number, and at EC 6 and 8 dS/m for seed yield where accumulation of Na^+ and Cl^- was also found commensurately increased though K^+ decreased, *vis-a-vis* non-significant decrease in N and P levels, however, seed oil content remained unaffected irrespective of salinity levels though total oil yield was reduced by 50 per cent at 8 dS/m and with a decrease in the yield of stearic and oleic acids and significant increase in the linolenic acid (omega-3) at high salinity levels. Demiral (2003) in greenhouse applied four concentrations of NaCl + $CaCl_2$ to *Matthiola tricuspidata* plants and though conducting Na, Cl and EC analyses of soil, plant and leached samples, *vis-à-vis* plant growth & development, plant dry matter content and yield, he recorded that increasing salinity increased translocation of Na and Cl in plants with significant effect on dry matter content of shoots and leaves, flower numbers and seed pod numbers; though in the second treatment, he recorded highest seed pod number under 2.0 dSm^{-1}, and increased dry matter up to 6.0 dS^{-1} in shoots & up to 8.0 dS^{-1} in the leaf which suggests that stock can be grown with saline water. Carter and Grieve (2006) investigated into the salinity tolerance of cut flowers of *Matthiola incana* 'Cheerful White' and 'Frolic Carmine' at the U. S. Salinity Laboratory to determine marketability based on stem length, where plants were exposed to differing water ionic compositions and salinity levels which though caused reduction in plant height with increase in the levels but the flowers were of marketable quality at moderate salinity levels. Grieve *et al.* (2008) subjected the stock cv. 'Cheerful White' with four irrigation water salinities (2, 5, 8, and 11 dS.m-1) and four nitrate concentrations (2.5, 3.6, 5.4, and 7.1 $mmol.l^{-1}$; N: 35, 50, 75, and 100 ppm), ammonium nitrogen being included in the nutrient solutions, where they recorded salinity significantly delaying initiation of the exponential growth phase, shortening its duration, and reducing the rate of plant development, being more pronounced at higher salinity levels so harvesting of marketable stems was delayed though in all the treatments the stems produced were of marketable size, *i.e.* ~60 cm. Increasing salinity increased leaf Na, Mg and Cl levels but decreased Ca and K, and increase in N levels (whose requirement in stock is very low) increased the Na level, decreased the K and Cl though Ca and Mg were unaffected. In a greenhouse experiment for determining effect of salinity (1.0, 2.0, 4.0, 6.0 and 8.0 dS/m) on stock, Heuer and Ravina (2004) recorded increase in soil salinity with saline water irrigation though relatively a constant soil salinity was found maintained in the root zone. For the first 80 days after initiation of salinity treatment (DAS), plant height was similar in all treatments, after which 2.0 dS/m yielded the tallest plants and 8.0 dS/m the shortest but after 125 DAS, plant height increased by 8 per cent at 2.0 dS/m with highest number of flowers and decreased by 14 per cent at 8.0 dS/m with lowest harvest index.

Weed Control

Since stock is grown as a winter annual in India so it encounters a large number of weeds in the field plantings and hand-weeding is a labour-oriented expensive venture. Therefore use of effective weedicide is recommended for taking good cut flowers, *vis-à-vis* seed production. Lenacil at 2.2-4.4. kg/ha applied one week after planting *Matthiola incana* in autumn, was safe and controlled autumn-germinating annual broad-leaved weeds (Anon., 1972). Tenoran 7 kg/ha, Ramrod (propachlor 65 per cent) 7 kg, Amiben-Granulat (chloramben), Kerb 50W 3 kg, and Legurame flussig [(carbetamide (70 per cent)] 10 litres and Ronstar (oxidiazon [75 per cent]) 8 litres were tested for their selectivity in *Matthiola* spp. where Tenoran though gave best results but caused some severe necrosis; Kerb 50W and Legurame gave satisfactory weed control; Amiben-Granulat, Kerb 50W and Legurame flussig were well tolerated; Ramrod lacked adequate herbicidal activity and Ronstar caused severe necrosis (Anon., 1974). Lamont (1986) in the plantings of stocks found Alachlor, Napropamide, Oryzalin and chlorthal dimethyl quite effective in controlling 80 per cent of the grasses, Alachlor giving the best results in case of broad-leaved weeds (80 per cent control compared with 30-75 per cent of other herbicides), though Chloroxuron was not found effective. Napropamide (2.25 or 4.5 kg/ha) and Alachlor (1.125 or 2.25 l/ha) were found quite safe in direct-sown stock plants where former controlled 50 per cent though latter 70 per cent broad-leaved weeds and both up to 90 per cent grasses, latter being effective both in summer and winter, though chlorthal dimethyl (7.5 or 15.0 kg/ha) was not found satisfactory in direct-sown stocks (Lamont and O'Connell, 1986). Duczmal (1989) when applied Aziprotryne (Mesoranil 50 WP) at 2.5 kg/ha post-emergence, it gave 47 per cent weed control in *Matthiola bicornis* (*M. longipetala*).

Growth and Flowering

Seeds of *Matthiola incana* cv. 'Brilliant' were soaked for 24 h in solutions containing 50-200 mg GA_3/l, 50-200 mg GA_{4+7}/l, 5-20 mg NAA/l, 50-200 mg PP_{333} (paclobutrazol)/l, 250-100 mg ethephon/l, 25-400 mg RSW 0411 (triapenthenol)/l or 100-400 mg Sumi-7 (diniconazole)/l but soaking in 100 mg GA_3 or 100 mg GA_{4+7}/l only increased germination percentage, though triapenthenol and diniconazole markedly reduced it (Grzesik, 1989).

In Japan, Fujita (1978) had sown eight stock cultivars in March, and the seedlings from 2-4 unfolded-leaf stage were kept in natural days at 15, 20 or 25 °C, which caused no flowering in four of the mid-season cultivars but at 25 °C there in the extremely early flowering cvs 'Christmas Rose', 'Senshonoyuki' and 'Christmas Blue' there were 100, 100 and 72 per cent blooming, respectively; and in the trial only with the extremely early cultivars, flowering percentage of 'Christmas Rose', 'Akinobeni', 'Akinomurasaki' and 'Kanshio' was 100 per cent at 23 °C, those of 'Benihime' and 'Tsukinoyosooi' was 98 per cent, of 'Christmas Blue' and 'Christmas White' were quite lower at 23 °C and only 70 per cent at 18 °C, though the early cv. 'Zuisei' flowered only at 13 °C. Seeds of 25 non-branching stock cultivars sown on August 25 in an unheated greenhouse, flower initiation in very early cultivars occurred by mid-October (mean temperature 21.5 °C) whereas in early, mid-season and late cultivars it occurred from mid-November to early December at 14.4, 13.4 and 12.4

°C, respectively; in a second trial where plants of seven cultivars were sown on February 17 and December 21 in greenhouses with minimum temperatures at 0-5, 8-10 and 13-15 °C, the very early and 'Seikai' (mid-season) cultivars flowered at all temperatures but in the remaining two cultivars the flowering percentage fell at 15 °C; and in a third trial the three very early cultivars flowered 100 per cent at 8, 13 and 18 °C, but the highest temperature which could induce 100 per cent flowering in early and mid-season cultivars was 13 °C and for a late cultivar 8 °C (Fujita and Nishitani, 1978). Flowering in very early, early and mid-season cultivars was hastened by increasing temperatures up to 15 °C but higher temperatures delayed it. In all the cultivars the number of nodes below the inflorescence varied little at temperatures below 8 °C but increased as the temperature rose and in all very early, early and mid-season cultivars of Japanese origin the number of nodes below the inflorescence was in the narrow range (27-40) at 0-8 °C suggesting that the length of the juvenile phase was the same in all the cases. Fujita and Nishikami (1979) when grew the seedlings of several branching stock cultivars under glass at 3 °C minimum temperature and then at 8, 13, 18 or 23 °C, the minimum temperature for inducing flower bud initiation was found as 23 °C for the very early-flowering cultivars, 'Iwaiaka No.1' and 'Seven-week Trisomic Dwarf Double', 18 °C for early cultivars 'Iwaiaka No.2', 'Wakazakura' and 'Dwarf Ten-week', 13-18 °C for the mid-season cv. 'Kanchidori' and 13 °C for the late cvs 'Matsudoaka' and 'Trisomic Hi-Double Lilac-Lavender'. For the production of cut flowers with stalks of sufficient length, it was necessary to keep the temperature above the minimum for flower bud initiation for a certain period. Fujita (1979) when grew branching and non-branching stocks under glass with a daily minimum temperature of 0 °C from the 5th day after sowing until flowering, the number of nodes to the inflorescence was lower in early-flowering cultivars than in late ones of branching stocks, but with non-branching stocks there was little difference among extremely early-, early- and medium-flowering cultivars, and where the node numbers were lower these took less time to floral bud appearance. Both types of stocks when were grown initially at a temperature above 18 °C and then subjected to a low-temperature regime at various stages with a daily minimum temperature of 1 °C or 4 °C, earlier start of the low-temperature treatment produced lower number of nodes with exception of very early cultivars. Stocks grown at a minimum daily temperature of 18 °C until the 10th unfolded leaf stage, then transferred to a low-temperature regime with a minimum of 4 °C for 0, 10, 20, 30 or 40 days and then returned to the high-temperature regime, medium-flowering branching type and early to medium-flowering ones of non-brnaching cultivars gave 100 per cent flowering under 10 days, branching type late-flowering cultivars 20 days and non-branching type late cultivars 40 days. Good flowering occurred only when the flower buds had been well differentiated before the end of the low-temperature treatment. Weekly foliar applications of GA_{4+7} at 25 ppm during seedling growth promoted reproductive development of *Matthiola incana* whether grown in short days (8 h) or long days effected by night-breaks (NB), both GA and NB favourable for early flowering and associated stem elongation, inhibited leaf growth, though GA applications substituted the long-day requirement for flowering (Peat and Summerfield, 1977). When GA_4 and three gibberellin biosynthesis inhibitors, *viz.* uniconazole (UCZ), prohexadione-calcium (PCa) and trinexapac-ethyl (TNE) were studied on the growth and flowering of *Matthiola incana*, in the early flowering 'Sourei', UCZ inhibited stem elongation and flowering, but at 20/15 °C (day/night) and 25/20 °C temperatures, GA_4, PCa and TNE promoted flowering in 'Sourei' though only PCa and TNE at 20/15 °C in late-flowering 'Banrei' under 12-h photoperiods; and GA_4 and PCa at all the temperatures tried, caused stem elongation in both the cultivars but no induction of flowering (Hisamatsu *et al.*, 1998), whereas during winter in greenhouse conditions, PCa promoted flowering in 'Banrei' but neither GA_4 nor PCa in summer induced flowering. Hisamatsu *et al.* (1999) stated that in stock, low temperature requirement for flowering is more stringent for late flowering cultivars than early flowering ones, and by treating with 10 or 100 ppm prohexadione-calcium (PCa) or 100 ppm GA_3 the flowering season of cvs 'Sourei' (very early), 'Pearl Pink' (early) and 'Fujimusume' (intermediate) was significantly accelerated with no change in spike length and stem diameter by 10 or 100 ppm PCa though decrease was recorded significantly by GA_3 application, and in the late cv. 'Banrei' the flowering date was promoted by 100 or 1,000 ppm PCa though 1,000 ppm decreased cut flower length. While studying physiological mechanisms of gibberellin effect on the growth and development of *Matthiola incana*, Hisamatsu (2001) stated that GA was necessary for stem elongation and flowering: GA_4 in the non-13-hydroxylation pathway being the most active component for stem elongation and flowering; GA biosynthesis inhibitors of the acyclohexanedione type could promote stem elongation and flowering; and GA-biosynthesis pathway prior to GA_{53} might be blocked in the plants with rosette formation. Now it is known that in stock, stem elongation and flowering are promoted by exogenous GAs, including GA_4, and also by acylcyclohexanedione inhibitors of GA biosynthesis, such as PCa and trinexapac-ethyl (TNE), and as it was unclear how GA biosynthetic inhibitors could promote stem elongation and flowering, on examining their effect on GA biosynthesis through quantifying endogenous GA levels, *vis-a-vis* the sensitivity of stem elongation and flowering to various GAs in combination with the inhibitors, Hisamatsu *et al.* (2000) found that stem elongation and flowering were most effectively promoted by GA_4 when combined with PCa, and next in order by 2,2-dimethyl-GA_4, PCa, GA_4+TNE, TNE, GA_9+PCa and GA_4, though little or no promotion by GA_1, GA_3, GA_9, GA_{13}, GA_{20} and 3-epi-2,2-dimethyl-GA_4. Both the promotive effects of the acylcyclohexanediones on stem elongation and flowering, particularly when applied with GA_4, and the fact that TNE caused a build-up of endogenous GA_4 imply that one effect of TNE at the lower dose involved an inhibition of 2 beta-hydroxylation of GA_4 rather than an inhibition of 2-oxidation and 3 beta-hydroxylation of GAs which were precursors of GA_4. Overall, these results indicate that (i) GAs with 3 beta-OH and without 13-OH groups (*e.g.* GA_4) are the most important for stem elongation and flowering in stock, and (ii) growth promotion rather than inhibition can result if an acylcyclohexanedione acts predominantly to slow 2 beta-hydroxylation and so slows down inactivation of active gibberellins, including GA_4. It follows that a low dose of an acylcyclohexanedione can be a 'growth enhancer' for any applied GA that is liable to inactivation by 2 beta-hydroxylation. Hisamatsu and Koshioka (2001a, b) studied the role of GAs in controlling growth and development in stock by regulating through manipulation of GA-biosynthesis and stated that GA_4 is the most active component than $GA_{1,3,9,13,20}$ in the non-13-hydroxylation pathway of GA biosynthesis for stem elongation and flowering due to GA biosynthesis inhibitors of acylcyclohexanedione type, and they further stated that development of responsiveness to active GAs for flowering may be one of the essential factors in the flower bud initiation process as the flowering in summer-sown stock was shown to accelerate the flowering time due to PCa treatment under polyhouse conditions. Hisamatsu and Koshioka (2000) reported optimum temperature for induction of flower bud initiation in stock as 10 °C though they recorded minimum duration when the late flowering cv. 'Banrei' was subjected to 20 °C. They mentioned that stem elongation was markedly promoted by cold treatment regardless of flower bud initiation, though GA_4 level necessary for flower bud initiation was found lower in 10 °C treatment than in the 15 °C, and it became lower with longer durations of cold treatment, which indicate that the cold treatments enhance responsiveness to GA_4 not only in the stem elongation process but also in the flower bud initiation process. Since in stocks there is close relationship between flowering and gibberellins biosynthesis, cyclohexadione-type growth retardants were when tried (Takami *et al.*, 2004) in the open with stock cvs 'Banrei' (late flowering) and 'White Beach' by applying PCa at 10, 15 and 30 leaf stages, 'White Beach' flowered 5-42 days earlier depending upon stage of application as earlier application brought about earlier flowering, however, 'Banrei' flowered 11-13 days earlier with no significant differences between stages of application.

Roh (1978) stated ancymidol as the best chemical for suppressing main shoot growth, inducing development of lateral shoots and allowing good flowering; dikegulac inhibiting main shoot growth as well as laterals and ethephon and dikegulac (Atrinal) both inhibiting flowering. Gibberellins improved seed germination and the seedlings from soaking of seeds with GA_3 or GA_{4+7} (50-200 mg/l) and NAA (5-20 mg/l) were taller and started to flower earlier than plants from the other treatments with no effect on final plant height and seed yield or quality either through seed soaking or plant spraying, though there was no clear effect of ethephon (250-1,000 mg/l), paclobutrazol (50-200 mg/l) or uniconazole (100-400 mg/l) at any stage (Grzesik, 1995). Treatment with GA at 150 mg/l or CCC (chlormequat) at 5 g/l, GA at 150 mg/l advanced growth and flowering and increased the leaf content of polyphenols (chlorogenic and sinapic acids) and active indolic compounds (IAA, glucobrassicin and indoleacetonitrile), and accelerated the synthesis of indoles from tryptophan in stock whereas CCC at 5 mg/l retarded the growth, flowering and other processes (Runkova, 1977). Sprays of B-Nine, Cycocel (chlormequat) and Ethrel; and foliar spray and soil drench with ancymidol (as EL 531 and EL 558 formulations) greatly reduced plant height and extended the saleable life by 3-13 days in stocks (Powell, 1974). Foliar spray of B9 to stock plants though markedly reduced plant height and was temporarily phytotoxic at 5,000 and 6,000 ppm as these rates did offset the response to N, particularly at the highest rates (450 and 600 g N/m^3) when there was a mild negative N × B9 interaction on foliar dry weight (Thomas and Leong, 1980). Chlormequat (500-1500 mg/l) in stock decreased plant height, inflorescence and peduncle lengths; increased number of lateral shoots, inflorescences and flowers; caused the accumulation of N, P and K in the shoots and hastened flowering; and GA_3 (60 mg/l) though increased shoot growth but with no effect on number of inflorescences and flowers, *vis-a-vis* nullified the effect of chlormequat on shoot growth, time of flowering and mineral constituents; and showed an additive effect to that of chlormequat in respect of NPK accumulation (Hamza and Helaly, 1983). For growing compact pot plants of stock, Ecker *et al.* (1992) potted 20 days old seedlings of cvs 'Lavender' (medium growing) and 'Midget Red' (small growing) in 1.2-litre pots outside, and after 20 days applied paclobutrazol drench at 0.25, 0.5 or 2.0 mg/pot (total 50 ml/pot), which increased time to flowering in 'Lavender'with every increase in concentration though not in 'Midget Red', and decreased stem length in 'Lavender' *vis-à-vis* inflorescence in both the varieties and its 0.5 or 2.0 mg/pot produced a thicker and more compact plant with decrease in perecentage of senesced flowers in 'Midget Red', therefore he recommended 0.5 mg/pot for commercial pot plant production. The plug-grown *Matthiola incana* 'Midget Red' were transplanted into jumbo six packs and when the seedlings had attained ~ 5 cm in height or width, Kuehny *et al.* (2001) sprayed them with 1000/800, 1250/1250 or 1500/5000 mg l^{-1} of chlormequat/daminozide solutions which reduced the plant height significantly.

Postharvest

Stocks when start opening of a few flowers (around 6) in the raceme are severed from the plant with maximum of stalk and cut ends are immediately put in plain water. The vase life of glasshouse stock cut flower could be greatly improved by the following practices, *i.e.* a minimum of six open flowers, by transporting the flowering stems with roots still attached or pre-treating the bunches in a solution of the commercial preparations Anjer VB or Rosal and keeping the bunches upright to prevent the stems growing crooked, by cutting the stems above the root collar in the retail store, and by ensuring that the buyer uses a solution of a cut-flower preservative in the vase water.

Leaf pigments, such as chlorophyll and carotenoids, apart from being decorative plant parameters which in fact define postharvest life, are in fact, essential plant molecules providing carbohydrate and energy during entire plant life, but due to internal hormonal imbalance such as lack of cytokinins which cause leaf yellowing and petal senescence (Ferrante *et al.*, 2004) for which pulsing with 5 or 10 μM thiadiazuron (TDZ), 150 mgl^{-1} 8-HQS and combinations of TDZ with 8-HQS on total carotenoids and ABA concentrations were when studied in cut stock leaves, it was found that total carotenoids decreased drastically from 1548 μg cm^{-2} until reaching 565 μg cm^{-2} at the end of vase life and the endogenous ABA increased strongly at the same time starting from 167 ng g^{-1} DW at start of the study to 1322 ng g^{-1} DW at the end of vase life, where TDZ inhibited carotenoid degradation but not the ABA while HQS did not prevent carotenoid degradation but only slightly affected ABA concentration. Stock cv. 'Rubin' cut flowers were when subjected to 'Proflovit-72' (0.3 g 8-HQS + 0.05 g CCC + 50 g sucrose/l) and compared with $AgNO_3$ (20-50 ppm), $KMnO_4$ (4 ppm), streptomycin (200 ppm), or 8-HQS (200 ppm) and in some cases, 5 per cent sucrose and/or CCC at 50 ppm, 'Proflovit-72' was found to be the best for flower quality and vase life (Nowak and Rudnicki, 1975). Immediate placing of cut flowers after cutting in water and in various conditioning solutions before marketing, Kalkman (1983) recorded little improvement in its vase life, albeit Chrysal-avb proved even phytotoxic. Cut stock flowers when were placed in light and in dark conditions after pulsing for 24 hours with water (control) or vase solution (5-10 μM TDZ, 50-100 μM 1-NAA, 150 mg l^{-1} 8-HQS, or combinations of NAA or 8-HQS with TDZ), Ferrante and Mensuali Sodi (2006) observed that TDZ delayed leaf yellowing in light in the course of whole study period (30 days) but under dark it delayed chlorophyll losses until 10-12 days though other combinations in dark or in the light did not make any impact. Celikel and Reid (2002) stated that respiration of stock flowers has a Q10 of 6.9 at 0-10 °C by which simulated transport for 5 days resulted in marked reduction in the vase life transported at 10 °C or above, however, water uptake, flower opening and vase life increased somewhat in a vase solution containing 50 ppm NaOCl, and considerably in a commercial preservative containing glucose and a bactericide, though exposure to exogenous ethylene resulted in rapid desiccation and abscission of the petals, the effects that were prevented by pretreatment with STA or 1-methylcyclopropene (1-MCP), STS being more effective. Stock is an ethylene sensitive cut flower so pretreatment with low doses of 1-MCP gives results similar to STS in negating ethylene effects and prolongs the vase life in an ethylene-free environment and in presence of ethylene at 1 μl/l, the vase life of 1-MCP-treated flowers increases up to four times than control (Serek *et al.*, 1995).

Song CheonYoung *et al.* (1996) conducted pretreatment studies on the vase life of stock cv. 'Chohong' with water, sucrose (7 per cent) + STS (0.5 or 1.0 mM) or sucrose + STS (0.5 mM) + GA_3 (20 mg/l) for 2 h; cold storage in distilled water or newspaper at 4 °C for 1 or 2 weeks; and various preservative solutions [water, carbonated water (30 per cent) or 8-HQS (50-200 mg/l) alone or in combination with sucrose (1 or 7 per cent), GA_3 (5 mg/l) or $CaNO_3$ (0.5 or 1 mM)], where pretreatment with STS + sucrose or STS + sucrose + GA_3 extended vase life by about 2 days irrespective of storage durations and methods; preservative solutions containing HQS at 100-200 mg/l + sucrose at 1 per cent increased the vase life by 4 days with improved flower quality; and preservative solutions containing $CaNO_3$ at 0.5 or 1 mM decreased tip bending of flower stalks and increased vase life; however, the vase life of flowers held in HQS + sucrose after pretreatment with STS + sucrose was increased by about 5 days; the quality of flowers following 1 or 2 weeks dry or wet storage was good following these treatments. In Iran, stock is the most important ornamental plant due to its various shapes, colours, fragrance and ease of cultivation, therefore, an experimeant on stock in Iran carried out by Arab *et al.* (2006) on the effects of water (control) and 8-HQS (150 mg/l), alone or in combination with sucrose (2 per cent) and cold storage temperatures (4, 8 and 12 °C), revealed that high temperature causes early flowering but with short vase life, 8-HQS alone or with sucrose increases the vase life and stability of hydraulic conductance of flowers stored at the three temperatures tested, although increase in the vase life of flowers stored at 4 °C was higher with 8-HQS where increase in fructose, glucose and sucrose contents in the flowers, stems and leaves of cut flowers were also noted. The effects of STS and 1-MCP were studied by Dole *et al.* (2005) on cut stock by treating the unpacked cut stems and placing them either in deionized water (DI) and subjected to 1-MCP (740 nl l^{-1}) or ambient air for 4 h or DI + STS at 0.2 mM for 4 h. After treatment, stems were removed, placed in polyethylene sleeves and stored either wet in DI water or dry in plastic-lined floral boxes at 5 °C in the dark for 4 days. After storage, bunches were placed in DI water under 12 h light (76 to 100 μmol $m^{-2}s^{-1}$) per day where it was found that STS increased the vase life though 1-MCP did not. Using 1-MCP in stock flowers, Muneto *et al.* (2007) recorded enhanced vase life as compared to control and STS treated ones and this treatment decreased ethylene concentrations whereas 1-aminocyclopropane-1-carboxylic acid (ACC) synthase and ACC oxidase (ACO) activities were not clearly different among the treatments. Freshly harvested stock flowers were when held in tap water (control) or in a 20, 100, 200 or 1000 ppm benzalkonium chloride (surfactant), the latter at 200 ppm and higher increased the fresh weights irrespective of solution depth and when its treatment was for 4 hours and then packed in corrugated cardboard box, left for 18 hours and then held in a vase, the fresh weights of cut flowers increased even without recutting the cut ends, though in water without cutting of the cut ends there was no increase in fresh weight (Murahama, 2007). Macnish *et al.* (2008) when added aqueous chlorine dioxide (ClO_2) at 2 or 10 μl l^{-1} in clean deionized water, the build up of bacteria in vase solution was prevented and the vase life was found extended in cv. 'Ruby Red', and they recorded increased vase life even when ClO_2 was added in water already contaminated with 1011 CFUL-1 aerobic bacterium.

Insect-Pests, Diseases and Physiological Disorders

Parrini and Rumine (1989) recorded *Epichoristodes acerbella*, *Lipaphis erysimi*, *Myzus persicae*, *Nezara viridula* and *Tetranychus urticae* insect-pests feeding on stocks. *Plutella xylostella* has been found infesting various members of Brassicaceae, Solanaceae, Alliaceae and certain others, including *Matthiola incana* in China (Wu, 1993). Members of Brassicaceae are more vulnerable as mustard oil is a common component of the food plants of this plutellid. Fos *et al.* (1986) collected and identified *Neoaliturus haematoceps* leafhoppers from stocks and certain other plants in France

(Corsica), Morocco, Syria and Turkey which is a vector for *Spiroplasma citri*, and this is the first record as vector of *Spiroplasma*. *Circulifer tenellus* cicadellid collected from Israel is polyphagus and also transmits *S. citri* on stocks (Klein *et al.*, 1988). *Circulifer haematoceps* (syn. *Neoaliturus haematoceps*), a polyphagous leafhopper was also recorded from Israel which feeds on many plant species including stocks (Klein and Raccah, 1991). Carrillo (1974) in Chile recorded *Lipaphis erysimi* aphids infesting on stock plants. In the United Kingdom, *Delia radicum* fly maggots were recorded feeding on stocks at all growth stages (Lane and Wheatley, 1984). Almeida *et al.* (1997) in Israel collected several leafhopper (*Circulifer tenellus*) variants and injected with *Spiroplasma citri*, out of which two [variant from *Atriplex halimus* designated as *C. tenellus*-A (CTA) and the other from *Portulaca oleracea* as *C. tenellus*-P (CTP)] were found transmitting the disease to stock plants within one hour and symtoms were visible from 6-32 days of feeding. Chandra and Lal (1973) in Kullu Valley (H.P.) recorded *Thrips flavus* and *T. kodaikanalensis* feeding on stocks. Vikas Kumar *et al.* (2005) collected *Coleothrips mongolicus* from the flowers of stock plants. Takano *et al.* (2004) when irradiated (400 Gy) stocks, two-spotted spider mite (*Tetranychus urticae*) was found more tolerant than other tested species of cut flower pests such as American serpentine leafminer (*Liriomyza trifolii*), the Comstock mealybug (*Pseudococcus comstocki*), the melon thrips (*Thrips palmi*), the onion thrips (*Thrips tabaci*), the green peach aphid (*Myzus persicae*) and the common cutworm (*Spodoptera litura*) and in most cases the eggs and other stages were found sterilized.

Suatmadji (1988) in Australia recorded *Rotylenchus robustus* nematode infesting stocks. Ibrahim and Al-Yahya (2002) through a survey in Riyadh (Saudi Arabia) identified the nematode *Helicotylenchus* infesting on the roots of stocks. Nematodes can be controlled through use of Carbofuran granules in the field.

Niebisch (1985) found that **seed dressing** and substrate disinfection of stocks with Captan reduced damping off and foot rot diseases and therefore increased seedling emergence. Lorenzini *et al.* (1983) described *Uromyces limonii* and *Oidium matthiolae* attacking stocks in Italy. Parrini and Rumine (1989) recorded *Botrytis cinerea*, *Oidium matthiolae*, *Peronospora matthiolae*, *Pythium*, *Rhizoctonia solani*, *Sclerotinia sclerotiorum*, *Xanthomonas campestris* pv. *incanae* infecting stock plants in Italy. Pasini *et al.* (1996) in Italy isolated *Rhizoctonia solani* from stocks, causing collar or foot rot. ***Sclerotinia matthiolae*** isolated from stocks differs from ***S. sclerotiorum*** in its smaller apothecia, asci and ascospores, and particularly in its greater pathogenicity stock plants wither completely though *S. sclerotiorum* causes only mild infection in wounded plants, and to control these, Tecnazene, Quintozene, Cycloheximide and phenylmercury acetate have been found quite effective fungicides (Tapio, 1972). Radish **black spot** (*Alternaria japonica*) is a global problem on stocks producing lesions on leaves, petioles, stems, inflorescence, siliquae and the fungus also infects the seeds (David, 2000). ***Phytophthora megasperma*** var. *megasperma*, a soil-borne pathogen favoured by wet conditions with temperatures around 15-24 °C and which survives in soil for over one year primarily due to delayed germination of oospores, has been recorded on stocks (Waterhouse and Waterson, 1966). Stocks appear to be particularly prone to ***Rhizoctonia*** sp. and Quintozene is recommended as a preventive treatment, being added to the compost as a dust or wettable powder, though benomyl at 0.2 per cent applied as a drench to the top 6 mm of compost appears to arrest the spread of the fungus as well as combats ***Botrytis*** infection (Hendy and Roberts, 1974). Gerlach (1976) from Germany recorded ***Fusarium oxysporum*** f. sp. *conglutinans* for the first time on stock. Garibaldi and Gullino (1976) made new records of *Fusarium oxysporum* f. sp. *conglutinans* and *Sclerotinia sclerotiorum* on stocks from Italy. Bosland and Williams (1987) through a global collection of 123 putative isolates of *Fusarium oxysporum* collected in Canada recorded *F. oxysporum* f. sp. *matthioli* from stock plants. The pathogenicity of four isolates of *Fusarium oxysporum* f. sp. *raphani* when was tested produced typical symptoms on stocks (Garibaldi *et al.*, 2006). *Fusarium oxysporum* has been observed causing one-sided stock wilt in the greenhouse, where wilting progresses from the base upwards, growth stunts, leaves bleach and the vascular tissues become stained dark brown (O'Neill *et al.*, 2004). *Fusarium oxysporum* f. sp. *matthiolae* in UK causing widespread stock wilt since 2003 could not be controlled even through steaming of the soil in the glasshouses previously encountered with severe outbreaks but when O'Neill *et al.* (2005) cultivated the soil and used formaldehyde as formalin at 0.15 l/m^2 directly to the soil, followed by overhead sprinkler watering (3.9 l/m^2), they recorded complete elimination of *Fusarium* from stem pieces on the soil surface. When *Penicillium simplicissimum* isolated from non-sterilized Kureha soil in Japan, was introduced to sterilized Kureha soil, the suppressive effect of B-143 (4-allyl-2-azetidinone) was reproduced even to control *F. conglutinans* on stock cv. 'Tokoharu' (Kobayashi *et al.*, 1997). To control *Pythium splendens* on stock seedlings, when Niebisch and Kelling (1986) used Previcur N (propamocarb) and Ridomil (metalaxyl) + zineb, they recorded Ridomil superior over Previcur N, and application of Captan 80 or Thiram FW on seeds and seedlings, they recorded increased yield of cut flowers.

Foliar disease of stocks caused by *Peronospora parasitica* was in California first reported in 1999 and when the infected leaves were pressed into healthy plants and incubated, the disease symptoms appeared after 10 days (Koike, 2000). In Tuscany (Italy) when stocks were grown for late winter and early spring flower production, they were liable to be infected with downy mildew (*Peronospora matthiolae*), particularly at the young plant stage, more serious being in the crop meant for spring-flowering, where upper leaves were severely damaged, and for this with Fongarid (systemic fungicide), Trimboli and Hampshire (1978), with two applications of 0.1 per cent or 0.2 per cent Fongarid w.p. (furalaxyl) or 0.15 per cent Previcur (prothiocarb) at weekly interval, Dirske (1981) and with metalaxyl, benalaxyl, oxadixyl, furalaxyl and fosetyl-Al, Parrini and Rumine (1987) obtained excellent control of *Peronospora matthiolae*.

Bacteriosis on stocks is caused by *Xanthomonas campestris* and appears as leaf yellowing *vis-à-vis* black rot, browning to blackening of leaf veins, and black rot of stems and seeds and carried through the seeds to the next season and spreads in the field through rain and irrigation water by hydathodes of the leaf margins or by stomata or by wounds to the root system (Hayward and Waterson, 1965). Bazzi *et al.* (1987) reported bactereium *Xanthomonas campestris* infecting the stocks in Italy. In Tuscany (Italy), there was a serious epidemic of bacterial blight (*Xanthomonas campestris* pv. *incanae*) causing acropetal discolouration and leaf withering, *vis-à-vis* longitudinal cracks in stems, appearing at different stages from emergence to flowering on greenhouse-grown stocks during 1985 and 1986 causing up to 90 per cent loss, for which Minardi *et al.* (1988) suggested seed treatment at 53-55 °C for 10 minutes, followed by rapid cooling as a preventive measure. In Iran, Rahimian and Okhoyatian (1989) recorded *Xanthomonas campestris* pv. *incanae* as the causal agent of stock bacterial blight. Potted plants of 10 genotypes of stocks, *viz.* two accession each of *Matthiola tricuspidata*, *M. aspera*, *M. longipetala* and four breeding lines of *M. incana* were evaluated against bacterial blight (*Xanthomonas campestris* pv. *incanae*) by spraying with bacterial cell suspension, adjusted to 108 cells/ml, where Ecker *et al.* (1995) found that accessions as well as *M. incana* × *M. tricuspidata* cross exhibited high resistance to the disease while other breeding lines were found severely infected. When the seeds or rooted cuttings of stocks were inoculated with *Pseudomonas fluorescens* strain E6, Yuen and Schroth (1986) recorded increased growth and fresh weight (18-41 per cent greater) after 3-4 weeks of treatment. Through a survey of eight cruciferous crops including stocks, Butcher *et al.* (1976) recorded postitive correlation between presence of indole (neutral and acidic auxin) glucosinolates [glucobrassicin, neoglucobrassicin, or glucotropaeolin (a potential auxin precursor)] and susceptibility to *Plasmodiophora brassicae* clubroot bacterium (a large cancerous outgrowth covering and destroying whole root system), and stated that the concentrations of glucobrassicin in certain crop-tissues increased during infection by the pathogen. Clubroot (*Plasmodiophora brassicae*) in stocks affects even other members of the Brassicaceae, the soil remains infected for decades and its infection is favoured by wet, poorly drained and acidic soils where liming is a key control measure, boron and nitrate promote populations of soil bacteria antagonistic, and as there is no any specific pesticide to control this so the alternative left is only effective control of all the cruciferous weeds, autumn digging, winter fallowing and rotational cropping (Dixon, 2003).

Esfahan province (Iran) on stock experienced the incidence of a virus, *i.e.* a strain of TuMV (Bahar *et al.*, 1985). Rosciglione and Cannizzaro (1983) recorded TuMV on stock from Sicily (Italy). Foddai *et al.* (1991) for the first time from Sardinia (Italy) recorded turnip mosaic potyvirus (TuMV) causing leaf marbling and flower breaking on stock plants. In 1993 at Naples (Italy), polyhouse stocks were showing symptoms of mosaic, stunting, severe leaf malformation, breaking and greening of flowers and abortion of the inflorescence in 50-100 per cent of plants due to the mixed infection of cauliflower mosaic caulimovirus and turnip mosaic potyvirus (Alioto *et al.*, 1994). Mosaic symptoms on leaves and colour breaking of flowers on stocks in Korea were observed in 1995 due to attack of CMV and TuMV (Yoon JuYeon *et al.*, 1998). In China, Duan YongJia and Cai Hong (1998) observed brilliantly variegated stock flowers infected with tulip breaking potyvirus. From Markazi (Iran), Farzadfar *et al.* (2005), for the first time in 2003, recorded TuMV causing mosaic and colour breaking on stocks. Gera *et al.* (2006) recorded phytoplasma and/or spiroplasma disease on stocks in Israel. In Sicily (Italy) in April 2007 in a glasshouse cultivation of stock cv. 'White Beach', Davino *et al.* (2007) recorded an outbreak of a phytoplasma disease belonging to ribosomal subgroup 16SrII-A which caused stunting and rosetting with smaller flower size characterized by virescence symptoms.

References

Alioto, D., L. Stavolone and B. Aloj, 1994. Serious alterations induced by cauliflower mosaic virus (CaMV) and turnip mosaic virus (TuMV) on *Matthiola incana* (Italian). *Informatore Fitopatologico*, **44**(6): 43-46.

Almeida, L.de, B. Raccah and M. Klein, 1997. Transmission characteristics of *Spiroplasma citri* and its effect on leafhopper vectors from the *Circulifer tenellus* complex. *Ann. Appl. Biol.*,**130**(1): 49-59.

Anonymous, 1972. Weed control in fall-planted annual flowers. Research Branch Report 1971, Canada Department of Agriculture, p. 337.

Anonymous, 1974. *Ann. Report German Pl. Prot. Serv.*, No. 21, pp. 221-223.

Arab, M., A. Khalighi, K. Arzani and R. Naderi, 2006. Influence of cold storage, 8-hydroxyquinolin sulfate and sucrose on vase life and quality of cut stock flowers (*Matthiola incana* L.) cv. Asanami (Persian). *Iranian J. agric. Sci.*, **37**(1): 83-92.

Ardelean, M., R. Sestras, D.A. Pui, M. Cordea and V. Mitre, 2008. General (GCA) and specific (SCA) combining ability in several intraspecific hybrids of gillyflower (*Mattihola incana* L.) and their value for breeding. *Bull. Univ. Agric. Sci. Vet. Med.* (*7th int. Symp. on Prospects for the 3rd Millennium Agriculture*, sections Horticulture and Forestry- Economics and Management, Miscellaneous; held at Babes-Bolyai Univ., Mihail Kogalniceanu Street, Cluj Napoca Horticulture, Romania on Oct. 2-4), **65**(1): 44-47.

Bahar, M., D. Danesh and M. Dehghan, 1985. Turnip mosaic virus in stock plant (Persian). *Iranian J. Pl. Path.*, **21**(1/4): 11-12, 33-39.

Bazzi, C., P. Minardi and U. Mazzucchi, 1987. Bacterial diseases of flower and ornamental plants in Italy (Italian). *Informatore Fitopatologico*, **37**(6): 15-24.

Bondt, K.de, 1978. *Matthiola annua* (*incana*) (Dutch). *Verbondsnieuws voor de Belgische Sierteelt*, **22**(21): 767.

Bosland, P.W. and P.H. Williams, 1987. An evaluation of *Fusarium oxysporum* from crucifers based on pathogenicity, isozyme polymorphism, vegetative compatibility, and geographic origin. *Canadian J. Bot.*, **65**(10): 2067-2073.

Brinker, A.M. and G.F. Spencer, 1993. Herbicidal activity of sulforaphene from stock (*Matthiola incana*). *J. Chem. Ecol.*, **19**(10): 2279-2284.

Butcher, D.N., L.M. Searle and D.M.A. Monsdale, 1976. The role of glucosinolates in the club root disease of the Cruciferae. *Mededelingen van de Faculteit Landbouwwetenschappen, Rijksuniversiteit, Gent*, **41**[2 (1)]: 525-532.

Carrillo, L.R. 1974. Aphidoidea of Chile I (Spanish). *Agro-Sur.*, **2**(1): 34-40.

Carter, C.T. and C.M. Grieve, 2006. Salt tolerance of floriculture crops. In: *Ecophysiology of High Salinity Tolerant Plants*, pp. 279-287(eds Almal Khan, M. and D.J. Weber). Springer Science +Business Media, Dordrecht, Netherlands.

Cavins, T.J., J.M. Dole and V. Stamback, 2000. Unheated and minimally heated winter greenhouse production of specialty cut flowers. *HortTech.*, **10**(4): 793-799.

Celikel, F.G. and M.S. Reid, 2002. Postharvest handling of stock (*Matthiola incana*). *HortSci.*, **37**(1): 144-147.

Chandra, J. and O.P. Lal, 1973. Record of thrips on some vegetables and ornamental plants from Kulu Valley, Himachal Pradesh. *Indian J. Ent.*, **35**(2): 164-166.

Chandrasekar, K., S.K. Srivastava, D.K. Singh and R.D. Gaur, 2004. Additions to the flora of Himachal Pradesh from Pin Valley National park. *Indian J. Forest.*, **27**(3): 317-319.

Dansereau, B., Y. Zhang, S. Gagnon and H.L. Xu, 1998. Stock and snapdragon as influenced by greenhouse covering materials and supplemental light. *HortSci.*, **33**(4): 668-671.

David, J.C. 2000. *Alternaria japonica*. IMI Descriptions of Fungi and Bacteria, No. 144 (Sheet 1432), CABI Bioscience, CAB International, U.K.

Davino, S., A. Calari, M. Davino, M. Tessitori, A. Bertaccini and M.G. Bellardi, 2007. Virescence of tenweeks stock associated to phytoplasma infection in Sicily. *Bull. Insectol.* [(*1st Int. Phytoplasmologist Working Group Meeting*, eds Bertacinni, A. and S. Maini), held at Bologna, Italy, on November 12-15, 2007], **60**(2): 279-280.

Demiral, M.A. 2003. Determination of salt tolerance of stock (*Matthiola tricuspidata*) as a potential oil crop. *Turkish J. Agric. & For.*, **27**(4): 229-235.

Dirkse, F.B. 1981. Downy mildew on stocks can be effectively controlled (Dutch). *Vakblad voor de Bloemist.*, **36**(3): 35.

Dirmenci, T., F. Satl and G. Tumen, 2006. A new species of *Matthiola* R. Br. (Brassicaceae) from Turkey. *Bot. J. Linnean Soc.*, **151**(3): 431-435.

Dixon, G.R. 2003. Clubroot, a case for integrated control. *Plantsm.*, **2**(3): 179-183.

Dohi, M. and T. Fujiura, 1999. Stock seedling selection and transplanting robot - selection method using GA and end effect or for transplanting. *ASAE/CSAE-SCGR Ann. Int. Meet.* on July 18-21, 1999, at Toronto, Ontario, Canada (ASAE Paper No. 993070), p. 9.

Dole, J.M., W.C. Fonteno and S.L. Blankenship, 2005. Comparison of silver thiosulfate with 1- methylcyclopropene on 19 cut flower taxa. *Acta Hort.* (Proceedings of the Fifth International Postharvest Symposium, Verona, Italy, 6-11 June, 2004; eds Mencarelli, F. and P. Tonutti), No. 682 (vol. 2), pp. 949-956.

Duan YongJia and Cai Hong, 1998. Studies on the variegated flowers of flowering plants (Chinese). *Acta Phytopath. Sinica*, **28**(1): 85-89.

Duczmal, K.W. 1989. Weed control in ornamental plants grown for seed (Polish).*Biuletyn Instytutu Hodowli I Aklimatyzacji Roslin*, No. 169, pp. 59-76.

Durlak and W. Martyn, 1996. Preliminary studies on the possibility of utilizing acrylic gels for seed dressing of some ornamental species (Polish). *Zeszyty Problemowe Postepow Nauk Rolniczych*, No. 429, pp. 127-132.

Ecker, R., A. Barzilay, L. Afgin and A.A. Watad, 1992. Growth and flowering responses of *Matthiola incana* L. R. Br. to paclobutrazol. *HortSci.*, **27**(12): 1330.

Ecker, R., A. Barzilay and E. Osherenko, 1994a. The inheritance of flowering time in garden stock (*Matthiola incana* R.Br.). *Euph.*, **72**(1-2): 153-156.

Ecker, R., A. Barzilay and E. Osherenko, 1994b. Linkage relationships of genes for leaf morphology and double flowering in *Matthiola incana*. *Euph.*, **74**(1-2): 133-136.

Ecker, R., D. Zutra, A. Barzilay, E. Osherenko and D. Rav David, 1995. Sources of resistance to bacterial blight of stock (*Matthiola incana* R. Br.). *Genet. Res. & Crop Evol.*, **42**(4): 371-372.

Ecker, E., Z. Yaniv, M. Zur and D. Shafferman, 1991. Embryonic heterosis in the linolenic acid content of *Matthiola incana* seed oil. *Euph.*, **59**(1): 93-96.

Epping, B., M. Kittel, B. Ruhnau and V. Hemleben, 1990. Isolation and sequence analysis of a chalcone synthase cDNA of *Matthiola incana* R. Br. (Brassicaceae). *Plant Mol. Biol.*, **14**(6): 1061-1063.

Erwin, J.E., C. Rohwer and E. Gesick, 2006. Red:far red and photosynthetically active radiation filtering by leaves differs with species. *Acta Hort.* (Proc. 5th int. Symp. on Artificial Lighting in Horticulture; ed. Moe, R.), held at Lillehammer, Norway on June 21-24, 2005, No. 711, pp. 195-199.

Farzadfar, S., K. Ohshima, R. Pourrahim, A.R. Golnaraghi, S. Jalali and A. Ahoonmanesh, 2005. Occurrence of turnip mosaic virus on ornamental crops in Iran. *Pl. Path.*, **54**(2): 261.

Fehleisen, S.O. and B. Schnack, 1972. The production of double flowers in the stock, *Matthiola incana*, using a modification of Kappert's method (Spanish). *Revista de la Facultad de Agronomia* (Universidad Nacional de la Planta), **68**: 125-130.

Ferrante, A. and A. Mensuali Sodi, 2006. Studies on flower and leaf senescence of cut stock flowers (Italian). *Italus Hortus* (Atti del Workshop del Gruppo di Lavoro SOI sul Postraccolta "Stato della ricerca italiana sul postraccolta dei prodotti ortoflorofrutticoli", Sassari, Italy, 7 October 2005; eds Mulas, M. and M. Schirra), **13**(5):92-95.

Ferrante, A., P. Vernieri, G. Serra and F. Tognoni, 2004. Changes in abscisic acid during leaf yellowing of cut stock flowers. *Pl. Gr. Reg.*, **43**(2): 127-134.

Foddai, A., M. Cugusi and G. Idini, 1991. First detection of a virus disease of radish in Italy: a mosaic caused by turnip mosaic virus (Italian). *Phytopath. Medit.*, **30**(3): 182-186.

Fos, A., J.M. Bove, J. Lallemand, C. Saillard, J.C. Vignault, Y. Ali, P. Brun and R. Vogel, 1986. The leafhopper *Neoaliturus haematoceps* is a vector of *Spiroplasma citri* in the Mediterranean area (French). *Annales de l'Institut Pasteur Microbiologie*, **137A**(1): 97-101.

Fujita, M. 1978. Studies on the establishment of cropping systems for common stocks (*Matthiola incana*). II. Flowering at high temperature in extremely early flowering cultivars of non- branching stocks. *J. Japanese Soc. hort. Sci.*, **47**(2): 209-216.

Fujita, M. 1979. Studies on the establishment of cropping systems for common stocks (*Matthiola incana*). VI. Effect of low temperature for young plants on flower initiation and flowering (Japanese). *J. Japanese Soc. hort. Sci.*, **48**(3): 327-335.

Fujita, M. and T. Nishitani, 1978. Studies on the establishment of cropping systems for common stocks (*Matthiola incana*). III. Varietal differences of flowering in non-branching stocks grown in various temperature regimes. *J. Japanese Soc. hort. Sci.*, **47**(2): 217-226.

Fujita, M. and T. Nishitani, 1979. Studies on the establishment of cropping systems for common stock (*Matthiola incana*). V. Varietal differences of flowering in branching stocks grown in various temperature regimes (Japanese). *J. Japanese Soc. hort. Sci.*, **48**(2): 213-223.

Garibaldi, A. and G. Gullino, 1976. Diseases of flower and ornamental plants new or little known in Italy. III. Root and vascular diseases (Italian). *Agricolt. Ital.*, No. 105, pp. 273-284.

Garibaldi, A., G. Gilardi and M.L. Gullino, 2006. Evidence for an expanded host range of *Fusarium oxysporum* f.sp. *raphani*. *Phytoparas.*, **34**(2): 115-121.

Gautam, V.K., A. Mittal, K. Nanda and S.C. Gupta, 1983. *In vitro* regeneration of plantlets from somatic explants of *Matthiola incana*. *Plant Sci. Lett.*, **29**(1): 25-32.

Gera, A., L. Maslenin, A. Rosner, M. Zeidan and P.G. Weintraub, 2006. Phytoplasma diseases in ornamental crops in Israel. *Acta Hort.* (*Proceedings of the XIth Int. Symp. on Virus Dis. of Ornam. Plants*, held at Taichung, Taiwan, on March 9-14, 2004; ed. Chang, C.A.), No. 722, pp. 155-161.

Gerlach, W. 1976. The first case of *Fusarium* wilt on garden stock (under glass) in Germany (German). *Nachrichtenblatt des Deutschen Pflanzenschutzdienstes*, **27**(2): 17-20.

Gladunov, I.M., N.F. Maslova and E.N. Krasnobaev, 1974. Determining doubleness in summer stock (Russian). *Vopr Intensifik Dekor Sadovodstva, Materialy Simpoz*, USSR, pp. 58-61.

Grieve, C.M., J.A. Poss, P.J. Shouse and C.T. Carter, 2008. Molecular growth of *Matthiola incana* in response to saline waste waters differing in nitrogen levels. *HortSci.*, **43**(6): 1787-1793.

Groote, A.De, 1980. Assimilation lighting of stocks (Dutch). *Verbondsnieuws voor de Belgische Sierteelt*, **24**(18): 714.

Grzesik, M. 1989. Effect of growth regulators on the seedling-growth of *Lathyrus odoratus*, *Zinnia elegans*, *Matthiola incana* and *Antirrhinum majus*. *Acta Hort.* (Proc. Symp. on Growth Regulators in Ornam. Hort., held at Skierniewice, Poland on September 5-10, 1988), No. 251, pp. 71-74.

Grzesik, M. 1995. Effect of growth regulators on plant growth and seed yield of *Matthiola incana*, 'Brilliant Barbara'. *Seed Sci. & Tech.*, **23**(3): 801-806.

Hammami, S., I. Khoja, H.B. Jannet, M.B. Halima and Z. Mighri, 2006. Flowers essential oil composition of Tunisian *Matthiola longipetala* and its bioactivity against *Tribolium confusum* insect. *J. Essent. Oil Bear. Pl.*, **9**(2): 156-161.

Hamza, A.M. and M.N.M. Helaly, 1983. Interaction between chlormequat (CCC) and gibberellin (GA_3) on growth, flowering and mineral constituents of some ornamental plants. *Acta Hort.*, No. 137, pp. 197-209.

Handreck, K.A. 1986. Critical concentrations of sulfur in liquid feeds for plants in containers. *Sci. Hort.*, **30**(1/2): 1-17.

Hayward, A.C. and J.M. Waterston, 1965. *Xanthomonas campestris*. IMI Descriptions of Fungi and Bacteria, No. 5 (Sheet 47). CABI Bioscience, CAB International, U.K.

Heller, W., G. Forkmann, L. Britsch and H. Grisebach, 1985. Enzymatic reduction of (+)-dihydroflavonols to flavan-3,4-cis-diols with flower extracts from *Matthiola incana* and its role in anthocyanin biosynthesis. *Planta*, **165**(2): 284-287.

Hemleben, V., A. Dressel, B. Epping, R. Lukacin, S. Martens and M.B. Austin, 2004. Characterization and structural features of a chalcone synthase mutation in a white-flowering line of *Matthiola incana* R. Br. (Brassicaceae). *Plant Mol. Biol.*, **55**(3): 455-465.

Hendy, A. and G. Roberts, 1974. Recognising the diseases in bedding plants. *Comm. Gr.*, No. 4091, pp. 925, 927-928.

Hetman, J., H. Laskowska, W. Durlak and W. Martyn, 1996. Preliminary studies on the possibility of utilizing acrylic gels for seed dressing of some ornamental species (Polish).

Zeszyty Problemowe Postepow Nauk Rolniczych, No. 429, pp. 127-132.

Heuer, B. and I. Ravina, 2004. Growth and development of stock (*Matthiola incana*) under salinity. *Australian J. agric. Res.*, **55**(8): 907-910.

Heuer, B., I. Ravina and S. Davidov, 2005. Seed yield, oil content, and fatty acid composition of stock (*Matthiola incana*) under saline irrigation. *Australian J. agric. Res.*, **56**(1): 45-47.

Heuer, B., Z. Yaniv and Israela Ravina, 2002. Effect of late salinization of chia (*Salvia hispanica*), stock (*Matthiola tricuspidata*) and evening primrose (*Oenothera biennis*) on their oil content and quality. *Indust. Crops & Products*, **15**(2): 163-167.

Hisamatsu, T. 2001. Studies on the role of gibberellin biosynthesis in the control of growth and development in *Matthiola incana* (L.) R. Br. and *Eustoma grandiflorum* (Raf.) Shinn (Japanese). *Bull. Nat. Res. Inst. Veget., Ornam. Pl. & Tea*, No. 16, pp. 79-133.

Hisamatsu, T. and M. Koshioka, 2000. Cold treatments enhance responsiveness to gibberellin in stock (*Matthiola incana* (L.) R. Br. *J. hort. Sci. & Biotech.*, **75**(6): 672-678.

Hisamatsu, T. and M. Koshioka, 2001a. Regulation of flowering in stock [*Matthiola incana* (L.) R. Br.] by manipulation of gibberellin biosynthesis. *JARQ (Japan Agric. Res. Quart.)*, **35**(4): 263-269.

Hisamatsu, T. and M. Koshioka, 2001b. The role of gibberellin in the control of growth and flowering in stock (Japanese). *Regul. Pl. Growth & Dev.*, **36**(1): 85-90.

Hisamatsu, T., M. Koshioka, S. Kubota and R.W. King, 1998. Effect of gibberellin A_4 and GA biosynthesis inhibitors on growth and flowering of stock [*Matthiola incana* (L.) R. Br.]. *J. Japanese Soc. hort. Sci.*, **67**(4): 537-543.

Hisamatsu, T., M. Koshioka, S. Kubota, Y. Fujime, R.W. King and L.N. Mander, 2000. The role of gibberellin biosynthesis in the control of growth and flowering in *Matthiola incana*. *Phys. Plant.*, **109**(1): 97-105.

Hisamatsu, T., S. Kubota and M. Koshioka, 1999. Promotion of flowering in stock [*Matthiola incana* (L.) R. Br.] by prohexadione-calcium in plastic-film greenhouse conditions. *J. Japanese Soc. hort. Sci.*, **68**(3): 540-545.

Hisamatsu, T., N. Oyama-Okubo, K. Ichimura, S. Esaki, R. Oi and M. Koshioka, 2002. Interactions of red and far-red light modification with temperature on shoot extension and flowering in stock (*Matthiola incana* (L.) R. Br.). *J. hort. Sci. & Biotech.*, **77**(1): 1-8.

Holmes, R.L., G.D. Coorts and T.L. Rosenfeld, 1981. Potassium nutrition of stocks in bark, peat-lite, and soil media. *HortSci.*, **16**(4): 560-561.

Houbowicz, R., K. Tylkowska and E. Jankowska, 1997. Methods of investigating seed vigour of selected ornamental plant species. *Folia Hort.*, **9**(1): 15-23.

Ibrahim, A.A.M. and F.A. Al-Yahya, 2002. Phytoparasitic nematodes associated with ornamental plants in Riyadh Region, Central Saudi Arabia. *Alexandria J. agric. Res.*, **47**(3): 157-167.

Ishak, I.F. 1989. Effect of temperature on the germination of seeds of some species of the Brassicaceae. *Egyptian J. Bot.*, **32**(3): 191-199.

Ito, A., T. Hisamatsu, N. Soichi, M. Nonaka, M. Amano and M. Koshioka, 1997a. Effect of diurnal temperature alterations on the growth of annual flowers at the nursery stage. *J. Japanese Soc. Hort. Sci.*, **65**(4): 809-816.

Ito, A., T. Hisamatsu, N. Soichi, M. Nonaka, M. Amano and M. Koshioka, 1997b. Effect of altering diurnal fluctuations of day and night temperatures at the seedling stage on the subsequent growth of flowering annual. *J. Jaspanese Soc. hort. Sci.*, **65**(4): 817-823.

Jana, S. and W. Seyffert, 1972. Simulation of quantitative characters by genes with biochemically definable action. IV. The analysis of heritable variation by the diallel technique. *Theor. Appl. Genet.*, **42**(1): 16-24.

Kalkman, E.C. 1983. Pre-treatment improves the quality of summer cut flowers (Dutch). *Vakblad voor de Bloemist.*, **38**(50): 26-29.

Kasten, P., K. Sommer and A. Papenhagen, 1990. Ammonium placement fertilizing of cut-flowers. (German). *Deutscher Gartenbau*, **44**(26): 1710-1713.

Kawabata, S., Li YuHua, M. Adachi, H. Maruyama, E. Ozawa and R. Sakiyama, 2002. Role of photoreceptor- and sugar-mediated reactions in light dependent anthocyanin production in lily and stock flowers. *J. Japanese Soc. hort. Sci.*, **71**(2): 220-225.

Khosravi, A.R. and A.A. Maassoumi, 1998. Contribution to the cytotaxonomy of some Cruciferae from Iran. *Iranian J. Bot.*, 7(2): 193-206.

Klein, M., P. Rasooly and B. Raccah, 1988. New findings on the transmission of *Spiroplasma citri*, the citrus stubborn disease agent in Israel, by a beet leafhopper from the Jordan valley (Hebrew). *Hassadeh*, **68**(9): 1736-1737.

Klein, M. and B. Raccah, 1991. Separation of two leafhopper populations of the *Circulifer haematoceps* complex on different host plants in Israel. *Phytoparasitica*, **19**(2): 153-155.

Kobayashi, Y., T. Arie, M. Shibasaki and I. Yamaguchi, 1997. 4-Allyl-2-azetidinone and *Penicillium simplicissimum* cooperate to control soilborne *Fusarium* diseases. *J. Pesticide Sci.*, **22**(2): 113-118.

Koike, S.T. 2000. Downy mildew of stock, caused by *Peronospora parasitica*, in California.*Pl. Dis.*, **84**(1): 103.

Koku, K., N. Aoki, M. Dohi, T. Fujiura and K. Takeyama, 1999. Studies on automatic discrimination of double flowering stock from single flowering seedling by fuzzy theory. *J. Japanese Soc.Hort. Sci.*, **68**(1): 49-54.

Kuehny, J.S., A. Painter and P.C. Branch, 2001. Plug source and growth retardants affect finish size of bedding plants. *HortSci.*, **35**(2): 321-323.

Lamont, G.P. 1986. Herbicide evaluation trials on cut flowers. *Australian Hort.*, **84**(7): 89-91.

Lamont, G.P. and M.A. O'Connell, 1986. An evaluation of pre-emergent herbicides in field-grown cut flowers. *Plant Prot. Quart.*, **1**(3): 95-100.

Lane, A. and G.A. Wheatley, 1984. Cabbage root fly. Leaflet No. 18, p. 10. Ministry of Agriculture, Fisheries and Food, UK.

Lan ZeQu, Luo Peng, Zhou SongDong, Xu HuaiLiang and Zhou HongFen, 1999. A study on karyotype and meiosis of *Matthiola incana* - an oil plant for special purposes (Chinese). *Chinese J. Oil Crop Sci.*, **4**: 37-40.

Lesley, M.M. 1973. An unstable S chromosome in diploid *Matthiola incana*. *J. Hered.*, **64**(6): 359-362.

Lin LiangBin, Ma ZhanQiang, Zhang ChuanLi, Liu YaTing, Mao XiaoQiang, Xia-GuoYin and Ma ZhongQin, 2008. Studies on somatic hybridization between *Brassica napus* and *Matthiola incana* B. Br. (Chinese). *J. Yunnan agric. Univ.*, **23**(3): 295-301.

Loeser, H. 1986. Comparison of stock cultivars at Heidelberg (German). *Deutscher Gartenbau*, **40**(32): 1474-1475.

Loeser, H. 1991. *Matthiola*, *Eustoma*, *Trachelium* and *Callistephus* in a closed system (German). *Gartenbau Mag.*, **38**(10): 51-53.

Lorenzini, G., E. Triolo and G. Scaramuzzi, 1983. Further contribution to the knowledge of new or little known diseases of ornamental plants in Tuscany (Italian). *Rivista della Ortoflorofrutticoltura Italiana*, **67**(1): 23-34.

Macnish, A.J., R.T. Leonard and T.A. Nell, 2008. Treatment with chlorine dioxide extends the vase life of selected cut flowers. *Postharv. Biol. & Tech.*, **50**(2/3): 197-207.

Magnani, G., F. Falleri and G. Vedrani, 1997. Foamed' film, a new covering material. *Plasticult.*, No. 113, pp. 2-10.

Marzouk, M.M., S.A. Kawashty, L.F. Ibrahium, N.A.M. Saleh and A.S.M. Al-Nowaihi, 2008. Two new kaempferol glycosides from *Matthiola longipetala* (subsp. *livida*) (Delile) Maire and carcinogenic evaluation of its extract. *Natural Prod. Comm.*, **3**(8): 1325-1328.

Mensuali Sodi, A., M. Brea, M. Panizza, G. Serra and F. Tognoni, 1994. *In vitro* regeneration of shoots in *Matthiola incana* L. from seedling explants of different age. *Gartenbauwissenschaft*, **59**(2): 77-80.

Minardi, P., U. Mazzucchi and C. Parrini, 1988. Epidemics of bacterial blight of stock (*Matthiola incana* R. Br.) caused by *Xanthomonas campestris* pv. *incanae* in Tuscany (Italian). *Informatore Fitopatologico*, **38**(1): 43-46.

Mizozoe, M., N. Inagaki, M. Okano, M. Kanechi and S. Maekawa, 1992. Effects of shading and fertilizing of mother plants, and NH4+:NO3- ratio in the medium on *in vitro* organogenesis of stock (*Matthiola incana* R. Br.) (Japanese). *J. Japanese Soc. hort. Sci.*, **61**(3): 625-633.

Muneto, S., H. Itamura, A. Nakatsuka and K. Ohta, 2007. Control of 1-MCP on vase life in stock (*Matthiola incana* R. Br.) flowers (Japanese). *Bull. Fac. Life & environ. Sci.* (Shimaane Univ.), No. 12, pp. 11-14.

Murahama, M. 2007. Effects of benzalkonium chloride on hydration of cut stocks (*Matthiola incana* (L.) R. Br.) (Japanese). *Hort. Res. Japan*, **6**(3): 487-490.

Niebisch, R.M. 1985. Use of Bercema-captan 80 for improving seedling emergence in ornamentals (German). *Nachrichtenblatt für den Pflanzenschutz in der DDR*, **39**(9): 190-191.

Niebisch, R.M. and K. Kelling, 1986. Results of chemical control of fungal diseases in ornamental plant production (German). *Gartenbau*, **33**(7): 215-218.

Nowak, J. and R.M. Rudnicki, 1975. The keeping quality of stocks can be nearly doubled. *Prace Instytutu Sadownictwa w Skierniewicach, B*, **1**: 173-179.

Okuse, I., Y. Satoh and K. Saga, 1996. Effects of cytokinins and leaf explants from single and double flowering plants on callus growth and organogenesis in stock plants (*Matthiola incana* R. Br.) (Japanese). *Bull. Fac. Agric. Hirosaki Univ.*, No. 59, pp. 71-78.

O'Neill, T.M., A. Shepherd, A.J. Inman and C.R. Lane, 2004. Wilt of stock (*Matthiola incana*) caused by *Fusarium oxysporum* in the United Kingdom. *Pl. Path.*, **53**(2): 262.

O'Neill, T.M., K.R. Green and T. Ratcliffe, 2005. Evaluation of soil steaming and a formaldehyde drench for control of *Fusarium* wilt in column stock. *Acta Hort.* (*Proc. of the VIth Intern. Symp. on Chemical and Non-chemical Soil and Substrate Disinfestation*; ed. Vanachter, A.), held at Corfu, Greece, on October 4-8, 2004; No. 698, pp. 129-133.

Parrini, C. and P. Rumine, 1987. Downy mildew of stocks (Italian). *Colt. Prot.*, **16**(4): 81-84.

Parrini, C. and P. Rumine, 1989. Diseases and pests of greenhouse-grown snapdragon and stock (Italian). *Colt. Prot.*, **18**(12): 29-38.

Pasini, C., T. Berio, P. Curir and F. D'Aquila, 1996. Further characterization of *Rhizoctonia solani* isolated from carnation and other ornamental plants. *Informatore Fitopatologico*, **46**(6): 33-36.

Peat, J.R. and R.J. Summerfield, 1977. Environmental and cultural effects on vegetative growth and flowering of selected "bedding" ornamentals. II. Night-break lighting and gibberellic acid applications. *Sci. Hort.*, **7**(1): 81-89.

Powell, L. 1974. Growth regulators can extend plant shelf life. *Commercial Grower*, No. 4041, p. 465.

Pui,D.A. and M. Ardelean, 2007. Heterosis manifestation of several quantitative characters of ornamental interest in gillyflower (*Matthiola* sp.) breeding. *Contributii Botanice*, **42**: 99-104.

Pui, D.A., M. Ardelean, M. Cordea, V. Mitre and A. Nicoara, 2007. Heritability of several quantitative characters of interest in gillyflower (*Matthiola incana*) breeding. *Bull.*

Univ. agric. Sci. and Vet. Med., Cluj Napoca, Horticulture, Romania, **64**(1/2): 54-57.

Rahimian, H. and H. Okhovatian, 1989. Bacterial blight of stock in Mazandaran (Persian). *Iranian J. Pl. Path.*, **25**(1-4): 11-12.

Rainbow, A. and N. Wilson, 1998. The transformation of composted organic residues into effective growing media. *Acta Hort.* (*Proc. Int. Symp. on Composting and Use of Composted Materials for Horticulture*, ed. Szmidt, R.A.K.), held at Auchincruive, Ayr, U.K. on April 5-11, 1997; No. 469, pp. 79-88.

Rall, S. and V. Hemleben, 1984. Characterization and expression of chalcone synthase in different genotypes of *Matthiola incana* R.Br. during flower development. *Plant Mol. Biol.*, **3**(3): 137-145.

Ramsay, N.A., A.R. Walker, M. Mooney and J.C. Gray, 2003. Two basic-helix-loop-helix genes (MYC-146 and GL3) from *Arabidopsis* can activate anthocyanin biosynthesis in a white-flowered *Matthiola incana* mutant. *Plant Mol. Biol.*, **52**(3): 679-688.

Roh, S.M. 1978. Lateral shoot induction on stock, *Matthiola incana*, and petunia, *Petunia hybrida*, by growth regulator and light treatments, and their flowering responses. *J. Korean Soc. hort. Sci.*, **19**(2): 154-166.

Rosciglione, B. and G. Cannizzaro, 1983. *Matthiola incana* R. Br., natural host of turnip mosaic virus in Sicily (Italian). *Tecnica Agricola*, **35**(3): 251-257.

Runkova, L.V. 1977. Interaction between exogenous and endogenous growth regulators as exemplified on ornamental plants (Russian). *Byulleten Glavnogo Botanicheskogo sada*, No. 106, pp. 62-70.

Saga, K. 1996. Effects of cytokinins and leaf explants from single and double flowering plants on callus growth and organogenesis in stock plants (*Matthiola incana* R. Br.) (Japanese). *Bull. Fac. Agric. Hirosaki Univ.*, No. 59, pp. 71-78.

Saito, N., F. Tatsuzawa, A. Nishiyama, M. Yokoi, A. Shigihara and T. Honda, 1995. Acylated cyanidin 3-sambubioside-5-glucosides in *Matthiola incana*. *Phytochem.*, **38**(4): 1027-1032.

Serek, M., E.C. Sisler and M.S. Reid, 1995. Effects of 1-MCP on the vase life and ethylene response of cut flowers. *Pl. Gr. Reg.*, **16**(1): 93-97.

Sheng XiaoGuang, Liu Fan, Zhu YueLin, Zhao Hong, Zhang Li and Chen Bin, 2008. Production and analysis of intergeneric somatic hybrids between *Brassica oleracea* and *Matthiola incana*. *Plant Cell, Tissue Org. Cult.*, **92**(1): 55-62.

Siemens, J. and M.D. Sacristan, 1995. Plant regeneration from mesophyll protoplasts of *Matthiola incana* (L.) R. Br. *Plant Cell Rep.*, **14**(7): 446-449.

Skiebe, K. 1985. Genetic problems of fully-double stock (*Matthiola incana* (L.) R. Br.) and the implications for breeding (German). *Archiv für Zuchtungsforschung*, **15**(2): 103-113.

Song CheonYoung, Bang ChangSeok, Huh KunYang and Song JeongSeob, 1996. Effects of preservatives and cold storage on vase life and quality of cut hybrid stock (*Matthiola incana*) (Korean). *RDA J. agric, Sci.: Hort.*, **38**(1): 598-603.

Suatmadji, R.W. 1988. *Pratylenchus penetrans* and *Rotylenchus robustus* on thirty herbaceous ornamental species. *Australian Pl. Path.*, **17**(4): 97-98.

Summerfield, R.J., E.M. Dawson and J.R. Peat, 1977. Environmental and cultural effects on vegetative growth and flowering of selected "bedding" ornamentals. I. Night temperature. *Sci. Hort.*, **7**(1): 67-79.

Takano, T., Y. Tsuchiya, T. Sakaguchi and S. Masaki, 2004. Irradiation effects on insect pests of cut flowers. *Res. Bull. Pl. Prot. Serv., Japan*, No. 40, pp. 25-32.

Takami, T., T. Hisamatsu and M. Koshioka, 2004. The utilization as a flowering promoter of growth retardant prohexadione-calcium (Japanese). *Regul. Pl. Growth & Dev.*, **39**(2): 257-259.

Tapio, E. 1972. *Sclerotinia* disease of stocks (*Matthiola incana* R. Brown). *Annales Agriculturae Fenniae*, **11**(2): 111-114.

Teusch, M. 1986. Uridine 5'-diphosphate-xylose:anthocyanidin 3-*O*-glucose-xylosyltransferase from petals of *Matthiola incana* R. Br. *Planta*, **169**(4): 559-563.

Thanos, C.A., K. Georghiou and P. Delipetrou, 1994. Photoinhibition of seed germination in the maritime plant *Matthiola tricuspidata*. *Ann. Bot.*, **73**(6): 639-644.

Thomas, M.B. and A. Leong, 1980. Effects of nitrogen, phosphorus, potassium, and B-nine on the growth of Brompton stocks (*Matthiola incana*). *Intern. Pl. Prop. Soc.* (Combined Proceedings), **30**: 156-164.

Thomas, M.B., M.R. Oates and M.I. Spurway, 1980. Evaluation of low-cost potting mixes for bedding plants and vegetable seedlings. *N. Z. J. exp. Agr.*, **8**(3/4): 281-286.

Trimboli, D.S. and F. Hampshire, 1978. Control of downy mildew of stocks with Fongarid. *APPS N.L.*, **7**(1): 9-10.

Vikas Kumar, Kaomud Tyagi and J.S. Bhatti, 2005. On some new records of *Thysanoptera* (Insecta) from India. *Entomon*, **30**(3): 249-254.

Waterhouse, G.M. and J.M. Waterston, 1966. *Phytophthora megasperma*. IMI Descritions of Fungi and Bacteria, No. 12 (Sheet 115). CABI Bioscience, CAB International, U.K.

Wu, W.J. 1993. Study on the host range of *Plutella xylostella* (L.) (Chinese). *Entom. Knowl.*, **30**(5): 274-275.

Xu HuaiLiang, Luo Peng, Li XuFeng, Chen YanHong, Yang Fan, Yang JiangYi, Wang Jin and Yuan Xu, 2002. Callus induction and shoot regeneration of *Matthiola incana* (Chinese). *Chinese J. Oil Crop Sci.*, **24**(1): 22.

Yamanouchi, M. and Yang XiuZhen, 2000. The effects of nitrogen supply on the growth and quality of cut-flower stock (*Matthiola incana* R. Br.) (Japanese). *Japanese J. Soil Sci. & Pl. Nutrit.*, **71**(2): 204-209.

Yaniv, Z., D. Schafferman, M. Zur and I. Shamir, 1997. Evaluation of *Matthiola incana* as a source of omega-3-lenolenic acid. *Indust. Crops & Prod.* (Spl. Issue), **6**(3/4): 285-289. *Selected Papers from the Third European Symposium on Industrial Crops and Products*, held at Reims, France on April 22-24, 1996.

Yoon JuYeon, Choi HongSoo, Ryu HwaYoung, Harm YoungIl and Choi JangKyung, 1998. Colour breaking syndrome of *Matthiola incana* caused by double infection of cucumber mosaic virus and turnip mosaic virus.*Korean J. Pl. Path.*, **14**(3): 220-222.

Yoshimura, T., M. Nishiyama and K. Kanahama, 2002. Effects of red or far-red light and red/far-red ratio on the shoot growth and flowering of *Matthiola incana* (Japanese). *J. Japanese Soc. hort. Sci.*, **71**(4): 575-582.

Yoshimura, T., A. Sasaki, T. Moriyama, Y. Shibahara, K. Katsuta and K. Kanahama, 2006. Effects of various light sources for night irradiation and light intensity on the flowering of stock [*Matthiola incana* (L.) R. Br.] plant (Japanese). *Hort. Res. Japan*, **5**(3): 297-301.

Yuen, G.Y. and M.N. Schroth, 1986. Interactions of *Pseudomonas fluorescens* strain E6 with ornamental plants and its effect on the composition of root-colonizing microflora. *Phytopath.*, **76**(2): 176-180.

Zhang ZhiSheng and H. Dorna, 2005. The effects of ABA priming on germination of *Matthiola bicornis* seeds. *J. Trop.& Subtrop. Bot.*, **13**(3): 224-228.

27

Narcissus (Family: Amaryllidaceae)

Priyanka Thakur, R.L. Misra and Sanyat Misra

[**Common names:** Angel's tears (*Narcissus triandrus* and *N. t. albus*), Autumn-flowering group of narcissi (*N. autumnalis* syn. *N. elegans, N. viridiflorus*), *Bulbocodium* group of narcissi (*N. bulbocodium, N. cantabricus*), Campernelle jonquil (*N.* × *odorus*), Chinese sacred lily (*N. tazetta orientalis*), *Cyclamineus* group of narcissi (*N. cyclamineus*), Daffodil (*Narcissus* spp.), Hoop petticoat daffodil (*N. bulbocodium*), *Jonquilla* group of narcissi (*N. jonquilla, N. rupicola, N. watieri*), Lent lily/Wild daffodil/ Easter flower (*N. pseudonarcissus*), Pheasant's eye/*Poeticus* or Poet's group of narcissi (*N. poeticus*), Polyanthus daffodil/Bunch-flowered daffodil (*N. tazetta*), Rush-leaved jonquil (*N. juncifolius*), *Tazetta* group of narcissi (*N. canaliculatus, N. tazetta*), Tenby daffodil (*N. lobularis, N. obvallaris*), *Triandrus* group of narcissi (*N. triandrus*), *Trumpet* group of narcissi (*N. asturiensis, N. pseudonarcissus*), Wild jonquil (*N. jonquilla*), *etc.*]

Introduction and Origin

Misra (1989) gave a detailed account of its origin and history. The genus is primarily found in the Mediterranean area and the centre of origin is Europe (Spain, Portugal and the Iberian Peninsula). Species are also found in northern Africa, France and Greece. These have originated from Northern Hemisphere, *i.e.* Europe, especially Spain and Portugal, France, Switzerland, Yugoslavia and North Africa with some exceptions of bunch-flowered (*Narcissus tazetta* L.) narcissi which are found across Iran, China, Japan and West Asia. A narrow band of *N. tazetta* grows naturally into China and Japan. Later these were distributed, planted and spread along ancient trade routes (Bailey and Bailey, 1976; Hanks, 1993; Meyer, 1966). Authorities doubt whether *N. tagetta* also originates through Iran to China and Japan and they believe that it should be native of Portugal from where it escaped from the gardens through Portuguese settlers in the colonies. The most important species is probably *N. hispanicus*, the tallest among the trumpets (grown in the Bristish gardens for over 300 years as *N. maximum* (syn. *N. maximus superba*) which inhabits through Spain and SW France along the Pyrenees (Rees, 1972). *Narcissus* L., a Greek name is said to be derived from *narke*, 'numbness' or 'torpor', from its narcotic properties. The narcissus was named after *Narkissos,* the son of the river god Kephissos and the beautiful nymph Leiriope. Narkissos was handsome and the nymph Echo fell deeply in love with him. He did not, however, return her love and the angry nymph consequently sought revenge. She asked Nemesis to devise a punishment. When Narkissos wanted to drink from a clear pond, he saw himself and promptly fell in love with his own reflection. For days together, he gazed himself continuously in pond, became fascinated and finally drowned. On that spot a bloom opened, Linnaeus found the flower so beautiful that he named it *Narcissus*. Others were also impressed by this flower, and Prophet Mohammed claimed, 'He who has two loaves of bread sells one to buy a narcissus, because bread is a food for body but narcissus is the food for the soul'. Shakespeare mentioned daffodils as a common symbol of the English spring and there were many songs which were written in praise of this magnificent flower. William Wordsworth (1770-1850) wrote '*The Daffodils*', I wandered lonely as a cloud, That floats on high o'er vales and hills, When all at once I saw a crowd, A host, of golden daffodils; Beside the lake, beneath the trees, Fluttering and dancing in the breeze... And then my heart with pleasure fills, And dances with the daffodils. Daffodil stands for the message of regard. They are among the most admired garden plants in the world, exclusive due to their matchless beauty. These are considered one of the heralds of spring, planted in autumn season and burst up in spring season, also known as 'queen of spring flowers'. Narcissi add attractive bright colour to any landscape. These ornamental bulbous plants require very little care and flower consistently year after year. They are used mainly as cut flowers, though they are excellent for formal or informal beddings, in pots and bowls (for exhibition and instant decoration), for edging and borders along the paths and river banks, in lawns and rockeries and even their mass plantings naturalize the surroundings in woodlands under the trees, in wild gardens, in meadows, prairie and alpine gardens and pastures because of different shapes and sizes of perianth and corona and various colours such as white, yellow, orange, pink, red and combinations of these. The main flowering period is spring but flowering period extends from late autumn to early summer. Various *Narcissus* species are found growing naturally in areas with cool temperate climates in different habitats at and above 3,000 meters mean sea level. The bulbs of *N. jonquilla, N. poeticus, N. pseudonarcissus* and *N. tazetta* possess purgative, dermal and depurative properties (Pizzetti and Cocker, 1975). These are used for extraction of essential oil; about 165g of essence is obtained from one quintal of flowers.

Brief History

Narcissus is an old Greek name. The *Narcissus* flower is perceived quite differently in the east than in the west where the flower is seen as a symbol of vanity. In China, the same flower is seen as a symbol of wealth and good fortune. The ancient Greeks believed this plant originated from the vain youth, Narcissus. He died after becoming so obsessed with his reflection in a pool. The Greeks say that the gods turned his remains into the *Narcissus* flower. This also led to the daffodil being a symbol of unrequited love. In ancient China, there is a legend about a poor but good man, who received many cups of gold and wealth by this flower. Since the flower blooms in early spring, it has also become a symbol of Chinese New Year. *Narcissus* bulb carving and cultivation is even an art akin to Japanese Bonsai. If your *Narcissus* blooms on Chinese New Years, it is said to bring you extra wealth and good fortune throughout the year. On top of that, it has one of the sweetest fragrances of any flower. So it is highly revered in Chinese culture. In Hawaii, the Chinese Chamber of Commerce of Hawaii sponsors a Chinese cultural festival, called the 'Narcissus Festival', culminating with a beauty pageant whose winner is called the 'Narcissus Queen'. Narcissus was cultivated in Netherlands since 16th century and it was famous and occupied eighth position as a cut flower and sixth position for bulb production during 1998. It became one of the important commercial bulbous crops during 2005. British breeder Peter Barr (1826-1909) was known as 'daffodil king' because he collected 500 species, hybrids and cultivars of daffodils. First daffodil conference was conducted in 1884 on behest of RHS.

Botany, Cytology and Breeding

The genus consists of more than 50 species of bulbous plants which are natives of Europe, North Africa and West Asia. All the *Narcissus* species are herbaceous perennials, hardy or tender, growing through bulbs or seeds and flowering in the spring or autumn. The leaves are linear, strap-shaped, thread- or rush-like, parallel-veined and basal. Plant heights are 15-45 cm, stems green, generally grooved, erect and leafless bearing either solitary or umbellate and usually drooping flowers surrounded by membranous spathe or sheath (structure formed by the fusion of 2 prophylls) in bud, and in the development the spathe is a 2-keeled tube with a vascular strand along the median line of each keel (Chen, 1952). Flowers are small to large, regular, bisexual with tubular base with the 6-lobed corolla (3 modified sepals and 3 petals) which may be in various shades of white, yellow or orange and fused at the base with an extended central trumpet (cup, crown or corona) of various lengths, configurations and colours (white, yellow, orange or red in various shades) usually being darker than segments, style 1 and 3-pronged, stamens 6 (usually in two whorls) and attached to the perianth tube, ovary 3-celled and inferior with many ovules, and the fruit is an ellipsoid to almost spherical and syncarpous capsule with many black seeds, sometimes with an appendage.

All narcissi are daffodils and *vice-versa*, daffodil being the common name and *Narcissus* the Latin name of the family but those with central trumpets (corona or cup) as long as or longer than the surrounding petals, are usually known as **daffodils** (Figure 27.1) (Salinger,1985) such as 'Bridal Gown', 'Carlton', 'Cheerfulness', 'Delna Shaugh', 'Dick Wilden', 'Dutch Master', 'Flower Carpet', 'Fortune', 'Golden Harvest', 'Gigantic Star', 'Ice Follies', 'King Alfred', 'Magnificence', 'Melbourne City', 'Ptolemy', 'Titoki', 'Trumpet Mutant', 'Unsurpassable', 'War Cloud', *etc.* In **double narcissi** (pl.), the corona is inconspicuous such as 'Golden Ducat', 'Sir Winston Churchill', 'Tahiti', 'Texas', *etc.* whereas the common name **narcissus** includes the whole lot irrespective whether corona is larger, equal, smaller or even inconspicuous when compared with the segments, though loosely it is referred only to those having shorter corona. Most narcissi have flat leaves but a few cylindrical and rush-like, especially the jonquils. All narcissi have some fragrance, and the jonquils and tazettas (Division 8) have the strongest. Most daffodils bloom within 4 to 6 weeks after the first appearance of foliage in the very early spring. Depending on location and cultivar, the blooming season can last from two months to almost six months.

All *Narcissus* species have a central trumpet-, bowl-, or disc-shaped corona surrounded by a ring of six floral leaves called the perianth which is united into a tube at the forward edge of the 3-locular ovary. The seeds are black, round and swollen with hard coat. The three outer segments are sepals,

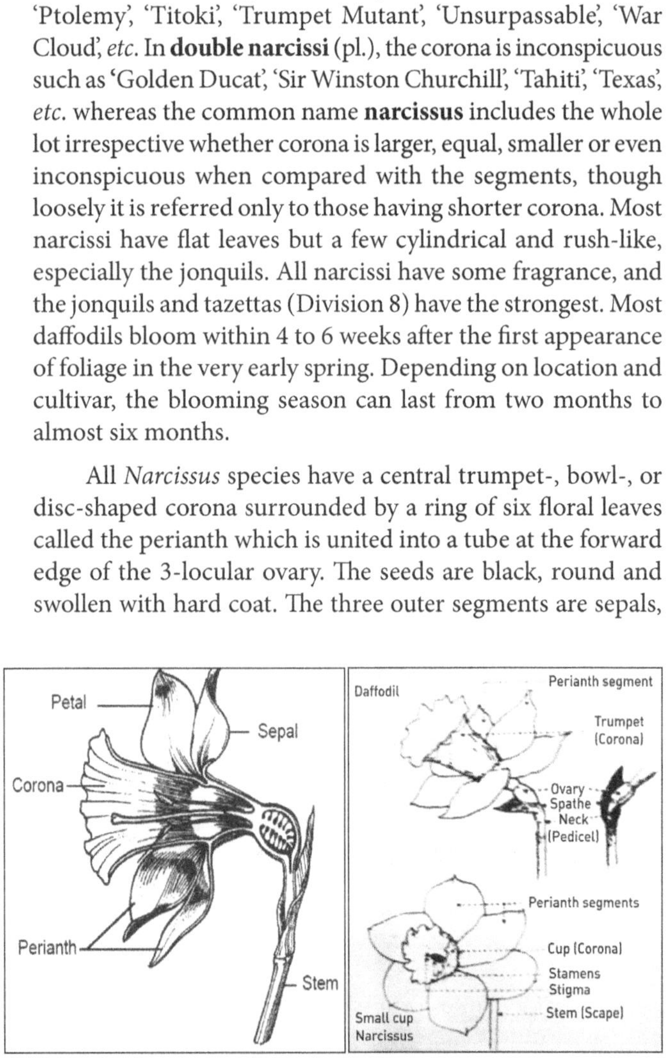

Figure 27.1: Narcissus and Daffodil.

and the three inner segments are petals. Though the traditional daffodil of folklore, poetry, and field may have a yellow to golden-yellow colour all over, both in the wild species and due to breeding, the perianth and corona may be variously coloured (Michael and Lerner, 2001). Breeders have developed some daffodils with double, triple, or ambiguously multiple rows and layers of segments, and several wild species also have known double variants.

Fernandes (1967) in Portugal has spent decades attempting to unravel relationships within the genus, based on chromosome numbers, morphology and geographical distributions. More recently, *The Flora Europea* has condensed the number of species severely (Webb, 1980), and the Royal Horticulture Society (RHS) is revising its checklist against the Fernandes scheme, with a few modifications. *The Flora Europea* groups the species into eight sections. Section I: Jonquillae with species of *Narcissus atlanticus, N. calcicola, N. cuatrecasasii, N. fernandesii, N. gaditanus, N. jonquilla, N. requienii, N. rupicola, N. scaberulus, N. viridiflorus, N. watieri* and *N. willkommii* from Africa. Section II: Narcissus with *N. poeticus*. Section III: Serotini with *N. serotinus*. Section IV: Tapeinanthus with *N. humilis*. Section V: Tazettae with *N. corcyrensis, N. dubius, N. elegans, N. pachybolbus, N. papyraceous* and *N. tazetta* from Africa. Section VI: Ganymedes with *N. triandrus*. Section VII: Bulbocodii with *N. bulbocodium, N. cantabricus, N. hedraeanthus* and *N. romieuxii* from Africa. Section VIII: Pseudonarcissi with *N. asturiensis, N. bicolor, N. cyclamineus, N. lagoi, N. longispathus, N. minor, N. obvallaris* and *N. pseudonarcissus*. Section IX: Aurelia with the single African species, *N. broussonetii*. Pigments in narcissi are mostly carotenoids and the red corona of some cultivars are a rich source of ß-carotene.

Narcissus with n (x) = 7 to diploid chromosome numbers are recorded most frequently as in case of *N. bulbocodium, N. calcicola, N. cyclamineus, N. gaditanus,* ancestral *N. jonquilla, N. juncifolius, N. marvieri, N. poeticus, N. pseudonarcissus, N. rupicola, N. scaberulus, N. serotina, N. triandrus, N. viridiflorus, N. watieri, etc.*, triploid (x = 21) in *N. jonquilloides, N. poeticus,* ancestral *N. tazetta, etc.*, 2n = 14 to 42 in case of *N. bulbocodium* and *N. pseudonarcissus*, tetraploid 2n = 28 in *N.jonquilla, N. juncifolius, N. serotinus, etc.* whereas 2n = 20 in *N. elegans* ssp *intermedius* and cv. Fallax, *N. tazetta, etc.*, 2n = 22 in *N. tazetta*, 2n = 30 in *N. serotinus*, 2n = 50 in *N. dubius*, but species in the sections Serotini and Tazettae have n (x) = 10 (*N. tazetta* being diploid with 2n – 20) and 11 chromosomes (*N. broussonettii* being diploid with 2n = 22) (Misra, 1989). Jefferson-Brown (1969) mentions that all the species possess basic chromosome number of 7 except a few (*N. broussonettii, N. elegans* and *N. tazetta*) where it may be either 10 (*N. elegans*) or 11 (*N. broussonettii*) or 10 and 11 both (*N. tazetta*) and the present day plants of specific status have probably evolved from extinct ancestral forms with diploid (2n = 14) chromosome number by gene mutation, hybridization, polyploidy and loss or alterations of chromosomes.

Many important modern cultivars were bred in Britain and in the Netherlands. Dean Herbert was first to effect deliberate hybridization in narcissi in 1820s (Gorer, 1970), followed by Blackhouse and Leeds in 1850s and then many others. Most cultivated species are derived from a few species, in particular *N. pseudonarcissus, N. poeticus* and *N. tazetta* in order of merit, crosses between the first two giving single flowered progeny and between the last two, multi-flowered hybrids, because the corona of *N. poeticus* is suffused with red on orange pigment which has been the ultimate source of all the brilliant red and orange shades available in modern garden clones. However, due to extensive interbreeding among them, this consideration is now no longer practical. Large-cupped narcissi (biggest commercial group), were produced by the crosses of trumpet forms and *N. poeticus.* Majority of cultivars are tetraploids. *N. tazetta* with basic number of 10 or 11 was when crossed with *N. poeticus* (diploid and tetraploid with basic number of 7) produced in 1923 'Laurens Koster' with chromosome numbers of 10 + 7 and 'Cheerfulness' with chromosome numbers of 10 + 14. The third prominent cv. 'Geranium' with chromosome number of 17 was created by J.B. van der Schoot in 1930 which is almost completely pollen-sterile. In 1860s in England, triploid form of *N. pseudonarcissus* with sterile pollen appeared which is now known as *N. bicolor* though this when crossed with others produces plenty of seeds. When this triploid was crossed with diploid *N. pseudonarcissus*, it resulted into tetraploids as well as other forms of even higher ploidy levels, *viz.* 2n = 41 (near hexaploid) in 1890s. *N. cyclamineus* with diploids and tetraploids from the trumpet and large-cupped groups resulted into fertile triploids. In Netherlands, area under narcissus production was 1639 hectares during 1990-91devoted to 'Carlton' (18 per cent), 'Ice Follies' (9 per cent), 'Dutch Master', 'Golden Harvest' and 'Tète-a-Tète' (8 per cent each). In United Kingdom, about 35 cultivars are considered to be important but six account for most of the areas. They are the first four of the Dutch list above, plus 'Fortune' and 'Cheerfulness'. Most cultivars are diploid, but some are triploid, the well known but little grown 'King Alfred' is tetraploid, and there are some hexaploids (Rees, 1972). Barrett *et al.* (2004) studied measures of selfing rates and levels of genetic diversity at allozyme loci in six populations of *Narcissus longispathus* (a self-compatible species) endemic to a few mountain ranges in south-eastern Spain. The populations were distributed among four distinct river valleys encompassing two main watersheds in the Sierra de Cazorla mountains. Selfing rates averaged 0.37 (range 0.23–0.46), resulting in significant inbreeding coefficients for the progeny (f=0.324). The observed patterns of genetic differentiation among populations are likely influenced by the mating system, and a combination of local topography, watershed affinities and gene flow. The studies on its mutation breeding is quite limited and nonsignificant.

Classification, Species and Varieties

Still there is controversy about number of true *Narcissus* species, some authorities putting the number as nine while others 40 to 60 with about half the number of natural hybrids. According to 'The American Daffodil Society' (ADS), all *Narcissus* have been divided into 13 horticultural divisions, which are based 50 per cent on flower form and 50 per cent on genetic background. Horticulturist Robert F. Gabella of

GardenOpus suggested replacement of the words 'cup' with 'corona' and the 'petals' with 'perianth segments' and the corona length and corona radius for cases where corona does not protrude outward. Gabella has further emphasized the prevalence of species phenotype comparable to the genetic lineage of ADS Divisions 5 through 10, and has also suggested garden and/or wild origin. The prevalent flower colours in daffodils are all shades and tones of green, white, yellow, orange, pink and red. They may be self-coloured, *i.e.* both perianth and corona identical in colour and shade or may differ widely. Some perianths and some coronas may also contain more than one colour or shade. Pinks vary from apricot to rose in shades from pale to deep, and some more recent cultivars have hints of lavender or lilac. Reds vary from orange-red to salmon-red to near scarlet. Pink, red, orange and green tones are mainly confined only to coronas. However, breeders are currently working against the genera's natural pigmentation and genetic barriers to create cultivars in which pink, rose, red, orange and green tones suffuse or bleed from the more highly coloured coronas onto the perianth segments of white or yellow. There are an increasing number of commercially available varieties which display this enhanced coloration. The flower's two regions (perianth and corona) are assigned colour somewhat differently. The perianth colours, in case of multi-coloured ones, are assigned from the outer edge of the segment inward to the base of the corona. The corona colours are assigned from the base of the corona outward to the rim. Thus, 'Actaea', a Poeticus (Division IX) daffodil, is officially classified as 9 W-GYR, while 'Accent', a Large-cupped (Division II) daffodil possessing a white perianth and a pink corona, is officially classified as 2 W-P, where G, P, R, W and Y stand for green, pink, red, white and yellow, respectively.

Division I (Daffodils or Trumpet narcissi)

Plants are of garden origin having one flower to a stem. Corona length is equal to or exceeds the length of the perianth segments. This division has four sub-divisions based on their colour combinations. **Sub-Division IA**: Height 35-45 cm and perianth and corona coloured but corona not paler than perianth such as 'Burgemeester Gouverneus' (golden-yellow), 'Dutch Master' (golden-yellow with central trumpet dentate), 'Garron' (primrose-yellow), 'Goldcourt' (lemon-yellow, cup deeper), 'Golden Harvest' (golden-yellow), 'Irish Luck' (golden-yellow), 'King Alfred' (golden-yellow), 'Kingscourt' (lemon-yellow), 'Moonrise' (lemon-yellow), 'Moonstruck' (lemon-yellow), 'Mulatto' (sulphur-yellow, central trumpet white inside), 'Unsurpassable' (golden-yellow), *etc.* **Sub-Division** IB: height 30-48 cm, perianth white and corona coloured (bicolour trumpets) such as 'Celebrity' (cream-white, trumpet yellow), 'Foresight' (milky-white, trumpet yellow), 'Music Hall' (perianth white, trumpet yellow), 'Oklahoma' (white, trumpet dark golden-yellow), 'Preamble' (perianth white and smooth, corona bright yellow), 'Queen of Bicolours' (white, trumpet canary-yellow), 'Spring Glory' (white, trumpet yellow), 'Trousseau' (white, trumpet pale-yellow), 'World's Favourite' (white, trumpet pale-yellow), *etc.* **Sub-Division IC**: Height 30-45 cm, perianth and corona white (sometimes yellow in bud) such as 'Angel's Wings' (large silver-white), 'Beersheba' (white, trumpet narrow), 'Broughshane' (cream but later white), 'Cantatrice' (white, trumpet narrow), 'Mount Hood' (perianth rounded and white, trumpet cream-white), 'Mrs. E.H. Krelage' (white, trumpet ivory-white), 'Vigil' (pure white), *etc.* **Sub-Division ID**: Height 40-45 cm. Any colour combination nor covered above such as 'Mrs. R.O. Backhouse' (ivory-white, trumpet dark pink), 'Pink Isle' (white, trumpet pink), 'Rose Caprice' (white, trumpet pink, base suffused green), 'Salmon Trout' (best pink, perianth white, trumpet salmon-pink), 'Spellbinder' (yellow, trumpet turning white), 'Toscanini' (white but reverse apricot-pink, trumpet ivory-white), *etc.*

Division II (Large-cupped narcissi)

Narcissi of garden origin with one flower to a stem, and corona more than one-third as long as the tepals. **Sub-Division II A**: Height 35-55 cm. Perianth segments coloured in shades of yellow, orange and red and cup usually darker such as (i) yellow-cupped 'Carlton' (yellow, broad-frilled cup), 'Galway' (deep yellow), 'Golden Torch' (golden-yellow), 'St Keverne' (rich gold), *etc.*; and (ii) orange- or red-cupped 'Armada' (cup orange), 'Aranjuez' (cup golden with bright red rim), 'Fortune' (orange to red cup) and 'Scarlet Elegance' (cup orange-red), *etc.* **Sub-Division IIB**: Height 33-50 cm. Segments white and corona coloured in varying shades of yellow and orange such as (i) yellow-cupped 'Chinook' (perianth white, corona yellow), 'Coverack Perfection' (petals star shaped and cream-white, cup rich yellow), 'Duke of Windsor' (petals white, cup apricot-orange), 'Green Island' (segments cream-white, corona variable green-yellow), 'Hyperion' (tepals and cup yellow), 'Kissproof' (segments and cup yellow), 'Polindra' (petals cream-white, cup mimosa-yellow), 'Tudor Minstrel' (segments white and overlapping, cup orange-yellow and frilled), *etc.*; and (ii) orange- or red-cupped 'Arbar' (tepals white, cup orange-red), 'Eddy Canzony' (tepals white, corona flat and bright orange but centre yellow), 'Fermoy' (petals white, corona orange-red), 'Flower Record' (tepals white, corona red), 'John Evelyn' (tepals white, corona orange and expanded), 'Kilworth' (tepals cream with yellow base, corona red), 'Roseworthy' (tepals white, corona pink), 'Salome' (tepals white, corona salmon); *etc.* **Sub-Division IIC**: Height 35-45 cm. Segments white, corona white but not paler than the segments such as 'April Snow' (white), 'Ave' (white, segments broad and pointed, corona rounded), 'Castella' (corona lemon, ageing cream), 'Gypsy Moth' (creamy-white), 'Ice Follies' (white, corona wide, flattened, yellow fading to white), 'Tenedos' (cream, corona yellow turning cream-white), 'Truth' (rounded and slightly frilled), *etc.* **Sub-Division IID**: Any other combination other than IIA, IIB or IIC such as 'Bethany', 'Binkie', 'Charter', 'Day-dream', 'Gleeful', 'Rushlight', *etc.*

Division III (Small-cupped narcissi)

Narcissi of garden origin, with one flower to a stem, and where corona is not more than one-third the length of the segments. **Sub-Division IIIA**: Height 43-45 cm. Segments and corona coloured such as 'Alcida' (yellow and orange), 'Ardour' (yellow, corona orange-scarlet), 'Chungking' (flowers circular, yellow, corona orange-red), 'Doubtful' (predominantly red),

'Edward Buxton' (yellow and orange), 'Irish Coffee' (yellow, corona brown-red), 'Perimeter' (red), 'Seraglio' (yellow, corona green-yellow with red rim), 'Varna' (lemon yellow, corona orange-red), *etc.* **Sub-Division IIIB**: Height 35-45 cm. Perianth white and corona coloured such as 'Barret Browning' (tepals white, corona orange-red), 'Bishmills' (white, corona red), 'Blarney' (cream-white pointed, corona open-fluted and rich apricot), 'La Riante' (white, corona vivid orange), 'Merlin' (white, corona orange), 'Rockhall' (white, corona orange-red), 'Valhalla' (white, corona orange-red), 'Verger' (flowers white, base yellow, corona flat and orange-red), 'Yalta' (cream, corona orange-yellow and rim red, *etc.* Other show cultivars without red colouration are 'Aircastle', 'Green Hills', 'Lady Kesteven', 'Silken Sails', 'Syracuse', *etc.* **Sub-Division IIIC**: Height 35-40 cm. Perianth white, corona short and white and not paler than perianth such as 'Angel' (white with a green and yellow eye), 'April Clouds', 'Chinese White', 'Cushendall', 'Polar Ice', 'Sacramento'Tern', 'Verona', etc. **Sub-Division IIID**: Any colour combination not described above but such varieties are very rare such as 'Reversa'.

Division IV (Double-flowered narcissi)

Height 25-45 cm. Any narcissus of garden origin in which more than one layer of perianth segments and/or more than one layer of corona segments are present. The combination of doubled perianth and corona segments can vary widely between cultivars, and there may be predominantly one, rarely more flowers per stem, also varying by cultivar such as 'Bridal Crown' (tepals white, centre yellow), 'Camellia' (soft yellow, no visible trumpet), 'Cheerfulness' (1-3 flower on a stem, outer segments cream-white, cup orange-rimmed and filled with more cream petals), 'Dick Wilden' (yellow), 'Golden Ducat' (deep yellow), 'Flower Dream' (yellow), 'Indian Chief' (sulphur-yellow, striped orange), 'Inglescombe' (primrose-yellow), 'Irene Copeland' (white and yellow), 'Mary Copeland' (cream-white, some central petals lemon and orange), 'Tahiti' (yellow-orange), 'Telemonius Plenus' syn. 'Van Sion' (golden-yellow, with or without a complete trumpet), 'Texas' (yellow and orange-red), 'Twink' (pale-primrose with orange centre), 'Valencia' (yellow and orange), 'White Cheerfulness' (blooms fragrant creamy white, 3-4 blooms per stem), 'Yellow Cheerfulness' (3-4 yellow fragrant blooms per stem), *etc.* Other show cultivars are 'Acropolis', 'Double Event', 'Gay Challenger', 'Gay Time', 'Hawaii', 'White Lion', *etc.*

Division V (Triandrus narcissi)

Height 20-38 cm. Narcissi of garden origin developed from *Narcissus triandrus* and its allies, so its characters, the pendent flowers having reflexed and back-swept petals are clearly evident. Plants most often bear two or more flowers per stem. **Sub-Division VA**: Corona not less than two-thirds the length of the perianth segments such as 'Liberty Bells' (creamy-yellow), 'Rippling Waters' (2-3 white flowers on a stem), 'Thalia' (pale-cream, corona white), 'Tresamble' (cream-white), *etc.* One of its varieties 'Arish Mell' is very showy cultivar. **Sub-Division VB**: Corona less than or equal to two-third the length of the perianth segments such as 'April Tears' (deep yellow), 'Hawera', 'Silver Chimes', 'Tincleton', 'Tuesday's Child', *etc.*

Division VI (Cyclamineus narcissi)

Height 20-38 cm. Characteristics of *Narcissus cyclamineus* and its allies clearly evident as there is only one nodding flower per stem and the perianth segments are often reflexed or wind-swept in appearance. Long trumpet-shaped corona varies but can sometimes exceed the length of perianth segment. **Sub-Division VIA**: Corona not less than two-thirds the length of the segments such as 'Baby Doll' (yellow with little deeper corona), 'Bartley' (golden reflexed flowers), 'Charity May' (yellow, corona little deeper), 'Dove Wings' (tepals white, corona yellow), 'February Gold' (yellow with slightly deeper corona), 'February Silver' (tepals white, corona yellow), 'Jack Snipe' (cream-white and a bit reflexed segments, corona orange-yellow with pale margin), 'Jumblie' (back-reflexed yellow tepals, corona turmeric-yellow), 'March Sunshine' (yellow), 'Peeping Tom' (shining yellow), *etc.* Other show cultivars are 'Jenny', 'Tête à Tête' (yellow), 'Woodcock', *etc.* **Sub-Division VIB**: Corona less than two-third the length of tepals such as 'Beryl' (cream and orange), 'Kitten', 'Little Imp', 'Lofty', 'Roger', *etc.*

Division VII (Jonquilla narcissi of garden origin)

Height 27-43 cm. Characteristics of *Narcissus jonquilla* and its allies clearly evident. Flowers are small to medium, perianth segments are flat, corona length varies but is usually short and semi-spherical, foliage may be rush-like and dark green as in the species but crossbreeding with other divisions has produced a range of foliage types. Fragrance is usually prominent. Flowers may be borne one to several to a stem, depending upon cultivar. Corona is short and sometimes flanged, petals are often flat, fairly broad and rounded. **Sub-Division VIIA:** Corona not less than two-thirds the length of the segments such as 'Desert Song' (golden), 'Golden Guinea' (golden), 'Golden Perfection' (golden), 'Golden Sceptre' (golden yellow, corona long), 'Shah' (golden), 'Sweetness' (canary yellow, corona small), *etc.* **Sub-Division VIIB**: Corona less than two-thirds the length of the perianth segments such as 'Chérry' (pale-cream, corona dull yellow turning pink), 'Finch' (yellow), 'Lanarth' (yellow), 'Lintie' (yellow, tepals rounded, corona orange-edged and small), 'Orange Queen' (orangish-yellow), 'Pipit' (deep yellow), 'Stratosphere' (orange, corona darker), 'Suzy' (segments yellow, corona deep orange), 'Trevithian' (buttercup yellow), *etc.* Other outstanding show cultivars are 'Nancegollan', 'Sweet Pepper', 'Tittle Tattle', *etc.*

Division VIII (Tazetta-Poetaz or bunch-flowered narcissi of garden origin)

Height 38-43 cm. Characteristics may be either true tazettas which are often known as polyanthus narcissi, and intermediate type poetaz narcissi developed between *Narcissus tazetta* and *N. poeticus*. The leaves are mid to dark green. Perianth segments are flat, corona length is usually short and semi-spherical. Fragrance is usually prominent. Flowers may be borne in clusters of a few to over a dozen per stem, depending upon cultivar. True tazettas though do not survive outdoors but are forced in bowls for winter flowering such as 'Avalanche' (segments white flushed yellowish, corona small and lemon-yellow), 'Grand Primo Citronière' (milky-white, corona lemon-yellow, strongly sweet scented), 'Minnow' (2-4 small creamy-yellow flowers per stem, corona lemon, and only 15 cm high), 'Paper White' (whiter, strongly scented), 'Pride of Cornwall' (white, corona small and deep orange), 'Scarlet Gem' (dull yellow, corona small and orange), 'Silver Chimes' (beautiful light yellow, corona little darker), *etc.* Poetaz which are hardy, bear fewer and larger flowers per stem but flower later than true tazettas and the varieties in this group are 'Cragford' (white, corona red), 'Geranium' (pale-cream, corona orange), 'Grand Soleil d'Or' (yellow, corona darker, strongly scented), 'Martha Washington', 'Soleil d'Or' (golden-yellow, deliciously and strongly scented), 'St. Agnes', *etc.*

Division IX (Poet's or Poeticus narcissi of garden origin)

Height usually 35-43 cm. Characteristics of *Narcissus poeticus* and its allies clearly evident. One flower one stem but rarely two, flowers medium-sized, perianth segments flat and almost always white, corona very small, flat, wrinkled, usually green-eyed and orange-to-red banded, and often with intermediate shades of yellow. Fragrance is usually prominent. Varieties are 'Actaea' (tepals glistening white, corona very small, yellow and red-edged), 'Cantabile' (white, corona green and edged red), 'Sarchedon' (tepals white, base yellow, corona yellow but edged red), 'Sonata' (white, corona flat, green-yellow and red-edged), *etc.* Other show cultivars are 'Milan', 'Sea Green', *etc.*

Division X (Species, wild variants and wild hybrids)

The genus *Narcissus* is complex taxonomically having about 50 species and three sub-genera, *i.e. Ajax, Corbularia* and *Narcissus.* All the narcissi described up to division IX are of garden origin, but all those narcissi and their naturally wild occurring forms (ecotypes and natural hybrids) are described under this division such as *Narcissus asturiensis* (*N. minimus*), *N. bulbocodium, N. b.* subsp. *bulbocodium, N. b. cantabricus, N. b. conspicuus, N. b.* var. *citrinus, N. b. monophyllus* (*N. cantabricus monophyllus*), *N. b. nivalis, N. b. obesus, N. b. romieuxii* (*N. romieuxii*), *N. b. tenuifolius, N. campernellii* (*N. canaliculatus, N.× odorus, N. tazetta*), *N. cantabricus* (*N. bulbocodium cantabricus, N. clusii*), *N. cantabricus foliosus, N. cyclamineus, N. elegans* (*N. autumnalis*), *N. jonquilla, N. juncifolius* (*N. assoanus*), *N. j. rupicola* (*N. rupicola*), *N. lobularis* (*N. obvallaris*), *N. minor* (*N. nanus, N. pumilus*), *N. moschatus, N. odorus regulosus, N. poeticus, N. p. verbenensis, N. pseudonarcissus, N. pseudonarcissus gayi* (*N. gayi*), *N. p. moschatus, N. p. obvallaris, N. poeticus* var. *recurvus, N. romieuxii, N. scaberulus, N. tazetta canaliculatus, N. t. papyraceus, N. triandrus, N. t. albus, N. t. concolor, N. viridiflorus, N. watieri, etc.* Most popular ones are being described here with.

Narcissus asturiensis (*N. minimus*) is a native in the mountainous regions of Spain growing to 7.5 cm high bearing weak stems, leaves minute and grass-like, and flowers deep golden-yellow with some 2.5 cm long, yellow and dentate trumpet. It belongs to trumpet-flowered narcissus in miniature. It fits the division I and is most suitable for growing in pots and rock gardens. Flowering occurs in February. It requires protection against rough weather. ***Narcissus alpestris*** is a native of Pyrenees grows only to 10 cm in height, bearing narrow and deeply channelled leaves, white flowers appearing in April with twisted segments which droop towards the slender and little cut-corona, little flanged corona is small and the anthers are bright yellow. ***Narcissus bulbocodium*** is a native of SW France, Spain, Portugal and NW Africa growing up to 15 cm high with green, slender, thread-like, cylindrical and 10-30 cm long leaves and bears solitary, all-yellow and funnel-shaped wide-open flowers some 2.0-3.5 cm long having 2.5 cm long, funnel-shaped and pale to deep yellow trumpets over narrow and pointed petals and curved stamens and style. It flowers from February to March. This is a very variable species. It has produced numerous varieties but do not fit to any of the divisions of the RHS. *N. b. obesus* is a deep yellow form bearing thick leaves and some 3.75 cm long flowers which appear in March and April and the corona is large and deep yellow. *N. b. romieuxii* is a pale-yellow form where flowers appear from December to February. *N. b. vulgaris conspicuus* is deep yellow and *N. b. citrinus* is lemon-yellow, both flowering in March and April. *N. b. monophyllus* (syn. *N. cantabricus*) from Spain, France and NW Africa grows up to 10 cm in height bearing thread-like leaves, 2-3 transparent-white and funnel-shaped flowers per stem, with widely expanded corona and very prominent stamens, flowering in December to January and is suitable for pot-growing. ***Narcissus calcicola*** is a native of Portugal growing up to 15 cm with linear leaves, jonquil-like clustered deep yellow and scented flowers, each some 2 cm across with flattened corona and appear in March and April. It is also sometimes considered as the taller form of *N. juncifolius.* It is suitable for pot-growing. It fits roughly to division VII. ***Narcissus canaliculatus***, a native of Canary Islands and S. Europe is a dwarf form of *N. tazetta.* It grows about 20 cm tall and has all the characteristics of that species though flowers later from February to April in bunches of up to six per stem. Its flowers are highly fragrant, the petals are white but corona is yellow. It is a most suitable species for rock gardens and for growing in pots. ***Narcissus cyclamineus*** is the only species of its group. A native of Spain and Portugal growing up to 20 cm in height with dark green narrow-linear leaves having a deeply grooved keel, slender scape and most beautiful rich gold pendent mini-flowers having some 2-4 cm long petal-coloured trumpets with crenate mouths though petals are entirely swept back similar to cyclamen. It can be grown in open or partial shade best in moist soils at a cool and shady location though is not frost-hardy. It is the parent of many hybrids and varieties fitting into division VI. ***Narcissus elegans*** (*N. autumnalis*), a native of Algeria, S. Italy and the Mediterranean islands is a non-frost hardy **autumn-flowering** (September-October) species which grows up to 25 cm in height. The leaves are linear and flat, some 25 cm long and 6 mm wide, channelled and appear with the flowers. The stems are erect and may bear solitary or more blooms some 5 cm across where petals are widely spaced, thin and wavy and greenish-white while central corona is pale-orange and almost flat. It likes sunny location. ***Narcissus jonquilla*** is a native of Algeria with distribution in Portugal, Spain, North Africa and Balearic Islands growing up to 45 cm in height, bearing cylindrical stems and dark green rush-like leaves up to 30 cm long and strongly scented deep yellow flowers in a cluster of 2-6, each measuring 3.7-5.0 cm wide, and small central corona. Its double variety is 'Queen Anne's Jonquil'. *N. jonquilla* with *N. poeticus* produced *N.* × *gracilis* growing up to 38 cm in height and bears linear mid-green leaves and scented bright pale-yellow flowers some 3.7-5.0 cm wide with short corona which appear in May though others flower in April. *N. jonquilla* with *N. pseudonarcissus* (a trumpet narcissus) has produced an April-flowering hybrid *N.* × *odorus* known since 16th century, growing up to 45 cm in height with linear leaves and 2-3 clustered, fragrant and pale-yellow flowers some 6.25 cm across with bell-shaped corona. *N.* × *o.* 'Campernelli Plenus' is its double form. These prefer warmer climate and deep rich soil with alkaline reactions. These all the varieties fit into division VII. ***Narcissus juncifolius*** is a native of SW Europe (Spain, Portugal and France) growing up to 15 cm in height with cylindrical leaves and May-borne deep yellow some 2.5 cm across flowers with flattened corona. This fits division VII. It is suitable for a cool location and in rock gardens. ***Narcissus minor***, a native of Portugal growing up to 20 cm in height with narrowly strap-shaped leaves, is a variable species flowering in March having pale-yellow flowers but slightly deeper corona which is 3.75 cm long. It fits into division I. ***Narcissus nanus*** is a native of Europe and grows to 20 cm or more in height. The leaves are grey-green, flowers are upright with segments and corona almost equal in size, *i.e.* some 3.5 cm long, perianth segments are flat, pointed and soft yellow though corona is deep yellow. ***Narcissus poeticus***, a native of Mediterranean region (S. Europe, Spain to Greece) is a variable species, growing to a height of 45 cm bearing 2-edged scape, glaucous green and narrow strap-shaped leaves, and solitary, 4.5-7.5 cm across, pure white and highly fragrant flowers in April with crisped and frilled, flat and bright red corona. This has many wild forms and named varieties found on wet mountain meadows so requires moist conditions for their growth. *N. p. ornatus* is one of its robust growing wild forms. In the development of the ancient varieties, out of the three ancestral species this occupies the second position and is parent of many small-cupped cultivars. ***Narcissus pseudonarcissus*** is a native of France, Switzerland, N. Italy, N. Spain and United Kingdom, growing up to 30 cm in height with grey-green, strap-shaped and 1.5 cm wide leaves. It is hardy and very variable in shape, size and colour of the flowers which are scented, nodding and near-white to light yellow with 5.0-6.25 cm long and lemon-yellow trumpets, which appear in April. This is found growing in the wild in the open and damp meadowlands. Among all the three ancestral species used in the breeding programme, this occupies the first position with many of the wild or developed varieties. *N. p. moschatus* is a white form, and *N. p. obvallaris* (syn. *N. obvallaris*) from Europe bears flat, up to 30 cm long and bluish-green leaves, uniform yellow flowers some 4-5 cm long having flat tepals and golden-yellow corona of about the same length but neither of them is as vigorous or self-propagating as the species. These fit to division I. ***Narcissus rupicola*** is a native of Spain and Portugal growing up to 13 cm high, bearing grey-green, narrow, triangular and some 20 cm long leaves, and fragrant, clear yellow solitary flowers some 1.8-2.5 cm across with nearly flat 6-lobed corona. It flowers in April and May and fits into division VII. ***Narcissus scaberulus*** is a native of Portugal, grows up to 13 cm high and bears slightly glaucous green and cylindrical leaves. Flowers up to 13 mm across are yellow (deeper than *N. rupicola*). It is best grown at cooler locations. It fits into division VII. ***Narcissus tazetta***, a native of Canary Islands, S. Europe (Spain) to Iran, Kashmir, China, Japan and North Africa, is half-hardy and grows up to 45 cm high bearing strap-shaped some 30 cm long leaves. This variable species is divided into three categories, *viz.* albae, bicolours and luteae. In albae the flowers are pure white, in bicolours the perainth is white and corona yellow while in luteae the flowers are yellow. *N. tazetta* has several (4-10) white flowers on one scape, each flower some 2.5-3.8 cm across and a small cup-like white to yellow corona, generally strongly scented and earlier to flower (January to February). It does not require cold for its growing (Bailey and Bailey, 1976). It is most suitable for growing in pots. Among all the three ancestral species involved in breeding programme, *N. tazetta* occupies the third place. With *N. poeticus* it has produced numerous poetaz varieties. These fit into division VIII. *N. t.* var. *orientalis* bears sulphur-yellow flowers with cup-shaped orange-yellow corona, having one-third the length of its segments. *N. t. papyraceus* ('Paper White') is native to S. France across to Dalmatia and grows to 52 cm high bearing strap-shaped leaves and all-white scented flowers some 3.8 cm across which appear in March outside but by Christmas under glasshouse in pots. ***Narcissus triandrus*** is a native of Spain, Portugal and Isles des Glenans. The scape is very slender and about 30 cm long. Its leaves are grey-green, few, rush-like, some 15 cm long and channelled. The flowers are white, nodding, 3 cm across, petals swept-back and in a bunch of 1-6. The corona is 1 cm long and broad, and cup-shaped. The seeds are produced freely. They also need plenty of moisture, but combined with a well-drained sandy, peaty soil which partly dries out in summer. It fits to division V. *N. t. albus* from Spain and Portugal grows up to 10 cm high bearing pendent cream-white flowers with reflexed and back-swept petals and cup-shaped corona some 1.25 cm long, and the flowers appear in March-April. *N. t. concolor* from Spain and Portugal is a golden-yellow and sweetly scented version of *N. t. albus*. ***Narcissus viridiflorus***, a native of Gibralter, S. Spain and Morocco is **autumn-flowering** (October-November) which grows up to 20 cm in height and prefers sun. Leaves are filiform, the flowers are 1-4 per stem of dull olive-green colour with about 5 cm in diameter. Petals are narrow, twisted, widely-spaced and star-like. ***Narcissus watieri***, a native of N. Africa and the Atlas Mountains is not hardy and grows to 15 cm high with narrowly strap-shaped, erect and some 15 cm long leaves, solitary pure white and sweetly scented flowers some 1.9 to 3.75 cm wide in April possessing an almost flat cup. It does well in pots in cold houses. It fits in division VII.

Division XI (Miscellaneous narcissi)

Plants are of garden origin and can represent any potential genetic background but not covered above. These may be the **collar** and **split-corona** groups having divided and petal-like coronas. The split can be tri-radial or hexi-radial of any length or orientation from the outer rim inward at more than half its natural length. In some cases the segments may be broad enough to underlap and overlap alternating perianth segments such as 'Cassata', 'Golden Orchid', 'Joli Coeur', 'Lemon Beauty' and 'Parisienne'. Though flowers are most often borne one to a stem, there are cultivars with multiple flowers per stem. In **collar** daffodils, the corona segments lie opposite the perianth segments and are usually in two whorls of three such as 'Ahoy', 'Evolution' and 'Gold Collar', but in **Papillon** daffodils the corona segments lie alternate to the perianth segments and are usually in a single whorl of six such as 'Donna Bella' and 'Elfhorn'. **Miniature daffodils** are not an official ADS Division but can occur in each of the other divisions and possess the same descriptive characteristics. However, the flowers are 3.8 cm or less in diameter, and are borne on proportionately smaller plants.

Propagation

Narcissi are propagated through vegetative means (bulbs, bulb offsets, bulb division, twin-scaling), tissue culturing and through seeds. Underground **bulb** splitting is common natural process but is very slow. Sanyat Misra and Misra (2013) have furnished a detailed account on the growth of narcissus bulb. In fact, narcissus bulbs do not renew themselves completely every year. A bulb consists of successive generations of bulb units, each unit comprising bulb scales, leaf bases and flower (in larger bulbs). The vegetative buds develop in the axils of the fleshy leaves. The apical meristems in the leaf axils of mature bulbs in each season produce 2, 3 or 4 scale leaves, 2-3 leaves and then a flower bud. These new bulbs enlarge and cause gradual disintegration of the previously intact outer bulb scales. With the growing of the buds in the axils of the fleshy leaves (Rees, 1972), the slabs of small and flattened bulbs are formed so when such bulbs are planted after separation develop into round, double-nosed and triple-nosed and mother bulb stages, the latter stage requiring about three years to develop from a slab. The life span of a bulb unit extends over about four years. Therefore the mother bulbs have a number of noses and offsets (bulbils), and are usually replanted after removing any loose offset. These provide a high proportion of rounds and many large bulb offsets. These are only rounds which are used for flowering and early forcing if they are large enough though second and third grade rounds are used for later forcing. Top size bulbs are known as 'wares' and these develop only when every year lifting and planting are effected. Double- and triple-nosed bulbs do not have proper-sized rounds, hence are used as planting stock for second or third year forcing after these once or twice have been grown after separation. Mother bulbs on replanting may produce large rounds, small rounds, double- or triple-nosed bulbs and many offsets. These **offsets** develop directly in rounds within shorter duration (1 year) as compared to the flat bulbs (2 years). Bulb production in a 2-year period is roughly 2.5 times. **Scooping** is a process where in vigorous varieties the whole basal plate (root disc) is carefully sliced out (not too deep into the scales) through a specially designed curved-bladed knife in a manner that neither the potential portion of the scale bases are removed nor any part of root disc is left after scoring as both the cases are harmful. The main aim is to kill the apically dominant bud and to provide full access to scale bases to form new bulblets after callusing (Rees, 1972). Scooping process should be initiated some 15-17 weeks before final planting. The scooped bulbs are immediately dipped in clean hydrated lime or in hypochlorite solution and then placed in propagating chamber at 21 °C, and this temperature may be raised up to 30 °C for two weeks but after about 15-20 days of scooping. When newly formed bulblets on the scale bases of the planted bulbs have started forming roots, these are planted out. Through scooping, one can obtain some 30 bulblets per parent bulb, and such bulblets will take up to five years to attain flowering size. **Chipping** (cutting or cross-cutting) is the conical incision on the basal plate of bulbs measuring 10-18 cm circumference, to kill the shoot for breaking apical dominance so that numerous daughter bulbs along the cut edges of the scales are formed. Chipping may be effected into halves, quarters, $^1/_8$ and $^1/_{16}$ and incubated in plastic bags and planted in a 50:50 grit + peat medium which he reported better than twin-scaling and quicker method for obtaining bulbs without wasting any bulb material. **Twin-scaling** is the further exploitation of chipping where many (8-10) such incision through apical portion to basal plate is made in 12-14 cm grade bulb, and then attached twin-scales are gently removed, disinfected in Thiram or Captafol, planted in moist sand or vermiculite and incubated at 15-23 °C for 8-10 weeks in darkness, these regenerate into daughter bulbs with better bulbil formation in the outer scales as compared to inner ones though bulbil size is better in the inner scales (Sanyat Misra and Misra, 2013). The effects on yield of propagating narcissus cv. 'Fortune' bulbs of different size grades by chipping and twin-scaling were studied in field and frost-protected greenhouse experiments. After the first year, the average number of bulbs per propagule cut from offset-derived bulbs varied between 0.51 and 1.50, with propagule size being a major determining factor. Total bulb weight yield derived from a single initial bulb was dependent on sectioning treatment. Up to an initial bulb weight of 25 g, uncut bulbs gave the highest yield, but sectioning above that by chipping gave higher total yields. There was a linear relationship between bulb weight per chip and initial chip weight; allowing target bulb weights to be achieved after the first year by adjusting the cutting rate relative to initial bulb grade. Further, bulb weight increase in the second year of growth was independent of sectioning method and could be described in terms of a jointed linear regression. Percentage weight increase was greatest in small bulbs. Bulb numbers in sectioning treatments did not increase during the second year since critical weights for offset production were not reached; uncut bulbs tended to produce offsets once a critical weight of about 35 g was reached. Bulb weight/cutting rate combinations were identified which approached the ideal of yielding just one, uniformly large bulb per propagule. Extrapolation to give estimated distributions of bulbs after 4 years of growth suggested that populations of bulbs derived from sectioning will approach a dynamic equilibrium of size grades characteristic of populations derived by natural increase (Fenlon, 1990).

For quick multiplication of elite material and to get disease-free planting material, narcissi can be propagated ***in***

vitro by taking scale bases, twin-scale bases, stems from the basal plate, leaf bases devoid of chlorophyll, floral stems and elongating shoots as explants and through this method up to 2,000 bulbils can be produced in 18 month duration (Sanyat Misra and Misra, 2013). In case of explants from twin-scales, bits some 1 cm wide and 1.5–2 cm high are generally taken. MS medium containing 2.0 mg 2,4-D/l was found to be most effective for callus induction through *in vitro* propagation of *N. pseudonarcissus* and when the medium was supplemented with NAA and BA resulted in a 1.5-2 times increase in plant regeneration compared with the medium 2,4-D and kinetin. However, through micropropagation the propagation rate of shoot clumps decreases with every cycle and bulblets do not grow rapidly after transfer to soil (Langens-Gerrits and Nashimoto, 1997). Narcissus bulbs consist of successive generations of bulb units, each comprising bulb scales, leaf bases and in larger bulbs a flower. They prepared the twin-scales noting the generation (1st, 2nd or 3rd year) and origin (bulb scale, leaf base or flower scape) of the twin-scale pieces. After incubation at 20 °C for 12 weeks, bulbil production was highest on propagules consisting of a 3rd year leaf base plus a 2nd year bulb scale; other types performed well, except those with 3rd or 1st year bulb scales or those including a scape. Bulbil production was followed during prolonged incubation of twin-scales cut from the outermost position (A) through to the innermost position (E). Twin-scales of position 'A' soon produced visible bulbil initials (mean 6.4 weeks) but their period of bulbil growth was limited (13.5 weeks) and the bulbils obtained were small (113 mm^3). Progressively more central propagules produced bulbils later but these had an extended growing period and became larger (corresponding values for position 'E' 12.7 weeks, 33.2 weeks, 366 mm^3). The attempt to modify bulbil production under a standard 12-week incubation, twin-scales cut after 0–10 weeks of prior bulb storage at 5–30 °C, and incubated with GA_3 (100 ppm) or ancymidol (1–10 ppm), the bulbil production was recorded maximum in positions 'B' and 'C', decreasing in the inner twin-scales, and was minimum in position 'A' (Sochacki and Orlikowska, 2005). In general, bulbil production decreased with delayed cutting. There was little effect of pre cutting storage temperature, although the highest numbers and sizes of bulbils were obtained after 30 °C storage. Bulbil production was reduced by GA_3, but was little affected by ancymidol. Sharma and Kanwar (2003) developed the technique for faster clonal multiplication of promising cultivars of tulips and daffodils through micropropagation. MS medium containing 1 mg l^{-1} NAA, 5 mg l^{-1} BA, 3 per cent sucrose and 0.8 per cent agar was found to be the best for shoot multiplication of tulip 'Recreado' and 'Christmas Marvel' and daffodils 'Carlton' and 'Dutch Master'. *In vitro* rooting was difficult for tulip. MS medium containing 9 per cent sucrose, 0.8 per cent agar and 1 mg l^{-1} NAA was found to be most effective for *in vitro* bulblet formation and rooting of daffodil. Sage *et al.* (2000) and Sage and Hammatt (2002) succeeded in producing somatic embryos (SEs) from leaf explants of *Narcissus pseudonarcissus* cvs 'Golden Harvest' and 'St. Keverne'. Initial experiments with cv. 'Golden Harvest' resulted in SEs from leaf lamina, leaf base, bulb scale and stem (scape) explants. Embryogenesis was induced on media with a range of 2,4-D and 6-BAP concentrations. There were significantly more SEs with media containing 5 μM 2,4-D and 0.5 or 5 μM BAP than any other growth regulator combination. Scape-explants produced SEs earlier than other explant types, and when oriented with their basipetal surface away from the medium, they produced significantly more advanced SEs than those with their surface in contact with the MS based medium. SEs converted to plantlets most efficiently with a pre-treatment at 24 °C for 10-20 weeks on a medium with IBA, followed by transfer to 4 °C. Leaf explants from shoot cultures of 'Golden Harvest' produced SEs on medium with BAP combined with 2,4-D or 1-NAA, but 4-amino-3, 5, 6-trichloro-2-pyridinecarboxylic acid (picloram) was ineffective. SEs converted to plantlets efficiently following a 4 °C treatment in addition to 4.9 μM IBA. These plantlets were readily transferred to *ex vitro* conditions. Nodular callus was produced from scape explants of 'Golden Harvest' and 'St. Keverne'. This regenerated most successfully on a medium with TDZ in combination with either NAA or 2,4-D. Preliminary studies of the growth of callus in a liquid medium were promising. Transformation has been achieved via *Agrobacterium tumefaciens*, both with a wild type and an engineered strain.

Sage (2005) opined that propagation and protection are two major considerations in the commercialisation of new flower bulb crops. Many propagation techniques are used with bulbous crops but the use of bioreactors might be necessary for the rapid and economical commercialisation of new cultivars. Some commercial bioreactors are now being developed for flower bulbs. Nodular callus was induced for two narcissus cvs 'Golden Harvest' and 'St. Keverne'. After regeneration trials on semi-solid media, up to 519 shoots g^{-1} callus for 'Golden Harvest' and 309 g^{-1} for 'St. Keverne' were achievable in a year. Trials with nodular callus in RITA bioreactors were promising, but further work is needed to speed up the system. Protection of bulb crops usually involves an integrated approach utilising multiple control methods, frequently including the use of agrochemicals. A desire to reduce agrochemical use has led, in some cases, to the use of transformation technology for pest and disease resistance. Where not appropriate, or politically unacceptable, marker assisted breeding has sometimes been investigated for more rapid and economical cultivar improvement than conventional breeding could allow. Marker assisted breeding, using AFLP technology is now being developed for narcissus to move resistance to *Fusarium oxysporum* f. sp. *narcissi* from wild species/resistant cultivars into susceptible/new ones. *Fusarium* is the worst problem faced by narcissus growers in Great Britain.

Except a few true breeding species, the narcissus crop is raised through **seeds** only for evolution of new varieties, so this is the work only of a breeder as it takes 4-7 years to produce flowers from seeds and where in the first year the flowers are also usually quite freakish (Hay and Beckett, 1971), *vis-a-vis* have different flowering characteristics from those of their parents. After pollination and fertilization, seed capsules are allowed to remain on plant for ripening which takes 4-6 weeks after capsule formation, and when capsules become brown or after shaking when these produce rattling sound, *i.e.* usually from April to June, the seeds are collected and after a few days are imbibed and sown in seed-beds, pots, pans or flats outside or in frames in the mixture containing sifted sterilized loam 2 parts, peat 1 part and coarse sand 1 part and to each cubic metre of the medium 1 kg of superphosphate and ½ kg of powdered chalk should be mixed. These are sown leaving 3 cm each way and then are covered thinly with soil mixture. At this time, 23-27 °C temperature should be provided for a week to 10 days and then these require a temperature regime of 5-16 °C for quick germination. Each evening the pots are watered lightly with fine spray. Each seed produces a thin rush-like leaf which dies down late in the season but the next season the seedlings appear stronger with 1-2 rush-like or flat leaves (depending upon the type of cultivar). The third season, the young bulbs are taken out and planted at 20 × 10 cm distance, followed by proper cultural operations until flowering (Sanyat Misra and Misra, 2013).

Cultural Practices

Narcissus can be grown in many **soil** types with varying textures. A well drained deep, fertile soil containing abundant organic matter, however, results in the best performance. Also a report from Poland mentions that black soils are best for narcissus cultivation. Heavy clay soils can be loosened by adding large amount of organic matter. Field soil along with bark, peat, perlite, sand, vermiculite can also be used for planting bulbs. The soil should be of average fertility because high nitrogen content promotes excessive vegetative growth at the expense of flower production. Most cultivars are tolerant of either acid or alkaline soils. Species will, of course, favour a soil type close to that of their habitat, *e.g.* acid to neutral soils having a pH range of 5.5-7.0 for most forms of *Narcissus bulbocodium*, *N. triandrus*, *N. asturiensis* and *N. cyclamineus*, and the slightly alkaline pH, *i.e.* 7-8 for *N. jonquilla* and *N tazetta.* Also there are reports that narcissi can be successfully grown at a pH as low as 4.9. **Temperature** plays a vital role in bulb sprouting, growth and flowering in narcissus. Medium and high hills having average temperature of 18-23 °C is optimum for better performance of daffodils. Narcissus requires low temperature which can be fulfilled by growing bulbs outdoors in temperate climate (Stevens, 1977). A low temperature treatment for 12-14 weeks is required after planting for successful flowering in narcissi. Locations having more than 6 h **sunlight** is suitable for quality flower production in majority of the cases (Miles, 1963). Planting in partial shade will ensure a longer flowering period. Flower regulation in narcissus is not dependent on photoperiod because flowers can be initiated at any time of the year (Rees, 1972) though for harvesting long-stemmed flowers, long days are essential (Stevens, 1977). However, the overall performance with regard to bulb and flower production is certainly better under full sunlight.

The field used for planting of daffodil bulbs should be ploughed to a depth of 30 cm to get a fine tilth. Jefferson-Brown(1969) suggested that yield and quality of bulbs and flowers were not affected by hand digging, deep ploughing up to 30 cm or rotavating to 17.5-22.5 cm deep. However, in any case the clods should be properly broken and plastic sheets, weeds as well as stones or any other hard material removed. The soil should be loosened with fork and also by digging to a depth greater than the planting depth so as to allow easy penetration of roots. At least before two days of planting, the bulbs should be dipped for one hour in 0.2 per cent Captan or in its other formulations, or in Thiram, shade-dried and then planted. **Planting** is done at any time after the bulbs have been lifted, but it is usually from mid-August to full October, however, the October planting generally results into best performance. This enables the bulbs to develop strong roots in the fall season which supports flowering in the spring. Delayed planting results in poor quality flowers and low bulb yield (Jefferson-Brown, 1969). More number of flowers, good quality blooms and long stems were observed by planting daffodils in August in England (Horton, 1956). Stevens (1977) recommended bulb planting 10-15 cm deep and 10-15 cm apart in single, double or triple rows separated by a 60 cm wide path. Proper **planting depth** for narcissus bulbs is governed by the variety to be grown, the size of the bulbs and the soil type. Bulbs should be planted one and half times the size of bulbs into the soil. Hay and Beckett (1971) recommend 3-times the depth of the polar diameter of the bulbs to be planted. Jefferson and Brown (1969) suggested planting of bulbs to 2½ times of its own depth. For good flowering in var. 'Carlton', the bulb weight should be 45-55 g to get flowering in more than 80 per cent plants, however, in case of the bulbs weighing 25-35 g the flowering may come down to less than 10 per cent, though each variety has its own standard bulb size for flower production. The rounds produce better cut flowers than the flat ones obtained from double- to triple-nosed bulbs. Planting the bulbs 15cm deep and 10 to 15 cm apart will be appropriate if they are the larger daffodil varieties though for smaller and miniature types, these should be planted 10 cm deep and 8 cm apart. Jefferson-Brown (1969) suggested 10 cm distance in rows and 30 cm apart in case of small bulbs, whereas 15 cm in the rows spaced 37.5 cm apart for the larger ones. In light soils, they require to be planted rather deeply. Smaller-sized but more number of bulbs are produced by shallow planting though deeper planting produces larger bulbs but with reduced number. Flower number per unit area in the first year crop is directly related to planting density, but subsequently it is found reduced at higher planting densities due to splitting of bulbs underground. In case planting density is not too much, maximum number of flowers is produced during second, third and fourth years and when plants have become overcrowded and the flower production is towards declining trend, the bulbs should be lifted and replanted. Best flowers are produced normally in the second year of planting whereas, largest number in the third year (Hay and Beckett, 1971).

For taking a good crop, at the time of soil preparation for planting 3 kg/m^2 (300 quintals/ha) of farmyard manure should be incorporated in the soil. Narcissus being highly sensitive to chemical **fertilizers** and as neither it roots nor sprouts if before planting any fertilizer in the soil has been incorporated, so only after sprouting or at the time of soil preparation the basal dose of fertilizers can be given (Sanyat Misra and Misra, 2013). This applies even when narcissi are being forced. High nitrogen level in the soil promotes bulb growth and produces longer flower stems but delays flowering (Stevens, 1977). Nitrogen 120 kg/ha in the form of ammonium sulphate or calcium ammonium nitrate together with P_2O_5 in the form of single superphosphate and K_2O in the form of muriate or sulphate of potash, each at 120 kg/ha should be incorporated in the bed at the time of soil preparation, and 60 kg N again after 3½ months of planting, *i.e.* just after sprouting. Deficiency of P_2O_5 and K_2O reduces growth, delays flowering, shortens flower stalks and induces early senescence of foliage. Ca and Mg deficiency in the soil causes reduced growth while Mn deficiency causes early senescence of the foliage (Sanyat Misra and Misra, 2013). Waters (1977) reported increase in number of flowers and bulb weight with application of 625 kg/ha each of single superphosphate and muriate of potash. In addition, NPK (19:19:19) @ 1ml/l at weekly intervals before bud formation and Multi-K @ 1ml/l fortnightly during bud formation can also be applied to the plants for better results. Moisture is required at all the stages of plant growth (De Hertogh, 1996) for better flowering and bulb production so field has to be **irrigated** uninterrupted depending upon requirement of plants to ensure growth otherwise bulbs may result into split scales to uneven growth. Since bulbs at planting have already initiated flowers and leaves so just for expansion of their tissues these require moist soil during all the stages of growth and flowering, however, waterlogged situation should always be avoided. At the time of planting the soil should have sufficient moisture to sustain the crop till sprouting. After first irrigation, the field is to be **mulched** 5 cm thick with the help of chopped straw, crushed dry leaves, rice husk, chopped bagasse, bark broken in small pieces, sawdust, coarse peat, compost or farmyard manure, *neem* cake or any other organic waste which will check weed growth, conserve moisture, facilitate air movement, will protect the bulbs against frost injury, will provide warmth to the bulbs and together will induce earlier flowering and in the end will decompose and provide organic matter to the soil. The overall effect will positively reflect on flower and bulb production. Though all these benefits can also be achieved by mulching with perforated black polyethylene sheets, perforations being only at the point of bulbs planted, but this is uneconomic and instead of providing organic matter in the end, these hinder with the plant growth and development. During spring, leaving some 2.5 cm of mulch in the field, rest is removed after the danger of frost is over. The left out mulch, in summer, apart from conserving the moisture and controlling the weeds will also facilitate better aeration and in the end after decomposition will add organic matter in the soil. Daffodils naturalise best so these need not be lifted each year in case the crop is not meant for commercial cut flower production. During growth and development, the soil should never be allowed to dry out as its impact will negatively be reflected on flower as well as bulb formation. After removal of the flowers from the daffodils for cut flower use, since flower initials for ensuing season are being formed inside so it should be ensured that these are provided at least with one dose of phosphatic fertilizer in the soil, which will provide the bulbs with the nutrients that they need for next year's growth. Faded flowers should immediately and regularly be removed so that energy towards growth of capsule formation and seed setting is saved for bulb development.

It is essential at this point that foliage is not removed until they become pale or brown as these help manufacturing the food for bulbs through photosynthetic activities which helps in bulb development and in next year flowering.

To sustain the growth and development, **weeds** compete with the main crop for nutrients, light and water, and as their growth is normally faster and usually horizontal so these mask the main crop quickly. In case of their vertical growth, these grow taller than the main crop to compete with them. If the weeds are not controlled timely and effectively, the crop stand will be utterly poor and there will be almost complete loss. If the beds have been mulched these will certainly hinder their germination and growth but in case when mulching has not been used, the field will have different types of weeds during winters. As soon as winter is gone and spring appears, the summer weeds start coming up. All these weeds are to be controlled either manually or through application of Stomp (pendimethalin) which remains effective for 70 days from the date of application. After 70 days, if there in the field any weed has grown up that should be removed and then again pendimethalin should be sprayed in whole planting by hiding the main crop with alkathene sheets otherwise the crop will also be injured which will affect flower and bulb production. Pendimethalin at the rate of 1.5 l/ha will be sufficient and it should be applied through spraying by dissolving it in water. While spraying, the whole land surface should be covered, followed with gentle flood-watering without entering the main beds. Atrazine @ 1 kg/ha was found effective in Japan in controlling weeds by 64 per cent in narcissus crop during autumn and spring without any adverse effect on flowering (Rusalenko *et al.*, 1970).

Pot Culture

Most of the smaller species, *e.g. N. assoanus, N. cantabricus, N. romieuxii, N. rupicola* and *N. watieri* are really best grown in pots in an unheated glasshouse or bulb frame where their miniature beauty can be appreciated and there is more control of their growing conditions. *N. serotinus* needs a real baking to induce them to flower. *N. bulbocodium* variants are also successful in pots but they can be tried in the open ground on a rock garden or in fairly fine grasses not forming tufts. These are usually seen on sloppy sandy acidic soils. Some clustered species are less frost-hardy and require a warm dry summer for flowering such as *N. papyraceus, N. tazetta* and their relatives. In cold areas these will certainly require to be planted against a warm sunny wall. The selections and hybrids of these types are used for early forcing indoors. The varieties which are best for pot growing are miniatures such as 'February Gold', 'Jetfire' and 'Tète a Tète'. When growing in pots, it should be ensured that there is plenty of crocks in the bottom of the pot for good drainage. Bulbs can of course be planted touching each other if required. Daffodils are well suited to pot cultivation, either for exhibition or for any other decorative purpose. Bigger and deep pots should be used to provide enough space for the roots. Either soil-based or soil-less potting media which is free draining can be used. In 25 cm pot size, 4-5 double-nosed while in 23 cm pots 3 double-nosed or 4-5 single nosed bulbs can be planted. Bulbs and medium in and around bulbs should be pressed firmly. Watering should be done with watering can and soil should be kept moist but not too wet. If plastic pots are used, care should be taken that pots do not get overheated due to direct sunlight. In no case the medium should be allowed to dry out at any time from planting to harvesting. Two parts high fertility soil-based medium mixed with one part coarse grit or fine chippings is suitable for most species. Small bulbs can be planted 7 cm deep in the 12-cm pots.

Growth and Development

The growth cycle of narcissus is dependent upon temperature and moisture cycles. It emerges quickly in the spring, flowers and survives for lengthy period in hot summers in vegetative stage. *N. cyclamineus* and *N. pseudonarcissus* require an absolute cold treatment for further floral differentiation, development and rapid emergence though *N. tazetta* does not require cold. All these three narcissus species require warm temperature for floral initiation and differentiation which occur before harvest and then continue afterwards. With fluorescent lamps emitting white, blue, red, yellow and green light having photosynthetic flux density (PPFD) of 25 ¾·m-2·s-1 to four narcissus varieties already getting 12 h sunlight, it was found that white and blue lights bring about stem stiffness in the cultivars 'Ice Follies', 'Johann Strauss' and 'Unsurpassable' though red light is at all not suitable as it gives poor results (WoŶny and Jerzy, 2007). Daffodil bulbs are used for glasshouse **forcing** in North America and Western Europe, mainly for cut flowers together with potted plants though they are also usually grown outdoors. In India, however, the crop is grown outdoors for bedding and cut flower production both under temperate and sub-temperate climates. At harvest, narcissus has an almost completely initiated flower (Hartsema, 1961) inside which had started taking place shortly after flowering of mother bulbs in April-May. The bulbs of *Narcissus* require warm-cool-warm temperature sequence for growth and development (Rees, 1972). Normally, round or double-nosed bulbs are used for forcing having bulb weight of at least 25 g (De Hertogh, 1977). When the bulbs are lifted from the field, they have partially formed flower parts. Hence, while the initial warm temperature helps to complete flower development, the subsequent low temperatures are required for stalk and leaf elongation and warm temperature for flowering (De Hertogh, 1974). For early forcing, the bulbs are harvested in July and kept at 34 °C for one week, followed by 17-20 °C until pre-cooled in August at 9 °C (Gordon *et al.*, 1984). Bulbs for cut and potted daffodils are handled separately after planting in early October. Cut daffodils are rooted and cooled continuously at 9 °C, which is optimum temperature for shoot elongation. After accumulation of 15-16 weeks of cold period, forcing takes place at a 13-15 °C, whereas pot daffodils are rooted at 9 °C but as soon as rooting takes place the bulbs are shifted to 5 °C. They are forced in the greenhouse at 16-18 °C after 15 cold weeks. The bulbs of daffodils are harvested in late July or August and stored at 17-20 °C until planting for mid-season and late forcings. Bulbs are planted by providing 17-18 cold weeks for cut daffodils and 14-16 cold weeks for pot daffodils. These are subjected to rooting at 9 °C and then cooled at 5 °C or 2 °C. The growth of the bulb shoot should not be more than 10 cm in the rooting room. Forcing of cut and potted daffodils is done at 13-15 °C or 16-17°C in the greenhouse.

'Paper White' tazetta narcissus leads bulb production in Israel, with more than 20 million bulbs exported annually for dry sales and flower production. Recent release of new narcissus cvs 'Ariel', 'Inbal' and 'Nir' suitable for pot-plant production, developed by Agricultural Research Station, Volcani Center of USA necessitates to device the methods for flowering manipulations. Under natural conditions in Israel, 'Paper White' flowers in December-January. In order to advance flowering to October-November, three forcing strategies were applied to 'Ariel', 'Inbal' and the well-known cultivar 'Ziva'. Following the determination of the stage of flower development, the bulbs kept in different storage regimes, *viz.* open shed, temperatures 18-22/25-30 °C (night/day), constant 25 °C, 25 °C followed by 2 weeks at 13 °C, and 25 °C followed by 2 weeks at 9 °C, storage at 25 °C accelerates intra-bulb flower differentiation and shortens forcing time by 2 weeks in 'Ziva' and 'Inbal' and by 4 weeks in 'Ariel', and storage at 9 °C and 13 °C for 2 weeks prior to planting causes stalk elongation inside the bulbs. In all the cultivars, storage at 25 °C followed by a short period at lower temperatures of 9 °C or 13 °C results in anthesis after 37-40 days of growth. Thus the time to anthesis is reduced by 16-18 days. Storage at 25 °C advances anthesis by 9-10 days. In 'Ariel' and 'Inbal', lower storage temperatures significantly increase the number and diameter of individual flowers, while the opposite result is obtained for 'Ziva'. Storage at 25 °C significantly increases the number of flower stalks per bulb in 'Inbal', but in 'Ziva' the increased number of flower stalks is observed after storage in the open shed. It is so inferred that physiological requirements for 'Ariel' and 'Inbal' are different from those of 'Ziva', and that a precise pre-planting protocol is needed for each cultivar.

There are two basic narcissus groups, the 'hardy narcissus' and the 'non-hardy Paper White narcissus'. Hanks (1993) reviewed the physiology, pests, and other aspects of these groups. With the exception of the hot water treatment (HWT), non-planted hardy narcissus bulbs are never stored > 34 °C, while 'Paper Whites' are not stored > 30 °C (Hanks, 1993; Cohen *et al.*, 2011). Neither group of bulbs is stored < 0 °C (Van Aartrijk, 1995). The largest group of narcissus is comprised of the hardy cultivars and there are hundreds grown in many countries. The major storage diseases are *Fusarium* and *Penicillium* (Byther and Chastagner, 1993; Van Aartrijk, 1995). To control *Fusarium* in the planting stock, one approved contact fungicide is added during HWT for 1-2 h at 43.5 °C. This treatment also controls nematodes and the maggots of bulb flies present inside. *Penicillium* can be controlled by use of proper ventilation conditions and a RH of 85 to 90 per cent. Except for the HWT, planting stock bulbs are stored at 17-20 °C. Commercial bulbs for early forcing are given one week treatment at 34 °C followed by 17 °C. All other bulbs are provided 17-20 °C temperature. De Hertogh (1996) has provided forcing programmes for hardy narcissus cultivars in North America. Transportation of hardy daffodils either for forcing or garden use should be at 17 °C under highly ventilated conditions. Most cultivars of 'Paper White' narcissus bulbs are produced in Israel. They are not considered to be hardy bulbs, but can be planted outdoors in USDA Climatic Zones 9 to 11 (Cathey, 1990). After harvesting, the planting stock and commercial bulbs are stored at 25 to 30 °C. Commercial bulbs should be shipped at 25 to 30 °C under highly ventilated conditions. After arrival, bulbs should be stored at 25 to 30 °C until shoots begin to emerge. Subsequently, they should be placed at 2 °C. Prior to planting, they require 2 weeks at 9 to 17 °C. Narcissi bulbs can directly be forced by treating the bulbs for 10-12 weeks at 9 °C and then forcing at 16 °C in the polyhouse (Sanyat Misra and Misra, 2014).

GA 100-1,000 ppm treatment for 2-4 weeks of narcissus bulbs for outdoor planting gave about one fortnight earlier flowering in shorter treatment though flower production was not found affected. For shipping the bulbs from one continent to another, the bulbs can be **retarded** by subjecting them to continuous warm temperature storage at 25-28 °C but in this case bulb quality is adversely affected, however, treating the just lifted bulbs for 12 weeks at 30 °C, for 9 weeks at 0 °C, and then for 8 weeks at 25 °C will retard anthesis by more than six months without affecting the bulbs adversely (Sanyat Misra and Misra, 2013).

Miller and Finan (2206) observed that ethanol reduces unwanted elongation of floral scape and leaves in cv. 'Ziva'. When they grew *N. tazetta* plants with traditional pebble culture, the root zone ethanol concentration of 1-5 per cent (v/v) reduced plant height without any visible phyto-toxicity to the roots. Other ethanol sources including gin, vodka, whisky, schnapps, rum and tequila at 4 per cent were equally effective in reducing growth though beer and wine (white or red) with same level of alcohol were found unsuitable.

Postharvest

Yield of narcissus flowers will depend upon size of the bulbs, soil type and its fertility, growing season and the genotype. One large bulb produces 1-2 flower stems depending upon the number of noses present. Trumpet and cupped narcissi should be cut at the fat goose-neck stage when the perianth segments are just showing colour (Stevens, 1977) and polyanthus types ('Paper White', 'Soleil d'Or', *etc.*) at the time when the first two flowers in the bunch are opening (Waters, 1977; De Hertogh, 1996). The cut scapes can be stored dry for local markets but immediately after reaching there these should be kept with their cut ends in water. Wet storage for 2-4 hours or overnight in water at 0-2 °C immediately after cutting will help in proper opening and prolonging the life. Cut stems can be held for two weeks at 1-2 °C and 90 per cent RH, and then can be rehydrated in warm water. Tulips and daffodils should not be kept in same vase solution because cut stems of narcissus exude a poisonous sap. These are graded for quality and packed in the bundles of tens just by putting one rubber band below the flowers and one near the cut ends. These are now stored either in water or dry in polythene bags or in newspapers and held vertically in wooden or metallic boxes but for distant marketing, these are kept flat in wooden or corrugated boxes. Whether these are stored dry or wet but they are to lose less than a quarter of their potential life. Lower temperatures prevent flower opening while higher storage temperatures reduce the vase life, but a temperature regime of 10-16 °C is best for flower opening. Vase life of narcissus varies greatly with the cultivar, and it ranges from 5-11 days, with an average life of 7 days at 15.6-16 °C and 65 per cent RH. 8-HQC at 200 ppm and sucrose at 2 per cent enhance the vase life. Shipment of potted trumpet narcissus is done when buds are in the pencil stage of development, *i.e.* when scape is erect.

Lifting, Curing and Storage of Bulbs

Lifting of narcissus bulbs is done when entire foliage have senesced and 95 per cent have died down and that is from mid-May to late July, or when required especially for forcing or for export, the desiccant spray generally of diluted sulphuric acid will kill the non-senesced foliage. In New Zealand they lift these during December-January (Stevens, 1977). The dormancy in narcissi is little and of tulip-type, and when leaves have completely senesced, the bulbs seem to be dormant, but in fact, inside there is a very hectic activity of primordial initiation and growth of daughter bulb units, shoots and roots. Poeticus and Poetaz have short resting period and in these cases generally roots begin appearing before senescence of leaves. Through fork the soil is loosened and gently the bulbs are cleaned, roots removed and bulbs collected without injuring them and giving any chance to narcissus fly to lay its eggs over them. Any loose scale if is there, that should be taken out. After lifting the bulbs are divided, graded and kept in trays at a ventilated place for drying and curing. Two flat bulbs and only a few bulblets are produced from one large bulb. On an average the bulb yield each year results into more than one and half times than planted. Normally, per hectare plant density is from 300,000-500,000. In the store, routine checking of bulbs should be conducted for any disease infection and it would be better if these are given frequent fungicidal sprays. Preferred temperature from lifting to grading is 18 °C as this will ensure no harm to flower initials inside the bulbs due to HWT and no infection of basal rot pathogen which is active at 21-30 °C. Bulbs meant for sale are usually 14 cm grade which are stored at 18 °C with proper ventilation to prevent premature rooting while for forcing the grade should be 12-14 cm in circumference.

Insect-Pests, Diseases and Physiological Disorders

Many insect-pests cause severe damage to narcissus crop, *viz.* large narcissus fly (*Merodon equestris*, syn. *Lampetia equestris*), small narcissus fly (*Eumerus tuberculatus* and *E. strigatus*), bulb scale mite (*Steneotarsonemus laticeps*, syn. *Tarsonemus laticeps*), bulb mite (*Rhizoglyphus echinopus*), stem and bulb eelworm (*Ditylenchus dipsaci*), bulb and leaf nematode (*Aphelenchoides subtenui*), root lesion nematode (*Pratylenchus pratensis*, syn. *Ditylenchus pratensis*), and slugs and snails, *etc.* **Large narcissus fly** is active during summer and resembles a small humble-bee about 1.2 cm long with black, yellow or ginger hairs. Females lay single eggs in the neck of the bulbs, and after hatching the maggots crawl down to basal plate to enter the bulbs but pupate in the soil in the following spring. The infested bulbs become soft. Dieldrin 0.1 per cent with HWT for 3 hours is effective to control this pest. **Small narcissus flies** resemble the large narcissus or house fly and is about 6 mm long. Their females lay eggs on the injured parts of the bulbs, and the maggots which are also some 6 mm long, after hatching enter the bulbs normally through the neck and feed inside them. Each season these flies make a number of generations. Their damage and control measures are similar to large narcissus fly. HWT at 43.5 °C for one hour or 42.5 °C for 1½ hours will control these flies. **Bulb scale mite** is highly devastating in the warmer parts, especially in greenhouse-forced narcissi. These are 0.2 mm long, enter the bulb scales and feed around the neck portion surviving on emerging flowers and foliage but when bulbs sprout these move upward and start feeding on the foliage. The infested leaves, after emergence appear with yellow stripes, peduncles elongate with saw-toothed edges and flowers deform. HWT at 44 °C for 3 hours to the dormant bulbs will kill this pest. **Bulb mite** enters the bulbs through ailing portion as it is a sporofungus-pest, and there it multiplies rapidly and extends the injured area. They can be detected as a grey powder on a brown patch. These can travel far and wide by clinging to the narcissus flies. Control is the same as for bulb scale mite. Wireworms and other lepidopterous pests will automatically be controlled while controlling other insects. **Stem and bulb nematode** is a highly devastating pest which enters the bulbs through soft tissues and feed surrounding the neck region and by entering the rapidly growing leaves and stems causing distortion and discolouration, especially in poorly drained soil where its spread is quite fast. At the base of the bulbs these also make dry eelworm wool. Its presence can be realised through the swellings and the light coloured speckles. Such bulbs, especially at the neck region become soft and on cutting these across these present concentric browning. **Bulb and leaf nematode** infest only a few of the selected varieties and its infestation causes scarification with blistering and crinkling on the outer scales of the infested bulbs, and when the bulbs are cut transversely, these show grey discolouration. Such plants show premature leaf yellowing. Its complete control can be obtained through use of nethyl bromide at 100 ml/m^2. **Root lesion nematode** infests only the roots and incites secondary root-rot infection of *Cylindrocarpon radicicola* which causes premature leaf yellowing. Bulb storage first at 32.2 °C for 1 week, followed by 3 h soaking in water at 26.7 °C and then HWT at 44.5 °C for 3 hours will control all these nematodes. **Slugs and snails** feed on the crop during humid weather conditions and enter the bulbs from their bases so leave the chances for attack of various other insect-pests and pathogens. These can be controlled by mixing 6 meta fuel pellets (metaldehyde base) in the powdered form to 1 kg of bran and left in the form of small and scattered heaps in the field infested with slugs. Baits prepared by mixing 1 part metaldehyde and 40 parts bran (v/v) causes the slugs in dry weather to produce more slime and eventual death.

Basal rot (*Fusarium oxysporum* f. sp. *narcissi*), smoulder (*Sclerotinia narcissicola*, syn. *Botrytis narcissicola*), green mould (*Aspergillus niger*), *Penicillium* bulb rot (*Penicillium corymbiferum*), white root rot (*Rosellinia nectarix*), white mould (*Ramularia vallisumbrosae*), leaf scorch (*Stagonospora curtisii*), *Narcissus* fire (*Botryotinia polyblastis*, syn. *Sclerotinia polyblastis*), *etc.*are the fungal diseases which infect narcissi. **Basal rot**, a serious disease is characterised by decay in the root plate and base of the scales in storage and in the field when temperature is above 14 °C, and becomes more serious when temperature is above 21 °C. It is caused normally in storage and transit where bulb tops turn yellow and die before maturity, and the roots as well as scales become purplish in colour and die. Infected bulbs when planted appear with yellowing of leaves. Methyl bromide fumigation at 50 g/m^2 before planting or Aldicarb 10 per cent a.i. at 4 g/m^2 twice during growing period will control this problem. The disease can also be controlled through bulb dipping in Captan 0.5 per cent, zineb 0.5-0.7 per cent, Captan + formaldehyde 1 per cent or by HWT at harvest. **Smoulder** is most prevalent during cold-wet seasons in field as well as in storage. Masses of small black and flat sclerotia are seen under the outer papery scales of affected bulbs, and on the rotting spots of leaves and flowers. Shoots become distorted and yellow with withered dark brown or black leaf tips. The rains and wind help spreading this disease far and wide. Proper field and store-room sanitation and avoidance of high humidity in growing area or in storage are the practices to minimize this disease. Within two days of lifting, the bulbs should be dipped for 30 minutes in a 0.2 per cent carbendazim or thiophanate methyl, followed by quick drying for storage will control this problem provided none of these chemicals are tried again in the following season. Effective control of this disease can be had through fortnightly sprayings of benomyl 0.1 per cent alternate with carbendazim 0.2 per cent. **Green mould** develops on the wounded or bruised bulbs under humid and non-airy conditions in storage which causes fluffy- or spongy-rotting in the bulbs. Injured bulbs should be discarded from storage and the healthy ones should be treated with Thiride or Captan which apart from controlling green mould will also control the ***Penicillium* bulb rot**. **White root rot** develops under moist conditions in the form of root rot and black rotting of the outer scales.White fungal strands are visible in the form of woolly mass near the basal plate. Such bulbs should be destroyed and others treated with Captan or thiride. **White mould** or ***Ramularia* blight** spreads through water or winds in moist-warm weather.It producs grey or white powder and its infection destroys whole of the foliage, *vis-a-vis* stalk. This disease may be eliminated by 3-4 weekly sprayings of 4:4:50 Bordeaux mixture or tank-mix zineb when leaves are 5-10 cm high. **Leaf scorch** pathogen survives in the dry papery scales of the bulbs but infects the leaves on their emergence making them scorched and burnt at the tips but this attacks buds and flowers only during damp weather condition. After senescence of the foliage the spores harbour the neck of the bulbs. It is highly devastating in the cool-stored and late-planted narcissi which flower early. In the first year the disease is controlled through HWT and formalin but in the second year by spraying with 0.2 per cent zineb at 10-14 days intervals after removing and burning diseased tips. The measures adopted to control white mould will control even this disease. **Narcissus fire** is characterized by quickly spreading pale-brown lesions on the leaves, which extend even to perianth segments. Premature death of foliage occurs due to its infection. Moist weathers are highly conducive for spread of this disease. Its infection causes killing of the first flowers and then pathogen moves to foliage where it overwinters. Removal of infected flowers and foliage, proper sanitation and fortnightly spraying with 4:4:50 Bordeaux mixture and tank-mix zineb will control this problem.

There are several **viruses** which attack narcissi. 'Cucumber mosaic virus' causes lighter striping and mottled pale-yellow discolouration of the foliage and floral stalks though not of much significance, 'narcissus yellow stripe' spreading through aphids is the highly devastating viral disease causing bright yellow streaks or stripes from emergence to flowering and thereby reducing yield and quality of flowers and bulbs, 'narcissus white streak' or 'silver leaf' appears during flowering and causes silvery leaf or white striping, thereby early senescence of leaves occurs with poor quality of bulbs, 'narcissus latent virus', 'jonquil mild mosaic' and 'Grand Soleil d'Or' viruses are transmitted through aphids but are not of much significance. 'Narcissus mosaic' is transmitted mechanically. 'Arabic mosaic virus', 'raspberry ring spot', strawberry latent ring spot', tobacco rattle virus', 'tobacco ring spot', and 'tomato black ring or chocolate spot virus' show large dark brown spots, all transmitted through nematodes. Regular spraying with contact insecticides will control aphids, and nematicides the nematodes and other vectors. Infected plants should also be rogued out and burnt to check further spread.

Hot water treatment (HWT) is an essential practice to control narcissus flies, the mites, nematodes, *Fusarium* basal rot, leaf scorch, *etc.* However, the treatment should not be detrimental to the bulbs, especially the flower initials present inside the bulbs, though should kill all the germs and insect-pests present there on and there in. Therefore, HWT should be given to the bulbs at a time when no root initials have formed and when there is complete resistance of the flower buds to heat of the HWT. Controlled (warm) storage of narcissi bulbs, followed by HWT at 44.4 °C along with 1 per cent formalin for three hours, within three weeks of bulb lifting have been found quite satisfactory. Delayed treatment may cause nematodes making protective wool around them.

After flowering for a number of years, sometimes it is noticed that daffodils produce lots of leaves but no flowers due to overcrowding, and this is referred to as **blindness**. Blindness also occurs due to unfavourable temperature at the time when floral primordia initiate inside the bulb before harvesting, improper forcing temperatures or due to prolonged or high HWT. Such clumps require lifting, division and replanting after addition of nutrients in the soil. Forcing of trumpet narcissi at 18 °C or higher causes **bull-nosing** where neither flowers expand nor open. HWT causes aborted or small starry flowers, dead buds, leaf mottling, roughening, thickening, various other types of distortion, grey rings in the scales and grey spots above the basal plate, *etc.* Brown corky areas around the periphery of the basal plate, reduction in flower number and deformed flowers are noticeable due to formaldehyde injury. Herbicides may also cause various abnormalities in narcissi. Paraquat causes leaf scorch, and when it or triazoles enter the bulb, the plants show chlorosis, delapon or trichloroacetic acid causes distorted flowers, and chlorpropham, diuron, fenuron or simazine causes reduced growth. MCPA and 2, 4-D damage the basal meristem and cause flaccidity.

References

Bailey, L.H and E.Z. Bailey, 1976. Narcissus. In: *Holland Bulb Forcer's Guide* (5th ed.), pp. B 69-93, C123-132. International Flower bulb Information Centre, Hillegom, The Netherlands.

Barrett, S.C.H., W.W. Cole and C.M. Herrera, 2004. Mating patterns and genetic diversity in the wild daffodil *Narcissus longispathus* (Amaryllidaceae). Heredity, **92**: 459–465.

Byther, R. and G. Chastagner, 1993. Diseases. In: *The Physiology of Flower Bulbs* (eds De Hertogh, A. and M. Le Nard), pp. 71-99. Elsevier Science Publishers, Amsterdam, The Netherlands.

Cathey, M. 1990. USDA Plant Hardiness Zone Map. USDA Agricultural Research Station (Miscellaneous Pub. No. 1475). U.S. Govt. Printing Office, Washington, D.C., USA.

Chen, T.T. 1952. Daffodil and Tulip Yearbook (U.K.), **17**: 3-31.

Cohen, D., D. Ziv, C. Sandler, A. Fintea, A. Ion, A. Cohen and R. Kamenetsky, 2011. Forcing of new cultivars of Paper White narcissus for early flowering. *Acta Hort.*, No. 886, pp. 427-434.

De Hertogh, A.A. 1974. Principles for forcing tulips, hyacinths, daffodils, Easter lilies and irises. *Scient. Hort.*, **2**: 313-355.

De Hertogh, A.A. 1977. Holland Bulb Forcer's Guide. Netherlands Flower bulb Institute, The Netherlands.

De Hertogh, A.A. 1996. Holland Bulb Forcer's Guide (5th ed.). International Flower Bulb Centre, Hillegom, The Netherlands.

Fenlon, J.S., S.K. Jones, G.R. Hanks and F.A. Langton, 1990. Bulb yields from narcissus chipping and twin-scaling. *J. Hort. Sci.*, **65**(4): 441-450.

Fernandes, A. 1967. Keys to the identification of native and naturalized taxa of the genus *Narcissus*. RHS Daffodil and Tulip Yearbook (published 1968), pp. 37-66.

Gordon, R., G.R. Hanks and A.R. Rees, 1984. Early forcing of narcissus: The effects of lifting date and stage of floral development at the start of cooling. *Scient. Hort.*, **23**(3): 269-278.

Gorer, R. 1970. *The Development of Garden Flowers*. Eyre and Spottiswoods Ltd., London, U.K.

Hanks, G.R. 1993. Narcissus. In: *The Physiology of Flower Bulbs* (eds De Hertogh, A. and M. Le Nard), pp. 463-558. Elsevier Science Publishers, Amsterdam, The Netherlands.

Hartsema, A.M. 1961. Influence of temperatures on flower formation and flowering of bulbous and tuberous plants. In: '*Handbuch der Pflanzenphysiologie*' (ed. Ruhland, W.), vol. XVI, pp 123-167. Springer-Verlag, Berlin and New York.

Hay, R. and K.A. Beckett, 1971. *Reader's Digest Encyclopaedia of Garden Plants and Flowers*, pp. 454-462. Reader's Digest Association Ltd., London, England.

Horton, D.E. 1956. *Bulbs*, pp. 18-36. A.R. Rosewame Exp. Hort. Stn., England.

Jefferson-Brown, M.J. 1969. Daffodils and Narcissi. Faber & Faber, London, England.

Langens-Gerrits, M. and S. Nashimoto, 1997. Improved protocol for the propagation of narcissus *in vitro*. In: *VII ISHS International Symposium on Flower Bulbs. Acta Hort.*, No. 430, pp. 311-314.

Meyer, F.G. 1966. Narcissus species and wild hybrids. In: *Daffodil Handbook* (ed. Lee, G.). *The Amer. Hort. Mag.*, **45**(1): 47-76.

Michael N. Dana and Rosie B. Lerner, 2001.The Narcissus, 4 p. Purdue University Cooperative Extension Service. Department of Horticulture, West Lafayette, USA.

Miles, B. 1963. *The Wonderful World of the Bulbs*. D. Van Nostrand Co., Inc, London, England.

Miller, W.B. and Erin Finan, 2006. Root zone alcohol is an effective growth retardant for Paper White *Narcissus*. *Hort Technol.*, **16**(2): 294-296.

Misra, R.L. 1989. *Narcissus*. In: *Commercial Flowers* (eds Bose, T.K. and L.P. Yadav), pp. 733-787. Naya Prokash, Calcutta, India.

Pizzetti, I. and H. Cocker, 1975. Flowers: A Guide for Your Garden. Harry N. Abrams, Inc., New York, USA.

Rees, A.R.1972. The Growth of Bulbs. Academic Press. London and New York.

Rusalenko, V.G., Zh.A. Rupasova, N.F. Murashova, L.I. Yanitskaya, R.N. Rudakovskaya, T. Saigusa and T. Yoshihara, 1970. *Res. Bull. Plant Prot. Serv. Japan*, No. 8, pp. 21-29.

Sage, D. and N. Hammatt, 2002. Somatic embryogenesis and transformation in *Narcissus pseudonarcissus* cultivars. *Acta Hort.*, No. 570, pp. 247-249.

Sage, D. O., James Lynn and Neil Hammatt, 2000. Somatic embryogenesis in *Narcissus pseudonarcissus* cvs. Golden Harvest and St. Keverne. *Plant Sci.*, **150** (2): 209-216.

Sage, D.O. 2005. Propagation and protection of flower bulbs: Current approaches and future prospects, with special reference to narcissus. In: *IX International Symposium on Flower Bulbs. Acta Hort.*, No. 673, pp. 323-334.

Salinger, J.P. 1985. *Commercial Flower Growing*. Butterworths, New Zealand.

Sanyat Misra and R.L. Misra, 2013. *Narcissus*. In: *Commercial Ornamental Bulb Science*, pp. 109-124. Westvelle Publishing House, New Delhi.

Sharma, Y.D. and S.B. Kanwar, 2003. Studies on micropropagation of tulips and daffodils. *Acta Hort.* (ISHS), No. 624, pp. 533-540.

Sochacki, D. And T. Orlikowska, 2005. Factors influencing micropropagation of *Narcissus*. In: *IX International Symposium on Flower Bulbs. Acta Hort.*, No. 673, pp. 669-673.

Stevens, R.B. 1977. *Introduction to Horticulture* (Bull. No. 30). Min. Agric. & Fish., Levin, New Zealand.

Van Aartrijk, J. 1995. Ziekten en Afwijkingen bij Bolgewassen. Deel 2: Amaryllidaceae, Araceae, Begoniaceae, Cannaceae, Compositae, Iridaceae, Oxalidaceae, Ranunculaceae. Tweede Druk. Laboratorium voor Bloembollenonderzoek, Informatie en Kennis Centrum Landbouw, Lisse, The Netherlands.

Waters, R.A.S. 1977. *Narcissus Production*, 11 pp. Ministry of Agriculture and Fisheries, New Zealand.

Webb, D.A. 1980. *Narcissus*. In: *Flora Europea* (eds Tutin, T.G., V.H. Heywood, N.A. Burges, D.M. Moore, D.H. Valentine, S.M. Walters and D.A. Webb), vol. 5, pp. 78-84. Cambridge University Press, England.

WoŸny, A. And M. Jerzy, 2007. Effect of light wavelength on growth and flowering of narcissi forced under short-day and low quantum irradiance conditions. *J. Hort. Sci. Biot.*, **82**(6): 924-928.

28

Orchids (Family: Orchidaceae)

Sanyat Misra and R.L. Misra

[**Common names**: Autumn lady's tresses (*Spiranthes spiralis* syn. *S. autumnalis*), Blue *Vanda* (*Vanda caerulea*), Beard orchid (*Calochilus robertsonii*), Bee orchid/Brown bee orchid (*Ophrys fusca*), Burnt-tip orchid (*Orchis ustulata*), Buttonhole orchid (*Bulbophyllum leopardinum*), Chien lan/Fukien orchid (*Cymbidium ensifolium*), Common twayblade (*Listera ovata*), Corkscrew orchid (*Trichopilia*), Dancing lady orchid (*Oncidium*), Dove orchid (*Peristeria*), Eastern underground orchid (*R. slateri*), Early spider orchid (*Ophrys sphegodes*), Eyelash orchid (*Epidendrum ciliare*), Flor de San Miguel (*Laelia anceps*), Flying duck orchid (*Caleana major*, syn. *Caleya major; Caleana nigrita*), Foxtail orchid (*Aerides, Rhynchostylis*), Frog orchid, Gardner's/Western underground orchid (*Rhizanthella gardneri*), Jewel orchid (*Anaectochilus roxburghii, Haemaria*), Lady's slipper/Venus slipper orchid (*Cypripedium* spp., *Paphiopedilum fairieanum, Phragmipedium* spp.), Lady of the night (*Brassavola nodosa*), Lamington's underground orchid (*R. omissa*), Leopard orchid (*Ansellia gigantea* var. *azanica*), Little bog orchid (*Malaxis paludosa* syn. *Hammarbya paludosa*), Lizard orchid, Kinta weed/bone plant (*Vanda hookeriana*), Man orchid (*Aceras anthropophorum*), Monkey orchid (*Orchis simian*), Moth orchid (*Phalaenopsis*), Pansy orchid (*Milfonia*), Purple enamel orchid (*Elythranthera brunonis*), Pyramidal orchid (*Anacamptis pyramidalis*), Red spider orchid (*Caladenia filamentosa*), Scorpion/Spider orchid (*Arachnis;* common scorpion orchid-*Arachnis flos-aeris*), Smaller duck orchid (*Paracaleana minor*), Spider orchid (*Brassia*), Tiger orchid, Tongue orchid, Tulip orchid (*Anguloa clowisii*), White butterfly orchid (*Phalaenopsis amabilis*), White helleborine (*Cephalanthera damasonium*), etc.]

Introduction and Origin

Orchids on the earth seem to have been originated 130 million years back. Around 2,800 B.C. in China, Emperor Sheng Nung emphasized the medicinal properties of a *Dendrobium* species. Confucius (551-479 B.C.) mentioned '*lan*', the Chinese name for orchids, in his writings as Chinese used them to decorate their homes. Theophrastus (370-285 B.C.), whom many considered as father of botany, through his treatise '***Inquiry into Plants***' coined the name '*orchis*' (a Greek word for orchid, *orchis* means testicle owing to paired pseudobulbs in some species) to the group of bizarre plants with unusual and interesting flowers, named so due to the appearance of underground tubers of these plants to masculine anatomy (the testes). '*Vedic*' scriptures of Hindus also mention *Vanda*, which later on was adopted as the generic name for one of its most beautiful groups. But only in the 17th century the aesthetic interest in orchids started developing. The first mention of a tropical orchid to be seen in the European greenhouse was *Bletia verecunda*, sent from Bahamas in 1731 by Peter Collinson to Mr. Wager in England. Afterwards *Epidendrum* and *Vanilla* were introduced into England. In 1778, *Phaius tankervilliae* and *Cymbidium ensifolium* were introduced from China by English physician John Fothergill, and afterwards *Encyclia cochleata* and *E. fragrans* from West Indies. By 1794, in Royal Botanic Gardens at Kew in London, there were 15 orchid species in cultivation. Throughout 19th century, England had been the principal importer of orchids, followed by Holland and Belgium. In 1821, 'Conrad Loddiges and Sons' grew orchids in their nursery at Hackney, and afterwards John Daminy, head gardener of the orchid section of the firm 'J. Veitch & Sons Nursery' of Exeter in England produced in 1852 the first orchid hybrid, named as *Calanthe dominyi* by John Lindley by crossing two *Calanthe* species, *viz. C. furcata* and *C. masuca*, which flowered in 1856. In 1863 the first intergeneric hybrid *Laeliocattleya schilleriana* by crossing *Cattleya mossiae* with *Laelia crispa*, and in 1892 the first intergeneric hybrid was created by crossing *Sophrinitis grandiflora* with *Laeliocattleya schilleriana* (Fanfani, 1989). However, for cut flower, in 1913, the 'Sun Kee Nursery' was established in Singapore to produce spray-type orchids of *Arachnis, Aranda* and *Aranthera*. Singapore, Hawaii and Thailand grow mainly tropical *Dendrobium* hybrids. Malaysia also grows other cut orchid sprays, apart from Singapore and Thailand, and they export these to Europe, the major one being West Germany. Apart from *Dendrobium*, Thailand is the largest producer of *Vanda* also. Countries like Thailand, Singapore, Malaysia, Indonesia, Japan, Sri Lanka, Australia and New Zealand are growing orchids commercially. The major importing countries of Indian orchids are USA, the Great Britain, Japan, Italy, Canada, Holland, Germany, etc. Some 60 gernera are involved in the export list of orchids from India where *Pleione* tops the list whereas *Dendrobium* is second, followed by *Paphiopedilum, Coelogyne, Cymbidium, Phalaenopsis, Vanda, Ascocentrum, Aerides* and *Rhyncostylis*. *Cymbidium* is being grown in the Southern Hemisphere. The flower symbolism associated with the orchid is love, beauty, refinement, bearing many children, thoughtfulness and mature charm.

Orchidaceae (monocotyledonous, order: Asparagales), the second largest family among the angiosperms, is the most highly evolved family of the entire plant kingdom. Orchid flowers occupy top position among all the flowering plants and are highly priced in the international florist trade as a cut flower, as potted plants and for corsages. The cut flowers are very beautiful and long-lasting, some even more than two months. The species most suitable as cut flowers are *Arachnanthe clarkei, Coelogyne barbata, Cymbidium aloifolium, C. giganteum, C. lowianum, C. tigrinum, Dendrobium heterocarpum, D. moschatum, D. nobile, D. primulinum, D. transparens, Paphiopedilum fairieanum, P. insigne, P. spicerianum, Phaius tankervilliae, Renanthera imshootiana, Vanda bicolor, V. coerulea, V. stangeana, V. tessellata*, etc. They are also used as potted flowering plants, in window boxes, in hanging baskets and as bedding flowering plants. Most suitable orchids for indoor culture are *Anoectochilus sikkimensis, Cymbidium giganteum, C. lowianum, C. tracyanum, Dendrobium chrysotoxum, D. infundibulum, D. nobile, Paphiopedilum fairieanum, P. insigne, P. villosum, Pleione maculata, P. praecox, Renanthera imschootiana, Vanda coerulea, V. pumila*, etc. The orchids suitable for other landscape beauty are *Aerides multiflorum, A. odoratum, Arundina graminifolia, Coelogyne cristata, C. nitida, Cymbidium devonianum, C. elegans, C. simonsianum, Dendrobium aphyllum, D. chrysanthum, D. densiflorum, D. lituiflorum, Paphiopedilum venustum, Pleione maculata, P. praecox, Rhynchostylis retusa, Vanda parviflora, V. teres*, etc. Among all the flowering plants on the globe, orchids are most exquisite in nature for their range of delicate colours (white, light to golden yellow, pink, scarlet, salmon, crimson, red, brown, violet, blue, with some of them being sweetly fragrant), the size and bizarre flower forms, their habits and wide geographical distribution, and to some extent for their pharmaceutical uses and vanilla production, aromatic uses and food. In various parts of NE region various orchids are used as food. The leaves and pseudobulbs of *Cymbidium* being of great nutritive value are used as food and in Arunachal Pradesh, *Anoectochilus* leaves are used as vegetables, *Dendrobium* dried leaves when cooked with rice give delicious food with exotic flavour, and the fragrant fruits of *Cephalanthera pallens, Eria pannea, Gastrodia elata, Jumellea fragrans, Nigritella nigra, Orchis fusca, Spiranthes cernua, Stanhopea tigrina, Thrixspermum malayanum*, climbing *Vanilla fragrans* and *V. planifolia* are the source of essence, *i.e.* vanillin which is the major constituent of vanilla. Flavouring agent vanilla is normally extracted from three species, *viz. Vanilla planifolia, V. pompon* and *V. tahitensis*, though vanilla for commerce is extracted only from *V. planifolia*. In Europe during 16th century, vanilla was introduced by the Spanish chocolate making factories. Tribal people of NE region use orchids as medicines in various ways; *Vanda coerulea* flower juice is used as eye drops against glaucoma, cataract and blindness, *Dendrobium nobile* flowers for various eye troubles, *Phaius tankervilliae* pseudobulbs along with wild ginger against dysentery and fracture, *Paphiopedilum insigne* whole plant against amoebic dysentery, dried pseudobulbs of some 30 orchids are said to have aphrodisiac properties and are known in the trade as 'SALEP', *Dendrobium nobile* contains alkaloids such as 'dendrobine' and 'nobiline', chewing of seeds of a few orchids mixed with *Hibiscus* leaves works as oral contraceptive in women, and some orchids are reported inducing sterility in the women. Ayurvedic tonic 'Chyavanprash', apart from having many other ingrediants, it also contains four orchids, *viz. Habenaria edgeworthii, H. intermedia, Malaxis acuminata* and *M. muscifera*.

Majority of the orchids are native to tropical countries, found especially in the humid tropical forests of South and Central America, Mexico, India, Sri Lanka, Myanmar, South China, Thailand, Malaysia, Philippines, New Guinea and Australia. The species such as *Calypso bulbosa, Microstylis monophyllos, Liparis loeselii* and *Orchis* are native to North temperate and the subarctic zones though now are widely distributed throughout the northern regions of both hemispheres (Equator to Arctic Circle). No orchid is native to Polar Regions and the great deserts. Brazilean *Cattleya*, Mexican *Laelia* and Indian *Dendrobium, Cymbidium* and *Vanda* have maximum contribution in the development of global orchid industry. Terrestrial orchids are common in NW India, epiphytic in NE India and the small-flowered ones in Western Ghats. The largest terrestrial genus is *Habenaria* with over 100 species, and the largest epiphytic genus is *Dendrobium* with over 70 species.

The orchids have 1,000 described genera with over 35,000 species, and perhaps another 1.5 lakh manmade hybrids and cultivars, and a worldwide distribution [India accounting some 1,331 belonging to more than 186 genera (Misra, 2007)], out of which over 900 species (Chowdhery, 2009) are found only in NE region including Sikkim and Arunachal Pradesh where climatic conditions are highly variable having temperate, sub-tropical and tropical, the richest in the world, *i.e.* Arunachal Pradesh 558 (Rao, 2010), Sikkim 557 (Lucksom, 2007) and Meghalaya 439 (including intraspecific taxa). Rao and Singh (2015) listed all the 439 species and interspecific taxa such as

Acampe ochracea, A. papillosa, A. rigida, Acanthephippium striatum, A. sylhetense, Aerides multiflora, A. odorata, Agrostophyllum brevipes, A. callosum, A. flavidum, A. planicaule, Anoectochilus brevilabris, A. roxburghii, Anthogonium gracile, Aphyllorchis montana, Appendicula cornuta, Arachnis labrosa, Arundina graminifolia var. *graminifolia* syn. *Bletia graminifolia, A.g.* var. *revoluta, Ascocentrum ampullaceum, Biermannia quinquenquecallosa, Brachycorythis galeandra* syn. *Platanthera galeandra, B. helferi* syn. *Gymnadenia helferi, Bulbophyllum affine, B. ambrosia, B. bisetum, B. blepharistes, B. candidum* syn. *Ione candida,*and *Sunipia candida, B. careyanum* syn. *Anisopetaloncareyanum, B. cariniflorum, B. caudatum, B. cauliflorum, B. cherrapunjeensis, B. chyrmangensis, B. crassipes, B. cylindraceum, B. delitescens, B. depressum, B. elatum, B. forstii, B. gamblei, B. griffithii, B. guttalatum, B. gymnopus, B. gyrochilum, B. helenae, B. hirtum, B. hymenanthum, B. jejosephii* syn. *Trias pusilla* and *Jejosephia pusilla, B. khasyanum, B. leopardinum* syn. *Dendrobium leopardinum, B. leptanthum, B. manabendrae, B. moniliforme, B. oblongum* syn. *Trias oblonga, B. odoratissimum* syn. *Stelis odoratissima* and *Bulbophyllum congestum, B. ornatissimum* syn. *Cirrhopetalum ornatissimum, B. penicillium, B. piluliferum, B. polythizum,B. protractum, B. repens, B. reptans* syn. *Tribrachia reptans, B. retusiusculum* syn. *Cyrrhopetalum retusiusculum, B. roseopictum* syn. *Ione bicolor, Sunipia bicolor, Dipodia khasyanum* & *Ione khasyana, B. roxburghii* syn. *Cirrhopetalum roxburghii, B. sarcophyllum* syn. *Cirrhopetalum sarcophyllum, B. scabratum* syn. *B. confertum, B. secundum, B. sikkimense* syn. *Cirrhopetalum sarcophyllum, B. spathulatum* syn. *Cirrhopetalum spathulatum* and *Rhytionanthos spathulatum, B. striatum* syn. *Dendrobium striatum, B. sunipia* syn. *Sunipia acariosa, B. tortuosum, B. trichocephalum* syn. *Cirrhopetalum trichosephalum, B. tricorne, B. triste, B. umbellatum, B. viridiflorum* syn. *Cirrhopetalum viridiflorum, B. wallichii* syn. *Cirrhopetalum wallichii, Calanthe alismifolia, C. anthropophora, C. clavata, C. densiflora* syn. *Phaius epiphyticus, C. herbacea, C. mannii, C. masuca, C. odora, C. puberula, C. triplicata* syn. *Orchis triplicata, Callostylis rigida, Cephalantheropsis longipes* syn. *Calanthe longipes, C. obcordata* syn. *Bletia obcordata, Ceratostylis himalaica, C. subulata* syn. *Ceratostylis teres, Chamaegastrodia asraoa, C. vaginata* syn. *Aphyllorchis vaginata, Cheirostylis griffithii, C. pusilla, Chilochista lunifera* syn. *Thrixspermum luniferum, Cleisocentron pallens, Cleisostoma appendiculatum* syn. *Aerides appendiculata, C. aspersum* syn. *Sarcanthus asperses* and *Stereochilus bicuspidatus, C. filiforme* syn. *Sarcanthus filiformis, C. linearilobatum* syn. *Sarcanthus linearilobatus* and *Cleisostoma sagittiforme, C. paniculatum* syn. *Aerides paniculata, C. racemiferum* syn. *Saccolabium racemiferum, C. simondii* syn. *Vanda simondii* and *Echioglossum simondii, C. striatum* syn. *Echioglossum striatum, C. subulatum, Coelogyne corymbosa, C. cristata, C. fimbriata, C. flaccida, C. fuliginosa, C. fuscescens, C. holochila, C. longipes, C. micrantha, C. nitida* syn. *Cymbidium nitidum, C. occultata, C. ovalis, C. prolifera* syn. *C. flavida, C. punctulata, C. raizadae, C. rigida, C. schultesii, C. stricta* syn. *Cymbidium strictum, C. suaveolens* syn. *Pholidota suaveolens, C. viscosa, Corybas himalaicus* syn. *Corysanthes himalaica* and *Corybas purpureus, Corymborkis veratrifolia* syn. *Hysteria veratrifolia, Crepidium acuminatum* syn. *Malaxis acuminata* and *M. a.* var. *biloba, C. biauritum* syn. *Microstylis biaurita* and *Malaxis biaurita, C. calophyllum* syn. *Microstylis calophylla, C. josephianum* syn. *Microstylis josephiana* and *Malaxis josephiana, C. khasianum* syn. *Microstylis khasiana* and *Malaxis khasiana, C. maximowiczianum* syn. *Microstylis maximowicziana, Cryptochilus sanguinea, Cryptostylis arachnites* syn. *Zosterostylis arachnites, Cymbidium aloifolium* syn. *Epidendrum aloifolium, C. bicolor, C. cochleare, C. cyperifolium, C. devonianum, C. eburneum, C. elegans* syn. *C. longifolium, C. ensifolium* var. *ensifolium, C. ensifolium* var. *munronianum* syn. *C. munronianum, C. erythraeum, C. hookerianum* syn. *C. grandiflorum, C. iridioides* syn. *C. giganteum, C. lancifolium, C. macrorhizon, C. mastersii, Dendrobium acinaciforme, D. amoenum, D. anceps, D. aduncum, D. aphyllum* syn. *Limodorum aphyllum, D. bicameratum, D. cariniferum, D. chrysanthum, D. chrysotoxum, D. chryseum, D. crepidatum, D. cumulatum, D. denneanum, D. densiflorum, D. denudans, D. devonianum, D. eriiflorum, D. falconeri, D. farmerei, D. fimbriatum, D. formosum, D. gibsonii, D. heterocarpum, D. hookerianum, D. infundibulum, D. jaintianum, D. jenkinsii, D. khasianum, D. lindleyi, D. lituiflorum, D. longicornu, D. moniliforme* syn. *Epidendrum moniliforme* and *D. candidum, D. moschatum* syn. *Epidendrum moschatum, D. nobile, D. ochreatum, D. polyanthum* syn. *D. cretaceum, D. porphyrochilum, D. pulchellum, D. praecinctum* syn. *D. pauciflorum, D. primulinum, D. ruckeri* syn. *D. ramosum, D. salaccense* syn. *D. cathcartii* and *Grastidium salaccensse, D. stuposum, D. sulcatum, D. terminale, D. transparens, D. wardianum, D. williamsonii, Didymoplexis pallens, Dienia ophrydis* syn. *Epidendrum ophrydis, Gastroglottis ophrydis* and *Malaxis latifolia, Diplomeris pulchella, Diploprora championii* syn. *Cottonia championii, Epigeneium fuscescens* syn. *Dendrobium fuscescens, E. naviculare* syn. *Katherinea navicularis, E. rotundatum* syn. *Sarcopodium rotundatum* and *Dendrobium rotundatum, Epipogium roseum* syn. *Limodorum roseum, Eria acervata* syn. *Pinalia acervata, E. amica* syn. *Pinalia amica, E. apertiflora, E. bambusifolia* syn. *Callostylis bambussifolia, E. biflora, E. bipunctata* syn. *Pinalia bipunctata, E. bractescens* syn. *Pinalia bractescens, E. carinata, E. clavicaulis* syn. *E. khasiana, E. coronaria* syn. *Coelogyne coronaria, E. crassicaulis, E. excavata* syn. *Pinalia excavata, E. ferruginea, E. glandulifera, E. javanica* syn. *Dendrobium javanicum, E. laniceps, E. lasiopetala* syn. *E. pubescens* and *Aerides lasipetala, E. muscicola* syn. *Dendrobium muscicola* and *Conchidium muscicola, E. paniculata* syn. *Mycaranthes floribunda, E. pannea* syn. *Mycaranthes pannea, E. pudica* syn. *Bryobium pudicum, E. pumila* syn. *Pinalia pumila, E. pusilla* syn. *Conchidium pusillum, E. spicata* syn. *Octomeria spicata* and *Pinalia spicata, E. stricta* syn. *Pinalia stricta, E. tomentosa* syn. *Epidendrum tomentosum, E. vittata, Eriodes barbata* syn. *Eria barbata* and *Tainia barbata, Erythrodes blumei* syn. *Physurus blumei, Esmeralda clarkei* syn. *Arachnanthe clarkei, Esmeralda cathcartii* syn. *Vanda cathcartii* and *Arachnanthe cathcartii, Eulophia bicallosa, E. bracteosa, E. graminea, E. spectabilis* syn. *E. nuda* and *Wolfia spectabilis, E. zollingeri* syn. *E. sanguine* and *Cyrtopera zollingeri, Flickingeria fugax* syn. *Dendrobium fugax, F. macraei* syn. *Dendrobium macraei, F. ritaeana* syn.

Dendrobium ritaeanum, Galeola falconeri, G. lindleyana, Gastrochilus acutifolius syn. *Saccolabium acutifolium, G. calceolaris* syn. *Aerides calceolaris, G. distichus* syn. *Saccolabium distichum, G. inconspicuous* syn. *Saccolabium inconspicuum, Luisia inconspicuua* and *L. macrantha, G. intermedius* syn. *Saccolabium intermedium, Gastrodia exilis, Geodorum densiflorum* syn. *Limodorum densiflorum, Goodyera foliosa* syn. *Georchis foliosa, G. hispida, G. procera* syn. *Neottia procera, G. recurva, G. schlechtendaliana* var. *robusta* syn. *G. robusta, G. s.* var. *schlechtendaliana, G. viridiflora* syn. *Neottia viridiflora, Habenaria acuifera, H. arietina, H. dentata* syn. *Orchis dentata, H. khasiana, H. malleifera* syn. *H. furfuracea, H. marginata, H. pantlingiana* syn. *H. stenopetala* var. *polytricha, H. pectinata* syn. *H. ensifolia* and *Kryptostoma pectinatum, H. reniformis* syn. *Listera reniformis, Herminium lanceum* syn. *Ophrys lancea, Herpysma longicaulis, Hetaeria affinis* syn. *H. rubens* and *Goodyera affinis, Liparis acuminata, L. assamica, L. bistriata, L. bootanensis, L. caespitosa* syn. *Malaxis caespitosa, L. cordifolia, L. deflexa, L. delicatula, L. elliptica, L. luteola, L. mannii, L. nervosa* var. *khasiana* syn. *L. bituberculata* var. *khasiana, L. nervosas* var. *nervosa* syn. *Ophrys nervosa* and *Liparis macrocarpa, L. odorata* syn. *L. paradoxa* and *Malaxis odorata, L. petiolata* syn. *L. pulchella*,and *Acianthus petiolatus, L. plantaginea, L. resupinata, L. rupestris, L. stricklandiana, L. torta, L. vestita, L. viridiflora* syn. *L. longipes* and *Malaxis viridiflora, Luisia brachystachys* syn. *Mesoclastes brachystachys, L. filiformis, L. psyche, L. teretifolia, L. trichorrhiza* syn. *Vanda trichorrhiza, L. volucris, Micropera mannii* syn. *Sarcochilus mannii, Micropera obtusa* syn. *Camarotis obtusa, M. pallida* syn. *Aerides pallida, M. rostrata* syn. *Aerides rostrata, Monomeria barbata, Neogyna gardneriana* syn. *Coelogyne gardneriana, Nephelaphyllum pulchrum, N. cordifolium* syn. *Cytheris cordifolia, Nervilia aragoana, N. khasiana, N. macroglossa* syn. *Pogonia macroglossa, N. plicata* syn. *Arethusa plicata, Oberonia acaulis* syn. *O. myriantha, O. brachystachys, O. caulescens, O. clarkei, O. ensiformia* syn. *O. iridifolia* and *Malaxis ensiformis, O. falconeri, O. jenkinsiana, O. mannii, O. mucronata* syn. *Stelis mucronata, O. obcordata* syn. *O. orbicularis, O. pachyracis, O. pyrulifera, O. recurva, O. ritaii, O. rufilabris, O. teres, Odontochilus crispus* syn. *Anoectochilus crispus, O. elwesii, O. grandiflorus, O. lanceolatus* syn. *Anoectochilus lanceolatus, Ornithochilus difformis* syn. *Aerides difforme, Otochilus albus, O. fuscus, O. lancilabius, O. porrectus, Pachystoma pubescens* syn. *P. senile, Panisea demissa* syn. *Dendrobium demissum, P. uniflora* syn. *Coelogyne uniflora, Paphiopedilum hirsutissimum* syn. *Cypropedium hirsutissimum, P. insigne* syn. *Cypripedium insigne, P. venustum* syn. *Cypripedium venustum, Papilionanthe teres* syn. *Dendrobium teres, P. uniflora* syn. *Mesoclastes uniflora, P. vandarum* syn. *Aerides vandarum, Pecteilis susannae* syn. *Orchis susannae, P. triflora* syn. *Habenaria triflora, Pelatantheria insectifera* syn. *Sarcanthus insectifer, Pennilabium proboscideum, Peristylus affinis* syn. *P. goodyeroides* var. *affinis* and *Habenaria affinis, P. constrictus* syn. *Herminium constrictum, P. cubitalis* syn. *Habenaria cubitalis* and *Plantanthera cubitalis, P. densus* syn. *Coeloglossum densum* and *Habenaria stenistachya, P. gracilis, P. goodyeroides* syn. *Habenaria goodyeroides, P. hamiltonianus, P. lacertifer* syn. *Coeloglossum lacertiferum, P. mannii* syn. *Coeloglossum mannii, P. perishii, P. richardianus, Phaius flavus* syn. *Limodorum flavum, P. mishmensis* syn. *Limatodes mishmensis, P. tankervilliae, P. mannii, P. taenialis* syn. *Aerides taeniale* and *Kingidium taenialis, Pholidota articulata* syn. *P. griffithii* and *P. khasyana, P. convallariae* syn. *Coelogyne convallariae, P. imbricata* syn. *Coelogyne convallariae, P. imbricata* syn. *P. calceata, P. pallida* syn. *P. imbricata* var. *sessilis, P. recurva, P. rubra, Phreatia elegans, P. laxiflora* syn. *Dendrobium laxiflorum, Platanthera concinna* syn. *Habenaria concinna, P. dyeriana* syn. *Habenaria dyeriana, Pleione maculata syn Coelogyne maculata, P. praecox* syn. *Epidendrum praecox, Podochilus cultratus, P. khasianus,Polystachya concreta* syn. *Epidendrum concretum, Porpax elwesii* syn. *P. meirax* and *Eria elwesii, P. gigantea, Pteroceras teres* syn. *P. suaveolens* and *Dendrocolla teres, Rhomboda lanceolata* syn. *Dossinia lanceolata, R. pulchra* syn. *Zeuxine pulchra, Rhynchostylis retusa* syn. *Epidendrum retusum, Robequetia succisa* syn. *Sarcanthus succisus, Sarcilabiopsis pusilla* syn. *Oeceoclades pusilla, Satyrium nepalense, Schoenirchis gemmata* syn. *Saccolabium gemmatum, Smitinandia micrantha* syn. *Saccolabium micranthum, Spathoglottis pubescens* syn. *S. p.* var. *parvifolia, Spiranthes sinensis* syn. *Neottia sinensis, Staurochilus ramosus* syn. *Saccolabium ramosum, Stereochilus hirtus, Stigmatodactylus serratus* syn. *Panthingia serrata, Taeniophyllum crepidiforme* syn. *Sarcochilus crepidiformis, T. glandulosum* syn. *T. khasianum, Tainia latifolia* syn. *T. khasiana* and *Ania latifolia, T. minor, T. viridifusca* syn. *Calanthe viridifusca, Thelasis bifolia, T. khasiana, T. longifolia, T. pygmaea* syn. *Euproboscis pygmaea, Thrixspermum centipeda* syn. *T. arachnites, T. musciflorum, T. pauciflorum* syn. *Sarcochilus pauciflorus, T. pygmaeum* syn. *Sarcochilus pygmaeus, Thunia alba* syn. *Phaius albus, Trichotosia dasyphylla* syn. *Eria dasyphylla, T. pulvinata* syn. *Eria pulvinata, Tuberolabium coarctatum* syn. *Saccolabium coarctatum* and *Trachoma coarctatum, Uncifera acuminata, U. obtusifolia, Vanda alpina* syn. *Luisia alpina, V. coerulea, V. cristata, V. jainii, V. pumila, Vandopsis undulata* syn. *Vanda undulata, Zeuxine affinis* syn. *Monochilus affinis, Z. agyokuana* syn. *Z. bidupensis, Z. flava* syn. *Monochilus flavus, Z. goodyeroides, Z. gracilis* syn. *Psychechilos gracilis, Z. nervosa* syn. *Monochilus nervosus* and *Z. strateumatica.*

Status of species in other orchid states is also very encouraging, *viz.* Assam 350, Manipur 251, Mizoram 244, Nagaland 241 and Tripura 100; though many species overlapping; the other regions in India being north-western Himalayas with some 300 species, and southern parts, especially Western Ghats some 300 species). In India, though some genera, *vis-a-vis* species are common for various zones but Eastern India has 107 genera, Western India 44, Southern and Central India 66 and Andaman & Nicobar Islands 18 genera with some 70 species. Bose and Bhattacharjee (1980), as per presence of various important orchid genera in various zones of India, gave an account that *Acampe, Aerides, Agrostophyllum, Ania, Arundina, Bulbophyllum, Calanthe, Coelogyne, Corymbis, Cymbidium, Cyperipedium, Dendrobium, Diplomeris, Doritis, Eria, Aulophia, Galeola, Goodyera, Habenaria, Microstylis, Nephelaphyllum, Nervilia, Oberonia, Odontochilus, Orchis, Ornitharium, Paphiopedilum,*

Peristylis, Phaius, Phalaenopsis, Pholidota, Renanthera, Rhyncostylis, Sarcanthus, Sarcopodium, Spathoglottis, Tainia, Vanda and *Vanilla* belong to **Eastern India;** *Aerides, Anoectochilus, Bulbophyllum, Calanthe, Cephalanthera, Dendrobium, Epipogum, Eria, Eulophia, Galesta, Habenaria, Liparis, Luisia, Microstylis, Oberonia, Orchis, Ornithochilus, Pachystoma, Phaius, Rhyncostylis, Saccolabium, Sarcanthus* and *Spiranthes* to **Western India**; *Acampe, Acanthephippium, Aerides, Anoectochilus, Aphyllorchis, Arundina, Bulbophyllum, Calanthe, Cheirostylis, Chiloschista, Chrysoglossum, Coelogyne, Cymbidium, Dendrobium, Desmotrichum, Disperis, Epipactis, Eria, Eulophia, Goodyera, Habenaria, Liparis, Luisia, Malaxis, Microstylis, Oberonia, Odontochilus, Paphiopedilum, Peristylis, Phaius, Rhyncostylis, Spiranthes, Tainia, Thelasis, Thunia, Vanda* and *Vanilla* to **Southern India**; *Aeides, Bulbophyllum, Cleisostoma, Cymbidium, Dendrobium, Eria, Eulophia, Geodorum, Luisia, Microstylis, Oberonia, Phalaenopsis, Pholidota, Pomatocalpa, Sarcochilus* and *Vanda* to **Andaman and Nicobar Islands**. Areas with maximum orchid diversity in India are Kerala and Tamil Nadu (Agastyavanam, Attapadi, Wynad hills, Nilgiris); Andhra Pradesh (Eastern Ghats); Maharashtra (Panchgani-Mahabaleswar); Arunachal Pradesh (Kameng, Tirap); Meghalaya (Shella and Garo hills); Nagaland (Jharnapani); Sikkim (Mangan, Rabong); Odisha (Mahendragiri, Sigaraja); West Bengal (Darjeeling); Andaman & Nicobar Islands; and Uttarakhand (Kaflani, Dafia Dhoora).

Most of the species of *Paphiopedilum* are restricted to NE Himalayas with exception of *P. druryi* which has been recorded from Kerala though is now almost extinct. Many species in the genera like *Arundina, Cymbidium, Coelogyne, Dendrobium, Paphiopedilum, Renanthera* and *Vanda* are also almost extinct from India due to their over-exploitation. *Acanthephippium sylhetense, Aerides bisvasianum, A. roxburghii, A. sikkimensis, A. tetropterus, Anoectochilus griffithi, A. roxburghii, A. sikkimensis, Aphyllorchis montana, Arachnanthe clarkei, Arundina graminifolia, Bulbophyllum piluliferum, Calanthe brevicornu, C. cristata, C. masuca, C. ochracea, C. whiteana, Coelogyne albolutea, C. rigida, Corallorrhiza innata, C. giganteum, C. grandiflorum, C. pendulum, C. sikkimense, Cymbidium elegans, C. giganteum, C. macrorhizon, C. sikkimense, C. whiteae, Dendrobium aphyllum, D. arachinites, D. candidum, D. chrysotoxum, D. crassinode, D. densiflorum, D. falconeri, D. formosum, D. hookerianum, D. nobile, D. wardianum, Didicea cunninghamii, Diplomeris pulchella, Eria crassicaulis, Galeola cathcartii, G. falconerii, G. lindleyana, Gastrodia exilis, Habenaria cristata, H. pubescens, Oberonia lobulata, Orchis chusua, Paphiopedilum cordigerum, P. druryi, P. fairieanum, P. hirsutissimum, P. insigne, P. venustum, P. villosum, P. wardii, Phaius tankervilliae, Phalaenopsis decumbens, P. mannii, Pleione humilis, Renanthera imschootiana, Spathoglottis ixioides, Thunia alba, Vanda alpina, V. amesiana, V. coerulea, V. pumila, V. roxburghii, Vanilla walkeriae, Yoania prainii, Zeuxine pulchra*, etc. are among some 150 species listed provisionally as the endangered or extinct orchids in India (Arora and Mukherjee, 1983; Foja Singh, 1986; Misra *et al.*, 2005). *Anthogonium, Cleisocentron, Cryptochilus, Didicea, Proteoceras, Risleya* and *Sirhookera* are endemic to India. Indian orchids of high ornamental values are *Anaectochilus roxburghii, Arachnis clarkei, Aerides crispum, A. fieldingii, A. multiflorum, A. odoratum, Arundina graminifolia, Bulbophyllum leopardinum, Calanthe masuca, Coelogyne devonianum, C. elata, Cymbidium longifolium, C. munronianum* (fragrant), *C. pendulum, Dendrobium aggregatum, D. aphyllum, D. barbatulum, D. chrysanthum, D. densiflorum, D. devonianum, D. farmeri, D. fimbriatum, D. jenkinsii, D. moschatum, D. nobile, Paphiopedilum faireanum, P. hirsutissium, P. insigne, P. venustum, Phaius wallichii, Pleione praecox, Rhynchostylis retusa, Thunia alba, Vanda coerulea, V. coerulescens* and *V. cristata* (Foja Singh, 1986). The most commercial orchids fall under epiphytic types, followed by terrestrial ones, *Cymbidium* and *Dendrobium* both being sympodial rank high in the export market. Apart from these two sympodial orchids, *Cattleya* and *Oncidium* are grown exclusively in Kerala, along with monopodial genera *Arachnis, Phalaenopsis* and *Vanda* together with inter-generic monopodial hybrids ~ *Aranda, Ascocendra* and *Mokara* (Rajeevan, 2001). Reddy *et al.* (1992) stated that a net profit of Rs. 6,000 per month may be earned by growing 100 m^2 area accommodating 1,750 potted orchid plants under shade-house conditions, and now this profit would have been trebled due to increased inflation in the country.

Botany, Breeding and Growth Habit

Most orchids are perennial herbaceous plants with simple leaves, quite thick to almost grasslike having parallel veins. In majority of cases the **leaves** are distichous, *i.e.* alternate or opposite on the stem, more clearly in the sympodials. In species where pseudobulbs are absent the leaves transform them to store water and food as in case of many species of *Phalaenopsis* where leaves are slightly fleshy, to those of intermediate types such as *Angraecum distichum* and *Dendrobium leonis* where leaves are clearly fleshy and to those leaves are transformed into true reserve organs such as *Dendrobium cucumerinum* and *D. linguiforme*. Most orchids have leathery leaves but in some cases where the pseudobulbs are large the leaves may be even membranous. Though, all the orchids are similar in tripartite flower structures conforming to a standard plan, their vegetative organs exhibit greater variation, apart from being **epiphytes** (growing above the ground, supported nonparasitically by another plant but deriving the nutrients and water from the surrounding air), **lithophytes** (growing on the surface of rocks), **saprophytes** (growing on dead or decaying organic matter), **terrestrial** (growing in ground), and **subterranean** (growing beneath the surface of growing medium), and some being **without leaves** such as *Chiloschista lunifera* (Larson, 1980). The structure of orchids corform to two basic models ~ **monopodial** (having neither rhizomes nor pseudobulbs and grow from single vegetative apex). The stem of varying length is swollen, erect or drooping and may have leaves well spaced or close together along its entire length, or only near the apex where lower leaves have fallen as in case of all the vandaceous orchids, or no leaves at all so the photosynthetic functions are performed through roots such as *Chiloschista* and *Microcoelia*. These orchids sprout new growth through axillary shoots growing into new plants and many of the species under monopodials produce thick, round or flattened roots also from stem's aerial part; and **sympodials** grow from a number of vegetative apices situated at varying intervals on the creeping rhizome (sometimes underground), which is often much branched and produces more or less erect or in some cases quite absent (subfamily Cypripedioideae) nonsecondary other stems, and these may swell into pseudobulbs (absent in case of subtribe Pleurothallidinae) containing reserve foods. The deciduous or persistent leaves in sympodials are produced either at the base, the apex or along the entire length of pseudobulbs, and the roots from the rhizomes, though in case of *Dendrobium* and in certain others new shoots may also arise from the nodes of the pseudobulbs, *vis-à-vis* roots from the base of these shoots whereas in case of *Maxillaria* the pseudobulbs come out from a secondary stem that has leaves along its whole length, and in *Scaphyglottis* the pseudobulbs may be produced at the apex of the old ones. Pseudobulbs may vary in shape and size in the same genus, from 1-2 mm in diameter as in *Bulbophyllum globuliforme* and *B. minutissimum* or as much as 5 m as in *Gramatophyllum papuanum* and *G. speciosum* (giant orchids). Likewise, rhizome even may be very short from one pseudobulb to another to pack these closely together as in case of *Cymbidium* and *Stanhopea* though quite long to put out new growth freely to make a large clump as in case of *Cattleya* and *Bulbophyllum*.

The **pseudobulb** is a single swollen internode appearing as bulb in epiphytic genera such as *Calanthe, Coelogyne* and *Maxillaria*. These may be round, ovoid, fusiform, clavate, elongated or cylindrical, depending upon the genera. These pseudobulbs produce a tuft or fan of strap-shaped or elongated lanceolate leaves, while in other genera such as *Cattleya, Dendrobium* and *Epidendrum*, the pseudobulb is stem-like with a few to many joints, sometimes swollen and the leaves are strap-shaped or lanceolate to oblong, and one being formed at each joint usually. The *Vanda* group of epiphytes, *viz. Aerangis, Aerides, Angraecum, Phalaenopsis* and *Vanda* do not form pseudobulbs but elongated stems with pendulous aerial roots and usually with thick and strap-shaped leaves. Cymbidiums initially grow from small rhizomes which develop shoots with green leaves, and on maturity of these shoots the bases swell forming the pseudobulbs on whose bases the vegetative leads and spikes appear. Senescence of the leaves on the pseudobulbs takes place in about three years with a distinct abscission layer, and this stage is called the **back bulb**. These start growth only when detached from the parent plant. In general, the function of pseudobulb or thickened stem in the orchids is like the bulb, which stores water and nourishments for the new growing shoot. Unlike certain orchids such as *Dendrobium*, the *Cymbidium* grows continuously and does not have a quiescent period.

In orchids, the **root system** is very strong and prolific arising from the shoot axils of basal internodes. Similar to bulbous plants, every lead in the orchids also produces roots at the base. They are fleshy and the **velamen** (the outer spongy and absorbent tissue layer around the true roots) is well developed not only in epiphytes and lithophytes but also in terrestrial orchids, and that absorbs water and nutrients quickly either from the growing media (terrestrial orchids) or from air (epiphytes), especially designed for epiphytes, so the function of velamen is similar to root hairs as in dicotyledonous root system though in *Paphiopedilum* root hairs are also present. The function of the velamen is also the conservation of moisture and protection of the plants from strong light. In certain species where these lack leaves, photosynthesis is carried out in the entire root structure. Terrestrial orchids such as *Cypripedium*, and all native British orchids, have either a tuft of fleshy roots at the base of the plant or underground tubers. The roots in orchids can persist for many years and senesce gradually and while repotting these old roots are removed. The leaves that are usually strap-shaped but more lax than epiphytic species appear only in the basal tufts.

The **inflorescence** (a raceme, spike, a panicle, a fascicle, an umbel or a cyme) in orchids is formed in the leaf axils, at the base of the pseudobulb, from the pseudobulb itself or at the apex of the stem. The flower numbers are highly variable, *i.e.* from a few as in *Cattleya* and *Paphiopedilum*, to hundreds as in *Epidendrum diffusum*. Leffring (1976) reported that in *Cymbidium*, floral spikes initiate before the leads in early cultivars, while well developed leads before spike emergence are found in mid-season and late-flowering cultivars. Timmerman (1978) stated that irrespective of the *Cymbidium* being early, mid-season or late, the spike initiation is at the same time in all the cultivars but it is the subsequent development that decides the actual time of flowering. **Flowers** in orchids may be insignificant or showy, long-lasting or fading soon, large or small and scented or nonscented, normally with bilateral symmetry but in certain cases such as *Ludisia, Mormodes,* etc. as the column or lip is rotated so such structures are not evident and the flowers appear asymmetrical. Flowers are stalked and enclosed in a bract. Ovary is always inferior and not fully developed at the time of flowering but is developed fully after fertilization, especially in the epiphytes. Orchid flowers comprise of two whorls of three parts, *e.g.* three sepals (the outer whorl) and three petals (the inner whorl), the third petal (often larger) transforming into a lip (labellum) which varies from genus to genus but is often spurred and variously lobed, pouched or keeled and is highly decorative. It is the lip in the flowers that attracts the pollinators. The second unusual characteristic feature is the column (fused style and stamens which is technically called as gynostemium) which has combination of pollen being formed at the top and the ovary just below. The pollen is not powdery but is found in the form of two, four, six or eight spherical masses or pollinia which are easily carried to the ovary by insects. The flowers in most orchids are hermaphrodite but a few species in the genera like *Catasetum, Cycnoches* and *Mormodes* bear flowers of separate sexes on different inflorescences of the same plant but normally at different times to rule out self-fertilization. They hybridize easily where multi-genera crosses are possible, and up to five genera have yet been combined in a single hybrid plant. Embryo is composed of relatively undifferentiated, mostly isodiametric cells with dense granulated cytoplasm and loses its viability very fast. The fruit is a capsule containing exceedingly small and dust-like 1,300 to 4,000,000 seeds (the colour being white, cream, pale-green, orange, reddish and dark brown) per capsule. Seeds have very little or no endosperm, therefore, these have symbiotic association with mycorrhizal fungi for their germination.

Pollen studies of *Dendrobium* varieties and hybrids, *e.g.* 'Emma White', 'Hawaiian Beauty' × 'Kasem Pink', 'Hieng Beauty', 'Kasem White', 'Kiomi Beauty' × 'Banyat Pink', 'New Pink', 'Sonia', 'Sonia # 28", 'Sonia 28 Mutant B' and 'White Nern' revealed that pollen grains are agglutinated in masses called pollinia where each pollinia contained 38,282 to 1,93,750 pollen with fertility percentage varying from 4.42 to 73.98 (Varghese, 1995). At 4 °C storage of pollen caused longest viability, best pollen germination occurred in medium containing 2 per cent sugar + 1 per cent agar, and the medium supplemented with 75 ppm boric acid proved best for pollen tube elongation (Varghese, 1995). Evaluation of 14 of the *Dendrobium* genotypes including two species for 21 characters revealed high heritability combined with high GCV and genetic advance for inflorescence length, scape length, flower number per inflorescence and nodes per cane, and significant positive inter-correlation in all pairwise combinations at genotypic and phenotypic levels was observed among seven characters, *viz.* leaf number per clump, leaf area per cane, cane height, age at first flowering, cane to flower first, length of inflorescence and vase life (Lekha Rani, 2002). Ninitha Nath (2003) also observed high genotypic and phenotypic coefficients of variation for aerial root number, leaf area and leaf width; the number of flowers per inflorescence had high heritability coupled with high genetic advance; and significantly positive correlation at genotypic and phenotypic levels was observed for length and width of flower with number of leaves per cane, number of aerial roots, number of spikes per cane, inflorescence length and scape length. Fifteeen *Dendrobium* varieties differing for various characters from each other, recorded high GCV and PCV for petal thickness, followed by flower number while high heritability was observed for petal thickness, *vis-a-vis* floral length and width (Padmanabha Pillai, 2003), though petal thickness, stomata number on the abaxial surface of leaves and number of flowers exhibited high heritability along with genetic advance, and so using Mahalanobis D^2 analysis, the varieties were grouped in three clusters. Through RAPD technique, the similarity coefficient values ranged from 0.2667 to 0.8824 and genetic distance ranged from 0.1176 to 0.5806, thus putting 15 varieties into six clusters. Krishnapriya (2005) studied 12 *Dendrobium* varieties and recorded high GCV and PCV for shoot length, followed by leaf area, internodal length, inflorescence length and flower number per inflorescence, highest heritability values were recorded for stomata number on the upper surface of leaves, and highest genetic advance percentage was recorded for flower column length. He recorded high positive phenotypic correlation between per inflorescence flower number and inflorescence length, and the column length had significant positive correlation with leaf number, inflorescence length and thickness and length of flower pedicel. Varghese (1995) recorded high rate of self- and cross-incompatibily in *Dendrobium*. 'Emma White', 'New Pink' and 'Kiomi Beauty' × 'Banyat Pink' which were found self-compatible; 'Emma White', 'Hieng Beauty' and 'White Nern' acted as good female parents; 'Hawaiian Beauty' × 'Kasem Pink' and 'New Pink' proved as better male parents; while 'Kasem White' and 'Kiomi Beauty' × 'Banyat Pink' were suitable as male as well as female parents (Varghese, 1995). Sobhana (2000) while studying 10 varieties and six species of *Dendrobium*, recorded anthesis from 7.30 AM to 2.30 PM, stigma receptivity from 2nd to 10th day of flower opening, pollen fertility ranging from 46.77 per cent (*D. chrysanthum*) to 91.93 per cent ('New Pink'), the var. 'Sabine' exhibiting maximum pollen size (47.46 μ) and 'Emma White' maximum pollen germination (80.63 per cent), and maximum pollen germination (80.33 per cent) occurring in 3 per cent sucrose + 1 per cent agar or in 2 per cent sucrose + 2 per cent agar. She recorded all the varieties studied as self compatible except 'Hieng Beauty', as a female parent 'Emma White' showed maximum cross compatibility while 'Sabine' the least, and 'Emma White', 'New Pink' and 'Pink Tips' performed as good male parents, and out of the six species tried, it was *D. fimbriatum* which gave maximum cross compatibility. Lekha Rani (2002) crossed 14 *Dendrobium* genotypes in all the possible combinations (88 crosses, 88 reciprocals and 14 selfs) excepting the six due to nonsynchronisation of the flowering season and got established the progenies from 67 combinations (62 crosses and 5 selfs) in the *ex vitro* greenhouse conditions where 123 combinations succumbed to incompatibility. She observed percentage capsule yield from 8 to 33, percentage seed filling in the capsules from 10.79 to 75.93, and percentage seed germination from 8.00 to 70.73 in ½ strength basal MS medium. Transplanting of these seedlings in potting media, best being broken tiles + charcoal + soilrite (2:2:1) under 85 to 95 per cent relative humidity where 94 per cent seedlings survived which grew well and out of which 16 hybrid combinations flowered with 25 new promising hybrids two years after transplanting recording significant differences for all the 12 floral characters studied. Ninitha Nath (2003) when crossed 12 genotypes in all possible combinations, observed incompatibility reaction at different stages and thus 58 combinations succumbed, however, 58 other combinations which produced green capsules ranging from 8 to 38 per cent with 18 to 76 per cent filled seeds (capsules in 12 combinations did not contain seeds and capsules from 10 combinations did not germinate while protocorms of developing seedlings from 12 combinations aborted *in vitro* at various stages) were cultured on ½ strength MS medium, where on growth seedlings showed variations for various characters. Stem nodal segments out of various explants tried, responded well and showed early PLB differentiation and so was the case with various media used as only VW medium gave early culture establishment with rapid PLB development (Sivamani, 2004). Out of various combinations tried, BA + NAA each at 2.0 mg l^{-1} gave earliest plantlet development, BA + NAA each at 8.0 mg l^{-1} produced more number of shoots, BA at 2.0 mg + IAA 4.0 mg, each per litre of medium produced plantlets quite earlier while BA and IAA each at 8.0 mg l^{-1} produced maximum number of shoots. Likewise, KIN and NAA each at 2.0 mg l^{-1} caused earlier plantlet development, 8.0 mg l^{-1} each produced highest number of shoots, while 4.0 mg l^{-1} each of KIN and IAA caused rapid plantlet development though KIN 6.0 mg and IAA 2.0 mg, each per litre of medium produced highest number of shoots. He recorded best *in vitro* rooting of microshoots in ½ strength MS medium supplemented with IBA 2.0 mg l^{-1}. Sobhana (2000) subjected the protocorms, PLBs and callus to *in vitro* mutagenesis with 60 Gy ^{60}Co gamma rays and recorded smaller and broader leaves with dark green colouration in higher doses.

Classification, Species and Varieties

The Swedish botanist Carl von Linne, better known as Linnaeus is the first to put system of classification of all the plants then known and he arranged them into several groups as per similarities in stamens and pistils as mentioned in his treatise '*Species Plantarum*' published in 1753. He described 8 genera and 69 orchid species under the group XX-Gyanandria (stamens adnate to the pistil) and Diandria (two stamens). Olof Swartz (1800) published an exclusive article on orchids, describing 25 genera under two divisions: (i) orchids with one anther, and (ii) orchids with two anthers, and thus he was honoured as the first Orchidologist. These two divisions today are recognized as Monandrae and Diandrae. John Lindley (1799-1865), an Englishman published 'Genera and Species of Orchidaceous Plants' during 1830-1840, based on all the orchids collected world over till then, and he put orchids in eight tribes as following:

I. Monandrae (one anther)

A. Pollen masses waxy

- ☆ Tribe 1: **Malaxideae:** No caudicle or separate stigmatic gland
- ☆ Tribe 2: **Epidendreae:** A distinct caudicle but no separate stigmatic gland
- ☆ Tribe 3: **Vandeae:** A distinct caudicle attached to a deciduous stigmatic gland

B. Pollen masses powdery, granular or sectile

- ☆ Tribe 4: **Ophrydeae:** Anther terminal, erect or incumbent, adnate to top of column
- ☆ Tribe 5: **Aretheuseae:** Anther terminal, operculate
- ☆ Tribe 6: **Neottieae:** Anther dorsal

II. Diandrae or Triandrae (two or three anthers)

- ☆ Tribe 7: **Cypripedieae:** Ovary one or three-celled
- ☆ Tribe 8: **Apostasieae:** Ovary three-celled

Heinarich Gustav Reichenbach (popular name Reichenbach fil), a German, published an article in form of a synopsis in ***Walper's Annales Botanices Systematicae,*** during 1861-1866. On the system of classification, there had been a lot of advocacies, the latest being by Holttum (1964), who outlined a system of classification based on orchids of Malaysia, and published **'A Revised Flora of Malaya (vol. I) – Orchids of Malaya',** which classifies orchids (in all, having 28 tribes) as ~ family **Orchidaceae** having 2 subfamilies: (i) **Pleonandrae** (Apostasia tribe, more than one anther, *e.g.* slipper orchids), (ii) **Monandrae** (one anther) which has two types, *viz.* **Basitonae** (anther attached by its base, Habenaria tribe) and **Acrotonae** (anther attached by the top). **Acrotonae** is further divided into two: (i) pollinia granular [saprophytes and their allies (4 tribes), Goodyera tribe, Corymborchis tribe], and (ii) pollinia waxy which is further divided into (i) **Monopodial** (*Vanda-Arachnis* tribe), and (ii) **Sympodial orchids** which comprise, (a) **mainly terrestrial** (pollinia without disc, *viz. Phaius* tribe, *Nephelaphyllum* tribe, *Arundina* tribe, *Sobralia* tribe, *Liparis* tribe), (b) **Epiphytes** (pollinia with no disc, *viz. Coelogyne* tribe, *Dendrobium* tribe, *Bulbophyllum* tribe, etc.), and (c): **Epiphytic and Terrestrial** (pollinia with disc, sometimes with a stripe, *viz. Agrostophyllum* tribe, *Appendicula* tribe, *Thelasis* tribe, *Eulophia* tribe, *Bromheadia* tribe, *Cymbidium* tribe, etc.). Though this system is very simple and easy to follow but still the **Classification of Orchids** has to evolve finally to make it more understandable. Leslie Andrew Garay (1972) classified orchids into five subfamilies, *viz. Apostasioideae, Cypripedioideae, Orchidoideae, Neottioideae* and *Epidendroideae*, however, Robert Dressler (1981) gave a more comprehensive classification system putting Orchidaceae into six subfamilies such as *Apostasioideae; Cypripedioideae; Spiranthoideae* with 2 tribes namely Erythrodeae and Cranichideae; *Orchidoideae* with 4 tribes namely Neottieae, Diurideae, Orchideae and Diseae; *Epidendroideae* with 9 tribes namely Vanilleae, Gastrodieae, Epipogieae, Arethuseae, Coelogyneae, Malaxideae, Cryptarrheneae, Calypsoeae and Epidendreae; and *Vandoideae* with 4 tribes namely Polystachyeae, Vandeae, Maxillarieae and Cymbidieae, while further in 1990, he made only five subfamilies abolishing *Vandoideae* by placing all its tribes in the subfamily *Epidendroideae* (Arditti, 1992).

The orchids have over 1,000 genera with some 35,000 species, out of which India is native of some 1,300 species. Almost half the species are (a) **terrestrial** (growing on the ground of the forest) and remaining ones (b) **epiphytes** [growing non-parasitically on the host plants, *i.e.* tree branches, getting their nourishment from air, from humus in bark crevices through specially adapted roots, and through rain water, and normally consists of an horizontal rhizome from which at intervals arise usually erect and swollen stems or pseudobulbs, and these are of two types: (i) **monopodials** growing from a single vegetative apex as they sprout new growth by developing axillary shoots which eventually grow into new plants, the main stem growing continuously year after year and has a dominant apical meristem from where in the leaf axils inflorescence arises, and have neither rhizome nor pseudobulbs *e.g. Arachnis, Chiloschista, Microcoelia, Phalaenopsis, Renanthera, Vanda, Vanilla,* etc. where *Chiloschita* and *Microcoelia* are devoid of leaves and the roots have a photosynthetic action, and (ii) **sympodials** growing from a number of vegetative apices present at intervals on the normally branched rhizomes with or without pseudobulbs, where main stem terminates its growth at the end of each season and produces a lead or shoot from the base in the growing season by forming its own bulbous stem (pseudobulb) which eventually flowers, *e.g. Cattleya, Cymbidium, Dendrobium, Oncidium,* etc. Epiphytic species are rather easy to grow. Most of the temperate zone orchids are terrestrial, and most of the tropical orchids are epiphytes. Majority of the orchids are tropical and others sub-temperate to temperate, starting from the sea level to 3,500 m altitude. Apart from (a) **terrestrial** and (b) **epiphytic** orchids, there are: (c) **lithophytes** or **epilithic** orchids growing on the surface of the rocks, (d) **saprophytic** growing on dead or decaying organic matter, (e) **subterranean** (underground) growing beneath the surface of the growing medium, and (f) **semi-aquatic** orchids growing in water.

The genera of orchids which could be commercially grown under the climatic conditions prevailing in this country are:

Aerides (Foxtail orchid)**:** This is monopodial and epiphytic orchid (a few lithophytes), closely alied to *Vanda* and may be grown in temperate as well as in tropical areas up to 700 m elevation, depending on the species. The basic chromosome number in *Aerides* is n=19. There are some 60 species and the important species are *A. crispum, A. fieldingi, A. maculosum, A. odoratum, A. racemeformis, A. vandarum,* etc. Hybrids are *Aeridachnis* (*Aerides* x *Arachnis*), *Aeridocentrum* (*Aerides* x *Ascocentrum*), *Aeridofinetia* (*Aerides* x *Neofinetia*), *Aeridopsis* (*Aerides* x *Phalaenopsis*), *Aeridostylis* (*Aerides* x *Rhynchostylis*), *Aeridovanda* (*Aerides* x *Vanda*), *Renades* (*Aerides* x *Renanthera*), *Tanakara* (*Aerides* x *Phalaenopsis* x *Vanda*), *Vandopsides* (*Aerides* x *Vandopsis*), etc. Pink Glory, Crestwood, Malibu Gold, Yip Sum Wah, Cherry Blossom, Frank Johnston, Hawaii, Tsuruko Iwasaki, etc. are the varieties developed using *Aerides.*

Angraecum eburneum. *Angraecum* derives from the Malaysian name *angurek*. This genus of 220 epiphytic orchid species native to tropical and sub-tropical areas of Africa, Malagasy, Indian Ocean islands and the Philippines is without pseudobulbs. These orchids grow erect with stems bearing two rows of fleshy and narrowly ovate to strap-shaped leaves. Starry, usually quite large and white & green or rarely yellow flowers appear in racemes or sometimes even solitary, having a lip with slender spur which elongates mostly to a great length. These plants requiring more humidity are most suitable for growing in conservatories but are brought home when to flower. These can be propagated through cuttings of side shoots. ***A. eborneum*** (*A. superbum*) from Mascarene Islands grows up to 2 m high with single stem, bearing up to 90 cm long ovate-lanceolate leaves which are bi-lobed on the tips. Green to greenish-white, fragrant and up to 7 cm long flowers with a white labellum having some 10 cm long spur appear during autumn to winter in two ranks on a stem growing up to or more than 90 cm long. ***A. eichlerianum*** from tropical West Africa grows 1.2 m or more in length with pendulous or climbing stems, bearing leathery and oblong-elliptic leaves which are notched at the tips. Strongly fragrant and 8-9 cm across yellow-green flowers appear 1-3 during autumn to winter having a green-marked white labellum with a spur about 4.5 cm long. ***A. infundibulare*** from tropical West Africa has more than 1.2 m long climbing to pendulous stems, bearing up to 10 cm long oblong-elliptic and leathery leaves. Fragrant yellow-green and up to 9 cm across solitary flowers appear during autumn to winter with a white labellum. ***A. rhodostictum*** (*Aerangis rhodostictum*) from Cameroons, Ethiopia, Kenya and Tanganyika generally grows pendent with short stems, bearing bright green and strap-shaped leaves up to 15 cm long and bi-lobed at the tip. Arching inflorescences grow to about 35 cm long bearing two parallel ranks of up to 20 white to cream or palest-yellow flowers during winter to Spring, each flower about 2.5-3.0 cm across with a red column and green-tipped spurs. ***A. sesquipedale*** from Malagasy is a robust plant with more than 60 cm long and erect stems, bearing blue-green leaves some 30 cm long which are unequally bi-lobed at the tips. In winter its 18 cm across, fragrant and fleshy white flowers appear with a cordate lip and up to 25 cm long spur.

Anguloa clowesii (Tulip orchid). *Anguloa* was named for Don Francisco de Angulo, a Spanish botanist of the 18th century. A native to tropical South America, especially in the Andean forests, the genus consists of 10 epiphytic or terrestrial clump-forming species, bearing ovoid to oblong pseudobulbs. Its leaves are lanceolate and prominently veined. It produces solitary erect flowers with overlapping tepals and appear as flowers of tulips. Propagation is through division. Though it is a greenhouse plant but can be brought inside the house for flowering. ***A. clowesii*** from Colombia and Venezuela produces 11-15 cm tall pseudobulbs, broadly oblanceolate leaves up to 60 cm long, pale to golden-yellow and waxy flowers about 8 cm long but hinged, boat-shaped and lobed labellum is white to orange-yellow. ***A. ruckeri*** from Colombia though smaller but is similar to *A. clowesii.* It produces up to 9 cm long fragrant flowers which in the outside are greenish-brown and inside yellow with dense red, blood-red or white spots. ***A. uniflora*** from Colombia to Peru produces a little angular pseudobulbs 10-18 cm tall with 45-60 cm long, folded and broadly lanceolate leaves, and the cup-shaped and unpleasantly fragrant cream to white and waxy-textured flowers, which are sometimes brown-spotted externally and pink-spotted internally.

Anoectochilus (syn. ***Haemaria***, Jewel orchid)**:** It is temperate terrestrial orchid which has some 20 species. It is valued for its most colourful leaves and not for the flowers which are normally insignificant. Important species are *A. geniculatus, A. regalis, R. roxburghii, A. setaceus, A. sikkimensis,* etc. *Anoectomaria* Dominyi is a beautiful hybrid between *Anoectochilus roxburghii* and *Haemaria discolor.*

Arachnis (*Arachnanthe*; Scorpion or spider orchid)**:** A genus of about 17 epiphytic, lithophytic or terrestrial species. This is a monopodial tropical orchid, flowering almost year round, and prefers bright sunlight and high humidity for its growth and flowering. Spikes are 75 to 100 cm or more in length and bear 8-15 flowers. Important species are *A. annamensis, A. cathcartii, A. clarkei, A. flos-aeris,* hybrids are *Aeridachnis* (*Arachnis* x *Aerides*), *Arachnopsis* (*Archnis* x *Phalaenopsis*), *Aranda* (*Arachnis* x *Vanda*), *Aranthera* (*Arachnis* x *Renanthera*), *Armodachnis* (*Arachnis* x *Armodorum*), *Limara* (*Arachnis* x *Renanthera* x *Vandopsis*), *Ridleyara* (*Arachnis* x *Trichloglottis* x *Vanda*), *Trevorara* (*Arachnis* x *Paraphalaenopsis* x *Vanda*), *Ascoarachnis* (*Arachnis cathcartii* x *Ascocentrum curvifolium*, an intergeneric hybrid), etc. and the prominent varieties are Catherine, Ishabel, Maggie Oei Red Ribbon, Maggie Oei Yellow Ribbon, etc.

Ascocentrum. It is a genus of nine orchid species allied to *Ascoglossum* and *Vanda*, native in the Himalayas to SE Asia to Philippines, Borneo, Java, S. China and Taiwan. Once the genus was included in *Saccolabium*. These dwarf, compact, erect and leafy-stemmed monopodial epiphytes (mostly) to lithophytes (rarely) bear stem -enseathing basally, alternate, carinate, fleshy, stiff, linear, apex with two teeth and up to 30 cm long leaves, many-flowered erect to spreading inflorescence which may be sometimes more than one with closely set flowers and as long as the leaves, and small (1-3 cm), closely set, erect, flat, often showy, long-lasting and wide-opening flowers coming up in succession with similar petals and sepals, which appear in lateral upright racemes. The tepals are five, lower two sometimes larger than others, the top one is hooded and the labellum is small with long narrow nectar spur. Their colour is yellow, orange-red, scarlet or purple with a spur. They are best all timer for indoor cultivation. They are propagated in spring after flowering is over, through stem tips having aerial roots. Important species are *A. ampullaceum, A. curvifolium* and *A. miniatum*. ***A. ampullaceum*** (*Aerides ampullaceum, Gastrochilus ampullaceum, Saccolobium ampullaceum*) from Himalayas to Myanmar produces up to 25 cm long stems, bearing leathery, linear, irregularly toothed at the tips and about 13 cm long leaves. Rose-red to rose-purple flowers appear from spring to summer with labellum sometimes toned paler and column white. ***A. curvifolium*** from NE India and Indo-China Peninsula also bears erect spikes bearing large scarlet florets with yellow blotched-lip. ***A. miniatum*** (*Gastrochilus miniatum, Accolobium miniatum*) from Himalayas to Borneo and Malaysia produces thick, woody and 10 cm tall stems, bearing firm-textured, fleshy, linear and 8-20 cm long leaves. Bright orange-red, sometimes orange-yellow to vermillion, and 2 cm across flowers appear from spring to early summer.

Brassavola nodosa (Lady of the night). *Brassavola* was named for Antonio Musa Brassavola (1500-1555, an Italian doctor *cum* botanist & professor at Ferrara). This genus comprises of 15 evergreen, epiphytic and clump-forming species producing stem-like cylindrical and slender pseudobulbs which terminate into a single and fleshy leaf. The showy flowers are single or racemose with similar sepals and petals (tepals) with a larger base-rounded labellum which has a slender tail-like tip. These can easily be grown as basket plant, on tree ferns or on the bark of the trunks. They are propagated through division of densely-formed clumps. ***B. cucullata*** (*B. appendiculata, B. odoratissima, Epidendrum cucullatum*) from Mexico to South America bears up to 13 cm long pseudobulbs, arching to pendulous 60 cm long slender leaves, and 1-3 fragrant and white to cream or greenish-white flowers which are drooping with up to 9 cm long-tailed and fringed tepals. This flowers during winter. ***B. digbyana*** (*Rhyncholaelia digbyana, Laelia digbyana*) from Mexico to Guatemala produces up to 15 cm or more long, a little flattened and club-shaped pseudobulbs which bear grey-green, up to 20 cm long, tough, fleshy and linear to narrowly elliptic leaves, and night fragrant creamy-white solitary flowers up to 18 cm across having greenish throat. ***B. glauca*** (*Rhyncholaelia glauca, Laelia glauca*) from Mexico to Panama produces spindle-shaped pseudobulbs about 10 cm long, bearing grey-green, leathery, stiff, up to 12 cm long and oblong-elliptic leaves and long-lasting, white to lavender or green fragrant flowers 12 cm across with pink spots or lines in the lips. ***B. nodosa*** (*B. grandiflora, Epidendrum nodosum*) from Mexico to Peru bears up to 15 cm tall stem-like pseudobulbs. Leaves are very fleshy, erect, linear and up to 30 cm long. White to yellowish-green and long-lasting night-fragrant flowers 9 cm across appear in racemes of 1-6 though continue appearing intermittently throughout the year, especially during autumn and winter.

Brassia (Spider orchids). *Brassia* was named for William Brass (*d.* 1783), who in West Africa collected many plants for Sir Joseph Banks. This genus contains some 50 evergreen and epiphytic species from tropical America, extending from Mexico to Brazil, and the West Indies thriving in intermediate temperatures, and possessing large flattened pseudobulbs. The long-tailed sepals and petals are the characteristic feature of *Brassia*. They are borne by creeping rhizomes and large, oblong (egg-shaped), flattened and crowded pseudobulbs where each one is topped with dark green, 1-3, thick, leathery and narrowly elliptic leaves. Long, stiff and arching floral stalks in the form of racemes arise from the base of the pseudobulbs and bear closely spaced florets on the upper part where each flower has three outer tepals much larger than the inner. Quite long sepals and petals with strange shape emerge from a narrow base, *vis-a-vis* taper to a narrow point in such a manner that both appear as a tail and the short lip often hangs down between the lower sepals like a tongue. They are best grown in baskets or on slabs of bark or tree fern, as well as in pots. They are propagated by division. ***B. brachiata*** from Guatemala produces oblong (egg-shaped), compressed and 7.5-12.5 cm long pseudobulbs, some 30 cm long leaves and 6-12 flowered racemes. The sepals and petals are light yellowsish-green with a few purple spots towards the base and the sepals are longer than the petals. Centrally constricted light yellow lip has orbicular base, upper part ovate-triangular and acuminate, and bears dark green warts. ***B. caudata*** is native to West Indies. Its pseudobulbs are yellowish-green, cylindrical, 7.5-15.0 cm high and some 2.5 cm wide, bearing 2-3 leaves some 17.5-23.0 cm long and 6.3 cm wide and up to 46 cm long floral stalk which may bear up to 12 fragrant flowers some 12.5 cm long and 7.5 cm across. Sepals and petals are light greenish-yellow with brown spots and bars near the base and the lip is triangular, light yellow and with reddish-brown spots. ***B. gireoudeana*** is from Costa Rica. Its pseudobulb is much compressed, 7.5-10.0 cm long and 3.75-5.0 cm wide, 1-leaved and leaf some 30 cm long, racemes 6-12-flowered, sepals and petals yellowish-green but sepal base brown-spotted and petal base brown, lateral sepals 12.5 cm long and little longer than dorsal sepals, petals half as long as dorsal sepals, and the clawed lip is yellow with brown spots where upper part is almost orbicular and acute. ***B. lanceana*** from Guiana is very similar to *B. lawrenceana* with regard to shape, size and colouration of pseudobulb, leaves, sepals and petals except that the sepals may be slightly smaller and petals its half, lip oblong, wavy, acute, yellow and without or with a few basal brown spots. ***B. lawrenceana*** is from Brazil with much compressed, ribbed and 7.5-12.5 cm long pseudobulbs, bearing two leaves which measure some 30 cm long. Its racemes are 7-12-flowered with light yellow sepals and petals which are basally brown-spotted, sepals some 7.5 cm long, petals 3.75 cm long and the lip 3.75 cm long, elliptic wavy and acute. ***B. longissima*** (*B. lawrenceana* var.

longissima) is from Costa Rica. Its pseudobulb is compressed, oblong-elliptic, 7.5-15.0 cm long and 1-leaf or sometimes paired with about 20-60 cm length and elliptic. The raceme comprises of numerous deep orange-yellow to greenish-yellow flowers with a few large basal blotches, lateral sepals some 17.5-30.0 cm long, *i.e.* almost double the length of dorsal sepals, 6 mm wide at base, petals 5.0-15.0 cm long and the lip cream to pale-yellow with purple spots at the base, elliptic, acuminate and some 7.5-15.0 cm long. ***B. maculata*** is from Jamaica. Its 7.5-10.0 cm long pseudobulb bears solitary leaf some 23 cm long and 5-10-flowered raceme. Tepals (sepals and petals) yellowish-green with lower part brown-spotted. The length of petals is $^{2}/_{3}$rd of sepals and sepals are some 7.5 cm long. Lip is creamy-white with purple dots, where upper part is broadly ovate and acute and the claw broad. Var. *Guttata* from Guatemala bears small and green flowers. ***B. verrucosa*** is from Guatemala. Its much compressed pseudobulbs are ovoid (egg-shaped), little furrowed and some 10 cm long with two elliptic leaves some 35 cm long. Its up to 60 cm long racemes have 8-15-waxy flowers with light yellowish-green tepals (sepals and petals) which are spotted with darker green or red basally. The sepals are up to 10 cm long with petals being half to their length, and the lip is usually smaller but sometimes up to 20 cm long, white with many dark green warts at the base, claw broad and dilated, upper part almost orbicular and acute. Some of the intergeneric hybrids are *Brapasia* (*Brassia* × *Apasia*), *Brassidium* (*Brassia* × *Oncidium*), *Miltasia* (*Brassia* × *Miltonia*), *Odontobrassia* (*Brassia* × *Odontoglossum*), *Rodrassia* (*Brassia* × *Rodriguezia*), etc.

Bulbophyllum: This temperate orchid is the largest genus comprising some 2,000 species, having roundish pseudobulbs and basal inflorescence. Basic chromosome numbers in *Bulbophyllum* are n=19, 20. Important species are *B. affine, B. barbigerum, B. biflorum, B. bracteolatum, B. campanulatum, B. cauliflorum, B. chinense, B. comosum, B. crassipes, B. cylindricum, B. falcatum, B. fascinator, B. hirtum, B. leopardinum, B. lobbii, B. longissimum, B. macranthum, B. maximum, B. medusae, B. ornatissimum, B. parviflorum, B. reticulatum, B. robustum, B. sikkimens, B. umbellatum, etc.* Interspecific hybrids are Fascination (*B. fascinator x B. longissimum*), Louis Sander (*B. longissimum* x *B. ornatissimum*) and David Sander (*B. lobbii* x *B. virescens*).

Calanthe vestita. *Calanthe* derives its name from the Greek *kalos* for 'beautiful' and *anthos* for 'a flower'. This evergreen to deciduous genus of 120 species of mainly terrestrial orchids, is native to the tropics and sub-tropics from South Africa eastwards to SE Asia and northern Australia, and one species in the Americas. Evergreen species which have large conical to ovoid pseudobulbs, rarely constricted close to the middle are suitable for cool houses while deciduous ones which have smaller and more globular pseudobulbs nearly hidden by the leaf bases are suitable for growing at warmer situations, especially in the home, and both the types have pleated and elliptic to lanceolate leaves. Flowers without any distinction between sepals and petals are spurred with large and lobed labellum. These are propagated through division at potting time. ***C. brevicornu***, an evergreen temperate species from Nepal to Assam bears up to 30 cm long oblong to elliptic leaves. It bears purple-brown flowers 2-3 cm across on up to 60 cm long racemes during spring and these flowers shade to yellowish-buff in the centre, and the labellum is red-purple with a white edge and a yellow keel. ***C. masuca*** (*Bletia masuca*) from Himalayas, an evergreen temperate species which blooms during summer and autumn, bears strongly pleated, up to 60 cm long and narrowly elliptic leaves. Deep violet-purple flowers up to 4 cm across with labellum darker and having golden-yellow patch, appear on up to 75 cm long racemes. ***C. vestica*** from SE Asia is a deciduous tropical species producing silvery-green, ovoid-conical, centrally constricted and15-20 cm tall pseudobulbs which bear up to 90 cm long lanceolate leaves. Flowers appear during winter before expansion of the leaves, they are white though in some cases the labellum is flushed pink,and moré than 4 cm across flowers appear on up to 90 cm long arching inflorescences.

Cattleya: *Cattleya* was named for William Cattley (*d.* 1832), a collector and grower of rare plants. This epiphytic mostly tropical or sometimes temperate orchid, loving partially shaded conditions, has sympodial growth and originates in South America. It is related to *Brassavola, Encyclia, Laelia, Sophronitis, etc.* It is the most popular and widely grown in the entire Orchidaceae. It has some 65 species, several thousand hybrids and numerous varieties. These are native to tropical Americas and at one time these were very popular for corsages. New hybrids are available in a variety of sizes and colours from white to purple and make attractive potted plants. Grow best in greenhouses. Hager (1957) stated that by proper selection of hybrids and cultivars the flowers can be harvested daily. Through photoperiodic adjustments some cultivars may be flowered twice a year, and once the flower buds have been formed their development can be retarded or hastened through temperature adjustments. Important species are *C. aurantiaca, C. bicolor, C. bowringiana, C. citrina, C. dowiana, C. intermedia, C. labiata, C. lawrenceana, C. leopoldii, C. luteola, C. maxima, C. mossiae, C. multiflorum, C. schilleriana, C. skinneri, C. violacea, C. walkeriana, C. warscewiczii*, etc. A few of the large number of hybrids are Hardyana (*C. dowiana aurea* x *C. gigas*), Guatemalensis (*C. skinneri* x *C. aurantiaca*), Intricata (*C. intermedia* x *C. leopoldii*), Victoria Reginae (*C. labitata* x *C. leopoldii*), Ballantineana (*C. trianaei* x *C. warscewiczii*), Clotho (*C. enid* x *C. trianaei*), Fabia (*C. dowiana* x *C. labiata*), Iris (*C. bicolor* x *C. dowiana*), Omar (*C. enid* x *C. leda*), Whitei (*C. warneri* x *C. schilleriana*), etc. *Laeliocattleya* is bigeneric hybrid with *Laelia*. With the genera like *Brassavola* (*Adamara*, a hybrid of *Brassavola, Cattleya, Epidendrum* and *Laelia; Brassolaeliocattleya*, a trigeneric hybrid of *Brassavola, Laelia* and *Cattleya;* and *Potinara*, a hybrid of four genera *i.e. Brassavola, Cattleya, Laelia* and *Sophronitis*); *Broughtonia, Diacrium, Epidendrum, Laelia* (*Saphrolaeliocattleya*, a trigeneric hybrid of *Cattleya, Laelia* and *Sophronitis*), and *Sophronitis* (*Lowiara*, a hybrid from *Sophronitis grandiflora* and *Brassolaelia* 'Helen'), this has produced intergeneric crosses. Some popular varieties are Bob Betts, Bow Bells, Bowrevel, Chickamauga, Diane Salo, Doty Sykora, Edith Bow Bells, Empress Bells, Estelle, Margaret Stewart, North Star, Suzanne Hye, White Christmas, White Emblem, *etc.*

Coelogyne: This orchid is epiphytic or lithophytic and has uninodal pseudobulb. There are some 200 species under this genus, some suitable for the temperate areas while others for tropical conditions. The species can be identified by characteristic positioning of the inflorescence. The basic chromosome number in *Coelogyne* is n=20.The important species are *C. barbata, C. corymbosa, C. cristata, C. dayana, C. elata, C. fimbriata, C. flaccida, C. fragrans, C. fuscens, C. nitida, C. ochracea, C. odoratissima, C. peciosa, C. tomentosa, C. uniflora,* etc. Hybrids are Mem. William Micholitz (*C. lawrenceana* x *C. mooreana*) and Stanny (*C. asperata* x *C. pandurata*).

Cymbidium: By and large, this is an evergreen terrestrial orchid though some are epiphytes and still a few others as lithophytes, but in cultivation all are grown only terrestrially, most suitable for growing in NE Regions of India, most of them preferring cool conditions with frost-free winters and warm sunny summers (*C. giganteum, C. longifolium, C. lowianum,* etc. above 1,500 m elevation though *C. elegans* around 700 m) though a few are also suitable for tropical regions. This is a sympodial orchid with basic chromosome number n=20 or 22. These plants are found from Sri Lanka to India and Japan, and throughout Malaysia to Australia. These are mainly used as cut flowers though dwarf ones as potted plants. The plants in flower can be placed in temperature controlled environments for forcing or placed in cold storage for up to one month (Dole and Wilkins, 1999). It has some 70 species, and forms abundant fleshy roots, and short and stout pseudobulbs. Cymbidiums are generally easy to grow which require cool or intermediate temperatures, *i.e.* warm temperate regions. The flower spikes are arching or drooping and display a wide variety of colours. The plants under this genus bear loveliest and long lasting (sometimes even more than two months) inflorescences, and therefore is most popular and is being grown on a commercial scale in many parts of the world, for supply to northern hemisphere to the USA, Western Europe and Japan. Important species are *C. aloifolium, C. bicolor, C. dayanum, C. devonianum, C. eburneum, C. elegans, C. ensifolium, C. giganteum, C. grandiflorum, C. insigne, C. longifolium, C. lowianum, C. madidum, C. mastersii, C. pendulum, C. simonsianum, C. tigrinum, C. tracyanum,* etc. Though there are several thousand hybrids but a few most popular ones are Angelica Advent, Alexanderi Westonbirt (*C. eburneum* x *C. lowianum* x *C. insigne* Westonbirt), Arabian Nights, Arunta Amber, Arunta Lovely Lips, Babylon Castle Hill, Balkies, Baltic, Bengal Bay, California Cascade, Cariga, Carisbrook, Comte de Hemptirne, Dingwall Lewes, Esmerelda 'S', Flirtation, Golden Fleece, Great Day, Karachi Poetic, King Arthur, Mahogany, Marcia, Moeke, Molly, Miretta, Mission Bay, Moriah Hindu, Nederhorst, No. 60, Oriental Legend, Pearl Magnificum, Peter Pan Greensleaves, Prince Charles, Princess Elizabeth, Promona, Pumander (Louis Sander x *C. pumilum*), Putana Showpiece, Queen of Picardy, Ramley Erin, Red Beauty, Rosanna Pinkie, Rose Beauty, Show Girl, Swallow, Vieux Rose, York, Yule Log, etc. This genus has intergeneric hybrids with *Phaius* and *Cyperorchis.* The top selling varieties in the Dutch Auctions are Cascade, Golden Fleece, Mieke, Molly, Nederhorst and No. 60.

Cypripedium **(Lady's slipper):** Commonly it is known as 'lady's slipper' because of its pouch like leaves, appearing as true lady's slipper. Generally they are deciduous and rhizomatous. Some 50 species are found in this genus which are mainly terrestrial with only a few lithophytic or epiphytic. It is widespread in North Temperate Zone with a few species extending in warm areas such as Mexico and parts of tropical Himalayas. These like tree shade for their growing. Important species are *C. bellatulum, C. calceolus, C. candidum, C.curtisii, C. insigne, C. parishii, C. parviflorum, C. pubescens, C. reginae, C. spicerianum, C. venustum, C. villosum, etc.* Andrewsii Fuller (*C. candidum* x *C. parviflorum*) and Favillianum Curtis (*C. candidum* x *C. pubescens*) are the interspecific hybrids.

Dendrobium: In the entire Orchidaceae, this is the second largest genus with some 1,340 species, widely distributed in tropical Asia from Sri Lanka to the Samoan and Tongan islands, Pacific Islands and across to Japan in the north and to New Zealand in the south. These are used primarily as cut flowers and are suitable for growing in greenhouses. Rotor (1952) stated that *Dendrobium nobile* can be timed by maintaining 18 °C and above temperatures, followed by LD and 13 °C where flowering occurs four months after lowering of the temperatures. The temperate species are *Dendrobium nobile, D. transparens,* etc. growing above 1,500 m while *D. chrysanthum, D. fimbriatum* var. *oculatum, D. lituifolium, D. moschatum,* etc. around 700 m elevation though many others are tropical in nature. These are non-bulbous producing slender canes to tiny pseudobulbs, out of the underground rhizomes, with a great majority being deciduous epiphytes while a few the more tropical ones are evergreen. For their growing, these prefer partial shade and high humidity coupled with good air circulation. Basic chromosome number in *Dendromium* is n=19 in most species but *D. amoenum* has n=20, *D. densiflorum* with n=20+0-2B, *D. parishii* being from diploid to tetraploid with 2n=38 and 76. The important species are *D. aggregatum, D. amplum, D. aphyllum, D. aureum, D. bigibbum, D. bellatulum, D. bensoniae, D. boxali, D. brymerianum, D. canaliculatum, D. chrysanthum, D. chrysotoxum, D. clavatum, D. crepidatum, D. densiflorum, D. devonianum, D. draconis, D. falconeri, D. farmeri, D. fimbriatum, D. formosum, D. gibsoni, D. gouldii, D. grantii, D. jamesonium, D. kingianum, D. loddigesii, D. longicornu, D. moschatum, D. nobile, D. ochreatum, D. parishii, D. phalaenopsis, D. pierardii, D. primulinum, D. speciosum, D. sulcatum, D. thyrsiflorum, D. undulatum, D. verartifolium, D. victoriae-reginae, D. wardianum, D. williamsoni,* etc., and the varieties are B.B. White, Boonchoo Gold, Caesar, Dorine White, Ekapol Panda, Emma White, Hieng Beauty, Intuwong, Jaquelyn Thomas, Jiad Gold, Joan Kushima, Karen No. 4, Kasem Gold, Kasem White, Lady, Laura, Lowana Nioka, Mme Vipar, Mary Mak, Nanatavarn, PM 11, Nurashikin, Pampadour, Pramot Sabine, Renapa, Royal Pink, Sabin, Sakura Pink, Sarifa Fatimah, Sonia No. 16, Sonia 17, Sonia 17 Mutant, Sonia 28 Mutant, Venus, Waipahu Beauty, Walter Oumae, Yupadeevam, etc. These varieties are being grown commercially in Kerala.

A few of the thousands of other varieties are Album, American Beauty, Banyat Pink, Boonchoo Gold, Candy Stripe,

Earsakul, Emma White, Gold Flush, Hawaian Beauty, Hieng Beauty, Jaquiline Thomas, Joe and Jim Biden (Anon., 1913; named in honour of US Vice President and his wife when they visited Indonesia's 63-hectare orchid garden on July 26, 2013), Jurie Red, Kanjana Green, Kasem Gold, Kasem White, Lady Hamilton, Mme Vipor, Nette White, New Pink, New Wanee, Pink Tips, Pramott-II, Renappa, Sabine, Sabine Red, Sakura Pink, Sonia Bom Jo, Sonia 17, Sonia 28, Snow White, Theodore, Tongchai Blue, White Nern, etc.

Epidendrum: This is among the largest genera of the Orchidaceae, containing more than 1000 species, is epiphytic (rarely terrestrial) in nature, and most species behave as weeds in the orchid fields, hence, most of the species are of minor importance horticulturally. The genus contains highly varied forms. Some prefer hot while certain others prefer cool conditions for their growing. Generally, they originate in Central America *i.e.* New World tropics. Important species of horticultural importance are *E. alatum, E. anceps, E. arachnoglossum, E. ciliare, E. cochleatum, E. difforme, E. endresii, E. fragrans, E. ibaguense, E. parkinsonianum, E. polybulbon, E. prismatocarpum, E. radicans, E. radiatum, E. vitellinum, E. wallisii, E. xanthinum*, etc., the interspecific hybrids are Endresio-Wallisii, Burtoni, Charlesworthii, Dellense, etc., and the intergeneric hybrids are *Epitonia* with *Broughtonia, Epidiacrium* with *Diacrium, Epigoa* with *Domingoa, Epilaelia* with *Laelia, Epilaeliopsis* with *Laeliopsis, Epicattleya* with *Cattleya, Epiphronitis* with *Sophronitis, Epiphronitella* with *Sophronitella, Schomboepidendrum* with *Schomburgkia, Kirchara* with *Cattleya* x *Laelia* x *Sophronitis*, etc.

Eria: This genus contains some 550 species, and is native to Tropical Asia, extending from the Himalayas, China and Sri Lanka to the Fiji Islands and Samoa, with greater concentration in Indonesia and New Guinea. This is allied to *Dendrobium* but has eight pollinia. These are either epiphytic or lithophytic. The flowers are small of brown and white colours. Important species are *E. bambusifolia, E. carinata, E. coronaria, E. dalzellii, E. fragrans, E. nana, E. mysorensis, E. pauciflora, E. polystachya, E. pseudoclavicaulis, E. reticosa, E. spicata, etc.*

Eulophia: *Eulophia* in Greek means 'handsome crest'. It is a terrestrial genus of small plant with 50-60 species from the tropics of both the hemispheres and is cultured similar to *Calanthe*. It develops from the pseudobulbs with membranous leaves, Scapes form on the base with several flowers where sepals and petals are spreading type, labellum is 3-lobed and pollinia 2. ***E. maculata*** from Brazil bears ompressed and ovate pseudobulbs, spotted or blotched leaves and small flowers where upper sepal is hooded while laterals acuminate and reddish-brown, petals wide, white or pale-rose, labellum cordate with two crimson spots near the base otherwise white. ***E. quartiniana*** from Central and East Africa, Sudan and Ethiopia was introduced into Europe round 1895. It was discovered by French botanist and explorer Richard Quartin-Dillon who died in Ethiopia in 1841. It is an erect growing species with long, wide and green overlapping leaves, the inflorescence sralk growing to more than 50 cm in length with about 20 cm rachis, bearing mauve drooping flowers. Greenish-brown sepals are back-reflexing. The species is very attractive and most suitable as cut flowers. ***E. scripta*** from Madagascar produces linear and subdistichous leaves, purple and yellow flowers, sepals and petals are linear-oblong, labellum is 3-lobed with lateral lobes rotund at the apices.

Miltonia (Pansy orchid). *Miltonia* was named for Viscount Milton (1786-1851), a patron of horticulture. It is an evergreen epiphytic genus of 25 clump-forming rhizomatous orchid species from central and tropical South America, producing greyish-green, shiny & smooth and flattened-ovoid pseudobulbs about 5-10 cm high, 3.75 cm wide and 1.25 cm thick, bearing normally 1-3 pale-green, long, narrow, gracefully arching and almost transluscent leaves at the tips, and 1-10 short-stalked, large, flat, richly coloured and often fragrant flowers 5-10 cm across, resembling pansies that emerge on erect stalks 15-45 cm long and these stalks rise from the base of pseudobulbs. Flower texture may be, especially of the hybrids, velvety. Tepals are oblong so giving overlapping appearance and their colour contrast with lip colour as in one of the two larger lobes in the upper part there is a contrasting blotch. Flowering is usually during late spring through summer but sometimes there is repeat in the fall. Each bloom lasts 4-5 weeks. They may be given direct morning or late afternoon light only for 2-3 hours daily during winter but for rest of the year only medium light. For their effective cultivation, the daytime tempearature should be around 21 °C and nighttime 18 °C with high humidity. In no case the temperature should rise above 24 °C and fall below 16.7 °C. They are propagated through division of the rhizomes, each segment having at least two pseudobulbs. Important cultivated species are *M. roezlii, M. spectabilis, M. vexillaria* and *M. warscewiczii*. ***M. roezlii*** (*Odontoglossum roezlii*) from Colombia is closely allied to *M. vexillaria* and produces 2.5-5.0 cm high and narrowly ovate pseudobulbs, bearing numerous slender, nerved and narrowly linear-lanceolate leaves 20-30 cm long. Flower stalks about half the length of the leaves, each stalk with 2-3 white, large and flat flowers 7.5-8.25 cm across and having a purple band at the base of the petals, and large and broadly obcordate labellum at its base is more or less marked reddish-brown. Tepals are ovate-oblong and acute. Labellum in the sinus has a tooth and a spur-like horn projecting backward on each side of the column. It flowers twice, in winter and spring. ***M. spectabilis*** from Brazil is a robust plant, producing 7.5-10.0 cm tall pseudobulbs, each with two strap-shaped and up to 15 cm long leaves. Floral stalks are up to 20 cm tall, topped with solitary flowers 7.5-10.0 cm across from July to September. The tepals are white to cream and tinged pink towards the bases. The labellum is bright rose-purple marked with a central blotch, margins pale-rose & wavy and veins deeper. Its form 'Moreliana' bears purple tepals, wine-purple lip and more heavily veined. ***M. vexillaria*** (*Odontoglossum vexillarium*) from western slopes of Andes, Colombia produces 3.75-6.25 cm high pseudobulbs, each with 3-5 strap-shaped and 20-30 cm long leaves. several floral stalks 45-50 cm tall appear at one time from each pseudobulb, each bearing 3-10 pale-rose-mauve (rarely white), flattened and fragrant flowers 8.75 cm across from May to June. Labellum is large (up to 6.25 cm across) and veined bright yellow. ***M. warscewiczii*** (*M. weltonii, Odontoglossum weltonii, Oncidium* *fuscatum, O. weltonii*) from Peru produces much flattened and 7.5-12.5 cm long pseudobulbs. The leaves are linear-oblong, obtuse and 12.5-15.0 cm long. Panicles appearing in March are nodding, branched and bear numerous pale-reddish-brown flowers 5.0 cm across with whitish tips. Tepals are cuneate-obovate, waving and crisped and white labellum is oblong, fan-shaped, bifid having a large rose-purple disc where in the centre there is a large brownish-yellow blotch.

Odontoglossum: This very popular orchid is among the loveliest orchids, hence is being widely cultivated in many tropical parts of the world. The genus comprises some 300 species and is allied to *Brassia* and *Oncidium*. This has oval to round pseudobulbs compressed at the sides. The plants of this genus are mostly epiphytic but a few are lithophytic or terrestrial. Important species are *O. bictoniense, O. cirvantesii, O. cirrhosum, O. citrosmum, O. cordatum, O. crispum, O. grande, O. harryanum, O. lindleyanum, O. luteo-purpureum, O. maculatum, O. nobile, O. rossii, O. stellatum, O. triumphans*, etc., a few of the thousands of interspecific hybrids are Adrianae, Alispum, Anneliese Rothenberger, Aspermum, Brandtii, Crispania, Crispelus, Elegans, Elise, Elle's Triumph Dahlenburg, Eudora, Himachal, Humeanum, Maronius, Nisus, Ophoon, Perryanum, Princess Elizabeth, Wilckeanum, etc., and intergeneric hybrids are *Odontioda* with *Cochlioda, Odontocidium* with *Oncidium, Odontonia* with *Miltonia, Odontobrassia* with *Brassia, Adaglossum* with *Ada, Symphodontoglossum* with *Symphyglossum, Beallara* with *Brassia* x *Cochlioda* x *Miltonia, Burrageara* with *Cochlioda* x *Miltonia* x *Oncidium, Sanderara* with *Brassia* x *Cochlioda, Vuylstekeara*, a trigeneric hybrid of *Cochlioda, Miltonia* and *Odontoglossum, Wilsonara* with *Cochlioda* x *Oncidium, Colmanara* with *Miltonia* x *Oncidium*, and *Lagerara* with *Aspasia* x *Cochlioda.*

Oncidium (Dancing lady or dancing girl)*:* By and large, these are sympodial evergreen epiphytic (a few terrestrial or lithophytes) orchids with some 750 species found in the American sub-tropics, having compressed pseudobulbs (certain are mule eared, *i.e.* devoid of pseudobulbs) and are highly variable. It likes partially shaded conditions. The elegant flowers are borne in drooping and branched spikes. Common characteristics are the petal-like wings on the column, a bump below the stigma and a large and many-toothed crest at the base of the lip. Usually the flowers are yellow or brown with dark crimson central lobe, bright and showy but small. Important species are *O. altissimum, O. ampliatum, O. auriferum, O. cavendishianum, O. cheirophorum, O. crispum, O. divaricatum, O. forbesii, O. lanceanum, O. macranthum, O. marshallianum, O. ornithorbynchum, O. papilio, O. sarcodes, O. varicosum, O. variegatum*, etc., important interspecific hybrids are Burzeffianum, Haematochilum, Kaiulani, Mantinii, Punctatum, Vizer, Wheatleyanum, etc. and intergeneric hybrids are *Brassidium* with *Brassia, Charleswortheara* with *Cochlioda* x *Miltonia, Miltonidium* with *Miltonia, Oncidioda* with *Cochlioda, Rodricidium* with *Rodriguezia, Trichicidium* with *Trichocentrum*, etc. Some other popular varieties are Album, Esther, Folksy Misty, Golden Shower, Golden Sunset Robsan II, Goldiana, Gower Ramsay, Irina Murray Orchidglade, Josephine, Potpourri Scarlet, St. Anne, Tiny Tim, Wikki Sunset, Willow Pond, etc.

Paphiopedilum (Lady's Slipper)**:** A genus of 60 evergreen species with many of the tropical Asiatic orchids having no pseudobulbs, and the flowers are quite similar to *Cypripedium*. These are used primarily as potted flowering plants and do well in homes, especially in cool and humid areas. Many species are also temperate in nature such as *P. fairieanum, P. hirsutissimum*, etc. which grow above 1,500 m elevation. They are mostly terrestrial, a few lithophytic or rarely epiphytic. The spectacular and the long-lasting flowers are produced singly or in few-flowered racemes. The basic chromosome number in *P. fairieanum, P. hirsutissimum, P. insigne, P. spicerianum*, and *P. villosum* all with n=13 whereas *P. venustum* possesses n=20+1B. Important species are *P. appletonianum, P. barbatum, P. bellatulum, P. callosum, P. concolor, P. fairieanum, P. hirsutissimum, P. insigne, P. lawrenceanum, P. niveum, P. parishii, P. rothschildianum, P. spicerianum, P. venustum, P. villosum*, etc. A few of thousands of the interspecific hybrids are Aylingii, Arthurianum, Bion, Charles Richman, Demura, Gloriosum, Goultenianum, Harrisonianum, Holdenii, Kowloon, Le Douxiae Leeanum, Lethamianum, New Era, Maudiae, Tesselatum, Vinter's Treasure, Winnipeg Shillington, Wottonii, etc. Two bigeneric hybrids are *Phragmipaphiopedilum* and *Selenocypripedium* which were evolved by crossing with *Phramigpedium.*

Peristeria (Dove orchid)**:** A group of stately South American pseudobulbous warm house orchids. Some five species are reported under this genus where only three are cultivated, *viz. P. cerina, P. elata* and *P. pendula*, which bear 6 to 20 flowers per raceme.

***Phaius*:** This tropical genus consists of 30 deciduous species of terrestrial (a few species also epiphytes) orchids, having elongated pseudobulbs and typically erect inflorescence with beautiful long-lasting flowers. *Phaius tankervilliae* has been found growing up to 700 m elevations. This possesses basic chromosome number as n=19. Important species are *P. amboinensis, P. callosus, P. flavus, P. humblotii, P. longipes, P. maculatus, P. mishmensis, P. pauciflorus, P. sanderianus, P. schnoebrunnensis, P. simulans, P. tankervilliae* (syn. *Bletia tankervilliae, P. grandifolius Lour.*), *P. tuberculosus, P. wallichii* (syn. *P. bicolor, P. blumei, P. grandifolius* Lindl. *or P. grandiflorus*) etc. Some of the interspecific hybrids are Cooksonii, Marthae, Norman, Phoebe, etc. while intergeneric hybrids are *Gastophaius* with *Gastorchis, Phaiocalanthe* with *Calanthe, Phaiocymbidium* with *Cymbidium*, and *Spathophaius* with *Spathoglottis.*

Phalaenopsis (Moth orchid)*:* Some 55 epiphytic (tree trunks and on the sides of the rocks) species that require partially shaded conditions, are reported in this genus which are generally evergreen and native of tropical India and Indonesia to the Philippines, New Guinea and northern Australia, mostly of tropical and subtropical nature, and are related to *Doritis*. These are primarily most suitable as potted flowering plants, and as cut flowers for design works, especially during weddings due to its predominantly pure white flowers though these have other colours from yellows to pinks. These orchids are eeasy to maintain at home towards the sunny side. The plants of this genus are monopodial in habit, have short rhizomes instead of pseudobulbs and tightly clinging long aerial roots. The inflorescences are lateral, short or long, erect or pendulous and sometimes branched. The flowers of some species are excellent for cutting as they are very attractive and long-lasting. This has been recorded with a basic chromosome count of n=19. Important species are *P. amabilis, P. fuscata, P. lueddemanniana, P. mannii, P. parishii, P. rosea* syn. *P. equestris, P. schilleriana, P. speciosa, P. stuartiana, P. violacea*, etc. Interspecific hybrids are Alice Gloria, Ann Marie, Artienne, Bride's Maid, Canary, Carrier Girl, Diana Pinky, Encantadora, Eye Dee, Golden Liz, Golden Pride, Irene Dobkin, Hennessy, Keith Shaffer, Malaysian Beauty, Natasha, Peppermint, Rose Parade, Ruby Lip Tip, Samba, Spotted Moon, Spring Shower, Star of Rio, Tammy, Texas Star, Reve Rose, Riea, Temple Cloud, Tiger Butter, Tiger Butter Larkin Valley, White Goddess, Zada, etc. and intergeneric hybrids are *Aeridopsis* with *Aerides, Arachnopsis* with *Arachnis, Doritaenopsis* with *Doritis, Phaladopsis* with *Vandopsis, Renanthopsis* with *Renanthera, Rhynchonopsis* with *Rhynchostylis, Rhynodoropsis* with *Doritis x Rhynchostylis, Tanakara* with *Aerides* x *Vanda* and *Vandaenopsis* with *Vanda*. Certain intergeneric varieties are George Molar, Golden Shower, Jerry Sue King, etc.

Phragmipedium (syn. *Selenipedium*, Lady's slipper orchid): It is native to Colombia, Costa Rica, Ecuador and Panama, and consists of some 12 species. These are terrestrial (sometimes epiphytic) tropical glasshouse orchids without pseudobulbs, rarely doing well under typical home environment, and are used as the potted flowering plants. Growth is clumpy through rhizomes. Leaves are leathery, linear, 40-50 cm long and 3-4 cm wide, pointed and curved. Apical inflorescence erect and consist of a number of flowers, opening one by one. Flowers are white, green or purple, large (4 cm in diameter), attractive and long-lasting with pendulous petals. Some of the hybrids and varieties are 'Albiflorum', 'Aureum', 'Dominianum', 'Giganteum', 'Grande', 'Grandiflorum', 'Latifolium', 'Luxembergense', 'Magniflorum', 'Nigrescens', 'Roseum', 'Rubrum', 'Schroderae', 'Sedenii', 'Seegerii', 'Splendens', 'Superbum', 'Wallisii', etc.

Pleione. *Pleione* derives from the Greek goddess of that name, mother of the seven Pleiades. It is native to SE Asia, from India to China, closely allied to and once included in *Coelogyne*. It is a genus of 10 temperate deciduous terrestrial or semi-epiphytic orchid species with pseudobulbs, and in the wild these are often found growing on mossy rocks, tree trunks and branches. These are suitable for growing in a cool greenhouse or in an alpine house or for growing as house plants except the *P. bulbocodioides* and its varieties which survive outdoors in sheltered rock gardens. Their pseudobulbs are small, each producing mid-green, 1-2, a little pleated, elliptic to lanceolate and prominently ribbed leaves which appear after the flowering rising from the bases of the short and squat pseudobulbs and survive one season only. The flowers often large for the size of the plant have spreading similar tepals and a 3-lobed trumpet-shaped to tubular labellum fringed at the mouth are borne singly (rarely 2) on the stems arising from the pseudobulbs. They are available in shades of white, yellow, pink or mauve often heavily marked with other contrasting colours and are most suitable for growing in shallow containers. The medium is kept dry since the leaves start yellowing till new foliage starts appearing in spring. They are propagated by dividing the clumps or by potting individual pseudobulbs separately which have been formed on top of the clump. These require filtered but bright sunlight, however, during short days supplementary fluorescent lighting may be helpful, and a temperature 10-16 °C. Important species are *P. bulbocodioides, P. forrestii, P. humilis*, and *P. praecox*. ***P. bulbocodioides*** is native to Tibet to Formosa *via* China and Taiwan. This is the most common species in which *P. formosana* (blush of dawn), *P. delavayi, P. limprichtii, P. pogonoides* and *P. pricei* are now included after the revision. This is suitable for growing in a cold frame, on a sheltered rock garden and in a living room. It produces dull green (sometimes partly blackish-purple) and nearly rounded pseudobulbs nearly 2.5 cm high. Flowering stalks nearly 15 cm long and the leaves about 20 cm long on maturity and 3-6 cm wide appear together from January to May. Pure white to deep mauve-pink flowers 7-10 cm across are borne with white to paler labellum which are fringed, crested and spotted with red, purple or yellow. This has produced many cultivars such as 'Alba' (graceful pure white), 'Blush of Dawn' (*P. formosana*; pale lilac-pink tepals and white labellum tinged with pale-mauve), 'Limprichtii' (tepals magenta-purple, labellum paler with dense red spots and streaks), 'Oriental Splendour' (*P. pricei*; pseudobulbs dark purple and more flask-shaped, flowers pale-violet, labellum white but finely lined with orange), 'Polar Sun' (flowers pure white), *etc*. ***P. forrestii*** from SW China and Myanmar bears pseudobulbs gradually narrowing towards the top, blooms opening after the death of the leaves in summer, flowers are yellow to orange-yellow, and marked with reddish-brown on the labellum. ***P. humilis*** from North India and Nepal is a half-hardy dwarf species with 2.5-3.75 cm high pseudobulbs. Solitary white to pale-mauve flowers about 7.5 cm across appear on a 10 cm long stalk from January to May where lip is frilled, white-keeled and heavily spotted with lines of purple, crimson or yellow-brown spots. ***P. praecox***, a native of Nepal, China and India, is a tender species which produces about 3.75 cm high, deep green and barrel-shaped pseudobulbs spotted with maroon. Its floral stems are 15 cm long, carrying a single 7.5 cm across flower between November and January. Tepals are deep rose-pink, and labellum is pale-pink with a pale-yellow disc and five keels.

Renanthera (Sun loving orchid)**:** This tropical monopodial orchid possesses some 15 epiphytic, tall (0.6-3.6 m) and climbing species which has good commercial value. Stems are branched and it climbs by means of fleshy roots. Its temperate species *R. imschootiana* grows above 1,500 m elevations. Inflorescence is quite large and drooping. Important species are *R. coccinea, R. elongata, R. imschootiana, R. pulchella* and *R. storiei*. The basic chromosome number recorded in *R. imschootiana* is n=19. Some of the bigeneric hybrids are *Aranthera* with *Arachnis, Leathers* with *Renantherella, Pararenanthera* with *Paraphalaenopsis, Renades* with *Aerides, Renanopsis* with *Vandopsis, Renantanda* with *Vanda, Renanthoglossum* with *Ascoglossum, Renanthopsis* with

Phalaenopsis, Renantylis with *Rhynchostylis, Renencentrum* with *Ascocentrum* and *Sarcothera* with *Sarcochilus.* The trigeneric hybrids of *Renanthera* are *Hawaiiara* with *Vanda* x *Vandopsis, Holttumara* with *Arachnis* x *Vanda, Limara* with *Arachnis* x *Vandopsis, Moirara* by crossing *Renanthera* with *Paraphalaenopsis* x *Vanda,* etc. The interspecific varieties are Brookie Chandler, Higgins, Kilauea, Mauricette, Mok York-Seng, Poipu, Tom Thumb, etc.

Rhynchostylis (Fox-tail orchid): This monopodial epiphytic orchid is quite robust and has some four species, distributed over tropical Asia from the Indo-Myanmar region to Thailand. *R. retusa* has been found growing in the wild above 1,500 m elevations. Flowers appear from the axils of the leaves, in dense racemes and gracefully drooping. *R. retusa* has been recorded with a basic chromosome number of n=19. There are four species where *R. retusa* is native to India, *i.e.* Western Ghats. Its various varieties are in trade, *viz. alba, dayi, gigantea, holdfordiana, majus, superba,* etc. *R. coelistis,* and *R. violacea* (flowers have disagreeable flavour) are the other two commercial species, latter having a var. *harrisonianum.* Its intergeneric hybrids are *Aeridostylis* through *Aerides, Renanstylis* through *Renanthera, Rhynchocentrum* through *Ascocentrum, Vandacostylis* through *Vanda,* etc. Gigantea Red, Gigantea White, etc. are a few most important varieties.

Spathoglottis. In Greek, *Spathoglottis* means 'spathe' and 'tongue', owing to the shape of the lip. It is a terrestrial orchid with about 40 species from Asia, Australasia, and the Malay Islands. It has broadly conical pseudobulbs some 7 cm tall and 5 cm across which are 3-leaved, numerous apical leaves are long-petioled, narrow, linear-lanceolate, plicate, membranous, pointed, and sometimes about 1 m long and 2-7 cm wide. Scape is lateral and bears large flowers in terminal raceme, sepals are free and subequal, petals similar or broader and longer, labellum is not spurred, lateral lobes are little convolute, middle one clawed, column is slender and pollinia is eight. Important species are *S. affinis, A. aurea, S. ixioides, S. plicata, S. grandifolia, S. hardingiana, S. kimballiana, S. vieillardii, etc.* With *Phaius,* this had produced *Spathophaius* hybrid. Two most important ones are ***S. plicata*** (*S. lilacina, Bletia angustata*) from Philippines, India, Indonesia, New Guinea and Indo-Chinese Peninsula is a medium-sized terrestrial plant having ovoid, 7.5 cm tall and 5 cm across pseudobulbs. There are numerous stalked, linear-lanceolate, pointed, membranous and plicate apical leaves up to 1 m long and 2.5-7.5 cm wide. Erect floral stalk from 20 cm to 1 m long, bears a dense cluster of about 25 very attractive purple to white flowers opening in succession several times a year. ***S. vieillardii*** (*S. augustorum*) from New Caledonia bears 30-60 cm long, lanceolate and acuminate leaves. It produces a robust scape up to 45 cm long with about 15 cm long, broad and corymb-like raceme. Quite pale-lilac to nearly white flowers are about 5 cm across with tepals being ovate-oblong and subacute, labellum as long as the tepal, lateral lobes orange-brown and middle lobe narrow.

Stanhopea. *Stanhopea* was named for Philip Henry Stanhope (1781-1855), 4th Earl of Stanhope and president of the Medico-Botanical Society of London. This genus is distributed from Mexico to Brazil with about 25-45 epiphytic orchid species bearing ovoid and clustered pseudobulbs 5-7 cm high, each carrying one pleated leaf 30-60 cm long and 2-6 fragrant flowers usually on pendent stalks. The tepals are reflexed, outer ones (sepals) wider than inner ones (petals), the labellum is round to shoe-shape, terminating into a tongued-tip or two horn-like projections near the base, and the long column arches over the tip of the labellum. They are most suitable for growing on bark slabs and in hanging baskets indoors. They grow best at 16-21 °C temperatures and high humidity. These are rested after completion of the growth, i.e in October-November and then temperature is also lowered. They are propagated by dividing the psudobulbs in July or August. Important species are *S. costaricensis* (*S. graveolens*), *S. devoniensis, S. eburnea* (*S. grandiflora*), *S. ecornuta, S. oculata, S. tigrina* (*S. hernandezii*) and *S. wardii.* ***S. costaricensis*** (*S. graveolens*) from Mexico to Brazil bears ovoid and clustered pseudobulbs about 3.75 cm high. In late summer to early autumn, each pseudobulb produces 12.5 cm long sheathed floral stalks with two flowers up to 12.5 cm across from spring to early summer. The flower 10-13 cm long comprises of buff-yellow, concave and 7.5 cm long sepals patterned with light red ring-spots, narrow and buff-yellow petals with wavy margins and small but bold red spots, and a pair of dark red patches at the base of the yellow-white lip. ***S. devoniensis*** from Mexico produces pale-yellow flowers 10 cm across with red-brown to deep red-purple spots, and the cream labellum blotched and flushed with purple, along with two horns. The flowers appear during summer. ***S. eburnea*** (*S. grandiflora*) from Brazil bears 3.75 cm high pseudobulbs, and the floral stalks which appear from August to October are 15-25 cm long with 1-2 flowers 10 cm across having long and narrow white tepals, and the labellum is narrow and white with a purple base but lacks the usual long curving horns. ***S. ecornuta*** from Central America produces 10-12 cm across flowers in summer bearing creamy-white tepals having purple spots basally, and the yellow labellum is without horns and shade to orange at the base with age. ***S. oculata*** from Mexico bears 5 cm high pseudobulbs, and the inflorescence which appear during July to October is more than 30 cm long with 3-10 fragrant and 10 cm across flowers. Light yellow tepals have red spots, and the labellum is narrow, base orange-yellow with 2-4 large black spots and the tip of the labellum is white with red to purple spots. ***S. tigrina*** (*S. hernandezii*) from Mexico produces in summer the orange-yellow flowers 20 cm across bearing purple-brown blotches and horned labellum. ***S. wardii*** from Mexico to Panama *via* Guatemala and Venezuela bears 5 cm high clustered pseudobulbs, producing slender floral stalks 23-30 cm long at any time from July to November. Stalks bear 6-10 strongly fragrant flowers 7.5-13.0 cm across with tepal colour being golden-yellow but spotted purple, and the deep yellow labellum is horned and patterned with two circular and velvet-purple blotches.

Trichopilia (Corkscrew orchid). *Trichopilia* in Greek stands for 'hair and cap', as the anther is concealed under a cap surrounded by three tufts of hairs. This epiphytic orchid comprises of 20 species ranging ftrom Mexico to South America.This is very handsome genus grown in pots and most suitable for indoor cultivation. Its short rhizomes are crowded with flattened and elongated pseudobulbs, each with one leaf and a tuft of dry scales at the base. Leaves are solitary, large, erect, fleshy and keeled. Scapes are short, decumbent or nodding and bear abundance of flowers. Tepals are narrow but spreading and twisted, labellum is large & most conspicuous and united with the column below. Its flowers are long-lasting. For its growing it requires an intermediate temperature. Propagation is through division of rhizomes. Important species are *T. crispa, T. fragrans, T. galeottiana, T. marginata, T. nobilis, T. suavis* and *T. tortilis*. ***T. crispa*** from Costa Rica is closely related to *T. marginata*. It produces dark green, ovate, flattened, 5.0-7.5 cm long and 1-leaved pseudobulbs. Leaves are fleshy, acute-pointed, keeled, 15 cm long and 5 cm wide. Drooping but short floral stalks emerge from the base in May or June bearing three flowers 5 cm long, pedicel included. Tepals are brownish-yellow, spreading, twisted & wavy-edged, 6.25 cm long and 1.25 cm wide, and labellum is deep crimson with a white margin, 3.75 cm across, folded over the column but spreading in the front. ***T. fragrans*** (*Pilumna fragrans, T. bachhousiana*) from Colombia bears 7.5-12.5 cm high, flattened and clustered 1-leaved pseudobulbs. Leaves are oblong-lanceolate, acute and 15-20 cm long. Scape some 30 cm long is pendent with six flowers having 7.5 cm long pedicels and appear in summer. Greenish-white tepals are linear-lanceolate, twisted & undulated, spreading and 6.25-7.5 cm long, and a little lobed labellum is white with yellow blotch over throat, folded over the column and spreading in the front. ***T. galeottiana*** from Costa Rica and Mexico bears flattened and narrow pseudobulbs about 12.5 cm long. Leaves are about 15 cm long, oblong and acute. Scapes are short, usually 1-flowered, flowers yellowish-green with sometimes banding below the middle, tepals cuneate-lanceolate, and the labellum is cup-shaped and whitish patterned with purple blotches (dots and streaks) in the centre. ***T. marginata*** (*T. coccinea, T. crispa* var. *marginata*) from Central America bears oblong, compressed and clustered pseudobulbs. Leaves are subauriculate at the base, broadly lanceolate and suddenly acuminate. Scape is generally small and with three large flowers appearing during May and June and these are whitish externally and reddish-purple internally, tepals are linear-lanceolate having white margins with slightly twisted petals, and the labellum is large, wavy, rounded and a cup-shaped 4-lobed blade. ***T. nobilis*** (*Pilumna nobilis, T. candida*) from Venezuela produces large pseudobulbs with broadly oblong-acute leaves. Flowers are white, tepals are little twisted, linear-oblong, acute and 5 cm long, and the labellum is large and white having a yellow spot in the throat. ***T. suavis*** from Central America produces thin and compressed pseudobulbs 5 cm high, bearing 20 cm long and broadly oblong leaves. Stalks are long, arching or pendent and with three creamy large flowers appearing in May and June where lanceolate-acuminate tepals are straight but wavy and about 5 cm long, and the labellum is also cream to white with pale-purple or yellow spots in the throat and the limb is large-lobed, wavy and crenate. ***T. tortilis*** from Mexico produces little curved, compressed and oblong pseudobulbs 5-10 cm high. Leaves are oblong, acute, solitary and 15 cm long. Stalk is decumbent, shorter than the leaves and 1-flowered and flowers about 4.5 cm across being borne in profusion during summer and again in winter. Flowers are brown with yellowish margins, tepal is 5 cm long and spirally twisted, and the labellum where upper portion is expanded is 4-lobed, white with crimson spots outside & within entirely crimson, and this forms a tube around the column.

Vanda: This tropical orchid is related to *Aerides* and *Rhyncostylis* genera. *Vanda coerulea* has been recorded growing above 1,500 m height though *V. teres* and *V. tessellata* up to 700 m. This sun-loving monopodial evergreen epiphytic orchid, in habit being similar to *Arachnis*, has highly attractive large flowers in all the species, has some 70 species spread in East India and the Malay Islands with outlying species in China and New Guinea. The species may be short-stemmed and erect (compact) or tall and branched (sometimes climbing to some height) and without pseudobulbs. The inflorescence is erect or pendulous and arises from the leaf axils. Based on leaf shape, it may be terete (cylindrical-leaved), semi-terete and strap-leaved. There are two types: (i) with strap-shaped leaves, and (ii) with cylindrical leaves. *Vanda* possesses basic chromosome number of n=19. Important species are *R. alpina, V. amesiana, V. bensonii, V. coerulia, V. coerulescens, V. cristata, V. denisoniana, V. hookeriana, V. parishii, V. parviflora, V. roxburghii, V. stangeana, V. teres, V. tricolor, V. tessellata*, etc. *V. coerulea* is used for breeding blue *Vanda* hybrids, vars 'Bhimayothin' for pink colour, 'Madame Rattana' for pink and blue vandas, 'Pranerm Ornete' for brownish yellow, 'Piyaporn' and 'Prakpetch' for purplish red and blue black, 'Gordon Dillon' and 'Siam Ruby' for pink, and purplish red 'Rassi' for yellow vandas. Two different plants of *Vanda coerulea* when crossed together resulted into a clone 'Buranasin'. 'Miss Joaquim', 'Deva', 'Nam Phung', 'Rasri', 'Kultana', 'Prakpetch', 'Sumon Sophonsiri', 'Kasem's Delight', 'Pissamai', 'Wirat', etc. are the interspecific hybrids. Some other popular varieties are Ellen Noa, Emily Notley, Evening Glow, Bill Sutton, Hilo Blue, Honolulu, Honomu, John Club, Josephine van Brero, Norbert Alphonso, Onomea, Ruby Prince, Wirat Uchida, etc.

Zygopetalum. *Zygopetalum* derives from the Greek *zycos* meaning 'a yoke' and *petalon* for 'a petal', a swelling at the base of the labellum seems to join or yoke the petals together. The evergreen genus from central and tropical South America (Brazil and Guianas) comprises 20 terrestrial and epiphytic orchid species, producing clump-forming ovoid pseudobulbs, each bearing two or more mid to deep green, distichous, narrow lanceolate and terminal leaves sheathing a short stem, and often racemes (sometimes even solitary) of large blooms which are often very attractive. Usually purple-spotted green flowers bear similar five radiating tepals where sepals and petals are often united to each other at the base, the sepals forming a short chin with the foot of the column and a large fan-shaped labellum of contrasting pattern. After flowering is over, these are propagated by division in spring. Most promising and common species are *Z. crinitum, Z. intermedium, Z. mackayi* (*Z. mackaii*), *Z. maxillare* and *Z. perrenoudii*, all from Brazil. ***Z. crinitum*** is similar in habit to *Z. intermedium*. Its leaves are broadly linear-lanceolate. Floral scape is stout and long which bears green with sparsely brown-blotched tepals that are 5.0 cm long and oblong-lanceolate.

Labellum is white with purple veins which radiate from the thick crest, scarsely emarginate, 5 cm across, spreading & wavy, and disc hairy, For flowering not particular about the time. *Z. mackaii* var. *crinitum* has comparatively fewer brown blotches on the tepals than *Z. intermedium*. The veins are dark blue in the var. *caeruleum*. This species has several varieties with vein colour of the labellum being pink, blue or almost colourless. ***Z. intermedium*** produces bright green pseudobulbs about 9 cm tall and with a basal sheath. The leaves are bright green, lanceolate-elliptic to nearly strap-shaped, 3-5-clustered or fan-shaped, arching and 30-45 cm. Fragrant, waxy, long-lasting, 8 cm across with almost similar size of tepals, each tepal being usually yellow-green but blotched red-brown while labellum is white patterned with branched purple-red radiating veins where margins are wavy or crimped, are borne during autumn and winter. ***Z. mackayi*** is the most popular evergreen epiphytic orchid species in the genus, growing up to 30 cm in height with broad, ovoid and 5.0-7.5 cm high pseudobulbs.Its leaves are ribbed, narrowly oval and about 30 cm long. During autumn (November to February) on a 30-60 cm long flowering stalk, the plants produce fragrant and purple-brown-blotched yellow-green flowers 5-8 cm across with spreading tepals where lips are white veined with violet-purple and the crest is ridged. ***Z. maxillare*** produces 5 cm high pseudobulbs with 30 cm long lanceolate leaves. Floral stalk (scape) is 20-25 cm long with 6-8 green but small flowers 1.25 cm across which are borne out during winter. Tepals are ovate-oblong and acute with brown transverse bands. Purple labellum is horizontal with a quite large glossy purple notched horse-shoe-shaped crest where central lobe is waved and roundish. ***Z. perrenoudii*** is an evergreen epiphytic orchid for a cool house. This grows up to 30 cm tall, bearing ribbed, 30 cm long and narrowly oval leaves. The spikes bear violet-purple-lipped fragrant dark brown flowers 8 cm across in winter.

Some of the intergeneric hybrids are ×*Alexanderara* (*Brassia* × *Cochlioda* × *Odontoglossum* × *Oncidium*), ×*Aliceara* (*Brassia* × *Miltonia* × *Oncidium*), ×*Angulocaste*, ×*Aranda* (*Arachnis* × *Vanda*), ×*Aeridovand* (*Aerides* × *Vanda*), ×*Ascocenda* (*Ascocentrum* × *Vanda*), ×*Ascofinetia* (*Ascocentrum* × *Phalaenopsis*), ×*Brassocattleya* (*Brassavola* × *Cattleya*), ×*Brassoepidendrum* (*Brassavola* × *Epidendrum*), ×*Brassolaeliocattleya* (*Brassavola* × *Cattleya* × *Laelia*), ×*Cattleytonia*, ×*Christieara* (*Aerides* × *Ascocentrum* × *Vanda*), ×*Dialaelia*, ×*Doritaenopsis* (*Doritis* × *Phalaenopsis*), ×*Epicattleya* (*Epidendrum* × *Cattleya*), ×*Epiphronitis*, ×*Laeliocattleya* (*Laelia* × *Cattleya*), ×*Luisanda* (*Luisia* × *Vanda*), ×*Maclellanara* (*Brassia* × *Odontoglossum* × *Oncidium*), ×*Miltonidium* (*Miltonia* × *Oncidium*), ×*Miltonioda* (*Miltonia* × *Cochlioda*), ×*Mokara* (*Arachnis* × *Ascocentrum* × *Vanda*), ×*Nakamotoara* (*Ascocentrum* × *Neofinetia*), ×*Odontioda* (*Odontoglossum* × *Cochlioda*), ×*Odontobrassia* (*Brassia* × *Odontoglossum*), ×*Odontocidium* (*Odontoglossum* × *Oncidium*), ×*Odontonia* (*Odontoglossum* × *Miltonia*), ×*Odontorettia* (*Odontoglossum* × *Comparettia*), ×*Oncidiola* (*Cochlioda* × *Oncidium*), ×*Opsisanda* (*Vandopsis* × *Vanda*), *Opsistylis* (*Vandopsis* × *Rhyncostylis*), ×*Potinara* (*Brassavola* × *Cattleya* × *Laelia* × *Sophronitis*), ×*Renanopsis* (*Renanthera* × *Vandopsis*), ×*Renantanda* (*Renanthera* × *Vanda*), ×*Renanthopsis* (*Renantheran* × *Phalaenopsis*), ×*Rhynchovanda* (*Rhynchostylis* × *Vanda*), ×*Sarconopsis* (*Sarcanthus* × *Phalaenopsis*), ×*Sophrocattleya* (*Sophronitis* × *Cattleya*), ×*Sophrolaelia* (*Sophronitis* × *Laelia*), ×*Sophrolaeliocattleya* (*Sophronitis* × *Laelia* × *Cattleya*), ×*Stewartara* (*Ada* × *Cochlioda* × *Odontoglossum*), ×*Trichovanda* (*Trichoglottis* × *Vanda*), ×*Vandanthe* (*Vanda* × *Euanthe*), ×*Vandofinetia* (*Vanda* × *Neofinetia*), ×*Vascostylis* (*Ascocentrum* × *Rhyncostylis* × *Vanda*), ×*Vuylstekeara* (*Cochlioda* × *Miltonia* × *Odontoglossum*), ×*Wilsonara* (*Cochlioda* × *Odontoglossum* × *Oncidium*), etc. The basic chromosome number in *Calanthe masuca*, *Pleione praecox* and *P. maculata*, *Pholidata imbricata*, and *Thunia alba* have been recorded as n=20.

Some other important orchid genera are *Acacallis, Acampe, Acanthephippium, Aceras, Acianthus, Acineta, Acriopsis, Ada, Aerangis, Aeranthes, Aganisia, Amblostoma, Amesiella, Amitostigma, Anacamptis, Ancistrochilus, Ancistrorhynchus, Angraecopsis, Ansellia, Anthogonium, Aphyllorchis, Aplectrum, Arachnanthe, Arethusa, Arpophyllum, Arthrochilus, Arundina, Aspasia, Barbosella, Barbrodria, Barkeria, Barlia, Bartholina, Batemannia, Bifrenaria, Bletia, Bletilla, Bollea, Bolusiella, Bonatea, Brachycorythis, Bromheadia, Broughtonia, Cadetia, Caladenia, Caleana, Calochilus, Calopogon, Calypso, Calyptrochilum, Camarotis, Capanemia, Catasetum, Cattleyopsis, Caularthron, Camaeangis, Cephalanthera, Chaubardia, Chiloglottis, Chondrorhyncha, Chysis, Cirrhaea, Cirrhopetalum, Cleisocentrum, Cleisostoma, Cleistes, Clowesia, Coelia, Coeliopsis, Coeloglossum, Cochleanthes, Cochlioda, Coelia, Colax, Collabium, Comparettia, Comperia, Constantia, Corallorhiza, Coryanthes, Corybas, Corymborchis, Corysanthes, Cribbia, Cryptochilus, Cryptophoranthus, Cryptostylis, Cuitlauzina, Cycnoches, Cycorchis, Cymbidiella, Cycnoches, Cynorchis, Cyperorchis, Cyrtidium, Cyrtopodium, Cyrtorchis, Cyrtostylis, Dactylorrhiza, Dendrochilum, Dendrolobium, Diacrium, Diaphananthe, Dichaea, Didymoplexis, Didiciea, Dimerandra, Dimorphorchis, Diplocaulobium, Diplomeris, Dipodium, Dipteranthus, Disa, Disperis, Dissinia, Diuris, Domingoa, Doritis, Dracula, Dressleria, Dryadella, Dyakia, Elleanthus, Elythranthera, Encyclia, Epigeneium, Epipactis, Eriochilus, Eriopsis, Erycina, Esmeralda, Euanthe, Eulophiella, Eurychone, Flickingeria, Galeandra, Galeola, Gastrodia, Gastrochilus, Geodorum, Glossodia, Gomesa, Gongora, Goodyera, Govenia, Grammangis, Grammatophyllum, Graphorkis, Grobya, Grosourdya, Gymnadenia, Habenaria, Haemaria, Hartwegia, Helcia, Herschelia, Hexadesmia, Hexisea, Himantoglossum, Holcoglossum, Holothrix, Houlletia, Huntleya, Huttonaea, Ibidium, Inobulbon, Ionopsis, Isabelia, Isochilus, Jacquiniella, Jumellea, Kefersteinia, Kegeliella, Kingidium, Koellensteinia, Lacaena, Laelia, Laeliosis, Lanium, Lemboglossum, Leochilus, Lepanthes, Lepanthopsis, Leptotes, Leucohyle, Limatodes, Limodorum, Linneara, Liparis, Lissochilus, Listera, Lisrostachys, Lockhartia, Ludisia, Lueddemannia, Luisia, Lycaste, Lyperanthus, Macodes, Macradenia, Macroplectrum, Malaxis, Masdevallia, Maxillaria, Megaclinium, Meiracyllium, Mendoncella, Mesospinidium, Mexicoa, Microstylis, Microterangis, Microtis, Miltoniopsis, Monomeria, Moorea, Mormodes, Mormolyca, Mycrostylis, Mystacidium, Nageliella, Nanodes, Neobathiea, Neobenthamia,*

Neocogniauxia, Neofinetia, Neogardneria, Neogyne, Neolauchea, Neomoorea, Neottia, Nephelaphyllum, Nervilia, Neuwiedia, Nidema, Nigritella, Notylia, Oberonia, Octomeria, Oeceoclades, Oeonia, Oeoniella, Oerstedella, Ophrys, Orchis, Ornithidium, Ornithocephalus, Ornithochilus, Ornithophora, Osmoglossum, Otochilus, Otoglossum, Otostylis, Pabstia, Pachystoma, Palumbina, Panisea, Paphinia, Papilionanthe, Papperitzia, Paraphalaenopsis, Pecteilis, Penthea, Pescatoria, Pholidota, Physosiphon, Physurus, Platanthera, Platyclinis, Platylepis, Playstele, Plectorrhiza, Plectrelminthus, Pleurothallis, Podandria, Podangis, Pogonia, Polycycnis, Polyrrhiza, Polystachya, Pomatocalpa, Ponerarchis, Porpax, Porroglossum, Prasophyllum, Promenaea, Psychopsis, Pterostylis, Rangaeris, Restrepia, Restrepiella, Rhinerrhiza, Rhisalidopsis, Rhyncholaelia, Risleya, Ritaia, Robiquetia, Rodriguezia, Rodrigueziella, Rossioglossum, Saccolabium, Sarcanthus, Sarcochilus, Sarcopodium, Satyrium, Scaphosepalum, Scaphyglottis, Schlimmia, Schoenorchis, Schomburgkia, Scuticaria, Sedirea, Seidenfadenia, Selenipedium, Serapias, Sievekingia, Sigmatogyne, Sigmatostalix, Sirrhookera, Smitinandia, Sobralia, Solenangis, Solenidium, Sophronitella, Sophronitis, Spiranthes, Stauropsis, Stelis, Stenia, Stenoglottis, Stenorrhynchos, Summerhayesia, Sunipia, Syphyglossum, Taeniophyllum, Tainia, Tainiopsis, Telipogon, Tetramicra, Teuscheria, Thecostele, Thelymitra, Theodorea, Thrixspermum (Sarcochilus), Thunia, Ticoglossum, Tipularia, Trevoria, Trias, Trichocentrum, Trichoceros, Trichoglottis, Trichosalpinx, Trichosma, Trichotosia, Tridactyle, Trigonidium, Triphora, Trisetella, Trudelia, Vandopsis, Vanilla, Warmingia, Warrea, Warreella, Warscewiczella, Xylobium, Ypsilopus, Xeuxine, Zootrophion, Zygosepalum, Zygostates, etc.

Some highly curious looking orchids are epiphytic monkey orchids (*Orchis simian*) having highly foul odour similar to faeces, a native to Europe, the Mediterranean, Russia, Asia Minor and Iran which has lobed lip mimicking the general shape of the monkey's body hence its common name monkey orchid and it has grace pink to reddish flowers; flying duck orchid (*Calcana major*, syn. *Caleya major*; named after George Caley; a native to ES Australia from Queensland to S. Australia to Tasmania; and smaller duck orchid, *Paracaleana minor*) is a terrestrial and marshland orchid first recorded in 1803 but first featuring on an Australian postage stamp in 1986, which is a small orchid being pollinated by sawflies through pseudocopulation, and whose flowers are reddish-brown, rarely greenish with dark spots; and tongue orchids of Australia closely resemble in shape and colour to certain female wasps and the imitation is so realistic that some male wasps mate only with the flowers and in doing so the pollens stick to their abdomens which are delivered to other flowers when these wasps visit them. A strange and wonderful subterranean orchid genus (*Rhizanthella gardneri*, Gardener's underground orchid) which is saprophytic and leafless, is reported near to the city of Corriginand from locations close to the south coast of Western Australia (Etienne Delannoy, 2011) which lives its entire life cycle, including flowering several centimetres underground. This orchid belonging to the subfamily *Epidendroideae*, tribe *Diurideae* and subtribe *Rhizanthellinae*, was discovered by Jack Trott in 1928 in the wheat field through an odd soil crack which was emitting a sweet fragrance (Wikipedia). Shoots of this white orchid emerge from the deeply buried bulb, and the floral head consists of 8-150 tightly packed tiny flowers of dark maroon colour about 1.25 cm across. This beautiful and bizarre orchid is an endangered species, with only 50 known plants in the wild found in five locations in Western Australia. It is a parasite taking its food (nutrients and CO_2) from the symbiotic mycorrhizal fungus (myco-heterotroph) named *Thanatephorus gardneri* which lives with the roots of the broom brush (broom honeymyrtle) shrub in Western Australia. Usually one plant produces three daughter plants vegetatively. For sexual reproduction, the underground insects such as gnats and termites are attracted to its flowers for its fragrance and so the flowers are pollinated. The pollinated flowers take some six months to mature the seeds. Its fruit is a fleshy indehiscent drupe with about 150 minute and fleshy seeds (unique to orchids) and eaten by rats but still germinate when the seeds are infected by this fungus. Though it is unable to photosynthesize its own food, still it retains its chloroplasts. Two other species, *Rhizanthella omissa* (Lamington's underground orchid) and *R. slateri* (Eastern underground orchid) were also recorded together. However, some other 400 genera are of not much commercial importance.

Propagation

Orchids can be propagated through **division**, **offset** (**offshoots** or **keikis**), **cuttings** or **seeds** but the division is the most simple and popular method.

Monopodial orchids are though difficult to propagate vegetatively but such as *Aerides, Arachnis, Aranda, Dendrobium, Epidendrum, Phalaenopsis, Renanopsis, Renanthera* and *Vanda* are propagated by **stem cuttings** (mostly with 40-50 cm long top cuttings with at least two well developed aerial roots are ideal but smaller cuttings of 10-12 cm length with 2-3 nodes will also serve the purpose though flowering will be considerably delayed). **Flower stalks cuttings** (flowered spikes of certain genera like *Calanthe, Phaius, Phalaenopsis* and *Thunia* sometimes produce vegetative shoots) from the flower stalks can be taken and planted horizontally in moist media (sphagnum moss or coconut husk bits). In *Phaius*, the new shoots can easily be induced by placing the floral stalks on moist sand bed (Das and Bhattacharjee, 2006). Mukherjee (1979) successfully raised shoots through *Phaius* floral stalks, and through use of growth regulators on the cuttings, Bhattacharjee (1995) stimulated rooting. **Layering** (*Vanda* and certain other monopodials) by striking a slanting cut some 20-30 cm below the apex of the stem and wrapping the wound with some moist medium akin to other layers and these should be separated and planted when these strike roots. *Anoectochilus* is though sympodial orchid but also responds well to cuttings than any other method of vegetative propagation. The genera like *Aerides, Arachnis, Vanda*, etc. can directly be planted in pots for rooting though nobile type *Dendrobium* and *Phalaenopsis* require special care and therefore should be potted in propagation beds. To get uniform plants without any genetic variation, the propagation through cuttings is still being followed by many to maintain true to type plants, and there are certain orchid genera such as *Anaectochilus* that respond well only to propagation through cuttings (Foja Singh, 1986). The rooting in cuttings can also be promoted through the use of different concentrations of IBA, NAA and IAA. Patil (2003) studied the 3-node stem cuttings of *Dendrobium chrysotoxum, D. crepidetum, D. densiflorum, D. farmeri, D. moschatum, D. nobile, D. pierardii* and *Epidendrum radicans* by using 6,000 ppm IBA drenching in a 1:1:1 rooting medium containing brick pieces, coal pieces and sphagnum moss, and obtained 50 per cent success in *D. nobile*, followed by 28 per cent in *D. pierardii* and 4 per cent in *D. chrysotoxum*, and where IBA enhanced earlier and twice the normal rate of sprouting. Barman and Rajni (2003) studied IBA (200-1,000 mg/l) on the rooting of young, 1-year old and 2-year old keikis of *Arundina graminifolia* where response in 1-year old keiki was found better than others, however, every increase in IBA concentration increased the rooting percentage, root number, root length and root diameter.

Most sympodial orchids such as *Cattleya, Coelogyne, Cymbidium, Dendrobium, Epidendrum, Laelia, Oncidium, Paphiopedilum, Zygopetalum*, etc. are propagated through **division** (division of large clumps into smaller units) where without injuring the roots the rhizomes of well established plants are halved, each half having 4-5 shoots (growing points) or in many pieces, each having 2-3 pseudobulbs but for a large number of plants even single pseudobulb sections on the rhizomes may be made from where the older ones may produce new growth, however, bigger the size of the division earlier is the establishment, *vis-a-vis* flowering. Vij and Sembi (2001) stated that *Cymbidium* and *Cattleya* will not flower if a clump has a less than 4 pseudobulbs. The division with some new growth having a portion of backbulb will establish with ease due to supply of food. **Offshoots** (**offsets,** or **keikis** meaning babies) as in case of *Dendrobium* and *Pleione* which produce small bulbils or offsets, which when detached and potted separately flower within 2-3 years but when these are removed from the mother plant after these have once flowered there these respond well. These are produced from the nodal region of the stem or inflorescence axis. These in fact are tiny plantlets which are not desired as these suppress flower production, show that plants are not healthy and also show that plants have become old. One or two sprays of 750 or 1,000 ppm PBA in *Ascocendra* induced keikis (Kunisaki, 1975). *Ex vitro* exogenous application of 50 mg/ml transcinnamic acid and 5 mg/ml BA to nodal inflorescence buds stimulated on an average 4 shoots per bud in *Phalaenopsis* (Griesbach, 1984). Vij and Sembi (2001) stated that formation of keikis in some *Phalaenopsis* species can be encouraged through wrapping moist moss on the basal nodes, and/or through use of some growth regulators. **Backbulbs** as in case of sympodial orchids such as *Cattleya* (the physiologically lesser active older shoots or canes of tall and climbing orchids which are sometimes devoid of foliage and are green in colour) which do not have visible eyes and may have only a few roots, are separated (lengths being 10-15 cm) and planted horizontally in moist medium and when these sprout from the nodal region and send out roots, the plantlets are separated and planted in individual pots. Delizo (1973) reported 71 and 86 per cent rooting in *Cattleya* hybrid under 60 and 90 ppm NAA pseudobulb dipping treatments, respectively. Nagabhushana (1982) reported best rooting in *Dendrobium aggregatum* after treating the pseudobulbs with IBA at 1,000 or 2,000 ppm. Through injecting BA 10^{-3} M + GA 10^{-4} M in excised mature pseudobulb of *Dendrobium*, Goh (1984) induced as many as seven buds which produced roots soon from below the point of injection. *Dendrobium* pseudobulb preplanting soaking in 250 mg/l IBA promoted better shoot and root growth (Barman *et al.*, 2001). Wadasinghe and Hew (1995) with *Dendrobium* found that leaves on back shoots supplied significantly more ^{14}C assimilates to the shoot bearing inflorescence than to other plant parts within the back shoot itself. The backbulbs are normally planted flat (horizontally) on sand or moss but Das *et al.* (2001a, b) with *Cymbidium traceyanum* backbulbs reported that though sand is good medium for quick sprouting but after development of shoots the backbulbs should be shifted to medium containing FYM for nutritional support, however, placing of the backbulbs in plastic bags with moist sphagnum moss at the base, sealed or loosely tied to maintain proper supply of CO_2 has been found a better proposition. **Offshoots** are also produced on the main stem often in certain monopodial orchids such as *Ascocenda, Phaius* and *Phalaenopsis* if their inflorescences are cut off and laid horizontally in moist sand and sphagnum medium or when their apices have lost the capability of suppressing the emergence of axillary buds. *Cymbidium* and *Paphiopedilum* also produce offshoots. Use of cytokinins can induce keikis even in dormant buds. *Paphiopedilum*, a difficult to propagate genus, though does not respond much to tissue culture technique, but can be forced to produce more number of lateral buds through single application of BA at 1 mg/litre (Foja Singh, 1986). Most of the dendrobiums produce **aerial shoots** or bulbs on old backbulbs devoid of leaves, normally at the upper part of the backbulb growing slowly and take about 90-120 days to develop roots so at this stage these can be detached with a portion of backbulb and planted as separate plants. Abraham and Vatsala (1981) also reported that in genera like *Goodyera*, the rhizomes produce special lateral branches in the form of aerial shoots which send out their own roots to establish themselves to grow independent of the mother plants. They further mentioned that tubers in genera like *Nervillia* and *Peristylus* produce roots from above the tubers, each forming a tubercle which produces a new plant after one year. Even through **air layering**, *Vanda* and many other monopodial orchids can be propagated. Monopodial orchids due to their indeterminate growth habits grow very tall producing aerial roots along the stem where apical region bearing 2-3 aerial roots can be severed and planted in the growing media. Even if there is no root, a cut through the stem below 20-25 cm of the apex may be made and wrapped with moist sphagnum moss and this wrapped part is kept moist throughout. Whenever the roots are out these can be detached and planted as a separate plant. Removal of apical region does not affect the basal part and instead it produces a new shoot as usual. This method allows making only a few plants. However, a partially slanting cut through the stem 20-30 cm below the apex and then wrapping this part with moist sphagnum moss or osmunda fibre which will induce rooting after sometime when these should be removed and planted in the medium (Bhattacharjee and Das, 1975). Incorporation of rooting hormones will encourage root formation early and for this Goh (1983) recommended NAA 10^{-4} M at third node or TIBA 10^{-4} M at seventh node to initiate roots within two weeks in *Aranda*.

In every case, the wounds are treated with effective fungicides if not separated, and are planted under aseptic conditions.

Commercially, orchids are propagated either through asexual mericlonal technique or through sexual methods (seeds are devoid of endosperm which store the growth promoting factors, therefore these do not grow like seeds of other plants). Orchids are the first plants which used to be propagated through tissue culture on commercial scale (Griesbach, 1986; Goh, 1990) and this meristem culture technique was initially developed by George M. Morel (1960). Through this technique up to 2,00,000 plants can be regenerated within a year (Vij and Sharma, 1996). The most recent discovery in this field is thin cell layer culture (Prakash *et al.,* 1995, 1996) through which innumerable number of plants can be regenerated without any damage to the mother plant. Some 69 genera and their hybrid varieties are being multiplied through tissue culture method globally. To minimize the young explant losses and improvement of growth in case of *Dendrobium* and *Phalaenopsis*, Keithly *et al.* (1991) recommended dipping of young seedlings in a 10 ppm DCPTA [2-(3, 4-dichlorophenoxy) triethylamine] solution after removing them from tissue culture chamber. Some Asian countries like Thailand, Singapore, Japan, Korea, Taiwan and Malaysia multiply some 32.0, 2.0, 2.5, 2.5, 1.5 and 1.7 million tissue cultured plants annually, respectively. The shoot tips of *Aranda, Arachnis, Cymbidium, Lycaste, Miltonia, Odontoglossum* and *Oncidium*; the shoot tips, leaf tips and axillary buds of *Cattleya*; through root tips and shoot tips in *Vanda*; and shoot and stem tips, root tips, leaf tips and inflorescence tips in case of *Phalaenopsis* are used as explants. In sympodial orchids like *Cattleya, Cymbidium* and *Dendrobium*, a young shoot arising from backbulb is most suitable as explant while in monopodial orchids such as *Aerides, Epidendrum, Phalaenopsis, Thunia* and *Vanda*, the floral stalks, the nodal sections, shoot apices and keikis are potent explants (Foja Singh, 1986). On commercial level, Knudson C (1946), Vacin and Went (1949), Murasige and Skoog (1962), Wimber (1963), Morel (1964), or Arditti (1977) media are used for orchid tissue culturing. Looking into the requirements of the individual species, the nutrients are adjusted in the medium. Whether it is asexual or sexual propagation method, aseptic conditions are required under both the circumstances. Their growth may be promoted by use of certain hormones such as DCPTA [2-(3, 4-dichlorophenoxy) triethylamine], 2, 4-D, IBA, NAA, IAA and 2, 4, 5-T; and cytokinins such as BA, kinetin and 2iP. The concentration differs from species to species, however, the pH of the medium should be slightly acidic *i.e.* 5.8. Lakshmi Devi (1992) with *Dendrobium fimbriatum, D. moschatum* and *D. nobile* axillary buds in April for maximum survival percentage found ideal surface sterilization using 0.1 per cent mercuric chloride for 10 minutes; for *D. moschatum* MS medium and for *D. fimbriatum* & *D. nobile* VW medium each with NAA 1.5 ppm + BA 1.0 ppm were found to be effective for early bud initiation and elongation, and for all the three species studied, VW medium with NAA 2 ppm + BA 3 ppm gave maximum number of shoots, and with addition of CW 15 per cent further increase in the shoots was noted. She found best *in vitro* rooting in full strength MS medium with IBA 4 ppm which was further boosted by addition of 1.5 per cent sucrose and 0.10 per cent AC. Of the various explants tried for callus mediated somatic organogenesis, root segments (aerial or from culture) initiated best callusing. Out of half strength MS and UW media tried, Sudeep (1994) recorded ½ strength MS medium best when added with casein hydrolysate for multiple shoot and leaf production and addition of coconut water further boosted this in both the media, however, in UW medium even shoot length and leaf number were also significantly influenced. Maximum survival of the *in vitro* plantlets in open in coconut husk potting medium, best multiple shoot production occurring in UW medium with 40 ppm peptone + 2.0 ppm NAA + 5.0 ppm BA. Jyothi Bhaskar (1996) for tissue-culturing with field-grown *Phalaenopsis* obtained success with node and stalk tip of inflorescence and the pollinia, and with plantlets grown *in vitro* through apical bud, shoot node, basal portion and plantlet leaf and root, and she recorded maximum survival of nodal explants with 0.01 per cent mercuric chloride for 30 minutes and 0.01 per cent streptomycin + penicillin for 90 minutes. Half-strength MS liquid medium with 5 ppm BA + 2 ppm NAA + 2 ppm 2, 4-D + 15 per cent CW recorded minimum number of days for nodal swelling and bud development. Sucrose 1.5 per cent or thiamine-HCl at 20 ppm in the medium produced maximum shoot and leaf numbers after 8 weeks of culturing. Production of shoot and leaf numbers was maximum in ¼ MS medium with BA 20 ppm + 2, 4-D 2.5 ppm, though of roots was at BA 20 ppm + 2, 4-D 5 ppm, however, the combined effect of BA 5 ppm +NAA 2 ppm + 2, 4-D 2 ppm recorded maximum number of shoots and leaves. Callusing time in pollinia was minimum (2 days) in ½ MS medium containing 3 per cent sucrose when added with BA 5 ppm + NAA 2 ppm + 2, 4-D 2 ppm. Sherly Kuriakose (1997) tried explants of shoot apices, root and leaf segments, keikis, and inflorescence stalks in a monopodial orchid *Aranthera* 'Annie Black' and 'Sonia-17' of the sympodial orchid *Dendrobium* and found enhanced release of axillary buds only from shoot apex and that too in VW (Vacin and Went) basal medium, and maximum shoot proliferation (35.33 shoots/culture) in 'Sonia-17' in ½ MS basal medium added with BA 2.5 mg, NAA 1.0 mg, sucrose 3.0 g, boiled *cum* filtered coconut water 150.0 ml and agar 6.0 g, each per litre of medium. Direct organogenesis was obtained in 'Sonia-17' from the leaf base from culture and in 'Annie Black' in ½ MS basal medium in presence of light with BA 3.0 mg, NAA 2.0 mg, sucrose 30.0 g, coconut water 150.0 ml and agar 6.0 g, each per litre of medium, and when 'Sonia-17' was left unsubcultured for 3-4 months, *in vitro* flowering was obtained in ½ strength MS medium supplemented with BA 2.5 mg and sucrose 30.0 g, coconut water 150.0 ml and agar 6.0 g, each per litre of medium, but *in vitro* root regeneration in 'Sonia-17' could be obtained in ½ solid MS medium added with NAA 1.0 mg, sucrose 30.0 g, and agar 6.0 g, each per litre of medium in presence of light. With *Dendrobium* 'Sonia 17',

Samasya (2000) obtained maximum survival of the *in vitro* plantlets, *vis-à-vis* higher leaf area, root length, total fresh weight and dry weight of the plantlets along with increase in contents of total chlorophyll, chlorophyll a, Chlorophyll b and carotenoids, as well as maximum protein and carbohydrate contents when 40 g/l sucrose was added in rooting medium. Triazole treatment of plantlets during planting out produced more number of roots and leaves with higher leaf area after 30 days of treatment, especially with 5 mg/l, thereby with increased photosynthetic rates and decreased respiration rates, favoured with 50 per cent light intensity and 70 to 90 per cent RH. In *in vitro* plantlets, the stomata remained open but with less number per unit area, cuticle layer was absent and the mesophyll layers were also found less compared to normal plantlets. Sobhana (2000) found *Dendrobium* 90-110 days old green pods giving best results for micropropation, and there occurred early germination and protocorm formation, *vis-à-vis* shoot and root production in medium containing kinetin/BAP 4-8 mg l^{-1} and IBA/NAA 3-6 mg l^{-1}. She reported that callusing was favoured by NAA 4 mg l^{-1} + 2, 4-D 2 mg l^{-1} along with kinetin 2 mg l^{-1}. *In vitro* leaf culturing produced PLBs in medium containing BAP 25 mg l^{-1} + NAA 2 mg l^{-1}.

Orchid seeds are very tiny and numerous per capsule (1,300-4,000,000) but are devoid of endosperm, hence, are unable to grow on their own. Through seeds, to attain flowering stage it takes some 5 years for *Cymbidium*, 5+ years for *Cattleya, Dendrobium*, 5-6 years for *Phalaenopsis* and 6 years for *Paphiopedilum* and *Phragmipedium* (Goh and Arditti, 1985). For their germination and survival, these require the presence of a symbiotic fungus (orchid mycorrhiza, *viz. Rhizoctonia languinosa, R. mucoroides* and *R. repens* and some 50 or more others) which brings about certain chemical changes in the tissues of the seeds and supplies some nutrients including simple sugars for the germination after the protocorm stage. These symbiotic fungi were first reported by Frank, the German botanist in 1885, and Noel Bernard, a French biologist in 1889.The independent studies of Bernard and Burgeff published separately in the form of various papers in the first decade of 19th century on the association of mycorrhizae fungi with germination of orchid seeds, laid the foundation stone for orchid breeding and the germination of orchid seeds. Knudson in 1922 developed an *in vitro* medium (various mineral salts) in place of the fungi where he substituted the starch with simple sugars, and this medium was later improved by Knudson himself which proved a boon for germinating seeds of various kinds of orchids. High germination rates, sometimes even up to 100 per cent, are obtained by sowing the seeds in flasks of agar and mineral nutrients under sterile conditions. Seeds are also sown by sprinkling them around the base of a mature orchid plant of the same species or the allied ones where this root fungus naturally occurs. When these germinate and reach the stage of 3-4 leaves, these are gently shifted singly in 5-cm pot filled with well-chopped compost. Varghese (1995) when planted out well developed seedlings in coconut husk pieces for *ex vitro* establishment after 270 days of *in vitro* planting, recorded best success. Though seeds were quite minute but while seeing microscopically, the fertile ones were browner than infertile ones.

Cultural Practices

Orchids are perennial plants of varying growth habits, *viz.* terrestrial, epiphytic (monopodial or sympodial, and even climbing types), lithophytic (epilithic), saprophytic, subterranean and semi-aquatic; sun-loving or shade-loving; pseudo-bulbous [(0.75 mm, too small) to 5.0 metres in diameter] or non-pseudo-bulbous; leafy or leafless; and temperate or sub-tropical and tropical, hence there are varying requirements of their growth factors, *viz.* growing medium, nutrition, watering, ventilation, temperature, light and humidity as defined here under.

After planting, the species and varieties in most genera attain the stage of flowering in more than a year so such plants are given proper spacing; however, a little grown (young) plant may take more than six months which can be planted at closer spacing and grown at 15-25 °C temperatures in full sunlight, though these require shading only against fierce sunlight to reduce temperatures. Young plants may be grown on capillary benches or through microtube irrigation system.

The true orchid species as well as their natural hybrids require a **temperature** regime similar to their natural habitats. Some have come from the temperate habitats, some others from sub-tropical or sub-temperate regions and the remaining ones from the tropical regions so such conditions are to be created. Therefore, orchid growers throughout the world have created such conditions by constructing **orchid houses/orchidaria.** Orchidarium can be made from the materials like split bamboo, glass, fibre glass, etc. Lath house with inclined or flat-roof is constructed with wooden posts embedded in a concrete base. The floor is made of paved brick, porous gravel, pumice, cinder or similar absorbent material to maintain high humidity inside the house. The roof is made with split bamboos or thin wooden slates but in high rainfall areas, the slate is covered with polythene. Long wooden benches should be provided in the house for displaying potted orchids but in the centre there should be water reservoir or lily pool or small tanks for water spraying and for maintaining humidity inside. Fibre glass houses are just like the lath house except the fibre glass roofing which provides uniform and diffused sunlight. The short walls are there with fixed glass windows all around. In such houses all sorts of tropical and sub-tropical orchids can be grown. Controlled glasshouses are the best as there temperature, humidity, aeration, light and nutrition can be provided as per the specific requirements of the orchid species. Precisely, controlled glasshouses can be categorized in **(i) cool, (ii) intermediate,** and **(iii) summer houses.** The **cool houses** where all the thin leaved orchids such as *Calanthe, Cymbidium, Cypripedium,* some *Miltonia, Odontoglossum, Paphiopedilum* with ordinary leaves, *Pleione* and many others collected from higher altitudes are grown, should have winter night temperature of 7-14 °C (preferably not below 12 °C) and day time temperature of 20-22 °C (preferably not to exceed 24-25 °C); **intermediate houses** where night temperature should not fall below 15 °C and the day temperature not beyond 30 °C where *Cattleya, Coelogyne, Dendrobium* species, *Eria, Laelia, Oberonia, Oncidium,* mottled-leaf *Paphiopedilum,* and several others are grown; and the **warm** or **hot houses** where night temperature should not fall below 18 °C and day temperature should approximate 25-32 °C or even more and here thick-leaved *Arachnis, Aerides, Aranda, Cymbidium* (a few tropical types), *Dendrobium* hybrids, *Mokara, Phalaenopsis, Rhynchostylis, Vanda,* etc. can be grown. Minimum winter night temperature should range from 7-14 °C, and the minimum summer night temperature from 14-22 °C. They can tolerate a good deal of additional heat as up to a rise of 11 °C temperature over the minimum given is safe for orchid growing but the temperature rising beyond this limit should be controlled through ventilation. On cold winter days (especially the temperate regions) when sunshine is to a minimum, in all the three types of greenhouses the heating system is set to create a contrast of at least 5 °C between day and night temperatures. Only if the orchids are given proper temperature range they may flower because temperature is not that crucial for growth and survival as that of flowering. In case of cymbidiums, the optimum summer temperature for flower initiation is 21 °C day and 14 °C night (Leffring, 1976), however, at the time of extension of floral spikes and floret opening, the temperature range of 10 °C to 20 °C is favourable, though below the minimum and above the maximum defined here, it may be unfavourable (Anon., 1981). Sarkar *et al.* (2009) stated that cool-growing orchids such as *Calanthe, Cymbidium, Odontoglossum, Pleione* and several other high altitude orchid species grow well at 10 °C night and around 10-12 °C day temperatures, however, *Cymbidium* requires 11 °C to 25 °C growing temperatures with a temperature difference of about 10 °C between day and night during flower bud emergence. In fact, tropical orchids such as *Arachnis, Aranda, Dendrobium, Mokara, Phalaenopsis* hybrids and *Vanda* —are comfortable at 18 °C and above night temperatures and 25 °C or above day temperatures, while *Cattleya, Coelogyne,* few *Dendrobium* species, *Oncidium, Paphiopedilum,* etc. do well at 15 °C and above night temperatures and around 30 °C day temperatures. Bhattacharjee (1979a,b,c) reported for *Vanda* that a day temperatures above 21 °C and night in the range of 15.5 21 °C, for *Oberonia* a range of 16-21 °C and for *Microstylis* which is deciduous a winter night temperature of 15.5 °C are quite favourable for flowering. In case of *Phalaenopsis,* Fouche *et al.* (1997) recorded 3.5 months delayed flowering when the plants were passed on to a 22/17 °C day/night temperatures from a 24/22 °C regime. Matsui and Yoneda (1997) in Japan when grew *Miltonia* at sea level and at 700 m above sea level, recorded 10 days earlier flowering at 700 metres height, and when prior to planting the plants were artificially chilled for 40 days in the growth chamber, they recorded accelerated growth and flowering, which suggests that through proper selection of species and cultivars, by growing these at lowland and at upland locations and with or without chilling the flowering can be altered and prolonged in *Miltonia.* Initiation of reproductive growth in *Anoectochilus formosanus* occurred when the plants were subjected to high temperature treatment (Shiau *et al.,* 1996).

Cymbidium plants at tight (puffy) bud or at first open flower stage can be cold-stored for up to 30 days at 4 °C and then brought back to growing **temperatures** with no any ill effect, and after opening of the first flower it may again be stored for up to 35 days (Dole and Wilkins, 1999). Just by selecting early, medium and late cultivars and through aforesaid storage technique, production may be generated throughout the year. *Cymbidium* usually flowers from winter to early summer. *Dendrobium* flowers from late autumn to early spring but through proper scheduling and selection of cultivars, it can also be flowered year round. *D. nobile* can be grown at or above18 °C temperature and then subjecting at 13 °C temperature and long days, it can be flowered four months after LD. Tissue cultured plants of *Phalaenopsis* flower in 18 months. This flowers in 104 days after 25/20 °C (day/night) temperature treatment for four months, followed by 28/18 °C (day/night) temperature until flowering. *Paphiopedilum* and *Phragmipedium* flower naturally during autumn. In this, flower initiation occurs within six months if grown around 13 °C temperature. When inflorescence heads the development can be hastened by forcing at 18 °C. Around 17 °C temperature and 8 h days photoperiod hasten flowering in *Cattleya.* LD delays and high light induces flowering.

Shading (50 to 70 per cent) of the greenhouse prevents too much direct **sunlight,** and controls temperature generated through direct sunrays, especially during summer. Indirect light is always ideal for orchids; therefore, these require to be grown under greenhouse conditions where light intensity akin to their requirement may be given as orchids need less light intensity. Seedlings require less **light** than adult plants. The standard commercial orchids are normally day neutral with exception of *Cattleya* (2.6-3.9 klx) and *Phalaenopsis* (1.6-1.9 klx) which require low light levels during their entire growth cycle, and where LD delays flowering and high light encourages flower induction whereas *Cymbidium* requires high light in the winter and reduced light in summer (Poole and Sealey, 1978; Sheehan and Sheehan, 1979). However, this is not true with many of the tropical orchids which respond well to small fluctuations in day length and temperature. The maximum CO_2 uptake (day and night) is found with 650 fc, with no difference between juvenile and older leaves, but in greenhouses, to control temperatures, light is reduced to 1,200-2,000 fc (Wang and Lee, 1994a). The growing of *Cymbidium* first under long days and warm conditions and then shifting to a night temperature of 10 °C make the plants generative, and maintenance of plants continuously under 16 h photoperiod and at 20 °C day and 14 °C night temperatures, all the developing buds on entire plant turn out to be inflorescences and this way it may be flowered throughout the year (Went, 1990). The growth in *Cattleya* and *Phalaenopsis* is 2-4 times more in controlled growth chambers than in greenhouses. Recommended light levels in case of *Cattleya* is 1,500-3,000 fc, in case of *Cymbidium* it is 2400-3600 fc (Yoneda and Sasaki, 1978), in case of *Dendrobium* it is 2,400-3,600 fc, in *Phalaenopsis* it is 1,200-2,000 fc, and in *Paphiopedilum* & *Phragmipedium* it is 700 fc. *Cypripedium* and *Phalaenopsis* require only 200-300 foot-candles light while genera such as *Aranda* and *Vanda* feel more comfortable under 800 foot-candles (Sarkar *et al.*, 1909). Both, the quantity and quality of light determine potential of blooming in most of the orchids and if a healthy plant is not initiating blooms it means the light received by the plant is improper. Every orchid species and natural hybrids should get optimum light conditions as to that of its natural habitat. Persisting very long days (more than 16 h) are detrimental to most of the orchids, though Perata (1981) recorded fluorescent lighting for 16 hours responding best to *Phalaenopsis* and *Wilsonara* though Keithly *et al.* (1994) advocated that *Phalaenopsis* and *Dendrobium* should be produced under 40 per cent shade cloth and a 12 h photoperiod through HID lighting. During propagation, aseptic cultural conditions are maintained where fluorescent or incandescent lamps are used in growth chambers. Sudden exposure to high light condition is also detrimental, though more light is not a problem in most cases if it is gradual. Ding *et al.* (1980) stated that shade-loving orchids such as *Dendrobium, Oncidium* and *Phalaenopsis* can not tolerate full exposure of tropical sun, and *Oncidium goldiana* when was exposed even to partial shade, recorded negative correlation. Sheron Fernandez (2001) with *Dendrobium* recorded maximum plant height, leaf production, internodeal length, spike length, spike longevity on plants, floral number per spike, vase life, total nitrogen and phosphorus content in the plants, anthocyanin content in the flowers and chlorophyll 'a' in 50 per cent double-level shading; longest spike with increased spike longevity in single-level 50 per cent shading; maximum shoot production as also to that of 25 per cent double-level shading, leaf production as to that of 50 per cent double-level shading, total chlorophyll and chlorophyll 'b' under 35 per cent double-level shading; maximum potassium content in plants in 35 per cent single-level shading; maximum shoot as well as spike production and dry matter accumulation in double-level of 25 per cent shading; and earliness in flowering and high vase life in 25 per cent single-level shading. However, Radha *et al.* (1994) found out that *Bulbophyllum nilgherrense* and *Vanda spathulata* plants naturally adapted to sunny or shade conditions, the performance of the plants under sunny condition was better though net photosynthesis became saturated in shade (low light intensity) than under sun.

Looking into the requirement of **light intensities,** particularly from summer to autumn by various orchid species, they can be put under three groups: (i) light shade (20-30 per cent of outdoor light), (ii) medium shade (15-20 per cent light), and (iii) heavy shade (10-15 per cent light). *Cattleya* and *Cymbidium* can be grown successfully under fluorescent lamps than natural or high density discharge light but *Phalaenopsis* under almost all types of fluorescent lamps. Tran-Thanh and Van (1974) reported that reduction in day length and light intensity in most orchids accelerated flowering. Krizek and Lawson (1974) reported better performance of *Cattleya* under high light intensity. In *Aerides multiflorum, Renanthera imschootiana* and *Rhynchostylis retusa*, Bose and Mukhopadhyay (1977) reported 42-49 days earlier flowering under 9 h short day condition, with increased flower number and longevity in all the three species, *vis-a-vis* spike length in *Aerides*. Long days inhibited complete flowering in *Phaius* though it (16 h) only delayed flowering by 21-30 days in *Dendrobium phalaenopsis, Phalaenopsis amabilis* and *P. schilleriana*, whereas short days of 8 h the flowering in all these three species was found advanced by 50-59 days (Bhattacharjee, 1979d). Muira (1981) recorded highest photosynthetic rate in *Cymbidium, Dendrobium* and *Oncidium* under 20-30 klx light intensity. In *Cymbidium goeringii*, the growth was poor at light intensity of 80,000 lux though 70 per cent shading which reduced the intensity to 7,000 lux, the growth of the plants under this shade responded favourably by lowering the temperatures (Cho and Kwack, 1996). Likewise, with *Dendrobium*, Rajeevan (1997) recorded better performance after 18 months through 25 per cent shading, *i.e.* at 4.5 m height and at 2.5 m height. Goh and Wan (1974) in Singapore reported that inflorescence production was dependent on length of exposure to direct sunlight, those plants exposed longer flowered more profusely while those exposed only to 4 h of direct sunlight daily did not flower. Goh (1984) reported that *Arachnis, ×Aranda, ×Aranthera, Ascocentrum, Renanthera* and *Vanda* require full sun for free-flowering whereas any shading delays or suppresses it. With *Arachnis* 'Maggie Oei Red Ribbon', Sabina George (1996) reported greater inflorescence production and vase life under 100 and 75 per cent light and the more inflorescence branching under 75 per cent light; in *Dendrobium* 'Sonia 16' the number and length of inflorescence were greater under 75 per cent light. Poole and Seeley (1977) reported fluorescent light sources better than high density discharge lamps (HID) for growth of *Cattleya, Cymbidium* and *Phalaenopsis* when spaced 7.5 cm from light, *Cattleya* and *Phalaenopsis* growing best when spaced 7.5-22.5 cm in former case and from 22.5-37.5 cm in latter case. Fluorescent or incandescent (70:30 per cent of wattage) lamps are necessarily used in aseptic culture conditions for propagation. Direct full sunlight exposure to *Bulbophyllum* plants causes sunburn injuries (Bhattacharjee, 1977a). Growth and flowering were observed adversely affected under more than 50 per cent shade in *Epidendrum radicans* (Yoneda and Sasaki, 1978). Light intensity in summer before flowering is a critical factor for flowering in *Cymbidium* (Roost, 1978).

When various orchid genera were grown in phytotron with controlled atmosphere (temperature, light and RH) and compared with those grown under protected greenhouses, Tran and Van (1974) obtained optimal flower production in *Phalaenopsis* at day/night temperatures of 24 °C/17 °C, in *Odontonia* the time period of 3-4 years for flowering could be reduced to less than one year, and at 17 °C *Miltonia* flowered maximally.

Proper **ventilation** should be ensured in the greenhouse. A rise of 11 °C temperature over the minimum given is safe for orchids. However, temperatures higher than this should be controlled through ventilating the polyhouse, even by installing an electric fan. Since epiphytes draw their nutrients from air as this consists of oxygen, nitrogen, carbon dioxide and water vapour, CO_2 being most important as food for the plants, therefore also fresh air movement in the polyhouse should be ensured. Epiphytes require constant and free movement of air in the greenhouse throughout the growing area, especially during summer months. Direct contact with too dry and warm winds even to the tropical orchids may kill them or too cold wind entering the warm house will injure the plants. Hot air contains negligible moisture so it absorbs moisture from the tender parts of the plant, and that is why during summer in northern plains of India, orchids either die or highly scorched. However, during cold weather also, the air movement is necessary to avoid *Botrytis cinerea* infection. The constant movement of the air in the orchidarium or greenhouse (cool, intermediate or hot) ensures good health.

Most terrestrial orchids can be grown in ordinary flower **pots** (porous earthen or plastic, earthen being the best) having holes on their sides and bottom but these should be clean, well-drained and crocked. In fact, earthen pots do a lot of breathing while plastic pots do not at all, so earthen pots are more suitable as roots have a much better access to the air for providing good health, though watering to the plants in the plastic pots is less frequent. Baskets for the purpose should be made of hard woods. Epiphytes feel comfortable in specially perforated pots, on pieces of tree trunk, bark, on osmunda fibre rafts, on tree-fern stem bits or in the wooden basket by fixing the plants through wire and such slabs are suspended at a suitable place. The upper portion of the pot should be filled with fresh broken bricks (5 parts small pieces) and fern fibre (1 part). The roots of the Vandaceous plants should not be buried in the compost and instead should be exposed to the air. **Potting** and **repotting** are done when the plant has overgrown its container, preferably just after the flowering or when new growth is appearing. While potting a young plant of an epiphyte, the rhizome is stationed on the drainage material and then all around it the compost is pressed using a strong but blunt stick, putting crown of the plant at the level of the pot. For repotting an established epiphyte, the plant is taken out of container, the soft and rotted compost is teased off, the dead roots are removed, the plant with intact rhizome is then held upside down and the compost is pressed in between and around to make it a ball larger than the size of the pot, and then it is squeezed and pressed firmly in the pot, and then finally some further compost is pressed in the pot to fill in the gaps. While dividing the plant, each division is treated the same way as for the whole plant. It should be done every three years.

At the start, the inflorescences are very poor so these should be pinched off to force the plants attaining good growth so that emerging inflorescences afterwards may be quite stout and of proper size. A number of **growing media** are used for orchid growing as per the specification of the individual genus or species; consistent with their place of origin but the pH of the medium should range in between 5.0-6.0 (Griesbach, 1985) and the E.C. not over 2.0. In hydroculture, some 70 per cent of ornamentals suffer due to excess pH (Molitor, 1992). The medium should provide support to the plants, should be highly porous for proper aeration, should have good drainage facilities, should have adequate water holding capacity and should be rich in nutrients so that roots may not starve. Exchange of gases in the root zone is vital for growth and survival of orchids so the media are designed to hold both, *i.e.* water and air, and in case when persisting scarcity of the one occurs the other occupies its place which becomes dangerous. If medium becomes once dry, it is difficult to rewet and at this stage if it is watered the water just runs through the medium without getting absorbed into it, and scarce results into increased salt concentration which is toxic to the roots and plants. For growing epiphytes, the medium should have adequate capacity for holding moisture but should not remain soggy and wet. Sphagnum moss or leaf mould in the plains decomposes soon and invites fungal diseases, *vis-a-vis* becomes slimy, hence is not preferred but under cooler climates sphagnum moss is preferred being acidic in nature, also as it contains certain amount of iodine which keeps the compost fresh for long period. Osmunda fern root (Post, 1949) or polypodium fern fibres alone or in combination with sphagnum moss (2:1 by volume) for improving water holding capacity is the best medium for most epiphytes as this lasts longer and decomposes slowly releasing 2-3 per cent nitrogen but due to its scarcity now root and trunk fibres of various tree fern species are being used. Apart from these, peat + dried fern + sphagnum moss, the beaten redwood/fir bark chips (Wang and Lee, 1994a) or fir bark + peat + perlite, coconut husk + pieces of bricks and charcoal, moss peat, saw dust, rice husks, peanut shells and cork are also used for both, *i.e.* epiphytes and terrestrials. Wu *et al.* (1994) did not succeed effectively when grew *Phalaenopsis* in sugarcane bagasse, however, increased use of fertilizer improved the growth significantly. Paul *et al.* (1992) reported that brick + gravel medium is best for *Dendrobium farmeri, D. fimbriatum, D. moschatum and D. nobile*, however, they further reported gravel + coconut fibre, gravel + coconut husks, brick + coconut fibre, brick + coconut husk, and charcoal + brick + gravel + coconut husk as equally good media. Kumar (1992) while studying 10 different media against 1-year old hybrid *Dendrobium* seedling, reported charcoal, followed by fern roots and rubber seed husks as the best whereas coconut husks, coconut shells, wood shavings, broken tiles, moss, gravel, and grass roots were least effective. The old coconut husks are used in the high rainfall tropics for growing *Phalaenopsis* and Vandaceous plants having no pseudobulbs, and *Phalaenopsis* plants when fixed to mango and other trees grow very well. Perlite: Metromix 700: charcoal (1:1:1) or fir bark: peat moss: perlite grade 3: perlite grade 2: vermiculite: rock wool (4:2:1:1:1:1) provide good yield in *Phalaenopsis. Cattleya* and its hybrids have been found responding well to coarse peat moss + dried and undecomposed fibre both in equal parts by volume (Davidson, 1960). Elle (1962) reported that most of the orchids can successfully be grown in 40 per cent each of pine bark pieces and sphagnum along with 20 per cent dry birch or oak leaves or that of 30 per cent each of polypodium, sphagnum and pine bark along with 10 per cent beech or oak leaves. Bhattacharjee (1977a, b) stated that *Bulbophyllum* species are comfortable on soft tree fern stem with a little sphagnum moss around the root whereas terrestrial *Coelogyne* grows best in mixture containing equal parts of sphagnum moss, leaf mould, white sand and loamy soil. Bhattacharjee (1978a, b) recommended 1:1:1 osmunda fibre + fibrous loam + sphagnum moss for majority of *Eria* species and sandy loam + leaf mould + chopped tree fern + white sand + well decomposed cow dung manure for *Calanthe*. Frei *et al.* (1975)

recommended incense cedar bark for *Encyclia tampensis* seedling. However, Poole and Sheehan (1982) suggested that both terrestrial and epiphytic orchids can be grown in 1:1 mixture of peat and perlite though reported that *Aerides midiflorum* grows well in a substrate of hardwood charcoal. Yadav and Bose (1985) found the potting medium with equal parts of charcoal, brick pieces, coconut fibre and leaf mould giving best growth and flowering in several epiphytic orchids such as *Aerides, Cattleya, Dendrobium, Vanda*, etc., and the compost containing equal parts of loamy soil, river sand, leaf mould, charcoal dust and old mortar best for improving growth and flower production in terrestrial orchids such as *Calanthe, Cymbidium, Phaius, Thunia*, etc. Boodley (1988) stated that *Cymbidium* orchids are being routinely grown in an oasis medium. Pradhan and Pradhan (1997) recommended coconut husk, polystyrene granules, broken bricks, charcoal, tree fern, sphagnum moss and cork suitable for epiphytic orchids whereas chopped sphagnum moss, garden topsoil, polystyrene granules, peat, fibrous loam, leaf mould and fine grade charcoal for terrestrial orchids. Gordon (1997) described the use of rubber pieces of old tyres as potting medium. Wilford (2000) suggested sphagnum moss for terrestrials, *Phalaenopsis* and some *Odontoglossum* hybrids for seedling orchids. Jawaharlal *et al.* (2001) while trying various media comprising of brick pieces, charcoal, coir dust, brick pieces + charcoal, brick pieces + coir dust, coir dust + charcoal, gravel pieces, coir dust + charcoal + brick pieces and tree fern bark + moss, reported the brick pieces + coir dust as the best medium for growing *Vanda rothschildiana* at Yercaud. Indhumathi *et al.* (2003) while trying various media reported that after pre-hardening the rooted plantlets when were maintained in pots containing charcoal + brick + cocopeat, the plantlet survival, plant height, leaf size and root length were observed best, but root number was found highest under charcoal only. Wu-RongZhe *et al.* (2005) recommended 3:1 perlite + rockwool for better results. *Cypripedium* and *Cymbidium* grow well when $^{1}/_{4}{}^{th}$ teaspoon of hoof and horn meal, coarse bone meal and leather meal are sprinkled over the drainage layer among the crocks. Borg (1965) stated that medium containing fine pine bark + 30 per cent coarse poultry peat or a mixture of 30 per cent oak leaves, 50 per cent bark and 20 per cent palco wool gives good results with *Cymbidium*. For *Cattleya* and *Cymbidium*, Penningsfeld (1971) suggested that plants should be grown in an inner perforated container filled with expanded clay which is stood in an outer pot containing nutrient solution to be drawn up into the clay by capillary action. For *Cymbidium*, in Holland most often only peat is used with desired nutrient status, *i.e.* 2.0:0.8:1.5 NPK at 5.5 to 6.0 pH (Timmerman, 1978). *Bulbophyllum* does best on soft tree fern stem with a little sphagnum moss around the roots. Terrestrial orchids require loamier compost *i.e.* equal parts (all by volume) coarse peat or rough leaf mould, fibrous loam, sand and moss. Terrestrial orchids such as *Calanthe, Phaius*, etc. have luxurious growth when grown in a mixture of leaf mould, chopped tree fern, white sand, sandy loam and well rotten cow dung manure or charred cow dung. *Coelogyne* can be grown successfully in a mixture containing equal parts of sphagnum moss, leaf mould, sand (white) and loamy soil. A porous medium rich in organic matter is excellent for *Habenaria*, osmunda fibre, fibrous loam and sphagnum moss in equal proportion for majority of *Eria* species, a mixture of chopped tree fern fibre, charcoal pieces and sphagnum moss for *Ione* and *Oberonia* species, osmunda fibre, sphagnum moss, charcoal and very small pieces of cork in shallow pans for *Liparis*, a mixture of equal parts of shredded tree fern fibre, leaf mould, sandy loam soil and white sand for *Microstylis*, and small brick chips, white sand, chopped tree fern fibre, leaf mould and a little quantity of charcoal and bone meal in a pot filled to its third with cork is good for *Goodyera*.

Adult orchid plants require balanced **nutrition** for about 10 months in a year except in the period when they are in new growth. Like other plants, they also require NPK and minor or trace elements (Mn, Mg, Ca, Fe, Zn, Cu and boron) for their healthy growth. Balance nutrient application promotes rapid plant growth with more number of leads and larger size of pseudobulbs, and may save one complete growing season as has been reported by Brundell (1981). Adequate fertilization is most essential in epiphytic orchids since the media in which they grow are normally devoid of nutrients. Under natural conditions, orchids draw these inorganic nutrients from the bark of the tree where they are growing, *vis-a-vis* from atmosphere and decaying organic matters (plant parts and bird droppings) but under controlled conditions these are to be supplied separately. During the period of new growths these require nitrogen rich fertilizers but not in excess to avoid salt deposition in the pots as orchids are highly sensitive to excess salt which may result in blackening of roots, burning of leaf tips and white encrustation on the media in the pots (Popenhagen, 1980). Intake of nutrients in orchids is very little through leaves, *i.e.* 1/8th as compared to roots (Benzing and Pridgeon, 1983) though contrary to this, Poole and Sheehan (1982) recommended foliar application. Hybrid plants of *Cymbidium* and *Phalaenopsis* respond well to 100 ppm N, 50-100 ppm K and 25 ppm Mg though young *Phalaenopsis* plants require heavy feeding, *i.e.* 200 ppm N through 20-20-20 fertilizer with each irrigation whereas *Cattleya* with 50 ppm of N, K and Mg and the levels of P, Ca, Fe, Mn, B, Zn, Cu and Mo are 20, 200, 3, 1, 0.25, 0.20, 0.025 and 0.0001 ppm, respectively (Poole and Sealey, 1978). Keithly *et al.* (1991) reported that *Dendrobium* and *Phalaenopsis* seedlings fed every 10 days with 50 ppm N, 4.5 ppm P and 8.2 ppm K + iron chelate and micronutrients which responded well. Sabina George (1996) recorded increased length of inflorescence, flower number per inflorescence and floral span area when *Dendrobium* 'Sonia 16' plants were fed with 500 ppm nitrogen or 400/500 ppm N, 400/500 ppm K and 500 ppm P where number of inflorescences was greater. A nutrient dose of 2.0 mg each of N, P and K from 3 to 6 months age of *ex vitro* stage; 6.0, 2.0 and 2.0 mg of N, P and K from 6 to 9 months age; 6.0 mg of N, 2.0 to 6.0 mg of P and 2.0 mg of K from 9 to 12 months of age; and 6.0 mg, 6.0 mg and 2.0 mg of N, P and K, respectively to the plants of *Dendrobium* 'Sonia 17' above one year of age are quite beneficial for increasing stem girth, plant height, leaf number and area, number of backbulbs, length of roots and flowering (Uma Maheswari, 1999). NPK 30:10:10 at 0.2 per cent twice per week on *Dendrobium* 'Sonia 17' produced superior plant height though number of shoots and leaves, and leaf area were found superior with NPK 30:10:10 at 0.2 per cent + GA_3 200 ppm twice a week (Swapna, 2000), however, NPK 30:10:10 0.2 per cent twice a week + BA 200 ppm followed by NPK 20:10:10 gave earlier flowering, *i.e.* 300 days after planting. Greencare with 13:27:27 NPK when applied at 0.2 per cent to the two *Dendrobium* varieties, *i.e.* 'Sonia 17' and 'Sonia 28', Sanjeev Nair (2001) found improved leaf number. In 'Sonia 17', the total phenolic content at shoot emergence was produced by 1:10 groundnut oil cake + 0.1 per cent Greencare + 250 mg l^{-1} BA, and during spike emergence by a combination of 0.1 per cent Greencare + 250 mg l^{-1} BA. Studies conducted using ^{32}P revealed that translocation of ^{32}P occurred from the backbulb to the younger shoots. Bhattacharjee (1978a) reported *Eria* species responding well to light doses of liquid cow dung manure at monthly intervals during active phases of growth. Application of chicken manure @ 50-100 g per plant on the ground to *Arachnis, Aranda, Aranthera, Dendrobium* and *Oncidium* provides increased inflorescence yield. In case of *Dendrobium,* neither medium nor fertilizer frequency brings the flower initiation of young pseudobulbs but flowering is hastened with increased number when grown in fir bark (small or large grades) amended with 30 per cent peat moss by volume and regular fertilizer application with 200 ppm N from an NPK fertilizer. Esser (1973) while testing various media stated that peat has capacity for high water absorption and retention, and that peat and a 2:1 mixture of dried fern + sphagnum moss has greatest nutrient salts whereas nutrient solution remains stable in peat and perlite, and addition of 100 g per plant chicken manure in case of *Derndrobium* whereas 50 g per plant for *Oncidium* also increases top shoot production, inflorescence length, flower yield and number of flowers per spray, though higher doses are sometimes harmful (Chua, 1974). Liquid manure prepared through cow dung or oil cake @ 1 per cent strength improves growth and flowering in *Aerides multiflorum* and *Dendrobium moschatum*. Application of 5 mmol K l^{-1} results in quality and quantity blooms in *Cymbidium sinense.* In fact, orchids prefer frequent weak solutions of fertilizers than the strong ones less often, very little nitrogen during flowering, no high N fertilization to seedlings and no fertilizer to sick plants. Nitrogen deficient plants remain stunted, mature earlier where older leaves turn yellow and fall off whereas excess of nitrogen accelerates excessive vegetative growth, delays flowering and causes rotting of roots and bulbils. Phosphorus deficient plants become stunted but with dark green leaves, sometimes foliage starts decaying. Potassium deficiency also causes stunting but here leaves are frequently scorched and dead on the edges. Calcium is necessary for regulation of cell activities and cell wall formation and its deficiency causes distortion of the plants, discolouration of leaves from margin to midrib due to loss of chlorophyll, purple tinting of the veinlets, mature leaves turning black and dropping off and developing pseudobulbs may also turn black though most orchids require calcium-less water spraying. Deficiency of sulphur may affect root growth. Magnesium promotes uptake of phosphorus and certain other elements whereas its deficiency causes chlorosis in between the veins of the older leaves which may cause ultimate death of the plant. Iron maintains the normal growth of the plant as it is essential for chlorophyll formation but its deficiency causes yellowing of the younger leaves. The role of boron, copper, manganese, molybdenum and zinc is not properly understood in case of orchid cultivation. Jyothi Bhaskar (1996) recommended 30:10:10 NPK at 0.5 per cent to micropropagated *Phalaenopsis* plantlets and recorded highest survival and growth after 12 weeks of planting. Sobhana (2000) found 30:10:10 NPK solution at 0.1-0.2 per cent at alternate days on *Dendrobium* hybrid seedlings giving best results in terms of number of shoots and leaves. With *Dendrobium* 'Sonia 17', when Binisha (2003) tried biofertilizers in combination with inorganic fertilizers and recorded highest plant height, number of leaves, number of leafy shoots, shoot girth and internodal length in NPK (10:5:10) along with *Azospirillum*, however, the number of pseudobulbs was found maximum under NPK (10:5:10) inoculated with *Azospirillum* and phosphobacteria at the time of planting. At the time of planting, the plants inoculated with *Azospirillum* and sprayed with NPK 20:10:10 at 0.2 per cent concentration bloomed earlier with longer spikes bearing larger and more number of flowers per spike. Root characters and dry matter production were markedly influenced due to biofertilizer treatment.

Watering of orchids depends upon the type of species, the plant size and stage (old, adult or young), the type of container [pot (wooden basket, plastic, cemented or earthen), raft or direct in the soil] and/or medium used, existing temperature, the season and the prevailing weather conditions, and the conditions from where these have been originally collected. The watering should be adequate to leach the soil mix periodically to avoid soluble salt build up and to keep the plants uniformly moist. The plants grown on a raft require the spraying of the constituent material several times a day in the growing season and watered every second day. Osmunda fibre (*Osmunda regalis* fern roots), tree-fern fibre (various species of *Dicksonia*), and sphagnum retain moisture much longer than coniferous bark, charcoal and perlite. Young plants of even same species with more delicate growth but less developed root system require watering more often than adults. Plastic pots retain moisture better and longer than the cemented, clay or wooden baskets though there is very poor root aeration in plastic pots. There should be copious atmospheric humidity and moisture in the media but these should be watered only 7-10 days after potting, and afterwards sparingly for a fortnight until the young new roots are well established. However, from April to September, *i.e.* during growing season all the plants should be kept moist if there is no rain. For a few minutes the pots of epiphytes are submerged in water every week in spring and autumn while twice a week in hot months and the terrestrials can be watered by can *in situ. Phalaenopsis* and *Paphiopedilum* are watered every 2 days, *Cymbidium* once every 3 days and *Dendrobium* every 7 days (Dole and Wilkins, 1999). Most orchids like water less in calcium, magnesium and sodium. A high chlorine level in the water hampers the functioning of the root system. However, good irrigation water is that which has 5.5 to 6.5 pH reaction (though *Cattleya* may be grown from 4 to 9 pH range) and contains soluble salts of less than 125 ppm while that containing 125 to 500 ppm is tolerable, and 500 to 800 ppm should be used with caution but above 800 ppm it is at all not suitable. Collection of water in the cities after the first rain is also ideal for orchids, however, where there is no industrial pollution it can always be collected from the rains. Drinking water is also suitable if stored in the tank for a day or so. Watering in the morning is preferred so that standing water may disappear by the evening. Water scarcity causes decreased CO_2 uptake (Goh and Arditti, 1985). *Dendrobium* and *Calanthe* are watered every seven days, *Cattleya, Oncidium* and *Odontoglossum* every five days, *Cymbidium* every three days, and *Cypripedium, Paphiopedilum* & *Phalaenopsis* every two days as these require moist medium throughout. Poole and Sealey (1977) recommended that best growth can be obtained in *Cattleya* plants when watered 5-7 times per week, *Phalaenopsis* three times a week though *Cymbidium* is not so sensitive. Syrovatka (1980) suggested use of fibre glass netting in wicks or some other form for capillary watering of pot-grown orchids. Sinclair (1990) reported that water vapour can not be a significant source of water for orchids, and the condensation of fog on the plant surface may provide adequate moisture to the epiphytes but not to terrestrial orchids. Pan *et al.* (1994) stated that *Cymbidium* plants with older leaves are less affected with moisture stress than the younger ones. Thomas (1996) recommended watering twice the plants per day during flowering as compared to other times.

A rise of temperature up to 32 °C or even more in the warmhouse during the summer growing season is safe if **humidity** in the polyhouse is around 100 per cent. Moreover, higher temperature undoubtedly encourages the growth provided commensurate humidity is present. Humidity in the orchidarium can be maintained by constructing a pool or water tank in the centre to ensure at least 30 per cent humidity in the night and 80 per cent during day time. Misting units or foggers (in the small units) and humidifiers (in the large units) will ensure adequate humidity, and the frequency of the humidifier may be timed depending upon the prevailing weather conditions. The orchids that originally come from tropical zones should in no case have less than 50 per cent humidity (preferably 50-70 per cent) in the surrounding with good air circulation to prevent fungal pathogen build up. Monopodials require higher humidity than the sympodials. In the hot climate, the plants may be watered twice a day as a rise in temperature will lower the humidity. Alternatively, the greenhouse should have provision of cooling system, cladding material made of porous articles with a closed circle supply of water fitted to one end of the greenhouse and an exhaust fan of the size commensurate to the size of the greenhouse and the area of cladding material fitted to another end. This will circulate cool and moist air throughout the greenhouse. In small greenhouses, the cooling effect may be generated through frequent water spraying (twice or thrice a day during summer while once in winter, depending on the type of houses because warm houses require more frequently than the cool houses) on the path and floor which may preferably be of clay, sand or gravel or through open vents on the shady sides.

Control of Growth and Flowering

For successful cultivation of orchids, conditions identical to their natural environment are provided. Optimum growing conditions are created only through protected structures under which unblemished flowers of high quality can be obtained. Structures protect the plants from fierce heat during summer months and from frosty winters. Such structures should have cooling facilities for summer months to maintain the maximum day temperature of 28 °C, and heating facilities during winters to maintain a minimum night temperature of 10 °C. The soil/potting mixture in the structure should be highly porous and must have proper drainage facilities as during summer it requires a lot of water and humidity for growth and survival. It would be better using containers for planting the orchids inside the structures for proper management and so low benches should be used. It is possible to produce 100 spikes m^2 in case of cymbidiums through choosing proper cultivars and by applying optimum growing environment.

The factors which influence their growth and flowering are the genotype (the major factor) and environmental conditions (the key factors) including growing media, *vis-a-vis* nutrients as have already been elaborated in the second paragraph of **Cultural Practices.** Several tropical species are free-flowering apart from having a peak flowering season which is due to their genetic makeup. *Dendrobium phalaenopsis* and its hybrids can be flowered throughout the year in the tropical areas regardless of day lengths, main flowering period being from late autumn to early spring. Many species have definite growth and flowering seasons, especially the temperate ones. *Cymbidium* requires cool nights (10-21 °C night-day temperatures) and partial shade to full sun for flower initiation in the spring. At stage 1, *i.e.* young plants should be given 15/25 °C night/day temperatures from planting to flowering, at stage 2, *i.e.* flower initiation 14/21 0C night/day temperatures, and at stage 3, *i.e.* spike development 10/20 °C night/day temperatures, though *Dendrobium* 22-30/16-22 °C day/night temperatures along with 65-80 per cent RH, 25-30 klux light intensity and good air circulation (Anon., 1997). High light intensities, *i.e.* 50 per cent shade and 15 °C day temperatures are conducive to good initiation in *Cymbidium,* though slight reduction in light intensity during flower bud development results in better flower colouration and reduces bud blasting in *Cymbidium.* Long days, high light and cool temperatures provide best flowering in *Cymbidium,* though short days result in weak growth and long days stimulate vegetative growth. To extend the flowering span beyond winter and spring in *Cymbidium,* summer temperature is controlled. *Cymbidium* hybrids flower in winter and early spring if flower initiation has occurred in the previous autumn. Mature pseudobulbs usually produce only vegetative shoots which may also produce flowers if climatic conditions are quite favourable but the young developing pseudobulbs of the current growth produce spikes. Dole and Wilkins (1999) describe that flower initiation is low temperature dependent for *Cymbidium, Dendrobium, Paphiopedilum* and *Phalaenopsis,* ranging from 5-8 °C for *Cymbidium,* 13-18 °C for Dendrobium, 13 °C for *Paphiopedilum* and 15-20 °C for *Phalaenopsis* as these temperatures are related to their native haunts. Keithly *et al.* (1991) described optimum production temperatures for young seedlings of *Dendrobium* and *Phalaenopsis* as 28

°C ± 4° day and 20 °C ± 2° night, however, Yoneda *et al.* (1991) stated that no flowering occurred either with juvenile or senile *Phalaenopsis* plants at a constant temperature of 28 °C. Wang and Lee (1994a, b) stated that *Phalaenopsis* flowers within 104 days after subjecting the plants for 4 weeks at 25/20 °C day/night temperatures, followed by 28/18 °C until flowering. Through tissue culture, flowering in *Phalaenopsis* occurs in about 18 months (Dole and Wilkins, 1999). Ota *et al.* (1991) stated that for CO_2 uptake in *Phalaenopsis*, a constant temperature of 20 °C is optimum, and after flower initiation, Wang and Lee (1994b) stated the optimum temperatures for reproductive growth and development as 25/20 °C day/night, *i.e.* for energy conservation the greenhouse temperature may be maintained at 18-28 °C day and 15-20 °C at night, however, any temperature below 25 °C day or night (15-20 °C being optimum where initiation occurs within 4-5 weeks) results in initiation, though when compared to older plants newly mature plants required lower temperatures or longer durations to become reproductive. They further described that with a day of 25 °C, each 1 °C drop in night temperature between 15 and 25 °C will delay flowering by one day. Vegetative growth in *Dendrobium* is best at 20 °C night temperature but for flower initiation it requires lower temperature of 17 °C (Powell *et al.*, 1988), and for *Cattleya* hybrids, they recommended 16-18 °C day and 16 °C night temperatures, and 26 °C as optimum day or night temperatures for maximum number of flowers per inflorescence in case of *Cymbidium* after flower initiation, however, Rotor (1952) and Goh and Arditti (1985) suggested in case of *Paphiopedilum* and *Phragmipedium* a day temperature range of 18-21 °C for active vegetative growth but afterwards lowering it to 13 °C for rapid flower induction though still these may require 6 months for initiation and development of flowers, and once the inflorescence becomes visible the development can be hastened by forcing at 18 °C (Dole and Wilkins, 1999). *Arachnis* and its hybrids require 18 °C cultural temperature and 2,400-3,600 fc light for their optimum blooming. *Oncidium* and *Paphiopedilum* for their optimum flowering require 2,400-3,600 fc light, and 15-21 °C and 18-21 °C growing temperatures, respectively. *Paphiopedilum* hybrids generated through extensive hybridization with other genera flower throughout the year though the flowering season of this species is winter and spring. By proper selection of the cultivars, flowering may be obtained for six months, *i.e.* from November to May or even longer. Short days and low temperatures in temperate areas induce flowering in some species of *Cattleya, Dendrobium* and *Phalaenopsis.* In summer, *Dendrobium crumenatum* and *D. flavellum* come in spectacular bloom after two weeks following low temperature and heavy rains. *Cattleya* hybrids also produce abundance of blooms under short normal day lengths. *Vanda* and its hybrids bloom profusely under full sun (at least 10 h of full sunlight). It is possible to have flowers all the year round just by selection of species, hybrids and the cultivars as in case of *Cattleya, Phalaenopsis amabilis* & its white hybrids. However, *Vanda. Phalaenopsis amabilis* and its hybrids can give 3-4 crops per plant in a year, primary spike being produced in the fall which may last for over two months and after the harvesting of the flower the top of the spike is removed from below the first flower bud and so within eight weeks secondary spike will appear, and likewise when this spike is also removed from below the first flower, tertiary spike will also appear.

The variation in the juvenile phase of orchids is enormous among the species as well as cultivars, *e.g.* some of the *Phalaenopsis* hybrid seedlings of the same cross flower within 15 months while certain others require two years or more. Normally, it takes 4-7 years for flowering from seeds for many orchids. After passing of the juvenile period, two primary environmental factors, *viz.* low temperature and light (photoperiod and light intensity) control the flowering as is elaborated under **temperature** and **light**. Almost standard commercial orchids are day neutral except *Cattleya* which responds well under SD, however, many of the tropical orchids found close to the equator react even to small changes in day length and also to temperature (Dole and Wilkins, 1999).

Vanda 'Miss Jonquim' shows inverse correlation of the auxin level in the shoot apex at the time of flowering, more the flowering less is the auxin level, however, at the vegetative phase the auxin level is found increased. *Phalaenopsis schilleriana* is poor in flowering when sprayed with auxin solutions due to excess of it. Induced flowering is obtained in *Aranda* 'Deborah' when treated with anti-auxins and growth retardants or to decapitation but the effect is negated through continuous supply of 10^{-4}M IAA solution. Benzyl adenine causes multiple inflorescences in *Aranda. Cymbidium* responds well to GA_3 treatment from 1,000-10,000 ppm at an interval of 2, 4 or 6 weeks where flower size and spike length are found increased. Similar to the forced *Cymbidium* plants through light and low temperature treatment, BA 400 ppm also prevents decrease and increases the flower number in non-forced plants due to increase in pseudobulbs, as has also been reported with *Dendrobium* ('Lady Hochoy' and 'Buddy Shepler' × ' Peggy Shaw') by Goh and Yang (1978) when they treated mature pseudobulbs with BA, however, though alone GA was ineffective but together with BA it enhanced the effect whereas IAA suppressed the promotive effect of BA. They further stated that when mature pseudobulb was decapitated, a new basal pseudobulb was found formed but this did not have any effect on flowering. Swapna (2000) with *Dendrobium* 'Sonia 17' recorded maximum number of spikes per year, *i.e.* 12.98, and floret number per spike, *i.e.* 10.50 through use of BA 100 ppm + NPK 10:20:10 at 0.2 per cent twice a week; and significantly supereior spike length, i.e 55.40 cm through GA_3 20 ppm + NPK 10:20:20 at 0.2 per cent twice a week. Zheng *et al.* (1998) through application of 5 to 20 mg of ABA and GA_3 obtained root growth which in turn promoted shoot development in three *Cymbidium* cultivars during summer season. Goh (1979) obtained flowering only when vegetative growth on the pseudobulb of *Dendrobium lousiae* cv. 'Dark' terminated, when the plants were placed under shade intermediate shoots with both vegetative and reproductive growth appeared, decapitation did not affect floral initiation, and in mature pseudobulbs BA stimulated flowering, GA though itself was not effective but enhanced the effect of BA. Goh (1984) when injected BA 10^{-3}M+GA 10^{-4}M in the excised mature pseudobulb of *Dendrobium* 'Mary Mak', more (7) bulbs were formed while in BA alone it was normal. In *Dendrobium* hybrids, 4,000 ppm BA treatment increases the flower number, and also improves the decreased flower number caused by forcing through temperature and lighting. Spraying thrice with tricantanol at 0.01 g/900cm^2 on *Dendrobium* seedlings induces earlier flowering, *i.e.* 18 months after spraying, and 0.01 mg or 0.05 mg/900cm^2 spraying four times induces flowering after 21 months. The roots of *Dendrobium nobile* when soaked for two hours with various concentrations of growth regulators, 6-BA, NAA and paclobutrazol, especially with 10 mg/litre BA increased the tillering rate though it was found inhibited by GA but it promoted bud development (Xia *et al.*, 1999). Fortnightly spraying with BA at 50 mg l^{-1} to *Dendrobium* seedlings produced maximum number of shoots and leaves, whereas GA 10 mg l^{-1} provided maximum plant height (Sobhana, 2000). Ethrel spraying at 1,000-4,000 ppm once to thrice in *Oncidium* 'Golden Shower' promotes flower induction. Peres *et al.* (1999) when studied the stimulatory effect of exogenous cytokinins on root-to-bud conversion in *Catasitum fimbriatum*, they recorded that it was probably mediated through establishment of an endogenous auxin: cytokinin balance favourable to shoot formation.

Harvesting of Flowers and Postharvest Management

Proper harvesting stage for majority of the commercially important orchids is fully opened and mature flowers. Mostly orchid flowers do not open properly even in floral preservatives if harvested early. *Cattleya, Cymbidium, Paphiopedilum* and *Phalaenopsis* mature only after opening of 3-4 flowers, therefore these should not be **harvested** until 3-4 days of their opening but *Dendrobium* may be harvested 1-2 days before full opening. On the inflorescence of *Phalaenopsis*, the longevity of individual flowers on the plant is from 30 days (*P. violacea*) to 120 days (*P. amabilis*) (Goh and Arditti, 1985) and for cut flower use either individual flower or the entire inflorescence is harvested (Dole and Wilkins, 1999). Rampal and Dayamma (2001) recommended harvesting of *Cattleya* inflorescences 3-5 days after bud split, pulsing of all types of orchids at 20-27 °C and about 20,000 lux light intensity, packaging under cool atmosphere and use of ethylene scrubbers in the boxes. Pollinia removal will adversely affect the life of the concerned flower as well as of the others on the inflorescence as ethylene production increases (Porat *et al.*, 1995; Suh *et al.*, 1998) and orchid cut flowers are highly susceptible to ethylene even to 3 ppm at 20 °C (Woltering, 1984) for one day. Insect-pest damage, mechanical injury on the inflorescence and/or night chilling increases ethylene production (Gregg, 1984). Sarkar *et al.* (2009) stated that *Dendrobium* should be harvested at 75 per cent bloom stage and *Cattleya* 3-5 days after bud split. In spray-type orchids, each flower on the inflorescence opens 1.5-2.0 days apart so when three or more are open these may be cut and taken dry to the market. However, the spikes with all the flowers open may also be harvested preferably in the evening, only when the flowers are required, so that these may give full instant display, suitable for marriages and other parties or occasions and moreover as also the flowers are long-lived on the plants themselves so one should not get panicky in getting them harvested when these are not required. While harvesting, only a few centimeters of stem length should be left on the plant and the wounded part should be treated with some contact fungicide to prevent possible infection of the pathogens through the wounds. The yield of inflorescence varies from genus to genus, species to species and variety to variety. On an average, a commercial *Dendrobium* variety produces 6-8 spikes while 'Sonia 17' even more under Kerala conditions though during monsoon, the spike production in most varieties is lessened.

Immediately after cutting the cut ends are held in salt-less water and taken to the processing room where these are graded and packed with peduncles inserted into a tube (plastic or rubber) of water or wrapped around with a little moist cotton and then taped to the bottom of the box. Shredded wax paper should be placed around the flowers to protect sepals and petals from physical injury during transit, carried preferably in a chamber having above 10 °C temperature as majority of the orchids suffer from chilling injury below 5 °C, though *Cymbidium* and *Paphiopedilum* can tolerate even a storage temperature of -0.5 °C. *Cymbidium* flowers in small tubes are **packed** 6, 8, or 12 in glassine-fronted boxes, for instant corsages. Japan demands relatively short spikes with not more than 10 flowers where miniature cymbidiums (polymins) work though in USA individual flowers are sold so wholesalers purchase only large spikes to remove individual florets for prepackaging or for corsages (Salinger, 1985). The spikes of cymbidiums should conform to the specifications of straightness, all florets fully open with firm sepals and petals, flowers appearing circular in outline and clear in colouration, blemishfree, and though not always possible but preferred when facing one way, and when longevity of whole spike is more than 2 to 4 weeks. Some 100 spikes of *Cymbidium* are packaged in a box and *Cattleya* spikes are also packaged in standard florist boxes. Four dozen sprays are packaged in a standard box (size 75 × 25 × 17.5 cm) in case of *Dendrobium.* The containers should be perforated so that ethylene build up due to removal of pollinia in certain orchids, especially from *Vanda* 'Miss Joaquim' may be avoided as ethylene bleaches the flowers of 'Miss Joaquim' from lavender to dirty white. Ethylene scrubbers with $KMnO_4$ or Purafil (activated alumina pellets) should be put in the packaging box (Sarkar *et al.*, 2009) and the boxes should have perforations all around for proper ventilation during transit. **Storage** temperature of -1 °C is highly detrimental as within three days flowers start turning brown. However, most orchids can be stored safely at 7-10 °C for 3-4 weeks (Rampal and Dayamma, 2001). During transportation, the packed flowers are kept in refrigerated vans or trucks, the temperature being akin to the requirement of the type. After receipt of the consignments the spikes are taken out and their cut ends are held in water but only after removing 0.75 cm of the stem. This practice should be repeated after every 2-3 days for better absorption. The longevity differs from genus to genus, species to species and variety to variety. *Cattleya* inflorescence may last in the field from 9-60 days (normally 3-4 weeks) and the flowers and plants can be stored from 10-16 °C, *Cymbidium* from 15-90 days at 10 °C though flowers are tolerant to -1 to 4 °C temperatures but are highly sensitive to ethylene whose action may be negated through STS or GA_3, *Dendrobium* from 19-55 days at 10-13 °C with no ethephon problem but microbial blockage of xylem is the major cause of poor vase life though just after cutting the dipping of the cut ends in hot water with a floral preservative and anti-microbial chemical is effective, and *Paphiopedilum* and *Phalaenopsis* from 30-120 days though flowers are highly sensitive to ethylene. In case of *Paphiopedilum*, the flowers on potted plants can last for 90 days and the flowers can be stored up to three weeks in water at -1 to 4 °C. *Phalaenopsis* can be stored for 14 days, the vegetative plants at 25 °C and the flowers at 7-10 °C. Being highly sensitive to ethylene, just after cutting, the cut ends should be dipped for 10 minutes in STS at 1,000 ppm to negate the effect of ethylene. Klougart (1984) recommended the dipping of cut ends in STS for one hour to increase the vase life of *Paphiopedilum* cut flowers; however, spraying or whole flower dipping in STS will also improve the life. Bhattacharjee (1995) suggested placing of the cut flowers in 250 ppm boric acid or standard HQC solution. Ketsa *et al.* (1995a, b) reported higher water uptake when inflorescences had open and unopen buds, and when they kept *Dendrobium* inflorescences in HQS + $AgNO_3$ + glucose, they obtained maximum vase life of 51.51 days with 89.46 per cent opening compared with only 7.25 days in distilled water with 18.82 per cent opening. *Dendrobium* 'Sonia 17' spikes were when pulsed in HQ 500 ppm + sucrose 5 per cent for 6 h, there was minimum loss of weight and pulsing for 12 h prolonged the life of first floret considerably, *i.e.* 19.92 days though last floret wilted in 25.00 days; however, holding treatment with $AgNO_3$ 50 ppm + HQ 400 ppm + sucrose 5 per cent recorded maximum vase life of 24.73 days (Swapna, 2000). Vase life in *Cymbidium* is not limited to problems related to water uptake as this contains almost no stomata so transpiration is very low, however, in *Phalaenopsis* the vase life is low because of vascular occlusion due to exudation of mucilage at the cut end in the vase water (Doorn *et al.*, 1999). Jomy *et al.* (2000) studied the effect of holding solutions on pulsed and packed inflorescences of *Dendrobium* 'Sonia' and recorded highest vase life of 22 days with 2 per cent sucrose +400 mg 8-HQ + 30 mg $AgNO_3$/litre compared to only 8.33 days in tap water. Ketsa and Kosonmethakul (2001) reported $Al_2(SO_4)_3$ 50 mg/l and $CoCl_2$ 200 mg/l as holding solution effectively increased the vase life and bud opening in *Dendrobium* 'Sonia Bom'. Ketsa *et al.* (2001) recorded 225 mg/l 8-HQS, 30 mg/l $AgNO_3$ and 4 per cent glucose solution increasing bud opening and delaying wilting of the open florets in *Dendrobium* cv. Caesar. Pulsing of *Dendrobium* cv. Sonia from 2-6 per cent sucrose + 200-600 ppm 8-HQ exhibited significantly positive effect, however, it was 4 per cent sucrose + 400 ppm 8-HQ which exhibited highest vase life of 21.33 days (Jomy and Sabina, 2002). With regard to *Dendrobium* 'Sonia-17' inflorescences, Jawaharlal *et al.* (2002) when used holding solutions of BA 25 ppm, BA 25 ppm + Bavistin 0.2 per cent, 8-HQ 100 ppm + sucrose 2 per cent, STS 1.0 mM + sucrose 2 per cent, $NiCl_2$ 300 ppm + sucrose 2 per cent, acetylsalicylic acid 100 ppm, coconut water 50 per cent and distilled water, it was 8-HQ 100 ppm + sucrose 2 per cent which gave best results. Likewise, Dineshbabu *et al.* (2002) with *Dendrobium* cv. Sonia-17 also recorded best life of 30.66 days only with HQ treatment. Rattanawisalanon *et al.* (2003) with inflorescences of *Dendrobium* 'Jew Yuay Tew' recorded no change of vase life when aminooxyacetic acid (AOA), glucose or sucrose alone was used; however, at pH 3.0-3.2 when AOA 0.5 mM + sucrose 4 per cent were added, the life with better quality was obtained. At consumer level, immediately after receiving the consignments, the tubes of individual flowers are removed, 0.75 cm of the peduncle is cut off and the stem is put in palatable water with preservative. Spray orchid's cut ends are removed at 2.5 cm upon arrival and placed at 38 °C warm water with a floral preservative and then hardened off at 5 °C. For display, sprays should be kept in water with a preservative for enhancing their shelf life. 8-HQC (100-200 ppm) with sucrose (3-4 per cent) and boric acid (0.1 per cent) in the holding water improves the life.

Insect-Pests, Diseases and Physiological Disorders

If proper sanitation and care have been taken, there is little scope for the pests to build up in the orchidarium. Environmental conditions suitable for orchid growing though are themselves congenial for harbouring various insects and diseases. Therefore it is essential to regularly monitor orchid planting for any sign of **insect-pest** infestation and **disease** infection.

Insect-pests attacking orchids are **aphids** (*Cerataphis lataniae, Macrosiphum luteum, Aphis gossypii, Neomyzus circumflexus, Myzus persicae*), **thrips** (*Chaetanaphothrips orchidii, Dichromothrips corbetti, D. fenestratus, D. spadix, Frankliniella occidentalis, F. schultzei, Haplothrips* sp., *Microcephalothrips abdominalis, Thrips palmi, T. simplex, T. sumatrensis, T. tabaci*, etc.), **mites** (*Phalaenopsis* mite *i.e. Teniupalpus pacificus*, two-spotted mite *i.e. Tetranychus urticae* and others *viz. Tyrophagus longior, T. neiswanderi, T. curvipenis, Amblyseius longispinosus* and *Breialpus phoenicts*), **scales** (*Asterolecanium russellae*, common armored scale *i.e. Diaspis boisduvalii*, orchid scale *i.e. Furcaspis biformis* or *Conchaspis angraecum, Vanda* orchid scale *i.e. Genaparlatoria pseudaspidiotus*, and brown soft scales *i.e. Coccus hesperidum*), **bugs** (*Tenthecoris bicolor, T. figueiredoi, T. orchidearum, T. vestitus*, etc.), **mealy bugs** (long-tailed mealy bug *i.e. Pseudococcus adonidum*, Baker mealy bug *i.e. P. maritimus, Ferrisia virgata, Nipaecoccus nippae* and *Planococcus citri*), **cattleya fly** (a tiny wasp, *Eurytoma orchidearum*), **cattleya midge** (*Parallelodiplosis cattleyae*), **root miner** (*Agromyza orchidearum*), **black twig borer** (*Xylosandrus compactus*), **orchid weevils** (*Orchidophilus aterrimus*), **orchid beetle** (*Stethopachys formosa* and *Lema pectoralis*), **grasshoppers, ants, cockroaches** (*Periplanata* spp.), **nematodes** (*Aphelenchoides aligarhiensis, A. besseyi, A. composticola, Helicotylenchus pseudorobustus*), **slugs** (*Slytommatophora* sp., *Deroceras laeve*), **snails** (*Achatina fullica* and *Zonitoides nitidus*), and rats. A *Cymbidium* hybrid in a farmer's field in Sikkim exhibited severe necrosis, swelling and fluffy root system, leaf bending from the base, its curling, twisting and unusual enlargement which was due to infestation of a nematode *Helicotylenchus microcephalus* (Anon., 2009). Sajitha Kumari (1998) through a survey in Kerala recorded grasshopper (*Oxya chinensis*), spiraling whitefly (*Aleurodicus disperses*) and Bihar hairy caterpillar (*Diacrisia obliqua*) from the leaves of *Spathoglottis* spp., the larvae and adult of thrips (*Megalurothrips distalis*) on the buds and flowers of *Dendrobium* spp. and *Spathoglottis* spp. causing their distortion, tobacco caterpillar (*Spodoptera litura*) on the flowers of *Spathoglottis* and *Dendrobium* spp., the ant (*Monomorium indicum*) causing root damage of orchids, *Lema* spp. on *Spathoglottis* flowers, banded blister beetle (*Mylabris pustulata*) sometimes infesting on *Spathoglottis* spp., sowbug (*Oniscus asellus*), the grubs and adults of an unidentified curculionid damaging the pseudobulbs of *Dendrobium* spp., land snail (*Ariophanta* sp.), black slug (*Arion* sp.), and grey slug (*Limax* sp.).

Ants transport aphids and scale insects, and these can be controlled through use of chlorpyrifos in the burrows. **Grasshoppers** feed on foliage and flowers which can be controlled through sanitation and through spraying with carbamate or organophosphate. **Rats** cause damage to ground (terrestrial) orchids by eating on foliage, rhizomes and pseudobulbs, and by burrowing the plantings. It can be controlled through poison baits. Attack of **slugs** and **snails** is cumbersome for outdoor plantings during night and these create holes in stems, leaves and spikes and also eat on root tips, and through their movement these leave silvery trails. Its menace can be controlled by spreading 5 per cent metaldehyde bait and metaldehyde mixed carbaryl bait (2.5-2.5 per cent), methylated pellets around the orchid racks or through chemicals like Slugit, methiocarb and aluminium sulphate. Phorate 1 g treatment is also quite effective in controlling the slugs. These can also be trapped and killed during night. **Aphids** (nymphs and adults) suck the sap of the tender parts, transmit viruses and secrete honeydew which attracts ants and fungus. These should be destroyed before a colony develops, by spraying with any insecticide, *viz.* dimethoate or Malathion. **Thrips** (adults and nymphs) lay their eggs in the tissues of the petals, nymphs and adults feed on buds, flower parts and other tender parts which cause flower withering. This can be controlled by spraying with abamectin 0.15 EC, aldicarb, carbaryl, carbofuran, dimethoate, Metasystox, etc. **Scales** are small and brown and suck the sap from the leaves, stems and roots and induce intoxicants which hamper the plant vigour. These can be controlled through scrubbing with soft brush or by wiping with cotton swabs soaked in methylated spirit, and through organophosphate spraying. **Bugs** suck sap and cause pale spots on the leaves. **Mealy bugs** have soft and filamentous body coated with white powdery wax and they look pink to yellow without coating. Mealy bugs are sucking pests, secrete honeydew, attract ants, and sooty mould may develop on the leaves in their heavy infestation. These can be controlled by spraying with 0.02 per cent Parathion. **Mites** also suck sap from leaves; produce fine webs and silvery marks (which afterwards turn brown or black) on underside of the leaves and cause flower damage due to purple brown spots, size reduction and sometimes twisting of petals. Azocyclotin, bromopropylate, cyhexatin, dicofol, ethion, fluvalinate and propargite acaricides can control this pest. *Phytoseiulus persimilis* may also be used to predate upon them. **Black twig borer** feeds on plant twigs which can be controlled with chlorpyrifos at preliminary stage. **Orchid weevil** punctures flower buds, forms white streaks there and lays eggs on depressions caused by adult feeding, and the larvae mine the pseudobulb and pupate there. This can be controlled through chlorpyrifos or methyl parathion at 60 g/100 litres of water or acephate or bendiocarb at 1.2 g/l water. Adults of **orchid beetle** feed on leaves, pedicels, cut the flowers from peduncle or feed on flowers making hole-like structures. Control measures are the same as in orchid weevil. **Nematodes** produce leaf blotch symptoms which can be kept under check through oxamyl and fenamiphos drenching in the pot-mix. **Cockroaches** are very harmful and feed on root tips, buds and newly opened flowers during night. These are controlled with boric acid + sugar pellets. **Root miner** causes wilting of leaves and pseudobulbs which can be controlled through fumigation with methyl bromide 30 g/m^3 for four hours in a chamber. The larvae of ***Cattleya* midge**, gall on the *Cattleya* roots which can be controlled with benzene hexachloride. The larvae of ***Cattleya* fly**, a tiny wasp burrow within the small flower buds or pseudobulbs which can be controlled by spraying with 0.1 per cent Malathion.

Orchids are prone to numerous fungal, bacterial or viral **diseases,** certain being highly devastating. **Leaf spot** is caused by *Cercospora, Colletotrichum, Gloeosporium, Glomerella, Haplosporella* and *Phyllostictina* which is not a serious problem in most cases but within a few days of infection sunken spots on leaves appear which afterwards turn brown. Bordeaux mixture (4-4-50) or yellow oxide of copper controls this problem. **Petal blight** is caused due to continuous low temperatures coupled with high humidity, mainly by *Botrytis cinerea* but also with *Cladosporium oxysporum* and *Curvularia geniculata* on *Cattleya, Cymbidium, Dendrobium, Oncidium, Phalaenopsis* and *Vanda,* which initially causes spots on flowers but afterwards blights the whole of the flower. Blight of new shoots and inflorescence has also been observed on *Coelogyne dayana* due to *Fusarium oxysporum* (Anon., 2009). Proper air circulation among the plants and keeping water off the flowers during cool weather, *vis a vis* safe destroying of affected flowers will keep this problem under check. Dithane M-45 is also very effective against this disease. Certain saprophytic fungi grow on nectar secreted by the perianth, showing grey or black spots (Salinger, 1985). **Black rots/damping off** are caused by *Phytophthora cactorum* and *P. palmivora* in cool weather and are slow in attack while *Pythium ultimum* during warm humid weather on seedling plants which is highly dangerous as infection spreads throughout the plant at a very rapid rate. The symptoms of attack by both the pathogens are almost similar with infected parts (leaves) turning black, sometimes with a yellow margin, and the stems and pseudobulbs also rot. The humid conditions in the greenhouse should be avoided to keep the foliage dry when these fungi are present. Infected plant parts should be removed and destroyed and should be repotted in a sterilized medium. Affected plants should be sprayed with Ridomil (metalaxyl) at 0.1 per cent a.i. or Ridomil MZ (metalaxyl + mancozeb) at 0.15 per cent a.i. or zineb (Captan) at 0.2 per cent. **Orchid wilt/basal rot** (*Sclerotium rolfsii* and *Marasmiellus inoderma*) is symptomized by yellowing of plants, rotting of plants, pseudobulbs and roots and ultimate death of the plant; and **collar rot** (*Sclerotium rolfsii*) is symptomized by yellowing or browning of leaf and stem bases, and eventual rotting and death. In the former case the infection spreads through leaf mould or potting mixture so potting media should be heat sterilized to kill the sclerotia, the infected plants should be destroyed, and the benches along with contaminated pots and potting mix should be sterilized with 2 per cent formalin. In case of collar rot, the plants are dipped in carboxin and repotted in the sterilized media. **Root rot** (*Rhizoctonia solani, Fusarium* sp. and *Pellicularia filamentosa*) initially causes brown root rot which afterwards spreads to rhizomes and eventually plants wither. After removing infected roots, the plants are dipped in thiram and repotted in fresh medium. **Slimy rot** (*Fusarium oxysporum, F. moniliforme* and *F. subglutinans*) is symptomized as slimy rot of entire shoot, but later on lesions at leaf bases and brown lesions on pseudobulbs which can be controlled through proper sanitation coupled with fortnightly sprayings with bavistin 0.1 per cent alternate with captan 0.2 per cent. **Rust** (*Coleosporium bletiae, Hemileia americana, Sphenospora mera, Uredo* sp.) is characterized by small raised blister-like pustules of yellow/rust colour on the underside of young leaves, initially appearing as small dots but at a later stage acquires larger dimension of dark brown colour. NRC for Orchids at Pakyong (Sikkim) recorded *Uredo* sp. on *Phaius tankervilliae* and *P. triplicata, Coleosporium* sp. on *Anthogonium gracile* and *Herminium angustifolium*, and also the rust symptoms on *Calanthe discolor, C. plantaginea* and *C. trulliformis*. *Puccinia* rust was recorded on *Satyrium nepalense* (Anon., 2009). Affected leaves should be removed and copper fungicide should be sprayed.

Bacterial brown spot (*Xanthomonas cattleyae*) infects young as well as adult plants. An initially water-soaked spot appears on leaves, enlarges afterwards and soon leaves become soft and die. The disease may progress further infecting the growing point which may kill the plant. Overhead misting should be avoided and proper sanitation should be maintained. **Bacterial soft rot** (*Erwinia carotovora*) is a very serious disease of *Cattleya* which appears from the leaf tip as a small water-soaked darker spot. The pseudobulbs of such plants also turn soft, pulpy (pulp of yellow colour) and foul smelling. Agrimycin spraying is beneficial. **Bacterial brown rot** (*Erwinia cypripedii*) infects the plants through wounds and cracks by which leaves turn yellow and brown and its infection to the crown kills the plant. Its infection may be minimized by submerging the plants in 1:2000 solution of 8-quinolinol benzoate or natriphene for 1-2.5 hours.

Viruses are the worst enemies of orchids. The symptoms produced by one virus on different varieties, species under the same genus and various genera of orchids differ in appearance which gives impression as if all these symptoms have appeared from different viruses. Most prominent ones are *Cymbidium* mosaic virus (CyMV), *Odontoglossum* ring spot virus (ORSV) and tobacco mosaic virus-o (TMV-o). These appear in uneven and unpredictable colour patterns, malformation of leaves and flowers, crinkling and twisting, and reduced quantity and quality of flowers. CyMV produces mosaic patterns on young leaves of *Cymbidium, Cattleya, Phalaenopsis, Spathoglottis* and many others which afterwards turn necrotic and black. ORSV infects *Odontoglossum* and many other cultivated orchids and the characteristic symptoms of its infection are single or concentric necrotic rings of green to black tissue appearing on leaves, on *Cymbidium* leaves it produces diamond mottle while in *Cattleya* a mild colour break. TMV-o produces chlorotic streaks on the leaves of *Cymbidium* while colour break in flowers of *Cattleya*. Viruses are controlled through proper sanitation, careful cutting of the spikes in a way that sap of one cut spike should not touch the other plants through cutter or knife. Therefore the cutting tools should be dipped in a solution of trisodium phosphate or a saturated lime solution (pH 12) to disinfect them after cutting of every flower. There should be regular spray of some insecticides to control virus vectors. However, virus infected plants should immediately be lifted and buried in the soil to check further spread through such plants.

There can be certain abnormalities other than diseases and insect-pests which in fact are **physiological disorders.** Water stress and high temperature, in general for all the orchids, cause flower bud abortion. In excessive cold temperature, internodes may not elongate, flowers become malformed and may also abort at bud stage. *Phalaenopsis* after inflorescence formation becomes vegetative if temperature is below or above 25-30 °C. Rapid change in growing temperature and high temperature coupled with dry air causes floral wilting if ventilation is poor. *Phalaenopsis* exhibits mesophyll cell collapse most frequently during winter months below 7 °C temperature and one or a few leaves (other leaves may look normal) on the plant become pitted. The pitted portion may turn yellow to tan and finally black. This malady can be prevented by raising the growing temperature at or above 15 °C at night.

Excessive light causes sun burn so proper shading as per requirement of the species should be provided. Ethylene gas generated from industries, automobiles, even heaters kept inside the house, etc. is very dangerous to orchids as they are highly sensitive to ethylene. Other air pollutants also affect the flower buds adversely.

Orchids do not feel comfortable in too large a pot as their roots will not be able to cling on the inside surface of the pot, the pot will have more potting material, and the water in large pots dries off quite slowly.

Use of disproportionate mix of plant protection chemicals injures the plants. Nutrient deficiency or excessiveness also causes various abnormalities. During active growth of certain orchids during warm weather, Ca deficiency causes blackening of young leaves and new leads starting from the tip towards the stem in *Cattleya*. In severe deficiency, the young pseudobulbs also turn black. Ca application at this stage prevents this malady.

References

Abraham, A. and P. Vatsala, 1981. *Introduction to Orchids.* Tropical Botanic Gardens and Research Institute, Trivandrum.

Anonymous, 1981. *Cymbidium* orchid culture. Seminar Proc. Ministry of Agriculture and Fisheries, Auckland, New Zealand.

Anonymous, 1997. *Production Manual Orchids.* Agricultural and Processed Food Products Export Development Authority, New Delhi.

Anonymous, 2009. *Report of the Quinquential Review Team (2003-2008).* NRC for Orchids (ICAR), Pakyong, East Sikkim.

Anonymous, 2013. Bidens have orchid named after them in Singapore. *Hindustan Times.* New Delhi (Saturday, July 27), **LXXXIX**(178): 19.

Anthony Huxley (ed-in-chief), Mark Griffiths (ed.) and Margot Levy (Man. ed.), 1992. The New Horticultural Society Dictionary of Gardening (vol. I ~ A-C, vol. II ~ D-K, vol. III ~ L-Q, vol. IV ~ R-Z). The Macmillan Press Ltd., London.

Arditti, J. 1977. *Clonal Propagation by Means of Tissue Culture ~ A Manual.* In: *Orchid Biology: Reviews and Literature* (ed. Arditti, J.), pp. 203-293. Cornell University Press, Ithaca, USA.

Arditti, J. 1992. *Fundamentals of Orchid Biology.* John Wiley and Sons, New York, USA.

Arora, Y.K. and A. Mukherjee, 1983. *Ornamental Orchids of North Eastern india* (Tech. Bull. No. 5). ICAR research Complex for NEH Region, Shillong (Meghalaya).

Barman, D. and R. Rajni, 2003. Studies on the rooting of the keikis in bamboo orchid (*Arundina graminifolia* L.). *J. Ornam. Hort.*, **6**(3): 260-263.

Barman, D., S.P. Rajni Das, V. Nagaraju and R.C. Upadhyay, 2001. Growth regulators and regeneration of *Cymbidium* pseudobulbs. In: *Souvenir and Abstracts, Orchid Diversity in India. I. Science and Commerce 6th Natl. Seminar,* TOSI, Palampur.

Benzing, D.H. and A.M. Pridgeon, 1983. Foliar trichomes of Pleurothallidinae (Orchidaceae): functional significance. *Amer. J. Bot.*, **70**(2): 173-180.

Bhattacharjee, S.K. 1977a. Native *Bulbophyllum* of India. *Amer. Orchid Soc. Bull.*, **46**(12):1094-1100.

Bhattacharjee, S.K. 1977b. Indian treasure house, Indian coelogynes. *Orchid Rev.*, **85**: 13-16.

Bhattacharjee, S.K. 1978a. Native *Erias* of India. *Orchid Rev.*, **86**: 246-249.

Bhattacharjee, S.K. 1978b. Some light on Indian *Calanthe. Orchid Rev.*, **86**: 332-334.

Bhattacharjee, S.K. 1979a. The native vandas of India. *Hawaii Orchid J.*, **8**(1): 7-9.

Bhattacharjee, S.K. 1979b. The native oberonias of India. *Orchid Rev.*, **87**: 306-312.

Bhattacharjee, S.K. 1979c. Regulation of growth and flowering in *Cattleya* orchids by altered day lengths. *Singapore J. Prim. Ind.*, **7**(2): 90-92.

Bhattacharjee, S.K. 1979d. Photoperiodism effects on growth and flowering in some species of orchids. *Sci. & Cult.*, **45**(7): 293-295.

Bhattacharjee, S.K. 1995. Cultural Requirements of Orchids. In: *Advances in Hortriculture* (vol. II, eds Chadha, K.L. and S.K. Bhattacharjee), pp. 673-701. Malhotra Publishing House, New Delhi.

Bhattacharjee, S.K. and P. Das, 1975. *Vanda*: Spectacular orchids. *Indian Hort.*, **19**(4): 17-19, 23.

Binisha, S. 2003. Supplementary Effect of Biofertilizers in *Dendrobium* (M.Sc. Thesis). Kerala Agricultural University, Vellanikkara, Thrissur, Kerala.

Boodley, J.W. 1988. *Directions for Growers Use of Oasis Orchid Medium.* Smithers Oasis, Kent, Ohio.

Bose, T.K. and S.K. Bhattacharjee, 1980. *Orchids of India.* Naya Prokash, Calcutta, India.

Bose, T.K. and T.P. Mukhopadhyay, 1977. Effects of day length on growth and flowering of tropical orchids. *Orchid Rev.*, **85**(1010): 245-247.

Brundell, D.J. 1981. Orchid production and potential. *Proc. N.Z. Commercial Flower Growers Conf.*, Hastings, pp. 22-34.

Cho, K.H. and B.H. Kwack, 1996. Effect of environmental factors on growth of *Cymbidium goeringii* in summer. *J. Korean Soc. hort. Sci.*, 37(6): 805-809.

Chowdhery, H.J. 2009. Orchid diversity in North-eastern states of India. *J. Orchid Soc. India*, **23**: 19-42.

Chua, S.E. 1974. The effects of different levels of dried chicken manure on growth and flowering of *Oncidium* Golden Shower (var. Caldwell) and *Dendrobium louisae* Dark. *Singapore J. Prim. Ind.*, **4**(1): 16-23.

Das, S.P., D. Barman, T.K. Bag, P.C. Bhutia and R.C. Upadhyay, 2001a. An alternative method of regeneration from back bulbs in orchids. In: *Souvenir and Abstracts, Orchid Diversity in India.* Science and Commerce 6th Natl. Seminar, TOSI, Palampur.

Das, S.P., D. Barman, V. Nagaraju, P.C. Bhutia and R.C. Upadhyay, 2001b. Influence of growing media on shoot initiation and growth from *Cymbidium traceyanum* back bulbs. In: *Souvenir and Abstracts, Orchid Diversity in India.* Science and Commerce 6th Natl. Seminar, TOSI, Palampur.

Das, S.P. and S.K. Bhattacharjee, 2006. Orchids. In: *Advances in Ornamental Horticulture.* Vol. 2. *Herbaceous Perennials and Shade Loving Foliage Plants* (ed. Bhattacharjee, S.K.), pp. 1-65. Pointer Publishers, Jaipur.

Davidson, O.W. 1960. Principles of Orchid Nutrition. In: *Proc. World Orchid Conference*, London, pp. 224-233.

Delizo, T.C. 1973. The vegetative propagation of orchids belonging to the genus *Cattleya* using alpha naphthalene acetic acid. *Araneta Res. J.*, **20**(2): 153-164.

Dineshbabu, M., M. Jawaharlal and M. Vijayakumar, 2002. Influence of holding solutions on the postharvest life of *Dendrobium* hybrid Sonia-17. *South Indian Hort.*, **50**(4-6): 451-457.

Ding, T.H., H.T. Ong and H.C. Yong, 1980. *Proc. 3rd Asian Orchid Congr.*, pp. 65-78. Organized by Min. of Agric., Malaysia,

Das, S.P. and S.K. Bhattacharjee, 2006. Orchids. In: *Advances in Ornamental Horticulture*, Vol. 2 (ed. Bhattacharjee, S.K.), pp. 1-65. Pointer Publishers, Jaipur.

Dole, J. M. and H.F. Wilkins, 1999. Orchidaceae. *Floriculture-Principles and Species*, pp.438- 445. Prentice Hall, New Jersey.

Doorn, W.G.van, G. Fischer (ed.) and A. Angarita, 1999. Water relations of cut flowers. II. Some species of tropical provenance. *Acta Hort.*, No. 482, pp. 65-69.

Elle, A. 1962. Pine bark as a substrate for orchids. *Die Orchidee*, **13**: 48.

Esser, G. 1973. Substrates and fertilization for orchids ~ results from orchid growing in Lemforde. *Orchidee*, **24**(3): 101-107.

Etienne Delannoy, 2011. (http://earthsky.org/earth/the-odd-life-of-an-underground-orchid).

Fanfani, A. 1989. *The Macdonald Encyclopedia of Orchids*. Macdonald & Co. Ltd., London.

Foja Singh, 1986. Orchids. In: *Ornamental Horticulture in India*, pp. 127-153. ICAR, New Delhi.

Fouche, J.G., L. Jowe, J.F. Hausman, C. Kevers and T. Gaspar, 1997. Are temperature induced early changes in auxin and polyamine levels related to flowering in *Phalaenopsis*. *J. Pl. Physiol.*, **150**(1-2): 232-234.

Frei, J.K., R.C. Fodor and J.L. Haymick, 1975. The orchids of Krakatau ~ evidence for a mode of transport. *Ann. Bot.*, **44**(1): 51-54.

George M. Morel, 1960. Producing virus-free cymbidiums. *Amer. Orchid Soc. Bull.*, No. 31, pp. 473-477.

Goh, C.J. 1979. Harmonal regulation of flowering in a sympodial orchid hybrid *Dendrobium lousiae*. *New Phytologist*, **82**(2): 375-380.

Goh, C.J. 1983. Aerial root production in *Aranda* orchids. *Ann. Bot.*, **51**(1): 145-147.

Goh, C.J. 1984. Root production in orchids. *Orchid Rev.*, **92**(1985): 88-89.

Goh, C.J. 1990. Orchids – Monopodials. In: *Handbook of Plant Cell Culture*. Vol. 5. *Ornamental Species* (eds Ammirato, P.V., D.A. Evans, W.R. Sharp and Y.P.S. Bajaj), pp. 598-637. McGraw Hill, New York, USA.

Goh, C.J. and J. Arditti, 1985. Orchidaceae. In: *Handbook of Flowering*, vol. I (ed. Halevy, A.H.), pp. 309-336. CRC Press, Boca Raton, Florida.

Goh, C.J. and H.Y. Wan, 1974. *Plant Growth Substances* (ed. Sumiki, Y.). Hirokawa Publishing Company, Tokyo.

Goh, C.J. and A.L. Yang, 1978. Effects of growth regulators and decapitation on flowering of *Dendrobium* orchid hybrids. *Pl. Sci. Letters*, **12**(3-4): 287-292.

Gordon, S.W. 1997. Treating new ground. A tyre-crumb mix is suggested as a new orchid potting media. *Orchids*, **66**(11): 1168-1169.

Gregg, K.B. 1984. Stress-induced ethylene production by developing racemes of *Catasetum* and *Cycnoches* ~ how orchids say "Ouch"! *Amer. Orchid Soc. Bull.*, **53**(1): 50-55.

Griesbach, R.J. 1984. The *in vitro* propagation of *Phalaenopsis* orchids. *Amer. Orchid Soc. Bull.*, **53**(12): 1303-1305.

Griesbach, R.J. 1985. An orchid in every pot. *Florists' Rev.*, **176**(4548): 26-30.

Griesbach, R.J. 1986. Orchid Tissue Culture. In: *Tissue Culture as a Plant Production System for Horticultural Crops* (eds Zimmerman, R.H., R.J. Griesbach, F.A. Hammerschlog and R.H. Lawson), pp. 343-349. Martinus Nijhoff Publishers, Dordrecht, The Netherlands.

Hager, H. 1957. Control of flowering in *Cattleya*. *Proc. of the 2nd World Orchid Conf.*, pp. 130-132., Honolulu Orchid Society, Honolulu, Hawaii.

Hay, Roy 1971. Orchids. *Reader's Digest Encyclopaedia of Garden Plants and Flowers*, pp. 484-487. Reader's Digest Asscn, London.

Indhumathi, K., M. Kanna, M. Jawaharlal and Veena Amaranth, 2003. Standardization of prehardening and hardening techniques for *in vitro* derived plantlets of *Dendrobium* orchid hybrid Sonia-17. *J. Ornam. Hort.*, **6**(3): 212-216.

Jawaharlal, M., K. Rajamani, D. Muthumanickam and G. Balakrishnamurthy, 2001. Potting media for *Vanda*. *J. Ornam. Hort.*, **4**(1): 55-56.

Jawaharlal, M, D. Babu, T. Arumugam and M., Vijayakumar, 2002. Effect of holding solutions on the postharvest characters of dendrobium. *Proc. Nat. Symp. Indian Floriculture in the New Millennium*, held at Lal Bagh, Bangalore, under the auspices of ISOH, New Delhi, pp. 271-272.

Jomy, T.G. and G.T. Sabina, 2002. Effect of conditioning and pulsing on vase life of *Dendrobium* Sonia inflorescences. *J. Ornam. Hort.*, **5**(1): 80-81.

Jomy, T.G., G.T. Sabina and S.R. Nair, 2000. Effect of holding solutions on vase life of pulsed *Dendrobium* cv. Sonia inflorescences. *J. Tropical Agric.*, 38(1-2): 41-45.

Jyothi Bhaskar, 1996. Micropropagation of *Phalaenopsis* (Ph.D. Thesis). Kerala Agricultural University, Vellanikkara, Thrissur, Kerala.

Keithly, J.H., D.P. Jones and H. Yokoyama, 1991. Survival and growth of transplanted orchid seedlings enhanced by DCPTA. *HortSci.*, **26**: 1284-1286.

Ketsa, S. and N. Kosonmethakul, 2001. Prolonging vase life of *Dendrobium* flowers: the substitution of aluminium sulfate and cobalt chloride for silver nitrate in holding solution. *Acta Hort.*, No. 543, pp. 41-44.

Ketsa, S., Y. Piyasaengthong and S. Prathuangwong, 1995b. Mode of action of $AgNO_3$in maximizing vase life of *Dendrobium* 'Pompadour' flowers. *Postharvest Biology & Techn.*, **5**(1-2): 109-117.

Ketsa, S., A. Uthairatanakij and A. Prayurawong, 2001. Senescence of diploid and tetraploid cut inflorescences of *Dendrobium* 'Caesar'. *Scientia Hort.*, **91**(1-2): 133-141.

Ketsa, S., C. WongsAree, T. Fjeld (ed.) and E. Stromme, 1995a. The role of open florets in maximizing flower bud opening of *Dendrobium* held in preservative solution. *Acta Hort.*, No. 405, pp. 381-388.

Klougart, A. 1984. The vase life of orchids can be increased. *Gartner Tidende*, **100**(4): 86-87.

Knudson, L. 1922. Non-symbiotic germination of orchid seeds. *Bot. Gaz.*, **73**: 1-25.

Knudson, L. 1946. A new nutrient solution for germination of orchid seed. *Amer. Orchid Soc. Bull.*, No. 15, pp. 214-217.

Krishnapriya, M. 2005. Morphological and Cyto-molecular Characterization of *Dendrobium* Sw. Cultivars (M.Sc. Thesis). College of Agriculture (K.A.U.), Vellayani, Kerala.

Krizek, D.T. and R.H. Lawson, 1974. Accelerated growth of *Cattleya* and *Phalaenopsis* under controlled environment conditions. *Amer. Orchid Soc. Bull.*, No. **43**(6): 503-510.

Kumar, P.K.S. 1992. Potting media and post-transplantation growth of *Dendrobium* hybrid seedlings. *J. Orchid Soc. India*, **6**(1-2): 131-133.

Kunisaki, J.T. 1975. Induction of keikis on ascocendras by cytokinin. *Amer. Orchid Soc. Bull.*, **44**(12): 1066-1067.

Lakshmi Devi, S. 1992. Standardisation of Explant for *in vitro* Propagation in *Dendrobium* spp. (M.Sc. Thesis). Kerala Agricultural University, Vellanikkara, Thrissur, Kerala.

Larson, R.A. 1980. *Introduction to Floriculture*. Academic Press, USA.

Leffring, L. 1976. Influencing flowering in *Cymbidium. Proc. 4th European Orchid Congress*, pp. 404-411.

Lekha Rani, C. 2002. Intra and Interspecific Hybridization in *Dendrobium* spp. (Ph.D. Thesis). College of Agrticulture (K.A.U.), Vellayani, Kerala.

Leslie Andrew Garay, 1972. On the origin of Orchidaceae. II. *J. Arnold. Arboretum*, **53**: 202-215.

Lucksom, S.Z. 2007. *The Orchids of Sikkim and North East Himalaya*. S.Z. Lucksom, Sikkim, India.

Matsui, N. and K. Yoneda, 1997. Effect of summer temperatures on growth and flowering of *Miltonia*. *J. Japanese Soc. hort. Sci.*, **66**(3-4): 597-605.

Misra, R.L.,C.T. Sakkeer Hussain, N.S. Pathania, Sanyat Misra and Anjula Pandey, 2005. *Ornamental Plants*. In: *Plant Genetic Resources: Horticultural Crops* (eds Dhillon, B.S., R.K. Tyagi, S. Saxena and G.J. Randhawa), pp. 309-327. Narosa Publishing House, New Delhi-110002.

Misra, S. 2007. *Orchids of India*. Bishen Singh Mahendra Pal Singh, Dehradun, India.

Molitor, H.D. 1992. Nutritional disorders in plants grown in hydroponic culture. *Gartenbau Mag.*, **1**(1-2): 73-75.

Morel, G.M. 1964. A new means of clonal propagation of orchids. *Amer. Orchid Soc. Bull.*, No. 31, pp. 473-477.

Muira, Y. 1981. Studies on establishment of orchids cultivated on the basis of their photosynthesis properties. I. Influence of temperature, light intensity and air humidity on photosynthetic rate. *Bull. Kanagawa hort. Exp. Stn*, No. 28, pp. 64-72.

Mukherjee, A. 1979. Ornamental *Phaius* as house plant. *Indian Hort.*, **24**: 23-28.

Murasige, T. and F. Skoog, 1962. A revised medium for rapid growth and bioassays with tobacco tissue culture. *Physiol. Plant.*, **15**: 473-497.

Nagabhushana, S.R. 1982. Propagation and cytotaxonomic studies in some Indian orchids. *Thesis Abstracts*, **8**(4): 383-384.KAU, Thrissur

Ninitha Nath, C. 2003. Compatibility Studies in Monopodial Orchids (M.Sc. Thesis). College of Agriculture (K.A.U.), Vellayani, Kerala.

Ota, K., K. Morioka and Y. Yamamoto, 1991. Effects of leaf age, inflorescence, temperature, light intensity, and moisture conditions on CAM photosynthesis in *Phalaenopsis*. *J. Japanese Soc. Hort. Sci.*, **60**: 125-132.

Padmanaba Pillai, N. 2003. Morpho-anatomical and Molecular Characterization of *Dendrobium* Sw. Cultivars (Ph.D. Thesis). College of Vellayani (K.A.U.), Vellayani, Kerala.

Pan, Rui Chi, Jian Yuau Chen, Zhao Qing Wen, R.C. Pan, J.Y. Chen and Z.Q. Wen, 1994. Influence of different potassium levels on growth, development and physiology in *Cymbidium sinense* following potassium starvation. *J. Tropical & Subtropical Bot.*, **2**(3):46-53.

Papenhagen, A. 1980. Suitability of irrigation water for ornamental plants under glass. *Deutscher Gartenbau*, **34**(6): 238-240.

Patil, P.V. 2003. Exotic orchid propagation through stem cuttings. *J. Maharashtra agric. Univ.*, **28**(1): 90-91.

Paul, C.A., P.K. Rajeevan and C. Anitha Paul, 1992. Influence of media on growth parameters in *Dendrobium*. *J. Orchid Soc. India*, **6**(1-2): 125-130.

Penningsfeld, F. 1971. The nutrient requirements of *Cattleya meleagris*. *Orchidee*, **22**(3): 95-100.

Perata, P. 1981. Growing orchids under artificial light. *Informatore di Ortoflorofrutticoltura*, **22**(8): 15-18.

Peres, L.E.P., S. Amar, G.B. Kerbauy, A. Salatino, G.R. Zaffari and H. Mercier, 1999. Effects of auxin, cytokinin and ethylene treatments on the endogenous ethylene and auxin-to-cytokinins ratio related to direct root tip conversion of *Catasetum fimbriatum* Lindl. (Orchidaceae) into buds. *J. Pl. Physiol.*, **155**(4-5): 551-555.

Poole, H.A. and J.G. Seeley, 1977. Effects of artificial light sources, intensity, watering frequency and fertilization practices on growth of *Cattleya, Cymbidium* and *Phalaenopsis* orchids. *Amer. Orchid Soc. Bull.*, No. **46**(10): 923-928.

Poole, H.A. and J.G. Seeley, 1978. Nitrogen, potassium and magnesium nutrition of three orchid genera. *J. Amer. Soc. hort. Sci.*, **103**(4): 485-488.

Poole, H.A. and T.J. Sheehan, 1982. Mineral Nutrition of Orchids. In: *Orchid Biology ~ Reviews and Perspectives* (vol. II, ed. Arditti, J.), pp. 195-212. Comstock Publishing Associates, Ithaca,USA.

Porat, R., A. Borochor, A.H. Halevy, T. Fjeld (ed.) and E. Stromme, 1995. Is jasmonim acid involved in the endogenous regulation of *Dendrobium* orchid flower senescence? *Acta Hort.*, No. 405, pp. 314-319.

Post, K. 1949. Orchids. In: *Florist Crop Production and Marketing*, pp. 663-719. Orange Judd Publishing, New York.

Powell, C.L., K.I. Caldwell, R.A. Littler and I. Warrington, 1988. Effect of temperature regime and nitrogen fertilizer level on vegetative and reproductive bud development in *Cymbidium* orchids. *J. Amer. Soc. hort. Sci.*, **113**: 552-556.

Pradhan, U.C. and S.C. Pradhan, 1997. 100 Beautiful Himalayan Orchids and How to Grow Them. Primulaceae Books, Kalimpong, India.

Prakash, L., C.L. Lee and C.J. Goh, 1996. *In vitro* propagation of commercial orchids: an assessment of current methodologies and development of a novel approach – thin section culture. *J. Orchid Soc. India*, **10**(1-2): 31-41.

Prakash, C.S. Loh and C.J. Goh, 1995. An *in vitro* method for rapid regeneration of a monopodial orchid hybrid *Aranda* 'Deborah' using thin section culture. *Plant Cell Rep.*, **14**: 510-514.

Radha, R.K., V.S. Menon and S. Seeni, 1994. Physiological analysis of sun and shade plants of selected orchids from the Western Ghats. *J. Orchid Soc.* India, 8(1-2): 55-59.

Rajeevan, P.K. 1997. An eco-compatible design for growing dendrobiums in Kerala. *J. Orchid Soc. India*, **11**(1-2): 47-50.

Rajeevan P.K. 2001. Orchids. In: *Handbook of Horticulture* (ed. Chadha, K.L.), pp. 573-577. ICAR, New Delhi.

Rajeevan, P.K., A. Sobhana, J. Bhaskar, S. Swapna and S.K. Bhattacharjee, 2002. Orchids, 62 p. AICRPF, IARI, New Delhi.

Rampal and M. Dayamma, 2001. Post production care and handling of orchids. In: *Souvenir and Abstracts, Orchid Diversity in India. Science and Commerce 6th Natl. Seminar*. TOSI, Palampur.

Rao, A.N. 2010. Orchid Flora of Arunachal Pradesh: An update. *Bull. Arunachal For. Res.*, **26**(1-2): 82-110.

Rao, C.S. and S.K. Singh, 2015. *Wild Orchids of Meghalaya ~ A Pictorial Guide*, 203 p. Meghalaya Biodiversity Board, Shillong.

Rattanawisalanon, C., S. Ketsa and W.G. van Doorn, 2003. Effect of aminooxyacetic acid and sugars on the vase life of *Dendrobium* flowers. *Postharvest Biol. Tech.*, **29**(1): 93-100.

Reddy, P.V., C.V. Ramana aand C.S. Reddy, 1992. Prospects of commercial cultivation of orchids in India. *J. Orchid Soc. India*, **6**(1-2): 15-20.

Robert L. Dressler, 1981. *The Orchids ~ Natural History and Classification*. Harvard University Press, Cambridge, USA.

Roost, V.V. 1978. Factors related to growth and development in hybrid cymbidiums. *Byulleten Glavnogo Botanicheskogo* Sada, No. 107, pp. 93-97.

Rotor, G.B., Jr. 1952. Day length and temperature in relation to growth and flowering of orchids. Bull. No. 885. Cornell Univ. agr. Exp. St., Ithaca, New York.

Sabina George, T. 1996. Performance of Selected Orchids under Varying Light Regimes, Culture Methods and Nutrition (Ph.D. Thesis). College of Agriculture (K.A.U.), Vellayani, Kerala.

Sajitha Kumari, 1998. Biology and Management of Orchid Pests (M.Sc. Thesis). Kerala Agricultural University, Vellanikkara, Thrissur, Kerala.

Salinger, J.P. 1985. *Cymbidium*. In: *Commercial Flower Growing*, pp. 180-191. Butterworths of New Zealand Ltd., Victoria Street, Wellington, New Zealand.

Samasya, K.S. 2000. Physiological Aspects of *ex vitro* Establishment of Tissue Cultured Orchid *Dendrobium* var. Sonia 17 Plantlets (M.Sc. Thesis). College of Agriculture (K.A.U.), Vellayani, Kerala.

Sanjeev Nair, U. 2001. Endogenous and Exogenous Regulation of Growth and Development in *Dendrobium* var. Sonia 17 and Sonia 28 (M.Sc. Thesis). Kerala Agril. University, Vellanikkara, Thrissur, Kerala.

Sarkar, I., T. Mandal, P.N. Kumar, R. Kumar, Sanyat Misra, R.L. Misra and K.P. Singh, 2009. Temperate Orchids, 77 p. AICRPF, IARI, New Delhi.

Sheehan, T. J. 1980. Orchids. In: *Introduction to Floriculture* (ed. Larson, R. A.), pp. 133-164. Academic Press, USA.

Sheehan, T.J. and M. Sheehan, 1979. *Orchid Genera Illustrated*. Van Nostrand Reinhold Co., New York.

Sherly Kuriakose, 1997. Standardisation of *in vitro* Techniques for Mass Multiplication of *Aranthera* and *Dendrobium*

(Ph.D. Thesis). College of Agriculture (K.A.U.), Vellayani, Kerala.

Sheron Fernandez, 2001. Standardisation of Shade Requirement in *Dendrobium* (M.Sc. Thesis). Kerala Agricultural university, Vellanikkara, Thrissur, Kerala.

Shiau, Yih Juh, Uei Chin Chen, Hung Min, LinChih Wen, Hsin Tsay and H.S. Sheng, 1996. Studies on flowering adjustment and pollination of *Anoectochilus formosanus* Hayata. *J. agric. Res. China*,**45**(3): 250-259.

Sinclair, R. 1990. Water Relations in Orchids. In: *Orchid Biology ~ Reviews and Perspectives* (vol. V; ed. Arditti, J.), pp. 63-120.

Sivamani, S. 2004. Micropropagation of *Dendrobium* Hybrids (M.Sc. Thesis). College of Agriculture (K.A.U.), Vellayani, Kerala.

Sobhana, A. 2000. Improvement of *Dendrobium* through Hybridization and *in vitro* Mutagenesis (Ph.D. Thesis). Kerala Agricultural University, Vellanikkara, Thrissur, Kerala.

Sudeep, R. 1994. Standardisation of Medium Supplements for Shoot Proliferation in *Dendrobium* (M. Sc. Thesis). Kerala Agricultural University, Vellanikkara, Thrissur, Kerala.

Suh, Jung Nam, Chun Ho Pak, Beyoung Hwa Kwack, J.N. Suh, C.H. Pak and B.H. Kwack, 1998. Ethylene production of cut *Cymbidium* flowers as affected by temperature, vibration and plant growth regulators. *Korean J. hort. Sci. Tech.*, **39**(1): 103-106.

Swapna, S. 2000. Regulation of Growth and Flowering in *Denbdrobium* var. Sonia 17 (Ph.D. Thesis). Kerala Agricultural University, Vellanikkara, Thrissur, Kerala.

Syrovatka, T. 1980. A new method of continuous irrigation for orchids. *Amer. Orchid Soc. Bull.*, No. **49**(11): 1258-1259.

Thomas, S. 1996. *Odontoglossum harryanum. Curtis's bot. Mag.*, **13**(3): 134-138.

Timmerman, D.A. 1978. *Cymbidium Culture, Management and Marketing* (Dutch). Netherlands Horticultural Advisory Service Bulletin.

Tran-Thanh van, M. and M.T.T. Van, 1974. Methods of acceleration of growth and flowering in a few species of orchids. *Amer. Orchid Soc. Bull.*, No. 43(8), pp. 699-707.

Uma Maheswari, R. 1999. Nutrition of Tissue-cultured Plants of *Dendrobium* Sonia-17 (M.Sc. Thesis). College of Agriculture (K.A.U.), Vellayani, Kerala.

Vacin, E.F. and F.W. Went, 1949. Some pH changes in nutrient solution. *Bot. Gaz.*, **110**: 605-913.

Varghese, S. 1995. Floral Biology and Compatibility Studies in *Dendrobium* (M.Sc. Thesis). Kerala Agricultural University, Vellanikkara, Kerala.

Vij, S.P. and J.K. Sembi, 2001. Orchids and their propagation. In: *Souvenir and Abstracts, Orchid Diversity in India. Science and Commerce 6th Natl. seminar*, pp. 34-38. TOSI, Palampur.

Vij, S.P. and V. Sharma, 1996. Regerative competence of *Vanda cristata* perianth segments ~ a Study *in vitro*. *J. Orchid Soc. India*, **10**(1-2): 25-29.

Wadasinghe, G. and C.S. Hew, 1995. The importance of back shoots as a source of photoassimilates of growth and flower production in *Dendrobium* cv. Jashika Pink (Orchidaceae). *J. hort. Sci.*, **70**(2): 207-214.

Wang, Y.T. and N. Lee, 1994a. Potted blooming orchids. *Greenhouse Gr.*, **12**(1): 79-80.

Wang, Y.T. and N. Lee, 1994b. Potted blooming orchids, Part 2. *Greenhouse Gr.*, **12**(2): 36-38.

Went, E.W. 1990. Orchids in my life. In: *Orchid Biology ~ Reviews and Perspectives* (ed. Arditti, J.), pp. 21-36.

Wilford, N. 2000. Trade secrets, culture guide to growing award winning orchids. *Orchids*, **69**(10): 938-951.

Wimber, D.E. 1963. Clonal multiplication of *Cymbidium* through tissue culture of shoot meristem. *Amer. Orchid Soc. Bull.*, No. 32, pp. 105-107.

Wotering, E.J. 1984. Influence on ornamentals tested. Many species susceptible to ethylene. *Vakblad-voor-de-Bloemisterij*, **39**(4): 112-113.

Wu, G.D., W.H. Chen, J.B. Chen, M.S. Chyou and Y.Y. Cheng, 1994. Effect of applying nitrogen and organic fertilizer to bagasse medium on growth of *Phalaenopsis*. *Report Taiwan Sugar Res. Instt.*, **146**: 18.

Wu-RongZhe, Debasis Chakrabarty, Hahn EunJoo and Paek KeeYoeup, 2005. Growth of *Doritaenopsis* in peat-substituted growing medium. *J. Korean Soc. hort. Sci.*, **46**(1): 76-81.

Xia, H.X.L.Q., M. Zhang, Y.Z. Zhang and Z.L. Bie, 1999. The effect of different plant growth regulators on growth of *Dendrobium nobile*. *Acta Hort. Sinica*, **26**(4): 275-276.

Yadav, L.P. and T.K. Bose, 1985. Possibility of commercial orchid culture in the plains of West Bengal. *Farmer & Parliament*, **20**: 19-21, 40.

Yadav, L.P. and T.K. Bose, 1989. Orchids. In: *Commercial Flowers* (eds Bose, T.K. and Yadav, L.P.), pp. 151-265. Naya Prokash, Calcutta.

Yoneda, K. and H. Sasaki, 1978. Studies on the flowering of orchids. I. The effect of shading and defoliation on the growth and flowering of *Epidendrum radicans*. *Bull. College Agric. & Vet. Medicine, Nihon Univ.*, No. 35, pp. 80-87.

Yoneda, K., H. Momose and S. Kubota, 1991. Effect of daylength and temperature on flowering in juvenile and adult *Phalaenopsis* plants. *J. Japanese Soc. hort. Sci.*, **60**: 651-657.

29

Polianthes tuberosa (Family: Agavaceae, formerly Amaryllidaceae)

Sanyat Misra and R.L. Misra

(**Common names**: Tuberose, Hindi-**Rajanigandha**, and Malayalam-**Nishigandha**)

Introduction and Origin

In 1738, the generic name *Polianthes* was given to tuberose by Linnaeus and then he described it *Polianthes floribunda alternis in his* **Hortus Cliffortianus** but later on in 1753 through his treatise **Species Plantarum**, he grouped this and *Hyacinthus indicus tuberosus* of Gaspard Bauhin (1671) under the genus *Polianthes* and species *tuberosa. Polianthes* takes its name from the Greek *polios* meaning 'white' or 'bright' and *anthos* meaning a 'flower', and contains one of the largest cultivated species of cut flowers, *Polianthes tuberosa*, which was known to Clusius in the 16th century, and the double form was known in the early part of 18th century. It is thought to be native of Mexico where it was cherished and cultivated before the Spanish conquest in 1522 though there is no record that where from it was first collected from the wild, however, as it has a long history for its domestication and may have spread from Mexico to other countries as an ornamental and for use in the industry (Zeven and Zhukovsky, 1975) but Bailey (1930) mentions it being probably native to the Andes Mountains of South America. It was introduced to European countries towards the end of 16th century but how and when it reached India and elsewhere in the orient is not documented (Trueblood, 1973). At one time, the genus *Polianthes* contained red- and orange-flowered species which have now been placed under *Bravoa*. Tuberose is a sun-loving plant. Flowering occurs throughout the year in tropical climate, from summer to full autumn in the sub-tropical regions and during summer to rainy seasons in the temperate regions of the country. If watering is not the limiting factor and there is protection against hot summer sun and chilly winters in sub-tropical conditions, in the tropical and subtropical regions it flowers throughout the year, peak being from late-rainy season to full autumn, *i.e.* September to November. Flowering is scarse during hot summers and chilly winters though 'Calcutta Single' is exception. The common name tuberose is not well understood but surely it has nothing to do with rose but seems to be a corruption of the word "tuberous" (tuber-like), referring to the plant's tuberous rootstock. Tuberose is the most homely flower of India. It is being grown in every part of the country irrespective of the climatic conditions prevailing there. Double types are being cultivated only for decoration purposes in India and elsewhere. In India, its commercial cultivation is confined to Delhi and its adjoining districts of Haryana (Faridabad and Sonipat), Uttarakhand (Udham Singh Nagar) and Uttar Pradesh (Meerut, Baghpat and Muzaffar Nagar); Maharashtra (Pune and Thane); Andhra Pradesh (East Godavari, Guntur, Chittoor and Krishna); Karnataka (Devanahalli, Tumkur and Mysore); Tamil Nadu (Coimbatore); and West Bengal (Ranaghat, Kolaghat and Panskura); *etc.* but only for cut and loose flower purposes and not for distillation. Guenther (1952) stated that it is grown in the Grasse region of South France and in Morocco for extraction of essential oil from its flowers and in the Grasse region then more than 75 metric tonnes of flowers from the single types were being harvested annually for processing by the essential oil distilleries though now there is decline due to more land being utilized for vegetable growing, however, in Morocco the area is being expanded. Though its flowers are white but they have intense pleasing fragrance and are available in two forms: single and double.

Its 1150 g of flowers yield 1 g of oil which is used extensively in perfumery. The intensity of fragrance is more in singles than in doubles. On the temperate region of the country as well as on the lower hills, the doubles are also quite fragrant. It is used as cut flowers, as loose flowers, as border plants, near windows for night fragrance, and for extraction of essential oils. Ali *et al.* (1981) isolated two sapogenins, hecogenin and 9, 11-dehydrohecogenin, and two saponins, becogenin-3-0-glucoxyloside and 9, 11-dehydrohecogenin-3-0-glucoxylogalactoside from whole plant of *Polianthes ruberosa.*

Botany

Its rootstock is a bulb though has a rhizomatous base with thickened roots, leaves are succulent, basal, very dense, long, narrow and strap-shaped, the inflorescence stalk is strong, erect, bracteate and up to 75-100 cm, flowers white, sweetly fragrant, waxy-textured, produced in a terminal raceme, long, cylindrical and tubular, bent near base and then flaring in funnel-form, and are borne in pairs in the axils of the bracts. Segments 6, unequal but short, stamens 6 and filaments thread-like, short, anthers dorsifixed and erect, style thread-like, stigma sickle-like, ovate and 3-lobed, ovary inferior and 3-chambered, fruit a capsule with black and flat seeds.

There are some 12-15 species under the genus *Polianthes,* nine having only white flowers, but except *Polianthes tuberosa* others seem to be obsolete. Almost all the *Polianthes* species are fertile except the double form 'The Pearl' under *P. tuberosa* which possesses n = 25 chromosomes and it is hypothesized that double forms are derived from the single form by the loss of the 5 pairs of smallest chromosomes possibly due to spontaneous abnormal meiotic or mitotic divisions. The chromosome numbers in single tuberose is recorded n = 30 which falls under four groups, *i.e.* 5 large pairs, 4 medium pairs, 7 medium-small pairs and 14 small pairs. Single forms are fertile whereas doubles are sterile, therefore, in a breeding programme double form is normally not used as male parent. There is self-incompatibility whether it is single or double form. Krishnan (2003) through 15 Gy and 20 Gy ^{60}Co gamma rays and 1.0 and 2.0 per cent EMS induced nine vM_2 stable mutants which revealed the differences in banding pattern of esterase, peroxidase and catalase between parents and mutants through isozyme analysis. Rajvir singh *et al.* (2008) gave pre-planting treatment of EMS at 0, 0.25, 0.50 and 0.75 per cent to the bulb tips of tuberose cv. Double and obtained delayed sprouting in all the treatments, in vM_1 increased duration of flowering under 0.25 per cent EMS, and in vM_1 and vM_2 both longest rachis, maximum spike weight and maximum bulb weight with 0.25 per cent EMS, but no any genetical modification was observed.

Classification, Species and Varieties

Except *Polianthes tuberosa,* other 11 species are not of any commercial importance. All the varieties available in the country or elsewhere fall under only one species, *P. tuberosa.* Though there is no any special classification of tuberose but on the basis of the flower form, it is classified as 'Single' or 'Double'. Initially only two varieties were introduced in the country *viz.* 'The Pearl' (double) and 'Mexican Single'. Afterwards, two varieties, 'Swarna Rekha' (double) and 'Rajat Rekha' (single) were developed through ^{60}Co gamma treatment from N.B.R.I., Lucknow. These varieties are only collector's item and do not have any commercial value as both flower seldom. However, there are golden stripes on the leaves of 'Swarna Rekha' and silver stripes on the leaves of 'Rajat Rekha'. Like to that of 'Swarna Rekha', one more variety was developed in Sikkim which also has golden stripes on the leaves but slightly different, and this is known as 'Sikkim Selection'. Its spike is taller than others and grows more than one metre in length. I.I.H.R., Bangalore developed four varieties *viz.* 'Prajwal' (quite sturdy, tall and vigorous and the best among all the singles existing in the country), 'Shringar' (single), 'Suvasini' (double) and 'Vaibhav' (double). One variety has been developed by A.N.G.R.A.U., Hyderabad, probably a selection and has been named as 'Hyderabad Single' which is very similar to 'Shringar'. One variety as 'Phule Rajani' has been developed by M.P.K.V. Pune centre. This single is also quite compact and beautiful. This variety at bud stage is green. Many states in the country grow single and double forms except those described above, which they claim their own but, in fact, these are one, the single one is 'Mexican Single' and the double one 'The Pearl'. All the varieties at the bud stage are pinkish except 'Phule Rajani', 'Mexican Single' (Calcutta Single) and 'Suvasini' which are green at bud stage. Thus in real sense only 11 varieties of tuberose exist in the country. The singles being 'Hyderabad Single', 'Mexican Single'(Calcutta Single), 'Phule Rajani', 'Prajwal', 'Rajat Rekha', 'Shringar' and 'Sikkim Selection' while doubles are 'Suvasini', 'Swarna Rekha', 'The Pearl' (Calcutta Double), and 'Vaibhav'. **The Dutch-based 'Ludwig & Co. Holland', in 2013, released coloured tuberoses through genetic engineering, and now this firm is licensee for these five varieties which have high fragrance. Out of these, the boldly coloured single-flowered varieties are 'Cinderella' (lavender-pink) with stalk length of 35 cm, 'Sensation' (pink) with stalk length of 45 cm, 'Super Gold' (deep yellow) with stalk length of 65 cm and 'Yellow Baby' (yellow) with stalk length of 35 cm; and the double-flowered one is 'Double Pink' (pink) with stalk length of 65 cm (www.floraculture.au). Also in Taiwan the work in the direction of conventional breeding is going on and they have achieved success in developing pink, yellow and orange variants for multi-purposes such as bedding, cut flower and pot-growing.**

Propagation

Tuberose is propagated vegetatively through bulbs, lateral bulblets forming on the side of the main bulb (also known as offsets), through division of bulbs (bulb segments) and through micropropagation. The most common method for tuberose multiplication is through **bulbs** and **bulblets**. Spindle-shaped bulbs having diameter of 2 cm are flowering size but other smaller ones (bulblets) are only planting stocks. The basal plate of tuberose is so formed that one confuses it with a rhizome and some also call these as tubers though these are bulbs. Though one year planted bulb is not so confusing but the bulbs taken out from perennial cropping give this impression. The formation of lateral buds is so fast that basal plate extension is necessitated to accommodate these bulblets and on lifting when bulbs and bulblets are taken out one by one, a big chunk of the disc (basal plate) is left in the end. In a two-year cropping, a plant of certain varieties produces about 5-10 large (flowering size) bulbs and about or more than 50 bulblets. On planting, in one year, 2-3 large bulbs and 15-30 small bulbs (bulblets) are formed. If cared well, bulblets also grow full flowering size bulbs in one year, if planted singly after separating these from the clump. Since tuberoses are, by and large, grown as perennial crop, *i.e.* 2-4 years so because of their dependency on the mother bulb as these are sharing the same disc for years, the bulblets get only a small share of nutrients hence their development is very slow. Four-year perennial cropping is not profitable as quality of the spike becomes very poor and so is the bulb and bulblet formation. After planting the large-sized bulbs, first year one gets only 1-3 spikes, but second year 4-6 with slightly poor quality of spikes, while the third year though there is increase in the number of spikes, but quality is obviously poor. Therefore, only two-year cropping should be adopted. It would be better if large bulbs are separated every year for flowering and bulblets for raising the stock. Removal of the old basal-rhizomatous-bulbs from offsets before planting lowers the number of new-forming offsets (Cirrito and De Vita, 1982). If cared well, many of the large bulblets will also provide quality spikes the first year.

Full-sized bulbs and large bulblets may also be **fractionated** into ½-$^1/_8$, each piece having a part of the disc as well as fleshy scales, treated with some effective contact fungicide and planted in sandy loam soil rich in humus. Each segment will provide a new plant. Sharga (1980) selected the tuberose bulbs ranging from 0.5 to 3.0 cm in diameter and divided these into six grades, each grade bulb fractionated longitudinally into two halves but when compared with entire bulb the flowering was considerably delayed, the bulb production was comparatively better though bulblet production was poor, and the segments only from the large diameter bulbs (>2 cm) regenerated well.

Micropropagation may also be adopted for quick multiplication. Bulb segments on MS medium, with addition of growth regulators and vitamins and incubating at 22-25 °C using continuous fluorescent light of 3,500 lux, may provide some 500-800 plants in one year. Narayanaswamy and Prabhudesai (1979) cultured tuberose explants of bulb scale, bracts and petals on a medium containing 2, 4-D (2 ppm) or kinetin (1 ppm) and the explants differentiated into numerous embryoid-like structures though these embryoids developed only a primary root and were devoid of shoot apical meristem. Bose *et al.* (1987) cultured tuberose cv. 'Single' explants from bulb scales, on MS medium containing 2, 4-D at 0.1 mg/l and obtained good callusing which regenerated into plantlets, the medium containing kinetin at 0.05-0.2 mg/l initiated shoot appearance before the rooting, and when the callus was subcultured on a medium containing NAA at 2 mg/l, this resulted in root and shoot initiation in 15-20 days with good plantlet production. Krishnan (2003) found scale stem section of the bulb the best as the ideal explant source and in somatic embryogenesis the inflorescence segments containing immature flower buds, and he found MS medium supplemented with BAP 6.0 mg l^{-1} + KIN 4.0 mg l^{-1} as the best for high rate of multiple axillary buds though for maximum root initiation MS medium supplemented with IBA 4.0 mg l^{-1} + 0.2 per cent activated charcoal. However, direct organogenesis was obtained from inflorescence segments in MS medium supplemented with NAA 0.2 +BAP 2.0 + KIN 1.0-3.0 mg l^{-1}.

Growers should not be concerned with seed propagation as it is a tedious job where there is only little success and that too with time constraints. Moreover, at flowering seed-raised plants may not come true to the type. 'Hyderabad Single', 'Shringar', 'Mexican Single' (Calcutta Single), 'Phule Rajani', and 'Sikkim Selection' all singles produce seeds which are small, shining black, light and flat. Immediately, when capsules mature, these should be taken out and sown on raised beds having well workable sandy-loam soil rich in organic matter. These beds should lightly be watered daily. Within a month or two, these will germinate.

Cultural Practices

Reasonably fertile sandy-loam **soil** which is light and well-drained but with good water-retention capacity, having pH range of 6.5-7.5, is most suitable for tuberose growing, though tuberoses can successfully be grown up to 8.5 pH provided the field is rich in organic matter. This prefers a sunny situation. Even partial shade hampers the growth. It grows very well from 18-35 °C temperatures.

Its proper **planting time** in the subtropical regions is February and in temperate regions when danger of frost is over, *i.e.* March-end to early-April. In tropical regions, it can be planted any time, preferably from October to January. If planted at proper time, these require 20 to 50 days for sprouting, the smaller ones sprouting earlier and the larger ones later as was experienced by Pathak *et al.* (1980) when they planted small (5-10 g), medium (11-15 g) and the large bulbs (19-35 g), though for flower emergence from emergence these took 70 days for small ones, 58 days for medium ones and 45 days for larger ones. In Italy, Farina and Patertniani (1986) recommended tuberose planting in early June for highest flower production when they studied the varieties 'La Perla' (The Pearl, double), 'Fiorentina' (semi-double) and a single-flowered type; however, Cirrito *et al.* (1982) stated best planting time in Italy from 14th to 29th June for bulb enlargement when they planted the bulbs having 6.0-8.0 cm circumference. They are planted shallowly just to cover the tips of the bulb with the soil. For its planting the **soil preparation** should be done as for potato planting. Soil should be well dug up to 30 cm deep thrice, every time followed with planking. The soil should have sufficient moisture at the time of third ploughing and planking so that after planting the field may not require immediate watering. Now the beds of convenient sizes are made after leaving the space for bunds and channels. **Bulb size** for the purpose of cutting should be 8.0-9.0 cm in circumference (Farina and Paterniani, 1986) or at least 2.5 cm diameter (Brundell and Steenstra, 1986), though Curir *et al.* (1984) reported flowering in Italy with 4-7 cm circumference of bulbs provided such bulbs attain 6.0-6.5 cm of circumference at flowering but still the larger bulb size is required for production of commercial flower heads. Brundell and Steenstra (1986) reported increased yield of inflorescences and lateral bulbs by planting large size of bulbs, clumped bulb giving greater response than individual bulbs though bloom quality increases with individual bulbs. Lopes (1971) planted the bulb-transplants weighing 70, 200 or 340 g at population levels of 30, 40 or 50 thousand per hectare and recorded no effect of planting density on flower production, length of floral stalk or plant weight though larger transplant size was associated with greater flower production and plant weight only. However, Sharga (1982) when used bulbs weighing from 3.0 to 49.5 g with single-flowered variety, obtained satisfactory results when the size was 19.4 g and above. Yadav *et al.* (1984) with tuberose var. 'Single' obtained 2.6-3.0 cm bulb diameter with highest yield of spikes and flowers when planted 6 cm deep. For 1-year crop, the proper distance found is 15 × 15 cm and for 2-year crop 17 × 17 cm, planting only one bulb per hill in both the cases (Patil *et al.*, 1980). Patil *et al.* (1987) when planted bulbs of the diameter of 1.5-2.5 or 2.6-3.0 cm at 15 × 20 cm, 20 × 20 cm or 25 × 20 cm spacings for a 3-year cropping, they obtained highest top quality flower yield from smaller bulbs planted at 15 × 20 cm distance. **Planting** is done at 30 cm apart in the rows if it is to be made a two-year crop. However, for a three-year cropping, it requires more distances and that is 40-45 cm apart. After planting, pendimethalin (stomp) pre-emergence **weedicide** at the rate of 3 litre/ha after mixing the chemical in 5,000 litres of water should be applied evenly in the planted area just to wet the surface. This will prevent germination of annual weeds up to 75 days. Then after, only stray weeds are to be removed manually because the main crop covers the in-between area with its own growth of leaves which normally does not permit other weeds to grow. In case, weedicide has not been used, organic or black polyethylene mulching is used. In Italy, Cirrito *et al.* (1980) obtained control of almost all the weeds except *Cyperus longus* with use of Treflan (trifluralin) at 2 kg/ha in tuberose cv. 'Perla', though other herbicides they used were Alipur (chlorbufam), Prevenol (chlorpropham) and Tok E25 (nitrofen). Pre-plant application of atrazine at 3.0 kg/ha caused maximum weed control in tuberose cv. 'Single' and yielded 10.2 tonnes/ha of good quality flowers compared with 5.0-10.0 tonnes in unweeded control (Yadav and Bose, 1987). Panwar *et al.* (2010) encountered weeds in the cultivation of tuberose cv. 'Double', some of them being *Amaranthus* sp., *Bracharia reptens, Corchorus* sp., *Cynodon dactylon, Cyperus rotundus, Euphorbia* sp., *Sesbania aculeata, Solanum nigrum, Sorghum halepense, Trianthema portulacastrum, etc.*, for which they used Atrazine and pendimethalin at 1.0, 1.5, 2.0 kg/ha as pre-emergence, while glyphosate at 1.0, 1.5 & 2.0 kg/ha and XL-71AG at 0.6, 0.9 & 1.2 per cent 30 days after emergence by dissolving in 625 l water for one hectare, and recorded post-emergence treatment quite injurious to the crop.

In case weedicide is not used, **hoeing** and earthing up should also be carried out as and when required. Hoeing will keep down the weeds and will permit aeration while **earthing** will give good support to the inflorescences to stand erect against winds and rains.

Irrigation and Feeding

First light **irrigation** is followed immediately after, if pendimethalin weedicide has been applied in the field. Whether it is temperate or subtropical region but its growing season passes through spring to autumn hence it requires copious watering but shallowly every time, starting from its sprouting to leaf senescence. In completely tropical region also, the watering frequency is the same. In light soils, it should be watered every 6 days while in heavy soils 8 days. Watering very much depends on the prevailing weather conditions. During rains, it does not require watering. If it is watered well regularly, one can get the flowers throughout the year in the tropical regions, from May-June to October-November in the sub-tropical regions, and from June end to October in the temperate regions. In Egypt, El-Roumi (1981) provided 40 litres water/m^2 at every 7 or 14 days and found better tesults where tuberose 'Pearl' plants were irrigated every week. In Italy, Cirrito *et al.* (1981) when drip-irrigated at 2,000-8,000 m^3/ha/year in a 3-year cropping the tuberose cv. 'La Perla' planted in late April at a density of 36 plants/m^2, recorded best yield of commercial-calibre bulbs, *i.e.* 30-32/m^2. Erection of polyhouses over the plantings during winter season, *i.e.* from November to March in India in sub-tropical and temperate regions will make the flowers available regularly for all the 12 months. Even otherwise, the variety 'Mexican Single' which is good only as a loose flower blooms throughout the year under Delhi condition if the crop is not damaged due to frost in winter. If it does not occur, to some extent leaves even of other varieties remain green even during winter otherwise senesce, and new growth starts only by April. Under Delhi condition, there is profuse flowering from September to November.

At the time of preparation of soil, well rotten **compost** or farmyard manure at the rate of 300 quintals/ha should be incorporated in the soil by mixing thoroughly. Being a perennial crop it requires heavy feeding. For tuberose crop, Cirrito (1972) recommended 24 kg each of N and P_2O_5 and 36 kg/1,000 m^2 of K_2O in Italy. In a 3-year trial with tuberose cv. 'La Perla', at Palermo (Italy) on alkaline sandy soils. Cirrito and De Vita (1981) when applied 500 kg/ha of chicken manure as pre-planting and 800 kg/ha ammonium sulphate, 1,000 kg/ha rock phosphate and 180 kg/ha K_2SO_4 as post-planting, they obtained highest production of new lateral bulbs measuring 4-6 cm out of which some 92 per cent were ready for flowering. Cirrito (1975) in Italy stated that tuberose bulbils for getting saleable bulbs should be applied with 20 kg N, 40 kg P_2O_5 and 60 kg K_2O/1000 m^2 at planting. Tuberose bulbs of 12-18 g size or those weighing 19-25 g planted in late April and applied with PK basal dressing though nitrogen in the form of ammonium sulphate at two split-doses, one at 2-3 leaf stage and the other before flowering, resulted into earliest and highest percentage of flowering and with highest number of flowers per spike with those receiving 75 kg N/ha with large bulbs (Mitra *et al.*, 1979). A soil deficient in N and P and high in K, with cv. 'Single', Nanjan *et al.* (1980) recorded highest flower production when applied 200 kg N, 60 kg P_2O_5 and 0 kg/ha K_2O. With cv. 'Single', application of 300 kg/ha N in two split doses, half at planting and the other half 40 days later, and 200 kg/ha P_2O_5

gave highest flower yield and plant growth (Yadav *et al.*, 1985). Bankar and Mukhopadhyay (1985) with tuberose cv. 'Single' when applied 25-35 g N, 60-80 g P_2O_5 and 60-80 g/m^2 K_2O; P_2O_5 improved only floral quality though N had a significant beneficial effect on all the parameters studied including spike durability in the field (22.8 days) which is longest in the highest N rate. Mukhopadhyay and Banker (1985) when applied 20 g/m^2 N in one dose or in 2-4 split doses to tuberose cv. 'Single', they recorded greatest plant height and number of spikes per plant with application of 1/3rd N before planting, 1/3rd after 30 days and remaining 1/3rd at 60 days after planting. By applying 30 g/m^2 N, Muralee Manohar (2000) recorded longest rachis length in tuberose, though 20 g/m^2 N and 15 g/m^2 P_2O_5 application was responsible for increased weight of bulbs and bulblets. In a 2-year tuberose cv. 'Single' crop, Mukhopadhyay and Bankar (1986) when applied 0-20 g N, and 0-40 g/m^2 each of P_2O_5 and K_2O, they found only N at the highest rate improving plant growth, spike yield, flower quality and production of daughter bulbs but not the main bulb. In a 2-year tuberose crop of cv. 'Double' when N at 0, 5, 10, 15 or 20 g/m^2, half at planting and remaining half as top-dressing at flower emergence; and P_2O_5 and K_2O each at 0, 20 or 40 g/m^2, both as basal dose, N improved vegetative growth, flowering and bulb production, and advanced flowering, and the highest N rate produced highest number of flower spikes (20.09)/m^2, however, the effect of P and K with respect to duration of flowering, spike number and rachis length could be seen only in the second year (Bankar, 1988; Bankar and Mukhopadhyay, 1990), and in the end there was recommendation that per square metre the 20 g each of N, P_2O_5 and K_2O should be applied for best results. In a 2-year pot-culture of tuberose, Sharga and Motial (1983) when applied foliar spray of N as urea at 2.14, 4.28 or 6.42 g/l; orthophosphoric acid at 0.9, 1.8 or 2.6 ml/l; and/or potassium citrate at 2.2 or 4.42 g/l, they obtained highest yield of essential oil from plants receiving medium N rate, highest P and lower K rate when sprayed 50 ml of solution every 15 days starting on June 2. In a sand culture experiment, Jana *et al.* (1974) found increased number of leaves, bulbs, spikes and flowers with high N and P though their deficiencies suppressed flowering whereas low K reduced the number of flowers on the spike. Sharma *et al.* (2008) recommended 200 kg N, 70 kg P_2O_5 and 40 kg K_2O per hectare as optimum for tuberose cultivation under Haryana conditions. They found 200 kg/ha nitrogen or 70 kg P_2O_5, followed by 40 kg K_2O increasing plant height, spike length, flower yield & quality and bulb & bulblet yield. Potassic fertilizers should only be incorporated in the soil if soil is deficient in potash. Therefore, it would be better if soil analysis has already been done so that only recommended dose of NPK or any other nutrients may be given. Nitrogen 200 kg and P_2O_5 60 kg/ha will give better yield of flower and bulbs. In potash deficient soils, 50 kg/ha K_2O will benefit the crop. Whole of the potassium and phosphorus and half of nitrogen may be applied in the soil while preparing the land for tuberose cultivation. Half nitrogen may be applied when inflorescence starts emerging. In the standing crop, always after fertilizer application, irrigation should be carried out. Another year also the same quantity of fertilizer may be used if it is a two year crop, and subsequently in the third year also the same quantity if it is three-year crop. Application of *Azospirillum*, *Azotobacter*, and phosphobacteria alone or in combination, and also when combined with inorganic fertilizers give excellent results in terms of number of spikes per plant, number of flowers per spike, number of bulbs and bulblets and their sizes per plant. Application of granular superphosphate at 60 g/m^2 combined with phosphobacteria increased the number of flowering plants, number of flowers per spike and floral durability (Aikincai, 1960). Chaudhary (2008) used VAM, nitrogen at 50, 100 and 200 kg/ha, phosphorus at 25, 50 and 100 kg/ha, and their combinations in tuberose 'Double' planting, and recorded increased growth and flowering parameters with increase in N and P applications, and the results in VAM was found to be at par to phosphorus at 50 and 100 kg/ha. Tuberose responds well to certain micronutrients. Application of zinc @15 kg/ha or spraying the crop with 0.25 per cent $ZnSO_4$ results in better production of flowers, bulbs and bulblets. $CaCl_2$ and NaCl salinity (EC 1.5-15.0 mmhos/cm) has detrimental effect on bulb sprouting, flowering time, spike length, flower number and flower size, *vis-à-vis* essential oil content, the effect being quite drastic at early stages of growth though later stage tolerates quite high salinity levels, however, the uptake of Na^+, K^+, Ca^{++} and Mg^{++} is also badly affected (Singh *et al.*, 1974). Malini and Khader (1989a, b) when provided 6,000, 8,000 or 10,000 ppm $CaCl_2$ or NaCl to tuberose 'Single' with irrigation water and assessed their effect at 80 and 145 days after planting, they recorded that plants were more tolerant to NaCl than $CaCl_2$ and where both were added even at lower level the growth was utterly poor. Control flowered in 122 days though at highest salinity level it was delayed up to 165 days, and the flower yield in control per plant was found to be 24.7 g though at highest salinity level it as 9.8 g per plant.

Growth, Development and Flowering

The bulbs have vegetative apex protected by a layer of 3-4 sheath leaves, and after a certain growth these change to generative phase and from then it takes hardly 25 days to form a complete flower inside. When such bulbs are planted in field, they initiate flowers in 90 days. Tuberoses do not have true dormancy and if the plants are getting irrigation water regularly these remain green throughout the year and flower though severe winter temperatures force them to senesce but soon after weather improves the bulbs remaining in the field start sprouting. However, for their forcing in greenhouses the dormancy of the bulbs can be broken by 4 per cent thiourea dip for one hour, ethylene chlorhydrin is effective only in case of large bulbs, and carbon tetrachloride dip is partially effective in case of large bulbs but is harmful to small ones (Hakim *et al.*, 1958). With respect to floral bud differentiation in tuberose, Kosugi and Kimura (1961) in Japan when planted the bulbs on April 1, through microscopical examination recorded floral bud initials on June 11 when plants were 19.5 cm high with 19.8 leaves, and when mean minimum temperature was 13.4 °C, and by June 18 the flower bud primordia had differentiated into two parts, on June 25 the sepals were first observed on the two primordia when mean minimum temperature was 19.3 °C and then the plant height was 29.2 cm with 23.3 leaves. Under protected cultivation of tuberose cv. 'Pearl' with no heating arrangement, Brundell and Steenstra (1986) in New Zealand obtained 10 days advance flowering with improved flower quality, increased lateral tuber production and 150 per cent increase in the stem yield. They recorded 50-60 per cent overall increased floral-stem yields when planted pre-heat-sprouted bulbs. The growth and differentiation of the apex depends on the supply of soluble carbohydrates from the scales. Storage temperature influences the number of bulbs lifted and the quality of flower spike. Storage at higher temperature (30 °C) for longer duration, though advances flowering but reduces the quality of the flower and bulb yield. However, storage at 10 °C for one month is optimum which improves the yield of flowers, but prolonged storage decreases bulb yield. Planting of the fresh bulb will lead only to profuse vegetative growth and little flowering, hence, for sometime the bulbs should be rested to encourage better growth and flower production.

For maximum root growth, soil temperature should be above 20 °C. For maximum bulb production, atmospheric temperature should be in between 20-30 °C. If bulb growth does not terminate soon after flowering, new lateral buds will initiate flowers. After senescence of aerial parts due to frost, flower and bulb growth stops. In tuberose, the lateral bulb production is prolific, and it takes 2-3 seasons to attain the flowering size. In perennial cropping after senescence of the foliage (not chopping of the foliage), certain lateral buds mature by attaining proper growth and start flowering later in the season.

Application of GA_3 and chlormequat stimulate the enlargement of laterals as well as the central bud by reducing the number of laterals. Muralee Manohar (2000) when used GA_3 on tuberose, he recorded longest rachis and its highest weight in 50 ppm, and enhanced floret opening, *vis-a-vis* maximum number of floret-opening at a time in 100 ppm; though IAA 25 ppm enhanced the vase life. By spraying the plants of tuberose cv. 'Single' at 4-5 leaf stage and 30 days later with 10-1,000 ppm GA_3, B_9 and ethephon, and 500-2,000 ppm CCC, Jana and Biswas (1979, 1982) recorded maximum number of flowers (35.5) per plant with 2,000 ppm CCC, 1,000 ppm B_9 or 10 ppm GA_3, B_9 being more effective though 10 ppm GA_3 brought about earliest floret opening. Basu and Bose (1980) reported advanced flowering by 9 days and increase in flower number from 35.4 to 44 per plant when treated the plants of tuberose cv. 'Single' with 200 ppm ancymidol at 5-6 leaf stage. Pre-treatment of tuberose bulbs with 300 ppm GA_3 or kinetin, or 10 or 50 ppm ABA showed that GA_3 was not found effective but kinetin and ABA stimulated the sprouting appreciably though both inhibited flowering, and the ABA being very effective for sprouting in case of small bulbs, *vis-à-vis* in stimulating flowering in large bulbs (Pathak *et al.*, 1980). Spraying of tuberose 'Single' plants 40 days after planting and twice more at fortnightly intervals with 25-100 mg/l GA_3 or 500-2,000 mg/l ethephon, Mukhopadhyay and Bankar (1983) recorded reduced plant height, bulb and bulblet production with increased ethephon concentrations though GA_3 increased spike length and number of florets per spike at 75 and 100 mg/l but inhibited bulb production at all the concentrations. Soaking of freshly harvested bulbs of tuberose 'Single' for 24 hours in GA_3 at 10, 100, 250 and 500 ppm and planting in 30-cm pots resulted into delayed sprouting from 8.1 days in control to 17.4 at 500 ppm, reduced plant height from 42 to 34 cm and number of flowering bulbs per plant but increased the rachis length by 4.85 cm at 100 ppm (Mukhopadhyay and Sadhu, 1985). Storage of bulbs (1.5-2.0 cm diameter) at 4-10 °C for 10-30 days and then soaking in 200 mg/l GA_3 or 2,000 mg/l thiourea for 6 hours improves plant growth, causes early emergence of flower spikes with improved quality and increases the yield of spikes and flowers (Dhua *et al.*, 1987). Choudhary (1987) when used 500, 1,000 and 1,500 ppm MH, 500-1,500 ppm CCC and 50-150 ppm GA_3, he recorded MH decreasing plant height at all the levels, though 500 ppm MH, 100 ppm ethephon, 1,000 ppm CCC or 50 ppm GA_3 gave best results with respect to number of flowers per plant. Planting of dormant bulbs of tuberose after dipping for one hour in solutions of 50-400 ppm GA, and 500-5,000 ppm ethephon and CCC, Ramaswamy *et al.* (1977/1978) recorded 100 ppm GA and 500 ppm CCC advancing flowering for more than a fortnight though flower yield was highest only with 1,000 ppm CCC. Hassan and Agina (1980a, b) through summer (15th March) and autumn (early September) plantings of tuberose cv. 'Double' and treating by spraying and through drenching on 19th April (repeated thrice at monthly intervals) to summer planted one and spraying in early October in autumn planted one with CCC at 1,000, 2,000 or 3,000 ppm, and recorded decreased fresh bulb weight, delayed flowering and slightly increased number of florets with spraying which later on was found significantly decreased after drenching with 3,000 ppm in summer planted crop, though in autumn planted crop the treatments extended the flowering period, and increased the number of floral stalks & the florets with increase in the concentrations and brought about some cold resistance in the plants. Itty (2001) got maximum height reduction in tuberose plants when used paclobutrazol at 100 ppm at 60 days after planting or 200 ppm at 30 days of planting and the latter treatment induced earlier flowering but when applied at flowering it produced longest rachis with larger size of floret. Paclobutrazol at 100 ppm applied at 30 days after transplanting brought size increase of bulblets. Highest plant spread was recorded when bulbs were dipped in CCC at 1,000 ppm, and its 500 ppm recorded more number of small sized bulbs and bulblets. BA 50 ppm treatment given at 60 days after planting produced more tillers and delayed floret opening but longer spikes, and its 50 ppm applied at flowering increased bulb size and weight.

Though tuberose is photo-insensitive, long-day exposure promotes vegetative growth and accelerates flowering with increased spike length while low light intensity and duration promote leaf length.

Postharvest

Spikes are harvested preferably in the morning when 2-3 florets have opened. After harvesting, immediately their cut ends are placed in a bucket containing water. Spikes should be harvested from the ground level with a sharp knife or should be taken out by giving a gentle push from the sides at ground level from where it breaks from the plant. Picking for loose flowers may also be done in the morning only for 2 hours and only opened and opening buds are picked. One person can harvest some 15 kg of loose flowers in 2 hours duration. The yield of loose flowers may vary from 12,000-30,000 kg/ha/year, depending upon the varieties chosen and the cropping year. Care should be taken that picking should not injure the spikes because other opening flowers from the same spike are to be harvested later. These loose flowers are used for various floral decorations, hair adornments and for making garlands. The var. 'Mexican Single' (Calcutta Single) is meant only for loose flower production. Loose flowers are packed in hessian cloth lined bamboo baskets of various sizes, holding 5-20 kg of flowers and are transported the same day to the designated market where it is sold by weight. If these flowers are marketed the next day, there will be weight loss up to 40 per cent.

After cutting, the flower spikes are also graded (on the basis of flower quality, spike thickness, spike length and rachis length), dressed and bundled in fifties or hundreds, and then wrapped in newspapers, and then finally packed in hessian cloth which may accommodate some 600-1000 spikes, depending on the varieties. The refined way of packing them is to wrap the spikes in tissue paper or corrugated brown paper and then with alkathene papers, and then packed in corrugated boxes, specially prepared for the purpose or in rectangular bamboo baskets lined with hessian cloth. These can easily withstand 48 hours of dry storage during transit.

Cut spike postharvest life depends on the variety, the cropping season, the cultural practices adopted, soil moisture status during whole cropping season, carbohydrate reserves and endogenous growth regulators present, the growing season (status of temperature), photoperiod and light intensity, relative humidity, bulb size and spacing. High nitrogen levels administered to the crop produces tall and succulent spikes which may easily break even without winds. Low light intensity but high temperature and excessive moisture reduce the cut flower life. Attack of pests and diseases also affect flower life adversely. Cut flower life improves when bulbs at planting are treated with daminozide or chlormequat 1,000-5,000 ppm, MH 1,000-2,500 ppm or GA_3 10-1,000 ppm. Panwar *et al.* (2008) when sprayed tuberose cv. Single plants at 4-5 leaf stage with GA_3 (25, 50, 75 and 100 ppm) and placed the spikes in vase solution of 4 per cent sucrose +200 ppm cobalt sulphate, they recorded that increasing concentrations of GA_3 increased the fresh weight and vase life of spike as well as fully opened florets. Benzimidazole or GA_3 (0.01-0.03 per cent) in combination with daminozide, ascorbic acid, glucose and hydroxyquinoline (0.001-0.003 per cent) improve the opening of young green unopened buds or matured half-bloomed flowers and increase the vase life. Calcium nitrate 0.01 per cent + citric acid 250 ppm + sucrose 3 per cent provide maximum vase life (14 days) to tuberose cut flowers. Normal vase life of tuberose cut flower is 7-8 days. Pulsing with 10 per cent sucrose + 250 ppm aluminium sulphate for 12 hours and the cut spikes stored for 5 days, prolongs life of the cut spikes and opening of all the florets are obtained. Khondakar and Mazumdar (1985) when subjected the 30 cm long cut spikes of tuberose in different solutions for up to 96 hours, they found 3 per cent sucrose + 0.03 per cent 8-HQC + 0.01 per cent $AgNO_3$ quite effective. Naidu and Reid (1989) stated that pre-treatment of cut spikes with 20 per cent sugar for 15-20 hours and 1.5 per cent sugar as vase-solution improves display life of stems before or after storage. The response of the sucrose pulsing and the presence of open flowers on the spike increases solution uptake, and as there is very little ethylene evolution by the florets so exogenous ethylene exposure is not injurious to these spikes. Gowda (1990) recorded longest vase life of 12 days when the cut flowers were held in 1 per cent sucrose + 200 ppm aluminium sulphate or 2 per cent sucrose + 400 ppm aluminium sulphate. Itty (2001) recorded maximum days for complete opening of florets when tuberose spikes were pulsed with 250 ppm 8-HQS and its 400 ppm gave maximum opened florets at a time though pulsing with $AgNO_3$ 50 ppm exhibited maximum vase life when held in 0.25 mM $AgNO_3$ solution. Citric acid 250 ppm + sucrose 4 per cent, followed by aluminium sulphate 250 ppm also improved postharvest life of tuberose cut flowers (Vijayalaxmi *et al.*, 2011).

Tuberoses have only white flowers. Hence, for obtaining flowers of various colours, *viz.*, yellow, rose, red, scarlet and blue, one will have to dissolve chemicals like bromocresol green and bromocresol blue for various shades of blue, phenol red for yellow, erythrocin red for rose, ammonium purpurate for rhodomine red and eosin for scarlet colour when pulsed at 0.1 per cent each for five hours as water solution. This process is known as **tinting**. Sambandhamurthi and Appavu (1980) used tuberose var. 'Mexican Single' with stalk lengths of 30, 45 or 60 cm with two basal flowers open and stood in solutions containing bromocresol green, bromophenol blue, eosin, erythrocin red, phenol red or ammonium purpurate, each at 0.025-0.2 per cent for different durations and found colour changes ranging from blue, red, scarlet, rose to yellow.

The yield of concrete from fresh flowers is 0.08-0.11 per cent, of which 18-23 per cent constitutes alcohol-soluble absolute (Guenther, 1952). Its oil is used in costliest perfumes or as modifiers in other absolutes such as jasmine or rose, and to lesser extent in flavouring candy, beverages and baked foods. Bhattacharjee (1984) stated that on an average yield of 10,000 kg/ha of fresh loose flowers of a single type yields 9.17 kg/ha of concrete and subsequently 1.83 kg/ha of absolute. Tuberose essential oil extraction may be carried out by enfeurage or by use of volatile solvents with petroleum ether where in the former case it yields 15 times more volatile oil but this system of extraction is almost abandoned in India and only petroleum ether extraction is used. For extraction, flowers are collected early in the morning when these have started opening. These are extracted with n-hexane (B.P. ranges 70-75 °C) in cold, miscella is separated and distilled to recover solvent, traces of solvent are removed under vacuum and floral concrete is obtained which is treated with chilled ethanol, waxes are filtered and the filterate is distilled under vacuum and then the floral absolute is obtained (Ramachandraiah *et al.*, 1987). Through petroleum ether extraction, in tuberose it yields 0.08-0.11 per cent concrete which gives 18-23 per cent absolute when treated with alcohol. One kg of concrete is obtained from 1,150 kg of tuberose flowers.

Lifting and Storage of Bulbs

After flowering is over, the growth of the plant also ceases and in the sub-tropical region of Delhi as well as whole of North India, the winter also sets in. By 20th of November, it becomes very cold (temperature going below 10 °C, and sometimes even 5 °C, while in January it may go even below 5 °C) which forces the leaves to die down though October-November is the peak flowering time. It is therefore advisable to lift the bulbs (clumps) by December to save the crop from rotting. These, if not lifted, will start sprouting by April provided the field has been watered from March onwards. Even after collapse of the foliage, the field should be cleaned of the weeds. In the temperate climate of India, the clumps should certainly be lifted by November as winter sets in there and continues till mid-March. In tropical regions of the country though all the weathers exist but winter is very mild so lifting may be carried out when peak flowering is over.

After harvesting (lifting), the clumps should be dried in single layer under a shed which is well aerated. These clumps are now separated into bulbs and lateral bulblets, are graded and then again dried. Those bulbs which are spindle-shaped and at least 2 cm (>6/8 cm circumference) or above in diameter are flowering size and those less than 2 cm (<6/8 cm in circumference) are planting stock. The grading may be done on the basis of diameter, circumference or weight. The bulbs having circumference of 8/10, 10/12 and 12-14 cm are the commercial grades. Through good cultural practices, below 2 cm diameter bulbs may also produce flowers of commercial grade. However, <1.5 cm bulbs may not produce commercial spikes. Large bulbs produce more laterals than the laterals themselves.

The yield of bulbs per year is roughly 20 tonnes per hectare, if raised through bulbs having diameter of 2.5-3.0 cm. However, the collective yield in a 3-year cropping period will come to about 50 tonnes.

Flowering may be advanced by storing the bulbs at higher temperature *i.e.* 30°C for longer duration, but the quality of the flower will be poor and bulb yield will be reduced. Lower temperature (10 °C) for longer duration also hampers the bulb production, but if stored at this temperature only for one month, more flowers with better quality are obtained. Storage temperature of at least 18 °C for 4-6 weeks, or 30 °C for six weeks has been advocated for production of more commercial size bulbs. Initiation and development of the flowers occur after planting of the bulbs. Initiation is affected by physiological stage of the plant (bulb size, number of leaves, soil and air temperature). Earlier planting provides longer harvest duration of spikes, with more number of spikes and better quality blooms. In Delhi, the optimum planting season for tuberose is February when there exists low soil and air temperatures that is why flower harvesting period is prolonged though flowering starts slightly late.

Insect-Pests and Diseases

Tuberose is a very sturdy and easy growing crop which is normally not affected by any serious insect-pest or disease. However, sometimes a few observed are described here under.

Grasshoppers (*Hieroglyphus* spp.) feed on the young leaves and flower buds, especially during rainy season which may be controlled by spraying 0.2 per cent methyl parathion. The adult **weevils** feed in darkness on the leaves making notches on the edges which may be trapped and killed as they are only a few. Their larvae feed on the roots and tunnel into the bulbs whose attack may be prevented by applying Furadan granules @ 3-4 kg/ha at the time of bed making. **Aphids** (*Aphis* spp.) are very small in size and multiply rapidly. They feed on the growing points and floral buds which may be controlled by spraying nicotine solution. **Thrips** (*Taeniothrips* spp.) are minute insects which suck the sap of the leaves, buds and flowers, and act as a carrier agent in causing 'bunchy top' disease where the inflorescence is malformed. While controlling the grasshopper, this will also be controlled. **Red spider mites** suck the sap on foliage causing yellow stripes and streaks, and in severe cases leaves turn yellow, silvery or bronze and finally deformed. Kelthane spraying will prevent its infestation.

Two **nematode**, *Aphelenchoides besseyi* causes greasy streak on foliage and *Meloidogyne* spp. (*M. incognita, M. javanica, M. arenaria, M. acritata*) cause root galling, poor growth of plants with leaf tip burn and yellowing, and suppress spike emergence in their severe infestation to tuberoses. *A. besseyi* was first reported as foliar nematode of tuberose by Holtzmann (1968), whose control was suggested by Trujillo (1968) as HWT, bed fumigation and post-planting drenches with methyl bromide. The treatment advocated for controlling larvae of the weevil will keep these pests also under check.

Stem rot (*Sclerotium rolfsii*) is caused at the soil level. Leaves loose their greenness and whole leaf rots and gets detached from the plant. Due to its infection, round and brown sclerotia are formed on and around the infected leaves. Ultimately the plants become too weak to produce flowers. Mercuric chloride 0.1 per cent and commercial formalin 0.2 per cent have excellent control of the disease. Dusting 20 per cent brassicol has been found quite effective. **Botrytis spots and blight** (*Botrytis elliptica*) is a problem under cool cum moist growing conditions. Initially when noticed, should be controlled by spraying with 0.2 per cent maneb weekly and the field humidity should be kept under check. Before planting, the bulbs should also be treated by dipping them for one hour in 0.2 per cent Bavistin or benomyl. ***Alternaria* leaf spot** (*Alternaria polyanthi*) is also seen on tuberose leaves. **Flower bud rot** (*Erwinia carotovora*) sometimes infects the floral buds which may be controlled through mercuric chloride or streptocyclin. Sometimes, leaf mottling is caused due to a virus. Homer and Pearson (1988) described a partial purification procedure for a flexuous filamentous virus from leaves of tuberose exhibiting mottle symptoms. Particular problems associated with the purification included particle aggregation and the highly viscous nature of the plant sap. The former difficulty was partially alleviated by addition of 1.0 M urea to the extraction buffer and the latter by treatment of the extract with cellulose. The mean virus particle length from negatively stained leaf dips was 742 nm which is characteristic for potyviruses. In addition, TEM of thin sections of infected leaf tissue demonstrated the presence of several types of inclusion bodies similar in appearance to inclusions previously described in association with potyvirus infections in other plants. Such plants should be destroyed to check further spread.

References

Ailincai, N. 1960. The influence of some fertilizers on the flowering of *Polianthes tuberosa* (Italian). *Lucr. Sti. Inst. Agron. Isasi*, pp. 355-360.

Ali, A.A., S.A. Ross, A.M. Moghazy and M.A. El-Shanawany, 1981. Steroidal saponins from *Polianthes tuberosa*. *Fitoterapia*, **52**(3): 137-139.

Bailey, L.H. 1930. *Polianthes. The Standard Cyclopedia of Horticulture*, vol. III, pp. 2731-2733. Macmillan Publishing Co. Inc., New York.

Bankar, G.J. 1988. Nutritional studies in tuberose (*Polianthes tuberosa*) cv. 'Double'. *Progressive Hort.*, **20**(1-2): 49-52.

Bankar, G.J. and A. Mukhopadhyay, 1985. Response of *Polianthes tuberosa* L. cv. 'Single' to high doses of NPK. *South Indian Hort.*, **33**(3): 214-216.

Bankar, G.J. and A. Mukhopadhyay, 1990. Effect of NPK on growth and flowering of tuberose cv. Double. *Indian J. Hort.*, **47**(1): 120-126.

Basu, T.K. and T.K. Bose, 1980. Uses of ancymidol on flowering bulbous plants. ***Sci. Cult.***, **46**(9): 323-325.

Bhattacharjee, S.K. 1984. You can farm fragrance. *Farmer's J.*, **3**(9): 32-34.

Bose, T.K., B.K. Jana and S. Moulik, 1987. A note on the micro-propagation of tuberose from scale stem section. *Indian J. Hort.*, **44**(1/2): 100-101.

Brundell, D.J. and D.R. Steenstra, 1986. The effects of protected cultivation, pre-sprouting and lateral tuber removal on tuberose production. *Acta Hort.*, No. 177 (vol. II), pp. 361-367.

Choudhary, M.L. 1987. Effect of growth regulatants on growth and flowering of tuberose (*Polianthes tuberosa* L.). *Ann. Agric. Sci., Ain shams Univ.*, **32**(2): 1451-1457.

Chaudhary, S.V.S. 2008. Influence of VAM, nitrogen and phosphorus on growth and flowering of tuberose (*Polianthes tuberosa* Linn.). *Haryana J. hort. Sci.*, **37**(3/4): 289-290.

Cirrito, M. 1972. Preliminary studies on fertilizing to increase the size of tuberose bulbs (*Polianthes tuberosa* L.). *Annali dell'Istituto sperimentale per la Floricoltura*, **3**(1): 1-11.

Cirrito, M. 1975. The effect of manuring and bulbil circumference on the enlargement of bulbs of Tuberose (Italian). *Annali dell'Istituto sperimentale per la Floricoltura*, **6**(1): 27-43.

Cirrito, M. and M. De Vita, 1981. A comparison between different types of manuring in the culture of tuberose for rhizome enlargement. *Annali dell'istituto Sperimentale per la Floricoltura*, **12**(1): 41-51.

Cirrito, M. and M. De Vita, 1982. A preliminary trial on the prevention of lateral development of offsets on tuberose (*Polianthes tuberosa*) cultivated for rhizome enlargement (Italian). *Annali dell'istituto Sperimentale per la Floricoltura*, **13**(1): 43-51.

Cirrito, M., M. De Vita and G.V. Zizzo, 1980. Results of triennial experiments on weed control under gladiolus and tuberose (Italian). *Annali dell'Istituto sperimentale per la Floricoltura*, **11**(1): 123-148.

Cirrito, M., M. De Vita and G.V. Zizzo, 1981. Different watering rates and mulching with culture of tuberose for rhizome enlargement (Italian). *Annali dell'Istituto Sperimentale per la Floricoltura*, **12**(1): 53-65.

Cirrito, M., G.V. Zizzo and M. De Vita, 1982. The effects of different planting times of tuberose offsets of circumference 6-8 cm on tuberose rhizome enlargement (Italian). *Annali dell'Istituto Sperimentale per la Floricoltura*, **13**(1): 33-41.

Curir, P., A. Fusconi, S. Sulis and L. Parisi, 1984. Aspects of flower differentiation of the meristematic apex of *Polianthes tuberosa* L. cv. Perla during the bulb enlargement phases (Italian). *Ann. Instituto Sperimentale Floricoltura*, **15**(1): 73-88.

Dhua, R.S., S.K. Ghosh, S.K. Mitra, L.P. Yadav and T.K. Bose, 1987. Effect of bulb size, temperature treatment of bulbs and chemicals on growth and flower production in tuberose (*Polianthes tuberosa* L.). *Acta Hort.*, No. 205, pp. 121-128.

El-Roumi, M.K. 1981. Effect of irrigation frequency on the growth and flowering of tuberose *Polianthes tuberosa* L. *Agricultural Res. Rev.*, **59**(3): 305-312.

Farina, E. and T. Paterniani, 1986. The cultural programming of tuberose for cut flowers. Results of two years of trials in the region of western Liguria (Italian). *Annali dell'Istituto Sperimentale per la Floricoltura*, **17**(1): 49-63.

Gowda, J.V.N. 1990. Effect of sucrose and aluminium sulphate on the postharvest life of tuberose Double. *Curr. Res. Univ. agric. Sci.* (Bangalore), **19**(1): 14-16.

Guenther, E. 1952. Concrete and absolute of tuberose. *The Essential oils*, vol. 5, pp. 343-348. D. Van Nostrand Co. Inc., New York.

Hakim, S., A. El-Gamassy and M.R. Shedid, 1958. A study of the rest period of *Polianthes tuberosa*. *Ann. agric. Sci., Cairo*, **3**(1): 101-113.

Hassan, A.H. and E.A. Agina, 1980a. The effect of chlormequat (CCC) on the growth and flowering of tuberose. I. Autumn planting. *Ann. agric. Sci., Moshtohor*, **12**: 211-221.

Hassan, A.H. and E.A. Agina, 1980b. The effect of chlormequat (CCC) on the growth and flowering of tuberose. II. Summer planting. *Ann. Agric. Sci., Moshtohor*, **13**: 145-152.

Holtzmann, O.V. 1968. A foliar disease of tuberose caused by *Aphelenchoides besseyi*. *Plant Dis. Reptr*, **52**: 56.

Horner, M.B. and M.N. Pearson, 1988. Purification and electron microscopy studies of a probable potyvirus from *Polianthes tuberosa* L. *J. Phytopath.*, **122**(3): 261-266.

Itty, Shiji K. 2001. Regulation of Flowering and Postharvest Behaviour of Tuberose (*Polianthes tuberosa* L.). M.Sc. Thesis. Kerala Agricultural University, Vellanikkara, Thrissur (Kerala).

Jana, B.K. and S. Biswas, 1979. Effect of growth substances on growth and flowering of tuberose (*Polianthes tuberosa* L.). *Haryana J. hort. Sci.*, **8**(3/4): 216-219.

Jana, B.K. and S. Biswas, 1982. Effects of growth regulators on growth and flowering of tuberose. *South Indian Hort.*, **30**(2): 163-165.

Jana, B.K., S. Roy and T.K. Bose, 1974. Studies on the nutrition of ornamental plants. III. Effect of nutrition on growth and flowering of dahlia and tuberose. *Indian J. Hort.*, **31**(2): 182-185.

Khondakar, S.R.K. and B.C. Mazumdar, 1985. Studies on prolonging the vase life of tuberose cut flowers. *South Indian Hort.*, **33**(2): 145-147.

Kosugi, K. and Y. Kimura, 1961. On flower bud differentiation and flower bud development in *Polianthes tuberosa* L. *Tech Bull. Fac. Agric. Kagawa*, No. 12, pp. 230-234.

Krishnan, Anu G. 2003. *In Vitro* Multiplication and Genetic Improvement of Tuberose (*Polianthes tuberosa* Linn.). Ph.D. Thesis, Kerala agricultural University, Vellanikkara, Thrissur (Kerala).

Lopes, L.C. 1971. The effect of transplant size and population on the flowering and growth of *Polianthes tuberosa* (Brazilian). *Revista Ceres*, **18**(99) 399-405.

Malini, G. and M.A. Khader, 1989a. Effect of NaCl and $CaCl_2$ on the growth attributes of tuberose (*Polianthes tuberosa* L.) var. Single. *South Indian Hort.*, **37**(4): 239-241.

Malini, G. and M.A. Khader, 1989b. Effect of different salinity levels on flowering and yield of tuberose (*Polianthes tuberosa* L.) var. 'Single'. *South Indian Hort.*, **37**(6): 357-358.

Mitra, S.N., P.S. Munshi and S. Roy, 1979. Effects of different levels of nitrogen and bulb size on growth and flowering of tuberose (*Polianthes tuberosa* Linn.). *Indian Agric.*, **23**(3): 185-188.

Mukhopadhyay, A. and G.J. Bankar, 1983. Regulation of growth and flowering in *Polianthes tuberosa* L. with gibberellic acid and Ethrel spray. *Scientia Hort.*, **19**(1/2): 149-152.

Mukhopadhyay, A. and G.J. Bankar, 1985. Effect of split application of nitrogen on growth and yield of *Polianthes tuberosa* L. cv. 'Single'. *South Indian Hort.*, **33**(1): 60-62.

Mukhopadhyay, A. and G.J. Bankar, 1986. Studies on nutritional requirement of tuberose. *South Indian Hort.*, **34**(3): 167-172.

Mukhopadhyay, A. and M.K. Sadhu, 1985. Studies on the effects of preplanting treatments of bulbs of *Polianthes tuberosa* L. with gibberellic acid. *Progressive Hort.*, **17**(2): 125-128.

Muralee Manohar, 2000. Regulation of Growth and Flowering in Tuberose. M.Sc. Thesis. Kerala Agricultural University, Vellanikkara, Thrissur.

Naidu, S.N. and M.S. Reid, 1989. Postharvest handling of tuberose (*Polianthes tuberosa* L.). *Acta Hort.*, No. 261, pp. 313-317.

Nanjan, K., K.M.P. Nambisan, D. Veeraraghavathatham and B.M. Krishnan, 1980. The effect of nitrogen, phosphorus and potash on the yield of tuberose (*Polianthes tuberosa* L.). In: *Natl Sem. on Prod. Techn. for Commercial Fl. Crops*, Coimbatore (India), pp. 76-78.

Narayanaswamy, S. and V.R. Prabhudesai, 1979. Somatic pseudoembryogeny in tissue cultures of tuberose *Polianthes tuberosa*. *Indian J. exp. Biol.*, **17**(9): 873-875.

Panwar, R.D., S.S. Sindhu, J.R. Sharma and R.B. Gupta, 2008. Spraying effect of gibberellic acid on vase-life of tuberose (*Polianthes tuberosa* L.). cv. Single. *Haryana J. hort. Sci.*, **37**(1/2): 87-88.

Panwar, R.D., S.S. Sindhu, J.R. Sharma and R.B. Gupta, 2010. Weed management studies in tuberose (*Polianthes tuberosa* L.) cv. Double. *Haryana J. hort. Sci.*, **39**(3/4): 304-307.

Pathak, S., M.A. Choudhuri and S.K. Chatterjee, 1980. Germination and flowering in different-sized bulbs of tuberose (*Polianthes tuberosa* L.). *Indian J. Pl. Physiol.*, **23**(1): 47-54.

Patil, B.A., D.P. Bhore, J.D. Patil and A.V. Patil, 1980. Effect of planting density on production of flower stalks and flowers in tuberose (*Polianthes tuberosa*). In: *Natl Sem. on Prod. Techn. for Commercial Flower Crops, Coimbatore*, pp. 71-72.

Patil, J.D., B.A. Patil, B.B. Chougule and N.R. Bhat, 1987. Effects of bulb size and spacing on stalk and flower yield in tuberose (*Polianthes tuberosa* L.) cv. Single. *Curr. Res. Reporter, MPAU*, **3**(2): 81-82.

Rajvir Singh, N.R. Godara and R.K. Goyal, 2008. Effect of ethylmethane sulphonate on germination, floral attributes and bulb production in tuberose (*Polianthes tuberosa* L.). *Haryana J. hort. Sci.*, **37**(3/4): 283-284.

Ramachandraiah, O.S., O.S. Gautama, G. Azeemoddin and S.D. Thirumala Rao, 1987. *Proc. VIII Pafai Seminar*, pp. 116-118.

Ramaswamy, N., C. Paulraj and P. Chockalingam, 1977/1978. Studies on the influence of growth regulators on flowering and yield of tuberose (*Polianathes tuberosa* L.). *Annamalai Univ. agric. Res. Ann.*, **7/8**: 29-33.

Sambandhamurthi, S. and K. Appavu, 1980. Effect of the chemicals on the colouring of tuberose. In: *Natl Sem. On Production Techn. for Commercial Fl. Crops*, Coimbatore (India), pp. 73-75.

Sharga, A.N. 1980. Maximising tuberose (*Polianthes tuberosa* Linn.) multiplication by using bulb segments. *Progressive Hort.*, **12**(2): 71-77.

Sharga, A.N. 1982. Effect of bulb size on vegetative growth and floral characters of tuberose (*Polianthes tuberosa* Linn.). *Progressive Hort.*, **14**(4): 258-260.

Sharga, A.N. and V.S. Motial, 1983. Studies on the effects of foliar nutrition on essential oil content of tuberose (*Polianthes tuberosa* Linn.). *Indian J. Hort.*, **40**(3/4): 260-265.

Sharma, J.R., R.D. Panwar, R.B. Gupta and Surender Singh, 2008. Nutritional studies in tuberose (*Polianthes tuberosa* L.). *Haryana J. hortic. Sci.*, **37**(1/2): 85-86.

Singh, R.S., V.S. Motial and L.B. Singh, 1974. Studies on salt tolerance of tuberose (*Polianthes tuberosa* Linn.). *Plant Sci.*, **6**: 80-88.

Trueblood, E. 1973. Omixochitl ~ The tuberose (*Polianthes tuberosa*). *Economic Bot.*, **27**: 157-173.

Trujillo, E.E. 1968. Diseases of tuberose in Hawaii. *Circ. Univ. Hawaii Coop. Ext. Serv.*, No. 427, p. 13.

Vijayalaxmi, M., A. Manohar Rao, A.S. Padmavatamma and A. Sivashankar, 2011. Influence of different holding solutions on vase life of tuberose. *J. Ornam. Hort.*, **14**(3/4): 106-111.

Yadav, L.P. and T.K. Bose, 1987. Chemical weed control in tuberose and gladiolus. *Acta Hort.*, No. 205, pp. 177-185.

Yadav, L.P., T.K. Bose and R.G. Maiti, 1985. Response of tuberose (*Polianthes tuberosa* L.) to nitrogen and phosphorus fertilization. *Progressive Hort.*, **17**(2): 83-86.

Yadav, L.P., T.K. Bose and R.G. Maiti, 1984. Effect of bulb size and depth of planting on growth and flowering of tuberose (*Polianthes tuberosa* L.). *Progressive Hort.*, **16**(3/4): 209-213.

Zeven, A.C. and P.M. Zhukovsky, 1975. Central American and Mexican Centre. *Dictionary of Cultivated Plants and their Centres of Diversity*, pp. 162. Centre of Agriculture, Publishing and Documentation, Washington.

30

Proteaceous Ornamentals

Kalkame Ch. Momin, Y.C. Gupta, S.R. Dhiman, R.L. Misra and Sanyat Misra

Introduction

Floriculture research has many levels of enquiry. Scientific research furthers our understanding of plant growth, delving into biochemical pathways to reveal patterns in growth and flowering regulation. Fresh flowers play a unique role in our lives, both in celebration and in sorrow. Unlike most other horticultural crops, demand for fresh flowers is often related to fashion trends. Our growing multicultural population has seen an increased use of flowers outside of the traditional peaks of Valentine's Day, Mother's Day and Christmas. For other festive occasions like Chinese New Year and Orthodox Easter, fresh flowers are used extensively. New flower lines and cultivars are being introduced all the time, either in new colours and shapes of traditional flowers like roses and chrysanthemums, or as new Australian native crops. Over the last decades, utilization of the members of the Proteaceae for cut flowers has intensified and changed from wild harvesting to cultivation for improved quality of the floral product.

The family name Proteaceae is given due to genus *Protea*. Plant life research indicates that Proteaceae probably originated in South Africa along the southern coastal mountain ranges. Proteaceae occurs in region where rainfall varies from as low as 180 mm to 2,500 mm per annum. The family includes over 1,400-1,500 species in over 60 genera of trees, shrubs and herbs from the southern hemisphere, especially Australia (30°0' S, 135°0' E) and South Africa (33°55' S, 18°22' E). Out of these species, over 800 in 45 genera are from Australia; Africa is home for about 400 species, including 330 in 14 genera from the Western Cape; about 90 species including *Embothrium* occur in CS America; 80 on the islands of New Guinea; and 45 in New Caledonia; though Madagascar, New Guinea, New Zealand (*Knightia, Persoonia, etc.*) and SE Asia host only small numbers of species (Rebelo, 1995). Today, the members of Proteaceae are cultivated for their flowers or colourful involucral bracts largely in South Africa and also in other countries like Australia, USA, Chile, Israel, New Zealand, Portugal, Spain and Zimbabwe.

The export of Proteaceae cut flowers provides an important source of foreign exchange and livelihood for these countries. Total area of Proteaceae under cultivation is now more than 6,000 ha, out of which 3,000 ha is in South Africa and 2,000 ha in Australia. The area under cultivation of Proteaceae in South America is approximately 8 ha, with 0.5 ha in Chile and 7.5 ha in El-Salvador. Spain and Portugal have approximately 30 ha of cultivated Proteaceae, located mainly on the islands of Madeira and Tenerife. The *Protea* cut flower industry started by using indigenous genetic material from both South Africa and Australia in 1940s (Parvin *et al.*, 2003). The products were initially harvested only from wild growing plants in their natural habitat, and then after when the markets were fascinated by the range and novelty of protea products, demand exceeded the supply, it was only then that its commercial cultivation started. Growers shifted from collection in the wild to cultivation, from seed-growing where plants do not come true to the type to clonal propagation to perpetuate desired types, assessment of the product for use in the cut flower industry and to develop speciality crop novelties for pot culture, keeping in view the growing floral industry based on Proteaceous products. This also caused improvement in its packaging, in its shipment through sea routes and air, *vis-à-vis* development of new markets so that higher income is generated.

Banksia ericifolia, B. robur, B. spinulosa, Grevillea thelemanniana, Isopogon formosus, Leucadendron adscendens, L. argenteum, L. comosum, L. eucalyptifolium, Leucospermum catherinae, L. conocarpodendron, L. cordifolium, L. cuneiforme,

Protea cynaroides, P. eximia, P. grandiceps, P. longiflora, P. obtusifolia, P. repens, etc. are quite suitable for use in **landscaping**. *Leucospermum catherinae, L. conocarpodendron, L. cordifolium, L. reflexum, Protea barbigera, P. compacta, P. cynaroides, P. eximia, P. lacticolor* hybrid, *P. pulchra, P. repens, Serruria florida, Telopea speciosissima, etc.* are suitable as **cut flowers**.

Typical Features of Proteaceae

Proteaceous plants tend to be sclerophylls, and they have leathery leaves which implies that these do not require a high atmospheric moisture level, *vis-à-vis* can tolerate moisture stress to a great extent. Their leaf buds are not protected by scale leaves which make them susceptible to frost-damage. Mild climates are quite congenial for their growth throughout if the soil remains sufficiently moist, and lack of water coupled with warm conditions causes these plants to go dormant for a while in summer. Many genera have proteoid roots (*Coprosma* or *Coriaria*) which enable them to draw nutrients from the soil even when nutrient level there is quite low (Thomas, 1979; Matthews and Carter, 1983). However, in nature, Proteaceous plants grow generally on infertile soils, and some even in sandy soil. Certain members of the family produce lignotubers, a short of swelling at trunk base which throws out new shoots, especially when main trunk is damaged. One more feature is quite prevalent in such plants that within the same species, very local forms have developed differing in flower colour, flowering time and growth habits. In *Grevillea* and *Protea*, bracts and floral parts both are attractive but in *Leucospermum* the bracts are small though styles and stigma are centre of attraction whereas in *Leucadendron* and *Telopea* the leaves surrounding the flowers are point of attraction (Salinger, 1985). Except for some *Grevillea*, the flowers grow mostly in clustered heads as in Asterceae.

Researchable Issues

Protea has become an established horticultural crop, with a world sale of approximately 8 million flowering stems. In South Africa 3.01 million stems are exported, 1.14 million sold through the formal market, and 1.01 million sold by the informal sector. Other producing countries do not have figures for sales of *Protea*, but total sales are estimated at 3.0 million. Less than 1.5 million stems are sold through the Dutch auction system annually. Proteas create great interest wherever they are seen and offer not only a new exotic appeal, but a good value as well. Protea industry has come a long way- from harvesting wild flowers in South Africa and Australia to cultivated plantations of superior cultivars in many different parts of the world. The continued international interest in Proteaceous cut flower crops provides the challenge for the plant breeders to supply new cultivars. This goal will be facilitated by a number of developments which require action over the next decade. One of the most important factors is the international co-operation in all aspects of cultivar development. In the last decade, research efforts enabling floriculture of Proteaceae have been concentrated on species selection, breeding, gene bank maintenance, propagation techniques, plant protection and postharvest techniques. The basics of plant nutrition and root growth aided the establishment of protea plantations. Knowledge of pests and diseases found in natural stands gave an idea of what to expect in cultivation, often enabling early control. To support the change from wild harvesting to cultivation, problems associated with the seed germination and plant establishment were studied. Research topics then began to explore ways to intensively cultivate proteas to achieve higher yields of acceptable quality at times of peak market demand. To enable greater control of plant growth and to provide uniformity, protea plantations were established using clonal material, rather than the seedlings. Propagation research provided protocols detailing propagation media, climatic conditions and treatments essential to successful root formation. The use of rootstocks was investigated to improve plant adaptability, but the higher costs associated with these techniques precluded widespread adoption of these results at this time. Plantations established using clonal material formed a monoculture in which pest and disease problems were enhanced. Studies on the susceptibility of species and cultivars to diseases allowed matching of cultivars to climatic areas to assist disease control. To offset the increased costs associated with the commercial production in plantations high productivity became essential. Despite the fact that proteas characteristically grow in nutrient-poor soils in the wild, the increased demands made on plants in cultivation depleted nutrient reserves. Research on nutrition management contributed to maintaining plant vigour. Investigations into flowering control aimed at matching supply of product to periods of high market demand. Global concerns about the impact of wild picking on conservation of natural flora put pressure on the industry to harvest fewer species from the wild. Consequently new products became economically viable to grow commercially and current management regimes, including propagation, were tested to validate their applicability. Cultivar registration of new products facilitated differentiation in the market between cultivated and wild-picked products, protecting market share for superior products. As prices increased research focused on aspects of quality to help industry attain a higher income per unit area. Postharvest problems such as leaf blackening were investigated. Cultivation of disease resistant or tolerant cultivars reduces production costs while providing better quality product. In addition to improved quality, higher income was attained by preferential market share captured by novelty products. Development of new products was supported by breeding and selection trials. As the production of cut flowers and plant material expanded, worldwide distribution costs came under close scrutiny. While growers increased packing efficiency, research sought reliable methods to use alternative to air transport. The high costs of air freight also forced attention onto reducing wastage in consignments. Improved fumigation techniques and processes have contributed to reduction in shipment rejection due to failure to comply with phytosanitary measures.

Crop Improvement

The main advantage of proteaceous blooms is their conspicuous and showy nature, which renders them ideally suited as standard blooms which form the focus or centerpiece of large floral arrangements. This prominence also means that the quality of the bloom is paramount and achieving this quality is one of the main challenges facing the plant breeders. The cut flower industry is subjected to changes in consumer preferences and plant breeding is by nature a long term undertaking. Sensitivity to and appreciation of potential future trends in the industry must be taken into account in the development of the selection criteria. Novelty is an important factor in success and the breeder must exploit available germplasm to maintain the flow of new and interesting cultivars. For most Proteaceous crops, the important selection criteria should be number of blooms per bush or per unit area; flower quality should qualify the international standard so there should be long straight stems with large, attractive and durable blooms; attractive leaf shape, texture and colour; resistance or tolerance to devastating root rot fungus *Phytophthora cinnamomi*; *etc.*

The Species

Protea is the most widely known genus of the Proteaceae, and other genera in this family that are widely used in floriculture are *Leucospermum* (Criley, 2001), *Banksia* (Sedgley, 1995) and *Leucadendron*. Other important genera, in addition to four mentioned above, are *Aulax*, *Dryandra*, *Grevillea*, *Hakea*, *Isopogon*, *Mimetes*, *Serruria* and *Telopea* which are being described here under. Collectively, all these genera would henceforth be called as Proteaceous due to the family Proteaceae or under the umbrella of *Protea* due to this being the most dominating genus.

Aulax. This is a South African genus of small- to medium-sized shrubs. This genus and *Leucadendron* are the only dioecious genera (separate male and female plants) of Proteaceae. Seeds of all three species, *viz. Aulax cancellata*, *A. pallasia* and *A. umbellata* are available but only *A. cancellata* is commonly planted. It grows to 1.5-2.0 m × 1 m and has fine needle-like leaves. In spring, female plants produce red edged yellow flowers that develop into red seed cones. The catkin-like male flowers are yellow, as are those of *A. pallasia* and *A. umbellata*, the female flowers of which are not very showy. *A. pallasia* grows to about 3 m and *A. umbellata* some 1.5m high. All are hardy to about -5 °C and are usually raised from seeds.

Banksia. An evergreen genus with 60 species, having handsome foliage, is found throughout Australia and into New Guinea, but only a few species are suitable as cut flowers either due to by-passing the flowerheads or suitable stem lengths are not available. The genus *Banksia* is named so in honour of the famous English naturalist and patron of science Sir Joseph Banks (1743-1820) who sailed with Captain Cook on his first voyage to the Pacific. Banksias come true through seeds with only a little variation in growth habits or flower colours. Its species range widely from the ground cover (low prostrate) to medium and tall trees, and so different species prefer different climates and soil conditions but planting these in the garden will attract birds of different kinds. Their foliage is very attractive, ranging from fine needle-like to wide serrated leathery, and the colourful blooms or brushes range in size. The flowering season is primarily from late winter to late spring and most species have cylindrical cone-like flowerheads composed of densely packed filamentous styles radiating from a central core. Creamy-yellow to light golden-yellow is the predominant colour range, although a few species, such as *B. ericifolia* (heath banksia) and *B. praemorsa* have golden-orange flowers and those of *B. coccinea* are red. Flowers of *Banksia* are suitable for drying of their flowerheads. The large leaves of *B. grandis* or *B. speciosa* are used by florists and the mature fruiting heads which are known as cobs or cones are used as dry flowers. Most species have narrow serrated leaves that are mid to deep green above and silvery-grey below but *B. ericifolia* has fine needle-like dark leaves. *B. ericifolia* is a dense shrub growing about 3.0 m, bearing long russet-coloured flowers and is most suitable for planting in the coastal gardens. In *Banksia*, leaf size varies from very small up to 50 cm long as in *B. grandis*. Hardiness varies with the species, some are quite frost-tender but some others can bear even -10 °C. From New South Wales, there is a species *B. serrata* which was discovered by Banks and Solander in 1770 from Botany Bay, as well as from Victoria and Tasmania. It grows up to 6.0 m high with tomentose young branches, leaves are 7.5-15.0 cm long, oblong-lanceolate, coriaceous and with deeply cut regular serration. Exterior of the broader tips of the perianth lobes are often hairy and bluish-grey but with opening of the flowers further when style predominates, it shows yellow colouration. *B. baueri* grows to about 2.0 m tall and is a dense-growing shrub. The flowers are large furry, yellowish inside and grey outside and appear during winter and spring. *B. candolleana* is a low shrub with yellow flowers. The flowers of *B. coccinea* are striped-red. *B. collina* (syn. *B. spinulosa*) is an open shrub growing up to 3.0 m tall, bearing narrow and notched foliage and honey-coloured brushes with black pins, and it grows well under filtered light positions. *B. grandis* grows up to 7.0 m high with yellowish flowers. *B. littoralis* from swampy areas of Western Australia is a tree some 18 m tall. Its inflorescence is charming yellow. *B. integrifolia*, a hardy species grows 3.0-4.0 m or sometimes even up to 12.0 m high, bearing 15 cm long and 2.5-4.0 cm wide, entire or occasionally slightly dentate leaves whose upper side is dark green but lower silvery-white, and inflorescence 7.5-15.0 cm long are produced during autumn, winter and spring, with about 2.5 cm long and greenish-yellow perianths. *B. menziesii* is a beautiful small tree growing up to 8.0 m tall and bears red flowers. *B. occidentalis* grows up to 3.0 m tall with red flowers. *B. paludosa* grows only up to 3.0 m tall and bears yellow flowers. *B. prionotes* is an attractive tree growing up to 8.0 m tall with long deeply toothed foliage, and the large orange flowers are produced during autumn and winter. *B. marginata* reaches to a height of up to 5.0 m, bearing narrow foliage which is white underside, and the small yellow flowers are produced during summer, autumn and winter. *B. robur* which grows up to 3.0 m tall, bears large foliage and the flowers appear as if these have developed along the limbs and up the trunk, initially appearing as bottle-green but turning yellow, bronze and brown as they mature. *B. speciosa* grows up to 4.0 m tall and bears greenish flowers.

Dryandra. An evergreen Australian genus of around 60 species of shrubs ranging in height from about 1.0-4.0 m in height and are found only in the south-western regions of

Western Australia. The genus commemorates Jonas Carlsson Dryander (1748-1810), a Swedish botanist and one time Librarian to Sir Joseph Banks. The leaves are narrow, mid to deep green, that are often very long and narrow with sharply toothed edges. Its flowers are similar to those of the related genus *Banksia* though differs in having less woody seed capsules and more dense heads of flowers surrounded by basal bracts. The rounded flowerheads, which appear from mid-winter, are usually light to bright yellow. The flowerheads of *Dryandra* are suitable for drying. *D. drummondii* which is similar to *D. calophylla*, both are found wild in the Swan river area of SW Australia, bearing pale-yellow flowers inside while ouside has covering of gingery perianth hairs. In the flowerhead, outer flowers open first, the gingery perianth lobes curl back to expose the yellow style and stigma and then the process proceeds to the inner side of the head up to the centre. Its young leaves are almost covered with hairs that are lost with maturity. It differs from *D. calopylla* as it does not have lateral subterranean shoots and small flowering heads. *D. formosa*, which grows to about 3.0 m and is hardy to around -5 °C once established, though most other species are less hardy. Dryandras are superb long-lasting cut flowers and some even dry well. They will grow on extremely poor soil and generally react badly to most fertilisers.

Grevillea. With some 250 species, this is the largest of the Australian Proteaceous genera, with a few found in New Guinea, New Hebrides and New Caledonia. Most of the common garden species and cultivars are ground covers to medium-sized shrubs (up to 3.0 m) with needle-like foliage. However, some species are far larger and attain the height of a complete tree. Its flowers resemble *Hakea* though differs in fruits. *G. alpina* (syn. *G. alpestris*) is found in Tasmania, Victoria and New South Wales and was first grown in 1857 in the Australian House in Kew. Its flowers have combination of yellow and red or pink and red. *G. australis*, a frost-hardy species grows prostrate and bears fragrant white flowers. *G. anethifolia* grows up to 2.0 m, is hardy and bears fine-pointed foliage, and creamy and greenish flowers. *G. banksii* has pinnate leaves similar to *G. robusta* but grows only to about 3.5 × 3.0 m in height. *G. baueri* is a prickly plant which grows up to 2.0 m tall, doing well in damp soils and bears deep red spider flowers twice in the year. *G. gaudichaudi* bears lobed foliage and burgundy tooth-brush flowers during spring or summer. *G. hookerana* is a frost-resistant, grows up to 3.0 m high and bears red toothbrush flowers in spring. Some of the variants of *G. juniperina* are nearly prostrate to semi-prostrate, ideal for embankments though others may be spreading tall shrubs bearing flowers in the colour range of cream, yellow, apricot or red. *G. juniperina* 'Molonglo' bears spreading arching branches with numerous apricot-coloured flowers. *G. juniperina* f. *sulphurea* (syn. *G. sulphurea*) is a rounded bushy shrub with almost needle-like leaves, above dark green and recurved though silky-haired below. It produces clusters of small spider-like pale-yellow flowers during spring-summer. *G. lanigera* produces a massed display of cream and pinkish-red flowers. *G. laurifolia* is frost-hardy and spreads above 4.0 m across. *G. punicea* (red spider flower) was first introduced into western Europe around 1857 from its native place of New South Wales. The species bears both large- and small-leaved plants and apricot or red flowers. *G. repens* (rock grevillea) grows prostrate and does well in demi-shaded conditions but in well-drained soil. Its leaves are holly-like and the flowers toothbrush-shaped red and green. *G. robusta* (silky oak), which is often seen in mild areas, can grow to 20 m and is common with most of the larger species. It has large pinnate leaves. The more densely-foliaged plants, especially *G. juniperina* and *G. rosmarinifolia* are often used as hedging plants. The plants of *G. rosmarinifolia* are frost-tolerant, grow 1.5-3.0 m high and bear green needle-like leaves and small red and cream flowers appearing from spring. For brightness and profuse flowering the forms of *G. speciosa* are highly suitable, and its ssp. *dimorpha* bears brilliant red flowers during winter. *G. sericea* (pink spider flower) grows about 2.0 m tall, is hardy and bears narrow pointed foliage and pink flowers most of the year. *G. thelemanniana* (*G. preissii*) is a spreading shrub having soft-tomentose young growths, and the leaves are pale or glaucous, 2.5-5.0 cm long, pinnatifid, lower pinnae divided with linear segments. Racemes are up to 4.0 cm long, dense and terminal, and bear pink flowers with green tips. *G. triloba* is hardy, grows up to 2.5 m high and bears grey spiky foliage and fragrant white flowers. There are many attractive *Grevillea* hybrids introduced over the years, especially in Australia, and some of these are 'Robin Gordon' with large reddish heads at the ends of the branches throughout the year, which grows to 2.0 m with spread of 3.0-4.0 m and responds very well to pruning; 'Misty Pink' grows up to 3.0 m, bearing heavily divided silvery foliage and large pink spidery flowers for most of the year; 'Pink Surprise' is much taller and its soft and pink bushes are produced mainly during autumn and summer; 'Poorinda Tranquility' is 1.0 m tall with serrated foliage and mauve flowers in spring; *etc. Grevillea* flowers are often described as 'spider flowers'. This refers to the styles of some species, which tend to radiate from the centre like a spider's legs. Some species have 'toothbrush' flowers; the styles are all on one side like the bristles of a toothbrush. The best known example of this type of flower is the common red-flowered cultivar 'Robin Hood'.

Hakea. *Hakea* was named so to honour Baron von Hake, a German botanist. This evergreen Australian genus includes about 130 species, few of which are widely cultivated even indoors. They are drought-resistant. The most common is probably *H. laurina* (*H. eucalyptoides*), the 'pincushion or sea urchin hakea', grows up to 9.0 m high with reddish-brown bark, bearing up to 15.0 cm long and some 2.5 cm wide, bluish-green, elliptic or lanceolate and sickle-shaped leaves tapering to a petiole, and produces strikingly handsome flowers of crimson colour in a globular involucre heads, from which numerous golden-yellow styles protrude in every direction. When not in flower, this species could easily be mistaken for a small eucalyptus. Its mature trees have a slightly weeping habit. The flowers appear in late autumn and early winter, opening cream and turning to orange and red as they age. This shrub is hardy to about -5 °C once well established and is easily grown in most well-drained soils. *H. elliptica* is a shrub which bears 5.0-9.0 cm long, spiny, oval or elliptic with undulated margins, almost sessile and attractive bronze foliage when young, and its flowers appear in white globose cluster. *H. saligna* is a glabrous and pale shrub (younger shoots hairy), growing up to 2.5 m high. Its leaves are 7.5-15.0 cm long, oblong or lanceolate, tapering to a short petiole and pinnately veined. Numerous small white and dense flowers are borne in clusters with recurved corolla. *H. ulicina* bears erect branches, with 2.5-20.0 cm long and hardly 3 mm wide, dense, narrowly linear, entire and acute foliage in appearance to ulex, and the flowers are very small. *H. suaveolens* (*H. pectinata*) is a 4.5 m high rounded shrub bearing 5.0-10.0 cm long, 1-5 irregularly terete-lobed leaves with spiny tips, and the flowers are fragrant white. Of the other species, the most common ones are *H. salicifolia*, *H. prostrata* and *H. sericea* (syn. *H. lissosperma*). They are hardy to about -8 °C or slightly lower and are easily grown in most soils. *Hakea salicifolia* has narrow, willow-like leaves; and spidery, white flowers that are produced in spring. It grows up to 5.0 m high and will tolerate poor drainage. *H. prostrata* and *H. sericea* have fine needle-like leaves and white or pale-pink flowers in winter and early spring. It grows to about 3 × 2 m high. All the members of this genus are usually raised from seeds, some can be grown from even cuttings, and a few, such as *H. franciscana*, are weak growers that often perform better when grafted onto more vigorous stocks, such as *H. salicifolia*.

Isopogon. The genus is commonly referred to as 'drumsticks' due to shape of the floral stems and unopened buds, though the name is often used for *Isopogon anemonifolius*. It is an Australian genus having 34 species of small to medium sized shrubs, most of which grow 1.0 to 2.0 m high and wide. It has a preference for poor but well-drained soil and collapses quickly if over-watered or overfed. Most species have about 7.5 cm long and narrow lanceolate leaves, though some such as common *I. anemonifolius* which is hardy, grows up to 2.0 m high, and bears finely cut foliage reminiscent of marguerite daisy or anemone leaves. The floral cones open in spring mainly to white, yellow or pink, predominantly yellow. The two most widely grown species, *I. anemonifolius* and *I. anethifolius* are hardy and can tolerate up to -5 °C, but many other species, such as *I. cuneatus* and the temptingly beautiful pink and yellow-flowered *I. latifolius* are damaged at temperatures below -2 °C. *Isopogon* species are usually raised from seeds. *I. dawsonii* (conebush) grows only up to 1.0 m high, does well in moist soils and bears yellow-cone-like flowers in spring.

Leucadendron. In Greek, *Leucadendron* means white tree. *Leucadendron* and *Aulax* are the only dioecious genera (separate male and female plants) of Proteaceae. Species of this genus are most widely grown of the South African Proteaceae and many are valued for their long-lasting qualities of flower bracts once cut. Most are medium-sized shrubs around 1.0-2.5 m high. However, one of the best known species, the silver tree (*Leucadendron argenteum*), can grow up to 10.0 m high though is less widely grown. Many species and cultivars are grown, but probably the most widely planted is 'Safari Sunset'. It is a hybrid between *L. laureolum* and *L. salignum* and is fairly typical of the genus. It has narrow, lanceolate leaves some 10.0 cm long. Some species, such as *L. argenteum* have tomentose foliage but 'Safari Sunset' does not. The upward-facing foliage densely covers the narrow, upright branches and develops deep red tints at the flowering tips. As the insignificant flowers near maturity, the bracts become intensely coloured. 'Safari Sunset' has red bracts but others appear in cream, yellow, pink or orange tones. The species and hybrids vary considerably in hardiness but most will tolerate frosts of at least -3 °C provided they have good drainage with proper humidity as it does not like excess humidity. 'Safari Sunset' is hardy to about -8 °C as to those of *L. salignum* and *L. laureolum*. North Island leucadendrons, generally thrive in all but the coldest central areas and they can be grown with varying degrees of success in all coastal areas of the South Island. Both, *Leucadendron* and *Protea* have similar requirements for their cultivation but nutritional requirement in case of *Leucadendron* is quite limited.

Leucospermum. A South African genus of about 50 species, most of which are medium to large shrubs that grow to about 1.5-3.0 m high. *Leucospermum* has two classes based on the type of flowerhead. The first group includes those species which have single flowerheads such as *L. cordifolium*, *L. lineare* and *L. tottum*. The second group has multiple flowersheads as seen in *L. cordifolium*, *L. muiri* and *L. mundii*. Some, such as *L. reflexum* have strongly upright growth habits but most including the commonly cultivated species *L. cordifolium*, are dense and bushy. Both of these species have tomentose greyish-green leaves that are usually broadly oval-shaped, often with small red-tipped lobes. The flowers are variously described as 'Catherine wheels', 'pincushions' and 'sky rockets', all of which refer to the numerous radiating styles. These are often incurved, creating a cupped effect. The flowers usually appear in late spring and continue for about two months. They are attractive when fresh, but often become unsightly once they die off. Most garden leucospermums are cultivars of *L. cordifolium* and are hardy to occasional frosts of about -5 °C, but they resent wet or humid winter conditions, which can often lead to tip die back. *L. reflexum* is an erect shrub with ascending branches. Leaves are dense, small and grey- to blue-green. Tubular, slender and crimson flowers with long styles appear in tight rounded heads during spring and summer.

Mimetes. This South African genus includes 16 species with faintly coloured leaf-like bracts, and all the species are more or less similar to *M. hottentoica*. *Mimetes* is easily recognized by the compact clusters of often brightly coloured flowers embedded on a sessile flowerhead around which the subtending bracts are also occasionally coloured. This is a bizarre genus bearing softly hairy silver leaves, withered and reflexed perianth segments of the yellow hue, and red and yellow styles that contrast with almost black stigma. Anthers come out from the spoon-shaped tips of the perianth lobes. *M. stokoei* bears tinged-red bright yellow styles, almost black stigmas, pinkish bracts subtending the flower clusters and satiny leaves. *M. cucullatus* is widely grown species, growing to about 1.5 × 1.5 m and is hardy to around -3 °C, having 4.0 cm long oblong leaves with small lobes at the tips, that densely cover the branches like upward facing scales, and the plant prefers moist, well-drained soil and is not very drought resistant. The small white flowers are enclosed within leaf bracts that change colour to a bright red as the flower buds mature. This species is usually raised from seeds. *Mimetes* may flower throughout the year but is usually at its best in late-spring when the new growth appears, as this is also red.

Protea. *Protea* was named so by Linnaeus in 1753, referring to the Greek mythical god, Proteus, who could change his shape at will, and this is an apt name due to the wide diversity of the genus. It is a large evergreen genus with 136 species of which 70 are distributed in the temperate zones of the southern hemisphere and the balance distributed in the sub-tropical to tropical zones of southern hemisphere. The natural habitat of *Protea* ranges at elevations from sea level to over 2,000 m and inhabits the well-drained, moderately acid and low fertility areas but not the humid ones. Once planted, they are low-maintenance plants. *Proteas* are suitable for drying of their flowerheads. The most widely recognized species in the genus is *Protea cynaroides* (meaning 'like a globe artichoke'), the 'king protea', the national flower of South Africa. It has flowerheads up to 30.0 cm across with widely spaced bracts arranged around a peak of flowers that vary in colour from near white to soft silvery-pink to deep rose-pink to crimson. *Protea magnifica*, the 'queen protea' is grown for its large 15-20 cm flowerheads of white to rose-pink to salmon colours. *Protea compacta* has lanky flower stems on a stiffly upright, sparsely branched shrub that grows to 3.5 m tall. The prominent flowerheads, unobscured by foliage, make fine winter cut flowers. *P. obtusifolia* is found growing in the white sandy soils in the coastal region of southern Cape Province with one of its white bracted variety occurring in the Albertinia district of South Africa. It bears smooth, shiny and brittle chaffy-textured floral bracts. *P. amplexicaulis* from southern Cape Province being found at altitudes between 300-610 m, is named so as the leaves clasp the stem from which they arise. Its remarkably coloured red-brown inflorescences are placed near the ground on one of the aerial branches, bracts are velvety exterior and paler interior. *P. grandiceps* is recorded from Table Mountain, Devil's Peak (near Cape division), Langeberg (near Swellendam), the Jonkershoek Mountains (near Stellenbosch) and the Cockscomb Mountains (near Port Elizabeth). It is a bush growing up to 1.5 m high. Its leaves only near the inflorescence are edged red though leaves below are entirely green. It is slow-growing and flowers only when very old. Its red bracts never open fully but even look very beautiful because of the pure white, grey or brownish beards on the top of the terminal inflorescence bracts. Other important species include *P. eximia, P. lacticolor, P. longiflora, P. neriifolia, P. repens* and *P. susannae*. Proteas prefer acidic soil having a pH range of 5.0-5.5, requiring plenty of nutrition for best growth and flowering but at planting no fertilizer should be used but soil should have sufficient humus, and afterwards only compost gives better results and not the animal manures. Its requirement for potash is very limited. After planting in spring or autumn, these do not require frequent disturbance so only little cultivation should be done only when necessary. Since its plants grow haphazard, so when needed these may be pruned to keep in shape. Apart from cut flowers when these last for a few weeks, its flowers can also be used after drying. Seeds can be sown 4-6 months after the flowering is over and the flowersheads have been allowed to mature on the plants. Seeds are sown deeper to their width and soils are kept moist until germination, and then seedlings should be grown in the 15-cm pots before planting these. Through seeds these take 3-4 years to bloom.

Serruria. *Serruria florida* (blushing bride) is very popular with florists because its nigella-like papery white bracts are very delicate and last well as cut flowers. The bracts, which are surrounded with finely cut lacy leaves, are produced freely in winter and spring. Blushing bride can be difficult to grow, because not only it is frost-tender (it tolerates only occasional exposure to -2 °C), it must also have full sun and absolutely perfect drainage. *Serruria* is a genus of 44 species from South Africa, of which the only other species commonly grown is *S. rosea*. It is a densely foliaged 70 × 90 cm bush with small pink bracts and is slightly hardier and definitely easier to grow than *S. florida*. *Serruria* species should be raised from seeds.

Telopea. Commonly known as waratah, native to Australia, this evergreen genus includes just four species. The New South Wales waratah (*Telopea speciosissima*) is most commonly grown species, having oblong, finely serrated leaves that are up to 12.5 cm long with small notches or lobes at the tips. It develops into a large shrub or small tree up to 5 × 5 m. The flowers, which are produced in spring and carried at the tips of the branches, are impressively large, bright red, and composed of numerous incurving styles surrounded by red foliage bracts. Several cultivars, such as the semi-dwarf 'Forest Fire' (2 × 2 m) are commonly available. The 'Victorian Waratah' (*Telopea oreades*) is a similar plant with slightly lighter coloured leaves and flowers. Both of these species and the cultivars are hardy to around -8 °C. *T. truncata* (Tasmanian waratah) is a upright-growing shrub which becomes bushy with age. The leaves are densely formed and deep green. The flowerheads are rounded bearing small, tubular and crimson flowers in late spring and summer.

Propagation

Seed is still widely used for the establishment of varieties not represented by cultivars and increasingly for the establishment of foliage plantations. The decline in research reports on seed germination techniques for commercial Proteaceaous varieties probably reflects the increased availability of cultivar material and/or that agriculturally satisfactory seed dormancy breaking techniques had been established. Dormancy seems to be imposed by a low temperature requirement and by the action of the pericarp, which prevents simultaneous germination of all achenes. Scarification, stratification, and incubation in pure oxygen improved the germination of *P. compacta*. Imbibition of *P. eximia* and *P. nerifolia* seeds to 100 ppm GA_3 for 24 h improved significantly the percentage and rate of germination (Rodríguez Pérez, 1995). *Leucadendron* species germinated optimally when temperature fluctuations between 20 °C (daytime) and 10 °C (night) were used (Sedgley *et al.*, 2001). Many species only required a low temperature to germinate optimally (11 °C). *Leucadendron* can be divided into two sections based primarily upon characteristics of the seed. Section A 'Leucadendron' has rounded nut-like flattened seeds, while section B has seeds that are flattened and have small wing. Flattened seeds from section B are reported to be easier and faster to germinate than the round nut-like seeds of section A, requiring no pretreatment prior to being sown in a 1:1 sand-perlite mix.

Most commercial Proteaceous species are propagated by using approximately 20 cm long terminal **semi-hardwood cuttings**. In general, a 5 second basal dip in 1,000 to 4,000 ppm IAA is followed by setting the cuttings in well aerated medium with intermittent mist and bottom heat at 22 to 25 °C (Malan, 1995). Rooting generally occurs within 6-16 weeks. Auxin concentration, auxin carrier, and hormone mixtures all influence rooting success. Specific requirements have to be adapted for each cultivar for optimum results. The time of taking the cuttings is important in *Protea,* where growth flushes are not always well synchronized (Malan 1995), because the physiological status of the new growth flushes may not be consistent. Scarring of the base of the cutting is effective in promoting rooting of some *Protea* cultivars. The propagation of *Protea obtusifolia* by stem cuttings following the standard technique, can be improved upon, when prior to hormone treatment, four longitudinal cuts, 2cm long, equally spaced are made in the bark of the cutting-base. Control of diseases while plants are rooting is important to ensure success and includes proper sanitation in the mother plants. Rodriquez *et al.* (2001) recommended the terminal cuttings for the vegetative propagation of *Leucospermum* spp., although some commercial nurseries also use basal cuttings. In some plant species, the use of basal wounding technique alone or combined with hormonal treatments (IBA) has stimulated root formation in stem cuttings. Use of this technique has improved the rooting in *Protea obtusifolia, Leucadendron* 'Safari Sunset' propagated in spring when rooting is difficult and in *Leucospermum* 'Sunrise'. Wounding followed by treatment with 4,000 ppm of IBA is quite effective for propagation of *Leucospermum cordifolium* 'California Sunshine' by terminal stem cuttings in order to shorten rooting process, although unwounded cuttings treated with 2,000 ppm of IBA were not successful.

Tissue culture techniques for propagation of *Protea* (Rugge, 1995) have been developed. The major factors that limit the tissue culture of the Proteaceae apparently relates to obtaining sterile explants from the field grown plants, phenolic browning of the medium and explants as well as difficulties in getting axillary buds on explants to sprout. However, shoot proliferation has been obtained in *P. cynaroides, P. obtusifolia* and *P. repens.* Successful transplanting of rooted shoots to soil has not been achieved. Rugge (1995) studied the micropropagation of *Protea repens* cv. 'Embers' and concluded that treatment of the actively growing axillary shoots on field grown mother plants with 200 mg/l BA significantly reduced browning and promoted bud sprouting *in vitro*. Through the use of GA_3 *in vitro* on the sprouting of multinodal explants of *Protea repens* cv. 'Embers', Rugge (1995) reported maximum axillary bud sprouting with 3-6 mg/l though increase in the GA_3 concentration to 9 mg/l did not result in the number of sprouted shoots. Croxford *et al.* (2006) obtained multiplication of *Leucadendron* hybrids on MS medium containing 20 g/l sucrose and 3 g/l Phytogel. They also studied the effect of rooting substrate and environment on the root strike and survival of *in vitro* grown shoots of *Leucadendron* genotype 007.

Cultural Practices

Small-leaved Proteaceous plants such as *Grevillea* are not so specific in their requirements though larger-leaved plants such as *Leucospermum* are very specific. Normally, the Proteaceous plants require acidic medium with a **pH** of 5.0 to 5.5 (Eliovson, 1965) for growing and flowering. The plants in case of most of the genera and species are raised in pots first and then planted at their permanent positions when these are 30-60 cm long. **Soils** for their planting should be free-draining, especially during winter months when evaporation and transpiration are very low. In water-logged soil, these are infected with *Phytophthora cinnamomi,* the root rot fungus, collapsing entire plant. However, some genera such as *Hakea* and some *Banksia* species feel comfortable in clay soils. **Planting** is carried out from late summer to early autumn or late spring to early summer in the sub-tropical and sub-temperate areas as during both the seasons plants experience warmer days when roots are encouraged to appear, however, planting in mid-summer is avoided as high temperatures irrespective of the soil moisture, cause stress to the plants thereby resulting into leaf scorching. Planting in Proteaceae is effected closely, and var. 'Safari Sunset' is planted in double-row system in a triangular placing at one metre distance from plant to plant, and so is the case with *Leucadendron xanthoconus* with one metre distance though the plants grow to a two metre high. Close spacing encourages upright growth. However, large plants such as *Banksia* and those being planted in high humidity areas especially hairy-leaf types which have tendency to hold moisture should be given more spacing. Planting distances vary as per genera and species as some are prostrate, certain grow only 30-60 cm tall, many others are shrubs growing 1-7 metres tall, some being small trees while others are tall trees. Those growing below one metre in height or are prostrate-growing, are planted very close, *i.e.* 30-45 cm distance, while shrubs growing taller than one metre and for trees, distances range from one metre to five metres. The planting may be carried out in lines or in the pots. Planting in pots may be done at any time and during inclement weathers may be brought inside the greenhouse while for liners which are planted outside, should be planted only during proper season. The liners are planted in deep furrows drawn by tractors. Planting should be done at a place (coastal areas; sandy or sandy-loam, heavy or light, moist or dry soils; at temperate, sub-tropical or tropical conditions;and in open sun, semi-shaded or shaded locations) as per requirement of a particular species to be planted. After planting the plants are looked after properly till these are properly established. Through seeds it may take many years to bloom though cutting-raised and even the tissue-cultured plants may start blooming the second year.

In Proteaceous plantings the **weeds** require to be controlled properly and timely. Spreading of black polythene of 900 mm width as mulch will control almost all the weeds and the film will last at least for two years. Alternatively, organic wastes, bark splitting, dry grasses and straw, dry leaves, husks of rice and groundnut, sawdust, *etc.* can effectively be used as mulches. Glyphosate herbicide is also effective in controlling weeds if the herbicide does not touch the leaves.

Flowers of the Proteaceae constitute a considerable proportion of the market, both locally and overseas. Proteas grow normally on leached and acidic soils which are poor in available minerals. Soil texture plays an important role in protea development. Surprisingly, the **nutritional** requirements of Australian and South African natives used in cut flower production are very poorly understood, particularly with respect to phosphorus. For example, advice from growers ranges from advocating the use of superphosphate to the use of no-phosphorus fertiliser. Because of the lack of information on nutritional requirements, fertiliser management is not scientifically based, for example, red/purplish coloured new leaves are commonly cited by growers as a sign of good health. In fact it is well known that such symptoms are a sign of phosphorus deficiency. Observations of plantations and descriptions from growers indicate that many plants in commercial plantations are growing sub-optimally due to either nutrient deficiency or toxicity. It is clear that growers need guidelines, based on hard facts, for nutrition management of their crops. Many Proteaceous plants grow poorly and develop necrotic and chlorotic leaf symptoms when given fertiliser at rates considered normal for other cultivated plants. This response has been attributed to phosphorus toxicity. The growth of Proteaceous plants in their native habitat gives a good indication of their cultural requirements. Nutritional status of *Protea, Leucadendron* and *Leucospermum* using soil test and plant analysis (Anon., 2001) is presented in Table 30.1.

Table 30.1: Soil and Leaf Analysis

Analyte	*Unit*	*Range*
	Soil	
pH	pH	5.0 - 5.5
Olsen phosphorus	mg/l	5.0 – 20
Potassium	me/100	0.40 - 0.80
Calcium	me/100	4.0 – 10
Magnesium	me/100	0.70 - 3.0
Sodium	me/100	0.0 - 0.50
CEC	me/100	12 – 25
	Leaf	
Nitrogen	per cent	0.75 - 1.5
Phosphorus	per cent	0.050 - 0.14
Potassium	per cent	0.35 - 0.60
Calcium	per cent	0.50 - 1.0
Magnesium	per cent	0.10 - 0.25
Sulphur	per cent	0.10 - 0.25
Sodium	per cent	0.0 - 0.40
Iron	mg/kg	20 – 70
Manganese	mg/kg	50 – 400
Boron	mg/kg	8.0 – 25
Zinc	mg/kg	10 – 50
Copper	mg/kg	3.0 - 8.0

Proteaceous species have both a lower requirement, *vis-a-vis* lower tolerance to fertilizer compared with plant species from other families. However, all the species grow better with applied fertilizer than without fertiliser, and large differences are found in responsiveness and sensitivity even within the same genus. It is also clear that these plants cannot be grown effectively without use of fertilisers. The lower requirement for fertiliser is based on the ability of these Proteaceous species to utilise nutrients efficiently. These features of the group are important for survival on the low fertility soils in which they have evolved. Specific knowledge on optimal nutrition will aid the consistent production of quality plants and cut flowers (Tables 30.2 and 30.3).

Parvin *et al.* (1973) stated that desired nutrient levels in *Protea neriifolia* leaves should be 0.5-1.0 per cent Ca, 0.03-0.06 per cent phosphorus, 0.3-0.7 per cent potassium and 0.1-0.3 per cent magnesium. Compost is preferred over animal manure and that too in autumn or spring when plants have started taking new growth. However, at the time of planting also the soil should be mixed with well rotten compost or any other manure. If the soil is properly mixed with compost, further there is no any need of fertilizer application, and even otherwise fertilizers should be discouraged, especially

Table 30.2: The Range of Fertiliser Rates and Mature Leaf Macronutrient Contents Associated with Optimum Growth

Species	*NPK (kg/m³)*	*N (per cent)*	*P (per cent)*	*K (per cent)*	*Ca (per cent)*	*Mg (per cent)*
Banksia hookeriana	1.75–10.3	2.60–3.55	0.09–0.32	0.80–1.20	0.58–1.00	0.31–0.32
Protea cv. Masquerade	1.50–6.75	0.90–2.30	0.06–0.17	0.75–1.40	0.50–0.70	0.13–0.30
Protea cv. Clarks Red	1.75–4.25	1.24–1.93	0.09–0.14	1.23–1.28	0.40–0.60	0.10–0.13
Leucodendron cv. Sundance	0.50–4.00	1.00–2.20	0.05–0.13	0.60–1.00	0.70–0.75	0.20–0.22

Table 30.3: The Range of Fertiliser Rates and Mature Leaf Micronutrient Contents Associated with Optimum Growth

Species	*Al (ppm)*	*Cl (per cent)*	*Cu (ppm)*	*Fe (ppm)*	*Mn(ppm)*
Banksia hookeriana	62–71	0.52–0.80	7 – 8	68–111	345 – 500
Protea cv. Masquerade	30–40	0.35–0.37	<10 -<10	20–25	170 – 280
Protea cv. Clarks Red	15–20	–	<10 -<10	20–22	180 – 280
Leucodendron cv. Sundance	30–30	0.55–0.55	<10 -<10	30–35	350 – 380

the phosphatic types. In the first year of planting, the plants require utmost care in **watering**. From mid-June to September, no watering is required because of frequent rainfall during the period, from November to February there is only nominal evaporation when also watering is generally not required, in October and from March to mid-June the weather is dry so during this period light weekly irrigation in the first year should be given so that plants may not die due to desiccation. After establishment, there is, generally no requirement of watering. Trickle irrigation is most economical and convenient.

Most members of Proteaceae produce flowers on new year's growth so after flowering, if seed is not required, these should be **pruned**. Pruning should be carried out to bring the plants into shape and also for better flowering the next season so the stems on which the flowers have appeared should be cut back to allow new fowering canes to appear.

Postharvest

Flowering stems are harvested at any stage between soft-bud, or anthesis of the outer ring of florets. The stems are best placed immediately in water, with cooling to 2 to 5 °C within 60 minutes after harvest. Thereafter the cool chain should be maintained until the stems are sold to the florist or consumer. In exporting countries, the cold chain is of necessity though broken during air transport. The stem length categories for export standards from South Africa starts at a minimum of 40 cm, with an increase in length of 10 cm for the next category. Maximum allowable blemishes, either physical or due to disease, on the involucral bracts and leaves are also defined, but each importing country sets its own phytosanitary restrictions. Malan (1995) specified the harvesting stages for different genera of Proteaceae. *Leucospermum* flowers can be harvested at 33-50 per cent styles reflexed stage, resulting in effective style opening and optimum travel ability and shelf life. *Protea* is harvested at soft-tip stage, except for *P. aurea, P. lacticolor* and *P. mundii* which are harvested at the emergence of the styles. Single stem of *Leucadendron* is harvested only during the period when it satisfies the consumer. Poorly coloured *Leucadendron* 'Safari Sunset', for instance, will harm the product's market position. *Leucadendron* foliage is harvested at any time subsequent to leaf maturity. *Serruria* flower is harvested when anthesis of second inflorescence occurs.

Stephens *et al.* (2003) pulsed the *Leucospermum* stems in either 0 or 2 per cent glucose solution for 20 hours at 18±1°C before being transferred to individual vases with tap water (Table 30.4). Vase life was assessed in a temperature controlled room subjected to natural light, and the number of leaves with at least 10 per cent blackened leaf area was determined daily.

From the Table 30.4 it can be seen that there was a significant increase in vase life of 'Cordi', 'Gold Dust', 'High Gold' and 'Succession' stored at 1°C for 3 days, followed by 2 per cent glucose pulse. No significant improvement in 'Scarlet Ribbon' or 'Tango' was found with 2 per cent glucose pulsing. Further, Stephens *et al.* (2003) reported that the glucose pulsing solutions significantly extended the vase life of 'Brenda', 'Carnival', 'Pink Ice', 'Susara' and 'Sylvia' cultivars of *Protea*.

Table 30.4: Effect of Glucose Pulsing Solution on Vase Life of *Leucospermum* Cultivars

Cultivar	*Control (Water) Vase Life (Days)*	*2 per cent Glucose Pulse*
Cordi	12	17
Gold Dust	15	24
High Gold	12	14
Scarlet Ribbon	16	16
Succession	10	15
Tango	20	18

Table 30.5: Effect of Glucose Pulsing Solution on Vaselife of Cut Flowers of Protea Cultivars

Cultivar	*Glucose Concentration (per cent)*						
	0	*1*	*2*	*3*	*4*	*5*	*10*
	Vase Life in Days						
Brenda	5	6	5	7	8	9	1
Cardinal	4	4	4	3	4	3	1
Carnival	6	6	7	7	8	10	1
King Protea	12	12	13	12	14	14	1
Pink Ice	5	6	8	10	1	1	1
Susara	6	6	7	8	9	12	14
Sylvia	5	5	6	8	9	14	12

From Table 30.5, it can be concluded that significant differences exist in the response of various *Protea* cultivars to glucose supplementation. Five of the seven cultivars responded positively to glucose supplementation. Increasing glucose concentration was associated with a significant improvement in the vase life of *Protea* cultivars. Phytotoxicity was observed in all proteas pulsed with 10 per cent glucose solution.

Gibberellic acid spray is usually known to cause elongation of the stem internodes. There are rare cases where GA spray enlarges the flower size. Sahebat and Zieslin (1995) reported an increase in weight of the detached rose petals or petal discs when imbibed in GA_3 containing sugar. Four weekly sprays of GA_3 at 1,000 ppm when the flowering buds of *Protea* cv. 'Pink Ice' were 4.0 cm long, caused an elongation of the involucres of bracts and thus increased the inflorescence size of this cultivar (Ben-Jaacov, 2006). The treated inflorescences were not only larger but had better appearance, *vis-a-vis* longer shelf life.

Large flowers such as *Hakea* & king and queen *Protea* are packed individually with rolled paper or cellophane though in many of the other cases where flowers are smaller are packed in fives in dry conditions in cartons lined with clean paper, however, for long storage the lining of the cartons should be done with polythene films to limit water loss. *Leucospermum* and *Protea*, after being sprayed with some botrycide can be stored dry for 2-3 weeks at 2 °C with 90-95 per cent relative humidity. Storage avoids glut in the market because there are certain species such as *Leucospermum cordifolium* that tend to flower at one time. For cool storage of the Proteaceous flowers in water for more than a few days, it is desirable to have a low light levels of about 400 lux during the daytime to break the continuous darkness. Blackening of leaves after harvest is also most prevalent in Proteaceous plants, especially *Protea* (*P. eximia, P. magnifica* and *P. neriifolia*) and *Leucospermum*. Postharvest life of the flower is dependent upon the leaf and stem carbohydrates for complete opening and quality development, though it is more complex morphologically and physiologically. As fresh flowers, *Protea* is limited by leaf blackening and loss of fresh flower appearance. Often, leaf blackening occurs within a week of harvest and occasionally upon removal from the shipping carton less than three days from harvest. Leaf blackening leads to loss of decorative value, loss of market and possible rejection of the consignment. Blackening after cutting occurs due to enzyme activity, the production of phenolic compounds and the breakdown of the cell tissues and this occurs particularly during transport when surface moisture of cut flowers was not dried before packing (Paull *et al.*, 1980).

Insect-Pests, Diseases and Physiological Disorders

These are not the subject to many pests and diseases. **Root rot** is most serious disease of Proteaceous crops caused by *Phytophthora cinnamomi* which can be avoided by managing proper drainage and plant-soil water relations. Since rooting in Proteaceous crops is almost very slow so use of chemicals will not prove effective. Other diseases are leaf spots as well as floral rots which can be controlled by spraying with 0.2 per cent Dithane M-45 or Dithane Z-78. ***Botrytis cinerea*** is the most common pathogen which affects flowers at low temperatures when humidity is high for which Dithane is very effective. Shoot blackening occurs when water is lodged on large hairy leaves of many of the Proteaceous plants, and in case of *Hakea* and *Telopea* which have flowerheads full of florets and nectors, so water lodges there in the heads causing development of *Botrytis*, ***Rhizopus*** and a few other pathogens including moulds that make rotting of the flowers. Silver leaf disease (*Chondrostereum purpureum*) seriously infects *Leucadendron* 'Safari Sunset' and greys the normal red-purple shoots. Though there is no any effective control but Difolatan sprayings reduce its occurrence. After cutting when surface moisture is not dried, the leaves in *Protea, Leucospermum* and in certain other Proteaceous leaves turn black within a week of harvest or in three days when taking out from the cartons which is due to enzyme activity, the production of phenolic compounds and the breakdown of the cell tissues. One can get rid of leaf blackening by drying the surface-moisture of cut material before packing, removing field heat and storage at cool temperatures (Salinger, 1985). Insect-pests are the same as the other plants such as aphids, thrips, bugs, grasshoppers, humble bees and caterpillars though not of serious nature. **Leaf roller caterpillars** feed on the tips of the shoots during active growth and tie the leaves together. This may be controlled by applying systemic organophosphates or pyrethrins.

Future Prospects of Protea Cultivation in India

India has an ancient heritage when it comes to floriculture. It has grown flowers for various purposes ranging from aesthetic to social and religious. A consistent increase in demand for cut and potted flowers has made floriculture one of the most important commercial trades in Indian agriculture. A mild winter, abundant sunlight, suitable agro-climatic conditions, low labour costs, availability of skilled manpower are factors that are beneficial for the growth and development of this sector into a potential earner of foreign exchange. Moreover, the thinking that only rose is the best exchange earner will have to be changed and one should go for various other options through product diversification. Various Proteaceous crops can be introduced into India for commercial cultivation. Places like Bengaluru, Pune, Nilgiri hills, lower Darjeeling hills, Koraput region of Odisha are some of the best places for introducing various Proteaceous crops. After introduction, the germplasm can be assessed for their performance and then the cultivation on commercial scale of the most promising ones should be taken. Afterwards, the improvements in the agro-technology may be undertaken along with the genetic improvement for developing novel hybrids and varieties to match with international competitors. A private company called 'Casablanca' in Bengaluru is evaluating certain Proteaceous plants for its suitability. Likewise, places having a mild climate, low rainfall with good amount of humidity need to be identified for popularizing their cultivation.

Efforts within the individual producer countries to develop quality standards based on the requirements of the consumer countries, should be coordinated internationally. These standards can then be used to set the selection criteria or individual breeding programmes. A programme on gene mapping of *Banksia* is already in progress so other Proteaceous crops should also be assessed for varietal identification. Detailed research into breeding systems of Proteaceous genera will assist in development of hybridization methods to produce a wider range of cultivars. Genus *Leucospermum* has so far proved to be the most flexible in terms of cultivar production *via* interspecific hybridization. Many of the other commercial genera are more difficult to manipulate, but are equally spectacular in terms of the cut flower market. Through genetic engineering, transfer of useful genes across taxonomic barriers can be accomplished. While research into gene isolation and control is quite advanced in a number of organisms, there is a major bottleneck with many plants as regeneration *in vitro* is generally required for transformation. Thus, an important goal of Proteaceous research should be *in vitro* regeneration of whole plants. When this has been achieved, then the path will be open for genetic manipulation and transformation of these genera. Refinement of cultivation practices, such as pruning, fertilization and irrigation are required to maintain the economic return of *Protea* as a crop and to ensure the delivery of quality blooms to a very competitive international market. The challenges of cultivating *Protea* differ from region to region, but the basic plant physiology controlling the plant's reaction to environmental stresses remains the same. The international flower markets are always searching for new and exciting products. *Protea* can fulfil this demand. A larger variety of cultivars with different forms and colours, longer vase life, exceptional quality, and extended availability during the year are needed to maintain and increase the market share. These goals will only be achieved by continued research.

References

Anonymous, 2001. http://www.hilllaboratories.com. Crop Guide to Proteas.

Ben-Jaacov J. 2006. Gibberellic acid spray increased size and quality of *Protea* 'Pink Ice' flowers- A preliminary experiment. *Acta Hort.*, No. 716, pp. 135-140.

Criley R A. 2001. Proteaceae ~ beyond the big three. *Acta Hort.*, No. 545, pp. 79-85.

Croxford, B., G. Yan and R. Sedgley, 2006. Micropropagation of *Leucadendron. Acta Hort.*, No. 716, pp. 25-33.

Eliovson, S. 1965. *Proteas for Pleasure*. Howard Timmins, Capetown, South Africa.

Malan, D. G. 1995. Crop Science of Proteaceae in Southern Africa ~ Progress and Challenges. *Acta Hort.*, No. 387, pp. 55-72.

Matthews, L.J.,and Z. Carter, 1983. South African Proteaceae in New Zealand. Matthews Publishing, Manakau, New Zealand.

Parvin, P. E., R.A. Criley and R.M. Bullock, 1973. Proteas ~ developmental research for a new cut flower crop. *HortScience*, **8**(4): 290-303.

Parvin, P. E., R.A. Criley and J.H. Coetzee, 2003. Proteas ~ A dynamic industry. *Acta Hort.*, No. 602, pp. 123-126.

Paull, R., T. Goo, R. Criley and P.E. Parvin, 1980. Leaf blackening in *Protea eximia*: Importance of water relations. *Acta Hort.*, No. 113, pp. 159-166.

Rebelo T. 1995. SASOL Proteas. In: *A Field Guide to the Proteas of Southern Africa*. Fernwood Press, Vlaeberg, South Africa.

Rodriguez-Prez, J. A. 1995. Effects of treatment with gibberellic acid on germination of *Protea cynaroides, P. eximia, P. nerifolia* and *P. repens. Acta Hort.*, No. 387, pp. 85-89.

Rodriguez-Prez, J. A., M.C. Vera-Batista, A. M. de Leon-Hernandez and P.C. Armas, 2001. Influence of cutting position, wounding and IBA on the rooting of *Leucospermum cordifolium* 'California Sunshine' cuttings. *Acta Hort.*, No. 545, pp. 171-175.

Rugge, B. A. 1995. Micropropagation of *Protea repens. Acta Hort.*, No. 387, pp. 121-125.

Sahebat, A. and N. Zieslin, 1995. Promotion of postharvest increase in weight of rose (*Rosa* × *hybrida* cv. Mercedes) petals by gibberellins. *J. Plant Physiol.*, **145**: 296-298.

Salinger, J. P. 1985. Proteaceae. In: *Commercial Flower Growing*. Butterworths of New Zealand, pp. 216-239.

Sedgley, M. 1995. Cultivar development of ornamental members of the Proteaceae. *Acta Hort.*, No. 387, pp. 163-169.

Sedgley, R., B. Croxford and G. Yan, 2001. Breeding new *Leucadendron* varieties through interspecific hybridization. *Acta Hort.*, No. 545, pp. 67-75.

Stephens, A. I., D.M. Holcroft and G. Jacobs, 2003. Postharvest treatments to extend the vase life of selected Proteaceae cut flowers. *Acta Hort.*, No. 602, pp. 155-159.

Thomas, M.B. 1979. Nutrition of container-grown plants with emphasis on the Proteaceae. Ph.D. Thesis, Canterbury Univ., New Zealand.

31

Rosa (Family: Rosaceae)

R.L. Misra, D.V.S. Raju and Sanyat Misra

[**Common names**: Albas (vigorous shrub roses deriving from *R.× alba*), Apple rose (*R. pomifera,* syn. *R. villosa*), Austrian briar/Austrian copper (*R. foetida*), Austrian copper (*R. foetida bicolor*), Autumn musk rose (*R. moschata*), Banksian rose/Banks's rose/Lady Bank's rose (*R. banksiae*), Blush Noisette (*R.* × *noisettiana*), Bourbon rose (*R.× bourboniana* and descendants of *R.* × *odorata* and *R. damascena*), Burgundy rose (*R. parvifolia,* syn. *R. burgundiaca*), Burnet rose (*R. spinosissima*), Burr/Chestnut rose (*R. microphylla*), Cabbage rose/Provence rose (descendants of *R.* × *centifolia*), Cherokee rose (*R. laevigata*), China/Bengal rose/ Monthly rose (*R. chinensis,* syn. *R. indica*)**,** Crimson Chinese rose (*R. semperflorens,* syn.*R. bengalensis*), Damask rose (originating from *Rosa damascena*), Dawson rose (*R. dawsoniana*), Dog rose/Wild briar (*R. canina*), Dwarf polyanthas (progeny of *R. multiflora* crossed with China tea and Noisette roses, have dwarf habits with profusion of blooms), Eglantine/Sweet briar (*R. rubiginosa*), 'Félicité et Perpétue' (*R.* ×*felicita*), Field rose (*R. arvensis*), Floribunda roses (derivatives from crosses between the Dwarf Polyantha and early Hybrid tea roses), French rose/Provins rose (*R. gallica,* syn. *R. rubra*), Green rose (*R. viridiflora*), Himalayan musk rose (*R. brunonii*), Hybrid perpetuals (descendants of Portland, Bourbon and China roses and predecessors of Hybrid teas, highly vigorous shrubs but less perpetual than the modern Hybrid teas, very popular in Victorian times), Hybrid Rugosas (some of the best varieties of *R. rugosa* within it or ouside with highly scented, sometimes with repeat-flowering), Hybrid musks (perpetual flowering shrubs derived from *R. moschata*), Hybrid tea (successors to the Hybrid perpetuals, and excellent as cut flowers), Gallicas (a largest group of old shrub roses and closest in appearance to the wild *R. gallica*), Macartney rose (*R. bracteata*), Manipur Tea rose (*R. gigantea*), Memorial rose (*R. luciae*), Miniature roses (derivatives of dwarf semi-double *R. chinensis* 'Minima' known as 'Roulettii' introduced into Switzerland by Dr. Roulet, have long flowering season with the beauty and form of the China rose, and 'Roulettii' though very beautiful, has further been improved by crossing with Hybrid tea and Floribunda), Moss rose (*R.* × *centifolia muscosa*), Noisette rose/Champney rose (*R.* × *noisette*), Penzance briars/Hybrid sweetbriars (those developed by Lord Penzance at the end of 19th century being fragrant and free-flowering shrubs often prone to mildew and appearing in abundance in small clusters), Portland roses (predecessors of the Bourbons and Hybrid perpetuals showing the influence of *R. chinensis*), Polyantha rose (*R. multiflora,* syn. *R. polyantha*), Pompon rose (*R. pulchella*, syn. *R. dijonensis*, 'De Meaux', 'Pomponia'), Scotch rose/Burnet rose (derivatives of *R. spinosissima,* syn. *R. pimpinellifolia*), Sweet briar/Eglantine (*R. rubiginosa*), Rose, Tea rose (*R.* × *odorata*), Threepenny-bit rose (*R. farreri persetosa*), White rose of York (*R.* × *alba*)]

Introduction and Origin

Rosa is ancient Latin name for a rose, and this is a genus of some 100-250 species, depending on the botanical authority but nearer to the first number seems to be more realistic as larger number of botanists have recognized over 100 species. They are native to colder and temperate regions of the northern hemisphere, in America extending to North Mexico, in north-east Africa to Ethiopia and in Asia to India, China and Japan. The garden roses (*Rosa* × *hybrida*, all non-species roses; Gudin, 2003) are most beautiful creations of nature and is universally acclaimed as 'queen of flowers'. Apart from being admired for its beauty as specimen plant, in the border, as hedges, as climbers for training on walls, pergolas and pillars and various other landscape uses, for loose flowers in worshipping, for making garlands and *rangolies*, as cut flowers and for making bouquets and various other decorations, as preserves, for making rose water and attar, and various other pharmaceutical uses. They are also grown for their beautiful fruits as well as for shining foliage. Because of diversified growth habit, exquisite shape, variation in size and form, attractive colour, delightful fragrance and numerous varieties exceeding more than 40,000 in number, roses have gained wide acceptability. Roses can fulfill the requirement of trees, shrubs, climbers, hedges and edges in the landscape planning and design of gardening. It occupies first position in the international cut flower trade. With the increasing demand of novel cultivars in the world trade, a large number of cultivars are being bred every year with more and more novel characters. Approximately eight billion rose stems, 80 million potted rose plants and 220 million garden rose plants are sold annually world over. About 31 per cent of all cut flowers traded at the European auctions are cut-roses with a total value of about 858 million euro (Heinrichs, 2008). In India, 4,330 hectare area (producing 874 million stems valued at Rs.44.00 crores) is under rose cultivation in the states of Delhi, Tamil Nadu, Karnataka, Maharashtra and West Bengal (www.nabard.org). Red roses symbolize love, passion, desire and respect; yellow means friendship, caring, happiness and freedom; coral roses symbolize desire; peach roses modesty; dark pink roses gratitude and appreciation; pale-pink roses grace, admiration and sympathetic thoughts; orange roses fascination and enthusiasm; white roses innocence, reverence, humility and truth; purple & lilac roses love at first sight, and enchantment; red blended with yellow gaiety and joviality; white with yellow blending harmony; and red roses with white blending symbolize bonding and harmony.

The rose cultivation in Asian gardens dates back to pre-Mahabharata period. In Asian countries, it has traditional use as loose flowers for worshipping, for making garlands for various social functions and festivities, for preparing rose water, *pankhuri, gulkand* and for extraction of essential oil, and for this the roses used are *R. bourboniana, R. centifolia, R. damascena* and many others which are fragrant. ***R. bourboniana*** locally in Kannauj known as *Chenia* though is being used for various industrial purposes but its importance is next to *R. damascena*. *R. bourboniana* is the perpetual flowering rose which is said to have originated through a cross between Quatre Saisons (*R. damascena bifera* the pink Autumn Damask, the Portland Rose or old Blush Rose) and Parson's Pink China (*R. chinensis* and *R. gigantea*) in the early days of queen Victoria's reign in Reunion, previously known as 'Isle de Bourbon' hence former name as'Ile-Bourbon' by Jacques, a gardener to Chateau de Nevilly in France who grew a few plants from the seeds of the original seedling (Steen, 1967; Edwards, 1975). Fairbrother (1965) mentions it reaching France in 1817 though Edwards (1975) mentions the year as 1819, into England in 1822 and in U.S.A. in 1828. Nacisse Desportes called it *R. bourboniana* and in neighbouring Mauritius it was known Rose Edward. Dr. Hurst said that Breon, curator of the botanical gardens in the Ile de Bourbon, sent the seeds of it in 1819 to France which in Paris were received and grown by Jacques, a gardener to king Louis Philippe, giving the French name 'Rosier de l'Ile Bourbon'. *Attar* based on sandal wood oil is prepared mainly from ***R. damascena***, locally in Kannauj known as *chaiti, phasli* or *gauraiya*. *Rosa damascena* (syn. *R. calendarum, R. polyanthus, R. belgica, R. gallica damascena*) is one of the Roman roses and Bunyard says it one of the 12 Pliny's roses (Edwards, 1975). Damask rose is believed to have originated in Asia Minor and is said to have entered Europe first in 1270 by Count de Brie who brought it back to France. Darlington cytologically found it a blend of *Rosa gallica* and *R. moschata*, and in Europe it is being cultivated since 16th century (Večeřa, 1971). *Rosa damascena triginipetala* is the form grown in Bulgaria and 'Gloire de Guilan' in Persia for production of *attar*. ***R. indica*** (syn. *R. chinensis, R. mutabilis*), a most important non-fragrant parent of all the modern cultivars, is being used as loose flower for all the religious and social purposes. *R. chinensis* and *R. gigantea* are the major contributors to recurrent roses (Zieslin and Moe, 1985), though *R.* × *centifolia, R. gallica, R. damascena* and *R.* × *borboniana* are the other species to have been involved in the evolution of modern hybrids (Post, 1945). ***R.* × *centifolia*** is an ancient rose of hybrid origin involving *R. canina, R. gallica, R. moschata* and *R. phoenicea*.

Rose water, rose attar, gulroghan, gulkand, rose dry petals (*pankhuri*) and rose oil are commercial products of rose for international and national market. Rose hips are one of the important sources of vitamin C. Rose water is used in eye problems, petals in nerve complaints or in preparation of tea in relieving brain fatigue. It is said that rose perfume (*ruh-e-gulab, attar gulab*) was first noticed in India by empress Nurjehan in her bathing tub, the knowledge of which was carried over by one of her personal attendants belonging to a family of mohalla Gandhiana in Kannauj (U.P.) which became the basis for the establishement of rose perfume industry in Kannauj (Misra and Chaturvedi, 1978). *Attar Gulab* in India is traditionally prepared through hydrodistillation which requires 3 to 4 hours and on the surface of rose water the *attar* floats which is the pure rose oil having 0.849 to 0.865 specific gravity, 1.452 to 1.466 refractive index at 25°, 63-84 per cent geraniol, citronellol/rhodinal 24-64 per cent, 7-25 per cent stearoptene, 1.5-3.8 acid value and 3.7-17.5 ester value. In *Rosa spinosissima* and *R. corymbifera* fruits were found yielding quercimeritrin, hyperoside, vitamin C and caffeic and chlorogenic acids (Zemtsova, 1977).

Botany, Genetics and Breeding

Roses are deciduous or semi-evergreen woody shrubs or scrambling climbers usually with prickly stems, though certain are thornless such as *Rosa banksiae albo-plena, R. banksiae lutea*, etc. Leaves are alternate, odd-pinnate (sometimes simple) and stipulate with 5-7 toothed leaflets having rounded or pointed tips. Flowers solitary usually with long stalks or corymbose at short stalks and often fragrant, petals and sepals 5 each, but 4 in case of *Rosa sericea*, the only 4-petalled rose, stamens and pistils numerous but sometimes only a few, enclosed in an urn-shaped receptacle which turns into a rounded to ovoid fleshy fruit called as hips or heps with maturity and these contain many bony achenes (seeds). The genus is divided into 4 sub-genera of which sub-genus *Rosa* includes nearly all the species (Gudin, 2000). The sub-genus *Rosa* is divided into sections of which the actual number is debatable as these are 100-250 with about 40,000 cultivars. The genome is complex comprising of 5 haploid chromosome sets each made of 7 chromosomes (Nybom *et al.*, 2005). The chromosome number of wild species, especially under *acicularis* group the number varies from 2n = 2x=14 to 2n= 8x=56, with most species being diploid or tetraploid (Nybom *et al.*, 2005) though high polyploidy is associated with the species found in the most northern parts of Siberia and Alaska. Several important species of Asian origin such as *Rosa chinensis, R. gigantea, R. moschata, R. multiflora* and *R. wichuriana* which are diploid with 2n=14 chromosomes were when crossed with European species such as *R. foetida* & *R. gallica* and their derivatives *R. centifolia* & *R. damascena*, these resulted into tetraploids with 28 chromosomes. However, the hybrids between European and Asian parents were by and large sterile triploids having 21 chromosomes, 14 from the European and 7 from the Asian but there had been certain spontaneous mutants with tetraploid structures having 2n=4x=28 chromosomes arising out of these triploids, and these fertile tetraploids became the basis towards the characters of Hybrid Perpetuals but with addition of *R. damascena* genes. The anthocyanidin (water soluble pigments) governs the floral colouration, *viz.* pelargonidins for orange-red to scarlet flowers, cyanidin for crimson to blush-red and delphinidin for blue and violet flowers. Delphinidin and myricetin responsible for true blue colouration are missing in roses though in early sixties of the 20th century the Japanese researchers reported the presence of even delphinidin pigments in roses, which with further breeding programme may fulfil the breeder's dream of getting true blues in roses through conventional system. In early eighties of the 20th century, the analysis of 1,200 progenies from 47 families revealed that cyanidin, peonin and pelargonin showing quantitative inheritance were found in 99, 52 and 31 per cent, respectively (Yadav *et al.*, 1989) and these pigments are reported to be mainly controlled by additive gene action. Apart from anthocyanidins, there are other pigments such as flavonols and carotenoids for white, yellow and brown pigmentation. The deep and bright yellow colouration is due to presence of pelargonidin, initially being inherited through *R. foetida* being crossed with others in France around 1930. The dominant colours in rose being magenta, pink or tyrian-rose to orange-red; pink over dark red, orange-yellow, yellow, white and scarlet; deep yellow over light yellow; and cream, light yellow and greenish-white over white. The inheritance of fragrance is governed by many genes and many a times it is due to complementary factors as by crossing non-fragrants sometimes results into frangrant seedlings. Duration of flowering is determined by a gene which restricts the flowering only once a year when in dominant form. Glossy leaves, climbing habits, flower doubleness and mildew resistance over dull leaves, dwarf or shrubby types, single flower type and mildew susceptibility are governed by single dominant gene though vigour, thorniness, leaf breadth, stalk length, neck strength, fragrance and shape of bud & flowers are dependent on interaction of many genes.

Recurrent flowering is one of the key characters crucial for the success of roses as ornamental crops as it leads to superior genotypes flowering throughout the whole growing season. This trait was most probably introgressed from *R. chinensis* in the early nineteeth century and is inherited as a single recessive gene. The inheritance of double (>5 petals) versus single (5 petals) flowers was shown to be caused by a single dominant gene (Debener, 1999). Moss phenotype originated from sports of *R. centifolia* in the 17th century and was shown to be inherited as a single dominant gene in crosses between tetraploid moss roses and tetraploid non-moss varieties (De Vries and Dubois, 1984). Several studies have shown that prickles on stems are inherited as single dominant genes (Debener,1999; Linde *et al.*, 2006) and that they are independently inherited from prickles on petioles (Lal *et al.*, 1982; Crespel *et al.*, 2002). Dwarf phenotypes probably introgressed from the diploid *R. chinensis minima* (Sims) which were used to breed miniature roses (De Vries, 2003). Wylie (1955) stated that there is a single dominant gene for glossy leaves, climbing habit and double flowers while polygenes govern characters like vigour, fragrance, thorniness, strength of neck, width of leaf, length of cut flower stock and shape and size of buds and flowers. The inheritance of fragrance is controlled by polygenes (Kaicker *et al.*, 1983).

Mr. B. S. Bhattacharjee of Deogarh (Bihar) is, in fact, the father of rose breeding in India. He released his first variety 'Rama Krishna Deva' in 1941. Rose improvement research was initiated at the Indian Agricultural Research Institute in 1956 in the Division of Floriculture and Landscaping (Division of Horticulture at that time). The work started with the organization of a National Rose Collection at the Institute. The breeding of roses started from the year 1957, keeping in view the fact that the roses bred under temperate climatic conditions may not perform well like those bred under tropical conditions. As a result of this, the first two rose varieties namely 'Rose Sherbet' and 'Delhi Pink Pearl' were evolved by late Dr. B. P. Pal in 1962. The development of new rose varieties at IARI has created a great deal of interest in rose cultivation in the country. Conventional hybridization and mutation breeding were employed for evolving new varieties. The main objectives of breeding are to evolve varieties with attractive flower colour and form, fragrance, floriferousness, long stems with long keeping quality, disease and pest resistance and suitability for growing under sub-tropical conditions.

Swarup *et al.* (1971) reported gamma ray induced mutants 'Pusa Christina' (pink) from 'Christian Dior' and 'Abhisarika' (pink striped) from 'Kiss of Fire'. IARI in 1978 developed a mutant 'Madhosh' from cv. 'Gulzar' through EMS (0.25 per cent for 8 hours) treatment of budwood. Mutagenesis in rose cv. 'Alliance', 'Folklore' and 'Suraga' under *in vivo* and only 'Folklore' under *in vitro* cultures (Wilson, 1993) recorded LD_{50} at 38 Gy under *in vivo*, 33 Gy under *in vitro* and two mutants in 'Folklore', *viz.* reddish-yellow mutant under 30 Gy and with increased petal number under 40 Gy under *in vivo*.

Steve Connor (2015) stated that scientists (Jean-Louis Magnard and co-workers) in France (University of Lyon in St. Etienne) have identified enzyme in the var. 'Papa Meillaand' that is key to fragrance in rose. A study into the chemistry of rose scents has found a new biochemical pathway in the petals of the plants which produce a sweet-smelling fragrance that could be reintroduced into rose varieties that have lost their smell. This particular enzyme playing the key role in producing sweet frangrance in roses has been named RhNUDX1, which is found in the cells of the flower petals and which generates the well-known fragrance substance called monoterpene geraniol, the primary constituent of rose oil.

Classification and Varieties

The *Rosa* genus is endemic to temperate regions of the northern hemisphere including North America, Europe, Asia and the Middle East, with the greatest diversity of species found in Western China. It is also distributed in the warmer areas such as New Mexico, Iraq, Ethiopia, Bengal (India) and Southern China (Nybom *et al.,* 2005). There is no endemic species in southern hemisphere. Among the many classification systems, ARS Classification Scheme 2000 is widely used (Cairns, 2003). There are 3 sub-divisions in this system, *viz.* species roses, old garden roses, and modern roses.

The genus *Rosa* contains approximately 150 species (Quest-Ritson, 2003), although the number varies according to authority. Wild roses (species and natural as well as man-made species-hybrids) from various parts of the world which share most of the characteristics of the parent species, and in the temperate regions bear single flowers generally in one flush in summer and the red or black hips in autumn though in the subtropical and tropical conditions a few may bear a flush during winter and the hips in summer, are categorized as **Species Roses**.

Basically a **species rose** (sub-division I) has a single flower of five petals only but some double or semi-double forms as sports have occurred in the wild. They are distinguished by often being resistant to pests and diseases. All species roses are almost hardy and deciduous. However, different taxonomy and cytological studies have concluded that only 11 species, *viz. R. canina, R. chinensis, R. foetida, R. gallica, R. gigantea, R. moschata, R. multiflora, R. phoenicea, R. rubra, R. rugosa* and *R. wichuraiana* (Hurst, 1941; Wylie, 1955; Maia and Vénard, 1976) were at the origin of the modern cultivars. As many as 16 species and four hybrid species were found growing wild in different phytogeographical zones of India (Rathore and Srivastava, 1992). The species endemic to Himalayas are *R. brunonii, R. gigantea, R. involucrata, R. longicuspis, R. leschenaultiana, R. macrocarpa, R. macrophylla, R. moschata, R. sericea,* and *R. webbiana* whereas the species growing wild are *Rosa × alba, R. banksiae, R. beggeriana, R. × bourboniana, R. × bracteata, R. × centifolia, R. chinensis, R. × damascena, R. ecae, R. foetida, R. gallica, R. indica, R. laevigata* and *R. multiflora* which perform well under sub-temperate climate and possess excellent vegetative and floral characters justifying their use as donor parents (Viraraghavan, 1991). Various localized rose species and their variants like *R. x bourboniana* (*gulkand* preparation) and 'Kakinada' rose (a red hybrid introducѐd in Kakinada, Andhra Pradesh which is highly floriferous and is most suitable and famous in South India as loose flowers) are potential candidates for geographical indicators. The cultivation of *R.× bourboniana* throughout the country has resulted in the evolution of many ecotypes, *viz. Desi, Cheenia, Kaithali* and *Meeruti* (Shrivastava and Chandra, 1985), *vis-à-vis Titree* in Rajasthan. The species roses are ***R. × alba*** of unknown origin which is thought either as an ancient form or hybrid of *R. gallica* and *R. canina* or *R. gallica* and *R. dumetorum*, grows some 1.8 m bearing sweetly scented single, semi-double or double white (even from white to carnation-pink) flowers 5.0-7.5 cm across in small clusters; ***R. banksiae***, a vigorous thornless fragrant climbing rose from CW China growing up to 15-20 metres if supported on trees, leaves glossy-green, flowers in large clusters and about 2.5 cm across, *R. banksiae alba* (*alba-plena*) is highly fragrant double not setting seeds under temperate and sub-temperate regions of India though is reported setting red, globose and small hips elsewhere, *R. banksiae lutea* (*luteo-plena*) is non-fragrant double, *R. banksiae lutescens* is a scented single yellow form of ancestral *R. banksiae normalis, R. banksiae normalis* is an original wild single species with scented white blooms from CW China discovered after the cultivars were named; ***R. beggeriana*** from N. Iran to Altai and Songaria is a dense shrub growing up to 1.5 m with paired prickles, corymbose white flowers; ***R. bracteata*** from China resistant to black spot disease, is an evergreen scrambling shrub growing up to 6.0 m on walls with solitary, single and fragrant white flowers 7.0-9.0 cm across and globose orange hips; ***R. brunonii*** (syn. *R. brownii, R. moschata nepalensis*) mostly erroneously listed as *R. moschata*, a climbing species native of Himalayas and W. China growing up to 5 m tall and bearing white flowers 3.75-5.0 cm across in large trusses during summer with musk fragrance, followed by red-brown small and globular hips in autumn; ***R. californica***, a native to British Columbia to California is a shrub growing up to 2.4 m with fragrant, single to semi-double and carmine-pink flowers 3.75 cm across in clusters, followed by globose-ovate red hips with prominent neck, var. *plena* has more showy flowers with sweet fragrance; ***R. canina*** is an upright shrub native of Great Britain and other parts of Europe, N. Africa and W. Asia, growing to a height of up to 3.0 m, used as a rootstock in England, flowers 1-3, white to pale-pink, single, 5.0 cm across, and hips ovate and orange-red or scarlet; ***R. × centifolia*** (syn. *R. gallica* var. *centifolia*), an ancient and sterile garden descendant of *R. gallica* quite similar to its parent, growing up to 1.5 m tall, originally recorded from E. Caucasus and bears highly fragrant and very double pink flowers on long and slender pedicels, a parent of numerous sports such as 'Bullata' with lettuce-like crisped leaves, 'Cristata'(syn. 'Chapeau de Napoleaon' with moss-like, resin-scented glands on exposed edges of sepals, 'Muscosa' (*R. centifolia muscosa*, moss rose) having resin-scented glands on sepals, stalks and sometimes even on leaves and rosy flowers, 'Alba Muscosa' (syn. 'White Bath') is a white sport of 'Muscosa', 'Cristata'with mossy excrescences at sepal's edges, 'Parvifolia' (syn. *R. parvifolia, R. burgundiaca*, 'Burgundy Rose') is a minirose with daintier rose-pink blooms 2.5 cm across, 'Pomponia' (syn. *R. pulchella, R. dijonensis*, 'De Meaux') a dwarf sport with light pink blooms, 'Variegata' with petals striped pink and white, etc., and through this and its varieties as well as *R. damascena* when crossed with *R. chinensis* and its varieties the Hybrid Perpetual or Remontant roses originated; ***R. chinensis*** (syn. ***R. indica***), a native of Central China, is a climbing single-flowered species of crimson or pink solitary flowers 5.0 cm across on glandular stalks, but now not in cultivation, however, the China roses of gardens which are 0.9-1.5 m tall shrubs are derived from crosses of *R. chinensis* and other species which have single deep red, crimson, pink or whitish flowers mostly in clusters, its var. *spontanea* is single deep red or pink and solitary, *semperflorens* (syn. 'Slater's Crimson', *R. semperflorens, R. bengalensis*) usually thornless with crimson or deep pink solitary flowers, *longifolia* (syn. *R. longifolia*) with deep pink single flowers, *minima* (syn. 'Roulettii', *R. lawrenciana, R. indica pumila*) growing up to 30 cm in height bearing pink or rose-red single or double flowers 2.5-3.75 cm across, and is the parent of the Miniature roses, and the 'Fairy Roses' belong to this variety, 'Old Blush' bears loose pink flowers 5.0-6.25 cm across, *viridiflora* with green flowers where petals are transformed into small narrow green leaves, *manetti* (syn. *R. manetti*) has vigorous upright growth with single or semi-double deep pink flowers which is easily multiplied through cuttings and is used as a rootstock for forcing roses, *R. chinensis mutabilis* is though a shrub but sometimes becomes a climber with coppery-flame at bud stage, paler-copper on opening of the size of 7.6 cm across in clusters, afterwards pink and finally fading to pale-crimson, etc.; ***R. clinophylla*** (syn. ***R. involucrata***) from tropical India, *viz.* E. Himalayas in N. Bengal, Sibsagar and Goalpara districts of Assam, the Garo Hills and Sylhet, in the vicinity of streams and marshy places, is an erect shrub with arching branches with solitary and single white flowers in small clusters and round, hairy and pale hips; ***R. damascena*** (syn. *R. bifera, R. calendarum*), a native to W. Asia is a shrub growing to 1.8 m and bears fragrant double pink flowers 7.5 cm across with incurved centre or green eye and flowers fading to white, its var. 'Semperflorens' (syn. 'Four Seasons') is a sport flowering in autumn, 'Trigintipetala' bears double but smaller red-pink flowers that are highly fragrant and is the variety being cultivated for rose oil in SE Europe, especially Bulgaria, 'Versicolor' (syn. 'York and Lancaster') bears semi-double white, pink or bicoloured blooms, etc.; ***R. ecae***, a native of Afghanistan grows up to 1.5 m with red stems, tiny leaves and shining single yellow flowers 2.5-3.0 cm across; ***R. farreri*** (syn. *R. elegantula*), a native of NW China is an attractive bristly shrub with arching branches which grows up to 1.8 m height and bears single pink flowers 3.75 cm across and small ovoid red hips, and this species under cultivation is represented by the var. *R. f. persetosa*; ***R. × felicita*** (*R. sempervirens × R. chinensis*) is a semi-evergreen vigorous rambler growing up to 3.0 m with double white flowers 2.5 cm across in clusters and its dwarf sport is 'Little White Pet'; ***R. filipes*** is a rampant climber native to W. China, grows to 12.0 m bearing single white fragrant flowers (sometimes pink-tinted) in large corymbs followed by ovoid to globose red hips, and its var. 'Kiftsgate' is a vigorous climber suitable for trees and for covering the banks; ***R. foetida*** (syn. *R. lutea*), a native of SW Asia is a shrub growing up to 1.5 m with 5.0-6.25 cm across bright yellow, single, fragrant and solitary flowers and this has introduced all modern bright yellow and bicolour roses through hybridization, *R. foetida bicolor* bears single orange flowers (upper side of the petals scarlet and lower yellow), 'Persiana' (syn. 'Persian Yellow') bears double yellow flowers; ***R. gallica*** (syn. ***R. rubra***), a native of W. Asia and S. Europe growing up to 1.0 m and bears solitary and attractive purple-pink flowers 5.0-7.5 cm across, followed by 2.5 cm wide round red hips, the var. 'Officinalis' (syn. 'Apothecary's Rose', 'Red Rose of Lancaster') bears large, loose, bright crimson and semi-double flowers, 'Versicolor' (syn. *R. mundi*) is a semi-double and striped (crimson streaked white) sport of 'Officinalis'; ***R. gigantea*** (syn. *R. odorata gigantea*), a native of Myanmar, SW China and India (Manipur, Nagaland and Sikkim at 1,800 m elevations) is a highly vigorous climber growing up to 15 m with prickly glabrous stems and single creamy-white to pale-yellow fragrant flowers 10-15 cm across, followed by yellow to orange globose hips about 3.5 cm in diameter and it has produced hybrids with *R. moschata, vis-à-vis* is an important ancestor of the Tea and Hybrid Tea roses through *R. × odorata*; ***R. × harisonii*** (*R. foetida × R. spinosissima*, syn. 'Harison's Yellow') is a shrub growing up to 2.0 m and flowering earlier with clear yellow semi-double flowers 6.25 cm across; ***R. × highdownensis*** is a vigorous hybrid of *R. moyesii* growing up to 3.0 m bearing single crimson flowers 5.0-6.25 cm across, followed by flask-shaped glossy red hips; ***R. hugonis***, a native of C. China is a shrub growing up to 2.7 m bearing solitary yellow flowers 5.0-6.25 cm across earlier in the season followed by round deep red hips in the late autumn; ***R. × kordesii*** (*R. rugosa × R. luciae*) developed in Germany is a climbing parent of the repeat-flowering Kordesii ramblers, growing up to 3.0 m with semi-double cupped pink flowers 5.0-6.25 cm across coming up in loose clusters; ***R. laevigata*** (syn. *R. camellia, R. cherokensis, R. nivea, R. sinica, R. ternata*) from China, Japan and Formosa is a climber growing more than 6.0 m on walls and railings with wicked thorns, prominently on new stems and large (6.25-8.75 cm) white (rarely rose) flowers, and this has produced hybrid *R. fortuneana* with *R. banksiae*, and Anemone rose with Tea rose bearing large single light pink flowers; ***R. leschenaultiana*** from India (Western Ghats, Nilgiri hills and Palni hills), Mediterranean regions, N. Africa and in the eastern Himalayas at an altitude up to 1,000 m, is an evergreen vigorous climber having purplish stems with slightly fragrant large white flowers appearing in clusters, followed by orange-red, rough and roundish hips; ***R. longicuspis*** (syn. *R. sinowilsonii, R. willmottiana*), a native of India (Assam), China and other temperate Himalayas is a vigorous climber growing up to 8.0 m on trees with single white flowers 5.0 cm across in clusters; ***R. luciae***, a native of E. Asia (Japan) is a semi-evergreen climber growing up to 3.6 m with trusses of single white flowers 1.8-3.1 cm across, *R. luciae wichuraiana* (syn. *R. wichuraiana, R. bracteata* Hort. not Wendl.) is prostrate-growing semi-evergreen creeping shrub with a few to many-flowered pyramidal corymbs bearing strongly fragrant white flowers 3.75-5.0 cm across and ovoid hips, a parent of Wichuraiana ramblers, most suitable for covering banks and rockeries, and it has produced many hybrids with Hybrid Teas, var. *rubra* Andre (*R. wichuraiana × R. multiflora* 'Crimson Rambler') with single carmine flowers 3.75-5.0 cm across, and *R. wichuraiana* has also produced a highly floriferous and beautiful crimson-flowered hybrid *R. × jacksonii* with *R. rugosa*, and other famed varieties are 'Alberic Barbier' and 'Dorothy Pertkins'; ***R. macrocarpa*** (syn. *R. xanthocarpa*) from India (Manipur) is also a climbing species growing some 15.0 m which some botanists consider synonym to *R. odorata* var.

gigantea (syn. *R. gigantea*) but bears yellow flowers and edible yellow fruits as large as a small apple; ***R. macrophylla*** (syn. *R. roxburghii*) from India, China and Japan (temperate Himalayas at an altitude ranging 1,300-3,000 m) is a shrub growing more than 2 m with single fragrant pale-pink and solitary flowers 6.0-7.5 cm across and orange-yellow, globular, flattened and densely prickly hips; ***R. moschata*** (syn. *R. ruscinonensis*), a native of India, Persia, S. Europe and N. Africa is a lax shrub growing up to 3.6 m with sarmentose (arching) branches with large trusses of highly fragrant (musk scent) single white (stamens yellow) flowers 5.0 cm across appearing late in summer, *i.e.* July to September though *R. brunonii* flowers earlier from May to June; ***R. moyesii*** (syn. *R. macrophylla* var. *rubro-staminea* Vilm., *R. fargesii* Hort.), a native of W. China is a 3.6 m tall shrub having red stems with solitary (rarely 2) single deep blood-red flowers 4.3-6.25 cm across having short pedicels and deep orange-red oblong-ovoid hips; ***R. multiflora*** (syn. *R. intermedia, R. polyantha, R. polyanthos, R. thyrsiflora, R. wichurae*), a native of Japan and China is a vigorus shrub growing up to 4.5 m with sarmentose or sprawling branches having many-flowered pyramidal corymbs with usually single white flowers 1.8-2.5 cm across and bright red & small round hips appearing in autumn, is a parent of the garden Polyanatha roses and an important rootstock, its varieties are 'Carnea' (syn. var. *plena* Regel., *R. florida* Poir; double pink), 'Cathayensis' (China; single pink; a parent of 'Carnea' and 'Platyphylla'), 'Goldfinch' (peach-yellow, semi-double, highly fragrant), 'Nana' (dwarf), 'Platyphylla' (syn. 'Seven Sisters'; rich purple fading to mauve-white, double; and this has also given 'Crimson Rambler', one of the best climbing roses with deep red and comparatively more flowers), 'Watsoniana' (crimped foliage like Japanese maple, flowers small), *R. multiflora* × *R. rugosa* resulted into *R. iwara* bearing small, single and white flowers, and *R. yedoensis* with small pink flowers, *R. multiflora* × *R. chinensis* resulted into *R. polyantha* Hort. not Roessig. (as a trade name for hybrids), *R. multiflora* × 'General Jacqueminot' resulted into *R. dawsoniana* (Dawson rose), and *R. multiflora* has also produced hybrids with *R. gallica, R. setigera* and *R. wichuraiana*; ***R.* × *noisettiana*** ('Blush Noisette', *R. moschata* × *R. chinensis*), developed in England is a lax shrub with climbing branches growing up to 3.0 m with profuse clusters of double fragrant flowers of white colour shaded pink and 3.75 cm across; ***R.* × *odorata*** (*R. gigantea* × *R. chinensis*; syn. *R. chinensis* var. *fragrans, R. indica* var. *odoratissima, R. thea*), one of the parents of modern Tea and Hybrid tea roses (Tea roses through intercrossing of this species and also with *R. chinensis*, and Hybrid tea through crossing Tea roses and other garden roses) developed in China is a shrub with sarmentose branches bearing solitary or 2-3 fragrant white, light pink, salmon-pink or yellowish semi-double flowers 5.0-7.5 cm across on short glandular stalks, and var. *ochroleuca* (syn. *R. indica* var. *ochroleuca*) with pale-yellow double flowers; ***R.* × *paulii*** (*R. rugosa* × *R. arvensis*) is a sprawling shrub growing to 1.2 m and most suitable as ground cover, especially on the banks as it roots while spreading, and bears white fragrant flowers in clusters, 6.25 cm across with crinkled petals, and its var. 'Rosea' bears pink flowers; ***R. phoenicea*** from Asia Miner and allied to *R. moschata*, is a straggling bush with single white flowers in pyramidal corymbs; ***R. pomifera*** (syn. *R. villosa*), a native of Europe and W. Asia is an upright shrub attaining 2.1 m height and quite similar to *R. canina*, with grey-green and resin-scented foliage, 1-3 (usually 1) pale-pink flowers 3.75-6.25 cm across on bristly and glandular pedicels, and fruits scarlet to dark red, ovoid or subglobose and covered with bristly hairs, its var. 'Duplex' is a double red-purple hybrid; ***R. primula***, a native of Turkestan to N. China is an upright shrub growing up to 2.4 m tall, foliage glabrous, double-toothed and aromatic when crushed, flowers pale-yellow, fragrant, solitary and 3.75 cm across; ***R. rubiginosa*** (syn. *R. eglanteria* Linn.), a native of Europe including England, very similar to *R. canina*, most suitable for a dense hedge, growing up to 2.4 m bearing sweet-pungent aromatic foliage, 1-3 bright pink single flowers 3.75-5.0 cm across and orange-scarlet hips, and has produced some hybrids of double forms with certain species; ***R. rubrifolia***, a native of CS Europe, preferring partial shade, growing up to 2.1 m and is related to *R. canina* but with glaucous-purple stems and grey-purple leaves, and single pink to red-purple flowers 3.75 cm across appearing in small clusters; ***R. rugosa*** (syn. *R. ferox, R. regeliana, R. coruscans*), a vigorous upright shrub native to N. China, Korea and Japan, most suitable for wild gardens and for hedging, grows up to 2.1 m with wrinkled foliage, solitary, deep pink and highly scented flowers 7.5 cm across and round orange-red hips, and its famed varieties are 'Hollandica' commonly known as *R. rugosa* is used as rootstock for standard roses, *R. rugosa* 'Alba' (white), 'Albo-plena' (double white), 'Plena' ('Roseraie de l'Hay', large double crimson-mauve), *R. rugosa* 'Frau Dagmar Hastrup' (single carnation-pink), *R. rugosa* 'Rubra' (large single magenta-purple), *R. rugosa* 'Scabrosa' (large single rose-magenta), 'Rosea' (pink), 'Rubro-plena' (double purple), etc.; ***R. sericea***, a native to India, China and other temperate Himalayas and growing straight up to 3.6 m with 7-11 narrow leaflets, cream-white to pink single flowers 2.5-5.0 cm across but with only 4 petals in each flower and small pear-shaped yellow or red hips, its varieties are *R. sericea omeiensis* with tapering hips, *R. sericea polyphylla* having up to 17 leaflets, *R. sericea pteracantha* having mahagony-red wing-like glowing prickles on new growths, etc.; ***R. spinosissima*** (syn. *R. pimpinellifolia*), a native of W. Asia to China and Europe including England is a low shrub adorned with dense bristles and fern-like foliage, growing to 0.3-0.9 m with solitary cream-white, pink or yellowish single flowers 2.5-3.75 cm across and black-purple round hips, this has produced bright yellow roses with *R. foetida, R. s. altaica* (tall, creamy), 'Andrewsii' (semi-double pink), *R. s. bicolor* 'Grahamstown' and 'Staffia' similar to the species but cream-white flowers are splashed pink, *R. s. hispida* (large sulphur-yellow), *R. s. lutea* (single yellow), *R. s. lutea* 'Plena' (double yellow), 'Nana' (small white, nearly double), *R. s. rosea* (single pale-pink), etc.; ***R. webbiana*** endemic to Ladakh region of Kashmir (temperate Himalayas) is a shrub growing to 2.0 m with arching stems and solitary pale-pink flowers 4.0-5.0 cm across and flask-shaped blackish-red hips; ***R. willmottiae***, quite different from *R. willmottiana* and *R. blanda* var. *willmottiae*, a native of W. China is a densely branched 3.0 m tall shrub with slender paired prickles, solitary rose-purple flowers 2.5-3.0 cm across on short lateral branchlets and hips subglobose orange-red; ***R. xanthina***, a native of Korea and N. China is similar to *R. hugonis*, having brown stems and triangular prickles with semi-double (original species with single flowers though extinct but the characters exist in *R.* × *spontanea* which in cultivation is known as 'Canary Bird') bright yellow flowers 2.0 cm across; etc.

Old Garden Roses (sub-divison II) are those deciduous shrub roses that arose as sports or hybrids of the species and were much in cultivation before introduction of hybrid teas (HT).The ARS has defined this category as those cultivated rose types prior to 1867 and are further divided into 21 types. By the 18th century, five broad old European classes had emerged such as Alba, Bourbons, Centifolia, Damask and Gallica (Marriott, 2003). However, Hay and Beckett (1971) stated that though precise classification is difficult due to much interbreeding but for reference 13 sub-sections based on their parentage where known and so a few varieties though not correctly Old Roses are also included if having the same characteristics. Here, several varieties have quartered flowers as appear in the flattened centre. These sub-sections are (i) **Albas** (*R.*× *alba*) which are hardy, vigorous and erect shrubs with only a few prickles, deriving from *R. alba* having white to pink sweet-scented flowers up to 7.5 cm across coming up in profusion. Most of these roses are suitable for training against trelliswork or as espaliers. The representative varieties being 'Céleste' (syn. *R. alba incarnata*; 1.8 m height and soft pink semi-double flowers), 'Félicité Parmentier' (1.2 m, blush-pink double), 'Königin von Dänemark' (1.5 m, blooms pink and quartered), 'Maiden's Blush' (1.5 m, scented blush-pink with paler pink margins), etc. (ii) **Bourbons** which are hardy vigorous perpetual flowering (June-October) shrubs deriving from *R.* × *odorata* and *R. damascena* with densely petalled incurved, globular or cup-shaped highly scented blooms 7.5 cm across. The representative varieties are 'Boule de Neige' (1.8 m, buds crimson opening to double white highly scented globular flowers), 'Bourbon Queen' (1.8-3.0 m, profusion of double, crimped and cup shaped pink and magenta flowers), 'Fantin Latour' (1.5 m, double pink and cup-shaped blooms), 'La Reine Victoria' (1.8 m, highly scented double pink cup-shaped flowers), 'Mme Isaac Pereire' (height 2.1 m ideal for walls, pillars and fences, very highly fragrant large quartered deep pink flowers), 'Mme Pierre Oger' (1.8 m, sport of 'La Reine Victoria', flowers double blush-pink), 'Souvenir de la Malmaison' (1.2 m and also with climbing form growing up to 4.5 m, flowers flesh-pink and quartered), 'Variegata di Bologna' (2.4 m with weak stems and prone to black spot, flowers globular and quartered with stripes of white and deep crimson), 'Zéphirine Drouhin' (a thornless Bourbon climber prone to mildew, growing up to 2.7 m, foliage tinted red, and heavily fragrant bright carmine-pink semi-double flowers blooming from summer to autumn), etc. (iii) **Cabbage/Centifolia/Provence** which are hardy and compact shrub roses descending from *R. centifolia* but as the branches are of drooping nature hence require support, and the fragrant double flowers 5.0 cm across appear in clusters. The representative varieties are 'Duc de Fitzjames' (1.8 m, semi-double deep pink to purple flowers with green centres fading to soft lilac), 'Petite de Hollande' (1.2 m, a compact miniature with pink flowers), 'Robert le Diable' (1.2 m, deep pink to purple flowers with green centre), 'Tour de Malakoff' (1.8 m vigorous shrub suitable as a pillar rose, purple-magenta flowers large, loosely double and fading to lavender and grey), etc. (iv) **Damasks** (*R. damascena*) which are lax and open hardy shrubs with thorny stems and short-lived fragrant double flowers 7.5 cm across on weak pedicels appearing in loose clusters, originating from *R. damascena*. The representative varieties are 'Blush Damask' (1.8 m dense shrub suitable for shrub border and hedging, short-lived but fully double nodding blooms in mauve-pink colours with paler petal edges), 'Celsiana' (1.5 m vigorous shrub with plenty of bright pink loosely semi-double flowers having yellow stamens), 'Mme Hardy' (1.8 m and is sometimes thought to be a hybrid of *R. centifolia*, with large clusters of well formed white quartered flowers having small green centres), 'Petite Lisette' (0.9 m with blush-pink rosette flowers having folded petals and button eye), etc. (v) **Dwarf Polyanthas** which are dwarf hardy progenies of *R. multiflora* × China tea and Noisette roses having slender hairy stems with profusion of small pompon-shaped large-clustered blooms 2.5-5.0 cm across appearing in summer as well as during autumn. Representative varieties are 'Cécile Brunner' (0.6 m but a form of it is a climber growing up to 3.0 m, and the rose-pink flowers are pale-yellow shaded and the centre is darker), 'Paul Crampel' (0.6 m with orange-red semi-double rosette-like flowers), 'Perle d'Or' (0.9 m robust shrub bearing rich yellow blooms in dense clusters), etc. In fact, the **dwarf miniatures** originated from Chinese breeding efforts during early days of 19th century. During 1930s, there had been tremendous progress in miniature rose breeding with crosses including dwarf *R. multiflora* in the Netherlands. Now many such cultivars are patented in Denmark, the Netherlands, United States and Canada (Pinney, 1964; Fitch, 1977). (vi) **Gallicas** (*R. gallica*) which are hardy, compact and strong shrubs with few thorns, and the largest group of the Old shrub roses close to *R. gallica*, bearing solitary and highly fragrant double pink to crimson and mauve flowers 5.0-7.5 cm across. The representative varieties are 'Belle de Crécy' (1.2 m having weak stems with fragrant fully double purple-red flowers changing to violet at maturity, and green button centre), 'Camaieux' (0.9 m arching shrub bearing semi-double and deep crimson-purple flowers splashed and striped white), 'Cardinal Richelieu' (1.2 m shrub excellent for hedging and bearing pale-pink fully double blooms changing to violet-purple), 'Charles de Mills' (1.2 m shrub excellent for hedging, bearing fully double deep maroon flowers shading to violet-purple), 'Complicata' (1.5 m low climber with single brilliant pink flowers appearing in clusters with white centre and 7.5-10.0 cm across), 'Due de Guiche' (1.2 m excellent for hedging with fully double crimson flowers splashed purple and green centre), 'Du Maître d'École' (2.4 m with arching branches and large double magenta-pink blooms changing to lilac and mauve), 'Francofurtana' (0.9 m vigorous shrub with slightly scented double rose-crimson cup-shaped flowers fading to pink), 'Old Velvet' (syn. 'Tuscany', growing to 1.2 m with semi-double rich purple flowers streaked white on a few petals inside), 'Président de Sèze' (1.05 m excellent as hedge, flowers double, rich maroon changing to purple), 'Sissinghurst Castle'

(0.9 m with semi-double maroon-purple blooms), 'Surpasse Tout' (1.2 to 1.5 m vigorous shrub excellent as hedge, with semi-double rose-crimson flowers), 'Tuscany Superb' (1.2 m vigorous with large, double and purple blooms), etc. (vii) **Hybrid Musks** (derivatives of *R. moschata*) which are hardy, sparsely prickly and perpetually flowering arching shrubs with large clustered scented flowers in the colour range of pink, mauve-crimson, deep crimson-scarlet, soft yellow and pure white, however, those differing the old types are included in the Modern shrub roses. Representative varieties are 'Ballerina' (1.2 m with large spray of pale-pink flowers), 'Bonn' (up to 1.8 m with clusters of orange-scarlet and slightly scented flowers), 'Buff Beauty' (up to 1.8 m with double tea-scented blooms of apricot colour), 'Cornelia' (1.5 m with coppery-apricot, richly scented, double blooms in large trusses suitable as cut flowers and blooming perpetually), 'Felicia' (1.35 m, free-flowering in clusters with deeply fragrant double flowers in silvery, salmon-pink colour), 'Pax' (up to 3.0 m being the tallest shrub in the group with sweetly scented cream-white and semi-double flowers appearing in mass), 'Penelope' (1.5 m vigorous shrub, continuously free-flowering in pink flushed apricot and fading to pale-yellow to cream-white, musk-scented and semi-double), (viii) **Hybrid Perpetuals** which are hardy and vigorous shrubs of upright habit, very popular during Victorian times, derivatives of the Portland, Bourbon and China roses and predecessors of Hybrid Teas but less perpetual than the modern HTs hence now replaced by these, excellent for training against walls and these produce cabbage-like rounded double flowers 7.5-10.0 cm across being borne singly or in small clusters from June to October, with the representative varieties as 'Baroness Rothschild' (1.2 m upright shrub with cupped flowers of rose-pink to white), 'Frau Karl Druschki' (1.5 m vigorous shrub excellent for making standard, light pruning causes profuse blooming, and the flowers are double, scentless and pure white), 'Général Jacqueminot' (1.2 m with highly fragrant, double and deep crimson flowers), 'Mrs John Laing' (1.5 m upright growing with sweetly scented, double and pink blooms being borne on straight stalk), 'Reine des Violettes' (1.8 m requiring hard pruning, the flowers are double mauve-purple), 'Roger Lambelin' (1.2 m with velvety deep crimson-purple and double fragrant flowers often with white petal margins), etc. (ix) **Hybrid Rugosas** which are vigorous and prickly, seldom repeat-flowering and the bowl-shaped flowers 7.5-10.0 cm across are mostly heavily fragrant and appear in small clusters. The representative varieties are 'Agnes' (1.8 m with fully double, pompon-shaped, yellow and fragrant flowers being borne in profusion), 'Blanc Double de Coubert' (1.8 m excellent as a hedge, with perpetual flowering habit bearing semi-double white flowers), 'F.J. Grootendoorst' (1.8 m with small double crimson flowers appearing in small clusters having fringed petals, and it has sported a pink sport as 'Pink Grootendoorst'), 'Sarah Van Fleet' (2.4 m sturdy, thorny and a free-flowering bush with abundance of highly fragrant, semi-double and pink flowers for longer duration), 'Schneezwerg' (syn. 'Snow Dwarf', growing up to 1.5 m with snow-white and anemone-like semi-double blooms), etc. (x) **Hybrid Sweetbriars/Penzance Briars** (raised by Lord Penzance at the end of the 19th century) which are hardy and vigorous free-flowering (3.75 cm across saucer-shaped flowers) fragrant shrubs with 5 petals, mostly suitable as hedges bearing aromatic foliage, often prone to mildew and blooming in abundance in small clusters, and in autumn these adorn the shiny crimson-scarlet hips. The representative varieties are 'Amy Robsart' (semi-double deep rose-pink flowers), 'Janet's Pride' (blooms single, pink with pale centre), 'Lady Penzance' (single with rich coppery-yellow tints), 'Lord Penzance' (flowers buff flushed pink), 'Meg Merrilees' (single crimson), etc. (xi) **Moss Roses** (*R. centifolia muscosa*, all being sports of *R. centifolia muscosa* or hybrids derived from these sports, hardy shrub roses closely allied to Provence or Cabbage roses but differing in having the stems, branches and petioles densely covered with bristles while the backs and edges of the sepals covered with resin-scented mossy glands) which are small (dwarf) to tall (pillar) sports of *R. centifolia* 'Muscosa' or hybrids of these sports, closely allied to the Provence or Cabbage roses though stems, branches and petioles are covered with bristles and sepal-edge backs with resin-scented mossy glands. The flowers are double or semi-double, being borne singly or in 2-3, heavily fragrant and 7.5 cm across. The representative varieties are 'Alfred de Dalmas' (syn. 'Mousseline', 1.2 m compact bush with small palest-pink flowers), 'Capitaine John Ingram' (1.5 m compact and free-flowering bush with double mottled crimson-maroon flowers), 'Comtesse de Murinais' (1.8 m vigorous grower with blush-white changing to pure white flowers with a green eye), 'Général Kléber' (1.5 m vigorous shrub fully covered with moss, flowers double pink), 'Gloire de Mousseux' (1.2 m stout shrub with lilac-pink flowers 12.5 cm across, fading to pale-pink), 'Henri Martin' (1.5 m graceful shrub with delicately charming foliage covered with moss, and semi-double flowers of crimson colour having petal-midrib white), 'Mme de la Roche-Lambert' (1.2 m having brown moss on the stem and flowers semi-double rose-purple), 'Nuirs de Young' (1.5 m wiry bush with dark moss and the flowers are double deep maroon-purple), 'Réné d'Anjou' (1.5 m shrub having bronze-coloured young leaves and brown-green moss, and double pink flowers changing to lilac-pink), 'Salet' (1.2 m shrub with continuous flowering, and flowers bright pink), 'William Lobb' (1.8 m heavily mossed shrub suitable for training against a wall or pillar, and large clusters of double purple-magenta flowers fading to lavender), etc. (xii) **Portland Roses** (being predecessors of the Bourbons and Hybrid Perpetuals and are among the earliest surviving roses showing the influence of *R. chinensis*) are hardy, compact and suckering shrubs bearing Damask-type flowers either solitary or in small clusters, 5.0-7.5 cm across with the representative varieties as 'Comte de Chambord' (1.2 m erect shrub with continuous flowering having heavy fragrance highly filled double pink flowers with lilac tones and rolled petals), 'Jacques Carrier' (1.2 m vigorous shrub with deep pink green-centred quartered flowers appearing intermittently in summer), 'Portland' (0.6 m shrub with semi-double bright crimson cupped flowers appearing in summer and again during autumn), etc. (xiii) **Scotch Roses** (hardy roses) which are descendants of *R. spinosissima* with vigorous suckering habit having straight bristles, singly or in small clusters saucer-shaped fragrant flowers 2.5-3.75 cm across appearing in May-

July, with the representative varieties as 'Falkland' (1.2 m with semi-double pale-pink flowers), 'Stanwell Perpetual' (a hybrid between *R. spinosissima* and *R. damascena*, a 1.2 m high lax shrub with fragrant blush-pink semi-double flowers fading to white), 'William III' (0.6 m shrub with semi-double velvey flowers of brilliant maroon colour but reverse of petals pale-lilac tinted), etc.

During the later part of the 18th century, the old European roses were crossed with the 'China rose'. The China group had distinct features that were lacking in the European roses such as low branching habit, recurrent flowering, crimson colouration, new fragrances, distinct flowers with high centre and slender flower buds that unfurled on opening (Higson, 2007) and thus a number of new classes like **Bourbon** (Autumn Damask × Parson's Pink China), **Hybrid Perpetual** (Portland × Bourbon × Hybrid China), **Noisette** (Parson's Pink China × *R. moschata*), **Portland**, and **Tea** (China × Bourbon or Noisette) (Marriott, 2003; Higson, 2007) emerged. Thus in all, these become Albas (*R.* × *alba*), Bourbons, China Roses, Hybrid Perpetuals, Centifolias (*R.*× *centifolia*), Musks (*R. moschata*), Moss Roses (*R.* × *centifolia muscosa*), Damasks (*R. damascena*), Gallicas (*R. gallica*), Noisettes, Portland and Tea Roses. **American Rose Society Classification Scheme** (Cairns, 2003) mentions old garden roses as Alba, Ayrshire, Bourbon, Boursalt, Centifolia, Damask, Hybrid Bracteata, Hybrid China, Hybrid Eglanteria, Hybrid Foetida, Hybrid Gallica, Hybrid Multiflora, Hybrid Perpetual, Hybrid Sempervirens, Hybrid Setigera, Hybrid Spinosissima, Moss, Noisette, Portland, Tea, and Miscellaneous. In 1979, the **World Federation of Rose Societies** including Royal National Rose Society (UK) and the American Rose Society discussed and decided that the terms HT and Floribunda have lost their original meaning due to immense hybridization and agreed to adopt a new classification system for clear understanding of the masses and thus the Hybrid Tea were named as "large-flowered bush roses", Floribunda as "cluster-flowered bush roses" and Patio roses in 1987 were named as "dwarf cluster flowered bush roses". After renaming them ~ in **Modern garden roses** the groups are **Large-flowered bush** (Hybrid Tea), **Cluster-flowered bush** (Floribunda), **Dwarf cluster-flowered bush** (Patio rose), **Miniature**, **Ground cover**, **Climbing**, **Rambler** and **Shrub**; in **Old garden roses** the groups are **Alba**, **Bourbon**, **China**, **Damask**, **Gallica**, **Hybrid Musk**, **Hybrid Perpetual**, **Moss**, **Noisette**, **Portland**, **Provence** (Centifolia), **Sempervirens** and **Tea**; and the **Species roses** include wild, or species, roses and species hybrids (which share most characteristics of both parent species) which are large and arching shrubs bearing one flush of single (5-petalled) flowers during spring or mid-summer and showy hips in autumn.

Commercially only four groups are recognized, *i.e.* (i) **cut flowers**, (ii) **miniature potted flowering holiday plants** grown from stem cuttings, (iii) **flowering pot-plants** produced from bare-root plants, and (iv) **potted garden plants** produced from bare-root plants (Dole and Wilkins, 1999). The cut flower varieties should be of short duration and requiring low light intensities so that dependence on supplementary high intensity discharge lighting (HID) is reduced to save the expenses. Zieslin and Moe (1985) stated that only 7-8 species are so far involved though we have a gene-pool of over 100 species. Miniatures are grown in the greenhouses in 10- to 15-cm pots. Several hybrid cultivar groups under Polyantha and Floribunda can be forced for holidays such as Christmas, Valentine's Day, Easter, Mother's Day, etc. Bare-rooted garden varieties are forced into leaf and flower for sale and these roses are large old fashioned shrubs which have low maintenance cost.

In broad term, **Modern Roses (1867+)** have originated from largely 10 species – *R. canina, R. chinensis, R. foetida, R. gallica, R. gigantea, R. multiflora, R. moschata, R. phoenicia, R. rugosa* and *R. wichuriana* (Gudin, 2000). The first of the modern roses were the Hybrid Tea roses (large flowered bushes) derived from crosses between Hybrid Perpetual and Tea roses (Marriott, 2003) and thus was having the germplasm of *R.damascena, R. moschata, R. chinensis, R. gigantea* and *R. gallica*. The first recognized Hybrid Tea was 'La France' discovered in 1867. In the early 1900's, a French breeder Joseph Pernet-Ducher made crosses between Hybrid Perpetuals and *R. foetida* to introduce yellow colour. Most HT rose cultivars are mainly upright mostly single well shaped flowers with high spiralling centres at the end of long stems; sturdy and often with shiny petals; characteristic pointed buds; large glossy leaves; and strong stems. The Hybrid Tea roses have contributed to many of the modern rose classes including Polyantha (*R. multiflora* × Hybrid Tea), Floribunda (Polyantha × Hybrid Tea, the cluster-flowering bushes), Grandiflora (Floribunda × Hybrid Tea, large flowered bushes), and Miniature (Polyantha × *R. chinensis minima*) (Marriott, 2003). Apart from these, other roses falling under modern garden roses are Patio (dwarf cluster-flowered bushes), Groundcover, Climber, Rambler and Shrubs. **American Rose Society Classification Scheme** (Cairns, 2003) mentions modern roses as Floribunda, Grandiflora, Hybrid Kordesii, Hybrid Moyesii, Hybrid Musk, Hybrid Rugosa, Hybrid Wichuriana, Hybrid Tea, Large Flowered, Miniature, Mini-Flora, Polyantha and Shrub.

Since the beginning of the nineteenth century, rose breeding has flourished, leading to the actual probable number of some 20,000 varieties cultivated in the world (Hareing, 1960). Some of the important **varieties** under **Hybrid Teas** (major glasshouse forcing type) with long stemmed (125 cm) single large bloom (>4 cm) are 'Aalsmeer Gold', 'Abhisarika', 'Ace of Hearts', 'Alinka', 'Adolf Horstmann', 'Adora', 'Alec's Red', 'Allegro', 'Anne Letts', 'Amalia', 'American Heritage', 'American Pride', 'Angel Bells', 'Angel Wings', 'Angelique', 'Angelus', 'Anuraag', 'Anvil Sparks', 'Arianna', 'Arjun', 'Arkansas', 'Arlene Francis', 'Athena', 'Atoll', 'Avon', 'Babylon', 'Baccara', 'Bacchus', 'Baden-Baden', 'Bajazzo', 'Ballet', 'Banco', 'Barbara Richards', 'Beauté', 'Bel Ange', 'Belami', 'Berolina', 'Beryl Bach', 'Bettina', 'Bewitched', 'Bhim', 'Birmingham Post', 'Black Beauty', 'Black Delight', 'Black Lady', 'Blanche Comete', 'Blood Stone', 'Blue Delight', 'Blue Moon', 'Blue Nile', 'Blue Ocean', 'Blue River', 'Blue Sky', 'Bonne Nuit', 'Bonsoir', 'Brandy', 'Brasilia', 'Bridal Robe', 'Brilliant', 'Brilliant Light', 'Buccaneer', 'Calico', 'Careless Love', 'Carina', 'Carmousine', 'Casanova', 'Catherine Denevue', 'Century Two', 'Champs Élysées', 'Charles de Gaulle', 'Charles Mallerin', 'Charlotte Armstrong', 'Chaten de Versailles', 'Cherry

Brandy', 'Chicago Peace', 'Christian Dior', 'Chrysler Imperials', 'City of Panjim', 'Cleopatra', 'Coalite Flame', 'Colour Wonder', 'Command Performance', 'Comtesse Vandal', 'Condesa de Sastago', 'Confidence', 'Crimson Glory', 'Crimson Halo', 'Cynthia', 'Delicia', 'Derek Nimmo', 'Diamond Jubilee', 'Diorama', 'Directeur Guerin', 'Dolly Parton', 'Doris Tysterman', 'Dorothy Anderson', 'Dorothy Peach', 'Double Delight', 'Dr. A.J. Verhage', 'Dr. B. P. Pal', 'Dr. Darley', 'Dr. Dick', 'Dr. F.G. Chandler', 'Duke of Windsor', 'Dutch Gold', 'Eden Rose', 'Eiffel Tower', 'Elaine', 'Elfe', 'Elizabeth Harkness', 'Ellinor le Grice', 'Ena Harkness', 'Ernest H. Morse', 'Esmerelda', 'Eterna Fascination', 'Ethel Sanday', 'Etoile de Hollande', 'Eve Allen', 'Evensong', 'Fandango', 'First Love', 'First Prize', 'Flaming Sunset', 'Folklore', 'Forgotten Dreams', 'Fragrant Cloud', 'Frau Karl Druschki', 'Fred Gibson', 'Friendship', 'Fritz Thiedemann', 'Funkühr', 'Gail Borden', 'Galia', 'Gallivarda', 'Ganga', 'Garden Party', 'Gavotte', 'Garvey, 'Gay Crusader', 'Gladiator', 'Gold Crown', 'Golden Dawn', 'Golden Giant', 'Golden Jubilee', 'Golden Melody', 'Golden Splendour', 'Gold Medal', 'Granada', 'Grand'mère Jenny', 'Grandpa Dickson', 'Grand Masterpiece', 'Grand Siecle', 'Grey Pearl', 'Guinevère', 'Happiness', 'Harmonie', 'Haseena', 'Hawaii', 'Headliner', 'Heart of England', 'Heidi Jayne', 'Helen Traubel', 'Helmut Schmidt', 'Hidalgo', 'Honor', 'Hot Pewter', 'Ico Ambassador', 'Ico Delight', 'Ico Trimurti', 'Inge Horstmann', 'Ingrid Bergman', 'Interflora', 'Isabel de Ortiz', 'Jacaranda', 'Jackino', 'Jadis', 'Jan Guest', 'Janina', 'Jennifer Hart', 'J.F. Kennedy', 'Jogan', 'John S. Armstrong', 'Josephine Bruce', 'June Bride', 'June Park', 'Just Joey', 'Kardinal', 'Karl Herbert', 'Kentucky Derby', 'King's Ransom', 'Kiss of Fire', 'Konrad Adenauer', 'Kronenbourg', 'Lady Belper', 'Lady Meilland', 'Lady Rose', 'Lady Seton', 'Lady Sylvia', 'Lady Vera', 'Lady X', 'Lady Zia', 'Lagerfeld', 'La Jolla', 'Lal', 'Lancome', 'Landora', 'Las Vegas', 'Laura', 'Lemon Sherbet', 'Lilac Time', 'Louisiana', 'Love', 'Lover's Meeting', 'Loving Memory', 'Lucky Champhorn', 'Mabella', 'Maid of Honour', 'Marcel Gret', 'Madelon', 'Margaret', 'Maria Callas', 'Marianne Powell', 'Mary Wheatcroft', 'Matterhorn', 'May Lyon', 'Madame Violet', 'McGredy's Ivory', 'McGredy's Sunset', 'McGredy's Yellow', 'Medallion', 'Meduse', 'Memorium', 'Message', 'Michèle Meilland', 'Milord', 'Mischief', 'Miss Ireland', 'Mme Butterfly', 'Mme Henri Guillot', 'Mme Luise Laperrière', 'Modern Art', 'Mojave', 'Moncheri', 'Monique', 'Montezuma', 'Montreal', 'Moonbeam', 'Mount Shasta', 'Mr. Chips', 'Mr. Lincoln', 'Mridula', 'Mrinalini', 'Mrs. Charles Lamplough', 'Mrs. Sam McGredy', 'Mullard Jubilee', 'My Choice', 'National Trust', 'Nefertiti', 'Night-'N'-Day', 'Night Time', 'Norma', 'Norman Hartnell', 'Numéro Uno', 'Nurjehan', 'Oklahoma', 'Olympiad', 'Only You', 'Ophelia', 'Oriana', 'Orient Express', 'Osiria', 'Pampa', 'Papa Meilland', 'Paradise', 'Paris Match', 'Pasadena', 'Pascali', 'Patsy Cline', 'Peace', 'Peach Beauty', 'Peer Gynt', 'Peggy Lee', 'Percy Thrower', 'Peter Frankenfeld', 'Perfecta', 'Peudouce', 'Phyllis Gold', 'Piccadilly', 'Picture', 'Pigalle', 'Pink Charming', 'Pink Favourite', 'Pink Panther', 'Pink Peace', 'Pink Spiral', 'Pink Supreme', 'Poinsettia', ' Polar Star', 'Polly', 'Portrait', 'Prelude', 'Prima Ballerina', 'Prima Donna', 'Princess Margaret of England', 'Princess Marina', 'Pristine', 'Pusa Ajay', 'Pusa Arun', 'Pusa Mansij', 'Pusa Shatabdi', 'Raja Surendra Singh of Nalagarh', 'Rajni', 'Raktagandha', 'Raktima', 'Red Ensign', 'Red Lion', 'Red Masterpiece', 'Red Planet', 'Red Recker', 'Rendez Vous', 'Rex Anderson', 'Rose Gaujard', 'Rose Mary Harkness', 'Rouge Meilland', 'Royal Highness', 'Royal Tan', 'Royal William', 'Ruksaar', 'Sabine', 'Sam McGredy', 'Sandra', 'Sarah Arnot', 'Savkar', 'Seiko', 'Sheerbliss', 'Shiralee', 'Shot Silk', 'Show Girl', 'Signora', 'Silver Jubilee', 'Silver Lining', 'Simon Dot', 'Sir Henry Segrave', 'Snow White', 'Sonia Meilland', 'Sophocle', 'Spek's Yellow', 'Stella', 'Sugandha', 'Sultane', 'Summer Fragrance', 'Summer Holiday', 'Summer Sunshine', 'Sundowner', 'Sunfire', 'Super Star', 'Supreme', 'Supriya', 'Suzon Lotthé', 'Sterling Silver', 'Sutter's Gold', 'Swarnrekha', 'Swagatham', 'Swarthmore', 'Sweet Afton', 'Sylvia', 'Taj Mahal', 'Tallulah', 'Tapestry', 'Tapti', 'Tarantella', 'Tata Centenary', 'Thäis', 'Tzigane', 'Teenager', 'Tendresse', 'The Doctor', 'Touch of Class', 'Tournament of Roses', 'Ulster Monarch', 'Vanamali', 'Velvet Fragrance', 'Verschuren's Pink', 'Via Mala', 'Victor Hugo', 'Vienna Charm', 'Violaine', 'Violinista Costa', 'Virgo', 'Voodoo', 'Wendy Cussons', 'Westminister', 'Westward Ho', 'Whisky', 'White Christmas', 'White Success', 'William Harvey', 'William Moore', 'Wind Sounds', 'Wisbech Gold', 'Woman and Home', 'Yonkee', 'Yorkshire Bank', etc.; under **Floribundas** (large and clustered flowering type with no disbudding) with shorter stems (<60 cm), flowers in bunches and smaller than HTs are 'Africa Star', 'Akito', 'Alain', 'Alamein', 'Alison Wheatcroft', 'Allgold', 'Allotria', 'Ama', 'Ambrosia', 'America's Junior Miss', 'Annabell', 'Anna Wheatcroft', 'Anne Cocker', 'Anne Harkness', 'Arabian Nights', 'Arthur Bell', 'Arunima', 'August Seebauer', 'Banjaran', 'Beaulieu Abbey', 'Belinda', 'Bellona', 'Betty Prior', 'Blessings', 'Blue Diamond', 'Bonica', 'Border Coral', 'Border King', 'Border Queen', 'Bridal Pink', 'Brown Velvet', 'Carol Amling', 'Cecile Brunner', 'Celebration', 'Centenary', 'Champagner', 'Chanelle', 'Charisma', 'Charleston', 'Charlotte Elizabeth', 'Charming Maid', 'Chinatown', 'Chorus Girl', 'Circus', 'City of Leeds', 'Clydebank Centenary', 'Cocorico', 'Confetti', 'Concerto', 'Cordula', 'Courtosie', 'Daily Sketch', 'Dainty Maid', 'Dearest', 'Deep Purple', 'Delhi Princess', 'Diamant', 'Donald Prior', 'Dorothy Wheatcroft', 'Drummer Boy', 'Drummer Marigold', 'Dusky Maiden', 'Elizabeth of Glamis', 'Else Poulsen', 'Escapade', 'Europeana', 'Evelyn Fison', 'Fantasia', 'Fashion', 'Faust', 'Fergie', 'Fervid', 'Festival Fanfare', 'Firecracker', 'Fire King', 'First Edition', 'Flamenco', 'Fred Loads', 'Frensham', 'Friedrich Heyer', 'Gene Boerner', 'Gipsy', 'Glacier', 'Glengarry', 'Golden Fleece', 'Golden Jewel', 'Golden Slippers', 'Golden Times', 'Goldgleam', 'Goldilocks', 'Gold Marie', 'Hakunn', 'Happy Anniversary', 'Happy Wonderer', 'Harkness Marigold', 'Highlight', 'Hunter', 'Iceberg', 'Ico', 'Impatient', 'Imperator', 'Independence', 'Inner Wheel', 'Innisfree', 'International Herald Tribune', 'Irene of Denmark', 'Irish Mist', 'Isle of Man', 'Ivory Fashion', 'Jan Spek', 'Jiminy Cricket', 'Judy Garland', 'June Bride', 'Karen Poulsen', 'Kim', 'King Arthur', 'Korona', 'Lady Sonia', 'Laminuette', 'Laughter Lines', 'Len Turner', 'Lilac Charm', 'Lilli Marlene', 'Lorena', 'Lubeck', 'Lutin, Madhura', 'Magali', 'Manx Queen', 'Märchenland', 'Margaret Merryl', 'Marielle', 'Marina', 'Marlena', 'Masquerade', 'Mercedes', 'Michelle', 'Milena', 'Mohini', 'Molly McGredy', 'Moonraker', 'Moulin Rouge', 'Neelambari', 'News', 'Nordia', 'Ohlala', 'Old Master', 'Orangeade', 'Orange Sensation', 'Orange Silk', 'Orange Special', 'Paddy McGredy', 'Pantomime', 'Papillon Rose', 'Paprika', 'Park Place', 'Pernille Poulsen', 'Picasso', 'Pink Parfait', 'Pinnochio', 'Poulsen's Bedder', 'Poulsen's Pink', 'Prema', 'Pusa Baramasi', 'Pusa Gaurav', 'Pusa Komal', 'Pusa

Muskan', 'Queen Elizabeth', 'Queen of Bermuda', 'Rare Edition', 'Red Dandy', 'Red Favourite', 'Red Gold', 'Red Wonder', 'Regensburg', 'Rimosa', 'Rob Roy', 'Rodeo', 'Rosemary Rose', 'Rose of Tralee', 'Rose Parade', 'Rudolph Timm', 'Rumba', 'Sadabahar', 'Salmon Perfection', 'Scania', 'Scarlet Queen Elizabeth', 'Scented Air', 'Sea Pearl', 'Shepherdess', 'Shepherd's Delight', 'Shocking Blue', 'Siren', 'Summer Snow', 'Sundance', 'Sun Flare', 'Sunny Maid', 'Suryakiran', 'Tambourine', 'Telstar', 'Tiki', 'Tip Top', 'Tombola', 'Travesti', 'Vera Dalton', 'Violet Carson', 'Vogue', 'White Junior Miss', 'White Queen Elizabeth', 'Woburn Abbey', 'Yellow Hammer', 'Yellow Pinocchio', 'Yvonne Rabier', 'Zambra', 'Zizi, Zorina', etc.; under **spray types** with 5-6 flowers per stem are 'Eviler', 'Joy', 'Nikita', etc.; under **Polyanthas** are 'Baby Faurax', 'Bharani', 'Cameo', 'Cécile Brunner', 'Chattilon', 'Chattilon Rose', 'Echo', 'Nartaki', ' Paul Crampel', 'Perle d'Or', 'Pink Showers', 'Swati', 'The Fairy', 'Vatertag', etc.; under **Modern shrub roses** are 'Constance Spry', 'Fred Loads', 'Fritz Nobis', 'Frühlingsgold', 'Frühlingsmorgen', 'Golden Chersonese', 'Golden Wings', 'Heidelberg', 'Joseph's Coat', 'Kassel', 'Kathleen Ferrier', 'Lavender Lassie', 'Lichterlohe', 'Marguerite Hilling', 'Munster', 'Nymphenburg', 'Queen Elizabeth', 'Réveil Dijonnais', 'Uncle Walter', 'Wilhelm', 'Will Scarlet', etc.; under **Miniatures** with very small buds are 'Baby Crimson', 'Baby Faurax', 'Baby Gold Star', 'Baby Masquerade', 'Blue Mist', 'Chandrika', 'Cherryade', 'Chipper', 'Cinderella', 'Coralin', 'Cricri', 'Dazzler', ' Deep Velvet', 'Dwarf Queen', 'Eleanor', 'Ernest H. Morse', 'Fire Princess', 'Honey Comb', 'Josephine Wheatcroft', 'Little Buckaroo', 'Little Flirt', 'Little Princess', 'Magic Wand', 'Maid Marion', *Rosa chinensis* 'Minima' (syn. 'Rouletti'), 'New Penny', 'Norwich Gold', 'Over the Rainbow', 'Peon', 'Perla de Alcanada' (syn. 'Baby Crimson'), 'Perla de Monserrat', 'Pink Cameo', 'Pixie', 'Pompon de Paris', 'Pour Toi', 'Pushkala', 'Red Imp', 'Rise-N-Shine', 'Rosina', 'Roulettii', 'Showoff', 'Starina', 'Strawberry Swirl', 'Sweet Dream', 'Sweethearts', 'White Madonna', 'Yellow Doll', etc.; under **Ground covers/prostrate** roses are'Grouse', 'Max Graf', 'Nozomi', 'Pheasant', 'Raubritter', 'Red Bells', *Rosa luciae wichraiana*, 'Rosy Cushion', etc.; under **Climbers/ramblers** are 'Albéric Barbier', 'Albertine', 'Alister Stella Gray', 'Allen Chandler', 'Aloha', 'American Pillar', *R. banksiae lutea*, *R. banksiae lutescens*, *R. banksiae normalis*, 'Bantry Bay', 'Blush Noisette', 'Bobbie James', 'Caroline Testout', 'Casino', 'Cécile Brunner', 'Chaplin's Pink', 'Cocktail', 'Copenhagen', 'Coral Dawn', 'Crimson Glory', 'Crimson Shower', 'Cupid', 'Danse des Sylphes', 'Danse du Feu', 'Delhi Pink Pearl', 'Delhi White Pearl', 'Dorothy Perkins', 'Dr. W. van Fleet', 'Easlea's Golden Rambler','Elegance', 'Emily Gray', 'Ena Harkness', 'Etoile de Hollande', 'Étude', 'Excelsa' (syn. 'Red Dorothy Perkins'), 'Flammentanz', 'François Juranville', 'Frühlingsgold', 'Gloire de Dijon', 'Golden Dawn', 'Golden Ophelia', 'Golden Showers', 'Guinée', 'Hamburger Phoenix', 'Handel', 'Heidelberg', 'High Noon', 'Iceberg', 'Joseph's Coat', 'Kathleen Harrop', 'Kiftsgate', 'Lawerence Johnston', 'Leverkusen', 'Little White Pet', 'Marechal Niel', 'Marigold', 'May Queen', 'Meg', 'Mermaid', 'Mme Abel Chatenay', 'Mme Alfred Carrière', 'Mme Butterfly', 'Mme Alfred Carrière', 'Mme Edouard Herriot', 'Mme Grégoire Staechelin', 'Mrs. Arthur Curtiss James', 'Mrs. Sam McGredy', 'Nevada', 'New Dawn', 'Norwich Pink', 'Norwich Salmon', 'Parkdirektor Riggers', 'Paul's Lemon Pillar', 'Paul's Scarlet Climber', 'Paul's Lemon Pillar', 'Pinata', 'Pink Perpetue', 'Polyantha Grandiflora', 'Prosperity', *Rosa* × *paulii*, 'Rose-Marie Viaud', 'Rosy Mantle', 'Royal Gold', 'Sander's White', 'Schoolgirl', 'Shot Silk', 'Thelma', 'The New Dawn', 'Veilchenblau', 'Wedding Day', 'Zephirine Drouhin', etc.; under **Moss** roses are 'Alfred de Dalmas', 'Capitaine John Ingram', *Rosa centifolia alba muscosa*, 'Comtessede Murinais', 'Géneéral Kléber', 'Gloire de Mousseux', 'Henri Martin', 'Mme de la Roche-Lambert', 'Mousseline', 'Nuits de Young', 'Réné d'Anjou', 'William Lobb', etc.; and under **Hybrid perpetuals** are 'Baroness Rothschild', 'Frau Karl Druschki', 'Général Jaqueminot', 'Mrs. John Laing', 'Reine des Violettes', 'Rogert Lambelin', etc. There are **glasshouse roses** for long-lasting cut flower use out of which the **long-stemmed** ones are 'Aalsmeer Gold', 'Bellona', 'Bianca', 'Black Magic', 'Carona', 'Carambole', 'Cora', 'Corvetti', 'Dallas', 'Diplomat', 'Dr. Verhage', 'Esacada', 'Femma', 'Film Stone', 'First Red', 'Gabriella', 'Golden Gate', 'Grand Gala', 'Grand Prix', 'Ilona', 'Jacaranda', 'Konfetti', 'Laser', 'Magic Moment', 'Maybella', 'Monte Carlo', 'Movies Star', 'Nicole', 'Noblesse', 'Osiana', 'Papillion', 'Parea', 'Passion', 'Pavrotte', 'Pink Sensation', 'Ravel', 'Red Velvet', 'Rodeo', 'Rossini', 'Sacha', 'Samura', 'Sandra', 'Sandy', 'Sangria', 'Skyline', 'Soledo', 'Sonia', 'Spinks', 'Star Light', 'Sublime', 'Susanne', 'Temptation', 'Texas', 'Tineke', 'Virginia', 'Vivaldi', 'Zéphirine Drouhin', etc.; the **medium-stemmed** ones are 'Aalsmeer Gold', 'Bellinda', 'Europa', 'Flirt', 'Frisco', 'Golden Times', 'Jack Frost', 'Jaguar', 'Kardinal', 'Kiss', 'Lambada', 'Mercedes', 'Motrea', 'Nordia', 'Ohio', 'Pink Prophyta', 'Prophyta', 'Sandokan', 'Souvenir', 'Tina', 'Vanilla', 'White Success', etc.; and the **small-flowered Floribundas** are 'Cecile Brunner', 'Garnette', 'Marimba', 'Sweetheart Rose', etc. There are many species and varieties of roses which are **fragrant**, a few of these being 'Abraham Darby', 'Alberic Barbier', 'Albertine', 'Alchymist', 'Alec's Red', 'Aloha', 'Angel Face', 'Anurag', 'Apple Jack', 'Apricot Nectar', 'Apsara', 'Arizona', 'Arthur Bell', *R. arvensis*, 'Austins Belle Isis', 'Banana Buff Beauty', *R. banksiae banksiae*, 'Belle Amour', 'Belle Storey', 'Blush Noisette', 'Bobbie James', *R. bracteata*, 'Break O'Day', *R. brunoni*, 'Buff Beauty', 'Cathedral', 'Celestine', the varieities of *R.* × *centifolia*, 'Cerise Bouquet', 'Charugandha', 'Cherish', 'Cherry Louis-Philippe', 'China Apricot', 'Chinatown', 'Chrysler Imperial', *R. cinnamomea*, 'City of Gloucester', 'Claire Rose', 'Command Performance', 'Compassion', 'Coquttes de Blanche', 'Coral Dawn', 'Cornelia', 'Cramoisie Superieur', 'Crimson Glory', 'Cuisse de Nymph', *R. damascena*, 'Daphne', 'Delhi White Pearl (clg.)', 'De Meaux', 'Distant Drums', 'Double Delight', 'Dr. M.S. Randhawa', 'Double Fragrance', 'Duchesse de Brabant', *R. eglanteria*, 'Electon', 'Etoile de Hollande (clg.)', 'Everest', 'Felicia', 'Fragrant Cloud', 'Fragrant Hour', 'Francesca', 'Friendship', 'Fritz Nobis','*R. gallica*, 'Garedenia', 'Gertrude Jakyll', 'Gloire de Dijon', 'Golden Afternoon', 'Grenada', 'Granny Grimmetts', 'Gulzar', 'Henry Hudson', 'Heritage', 'Honeysweet', 'Honorine de Brabant', 'Indian Princess', 'Intrigue', 'Jawahar', 'Jayne Austin', 'Jens Munk', 'Jude the Obscure', 'Jwala', 'Kamla Devi Chattopadhyay', 'Korresis', 'Lady Hillingdon', *R. laevigata*, 'L'Aimant', 'La Reine des Violettes', 'Lavender Crystal', 'Lavender Lassie', 'Love', 'Mme Gregoire Staechelin', 'Madame Hardy', 'Madame Issac Pereire', 'Madhura', 'Maggie', 'Marechal Niel', 'Marie Pavie', 'McCartney Rose', 'Monsieur Tillier', 'Margaret

Merrill', 'Max Graf', *R. moschata*, 'Mother Teresa', 'Mridula', *R. mulligani*, *R. multiflora*, *R. muscosa purpurea*, 'Nastarana', 'New Dawn', 'New Zealand', 'Nishada', *R. nitida*, 'Nurjehan', 'Nymphenburg', 'Old Blush', 'Old Pink Moss', 'Papa Meilland', 'Parson's Pink China', 'Penelope', 'Perception', 'Petite de Hollande', 'Perfume Delight', *R. phoenicea*, 'Preyasi', 'Prima Ballerina', 'Queen Margarethe', 'Rajhans', 'Ralph's Creeper', 'Ranjana', 'Rose à Parfum de L'Hay', 'Rose de Resht', 'Rosemary Harkness', 'Roseraie de l'Hay', 'Rose Sherbet', *R. rugosa rubra*, 'Safrano', 'Saratoga', 'Scented Air', 'Sceptred Isle', 'Sheer Bliss', 'Shreyasi', *R. sempervirens*, 'Sir C.V. Raman', 'Slater's Crimson', 'Sombreuil', 'Souvenir de Mme Boullet', 'Sugandhini', 'Sujata', 'Sundowner', 'Sunsprite', 'Sutter's Gold', 'Sweet Chariot', 'Sweet Surrender', 'Tiffany', 'Trier', 'Tuscany', 'Warwick Castle', 'Wendy Cussons', 'White America', 'White Dawn', 'White Maman Cochet (clg.)', *R. wichuriana*, 'Wife of Bath', 'Woburn Abbey', 'Zéphirine Drouhin', etc.

Propagation

Roses can be propagated through vegetative means, and for production of new varieties through seeds. **Seed** propagation is generally followed by breeders to produce new varieties as through this method usually seedlings show variation in colour though there are many species which breed true such as *R. glauca*. Seed propagation is also used to multiply rootstocks like *R. canina*. Seed is harvested from mature and ripe hips normally during autumn. The seeds of rose contain an immature embryo which results into dormancy, *vis-à-vis* a very hard seed coat. Due to its very hard seed coat, it is essential to soften it through acid scarification. However, stratification is an ideal and widely followed method for breaking dormancy. The seeds after extraction are stored in a dry, dark and airy place at 1-5 °C until their use. However, for raising them, these are first soaked in water for 24 hours and then placed in media like sphagnum moss or moist peat + sand, and wet-stratified at 1.6-4.4 °C for 2-6 months depending on the genotype. The treated seeds are sown to their own depth with sand or grit and placed at an airy place or in raised beds at 5 cm apart or in plugs in the month of October- November in fine and lightly pressed compost (having only a little ~ roughly one-quarter of the fertilizer contents of the potting compost and a fungicide to prevent damping off disease) which is capable of drawing water up to seeds placed near its surface. However, sowing during other months will require placing of the trays in cold frames. These may take about one year to germinate. After formation of the first true leaves, the seedlings being highly fragile are pricked out gently in individual containers. Up to their establishment, the seedlings are retained in the cold frame, and then these are hardened off gradually by moving them out of the cold frame during day time. After full acclimatization these seedlings are kept in the open, potted on as and when necessary and cared as other rose plants.

Vegetatively, initially it was **inarching** method of grafting through which roses were being prepared but now this method is obsolete, and is rarely followed in modern times. **Stenting** is a quick method of propagation of roses based on whip grafting. In stenting method a selected cultivar budded on an unrooted cutting of a rootstock results in a complete plant in 3-4 weeks. A piece of stem with a 5 leaflet-leaf and a single dominant axillary bud is used as a scion. A piece of stem consisting of a single internode without buds and/or leaves is used as rootstock. Scion and stock are held together with the help of a peg. The basal end of a rootstock is dipped in a talc with IBA at 0.4 per cent. The rooting medium is a mixture of peat and perlite in 1:1 ratio. The relative humidity is maintained at 100 per cent with intermittent misting and the temperature is maintained around 25 °C. There are roses that grow on their own roots and send out suckers which when root out should be **divided** and planted separately when these are dormant (Brickell, 1992). The roses such as some of the Gallicas and the cultivars of *R. pimpinellifolia and R. rugosa* can easily be propagated through this technique. Roses grown through cuttings and also through seeds can be propagated through this method. Propagation of roses through **layering** (ground and air) is now obsolete due to a very slow rate of multiplication though at one time it was an usual practice. Though it is usually practiced during spring, and is limited only to climbers and ramblers, *vis-à-vis* many of the shrub roses (Hessayon, 1981) such as Damask, Provence, Bourbon, Alba and species roses as well as all the ground-cover roses. In fact, the rose with long and flexible shoots can be propagated through simple ground layering. The shoots of such climbers or ramblers that are healthy and with mature woods when the flowering is over should be bent over, a little portion (some 3-5 cm) notched or left as such and covered with soil, and pegged into the ground to hold the layer at its position but with terminal end exposed, and in the next spring such layers should be severed from the parent plant just below the roots and planted. Earlier and best rooting may be induced through application of 25 ppm 2, 4-D, 500-2,000 ppm IBA, and 200 ppm NAA or 500 ppm IAA. These promoters may be variety or species specific. Rooting may occur within a month and then after a fortnight the layers may be severed from the parent plant and planted as the usual plant which will flower similar to that of its mother plant. In case of air-layering, the bark to a length of some 2.5 cm is removed without damaging the cambium layer and the wounded portion with a little adjoining bark is wrapped with moist sphagnum moss and then with polythene securely over the wound to both the sides, however, certain rooting hormones may also be added with the sphagnum moss for better results.

Hartmann and Kester (1972) stated that certain *Rosa* species such as *Rosa blanda, R. nitida* and *R. virginiana* can be propagated even through **root-cuttings**. Proximal (uppermost) ends of the root-cuttings should always be up and at the level of rooting media to maintain the polarity. **Stem cuttings** are used for clonal propagation of rootstocks, certain vigorous cultivars, polyanthas, climbers, miniatures and certain others that are related quite closely to wild species such as rambler roses. Through cuttings it takes about three years for the new plants to become established, the longest wait, except for miniature roses which root so easily to propagate commercially. IBA 500-2,000 ppm (Singh *et al.*, 2003) is used for rooting of cuttings. Cuttings taken during monsoon and spring respond well (Bose and Mukherjee, 1977). The length of hardwood cuttings should be 20-25 cm (in case of miniatures only 5-10 cm) and softwood (semi-ripe) cuttings 10-15 cm severed above a bud where the shoot is turning woody but the soft tips are trimmed off with the basal cut just below a node, atleast with 3 nodes and devoid of leaves at least the lower ones, out of which 2/3rd of the length is pushed down in the soil for rooting and 1/3rd above. When there is dearth of material, even those cuttings with 1-2 nodes can also be used, especially during high humid atmospheric conditions or under mist. However, miniature flowering pot-roses are exclusively propagated by single leaf-node stem cuttings (2 cm of stem above and below the node), best being from shoots from the second cutback or pinch of plants already in production. These cuttings can be stored moist in plastic bags at 2 to 5 °C for up to 7 days (Dole and Wilkins, 1999). The upper cut should be slant to distinguish it from the bottom cut so that while planting there may not occur any polarity mistake and also to facilitate rain water not stagnating on the wound, and trickling down easily. Cuttings are always planted in slanting position with upper cut facing downward to avoid any pathogenic infection. Rooting media are fresh sand, sand + peat moss, vermiculite or vermiculite + peat moss, perlite or perlite + peat moss or good garden soil. Media should be porous and should have good water holding capacity. The rooting and further growth of cuttings are influenced by type of cuttings (the age and the physiological condition) and their treatment, *vis-à-vis* environmental conditions. All the roses can be propagated through soft-wood, semi-hard wood or hardwood cuttings taken from current season's growth under mist and with treatment of IBA. Miniatures under mist respond well through soft-wood or semi-hardwood cuttings, though in case of rootstocks, & strong-growing climbers, ramblers and polyanthas hardwood cuttings are most suitable (Harstmann and Kester, 1972), though large-flowered bush roses and that too of complex parentage do not root so readily and when root also the growth may be very poor. However, to avoid damping off caused by *Cylindrocladium scoparium*, it would be better to steam-sterilize the media. During course of rooting and even afterwards, there should be proper illumination.

Roses are commercially propagated by 'T' **budding** or **shield budding**. In this case the plant material from two different roses are united to combine the virtues of both, one that is bud-grafted is known as scion and is meant for producing desired quality of flowers on a desired stock on which the scion is budded. Usually the stock (rootstock or understock) is chosen on the basis of its suitability to a particular agroclimatic condition, its compatibility with the scion, *vis-a-vis* vigorosity. The dormant bud from the scion is taken and inserted under the bark of the rootstock and then scion with the stock is securely wrapped and tied tightly around leaving only the portion of the bud for growth. In case of shield budding, the bud along with the rectangular bark around is gently removed and inserted through the incision made out on the stock at a desired height but below a node but in case of 'T' budding, an inverted 'T' cut is made onto the stock where the bud with a little portion of bark from desired scion is inserted. There are various stocks to suit varying conditions. Climate, compatibility, soil, pests and diseases, and utility play an important role in the selection of rootstocks. *Rosa indica* var. *odorata* is recommended for northern plains. This can withstand dry or wet conditions and salinity. This is also tolerant to powdery mildew and insect-pests. *R. indica major* (syn. *R. chinensis major, R.* × *odorata*) is easily propagated through cuttings and this bears pink, semi-double and fragrant flowers. This is most suitable rootstock for hot conditions, and especially for the 'Baccara' roses. *Rosa bourboniana* is widely used in North India for producing standard roses as it throws vigorous and strong shoots though it did not gain prominence due to being susceptible to powdery mildew. *Rosa multiflora* is widely used as rootstock in South and Eastern India, and in the temperate regions of the country where winters are very cold. It is very vigorous all through the season from summer to autumn and is most suited to poor and light soils though scions budded onto this are not long-lived. It is resistant to nematodes. *Rosa canina* and *R. c. inermis* are commonly used rootstocks in Europe as in latter case the plant remains less thorny, that is why, it has been named *inermis* which means thornless, however, *R. c. pollmeriana* rootstock possesses coloured green shoots. *Rosa canina* produces hardy plants and is popular where winters are severe or on heavy soils but it suckers heavily. It is propagated by seed and is resistant to drought and alkaline conditions. 'Dr. Huey' (syn. 'Shafter') is most suitable with areas having short-lived dormancy in roses. *R. ceriifolia froebelii* (syn. *R. dumetorum laxa*, popularly known as 'Laxa'), a moderately thorny or almost thornless rootstock producing no suckers, is highly vigorous and most suitable for budding the red varieties. This is reliable on most soils and climates and is replacing most other commercial rootstocks. *R. eglanteria* (syn. *R. rubiginosa*) as rootstock is very thorny, and slow-growing, hence is most suitable for pot-roses and miniatures. *R. manetti* (*R.* × *noisettiana*) and Natal Briar are also used as rootstocks in other countries. It is *R.* × *noisettiana* which is used in North America while *R. indica major* (*R. chinensis*) and *R. canina* are commonly used in Europe and Israel. Colombia, Ecuador and Mexico growers generally use *R.* × *noisettiana, R. chinensis* and *R. canina* as rootstocks. *R. canina* and *R. manetti* perform well in India also (Singh *et al.*, 2003). Under Delhi condition when various rootstocks (*R. bourboniana* strain, *R. canina, R. indica odorata, R. laxa, R. manetti*, 'Dr. Huey') were tried, the performance of *R. indica odorata* was recorded best (Singh *et al.*, 2003). Natal Briar is gaining prominence as rootstock for greenhouse roses. The ideal time for budding in North India is from January to March. In case of areas with mild climate, it can be done throughout the year, and in temperate areas from mid- to late summer when air is quite humid. When new shoots grow after bud take, the shoots are cut back 20-25 cm above the graft union and then for 4 weeks the plants are exposed to -0.5 to 0 °C and high humidity which will stimulate axillary bud activity after planting (Dole and Wilkins, 1999). Anitha (1989) when tried the rootstock *Rosa indica* for three scion varieties, *viz.* 'Ambassador', 'Pink Panther' and 'Princess' during different seasons, recorded 82-98 per cent success from second fortnight of August to first fortnight of October when there is downpour and atmospheric humidity is high, however, from first fortnight of February to second fortnight of March was found least favourable due to hot climate and high sunshine, and stated that critical periods for success in budding was recorded as the current fortnight, one preceding and the one succeeding. Swelling initiation of the bud at the union shows the success of the bud take, and when shoots grow out sufficiently from this bud, growth of the rootstock above the union is cut off.

For rapid multiplication, roses can also be propagated through **tissue culturing** the terminal buds, immature floral buds and petals. Uma (1991) studied chemical mutagenesis in rose cv. 'Alliance' under *in vitro* culture, and recorded 0.08 per cent mercuric chloride treatment for 15 minutes to be the best for surface sterilization of axillary buds, BAP 2 mg l^{-1} + 2, 4-D 1 mg l^{-1} for culture establishment, BAP 2 mg l^{-1} + GA_3 1 mg l^{-1} for shoot proliferation and the full strength MS medium when supplemented with 2 mg l^{-1} IAA for maximum rooting. For EMS treatment, 0.125 and 0.250 per cent were found to be the best for time taken for bud take, multiple shoot production and rooting for the buds excised at the time of flower harvest though 0.375 and 0.500 per cent EMS curbed multiple shoot production. Under *in vitro* studies with the cv. 'Folklore' (Wilson, 1993), mercuric chloride 0.08 per cent for 12 minutes for shoot and axillary bud explants and 0.06 per cent for 12 minutes for internodal segments and leaf disc explants minimized the contamination maximally. Axillary buds excised 4 days after flower opening had the best response in culture establishment. The MS medium supplemented with kinetin 2.0 mg l^{-1} + GA_3 1.0 mg l^{-1} recorded early multiple shoot induction and highest number of shoots per culture, though addition of BAP 2.0 mg l^{-1} + GA_3 0.75 mg l^{-1} proved best with respect to highest percentage of cultures with multiple shoots, however, floral bud initiation was observed under BAP 2.0 mg l^{-1} + GA_3 0.50 mg l^{-1}. *In vitro* rooting was best when the medium was supplemented with IAA and NAA each at 1.0 mg l^{-1} + activated charcoal 500 mg l^{-1}, however, successful hardening and *ex vitro* establishment of plantlets were achieved by surface inoculation of germinated spores of VAM mycorrhizae in liquid suspension. Callus induction was found best under BAP 0.5 mg l^{-1} + NAA 2.0 mg l^{-1} +2, 4-D 0.5 mg l^{-1} though callus proliferation was recorded best under BAP 0.5 mg l^{-1} + NAA 0.1 mg l^{-1} + ascorbic acid 5 mg l^{-1}.

Pre-culture dipping of explants for 3 hours in 2,000 mg/l carbendazim and 200 mg/l 8-HQC, and afterwards with 500 mg/l gentamycin to the establishment medium control the contamination to a greater extent. More than 90 per cent explant survival and its sprouting with dark green colouration is obtained in MS medium where sprouting occurs within 8 days, and addition of 3 mg/l BAP + 0.1 mg/l NAA and 1 mg/l GA_3 causes more than 95 per cent sprouting within 4 days producing more than 3 shoots per explant (Singh *et al.*, 2003). A combination of 0.2 mg/l BAP, 0.2 mg/l NAA and 0.5 mg/l GA was found for cvs 'Priyadarshni', 'Raktagandha' and 'Raktima'; and 1.0 mg/l GA instead of 0.5 mg/l in case of cv. 'Arjun' quite favourable for initial establishment through shoot tip culture though for maximum proliferation it was 1.5 mg/l BAP and 0.2 mg/l NAA in case of 'Priyadarshni' and 'Raktagandha' and 2.0 mg/l BAP and 0.2 mg/l NAA in case of 'Arjun' and 'Raktima' (Singh *et al.*, 2003).

Cultural Practices

Rose production is favoured by an extended growing season with a majority of sunny days. Roses in gardens should have **sunshine** for a minimum of six hours and prefer morning sunlight. **Temperature** has a significant effect on quality aspects such as stem length, bud and leaf size and stem diameter. The ideal temperature for rose cultivation is 15-27°C. Best commercial quality is obtained when the night temperature is around 16°C. Optimal rose production occurs at 12-15 mol/m^2/day. The ideal **humidity** is 60-65 per cent.

Roses prefer well drained **soil** with a pH between 6.0-7.0, though a pH below 5.5 or above 7.5 is not congenial for their growing. The ideal pH range is 5.5 to 6.0 (Dole and Wilkins, 1999). Alkaline soils can be improved through use of plenty of organic matter. Roses are not salt tolerant, so EC values should be less than 2.0 dsm^{-1}, preferably less than 1.5 ds/m as 1.5 ds/m level is critical. Soils of sandy loam are preferred for their rapid infiltration rates. Roses cannot stand water-logging and are susceptible to leaf fall under such conditions.

Planting of rose in gardens is done in beds of different shapes. The beds are prepared by incorporation of organic manure and basal dose of phosphorus in the soil. Planting should be done in well prepared beds. Planting should be done in such a way that the bud union remains 2.5-5.0 cm above the soil level. After planting the plants are to be watered copiously. In case of bare root roses, the roots along with the stem up to the bud union should be kept in water for six hours before planting. In north Indian plains and Eastern India, planting can be done from October to February. In Southern India October to December planting is preferable. In temperate areas spring and autumn planting is done. In areas with mild climate, planting can be done throughout the year. Planting distance for garden roses varies from 30-75 cm between plants and rows depending on the class, variety, soil and climate.

Generally, a rose plant requires **watering** at 8-10 l/m^2 a day. Drought stress leads to defoliation and sun-burn of canes and may contribute to spider mite. However, overwatering or poorly drained soils may lead to root disease or nutritional problems. Water stress at the stage of leaf primordium formation results into delaying the rose production cycle but does not have any negative effects on the quality of flower shoots or the flower buds. The stage prior to petal initiation is more sensitive. Water stress at this stage affects the quality of flower buds (reduction in number of well formed petals and flower bud length). Water stress imposed prior to stamen initiation seems to be the most damaging on rose development affecting the quantity (upto 70 per cent reduction of production) and the quality of flowering shoots and buds (reduction of stem length, fresh weight of flowering shoot, *vis-à-vis* bud diameter & length as well as well-formed petal number in the flower bud). Water stress after stamen formation and/or carpel formation does not have any negative effect on the number of flowering shoots that have reached the marketable stage or on the quality of flowers produced.

In roses, **pruning** starts right from planting and (i) if any root is broken that is removed and the cane height is also reduced leaving only 3-4 dormant axillary buds, (ii) when new shoots start appearing and the floral bud is either smaller than pea-size, its removal is termed as 'soft pinch' and sometimes it is repeated twice or in case when bud size is larger than a pea-size it is termed as 'hard pinch'. In both these cases, either of the two is practiced, and these new flowering shoots are removed down to the second 5-leaflet leaf from the main stem though depends upon the cultivar and the season. Through soft pinch, axillary buds grow soon though from hard pinch it takes more time and flower later. This pinching structures the plant in proper architectural form to increase future quality production (Langhans, 1987). At the time of flower harvesting it is pruning No. (iii) when we should be sure whether one, two or three nodes on the stem are to be retained or it should be cut below the knuckle (the junction of the bud to the stem) and then afterwards the height of the plant can be controlled or long-stalked flowers can be obtained by cutting the flowers at their knuckle though further flowering will be delayed, and the No. (iv) pruning, i.e 4-5 weeks (depending upon the cultivar) prior to date of requirement of the flowers developing shoots are pinched, and No. (v) pruning is drastic when demand of flowers is almost nil or little, maintaining the height of the plant from 45 to 90 cm and this is known as annual pruning. Rose requires annual pruning to revitalize the plant before it produces a new flush of blooms in the season. Pruning encourages new growth from bud union. This removal of old growth sometimes acts as a form of dormancy in warmer climates which is otherwise lacking. Pruning slows down the plant processes for a short period of time. Pruning in rose is done to (i) re-invigorate an old and unproductive plant, (ii) to remove old and diseased shoots, (iii) to shape the plant, (iv) to encourage the plants producing fewer but larger blooms or many but smaller blooms, (v) to manipulate the blooming time to avoid glut in the market and as per demand in the market, and (vi) to deadhead and to encourage repeat blossoms. **Hard pruning** is done to rejuvenate sickly and neglected plants, and in case of newly planted bush roses in HT, Grandifloras and Floribundas, and to produce show blooms for exhibition though it is not preferable for established plants. Here the shoots are cut back to three or four buds from the base or bud union to encourage short and sturdy shoots. **Light pruning** is usually not recommended as it produces spindly bushes, and its repetition year after year will result in an early bloom with poor quality flowers. However, only in very special cases such as very vigorous hybrid teas, climbers and shrub roses, this type of pruning is considered. **High pruning** is cutting back of shoots to about 2/3rd of their length. This means that after removal of unwanted wood the remaining stems are merely tipped. **Moderate** or **intermediate pruning** is usually done in established plants, most suitable in case of HT, Floribundas and Grandifloras where shoots are normally removed to its half though weaker shoots more depending on their location on the bush. **Hybrid Teas** require hard pruning to build up a strong root system and to stimulate fresh and sturdy shoots close to the base of the bush in newly planted roses. In established HTs only 4-5 shoots are retained through moderate pruning, each shoot with 3-5 buds, the uppermost bud in each shoot facing outside. However, hard pruning in established roses is required for exhibition blooms. In case of 4-5 years old bushes, annually the oldest branch along with the thinning of the congested ones should be carried out. **Floribunda** and **Polyantha** roses should be cut back similar to HTs when newly planted but pruned hard the second year leaving only 1-2 buds (only top half of all strong stems) and moderate with 5-7 cm of the flowered stems and the tips as these bloom in clusters on shorter stems, *vis-à-vis* this ensures a long period of continuous blooming at varying heights. **Standard** roses are, in fact, HTs, Floribundas or Polyanthas budded 60-120 cm above the ground onto *Rosa canina, R. bourboniana* or *R. rugosa* stems and so the pruning should conform to the scion type budded. In newly **standard** roses, hard pruning but in established standrads only moderate to induce plenty of blooms on a properly formed head, and on the top around 20-25 cm length of the stem should be retained. Established standards do not like hard pruning as this encourages vigorous canes that affect the shape of the plants adversely. **Miniatures** and **climbers** usually do not require pruning as such for blooming but to eliminate any unwanted shoot which is sick or broken and that too through scissors and not the pruning shears. The tips in the climbers with the spent blooms should also be removed. **Ramblers** as flower best on previous season's strong stems so pruning in this case is done to induce as many strong and 1-year old stems as possible. After initial planting these are cut back to 30 cm, next year every stem at 30 cm or up to the base and this practice is adopted year after year but only after flowering. After pruning, the cuts should be sealed with Bourdeaux mixture or Blitox paste, and just before sprouting of the buds an effective pesticide spray should be applied for healthy growth. Fertilizer can be applied only after three weeks of pruning. **Bending** which originated in Japan, in fact, is a modification of pruning to induce more flowering shoots with longer stems. This way one annual cut is prevented and instead the shoots are frequently bent whenever there is a need and to maintain the height at working level. In this case all weak shoots are bent down to fill any area void of foliage and thus attain a desirable leaf area index to optimize photosynthetic potential and to facilitate the sugar transport to the developing shoots. Following knuckle cuts, as the weak shoots are bent the remaining shoots thus arising on dormant buds are normally more vigorous due to less competition for food and better exposure to light so these produce superior grade flowers.

Roses express deficiency symptoms quickly though return to normalcy slowly. Generally, the basic **nutrient** concentration of N in ammonium form (5:1 nitrate:ammonium in summer and 10:1 in the winter when soil or water has high pH) at 150 to 200 ppm is recommended. Jørgensen (1992) stated that low EC (1.5 ds/m) is critical and he advocated N:P:K levels at 220:30:195 ppm fertigation. A fertilizer dose of 520 kg N, 868 kg P_2O_5 and 694 kg K_2O/ha per year has been recommended for Delhi conditions for high density planting (30 × 30 cm) for 'Super Star' roses under open field conditions (Bhattacharjee and Damke, 1994). Potassium application rates affect the number and quality of flowering stems and reduce the disease incidence. Potassium closer to 50 g/m^2 per year is quite effective to obtain maximum quantity of commercial stems regardless of the cultivar. Heavy dose of organic manure like FYM at the rate of 50-100 t/ha is recommended in rose. Bhattacharjee (1994) observed that the soil application of sulphur at 10 kg/ha, magnesium sulphate at 50 kg/ha and calcium sulphate at 50kg/ha significantly improves flower yield and quality in cv. 'Raktagandha'. Ushakumari (1986) worked out effect of N, P and K nutrition on bud take, vegetative growth and flowering of roses under Kerala conditions. The fertilizer application improves the girth of stock at the time of budding. A fertilizer combination of 0.75 g each of N and P_2O_5 and 0.25 g K_2O, per plant has been found effective in promoting early bud take; a dose of 1.0 g N, 0.75 g P_2O_5 and 0.25 g K_2O for early emergence of leaves and for maximum height of the plants; however, the best combination was 0.75 g each of N, P_2O_5 and K_2O. She further reported that nitrogen and potassium alone and in combination were effective in early induction of first floral bud but phosphorus was found effective only when combined with nitrogen and potassium. Nirmala George (1989) recorded maximum number of flowers with 10.0 g N, 30.0 g P_2O_5 and 10.0 g K_2O per plant when applied at 30 day's interval; the longest floral shoot with 20.0 g N, 30.0 g P_2O_5 and 5.0 g K_2O applied at 45 day's interval; maximum floral diameter with 20.0 g N, 15.0 g P_2O_5 and 10.0 g K_2O applied at 15 day's interval; and maximum floral life with 10.0 g N, 45.0 g P_2O_5 and 15.0 g K_2O applied at 45 day's interval.

Duczmal (1989) obtained satisfactory weed control in *Rosa odorata* by applying 3.0-4.0 kg Trifluralin (Treflan) before transplanting or 6.0-7.0 kg Propachlor as post-emergence or after transplanting. Rajamani *et al.* (1992) recorded better control of monocot **weeds** with glyphosate 1.0 kg a.i/ha and dicot weeds with oxyfluorfen 0.5 kg a.i/ha. The weeds can be kept under check through mulching with black polythene strips and through a thick layer of straw or colossal garden waste which after rotting will also provide nutrients to the plants.

Protected Cultivation

Greenhouse roses are usually day-neutral with recurrent flowering occurring throughout the year to sustain regularity in supply to the market or directly to the consumers so that reliability of the market or consumer is fully ensured. Zieslin and Moe (1985) stated that garden roses include both recurrent and nonrecurrent flowering cultivars. Market reliability is lost when only the roses are supplied to the market periodically. Therefore, it is necessary to grow them in controlled atmosphere where optimum conditions for their growing are met, and it is possible only when the grower is equipped with greenhouses. **Greenhouse** is a glass or transparent plastic structure, often on a wooden or metal frame, for growing high value plants by providing required heat and light (intensity and duration), *vis-à-vis* protection from adverse weather conditions. The structures for roses are, in fact, warm greenhouses with 13-18 °C temperatures as these continue blooming throughout the year at 15/27 °C night/day temperatures. In temperate areas, glasshouses are required only during winter months when temperature goes very low and plants tend to become dormant so to create an environment for growing plants in winter such greenhouses are erected though summers are quite congenial for growing plants outdoors without any protection. During sunny days of May-June when temperature is anticipated to go beyond 30 °C, roses may require to be roofed with some semi-transparent material to provide only filtered lighting from 10.30 a.m. to 2.30 p.m. so that the plants are protected from scorching sunlight and uncongenial temperatures, however the sides are left free for proper air circulation. During winters, *i.e.* from third week of December to second week of March, the greenhouses prepared from polyethylene will be damaged due to weight accumulation of snows on their roofs in snowfall areas, especially higher reaches of Uttarakhand, Kashmir and Himachal Pradesh. Hence in such areas only glass-made greenhouses will work though cost will be prohibitive. In the sub-tropical areas, the polyhouses are required from mid-March to mid-October (7 months) to reduce the temperature levels as roses are successfully grown at 15-27 °C temperature range though here electricity cost becomes prohibitive, however, from mid-November to mid-February the polyhouses without electricity device for temperature control will be effective as the erection of such structures will increase the temperature levels from 8-12 °C which will facilitate rose growing as usual temperature in December-January in sub-tropical region of Delhi remains in the range of 5/15 °C (night/day). In quite tropical areas, the polyhouse will work throughout the year to maintain the congenial temperatures for its growing, except in winter. It may be either (i) **conventional greenhouses**, *viz.* 'traditional span', 'Dutch light', 'three-quarter span', 'lean-to', and 'Mansard' or 'curvilinear', or (ii) **special greenhouses**, *viz.* 'dome-shaped', 'polygonal', 'alpine house', 'conservation greenhouse', 'mini-greenhouse' or 'polytunnel'. All of these may have aluminium or timber framework, and glass walls or part-solid walls, except Dutch light greenhouses. Glasshouses, now-a-days are quite expensive but polytunnels are much cheaper and have added benefits. Greenhouse culture leads to 10-15 times higher yield, blooming consistently throughout the year than that of outdoor cultivation where neither quality is good nor blooming is perpetual, though greenhouse cultivation depends upon the greenhouse design, availability of environment control facilities, cropping systems, greenhouse management and crop type. A modern greenhouse has three major components – structure, covering and environment control system (temperature, light and humidity). Of various designs, gutter connected house covered with heavy-duty transparent plastic sheets with roof ventilation is most efficient for wide range of conditions. These types of greenhouses are cheaper and most feasible to automate the single consolidated space inside a gutter connected greenhouse than the multiple equivalent space in Quonset greenhouses. Management of materials and products into and out of the greenhouse requires less labour in a single large space than in numerous small spaces. The heating cost is less in multispan (gutter) greenhouses, because there is a less expanded area. The height of the gutter above the ground is increasing over the years to accommodate the growing evolution of climatic control equipment and automation devices. The original gutter connected greenhouse typically has a 2.4 m gutter height, but today 4.3 m is becoming very common. Gutter may be constructed from galvanized sheet of aluminium. The distance between gutter rows depends on the greenhouse brand purchased. The distance ranges from 3.2 to 12.2 m. Greenhouse with spacing between gutters of 3.7, 5.2, 6.7 and 9.1 m can be covered by film plastic sheets 4.3, 7.3, 7.6 and 11 m wide. Greenhouses are now offered that have roll up side curtains and can be installed on two or all the four walls. Greenhouses with retractable roof are becoming very popular in USA and Canada. The purpose of side curtain plus roof-ventilator system or the retractable roof with or without roll-up curtains is to replace high energy consuming 'fan and pad cooling systems'. These passive cooling systems work well in hot or cold climates. The greenhouse frames should preferably be covered with double layering covering 0.10-0.18 mm thick UV stabilized plastic. Today, polythene film as well as rigid FRP, acrylic and polycarbonate panels are available with an antilog surfactant built into the film or panel. It is advisable to use an antilog product because in addition to water dropping, the condensation also reduces light intensity within the greenhouse.

Fresh flower crops are grown in either ground beds or raised benches in the greenhouses. Such beds are 1.1 or 1.2 m wide and 20 cm deep, but 30 cm is best for rose beds. Fresh flower beds are oriented along with the length of the greenhouse with 45 cm aisles between them. This arrangement of beds allows for 67-70 per cent of floor space for growing. Soil is the natural medium for growing greenhouse crops, though roses and gerbera do not respond well in soil. The **cocopeat** available in plenty in South India can be used to get manifold yield with better quality. This substrate provides better root spread with minimum resistance, sufficient porosity, maximum nutrient uptake, better drainage of excess water, *vis-à-vis* minimum pathogens and scarce algal growth, as the medium is quite independent of the soil. Cocopeat, a powdered coconut fibre, is decomposed for 3-4 months in a tank filled with water which takes out the excess salts from the dust, and weathers it to the extent it becomes effective for crop growing. When roses inside polyhouse are planted in pots, a grower has almost a different mixture than others but a mixture containing coarse turfy loam 180 parts: wellrotten farmyard manure 100 parts:dry bonfire ash 40 parts:old soot 5 parts:bone meal 1 part, all by volume, should be mixed thoroughly and stored inside the store room where it should have been turned over several times, and this at the time of planting should be filled in 25 cm pots leaving only 4 cm at the top and then the plants are planted firmly pressing the roots all around and watered. Temperate as well as sub-tropical climates of North India face heavy chilly weather, requiring heating to sustain crop growth for flower regulation on special events like Christmas, New Year, Easter, Mother's Day and Valentine Day. Greenhouses can be **heated** with the oil burners, hot water or steam and electric heaters. A central heating system can be more efficient than unit heaters. In this system, two or more large boilers are in single location from where heat is transferred in the form of hot water or steam pipe mains to growing areas. Their heat is exchanged from hot water in a pipe coil located in plant zone or through overhead pipe. **Cooling** in the greenhouse is very essential where outside temperature goes above 30°C. Cooling system generally consists of fan at one side and pad at the other where principle of evaporation cooling is facilitated by running water stream over pad and consequent withdrawal of air through it by fans on the opposite side. A reduction of maximum 10-15 °C difference in temperature could be achieved depending upon the system and outside climate. Greenhouse should be airtight during the running of fan-pad system and care must be taken to periodically clean the pads from salts and algae. Other active way of cooling alternative is fog cooling. Control can be achieved through analog machines or by computers through aspirated chambers. Top and side ventilation also adds in cooling along with maintaining relative humidity.

The flower production and its quality may deteriorate due to very high **light intensity** during summer. **Shading** reduces light intensity and cools the microclimate inside the greenhouse. Shade paints (lime or Redusol or Vari clear), agro-shade nets or retractable thermal screens are generally used or operated manually or through automatic devices. Several plant species flower only when they are exposed to specific light duration. Yield and quality of flower crops could be increased with artificial lighting during night hours. **Fertigation** varies from single broadcasting of fertilizers to use of soluble grade fertilizers over different operating systems. One of the most modern technologies is currently offered by Priva-Phillips Nutriflux or Van Vliet Midi Aqua Flexilene System. Both the systems have a nutrient recycling method translating plant demand of nutrients in relation to EC/pH of the media, temperature, RH, light intensity, crop growth, mineral deficiency, etc. **Water quality** is very important though often overlooked. Total salt content levels, alkalinity levels, the balance of Ca and Mg, and levels of individual ions such as boron and fluoride can all have serious bearing on crop success. The water source should be tested before a greenhouse is established. **Electrical conductivity** level should be 0.75-1.50 dS/m and a **pH** of 6-7. Automatic watering system through drips or over-head foggers is generally used. A manual or semi-automatic control system is less capital intensive but requires a lot of attention and care. Now computerized control systems are available which can integrate temperature, light intensity, RH, CO_2, plant moisture, nutrient requirement, and plant protection measures. There are certain unheated structures in the gardens which are used to advance or extend the growing season such as cloches used *in situ* over plants in the garden itself, and **cold frames** also used for storing 'resting plants' and for hardening off plants propagated in the greenhouse.

In polyhouses, rose is planted on raised beds having one metre width, 30 cm height and length as per the structure. Roses are planted in double lines at 30 × 15-20 cm, accommodating 7 to 13 plants per square metre. The soil may be sterilized with formaldehyde and thoroughly mixed with FYM, phosphatic and potassic fertilizers. Sterilized farmyard manure at the rate of 100 t/ha should be thoroughly incorporated in the soil if the roses are to be planted in the flat beds. Usually these are planted in strips one metre width with medium filled in double-layered polythene sheets. Medium used is usually cocopeat whose depth may be from 20 to 30 cm. Initially the plants are supported to stand properly. Roses grown at a minimum **temperature** of 12 °C with 2,000 lux supplementary **lighting** (total light period of 16 h/day) from

HPS lamps during November to February increases flower yield. There may be species that requires cold for dormancy release. There are certain cut flower roses that produce fewer leaves under low irradiation levels than under high, and there are facultative LD cultivars which flower even with fewer leaves under long days (Moe, 1970).

Rose plants respond to $\mathbf{CO_2}$ enrichment (1,000 to 2,000 ppm) of the greenhouse atmosphere (Mastalerz, 1987a, b) with a decrease in the rate of lower bud abortion, an increase in lateral sprouting and an increase in the fresh weight of flower stems (Mortensen and Moe, 1983). CO_2 (700 to 1,000 ppm) in miniatures produces increased number of flowering shoots and hastens flowering with increased shoot lengths (Clark *et al.*, 1993), whereas in bare-rooted potted plants, Mortensen and Moe (1983) recorded 19 per cent increased floral shoots in floribunda 'Garnet' roses at 950 ppm CO_2 level. Prevention of flower bud abortion can be a result of alteration in partitioning of assimilates between upper buds. An additional possibility is that the effect of the CO_2 on the flower bud retention is due to anti-ethylene action of CO_2. The increase in number of flowers and quality may be the result of CO_2 on photosynthetic process, partitioning of assimilates and the effect of CO_2 on ethylene. Increasing CO_2 levels during day time to 800-1,000 ppm are beneficial for rose production due to increased dry matter production.

Postharvest

Roses should be harvested at the tight bud stage when one or two petals begin to unfold. These have 6-13 days of vase life depending upon the varieties, the climatic conditions and the postharvest treatments. The stage of harvest depends on the variety, distance to market place, climate and consumer preference. Roses cut too early may develop bent neck. Flowers should be cut in the morning or evening looking into the market requirement though those cut at evening have built up more photosynthates. They should be cut leaving two 5-leaflet leaves on the stem. After cutting, they are immediately placed in a hydrating solution to maintain turgidity and then these are taken to the cooling chamber where grading, cooling and other postharvest treatments are given. While grading, the stems, leaves and flowers should be blemish- and infestation-free, the leaves should be clean, shining, dark green and healthy with no any discolouration whatsoever or due to micronutrient deficiencies, too open or tight buds, damaged and bullhead types and those not conforming to the varietal specifications are rejected. and as per stem lengths these are graded. Long-stemmed varieties are graded from 40 cm onwards with a difference of 10 cm though smaller ones from 40-65 cm with a difference of 5 cm each. Roses are wet-stored for 5-7 days at 2-5 °C with a relative humidity of 90-95 per cent. However, these can be dry stored in sealed polythene sleeves for about 7 days (Kushal Singh *et al.*, 2013). Macnish *et al.* (2008) when added aqueous chlorine dioxide (ClO_2) at 2 or 10 $\mu l\ l^{-1}$ in clean deionized water, the build up of bacteria in vase solution was prevented and the vase life was found extended in *Rosa* × *hybrida* cv. 'Charlotte'. Red flowers have blackening of the petals due to a very low temperature storage therefore these should be stored at a little higher temperature. Precooling removes the field heat, slows down the respiration rates, lowers water loss, arrests the opening of the bud and thus improves the postharvest life. The hydrating solution may be acidified with aluminium sulphate or citric acid 300 ppm or bleach solution (50 ppm chlorine) to improve the uptake of solution. Pulsing with 3 per cent sucrose + 300 ppm aluminium sulphate is also quite effective (Kushal Singh *et al.*, 2013). Other effective pulsing solutions are 8-HQC 200 mg + cobalt chloride 250 mg + sucrose 4 per cent per litre of water or 8-HQC 250 mg + citric acid 200 mg + $AgNO_3$ 25 mg + aluminium sulphate 50 mg all in one litre of water + sucrose 1 per cent, or $AgNO_3$ 30 mg l^{-1} + sucrose 4 per cent or DICA 5 mg l^{-1} + sucrose 2 per cent or DDMH 50 mg l^{-1} + sucrose 2 per cent are also quite effective (Kushal Singh *et al.*, 2013). Also the stems may be held in a solution containing aluminium sulphate 300 ppm or through supplementation of 4 per cent sucrose or up to 50 ppm chlorine (prepared from bleaching powder, *i.e.* calcium chloride and not sodium hypochlorite) alone or through addition of sucrose 1.5 per cent in the solution (aluminium sulphate or chlorine) for better results (Kushal Singh *et al.*, 2013). After cooling, the flowers are shifted to grading room. All the inferior stems and those infested with pests and diseases are relegated. The flowers are sorted out to different grades manually or through automatic graders. Long stemmed varieties are graded from 40 cm onwards with a difference of 10 cm. The short stemmed varieties are graded from 40-65 cm with a difference of 5 cm. The graded stems are made into bundles of 20 each. The buds are wrapped with corrugated paper. The leaves are removed from lower 5 cm portion of the stems. The bunches are packed in fibreboard boxes. The stems should be tightly packed to avoid movement during transport. For increase of the vase life, aluminium sulphate at 300 ppm is quite effective.

Dormancy and Control of Growth and Flowering

Dormancy is not found in commercial cultivars. Generally, the glasshouse roses are day-neutral with recurrent flowering on a year-round basis and where flower initiation is independent to environmental factors while garden roses include even nonrecurrent cultivars (Zieslin and Moe, 1985), however, irradiance, temperature and CO_2 in greenhouses have been found influencing growth and development as low day temperatures of ≤17 °C and low night temperatures of ≤14 °C produced greatest stem diameter, *vis-a-vis* fresh and dry weights of stems, leaves and floral buds though flowering was found delayed; whereas increased irradiance increased floral stalk quality and growth rate (Hopper and Hammer, 1991; Passion and Lieth, 1994; Lieth, 1996). Some cultivars behave as facultative LD plants, as these flower with fewer leaves under LD, as also certain cut flower cultivars produce fewer leaves under low irradiance (Moe, 1970). In cut flower roses, the transition from vegetative to reproductive phase, *i.e.* floral initiation starts when the axillary shoots are 3 to 4 cm long which is usually 4 to 21 days (Laurie and Bobula, 1938; Zieslin and Moe, 1985). During winter when light is very poor, supplemental light is provided whether these are cut roses, potted miniature roses or roses for Valentine's Day. HPS (high pressure sodium) lamps at 300 to 1000 fc for 8 to 24 h per day (duration depending on the season and intensity) may be given (Tsujita, 1987) though it is not necessary in naturally high light areas. In case of miniatures, if natural light in winter is not enough, supplemental light (670 fc for 24 h) is required for rapid rooting (Jørgensen, 1992) but for production these require 350 to 500 fc for 12 h or more per day during winter. In bare-rooted potted plants, Asaoka and Heins (1982) stated that supplemental lighting (250 to 300 fc for 12 h) will result in more flowering stems per plant and decreased bud abortion during forcing, however, these plants require shading when intensity is higher than 3000 fc.

A rose plant when subjected to 2 to 4 °C for 4 to 6 weeks will stimulate axillary bud development and adventitious shoots from stem bases (Post, 1949; Hanan, 1979; Zieslin and Moe, 1985). Zieslin and Byrne (1981) observed that axillary buds from the upper nodes flower earlier but with few leaves. Though optimum temperature may vary with the cultivars but normally it is 16/17 °C night, 18 to 22 °C during cloudy winter days and 24 °C during sunny days, therefore in cut roses during summer in conditions like tropical/subtropical India, cooling system in the polyhouses will reduce temperatures otherwise flower quality will be hampered (Mastalerz, 1987a). Miniatures root at 23 to 24 °C, and after rooting to first cut back these require 20 to 22 °C for 7 to 10 days, 20 to 22 °C for 2 to 3 weeks of growth after cutback, 19 to 22 °C for 2 to 3 weeks after the second cutback (only when done), 18 °C for the final 3 to 4 weeks prior to flowering (Jørgensen, 1992) and if there is large increase in temperature over a short time the malformed flowers may be produced (Pemberton *et al.*, 1997). Garden plants as well as bare-rooted flowering pot-roses, after the appropriate cold treatment may be forced at 10 to 13 °C night temperature until commencement of axillary bud activity and then temperature may be increased to 18 °C (Post, 1949; Heins, 1981; Zieslin and Moe, 1985).

Insect-Pests and Diseases

There are a wide variety of insects that affect roses. Insect-pests degrade the ornamental value of roses in the garden and decrease productivity in commercial rose ventures. In case of severe infestation, they can even cause death of plants. Vijayan Nair (1989) through a survey in Thiruvananthapuram recorded thrips (*Rhipiphorothrips syriacus, Scirtothrips dorsalis*), scales (*Aonidiella aurantii*), the leaf feeding beetles (*Adoretus* spp.), and the mites (*Tetranychus cinnabarinus, T. neocaledonicus*), out of which *Tetranychus neocaledonicus* was recorded as the most important pest in all the locations, followed by *Adoretus* spp., *Aonidiella aurantii, Rhipiphorothrips syriacus, Scirtothrips dorsalis* and *Tetranychus cinnabarinus*.

Aphids (*Macrosiphum rosae*) are small, slow-moving, soft-bodied insects with piercing-sucking mouth parts and damage the plants by sucking their sap from tender parts. These insects grow early in the season on tender parts like shoots, buds and flowers. The infested flowers become malformed. The aphids secrete sweet honeydew on which fungi grow to produce black sooty mould which attracts ants and degrades the ornamental value of rose. The aphid population is kept under control by natural enemies like ladybird beetles. Excessive use of some of the insecticides increases the outbreak of aphids. In home gardens, in case of smaller populations they can be managed by spraying jet water. Severe infestations can be controlled by spraying insecticides like Monocrotophos, Acephate, Imidacloprid, Malathion and Dimethoate at 0.1 per cent. **Thrips** (*Rhipiphorothrips cruentatus, Scirtothrips dorsalis*) are one of the most important insect-pests of roses whose attack is noticed on new growths appearing after pruning, the first one which appears blackish-brown in the open field and the latter one which appears creamy in polyhouse roses. There are many other species also, but flower thrips and western flower thrips are two of the most common ones. Immature thrips are usually light yellow to lemon-coloured and are spindle-shaped. In roses, thrips cause damage mainly by feeding on flowers. Their injury reduces the beauty of the buds and flowers, and heavy infestation can prevent buds from opening. Thrips feed by punchuring plant cells with their needle-like mandibles and sucking up plant juices. This results in silvery or bleached- damaged areas on flower petals that eventually turn brown. Because feeding is concentrated on young and actively growing tissues, the petals and leaves are often crinkled or distorted, *vis-à-vis* brown markings on the curled leaves and burnt margins on deformed buds. Low levels of thrips injury tend to be more obvious on white or light-coloured blooms than on darker blooms, but severe injury will damage blooms regardless of colour. Thrips can effectively be controlled by spraying insecticides like Dimethoate, Phophomidon, Metasystox or Chlorpyriphos, each at 0.05 per cent or Fenitrothion at 0.075 per cent. Spinosad is one of the most important insecticides that provides excellent control of thrips. **Leafhoppers** (*Edwardsiana rosae*) are small, light green or pale-green insects which suck cell sap. Both adults and nymphs suck sap from the undersides of leaves and tender stems, causing leaves to become spotted or turning yellow. A white or yellow stippling of the leaves is one of the most common symptoms in roses. Leaf hoppers can effectively be controlled by spraying Malathion, Carbaryl, Imidacloprid or Permethrin at 0.2 per cent. **Spider Mites (*Tetranychus urticae*)** are minute insects not visible to the naked eye. They feed on the underside of the leaves by sucking the fluid from plant cells. Heavy infestations can cause severe injury or even kill the plants. Feeding by individual mites causes localized cell death, resulting in light-coloured stippling. When mite populations are heavy, leaves give a bleached or bronzed appearance. Severely injured leaves may curl and drop from the plant. In the initial stage, mite infestation is only on the undersides of leaves. Under heavy infestation the mites produce webbing and occur on the tops of leaves and other plant parts. Populations of plant-feeding mites are often kept in check by naturally occurring predatory mites and other predators. Outbreaks of spider mites often occur following insecticide treatments targeted against other pests because these treatments destroy the predatory mites. Foliar applications of Carbaryl, Acephate, or Pyrethroid insecticides have a tendency to trigger mite outbreaks. Mites are favoured by hot-dry weather, especially if accompanied by dusty conditions. Keeping plants well watered during periods of drought helps reducing the potential for mite outbreaks.

Washing foliage, especially the undersides with a water spray can also help control or prevent mites. Mites can be controlled by spraying miticides like Dicofol, Milbemectin, Malathion, Bifenzate and Dimethoate from 0.5 to 1 ml per litre. Vertimec (Abamectin 0.5 ml/l), Cascade (Flufenoxuron 1 ml/l) and Apollo (Clofentezine 0.6 ml/l) control the mites completely in one fortnight of application. **Scale insects** (*Aulacaspis rosae*) can be easily detected by the reddish encrustations on the shoots. Their bodies are covered with a hard scale-like coating that may be round, elliptical, tear-shaped or oyster-shaped depending on species. The scale covers often blend in with the bark of the plant, making its appearance very difficult to detect. Scales suck the sap from the plant with their thread-like sucking mouth parts. Heavy infestations also cause tissue damage, death of shoots or whole plant as they probe and feed on plant cells. Its infestation can reduce plant vigour and growth. The infested parts can be pruned to prevent further spread. They can be effectively controlled by spraying insecticides like Malathion, Rogor or Dimethoate at 0.2 per cent. **Mealy bugs** (*Pseudococcus* spp.) are small insects with white cottony filamentous growth on the exterior. They damage the plants by sucking sap from the shoots. They can be effectively controlled by spraying Monocrotophos and Dimethoate at 0.2 per cent spraying. **Whiteflies** (*Trialeurodes vaporariorum, Bemisia tabaci*) are small insects that are covered with a white waxy powder. They most often occur on the undersides of leaves but clouds of adults will fly around infested plants when disturbed. Immature whiteflies are immobile, scale-like insects that feed on the undersides of leaves. Like aphids, whiteflies suck plant sap through piercing-sucking mouth parts. They are also similar to aphids in their tendency to build the high populations and their ability to produce large amount of honeydew, which eventually results in sooty mould. They can be effectively controlled by spraying Imidacloprid, Acephate or Malathion at 0.2 per cent or neem oil (20 ml/l). **Termites** (*Odontotermes obesus*) cause damage to the rose plants even before they are fully established. They destroy the underground parts. They are very difficult to control as they colonize under the soil. One of the most difficult thing with termites is that by the time they are evident much of the damage would have been done to a plant. They can be controlled by soil application of Chlorpyriphos at 0.5 per cent, or through Carbofuran granules at 5 g/m^2 followed by watering. The larvae (**caterpillars**) of various moths or butterflies have been found infesting roses through their chewing type of mouth parts and feeding on the leaves, resulting in leaf skeletonization or defoliation. They can be effectively controlled through 0.2 per cent Acephate, Carbaryl, Dichlorvos or Spinosad sprayings and also through use of *Bacillus thuringiensis*. **Chafer beetles** (*Oxycetonia versicolor, Adoretus* spp.) and **ash beetle** (*Myllocerus* spp.) feed on the growing points making irregular holes and punches on the leaves. The grubs feed on the roots. The insecticides like Carbaryl, Acephate or Chlorpyriphos at 0.2 per cent controls these beetles. **Digger wasps** (*Crabro* sp.) damage the plants after pruning. They dig into the pruned stem through cut ends. Their digging also facilitates the entry of weak pathogens that cause die back. The pruned ends should be effectively sealed with a proprietary fungicidal paste to prevent the damage. In case of small gardens a few drops of insecticides like Dimethoate may be dropped into the tunnel. **Leaf cutting bees** (*Megachile* spp.) build their nests in hollow stems, reeds or pipes. They stuff the nests with pieces of leaves cut from various plants and feed them to developing larvae. Roses are one of the preferred plants for cutting leaves. Adult bees cut semi-circular holes in rose leaves, usually affecting only a few leaves on any one plant. Their damage is minor. They need not be controlled as their control may be harmful to the bee population.

Vijayan Nair (1989) reported that 0.05 per cent spray of Monocrotophos, Dimethoate or Fenithion is quite effective for controlling all the rose pests he recorded in Thiruvananthapuram though in case of *Scirtothrips dorsalis* 0.1 per cent concentration of the sprays was found effective. For integrated insect control, oxymeton methyl 0.05 per cent 30 days after planting (DAP), *neem* kernel extract 4 per cent (40 DAP), methyl parathion 0.05 per cent (50 DAP), *neem* oil 2 per cent (60 DAP) and *neem* kernel extract 4 per cent (70 DAP), followed at 10 day's interval provides significant control of aphids, beetles, caterpillars, thrips, and many other pests of roses under field conditions (Tejaswini *et al.*, 2007).

Roses are also infected with many of the pathogens such as *Diplodia rosarum* (die back), *Diplocarpon rosae* (black spot), *Sphaerotheca pannosa* (powdery mildew), *Peronospora sparsa* (downy mildew) and *Phragmidium mucronatum* (rust). Most common disease is **die back** which starts from top to downwards, especially through wounds or pruned ends. In its infection, initially the affected stem-part becomes black but afterwards whole plant dies. The affected part should be cut immediately in slanting position with a sharp knife in one attempt and burnt, *vis-à-vis* cut ends should be sealed with Bordeaux paste or copper oxychloride. Slant cut will avoid water stagnation at the point of cut. Carbendazim 0.2 per cent spraying also controls this problem. **Black spot** disease appears as dark brown circular spots with fringed borders which after sometime expands and coalesces. Leaves also turn brown and start falling. The spread of the disease is checked by burning the plant debris and through spraying with 0.2 per cent Bayleton (Tridumefon), Captan, Chlorothalonil, Fenarimol or Benlate or 0.5 per cent Tilt (Propicenazole). *Rosa laevigata, R. multiflora, R. roxburghii, R. rugosa, R. virginiana* and *R. wichuriana* are highly resistant to black spot. **Powdery mildew** appears as powder-graying of entire plant, especially on the abaxial surface of young leaves and floral buds which ultimately causes unopening of the flower. The conidia spreads through wind and the disease is favoured by humid or excessive dry weather conditions and also when days are warm and nights are cool. Sanitation of the plant debris in the surrounding will prevent the spread of the disease. It can be controlled through spraying with 0.05 per cent Karathane followed by 0.2 per cent Sulfex or Topaz (Penconazole 0.05 per cent). **Downy mildew** is quite prevalent in polyhouse roses where its infection is noticed on young plants with purplish-red to dark irregular spots on leaves during extended period of cool and humid weather in early to mid-spring. Leaf falling is most common and mostly in its severe infection the entire plant defoliates completely. Downy mildew is controlled through 0.20-0.25 per cent spraying with Chlorothalonil, Fosetyl-aluminium, Ridomil, Cuman-L, Aliette or Aliette 0.25 per cent + Mancozeb 0.20 per cent. **Rust** is evident by reddish-orange pustules on broken part of stems, petioles and on leaflets only in the temperate regions during rainy seasons, and these pustules become blackish at later stage. Defoliation of the plant occurs only in its severe infection. This disease is controlled through use of sulphur and oxycarboxin. Bazzi *et al.* (1987) reported *Agrobacterium tumefaciens* infecting the rose in Italy.

References

Anitha, I. 1989. Effect of Season and Position of Bud in Budding of Rose (M. Sc. Thesis). College of Agriculture, Kerala Agricultural University, Vellayani, Kerala.

Asaoka, M. and R.D. Heins, 1982. Influence of supplemental light and preforcing storage treatments on the forcing of 'Red Garnet' rose as a pot plant. *J. Amer. Soc. hort. Sci.*, **107**: 548-552.

Bazzi, C., P. Minardi and U. Mazzucchi, 1987. Bacterial diseases of flower and ornamental plants in Italy (Italian). *Informatore Fitopatologico*, **37**(6): 15-24.

Bhattacharjee, S.K. 1994. Influence of secondary nutrients on *Rosa hybrida* cv. "Raktagandha". *Floriculture: Technology, Trades and Trends* (eds Prakash, J. and K.R. Bhandary), pp. 100-102. Oxford & IBH Publishing Co. Ltd., New Delhi.

Bhattacharjee, S.K. and M.M. Damke, 1994. Response of Super Star rose to nitrogen, phosphorus and potash fertilization. *Indian J. Hort.*, **51**(2): 207-213.

Bose, T.K. and D. Mukherjee, 1977. *Gardening in India.* Oxford & IBH Publishing Co. Ltd., Calcutta, Bombay and New Delhi.

Brickell, C. 1992. *The Royal Horticultural Society Encyclopedia of Gardening*, pp.116-136. Dorling Kindersley Publishers Limited, London.

Cairns, T. 2003. *Horticultural Classification Schemes.* In: Encyclopedia of Rose Science, vol. I.(eds Roberts, A.V., T. Debener, S. Gudin), pp. 117-124. Elsevier Ltd., Oxford, U.K.

Clark, D.G., J.W. Kelly and N.C. Rajapakse, 1993. Production and postharvest characteristics of *Rosa hybrida* L. 'Meijikatar' grown in pots under carbon dioxide enrichment. *J. Amer. Soc. hort. Sci.*, **118**: 613-617.

Crespel, L., M. Chirollet, C.E. Durel, D. Zhang, J. Meynet and S. Gudin, 2002. Mapping of qualitative and quantitative phenotypic traits in *Rosa* using AFLP markers. *Theor. Appl. Genet.*, **105**: 1207–1214.

De Vries, D. P. 2003. Breeding/Selection strategies for pot roses. **In**: *Encyclopedia of Rose Science* (eds Roberts, A.V., T. Debener and S. Gudin), pp. 41-48. Elsevier Ltd., Oxford, U.K.

De Vries, D. P. and L.A.M. Dubois, 1984. Inheritance of the recurrent flowering and moss characters in F_1 and F_2 Hybrid Tea x *Rosa centifolia muscosa* (Aiton) Seringe populations. *Gartenbauwissenschaft*, **49**: 97–100.

Debener, Th. And M. Linde, 2009. 'Exploring Complex Ornamental Genomes: The Rose as a Model Plant'. *Critical Reviews in Plant Sciences*, **28**(4): 267-280.

Debener, T. 1999. Genetic analysis of horticulturally important morphological and physiological characters in diploid roses. *Gartenbauwissenschaft*, **64**:14-20.

Dole, J.M. and H.F. Wilkins, 1999. *Rosa.* In: Floriculture Principles and Species, pp. 495-508. Prentice Hall, New Jersey, USA.

Duczmal, K.W. 1989. Weed control in ornamental plants grown for seed (Polish).*Biuletyn Instytutu Hodowli I Aklimatyzacji Roslin*, No. 169, pp. 59-76.

Edward, G. 1975. *Wild and Old Garden Roses.* David E. Charles, London.

Fairbrother, F. 1965. *Roses.* Penguin, Great Britain.

Fitch, C.M. 1977. *The Complete Book of Miniature Roses.* Hawthorn Books, New York.

Gudin, S. 2000. *Rose: Genetics and Breeding.* In : *Plant Breeding Reviews* (vol. 17), pp. 159-189. John Wiley & Sons, Inc., U.K.

Gudin, S. 2003. *Breeding Overview.* In: *Encyclopedia of Rose Science*, vol. I (eds Roberts, A.V., T. Debener and S. Gudin), pp. 25-33. Elsevier Ltd., Oxford, U.K.

Hanan, J.J. 1979. Observation of a low temperature effect on rose. *J. Amer. Soc. hort. Sci.*, **104**: 37-40.

Hareing, P.A. 1960. *Modern Roses* 9. American Rose Society, Shreveport, LA, USA.

Hartmann, H.T. and D.E. Kester, 1972. *Plant Propagation, Principles and Practices* (3rd ed.). Prentice-Hall of India Pvt. Ltd., New Delhi.

Hay, R. and K.A. Beckett, 1971. *Rosa.* In: *Reader's Digest Encyclopaedia of Garden Plants and Flowers*, pp. 597-631. The Reader's Digest Association Limited, London.

Heinrichs, F. 2008. International statistics of flowers and plants. *AIPH/Union Fleurs.* **56**: 16–90.

Heins, R.D. 1981. Forcing pot roses for Valentine's Day. *Florists' Rev.*, **169**(4376): 14-15.

Hessayon, D.G. 1981. *The Rose Expert.* PBI Publications, England.

Higson, H. 2007. *The History and Legacy of China Rose.* Quarryhill Botanical Gardens, Glen Ellen, California (www.quarryhillbg.org/page14.html).

Hopper, D.A. and P.A. Hammer, 1991. Regression models describing *Rosa hybrida* response to day/night temperature and photosynthetic photon flux. *J. Amer. Soc. hort. Sci.*, **116**: 609-617.

Hurst, C.C. 1941. Notes on the origin and evolution of our garden roses. *J. Royal hort. Soc.*, **66**: 73-82.

Jørgensen, E. 1992. Growing Parade-Roses. Pejoe Trykcenter A/S, Hilreød, Denmark.

Kaicker, U. S., A.P. Singh and R.S. Malik, 1983. Breeding roses for fragrance. Abst. No. 988 XV *International Genetic Congress*, p. 554.

Kushal Singh, Gunjeet Kumar, Tarak Nath Saha and Ramesh Kumar, 2013. *Postharvest Technology of Cut Flowers* (DFR Bull. No. 6). Directorate of Floricultural Research (ICAR), Pusa Campus, New Delhi-110 012.

Lal, S. D., J.N. Seth, J.P. Yadav and N.S. Danu, 1982. Genetic variability and correlation studies in rose. I. Phenotypic variability, heritability and genetic advance. *Progressive Hort.*, **14**: 234-236.

Langhans, R.W. 1987. Planting. In: *Roses~A Manual of Greenhouse Rose Production* (ed. Langhans, R.W.), pp. 57-59. Roses Incorporated, Haslett, Michigan, USA.

Laurie, A. and P.F. Bobula, 1938. A study of flowering rose shoots with reference to flower bud differentiation. *Proc. Amer. Soc. hort. Sci.*, **36**: 767-768.

Lieth, H. 1996. Modeling roses for optimum production. *GrowerTalks*, **60**(1): 42, 44, 47, 48, 50.

Linde, M., A. Hattendorf, H. Kaufmann and T. Debener, 2006. Powdery mildew resistance in roses: QTL mapping in different environments using selective genotyping. *Theor. Appl.Genet.*, **113**: 1081–1092.

Macnish, A.J., R.T. Leonard and T.A. Nell, 2008. Treatment with chlorine dioxide extends the vase life of selected cut flowers. *Postharv. Biol. & Tech.*, **50**(2/3): 197-207.

Maia, N., and P. Vénard, 1976. *Cytotaxonomie du genre Rosa et origine des rosiers cultivés*, pp. 7-20. In: *Travaux sur rosiers de serre*. FNPHP, Antibes.

Marriott, M. 2003. *Modern Roses* (Post 1800). In: Encyclopedia of Rose Science, vol. I (eds. Roberts, A.V., T. Debener and S. Gudin), pp. 402-409. Elsevier, Oxford, U.K.

Mastalerz, J. 1987a. Environmental factors light, temperature and carbon dioxide. In: *Roses: A Manual of Greenhouse Rose Production* (ed. Langhans, R.W.), pp. 147-169. Roses Incorporated, Haslett, Michigan, USA.

Mastalerz, J. 1987b. Low temperature dry storage. In: *Roses: A Manual of Greenhouse Rose Production* (ed. Langhans, R.W.), pp. 273-282. Roses Incorporated, Haslett, Michigan, USA.

Misra, R.L. and O.P. Chaturvedi, 1978. Roses in Aligarh and Kannauj for perfumery. *Delhi Gdn Mag.*, pp. 13-16.

Moe, R. 1970. Growth and flowering of potted roses as affected by temperature and growth retardants. Meldinger Norges Landbrukshøgskole, **49**: 1-16.

Mortensen, L.M. and R. Moe, 1983. Growth responses of some greenhouse plants to environment. VII. Effect of CO_2 on photosynthesis and growth of roses. *Meldinger Norges Landbrukshøgskole*, **62**(3): 1-11.

Nirmala George, 1989. Effect of Split Application of N, P & K on the Growth and Flowering of Rose cv. Happiness (M. Sc. Thesiss). College of Agriculture, Kerala Agricultural University, Vellayani, Kerala.

Nybom, H., G. Werlemark, D.G. Esselink and B. Vosman, 2005. Sexual preferences linked to rose taxonomy and cytology. *Acta Hort.*, No. 690, pp. 21-27.

Passion, C.C. and J.H. Lieth, 1994. Nondestructive dry matter estimation of rose shoot leaves, stem and flower buds using regression models. *HortSci.*, **29**: 162-164.

Pemberton, H.B., J.W. Kelly and J. Farare, 1997. Rose. In: *Tips on Growing Specialty Potted Crops* (eds Gasten, M.L., S.A. Carver, C.A. Irwin and R.A. Larson), pp. 112-117. Ohio Florists' Assocn, Columbus, Ohio, USA.

Pinney, M.E. 1964. *The Miniature Rose Book*. D. Van Nostrand Company, New York.

Post, K. 1949. *Rosa*. In: Florist Crop Production and Marketing, pp. 758-803. Orange Judd Publishing, New York, USA.

Rajamani, K., S. Thamburaj, T. Thangaraj and S. Murugesan, 1992. Studies on the effect of certain herbicides in rose cv. "Happiness". *South Indian Hort.*, **40**(2): 121-122.

Rathore, D. S. and U.C. Srivastava, 1992. Rosa Species (A bulletin), 113 pp. NBPGR, New Delhi.

Shrivastava, H. P. and V. Chandra, 1985. Roses for perfumery industry-*R. bourboniana*. *Indian Rose Annual*, **4**: 198-213.

Singh, A.P., K.V. Prasad and M.L.Choudhary, 2003. *Panorama of Rose Research*. AICRP on Floriculture (ICAR), IARI, New Delhi.

Steen, N. 1967. *The Charm of Old Roses*. Herbert Jenkins Ltd., Australia.

Steve Connor, 2015. Roses may soon smell much sweeter. *Sunday Times of India* (New Delhi), July 5, p. 23.

Swarup, V., U.S. Kaicker and H.S. Gill, 1971. Induced mutations in French bean and roses. *Proc. Int. Symp.on 'Use of Isotopes and Radiation in Agriculture and Animal Husbandry Research*, pp. 13-25; 30 November to 2nd December, New Delhi.

Tejaswini, P. Naveen Kumar, R.L. Misra, K. Sujata, Sangama, N. Ramachndran, Jhansi Rani and Suchitra Pushkar, 2007. *Rose* (AICRP on Floriculture Tech. Bull. No. 25). ICAR, New Delhi-110 012.

Tsujita, M.J. 1987. High intensity supplementary radiation of roses. In: *Roses: A Manual of Greenhouse Rose Production* (ed. Langhans, R.W.), pp. 171-186. Roses Incorporated, Haslett, Michigan, USA.

Uma, B. 1991. Induced Chemical Mutagenesis in Rose under *In Vitro* Culture (M.Sc. Thesis). College of Agriculture, Kerala agricultural University, Vellayani, Kerala.

Ushakumari, S. 1986. Effect of Nutrition on the Establishment and Bud Take in Budded Roses (M. Sc. Thesis). College of Agriculture, Kerala Agricultural University, Vellayani, Kerala.

Večeřa, L. 1971. *A Concise Guide to Colour Roses.* Hamlyn, Middlesex, England.

Vijayan Nair, V. 1989. Biology and Control of Pests of Rose (M.Sc. Thesis). College of Agriculture, Kerala Agricultural University, Vellayani, Kerala.

Viraraghavan, M. S. 1991. Rose breeding for the tropics. *Indian Rose Annual,* **9**: 45-60.

Wilson, D. 1993. Induced Mutagenesis in Rose under *In vivo* and *In vitro* Culture (Ph. D. Thesis). College of Agriculture, Kerala Agricultural University, Vellayani, Kerala.

Wylie, A. P. 1955. Master Memorial Lecture on History of Garden Roses II. *Jour. Roy. hort. Soc.*, **89**: 8-24.

Yadav, L.P., N.K. Dadlani and R.S. Malik, 1989. *Rose.* In: *Commercial Flowers* (eds Bose, T.K. and L.P. Yadav), pp. 15-150. Naya Prokash, Calcutta (India).

Zemtsova, G.N. 1977. Investigation on plants of the family Dipsacaceae and the order Rosales with a view to introducing them into cultivation (Russian). *Sostoyanie i Perspektivy Nauch Issled po Introduktsii Lekarstv Rast.*, USSR, pp. 176-177.

Zieslin, N. and T. Byrne, 1981. Plant management of greenhouse roses. Flower cutting procedures. *Scientia Hort.*, **15**:179-186.

Zieslin, N. and R. Moe, 1985. *Rosa.* In: Handbook of Flowering, vol. IV (ed. Halevy, A.H.), pp. 214-225. CRC Press, Boca Raton, Florida, USA.

32

Salpiglossis (Family: Solanaceae)

Sapna Panwar, Poonam Kumari, Thaneshwari and R.L. Misra

[(**Common names**: Painted tongue, Scalloped tube tongue, Velvet trumpet tongue)]

Introduction

Salpiglossis species are members of the mainly Neotropical tribe Cestreae of the Solanaceae family. The genus *Salpiglossis* seems to have puzzled botanist for its placing. It was first supposed to belong to Bignoniaceae because of its trumpet-shaped corolla, it was then placed in Solanaceae from its apparent alliance to *Petunia*, the resemblance between the genus being in fact so great that several species now considered as petunias were first supposed to be in *Salpiglossis*. The name *Salpiglossis* is of Greek origin, *salpinx* for 'trumpet' and *glössa* for 'tongue', owing to mainly the shape of the style, and also the corolla form. It is a genus of 5-18 species of annuals,biennials and perennials and prefers temperate regions to grow. The plants grow straight, often with sticky hairs, alternate leaves and colourful 5-lobed flowers and the *S. sinuata* hybrids are daintier and aristocrats of the plant kingdom and are among the choicest of annuals.

Salpiglossis owes its ornamental value to the wide range of corolla colours, patterns and degrees of colour saturation present in the species. *Salpiglossis* is not as popular as hybrid petunia which it resembles. *Salpiglossis* has received sparse scientific study in terms of genetics, breeding, biochemistry and physiology though the plant has great ornamental value. *Salpiglossis* plant attains the height of up to 90 cm in the garden. Clusters of *Salpiglossis* are the focal point in the borders and beds, and these make excellent cut flowers.

Botany and Breeding

The closest relative to the *Salpiglossis* is *Petunia*. It is a tender, tropical, and annual herbaceous plant. It possesses a tap root system. Stems bear short glandular hairs. The plants are sparse in the foliage, the leaves tend to be opposite especially toward the inflorescence-end of the stem. Devoid of stipules, the leaves are narrow, sometimes lobed and simple, each leaf being unclustered and independent on the stem, basally larger though gradually smaller towards the blossom-end. Cymose inflorescence at the apex branch out in three or more, each bearing a flower some 5 cm across with a long pedicel and flower stalk which can be cut along at any plane (actinomorphic) and is without bracts or bracteoles. There are five sepals at the base of the petal which are united or fused (gamosepalous). The corollas of these first-rate annuals consist of five velvety and gamopetalous petals having yellow, orange to golden-yellow, pink, scarlet to red and maroon, lavender and purple, fused together basally to form a funnel, and these petals are forked and bilobed at the end. The male part of the flower consists of five stamens where 2-4 stamens are fertile instead of usual 5. There are two carpels in the female part of the flower, both fused together and are syncarpous, and the ovary is superior. Base-rounded but tapering towards apex, glandular, pubescent, brown, oblong to ovoid and capsular fruits are some 9-11 mm long, 4.0-4.5 mm wide, the mature ones being enclosed in a tight fitting composed of papery calyx as long as the capsule and dehisce by two-bifid valves with opening at apex, each capsule having about 200 seeds, each seed being angular, cuboidal, light to dark brown, reticulate with the dimension of 0.8-1.2 × 0.4-0.5 mm. Chromosome number is 2n = 44 (Fedorov, 1969).

The significance of *Salpiglossis sinuata* protoplast is in its potential for cell fusion with other Solanaceous bedding plant species such as *Browallia*, *Nicotiana alata*, *Petunia hybrida* and other *Petunia* species. Various attempts at intergeneric sexual crossing between *Salpiglossis sinuata* and the three other genera have not succeeded. Thus, cell fusion is a potential means of gene transfer, with the initial interest focused on gene controlling flower pigments and flower morphology. Several wide hybrids between sexually incompatible species have been created by somatic hybridization, but only three have resulted into whole plants (Sink, 1984). The widest hybrid created were *Petunia hybrida* × *Salpiglossis sinuata* and *Petunia axillaris* subs. *parodii* × *Salpiglossis sinuata* (Lee *et al.*, 1994). Putative hybrids were identified by chromosome and isozyme analysis. Both hybrids had abnormal leaf development and very poor root growth. Neither of the hybrids produced flowers.

Propagation

Commercially *Salpiglossis* is propagated by **seeds**. Salpiglossis is a cool-season annual that flourishes best before dry and hot weather hits. The best way to sow and grow salpiglossis plant is to choose seeds that have dried on the mother plant. Many times seeds that are sold in the nurseries are induced to dry and such seeds have poor viability and may wither away once the sapling has produced a few leaves. The seeds are fine, so these are gently pressed into the soil left uncovered. These are sown in February or March in the temperate regions in the glasshouse at a temperature of 18 °C. If sowing *Salpiglossis* seeds directly outdoors, it should be only when soil temperature is warm as growing it at 16-18 °C (Hay and Beckett, 1971) will be better. Prepare a seedbed and add compost to the soil so that it becomes rich in nutrients. Spacing between plants must be 15 to 25 cm apart. Other than growing in the wild or on seed bed, seeds must be sown in a container, more or less toward the surface of the soil and covered with newspaper or gunny bag. Flats, shallow dishes or an 8 cm pot should be filled with moist seed compost. Seeds are sown and covered lightly with a very fine layer of compost. The seeds require darkness to germinate well so after sowing the containers should be shaded. It germinates within 15 to 20 days at 21 °C. Once the sapling emerges, the covering is removed and the container should be shifted to the light gradually.

Tissue culture study on *Salpiglossis* has been limited though success has been obtained on the initiation and maintenance of callus and regeneration of plant. Hughes *et al.* (1973) first reported tissue culture of *Salpiglossis sinuata* with the induction of callus from anther placed on Nitsch medium containing 1mg/l kinetin and 5mg/l IAA. Depending on stage of development at the time of culture, 27-46 per cent anthers formed callus. None of the calli however originated from pollen. Shoots occurred after five months culturing in the somatic-derived callus and these rooted on MS medium containing 0.04mg/l kinetin and 10mg/l IAA. Lee *et al.* (1977) placed young leaf discs of *S. sinuata* on MS medium and found optimum shoot production at 0.1mg/l NAA and 2mg/l kinetin in 15-20 days. The shoot rooted at 60 per cent efficiency in MS medium and NAA was found inhibitory to rooting. Boyes and Sink (1981) studied the leaf explants on MS medium + 2,4-D at various concentration which produced callus that varied in friability. More uniform dividing friable callus was obtained on Uchimiya and Murashige 1974 medium. Second subculture of callus growing on these media was tested on subsequent shoot regeneration. Shoots occurred within a 6 week period and growth was optimum on MS + 0.1 and 1mg/l 2ip (purine). As in other tissue culture studies on *Salpiglossis*, rooting was the most difficult stage. It was found that low level of 2,4-D (0.001mg/l) in MS medium gave 75 per cent rooting of regenerated shoots and subsequent transfer to soil and survival in the greenhouse was 85-90 per cent.

Species and Varieties

Four *Salpiglossis* species are popularly known, such as *Salpiglossis erecta* from Mexico and Texas, *S. sinuata* and *S. spinescens* from Chile and *S. anomalia* from Argentina. *S. sinuata* is the only species in cultivation which is described herewith. ***Salpiglossis sinuata*** (syn. *S. atropurpurea*, *S. barclayana*, *S. coccinea*, *S. flava*, *S. gloxiniifolia*, *S. grandiflora*, *S. picta*, *S. staminea*, *S. superbissima* and *S. variabilis*; Beckett, 1983) is almost a hardy and erect annual native to southern Chile, and is grown for its colourful petunia-like flowers. Chile is native for 7-8 *Salpiglossis* species. *S. sinuata* grows up to 90 cm tall with branching in the lower part. The leaves are alternate, elliptic, pinnatifid with little undulated to deeply dentate margins (sinuately toothed), downy sticky and basal leaves larger up to 10 cm long but decreasing up the stems. Funnel-shaped, wide-mouthed and 5-lobed flowers some 6.5 cm long and 5.0 cm wide in shades of cream, yellow, scarlet, pink, red, purple and blue, and often striped or with a distinctive contrasting vein colouration appear in clusters from the leaf axils during summer to autumn in terminal racemes. Various horticultural strains have larger flowers in a wider range of colours and shades such as cream, mahogany, yellow, chestnut, scarlet, pink, red, crimson, carmine, blue, violet, purple or black. Petals are normally covered with a longitudinal network of yellow, red, maroon and blue veins. Colour intensity increases with flower maturity. It has been introduced to Europe between 1820 and 1830. It comprises a minor portion of the various plant species grown by the bedding plant industry.

Certain popular horticultural varieties are *S.s.* var. *grandiflora* growing up to 75 cm tall, *S.s.* var. *grandiflora nana* growing up to 40 cm tall, *S.s.* var. *superbissima* (syn. var. *imperiale*) growing up to 75 cm tall with solitary quite large flowers, *etc.* (Pizzetti and Cocker, 1975). Private Compnies have developed F_1 series such as 'F_1 Royale' (dwarf with basal branching with dimension of 60 cm height and 30 cm width and most suitable for pot culture), 'Royale Series', 'Royale Chocolate', 'Royale Mixed F_1' (height up to 30 cm, selfs and bicolours from sunshine yellow through blues, purples and reds and many with yellow throats and veins), 'Royale Purple', 'Royale Purple Bicolour' and 'Royale Red Bicolour'. Certain seed selections available are 'Black Trumpets', 'Bolero Hybrid Mix' (F_2 hybrid with 60 cm height, a mixture with various colours but throat veined gold), 'Casino Glass', 'Casino Mixed' (flowers in shades of yellow, orange, rose, red and purple, strongly veined, suitable for bedding), 'Festival Mixed' (plants compact, bushy and up to 30 cm high, flowers large in shades of gold, lavender, scarlet, crimson, pink and rose), 'Friendship' (upward facing flowers in a range of colours), 'Kew Blue' (flowers rich blue, veins gold), 'Little Friends', 'Peacock Flame' (exotic appearance, flowers bright red, throat splashed with creamy-yellow), 'Splash' (plants bushy with profusion of brightly and multi-coloured blooms), 'Stained Glass', *etc.*

Cultural Practices

Salpiglossis requires slightly acidic **soil** with a pH range of 6.1-7.5, ideal being 6.5, though the plant grows even at 5.5. In addition to pH, the plants grow well when the soil is sandy- to clay-loam. Even otherwise, the soil can be improved through addition of farmyard manure up to 300 quintals per hectare and in heavy soils even the coarse sand. Dry and hot seasons are not suitable for its growing, however, where cool weather lasts longer but no frost is experienced, the humidity is medium, the soil is well-drained and the sunshine is in plenty it thrives well. The flowering season ranges from early summer to mid-fall with temperatures ranging from 10 to 24 °C. These make attractive pot plants for a cool greenhouse and can be had in flowering during winter and spring from the late summer to early autumn sowing if minimum winter temperature of about 16 °C is maintined (Beckett, 1983). The soil should be ploughed 2-3 times, followed by plankings, all the hard materials and rootstocks of the perennial weeds are removed and the beds of convenient sizes are made by laying out the main and subsidiary irrigation channels cum bunds. Barely covered seeds though dusted thinly with fine soil, should be sown in March-April at warmer places, especially in the greenhouse if provision is, and the seedlings should be pricked out in boxes, flats or trays and when large enough, *i.e.* 10-15 cm tall, these are **transplanted** in the beds or filled-in 12-cm pots individually and immediately watered. In its case for direct field planting, direct sowing is more advantageous which may afterwards be thinned out some 25 cm apart. When growing, its main stems require support of twiggy **stakes** to protect its plants from strong winds. ***Mulching*** is one of the most beneficial practices used to keep the roots cooler for longer periods. Its crops require continuous supply of **moisture** from seeding to flowering but over-watering should be avoided as this causes root rot. These are heavy feeder so it would be better to supply them with **liquid feed** or fertigation with balanced fertilizer once every fortnight. **Pinching back** the tips should be done once the seedlings develop two or more true leaves. This will aid in promoting thicker and healthier plants with more basal branching. Spent flower spikes are also removed so that more flowers are encouraged to form with side shoots bearing larger flowers. Pods are allowed to dry on plant; break open to collect seeds properly cleaned, seed can be successfully stored. For good seed production, its pods are allowed to mature and dry on the plant and then these are collected and stored.

Growth and Flowering

Salpiglossis produces flower colours in various shades such as yellow, orange, pink, red, blue of various depths with or without striping. At first flower anthesis, if the plants are not treated with chemical growth retardants, the hybrids grow up to 30-45 cm high which is commercially unacceptable. Thus, if height could be restricted to a size appropriate for a potted plant, the marketability of salpiglossis would be enhanced and the consumer's desires for new and unusual plants could be satisfied. When Needham and Hammer (1990) covered the salpiglossis inbred line 'P-5' seedlings with black cloth for 16 h period and then exposing to 18 h photoperiod, they recorded flowering occurring 40 days after transferring the plants to long days. Application of daminozide from 1,000 to 5,000 ppm, alone and in combination with chlormequat chloride retarded its plant height considerably, however, concomitant restriction of corolla diameter was frequently observed.

Postharvest

The stem lengths for cutting should be 40-60 cm and they are harvested when the flowers are starting to open (Vaughan, 1988), especially in the summer as it is a temperate ornamental though a few winter-flowering hybrids have also been developed to be cultivated in the greenhouses. Cold storing of its cut flowers is generally not recommended. It can be shipped at 2-5 °C temperature. Its life is approximately one week.

Insect-Pests and Diseases

Soft-bodied, yellow to pale- and dark green, and globular green peach **aphids** (*Myzus persicae*) and melon aphids (*Aphis gossypii*) with long thin legs and antennae, *vis-à-vis* a pair of projected cornicles emerging from the rear end of the body, feed by clustering on the tender parts of the plants and continue sucking the cell sap through their piercing and sucking type of mouth parts. Both adults and nymphs cause extensive damage. Their feeding on the plants make the plants dirty and viscid, stunt them, transmit viral diseases and attract sooty mould. Neem seed oils and soaps, *via-a-vis* their formulations with pyrethrins are quite effective to control these. Imidacloprid being systemic insecticide, is very effective when infestation is serious but it may have negative impacts on predators, parasitoids, and pollinators, so its use should be avoided where soaps and oils provide adequate control.

Rhizoctonia* root and foot rot** (*Rhizoctonia solani*) and **damping off and root rot** (*Phytophthora* sp., *Pythium* sp.) are the common diseases that typically cause most damage to seedlings, but can also damage older plants. Their infection at later stage of growth may cause superficial lesions or sometimes wilting of the foliage though initial stage infection can stunt and ultimately kill the plants, making a complete loss. Foot rot and root rot infection occurs mostly when the plants are about to flower. Rusty-brown, dry and sunken lesions on stems and roots near the soil line are a characteristic symptom of *Rhizoctonia* infection. Lateral roots may be decayed. Seedlings or older plants may develop these infections and become stunted, yellow and may wilt. ***Pythium* and *Phytophthora also cause rotting of the seeds prior to germination, apart from pre- and post -emergence damping off. It produces tan-brown and soft rotted tissues. At the primary leaf stage, infected stems appear bruised and soft, secondary roots are rotted, the leaves turn yellow, and plants frequently wilt and die. Seed and seedling diseases of *Salpiglosis* are difficult to manage, *vis-a-vis* different pathogens require different management approaches, therefore it is a must to diagnose the pathogens correctly before treatment. In general, these diseases can be reduced by planting good-quality seeds in well-drained and non-compacted fields. Delaying planting until soils are moderately moist towards drier side and >13 °C temperature, so that germination is rapid with accelerated growth. Seed treatments at the time of sowing with two or more active ingredients, especially mefenoxam (ApronXL) or metalayl (Allegiance) can be effective against *Pythium* and *Phytophthora*, and products containing fludioxonil (Maxim) or a strobilurin product (azoxystrobin, trifloxystrobin, or pyraclostrobin) may help to reduce damage from true fungi such as *Fusarium* and *Rhizoctonia*.

References

Beckett, K.A. 1983. *A Concise Encyclopedia of Garden Plants*, p. 365. Orbis Publishing Limited, London.

Boyes, C.J. and K.C. Sink, 1981. Morphogenetic responses of *Salpiglossis sinuata* L. leaf explants and callus. *Scientia Hort.*, **15**: 53-60.

Fedorov, A.A. (ed.), 1969. Chromosome number of flowering plants, 926 pp. Acadamy Sciences, U.S.S.R., Moscow.

Hay, R. and K.A. Beckett, 1971. *Reader's Digest Encyclopaedia of Garden Plants and Flowers*, p. 639. The Reader's Digest Association Ltd., London.

Hughes, H., S. Lam and J. Janick, 1973. *In vitro* culture of *Salpiglosis sinuata* L. *HortSci.,* **8**: 335-336.

Lee, C. H., K.Y. Paek and J.K. Hwang, 1994. Production and characterization of putative intertribal somatic hybrids between *Salpiglossis* and *Petunia*. *J. Kor. Soc. hort. Sci.*, **35**: 360-369.

Lee, C.W., R.M. Skirvin, A.I. Soltero and J. Janick, 1977. Tissue culture of *Salpiglossis sinuata* L. from leaf disc. *HortSci.*, **12**: 547-549.

Needham, D.C. and P.A. Hammer, 1990. Control of *Salpiglossis sinuata* height with plant growth regulators. *HortSci.*, **25**(4): 441-443.

Pizzetti, I. and H. Cocker, 1975. *Flowers A Guide to Your Garden*, pp. 1179-1180. Harry N. Abrams, Inc., Publishers, New York, USA.

Sink, K.C. 1984. Taxonomy. *In*: Petunia (ed. Sink, K.C.), pp. 3-9. Springer, New York.

Vaughan, M.J. 1988. *The Complete Book of Cut Flower Care*, p. 68. Timber Press, Oregon, USA.

33

Scabiosa (Family: Dipsacaceae)

M.K. Singh, R.L. Misra, Sanyat Misra, Sanjay Kumar and Ajit Kumar

[**Common names**: Caucasian scabious (*Scabiosa caucasica*); Mourning bride, Pincushion flower, Scabious; Small or Blue scabious/Dove pincushion (*S. columbaria*); Sweet scabious (*S. atropurpurea*); Yellow scabious (*S. ochroleuca*), *etc.*]

Introduction

Scabiosa is Latin and named so as scabies, a disease was once supposed to be cured by certain members of the genus. Authorities differ about the number of species under this genus, but there are some 80-100 species of annuals, biennials and herbaceous perennials originating from Europe (especially from Mediterranean regions, Caucasus and Southern Europe), Africa and Asia (Asia Minor). In European fields and meadows, the beautiful wild forms are still found growing and emitting their delightful fragrance. The flowers are beautiful having pleasing colours, forms, fragrance with long-flowering habit though not so common in our gardens. Its reference appears in one of the poems by an unknown author written around 1099 during the pontificate of Pope Urban II, though now the species, *Scabiosa succisa* referred to here is renamed as *Succisa pratensis*, and the introduction of *Scabiosa* to the European gardens was many years after Urban's time. *Scabiosa stellata* was introduced to British gardens from Spain in 1596, followed by *Scabiosa atropurpurea* in 1629, and then afterwards other species one by one. There may be some confusion about the origin of certain species because of their ecotypes and homoploids being found away from their original homes adapting to a quite distinct environment so Hagen *et al.* (2008) tested whether *Scabiosa columbaria* s. str. could have originated by recent homoploid hybrid speciation because it is of intermediate morphology and is distributed in areas under glacial influence though original species is native to Central Europe. Johnson *et al.* (2003) tested the hypothesis that two rare South African orchid taxa *Brownleea galpinii* ssp. *major* (nectar-producing) and *Disa cephalotes* ssp. *cephalotes* (non-rewarding) are mimics of the nectar-producing flowers of a relatively common species, *Scabiosa columbaria*, with which they always occur sympatrically, flowers in both (these orchids and *Scabiosa*) having similar dimensions and almost identical spectral reflectance though orchids were not fragrant, and the orchids were pollinated exclusively by long-proboscid flies (*Tabanidae* and *Nemestrinidae*) that feed mainly on nectar in *Scabiosa* flowers but these flies did not discriminate between flowers of these orchids and *Scabiosa* while alighting on their flat-topped inflorescences but not on other orchids dissimilar to *Scabiosa* or on artificially reconstructed spike-shaped *Brownleea galpinii* orchid, so the authors warned that the term mimic should not apply to species whose resemblance to another species is due entirely to plesiomorphic traits that, in all likelihood, evolved prior to the ecological association.

Scabiosa is undemanding and sun-loving perennial, mostly growing wild in dry grasslands. In the garden they thrive on hot slopes or on patches of open ground between the stones of a rock garden. Flowers come in round, pin-cushion like heads with pale blue, misty blue, deep blue, red, pink, white to mauve in colours. It is suitable for growing in beds, herbaceous and mixed borders and the daisy-like long-lasting flowers are most suitable for cutting. They are best set in groups on open rock garden terraces, where they develop into spreading mats of varied vegetation. Nowak (1995) presented a note on the characteristics, cultural requirements and decorative value of *Scabiosa stellata* suitable for use as dried flowers. Its dried seed heads may be used in floral arrangements.

Dargaeva and Brutko (1977) isolated umbelliprenin, bergapten and coumarin from aerial parts of *Scabiosa comosa*. Quercetin, diosmin, ochroside and luteolin-7-glycoside were isolated from *S. ochroleuca* (Zemtsova, 1977). Aqueous extracts of the inflorescences of *S. atropurpurea* yielded a fraction containing tannins, saponosides, anthocyanins, sterols and triterpenes which showed analgesic and antipyretic effects in rats and mice (Marhuenda-Requena *et al.,* 1987). Acute and subchronic anti-inflammatory activity and *in vivo* antibacterial activity of *S. atropurpurea* in male Wistar rats revealed that a 10 per cent extract decoction was active against acute inflammation, the percentage inhibition being 4.5 at an administration rate of 300 mg dry matter/kg and 36.3 at 600 mg/kg, and the antibacterial activity against *Staphylococcus aureus* was observed at doses above 300 mg/kg (Saenz-Rodriguez *et al.,* 1987). When Kowalczyk and Krzysanowska (1999) screened the extracts of *S. ochroleuca* for antifungal activity against *Candida albicans, Rhodotorula rubra* and *Aspergillus fumigatus* pathogenic fungi, they recorded good activity against these fungi due to presence of polyphenolic compounds especially flavonoids [luteolin, luteolin 7-glucoside, apigenin, quercetin, kaempferol and phenolic acids (chlorogenic, protocatechuic, p-coumaric and p-hydroxybenzoic)]. The rootstocks of *S. columbaria* yielded loganin and sweroside, the latter at 1.0 mg/l (minimum inhibitory concentration) showed moderate antibacterial activity against the Gram-positive *Bacillus cereus, B. pumilus, B. subtilis, Micrococcus kristinae* and *Staphylococcus aureus* and the Gram-negative *Escherichia coli, Klebsiella pneumoniae, Pseudomonas aeruginosa* and *Enterobacter cloacae* (Hom *et al.,* 2001) and it was further revealed that pure sweroside was converted to the sweroside aglycone 1-acetoxy derivative after hydrolyis of the glucose unit, and was transformed into a new seco-iridoid having the novel trans diaxial configurations for the protons C-1, C-9, and C-5. *S. hymettia* was when first investigated by Christopoulou *et al.* (2008), they recorded two flavonoids, three coumarins, three iridoids and two phenolic constituents, where they found the kaempferol derivative I with highest activity against three human pathogenic fungi, and six Gram-positive or Gram-negative bacteria. Zheng Quan *et al.* (2004) isolated 11 new triterpenoid saponins, scabiosaponins A-K (1-11), and hookerosides A (12) and B (13) from the whole plant extract of *S. tschiliensis* and they found that scabiosaponins E, F, G, I (5, 6, 7, 9), hookerosides A, B (12, 13), and prosapogenin 1b all exhibiting strong inhibition of pancreatic lipase *in vitro*. Javindra *et al.,* 2006) characterized the essential oil of the aerial parts of *S. flavida* with 43 components constituting 94.2 per cent of the oil, the main components of the oil being tricosane (15.5 per cent), rosifoliol (15.3 per cent), (E)-caryophyllene (10.7 per cent), and α-humulene (7.9 per cent).

Botany, Genetics and Breeding

Scabiosa is a frost-hardy genus of 80-100 species that are tufted to clump-forming, having opposite pairs of entire, dentate-lobate or pinnately dissected leaves and generally more or less woody base. They are annuals, biennials or herbaceous perennials some of them being evergreen. These are native to Europe, Africa and Asia, but rarely found in the tropics. Its heads are terminal, depressed subglobose or ovoid-conical, pedunculate or rarely sessile in a dichotomous inflorescence. Its sepals are bristle-like and the flowers are cushion-like with the same formation as a daisy of Asteraceae, where overlapping ray florets surround a large central disc with protruding styles. Each floret is 4-5-lobed, tubular to funnel-shaped, the outer ones generally quite longer than the rest like the ray florets, and the stamens 4 (sometimes even 2). The flowers may be white, yellowish, rose or blue. At the base of each floret there is an epicalyx or involucel (a ribbed funnel-shaped structure which is differently formed in different species) with an expanded membranous margin and after flowering this epicalyx enlarges to form a one-seeded achene (fruit) for further propagation. The epicalyx is structured into developing pits and grooves, *vis-à-vis* a larger membranous margin to aid in its dispersal by wind. Botanists find a strong base in its fruit form for describing various species.

The progenies through hand-pollination to produce selfed, within-population crosses (WPC) and between-population crosses (BPC) from the field-collected seeds showed no significant difference in germination, seedling-to-adult survival, flowering percentage and the number of flower heads, but severe inbreeding depression was demonstrated for biomass production, root development, adult survival and seed set, and on calculating multiplicative fitness functions, WPC progeny showed a more than 4-fold, and BPC progeny an almost 10-fold, advantage over selfed progeny, indicating that *S. columbaria* is highly susceptible to inbreeding, though no clear relationship could be noted between population size and level of inbreeding depression, suggesting that the genetic load has not yet been reduced substantially in the small populations (Treuren *et al.,* 1994a). In natural populations of *Scabiosa columbaria*, Treuren *et al.* (1994b) estimated the outcrossing rate in hermaphrodites which was close to 1 and ranged from 0.84 ± 0.07 to 1.12 ± 0.11, and on studying the effect of plant density on outcrossing rates in 2 experimental populations of 27 individuals, contrary to expectation, estimates of outcrossing rates in hermaphrodites were about 100 per cent for both the densities, however, in the sparse population the fraction of developed seeds of plants used to estimate outcrossing rates was significantly lower than of plants in the dense population (0.41 ± 0.06 and 0.68 ± 0.08, respectively). Artificial pollination of these plants in a greenhouse showed that the fraction of developed seeds was 0.60 ± 0.01 and 0.83 ± 0.05 after self pollination and cross pollination, respectively. The combined results suggested that the differential success of self pollination and cross pollination might have caused equalization of the outcrossing rates in the experimental populations, despite different plant densities. Waldmann (2001) raised the plants of *Scabiosa canescens* (collected from Hallestad, Skane, Sweden), a locally rare species, under uniform growth conditions to examine the effects of one generation of selfing and outcrossing on fluctuating asymmetry (FA) in floral morphology and obtained significantly higher $p=0.038$ for inbred progeny than for outcrossed offspring which supports that expression of deleterious recessive alleles are responsible for increase of FA, and estimation at the individual level of the five fitness-related traits revealed that there was no correlation between the two though a companion study found significant inbreeding depression for all fitness traits, and a negative association between FA and fitness could therefore be asserted at the treatment level (inbred/outbred progeny), hence, FA seems to be useful to predict inbreeding depression, but specific individuals with high fitness cannot be identified based on their FA levels. Pluess and Stocklin (2004) selected 89 seed families from 11 populations of various sizes of *Scabiosa columbaria* for a study of molecular diversity, and used the same material in a greenhouse experiment to measure variation in fitness-related traits and the ability of populations to cope with competition, and by using RAPD-PCR they detected 71 RAPD-phenotypes among 87 genotypes, *vis-à-vis* obtained variable and relatively high molecular diversity within populations, with an expected heterozygosity, He ranging from 0.09 to 0.24; He, the Shannon index (SI) and the percentage of polymorphic bands were not correlated with population size, but the smallest populations had the lowest molecular diversity (He, SI); population differentiation was moderate with 12 per cent of the molecular diversity among populations; and measures of fitness in the greenhouse differed among seed families ($P<0.001$), but not among populations, therefore these results suggest an increased risk of local extinction of *Scabiosa columbaria* in the Swiss Jura caused by a decreased viability and reduced phenotypic plasticity due to genetic erosion in small populations.

Zymograms of glucose-6-phosphate isomerase (GPI) and phosphogluconate dehydrogenase (PGD) revealed 3 isoenzymes for each enzyme in *S. columbaria*, and Treuren and Bijlsma (1992) stated that intergenic heterodimers are formed between the polypeptides coded by Gpi1 and Gpi2 and between those coded by Pgd1 and Pgd2, indicating that a GPI and a PGD locus have been duplicated in the past in *Scabiosa*. The ancestral genes assort independently with their duplicated gene, suggesting that the duplications have originated from a process of translocation. Linkage was found only between Gpi1 and Pgd2 and between Gpi2 and Pgd1, suggesting that the duplicated loci were located on the same translocated chromosomal segment. The combined results suggest that a chromosomal segment containing Gpi2 and Pgd1 was translocated before the divergence of *Scabiosa*.

Species and Varieties

Scabiosa *atropurpurea* (*S. major, S. maritima, S. calyptocarpa*): It is an evergreen, moderately fast-growing, upright-growing, hardy biennial to short-lived bushy perennial species though grown as annual, with branched stems, growing 45 (dwarf types) to 100 cm (taller ones) high, and is native to Southern Europe to Turkey. It is being grown in the gardens since 1629. It is a highly variable species but very common in cultivation for its delicious fragrance and excellent cut flowers. The species has an almost prostrate compact rosette of mid-green, radical and deeply lobed leaves, lower ones lanceolate-ovate, lyrate and coarsely dentate, and the upper stem leaves pinnately lobed, lobes cut or dentate and oblong. Peduncles long and wiry, and bear 3-5 cm across scented and pin-cushion-like floral heads of dark maroon to lavender, deep purple, rose, red and white colours, and the fruiting heads become ovate or oblong. Its 'Cockade Series' plants are fast-growing bushy annuals with lobed leaves and large, rounded, double flowerheads in shades of pink, red, purple or blue appear on slender stalks. Its popular varieties are 'Blue Moon' (quite large conical inflorescences and lavender-blue flowers), 'Candidissima' (white), 'Candidissima flòre-plèno' ('Grandiflora', large-flowered double form), 'Coral Moon' (coral-red), 'Dobie's Giant Hybrids' (large double flowers in shades of pink, scarlet, lavender, blue, maroon and white), 'Dwarf Double Mixed' (compact plant with flower colours in shades of pink, scarlet, mauve, blue and maroon), 'Major' (golden), 'Maxima' (muchbranched dwarf-growing), 'Tom Thunb' (scarlet), *etc.*

Scabiosa brachiata (*Callistema brachiatum*): An annual species native of Europe and Asia, growing up to 40 cm in height; where lower leaves are ovate-oblong, the upper pinnately cut and lyrate, the lower lobes decurrent, terminal large, obovate and oblong; and the flowers are quite attractive in light blue colour.

Scabiosa canescens (*S. suaveolens*): A native of Central and Western Europe, which grows up to 50 cm in height and is ideal for growing in dry sandy-loam soils. Leaves are grey-hairy, undivided, and lanceolate at base though stem leaves are pinnatifid. Its light blue flowers are scented.

Scabiosa caucasica: A clump-forming herbaceous perennial is native to Caucasus, Central and North Russia and North Iran. Though it is fully hardy but is not suitable for outdoor cultivation in exceptionally cold zones. Suitable for growing for calcareous (alkaline) soils. The original species was introduced into European gardens in 1803. It grows up to 75 cm in height bearing clustered light green, opposite, pinnatifid, glaucous or whitish leaves, upper part of each leaf being cut into narrow segments, lower being lanceolate-linear and acute though upper ones pinnately-divided and glaucous, heads some 5.0-8.0 cm across, roundish-flattish and pin-cushion-like, and flowers frilled, showy and lavender-blue with pin-cushion-like centres on long rigid and almost leafless stems. Plant blooms throughout the summer if old stems are removed regularly. Plants are sensitive to damp conditions, particularly in winters and do not tolerate extreme heat or cold. Its popular varieties are 'Alba Perfecta' (syn. 'Perfecta', white), 'Bressingham White', 'Cannon Andrews' (deep purple), 'Caucasica Blue' (blue), 'Clive Greaves' (lavender-blue), 'Floral Queen', 'Imperial Purple' (petals frilled and flowers dark lavender-mauve), 'Kompliment', 'Loddon White', 'Magnifica' (large-flowered, deep lavender-blue and foliage silvery-grey), 'Miss Willmott' (creamy-white), 'Moerheim Blue' (indigo), 'Moonstone' (light blue), 'Penhill Blue' (blue, shaded mauve), 'Perfecta' (pale-blue), 'Sally' (grey-blue with pink stamens), 'Silver King' (silvery-white), 'Souters Violet' (dark violet), 'Vincent' (cobalt-blue), 'Wanda' (deep blue), *etc.*

Scabiosa columbaria: A quite variable tufted hardy herbaceous perennial which grows up to 75 cm in height, is native to Europe, Western Asia, Siberia and North Africa. Its branching stems are nearly glabrous, radical leaves obovate to oblanceolate, simple and crenate to lyrate-pinnatifid, membranous and pubescent on both the surfaces, stem leaves glabrous, pinnately lobed, the segments linear, entire or little incised, and hispid, peduncles long and pubescent, floral heads 1.5-3.5 cm across, ovate-globular and flowers blue (lavender to lilac) though its var.'Alba' bears white while var. 'Alpina' bears yellow flowers, latter being ultra dwarf some 15 cm tall. Two beautiful English varieties 'Pink Mist' (purple-violet) and 'Butterfly Blue' (violet) and one New Zealand var. 'Samanthas Pink' (red-purple) have recently been developed (Anon., 1992; 2000).

Scabiosa cretcica: A rounded herbaceous perennial, tolerating up to 5 °C minimum temperature, is native to Balearic Islands, Italy and North Africa. It grows in rocky places preferably in dry sheltered position. Plant grows upto 25 cm in height bearing oblanceolate leaves that are 3.5 to 5.0 times longer than breadth. Flowering heads are above the leaves and the flowers are pale-bluish.

Scabiosa graminifolia: It originates in Southern Europe and is a frost-hardy, attractive, semi-shrubby, tufted or clump-forming, evergreen herbaceous perennial which is short-lived, with vigorous growth having stiff but wiry stems and somewhat woody base, growing up to 30 cm in height, bearing dense, mainly basal, almost prostrate, pinnate or bipinnate silvery grey-green, linear (grass-like) to narrowly lanceolate and acute leaflets with silky hairs and pale-blue (lilac or mauve to pink) flowers. Flowerheads are often solitary, pin-cushion-like, spherical and 2.5-4.0 cm across. It is grown in open ground with warm and sunny sites and is an ideal plant for dry-stone wall decoration and rock gardens but does not like disturbance. Its var. 'Pinkushion' bears pink flowers.

Scabiosa lucida: A native to Alps region (Central and Southern Europe) growing up to 40 cm high, is most suitable for cultivation in stony soils, and is similar to *S. columbaria* though usually shorter with smooth and shiny leaves and rose-lilac to light red-purple flowers that are borne on solitary involucrate flower heads. The bases of bristle-like calyx on the seed is winged.

Scabiosa japonica: It is a tufted herbaceous, dichotomously-branched perennial from Japan which grows up to 75 cm high, and is allied to *S. atropurpurea*. Leaves are decorative bright shade, pinnatisect with narrow lobes. Peduncles quite long, flower-heads some 5 cm across, terminal or axillary, and the flowers pale-lilac or violet-blue, and dissimilar involucral bracts formed in two rows are shorter than the flowers.

Scabiosa ochroleuca (*S. columbaria* var. *ochroleuca*): A clump-forming fully hardy herbaceous perennial growing up to 80 cm high is native to East-Central and Southern-East Europe, Western Asia and Siberia and prefers moist soils. It is almost identical to *S. columbaria* though flowers are frilled and bright pale-yellow with pin-cushion-like centres. Its highly branched stems are somewhat hairy, leaves whitish-pubescent, the radical crenate or lyrately pinnatifid (pinnate or bipinnate), tapering to a petiole and pubescent on both the surfaces, stem leaves 1-2-pinnately divided or cleft into oblong or linear lobes, peduncles are long and slender and the leaves of the involucres shorter than the flowers.

Scabiosa prolifera: A native of Asia Minor, this hardy annual species grows up to 75 cm in height. Its foliage are narrowly oblong and acute and measure some 3.8 cm in length. Flowers are round, flattish, cream or ivory-white and are borne on rigid stems. Though it is quite beautiful but rarely cultivated. It blooms continuously for a long time if the spent blooms are regularly removed.

Scabiosa pterocephala (*Pterocephalus parnassi, P. perennis*): It is a little woody-based perennial native to mountains of Albania and Greece and is suitable for growing in the sunny positions especially in screes, rock crevices or dry walls. While revising the genus, this has now been placed under the genus *Pterocephalus* which contains some 20-25 species. Its leaves are 2-5 cm long, oblong-spathulate to lyrate, usually crenate and densely grey-downy. Solitary floralheads are dense low-cushioned and scabious-like, 2.5-4.0 cm across, florets 5-lobed, small and tubular, pink to rosy-purple and appear just above the foliage.

Scabiosa reuteriana: It is an erect and branched annual species native to Asia Minor which grows up to 40 cm in height. Lower leaves are oblong-spatulate and entire though upper ones lyrate, lateral segments 1-2 to both the sides, short and lanceolate though terminal quite large. Calyx limb short-stipitate, corolla pale-violet with denticulate lobes and the fruiting head is ovate.

Scabiosa stellata: It is a simple or little-branched annual species native to Southern Europe, which grows up to 50 cm in height. Whole plant is hairy; the leaves are cut or little lyrate, obovate and dentate with large terminal lobe, and the upper ones on the stems are often pinnatifid; and the floral stalks are gracefully slender, 2.5 cm in diameter, long and bears lovely blue floral heads where corolla is 5-parted with radiating lobes.

Scabiosa ucranica: A native of Europe, Asia Minor and Iran, this species is an herbaceous biennial or perennial with erect and branched stems which grows up to 40 cm in height. The leaves are pilose-pubescent with appressed white hairs, the lower pinnatifid, segments oblong or oblong linear, entire or dentate, and the upper ones are often entire. Calyx limb short-stipitate, corolla creamy to yellowish-white, sometimes rose or even blue and the lobes are nearly entire or crenate. The fruiting heads are spherical.

A few other important varieties are 'Blauer Atlas' (blue), 'Blausiegel', 'Ivory Queen' (white), 'Nacht Flalter' (blue), 'Stafa' (blue), *etc.* In France when Mallait (1988) planted five *Scabiosa caucasica* cultivars outdoors in June 1966, all except cv. 'Kompliment' produced flowers in the same year with a total production (during Septembr 1986 - November 1987) for all the cultivars combined was 170 flowers/m^2, the best results being with 'Clive Greaves' which gave high yields of good quality flowers (21.6/plant in 1987) and flowered from May to November (with peak production in June).

Propagation

It is propagated directly through seeds, by division and through cuttings. The **seeds** of the annual species are sown in the flowering site during autumn and spring and to maintain proper spacing, *i.e.* 30 × 30, 30 × 45 or 45 × 45 cm depending upon the characteristic spread of the concerned species or varieties, these are thinned out only during spring to summer when these are large enough to handle. The perennial species and their varieties can be propagated from seeds sown in May-June inside the frame or in the seed-beds, pricking them out in the nursery beds and planting out in their permanent positions in autumn or during spring. Seeds of *Scabiosa atropurpurea, S. caucasica* require 18-21 °C for germination, the former requiring covering in beds and the latter light. It takes 10-12 days for germination of seeds in case of *S. atropurpurea* and 10-18 days in *S.caucasica* (Nau, 1993). Seeds should be sown in early winter, preferably in seed tray or shallow pots or alternatively in raised bed or in cold frame. The shoots of *S.caucasica* have very few roots of their own, making them difficult to propagate. Those to be propagated through **divisions** as in case of many named varieties of perennial scabiosas, their clumps can be divided during autumn or spring and planted in the prepared field. **Cuttings** are planted during spring in the sandy soil under glass or polytunnel for early rooting. In a nursery trial, Ackerman and Hamemik (1994) sprayed 1,500 each of *Scabiosa* cv. 'Butterfly Blue' and cv. 'Pink Mist' plants, all with single shoot in 10-cm pots with GA_3 (as ProGibb) at 2-week intervals where two weeks after the initial application, a spring-like flush of growth started and previously dormant nodal buds started to swell and terminal leaves started to expand, and subsequently two weeks after the second application, a sufficient growth rate had been attained to make further applications unnecessary, and so in less than two months, >25,000 cuttings of each of the *Scabiosa* cultivars were obtained from the treated plants.

Romeijn and Lammeren (1999) tested several factors including explant type, donor plant, cold pre-treatment and medium composition for ***in vitro* regeneration** on the explants of anthers and ovules of *S. columbaria* where they recorded varying callus induction frequency among the treatments, and histological analysis revealed numerous sites of callus induction, however, gametophytic cells did not proliferate. Stepwise removal of growth regulators and simultaneous lowering of sucrose in the nutrient medium resulted in initiation of embryogenesis or shoot organogenesis, and allowed plant regeneration but only in case of one donor under the conditions tested. The regenerants were diploid (except one mixoploid individual), but showed various types of flower heads and probably they were of sporophytic origin. On MS medium supplemented with BA, Hosoki and Nojima (2004) took adventitious shoots or node sections of *S. caucasica* var. 'Caucasica Blue' and obtained best success with nodal or intermodal explants and leaf blades with highest frequency of adventitious shoots occurring in the MS medium with 4.4 and 18 μM BA, respectively, and addition of 0.19 or 1.9 μM NAA to the BA-containing medium promoted callus formation and reduced shoot organogenesis. During micropropagation, shoot nodal explants derived from *in vitro* shoots cultured in the MS medium supplemented with 4.4 μM BA yielded 8.9 shoots per explant within 40 days after culture initiation.

Cultural Practices

Scabiosa needs full sun for its cultivation but in the cold climate, its growing under protected structure will provide quality blooms. Since it is a crop of temperate regions, best it is sown during spring to early summer. For its growing, *Scabiosa* prefers a fertile and well-drained garden **soil**, preferably in chalk and limestone alkaline soils though its response is not proper especially on a clayey soil. *Scabiosa stellata* cv. 'Sternkugel' was grown (a) under 8-h **daylight** (DL), (b) 8-h DL + 8-h white incandescent lamps, (c) 8-h DL + 8-h blue light (Philips LT18), or (d) 8-h DL + 8-h red light (Philips LT15), where Zimmer (1989) reported that *S. stellata* is a facultative LD plant, but in this plant both blue and red lights were effective in extending the photoperiod and promoted flowering, however, in the second experiment, *S. atropurpurea* was grown (a) under 10-h DL, (b) 10-h DL with a 4-h break (incandescent lamps) during the dark period, or (c) continuous light (DL + incandescent lamps) and reported that *S. atropurpurea* is an obligate LD plant. Maintaining the night temperatures of 10-13°C for *Scabiosa atropurpurea* and13-14°C for *S. caucasica* (Nau, 1993) at post-germination production give about maximum production with quality blooms. Hellmann (1991) had sown four *S. caucasica* cultivars in May in Germany in a mixture of high moor peat and perlite, pricked out in July into peat-based substrate or rockwool blocks and grown in normal beds or in a 'plant plane hydroponic system' from July-August where growth and production of flowering stems were monitored until the following April, and he obtained that plants in hydroponic system started production four weeks earlier than those grown in soil and reached maximum production after 6 weeks, compared with 7 weeks for plants in soil, however, in both the systems, plant diameter was lowest in mid-March and shoot buds began to form in late January, yield being highest in cv. 'Perfecta', followed by 'Fama', 'Kompliment' and 'Perfecta-Weiss'.

Field preparation is similar to other crops as before planting the soil should be well pulverized and freed of any foreign material (large pieces of polythene, stones, bricks, glasses, hard woods, *etc.*), *vis-à-vis* weeds. Sowing or planting should be done preferably in autumn (August to October), alternately during spring (March-April) or in the tremperate regions even during summer (May-June). Soil should be quite fertile so at the time of field preparation, the soil should be thoroughly mixed with 300 quintals of wellrotten farmyard manure, compost or any other manure and watered so that the same is perfectly set with the upper surface of soil. Fully cultivated field should be divided into beds of the convenient sizes, preferably 1.6 m in width though length as long as the level of field is even for proper flood irrigation, however, drip irrigation will save the water to a great extent. While preparing the beds, the appropriate places should be earmarked for main and subsidiary channels for the purpose of flood irrigation, and should be prepared accordingly. The subsidiary channels may be flat so that these may work as path for convenience of the people who work there in the field. The beds should be slightly raised so that water released in the subsidiary channels is sucked up by the beds from both the sides so that water does not come in direct contact of the root system and only that much water is sucked which is essential at that time. These paths *cum* channels will also facilitate the workers for taking out the weeds, for hoeing the beds, for staking or any other cultural operation including pesticide application and flower harvesting, *etc.* from both the sides. Looking into the plant spread of an individual variety, hybrid or species, the per **plant space** is provided which may be 30 × 30 cm, 45 × 30 cm or 45 × 45 cm per plant per square meter. Immediately after planting, the plants are irrigated to recover from the lifting shock. *S.caucasica* should be planted in warm, sunny sites with moderately dry to moist soil. Planting of 5 plants/m^2 will cover the ground completely within two growing seasons but for for quick covering some nine plants/m^2 should be planted, however, these in natural courses are normally planted 3 to 10 plants per square meter. When Zizzo *et al.* (1995) transplanted *Scabiosa stellata* at 5, 10 or 15 plants per square meter, the highest seed yield was obtained at 10 plants/m^2. In Germany, Aderhold and Kuhn (1991) divided *S. caucasia* plants in April or mid-August and planted the crop in deep and rich soil having the pH range of 6.4 to 6.8 and supplied with basal and supplementary fertilizer doses which resulted into flower harvesting 2-3 times per week between mid-June to late October with an expected yield of 50-60 stems per plant by the third year. They suggested the planting of cultivars such as 'Blausiegel', 'Clive Greaves' and 'Miss Willmott' for higher yield.

Thinning out is a must in seed-grown crop to maintain proper plant spacing, *vis-à-vis* it is also necessary after a few years to avoid overcrowding. *Scabiosa* plants respond well to dead-heading. Removal of spent flowering stems in *S. caucasica* and other species increases the flowering period from June to November. In case of taller strains, the plants require to be **staked** with bamboo sticks, or erect canes collected from branches and twigs of weeping willows and other trees which produce smooth and erect twigs. The erection of such stakes and tieing these with the plants or inflorescences will save the flowers from breaking down due to lodging or wind-strokes.

In a 3-year experiment, applications of 75, 150 or 300 kg N/ha/year increased stem length and the number of flowers/plot, greatest being from the highest rate, and here leaf N content was found correlated with yield, critical level being >3 per cent N when leaves were sampled 30 days before beginning of flowering and 2.4-3.0 per cent later in the season, however, application of 100 or 200 kg/K_2O/ha/year had no effect since soil K content must have been high (Alt, 1978). In Germany on heavy sandy loam soil the *Scabiosa caucasica* cv. 'Fama' plants were when **fertilized** with 0, 60, 120 or 180 kg N/ha, Sprau (1998) recorded number of stems/m^2 in the range 100-129 as highest, *vis-à-vis* increasing fresh weight up to 120 kg N/ha though a large variation was found between different experimental years, however, decreased fresh weight was noted at 180 kg N/ha.

Weed is a problem in any crop growth as these rob the crop of its nutrients, light and moisture and harbour various insect-pests and diseases which may prove harmful to the plants, therefore, from the very beginning of the crop the field should be clean. Manual labour is time consuming and expensive hence one resorts to chemical control. Kerb 50W (50 per cent pronamide) applied at 2-3 kg/ha on 31 October to *Scabiosa caucasica* plants gave good control of *Agropyron repens, Veronica persica* and *Lamium purpureum* and complete control of *Poa annua*, but did not control *Senecio vulgaris* though application of 3 kg/ha showed some growth depression (Hofmann, 1975).

Polyhouse Growing

Kneissl and Sollner (1978) successfully forced *Scabiosa caucasica* in plastic tunnels (50 holes/m^2), under flat plastic (500 holes/m^2) and in some cases in a slightly heated plastic house. *Scabiosa caucasica* cv. 'Clive Greaves' grown under perforated plastic film (500 holes/m^2), performed well and flowered earlier with quality blooms and better cut flower yield than those without protection (Penningsfeld *et al.*, 1979, 1980). Plants of *S. caucasica* in Germany were set in April or May 50 × 50 cm apart for forcing the following year and in the first year all flower stems were nipped but after snowing, a plastic cover on a frame was placed over the plants and heat was supplied, 12 °C by day and 4-6 °C by night, and then by mid-April when the plants were strong enough and the buds started showing colour the flowers were harvested and afterwards in mid-May the plastic cover was removed (Plomacher, 1980). This way the cv. 'Clive Greaves' gave highest production though the new cvs 'Nachtfalter', 'Prachtkerl' and 'Blauer Atlas' performed poorly.

Growth and Flowering

Seeley (1989) studied effect of photoperiod on *Scabiosa* cultivars and reported that long days induce flowering and stem elongation though short days promote compact flowering. Keever *et al.* (2001) studied the effect of night-interrupted (NI) lighting (beginning late winter, *i.e.* February 1, February 15, March 1, March 15, and a natural photoperiod) on *Scabiosa columbaria* cv. 'Butterfly Blue' produced outdoors in a southern nursery setting of United States, which increased 44-51 per cent flower and flower bud counts, *vis-à-vis* plant height compared to plants under natural photoperiod.

Friedrich and Sackmann (1985) through spraying of CCC on *Scabiosa caucasica* plants obtained reduction in the stem growth but neither flowering time was altered nor there was any change in the flower quality or vase life, however, the marginal leaf chlorosis and necrosis, and occasionally shorter flower pedicels were obtained, so they suggested that watering CCC on the soil may give better results as it is then less affected by weather conditions. CCC (chlormequat chloride) and B-Nine (daminozide) are quite effective on *Scabiosa atropurpurea* for keeping the plants compact and of desired shape. *Scabiosa columbaria* 'Butterfly Blue' was treated with various growth retardants [B-Nine 5,000 ppm; B-Nine 5,000 ppm + CCC 1,500 ppm; Florel (ethephon) 500 ppm (applied twice); B-Nine 5,000 ppm + Florel 500 ppm; Sumagic (uniconazole) 20 ppm; or B-Nine 5,000 ppm + Sumagic 10 ppm] where only Florel and Sumagic reduced plant growth though Florel delayed flowering by about 10 days, however, combined treatments of PGRs, in general, were not found effective (Banko *et al.*, 2001). The next year of study, when Florel was applied once at 0, 250, 500, 750 and 1,000 ppm; B-Nine at 5,000 ppm; and Sumagic at 20 ppm, Florel suppressed the height of scabiosa 'Butterfly Blue' linearly up to 52 days of transplanting without affecting the initial number of inflorescences, and when first set of inflorescences was removed and a second set developed, there was a linear increase in inflorescence numbers with increasing concentrations of Florel (Banko *et al.*, 2001).

Postharvest

Scabiosa flowers from early summer to early autumn are harvested at the stage when half to fully open, for better life when half open (Nowak and Rudnicki, 1990). Nowak and Rudnicki (1990) mention about *S. caucasica* that flowers are sensitive to ethylene, and floral preservatives effective in this species are Bactoflor, Chrysal-AVB, Chrysal-OVB, Florever, Florissant 100, Florissant 400 and Floran G.

Insect-Pests and Diseases

Four fungal diseases occur on Scabiosa are blight (*Pellicularia rolfsii*), powdery mildew (*Erysiphe polygoni*), root rot (*Phymatotrichum omnivorum*) and stem rot (*Sclerotinia sclerotiorum*). *Erysiphe knautiae* was for the first time reported in Washington on *Scabiosa columbaria* cv. 'Butterfly Blue' by Glawe and Grove (2005) and its infection causes white to greyish patches of dense and sporulating mycelium on adaxial and abaxial leaf surfaces. Dietrich (1989) reported *Ustilago intermedia* causing anther smut on *Scabiosa columbaria* in German Democratic Republic. Kinsey (2002) stated that the soil-borne fungus *Phoma leveillaei* infects scabiosa. *Peronospora knautiae* has been reported causing downy mildew in the form of sparse violet or brown spots on *Scabiosa* leaves (Hall, 1994; Anon., 1995). Takeuchi *et al.* (1995) in Japan recorded *Botrytis cinerea* causing grey mould on *S. atropurpurea*. Pasini *et al.* (1996) in Italy isolated *Rhizoctonia solani* from scabious causing collar or foot rot. Mycoplasma-like organisms (MLO) have been found infecting *Scabiosa columbaria* in Denmark which could be transmitted through leafhoppers *Euscelis plebeja* & *Macropsis fuscula* and through *Cuscuta reflexa* (Begtrup and Thomsen, 1984). Beet curly top virus and aster yellows virus also infect scabious causing curly top. Mildews can be controlled through the use of sulphur fungicides. Leaf spot diseases can be controlled by spraying with 0.2 per cent Dithane M-45 or Z-78 and rots through spraying and drenching with 0.2 per cent Thiride, Bavistin and Benomyl. However, the MLOs and virus-infected plants should be uprooted and burnt.

Insects attacking the *Scabiosa* are aster yellow leaf hopper (*Macrosteles fascifrons*), rose beetle (*Pantomorus cervinus*), chrysanthemeum lace bug (*Corythucha marmorata*), etc. *Macrosteles fascifrons* transmitting aster yellow virus can be controlled through regular spraying of Malathion or Monocil. *Euphydryas aurinia* butterfly larvae were found feeding on *Scabiosa comosa* in the Russian Republic of Buryatia, on which Wahlberg *et al.* (2001) reared five species of hymenopteran parasitoids and three species of dipteran parasitoids. Beiger (2001) in Poland recorded larvae of *Chromatomyia scabiosella* sp. nov. mining the leaves of *Scabiosa* spp. and *S. ochroleuca*. Its males are easily distinguishable from their yellow fore-knees though are darker in *Phytomyza scabiosae*. *Pterostichus madidus*, one of the most abundant carabid beetles in Cambridge (U.K.) has been recorded feeding on wild flower seeds including *Centaurea scabiosa* (Hurst and Doberski, 2003). All these insects can be controlled by spraying with 0.2 per cent Malathion or Sumithion.

References

Ackerman, R. and H. Hamernik, 1994. Gibberellic acid to extend shoots and bud break on *Heuchera* and *Scabiosa*. *Intern. Pl. Prop. Soc.* (Combined Proceedings), **44**: 545-546.

Aderhold, A. and J. Kuhn, 1991. *Scabiosa caucasica* - a profitable cut-flower perennial (German). *Gartenbau Mag.*, **38**(10): 48-49.

Alt, D. 1978. Influence of nitrogen and potassium on perennials for cutting. I. *Scabiosa caucasica* (German). *Gartenbauwissenschaft*, **43**(1): 28-33.

Anonymous, 1992. *Scabiosa, Scabiosa columbaria* (Blakedown Nurseries Ltd., Kidderminster, UK). *Pl. Var. J.*, **5**(4): 20-21.

Anoymous, 2000. Variety: 'Samanthas Pink' (Application no: 1999/238, Super Perennials Ltd., Auckland, New Zealand). *Pl. Var. J.*, **13**(1): 69-70.

Anonymous, 1995. *Peronospora knautiae*. IMI Descriptions of Fungi and Bacteria (Distribution Maps of Plant Diseases, CABI Bioscience, U.K.), Map 678.

Banko, T.J., M.A. Stefani and M.S. Dills, 2001. Growth and flowering response of 'May Night' salvia and 'Butterfly Blue' scabiosa to growth retardants. *J. env. Hort.*, **19**(3): 145-149.

Begtrup, J. and A. Thomsen, 1984. Mycoplasma-like organisms (MLO) and their distribution in Denmark. *Tidsskrift for Planteavl*, **88**(3): 299-310.

Beiger, M. 2001. *Chromatomyia scabiosella* sp. nov. (Diptera: Agromyzidae), a new mining fly from Poland. *Polskie Pismo Entomologiczne*, **70**(1): 55-57.

Bijlsma, 1992. Duplication of the structural gene for glucosephosphate isomerase and phosphogluconate dehydrogenase in *Scabiosa columbaria* and their phylogenetic implications in the Dipsacaceae. *Biochem. Gen.*, **30**(1-2): 99-109.

Bijlsma, N.J. Ouborg and van W. Delden, 1994. The significance of genetic erosion in the process of extinction. IV. Inbreeding depression and heterosis effects caused by selfing and outcrossing in *Scabiosa columbaria*. *Evol.*, **47**(6): 1669-1680.

Christopoulou, C., K. Graikou and I. Chinou, 2008. Chemosystematic value of chemical constituents from *Scabiosa hynettia* (Dipsacaceae). *Chem. & Biodiv.*, **5**(2): 318-322.

Dargaeva, T.D. and L.I. Brutko, 1977. Coumarins of *Scabiosa comosa* (Russian). *Rastitel' nye Resursy*, **13**(1): 78-80.

Dietrich, W. 1989. On the occurrence of the anther smuts *Ustilago succisae* and *U. scabiosae* in the GDR (German). *Boletus*, **13**(1): 15-17.

Friedrich, H. and G. Sackmann, 1985. Cycocel: not suitable for use on perennials for cut flowers. Outdoor spray treatments did not improve quality (German). *Gb+Gw.*, **85**(20): 751-753.

Garaev, E.A., I.S. Movsumov and M.I. Isaev, 2008. Flavonoids and oleanolic acid from *Scabiosa caucasica. Chem. Natural Comp.*, **44**(4): 520-521.

Glawe, D.A. and G.G. Grove, 2005. First report of powdery mildew of *Scabiosa columbaria* (dove pincushions) caused by *Erysiphe knautiae* in North America. *Pl. Heal. Prog.*,Oct. 2005, pp. 1-2.

Hofmann, K. 1975. Weed control in perennials (German). *Zierpflanzenbau*, **15**(4): 117.

Hagen, K.B. von, G. Seidler and E. Welk, 2008. New evidence for a postglacial homoploid hybrid origin of the widespread Central European *Scabiosa columbaria* L. s. str. (Dipsacaceae). *Pl. Syst. Evol.*, **274**(3/4): 179-191.

Hall, G. 1994. *Peronospora knautiae*. IMI Descriptions of Fungi and Bacteria (CABI Bioscience, U.K.), No. 120, Sheet 1196.

Hallmann, S. 1991. Hydroponic production of perennials for cut flowers (German). *Gartenbau Mag.*, **38**(8): 56-57.

Horn, M.M., S.E. Drewes, N.J. Brown, O.Q. Munro, J.J.M. Meyer and A.D.M. Mathekga, 2001. Transformation of naturally-occurring 1, 9-trans-9, 5-cis sweroside to all trans sweroside during acetylation of sweroside aglycone. *Phytochem.*, **57**(1): 51-56.

Hosoki, T. and S. Nojima, 2004. Micropropagation of *Scabiosa caucasica* Bieb. cv. Caucasica Blue. *In Vitro Cell. & Dev. Biol. Plant*, **40**(5): 482-484.

Hurst, C. and J. Doberski, 2003. Wild flower seed predation by *Pterostichus madidus* (Carabidae: Coleoptera). *Ann. Appl. Biol.*, **142**(2): 251-254.

Javidnia, K., R. Miri and A. Javidnia, 2006. Constituents of the essential oil of *Scabiosa flavida* from Iran. *Chem. Natural Comp.*, **42**(5): 529-530.

Johnson, S.D., R. Alexandersson and H.P. Linder, 2003. Experimental and phylogenetic evidence for floral mimicry in a guild of fly-pollinated plants. *Biol. J. Linnean Soc.*, **80**(2): 289-304.

Keever, G.J., R. Kessler, Jr. and J.C. Stephenson, 2001. Accelerated flowering of herbaceous perennials under nursery conditions in the Southern United States. *J. env. Hort.*, **19**(3): 140-144.

Kinsey, G.C. 2002. *Phoma leveillei. IMI Descriptions of Fungi and Bacteria*, No. 151 (Sheet 1502), CABI Bioscience, Bakeham Lane, Egham, Surrey, U.K.

Kneissl, P. and V. Sollner, 1978. Forcing of cut flowers from perennials under plastic (German). *Deutscher Gartenbau*, **32**(47): 1946-1947.

Kowalczyk, A. and J. Krzyzanowska, 1999. Preliminaary antifungal activity of some Dipsacaceae family plants. *HerbaPolonica*, **45**(2): 101-107.

Mallait, M. 1988. Diversification of cut flowers with perennial plants. The Caucasian scabious: a species to recommend for outdoor cultivation (French). *Revue Hort.*, No. 284, pp. 51-53.

Marhuenda-Requena, E., M.T. Saenz-Rodriguez, and M.D. Garcia-Gimenez, 1987. A contribution to the pharmacodynamic study of *Scabiosa atropurpurea* L. I. Analgesic and antipyretic activity (French). *Plantes Medicinales et Phytotherapie*, **21**(1): 47-55.

Nau, J., 1993. Ball Culture Guide. *The Encyclopedia of Seed Germination* (2nd edn). Ball Publishing, Batavia, Illinois, USA.

Nowak, J. 1995.Less known plants for dry arrangements (Polish). Materiay ogolnopolskiej konferencji naukowe Nauka Praktyce Ogrodniczej z okazji XXV lecia Wydziau Ogrodniczego Akademii Rolniczej w Lublinie, pp. 887-890.

Nowak, J. and R.M. Ridnicki, 1990. *Postharvest Handling and Storage of Cut Flowers, Florist Greens, and Potted Plants*, pp. 37, 160. Timber Press, Portland, Oregon, USA.

Pasini, C., T. Berio, P. Curir and F. D'Aquila, 1996. Further characterization of *Rhizoctonia solani* isolated from carnation and other ornamental plants. *Informatore Fitopatologico*, **46**(6): 33-36.

Penningsfeld, F., P. Kurzmann and P. Kalthoff, 1979. Plants for cut flowers under flat plastic (German). *Annual Report of the Tech. Univ. at Weilhenstephan*, Jehresbericht, FRG, pp. 14-30.

Penningsfeld, F., P. Kurzmann and F. Kalthoff, 1980. Cut flower perennials under plastic sheeting (II) (German). *Deutscher Gartenbau*, **34**(17): 774-778.

Plomacher, H. 1980. Plants grown under plastic for cutting. *Scabiosa caucásica* (German). *Zierpflanzenbau*, **80**(1): 17.

Pluess, A.R. and J. Stocklin, 2004. Genetic diversity and fitness in *Scabiosa columbaria* in the Swiss Jura in relation to population size. *Coserv. Gen.*, **5**(2): 145-156.

Romeijn, G. and A.A.M. van Lammeren, 1999. Plant regeneration through callus initiation from anthers and ovules of *Scabiosa columbaria. Pl. Cell Tiss. Org. Cult.*, **56**(3): 169-177.

Saenz-Rodriguez, M.T., M.D. Garcia-Gimenez and E. Marhuenda-Requena, 1987. Contribution to the study of pharmacodynamics of *Scabiosa atropurpurea* L. II. Anti-inflammatory and antibacterial acitivity (French). *Plantes Medicinales et Phytotherapie*, **21**(3): 203-208.

Seeley, J.G., 1989. Light. *In: Tips on Growing Bedding Plants* (2nd edn.), pp. 34-36. Ohio State University, Columbus, Ohio, USA.

Sprau, G. 1998. Nitrogen fertilizer for scabious (German). *TASPO Gartenbau Mag.*, **7**(2): 20, 22.

Takeuchi, J., H. Horie and T. Hirano, 1995. Gray mold of some garden plants in Tokyo is caused by *Botrytis cinerea*

Persoon (Japanese). *Proc. Kanto Tosan Pl. Prot. Soc.*, No. 42, pp. 105-107.

Treuren, R. van, R. Bijlsma, N.J. Ouborg and van W. Delden, 1994a. The significance of genetic erosion in the process of extinction. IV. Inbreeding depression and heterosis effects caused by selfing and outcrossing in *Scabiosa columbaria*. *Evol.*, **47**(6): 1669-1680.

Treuren, R. van, R. Bijlsma, N.J. Ouborg and M.M. Kwak, 1994b. Relationships between plant density, outcrossing rates and seed set in natural and experimental populations of *Scabiosa columbaria*. *J. Evol. Biol.*, **7**(3): 287-302.

Treuren, R. van and R. Bijlsma, 1992. Duplication of the structural gene for glucosephosphate isomerase and phosphogluconate dehydrogenase in *Scabiosa columbaria* and their phylogenetic implications in the Dipsacaceae. *Biochem. Gen.*, **30**(1-2): 99-109.

Wahlberg, N., J. Kullberg and I. Hanski, 2001. Natural history of some Siberian melitaeine butterfly species (Nymphalidae: Melitaeini) and their parasitoids. *Entomologica Fennica*, **12**(2): 72-77.

Waldmann, P. 2001. The effect of inbreeding on fluctuating asymmetry in *Scabiosa canescens* (Dipsacaceae). *Evolutionary Ecol.*, **15**(2): 117-127.

Zemtsova, G.N. 1977. Investigation on plants of the family Dipsacaceae and the order Rosales with a view to introducing them into cultivation (Russian). *Sostoyanie i Perspektivy Nauch Issled po Introduktsii Lekarstv Rast.*, USSR, pp. 176-177.

Zheng-Quan, K. Koike, Han-LiKun, H. Okuda and T. Nikaido, 2004. New biologically active triterpenoid saponins from *Scabiosa tschiliensis*. *J. Nat. Prod.*, **67**(4): 604-613.

Zimmer, K. 1989. Photoperiodic reactions of summer cut flowers (German). *Deutscher Gartenbau*, **43**(28): 1719-1721.

Zizzo, G.V., U.A. Roxas and G. Iapichino, 1995. Effects of planting density on *Scabiosa stellata* L. seed production (Italian). *Sementi Elette*, **41**(1): 27-29.

34

Solidago (Family: Asteraceae)

M. Kannan, S. T. Bini Sundar, P. Ranchana, Sanyat Misra and R.L. Misra

[**Common names**: Aaron's rod, Beach/Seaside goldenrod (*Solidago sempervirens*), European goldenrod (*S. virgaurea*), Goldenrod, Sweet goldenrod (*S. odora*), Wreath goldenrod (*Solidago caesia*)

Introduction and Origin

Solidago takes its name from the Latin *solido* [to make whole or sound (health)], as it was thought that the plant had the virtue of rejuvenate and heal (Beckett, 1985; Pizzetti and Cocker, 1975). The genus comprises about 100-130 species (Bailey, 1960; Beckett, 1985) of perennial herbs mainly from North America (USA and Canada with more than 60 species), with a few on the Pacific Coast, and a few in Mexico and South America, temperate Europe (*S. serotina* and *S. virgaurea*), North Africa (*S. virgaurea*) and Asia (*S. virgaurea*). Asters (blue and blush) and goldenrods (golden and yellow) are the two very common flowers, complementing each other in colour and beauty, and giving full enchantment during autumn in America in thin woodlands and alpine meadows. Being so common and colonizing there, people in America do not appreciate its planting also as sometimes these become

invading weeds. Goldenrods are planted on the borders, in rock gardens, for colonizing in wastelands, and for cut flowers.

Its yellow flowers are used for making an infusion for treating kidney inflammation, and its decoction in poultices prepared from the flowers of *S. virgaurea* is used to fester wounds. Triterpenoid glycocides (monodesmosidic and bidesmosidic glycosides of polygalacic acid) which are obtained in *S. virgaurea* exert fungicidal properties, especially against *Candida* sp. (Bader *et al.*, 1990). *S. gigantea, S. virgaurea* and certain other species exert anti-inflammatory, spasmolytic & diuretic properties (Leuschner, 1995), and the latter species is anti-oxidative (Meyer, 1995), possesses anti-hypertensive properties and is also used against urological problems (Bader *et al.,* 1995), and its fresh aerial parts possess some anti-rheumatic properties (Kruedener *et al.,* 1995). Goldenrods also possess anti-phlogistic, antimicrobial and immunomodulatory properties (Bader, 1999). *S. Canadensis* is also used in traditional treatment of kidney and bladder diseases and its essential oil is insecticidal (Weyerstahl *et al.,* 1993).

Though beautiful when in blooms but becoming so common in America and invasive as weeds, it was not given due recognition for a very long time until H. Walkden, through his 20 year's concerted efforts developed interspecific hybrid *Solidago* × *hjybrida*, whose beauty enticed people to plant it in their gardens. Further, the goldenrods were crossed with a perennial *Aster* which resulted into intergeneric group of hybrids named as *Solidaster* and the best among these had been the beautiful *Solidaster luteus* with narrow aster-like leaves and graceful yellow flowers, most suitable for herbaceous border. This grows to 75 cm in height and flowers during autumn. *S. canadensis* and *S. virgaurea* are seldom grown in the garden but those which are grown are, in fact, the hybrids between these two and certain others. The dwarf ones have *S. virgaurea* 'Brachystachys' and ×*Solidaster luteus* in their parentage.

Botany

Solidago is a clump-forming perennial herb with long to short rhizomes, very useful for borders and for colonizing. It is an erect multibranched plant with narrow, simple and alternate leaves and many small yellow (rarely whitish) heads in terminal spikes, compound, feather-like golden coloured terminal panicles or recemes. The tiny flower head is oblong or narrow-campanulate, containing few florets (a few disc and short ray florets). The disc florets are all perfect and the ray florets in one series and pistillate. The basic chromosome number of *Solidago* is 9, and in its native range it is most often hexaploid (2n = 54), but triploid (2n = 27) and tetraploid (2n = 36) cytotypes also occur.

Species and Varieties

Some 100-130 herbaceous perennial species (Bailey1960; Beckett, 1985; Hay and Beckett, 1971) out of which only a few are of horticultural importance though remainders colonize similar to those of invasive weeds. The ones of horticultural importance are *Solidago caesia* L., a native to E. USA and extending westwards to Texas with smooth stems growing up to 90 cm, leaves dentate, lanceolate and some 12-13 cm long, and flowering early with inflorescence appearing from the leaf axils; *S. canadensis* Linn., a native of EN America and naturalized in Great Britain. This species is hairy, 90-180 cm tall with spread of 75-105 cm, vigorous and erect and unbranched or sparsely branched stems bearing simple mid-green, alternate, sharply serrated and lanceolate leaves. Heads are broad and plumose with pretty golden yellow flowers which are produced during autumn in large terminal panicles for several months. Its varieties by crossing with *S. virgaurea* and certain other species are 'Bellardii' (vivid golden-yellow) growing to 60 cm high, much branched and flowering during August-October; 'Cloth of Gold' (deep yellow) growing tall (up to 45 cm), vigorous and sturdy; 'Crown of Rays' (bright yellow) growing tall (60 cm) with panicles spreading; 'Goldenmosa' (bright yellow, stalks yellow) growing 75 cm tall, compact and bushy; 'Goldstrahl' (yellow) growing up to 40 cm tall with erect and stout stems, and is susceptible to mildew; 'Golden Thumb' (syn. Queenie, Tom Thumb, yellow, and foliage tinted-yellow) growing 30 cm tall and bushy; 'Golden Wings' (deep yellow) flowering in September up to 1.8 m high; 'Lemore' (lemon to primrose-yellow) growing some 75 cm tall with sturdy habit; 'Leraft' is an early flowering variety with large heads on 50-100cm tall stem; 'Lesale' (light yellow) growing to 75-90 cm with erect stem and excellent as cut flowers; 'Lesden' 90cm tall, compact growth habit and bright yellow flowers; 'Little Lemon' grows some 25 cm in height with free branching habit and is the lightest yellow-flowering compact goldenrod flowering for long duration; 'Minuta' grows some 20 cm tall and is most suitable for rock garden planting; 'Peter Pan' grows to 50-60cm tall, having large flower heads and bright yellow flowers; etc.; *S. capulinensis* is hypothesized to be most closely related to and perhaps derived from *S. wrightii* (Nesom and Lowery, 2011), the plants are densely leafy sub-shrubs growing more than one metre tall with numerous ascending-erect, woody stems arising from the base and producing a greatly extended thyrsoid capitulescence, and stipitate-glandular vestiture in the capitulescence and leaves toothed; *S. nemorallis* Ait. is slender, hairy, 48-54 cm tall, leaves thick and coarse, lower petioled, oblanceolate, crenate, upper becoming smaller, linear oblong, acutish and entire, flowers showy but not large, one-sided panicle, and the bracts of the involucres linear, oblong and obtuse; *S. odora* Ait., a native of E. USA is good for growing for dry sandy open conditions, growing some 45-60 cm in height with slender and simple stems and the plant is anise (*Pimpinella anisum*)-scented, leaves entire, acute or acuminate, lanceolate and 6.25- 10.0 cm long, small clustered flowers persisting for many weeks though not very attractive, and bracts involucre with acute tips; *S. rugosa* Mill., a native of USA and Canada, growing some 2.1 m tall with stout and erect main stem having strong and long branching, hairy, leaves lanceolate to ovate-lanceolate, more or less rugose and sharply serrated, head pyramidal panicle; *S. sempervirens* L; a native to E. USA and extending westwards to Texas in the coastal and swampy locations, stems stout, grows up to 2.4 m tall, leaves elongated ovate, not dentate, thick and fleshy and up to 25 cm long, inflorescence a large branched terminal panicle, and most suitable for salty sandy soils, especially the seaside gardens; *S. speciosa* is stout, usually simple-stemmed, smooth below, often rough above, leaves flabrous, firm, the basal 7-15 cm long and 1-3 cm wide, size diminishing above and pinnately veined, heads in a large showy terminal thyrsus, the branches of which are ascending and often leafy and bracts very blunt; *S. rugosa* Mell. is stout, erect and mostly stiff and 1.5-2.0 metres tall, leaves crowned above, lanceolate to ovate-lanceolate, sharply serrate, more or less rugose, heads in a broad pyramid, panicle, closely arranged on one side of the curving branches; *S. virgaurea* Linn., a best garden plant in the group is the European golden rod with simple, rough and stout stems, growing some 80-90 cm tall, basal leaves 10-20 cm long and 5-10 cm wide, upper leaves sessile or narrowed into margined petiole, flowers in dense terminals, rather narrow which are often 20-25cm long and bracts of the involucres acute or acutish; *etc.*

Propagation

Solidago can be propagated through seeds, rhizomes or suckers and cuttings. Disease-free plants can be produced through meristem culture. **Seed** propagation is not followed commercially due to great deal of variation. However, it can be propagated through seeds for developing new varieties. The seeds of *Solidago* are very small in size as 1,000 seeds weigh approximately only 0.261 g and the germination is proper only when these are stratified for 10 weeks at 5°C as has been reported in *Solidago petiolaris* (Biswas and Mondal, 2006). Germination is further aggravated due to high salinity levels in the soil or in the irrigation water. Germination of seeds are further influenced even by temperature and light as in the dark no seed has been found germinating even up to 14 days of sowing and the seeds of *Solidago chilensis* germinated to the maximum percentage (up to 20 per cent) at 20 °C temperature (Correia *et al.,* 1997). The behaviour of seed-propagated plants is almost different than those grown vegetatively or through rhizomes (Schmid and Weiner, 1993).

In goldenrod, commercial propagation is through **division** of vigorous growing and disease-free clumps (suckers) and through this method the aerial part of the plant is cut back immediately after flowering to promote shooting. When many shoots of the size of 3-4 cm bearing 5-6 leaves have appeared from the ground, the clumps (stools/suckers) are lifted between October and March, divided and planted in the properly prepared field. It is essential to divide clumps every three years and replant them properly. **Apical cuttings** with 12 buds placed in substrates containing soil is another method of propagation (Correia *et al.,* 1998) and the plants obtained through such cuttings appear quite uniform for flowering (Francine *et al.,* 1999).

Cultural Requirements

Solidago grows in mild climate as high rains, severe drought and excessively low (below 10 °C) and high persisting **temperatures** (above 40 °C) adversely affect solidago growing (Biswas and Mondal, 2006). The flower production is adversely affected by too much rain or severe drought conditions and under such conditions one is likely to loose few collections. The effect of **day length** also influences flower production in goldenrod though does not have high impact. Under short days, isopentenyladenine content was found decreased (Biswas and Mondal, 2006) which may hamper branching. Generally under long days (16-20 h), floral induction but with delayed flowering is promoted with more number of lateral branches and leaves though under short days (8-12 h) some species show rosetting but with rapid advancement from bud development to anthesis (Biswas and Mondal, 2006). The rate of flower bud development to anthesis is most rapid under short photoperiod (8-12 hours) and delays under long (16-20 hours) photoperiods.

Solidago may be grown in all types of **soils** ranging from light to heavy. Soil should be adequately porous and well-drained but together moisture-retentive. The main field should be thoroughly ploughed to a depth of more than 30 cm after mixing compost or dung manure at the rate of some 200 quintals per hgectare. At least three ploughings should be carried out and every time the weed-rootstocks should be collected from the field manually through hands or through forks. Third ploughing should make the soil completely pulverized with no stones, grits, wooden or plastic pieces left in the field. Now the beds should be prepared 1.0 to 1.6 metres wide and to any length depending upon the topography of the land but by leaving a clearance of 40-60 cm between the beds for cultural operations, *vis-à-vis* irrigation. Planting is done during the month of May to June preferably on raised beds or on ridges. Suckers are planted on ridges at 30 to 60 cm spacing or in flat beds. For first year crop 30 x 30 cm spacing is adopted while for ratooning, it is 30 x 60cm or 45 x 60 cm on raised beds of convenient length but the width is maintained at one metre if drip fertigation system is provided. A **fertilizer** mixture containing 100-150 kg nitrogen, 100-150 kg phosphorus and 50-100 kg potassium per hectare is recommended as basal dose in furrows (Biswas and Mondal, 2006). Nitrogen should be applied in 3 to 4 split doses as top dressing to ensure continuous flowering. Kim *et al.* (1997) stated that N and K efficiencies were significantly correlated with flower yield. They further stated that 180 kg N/ha was found most effective when its 40 per cent was given as basal and remaining 60 per cent as top dressing in *S. virgaurea* var. Asiatica. Rajput *et al.* (2010) recommended 250 kg N/ha for recording earliest flowering with maximum plant height, plant spread, number of suckers per plant, and length and number of panicles though vase life was recorded best only where no nitrogen was applied. **Watering** should be done immediately after planting of either suckers or leaf cuttings. Watering quantity and frequency is dependent on prevailing weather conditions, the soil types, the soil moisture and air humidity During winter months once in a fortnight but during the crop period it requires watering at the intervals of 7-10 days, depending upon the existing weather conditions. Initially, **weeding** is a must though afterwards when crop is fully grown, it does not permit growing of any weed as the crop itself covers its surrounding which does not permit any weed to come in direct contact of sunlight. However, initially weeds may also be controlled through tested herbicides as manual weeding is at all not cost effective.

Harvesting and Postharvest Management

Though the length of floral stalk differs from species to species and varieties to varieties but the species and the varieties which we grow in India normally grow taller. Normally, goldenrods start flowering three months after planting producing 60-120 cm of floral stalks throughout the year in tropical parts of the country, *viz.* tropical South, East and West India, and during period of weather transition in sub-tropical North and Central India, *i.e.* spring and autumn and during summers in the temperate regions so these can be removed when roughly one-fourth of flowers at the base of rachis have opened. One plant produces more than 25 floral stalks which should be removed and the cut ends should be placed in a bucket having distilled or soft water. One spike lasts for about one week. Ryagi *et al.* (1996) when compared the tap water with those of sucrose, $AgNO_3$, $Al_2(SO_4)_3$ and $AgNO_3$+ $Al_2(SO_4)_3$ each in various concentrations, recorded a vase life of 9.27 days in 0.4 per cent $Al_2(SO_4)_3$as compared to 5.47 days in tap water, while Patil *et al.* (1997) recorded 2 per cent sucrose+1 mM HQS increasing the vase life up to 12.75 days.

Vase life of cut 'Golden Rod' largely depends upon the postharvest treatments. Generally it remains fresh for 6-8 days after harvesting. Postharvest treatment with 2 per cent sucrose + 1mM HQS increases vase life up to 12.75 days. STS and BA delay leaf yellowing in cut spikes during vase life and STS also inhibits flower senescence; combining STS and BA is beneficial on both leaf and flower survival.

Insect-Pests and Diseases

Lace bug (*Corythucha marmorata*) feeds on the leaves arresting development of the plant. It also infests on aster, chrysanthemum and statice. **Orange tortrix** (*Argyrotaenia citrana*) is a buff-white and brown-headed caterpillar which spins, rolls and webs the leaves it feeds. **Herbivore insect** (*Trirhabda virgata*) feeds on the leaves causing decreased quality of leaves and its tissues, hampering nitrogen availability to the plants (Uriarte, 2000). **Leaf beetle** (*Trirhabda canadensis*) adults prefer feeding on younger leaves which are rapidly converted into simple biomass (Brown and Weis, 1995). Larvae of *Dichomeris leuconotella* spin and roll, fold and tie around the leaves and rapidly damage the leaves which causes plant stunting and hampers the flowering severely (Loeffler, 1996). Doeden and Teerink (1997) in western USA observed that *Procecidochares anthracina* is monophagus on *Solidago californica*. All these pests can be controlled by spraying Diazinon, Malathion or Sevin at 0.2 per cent. *Eurosta solidaginis* larvae contain high concentration of cytokinin, IAA, etc. (Mapes *et al.*, 1998) and form galls on *Solidago altissima* so are very difficult to control. Such branches should be cut back and burnt. **Aphids** (*Myzus persicae, Uroleucon tissoti* and *U. nigrituberculatum*) have been found feeding on golden rods (Kaakeh, 1991; Pilson and Rausher, 1995) which can be controlled through Permethrin at 0.2 per cent spraying.

Root rot is caused by *Rhizoctonia bataticola* so water-logging conditions in its planting should never be allowed, soil should be porous and it would be better if the soil before planting has been sterilized. Thiride drenching will also control this disease. **Scab** (*Elsinoe solidaginis*) infection occurs when plants are young and it causes stunting where leaves become covered with a fungal growth that causes them to wither and die as they unfurl, stems become rusty and finally the plants die. The infected leaves and even the plants should be removed and destroyed. **Powdery mildew** (*Erysiphe cichoracearum*) is reported to cause severe damage to *Solidago* (Sharma, 1989). A powdery off-white and discrete coating develops on the leaves, stems and aboveground parts including flowers which later on enlarges and covers the entire plant. Finally the leaves are deformed, turn yellow and drop prematurely. Spraying of Sulfex (0.2 per cent) or benomyl (0.1 per cent) proves effective. **Leaf spot** disease is caused by *Sclerotium rolfsii* which can be controlled through spraying with 0.2 per cent zineb or maneb. Solidago is also attacked by **rust** which is caused by *Coleosporium asterum*. Excessive humid conditions should be avoided and the field should be properly cleaned so that alternate hosts may not come up nearby. Dithiocarmate spraying will also control the rust.

References

Bader, G. 1999. Goldenrod: plant components, pharmacology, clinical use and cultivation. *Zeitschrift für Phytotherapie*, **20**(4):196-200.

Bader, G., Y. Kulhanek and H. Ziegler-Bohme, 1990. Antifungal action of glycosides of polygalacic acid. *Pharmazie*, **45**(8): 618-620.

Bader, G., V. Wray and K. Hiller, 1995. The main saponins from the aerial parts and the roots of *Solidago virgaurea* subsp. *virgaurea*. *Planta Medica*, **61**(2): 158-161.

Bailey, L.H. 1960. *The Standard Cyclopedia of Horticulture* (vol. III), pp. 3187-3189. The Macmillan Co., New York.

Beckett, K.A. 1985. *The Concise Encyclopedia of Garden Plants*, pp. 386-387. Orbis Publishing Ltd, London.

Biswas, B. and T. Mandal. 2006. Solidago. In: *Advances in Ornamental Horticulture* (vol. 2), pp.174-181. Pointer Publishers, Jaipur.

Brown, D.G. and A.E. Weis, 1995. Direct and indirect effects of prior grazing of golden rod upon the performance of a leaf beetle. *Ecology*, **76**(2): 426-436.

Correia, E., F.L.A. Camara and L.C. Ming, 1998. Vegetative propagation of Brazilian-Arnica (*Solidago chilensis*) by rhizome cuttings. *Revista Brasileria de Plantas Medicinais*, **1**(1): 23-27.

Correia, E., L.C. Ming, F.L.A. Camara, G. Gilberti, L. Craker, M. Lorenz, A. Mathe and A. Guilietti, 1997. Aspects of the sexual reproduction of the Brazilian arnica (*Solidago chilensis* Meyen, var. *megapotamica* (D.C.) Cabrere-Asteraceae). *Proc. Second World Cong. Medicinal Aromatic Plants for Human Welfare* (eds Gilberti, G., L. Craker, M. Lorenz and A. Mathe), WOCMAP-2. Agricultural Production, Postharvest Techniques and Biotechnology, Mendoza, Argentina, November 10-15, 1977. *Acta Hort.*, 1999, No. 502, pp. 89-91.

Doeden, R.D. and J.A. Teerink, 1997. Life history and description of immature stages of *Procecidochares anthracina* (Doane) (Diptera: Tephritidae) on *Solidago californica* Nuttall in southern California. *Proc. Entom. Soc. Washington*, **99**(1): 180-193.

Francine, L. C., K. Minami and O. Abrahao, 1999. Apical cutting increases the uniformity of harvesting of 'Tango' flowers'. *Scientia Agricola*, **56**(4): 1009-1012.

Hay, R. and K.A. Beckett (eds), 1971. *Reader's Digest Encyclopaedia of Garden Plants and Flowers*, pp. 571-672. The Reader's Digest Association Limited, London.

Kaakeh, W. and H.W. Hogmire, 1991. Biology and control of green peach aphid, *Myzus persicae* (Sulzer) on peach in West Virginia, USA. *Arab J. Pl. Protection*, **9**(2): 124-128.

Kim-Changbae, Lee-Hyunsook, Kim-Changkil, Choi-Kyung Bai, Choi-Boosull, C.B. Kim, H.S. Lee, H.S. Kim and B.S. Choi, 1997. Effect of nitrogen split application methods on the growth and yield of *Solidago virgaurea* var. Asiatica Nakai. *RDA J. Agr. environ. Sci.*, **39**(2): 30-34.

Kruedener, S. von, W. Schneider, E.F. Elstner and S. Von-Kruedener, 1995. A combination of *Populus tremula, Solidago virgaurea* and *Fraxinus excelsior* as an anti-inflammatory and anti-rheumatic drug. *Arzneimittel Forschung*, **45**(1-2): 169-171.

Leuschner, J. 1995. Anti-inflammatory, spasmolytic and diuretic effects of a commercially available *Solidago gigantea* herb extract. *Arzneimittel Forschung*, **45**(1-2): 165-168.

Loeffler, C.C. 1996. Adaptive trade-offs of leaf folding in *Dichomeris* caterpillars on golden rods. *Econ. Entom.*, **21**(1): 34-40.

Mapes, C.C., P.J. Davies, G. Ksoka (ed.), W.J. Mattson (ed.), G.N. Stone (ed.) and P.W. Price, 1998. Plant hormones and gall formations by *Eustoma solidaginis* on *Solidago altissima*. *The Biology of Gall-Inducing Anthopods. Proc. of an International Conference*, Matrafured,

Hungary, August 14-19, 1997. General Tech Rep., North Central Res. Station, USDA Forest Service, 1998, No. NC-199, pp. 161-172.

Meyer, B., W. Schneider and E.F. Elstner, 1995. Anti-oxidative properties of alcoholic extracts of *Fraxinus excelsior, Populus tremila* and *Solidago virgaurea*. *Arzneimittel Forschung*, **45**(1-2): 174-176.

Nesom, G.L. and T.K. Lowery. 2011. *Solidago capulinensis* (Asteraceae: Astereae) redivivus. *Phytoneugton*, **24**: 1-22.

Patil, S.R., B.S. Reddy, B.S. Kulkarni and P.T.K. Reddi, 1997. Effect of 8-hydroxyquinoline sulphate on postharvest physiology of golden rod (*Solidago canadensis*). *Karnataka J. Agric. Sci*, **10**(1): 107-111.

Pilson, D. and M.D. Rausher, 1995. Clumped distribution patterns in golden rod aphids : genetic and ecological mechanisms. *Econ. Entom.*, **20**(1): 75-83.

Pizzetti, I. and H. Cocker, 1975. *Flowers-A Guide for Your Garden* (vol. II), pp. 1235-1237. Harry N. Abrams, Inc., Publishers, New York.

Rajput, S.T., Giriraj Jat, A.V. Barad, A.K. Tank and H.L. Kacha, 2010. Effect of different nitrogen levels on golden rod (*Solidago canadensis* L.) during summer season. *Haryana J. hort. Sci.*, **39**(3-4): 271-272.

Ryagi, Y.H., U.G. Nalawadi, M.B. Chetty and P.A. Sarangamath, 1996. Effect of different chemical preservatives on vase life of golden rod. *Karnataka J. agric. Sci.*, **9**(1): 177-178.

Schmid, B. and J. Weiner, 1993. Plastic relationships between reproductive and vegetative mass in *Solidago altissima*. *Evolution*, **47**(1): 61-74.

Sharma, S.P. 1989. Solidago. In: *Commercial Flowers* (eds Bose, T.K. and L.P. Yadav), pp. 861. Naya Prokash, Calcutta.

Uriarte, M. 2000. Interation between goldenrod (*Solidago altissima* L.) and its insect herbivore (*Trirhabda virgata*) over the course of succession. *Oecologia*, **122**(4): 521-528.

Weyerstahl, P., H. Marschall, C. Christiansen, D. Kalemba and J. Gora, 1993. Constituents of the essential oil of *Solidago canadensis* from Poland - a correction. *Planta Medica*, **59**(3): 281-282.

35

Strelitzia (Family : Strelitziaceae, formerly Musaceae)

Naveen Kumar, R.L. Misra, J.K. Ranjan, R.K. Dubey, Sanyat Misra, Pragya Ranjan and Ajit Kumar

[(**Common names:** Bird of paradise flower/Crane flower (*Strelizia reginae*), Great white Strelitzia (*S. alba*), Rush-like Strelitzia (*S. parvifolia* var. *juncea*), Small-leaved Strelitzia (*S. parvifolia*), Wild banana/Wilde pisang (*S. caudata*)]

Introduction

Strelitzia was named in honour of Queen Charlotte Sophia (1744-1818), a daughter of the Duke of Mecklenburg-Strelitz in north Germany, wife of King George III of England and mother of 15 children, and Sophia was also great patron of botany. It is a native of sub-tropical to southern Africa, especially South Africa. The plants of this genus are herbaceous evergreen perennials, prized for their unusually curiously-looking brilliantly coloured asymmetrical flowers that are most exquisite. For any garden it enjoys a place of pride and is an important choice for landscaping. Though in the garden these exotic flowers look quite spectacular and special, bearing a close resemblance to a crested bird's head, apart from planting in the background of a herbaceous border or in front of a shrubbery or along the side of a tank or lily pool, these are also most suitable as cut flowers lasting for many days. Certain small types are also most suitable for pot-planting or the suitable size of the pot may be used for their pot culturing. Due to their unusually attractive flower shape and colour, these are grown in many parts of the world in order to produce cut flowers for both domestic and export markets. For cut bloom production these are being grown commercially in California, Hawaii, Israel, South Africa and many of the large glasshouses in South America. Being emblem of Los Angeles city, it has special significance there for its extensive planting on the streets.

Botany

Strelitzia is the perennial evergreen herb (1.2 to 2.0 metres) or small trees (6.0 to 10.0 metres) of palm-like form, either virtually stemless and clump-forming or bearing multiple, stout and simple stems, all coming up from the rhizomes which are sometimes subterranean and sometimes an woody erect stem; having undivided, long and leathery leaves in two ranks and resembling *Musa* but more rigid, oblong to lanceolate, broad and entire; petioles long and usually grooved, bases partially sheathed or strongly overlapping in taller species; inflorescence axillary, long or short stalked, normally horizontal, simple or compound; flowers appear sequentially from waxy rigid and boat-like spathe; flowers bilaterally symmetrical, full of nectars, white or coloured, calyx with three narrow segments, and three petals, two petals being paired to form a projecting arrow- or tongue-shaped structure resembling a large anther, but in fact enclosing the 5 threat-like anthers, united petals enclose 5 stamens and style with shorter upper part, and the third much reduced petal forms a small nectar-producing pouch at its base; anthers maturing before the stigma, stigma swollen and pollination through sugarbirds or sunbirds (*Nectarinia afra*), ovary 3-celled and leathery after fruit formation; capsular fruits are brown, many seeded, and seeds black and each has an orange matting of aril or oil-body for enticing the birds. Seeds matted with aril also looks very beautiful and become an item for floral display.

In *S. reginae*, the embryo is an elongated, club-shaped organ, the bulk of which consists of a massive cotyledon, and the apex of the cotyledon corresponding to the leaf blade remains in the seed during the process of germination and functions as a haustorium. Germination being hypogeal, elongation growth of the basal portion of the cotyledon carries the plant axis into the soil. A short cotyledonary petiole separates the plant axis from the haustorium. Proliferation of the cells around the cotyledonary clefts results in the formation of a ligule while extensive elongation of the basal portion of the cotyledonary sheath is absent. The first leaf to emerge from the ligule is a tubular scale leaf while the second to appear is the fisrt laminate leaf.

Species and Varieties

The genus *Strelitzia* consists of five species ~ *S. reginae* (the popular 'bird of paradise'), *S. parviflora*, *S. alba*, *S. nicolai* (named so in commemoration of the Emperor Nicolas of Russia) and *S. caudata*, all native to South Africa.

S. reginae grows naturally along the river banks and in clearings among scrub in coastal areas of Cape Province where it reaches the height of 1.2-2.0 metres, though it is stemless, and the leaves are 25-70 × 10-30 cm, glaucescent, oblong to lanceolate, acute to rounded, sometimes marginate, base rounded or tapering, often wrinkled, petiole oval and about 1 m, inflorescence a long sheathed peduncle equalling leaves, the condensed spike of flowers is enclosed in a stiff horizontal green flushed purple and orange sheath or spathe, about 12 cm, glaucous, from where about 10 cm long flowers appear in slow succession one by one at about one week's intervals, the flowers having 3 orange-coloured sepals and 3 blue petals, 2 being united into an arrowhead-like structure resembling a large blue anther though 5 thread-like anthers are enclosed therein, and the third quite small petal forms a small nector-producing pouch at its base (Fisher, 1976). Its popular varieties are 'Farinosa' (leaves oblong with long petiole and truncate base), 'Glauca' (leaves large, glaucous, oblong-lanceolate and attractive), 'Humilis' (Pygmaea, dwarf and compact, closely clump-forming, short petioled, leaves ovate-oblong, and flowers disproportionally large with shart stalk), 'Ovata' (leaves ovate with cordate base, petiole short) and 'Rutilans' (leaves attractive, midrib red or purple, and flowers very brightly coloured). *S. reginae juncea* (Ker Gawl) H.E. Moore has robust, rush-like leaf stalks with/without terminal blades, *S.r. intermedia* is half-way between the species and var. *juncea* but with much reduced leaf blades, and *S. r. humilis* is quite dwarf than the species and *S, r.* var. *citrina* is also admired everywhere in the world.

S. parvifolia is similar to *S. reginae*, producing equally brilliant orange and blue flowers but has much narrower (linear-lanceolate) leaves and is much smaller in stature. Stalks tall and stiff. Probably it is not in cultivation. Its var. *juncea* has bladeless leaves or no leaves and forms a very spiky tuft of stems which grow about 1.2 m in height, spathe green edged with magenta, sepals orange and petals deep blue tipped white.

S. alba (*S. augusta*) is very similar to *S. nicolai* and while growing one confuses it with *Ravenala madagascariensis* as both have large fan-like leaves, found in Cape Province, growing up to 10 m, petiole up to 180 cm long, the leaves at the summit of the stem, 60-90 cm long, oblong and acute, peduncle short and appearing from the leaf axil, spathe deep-purple, 25-30 cm in length, bearing short and purple pedicels with pure white flowers arranged in solitary boat-shaped bracts and where petals are of 8.75 cm length, and round at the base.

S. nicolai is found in S. Natal and NE Cape Province, may grow as tall as *S. alba*, with palm-like, erect, clustered and woody stems up to 10 m in height, which at maturity usually branches only at ground level making a massive clump, leaves oblong-ovate, rounded or cordate at base, 90-120 cm or even more in length on as long stalks (petioles) which are glossy-green, inflorescence short-stalked and partly concealed in axils, spathes 3-5, pruinose, chestnut-red and 40-45 cm long, floral bracts boat-shaped, and several forming on a stout peduncle, and its 20 cm long flowers have white sepals, and mauve (pale-blue to light mauve or sometimes even white) and white petals, and the tongue boat-shaped (Fisher, 1976).

S. caudata is found in Transvaal and Swaziland and there locally it is called wilde pisang (wild banana), it is clumped and its usually solitary stem also becomes woody and growing up to more than 6.0 metres in height, leaves 1.75 × 0.8 m, ovate to oblong, cordate at base and petioles equalling the blade, inflorescence bracteate, spathe solitary and dark purple-red, slightly glaucous and 30 cm in length, flowers 20 cm long, calyx white, corolla also white or blue-purple at base and forming a boat-shaped tongue, basal lobes obtuse and margins undulated.

S.* × *kewensis (*S. alba* × *S. reginae*) is of garden origin growing to some 1.5 m tall trunk or quite absent, leaves like to that of *S. alba* but blades 60 × 45 cm, spathe solitary and long-stalked, flowers vertical, pale-yellow, corolla blue, lilac-pink at base and resembling mostly to *S. reginae* but the small hooded petal like to that of *S. alba* and with lilac-pink patches at the base of the sepals.

The study conducted at Hirehally-Tumkur (southern dry zone of Karnataka) on possibility of growing bird of paradise, 44 local collections were made and evaluated (Anuradha Sane *et al.*, 2011). They obtained it a promising crop of the region flowering within two years of planting, flowering span being throughout the year though peak being during December-March.

Propagation

Bird of paradise is propagated through **seeds** but such plants require more time to attain the flowering stage. However, since it is highly cross-pollinated crop as also it is protandrous, so seed propagation should be done only to evolve new varieties. Skead (1975) stated that weaver birds (*Ploceus cuculatus*) are good pollinators in *S. reginae* as these visit the flowers in search of nectar which has been found rich in glucose, fructose and sucrose (Kronestedt and Robards, 1987; Kronestedt *et al.*, 1989). To obtain maximum seeds one should not depend only on the bird pollinators but should resort to even hand pollination to ensure good number of capsule set as one capsule may contain approximately 50 seeds. Usually in spring the seeds are sown at around 20-25 °C atmospheric temperature in seed pans containing sandy media, kept very damp and screened where these may germinate within 50 days and when the seedlings attain almost 7.5 cm height, these are transplanted in a porous medium containing good garden soil, well rotten compost and fresh sand where the seedlings should be allowed to mature enough to be transplanted at a permanent place. Seed to flower it usually takes 5-6 years, however, with a better management practices these may also flower in three years. Seeds have dormancy which was got released when Besemer (1976) treated the seeds of *S. reginae* with hot water for 30 minutes at 61 °C and obtained 59 per cent germination though abraded seeds did not germinate. Venter (1974, 1976) reported that seeds of *S. reginae* and *S. r. juncea* germinated well, *i.e.* 62.5 per cent in former case and 25 per cent in latter case, following treatment with H_2SO_4 for 5 minutes. El-Kholy and Hasssan (1983) tried X-ray irradiation at 500 R and 1000 R and found advanced germination by 12 days in both the treatments, however, in control germination was found in 48 days. However, with growth substances they recorded 100 per cent germination in case of *S. r. juncea* by increasing the O_2 concentration up to 100 per cent in the incubation chamber. When they treated the seeds with Tryptophan at 5 ppm, the germination was 82.6 to 89.3 per cent as compared to 15.3 to 20.6 per cent in control. Meillon *et al.* (1990a) studied the effect of oxygen on the seeds of *S. reginae* on their alcohol dehydrogenase activity, cytoplasmic malate dehydrogenase activity and ethanol content.

Simplest means for increasing stock in 'bird of paradise' plants is through **suckers** or by **division** of the old crowns. Offsets normally take three years to attain the flowering stage. Rooted suckers (offsets) are separated from the parent plant in the spring and planted in a high fertility medium for quicker results. Some 10-25 suckers are produced in a year by one plant though it may vary with the mother plant species and varieties planted and its previous growth, the age of the plant and the medium in which the mother plant has been grown. *S. reginae* is also propagated through division of naturally developed rooted branches of the apical dome known as 'fans' but it has its own limitation as in this case the multiplication rate is very poor, only 1-2 divisions per branch each year. However, there is no branching from axillary buds. Pol *et al.* (1988) studied for one year the fans by shortening their roots to 20 cm and planting them into medium. The apex was reached by a triangular excision of a part of the plant with a transversal cut 8-12 mm above the basal plate throughout the basal leaf sheaths keeping the leaves in contact with the roots. The apex was removed and after 2-6 months lateral sprouts developed varying in number from 2-30 per fan depending on age and size of the fan. To obtain shoots with roots most of the old roots were removed and during the next six months the newly formed laterals rooted and could be divided into individual plants and these plants could be treated again after one year and so on.

Due to oxidative browning of the wounded explant, the micropropagation was almost impossible in *S. reginae*, but Ziv and Halevy (1983) succeeded in breaking this barrier by treating the terminal and axillary buds with antioxidant solution by total submergence for 24 h and then culturing these on an agar medium with charcoal or on paper bridges in liquid medium with dithiothreitol where growth initiated with further shoot proliferation. Bettaieb and Tissaoui (1994) did not obtain good success as they used low level of IBA and BAP for shoot establishment of bird of paradise but in a further study, Bettaieb and Tissaoui (1997) when cultured the shoot tips on modified MS medium with addition of charcoal at the initiation stage in the medium, it proved very effective while at proliferation stage when pH was adjusted to 4.5 and the media were supplemented with IBA and BA at 0.25 and 2 mg/l, respectively, best proliferation was obtained. To eliminate the apical dominance, *vis-à-vis* promoting growth, several incisions were made on the apices. Application of 1 per cent activated charcoal completely checked the oxidative browning, increased micropropagated seedling height and initiation of leaves (Hosni, 2001). Shreevatsa *et al.* (2004) used terminal rhizome buds where explant browning was controlled by dipping the explants in ascorbic acid at 150 mg/l, and after scrapping the dead tissues the explants were transferred at 3 days intervals to the MS medium containing 0.5 mg/l IBA and 7 mg/l BAP in addition to 200 mg/l ascorbic acid + 30 g/l sucrose, and *in vitro* the multiple shoots were rooted on medium containing 2.5 mg/l IBA.

Cultural Practices

S. reginae grows and flowers well in the garden under mild climatic conditions either in full sun or under filtered conditions. In cold regions, it requires to be grown either in pots or in border in warm greenhouses especially during winter season, and in the open under sub-tropical conditions again these require to be protected especially during May-June when weather is very hot and relative humidity is very poor, however, tropical conditions are most suitable for its growing whether it is summer or winter. These are most suitable for growing throughout Kerala, eastern parts of the country, in Karnataka, southern part of Andhra Pradesh and many parts of Tamil Nadu. Wherever it is to be grown, it requires good soil, abundant supply of water, considerable sunlight and frost-free areas where temperature does not go below 10-13 °C ever. A day termperature of 20-22 °C has been suggested most suitable for its successful cultivation (Auman, 1980). Besemer *et al.* (1981) when evaluated 20 clones of bird of paradise, recorded peak in flower production from October to December and again in spring, and obtained an average of 51 per cent increased flower production from one year to the next. However, apart from direct relation of temperature to its flowering, number of leaves also relate to the flowering behaviour as Haouala and Zouari (2001) found 51-55 leaves contributing to good flower production in bird of paradise.

Strelizias can be grown in a variety of soils whether it is coastal soil or dry inland soil.Soil should be well-drained, loamy and rich in organic matter. Sandy loam soil is also suitable for its growing if that has been improved through addition of good quantity of well decomposed organic manure. Flower production is seasonal in most of the *Strelitzia* growing areas. At its native haunts in South Africa, it blooms in the fall-winter-spring period, in Hawaii at high elevations from spring to fall and at low elevations from June to September (Criley and Kawabata, 1984; Kawabata *et al.*, 1984), in California from September to May with a peak during October to December (Besemer *et al.*, 1982) and in Israel with a double peak, once in March-April and the other in September (Halevy *et al.*, 1976). Though flowering in the bird of paradise is seasonal but not the leaf emergence. The flower buds in the leaf axils should develop at anthesis which is approximately 250 days after leaf emergence. Kawabata *et al.* (1984) observed in Hawaii that flower yield increases with age of the plant but due to annual waves as in December-March the yield is low though peak occurs during July to September. Halevy *et al.* (1987) have given in Table 35.1 the fraction that each month represents of the total yearly harvest.

This Table shows that flowering as well as production were high in Hawaii during summer months, in South Africa and California during fall-winter months, and in Australia during late winter-spring.

Though **light** is not very crucial in its flowering in certain cases such as two strains brought in California from San Diego and Santa Barbara but Halevy *et al.* (1976) demonstrated that low light intensity reduces flowering, probably due to effect on assimilate supply and distribution. **Temperature** has been recorded playing major role on growth and flowering of *Strelitzia* for which Kawabata *et al.* (1984) reported that flower buds fail to open due to high temperature and worked out that a critical air temperature of 27 °C is the potential cause. Low temperatures may retard the development but when congenial temperatures appear, a surge of flowering can result, and in case of California and South Africa the period of warm temperature may be too brief for over all development, hence, the final phases of flowering may occur during or after another period of low temperature. Halevy *et al.* (1987) reported that low temperatures were associated with slow development and suggested that a temperature range of 17-27 °C would provide optimum environmental conditions for rapid and uniform flowering.

For **planting** of *Strelitzia*, the pits of 90 × 90 cm should be dug out and then filled with organic soil, peat and coarse sand, all in equal ratio or good garden soil 1 part + well rotten farmyard manure 2 parts + peaty soil or leaf mould 2 parts + coarse sand 2 parts all by volume. It would be better to fill the pits with water to settle the mixture, and if need be even afterwards again a little filling may be done. The planting is carried out from late spring to early summer. At the time of planting, in the centre the plants should be settled in a manner that only that portion should be in the mixture of the pit which had been at the place from where these have been taken out. The plant density in field should be maintained two plants per square metre (Krogt, 1981). Nair *et al.* (2011) studied 6-months old tissue-cultured *Strelitzia* plants for five levels of spacing, *viz.* 2 m × 2 m, 2 m × 1 m, 1.5 × 1.5 m, 1.5 × 1 m and 1 m × 1 m and recorded that the spacing of 1.5 × 1 m (6667 plants/ha) was optimum for the first, second and third year of planting. Watering is must after planting. For pot planting, the earthen pots of 20-25 cm size should be selected and filled with potting mixture containing two parts sandy-loam soil and one part of leaf mould, and one part well decayed farmyard manure, and charcoal. After potting, watering should be given lightly but daily to ensure proper establishment. During active growth it requires ample watering as roots absorb a great deal of water from the soil. The plants being the gross feeders, should be top-dressed with rich organic matter in the intervening years. Pot plants are supplied with a dilute liquid fertilizer fortnightly, particularly when the plants are making new leaf growth. If planted in the greenhouses, and when temperature reaches above 20 °C, the ventilators should be opened. The glass should be partially shaded during summer but all possible light should be allowed to admit from September end to early April. It would be better if these are repotted every two years. The weed if any is emerging that should be taken out so that it may not rob the plants of its nutrients.

For **intercropping** only those plants should be grown which do not deprive the main crop of its due light~lilies, amaryllis direct in between, while *Cyclamen* and *Hedera* in pots, however, foliage plants should not be grown as these do not permit ample sunlight to the main crop and also rob the plants of their nutrients (Armbruster, 1973).

Feeding and Irrigation

It requires plenty of **water** during active growth in summer and also during flowering. During summer the beds should be drenched thoroughly. During off season only the soil should be kept moist. However, after planting and until fully established these require daily watering to maintain sufficient humidity around the plant.

At the time of pit or pot filling, the mixture should contain sufficient well decomposed organic matter. After the plants have established, these can be fed continuously for three months with liquid fertilizer at 10 day's intervals with 6 g superphosphate + 3 g potassium nitrate in one litre of water, and then again after a gap of one month (Dakshini, 1971). El-Shoura and Hosni (1996) with *S. reginae* in a potting mixture containing 3:1, 1:1 or 1:3 sand:clay or only

Table 35.1: Seasonality of Flower Production and per cent Total Yearly Harvest

Site	*Low Months*	*Percentage*	*High Months*	*Percentage*
Australia	April-May	18.7	August-October	41.5
South Africa	September-January	27.3	April-July	49.5
California	June-August	9.7	October (high)	16.0
			November-May	67.0
Hawaii	December-April	21.2	June-October	63.7

clay, when applied N, P and K at 0.49:0.19:0.33, 0.99:0.38:0.66 or 1.48:0.57:0.99 g/pot and observed best results under the growing medium containing sand:clay at 1:3 or 1:1 and fertilized with 0.99:0.38:0.66 g NPK/pot. Ali (1998) applied Singral (NPK, 16:9:2) and Nitrophoska (NPK, 18:18:5) in a sprinkler irrigation system and at 0, 1, 2 and 3 per cent through drip system to the greenhouse *S. reginae* plants and with 3 per cent Singral recorded highest inflorescence length (104.75 cm), number (3.17) and weight (93.6 g) in the second year. Vasantha Kumar *et al.* (2011) obtained more number of flowers in two year old *Strerlitzia* plants by applying 50 g N/m^2, 20 g P_2O_5/m^2 or 50 g K_2O/m^2 individually, and when applied these in combined form the interactive flower yield was almost 33 per cent more. Application of sulphur at 100 g/m^2 and NPK (18:18:5) 1 per cent through trickle irrigation, Ali (1998a) recorded better growth and flowering though S at 50 g/m^2 registered highest concentration of leaf N, and NPK at 0.5 per cent registered highest K, Ca, Mg and P concentrations. In another study, when Ali (1998b) applied 3 per cent Singral (16:9:2 NPK) through drip irrigation, he recorded significantly increased growth and yield characters with leaves having higher concentration of K and Fe and where 2 per cent Singral was applied higher concentrations of P, Ca, Mg and Mn were obtained.

Postharvest

Usually the flowers are severed from the plants early in the morning when the first flowers from the bracts have emerged out and the cut ends are immediately placed in fresh tap water. Dipping of stem bases of *S. reginae* in STS for 10 minutes for checking ethylene production and action, *vis-à-vis* checking bacterial growth in vascular tissuaes, and then pulsing with 25 ppm of GA_3 overnight for delaying senescence process and increase of petal size and then further keeping the flowers in a solution containing 10 per cent sucrose (as carbohydrate source necessary for growth and development) + 200 mg HPQS (for antimicrobial effect) + 150 mg citric acid (for enhancing the effectiveness of 8-HQPS) per litre at a pH of 3.6 causes 100 per cent floret opening with 38.5 days of vase life compared to 15.5 days in control where floret opening is only 40 per cent (El-Saka *et al.*, 1994). El-Mokadem *et al.* (1994) reported that either 20 per cent sucrose or 25-50 ppm GA_3 significantly increased the vase life of *S. reginae* flowers. Campanha *et al.* (1997) studied the *Strterlitzia reginae* inflorescences cut at one flower open stage and observed floret opening from 1.6 in control to 2.7 in the treatment with the respective vase life of 7.6 days and 11.3 days, the treatment being distilled water and cutting of 2 cm of the stem every 2 days. Halevy *et al.* (1978) for storage of *Strelitzia* cut flowers reported that flowers should be cut 4-5 days earlier to colour showing when inflorescence angle to the stem is 50-60°, and for protection against *Botrytis* the flowers are required to be conditioned for 40-50 h at 22 °C in 250 mg/l 8-HQC, 100 g/l sucrose and 150 mg/l citric acid, then the flowers should be wrapped in paper and packed in boxes lined with polythene foil, and thus these may be stored up to one month at 7-8 °C, however, before storage the blooms require to be dipped in a fungicide solution containing benomyl or 2-(4-thiasolyl)-benzimidazide at 200 ppm to sustain long term shipment. In transportation the temperature should be maintained at 7-8 °C under dry state of flowers. Auman (1980) recommended that such flowers may be successfully stored at 10 °C, however, the best storage temperature was recorded at 8 °C. Finger *et al.* (1999) recommended 40 per cent sucrose pulsing for 24 h causing more number of flower opening and more vase life. Finger and Barbosa (2006) stated that the inflorescences of tropical flowers such as *Heliconia* and *Strelitzia* should not be stored below 13 °C temperature due to intense tissue discolouration.

Insect-Pests and Diseases

The **insect-pests** infesting bird of paradise are mealy bugs (*Planococcus citri, Pseudococcus adomidum*), scales (*Aspidiotus carnolliae*), caterpillars (*Opogona subcervinella*), thrips (*Chaetanaphothrips signipennis*, reported by Denmark and Osborne, 1985), crawlers (*Pseudaulacaspis cockerelli*), the first two being more serious, especially under greenhouse grown crops. Pirone *et al.* (1960) recommended 0.2 per cent Malathion spraying to control both the pests, especially the young crawlers. Malathion spraying will also control the caterpillars of *Opogona subcervinella*. However, fumigation of the crop with methyl bromide or HCN or HWT at 49 °C for 6 minutes is very effective. *C. signipennis* causes leaf rolling and other foliar malformation which can be controlled through Vydate (oxamyl) in the greenhouse. The adults, nymphs and crawlers of *Pseudaulacaspis cockerelli* have been controlled successfully by exposing the flowers to 49 °C hot water for 5.0, 5.5 and 6.0 minutes, respectively (Hara *et al.*, 1993).

Diseases that infect bird of paradise are bud rot, floral rot and root rot caused by certain fungi while there are certain bacterial diseases which affect *Strelitzia* plants. Bud rot (*Fusarium* sp.) and the floral rot (*Botrytis cinerea, Fusarium culmorum* and *F. avenaceum*) can be controlled through spraying with Bavistin 0.1 per cent alternate with Captaf 0.2 per cent, one after the other at fortnightly sprayings from bud forming stage to completion of flowering, however, the floral rot caused by *Botrytis cinerea* can be controlled through 0.2 per cent weekly sprayings when the weather is chilly, windy and humid. Lorenzini *et al.* (1983) reported *Fusarium* sp. infecting inflorescences of *Strelitzia reginae*. Orlikowski (1976) reported that *F. oxysporum, F. moniliforme, F. culmorum* and *F. solani* cause root necrosis. Root rot (*F. moniliforme*) is a seed-borne disease that can be controlled by soaking the seeds and seedlings in water first at room temperature and then only seeds at 57.2 °C for 30 minutes (Pirone *et al.*, 1960). Root rot caused by *Armillaria mellea* can be controlled through soil fumigation with methyl bromide and heat sterilization. From the diseased sample collected from Bangalore and Chandigarh, a leaf spot disease caused by the fungus *Colletotrichum gloeosporides* which is symptomised as scattered, marginal rosy-buff to ochreous spots with wavy margins surrounded by dark brown haloes has been recorded, for which spraying of 0.1 per cent Bavistin has been recommended. Bacterial leaf spot and blight (*Xanthomonas campestris*) has been reported by Chase and Jones (1987) in Florida at the temperature range of 21-27 °C, optimum being 24 °C, where it was observed forming angular, yellow to reddish-brown lesions on leaves and petioles, lesions being frequently bordered by leaf veins and after coalescing these were giving blighted appearance of the plants. Bacterial wilt (*Pseudomonas solanacearum*) was though reported serious in Japan but its another race has also been observed in Australia and Hawaii with low incidence (Anon., 1988). El-Goorani and abo-El-Dahab (1973) reported about the root rot of *S. reginae* caused by *Erwinia carotovora*. Bacterial soft rot (*Pseudomonas syringae* pv. *delphiniii* and *Erwinia carotovora* ssp. *carotovora*) were reported by Abd-El-Khair and Nafal (2001) from Egypt causing necrosis and soft rot on flower tissues, *Pseudomanas* being more pathogenic than *Erwinia*. For their control, they recommended application of chloramphenicol and tetracycline, either at 100 ppm.

References

Abd-El-Khair, H. and M.A. Nafal, 2001. Flowers bacterial soft rot of Bird of Paradise (*Strelitzia reginae* Banks) in Egypt and its control. *Arab. Univ. J. agric. Sci.*, **9**(1): 397-410.

Ali, M.S.S. 1998a. The response of bird of paradise plant to different rates of sulphur and nitrophoska (NPK 18:18:5). *Alexandria J. agric. Res.*, **43**(3): 81-92.

Ali, M.S.S. 1998b. Effect of sigral and nitrophoska fertilizers on growth, flowering and mineral composition of bird of paradise (*Strelitzia reginae*) plants. *Indian J. Hort.*, **55**(3): 257-262.

Anonymous, 1988. Records of bacterial wilt on bird of paradise (*Strelitzia reginae*). *Bacterial Wilt Newsletter, ACIAR*, No.3, p. 5.

Anuradha Sane, T. Janakiram, A.K. Nair and M.A. Suryanarayana, 2011. Evaluation of bird of paradise (*Strelitzia reginae*) for southern dry zone of Karnataka. *J. Ornam. Hort.*, **14**(3-4): 59-64.

Armbruster, J. 1973. Advantages and disadvantages of intercrops for *Strelitzia reginae*. *Gartenwelt*, **73**(8): 168-169.

Auman, C.W. 1980. *Introduction to Floriculture* (ed. Larson, Roy A.). Academic Press, New York.

Battaieb, T. and T. Tissaoui, 1977. *In vitro* propagation of *Strelizia reginae*. *Annales de l'Institut National de la Recherche Agronomique de Tunisie*, **67**(1-2): 125-131.

Battaieb, T. and T. Tissaoui, 1994. *In vitro* propagation of *Strelitzia reginae* Ait. *Annales de Institut Nationale de la Recherche Agronomique de Tunisia*, **67**(1-2): 125-131.

Besemer, S.T. 1976. Germination methods for bird of paradise seed. *Flower and Nursery Rpt.* (May-June), No. 2.

Besemer, S.T., R.B. McNaim and A.H. Halevy, 1981. Studies on bird of paradise clonal selections. Part II. Flower production. *Flower and Nursery Report* (winter), pp. 1-3.

Besemer, S.T., R.B. McNairn and A.H. Halevy, 1982. Bird of paradise clonal selections. Part II. Flower production. *Flor. Rev.*, **170**(4402): 16-18.

Campanha, M.M., F.L. Finger, P.R. Cecon and J.G. Barbosa, 1997. Water relationship of cut bird of paradise. *Revista Brasiliera de Horticultura Ornamentale*, **3**(1): 27-31.

Chase, A.R. and Jones, J.B. 1987. Leaf spot and blight of *Strelitzia reginae* (bird of paradise) caused by *Xanathomonas campestris*. *Pl. Dis.*, **71**(9): 845-847.

Criley, R.A. and O. Kawabata, 1984. Development of the flower spike of bird of paradise and its flowering period in Hawaii. *J. Amer. Soc. hort. Sci.*, **109**: 702-706.

Dakshini, M.N. 1971. Forcing the 'bird of paradise'. *Indian Hort.*, **16**(3): 15, 19.

Denmark, H.A. and L.S. Osborne, 1985. *Entomology Circulation*. Division of Plant Industry, Florida. Deptt. of Agric. and Consumer Dervices, No. 274, p.2.

El-Goorani, M.A. and M.K. Abo-El-Dahab, 1973. Root rot of *Strelitzia reginae* caused by *Erwinia carotovora*. *Egyptian J. Phytopath.*, **4**: 65-70.

El-Kholy, S.A. and E.A. Hassan, 1983. The effect of irradiation and growth substances on seed germination of *Strelitzia reginae*. *Minubiya J. agric. Res.*, **6**(27):267-279.

El-Mokadem, H., M. Khatlab, M.R. Hassan and A.M.Tarabeih, 1994. Effect of some chemicals on the keeping quality of *Strelitzia reginae* cut flowers. *Alexandria J. agric. Res.*, **39**(3): 231-241.

El-Saka, M., A.E. Awad, B. Fahmij, A.K. Dowh, A. Ait-Dubahou (ed.) and M. El-Otmani, 1994. Trials to improve the quality of *Strelitzia reginae* flowers after cutting, *Proc. intern. symp* held at Agadir (Morocco), pp. 480-488.

El-Shoura, H.A.S. and A.M. Hosni, 1996. Growing *Strelitzia reginae* in improved sandy growing media under different levels of fertilization. *Annals agric. Sci., Cairo*, **54**(7): 578-592.

Finger, F.L. and J.G. Barbosa, 2006. Postharvest physiology of cut flowers. *Advances in Postharvest Technologies for hort. Crops*, pp. 373-393 (*http://129.200.9.213:8595/webspirs/* doLS.ws?ss).

Finger, F.L., M.M. Campanha, J.G. Barbosa and P.C.R. Fontes, 1999. Influence of ethephon, silver thiosulphate and sucrose pulsing on bird of paradise vase life. *Revista Brasileira de Fisiologia Vegetal*, **11**(2): 119-122.

Fisher, J.B. 1976. Development of dichotomous branching and axillary buds in *Strelitzia*. *Canadian J. Bot.*, **54**(7): 578-592.

Halevy, A.H., A.M. Kofranek and J. Kubota, 1976. Effect of environmental conditions on flowering of *Strelitzia reginae* Ait. *HortSci.*, **11**: 538.

Halevy, A.H., A.M. Kofranek and S.T. Besemer, 1978. Potharvest handling methods for bird of paradise flowers. *J. Amer. Soc. hort. Sci.*, **103**(2): 165-169.

Halevy, A.H., R.A. Criley, S.T. Besemer, H.A. Vinter Vande, M.E. McKay and O. Kawabata, 1987. A comparison of flowering behaviour of *Strelitzia reginae* at four locations as affected by air temperature. *Acta Hort.*, No. 205, pp. 89-96.

Haouala, F. and M. Zouari, 2001. *Strelitzia reginae* ~ Effects of vegetative development on inflorescence production in open-air culture. *PHM Revue Horticole*, No. 426, pp. 30-33.

Hara, A.H., T.Y. Hata, B.K.S. Hu and V.L. Tenbrink, 1993. Hot water immersion as a potential quarantine treatment against *Pseudaulacaspis cockerelli* (Homoptera: Diaspididae). *J. Econ. Ent.*, **86**(4): 1167-1170.

Hosni, A.M. 2001. Increasing the potential of *in vitro* propagation of *Strelitzia reginae* by controlling oxidative browning of explant tissues. *Arab. Univ. J. agric. Sci.*, **9**(2): 839-852.

Kawabata, O., R.A. Criley and S.R. Oshiro, 1984. Effects of season and environment on flowering of bird of paradise in Hawaii. *J. Amer. Soc. hort. Sci.*, **109**: 706-712.

Krogt, T.M.van der, 1981. Sortimentsverbetering by strelitziateelt mogelik. *Vakbl. Bloem.*, **36**(4): 44-45.

Kronestedt, E.C. and A.W. Robards, 1987. Sugar secretion from the nectar of *Strelitzia* ~ and ultra structural and physiological study. *Protoplasma*, **137**(2-3): 168-182.

Kronestedt, E.C., M. Greger and A.W. Robards, 1989. The nectar of the *Satrelitzia reginae* flower. *Physiol. Plant.*, 77:341-346.

Lorenzini, G., E. Triolo and G. Scaramuzzi, 1983. Further contribution to the knowledge of new or little known diseases of ornamental plants in Tuscany (Italian). *Rivista della Ortoflorofrutticoltura Italiana*, **67**(1): 23-34.

Meillon, S. de, H.A. van de Vinter, J.G.C. Small, S. de Mellon and H.A. van de Venter, 1990a. The effect of dormancy release of seeds of *Strelitzia reginae* by oxygen on their alcohol dehydrogenase activity, cytosolic malate dehydrogenase activity and ethanol content. *S. Afr. J. Bot.*, **56**(3): 342-345.

Nair, A.K., M.A. Suryanarayana and Anuradha Sane, 2011. Effect of spacing on growth and flowering of bird of paradise (*Strelitzia reginae* Ait) in maidan regions of Karnataka. *J. Ornam. Hort.*, **14**(3-4): 85-88.

Orlikowski, L. 1976. Certain fungal diseases of *Strelizia reginae*. *Prace Institutu Sadownictwa w Skierniwicach* B, No. 2, pp. 163-167.

Pirone, P.P., B.O. Dodge and H.W. Richet, 1960. *Diseases and Pests of Ornamental Plants*. The Ronald press Co., New York.

Pol, P.A. van de and T.F. van Hell, 1988. Vegetative propagation of *Strelitzia reginae*. *Acta Hort.*, No. 226 (vol. II),pp.581-586.

Shreevatsa, K.S., K.V. Jayaprasad, P. Narayanaswamy and B.N. Satyanarayana, 2004. Establishing *in vitro* cultures of *Strelitzia reginae* using shoot tip explants. *J. Ornam. Hort.*, **7**(3-4): 341-344.

Skead, C.J. 1975. Weaver bird pollination of *Strelitzia reginae*. *Ostrich*, **46**(2): 183-185.

Vasantha Kumar, R., B.S. Reddy, S.K. Balaji and Prasad Kumar, 2011. Influence of nitrogen, phosphorus and potassium fertilization on flower yield and economics of bird of paradise (*Strelitzia reginae* Ait.). *J. Ornam. Hort.*, **14**(1-2): 34-37.

Venter, H.A. van de, 1974. Dormancy in seeds of *Strelitzia*. *South Afr. J. Sci.*, **70**(7): 216-217.

Venter, H.A. van de, 1976. Notes on the morphology of the embryo and seedling of *Strelitzia reginae*. *J. South Afr. Bot.*, **42**(1): 63-69.

Ziv, M. and A.H. Halevy, 1983. Control of oxidative browning and *in vitro* propagation of *Strelitzia reginae*. *HortSci.*, **18**(4): 434-436.

36

Tulipa (Family: Liliaceae)

Ajit Kumar, Sanyat Misra, Jyoti Bajeli, Tripti, Sunder Pal, R.L. Misra and Manish Kapoor

[**Common anmes**: Candlestick tulip/Candy tulip/Lady tulip (*T. clusiana*, syn. *T. aichisonii*), Didier's tulip/Garden tulip (*Tulipa gesneriana*), Forentine tulip/Forest tulip/Wild tulip (*T. persica, T. sylvestris*), Horned tulip (*T. acuminata*, syn. *T. cornuta*), Splendens tulips (*Tulipa hageri*), Tulip, Waterlily tulip (*T. kaufmanniana*)]

Introduction and Origin

The mentioning of tulip appears first in western Europe in or around 1554, though it is thought to have been introduced into Europe about 1572, and to Britain in 1577, Clusius sent the bulbs of *Tulipa claremont*, a parent of the early-flowering tulips a few years later (Pizzettii and Cocker, 1968). Conrad Gesner gave the plant its name and made the first drawings of it, prior to 1559, after seeing a specimen in bloom at Augsburg in the garden of Councellor Herwath (Pizzetti and Cocker, 1968). It was first cultivated in Turkey. Pizzetti and Cocker (1968) mention that it is a flower of Oriental origin and in the East its cultivation was initiated about a thousand years

ago. Near Kabul, the Great Mughal Baber counted 33 different species. It is thought that it was Clusius who introduced tulips into Holland since in 1593 he left Vienna to take the chair of professor of botany at Leiden (Pizzettii and Cocker, 1968). In the 17th century the tulip became an extravagant cult in Holland, becoming known to historians as Tulipomania, and single bulbs of new cultivars changed hands for vast sums of money (Beckett, 1983). Tulip is derived from the Turkish word *tulbend* or Persian word *toliban*, meaning 'turban', because of similar appearance of the flower in shape and colour. The very earliest known tulip was called *Tulipa gesneriana* in honour of Gesner. It was Gesner who in 1561 first mentioned *Tulipa suaveolens*, native to southern Russia, and this species in succeeding crosses imparted something of its perfume to subsequent hybrids (Pizzetti and Cocker, 1968). Tulip in Turkish culture was considered as a symbol of paradise on earth having a divine status. According to Pavord (1999), the first introduction of tulips started to appear on tiles and decorated ceilings in the 11th century. They were introduced from Turkey into Europe in the 16th century (Killingback, 1990). Tulips are commonly associated with the Netherlands and this association began in 1594 causing the famous 'tulipomania' in the 1600s during the Ottoman Empire. The Sultan and his Court prized the tulip so much that they held lavish parties when tulips bloomed. This is where the tulip first became associated with wealth, power and prestige. Botanist Carolus Clusius is credited with bringing the tulip to Holland, where he studied it, developed new varieties and planted in impressive display in the Botanical Gardens at Leiden. When Clusius refused to sell the tulip bulbs from his stock, people there began stealing this as this was considered a symbol of prosperity. Today, tulip flowers are so popular that several tulip festivals are held around the world every year. As ranked one of the third most popular cut flowers sold worldwide (Podwyszynska and Sochacki, 2010) next to rose and chrysanthemum, the largest area under any true bulb crop in the world is that of *Tulipa*, followed by *Narcissus*, *Iris*, *Hyacinthus* and *Lilium*.

Centre of origin of tulips is Central Asia (Siberia, Mongolia and China), and the area of greatest diversity is south into the Kashmir (India) area and west into Afghanistan, Iran, Caucasus and Turkey (Bailey and Bailey, 1976; Rees, 1985; Le Nard and De Hertogh, 1993; Dole and Wilkins, 1999).Tulips are native of Middle East (from Turkey south to Yemen, collectively known as Arabian Peninsula) to the Pamir and Hindu Kush mountains in Asia. However, the Pamir Alai and Tien Shan mountain ranges in Central Asia including the countries such as Turkey, Afghanistan, Iran, Kazakhistan, Turkmenistan, Uzbekistan and Georgia are referred as the primary gene centres of the genus *Tulipa* L. A couple of species extend into northern India and far western China but on limited basis. One species, *Tulipa stellata* is found wild in the mid- and high hills of Himachal Pradesh and *T. aitchisonii* is native to Himalaya.

In the Netherlands, Dutch growers were the first to start tulip industry, and there, apart from providing enchantment, tulips represent briefness of life, and the total area under tulip cultivation in Holland is around 10,000 ha and they are the leading growers in the world. There is tremendous growth potential of this crop all over the world with an increase @ 6-8 per cent per annum in production. In the Netherlands, about 75 per cent of the bulbs are exported to various parts of the world with the highest share to United States, followed by Germany. Tulips are also gaining importance in India because of the favourable climatic conditions existing in the temperate areas of Kashmir Valley, Himachal Pradesh and Uttarakhand at an altitude of 1,300 to 3,000 m above msl.

Tulips appear almost in every hue and colour except pure blue. In the florist trade, tulips having the faint violet hue are sold as blue tulips. They are widely cultivated for cut flower production as well as for pot culture, in beds, borders, formal and informal gardens. They are the primary choice of growers to grow in the orchards, lawns, wild gardens and the dwarf and compact ones in the rockeries. Mass planted tulips provide an excellent view. They are being used as decorative motifs on tiles, mosques, fabrics, crockery, etc. Tulips not only add charm by way of beautiful flowers but in some species flowers and bulbs are also edible and have potential in various value added products. Edible tulips are classified into three categories, *viz.* scented, textured and colourful. Scented tulips add fragrance and make a sultry salad with lamb's lettuce, red cabbage and chicory or roast peppers, squash and pecorino. Tulip petals are large, crunchy, colourful, sweet and can be taken raw. Fresh flower petals add colour and texture to salad dressings and garnishes (Mlcek and Rop, 2011). The white bit present at the base of the petals may sometimes be bitter in taste and should be chopped off. Larger petals have a stiffer texture and more crunchy. They produce crunching noise while breaking when eaten. They taste fantastic with the mixed bright green leaves of oakleaf lettuce, sorrel, mint and feta. The tulip petals of various colours like yellow, red, white, dark purple and near black add pleasing brittleness to the salad with an initial tang and a sweet taste afterwards. The petals are used to serve various canapés such as chicken tikka salad, guacamole (a Mexican dish of crushed avocado mixed with tomatoes, onion, chilli, etc.), watermelon and feta, chocolate and raspberry cheesecakes, plum and almond bites, ricotts, and upside down rhubarb crumbles. Edible tulip bulbs are either cooked or taken raw after removing the outer skin and inner core. They are considered as a famine food as they can be used at the time of hardship. Tulip bulbs can be dried, powdered and added to flour or cereal products, and sometimes used as replacement to onions in cooking.

Botany

Tulip is herbaceous perennial geophyte with true bulbs as its underground rootstock. The bulbs have papery to coriaceous tunic, sometimes lined with 'hairs' or 'wool', especially those species which originate to extremely cold areas. In various species and varieties, there is great variability in plant height, starting from 10 cm to 70 cm. Stems are simple and sometimes tinged red, erect, glaucous or pubescent. Leaves are a few (2-6, or sometimes even up to 12), sessile, grey-green with waxy coating, basal or cauline, alternately placed on the stem, strap-shaped, linear lanceolate to broadly ovate, sometimes undulate or crispate and hairy or hairless. Flowering stalk is erect and arise from centre of the bulb, and individual flowers which are erect (rarely nodding) are borne at terminal ends in the form of bell or funnel with distinct and free petals in two whorls, often blotched white or yellow near the base. Flower colours may be white, yellow, orange, pink, red, purple and bicolours. Irregular or broken petal colouration is due to virus infections. Flowers bear 6 stamens, with often triangular, flattened and narrow filaments, basifixed oblong-linear anthers, 3-angled superior ovary, 3-lobed stigma, short or absent style, and spherical or ellipsoid capsule with numerous flat seeds. The light to dark brown seeds have very thin seed coat and endosperms that does not normally fill the entire seed (Botschantzeva, 1982).

Tulips are cross-pollinating in nature, and the pollination is through wind as well as insects, mainly bees which are attracted by the bright colour flowers. About 85 per cent of the species are diploid (2n=24), some 14 per cent species are triploid, and a few in nature are tetraploid to pentaploid. Basic chromosome number in tulips is n = 12. *Tulipa aitchisonii* and *T. Sylvistris* subsp. *australis* are diploid with 2n = 24, *T. Aleppensis* and *T. Lanata* triploid with 2n = 36, *T. saxatilis* with diploid 2n = 24 and triploid 2n = 36, *T. chrysantha* diploid 2n =24 or tetraploid 2n =48, *T. sylvestris* subsp. *sylvestris* either triploid 2n = 36 or tetraploid 2n = 48, *T. stellata* with 2n = 24, 36 and 48, *T. orphanidea* with 2n = 24, 36 and 60, and *T. clusiana* with 2n = 24, 36, 48 and 60 (Misra and Misra, 2013). This has also been recorded that *Tulipa sylvestris* is a tetraploid form of diploid *T. australis*; *T. turkestanica* a tetraploid form of the diploid *T. biflora*; and *T. praecox* is a triploid form of the diploid *T. oculus-solis*, and so on (Pizzetti and Cocker, 1968).

Breeding

Tulips are great favourites of every flower lover because of their diversity in flower colour and shape. Its breeding started some 400 years ago, and it takes some 20 years for developing a new tulip variety through conventional system. In tulips, major cultivar groups have originated from crosses of several species (Killingback, 1990). It is one of the ornamental crops having a large number of cultivars which are present worldwide along with different colours, vibrant spectrum, form and size. Various cultivars consist of 15 groups according to their flower form, including single early and late, double early and late, lily-flowered (pointed petals), parrot, triumph and fringed. Only few wild varieties are existing in the gardens. Till date, several thousand varieties of tulips are registered and old varieties are replaced by new ones every year. Modern day tulips are hybrids evolved through hybridization involving the species *T. gesneriana* which has played an important role in their evolution. A wide diversity is available in growth and vigour, floral characters with regard to flower shape, size & colour, floral stalk length and reaction to various diseases. In order to further improve ornamental traits and bulb productivity in tulip, there is a need to study various breeding techniques such as overcoming of interspecific barrier, embryo rescue of abortive embryos and induction of polyploidy (van Tuyl and van Creij, 2006). Studies have been carried out by van Creij *et al.* (1997) to find out pre-fertilization barriers in pollen tube growth in the pistil and pollen tube penetration of ovules involving crosses between cultivars from *Tulipa gesneriana* and 12 tulip species from eight sections of the genus *Tulipa*. Most of the species are grouped together and are referred to as *Tulipa gesneriana*. As many characteristics are almost not solely present in *T. gesneriana* assortments such as desired earliness, resistance against tulip breaking virus, and so on, hence, other *Tulipa* species are sometimes used for introducing these characters. In spite of a number of pollen tubes in the ovary, none or only a few per cent pollen tubes penetrated the ovules of some crosses while in other crosses the penetration was to a tune of up to 79 per cent of the ovules. As a result no or only a few hybrids were obtained after seed maturation on the plant. Apparently, various kinds of barriers such as temperature, low viability of the pistillate parents, hampered fertilization or normal embryogenesis occur in interspecific crosses though interspecific hybridization has always gained much attention due to its specific benefits (van Tuyl, 1996). Interspecific hybridization generally makes more genetic variation than intraspecific hybridization and valuable specific traits of the wild species such as disease resistance can be introduced into cultivars. Reproductive and germinative barriers and long generation times limit interspecific hybridization. These barriers include pre- and post-pollination incongruity. Use of techniques such as bud pollination, cut styles, grafted styles, placental pollination, and pollination of isolated ovules help in overcoming crossing barriers (van Tuyl and van Creij, 2006). Some popular cultivars exhibiting characters like earliness, and resistence against tulip breaking virus & *Fusarium* are 'Beau Monde', 'Come Back', 'Explosion', 'Lefeber's Memory', 'Pink Impression, 'Purple World' and 'Spring'. Embryo rescue techniques are considered beneficial in preventing interspecific incongruity (Okazaki *et al.*, 1994). Custers *et al.* (1995) made a comparison between two embryo rescue techniques in intraspecific crosses and suggested that ovule culture was possible by using smaller embryos than full embryos. Successful rescue of abortive embryos could be achieved by hampering embryo breakdown through direct ovule culture to the interspecific cross of *T. gesneriana* × *T. kaufmanniana*. Likewise, van Creij *et al.* (1999) reported another type of embryo rescue technique in which ovary-slice culture was followed by ovule culture. This technique was similar to that of direct ovule culture.

Haploidization is the process whereby diploid cells become haploid by progressively losing chromosome through non-disjunction as they fail to separate equally into the daughter cells during mitosis, whereas in polyploidy there is total non-disjunction of chromosomes during mitosis or meiosis both. The production of *in vitro* haploid plants using micropore culture is a technique used in many plant species in order to produce homozygous plants. Tulip is more promising than lily to develop haploid plants *via* this technique (van den Bulk and van Tuyl, 1997). Haploidization and molecular techniques result in homozygous plants which assist in marker assisted breeding. Young pollen through tissue culture result in the formation of embryo-like structures. These haploids can be doubled either spontaneously or by use of chemical methods to produce homozygous plants required for breeding programmes. There are several methods of obtaining **polyploid** tulips. Among them, mitotic polyploidisation is the primary method, where colchicine or oryzalin is used during a stem regeneration process for chromosome-doubling (van Tuyl and van Creij, 2007). Polyploidy in tulips can also be induced by treatment of pollinated ovaries and seed-bulbs with laughing gas (nitrous oxide). A large number of commercial Darwin hybrids are sterile triploids ($2n = 3\times = 36$) such as 'Apeldoorn', 'Ad Rem', 'Pink Impression', 'Van Eijk', and 'Worlds Favourite', while 95 per cent of the cultivar population are diploids (Marasek-Ciolakowska *et al.*, 2012). Polyploid tulips can also be obtained *via* interploidy crossing (Okazaki and Nishimura, 2000). Triploid tulips which have generally more vigour, can also be developed through crossing of tetraploids with diploids or *vice versa* (van Scheepen, 1996).

For induction of artificial **mutations**, van Harten and Broertjes (1989) suggested use of X-rays in tulips. Tulip is principally propagated through vegetative means and hundreds of spontaneous sports have been picked up (van Scheepen, 1996). Mutations can lead to differences in flower colour or flower shape such as parrots, fringed and double types, leaf edge, plant height and bulb production. Many natural mutants have been recognized from specific cultivars, *e.g.* 'Bartigon', 'William Copland', 'Murillo' and 'Apeldoorn'.

They are bred for developing new combinations of colours, shape and size of the flowers, *vis-a-vis* stalk lengths, for introducing disease resistance, for prolonging vase life, for shortening of the juvenile period and to introduce disease resistance.

Flower colour is directly correlated with the proportion and composition of different colour **pigments**. Though white flowers are almost without any colour pigment, the pigments responsible for yellow flowers are only carotenoids; carotenoids and cyanidin imparting orange colouration; pelargonidin and cyanidin for red colouration; while delphinidin and cyanidin for purple colouration. Highest variation in flower colour pigments is exhibited by pink flowers but the level is mostly lower compared to other colours. Apart from single-coloured tulips, the multi-coloured tulips such as pink-red, purple-red and break of white, yellow, purple, red and so on contrast the other base colours and make tulip a most attractive flower. Howevewr, the colour break in tulip is mostly due to certain viruses also. Tulips being highly heterozygous in nature with complex genetic constitution as it took some 400 years in its breeding to attain the present colour status, also as without giving it any chance to segregate freely, we start it propagating clonally so prediction of the flower colour of the progeny becomes highly cumbersome on the basis of the flower pigment composition of the immediate parents. However, it is possible only when the fresh crosses are started at specific levels or when a parent has been used repeatedly in the breeding programme, but mostly the progenies in such former cases are freak and of no use. So is the case even in the inheritance of flower shapes where too any prediction is very difficult to make, though of course, where very close breeding has been followed, the possibility of such predictions, to some extent, are possible. For inheritance of flower shape and certain other floral characters, the crosses made involving a lily-flowered 'Diyanto' cultivar with those of double-flowered, edge-coloured, single-coloured and base-coloured cultivars or when double-flowered 'Abodement' was used as one of the parents with various other types of cultivars, the former did not give any of the off springs having pure lily-flowered type and the latter also did not result into any of the pure double-flowered type seedlings. *T. acuminata* was when used as a parent, it resulted into a large number of beautiful lily-flowered seedlings.

Tulips are invaded by several **pathogens** causing various kinds of diseases resulting in huge losses qualitatively and quantitatively. Among them bulb-rot (*Fusarium oxysporum*), tulip fire (*Botrytis tulipae*) and tulip breaking virus (TBV) are most common. Some others though not so serious pathogens are *Pythium* spp., *Rhizoctonia tuliparum*, *R. solani*, and viruses [tobacco necrosis virus (TNV) and tobacco rattle virus (TRV)] which result in drastic economic losses to the growers. Host resistance is the best known preventive measure against such diseases. Use of chemicals to get rid of various kinds of diseases is not only costly but also un-ecofriendly due to degradation of soil, water and environment. Therefore, development of cultivars resistant to various kinds of diseases will be very useful. Inheritance of *Fusarium* resistance in tulips was studied by van Eijk *et al.* (1983) and Romanov *et al.* (1991) in which they made reliable screening tests both for clones and for juvenile seedlings at the time of pre-selection. Tulip bulbs were planted in *Fusarium*-infected soil by providing optimum package of practices, and when the harvested bulbs were tested for *Fusarium* infection, almost complete resistance was found within the *Tulipa gesneriana* assortments. Thus, to develop *Fusarium* resistant cultivars, use of atleast one resistant parent is necessary. Against TBV, Romanov *et al.* (1991) and Straathof *et al.* (1997) conducted screening tests both at the clonal and seedling levels through virus inoculation by viruliferous aphids in leaves of flowering plants which showed flower breaking after one year of inoculation. Several cultivars of *T. fosteriana*, such as 'Cantata' and 'Princeps' showed greater resistance to TBV. Crosses made through *T. gesneriana* × *T. fosteriana* resulted into highly resistant genotypes, which were referred to as Darwin hybrids against TBV. These hybrids were mostly triploid and sterile in nature. However, this problem can be worked out through artificial doubling of the chromosome (van Tuyl and de Jeu, 1997). Straathof *et al.* (2002) screened tulips for resistance to *Botrytis tulipae*. There was a complete resistance in *T. tarda*, but this species is not compatible with *T. gesneriana*. However, partial resistance against *Botrytis tulipae* was recorded in *T. gesneriana* and *T. kaufmanniana*. Breeders must focus on development of multi-resistant cultivars of tulips for revolutionizing its cultivation across the globe. For the future, resistance to three main diseases of tulip must be combined in order to have multi-resistant tulip cultivars.

Vase life is the another most important character for improvement and it is positively correlated to the durability of the flowers on the plant under field condition. Therefore, the most field-durable cultivars should be selected for such breeding programmes so that naturally their life is prolonged in the vase solution. Senescence is a gradual process and is expressed in terms of leaf yellowing, floral-stalk neck drooping, colour deterioration and tepal senescence & abscission. For this purpose healthy bulbs should be selected only of the cultivars with known reputed pedigree and then proper treatment throughout the cycle of growth and development should be provided. Anti-senescence genes may also be introduced in the most wanted cultivars.

Tulips have a long **juvenile period** so there is need to reduce their growth cycle from one year to eight months by growing in environment-controlled greenhouses and storing them at required temperature so that three growth cycles are regulated in two years. Crop juvenile period can be reduced from 4-5 years to 3-4 years by applying optimal field and storage conditions. However, higher growth cycles per annum with change in micro-climate such as additional lighting results into smaller bulbs than those with only two cycles of cultivation so these aspects require to be investigated thoroughly.

Genetic modification or transformation is the genetic alteration of a cell resulting from the direct uptake and incorporation of exogenous genetic material, *i.e.* exogenous DNA from its surrounding cell membrane. These tools of molecular biology can accelerate breeding programme of tulips for novelty as well as resistance against various biotic and abiotic stresses. Flower bulbs have been transformed using both *Agrobacterium*-mediated and microprojectile-acceleration methods. The flowers of the viridiflora tulip produce greenish stripes in the tepals in whorls 1 and 2 after incorporation of the genes *TGSQA* and *TGSQB* from wild tulip species. Use of molecular marker techniques as a method of pre-selection is being employed successfully in tulips as they exhibit a long juvenile period. Marker-assisted breeding has been identified by use of *AFLP* markers in lily and tulip (van Heusden *et al.*, 2002). Several AFLP markers have shown tulip breaking virus (TBV) resistance. Application of *PCR* derived markers can also be used to speed up the breeding techniques of tulips.

Classification, Species and Varieties

Tulipa includes 16 taxa (150-160 species), among them two are endemic to Turkey (Mlcek and Rop, 2011). Tulip is grouped into the following divisions as per Royal General Dutch Bulb Grower's Society in collaboration with RHS: Initially the classification ran to 23 divisions but further study reduced it to 15.

Division I (Single early tulips)

This is the oldest group of tulips in cultivation. As the name implies, these bloom earlier among all the tulip divisions. The average plant ranges from 15-38 cm. These are single-flowered cultivars which are early flowering in nature. Plants are short but very strong stemmed which make them extremely adaptable against wind and rains. Flowers though commensurately small, are available in many colours, *viz.* apricot, orange, pink, purple, red, white and yellow, both jewel tones and pastels. Cup-shaped flowers have pointed petals. They are good for bed, borders, pot culture, indoor forcing and rock gardens. Varieties are 'Brilliant Star' (scarlet), 'Couleur Cardinal' (orange-red), 'Crown Imperial' (mahagony-red, banded wide yellow), 'Cullinan' (pink), 'Diana' (white), 'Mainau' (dark red), 'Mon Tresor' (yellow), 'Pink Beauty' (deep pink and white), 'Prince Carnival' (scarlet striped yellow), 'Proserpine' (carmine), 'Silver Standard' (pink, red, white), 'Sunburst' (golden-yellow), 'Ursa Minor' (golden yellow), 'Van der Neer' (purple), 'Vermilion Brilliant' (vermilion-scarlet), 'White Beauty' (white), 'White Hawk' (white), 'Yellow Prince' (canary yellow), *etc.*, which are most suitable for early forcing as well as for outdoor growing. Late forcing varieties which though flower earlier outdoors are 'Bellona' (golden-yellow), 'General de Wet' (orange-red), 'Keizerskroon' (carmine-red and yellow, impressive large flower), 'Pink Perfection' (carmine-pink), 'Prince of Austria' (orange and scarlet), *etc.* The varieties most suitable for forcing are 'Crown Imperial', 'Cullinan', 'Fred Moore', 'General de Wet', 'Mon Tresor', 'Pink Perfection', 'White Hawk', 'Yellow Prince', *etc.*

Division II (Double early or paeony-flowered tulips)

Their flowers last a bit longer than **single flowered** cultivars, and the flowers look more like a paeony, are early in flowering, have large double flowers packed with so many petals with short and sturdy stems (25-35 cm) and are suitable for February forcing or outdoor growing in April. The flowers show less variety of colours (white, yellow, orange, pink and red) than other tulips but are completely open under sunny condition. Plants are ideal for growing in beds, borders, containers and rock gardens. The varieties are 'Abba' (red), 'Caliph of Baghdad' (dark yellow), 'Dante' (blood-red), 'David Teniers' (violet purple), 'Electra' (pink-mauve), 'Goya' (orange-red), 'Holland Ballet' (white), 'Largo' (red), 'Madame Testout' (deep rose-pink), 'Marechal Niel' (shining yellow), 'Margarita' (lavender), 'Mr. van der Hoef' (yellow), 'Mondial' (white), 'Monte Carlo' (yellow), 'Murillo' (soft pink), 'Murillo Maxima' (pink suffused white),'Orange Nassau' (orange-brown), 'Peach Blossom' (deep pink), 'Scarlet Cardinal' (scarlet), 'Schoonoord' or 'Purity' (snow white), 'Tea Rose' (primrose yellow stained pink), 'Triumphator' (deep pink), 'Vuurbaak' (orange-scarlet), 'Wilhelm Kordes' (orange and red), 'Willemsoord' (carmine, edged white), *etc.* all these varieties originated as mutants from 'Murillo' (pink and white). The varieties most suitable for forcing are 'Marechal Niel', 'Murillo', 'Peach Blossom', 'Tea Rose', *etc.*

Division III (Mendel tulips, hybrids between single early and Darwin tulips)

The firm of Kreelage offered first Mendel tulips in 1909. Outdoors, they flower from late April to early May, while under glass during February-March, earlier than **Darwin** tulips though later than **single early** and **double early** types. These are of medium height (38-50 cm) and are most suitable for bedding and forcing. The varieties are 'Athlete' (white), 'Brightling' (rose pink stained with orange), 'Emmy Peeck' (lilac-pink), 'Her Grace' (white and pink), 'Krelege's Triumph' (dark red edged yellow), 'Mimosa' (yellow), 'Marjoran' (carmine-red), 'Orange Wonder' (orange-bronze and scarlet), 'Pink Gem' (white margined with pink), 'Scarlet Admiral' (red with black centre), 'Sulphur Triumph' (sulphur-yellow), 'Van der Eerden' (red), 'Weber' (white edged with lilac rose), 'White Sail' (cream white changing to pure white), *etc.* The varieties most suitable for forcing are 'Athleet', 'Brightling', 'Her Grace', 'Hildegarda', 'Orange Wonder', 'Pink Gem', 'Weber', 'White Sail', *etc.*

Division IV (Triumph tulips)

Among various tulip divisions, this one is the largest and most important. Though its flower shape is similar to Darwin tulips but is less in height and flowers some 10 days earlier. This division of tulips arose through hybridization between cultivars of the single early division and single late one. They possess single robust flowers on a strong and sturdy stem of medium height (up to 50 cm) and flower in mid-season (mid-April). The flowers are of almost every shade including apricot, orange, pink, purple, red, white and yellow. They are highly suitable for beds, borders, indoor forcing and as cut flower due to having long vase life. The varieties are 'Adorna' (salmon-orange), 'Bandoengn' (dark red and yellow), 'Bruno Walter' (golden brown with purple tint), 'Crater' (bright red), 'Dutch Princess' (buff orange and gold), 'Edith Eddy' (red edged white), 'Elizabeth Evers' (carmine rose), 'Elmus' (cherry-red, edged white), 'Garden Party' (white, edged carmine), 'Golden Eddy' (an yellow edged sport), 'Johanna' (deep salmon pink), 'Kansas' (pure white), 'Korneforos' (red), 'Merry Widow' (red and white), 'Nova' (lilac), 'Orange Korneforos' (bronze-orange), 'Paris' (orange-red and yellow), 'Pax' (white), 'Piccadilly' (cherry red, margined white), 'Princess Beatrix' (orange-scarlet, edged gold), 'Purple Star' (purple-black), 'Red Giant' (bright red)', 'Rose Korneforos' (rose-red), 'Sulphur Glory' (yellow), 'Sweet Seventeen' (pink), 'Virtuoso' (lilac-rose), *etc.* The varieties most suitable for forcing are 'Elmus', 'Glory of Noordwijk', 'Kansas', 'Korneforos', 'Paris', 'Red Giant', 'Telescopium', *etc.*

Division V [Darwin hybrid tulips (Darwin tulips × *T. fosteriana*)]

Cultivars of this group evolved by crossing the cultivars of Darwin group with *Tulipa fosteriana* and other varieties and botanical species which appear similar to *T. fosteriana.* When compared with other divisions, the flowers in this group of cultivars are pyramidal, largest-flowered of all tulips (15-18 cm wide when fully expanded), single on tallest sturdy stems (50-70 cm), bloom during April and May (mid-season), and some of them are fragrant. Brilliantly coloured bright flowers come in various shades such as black, orange, pink, purple, red, white and yellow. They are excellent for growing in beds, borders and as cut flowers. Being tallest, they are susceptible to wind damage hence require shelter. The varieties are 'Apeldoorn'(orange-red), 'Beauty of Apeldoorn' (golden-orange with a black base), 'Dover' (scarlet), 'Elizabeth Arden' (rose-purple), 'General Eisenhower' (orange-red, base yellow), 'Golden Oxford' (pale-yellow, margins narrowly edged), 'Golden Springtime' or 'Santiago' (deep golden-yellow), 'Gudosnik' (yellow, spotted red and pink), 'Holland's Glorie' (scarlet), 'Jewel of Spring' (sulphur-yellow, base green-black, margin narrowly red), 'Oxford' (scarlet-red), 'Parade' (scarlet-red), 'Red Matador' (scarlet), 'Tender Beauty' (white, margin pink), *etc.*

Division VI (Darwin tulips)

The first Darwin tulips were created by Jules Lenglart in 1886. They have long flowers, which are square rather than pointed in outline. Of all the garden hybrids, Darwin tulips are most beautiful and quite ideal for bedding as well as cutting. These grow 55-75 cm in height having rounded and deeply cup-shaped flowers with a broad almost square base, 10-12 cm wide and flowering in May. Varieties are 'Ace of spades' (purple-black), 'Adoration' (pink), 'Afterglow' (rose-amber), 'Bartigon' (red), 'Bishop' (deep lavender), 'Bleu Aimable' (lavender-mauve), 'Breitner' (blood-red), 'Charles Needham' (brilliant red), 'City of Haarlem' (scarlet), 'Clara Butt' (bright rose-pink), 'Copland's Purple', 'Demeter' (violet-blue), 'Dorrie Overal' (petunia-violet), 'Dreamland' (rose, base white), 'Eclipse' (scarlet-crimson), 'Giant' (purple), 'Glacier' (pure white), 'Golden Age' (deep yellow), 'Golden Hind' (dark golden-yellow), 'Golden Niphetos' (yellow), 'Heather Hill' (lilac-purple, base white edged blue), 'Helen Madison' (pink), 'Joan of Arc' (ivory), 'King Harold' (blood red), 'La Tulipe Noire' (maroon black), 'Mahagony' (dark mahogany), 'Mark Antony' (purple and bronze), 'Mimosa' (buttercup-yellow), 'Niphetos' (lemon-yellow), 'Nobel' (deep red), 'Paul Richter' (geranium-red), 'Pink Supreme' (deep pink), 'Pride of Haarlem' (carmine rose), 'Princess Elizabeth' (dark pink), 'Prunus' (salmon pink). 'Queen of Bartigons' (pink), 'Queen of Night' (blackish dark purple), 'Rose Copeland' (pink), 'Scarlet O'Hara' (scarlet, base black and yellow), 'Smiling Queen' (pink), 'Snow Peak' (white), 'The Bishop' (purple), 'Union Jack' (ivory-white flamed raspberry red, base white edged blue), 'Venus' (silvery rose), 'White Giant' (white), 'William Copeland' (mauve lilac), 'William Pitt' (strawberry red), 'Yellow Giant' (yellow), 'Zwanenburg' (white), *etc.* The varieties most suitable for forcing are 'Adoration', 'Demeter', 'Glacier', 'Rose Copeland', 'Pride of Haarlem', 'The Bishop', 'Venus', etc.

Division VII (Lily-flowered tulips)

These tulips were obtained at the beginning of the 20th century by crossing *Tulipa retroflexa* with a cottage tulip. They bloom at the same time as cottage tulips and have pointed petals which curve outwards at the tips. These are single-flowered cultivars with lovely colours marked with stripes, flames, feathers or coloured margins, and sometimes with fragrance, which bear large, elongated, globlet-, lily- or urn-shaped slender blooms with petals tapering to thin points, and appear either mid or late in the season (April), the time matching with Darwin hybrid tulips though developed after them, having pointed curled back petals. The cultivars have an average plant height of 45-60 cm with thin stem and are excellent for bedding and cutting. They have a variety of colours such as white, green, yellow, orange, apricot, pink, indigo, purple, violet and red. However, the plants require shelter as are susceptible to wind damage. The Varieties are 'Aladdin' (orange-red, yellow within), 'Alaska' (yellow), 'Arkadia' (buttercup yellow), 'Astor' (bronze tinged pink), 'Capt. Fryatt' (ruby purple), 'China Pink' (pink, base white), 'Dyanito' (bright red), 'Ellen Willmott' (primrose yellow), 'Florestan' (crimson), 'Gisela' (soft pink stained with yellow), 'Golden Duchess' (golden yellow), 'Koruna' (red interior, base green, purple flushed exterior, edged yellow), 'Leica' (cream white), 'Mariette' (pink), 'Painted Lady' (yellow and orange red), 'Picotee' (white edged with rose), 'Queen of Sheba' (red, edged orange), 'Stanislaus' (orange), 'White Duchess', 'White Trumphator', 'Yellow Marvel', *etc.*

Division VIII (Cottage or single late tulips)

As the name implies, this group of cultivars is last one to bloom among all the tulip divisions, as flowers in May. This is a very old division of vigorous and robust tulips, developed by combining four classes, *viz.* Darwin, Cottage, Old Breeder and Scheeper hybrid tulips. They are single-flowered cultivars and mainly long-stemmed (35-65 cm). Flowers are large (10-13 cm across), rounded or ovoid and almost egg-shaped, and some varieties have fragrance. They come in a wide range of colours such as white, yellow, apricot, orange, pink, purple, red and almost black. They are suitable for naturalizing with no disturbance for many years, and for growing in beds, borders and as cut flowers. Varieties are 'Advance' (cerise-scarlet, base blue), 'Arethusa' (lemon yellow), 'Artist' (green & pink interior, purple and pink exterior), 'Belle Jaune' (deep yellow), 'Beverley' (orange flame), 'Carrara' (white), 'Dido' (pink, yellow and orange), 'Geisha' (pink), 'Golden Harvest' (lemon-yellow), 'Grenadier' (orange), 'Groenland' (green, edged pink), 'Henry Ford' (deep pink), 'Ivory Glory' (cream white changing to pure white), 'Marshal Haig' (scarlet), 'Mrs. Moon' (yellow), 'Mount Shasta' (pandion), 'Perseus' (orange), 'President Hoover' (orange-red), 'Queen of Spain' (cream-yellow), 'Rosabella' (pink), 'Rosy Wings' (salmon pink), 'Viridiflora' (green), *etc.* The varieties most suitable for forcing are 'Advance', 'Albino', 'Carrara', 'Dido', 'Lemon Queen' ('Mother's Day'), 'Mongolia', 'Rosabella', *etc.*

Division IX (Rembrandt tulips)

These tulips are named after the famous Dutch painter Rembrandt (1606-1669 A.D.). Though they are Darwin tulips but with broken flower colours, striped or marked with black, bronze, brown, pink or purple and red on a red, white or yellow background due to a virus infection. They are also long stemmed. Commercially these are not grown but are used only as collector's items from academic point of view due to being of historical importance. Varieties are 'Absalon' (coffee-brown and yellow), 'American Flag' (red marbled white), 'Clara' (pink), 'Cordell Hull (red and white), 'Kathleen' (pink and white), 'May Blossom' (purple and cream), 'Montgomery' (white and pink), 'Paljas' (scarlet, variegated white), 'Pierrette' (white and violet-purple), 'Refinement' (white and red), 'Zomerschoon' (salmon-pink and cream), *etc.*

Division X (Parrot tulips)

First parrot tulip appeared spontaneously in 1665. Parrot tulips can not be propagated by seeds as all have occurred by chance. Parrot tulips developed through mutations of Triumph and other **divisions** including those of the late flowering tulip species and cultivars, and as the species are involved in the evolution, they bloom in the middle or late in the season, *i.e.* April and May. Their cup-shaped brightly coloured flowers in the shades of white, yellow, orange, apricot, pink, purple and red are large, single, sometimes fragrant and widely open giving flattened appearance and are often bi- to multi-coloured with twisted to curled, deeply feathered and irregularly fringed, crisped or slashed petals which look more curious, hence, most favourite among the flower arrangers. **Viridiflora or green-flowered tulips** are also single-flowered mutants from the cultivars of 'single late division', having some greenish stripes which extend from the base to the tips of each petal. They have sturdy stem of variable length and are late in flowering, *i.e.* May with the height of 35-45 cm. Cup-shaped flowers come in various colours like mauve, orange, pink, red, white and yellow. The plants being low in height are suitable for growing in beds, borders as well as for cut flowers. The varieties are 'Arist' (terra-cotta and green but later changing to green), 'Chérie' (yellow and green), 'Formosa' (yellowish-green), 'Greenland' (green and dark red), 'Viridiflora' (green and yellow with undulated petals), Sometimes parrot tulips have colour-breaking virus. The plant height of parrot tulips is usually 40-60 cm though stems are weak. However, the plants are sensitive to cold so require protection. Wet conditions for their growing should also be avoided. Varieties are 'Apricot Parrot' (apricot), 'Black Parrot' (dark purple), 'Blue Parrot' (mauve-blue), 'Doorman' (cherry red), 'Double Fantasy' (double pink), 'Fantasy' (pink), 'Faraday' (white), 'Fire Bird' (scarlet and green, a 'Fantasy' sport), 'Flaming Parrot' (red, flamed yellow), 'Galdelan' (dark violet-mauve), 'Orange Favourite' (orange and green), 'Orange Parrot' (fragrant), 'Red Parrot', 'Salmon Parrot' (salmon), 'Sunshine' (bright yellow), 'Texas Flame' (yellow, flamed red), 'Texas Gold' (yellow with narrow red margin), 'Van Dijk' (pink), 'Violet Queen' (lavender-blue), 'Webers Parrot' (pink, base white), 'White Parrot', *etc.*

Division XI (Double late Paeony-flowered tulips)

These tulips are packed with so many petals, hence, are referred to as paeony-flowered tulips, which appear late in the season, *i.e.* May and persist for long, many of which emitting their sweet fragrance, hence are most suitable for bedding and cutting. The plants bear large double flowers on 40-70 cm long stems, therefore, sometimes these require staking against strong winds though stems are quite rigid. To some extent these tolerate partial shade better than many other tulips. The foliage is blue-green-golden and the flowers appear in a variety of shades such as white, green, orange, pink, red and blue. The plants have the ability to withstand drought conditions and require low maintenance. The varieties are 'Angelique' (soft pink), 'Blue Flag' (light violet), 'Brilliant Fire' (scented red), 'Clara Carder' (clear pink), 'Eros' (scented white), 'Gerbrandt Kieft' (carmine red), 'Golden Lion' (golden yellow), 'Gold Medal' (deep yellow), 'Grand National' (cream-yellow, centre gold), 'Lilac Perfection' (lilac-blue), 'Livingstone' (cherry pink), 'May Wonder' (pink), 'Mount Tacoma' (white), 'Nizza' (cream marked with crimson), 'Orange Triumph' (soft orange and brown), 'Pavo' (rose pink), 'Rocket' (scarlet-orange), 'Symphonia' (carmine-red), 'Uncle Tom' (maroon), 'Upstar' (pink), *etc.*

Division XII (Kaufmanniana or waterlily tulips)

These tulips are one of the earliest to flower (March-April) and were developed from *Tulipa kaufmanniana*, hence these appear to its parent though differ in colouring. These are low growing with a height up to 20 cm displaying mottled or striped foliage. The flowers are more like waterlily, hence also known as waterlily tulips. They have multi-coloured flowers in the shades of white, yellow, orange, pink, red and violet. The flowers are pointed and open almost flat, and its petals are slightly reflexed with their exterior having a bright carmine blush. The plants being low in height are suitable for growing in beds, borders, containers and rock gardens. The varieties are 'Alfred Cortort' (scarlet and black), 'Aurora' (deep yellow and red), 'Berlioz' red and lemon-yellow), 'Brilliant' (red), 'César Franck' (yellow and red), 'Fritz Kreisler' (salmon-pink), 'Gaiety' (ceam-white and orange), 'Johann Strauss' (white, cream and red), 'Lady Rose' (soft pink), 'Mendelssohn' (cream and red), 'Shakespeare' (salmon-red, base yellow), 'Solanus' (golden-yellow and orange), 'Stresa' (margins and interior yellow, base blood-red blotched, and exterior red), 'The First' (carmine-red, edged white), 'Vivaldi' (cream-yellow), *etc.*

Division XIII (Fosteriana or Emperor tulips)

These cultivars are derived mainly from *T. fosteriana* so basically resemble the species but with varying colours, and once planted it continues blooming at the same place year after year in April. These tulips are medium growing (25-50 cm), displaying mottled or striped foliage. Flowers are cup-shaped, bearing large and wide petals in bright colours of white, yellow, orange, pink and red, predominantly in yellow and red with some varieties emitting even fragrance. They are suitable for planting in beds, borders, wild gardens, for naturalizing effects, in the rock gardens and in containers. The varieties are 'Canopus' (red), 'Cantara' (red), 'Czardas' (orange-scarlet), 'Easter Parade' (yellow, base black with red rim), 'Feu Superb' (red with yellow base), 'Galatea' (orange-scarlet), 'Gold Beater' (golden-yellow), 'Golden Eagle' (yellow and red), 'Holland's Glory' (glowing orange-red), 'Pinkeen' (pinkish-orange), 'Princess' (red), 'Purissima' or 'White Emperor' (pure white), 'Red Emperor' or 'Mme LeFeber' (vermillion, a selected large form of the species), 'Red Matador' (orange-red), 'Red Riband' (bright red), 'Rockery Beauty' (brilliant red), *etc.*

Division XIV (Greigii tulips)

Derived mainly from *T. greigii*, therefore the varieties have maroon or purple-brown veins, mottling or striping on foliage taken from the parent, but with flowers of different colours. These tulips have rigid stems and can attain 15-40 cm plant height, flowering in April, having 2-toned or streaked flowers in shades of white, yellow, peach, pink and red, the flowers being cup-shaped some 12 cm across when fully open and have pointed petals. The plants being short in height are suitable for growing in front rows of the borders, in containers, for indoor forcing and for rock gardens. The varieties are 'Fatima' (scarlet-orange), 'Margaret Herbst', 'Oriental Beauty' and 'Red Riding Hood', all with red flowers, 'Oratorio' (pinkish-red), 'Pandour' and 'Yellow Dawn', both yellow and red, 'Lamba' (yellow and bronze-green), 'Oriental Splendour' (lemon-yellow, base green tinged red), 'Perlina' (rose, base yellow), 'Pride of the Caucasus' (orange), *etc.*

Apart from the ones described above, two more popular types still exist such as 'Dutch breeder tulips' and the 'multiflora tulips'. **Breeder tulips** are May-flowering, growing some 60 cm in height, single and enormous long oval blooms in a vast range of colours, marked or stained with other colours, and most tulips lack for such colours. Varieties are 'Astrakhan' (magenta and gold), 'Bacchus' (violet), 'Barcarolle' (blue purple), 'Chappaqua' (cherry pink), 'Cherbourg' (old gold and bronze), 'Dillenburg' (salmon orange), 'Dom Pedro' (coffee brown), 'Indian Chief' (red brown), 'Kathleen Truxton' (shades of coffee-brown), 'Limnos' (salmon pink and orange), 'Louis XIV' (purple and bronze), 'Papageno' (reddish, brown and scarlet), 'President Hoover' (orange scarlet), 'Southern Cross' (cherry- and orange-red), 'Tantalus' (light yellow shaded with violet), 'Velvet King' (royal purple), *etc.* Breeder tulips most suitable for forcing are 'Cherbourg', 'Indian Chief', 'Limnos', 'Tantalus', *etc.* Likewise, the class of **multiflora tulips** is very interesting with branched floral stems which are quite stout and strong and produce 3-6 blooms or sometimes even up to 12 in May from a single bulb, height 45 cm but flowers small though elegant and are ideal for bedding and cutting. The varieties are 'Claudette' (white, petals margined red), 'Emir' (red), 'Georgette' (yellow), 'Rainbow' (pink), 'Rose Mist' (white suffused pink), 'Wallflower' (mahagony-red), *etc.*

Division XV (Miscellaneous ~ all the wild species, their ecotypes and those plants introduced from various parts of Europe as escapes from cultivation, at first thought to be true species but now grouped together as *Neotulipae*, Nt)

Though total number of species is estimated to be 160 but only those of commercial importance and used in breeding programme, which number some 30 are being described here with. These species are *Tulipa acuminata, T. aitchisonii, T. aucheriana, T. australis, T. batalinii, T. biflora, T. chrysantha, T. clusiana, T. eichleri, T. fosteriana, T. gesneriana, T. greigii, T. hageri, T. humilis, T. kaufmanniana, T. linifolia, T. marjolettii* (Nt), *T. maxicowiczii, T. montana* (*T. wilsoniana*), *T. orphanidea, T. praecox, T. praestans, T. pulchella, T. saxatilis*

(T. bakeri), T. sprengeri, T. stellata, T. sylvestris, T. tarda (T. dasystemon), T. tubergeniana, T. turkestanica, T. urumiensis, T. violacea, T. whittallii, etc.

Tulipa acuminata (probably of hybrid origin and a form of *T. gesneriana*; syn. *T. cornuta, T. turcica*): Though its origin is doubtful, but most probably it is from Turkey. It is a very distinct and curious tulip, once believed to be a true species but now thought to be of garden origin, and so is being classified as *Neo-Tulipae*. It bears 4 lanceolate leaves that are undulated at margins. Its glabrous peduncles are shining, flowers though attractive with varied colours are more curious-looking than being attractive, sometimes being bicoloured, often twisted and with thin, long and horned petals, mostly yellow blotched with red lines. The petals are up to 11 cm long and 1.25 cm wide with edges rolling inside. It flowers in April and grows up to 38 cm in height.

Tulipa aitchisonii: A species from Kashmir (Ladakh) near to Tibetan border, growing up to 30 cm in height with grey-green, linear and 7-10 cm long leaves. Flowers white but mostly flushed crimson, bowl-shaped, appearing in mid-spring, some 10 cm long, and sometimes appearing in pairs. Its petals are narrowly oval. It can be planted at an sheltered site or in an alpine house. Its var. *cashmeriana* (quite similar to *T. clusiana*) is shorter with pale-yellow flowers.

Tulipa aucheriana (a form of *T. linifolia*): A species from Iran and Syria grows up to 20 cm high, bearing small bulbs, mainly basal leaves that are up to 10 cm long, grey-green and linear, and 1-3 star-shaped flat flowers in pink colour, blotched yellow-brown at the base, outer segments with a greenish-yellow stripe while the inner ones with 2 green or brown veins externally, flowers 3-5 cm long. The flowers appear during early to mid-spring.

Tulipa australis: A native generally from higher altitudes of Savoy, France, Spain, Poprtugal and Algeria and is quite similar to *T. sylvestris* as both throw out stolons freely. It grows to 30-45 cm high with ribbon-shaped, channelled and 10 cm long leaves. Flowers are solitary or rarely in pairs, scented yellow, suffused red externally, and petals some 5 cm long with pointed tips.

Tulipa batalinii: A native of Bukhara (Turkestan), it grows up to 15 cm in height and flowers appear in April. This seems to be creamy-yellow version of *T. linifolia* with more glaucous leaves and is perhaps albino of that species. In smaller species, it is most beautiful, and is suitable for planting in containers and in rock gardens. Its leaves are 5, mostly basal and some 13 cm long, light green, grass-like, linear-lanceolate, glaucous and slightly undulated. Flowers are bowl-shaped at the base, up to 6 cm long, yellow with greyish- or glaucous-yellow markings, petals are broadly oval, notched, some 3.75 cm long and tips are blunt and cream-yellow. This when crossed with *T. linifolia* and *T. maximowiczii*, produced beautiful pink and pinkish-yellow flowers.

Tulipa biflora (syn. *T. callieri, T. mariannae, T. polychroima*): This tulip from Caucasus is a dwarf species treated as an annual where summer is quite hot and long and winters are cold. Plants can grow as tall as 30 cm high. The leaves are smooth, spear-shaped, grey-green and are 15 cm in length. It is an early flowering tulip, producing 1-5 with very attractive, creamy white, star-shaped, fragrant flowers having creamy yellow centres. Flowers with red edgings have a yellow base and blush greenish-grey or greenish-pink on the exterior. It is vegetatively propagated as flowers are either sterile and do not set seeds or plants will not come true to type through seeds. They are highly suitable for planting in containers, rock gardens, forcing indoors and as cut flowers.

Tulipa chrysantha (syn. *T. clusiana chrysantha*): This tulip from mountains of Central Asia and the Mediterranean regions is referred to as botanical tulip. This has an average plant height of 20-36 cm. Flowers are some 3.75 cm long with magnificent rosy-red or brown flushing exterior and yellow interior. This is excellent for rockeries, borders, wild gardens and small group plantings and requires a well-drained sheltered sunny position where this may remain for a number of years undisturbed.

Tulipa clusiana: A native of Afghanistan, Iran, Iraq, Kashmir, western Himalayas and the north Mediterranean shores. It is one of the most graceful of all wild tulips. The plant height ranges from 25 to 30 cm. Plants bear 2 to 5, erect, narrow, channeled, linear-lanceolate, pale- to grey-green but often margined red, 12 cm long leaves which are both stem and basal-borne. Flowers appear singly or in bunch in April, are star-shaped and 10 cm in diameter. Petals are elliptic-lanceolate, pointed, 2.5-3.8 cm long, have an unique colour combination of pure white inside and brushed with red or deep pink outside, which is not found in other species, and dark purple or almost black centre. The species is cold-hardy. It is an ideal plant for rock gardens, containers, for forcing indoors and as cut flowers.

Tulipa eichleri (syn. *T. boeotica, T. undulatifolia*): A native of Transcaucasia and northwest Iran. It is regarded as one of the largest and showiest among the various wild species. This early (April) flowering tulip grows up to the height of 25-38 cm and bears bell-shaped solitary flowers with 10 cm long and 5 cm wide segments. Leaves are up to 20 cm long, broadly lanceolate, glaucous, pointed, sturdy, pale-green and wavy at the margins. It has some 7 cm long and 12.5 cm wide, shiny orange-red, bell-shaped flowers with a tinge of yellow on sturdy upright stems. The centre of flower is black with a yellow ring. Plants thrive well where summers are dry and winters are cold. They are quite suitable for rock gardens, containers, for forcing indoors, and as cut flowers.

Tulipa fosteriana: A native of Uzbekistan to Tadzhkistan and Central Asia. It is most attractive among all the species of *Tulipa* and by nature is early flowering (March-April), therefore, was extensively used by breeders. Plants are tall with an average height of 40-45 cm. Leaves are broad, glaucous green or grey-green to bluish-grey, occasionally mottled or striped and 3-5 in number. Stem is thick and pubescent which bears saucer-like scarlet flowers, centrally blotched either with black or gold or in combination of these colours and has dark violet anthers. Fully expanded flowers measure some 22.5 cm wide with a shiny and silky sheen. Plants are good for growing in beds, rock gardens and as cut flowers and requires to be planted year after year.

Tulipa gesneriana: It is a *neotulipae* from Turkey and Asia Minor, but naturalized in southern Europe. The very earliest known tulip was called *T. gesneriana* in honour of Gesner. Plant height of *T. gesneriana* is 30 to 60 cm with flowering in April-May. Leaves 3-7 in number, are linear or broadly lanceolate to spear-shaped, glaucous green, 15 cm long, 2-7 in number and are either smooth or lightly covered with fuzz. There is a single showy flower some 7 cm long on a strong stem. Sweet-scented flowers are red, orange, yellow or burgundy and sometimes marked yellow or black at the base, and the perianths are campanulate, 2.5-6.3 cm long. Bulbs are extremely resistant to frost, and can tolerate temperatures well below freezing. The bulbs may be dried and pulverised and added to cereals or flour. The plants are best planted in large drifts, rock gardens, containers, for forcing indoors and as cut flowers. It is considered the basic species which by crossing has given many of the popular garden cultivars, though it is not known truly wild and is found in varied forms and colours, therefore its right to species status is doubted by some botanists.

Tulipa greigii: A moderately hardy species native to Turkestan and other parts of Central Asia, grows to 30 cm in height, and is another magnificient large-flowered species used extensively in breeding programme and its cultivars resemble the species but height ranges from 15 to 60 cm. Though quite similar to *T. fosteriana* but the leaves are boldly streaked and mottled dark purple. This bears usually 4, glaucous green, undulated, 20 cm long and broadly ovate leaves. Flowers appear in early April, are cup-shaped, solitary, up to 8.5 cm long, blunt-pointed and the colour is bright orange-scarlet with each tepal at its base having a yellow-edged black blotch.

Tulipa hageri: A native of Greece and western Iran and is frost hardy. The plant is dwarf and grows some 12- 30 cm in height and flowers in mid-spring. Leaves are 2-7, glaucous and smooth, spear-shaped to lanceolate and some 20 cm long. Flowers are solitary or 3-5 per stem, scented, star-shaped, 3-6 cm long and 5-7 cm across, base-tapered, and the petals are soft red, oval and externally tinged-green. Plants grow well in dry summer and cool winter conditions. They behave like annuals where summers are hot and long. They are ideal for planting in large drifts, rock gardens, containers, forced indoors and as cut flowers.

Tulipa humilis (*T. pulchella*): A native of Asia Minor. These are one of the lowest growing tulips and hence are named as '*humilis*'. Plant height ranges from 10-15 cm and flower appears in early to mid-spring. Leaves mainly basal, linear with presence of red edges and 10-15 cm long. Flowers 1-3 per stem, small (3.5 cm long), star-shaped, interior yellow-purple with cream to yellow centres and have white tips. This species is most suitable for rock gardens. This is a highly variable species and is often considered to include *T. aucheriana, T. pulchella* and *T. violacea*.

Tulipa kaufmanniana: These hardy tulips are native of Turkey and central Asia and are quite early (early to mid-spring) in flowering. Plants are robust and grow up to an average height of 20-25 cm, bearing glaucous mid-green, mottled or red-striped, broadly oblong to lanceolate and wavy leaves of similar or more length. Though plants are small in height, still they are damaged by wind, rains and lodging. Flowers are often scented, cup-shaped and open flat, 8.75 cm long and 5-10 cm across and look like a hexagonal star, with slandered and pointed petals having creamy or yellow colour, with a blush of carmine, pink or red on the exterior. These tulips are excellent for naturalizing, for rock gardens, for containers or for planting in the front row of the border. Breeders used it extensively for developing many of the cultivars and hybrids.

Tulipa linifolia: A highly variable, dwarf to medium growing species native of S. Russia (Turkestan), Turkey and Pamir Mtn. ranges in Asia, which grows from 15 to 30 cm in height, bearing linear, slightly glaucous grey-green but sometimes margined red, mostly basal, wavy and 13 cm long leaves. It is often considered to include *T. batalini* and *T. maximowiczii*. The flowers are red, bowl-shaped, 2-6 cm long with broadly oval and pointed petals, interior centre blackish-purple narrowly edged with cream or yellow, and appear in April and May.

***Tulipa marjolettii* (Nt.)**: This is of garden origin from SE France (Savoy) and NW Italy, and grows to 60 cm in height. The leaves are grey-green and lanceolate. Flower which appears in May are bowl-shaped, 5-6 cm long, yellow and red externally but ageing almost white and the petals are pointed, broadly oval and marked with deep pink. It is considered probably another form or hybrid of *T. gesneriana*.

Tulipa maxicowiczii: A native of Afghanistan, Uzbekistan and USSR. It is quite similar to *T. linifolia* except that the grey-green leaves are arranged up the stem, and in the flowers the basal blotch is black edged-white, and flowering is earlier in April. It grows 10-30 cm high bearing bowl-shaped, 2-6 cm long, bright red flowers with black centres, and broadly ovate petals which are bordered white.

Tulipa montana (*T. wilsoniana*): A native of N. Iran. Plant grows 10-20 cm, bearing several mid-green, undulating, channeled, spear-shaped to lanceolate and red-margined leaves. Flowers appear in April, are solitary, brilliant deep red with a small black blotch in the centre that surrounds the greenish-yellow ovary, 5 cm long, bowl- or cup-shaped on sturdy upright stems and the petals are spoon-shaped. They are well suited for drifts, rock gardens, containers, for forcing indoors and as cut flowers. They thrive well where summers are dry and winters are cool.

Tulipa orphanidea: It is allied to *T. hageri* but smaller and more slender in all its parts. A frost-hardy species native of E. Mediterranean, Greece and Turkey. It grows from 10 to 40 cm in height bearing green leaves often with reddish margins, and 1-4 fowers some 3-6 cm long on each stem during mid-spring. Its flowers are tapered at the base, vermillion to brick-red in colour, outer tepals buff stained-green or purple externally, blotched olive to black at base, rarely edged yellow, and the petals are oval.

Tulipa praecox: A native of Italy, France and W. Asia (Iran) but due to being stoloniferous, it is naturalized throughout most of the Mediterranean regions though at hot and sunny sites, grows up to 45 cm in height and flowers in

March-April. Flower stems are glabrous, bearing glaucous and wide leaves, flowers are solitary, medium-sized, petals in external whorls broader and larger (6.25 cm) than internal one, scarlet and pointed, each at base with a black patch which is edged greenish-yellow though often variable. It is a very beautiful species.

Tulipa praestans (syn. *T. suaveolens* var. *sylvestris*): This species is a native of Tadzhikistan in the southern Pamir Alai and Central Asia and its flowering time is early spring (March and April). Plants are robust and very showy having 20-45 cm plant height, and are suitable for planting in the rock gardens, in containers, for forcing indoors and as cut flowers. Stems are white-hairy, with grey-green, hairy, 25 cm long, broad but narrowly lanceolate and channelled leaves which are folded upwards from midrib, and bear 2-5 cup-shaped, 5.5-6.5 cm long flowers of vivid orange-red, and flamed yellow at the base. The petals are pointed. These can be grown in rock gardens, containers, forced indoors, and as cut flowers.

Tulipa pulchella: A native of SW Asia (Asia Minor) which flowers during March and grows from 10 to 15 cm in height. This tulip has ranked in the Sir Daniel Hall's monograph 'The Tulip' (1929, Martin Hopkinson, London), but confusion has arisen as closely related species *T. aucheriana, T. pulchella* and *T. violacea* have been put by some authorities as geographical forms of *T. humilis*. Whatsoever its name may be but it is a really very attractive plant when in bloom. The flowers are urn-shaped but base-tapering, some 3 cm long, the exterior of outer petals are violet-green and interior bright violet though intensity varying from mauve to carmine-purple. Each petal is though dark blue but with white or yellowish markings at the edges. Leaves are grey-green, 2-3 per plant, glaucous green often red-margined, narrowly lanceolate, channelled, 10-15 cm long and nearly prostrate. It requires well-drained soil and is excellent for planting in the rock gardens.

Tulipa saxatilis (*T. bakeri*): A sterile species native of Crete, naturalized in or perhaps native to Italy, which is stoloniferous, unusually frost-hardy and flowers in early spring. It grows from 15 to 45 cm in height, bearing shiny green and ovate to lanceolate leaves some 20 cm long which appear some weeks in advance of the blooms, cup-shaped 1-4 scented flowers per stem of 4.0-5.5 cm length which taper at the base, and the petals are obovate-spatulate, bright lilac-pink with deep yellow centre. It is shy bloomer until planted at a hot dry spot. It is excellent for naturalizing effect at a sunny position.

Tulipa sprengeri (syn. *T. brachyanthera*): A native of N. Turkey and Asia Minor is frost-hardy and one of the excellent tulips for naturalizing which is highly tolerant to shade and which flowers late in the season, *i.e.* May. It grows to 75 cm in height though in garden only up to 45 cm and bears bright green and narrow leaves, and solitary, base-tapering, globose, scarlet or dark red, seldom brownish-crimson flowers which are lighter outside, and the segments are narrowly oval and reflexed, with back of the outer 3 being buff-yellow.

Tulipa stellata: A native of Afghanistan and temperate areas of North Indian Himalayas extending to Pak-occupied Kashmir. It is cold-hardy with 20-30 cm stalk length and the flowers appear in April. Sometimes this is listed as *T. clusiana stellata*. Leaves are some 25 cm long, narrow and linear, greyish-green often red-edged, and channelled. Flowers are solitary, 3.75 cm long, star-shaped with recurving segments, white with yellow blotch and the back of the outer segments are tinged red though within yellow-centred. The stamens are also yellow.

Tulipa sylvestris (*Tulipa persica*): A native of Europe including Great Britain, W. Asia and N. Africa which is free-bloomer and flowers start appearing in April. It is cold-hardy, drought-tolerant and commonly found wild in the woodlands, orchards and vineyards of western Europe. It can be successfully grown either in full sun or in partial shade. It grows to 15 to 40 cm in height, bearing pale-green, smooth, narrow-linear and 15-20 cm long leaves, and solitary, fragrant, star-shaped, self-fertile, and some 3.5-6.5 cm long and 10 cm across tapered flowers with oval and reflexed outer petals which are often tinged green outside.

Tulipa tarda (*T. dasystemon*): This hardy species is native of Central Asia and Turkestan (USSR) with an average plant height of 15 cm and flowering in early spring (March). This species is quite drought tolerant and cold hardy. Leaves are 2-7 or even more, almost prostrate, glossy bright green with reddish edges appearing in rosette at flowering time, 15-18 cm long, and linear to lanceolate, and mainly basal. Flowers are solitary or in a tight bunch of 3-7 per stem, 3-4 cm long and tapering to the base, star-shaped with a burst of white where inside half below is bright yellow, tinged green or sometimes red ouside, and the oval and pointed petals are white-tipped inside and suffused with red outside.

Tulipa tubergeniana: An April flowering species native of Central Asia which grows with some 30 cm stalk length. Leaves are grey-green, broad-lanceolate-acuminaate and appear slightly above the ground. Flowers are red with black basal blotches, 8.5-10.0 cm long and petal edges waving.

Tulipa turkestanica (syn. *T. biflora turkestanica*): An early spring-flowering (March) tulip from Turkestan (USSR), Central Asia and NW China which grows up to 30 cm in height. This spreads slowly but is long-lived. Its hairy stems bear 2-4 glaucous grey-green leaves with pinkish colour at the margins. Flowers up to 12 per stem, star-shaped and pointed, 1.5-3.5 cm long and 5.0 cm across, and tapered at the base, unpleasantly fragrant, ivory white to pinkish red, with a yellow to orange basal blotch, which extends to about a third of the flower. The bulb is leathery, bright reddish-brown and has a hairy tunic. They are best planted in large drifts, rock gardens, containers, for forcing indoors, and as cut flowers.

Tulipa urumiensis: A native of NW Iran, this species is similar to *T. tarda* but exterior of the flowers are olive-red while entirely buttercup-yellow inside. It is early spring-flowering, growing from 10 to 20 cm in height. Stem is mostly below soil level and the leaves are greyish-green or green. Flowers are 1-2, some 4 cm long, tapering at base, and the petals are narrowly oval, yellow, flushed with mauve or red-brown outside. It is excellent for rock gardens.

Tulipa violacea (syn. *T. pulchella violacea*): A native to Iran is early spring-flowering, grows up to 30 cm in height, and is suitable for planting on raised beds or in the rock gardens. It is allied to *T. hageri, T. humilis* and *T. pulchella*. The grey-green leaves are 3-5 and crowded, the flowers are fragrant, bright violet-pink or rosy-crimson flushed with purple with yellow or bluish-black centres usually bordered with white, 2-5 cm long and 5 cm wide, and tapered to the base, and the perianth is campanulate with a contracted base, and the segments are oval, subacute.

Tulipa whittallii: A frost-hardy species native of W. Turkey where blooms appear during mid-spring. It is much like and allied to *T. hageri* though not vigorous as much, and with more orange-hued flowers. In height it grows 30-35 cm with 1-4 flowers each stem. The flowers are 3-6 cm long, tapered at the base. Oval petals are bright brownish-orange.

Propagation

Propagation of tulip is done through seeds as well as through vegetative means, *i.e.* from bulbs, bulblets, offsets and micropropagation. However, propagation through bulbs is the commercial method. **Seed propagation** is practiced only by breeders to develop new varieties or to multiply *Tulipa* species. In nature, this being a cross-pollinated crop, through seeds the varieties do not appear true to the type. Moreover, the plants grown from seeds have long juvenile phase and often take 5 to 8 years to come into flowering. However, most of the modern hybrids are sterile though a few set seeds. Open seeds and those obtained through crossing, after maturity of the capsules when these start splitting, these can be collected and stored in paper bags until autumn and these are sown thinly in beds, pots or boxes filled with leaf mould, good garden loam, compost and coarse sand, all equally by volume. These are sown 5-6 mm deep or covered with sand, and 10-20 mm apart. If planted in containers, these are kept in frame at a temperature of 5-8 °C. The seed-sown beds or containers should be watered lightly and regularly. After the winter is over, the seed-sown containers should be shifted to a well ventilated cold house. In the fourth year the bulbs will be large enough to plant out as the planting stock. Through seeds it takes 5-7 years to produce flowers. Rouhi *et al.* (2012) advocated that after sowing if seeds are given 13±1 °C temperature, these germinate within 3-4 weeks.

Vegetative method of propagation is used to derive uniform plants for various vegetative and floral attributes of a particular variety. Commercially the tulips are multiplied only through bulbs. Tulip bulb is renewed annually by natural division of mother bulbs and has a tough brownish membranous scale all around except below at the base where a tiny scar is present from where it was attached with the mother bulb. After removing the membranous scale, fleshy scales appear with one or more small bulbs adhering to the base near the scar which got nurtured by the membranous scale in the previous season when it was also fleshy but for the formation of these small bulbs, this passed on its food reserves rapidly to the buds in its axil. A mature tulip **bulb** consists of a series of concentric fleshy scales which are modified leaves containing food reserves, each scale having a **bud** in its axil, outer scales having more. The innermost scale serves as a food to the central bud which has been clearly formed by September with a short thick stem bearing leaves in the form of rolled umbrella around the flower bud. All these fleshy scales, axillary buds and central bud are situated on the basal plate, the true stem. After planting, the roots come out from the base of the basal plate, and the central bud as a shoot. The buds present in the axil of the fleshy leaves swell drawing food from the fleshy scales and the aerial shoot, the development being more prominent after flowering. When leaves are senescing, the mother bulb is fully depleted of its food reserves developing into 5-7 **daughter bulbs** on an average, though depending on the cultivar. At senescence of leaves or at lifting the new leaves or flowers inside the new bulb is not fully formed though occurs within about a month at 18-21 °C when the bulbs are in store, and then these bulbs require a long period of chilling for proper development, especially stem elongation. Therefore, on planting a large basal leaf emerges first, and then subsequently the flower stem attached with other leaves. After the flowering, the senescence process of leaves and flowers is very rapid, especially in warm weather.

A large bulb produces several **daughter bulbs** in the axils of the bulb scales, the innermost being the largest while other **axillary bulbs** (daughter bulb **offsets**) smaller. The offsets are found either clustered at the base of the floral stems or on stolons or droppers growing outwards from the mature bulbs. Under favourable temperature conditions, the floral bud in the flowering size daughter bulb is completely differentiated by September which on planting will produce flowers. By and large the central bulb is flowering size while other axillary bulbs may or may not be. A flowering size bulb contains at least five scales including/excluding the tunic. It is only the size of the bulb (diameter, circumference or weight) which determines its capability for flowering. By and large, the bulbs weighing above 6 grammes or having above 6 cm circumference may initiate flower though differences among the varieties occur as in var. 'Aureola' the size should not be less than 13 cm. Smaller ones are the planting stocks, not producing the flowers, though in 1-2 season these may attain the flowering size rounded bulbs, however, depending on the cultural conditions and the cultivars.

Two to three foliar applications of CCC (chlormequat) @ 1,000 ppm with Tracel 4g/l immediately after removal of the flower buds at green-bud stage at an interval of 3-4 weeks is helpful in increasing number, weight and circumference of bulbs and bulblets. Mugge and Richter (1984a,b) reported that CCC though did not affect bulb size and plant health in the production of forcing bulbs but there was a 26 per cent increase in optimum-sized bulbs. There is a common belief that deflowering at bud initiation will increase bulb size to some extent, but bulb production was found adversely affected by removal of flower along with one or two upper leaves (Anonymous, 2001-02). At Solan (H.P.) condition, tulip bulbs subjected to 21 °C temperature for four weeks resulted into increased number and weight of bulbs harvested, however, pre-planting treatment of bulbs with BA at 50 ppm promoted early emergence and increased bulb weight (Sehgal, 1995).

However, bulblets formed around the daughter bulbs can also be used to multiply them but it takes years to produce flowers. Production of daughter bulb-offsets, *i.e.* **bulblets**, is from the vegetative axillary buds in the axils of tunicated scales. Normally, 2-3 bulblets can be obtained annually from each bulb. They are detached from the parent bulb, cleaned and graded into various sizes. These bulblets are planted 7-8 cm deep in the light and rich soil at a sunny situation during November. Being small in size, they are raised in the field for 2 to 4 years till they attain a commercial size of 10-12 cm for raising a commercial crop. **Offsets** are separated from the mother bulbs at maturity and are stored dry for replanting in the next season. However, offsets require a year or so for the plants to come into flowering.

Traditional vegetative propagation is a limiting factor for successful introduction of a new variety into the market (Maarten *et al.*, 1997). Therefore, faster multiplication of quality planting material can be achieved through **micropropagation** using bulb scales (Koster, 1993) and stalks as explants. However, the best results for regeneration through bulb scales can be obtained only during July-August under temperate conditions of India, and beyond that period there is death of the explants in the media and hence hindering the multiplication through this noble technique. Hulscher *et al.* (1992) developed a method for the micropropagation of tulips from stem and axillary bud explants cultivated on media containing α-naftylacetic acid (NAA), N6-[2-isopentenyl-adenine (2iP)] and 6-benzylaminopurine (BAP) or zeatine. The regenerated shoots were able to form bulbs after cold treatment on a medium supplemented with 70 g/l sucrose, and lacking growth regulators. Kuijpers and Langens-Gerrits (1997) have reported the regeneration of tulip shoots on a culture medium containing 5µM zeatine and 5µM α-naftylacetic acid, pointing out the significant increase of the rate of meristem formation on stem explants when silver thiosulphate, and especially, paclobutrazol and methyl jasmonate were used. Genetic stability of micropropagated shoots can be confirmed using molecular markers. A new tulip micropropagation method based on the **cyclic shoot multiplication** in presence of the thidiazuron (TDZ) has been reported by (Podwyszynska and Sochacki, 2010), which enables the production of virus-free stock plants, boosts up breeding, and provides new genotypes for the market. Cyclic shoot multiplication can be performed for 2–3 years by using TDZ instead of other cytokinins such as 6-benzylaminopurine (BAP) and N6-[(2-isopentyl) adenine (2iP)]. It is possible to produce 500–2,000 micro-bulbs from one healthy plant. Somatic embryogenesis is admired as the most convenient way of regeneration. **Somatic embryogenesis** in stem tulip culture was first described by Gude and Dijkema (1997) and by Bach and Ptak (2001) in the ovary culture of *Tulipa gesneriana*. These embryogenic cultures can further be propagated in bioreactor and this culture can be used for cryopreservation and genetic modification (Paek *et al.*, 2001). The steps involved in the process of production of tulip somatic embryos include initiation of embryogenic callus, development and maturation of the embryos and their subsequent conversion into plants. Ptak and Bach (2007) observed the highest number of somatic embryos on medium with 25 µM Picloram and 0.5 µM BA using the base of the lower part of the flower stem isolated from bulbs chilled for 12 weeks as explant. Picloram was more efficient in inducing the formation of embryogenic nodular callus than 2,4-D, whereas the latter stimulated formation of colourless callus.

Cultural Practices

Tulip thrives best in well-drained, fertile, sandy loam **soil** having low soluble salts (low EC) and with a **pH** of 6-7. Acidic media or soils are not considered good, as acidity suppresses the effectiveness of ancymidol when tulips are forced as pot plants. Heavy soils can be improved by enriching with well decomposed farmyard manure or leaf mould. There should be proper drainage in the field as tulip cannot withstand waterlogged conditions, as it may cause root and bulb rot. Drainage is problematic in heavy and clayey soils which can be rectified through incorporation of sand and organic manures. Rajaei and Onsinejad (2014) studied the effects of peat @ 90 per cent and 100 per cent, and municipal solid waste compost (MSWC) at different levels (0, 10, 20 and 30 per cent) on flowering time and quality of early flowering tulips. Observations revealed that MSWC at 10 per cent and peat at 90 per cent were more suitable for most traits than the other treatments. Treatments containing MSWC (10 and 20 per cent) resulted in highest amount of chlorophyll content in leaves, but high amount of MSWC caused soil salinity and decreased vegetative and reproductive growth in the plants. Under polyhouse condition, growing media should have 2 parts soil, 1 part coarse sand and 1 part poultry or sheep manure for optimum growth and higher yield of plants.

It is well known that tulips demand definite 'warm-cold-warm' **temperature** changes during entire growth period (De Hertogh *et al.*, 1983). Flower initiation and development take place at 17-20 °C provided the bulbs have attained proper circumference or weight which is normally 12- to 14-cm or even more, the 12-cm being for potted flowering plants while larger for cut flower production, though the size may vary with cultivars. Plants like direct exposure to sunlight during morning and evening hours and partial shade conditions during mid-day. Therefore, selection of site should be done accordingly for tulip growing. In greenhouse, a temperature of 18-20 °C is more suitable for its luxuriant growth. They require a long period of chilling. In Kashmir valley, chilling requirement of tulip is met automatically by keeping them under open field conditions. It is not possible to grow tulips in frost-prone areas without pursuing forcing schedule. Climatic conditions of hilly areas of Himachal Pradesh, Uttarakhand and Kashmir offer excellent opportunities to grow this crop.

Tulips require ample **light** conditions for better growth and development as shade can lead to delay in blooming. Forcing of tulips is done in artificial light because a well insulated store can save energy. In very late forcing, some shading is necessary in greenhouses. Light requirement for greenhouse forced plants is 36 W/m^2 (1,100 fc) for a 12 hour photoperiod for potted plants and 24 hours photoperiod for cut flowers.

Well prepared field with good drainage is very necessary for tulip cultivation. Soil should be dug out deep (up to 30 cm), well pulverized by cultivating it thrice followed by planking each time for better growth and development of the crop. Well rotten compost is mixed at the time of **field preparation** as fresh animal manure can develop *Botrytis* blight. Field should be well levelled and as per requirement of the dimension the beds are prepared having good drainage system.

Tulip is planted from October to December in mid-hills (1,000-1,800 m m.s.l.) whereas on higher hills (above 1,800 m m.s.l.) the **planting** is done from November-December to February. The bulbs having 10-12 cm girth should be treated with Carbendazim @ 0.2 per cent solution for 30 minutes prior to planting. As a general rule, **planting depth** should be two to three times the diameter of the bulb. Larger bulbs are planted at a depth of 20 cm and smaller ones 10 cm deep. Under polyhouse conditions, 15th November planting is most suitable to exploit the better growth and production of flowers and bulbs, followed by 30th October planting. Late-flowering cultivars are planted at 15 cm **spacing** from bulb to bulb, while for early flowering cultivars, the spacing is kept at 10 cm. Bulb density of 50-68 bulbs/m^2 is optimum. In the Netherlands, about 39-80 bulbs/m^2 are planted, whereas 26-38 bulbs/m^2 is followed in Germany. About 8-9 tonnes bulbs are sufficient for one hectare area. Khan *et al.* (2008) found maximum stem thickness and tepal diameter when tulips were planted on 30th October while maximum scape length, wrapper leaf area, vase-life and bulblet weight per plant were recorded when the bulbs were planted on 15th November followed by 30th October.

Mulching should be done immediately after planting as it controls weeds and retains soil moisture. Mulching also prevents dirt from splashing up on the flowers when it rains and improves the soil as it gradually decomposes over the seasons. The thickness of the mulch should be 2.5-4.0 cm. Various organic materials such as dried or chopped leaves, straw, shredded bark, wood chips, rice husk, chopped bagasse, sawdust, *neemcake, etc.* over the bed can be used as a mulch. Crop should be kept free from **weeds**. Timely removal of weeds is an important operation in tulip cultivation as weeds not only compete with the main crop for space, nutrients, light and water but also harbour various harmful insect-pests and diseases which affect the plant growth and flower production significantly. Weed intensity is more during spring season than winter season. Various intercultural operations like hoeing, weeding, *etc.* should be carried out to make the soil porous for better aeration and to keep the fields free from weeds. Herbicides like Carfentrazone, Sulfentrazone and Fluminoxazin can be used to reduce weed population up to 85 per cent in tulip crop. Weedicides like Diquat and Paraquat can control weeds but yield of bulbs is reduced. Timely **roguing** is extremely important for quality bulb production. The plants which differ from the normal plant population in being weak, sickly, bolters, groundkeepers, or dissimilar should be removed at the earliest possible. To maintain genetic purity, roguing should be carried out at vegetative, flowering and harvesting stages.

Most of the cultivars have sturdy stalks. However, certain tulips such as Darwin hybrids bear very large flowers so they need **staking** as winds and heavy rains may knock them over. Staking can be done by bamboo stakes or similar objects which should be inserted quite deep into the soil near each stem without injuring the developing bulb but before opening of the flower bud. The stem should be tied loosely with the help of green string, a coated wire twist or through strips of soft cloth. To obtain flowering in the next season, dead flowers should be removed before they set seeds. However, stems and leaves should be allowed to die back gradually so that the foliage continues to store nutrients in the bulb for the next season.

Tulips have low requirement of **manures and fertilizers**. For obtaining good vegetative growth and higher flower production, the soil should be incorporated with wellrotten FYM at 50-100 tonnes/ha, nitrogen at 140-150 kg, phosphorus at 40-50 kg, and if the soil is deficient in potassium then even potassium at 110-120 kg along with calcium at 100 kg/ha but only after soil analysis. It also responds well to the application of bone meal at a rate of 500-600 kg/ha. Calcium deficiency in terms of scape toppling, flower abortion, chlorosis, water soaked spots, cracking, exudation and purple pigmentation of the lower leaves is often observed in tulips. Nelson and Niedziela (1998) studied the effect of ancymidol, in combination with temperature, $Ca(NO_3)_2$ and cultivar on the prevention of Ca deficiency symptoms during hydroponic forcing of *Tulipa gesneriana*. Scape topple was prevented by $Ca(NO_3)_2$ in a low temperature regime (21 °C day-18 °C night). When forced in a solution of ancymidol, topple was prevented but all other symptoms of Ca deficiency persisted. Application of ancymidol @ 0.18 mg per pot in combination with 5 mM $Ca(NO_3)_2$ was sufficient to completely prevent Ca deficiency in tulips.Frequent moderate **watering** is necessary for growing media to attain a healthy root system. Irrigation should be given every alternate day for crop grown inside polyhouse, whereas in open field at weekly intervals. Growing media should be kept moist but over-watering should be avoided as it can be more harmful than under-watering. Relative humidity inside the rooting room should be around 95-100 per cent (De Hertogh, 1996). Overhead irrigation should be avoided as wetting of foliage can lead to *Botrytis* infection. Proper drainage is necessary for proper root development and to save the bulbs from rotting. Therefore, it is advisable to moist the soil thoroughly from root formation until blooming but not allowing it becoming persistently wet as this condition becomes dangerous. After commencement of flowering, beds or containers are watered just to keep the soil or media moist and to avoid cracks on the surface of media.

The **flower yield** in tulip varies according to cultivar, bulb size, planting density, management practices, prevailing climatic conditions, etc. Approximately, 4-5 lakh cut flowers can be obtained from an area of one hectare at a density of 50-60 bulbs/m^2.

Growth and Development

Its growth follows natural **temperature** cycles in the growing areas. By winter, the bulb is properly rooted with floral and leaf meristems present. Cold is required for floral shoot elongation for further differentiation of leaves and flowers, and development (Dole and Wilkins, 1999). In spring, the warm temperature causes rapid shoot elongation and flowering, *vis-a-vis* initiation of daughter bulbs inside the axillary buds, and in late spring and early summer the aerial parts mature, senesce, the old bulb tissues degenerate and growth of the daughter bulbs is complete which now appear as dormant but inside still the differentiation process in the reproductive and vegetative tissues is going on along with the differentiation in the rooting zone of the basal plate. When in early summer the bulbs are harvested, these activities are continued and when after meeting the appropriate spring temperatures these are planted in autumn at the appropriate cold temperature, the development of roots starts and scapes elongate and emerge out. Under temperate conditions of Srinagar and other temperate regions of Jammu & Kashmir, Himachal Pradesh and Uttarakhand, growing temperatures range between 15 to 25 °C (day) and -2 to 5 °C (night), during the entire growing period which are congenial for their growth and development. Le Nard and De Hertogh (1993) mentions that the dominant axillary bud typically initiates four vegetative leaves, followed by petals, stamens and finally the gynoecium, and before the commencement of cooling of the bulbs, it is critical that flower primordia reach the G_{3} stage of development, *i.e.* when gynoecium has developed a well-indented 3-lobed stigma (De Hertogh, 1996). In exceptionally cool summers bulbs may not reach the G-stage though quite warm (hot) weathers cause smaller-sized bulbs and increase risk of *Fusarium* (Dole and Wilkins, 1999). Precooling should start before planting just as rooting zone at the basal plate is fully developed and then these can be transported or stored at 17 °C until planting so that completion of floral development and attainment of G-stage is appropriated.

Forcing is a technique in which tulip bulbs are subjected to a period of low temperature prior to planting, and to artificial illumination, for uniform, proper and staggered *cum* regulated flowering. The varieties suitable for forcing under artificial lighting are 'Aladdin', 'Albury', 'Apeldoorn', 'Charles', 'Dix Favourite', 'Jimmy', 'Kingdom', 'K&M's Triumph', 'London', 'Orient Express', 'Oxford', 'Paul Richter', 'Prominence', 'Ruby Red', 'Stockholm', 'Tommy, *etc.* under **orange and reds**; 'Blenda', 'China Pink', 'Christmas Marvel', 'Palestrina', 'Preludium', 'Rosario', *etc.* under **pinks**; 'Golden Apeldoorn', 'Monte Carlo', 'West Point', *etc.* under **yellows**; 'Hibernia', 'Pax', 'Snowstar', *etc.* under **whites**; 'Invasion', 'Kees Nelis', 'Madame Spoor', 'Thule', *etc.* under **reds and yellows**; and 'Attila' under purple. Only a few cultivars are suitable for forcing. Forcing in trays is the most commonly used method. After planting, the bulbs are allowed to develop root and shoot at 9 °C for 13 to 20 weeks. For early forcing the bulbs are cooled in August or September and potted in early October, though for late forcing the bulbs are held at 17 °C before potting and then are held at 9 °C but in early November to December the temperature is reduced to 5 °C when plants have properly rooted. Once the bulbs are removed from the rooting room for forcing, only a small amount of **light** is required. In early forcing if emerging shoots are not shadowed to reduce light, plants may remain short so they require to be etiolated initially by providing only 1,000 to 2,500 fc light for greenhouse forcing (De Hertogh, 1996). Tulips forced with fluorescent lights (36 W/m^2 or 1,100 fc) for a 12-h photoperiod for potted plants and 24-h for cut flowers with longer stalk lengths are similar or even better than greenhouse forced plants (Dole and Wilkins, 1999). In ealy January the temperatures are lowered to 0-2 °C. This treatment results in longer stem indicating that very low temperature promotes faster growth of crop. It should be ensured that bulbs reach the G_3-stage, when gynoecium is clearly visible on dissection before forcing the bulbs, otherwise flowers may abort. The bulbs are thawed immediately after digging and it takes 14 to 18 days to force them. They are subjected to low storage temperature stepwise from 23 °C in mid- September to 17 °C before the onset of cooling at 9 °C from mid-October to mid-December. The bulbs are then either planted or packed in peat in polythene covered boxes and frozen at -2°C until they are again removed for forcing in greenhouse at 11-18 °C for planting. This all forcing procedure involves lots of labour and energy and thus increases the cost of tulip bulbs. For greenhouse forcing, application of NPK complex fertilizers (8:8:8) @ 1 kg/33 row meter as basal dose and second dose of same amount should be applied at shoot emergence stage. Application of calcium nitrate is helpful to reduce the stem topple disorder. Fertigation with calcium nitrate (25 g/10 litre of water) is beneficial and should be done twice a week or it may be applied at the time of each watering.

For growing tulips in pots, besides planting dwarf varieties such as 'Greigii' hybrids, application of **growth regulators** is extensive in various tulip growing areas of the world. Various chemicals like A-Rest, CCC and PP333 are commercially used as growth retardants for controlling height in potted tulips. Spraying of plants with cycocel 100 ppm after deflowering increases bulb yield, and GA_3 is beneficial in balancing cold requirement of bulb as well as to enhance flowering. Dipping bulbs in GA_3 at 100 ppm for 6-8 hours before planting promotes vigorous growth and early flowering. Kumar *et al.* (2013) studied the effect of plant growth regulators on growth, flowering and bulb production in tulip (Tables 36.1 and 36.2). Three different growth regulators; GA_3 at 100, 200 and 400 ppm, and 2-chloroethyl trimethyl ammonium chloride (CCC) and maleic hydrazide (MH) each at 100, 200 and 500 ppm along with control were applied as dip treatment and foliar spray. Earliest flowering, maximum plant height and vase life, *vis-a-vis* number of daughter bulbs per plant were recorded with 400 GA_3, followed by 200 ppm GA_3 where even blooming period was also found increased. However, flowering was found delayed under 500 ppm MH. Rajaei and Onsinejad (2014) when used GA_3 at 0, 50 and 100 mgl^{-1}, recorded highest number of flowers and bulbs in the plants treated with 100 mgl^{-1}. Application of BA in lanolin paste at 0.25 per cent around the basal plate results in the formation of parrot like perianth in cultivars such as 'Triumph', 'Red Sunshine' and 'Redshine' due to an increase in levels of cytokinin as compared to parent cultivars.

Table 36.1: Effect of GA_3, CCC and MH on Vegetative Attributes of Tulip cv. 'Apeldoorn'

Treatment	Days to Bulb Sprouting	Plant Height (cm)	No. of Leaves per Plant	Wrapper Leaf Area (cm^2)	Floral Stem Diameter	Field Life (Days)
GA_3 100 ppm	81.38	32.83	3.77	126.80	6.86	178.20
GA_3 200 ppm	80.25	34.13	4.31	131.43	6.32	81.05
GA_3 400 ppm	78.62	37.32	4.54	137.20	6.15	183.70
CCC 100 ppm	83.25	31.51	3.67	123.36	5.62	180.75
CCC 200 ppm	85.47	29.93	4.64	119.56	5.76	182.00
CCC 500 ppm	86.16	29.08	4.76	117.16	6.06	184.75
MH 100 ppm	83.32	30.09	3.77	122.80	5.86	181.50
MH 200 ppm	84.55	28.75	3.83	119.00	5.86	180.49
MH 500 ppm	86.30	27.48	4.21	117.50	5.81	182.39
Control	82.89	31.91	3.62	120.90	5.75	178.04

Kumar *et al.* (2013).

Table 36.2: Effect of GA_3, CCC and MH on Floral Attributes of Tulip cv. 'Apeldoorn'

Treatment	Days to Floral Bud Appearnace	Days to Colour-Break Stage	Days to Flowering	Flower Size (cm)	Floral Stalk Length (cm)	Vase Life (Days)
GA_3 100 ppm	135.23	141.03	143.50	5.67	24.50	9.40
GA_3 100 ppm	131.84	138.46	142.43	6.39	26.36	10.43
GA_3 100 ppm	130.72	136.20	141.30	6.70	31.96	11.26
CCC 100 ppm	140.21	142.43	148.83	5.39	23.90	8.90
CCC 200 ppm	141.89	143.80	150.83	5.25	21.86	9.33
CCC 500 ppm	143.81	146.86	151.75	4.99	20.50	10.33
MH 100 ppm	139.03	143.60	151.40	5.29	21.90	8.26
MH 200 ppm	140.02	145.33	151.93	5.28	20.10	8.53
MH 500 ppm	141.70	145.63	152.96	5.16	18.86	7.63
Control	139.03	142.30	148.93	5.40	22.86	7.30
CD (P=0.05)	1.85	1.66	1.33	0.23	1.22	0.44

Kumar *et al.* (2013).

By and large, to initiate flowering at the specific dates, the **scheduling** is cultivar-specific though preferably the bulb size should be 12/14 cm size or above at least for delayed planting, whether it is for potted plants or for cut flower production. Forcers should also be aware whether the bulbs are regular or precooled. Dole (1996) developed delayed potting method to reduce cooler time and cooling space by still holding the bulbs in the cooler for next date for required number of weeks, and at six weeks before moving bulbs to the greenhouse the bulbs are potted, watered and returned to the cooler at 5 °C for development of roots and shoots simutaneously at the same time and so as per requirement to the next important date the second crop can be programmed in the cooler space when for first date the crop has been moved to the greenhouse for forcing. In any case the cold dry storage should not exceed 112 days (16 weeks) or late January. After this period the bulbs are potted and cold-applied in the moist medium for six weeks from 1-9 °C, preferably 5 °C whether it is for dry or moist treatments, however, the temperature below 5 °C generally decreases the flowering percentage while above this but up to 9 °C it lengthens the floral stalk (Dole and Wilkins, 1999). To utilize the cooler-space to its full potential, when a part is taken out for planting, for a later forcing the next lot of dry potted bulbs can be accommodated, but all the bulbs should be potted by the end of January and if it is noticed that there is excessive shoot or root growth in the cooler especially for later forcing, the temperature is lowered below 5 °C (Dole and Wilkins, 1999).

Bulb Lifting and Storage

Tulips should be be-headed to prevent seed production. Bulbs should be harvested after six weeks of flowering when the foliage has turned yellow. The forced bulbs are never re-used so while cutting the flowers the whole plants are lifted, bulb dissected and the stalk is sheared off from above the basal disc and by this way the stalk lengthens at least by 2.5 cm in length. However, the bulbs are collected only from the unforced plants and from those maintained as planting stock as these are new and quite vigorous. Lifting of bulbs from soil should be done every year manually with the help of spade. However, dwarf species of tulips such as *Tulipa fosteriana*, *T. greigii* and *T. kaufmanniana* and their hybrids often re-flower without lifting and they should be lifted only when clumps get overcrowded. Adhering soil to the surface of bulbs should be thoroughly cleaned and any damaged or diseased bulbs should be discarded. Normally a flowering mother bulb produces 4-5 daughter (at the most up to 7) bulbs which after lifting are separated, cleaned, infected and injured ones discarded, graded into flowering and non-flowering sizes, treated with some aphidicide or placed with some fumigant such as dichlorvos strips and then stored under dry and airy condition at 18-20 °C temperatures (Misra and Misra, 2013). The large bulbs are used for sale or planting and smaller ones retained separately as planting stock. In the midst of lifting and storage period, the bulbs though appear normal but inside them the process of differentiation of flowers, vegetative buds and roots occur. Storage temperatures influence all these processes and their subsequent development. Even smaller than flowering-size bulbs, if immediately after lifting, are stored at 30 °C for several weeks, often these become capable of producing flowers. Flower bud differentiation is quickest at 17-20 °C as compared to its higher even up to 35 °C or as lower as 1.5 °C (Misra and Misra, 1913). A short duration of temperature of 30-35 °C, followed by 15-20 °C induces earlier flower bud differentiation and rooting, and if once the organogenesis has occurred, these bulbs are stored at low temperature so that after planting the plant growth is rapidly promoted (Misra and Misra, 2013). Bulbs should be thoroughly dried in partial shade before storing and store them in trays in a cool place (7-9°C) for 6-8 weeks during September-October. Cold storage helps in the development of flower primordia. Storing of bulbs in high humidity should be avoided as this will make bulbs susceptible to fungal attack, however, 70 per cent RH is optimum, and too dry storage atmosphere causes shrivelling and loss of viability. The Netherlands is the major producer of tulip bulbs and produces more than 3 billion bulbs annually mainly for export. On an average three bulbs per plant can be obtained under proper management practices.

Insect-Pests, Diseases and Physiological Disorders

Tulips are infested by various species of **aphids** such as 'tulip bulb aphid' (*Dysaphis tulipae*), 'lily aphid' (*Aulacortum circumflexum*), 'green peach aphid' (*Myzus persicae*) and 'black bean aphid' (*Aphis fabae*). *Dysaphis tulipae* infests the bulbs and the aerial parts and potato aphid (*Macrosiphum euphorbiae*) also infests on tulips. The infested bulbs have red or brownish spots which are waxy in appearance but later on turn black due to the development of sooty mould. Aphid attack is also seen on shoots if sprouting occurs during storage. Although, aphids are not so common pests of tulips but they transmit deadly viral diseases. Bulbs should be stored at low temperature (<15 °C). Application of Fipronil @ 40-50 ml a.i./ha or Imidacloprid @ 25-50 ml a.i./ha should be done to get rid of aphid problem.

Bulb mites (*Rhizoglyphus echinopus*) attack the healthy bulbs as well as bulb tissues damaged mechanically, environmentally, by fungi, bacteria, nematodes or by any other means. They are yellowish white in colour with a pink tinge, bead-like, shiny and crawling. Leaves also show symptoms of mite's infestation which can reach young flower parts in the shoot and when combined with ethylene, these cause bud necrosis. Damaged tissues appear brownish in colour. **Tulip gall mites** (*Acaria tulipae*) infest the bulbs in store so these should be controlled through monthly fogging with primiphos-methyl (Muller and Conijn, 1981). In field in case of severe infestation, effective miticide should be used to control them.

Tulips are seriously affected by a race of **stem and bulb nematodes** (*Ditylenchus dipsaci*) though another race of the same nematode infests narcissi. This infests the bulbs from the base and causes soft grayish or brownish patches at the base on the outermost scale. Initial symptoms of infestation are best examined at flowering stage which appear as pale or purplish lesions on stems just below the flower and stem bends. These lesions ultimately enlarge both in downward and upward directions above the petals. In Europe, legislative laws have been established for *Ditylenchus dipsaci* as a significant quarantine pest. Soil application of Furadan (Carbofuran) @ 20-30 kg/ha is quite effective. Application of neem cake and cultivation of marigold flowers also minimise nematode population. Warm storage of dormant bulbs for one week at 34 °C and then HWT at 53 °C for 30 minutes eradicate these nematodes as has been proved at Kirton Experimental Horticulture Station in Great Britain (Anon., 1975). Dipping of bulbs in phoxim, thionazim or Carbofuran (Damadzadeh and Hague, 1979) or use of 0.6 per cent a.i. of Aldicarb or 0.28 per cent a.i. of Oxamyl (Windrick, 1985) is effective to control nematodes.

Tulip fire (*Botrytis tulipae*) is the most common and highly devastating disease of tulips. Initial symptoms include appearance of light tan patches on the leaves before proper emergence and the patches are more prominent on light-coloured varieties. Infected leaves have small brown necrotic spots which coalesce to form large brown patches and ultimately result into blight of entire leaves. Fungus also attacks the flowers and represents flecked-brown to white spots. White spots appear on coloured petals but brown on white petals. High humidity leads to necrotic spots on leaves, stems and flowers which are accompanied by sporulation of the brown-gray fungus. Rapid infection occurs once the fungus begins sporulation; and hence it is named 'tulip fire'. Leaves, stems and flowers collapse due to severe infection. Careful inspection of the bulb before planting is required after removing the outer husks of the bulb and the discoloured or spotted ones should be discarded from planting. Regular inspection of the field is necessary as early detection and hand-roguing is essential for eradicating this disease. Infected plant parts should be removed and burnt when weather is dry as wet-removal will help spreading the fungal spores (conidia). Overhead irrigation and crowded plants also lead to faster spread of the disease. Prior to planting the bulbs should be dipped in benomyl, Thiabendazole and formalin @ 2 per cent for 30 minutes to reduce the carryover of fungus. Zamorski (1985) used benomyl, carbendazim and thiophanate-methyl dipping of bulbs and observed that though the incidence was reduced but there was no complete control, however, the secondary leaf infection was found reduced and the conidial sporulation was almost completely inhibited when spraying was done. Spraying with Benomyl (0.1 per cent) and Mancozeb (0.25 per cent) just after leaf development is recommended to check developmenat of the disease. Koster *et al.* (1992) reported to have controlled this disease with reduced fungicide inputs up to 70-82 per cent through low doses of chlorothalonil, prochoraz, fluazmam, dicarboximide or benzimedazole.

***Fusarium* bulb or basal rot** (*Fusarium oxysporym* f. sp. *tulipae*) develops as grayish-brown spots on infected bulbs during storage and sometimes concentric rings and a clear yellow rim appear. Infected bulbs emit a distinctive and acrid smell. Even healthy bulbs may carry over spores of such fungus. Bulbs produce symptoms of disease from the base of the root crown at an early stage of plant growth. In moderate infection, symptoms include retardation in growth, yellowing of flower tips and desiccation of flower buds while severely infected bulbs will not emerge at all. Infected bulbs release ethylene gas during storage which leads to gumming. To control the spread of the disease, removal of infected plants is necessary. Healthy stock should be maintained and stored in a well-ventilated place at or below 15 °C temperature so that apart from checking its sporulation even ethylene build up is also restricted. Treatments of tulip bulbs with carbendazim + Thiram proved effective against *Fusarium* and *Botrytis* but Funaben 50 (benomyl) proved most effective (Niebisch and Kelling, 1986). The bulbs should be dipped in Benzimidazole fungicide immediately after lifting (within 48 hours). Also, dipping the bulbs for 2 hours in a 0.5 per cent formalin provides good control. Planting should be done at a soil temperature of 9°C or less and not above 13°C to manage the occurrence of the disease. Spray of Benomyl and Thiophanate (0.05 per cent) provides excellent control of the disease.

Root rot (*Pythium ultimum* var. *ultimum, P. irregulare* and *P. spinosum*) occurs both during tray cultivation and in the greenhouse, the former being the most important one. Inoculum appears from the tulip remnants, especially the roots present in the soil from the previous year's tulip crop. Less severe symptoms include partially rotted root system where other plant parts are not damaged, whereas in severe cases flower blast occurs and plants remain short. The infected roots have a narrow brown stripe, appear glassy or watery and break off easily. As the infection progresses, the entire root system turns brown. The severity of disease is enhanced with increase in temperature and moisture content of the soil. Planting should be done in fresh soil while the infected soil with *Pythium* needs additional soil treatment. Soil sterilization with steam or formalin before planting provides good control. The drainage should be proper in the field and infected plants should be immediately removed and burnt. Soil drenching with Thiram and Captan at 2.5 g/m^2 prevents infection. Biological control is achieved either by mixing the bacteria through the soil or by dipping the bulbs in a bacterial suspension immediately before planting using *Pseudomonas* isolate E113 and *Trichoderma harzianum*, where former is much more effective (Greeff and Duineveld, 1991).

Rhizoctonia solani affects tulip bulb stored at 5 °C and also the standing crop in the field. In its infection the growth of leaves is hampered, these become pale, often having a dead streak along the margins, lose turgidity and may die whereas the daughter bulbs are misshapen and when dissected the concentric rings are visible inside the bulbs. The disease is transmitted from one tulip or chrysanthemum crop to the other. It is only soil sterilization which checks the spread of the disease further. Koster (1983) found better control of *R. solani* with use of fungicides such as tolcxlogos-methyl, furmecyclox and pencycuron.

Grey brown rot (*Rhizoctonia tuliparum*) is a soil-borne disease and its infection to the bulbs causes these not producing shoots on planting, *vis-a-vis* rotting in the field and also in storage. Some 10 per cent of sclerotia of this fungus remain viable in the soil for up to 10 years, therefore, Coley-Smith *et al.* (1979) suggested the same field for tulip growing should be used only after five years of rotation.

Storage rot (*Penicillium hirsutum*) causes green fungal growth on the bulbs due to insufficient aeration in the storage or due to injury on the bulbs coupled with high humidity and warm atmosphere, and such bulbs rot completely. Prince *et al.* (1987) when treated the bulbs with Prochloraz, Etaconazole or Captan before packaging, recorded significant reduction of the basal plate rot due to *Penicillium* infection.

'Tulip breaking virus' is the most common viral disease and causes colour breaking or streaking of the pink, purple or red coloured and late flowering cultivars. In China, Duan YongJia and Cai Hong (1998) observed brilliantly variegated tulip flowers infected with tulip breaking potyvirus (TBV). However, white and yellow flowered cultivars are not infected by this disease. Leaves become mottled or striped and there is loss of vigour coupled with poor flower production. Although colourful flowers are produced but plants get weakened and subsequently die. This viral disease is spread by various aphid species such as *Aphis fabae, Macrosiphum euphorbiae* and *Myzus persicae* which require immediate effective control as advocated earlier. Infected plants should be immediately removed and destroyed. Planting of tulips near lilies should be avoided. **'Tobacco necrosis virus'** (Augusta disease) is spread by a fungus *Olpidium brassicae* and the symptoms include appearance of brown necrotic streaks on leaves and stems. The disease is more damaging to early flowering cultivars. Control measures include removal of infected plants, changing the site of tulip cultivation and late planting in November which help to minimize the menace. **'Cucumber mosaic virus'** appears as grey-brown sunken spots on the bulb scales during storage. When these infected bulbs are planted in the field, they produce deformed plants with leaf chlorosis and damage or abortion of flowers. Late flowering cultivars are more susceptible to this disease. The disease is spread through aphids so aphids should be controlled whenever these are seen.

Physiological disorders are 'stem topple' ('water neck'), 'ethylene-related maladies' ('plant distortion', 'flower abortion', 'flower blindness', gummosis), 'retarded growth', and 'leaf-flower fusion'. **Stem topple** appears due to calcium deficiency which is symptomised by glassy stem appearance, with top part becoming dark green and watery leading to shrinking of tissue and drooping and falling of the stem. Watery spots appear on flowers and leaves from which water drops are often released. High relative humidity in the greenhouse or poor rooting of the bulbs are the main reasons of this disorder. Lower transportation of water through the plant because of high relative humidity causes calcium deficiency in the fast growing parts of the plants, which in turn affects nitrogen absorption. In such case, excessive cold treatment is very harmful. Cultivars 'Gander', 'Kees Nelis', 'Leen van der Mark', 'Princes' and 'White Dream' are found to be quite susceptible. High RH levels (over 80 per cent) in the greenhouse and excessive growth of plants should be prevented. Affected part should be treated in a solution of 1 per cent calcium nitrate. Tulips are highly susceptible to **ethylene** when its concentration is >0.1 ppm and its exposure at any stage of forcing, programming and transportation, being further aggravated due to *Fusarium* infection in the bulbs, and these result into poor rooting, short and/or thin plants, bulb blasting, gummosis, bud necrosis, and floral abortion & malformations (Rees, 1968; Le Nard and De Hertogh, 1993; De Hertogh, 1996; Gude and Dijkema, 2005). Ethylene also interferes with the physiological processes such as respiration, *vis-a-vis* inhibits organ initiation. Bulbs should be stored at a temperature lower than 13 °C to minimize damage. It should therefore be ensured that storage rooms and the greenhouses are completely free from ethylene. **Retarded growth** of plants occurs when either the bulbs are insufficiently matured or if matured ones have been stored under quite warm conditions. **Winter injury** occurs when bulbs are planted too late in the autumn in a compact and poorly drained soil. Roots fail to grow and shoots become abnormal and distorted. Ultimately, bulbs decay and death of plants occur. Sometimes the normal floral buds are formed but these become blind and do not open properly or fail to develop completely which is attributed to exceessive **high temperature exposure** of bulbs before planting (Charles-Edwards and Rees, 1975). The abortion of either whole flower or part of flower may take place at certain stages of plant growth after initiation. To some extent the maladies caused by excessive temperatures can be corrected through GA_3 and kinetin treatments. **Leaf flower fusion** is an abnormal fusion of the leaves and flower petals together. The real cause of occurrence is though not known yet but it is assumed that it occurs due to hormone-type weed killer used during the previous growing season.

Postharvest

Potted tulips are ready for marketing when petals start showing colour and so should be the case with cut tulips but in latter case the petal colour intensity and development should be half complete (Dole and Wilkins, 1999).Tulip flowers are **harvested** at proper flower development stage but still closed, as this stage will minimize the damage during transportation. However, Darwin hybrid tulips are harvested at partial colour development stage. The harvesting time in mid-hills and high hills is February-April and April-June, respectively. When the bulbs from the forced crop are being lifted, after the storage period is over, these are dissected, entire part of stalk which is about 2.5 cm long inside (Dole and Wilkins, 1999) is removed and then bulbs are dumped for composting. Entire plant harvesting facilitates upright storage of the flowers. While harvesting the flowers from unforced bulbs, the wrapper leaf on each plant should be left for further development of the bulbs and after leaf senescence these are also lifted (Misra and Misra, 2013). Harvesting of the flowers should be done in the cool hours of the day, *i.e.* either early in the morning or late in the evening to minimize field heat. Immediately after harvest, tulips should be stored at a cool place to remove field heat. Pre-cooling is a method to remove field heat which induces glucose accumulation in the elongation of internodes that may be accompanied by reserve mobilization in the scales (Lambrechts *et al.*, 1994). The practice of flower harvesting along with bulbs also knocks out the pathogens from soil which otherwise could have hibernated in the soil and damaged the next crop.

Tightness of flower buds prolongs the cut flower life. These scapes are then bunched and wrapped tightly before storage at a low temperature (0.5-0.8°C) and high RH to prevent desiccation of petals. The flowers are preferably dry-stored, especially when storing for longer, and in a vertical position to prevent geotropic bending. For long period storage (up to 2 weeks), flowers are stored at 0 to 2 °C along with their bulbs till their bunching process and are kept upright, however, without bulbs the storage is limited only to five days. Potted plants and cut flowers both are kept as cool as possible at 0 to 2 °C where potted plants can be retained up to one month at the most (Dole and Wilkins, 1999). Modified atmosphere packaging (MAP) is widely accomplished for both whole and fresh cut products in order to extend storage life as well as the shelf-life (Watada and Qi, 1999). In fact, it is endogenous GA which controls elongation of flowering scape, and this GA in plants is released due to cold treatment. McDaniel (1990) reported that use of paclobutrazol and ancymidol can suppress the scape elongation in potted 'Paul Richter' tulips under greenhouse and low light postharvest environments. Paclobutrazol pre-plant bulb soaks at 2.5 or 5.0 mg/l prevented excessive scape elongation during exposure to low light. Application of A-Rest causes primary reduction of internode lengths with the two basipetal nodes during forcing as their elongation is greater as compared to two acropetal nodes whose elongation is primarily controlled by auxin of the gynoecium which occurs in the home environment (Dole and Wilkins, 1999) and is accelerated by low light and warm temperatures and presently there is no known method to check the elongation of these acropetal nodes (Aung and De Hertogh, 1979; Hanks and Rees, 1980; Rees, 1985; Le Nard and De Hertogh, 1993), however, ethephon or ethylene at 50 ppm can slow down this elongation process but commercially this is not being applied (Nichols and Kofranek, 1982). Ferrante and Tognoni (2003) observed that TDZ was able to inhibit leaf yellowing significantly during vase life in cut tulip cvs 'Attila' and 'White Dream'. Leaf yellowing was delayed by over 20 days in cut tulips treated with TDZ while in control, symptoms were observed after 7-9 days. However, higher doses of TDZ reduced the stem length. The vase life of tulip flower is 4 to 11 days at 16 °C and from 3 to more than 6 days at 23.5 °C (Misra and Misra, 2013). The cut flower life is influenced due to cultivar and prevailing weather conditions. Vase solution in the form of only the distilled water is best, however, 8-HQC 200 ppm + sucrose 4 per cent with GA_3 or $AgNO_3$ 25 ppm + sucrose 1.5 per cent is to some extent effective in enhancing the vase life (Misra and Misra, 2013). Nowak and Rudnicki (1975) when treated tulip cv. 'Lustige Witwe' with Proflovit-70 (0.3 g 8-HQS+0.05 g CCC+50 g sucrose/l) and compared with $AgNO_3$ at 20-50 ppm, $KMnO_4$ at 4 ppm, streptomycin at 200 ppm or 8-HQS at 200 ppm, sucrose 5 per cent or CCC 50 ppm, they recorded Proflovit-72 giving best results with regard to vase life and flower quality.

References

Anonymous, 1979. Fifteenth Annual Report of Kirton Experimental Horticulture Station, England.

Anonymous, 1981a. Organische bemesting. *Bloembollencultuur*, **91**: 1252-1253.

Anonymous, 1981b. Die planthoedontsmetting. *Vakblad voor Bloembollencultuur*, **92**: 137-138.

Aung, L.H. and A.A. De Hertogh, 1979. Temperature regulation of growth and endogenous abscisic acid-like content of *Tulipa gesneriana* L. *Plant Physiol.*, **63**: 1111-1116.

Bach, A. and A. Ptak, 2001. Somatic embryogenesis and plant regeneration from ovaries of *Tulipa gesneriana* L. in *in vitro* cultures. *Acta Hort.*, No. 560, pp. 391-394.

Bailey, L.H. and E.Z. Bailey, 1976. *Tulipa*. In: *Hortus Third: A Concise Disctionary of Plants Cultivated in the United States and Canada*, pp. 1132-1134. Macmillan Publishing, New York, USA.

Beckett, K.A. 1983. *The Concise Encyclopedia of Garden plants*. Orbis Publishing Ltd., London.

Botschantzeva, Z.P. 1982. *Tulips: Taxonomy, Morphology, Cytology, Phytogeography and Physiology*. CRC Press, USA, 120 p.

Charles-Edwards, D.A. and A.R. Rees, 1975. An analysis of the growth of forced tulips. 2. Effects of low temperature treatments during development on plant structure at anthesis. *Scient. Hort.*, **3**: 373-381.

Coley-Smith, J.R., D.R. Humphreys-Jones and P. Gladders, 1979. Long term survival of sclerotia of *Rhizoctonia tuliparum*. *Plant Path.*, **28**(3): 128-130.

Custers, J.B.M., W. Eikelboom, J.H.W. Bergervoet and J.P. Van Eijk, 1995. Embryo-rescue in the genus *Tulipa* L.: Successful direct transfer of *T. kaufmanniana* Regel germplasm into *T. gesneriana* L. *Euphytica*, **82**: 253-261.

Damadzadeh, M. and N.G.M. Hague, 1979. Control of stem nematode (*Ditylenchus dipsaci*) in narcissus and tulip by organophosphate and organocarbamate pesticides. *Plant Path.*, **28**(2): 86-90.

De Hertogh, A.A. 1996. Tulips. In: *Holland Bulb Forcer's Guide* (5th ed.), pp. 51-54. International Flower Bulb Centre, Hillegom, The Netherlands.

De Hertogh, A.A., L.H. Aung and M. Benshop, 1983. The Tulip: Botany, Usage, Growth and Development. *Hort. Rev.*, **5**: 45-125.

Dole, J.M. 1996. Spring bulb production: The delayed potting method. *Ohio Florists' Asscn. Bull.*, No. 806 (1), pp. 3-4.

Dole, J.M. and H.F. Wilkins, 1999. *Tulipa*. In: *Floriculture Principles and Species*, pp. 537-545. Prentice Hall, Inc., New Jersey, USA.

Duan YongJia and Cai Hong, 1998. Studies on the variegated flowers of flowering plants (Chinese). *Acta Phytopath. Sinica*, **28**(1): 85-89.

Ferrante, A. and F. Tognoni, 2003. Treatment with thidiazuron for preventing leaf yellowing in cut tulips and chrysanthemum. *In*: *Elegant Science in Floriculture* (eds Blom, T. and R. Criley). *Proc. XXVI intern. hort. congr.*, held on August 11-17, 2002 in Toronto on behest of Canadian International Development Agency (CIDA). *Acta Hort.* (ISHS), No. 624.

Greeff, R. de and T. Duineveld, 1992. Towards biological control of *Pythium* root rot in forced tulips. *Bull.-OILB-SROP*, No. 15(1), pp. 45-47.

Gude, H. and M.H.G.E. Dijkema, 2005. The use of 1-MCP as an inhibitor of ethylene action in tulip bulbs under laboratory and practical conditions. *Acta Hort.*, No. 673, pp. 243–247.

Gude H. and M.H.G.E. Dijkema, 1997. Somatic embryogenesis in tulip. *Acta Hort.*, No. 430, pp. 275-300.

Hanks, G.R. and A.R. Rees, 1980. Growth substances of tulip: The activity of gibberellins-like substances in field-grown tulips from planting until flowering (German). *Zeitschrift für Pflanzenphysiologie*, **98**: 213-223.

Hulscher M., H.T. Krijksheld and P.C.G. Van der Linde, 1992. Propagation of shoots and bulb growth of tulip *in vitro*. *Acta Hort.*, No. 325, pp. 441-446.

Khan, F.U., A.Q. Jhon, F.A. Khan and M.M. Mir, 2008. Effect of planting time on flowering and bulb production of tulip under polyhouse conditions in Kashmir. *Indian J. Hort.*, **65**(1): 79-82.

Killingback, S. 1990. *Tulips-An Illustrated Identifier and Guide to Cultivation*. Apple Press, London, pp. 21-93.

Koster, A.T.J. 1983. A new material for control of cereal volunteers in narcissi, irises, crocuses and gladioli. *Bloembollencultuur*, **93**(39): 1008-1009.

Koster, J. 1993. *In Vitro* Propagation of Tulip. M.Sc. Thesis. University of Leiden, The Netherlands.

Koster, A.T.J., L.J. van der Meer, K. Verhoeff, N.E. Malathrakis and B. Williamson, 1992. Control of *Botrytis* spp. in tulip with reduced input of chemical crop protection. Recent advances in *Botrytis* research. *Proc. of the 10th intern. Botrytis Symp.*, held at Heraklion, Crete, Greece on April 5-10, 1992, pp. 277-281.

Kuijpers, A.M. and M. Gerrits-Langens, 1997. Propagation of tulip *in vitro*. *Acta Hort.*, No. 430, pp. 321-324.

Kumar, R.N., Ahmed, D.B. Singh, O.C. Sharma, S. Lal and M.M. Salmani, 2013. Enhancing blooming period and propagation coefficient of tulip (*Tulipa gesneriana* L.) using growth regulators. *African J. Biotech.*, **12**(2): 168-174.

Lambrechts, H., F. Rook and C. Kolloffel, 1994. Carbohydrate status of tulip bulbs during cold- induced flower stalk elongation and flowering. *Plant Physiol.*, **104**: 515-520.

Le Nard, M. and A.A. De Hertogh, 1993. *Tulipa*. In: *Physiology of Flower Bulbs*, pp. 617-687. Elsevier, Amsterdam, the Netherlands.

Maarten, W.P.C., van Rossum, M. Alberda and H.W. Linus van der Plas, 1997. Role of oxidative damage in tulip bulb scale micropropagation. *Plant Sci.*, **130**: 207-216.

Marasek-Ciolakowska, A., H. He, P. Bijman, M.S. Ramanna, P. Arens and J.M. Van Tuyl, 2012. Assessment of intergenomic recombination through GISH analysis of F_1, BC_1 and BC_2 progenies of *Tulipa gesneriana* and *Tulipa fosteriana*. *Plant Syst. Evol.*, **298**: 887-899.

McDaniel, G.L. 1990. Postharvest height suppression of potted tulips with paclobutrazol. *Hort. Sci.*, **25**(2): 212-214.

Misra, S. and R.L. Misra, 2013. *Tulipa*. In: *Commercial Ornamental Bulb Science*, pp.165- 180. Westville Publishing House, New Delhi.

Mlcek, J. and O. Rop, 2011. Fresh edible flowers of ornamental plants - a new source of nutraceutical foods. *Trends Food Sci. Technol.*, **2**: 561-569.

Mugge, A. and P. Richter, 1984a. Application of CCC for improvement of tulip bulb production. *Gartenbau*, **31**(9): 282-283.

Mugge, A. and P. Richter, 1984b. Studies on the effect of CCC on the propagation of tulip bulbs. *Archiv für Gartenbau*, **32**(3): 133-138.

Muller, P.J. and C.G.M. Conijn, 1981. New possibilities of controlling tulip gall mites during storage. *Bloembollencultuur*, **919**(51): 1350-1352.

Nelson, P.V. and C.E.J. Niedziela, 1998. Effect of ancymidol in combination with temperature regime, calcium nitrate, and cultivar selection on calcium deficiency symptoms during hydroponic forcing of tulip. *Sci. Hortic.*, **74**: 207-218.

Nichols, R. and A.M. Kofranek, 1982. Reversal of ethylene inhibition of tulip stem elongation by silver thiosulfate. *Sci. Hort.*, **17**: 71-79.

Niebisch, R.M. and K. Kelling, 1986. Results of chemical control of fungal diseases in ornamental plant production (German). *Gartenbau*, **33**(7): 215-218.

Nowak, J. and R.M. Rudnicki, 1975. The effect of "Proflovit-72" on the extension of vase-life of cut flowers. *Prace Instytutu Sadownictea w Skierniewicach, B*, **1**: 173-179.

Okazaki, K. and M. Nishimura, 2000. Ploidy of progenies crossed between diploids, triploids and tetraploids in tulip. *Acta Hort.*, No. 522, pp. 127-134.

Okazaki, K., Y. Asano and K. Osawa, 1994. Interspecific hybrid between *Lilium* 'Oriental' hybrid and *L.* 'Asiatic' hybrid produced by embryo culture with revised media. *Breeding Sci.*, **44**: 59-64.

Paek, K.Y., E.J. Hahn and S. Son, 2001. Application of bioreactors for large scale micropropagation systems of plants. *In Vitro Cell. Dev. Biol. Plant.*, **34**: 149-157.

Pavord, A. 1999. The Tulip. Bloomsbury Publishers, London, U.K.

Pizzetti, I. and H. Cocker, 1968. *Flowers A Guide for Your Garden*, pp. 1310-1331. Harry N. Abrams, Inc., Publishers, New York, USA.

Podwyszynska M. and D. Sochacki, 2010. Micropropagation of tulip: production of virus-free stock plants. *In*: Protocols for *In Vitro* Propagation of Ornamental Plants (eds Jain, M. and J. Ochatt), pp. 243-256. The Humana Press Inc. Totowa, NJ, USA.

Prince, T.A., R.C. Herner and C.T. Stephens, 1987. Fungicidal control of infection by *Penicillium* sp. of precooled tulip bulbs in a modified atmosphere package. *Plant Dis.*, **71**(4): 307-311.

Ptak, A. and A. Bach, 2007. Somatic embryogenesis in tulip (*Tulipa gesneriana* L.) flower stem cultures. *In Vitro Cell. Dev. Biol.-Plant*, **43**: 35-39.

Rajaei, N. and R. Onsinejad, 2014. Effect of municipal solid waste compost and gibberellic acid on morphological and physiological traits of tulip (*Tulipa* spp.) cv. Bright Parrot. *European J. exp. Biol.*, **4**(1): 361-368.

Rees, A.R. 1968. The initiation and growth of tulip bulbs. *Ann. Bot.*, **32**: 68-77.

Rees. A.R. 1985. *Tulipa*. In: *Handbook of Flowering* (vol. I), CRC Press, Boca Raton, Florida, USA.

Rouhi, H.R., F.A. Karimi, A.R. Shahbodaghlo, M. Sheikhalian, R. Rahmatabadi, M. Samadi and F. Karimi, 2012. Effects of sulfuric acid, stratification, phytohormone and potassium nitrate on dormancy breaking and germination of water lily tulip (*Tulipa kaufmanniana* Regel.). *Int. J. Agri. Sci.*, **2**(2): 136-142.

Romanov, L.R., J.P. Van Eijk, W. Eikelboom, A.R. Van Schadewijk and D. Peters, 1991. Determining levels of resistance to tulip breaking virus (TBV) in tulip (*Tulipa* L.) cultivars, **51**(3): 273-280.

Sehgal, O.P. 1995. Cultural Requirements of Other Bulbous Ornamentals. In:*Advances in Horticulture* (vol. 12, *Ornamental Plants*, eds Chadha, K.L. and S.K. Bhattacharjee), pp. 783-794. Malhotra Publishing House, New Delhi.

Straathof, T.P., J.J. Mes, W. Eikelboom and J.M. Van Tuyl 2002. A greenhouse screening assay for *Botrytis tulipae* resistance in tulips, *Acta Hort.*, No. 570, pp. 415-421.

Straathof, T.P., W. Eikelboom, J.M. Van Tuyl and D. Peters, 1997. Screening for TBV resistance in seedling populations of *Tulipa* L. *Acta Hort.*, No 432, pp. 391-395.

van Creij, M.G.M., D.M.F.J. Kerckhoffs and J.M. van Tuyl, 1997. Interspecific crosses in the genus *Tulipa* L.: Identification of pre-fertilization barriers. *Sex Plant Reprod.*, **10**: 116-123.

van Creij, M.G.M., D.M.F.J. Kerckhoffs and J.M. van Tuyl, 1999. The effect of ovule age on ovary-slice culture and ovary culture in intraspecific and interspecific crosses with *Tulipa gesneriana* L. *Euphytica*, **108**: 21-28.

van den Bulk, R.W. and J.M. van Tuyl, 1997. *In vitro* Induction of Haploid Plants from the Gametophytes of the Bulbous Crops Lily and Tulip. In: *In Vitro Haploid Production in Higher Plants* 5, pp. 73-88. Kluwer Academic Publishers, the Netherlands.

van Eijk, J.P., F. Garretsen and W. Eikelboom, 1983. Breeding for resistance to *Fusarium oxysporum* f.sp. *tulipae* in tulip (*Tulipa* L.). 2. Phenotypic and genotypic evaluation of cultivars, *Euphytica*, **28**: 67-71.

van Harten, A.M. and C. Broertjes, 1989. Induced mutations in vegetatively propagated crops. *Plant Breed. Rev.*, **6**: 55-91.

van Heusden, A.W., M.C. Jongous, J.M. van Tuyl, T.P. Straathof and J.J. Mes, 2002. Molecular assisted breeding for disease resistance in lily. *Acta Hort.*, No. 572, pp. 131-138.

van Scheepen, J., 1996. Classified List and International Register of Tulip Names. Royal General Bulbgrowers' Association KAVB, Hillegom, the Nethelands.

van Tuyl, J.M., 1996. Interspecific hybridization of flower bulbs: A review. *Acta Hort.*, No. 430, pp. 465-476.

van Tuyl, J.M. and M.G.M. van Creij, 2006. Tulip. *Flower Breeding and Genetics* (ed. Anderson, N.O.), pp. 623-641. Springer, Oklahoma, USA.

van Tuyl, J.M. and M.G.M. van Creij, 2007. *Tulipa gesneriana and Tulip Hybrids*. In: *Flower Breeding and Genetics* (ed. Anderson, N.O.), pp. 623-664. Springer, Oklahoma, USA.

van Tuyl, J.M. and M.J. de Jeu, 1997. Methods for overcoming interspecific crossing barriers. *In*: *Biotechnology and Crop Improvement* (eds Sawhney, K.R. and V.K. Shivanna), pp. 273-293. Cambridge University press, Cambridge.

Watada, A.E. and L. Qi, 1999. Quality of fresh-cut produce. *Postharvest Biol. and Tech.*, **15**: 201-205.

Windrick, W.A. 1985. Control of stem nematode, *Ditylenchus dipsaci*, in tulip bulbs with aldicarb and oxamyl. *Crop Prot.*, **4**(4): 458-463.

Zamorski, C. 1985. Benzimidazole fungicides in the protection of tulip from tulip fire [*Botrytis tulipae* (Lib.) Lind] and *Fusarium* disease [*Fusarium oxysporum* (Schlecht.) S. & H. f. sp. *Tulipae* Apt.]. *Rev. Pl. Path.*, **65**: 1920.

37

Zantedeschia (Family: Araceae, sub-family: Philodendroideae)

M.K. Singh, Sanyat Misra and Sanjay Kumar

[**Common names**: Arum lily/Calla lily/White lily (*Zantedeschia aethiopica*), Aroid lily, Black-throated calla (*Z. melanoleuca*, now placed under *Richardia*), Calla lily/Cape lily/Golden calla (*Z. elliottiana*), Cuckoo-pint, Jack-in-the-pulpit, Pink calla/Rose calla (*Z. rehmannii*), Lily of the Nile, Little Gem (*Z. aethiopica* var. *minor*), Pig lily, Richardia, Spotted calla (*Z. albo-maculata*)]

Introduction

Zantedeschia was named so by German botanist Kurt Sprengi (1766-1833) in honour of Italian botanist Giovanni Francesco Zantedeschi (1773-1846). Its foremost generic name was *Calla* due its water-loving nature but when this was assigned to a European swamp-loving species *Calla palustris* to form the monotypic genus *Calla*, the next assigned name was *Richardia* but afterwards to avoid confusion with a completely different *Richardia* genus of the family Rubiaceae, it was finally named as *Zantedeschia*. Robert Sweet through his *Hortus Britanicus* mentions that *Zantedeschia aethiopica* was first introduced into Europe in 1731while *Z. albomaculata* in 1860. Its various species originate in southern parts of Africa chiefly Cape Province, Natal, the Transvaal, Basutoland; Angola; Malawi; and Rhodesia. Wide use of its flowers was made at memorial ceremonies and funerals as *Z. aethioppica* has white flowers. Calla lilies are one of the most popular plants which are now cultivated for cut flower, as pot plants and for bedding. Plants look impressive with their spectacular bursts of colours such as white, yellow, golden, peach, pink, red, mauve, etc. White callas have advantage over other rhizomatous flowering plants as they can be successfully planted in the bog gardens, and along ponds and lakes as border plants to enhance their landscape values. Callas can be planted as a specimen plant in containers, in tubs and in conservatories. The leaves are very attractive and are used in floral arrangements. All the parts of calla lilies are poisonous due to the presence of calcium oxalate, a needle-like crystal (raphides) in the sap which produces irritation and swellings in the mouth and throat if swallowed, and may also cause acute vomiting and diarrhoea. *Z. aethiopica* can be used for treatment of waste water as it removes nitrogen from it through the microbial process of nitrification (Belmonte and Metcalfe, 2003). White calla also produces some antialgal compounds which inhibit growth of green algae, *Selenastrum capricornutum* (Greca *et al.*, 1998). The plant has an interesting mechanism which allows it to grow in wet conditions as the leaves contain water stomata which can discharge excess water through guttation, because of which these can grow in water-logged or in marshy conditions (Alice Aubrey, 2001). White calla lily can also be used for controlling sap-sucking insects because its mannose-binding lectin causes inhibitory effects on these insects (Chen *et al.*, 2004, 2005).

Botany and Breeding

They are rhizomatous and moisture- loving herbaceous plants. Rhizomes are fleshy, tuber-like, oblique and branched. Leaves are 15-45 cm long, arrow-shaped and emerge from rhizome's apices. They are usually dark green, often spotted with white flecks, evergreen or deciduous. Leaf lamina is sagittate to cordate, margin undulate, main lateral veins merging with marginal veins. Peduncles often longer than foliage and arranged in spiral form on a thick stalk. Flowers comprise of waxy, subcylindric to funnel-shaped spathes enclosing a prominent, tube like spadix, often yellow in colour.

Spadix is shorter than spathe. Spathes grow on 45–120 cm tall shoots. Flowers are unisexual where male (staminate) zone occupies larger area in upper part and female (carpellate) zone covers less area on the lower part of the stalk. It is a cross- pollinated crop because anthers of each flower mature before the ovaries. Cross pollination is done by honey bees and certain other insects. Ovary is 3-locular and contains 1-8 ovules per locule. Fruits are generally green to orange and contain leathery seeds.

All *Zantedeschia* species are diploid having 2n=32 chromosomes (Earl, 1957; Yao *et al.*, 1994b). Triploid *Zantedeschia* hybrids were produced from **crosses** between diploid and tetraploid plants by using an embryo rescue technique. In five different crosses, the endosperm degenerated and embryo development was restricted to the globular stage because of post-fertilization incompatibility, as these plants were triploids as found by counting chromosomes in root-tip cells. The triploid plants flowered in a greenhouse and produced no pollen. As hybridizations among diploid plants in the winter-dormant section are generally compatible, severe incompatibility in the diploid-tetraploid crosses may be the result of the imbalance of chromosome numbers (Yao and Cohen, 1996). The occurrence of a major breeding barrier prevents the introgression of genes from the section *Zantedeschia* into cultivars of the section Aestivae (Wolff, 1919; Traub, 1948; Yao *et al.*, 1994a, 1995). This barrier results in albino plants that can only survive heterotrophically which is caused by reciprocal incompatibility between plastomes and genomes (Yao *et al.*, 1994a). Plastid development is probably blocked in an early stage in these albino plants (Yao and Cohen, 2000). Plastome-genome incompatibility (PGI) exists between species of different sections of the genus *Zantedeschia* as well. The plastome of *Z. odorata* is incompatible to the genome of *Z. aethiopica*, resulting in albino plants. The plastome of *Z. aethiopica* is partially incompatible to the genome of *Z. odorata*, resulting in virescence, a condition where plants initially are pale-green or yellow, but turn green as they age (Yao *et al.*, 1994a). Although functional interspecific hybrids can be obtained after artificial hybridization among species of the section Aestivae (Wolff, 1919; Shibuya, 1956; Horn, 1962), it is possible that PGI exists. An indication for PGI was delivered by New and Paris (1967) who mentioned hybrid variegation and a non-Mendelian segregation of virescence in interspecific hybrids. If PGI exists among the section Aestivae, then it could have a major impact on genetic studies and breeding. Knowledge of breeding barriers and PGI within the section Aestivae is a prerequisite of genetic analyses and breeding. Directions of PGI between species of the section Aestivae must therefore be determined to identify the most promising parents and crossing directions.

Flavonoids being the main group of colour-giving compounds in flowers, the flavonoid enzymes lead to the formation of anthocyanins responsible for colours ranging from orange and red to lilac and blue, and the lack of activity of one or more enzymes of this pathway leads to acyanic flowers, as is there in the white spathe of *Zantedeschia aethiopica* cv. Nili. There early enzymes of flavonoid biosynthesis, *i.e.* chalcone synthase [naringenin-chalcone synthase], chalcone isomerase and flavanone 3-hydroxylase [naringenin 3-dioxygenase] remain active resulting in the formation of dihydrokaempferol (DHK), and this compound was 3'-hydroxylated to dihydroquercetin (DHQ) by flavonoid 3'-hydroxylase (F3'H), and the reduction of dihydroflavonols by DFR to leucoanthocyanidins was clearly detected in Nili which suggests a lack of anthocyanidin synthase (ANS) activity in this variety which converts leucoanthocyanidins to anthocyanidins, so particle gun bombardment of Nili spathe with plasmids containing ANS from *Matthiola incana* led to red spots, which supports this hypothesis (Seitz, 2003).

Cultures of nine *Zantedeschia* cultivars were treated *in vitro* in a medium containing 0.05 per cent colchicine for 1, 2 or 4 days to induce chromosome doubling. After colchicine treatment, only few shoots survived which were multiplied through several subcultures. Rooted *in vitro* shoots were then transferred to a greenhouse. These plants were screened after two months by measuring stomatal length, and 110 out of 565 plants were selected as putative tetraploids with a stomatal length significantly greater than in diploid control plants. Through chromosome counts carried out on root tips of 44 plants, 38 were found tetraploids, 2 chimeras (predominantly tetraploid with a few octoploid cells), and 4 diploids. Stomatal length was rechecked in mature tetraploid plants of cv. 'Black Magic' which advocated that stomatal length is a good indicator of ploidy level in *Zantedeschia*. This study has shown that multiplying colchicine-treated shoots *in vitro* for several subcultures prior to transfer to soil produced very few chimeras. The stomatal length measurements are non-destructive and allow the rapid screening of a population for tetraploids (Cohen and Yao, 1997).

Species and Varieties

The genus *Zantedeschia* is native to southern Africa and consists of two sections (Letty, 1973; Singh *et al.*, 1996). The *Zantedeschia* section comprises of two species, *viz. Z. aethiopica* L. and *Z. odorata* Perry each with white flowers and rhizomatous storage organ. The section *Aestivae* consists of around six species with coloured flower and tuberous storage organ, such as *Z. rehmannii* Engl., *Z. jucunda* Letty, *Z. elliotiana* (Watson) Engl., *Z. pentlandii* (Watson) Whittm., *Z. albomaculata* (Hook) Baill. and *Z. valida* (Letty) Y. Singh (Singh *et al.*, 1996). This section is frost-sensitive and requires a period of dormancy (Funnell, 1993; Naor and Kigel, 2002). The species ***Z. aethiopica*** (L.) Spreng (syn. *Calla aethiopica* L., *Richardia africana* Kunth), a native of tropical and South Africa, is found in the wetland and is most commonly grown species. It is hardy and widely adopted in subtropical regions. Under congenial growing conditions when there is plenty of moisture, it remains evergreen but under unfavourable conditions this becomes dormant. Its rhizomes are blackish, fleshy and robust, roots fibrous and white, leaves appear only from the base and are long-stalked, elongated heart-shaped and pointed, floral stems up to 90 cm and terminate into an inflorescence, spathe white and quite long (up to 23 cm), vase-shaped and curled at the base and curved at apex, and spadix yellow to yellowish-orange bearing true flowers though insignificant. Its cv. 'Crowborough' is cold-hardy and grows up to 90 cm with large white spathe, cv. 'Candidissima' (pure white, large spathe), cv. 'Devonensis' (dwarf, fragrant and free-bloomer), cv. 'Green Goddess' grows up to 90 cm having faded-green leaves with large green spathe while spadix is white, 'Devonensis', a dwarf variety, free-bloomer and with fragrant flowers, and var. *minor* (Little Gem) is identical to its parent but miniature growing only up to 35 cm; cv. 'Perle von Stuttgart', var. *gedesfreyana* (smaller and free-bloomer) and var. *childsiana* are excellent for pot culture, also there are curious-looking forms of double and triple spathes, and the varieties with large spathes are 'Weisser Hercules', *gigantea* (quite large plants) and *grandiflora* (quite large flowers); ***Z. albomaculata*** (Hook.) Baill. (*Z. oculata* Lindl.), widespread in S. Africa, is deciduous which is bushy and compact, and grows up to 75 cm tall bearing elongated-lanceolate, pointed, 15-38 cm long green leaves often with white spots and white to pale-yellow flowers having purple interior bases, known with 3 sub-species, *viz. albomaculata* with often spotted leaves, *macrocarpa* with sometimes unspotted leaves, and *valida* (Letty) always with unspotted leaves, and spathes often opening to 15 cm long, and a few varieties such as 'Elliottiana' (golden-yellow), 'Solfatare' (yellow blushed black), etc.; ***Z. elliottiana*** (Watson) Engl., a native of Transvaal and many other parts of southeast Africa growing up to 75 cm tall, usually grows 45 cm tall bearing oval or heart-shaped brilliantly green leaves with numerous transparent white marks, and 15 cm long funnel-shaped yellow to golden-yellow spathes; ***Z. jucunda*** Letty growing up to 60 cm bearing leaves with many large white spots and 10-16 cm long tube-shaped golden-yellow spathes having pulple spot at base; ***Z. melanoleuca*** Engl. (syn. *Richardia melanoleuca*), a native of Natal growing up to 40 cm in stony soils, foliage and stems silky, leaves 12-18 cm long having whitish covering, elongated-lanceolate and pointed and spathe straw-yellow with purple mark at the base; ***Z. odorata*** (Perry), similar to *Z. aethiopica* is a sweetly scented species bearing white spathe; ***Z. pentlandii*** (Watson), a native of Transvaal growing up to 60 cm bearing usually unspotted yellow-green leaves and 13 cm long lemon to chrome-yellow spathes with dark purple spot at base; ***Z. rehmannii*** (Engl.), a native of Natal and Transvaal growing from 30 60 cm bearing unspotted dark green leaves of irregular elongated-lanceolate form covered with transparent whitish marks, and 12 cm long rosy-pink to dark maroon spathe, etc., and it has produced hybrids with *Z. elliottiana* such as 'Cameo' (apricot), 'Mango' (orange), 'Black-Eyed Beauty' (dark yellow with black eye), etc. Hybrids between *Z. elliottiana*, *Z. rehmannii* and *Z. albomaculata* with various leaf patterns and yellow to red via orange to pink flowers are available for commercial cultivation. Important calla lily varieties are 'Apple Court' (dwarf with white spathe), 'Aztec Gold' (bright orange to red), 'Best Gold' (spathe gold and large), 'Black Eye Beauty' (spathe cream, throat black), 'Black Magic' (yellow, some blooms tallest of all coloured cultivars), 'Bridal Blush' (blooms armine-red to deep red, spathe cream washed pink), 'Cameo' (blooms peach to apricot), 'Chianti' (growing 90 cm tall, bloom size 5-8 cm and colour deep Pink), 'Childsiana' (a dwarf variety with large and white spathe), 'Crystal Blush' (flowers small and white), 'Dusty Pink' (blooms light pink), 'Florex Gold' (bloom 5-8 cm, flowers deep golden), 'Fire Light' (orange), 'Giant White' (blooms white, spathes up to 18 cm), 'Galaxy' (plants dwarf up to 35 cm, spathes lavender), 'Hot Flashes' (bicoloured, bright pink with white throat), 'Lady Luck' (spathe rich yellow, faintly washed orange), 'Lilac Mist' (light lavender), 'Majestic Red' (leaves spotted green, blooms deep crimson-red), 'Mango' (foliage spotted green, blooms tall and orange), 'Millennium Gold' (bright golden-yellow), 'Neon Amour' (blooms bright pink with wavy edges), 'Parfait' (same plant bears rose-red and cream-pink flowers), 'Peach Chiffon' (spathes peach but edges apricot), 'Regal' (purple), 'Twilight' (bicoloured, light purple and white), 'Picasso' (leaves green spotted, blooms deep purple with cream edges), 'Pink Persuasion' (deep rose-pink with cream throat), 'Red Galaxy' (growing to 30-90 cm, flowers deep red), 'Schwarzwalder' (intense purple with black look), 'Sensation' (creamy-yellow), 'Treasure' (rich tone of red), etc.

Propagation

The conventional method of propagating callas is through **division** of underground rhizomes for getting true to type plants. Misra and Misra (2013) stated that rhizomes have many growing points which should safely be removed before start of the growth in the spring with the relevant portion of the rhizome having normally 4 cm diameter and then dipped in 2 per cent formaldehyde or 0.2 per cent Captan or Thiram for one hour and then superficially dried for planting to produce the flowers the same season. Planting is usually done 6-10 cm deep depending on the cultivar and size of the rhizome. In case of rhizomes being planted of 1-2 cm diameter, these throw out a great number of suckers or offsets (10-40) which can be removed with knife and planted 10 cm across to produce the new plants which are lifted after the season is over and again planted the following season for blooms (Misra and Misra, 2013). It can also be propagated by **seeds** or through tissue culture. Some F_1 hybrid cultivars and species such as *Z. elliottiana*, *Z. rehmannii* and *Z. albomaculata* can be multiplied through seeds. For quicker germination, seeds are soaked in warm water for 24 hours at 21-27 °C and then sown in the containers in the greenhouse or in the open beds at 5 cm across, in November in the plains of India and in March in the temperate areas, however, those sown in the open require shading during summer. It takes a minimum of 2 years or 2 growth cycles to produce flowering size rhizome though under controlled environment the rhizomes attain flowering size in nine months in case of *Z. aethiopica* (Misra and Misra, 2013).

Cultural Practices

Callas grow best in temperate to subtropical climate. They prefer full sunlight, cool and moist conditions. Direct sunlight is essential for brighter blooms of coloured cultivars, however, in warmer areas they require some shade to prevent early fading of colours. *Zantedeschia* is a **day-neutral plant** as there is no effect of photoperiod on flower production (Greene *et al.*, 1932). White callas can tolerate a wide range of light intensities. High or low light intensities may reduce spathe colour intensities (Corr, 1988; De Hertogh, 1996). The effects of irradiance level, day length on growth and flowering of *Z. elliottiana* and *Z. rehmanni* were studied by Corr and Widmer (1990) in greenhouse crop. Plants grown with 45 per cent or 15 per cent of natural irradiance were taller than those grown under full natural irradiance but flowered at the same time and produced a similar number of flowers. Scape length and spathe width increased with increase of the shade level (Armitage, 1991). Most of the species need an average **temperature** of around 13-24 °C for satisfactory growth and flowering. Hay and Beckett (1971) mentioned that *Z. aethiopica* requires 5 °C, *Z. rehmannii* 7-10 °C and *Z. elliottiana* & *Z. pentlandii* 10-12 °C during winter. For all types of callas the day temperatures of 18-21 °C and night temperatures of 16 °C (coloured callas) or 13 °C (*Z. aethiopica*, the white calla) are recommended (De Hertogh, 1996). They are difficult to grow in plains where temperature rises above 40 °C. However, they can be cultivated under protected conditions by decreasing temperature and light intensities. They should be grown in frost-free environment. Corr and Widmer (1990) studied the effects of air and medium temperature on growth and flowering of *Z. rehmanni*. When it was grown at constant air temperature of 15 °C, this flowered in 85 days though it took only 67 days when grown at air temperature of 20 °C. A well- drained loam **soil** rich in organic matter and with a pH range of 6.0-6.5 is ideal for calla lily cultivation. Good drainage is vital to avoid root-rot and other fungal diseases though almost all the species like wet soils and heavy feeding and are adaptable to sun or shade. Callas can be planted in raised beds for cut flower production. For bedding purpose, dig about 30 cm of the soil and apply plenty of well decayed farmyard manure. The land should be thoroughly **prepared** by repeated ploughing and spading. To have a healthy crop of callas the soil should be sterilized properly through solarization or by using methyl bromide @ 100 g/m^2 or formaldehyde @ 2 per cent to reduce the population of harmful pathogens. After the chemical treatment the soil should be covered with polythene for 72 hours. **Planting** of rhizomes can be started a week after removal of the soil cover. Rectangular raised beds with 1 m width having at least 30 cm wide clearance between two beds offer certain advantage for cultural operations. Only healthy rhizomes, free from any infection, chalking or injuries of the best commercial varieties should only be grown. The flowering rhizomes are sold in different sizes starting from 3.8-6.3 cm. For the coloured hybrids, tuber grade 1 is 5.0-6.0 cm, grade 2 is 4.0-4.5 cm and grade 3 is 2.5-3.5 cm in diameter. A rhizome of less than 2.0 cm diameter across may produce foliage but is normally not able to bloom. Before planting, rhizomes must be dipped in 2 per cent Carbendazim + 2 per cent Mancozeb solution for an hour. Plant the rhizomes about 2.5-5.0 cm below the surface with their growing tips facing upright. Width of mature plants will reach about 60-75 cm and so it is important to give them enough room to grow and develop. Plant **spacing** varies from 10 × 10 cm to 50 × 60 cm (row to row and rhizome to rhizome) depending upon rhizome size and expected canopy of plant. Welsh and Baldwin (1989) suggested two spacing systems depending on whether or not tubers are lifted. When to lift annually, it should be planted at 4 rows per bed, the distance between rows and plants should be 20 cm (for 4 cm rhizomes). When lifting after 2 years, it should be planted at 2 rows per bed, 20 cm between plants and 40 cm between rows (for 4 cm rhizomes). At planting there should be sufficient moisture in the soil so that the crop may sustain until sprouting. In mid hills, rhizomes of white calla lily or *Z. aethiopica* can be planted in September-October (autumn) whereas rhizomes of coloured cultivars can be planted in February- March.

Callas do not need heavy **fertilization**. Incorporation of large amount of organic manure during preparation of bed will promote successful cultivation of calla lily. The pattern of development is closely related with nutrient requirement. Study on the fertilizer requirements and time of application in case of *Z. elliottiana* (Astivae section) indicate that nitrogen and potassium are the most required elements, especially from 6 to 12 weeks following tuber planting. Application of 300 kg nitrogen, 45 kg phosphorus and 400 kg potassium per hectare for production of *Z. elliottiana* is said to be optimum. This should be in the form of slow release or soluble fertilizer, applied either as a basal dressing or immediately after planting before the emergence of shoots. Only 90 kg nitrogen, 10 kg phosphorus and 30 kg potassium per hectare would be retained by tubers and roots at dormancy. However, Talukdar (2006) advocated 40 kg FYM, 1 kg calcium ammonium nitrate, 2.5 kg single superphosphate and 750 g muriate of potash for a 10 m^2 plot. She further stated that 150 g of urea and 100 g of muriate of potash as liquid manure per 10 m^2 in light soil will be quite effective. She advocated two fertigations per month of 20:10:20 NPK at the rate of 20 litres per square metre but warned that excess of nitrogen will encourage more leaf formation and will cause greening of spathe. Misra and Misra (2013) stated that with the final ploughing, a basal dose of 70 kg N preferably in the form of ammonium nitrate, 200 kg P_2O_5 and if need be 100 kg K_2O per hectare should be applied in the field and a further 70 kg of N in the form of ammonium nitrate may be applied just before flower emergence.

Callas require adequate moisture during active growth of the plant. Light **watering** is given as growth begins but after attaining full growth the frequency is increased. Overhead watering should be avoided as this damages the rhizomes so it is only drip irrigation which is preferable over sprinkler system. Coloured callas neither prefer over-watering nor water-logging as always there is danger of root damage though *Z. aethiopica* is kept moist at all the times of its growing. Frequency of irrigation is reduced during later stages when the flowers begin to fade. In fact, to avoid *Erwinia* infection, the crop is watered after planting and then withheld until absolutely needed so that the medium always remains to the dry side but also together it should be thought that at no stage the crop should starve of water otherwise the plants will result into reduced leaf area which will adversely affect the growth and development of plants and the rhizomes, however, at full leaf stage these are watered copiously (Misra and Misra, 2013).

Since this crop encounters a lot of **weeds** during its growth period, their control is necessary to save the crop which is normally done manually though this is labour intensive. Organic mulching conserves moisture in the field, works as air conditioner and saves the plants from frost or dryness, checks growth of the weeds and together supplies additional nutrients to the field after decomposition so leaf mould, compost, dry leaves, bagasse, straw, rice husks, etc. can be used for the purpose. Black polythene films are quite effective in controlling the weeds. Mackay and Harrington (1994) used various herbicide combinations to the seedlings of *Zantedeschia albomaculata* cv. 'Chromatella' where it was recorded that weed competition reduced the rhizome yields by up to 50 per cent. The highest yields of rhizomes were obtained from post-emergence herbicides with residual activity, particularly terbutryn (0.5 kg/ha) and methazole (1.1 kg/ha). Mixture of bentazone (1.0 kg/ha), methabenzthiazuron (1.6 kg/ha) and a phenmedipham + desmedipham (0.66 + 0.66 kg/ha) also gave high rhizome yield. Pre-emergence herbicide treatments which appeared effective included oryzalin (3.5 kg/ha), terbuthylazine (1.0 kg/ha), simazine (1.5 kg/ha) and propazine 1.0 kg/ha.

Lifting of Rhizomes and Storage

Rhizomes of coloured callas must have a rest period of 6-12 weeks prior to planting so that these regain energy for next year's growth and flowering. Callas can be left in the ground or **lifted** when the plants become crowded, and then division of rhizomes is necessary. Plants are allowed to senesce by withholding irrigation water. Rhizomes of coloured species are lifted in September-October when the foliage has turned yellow and the plants complete one growth cycle. Lifted rhizomes should be cleaned, graded and treated with fungicides. They can be divided to obtain more rhizomes. Lifted rhizomes should be checked for soft rot, chalking and any physical damage. Plants grown on soil beds produce larger rhizomes than those grown in pots. Rhizomes are air-dried at 20-30 °C for one week and then the coloured types are **stored** at 9-24 °C whereas *Z. aethiopica* types at 2-9 °C. Rhizomes are stored in sand, perlite or vermiculite till the next planting season. If tuberous rhizomes are not given a resting period of at least 6 weeks the flowering may be delayed. The factors affecting storage of rhizomes are relative humidity, temperature level and its duration. Tubers of coloured hybrids should be placed in dry storage at 8 °C and at 70 to 80 per cent relative humidity (De Hertogh, 1996). Tubers can be stored up to 6 months with the optimal storage temperature of 8 °C which restricts their respiration and transpiration rates considerably. Higher temperatures of 13 °C can be used for short term storage (Corr, 1988; Funnell, 1993). Vegetative growth in coloured callas such as *Z. elliottiana* and *Z. rehmanni* is greatly influenced by the storage temperature and period of treatment given to the rhizomes. Tubers stored at 22 °C for 3 or more weeks produced more shoots and leaves, and were taller than non-stored tubers (Corr and Widmer, 1987a, b). Coloured hybrids of *Zantedeschia* need partial dehydration during storage for release of vegetative dormancy. Freshly harvested tubers, without a period of drought were either immediately replanted or were stored dry for 6 weeks at 22 °C. Tubers planted immediately at 22 °C without being subjected to storage did not sprout though those planted after 6 weeks of dry storage sprouted. However, the freshly planted rhizomes which did not sprout, when dug up and after subjecting to storage replanted sprouted as usual, proves the release of vegetative dormancy by dry storage (Corr and Widmer, 1988).

Control of Growth and Flowering

Shoot growth of calla lily rhizomes hardened at 15±4°C for 4 weeks was initiated after 4 weeks at 18 °C or after 2 weeks at 24 °C storage. Flower initiation was not observed at any storage condition. The days to shoot emergence and flowering were shortened by increasing storage temperature and duration. The flower numbers increased when the rhizomes were planted just after shoot elongation but reduced when maintained in storage after shoot elongation. The best conditions for storage were 8 °C for 12 weeks (Choi *et al.*, 2003). The effect of low-temperature (LT) storage at 10 °C on dormancy relaxation and bud break was studied in rhizomes of coloured *Zantedeschia*. LT storage for 35 or 64 days enhanced bud break by about 80 days compared to tubers stored at 20 °C. Primary bud elongation did not occur at 10 °C for 143 days, but took place after the rhizomes were given prolonged LT storage at 20 °C. It is proposed that LT storage enhances dormancy relaxation but at the same time inhibits bud elongation in non-dormant buds. This dual effect can serve growers in programmed production of year-round forcing of coloured calla lily flowers (Naor *et al.*, 2006).

Harrison (1972) stated that flower initiation and development can occur under any condition if favourable climate for vegetative growth is provided. The traditional white *Z. aethiopica* lily is winter-flowering, producing flowers from late autumn through early spring where without a resting period the prolonged flowering all the year round can be obtained through adjustment of suitable temperature and moisture levels though coloured species and hybrids flower only once a year. Rhizomes of the coloured species are planted in spring, *i.e.* from February-March and they continue blooming throughout whole summer. Calla lilies start flowering after 55-90 days of planting though days vary with cultivars, rhizome storage duration and the ambient temperature. Callas can continue blooming for about 5-8 weeks, provided required moisture, nutrition and sunlight are met with during this period. Old and faded flowers should be removed to encourage new flush of blooms. Flower productivity in coloured calla lilies is not only affected by the low number of floral stems per rhizome (not exceeding 2-3 floral stems even in tubers weighing more than 80 g) but also by the short period (2-3 months) of flowering (Naor *et al.*, 2005). Flowering potential is also affected by rhizome size, large size producing more foliage and flowers than small size rhizomes of a cultivar (Corr and Widmer, 1991; Naor *et al.*, 2005).

Immediate GA treatment of the dry harvested rhizomes **forces** formation of numerous flowers with a few leaves though those harvested and stored for 6 weeks at 10 °C produce few flowers with many leaves, however, storage at 10 °C and GA treatment accelerates rapid emergence of leaves and flower with more number (Corr and Widmer,1987b; Corr, 1988). Dipping of *Z. rehmanni* cv. 'Galaxy' rhizomes in 25-100 ppm GA_3 or Promaline solution (a mixture of GA_{4+7} and BA in 1:1 ratio) for 30 seconds, GA_3 produced 3.4 flowers per bulb in comparison to 1.4 flowers in untreated controls while Promaline at 100 ppm produced 5.9 flowers per bulb. Both GA_3 and promaline increased the number of primary flowers rather than the number of secondary ones. It was observed that GA_3 and Promaline induced floral shoot formation at an early stage and that the greater effect of Promaline was due to formation of inactive buds to develop into shoots. Stimulation of flowers increased by dipping of rhizomes of *Zantedeschia* cvs 'Red Sunset' and 'Orange Sunset' in 25 and 50 ppm GA_3 solution (Dennis *et al.*, 1994). Dipping of rhizomes of *Z. albomaculata* cv. 'Black Magic' in GA_3 (1-1,000 mgl^{-1}) or in GA_{4+7} (10-1,000 mgl^{-1}) increased axillary shoot initiation but with reduced number of leaves. Foliar spray of GA_3 (0, 6 mM) on *Z. rehmanni* cv. 'Calla Gold' delayed bud dormancy and caused elongation of floral stem with extended flowering period (Naor *et al.*, 2005).

Drenching paclobutrazol (Bonzi) at 0.5 to 4.0 mg a.i. per pot before attaining the height of 2.5-5.0 cm controls height of taller plants and darkens the leaf colour, and if required, a second dose can be applied after two weeks (Beckman and Luknes, 1997), however, the solution should come in contact with stems and/or roots as it is not translocated to the apex through leaves, and this way this moves quickly throughout the plant system (Styer, 1997). Early application causes excessive plant stunting though late application is less effective. Similarly, rate of application depends on intensity of light, as lower intensity needs higher rate and higher intensity the lower. Its application on the bark or in the acidic medium is not effective (Auman, 1980; Corr and Widmer, 1987a; De Hertogh, 1996; Funnell, 1993).

Postharvest

Flowers are pulled (not cut to avoid disease incidence) when they are fully open but 1-2 days before pollen shedding or safely at half open stage, kept in palatable water, brought to the storage room having 5-9 °C temperature and placed in a 2 per cent sucrose solution with a disinfectant for 6-12 h to prevent stem splitting or rolling (Misra and Misra, 2013). In fact, proper stage for harvesting is suggested when spathe begins to turn downward, *i.e.* cigar-shaped. Tija (1989) and Tija and Funnell (1986) advocated 5-10 °C temperature of the storage room. Nowak and Rudnicki (1990) recommended pulsing in 0.1 mM sucrose for 12 h and 4 °C storage in water for one week. Cut flower life is greatly dependent on the stage of flower development at the time of cutting as flowers in *Z. aethiopica* 'Childsiana' cut at large bud stage and stored at 75 fc low light condition for 12 h at 20 °C lasted for 26 days though those with fully open spathe but before pollen shed lasted only for 11 days (Plummer *et al.*, 1990). Greening and regreening of the senescing spathe is not due to ethylene (Post, 1952; Funnell and Downs, 1987) but because of hormonal factor (Pais and Chaves des Neves, 1982-1983). For marketing, the cut stems are bunched in fives (large ones) or in tens (small ones), however, for single bloom marketing these should be packed in boxes laid with tissue papers between each row.

Potted callas for marketing are transported at 10 °C with proper ventilation when the first flowers are unfurling in case of *Z. aethiopica* or when first flowers are showing colours in case of coloured callas. Potted callas can be stored at 3-4 °C for one week which afterwards remain fresh for another week or even more. However, dark storage is not recommended as at 4-12 °C these start showing deterioration after 1-2 days.

Insect-pests, Diseases and Physiological Disorder

Bacterial soft-rot (*Erwinia carotovora* var. *aroideae*, syn. *Pectobacterium carotovorum* subsp. *carotovorum*), a worldwide prevalent disease is a major problem in *Zantedeschia* species (Kuehny, 2000; Snijder *et al.*, 2004). This bacterium infects the base of the stem and rhizome, spreads readily moving in free water in the soil from infected to non-infected tissues where it enters through wounds and natural openings (Toth *et al.*, 2003). An infected plant shows a soft rot at the base near the soil level. As the bacteria travel upwards the leaves become blighted and plants finally wilt and die. This disease occurs under high relative humidity, high temperature, or under stress due to low soil aeration (Wright and Burge, 2000; Toth *et al.*, 2003). Incidence of soft rot can be minimized by reducing over-watering. Termination of watering as early as possible causes though rapid senescence of foliage but at lifting the rhizomes in such plants are less affected. Proper sanitation and controlled irrigation minimize this disease. Rhizomes suspected of infection may be saved through dipping in bacteridcide and fungicides (Wright and Burge, 2000) and for this streptomycin may prove quite effective. Rhizome rot may also be prevented by discarding all decayed rhizomes at lifting, removing decayed parts of rhizomes and dusting with some effective bactericide,*vis-a-vis* Captan. Infected soil must be drenched with formaldehyde solution before planting. **Leaf-spots** in calla lily is caused due to infection of *Cercospora richardiaecola*, *Gloeosporium callae* and *Phyllosticta richardiae* fungi. *Phyllosticta* causes small, roundish, ash-grey and discoloured spots which coalesce together as the disease progresses, producing irregular decayed areas leading to extensive necrosis of the lamina, being more pronounced on the older leaves of the plants grown in shade and irrigated by spraying. Sometimes stalks and flowers also get infected. It can be controlled by removing badly infected leaves and other affected parts and by keeping low greenhouse R.H. and through properly aerated environment. Dithane M-45 and Dithane Z-78 at 0.2 per cent spraying may also prove effective. **Crown-rot** (*Pellicularia filamentosa* and *Sclerotium delphinii*) is soft decay of lower half of the main rhizome and thick feeding roots. To prevent this disease, plant the rhizome in steam-pasteurized soil or soil treated with PCNB (Terraclor). Drenching with Thiride can also save the plants. **Root-rot** (*Phytophthora richardiae*) is characterized by leaf yellowing, loosing stiffness, turning brown and finally wilting, especially the lower and the outer ones when plants begin to flower and on opening the flowers are malformed, *vis-à-vis* discolouration extending to the stalk. This disease does not make sudden kill of the plant but as new leaves and flowers appear, they quickly show its effects. Pulling of infected plants will show decay of the root system and the remaining ones will be softer and water-soaked though rhizomes remain almost healthy. Its infection can be prevented by dipping the dormant rhizomes in hot water at 50 °C for 1 hour. Dormant rhizomes can also be soaked for 1 h in a formaldehyde solution. From Italy, Pasini *et al.* (1996) isolated *Rhizoctonia solani* from calla lily causing **collar or foot rot. Storage rot** (*Pythium ultimum*) causes decay of rhizomes of pink (*Z. rehmanni*) and yellow callas (*Z. elliottiana*) in storage.

Main **viruses** infecting callas are CMV and 'tomato spotted wilt virus' (TSWV) though certain others as reviewed by Huang and Chang (2005) are also reported infecting *Zantedeschia*, such as 'Dasheen mosaic virus' (DsMV), 'bean yellow mosaic virus' (BYMV), 'Konzak mosaic virus' (KoMV), 'impatiens necrotic spot virus' (INSV), 'alfalfa mosaic virus' (AMV), 'Arabic mosaic virus' (ArMV), 'potato virus X' (PVX), 'tobacco rattle virus' (TRV), 'turnip mosaic virus' (TuMV), and 'zantedeschia mosaic virus' (ZaMV). A new potyvirus 'zantedeschia mild mosaic virus' (ZaMMV) has also been found infecting callas (Huang *et al.*, 2007). Infection of CMV shows irregular mottling and streaking of the leaves. TSWV usually infects white, yellow and pink callas producing chlorotic to pale-white streaks and circular lesions between veins of the leaves, some of which may later become necrotic, wrinkling and distortion of the leaves with upward or downward curling of the margins and the bleaching of the spathes. To avoid menace of this malady the virus-free stock should be used, and the use of 0.15-0.2 per cent spraying of Malathion or Metasystox will control the vectors. Infected plants should be uprooted and burnt.

Physiological disorder 'chalking' is most prevalent in *Zantedeschia aethiopica* cv 'Childsiana' (dwarf white calla) which occurs due to prolonged and improper storage, where the rhizomes develop a hard, chalky exterior resulting in poor sprouting and growth. Another physiological disorder in flower and leaf may be caused by the excessive concentration or treatment duration of GA used as a pre-plant dips in rhizomes. It causes excessive leaf and flower scape elongation and narrowing of leaf blades. If GA treatments are applied too soon after harvest; flowers emerge with little leaf development.

Several **insect-pests** such as long tailed mealy bugs, caterpillars, thrips, aphids and bulb mites have been observed infesting callas. These insects can be controlled by proper monitoring and regular spray of insecticide like Malathion or Metasystox at 0.15 to 0.2 per cent regular spraying.

References

Alice Aubrey 2001. *Zantedeschia aethiopia* www.plantzafrica.com.

Armitage, A.M. 1991.Shade affects yield and stem length of field-grown cut flower species. *HortSci.*, **26**: 1171-1176.

Auman, C.W. 1980. Minor Cut Crops. In: *Introduction to Floriculture* (ed. Larson, R.A.). Academic Press, New York.

Beckman, P. and T. Luknes, 1997. Simple steps for pot calla success. *Grower Talks*, **60**(12):49, 54.

Belmonte, M.A. and C.D. Metcalfe, 2003. Feasibility of using ornamental plants (*Zantedeschia aethiopica*) in subsurface flow treatment wetlands to remove nitrogen, chemical oxygen demand and nonylphenol ethoxylate surfactants-a laboratory scale study. *Ecolog. Engineering*, **21**: 233-247.

Chen, Z., X. Sun and K. Tang, 2004. Genomic cloning and characterization of a novel lectin gene from *Zantedeschia aethiopica. Biosci. Reports*, **24**: 225-234.

Chen, Z., Y. Pang, X.J. Liu, X. Wang, Z. Deng, X. Sun and K. Tang, 2005. Molecular cloning and characterization of novel mannose-binding lectin cDNA from *Zantedeschia aethiopica. Biocell*, **29**: 187-193.

Choi, SoRa., Nou Bog Park and MyungJun Kim, 2003. Growth responses of calla lily 'Black Magic' according to storage temperature and duration of tuber. *Korean J. hort. Sci. and Techn.*, **24**(3): 398-403.

Cohen, D. and J.L. Yao, 1997. *In vitro* chromosome doubling of nine *Zantedeschia* cultivars. *Plant Cell Tissue and Organ Cult.*, **47** (1): 43-49.

Corr, B.E.1988. Factors influencing growth and flowering of *Z. elliottiana* and *Z. rehmanni*. Ph.D. Thesis, University of Minnesota, St.Paul, USA.

Corr, B.E and R.E. Widmer, 1987a. Calla lilies: An up-coming floricultural crop. *Minn. St. Flor. Bull.*, No. 36(1): 6-9.

Corr, B.E and R.E. Widmer, 1987b. Gibberellic acid increases flower number in *Z. elliottiana* and *Z. rehmanni. Hortsci.*, **22**: 605-607.

Corr, B.E. and R.E. Widmer, 1990. Growth and flowering of *Z. elliottiana* and *Z. rehmanni. HortSci.*, **22**: 925-927.

Corr, B.E. and R.E. Widmer, 1991. Paclobutrazol, gibberellic acid, and rhizome size affect growth and flowering of *Zantedeschia. HortSci.*, **26**: 133-135.

De Hertogh, A. 1996. *Zantedeschia* (Calla lily). In: *Holland Bulb Forcer's Guide*, 5th edition, pp.163-172. The International Flower Bulb Centre, Hillegom, The Netherlands.

Dennis, D.J., J. Doreen and T. Ohteki, 1994. Effect of gibberellic acid 'quick dip' and storage on the yield and quality of blooms from hybrid *Zantedeschia* tubers. *Sci. Hort.*, **57**: 133-142.

Earl, G. 1957. Chromosome numbers and speciation in the genus *Zantedeschia. Plant Life*, **13**: 140-146.

Funnell, K.A. 1993. *Zantedeschia*. In: *The Physiology of Flower Bulbs* (eds De Hertogh, A. and M. Le Nard), pp. 683-704. Elsevier, Amsterdam, The Netherlands.

Funnell, K.A. and C.G. Downs, 1987. Effect of ethylene on spathe regreening of *Zantedeschia* hybrids. *HortSci.*, **22**: 1333.

Greca, M.D., M. Ferrara, A. Fiorentino, P. Monaco and L. Previtera, 1998. Antialgal compounds from *Zantedeschia aethiopica. Phytochem.*, **14**: 1005-1008.

Greene, L., R. Withrow and M. Richman, 1932. The response of greenhouse crops to electric light supplementing daylight *Indiana agric. Exp. Station Bull.*, No. 366, p. 20.

Harrison, R.E. 1972. *Zantedeschia* hybrids. *J. Roy. hort. Soc.*, **97**: 131-132.

Hay, R. and K.A. Beckett, 1971. *Reader's Digest Encyclopaedia of Garden Plants and Flowers*, pp. 741-742. The Reader's Digest Association, London.

Horn, M. 1962. Breeding research on South African plants. III. Intra and interspecific compatibility in *Ixia* L., *Sparaxis* Ker, *Watsonia* Mill and *Zantedeschia* Spreng. *J. South Afr. Bot.*, **28**(4): 269-277.

Huang,C.H. and Y.C. Chang, 2005. Identification and molecular characterization of *Zantedeschia* mild mosaic virus, a new calla lily-infecting potyvirus. *Arch. Virol.*, **150**: 1221-1230.

Huang, C.H., W.C. Hu, T.C. Yang and Y.C. Chang, 2007. *Zantedeschia* mild mosaic virus, a new widespread virus in calla lily, detected by ELISA, dot-blot hybridization and IC-RT-PCR. *Plant Path.*, **56**(1): 183-189.

Kuehny, J.S. 2000. Calla history and culture. *Hort Techn.*, **10**: 267-274.

Letty, C. 1973. The genus *Zantedeschia. Bothalia,* **11**: 5-26.

Mackay, T.J. and K.C. Harrington, 1994. Weed control in calla seedlings. *Proc. 47th N. Z. Pl. Prot. Conf.*, pp.163-167. Waitangi, New Zealand, 9-11 August, 1994.

Misra, S. and R.L. Misra, 2013. *Zantedeschia*. In: *Commercial Ornamental Bulb Science*, pp. 324-339. Westville Publishing House, New Delhi.

Naor, V. and J. Kigel, 2002. Temperature affects plant development, flowering and tuber dormancy in calla lily (*Zantedeschia*). *J. hort. Sci. & Biotech.*, **77**(2): 170-176.

Naor, V., J. Kigel, M. Ziv and M. Flaishman, 2005. A developmental pattern of flowering in coloured *Zantedeschia* spp.: Effects of bud position and gibberellin. *J. Pl. Gene Regul.*, **23:** 269-279.

Naor, V., J. Kigel and M. Ziv, 2006. Control of bud sprouting and elongation in colored *Zantedeschia* tubers by low-temperature storage. *HortSci.*, **41**(3): 685-687.

New, E.H. and C.D. Paris, 1967. Virescence in *Zantedeschia. Proc. Amer. Soc. hort. Sci.*, **92**: 685.

Nowak, J. and Rudnick,1990. *Postharvest Handling and Storage of Cut Flowers, Florist Greens and Potted Plants*, Timber Press, Portland, Oregon, USA.

Pais, M.S.S. and H.J. Chaves des Neves, 1982-1983. Regreening of *Zantedeschia aethiopica* Spreng. spathe induced by reapplied cytokinins. *Plant Gr. Reg.*, **1**: 233-242.

Pasini, C., T. Berio, P. Curir and F. D'Aquila, 1996. Further characterization of *Rhizoctonia solani* isolated from carnation and other ornamental plants. *Informatore Fitopatologico*, **46**(6): 33-36.

Plummer, J.A., T.E. Welsh and A.M. Armitage, 1990. Stages of flower development and post production longevity of potted *Zantedeschia aethiopica* 'Childsiana'. *HortSci.*, **25**: 675-676.

Post, K. 1952. *Zantedeschia* (calla). In: *Florist Crop Production*, pp. 847-851. Orange Judd Publishing, New York, USA.

Singh. Y, H. Baijnath and A.E. van Wyk, 1996. Taxonomic notes on the genus *Zantedeschia* Spreng. (Araceae) in Southern Africa. *S. Afr. J. Bot.*, **62**: 321-324.

Shibuya, R. 1956. *Intercrossing among pink calla, white-spotted calla and yellow calla*. Kasai Pub. & Print. Co., Cornell University, USA.

Snijder, R.C., H.R. Cho, M.M.W.B. Hendriks, P. Lindhout and J.M. van Tuyl, 2004a. Genetic variation in *Zantedeschia* spp. (Araceae) for resistance to soft rot caused by *Erwinia carotovora* subsp. *carotovora. Euphy.*, **135**: 119-128.

Seitz, C., N. Oswald, D. Borstling, G. Forkmann and S. Martens, 2003. Being acyanic: an unavoidable fate for many white flowers? *Acta Hort.* (Proc. int. Symp. on Classical vs. Molecular Breed. of Ornamentals, held on August 25-29, at München, Germany, eds Forkmann, G., B. Heuser and S. Michaelis), No. 612, pp. 83-88.

Styer, R.C. 1997. Putting the brakes on plug growth. *GrowerTalks*, **61**(7): 40-45.

Talukdar, M.C. 2006. *Zantedeschia*. In: *Advances in Ornamental Horticulture*, vol. 3, *Bulbous Ornamentals and Aquatic Plants* (ed. Bhattacharjee, S.K.), pp. 248-255. Pointer Publishers, Jaipur.

Tija, B. 1989. *Zantedeschia*. In: *The Handbook of Flowering*, Vol.6 (ed. Halevy, A.H.), pp. 697- 702. CRC Press, Boca Raton, Florida, USA.

Tija, B. and K.A. Funnel, 1986. Postharvest studies of cut *Zantedeschia* inflorescence. *Acta Horticulturae*, No. 181, pp. 451-459.

Toth, I. K., K.S. Bell, M.C. Holeva and P.R.J. Birch, 2003. Soft rot erwiniae: from genes to genomes. *Molecular Pl. Path.*, **4**: 17-30.

Traub, H. P. 1948. The genus *Zantedeschia. Plant Life*, **4**: 3-32.

Welsh, T.E and S. Baldwin, 1989. Calla lilies: A New Zealand perspective. In: *Proc. 2nd nat. Conf. Specialty Cut Flowers*, pp. 81-90. Athens, Greece.

Wolff, H.A. 1919. Advances in Richardias. *Weekly for Bulb's Culture*, **30**(45): 196.

Wright, P.J. and G.K. Burge, 2000. Irrigation, sawdust mulch, and Enhance® biocide affects soft rot incidence, and flower and tuber production of calla lily. *N. Z. J. Crop and hort. Sci.*, **28**: 225-231.

Yao, J.L. and D. Cohen, 1996. Production of triploid *Zantedeschia* hybrids using embryo rescue. *N. Z. J. Crop hort. Sci.*, **24** (4): 297-301.

Yao, J.L., and D. Cohen, 2000. Multiple gene control of plastome-genome incompatibility and plastid DNA inheritance in interspecific hybrids of *Zantedeschia*. *Theor. Appl. Genet*, **101**: 400-406.

Yao, J.L., D. Cohen and R.E. Rowland, 1994a. Plastid DNA inheritance and plastome-genome incompatibility in interspecific hybrids of *Zantedeschia* (Araceae). *Theor. Appl. Genet.*, **88**: 255-260.

Yao, J.L., R.E. Rowland and D. Cohen, 1994b. Karyotype studies in the genus *Zantedeschia* (Araceae).*S. Afr. J. Bot.*, **60**: 4-7.

Yao, J.L., D. Cohen and R.E. Rowland, 1995. Interspecific albino and variegated hybrids in the genus *Zantedeschia. Plant Sci.*, **109**: 199-206.

38

Zinnia (Family: Asteraceae)

Pragya Ranjan, J.K. Ranjan, M.K. Bag and Sanyat Misra

[**Common names**: Common annual zinnia/Youth in old age/Youth and age/Cut and come again (*Zinnia elegans,* syn *Z. violacea*), Haage's zinnia/Mexican zinnia/Narrow-leaved zinnia/Orange zinnia (*Z. haageana*), Prairie/Rocky mountain zinnia(*Z. grandiflora*)]

Introduction and Origin

Zinnia is the first flower to bloom in space for which International Space Station commander Scott Kelly on Jan. 17, 2016 tweeted 'First ever flower grown in space makes its debut'! The zinnias are the second plant to be tested in zero-gravity; last year it was lettuce.

~***The Times of India*** Newspaper, dated Jan. 18, 2016 (page 1, column 1).

Zinnia (*Zinnia elegans* Jacq.) occupies a pristine place in the garden for its bright colours and long flowering period. It is named after Johann Gottfried Zinn (1727-1759), a German Professor of Medicine at Gottingen University. Zinnia gained its popularity as garden annual only in late 19th century. Both annual as well as perennial species are available, but in India the annual one is popularly grown as a summer season annual. In U.S.A., zinnias rank first in seed packet sales (Ewart, 1981) but comprise only 1 per cent of the total bedding plants grown for sale (Voight, 1985). Zinnia is native to Mexico, and its area of diffusion extends from Texas and Colorado in the far north to Brazil and Chile in the far south. Zinnias are the first choice of gardeners due to its appealing colours and shapes though not always so. When the Spanish first saw *Zinnia* species in Mexico, they found its flower quite unattractive so they named it *mal de ojos* meaning 'sickness of the eye', but over the years, through breeding, it went to tremendous changes in its colour range and shape, plant size, and disease resistance, hence present day zinnias are very attractive and suitable for almost all the occasions and purposes such as edging, bedding, borders, rock gardens, hanging baskets, naturalizing effects, etc. These are also excellent as cut flower, which last long but their cut flower production is restricted to outdoor areas only during summers. The dwarf cultivars also produce good cut flowers. They are also suitable as pot plants as well as for growing under glass. However, the use of zinnia as a potted plant has been limited due to its obligate short day length type which fails to flower under continuous irradiance (Sawhney and Sawhney, 1976). The dwarf cultivars are especially suited for edging and window boxes though they stand adverse weather conditions better than their taller counterparts. Zinnias also have nematicidal properties. Flower extract caused the greatest juvenile mortality followed by extract of leaves, stems and roots (Bano *et al*, 1986). When zinnia is grown as a mixed-crop along tomato, the population of *Meloidogyne incognita* and *Rotylenchulus reniformis* was found to be the lowest (Yassin and Ismail, 1995). Thus, it can be used as a pollution-free and inexpensive control against nematodes. They also attract butterflies, moths, bees and birds including hummingbirds.

Botany

Zinnia is a half-hardy annual, perennial or sub-shrubby erect but highly branched plant, 15-90 cm high; with hairy stems; stalkless leaves are opposite, large, oval or heart-shaped, pointed, mostly entire, rough surfaced and light green; inflorescence a terminal head or capitulum, florets of two types, *viz.* fertile pistilate ray florets which are mostly brilliantly coloured, and bisexual fertile disc florets that are small, tubular, yellow or orange, and in the case of hybrids, completely absent; and achenes laterally compressed, bi-toothed at the summit but from the inner angle mostly 1-awned (rarely-2).

Genetics and Breeding

Zinnia elegans, the most widely cultivated species, has a basic chromosome number of x = 12 while *Zinnia linearis* has x = 11. Most of the species are diploid. Bhattacharya and Ghosh (1978) studied the cytology of 8 varieties of *Zinnia elegans* and detected B chromosome in all except one. Ramalingam *et al.* (1976) developed a hybrid (2n=23) by crossing *Zinnia linearis* and *Zinnia elegans.* Fertility of the hybrid can be restored by colchicine treatment.

In zinnia, both male sterility and sporophytic self-incompatibility system is present. Metcalf and Sharma (1971) found out the male sterile plant in *Z. elegans.* Flowers of the male sterile plants had degraded petals, *vis-a-vis* degraded stamens without pollen, and the male sterility of *Z. elegans* was controlled by three recessive genes or a single recessive gene. Sporophytic self-incompatibility system is being used for F_1 hybrid seed production. Samaha and Boyle (1989) suggested the presence of a linear dominance series of S alleles in the pollen and either a linear dominance series or a combination of dominance and independent action of S alleles in the pistil. When intra-specific and reciprocal inter-specific crosses involving *Zinnia angustifolia* clones and *Zinnia elegans* were tested, the intensity of self incompatibility varied among the clones and lines, and high self seed set was associated with a concomitant decrease in callose florescence in papillae and pollen tubes. Incomplete stigmatic inhibition of pollen germination and tube growth were observed in reciprocal inter-specific crosses and associated with callose synthesis, suggesting S-gene activity (Boyle and Stimart, 1986). Callose florescence intensity ranged from 4.7 to 62.6 per cent for incompatible and from 6.4 to 9.9 per cent for compatible crosses and was negatively correlated with per cent embryo set (Samaha *et al.,* 1989). Swarup and Raghava (1974) studied the inheritance of leaf curl virus disease in X-ray irradiated (20 kR) population of *Zinnia elegans* and obtained one resistant strain through rigorous recurrent selections. The resistance of leaf curl virus was found to be governed by two dominant genes, *viz.* R for resistance and S, a suppresser gene. The gene S inhibits the expression of the gene R, both in homozygous and heterozygous conditions. Thus, the plant with complete resistance will have the genotype ssRR, while the genotype ssRr and SsRr will be regarded as highly resistant and partially resistant, respectively. Jones and Strider (1979) suggested *Zinnia linearis* cv. 'Classic', *Z. haageana* cvs 'Chippendale Daisy', 'Old Mexico' and 'Persian Carpet', and *Z. elegans* cvs 'Cut and Come Again' and 'Tangerine' as sources of resistance to bacterial leaf spot caused by *Xanthomonas nigromaculans.* The Ruffles series cultivars (medium height) had the highest tolerance to bacterial leaf and flower spot and blight, while cultivars 'Peter Pan White' and 'Short Stuff Orange' (both short) had the highest tolerance to powdery mildew (Gombert *et al.,* 2001).

A lot of **breeding** work was performed in Germany, Holland and Italy. Heterosis breeding, mutation breeding and polyploid breeding are among the important breeding methods adopted in zinnia. 'Pumila Mixed' (precursor of the 'cut-and-come-again' zinnias) and two selections from that strain, 'Mammoth' and 'Striata', were brought to these countries where these enjoyed great success with gardeners. However, the major breakthrough took place around 1920 when Bodger Seeds Ltd. introduced the dahlia-flowered 'Giant Dahlia'. John Bodger discovered it as a natural mutation in a field of 'Mammoth' and within the next few years selected the large, flat-flowered 'California Giant' from the strain. It was available in separate colours and was considered to be a new trend in plant habit and flower form. It won a gold medal from the RHS, England. The first tetraploid zinnia, called 'State Fair', came from Ferry Morse Seed Company in the 1950's. Dwarf selections of *Zinnia haageana* such as 'Persian Carpet' and 'Old Mexico' were also introduced by this time. Since then a large number of F_1 hybrids in different groups are bred every year with improved characters. F_1 hybrids are much superior with regard to sturdiness, more number and larger flower size coupled with doubleness in medium and tall growing habits and in cactus-floral-forms. Though there had been a lot of improvement works in zinnia but still we lag behind in developing novelties, which requires further concerted efforts. Venkatachalam and Jayabalan (1991) through gamma irradiation induced a wide range of colours. The crimson colour was changed to magenta, yellow, red, and red with white spots at radiation doses of 2.5, 5.0, 7.5 and 10.0 kR, respectively. However, Kaicker *et al.* (1974) when irradiated *Zinnia elegans,* recorded flat ray florets changing to tubular ray florets. Nikolova and Vasileva (1979) and Seregina (1973) through 1-4 kR gamma irradiation recorded increased flower size through increase in the number of ligulate florets in the former case while increase in stalk length, flower size and improved flower quality in the latter case. Colchicine-treated plants show stunted growth, shorter internodes, thicker leaves, more laterals and more and larger flowers than the untreated ones (Srivastava, 1965). Polyploidy breeding is not only important for restoration of fertility but is also being used for disease resistance and for altering the morphology. Bose and Panigrahi (1969) selected two varieties of *Z. linearis* from induced polyploidy. Madhava Menon *et al.* (1969) through 0.05 per cent colchicine-treatment in *Z. linearis* var. 'Yellow' found that induced tetraploids were later to bloom with much prolonged blooming than the diploids. Gupta and Koak (1976) induced autotetraploidy in *Zinnia elegans.* Raman *et al.* (1976) reported that autotetraploids were taller with larger leaves and the ray florets, and capitula were also of greater diameter than diploids. The sterile hybrid (2n = 23) between *Z. angustifolia* (2n = 22) and *Z. elegans* (2n = 24) was partially restored by inducing amphiploidy by colchicines-treatment of axillary buds (Boyle and Stimart, 1982; Terry-Lewandowski *et al.,* 1984). Preliminary tests showed that *Z. angustifolia* and its all the hybrids appeared to be resistant to powdery mildew (Boyle and Stimart, 1982). Colchicine-induced amphiploids of *Zinnia* sp. revealed high level of resistance to *Erysiphe cichoracearum* and *Alternaria zinniae,* and moderately high levels of resistance to *Xanthomonas campestris* pv. *zinniae* (Terry-Lewandowski and Stimart, 1983). *Z.* × *marylandica* is an artificial hybrid between *Z. angustifolia* var. *angustifolia* (2n = 22) as female parent and *Z. violacea* (2n = 24) as male parent. It is resistant to *Erysiphe cichoracearum*, *Alternaria zinniae* and *Xanthomonas campestris* pv. *zinniae* (Spooner *et al.*, 1991).

Classification, Species and Varieties

Zinnia can be classified on the basis of its flower forms as well as height. Different flower forms ~ single, semi-double and double vary in size from 2.5 to 15.0 cm across. Tall cultivars (Giant or Mammoth strain) range from 75 to 90 cm with a maximum of 15 cm flower diameter, medium tall, that is 45-60 cm (*Zinnia pumila*, *Zinnia nana* and *Zinnia compacta* as per trade catalogues), and dwarf types that are 15-45 cm and are of two types (Pompons or Liliputians and Tom Thumbs). Pompons are taller growing with smaller flowers while Tom Thumbs are the dwarfest possible plants with the largest possible flowers on them. Zinnias also have an amazing range of colours; in fact, flowers come in almost every shade except blue, most being solid but some, especially *Z. haageana* is bicoloured with a contrasting colour at the tip of each petal.

Though the classification of varieties is very confusing as California Giants include dahlia-flowered and Liliputs include the Thumbelina, but for the convenience of growers and the trade, it is necessary to classify them. *Zinnia elegans* hybrid can be divided into numerous types or classes on the basis of their plant height and flower size as stated by Pizzetti and Cocker (1975) and these are:

Double: Plants 75 cm high, flower 7-10 cm in diameter, varieties are 'Blaze', 'Burpy Hybrids', 'Canary Bird', 'Orange King', 'Persian Carpet', 'Purity', 'Scarlet Flames', 'Super Giants', 'Violet Queen', etc.

Lilliput or Pompon: Plants bushy, compact, 45 cm high with 3-5 cm flower diameter. Varieties are 'Canary Yellow', 'Peach Blossom', 'Rose Gem', 'White Gem', etc.

Thumbelina: Relatively a new strain requiring sheltered position, early, ultra dwarf, compact, and 10-15 cm high bearing double or semi-double flowers with 3.5 cm diameter.

California Giant: Plant height 90 cm or more bearing largest flower among all the zinnia hybrids, and flower size 15 cm or more. Varieties are 'Brightness', 'Cherry Queen'. 'Lavender Gem', 'Orange Queen', 'Purity', 'State Fair', 'Super Giant', etc.

Scabiosa Flowered: Typical scabiosa flower form growing 80 cm high with flower diameter of 8 cm.

Cactus Flowered: Plants 75-90 cm high, flowers up to 10 cm in diameter with slightly recurved and reflexed petals. Varieties are 'Empress', 'Red Man', 'Snowman', 'Sun Gold', etc.

Pumila: Plants up to 45 cm high, and flowers very double and 7 cm wide. Varieties are 'Early Wonder', 'Pink Buttons', 'Red Riding Hood', etc.

Sombrero: Plant height 35-45 cm, single flower 6 cm wide, petals crimson-scarlet tipped with yellow and excellent for cutting

Two popular series for outdoor cut flower production are 'Dahlia Blue Point' and 'Oklahoma' which are productive throughout the year in mild climates and have moderate disease resistance (Dole, 1997; Dole and Wilkins, 1999). The important varieties of other groups are:

Giant Dahlia: 'Canary Bird', 'Dream Meteor', 'Oriole', 'Polar Bear', etc.

Cupid: 'Golbin', 'Pink Button', 'Pixie', 'Snow Drop', etc.

F_1 zinnias also have several strains such as:

Dahlia: These are 'Blaze Eskimo', 'Canary Bird', 'Orange King', 'Peter Pan Series', 'Purity', 'Riverside Beauty', 'Scarlet Flame', 'State Fair', 'Super Giants', 'Violet Queen', etc.

***Zinnia pumila*:** 'Cut and Come Again' is a popular strain. As bedding plants, some promising cultivars among F_1 zinnia are also identified such as 'Dasher Scarlet', 'Fantastic Light Pink', the Ruffles series such as 'Cherry Ruffles', 'White Ruffles', 'Yellow Ruffles' etc., and the Pulchino series. For cultivation under cover, the recommended varieties are 'Cherry Buttons', 'Envy and Snowman', 'Krasnaya Snapochka', 'Pink Buttons', etc.

Pizzetti and Cocker (1975) mention some 20 *Zinnia* species. Two more Mexican species, *viz. Z. acerosa* var. *guanajuatensis* and *Z. zamudiana* have further been reported occurring (Rzedowski and Rzedowski, 1996). Modern zinnia cultivars were developed primarily from the species *Zinnia violacea* (common zinnia). However, there are some cultivars developed from *Zinnia angustifolia* (narrow-leaf zinnia) and *Zinnia grandiflora* (rocky mountain zinnia). A notable interspecific hybrid is listed as *Zinnia hybrida* developed from *Z. angustifolia* × *Z. elegans*.

Some of the important species are ***Zinnia angustifola*** (syn *Z. linearis*) which is a dwarf (15 cm tall) annual species creeping in nature and useful for naturalizing in rock gardens, bearing narrow leaves, and is suitable for containers especially the dwarf types, as edging, as fillers, for mass plantings and for hanging baskets, but its taller cultivars grow up to 90 cm, the flowers have bold colours from yellow to red via orange, soft pastel shades of apricot, rose, cream, light yellow and lavender, appearing in single with numerous disc florets (Huxley *et al.*, 1992), sometimes double, solid or bicoloured, and streaked or spotted, and the interspecific hybrid of *Z. angustifolia* × *Z. elegans* combines the mounding growth habit and disease resistance of *Z. angustifolia* with the larger leaves and flowers of *Z. elegans*, especially in the Profusion series of cultivars; ***Zinnia elegans*** (syn *Z. violacea*) is the common annual zinnia showing wide variation in plant height (25-120 cm) and flower size (2.5 to 12.0 cm) though original species has purple or lilac-pink flowers that are single, terminal and borne on rigid stalks swollen at the summit, almost all the modern cultivars being developed primarily from this species, with varieties having almost all the shades, and its common names include 'Youth in Old Age' because of the long life of the flowers and 'Cut and Come Again' because removing cut flowers stimulates new flowering shoots; ***Z. grandiflora***, commonly known as rocky mountain or Prairie zinnia, is a clump-forming perennial zinnia that grows 10 to 15cm tall and 30 to 50 cm wide with narrow leaves up to 2.5 cm long and 0.3 cm wide, naturally growing on arid hillsides throughout much of the SW United States at elevations from 600 to 1,800 m, widely distributed from Colorado and Kansas in the western USA and Mexico, and flowering throughout summer and into the fall with yellow-orange, daisy-like flowers; ***Z. haageana***, one of the small flowered annual species growing some 25-30 cm tall with single and double flowers in a wide range of colours, ray petals tipped with yellow and lemon having a broad zone of maroon at the base, smaller than common zinnia but similar in disease resistance, and 'Persian Carpet' is the only cultivar series common in the trade and is important for home gardens; ***Zinnia pumila*** A. Gray is a completely different species than *Zinnia elegans* with a tufted habit and small greyish leaves. The cultivated plant in Britain known as *Zinnia pumila* is, in fact, a dwarf form of *Zinnia elegans*; etc.

Among the recently reported species, ***Zinnia acerosa*** var. ***guanajuatensis*** differs from the species type in its prostrate habit, wider inflorescence containing a larger number of flowers, and almost completely pappose achenes. ***Zinnia zamudiana*** appears to be related to ***Zinnia juniperifolia*** but differs in having prostrate habit, inflorescence with short peduncles and shorter and less intensely coloured ligules (Rzedowski and Rzedowski, 1996).

Propagation

Zinnia can be propagated through both sexual and asexual means. Commercial method of propagation in annual zinnias is through seeds while perennial zinnias can also be multiplied through cuttings. F_1 hybrids also exist but they are very expensive. Some 100 to 200 seeds are recorded per gramme of zinnia seeds. Germination rate in zinnia is affected by presence and absence of pericarp. Miyajima (1996) observed that time to germination and variance of time to germination are different between intact seeds produced by ray florets and tubular florets. For the seeds without pericarp, percentage germination is nearly 100 per cent. Time to germination was shortened through acceleration of seed's imbibing speed and final water absorption by pericarp removal. The structure of the pericarp is different between ray florets and tubular floret seeds and this may be considered to be the cause of difference in water absorption and germination between these seeds. Miyajima (2000) suggested that factors like temperature, vermiculite particle size and frequency of watering also affect germination. The effect of pericarp removal on the percentage of emergence was greatest at 15 °C than at 20 °C. The treatment had no effect at 25 and 30 °C. However, the effect of vermiculite particle size was relatively small compared to that of frequency of watering at any temperature. Germination is rapid and emergence at 21 to 24 °C is within 5 to 7 days. Sharan *et al.* (2011) reported that the soils amended with vermicompost provide considerable increase in germination. The entire texture of the plant was found to change to a healthier look in the presence of vermicompost. Better growth of the plant, larger number of rootlets, and bigger leaf area can be suggested to be additive role of vermicompost on growth and development of *Z. elegans*.

Hybrid seed production in zinnia involves the use of male sterile lines which is a three gene recessive trait and produces only 50 per cent male sterile progeny when propagated from seed. Since no method of clonal propagation is available, seed-produced female lines require labour intensive field-roguing to ensure removal of all normal flowered individuals. To solve this problem, Rogers *et al.* (1990) developed a protocol for micropropagation of male sterile lines.

Cultural Practices

Zinnia requires hot and moderately dry climate. Although, it can be grown throughout the year, summer is the most appropriate season for its cultivation. For growing zinnia under glass, minimum temperature should be 18-20 °C. The growing plants should be kept humid. Plants receiving soil heat produce more flowers. Plant morphology is greatly influenced by day/night **temperature** differentials (DIF). Direct seeding or seedling transplanting in field should be done only when soil temperature is not less than 18 °C. Dole and Wilkins (1999) suggest 21 °C day and 16 to 18 °C night temperatures as optimum. Today, DIF is used as an alternative to chemical growth regulators to control stem growth. Neily *et al.* (2000) reported that it is the temperature and not the **growth regulators** which influences diurnal stem elongation rhythms in zinnia. GA_3 increases instantaneous stem elongation throughout the diurnal period while daminozide decreases it. Neither GA_3 nor daminozide alters the basic instantaneous stem elongation pattern at any DIF (Neily *et al.*, 2000). Zinnia is basically a **facultative short day** plant as it flowers more rapidly under SD (Dole and Wilkins, 1999). Long days cause increased number of ray florets and flower diameter when compared with SD grown plants. SD followed by LD treatments cause long stems with largest flower diameter. *Z. angustifolia* is day neutral (Dole and Wilkins, 1999). Carpenter and Beck (1973) reported that supplemental HID lighting after transplanting reduces the time for flowering and incandescent lighting increases stem length. Thus proper understanding of its day length requirement is necessary for its year round production. Sawhney and Sawhney (1976) reported that the short day requirement is obligate in nature while Boyle and Stimart (1983) reported it as facultative short day plant for floral initiation and development. They found that plants grown under long days (14,16,18 and 24 hours) were greater in height, leaf size, flower diameter and node number and flowered later than plants grown under short days (8,10 and 12 hours). The greatest plant response occurred between 12 and 14 hours. They also found that more long days before short days resulted in increase in height, stem and flower diameter, node number, ray petal number and days to flowering. Both vegetative and reproductive developments are affected with seasonal variation. Zinnia flowers earliest in summer and latest in winter (Boyle *et al.*, 1986). Armitage *et al.* (1981) reported that the time of flowering, plant height and weight of several *Z. elegans* cultivars were reduced by 8 hour photoperiod for 4 weeks. Armitage (1983) stated that high irradiance forces juvenile zinnias to attain senile phase. Reduction in flower size in cvs 'Exquisite', 'Golden Dawn' and 'Bonanza F_1 hybrid' was caused by short day treatment (Reimherr, 1980), while long days increased the plant height of both tall and dwarf cultivars but the start of flowering in dwarf cultivars was not affected by day length (Han and Yeam, 1978). Armitage (1985) found incandescent lighting increasing stem length while Carpenter and Beck (1973) stated supplemental HID lighting reducing the time from transplanting to flowering.

A well-drained loamy **soil** rich in organic matter and with a pH of 6.0 to 7.0 is ideal for zinnia cultivation provided it is at an open situation with full sunlight. It is believed that in richly manured soil, zinnias are more susceptible to viral diseases; hence, the soil should be only lightly manured. Damp conditions are at all not suitable for its growing, especially the large-flowered cultivars, and the blooms are also easily damaged through heavy rains.

Only healthy seeds are used for zinnia growing. The seeds are **sown** in the nursery beds, boxes or seed pans during February-March and August-September in the plains. In the northern plains, seeds are sown either early or late to avoid viral infection. Seedlings should be **transplanted** at 4-leaf stage, *i.e.* normally one month after sowing. **Planting distance** varies with cultivars. The distance between plants may vary from 20 to 40 cm. However, Oberthova (1981) suggested a spacing of 60 × 40 cm for getting highest seed yield. He stated that seed quality varies with method of sowing, date of harvesting and growing season. A spacing of 10 × 10 cm and 30 × 30 cm may be followed in greenhouse and in the field, respectively. However, Oberthova (1981) achieved the highest number of zinnia flowers from 20 × 20 cm spacing after three years of experimentation.

The seedlings after 20 days of transplanting in taller types, should be **pinched** to induce branching otherwise normally these require no pinching or disbudding. The plants have a tendency to flower at a young stage. Pinching of the seedlings also helps in attaining sound vegetative growth, *vis-a-vis* in delaying the flowering. In most cases, the first bud is not of good quality, especially in case of large double flowering types. Therefore, it is always better to remove the first bud. For further cropping, it is always advisable to collect seeds only from the best flowers. This being the annual crop becomes a problem for control of **weeds** as a lot of summer weeds encounter this crop where hand weeding becomes a costly affair, hence, Duczmal (1989) recommended 3.0-4.0 kg Trifluralin (Treflan) before transplanting in *Zinnia elegans* cv. 'Jowita' for maximum seed yield (1.1-12.1 kg/ha). Tenoran 7 kg/ha, Ramrod (propachlor 65 per cent) 7 kg, Amiben-Granulat (chloramben), Kerb 50W 3 kg, Legurame flussig [(carbetamide (70 per cent)] 10 litres and Ronstar (oxidiazon [75 per cent]) 8 litres were tested for their selectivity in *Zinnia elegans* where Tenoran though gave best results but caused some severe necrosis; Kerb 50W and Legurame gave satisfactory weed control; Amiben-Granulat, Kerb 50W and Legurame flussig were well tolerated; Ramrod lacked adequate herbicidal activity and Ronstar caused severe necrosis (Anon., 1974). In Australia, Lamont (1986) recorded Alachlor, Napropamide, Oxadiazon, Oxyfluorfen and chlorthal dimethyl weedicides in zinnia plantings quite effective against up to 90 per cent of grasses and 67-100 per cent broad-leaved weeds. Pre-emergence application during summer of Alachlor (1.125 or 2.25 l), Oxadiazon (l, 2 or 4 kg), chlorthal dimethyl (7.5 or 15.0 kg), Oxyfluorfen (2 litres) and Napropamide (2.25 or 4.5 kg/ha) showed that only Oxadiazon and Oxyfluorfen provide 100 per cent control of all the weeds in *Zinnia elegans* plantings, and in direct-sown crop Alachlor and Napropamide were found quite safe controlling 70 per cent in former case and 50 per cent in latter crop of broad-leaved weeds and up to 90 per cent grasses during summer as well as in winter, however, chlorthal dimethyl was found injurious in direct-sown crop (Lamont and O'Connell, 1986).

Balanced **fertiliser** dose is very essential for good growth and flowering. Nitrogen and phosphorus application leads to improvement in its growth and flowering (Tsurushina and Das, 1971). The amount of NPK to be added varies with soil type. Dhaka *et al.* (1999) advocated the optimum dose as 90 kg N, 60 kg P_2O_5 and 25 kg K_2O/ha as this induced earlier flower bud formation and its anthesis, prolonged duration of flowering and floral longevity in the standing crop, being more pronounced under N and P application. However, recently Iftikhar *et al.* (2012) reported that application of nitrogenous fertilizer increases yield equally in zinnia, but had no effect on stem length as well as days to harvest. The seed yield of plants receiving NPK was 82 per cent higher than those grown without fertiliser (Spaldon and Oberthova, 1973). Deficiency of nitrogen and phosphorus causes reduction in growth and flower production. Flowering is delayed due to phosphorus deficiency. Deficiency of potassium leads to reduction in the number and quality of flowers. Once the seedlings are established, regular irrigation with 100 ppm N gives quality blooms (Armitage, 1993). Zinnia is very susceptible to B deficiency and this causes flower bud blast. It is characterised by blackening of the developed bud as a result of necrosis of the scales, unopened calyx and enclosed tissues. Redington and Peterson (1977) stated that lower levels of B, *i.e.* 0.0025 or 0.025 ppm sometimes cause occasional terminal bud blasting and slow side branch development, coupled with lower plant height, and root & bud weight but its higher dose, *i.e.* 2.5 ppm delays the flowering. Plants receiving boron at 0.25 ppm were the healthiest and most vigorous (Redington and Peterson, 1983). Recent study suggests that zinnia can accumulate substantially more silicon from the municipal water source and growing medium and thus may be benefitted from applications of silicon-based fertilizers to promote cold hardiness and/or plant resistance to fungal pathogens and insect-pests (Brian *et al.*, 2012). Soil inoculation with VAM fungus (*Glomus etunicatum*) leads to faster flowering and with an increased number of flowers in zinnia. These effects of mycorrhizal inoculation appeared to be independent of effects on the plant of P, K and Na concentrations (Aboul-Nasr, 1994, 1996). Even different strains of bacterium have been used to improve its growth and flowering. **Water** should never be a limiting factor, especially during summer when the weather is sunny and warm. Sandy soils require more frequent watering than the clayey ones. Shallow watering should immediately be given when needed. It should be watered every fourth day during persisting sunny and hot weathers. Irrigating *Zinnia elegans* plants with 1-2 per cent solution containing a photosynthetic bacterium stimulated root development, increased root absorption, promoted plant growth and increased resistance to drought (Xia *et al.*, 1994).

Growth and Flowering

Seeds of *Zinnia elegans* were soaked for 24 h in solutions containing 50-200 mg GA_3/l, 50-200 mg GA_{4+7}/l, 5-20 mg NAA/l, 50-200 mg PP_{333} (paclobutrazol)/l, 250-1000 mg ethephon/l, 25-400 mg RSW 0411 (triapenthenol)/l or 100-400 mg Sumi-7 (diniconazole)/l but only 100 mg GA_3 or 100 mg GA_{4+7}/l increased germination percentage, though triapenthenol and diniconazole markedly reduced it (Grzesik, 1989).

Several compounds have been used to induce flowering under non-inductive photoperiods. GA_3 has been successfully used to induce flowering under non-inductive photoperiods (Sawhney and Sawhney, 1976). GA_3 also enhanced extension of growth and leaf differentiation. However, Das *et al.* (1977) found that GA_3 failed to induce flowering under long days, while IAA at 100 mg/1 or kinetin at 50 mg/1 enhanced flowering with additional light. Application of SADH at 5,000 ppm reduced plant height and flower diameter but increased time of flowering without altering the fresh weight in zinnia. Ancymidol 100 ppm restricted the height and fresh weight but not the time to flowering (Armitage *et al.*, 1981). Single drench application at 0.5 or 1.0 mg a.i./pot and single spray applications at 250, 500 or 1,000 ppm were effective in reducing the height (Cox and Keever, 1988). Seedling treatment with 5,000 ppm daminozide and 200 ppm ancymidol reduced leaf expansion and stem elongation, and delayed flowering in *Z. elegans* (Latimer, 1991). SADH at 500 and 2,000 ppm increased seed yield in *Z. elegans* (Shedeed *et al.*, 1986). Growth characteristics of zinnia are also affected by irradiation. Dry zinnia seeds which were irradiated at 0.5 Krad, produced the largest root system, and at 0.5-1.0 Krad the tallest plants. Nikolova and Vasileva (1979) stated that irradiation increases flower size by increasing the number of ligulate florets.

Postharvest

Flowers should be harvested when they are fully open. Immediately after harvesting, the flowers should be kept in air-free water at 10 °C to reduce the water loss. They can be stored for 5 to 7 days at 4 °C temperature. The floral preservative solution consisting of 8-HQC 200 mg/l + sucrose 10 g/l was found to be most effective for enhancing vase life (Stimart and Brown, 1982), the longevity being 6 to 10 days in the home environment. De-ionised water is very effective in prolonging the life but the fresh weight of flower was found decreased (Stimart *et al.*, 1983). Pre-treatment with solutions of Chrysal, Rosal or Flora-2000 controls bacteria and fungi, *vis-a-vis* inhibits ethylene development. Chrysal Universal at 12g/1 has been recorded most effective in increasing zinnia vase life (Lemper, 1984). Mugge (1983) found best cut flower life with AKN preservative [$A1_2(SO_4)_3$ 0.08 per cent + KCl 0.3 per cent + NaCl 0.02 per cent + sucrose 1.5 per cent], though cold storage did not leave any impact on flower quality. To avoid rotting of lower leaves and polluting the solution it would be better to remove them before putting the cut flowers in vase solutions (Nowak and Rudnicki, 1990; Sacalis, 1993; Stimart *et al.*, 1983).

Diseases and Insect-Pests

A large number of diseases attack zinnias. *Alternaria alternata, Alternaria zinniae, Glomerella cingulata, Cochliobolus lunatus, Phoma exigua and Fusarium* sp. are the major seed-borne **fungi**. They cause rotting of seeds and if seedlings emerge, they will die after some days. *Alternaria zinniae* causes blight and can be isolated from under the seed coat, and between and within the cotyledons. Symptoms are reddish-brown spots with greyish centre. Flowers and stem also get spotted. For seed-borne diseases, hot water treatment (66 °C) for 30 minutes has been suggested. *Alternaria* can be controlled by CTP 80 (a uric acid derivative) at 90 g/kg seed and TMTD (Thiram) at 45 g/kg seed (Imre, 1974). Ferbam and Bordeaux Mixture can also be used for controlling this disease (Pirone *et al.*, 1960). However, Zamorshi and Milczarek (1977) suggested the use of Thiram (0.3 per cent) and Captan (0.5 per cent) as the best control measure for *Alternaria.* Dithane M-45 was also found to be the most suitable fungicide for seed treatment (Srivastava and Gupta, 1983). Biological control for *Alternaria* is also being tried. Wu and Chou (1995) found that *Bacillus* strain <hash> 18 was as effective as spraying Pyrifenose in controlling *Alternaria carthami* on zinnia in both the conditions, *i.e.* greenhouse and field trials. Powdery mildew (*Erysiphe cichoracearum*) is controlled by spraying with wettable sulphur or Karathane. Powell (1974) recorded complete control through Benlate, Topsin-in and karathane. Stem rot (*Phytophthora* sp. and *Sclerotinia sclerotiorum*), head blight (*Botrytis cinerea*) and leaf spot (*Cercospora zinniae*) diseases are also experienced by zinnias, whose attack may be minimised through routine and normal sanitation precautions. When the seeds of zinnia were inoculated with *Pseudomonas fluorescens* strain E6, Yuen and Schroth (1986) recorded increased growth and fresh weight (18-41 per cent greater) after 3-4 weeks of treatment, less root colonization by *Penicillium* spp.but higher by *Fusarium* spp., however, growth promotion by strain E6 was related to a change in the composition of root microflora and a reduction in the deleterious effects of minor pathogens. *Xanthomonas nigromaculans* f. sp. *zinniae* is the common **bacterium** infecting zinnia. Hernandez and Trujillo (2000) found *Xanthomonas campestris* pv. *zinniae* also causing leaf spot in zinnia. Strider (1979a) suggested a method for detecting the pathogen in zinnia seed. He made dense sowings of seed under warm and humid conditions in the glasshouse. Distinct lesions were observed on the cotyledons after 5-14 days. Considerable pre-emergence spread occurred so that a very low percentage of contaminated seed resulted in a high number of infected seedlings, thus facilitating this method of indexing. Observations of disease development had to be made during the first two weeks after seedling, because infected cotyledons (or entire seedlings) frequently withered during this period and symptoms did not always develop again on the leaves until after another infection period. Bacterial leaf spot is reduced by treating seeds with a 30-minute chlorine bleach soak of 1:2 (bleach:water). Excessive bleach treatment will injure the seeds. After rinsing, the seeds are sown. To eliminate this disease, streptomycin was also used. Strider (1979b) reported that streptomycin is effective in limiting the pathogen but germinating seedlings become chlorotic, slightly stunted and not uniform. However, sodium hypochlorite (10,500 ppm) soak for 30 minutes was very effective in eradicating this pathogen. A combination of trimming of diseased leaves followed by a foliar spray of streptocycline (0.01 per cent) + Carbendazim (0.025 per cent) was found to be the most effective against *Xanthomonas* as well as *Cercospora* leaf spot (Jindal and Meeta, 1991). Curly top, leaf curl, mosaic, spotted with white and yellows are among the most common **viral** diseases infecting zinnias. *Ruga verrucosans* causes curly top and *Letham australiense* causes spotted wilt. Zinnias are badly affected with spotted wilt. At first, the leaves get curled, and after some time the plants get wilted. This disease is very common in the northern plains and heavy rainfall areas. However, the hills are comparatively free from this disease. It is transmitted by thrips so this pest should be controlled instantly. A new disease known as 'tomato aspermy virus' has been identified in zinnia (Kameya *et al.*, 1996) causing chlorotic spots and mosaic symptoms. Sastri *et al.* (1973) reported mosaic disease of zinnia being transmitted by *Aphis gossypii* and *Myzus persicae.* However, zinnia mild mottle virus is transmitted by *Myzus persicae* (Sulz.), but not by *Aphis solanella* (Padmanabhan and Padmanabhan, 1978). Zinnia is a symptomless carrier of watermelon mosaic virus (Tewari, 1976). It may be of importance in its epidemiology. From Markazi (Iran), Farzadfar *et al.* (2005), for the first time in 2003, recorded TuMV causing mosaic and colour breaking on *Zinnia elegans.* Once the plants are inflicted with viruses these cannot be recovered, hence, the affected plant should be uprooted and burnt or buried in the soil to avoid further infection. Control of vectors through effective insecticides will control even viral diseases.

The major **insect-pests** infesting zinnia are black bean aphid (*Aphis fabae*), red-banded leaf hopper (*Graphocephala coccinea*), tarnished plant bug (*Lygus lineolaris*), the long-tailed mealy bug (*Pseudococcus adonidum*), four-lined plant bug (*Poecilocapsus lineatus*) and six-spotted leaf hopper. Sometimes, zinnia is also infested by root-knot nematode (*Meloidogyne incognita*) and *Aphelenchoides ritzemabosi* which cause angular leaf spot (Pirone *et al.*, 1960). *Aphis gossypii, Aphis solanella, Myzus persicae and Bemisia tabaci* act as the vectors of major viral diseases. Aphids are sucking insects and suck the sap from tender shoots, leaves and flower buds. This results in twisted and distorted shoots or foliage. The affected plants become weak and deformed. Aphids also secrete honeydew on the foliage, causing sooty-mould. Spraying of the infesting crop with some effective insecticides such as Malathion, Rogor, methyl parathion at 0.2 per cent and Nuvan or Dimecron at 0.02 per cent will control all these pests, however, it is only soil application of Nemacur or Furadan which will control the nematode.

References

Aboul-Nasr, A. 1994. Effect of vesicular-arbuscular mycorrhiza on *Tagetes erecta and Zinnia elegans. Alexandria J. Agric. Res.*, **39**: 453-470.

Aboul-Nasr, A. 1996. Effects of vesicular-arbuscular mycorrhiza on *Tagetes erecta and Zinnia elegans'. Mycorrhiza,* **6**: 61-64.

Anonymous, 1974. *Ann. Report German Pl. Prot. Serv.*, No. 21, pp. 221-223.

Armitage, A.M. 1983. Determining optimum sowing time of bedding plants for extended marketing periods. *Acta. Hort.*, No. 147, pp. 143-152.

Armitage, A.M. 1985. *Zinnia elegans and Z. angustifolia.* In: *Handbook of Flowering,* vol. IV (ed. Halevy, A.H.), pp. 548-552. CRC Press, Boca Raton, Fla, U.S.A.

Armitage, A.M. 1993. *Zinnia.* In: *Speciality Cut Flowers*, pp. 138-141. Versity Press/Timber Press, Portland, Or., U.S.A.

Armitage, A.M., R.E. Bass, W.H. Carlson and L.C. Ewart, 1981. Control of plant height and flowering of zinnia by photoperiod and growth retardant. *Hort. Sci.*, **16**: 218-220.

Bano, M., S. Anver, S.A. Tiyagi and M.M. Alam, 1986. Evaluation of nematicidal properties of some members of the family Compositae. *Intern. Nematology Network Newsletter*, **3**(10): 1.

Bhattacharya, G.N. and R. Ghosh, 1978. Chromosome in *Zinnia elegans* Jacq. *Cell & Chrom. Newsletter*, **1**: 3.

Bose, S. and U.C. Panigrahi, 1969. Studies on induced polyploidy in *Zinnia linearis* Benth. *Cytologia*, **34**: 103-111.

Boyle, T.H. and D.P. Stimart, 1982. Inter-specific hybrids of *Zinnia elegans* Jacq. and *Z. angustifolia* HBK: embryology, morphology and powdery mildew resistance. Euphytica, **31**: 857-867.

Boyle, T.H. and D.P. Stimart, 1983. Development responses of *Zinnia* to photoperiod. *J. Amer. Soc. hort. Sci.*, **108**: 1053-1059.

Boyle, T.H. and D.P. Stimart, 1986. Self-incompatibility and inter-specific incompatibility: relationships in intra-inter-specific crosses of *Zinnia elegans* Jacq. and *Z. angustifolia* HBK (Compositae). *Theor. Appl. Gen.*, **73**: 305-315.

Boyle, T.H., D.P. Stimart and M.S. McIntosh, 1986. Seasonal variation in vegetative and reproductive development in *Zinnia elegans* Jacq. *J. Amer. Soc. hort. Sci.*, **111**: 260-266.

Brian K. Hogendorp, Raymond A. Cloyd and John M. Swiader, 2012. Determination of silicon concentration in some horticultural plants. *HortSci.*, **47**(11): 1593–1595.

Carpenter, W.J. and J.R. Beck, 1973. High intensity supplementary lighting of bedding plants after transplanting. *HortSci.*, **8**: 482-483.

Cox, D. A. and G.J. Keever, 1988. Paclobutrazol inhibits growth of zinnia and geraniums. *HortSci.*, **23**: 1029-1030.

Das, P., T.K. Bose and P.K. Roy, 1977. Effect of growth regulators on the photoperiodic response of short and long day plants. *Orissa J. Hort.*, **5**: 1-6.

Dhaka, R.S., M.S. Fageria, S. Mohemmed, A.S. Faroda (ed.), N.L. Joshi (ed.) and S. Kathiju, (ed.), 1999. Effect of different levels of N, P and K on floral characters of zinnia. *Recent Advances in Management of Arid Ecosystem.* Proc. of a Symposium held in India in March 1997, pp. 379-382.

Dole, J.M. 1997. The 1996 ASCFG National Cut Flower Trials. *The Cut Flower Quart.*, **91**: 31-37.

Dole, J.M. and H.F. Wilkins, 1999. Zinnia. In: *Floriculture Principles and Species*, pp. 555-558. Prentice-Hall, Inc., New Jersey, USA.

Duczmal, K.W. 1989. Weed control in ornamental plants grown for seed (Polish).*Biuletyn Instytutu Hodowli I Aklimatyzacji Roslin*, No. 169, pp. 59-76.

Ewart, L. C. 1981. Zeroing in on zinnias. *Amer. Veg. Grower & Greenhouse Grower*, **29**: 8-9.

Farzadfar, S., K. Ohshima, R. Pourrahim, A.R. Golnaraghi, S. Jalali and A. Ahoonmanesh, 2005. Occurrence of turnip mosaic virus on ornamental crops in Iran. *Pl. Path.*, **54**(2): 261.

Gombert, L., M. Windham and S. Hamilton, 2001. Evaluation of disease resistance among 57 cultivars of zinnia. *Hort. Tech.*, **11**: 71-74.

Grzesik, M. 1989. Effect of growth regulators on the seedling-growth of *Lathyrus odoratus*, *Zinnia elegans*, *Matthiola incana* and *Antirrhinum majus*. *Acta Hort.* (Proc. Symp. on Growth Regulators in Ornam. Hort., held at Skierniewice, Poland on September 5-10, 1988), No. 251, pp. 71-74.

Gupta, P. K. and R. Koak, 1976. Induced autotetraploidy in *Zinnia elegans* Jacq. *Cytologia*, **41**: 187-191.

Han, I.S. and D.Y. Yeam, 1978. The effect of photoperiod on the growth and flowering of marigold, salvia, calendula, petunia and zinnia plants. *J. Kor. Soc. hort. Sci.*, **19**: 117-128.

Hernandez, Y. And G. Trujillo, 2000. *Xanthomonas campestris* pv. *zinniae* infecting zinnia plants (*Zinnia elegans* Jacq.) in Venezuela. *Revista de la Facultad de Agronomia, Universidad del Zulia*, **17**: 156-163.

Huxley, A., M. Griffiths and M. Levy, 1992. *Zinnia.* In: *The Royal Horticultural Society Dictionary of Gardening*, vol. 4, pp. 739-740. Stockton Press, New York.

Iftikhar A., M.D. John and N. Paul, 2012. Nitrogen application rate, leaf position and age affect leaf nutrient status of five specialty cut flowers. *Scientia Hort.*, **142**: 14-22.

Imre, K.H. 1974. Investigations on *Alternaria zinniae* infecting *Zinnia elegans* seed. *Kerteszeti Egyetem Kozlemenyei*, **38**(6): 249-258.

Jindal, K.K. and M. Meeta, 1991. Control of bacterial and *Cercospora* leaf spots of zinnia (*Zinnia elegans*). *Indian J. agric. Sci.*, **61**: 156-158.

Jones, J.J. and D.L. Strider, 1979. Susceptibility of zinnia cultivars to bacterial leaf spot caused by *Xanthomonas nigromaculans* f.sp. *zinniae*. *Plant Dis. Rep.*, **63**: 449-453.

Kaicker, U.S., H.P. Singh and K.S. Shukla, 1974. The expression of tubular gene in some members of compositae through the use of gamma radiation and in natural population (abstr. page No. 48). Symp. on *The Use of Radiation and Radioisotopes in Studies on Plant Productivity*, held in 1974 at GBPUAT, Pantnagar.

Kameya, M.I., Y. Suzuki, K. Hanada, S. Ito and S. Tanako, 1996. A new mosaic disease of zinnia caused by tomato aspermy virus. *Annals Phytopath. Soc. Japan*, **62**: 175-176.

Lamont, G. 1986. Herbicide evaluation trials on cut flowers. *Australian Hort.*, **84**(7): 89-91.

Lamont, G.P. and M.A. O'Connell, 1986. An evaluation of pre-emergent herbicides in field-grown cut flowers. *Plant Prot. Quart.*, **1**(3): 95-100.

Latimer, J.G. 1991. Growth retardants affect landscape performance of *Zinnia*, *Impatiens* and marigold. *Hort. Sci.*, **26**: 557-560.

Lemper, L. 1984. Measures for improvement in the development of harvested cut flowers. *Gb+Gw*, **84**: 158-162.

Madhava Menon, P., R. Sethupathi Ramalingam, S.R. Sreerangasamy and V.S. Raman, 1969. Induced autotetraploidy in *Zinnia linearis*. *Madras agric. J.*, **56**: 261-267.

Metcalf, H.N.and J.N. Sharma, 1971. Germplasm resources of the genus *Zinnia* L. *Economic Bot.*, **25**: 169-181.

Miyajima. D. 1996. Germination of zinnia seed with and without pericarp. *Seed Science & Tech.*, **24**: 465-473.

Miyajima, D. 2000. Effects of watering frequency, grain size of vermiculite and various temperatures on the emergence of zinnia (*Zinnia violacea*) seeds with or without pericarps. *J. Japanese Soc. hort.* Sci., **69**: 60-62.

Mugge, A. 1983. Vase life and storage of annual flowers. *Gartenbau*, **30**: 218-219.

Neily, W.G., P.R. Hicklenton and.N. Kristie, 2000. Temperature, but not growth regulators, influences diurnal stem elongation rhythms in *Zinnia*. *Hort. Sci.*, **35**(1): 39-42.

Nikolova, N. and R. Vasileva, 1979. Effect of gamma rays on growth characteristics and flower quality of *Zinnia elegans*. *Nauchni-Trudove, Vissh Selskostopanski Institut Vasil-Kolarov*, **24**: 117-120.

Nowak, J. and R.M. Rudnicki, 1990. *Postharvest Handling and Storage of Cut Flowers, Florist Greens and Potted Plants.* Timber Press, Portland, Oregon, U.S.A.

Oberthova, K. 1981. Methods of growing zinnia transplants and their effect on seed yields. *Acta Fytotechnica*, **37**: 55-70.

Padmanabhan, C. and G. Padmanabhan, 1978. A leaf curl disease of *Zinnia elegans* Jacq. *Cur. Res.*, **6**: 100-101.

Pirone, P.D., B.O. Dodge and H.W. Rickett, 1960. *Diseases and Pests of Ornamental Plants.* Constable & Co. Ltd., London, U.K.

Pizzetti, I. and H. Cocker, 1975. Zinnia. In: *Flowers*, vol. II, pp. 1393-1396. Harry N. Abrams, Incorporated, New York, U.S.A.

Powell, C.C. 1974. Chemical control of powdery mildew of zinnia. *Research Summary.* Ohio Agricultural Research and Development Centre, U.S.A., **79**: 57-58.

Ramalingam, R.S., S.R. Ramaswamy and V.S. Raman, 1976. The cytology of an inter-specific hybrid of zinnia. *Cytologia*, **36**: 522-528.

Raman, V.S., S.R. Sreerangaswamy and R.S. Ramalingam, 1976. Cytomorphology and stability of induced variants in *Zinnia linearis. Cytologia*, **41**: 201-206.

Redington, C.B. and J.L. Peterson, 1977. Zinnia bud blast and boron deficiency. *Proc. Amer. Phytopath. Soc.* **4**: 196.

Redington, C.B. and J.L. Peterson, 1983. Influence of boron on bud blasting and plant growth in *Zinnia elegans. Bull. Torrey bot. Club*, No. 110, pp. 77-79.

Reimherr, P. 1980. Why grow zinnias under glass at all? Problems with cut flower cultivation in the greenhouse. *Gb+Gw*, **37**: 828-830.

Rogers R B, M.A.L. Smith and R. Cowen, 1990. Clonal micropropagation of male-sterile *Zinnia elegans, HortSci.*, **25**(9): 361-362.

Rzedowski, G.C. De and K. Rzedowski, 1996. Two new species of *Zinnia* subgenus *Diplothrix* (Compositae: Heliantheae) from central Mexico. *Acta Botanica Mexicana*, **36**: 77-83.

Sacalis, J.M. 1993. *Zinnia elegans.* In: *Cut Flowers, Prolonging Freshness*, 2nd edn (ed. Seals, J.L.), pp. 103-104. Ball Publishing, Batavia, Ill., U.S.A.

Samaha, R.R. and T.H. Boyle, 1989. Self-incompatibility of *Zinnia angustifolia* HBK (Compositae). II. Genetics. *J. Heredity*, **80**: 368-372.

Samaha, R.R., T.H. Boyle and D.L. Mulcahy, 1989. Self-incompatibility of *Zinnia angustifolia* (Compositae). I. Application of visible light and florescence microscopy for assessment of self-incompatibility. *Sexual Plant Reproduction*, **2**: 18-26.

Sastry, K.S., K.S.M. Sastry and S.J. Singh, 1973. Studies on virus causing mosaic disease on *Zinnia elegans. Indian J. Mycology Plant Path.*, **3**: 165-168.

Sawhney, S. and N. Sawhney, 1976. Floral induction by gibberellic acid in *Zinnia elegans* Jacq. under non-inductive long days. *Planta*, **131**: 207-208.

Seregina, M.T. 1973. The effect of pre-sowing irradiation of zinnia seeds on their growth and development. *Byulleten Nauchno Tekhnicheskoi Informatsii Po Agron Fiz.*,No. 17-18, pp. 81-84.

Sharan A. K, M. Kumar, R Singh, Neha, A. Kishor, G. D. Sharma and C. Jee, 2011. Effect of vermicompost on manifestation of pesticide action on growth of *Zinnia elegans. African J. Biotech.* **10**(36): 6991-6996.

Shedeed, M R., K.M. El-Gamassy, M.E. Hashim and A.M.N. Almulla, 1986. Effect of some growth regulators on the growth, flowering and seed production of some summer annuals. *Annals Agric. Sci. Ain Shams University*, **31**: 677-689.

Spaldon, E. and K. Oberthova, 1973. Investigations on the nutrient consumption by zinnia seed plants grown with different fertilisation. *Acta Fytotechnica*, **27**: 21-36.

Spooner, D.M., D.P. Stimart and T.H. Boyle, 1991. *Zinnia marylandica* (Asteraceae: Heliantheae), new disease-resistance ornamental hybrid. *Brittonia*, **43**: 7-10.

Srivastava, V.K. 1965. Colchicine treated zinnia'. *Sci. & Cult.*, **31**: 200-201.

Srivastava, R.N. and J.S. Gupta, 1983. Pathogenicity of seed-borne infections of zinnia and their control. *Indian Phytopath.* **36**: 14-16.

Stimart, D.P. and D.J. Brown, 1982. Regulation of postharvest flower senescence in *Zinnia elegans* Jacq.' *Scientia Hort.*, **17**: 391-396.

Stimart, D.P., D.J. Brown and T. Solomos, 1983. Development of flowers and changes in carbon dioxide, ethylene, and various sugars of cut *Zinnia elegans* Jacq. *J. Amer. Soc. hort. Sci.*, **108**: 651-655.

Strider, D.L. 1979a. Detection of *Xanthomonas nigromaculans* f.sp. *zinniae* in zinnia seed. *Plant Dis. Reporter*, **63**: 869-873.

Strider, D.L. 1979b. Eradication of *Xanthomonas nigromaculans* f.sp. *zinniae in* zinnia seed with sodium hypochlorite. *Plant Dis. Reporter*, **63**: 873-876.

Swarup, V. and S.P.S. Raghava, 1974. Induced mutation for resistance to leaf curl virus and its inheritance in garden zinnia. *Indian J. Gen. Pl. Breed.*, **34**: 17-21.

Terry-Lewandowski, V.M. and D.P. Stimart, 1983. Multiple resistance in induced amphiploids of *Zinnia elegans* and *Zinnia angustifolia* to three major pathogens. *Plant Disease*, **67**: 1387-1389.

Terry-Lewandowski, V.M., G.R. Banchan and D.P. Stimart, 1984. Cytology and breeding behaviour of inter-specific hybrids and induced amphidiploids in *Zinnia elegans* and *Zinnia angustifolia. J. Genet. Cytol.*, **26**: 40-55.

Tewari, J.P. 1976. *Zinnia elegans* Jacq., the latent host of watermelon mosaic virus. *Cur. Sci.*, **45**: 37.

Tsurushima, H. and N. Das, 1971. The effect of nitrogen, phosphorus and potassium on the growth and flowering of bedding plants. *J. Jap. Soc. hort. Sci.*, **40**: 407-415.

Venkatachalam, P. and N. Jayabalan, 1991. Induction of mutation in *Zinnia elegans* Jacq. *Mutation Br. Newsletter*, **38**: 10.

Voight, A.O. 1985. 1984 season weather beaten but now bowed for 1985? Smallest increased schedule. *Bedding Plant News*, **16**: 1-8.

Wu, W.S. and J.K. Chou, 1995. Chemical and biological control of *Alteraria carthami* on zinnia'. *Seed Science & Tech.*, **23**: 193-200.

Xia, Y.P., S.M. Chen and Z.S. Qian, 1994. Effects of photosynthetic bacterium on plant growth and nutrition of rhizosphere in zinnia. *Advances in Horticulture*, **1**: 600-603.

Yassin, M.Y. and A.E. Ismail, 1995. Effect of *Zinnia elegans* as a mix crop along with tomato against *Meloidogyne incognita and Rotylenchulus reniformis. Tropenlandwirt,* **96**: 221-225.

Yuen, G.Y. and M.N. Schroth, 1986. Interactions of *Pseudomonas fluorescens* strain E6 with ornamental plants and its effect on the composition of root-colonizing microflora. *Phytopath.*, **76**(2): 176-180.

Zamorshi, C. and K. Milczarek, 1977. The effect of seed dressing on the germination of seeds of ornamental plants. *Acta Agrobotanica*, **30**: 335-340.

Glossary

Abaxial. Of the surface or part of a lateral organ facing away from the axis and towards the plant base.

Acaulescent. Stem of a plant being absent or appears to be absent, being very short or subterranean.

Achene. A one-seeded dry and indehiscent small fruit (nut) with a tight thin pericarp as in certain members of Ranunculaceae.

Acropetalous. Leaves or flowers that grow in order from the base of a plant or stem towards the apex.

Actinomorphic. Regular flowers with radial symmetry, developing equally on all sides.

Acute. Having a short sharp point.

Acuminate. Where the tips or the base of leaves and perianth segments taper gradually to a point.

Adaxial. Surface or a part of a lateral organ facing towards the axis and apex.

Adnate. Being attached by its whole length to the face of an organ.

Adpressed (appressed). Hairs (indumentum), leaves, etc. lying flat or close to the stem or leaves to which these are attached.

Adventitious. Originating from an uncommon place as viviparously produced plantlets, the roots from the stem or leaf axil than the radicles, or the buds originating along the stem than at leaf axils.

Alternate. Arrangement of branches on a plant or leaves on the stem alternatively to each other at different heights.

Amplexicaul. Leaf sessile at the base as in case of *Tradescantia*.

Androecium. The stamen or stamens (the male component of a flower as a whole.

Angular. The laterally projecting angles as in case of longitudinal ridging and angular stems.

Annual. A plant completing its entire life cycle (seed to seed *i.e.* germination, flowering and death) within one season or in a year.

Anther. The pollen-bearing organ of the stamen.

Appendix. The long and narrowed development of the spadix in Araceae.

Apiculate. Of leaf tips which terminate abruptly in a short, often firm and sometimes sharp point.

Aquatic. A plant growing naturally in water entirely or submerged.

Arborescent. Tree-like.

Arboretum. A man-made garden of planted trees as a museum or collection.

Areole. A small pin-cushion like sunken area on the body of cacti functioning as a point for throwing out spines, producing flowers and working as a generating organ.

Aril. Fleshy layer covering the seed of a plant.

Arillate. Fleshy appendage of the hylum or funiculus enveloping the testa of seed entirely or partially.

Aristate. Bearing bristles on awns, often appearing as if bearded.

Articulate. Jointed.

Asymmetrical. Informal (irregular) or lop-sided, usually of leaves which have one half larger than the other.

Attenuate. Gradually long and tapering, as in case of a leaf.

Auriculate. Eared as in case of leaves of some plant species in which leaf has two basal lobes.

Awl-shaped. Appearing an awl, a short and broad needle-like tool usually flattened on one face.

Axillary. Arising from an axil (a junction point of stem and leaf).

Baccate. Berry-like.

Back bulb. A pseudobulb in case of an orchid, produced at the back of a flowering stem and one which has borne a flower in the previous season.

Basipetalous. In case of inflorescence where flowers open from apex to base

Beard. Referring to a tuft or zone of hairs on or within a flower as in bearded irises, the tuft or line of hairs on the falls.

Bending. A method used for increasing the yield of flowers and/or fruits through arching or bending of prospective young branches.

Berry. A fleshy and pulpy many seeded indehiscent fruit with a soft covering.

Bidentate. The apex or margins with two teeth, the teeth may also be toothed.

Biennial. The plants that grow and die (seed to seed) within two seasons, producing leaves only in the first season or year, flowers and seed the next.

Bilabiate. Two-lipped.

Bipinnate. Compound leaves with both the primary and secondary divisions pinnate.

Bipinnatisect. Of leaves, bracts or stipules which are twice or doubly pinnatisect, *i.e.*with leaf lobes which are again lobed.

Bisexual. Flowers having both the sexes, *i.e.* stamens and pistils.

Blade. The extended portion of a leaf.

Bloom. A flower, or a thin layer of white waxy powder on certain plants on their stems, leaves, flowers and fruits.

Bract. A modified leaf usually associated with an inflorescence, in the axils of which flowers arise. Some bracts are scale-like and insignificant, others are large and coloured as in *Bougainvillea* and *Euphorbia pulcherrima*.

Bracteole. A little bract.

Break bud. A flower bud which eventually appears at the end of the solitary main stem if left to grow in natural manner, but before sending out side growths. Normally such buds shrivel and do not develop into flowers.

Bud. A shoot in embryonic form containing within it a miniature stem, leaves and sometimes flowers. Also it is applied to a flower before it opens.

Bulb. A modified usually subterranean bud (growing point) consisting of a short thickened and solid stem surrounded by one or many tightly overlapping fleshy scales (modified leaves) attached to the basal plate (stem) enclosing the growing point.

Bulbil. Diminutive bulbs being formed in the axils of the leaves or in the inflorescences.

Bulblet. Small bulbs being formed by parent bulb underground directly around it.

Burr. Fruits or seeds with hooked or barbed hairs or bristles which cling to animal fur or feathers and are thus dispersed well away from the parent plant.

Caducous. Abscising or falling very early.

Callus (pl. calli). In orchid a ridge-, bump- or club-shape thickening in the labellum.

Calyx. The external whorl of floral organs composed of sepals, mostly protect the petals but sometimes become petaloid.

Campanulate. A broad tube of corolla extending in a flared limb or lobes forming bell-shaped structure.

Candelabra. Flowers of an inflorescence arranged in tiers up the stem.

Capillary. Hair-like.

Capitate. A flower head composed of small flowers arranged in dense cluster as in Asteraceae, or terminating in a knob or in somewhat spherical tip.

Capitulum. A flower head that looks like a large single flower but consists of numerous sessile or subsessile small flowers clustered together on a disc.

Capsule. A fruit becoming dry after ripening and opening through one or more vertical valves.

Carnivorous (insectivorous). A group of plants which catch insects and other small animals such as *Darlingtonia, Dionaea, Nepenthes, Pinguicula* and *Sarracenia.*

Carpel. The floral organ (female sporophyll or a simple pistil) bearing ovules.

Cataphyll. A reduced or little developed leaf, bract-like and often sheathing emerging shoots.

Catkin. Slender inflorescence of small and unisexual, usually wind-pollinated flowers.

Caudate. The apex of a leaf or a perianth-segment tapering gradually to form a long tail-like appendage.

Caudex. A trunk of a tree that bears leaves only at its apex as in palm and tree ferns, or the swollen stem base of certain perennial plants from where new growths appear.

Cauliflorus. Flowers or inflorescences being produced directly from the trunks and larger branches of the trees.

Cauline. Arising from the stem.

Cephalium. The area of stem in cacti, often at or near the apex from which the flowers emerge year after year from a dense mass of woolly hairs or bristles as in case of *Espostoa* and *Melocactus.*

Channelled (candiculate). A narrow leaf or leaflet with up-turned margins and forming a channel or gutter-shape.

Ciliate. Fringed with hairs.

Cladode. A green stem or branch that has taken over the major part of a plant's photosynthetic function.

Clavate. Thickening to the apex from a tapered base. Club or basketball bat shape.

Cleistogamous. Self-pollination occurring in the closed flower.

Climber. A plant not having strong stem to grow erect and requires support to grow straight in terms of twining stems, tendrils, hooked thorns, clinging roots and the leaves with reflexed stalks.

Column. The central part of an orchid flower which combines ovary, stigma and pollinia.

Compound. A division into two or more subsidiary orders.

Connate. Joined together as in case of leaf bases or sometimes flowers.

Conservatory. A structure composed partly or entirely of glass attached to the house and within which a large number of plants are grown.

Contractile. These are roots contracting the plants deeper into the soil akin to their natural habitat.

Cordate. The heart-shaped leaves and leafy stipules which have rounded base and a deep basal sinus or notch where the petiole is inserted.

Coriaceous. Leathery and tough but smooth in texture.

Corm. An annual solid and swollen part of a stem which is bulb-like underground structure having growing point at the apex and the basal plate below which forms the roots, and all around the body it has rings as scars of the previous season's leaves. Every ring has a prominent node just opposite the buds of its adjoining rings and many invisible buds. Every part of the corm has regenerating capacity. Entire corm is covered with shredded tunic as dry leaf bases.

Cormel (Cormlets). A small corm developing from and around the new developing (daughter) corm which normally has a very tough and intact tunic. It is as senile as the corm and can develop the flowers similar to a corm under optimum conditions.

Cormule. Similar and smaller than corm but normally larger than the cormels, with shredded tunic, developing 1-4 laterally or terminally at the first nodes but on the cost of spike. No further spike development in case it is terminal.

Corolla. The collective name for petals and if there are two whorls the inner one.

Corona. An extra cup-like appendage or crown as a development of the perianth (*Narcissus*), of the staminal circle (*Asclepias*), or located between corolla and the stamens (*Passiflora*).

Corymb. An indeterminate flat-topped or convex inflorescence where outer flowers open first.

Costa-palmate. The leaves being halfway between palmate and pinnate as in case of palms.

Cotyledon. The primary or seed leaf appearing immediately after germination.

Creeper. A plant not having strong trunk for growing erect so spreads horizontally prostrate over the ground or on a wall or any other object, usually throwing out clinging aerial roots or adhesive pads.

Crenate. Margins of leaf blades and other flattened organs with blunt and rounded teeth.

Crest. A ridge formed on the falls as in *Iris*, which is most commonly orange or yellow in colour.

Crisped. Of hairs, leaves or petal margins being wavy or waved.

Cristate. Crested or less commonly 'crest-like' or ruffled.

Crown. A corona, or the branching pattern, the foliage and the shape (spreading, conical, dome, etc.) of a tree or a shrub, or a part of herbaceous plant where the roots (or underground structures) and the aerial parts meet, or a plane (*i.e.* the point on rhizome or root tubers) having strong terminal bud for propagation.

Crownshaft. The sheathing frond bases wrapped around one another to form a smooth, usually green cylindrical shape with an apparent continuation of the trunk as in case of *Roystonia regia*.

Cultivar. A cultivated variety that has originated and persisted under cultivation.

Cuneate. Wedge-shaped or inversely triangular.

Cuspidate. Apices terminating abruptly in a sharp inflexible point.

Cutting. An unrooted part of stem, leaf or root of a plant taken for producing similar plant.

Cyme. More or less flat-topped and determinate inflorescence being formed of main and secondary stem where the central or terminal flower opens first.

Dealbate. Covered with a white powder.

Deccusate. At right angles to one another.

Deciduous. Of leaves, not persistence or evergreen, and generally fall in the autumn.

Decumbent. Having prostrate stems with tips erect.

Deliquescent. Appearing to melt or become fluid as in petals of *Commelina*.

Dehiscent. The fruit, pollen or spore which spontaneously open when ripe.

Dentate. Margins of leaf blades and other flattened organs having shallow and equilateral triangular teeth.

Denticulate. Having many very small teeth.

Determinate. Inflorescences as in cymes that terminate in a bud where terminal or central flower opens first.

Dichotomous. A method of branching in which each axis bifurcates at the tip.

Dicotyledon. A plant having two primary seed-leaves when germination takes place.

Didynamous. Containing two pairs of stamens, one pair shorter than the other.

Digitate. A compound leaf palmately arranged in which the leaflets arise from the same point at the apex of the petiole

Dioecious. Indicating male and female flowers on separate plants.

Diploid. Plants having two matching sets of chromosomes.

Diplont. An organism whose cells (other than reproductive) have a diploid number of chromosomes in their nuclei.

Diplontic selection (extra somatic selection). In case of mutation in plants when once the exhibited mutation does not appear again in the successive generation(s) as mutated (abnormal) tissues are not allowed to develop further due to suppression by the surrounding normal tissues, and which results in the loss of the mutation, it is known as *diplontic selection.*

Disc (Disk). The central part of the flower head (capitulum) of a composite plant made up of many tubular flowers, or a fleshy (or raised) development of the torus within the calyx or within the corolla and stamens or composed of coalesced nectarines or staminodes and surrounding the pistil, or the basal plate (condensed stem) of a bulb or a corm.

Discoid. Leaves with round fleshy blades and thickened margins, or the capitula of Asteraceae formed entirely of disc florets.

Dissected. Divided into numerous segments.

Distichous. Arrangement of leaves on stems or its branches in two opposite ranks.

Dormancy. The state of cessation of growth of a plant during adverse environmental conditions, usually during winter.

Dorsal. The back of an organ or the surface turned away from the axis *i.e.* abaxial.

Drupe. A fruit consisting of a more or less succulent pericarp which encloses a single many-celled stone.

Ellipsoid. Midway between oblong and ovate with equally narrowed or rounded ends and widest across the middle but 3-dimensional.

Elliptic. Midway between oblong-ovate but with equally narrowed or rounded ends and widest across the middle, i.e oval but narrowing towards each of the rounded ends.

Emarginate. Apex shallowly notched with acute indentation, *i.e.* petals or leaves having a notched tip.

Embryo. It is part of the seed after fertilization but before germination and this develops into a new plant, *i.e.* the rudimentary plant within the seed.

Endemic. The natural haunts being confined only to a particular region and not elsewhere.

Endosperm. The tissue surrounding the embryo in a plant seed to nourish the embryo while germinating.

Ensiform. Sword-shaped and with an acute point, especially the foliage.

Entire. The leaf margins not toothed.

Epicalyx. A ring of bracts just below the calyx or sepals and resembling them.

Epigynous. The sepals, petals and stamens growing on top of the ovary or appearing so.

Epipetalous. Borne upon the petals of a flower.

Epiphyte. Plants that grow habitually in the wild on the tree trunk or its branches, poles, telephone wires, or rock, *etc.* but are not parasitic.

Equitant. The conduplicate leaves overlapping inside each other in two ranks in a strongly compressed fan as in *Iris* or *Gladiolus* and in other members of Iridaceae.

Evergreen. A plant that remains in leafy conditions through all the weathers.

Exotic. An introduced plant not native to that area.

Exserted. Projecting or extending beyond the organs surrounding it as stamens coming out of the corolla.

Eye. The centre of a flower, particularly when of a contrasting colour.

F_1, F_2. Symbols used to indicate first or second generations from a deliberate cross.

Falcate. Curved and tapering to a point like a sickle.

Falls. Outer parts of the flowers which seem to fall downwards or outwards, often bearded, as in *Iris.*

Family. A plant or a group of plants having certain common botanical characteristics, especially in the flowers, and contain a number of genera.

Farina. The mealy or powdery coating on stems, leaves and sometimes on flowers of certain plants.

Farinaceous (Farinose). Having a mealy or granular texture.

Fasciated. The malformation with flattened stems looking as if many stems have joined together as in *Amaranthus* and *Celosia.*

Fascicle. A cluster of roots, stems, leaves, racemes or flowers though arising always independent but appear as if coming out from one common point.

Fastigiate. Of erect habit with all or most stems steeply ascending, creating an overall shape taller than wide.

Filament. The thread-like stalk of a stamen.

Filiform. Thread-like long and very slender leaves, branches, etc. and rounded in cross-section.

Fimbriate. Bordered with a fringe of slender processes usually derived from the lamina rather than attached as hairs.

First crown bud. The first bud which appears at the end of a lateral growth from the main stem of the plant.

Flabellate. A fanilike structure.

Flore-pleno. Double flowered, usually with stamens and

sometimes with carpels transformed to petals or petaloids.

Floret. A small individual flower, generally one of a group or a cluster.

Flower. A part of the plant containing reproductive organs. It may either be unisexual or bisexual.

Flowerhead (capitulum). A dense cluster of flowers of more or less regular size.

Foliaceous. Leaf-like in appearance or structure.

Form. A variant of a wild sprcies not so distinctive to be called a subspecies or a variety.

Frond. A usually large, much divided or compound leaf which appears consisting of many leaves such as in case of ferns and palms.

Fruit. The seed-bearing organ of a plant.

Frutescent. Shrubby or becoming so.

Fruticose. Though without a single main trunk, but bearing stems or branches similar to shrubs.

Fugacious. Withering soon.

Fusiform. Spindle-shaped.

Funnel-form. Flowers shaped like the inverted cone or funnel where petals may either be fused or overlapping and usually joined into a short tube at the base.

Fusiform. Swollen in the middle and tapering at both the ends *i.e.* spindle-shaped.

Gamopetalous. Petals united by their margins to form a tubelike corolla.

Gamosepalous. Having sepals united by their margins.

Genotype. The underlying genetic make up of an individual plant or clone.

Genus. The next upper rank above species in botanical classification.

Geophyte. A plant growing with stem or bulb (bulbs, corms, rhizomes and tubers) below ground.

Gesneriad. A plant belonging to African violet (Gesneriaceae) family. The most popular gesneriads are *Columnea, Gloxinia, Saintpaulia* and *Streptocarpus.*

Glabrous. Smooth, without hairs.

Gland. Usually the base of tepals secreting normally a sweet substance to attract insects for pollination.

Glaucous. Coated with a fine bloom mostly grey or blue-green in colour being formed by the waxy coating of the foliage, which may easily be rubbed off.

Globose (Globular). Spherical or near spherical in shape.

Glochid. Tiny barbed spine or bristle growing from areole in cactus.

Grex. A name applied to a group of individual plants, sometimes difficult to separate individually, as they differ slightly or much from each other but with common origin. Usually applied to a batch of hybrid plants of the same parentage, particularly orchids.

Gynoecium. Female organ of a flower, the carpels collectively.

Gynophore. Stalk of a pistil which raises it above the receptacle.

Gynostegium. A covering of any kind of gynoecium, a perianth.

Habitat. The district or geographical area where a plant is found in the wild.

Haft. The narrow basal portion of the 'falls' as in *Iris.*

Halophyte. A plant tolerant or adapted to saline soil.

Half-hardy. A plant requiring a minimum temperature of 10-13 °C for healthy growth.

Hardy. A plant withstanding prolonged exposure to temperature at or below 10 °C.

Hastate (Sagittate). Arrow-shaped, in triangular fashion, both downward lateral points and one upward.

Heel. A strip of bark and wood remaining at the base of side shoot cutting pulled of a main shoot.

Helophyte. A plant growing seasonally or permanently in mud.

Herbaceous, A plant with no persistent woody stem above ground though it may be evergreen or deciduous.

Hermaphrodite. Flowers having in them both types of sex organs, *viz.* stamens and pistils.

Heterogamous. In Asteraceae where two types of flowers (ray and disc) are borne, or where flowers bear abnormally arranged sexeual organs, or the functions being transferred from one sex to another.

Heteromorphic. Where plants assume different forms at different stages of the life cycle, which is opposite to polymorphic where different forms occur at the same stage of plant development.

Heterophyllous. Where a plant bears two or more forms of leaves on the same plant at the same or at different times.

Heterostylous. Species where flowers differ in the presence or number of styles, some being unisexual while some others hermaphrodite.

Hirsute. Covered with long, coarse, spreading hairs.

Hispid. Covered with bristle-like hairs.

Hoary. Leaves and stems being covered with very short, usually branched or star-shaped white hairs.

Homogamous. Either the flowers are hermaphrodite or of the same sex.

Hyaline. The margins of leaves or bracts being transparent or translucent.

Hybrid. The offspring of crosses involving two plants of the same variety or race, or two different varieties, species or even genera.

Hydrophyte. The plants growing in water.

Hypogynous. All or any of the floral organs situated below the ovary.

Hysteranthous. Where leaves develop after the flowers.

Imbricate. Leaves and bracts overlapping in a regular pattern, sometimes even encircling the axis.

Imparipinnate. Pinnate leaf with an odd terminal leaflet.

Imperfect. When flowers become unisexual due to certain parts though present but are undeveloped.

Incised. Having the leaves or petals deeply cut at the margins, generally in an irregular manner.

Incomplete. Where a flower lacks one or more of the four organs.

Incurved. Inwards or upwards curving nature of leaves and petals.

Indehiscent. Fruit or seed pod and spores splitting self for dispersal of seeds or spores.

Indeterminate. An inflorescence not terminating by a single flower but bears many flowers where opening starts from the bottom.

Indumentum. A covering of hairs, scales or scurfs, used in the sense of hairs as in *Anigozanthos*.

Indisium. The membranous covering over the sporangia of ferns which gives each sorus its particular shape.

Inflated. Bladdery, frequently used to describe a type of fruit or seed vessel.

Inflorescence. It refers to the spike of individual flowers on a stem or a branch. This may be a corymb, cyme, panicle, raceme or spike.

Internode. The part of a stem between two nodes.

Introrse. Facing inward or towards the centre.

Involucre. A single, a pair or a ring (whorl or whorls) of small bracts or leaves subtending a flower or inflorescence.

Involute. Rolling inward towards the uppermost side.

Irregular. Asymmetrical or zygomorphic.

Juvenile. Sexually immature phase of a plant's life.

Keel. A prominent ridge running longitudinally down the centre of the under-surface of a leaf, petiole, bract, petal or sepal as a longitudinal ridge on the foliage of *Kniphofia*.

Keiki. An offshoot or offset of certain plants as in orchids.

Labellum. The lowest of three petals in the orchid flower commonly known as lip. It is modified in a variety of ways to aid insect-pollination.

Labiate. The calyx or corolla possessing a lip or lips.

Labium. One of the lip-like divisions of the calyx or corolla.

Laciniate. An irregular but finely cut as if slashed.

Laevigate. Appearing smoothly polished.

Lamina. Leaf blade, the extended part of a leaf.

Lanate. Covered with thick woolly hairs.

Lanceolate. Lance-shaped tapering towards the apex, three to six times as long as broad, broadest point before the middle.

Lateral. Shoots arising on the sides of main or leading stems.

Leggy. Abnormally tall and spindly growth.

Ligule. The strap-shaped petals or leaves.

Linear. Slender and elongated with margins parallel or nearly so.

Lip. A staminode or petal differentiated from the others, or in Labiatae, one of the two distinct corolla divisions, the upper often hooded and the lower often forming a flattened platform for landing of pollinators. In orchids, it is labellum.

Lithophyte. A plant growing on rocks or stony soil by deriving nourishments from the atmosphere.

Littoral. Growing on the sea-shore.

Lobe. Divided usually into rounded segments and separated by adjacent lobes with sinuses as in petals and leaves.

Loculicidal. Longitudinal and dorsal splitting through a capsule wall into the capsule.

Long day plant. A plant which requires light for a longer period than it would normally receive from daylight in order to induce flowering, such as *Saintpaulia*.

Lorate. Strap-shaped.

Lyrate. Pinnatifid with a large rounded terminal lobe.

Membranous. Thin, soft, more or less translucent.

Mericarp. A one-seeded carpel where one of a pair splits apart at maturity from a syncarpous or schizocarpous ovary.

Meristem. Tissues yet not differenatiated but capable of developing into organs.

Mesophyte. The vast majority of plants which are midway to hydrophytes and xerophytes.

Midrib. The thicker central vein of a leaf.

Monadelphous. Stamens fused by their filaments into a single bundle.

Moniliform. A cylindrical or terete organ with regular constrictions to appear as string of beads.

Monocarpic. Fruiting or flowering only once after several years of growth and then dying as in case of bromeliads and sempervivums. Technically annuals and biennials are monocarpic, but horticulturally the term is used for plants which live more than two years before flowering as in case of some *Agave* species.

Monocotyledon. Having only one cotyledon or seed leaf at germination.

Monoecious. Male and female flowers being borne on the same plant, with one flower of one sex.

Monopetalous. Only with one petal.

Monopodial. A stem or rhizome growing continuously year after year from the apical or terminal bud, with little or no secondary branching.

Monotypic. It describes a taxon, in particular a genus, family, order, class or division that has only one member at its next lower level.

Mucronate. An apex terminating suddenly with an abrupt spine or spur developed from the midrib.

Mutation. A sudden change in the genetic make-up of a plant, leading to a new feature which may be inherited.

Natural break. Sending out of side growths from the main stem after appearance of the break bud at the end of the main stem which has yet not been removed.

Natural 1st crown. It is, in fact, 1st crown bud but is termed so when the plant has made a natural break – as distinct from a 1st crown bud which appears on a plant that has been stopped.

Nectary. Often a projected or depressed gland which secretes or sometimes absorbs the nectar.

Netted. Net-veined or reticulate.

Node. The part of stem where the leaf is joined and a lateral shoot grows out.

Obconical. Of a leaf, roughly conical but with the stalk at the narrow end.

Obcordate. Like to that of cordate but with the sinus at the apex rather than the base.

Oblanceolate. Similar to lanceolate but with the broadest part above the middle and tapering to the base rather than the apex.

Oblong. Linear and 2-3 times as long as broad with almost parallel sides terminating obtusely at both ends.

Obovate. Ovate but broadest at half past so narrow at the base.

Obovoid. As ovoid but broadest before the half.

Obtuse. Apex or base terminating gradually in a blunt or rounded end.

Odd-pinnate. A pinnate leaf with odd number of leaflets, though also implying that there is a single terminal leaflet.

Offset. A lateral growth at base of the plant used for propagation.

Oleifeous. Those plants which have leaves and seeds producing oil.

Opposite. Of leaves and flowers being borne on different sides of axis with bases on the same level.

Orbicular. Circular or nearly so.

Ovary. Usually the swollen base of a pistil containing one or more ovules, which after fertilization become the seeds.

Ovate. Egg-shaped, 1.5-2.0 times as long as broad, broadest before half past and rounded at both ends.

Over-potting. Repotting a plant into a pot which is too large to allow successful establishment.

Ovoid. Egg-shaped ovate but 3-dimentional.

Ovule. One or more minute roundish structures occurring within the chamber of the ovary, and each such structure contains an egg cell which after fertilization develops into a seed.

Palmate. A leaf with several lobes attached to the stalk like the fingers of a hand.

Palmitifid. Palmate with the sinuses reaching about half-way down, *i.e.* rounded leaves lobed in a palmate fashion to half their length.

Palmatisect. Rounded leaves palmately lobed almost to the base.

Panicle. Inflorescence in the form of a branched or compound raceme.

Papillae (papillose). Small, usually soft nipple-like protuberances on a leaf, stem or fruit.

Pappus. A tuft or whorls of delicate bristles or scales on the place of calyx as in certain Asteraceous plants.

Paripinnate. Pinnate leaf with an even number of leaflets.

Parthenocarpic. Fruiting without fertilization.

Pathenogenesis. Development of seed without fertilization.

Peat. Partially decomposed sphagnum moss or sedge used in making composts. It is valuable for its pronounced air and water-holding capacity and its freedom from weeds and disease organisms.

Pedate. A palmatisect leaf with atleast the two basal lobes again lobed.

Pedicel. The stalk supporting an individual flower or fruit.

Peduncle. The stalk of entire inflorescence.

Peltate. A leaf whose stalk instead of being attached at the edge is attached inside its margin normally at the centre below as in *Nelumbium*.

Pendent. Hanging downwards.

Pentamerous. Organs in a group of five or multiples thereof.

Perennate. The survival from year to year by overwintering.

Perennial. Any plant that lives from several to many years, but in the strict sense applied to non-woody species which produce new stems annually from, at or near ground level.

Perfoliate. A leaf whose base clasps the stem as in *Papaver somniferum* or paired leaves which fuse around the stem.

Perianth. The calyx and corolla together.

Perigynous. The flowers in which the perianth and the stamens are basally united and borne on the margins of a cup-shaped rim, itself being borne on the receptacle of a superior ovary, neither above nor below it.

Persistent. An organ neither withering nor falling.

Petal. A part of flower which forms the corolla.

Petaloid. Various organs of a flower adapted to give the appearance of a petal.

Petiole. The stalk of a leaf.

Photosynthesis. The food-making process which occurs usually in the leaf. This process requires water, air and adequate light.

Phylloclad. A flattened stem or branch to function as a leaf.

Phylloclade. Another name for cladode.

Phyllode. An expanded petiole functioning as a leaf blade.

Picotee. Of petals with a narrow margin having a contrasting colour.

Pilose. Covered with long ascending hairs.

Pinching. The act of removing the tip of a growth, either of the main stem or/and of a side growth.

Pinna (pl. pinnae).One of the primary divisions of a pinnate leaf (leaflet).

Pinnate. In compound leaves the arrangement of pinnae (leaflets) in two rows along the rachis. In bipinnate leaves the pinnae are divided into rachillae aand petiolules bearing pinnules (leaflets). Applied also to feather-like arranged veins.

Pinnatifid. Deeply lobed to about half-way down or more with the lobes pinnately arranged.

Pinnatisect. A leaf divided pinnately to the midrib or almost so, but not rounded off into leaflets.

Pinnule. The secondary or more complex division of a fern or palm frond.

Pips. Corms, rhizomes or tubers obtained by ripening-off seedling plants a few months after flowering as in case of anemone and convallaria.

Pistil. Female seed-bearing organ made up of ovary, style and stigma, *i.e.* gynoecium.

Pitcher plants. Species of carnivorous plants in which the trapping mechanism is of pitcher-like form, with a pool of digestive liquid in the bottom, such as *Cephalotus, Heliamphora, Darlingtonia, Napenthes* and *Sarracenia*.

Plicate. Pleated or folded lengthwise, mainly of leaves.

Plumose. Feather-like with long fine hairs which further have fine secondary hairs.

Pollen. The male spores which are developed in the pollen sacs or loculi of anthers.

Pollinia. A group of waxy pollen grains found in orchids and in the members of Asclepiadaceae.

Polygamous. A species bearing both unisexual and bisexual flowers on the same or on different plants.

Pot-bound. A plant growing in a pot which is too small to allow proper leaf and stem growth.

Potting on. The repotting of a plant into a proper-sized larger container which will allow continued root development.

Pricking out. The moving of the seedlings from nursery, tray or pot in which they were sown to other receptacles where they can be spaced out individually.

Prickle. A sharp-pointed broad-based outgrowth of a stem, leaf or other organ as in roses, as opposed to a thorn or spine.

Procumbent. Loosely trailing on ground without rooting.

Prophyll. The bract-like leaves subtending the entire plant or inflorescence in monocots as in *Gladiolus*.

Prostrate. Lying flat on the ground.

Protandrous. In a flower where mature anthers release the pollen before stigma receptivity, but when it continues even after the stigma receptivity it is known as incomplete protandy which enables self-pollination.

Prothallus. A tiny leaf-like body which develops from the spore of a fern, on which are borne the sex cells or gametes which upon fusion give rise to a new fern plant.

Pseudobulb. The fleshy bulb-like thickened aerial stem found in many epiphytic orchids.

Pseudostem. The aerial stem composed of only overlapping sheaths and stalks of basal leaves.

Pubescent. Hairy. When an organ is covered with silky or downy hairs.

Punctate. Dotted with minute, translucent impressions, pits or dark spots.

Pyramidal. Conical in shape but with more angular sides.

Raceme. An indeterminate and unbranched inflorescence composed of pedicellate flowers. Here each flower is borne on an individual stalk along a central stem.

Rachis. The axis of a compound leaf or of an inflorescence.

Radical. The leaves rising directly from the roots.

Ray floret. A small flower with a tubular corolla having expanded and flattened limb in a strap-like ligule (blade), normally occupying the peripheral rings in a radiate capitulum as in Asteraceae.

Receptacle. The elongated and either flat, concave or convex end of the stem from which the perigynous zone (floral parts) originates.

Reniform. Kidney-shaped.

Repent (repens). Growing flat on the ground, usually of stems that root at intervals.

Reticulate. Netted like a fish-net. Network of veins, ribs and colouring. Reticulate tunic is composed of a lattice of fibres in certain bulbs and corms.

Retrorse. A minor organ attached to a larger part such as prickle on a stem, a barb on a leaf or a callus on a lip, and so on, and turned, curved or bent backwards or downwards, away from the apex.

Revolute. Margins rolling towards the dorsal surface.

Rhizome. A specialized slender or swollen, simple or branched subterranean (sometimes close to soil surface, but exposed in epiphytes) and normally creeping stem which produces roots, stems, leaves and inflorescences along its length and at its apex.

Rhomboid. In case of leaves and tepals being angularly oval, the base and apex with acute angles and both the sides having obtuse angles.

Rootstock. The base of a stem from which the roots emerge but in graftage the rootstock refers to the locally adapted plants on whose rooted stems the desired variety is grafted or budded.

Rosette. Leaves clustered together in circular pattern.

Rosulate. Bearing a rosette.

Rugose. Wrinkled, puckered or corrugated appearance of leaves.

Runner. A slender, prostrate stem bearing plantlets at the well-spaced nodes.

Sagittate. Spear- or arrow-shaped leaves where triangular basal lobes point towards the stalk or downward.

Salverform. Where corolla have long and slim tube with an abruptly expanded flattened limb.

Scabrous (scabrid). Rough, usually with short and hard hairs.

Scale leaves. Specialized leaves which are fleshy and scale-like and cover the buds or make the bulbs.

Scandent. Climbing or scrambling stems.

Scape. A leafless erect stalk bearing a terminal inflorescence or flower.

Schizocarp. A dry fruit splitting at maturity into two or more one-seeded carpels which remain closed.

Second crown bud. Removal (stopping) of the tips of the lateral growths (first breaks) from the main stem, forces appearance of further side growths (second breaks) in the leaf axils of these laterals, and these laterals grow longer terminating into a bud at the end, and this is termed as second crown bud.

Secund. Where all the parts of the leaves or flowers are borne along one side of the axis or appear so due to twisting of the stalk.

Segment. One of the parts of a compound leaf or the flowers where sepals and petals are similar.

Sepal. One of the separate parts of the calyx.

Sericeus. Silky-hairy.

Serrate. Leaf blades and other flattened organs having sharp saw-toothed margins curving forward.

Sessile. Without a stalk as in case of leaves amd flowers.

Short day plant. A plant which requires light for a shorter period than it would normally receive from daylight in order to induce flowering, *e.g. Dendranthema* and *Poinsettia*.

Shrub. A woody plant with a framework of branches and little or no central system.

Simple. One piece leaf, neither composed nor divided.

Sinuate. With undulated or wavy margins.

Sinus. The space between two lobes or divisions.

Solitary. Used in case of flowers when they are borne one to a stem of leaf axil.

Sorus. A group of sporangia usually of rounded or oval shape on the back of a fern frond. Each sorus may or may not be covered with an indusium.

Spadix. A spike of flowers on a thick and fleshy axis, often subtended by spathe as in Araceae.

Spathe. A modified, often papery bract or leaf enclosing the inflorescence.

Spathulate (Spatulate). An oblong and spatula-shaped organ which is narrow and tapering at base but rounded at apex.

Spicate. Borne in a spike-like inflorescence.

Spike. An indeterminate and unbranched inflorescence bearing sessile flowers.

Spikelet. A very small, usually compact spike, generally used for the basic unit of a grass inflorescence.

Spine. A hard and sharp pointed structure formed by a modified twig.

Sporangium. An asexually formed spore produced by ferns and certain fungi.

Spore. Minute reproductive bodies formed of one or a few cells together, which give rise to new individuals, either directly as in fungi or indirectly as in ferns.

Sport. A plant which shows a marked and inheritable change from its parent; a mutation.

Stalk. The petiole, peduncle, filament, stipe (stalk of the pistil), etc. which provide stem-like support to their respective organs.

Stamen. Anther and filament constituting the male pollen-bearing part of the flower.

Staminate. A male flower bearing stamens and no functional pistils.

Staminode. A stamen which is sterile and often enlarged and petal-like.

Standard. The perianth segments of inner whorl, often erect as in *Iris*.

Stellate. Star-shaped.

Stigma. The part of the pistil which is receptive when pollen over it is deposited.

Stipe. An alternative name for the leaf stalks of ferns. Also used for the stems of mushrooms and toadstools.

Stipule. A bract-like or leafy appendage, often in pairs, at the base of a petiole, shedding soon.

Stolon. An underground trailing stem giving rise to plantlets at its nodes and apex or cormlets and bulblets at its simple or branched apices.

Stool. It is the root of an old plant with a portion of the old stem and its surrounding young shoots.

Stopping. The act of removing the tip of a growth, either of the main stem and/or the side one(s).

Stove plant. A plant which requires warm greenhouse conditions in winter.

Strain. A selection of a variety, cultivar or species which is raised from seed.

Striate. Fine parallel lines or stripes on the surface.

Style. The area of ovary between stigma and ovary.

Sub-shrub. A plant which at the adult stage has woody stems near the base and a framework of soft green growth.

Succulent. Fleshy, juicy as in cacti and many other succulents.

Sucker. A rooted shoot developing from an underground stem or the plant's root.

Sympodial. In growth of a plant where terminal bud terminates into an inflorescence or dies, and successive growth starts in the ensuing season through secondary axes from lateral buds.

Synanthous. Where leaves appear with the flowers.

Syncarpous. Where two or more carpels fuse in a compound pistil.

Synflorescence. A compound inflorescence which possesses a terminal inflorescence and lateral inflorescences.

Syngenecious. Stamens united into a cylindrical form by the anthers.

Taking a bud (securing a bud). The act of removing all leaf axil shoots and unwanted buds except the bud which is to develop into a flower.

Tender. A plant which requires a minimum temperature of 15.6 °C. Occasional short exposure to temperatures below this level may be tolerated.

Tendril. A slender twining organ that enables a plant to climb.

Tepal. An unit of undifferentiated perianth which can not be distinguished as a petal or a sepal.

Terete. Slender and cylindrical, circular in cross-section.

Terminal bud. It is a bud which develops surrounded by other flower buds, and this marks end of further vegetative growth.

Ternate. Usually of leaves, in whorls or clusters of three.

Terrarium. A partly or entirely closed glass container used to house a collection of indoor plants.

Terrestrial. Growing on the ground.

Tessellated. Chequered or marked with a grid of small squares.

Tetramerous. With parts in fours or group in fours.

Throat. The opening between the tube and the limb of the tubular part of calyx, corolla or the perianth.

Thyrse. A panicle which is broadest in the middle and tapers to base and apex.

To break. The branching or sending out of side growths.

Tomentose. A downy covering of tiny, wooly and stiff hairs on leaves and other plant parts.

Topiary. Art of clipping and training woody plants to form geometric shapes or intricate patterns.

Toothed. Dentated. Bearing small teeth on the edges.

Torus. A receptacle of a flower, the region in which floral parts are inserted. An elongated torus is a gynophore, the stalk of a pistil which raises it above the pistil.

Transpiration. The loss of water through the pores of the leaf.

Tree. A woody plant with a distinct woody trunk.

Trifoliate. With three leaves or leaflets.

Trimerous. Having parts in threes.

Trumpet. When the tepals (perianth segments) are fused or appearing to be fused into the flared bell of a trumpet as in *Hymenocallis, Narcissus,* etc.

Truncate. Ending abruptly or bluntly.

Tuber. A generally subterranean fleshy and swollen stem with eyes over the surface as in tuberous *Begonia,* or fleshy and swollen roots having buds (eyes) only on the crown, the part which is attached with the aerial stem as in *Dahlia.*

Tubercle. A small wart or knob-like projection on a stem, leaf, fruit, *etc.*

Tuberous. Producing tubers or resembling a tuber.

Tunic. The papery or netted covering of corms derived from widely expanded leaf bases.

Tunicate. Enclosure of a tunic, the papery covering of a bulb or a corm.

Umbel. Umbrella-type inflorescence, normally flat-topped and corymb-like where all the flowered pedicels arise from the same point at the apex of the main axis.

Uncinate. Hooked as in the spines of certain cacti.

Undulate. The wavy margins of leaves, bracts, petals, *etc.*

Unilateral. Arranged only at one side.

Unisexual. Having single-sexed flowers, though both may be on the same plant.

Urceolate. Of flowers, urn-shaped calyx or corolla with an inflated tube narrowed in the mouth but slightly broadening again at the tip.

Vaginate. Enclosed by a sheath.

Variegated. Mottled, streaked, edged or striped with colours of leaves mostly white to yellow other than the normal green.

Velamen. A multilayered corky covering of dead cells on some aerial roots as in some orchids, which enable them to absorb moisture from the atmosphere.

Ventral. The inner (adaxial) of the upper side of a leaf or other surface that faces towards the stem.

Virescens. Appearance of green pigmentation on the tissues not ordinarily green.

Vernal. Appearing in spring.

Verrucose. Covered with small wart-like outgrowths, usually on stems.

Verticil. An alternative name for whorl.

Verticulate. Arranged in a whorl or in a series of whorls.

Vesicle. A small bladder or blister-like organ.

Villous. Covered somewhat shaggily with long soft hairs.

Viscid. Sticky, glutinous.

Viviparous. The buds or bulbs which sprout and become plantlets or the seeds which germinate, while still attached to the parent plant.

Whorl. A circular arrangement of foliage or flowers round an axis, at a node, as in cases of leaves of some lilies.

Woolly. For leaves and stems, covered with long soft hairs that are usually white or grey.

Xerophyte. A plant which is able to live under very dry conditions.

Zygomorphic. Capable of being divided into two equal parts in one plane (only along a vertical axis), *i.e.* bilaterally symmetrical.

Index

D

Q

R

S

T

9 789359 191034

Printed at Repro India Ltd.